y Guide

Hans-Jörg Bullinger (Editor)

Technology Guide

Principles – Applications – Trends

With 1092 illustrations and 37 tables

Springer

Editor
Prof. Dr.-Ing. Hans-Jörg Bullinger
Fraunhofer-Gesellschaft, Munich
Germany

Editorial staff
Editor-in-chief: Dr.-Ing. Lothar Behlau (Fraunhofer-Gesellschaft, Munich, Germany)

Dr. Sabine Korte
(VDI Technologie Zentrum GmbH, Future Technologies Consulting, Düsseldorf, Germany)
Dr. Meike Spengel (Fraunhofer-Gesellschaft, Munich, Germany)
Dipl.-Biol. Andrea Vidal (Fraunhofer-Gesellschaft, Munich, Germany)
Maria Hahn (freelance assistant)

Image editing and graphics
Alexander Jost (freelance assistant)
Katharina Prehn (freelance assistant)

Translation and editing
Burton, Van Iersel & Whitney GmbH, Munich, Germany

ISBN 978-3-662-50178-8 ISBN 978-3-540-88546-7 (eBook)
DOI 10.1007/978-3-540-88546-7
Springer Dordrecht Heidelberg London New York

Softcover re-print of the Hardcover 1st edition 2009

Cover design: eStudio Calamar
Production, reproduction and typesetting: le-tex publishing services oHG, Leipzig, Germany

Printed on acid-free paper

Springer is part of Springer Science+Business Media (www.springer.com)

Foreword

"Everything that can be invented has been invented," declared Charles H. Duell, commissioner of the US patent office, in 1899. As we all well know, this was a grand misconception, seeing that Albert Einstein only a few years later heralded the beginning of a century of revolutionary discoveries. These precipitated such fundamentally important innovations as the computer, the electron tube, the laser and the microchip, which have in turn initiated an abundance of other inventions. Today we are once again at the forefront of profound structural change, as we move from the industrial towards a knowledge society. Driven by new technologies, this change is characterized by a tendency to treat information as a product and knowledge as a strategic commodity. The Internet now serves as the catalyst that gives information global reach on a massive scale, linking media and communication systems into an ever more finely interwoven and efficient network. Mobile devices have set the stage for an unlimited exchange of data, independent of time and place. In much the same way, biological technologies have broken ground beyond the health and food sectors, gaining an ever greater foothold in the production of materials.

Although it has grown rapidly in volume and complexity, information on new technologies is now essentially available to everyone, thus doing away with the long-held assumption that such knowledge remains the preserve of experts. Which begs the question, then, of whether it is still appropriate nowadays to compile information on new technologies in a book? We believe so – more than ever before, in fact. In our opinion, the decisive factor is not the medium by which content is presented – i.e. in a book or in electronic form – but the fact that this technology guide competently und comprehensively reports on up-to-date technologies on a consistent level, with clear cross-references between the technologies. It is important to maintain a clear overview. What we most urgently need in today's age of unmanageable quantities of information is orientation and reliable selection. Information can only become retainable knowledge if it is presented to the user in such a way that it can be recognised as new and relevant, and can also be readily understood. But only rarely is it available in the desired condensed form presented here.

What alternative could better serve the inquisitive mind than that which has stood the test of centuries? We turn to experts we can trust. Just like a theatre-goer might browse a playbill or a tourist consult a guidebook, everyone who now works with technology, or takes an interest in it, can refer to this technology guide. It is neither a conventional encyclopedia, nor a study of the future, offering visionary scenarios of tomorrow's world. Numerous publications already fulfill that description. The technology guide is more a work of reference that also makes a good read. It equips voyagers into the future with all the information they need, helping them navigate through current technologies, illustrating their applications, and signposting new trends that give readers a bearing on where we are going.

The book intends to inspire readers, to peek their curiosity as they browse through the pages. In discussions about the development of Fraunhofer's technology portfolio, we have learned that communication between researchers of different disciplines is facilitated by their having an up-to-date overview of the latest technologies. Tabling the latest topics and trends inevitably furthers the development of constructive new ideas and discussions of where these might be interlinked. Suitable platforms need to be created to promote understanding in dialogue between different specialists. Nowadays, innovations mainly emerge wherever knowledge is pooled in an unusual way, i.e. at the interfaces between disciplines and fields of expertise. The philosopher Jürgen Mittelstrass introduced the term "transdisciplinarity" in order to stress how formative the problem-oriented approach has become in overriding an adherence to thinking in disciplines.

In order to stimulate and accelerate the process of transdisciplinary collaboration, it is necessary to structure the wide variety of technologies and their applications in a form that helps inquirers get their bearings. Our science system has become worryingly unclear and confusing. Since we, the publishers of this guide, are unwilling to resign to this complexity, we risk an attempt here at multi-dimensional integration. The material has been grouped into four essentially discipline-oriented cross-section technology categories (materials, electronics/photonics, information and communication technologies, and biological technologies) and into nine chapters covering application-oriented technologies such as mobility or health. Although the contents of the cross-sectional and application-oriented technology fields in this book sometimes overlap, they are presented from a different perspective and with a different focus in each case. Our method of grouping together 13 topic areas covering 100 subjects is only one of many possibilities for structuring the available material, but – in our opinion – a very logical one.

However, anyone who picks out their specialist topic in the technology guide expecting to find something completely new has misjudged the intention of the book. Naturally, they will miss several things that they themselves would have considered essential, or at least worth mentioning. Nevertheless, each specialist article provides even experts with an overview of their own technological domains – and each and every reader is invited on this basis to judge the fundamental quality of this book. The need to present all relevant current technologies correspondingly forced us to condense the material. It was more important to us to clearly portray several prominent aspects of each topic in the limited space available – in such a way that readers can grasp the technical principles behind them and retain an understanding of them in the long term. A complete representation of each specialist area on just a few pages would only have led to a tandem sequence of technical terms on a very abstract level, throwing up more questions than answers. It was not easy for the authors to find a stable middle ground between broad-band superficiality and narrow-lane expertise, and to separate the blurred technology fields into clean-cut, easily digestible chunks. The Internet references provided offer a deeper insight into each topic.

The technology guide makes a particularly exciting read if readers let themselves be guided to areas that lie beyond the knowledge horizon already familiar to them: establishing interconnections to their own areas of work can spark new ideas, and precipitating such inspiring moments seemed more important to us than deeply profound scientific explanations.

At the same time, the technology guide is a reference book that briefly and concisely describes all the important current technologies. It explains the basic fundamentals, portrays applications and comments on future trends. A detailed keyword index and cross-references between different topics help to establish relevant links.

The discourse on future technologies and the search for innovations concern us all. Large international companies and research establishments are not the only ones responsible for innovation – operators and users of technology, too, play an important role, since new ideas could never succeed were society not open to innovation. Anyone with a better understanding of the latest technologies and how they are interlinked can competently join in on discussions of how to shape the future. The technology guide is a suitable aid in gaining this understanding: it is aimed at entrepreneurs, politicians, teachers, students, and ultimately anyone with an interest in technology.

In conclusion, we would like to make another comment on the format of the book. This is an anthology with contributions from more than 150 renowned technology experts from both small and large companies, research establishments, universities, associations and authorities; even a Nobel Prize winner has contributed to this book. When such large numbers of participants are involved, the homogeneity of the end product tends to suffer. In order to avoid a "patchwork character", a dedicated team of editors had to adapt the contributed articles to the pre-defined concept of the Technology Guide, through intensive dialogue with the authors. The goal was to create a uniform standard in terms of technological depth and a homogeneous structure throughout all the articles. The finished book is the result of these thorough review and verification efforts. I would like to thank the editorial team and also the authors, who, throughout several iteration loops, remained open and tolerant towards the sometimes unusual representation of their field of expertise.

Why did we invest so much effort in this project? Because we are confident that the Technology Guide will contribute towards a broader understanding of today's technologies. But the most desirable effect this guide could achieve would undoubtable be to stimulate readers and to spark new ideas that lead to further innovations. After all, we should not leave it to others to shape our own future.

The present issue is a strongly revised version of the German "Technologieführer" printed for the first time in 2007. On account of its success, we decided to update the book's content, perfect its structure, and make it available to a wider circle of readers by publishing it in English.

PROF. DR. HANS-JÖRG BULLINGER
President of the Fraunhofer-Gesellschaft

Authors

PROF. DR. EMILE AARTS, Philips Research Laboratory, Eindhoven, Holland
PROF. DR. STEFAN ALTMEYER, Cologne University of Applied Sciences, Cologne, Germany
PROF. DR. GARABED ANTRANIKIAN, Institute of Technical Microbiology, Hamburg University of Technology, Hamburg, Germany
DR. JENS ASSMANN, Polymer Research Division, BASF SE, Ludwigshafen, Germany
DIPL.-BETRW. CHRISTIANE AUFFERMANN MBA, Fraunhofer Institute for Material Flow and Logistics (IML), Dortmund, Germany
DIPL.-ING. WERNER BAHM, Forschungszentrum Karlsruhe GmbH, FUSION Program, Karlsruhe, Germany
DR. RUDOLF BANNASCH, BIOKON e.V./EvoLogics GmbH, Berlin, Germany
DR.-ING. WILHELM BAUER, Fraunhofer Institute for Industrial Engineering (IAO), Stuttgart, Germany
PROF. DR.-ING. JÜRGEN BEYERER, Fraunhofer Institute for Information and Data Processing (IITB), Karlsruhe, Germany
DIPL.-WIRT.-INF. NADINE BLINN, Department of Business Administration, Institute of Information Systems (IWI), University of Hamburg, Germany
DIPL.-ING. MATTHIAS BOXBERGER, E.ON Netz GmbH, Bayreuth, Germany
DR. LEIF BRAND, VDI Technologiezentrum GmbH, Future Technologies Consulting, The Association of German Engineers, Düsseldorf, Germany
PROF. DR.-ING. KARLHEINZ BRANDENBURG, Fraunhofer Institute for Digital Media Technology (IDMT), Ilmenau, Germany
PROF. DR. FRANZ BRANDSTETTER, formerly Polymer Research Division, BASF SE, Ludwigshafen, Germany
PROF. DR. GÜNTER BRÄUER, Fraunhofer Institute for Surface Engineering and Thin Films (IST), Braunschweig, Germany
DR. ANNETTE BRAUN, Fraunhofer Institute for Physical Measurement Techniques (IPM), Freiburg, Germany
DR. MATTHIAS BRAUN, VDI Technologiezentrum GmbH, Future Technologies Consulting, The Association of German Engineers, Düsseldorf, Germany
PROF. DR. GERT-PETER BRÜGGEMANN, Institute of Biomechanics and Orthopaedics, German Sport University Cologne, Germany
PROF. DR. OLIVER BRÜSTLE, Institute of Reconstructive Neurobiology, Life & Brain Centre, University of Bonn, Germany
ANTONY T. BULLER, formerly StatoilHydro ASA, Stavanger, Norway
DR. DANIEL E. BÜRGLER, Forschungszentrum Jülich, Germany
DIPL.-GEOL. DETLEF CLAUSS, Institute for Sanitary Engineering, Water Quality and Solid Waste Management, Stuttgart University, Germany
DR. WOLFGANG CLEMENS, PolyIC GmbH & Co. KG, Fürth, Germany
PROF. DR.-ING. MANFRED CURBACH, Institute of Concrete Structures, Technical University Dresden, Germany
PROF. DR.-ING. FRANK DEHN, Leipzig Institute for Materials Research and Testing (MFPA Leipzig GmbH), Leipzig, Germany
HEINZ DEININGER, Cysco Systems GmbH, Stuttgart, Germany
DR. MARKUS DEMMEL, Institute of Agricultural Engineering and Animal Husbandry & Bavarian State Research Center for Agriculture, Freising, Germany
DIPL.-ING. KERSTIN DOBERS, Fraunhofer Institute for Material Flow and Logistics (IML), Dortmund, Germany
DR. CHRISTIAN DÖTSCH, Fraunhofer Institute for Environmental, Safety and Energy Technology (IUSE), Oberhausen, Germany
DR. JOCHEN DRESSEN, VDI Technologiezentrum GmbH, Division EINS - Electronics, The Association of German Engineers, Düsseldorf, Germany
PD DR. FRANK EDENHOFER, Institute of Reconstructive Neurobiology, Life & Brain Centre, University of Bonn, Germany

PROF. DR.-ING. MANFRED EHLERS, Institute for Geoinformatics and Remote Sensing (IGF), University of Osnabrück, Germany
DR. HEINZ EICKENBUSCH, VDI Technologiezentrum GmbH, Future Technologies Division, The Association of German Engineers, Düsseldorf, Germany
PROF. DR.-ING. PETER ELSNER, Fraunhofer Institute for Chemical Technology (ICT), Pfinztal, Germany
DIPL.-ING. THOMAS EUTING, Fraunhofer Institute for Technological Trend Analysis (INT), Euskirchen, Germany
DR. BIRGIT FASSBENDER, Bayer Schering Pharma AG, Wuppertal, Germany
PROF. DR. DIETER W. FELLNER, Fraunhofer Institute for Computer Graphics Research (IGD), Darmstadt, Germany
DR. GABI FERNHOLZ, VDI/VDE Innovation + Technik GmbH, Berlin, Germany
DR. TIMO FLESSNER, Bayer Schering Pharma AG, Wuppertal, Germany
DIPL.-ING. JOHANNES FRANK, Cologne University of Applied Sciences, Cologne, Germany
DR. TORSTEN GABRIEL, Fachagentur Nachwachsende Rohstoffe e.V. (FNR), Gülzow, Germany
PROF. DR.-ING. UWE GLATZEL, Dept. of Applied Sciences, University of Bayreuth, Germany
DIPL.-WIRT.-ING.THOMAS GOETZ, Fraunhofer Institute for Mechanics of Materials (IWM), Freiburg, Germany
DR. LARS GOTTWALDT, Volkswagen AG, Research Base, Wolfsburg, Germany
PROF. DR. PETER GRÜNBERG, Forschungszentrum Jülich, Germany
DR. MATTHIAS GRÜNE, Fraunhofer-Institute for Technological Trend Analysis (INT), Euskirchen, Germany
DR. MARCUS GRÜNEWALD, Bayer Technology Services GmbH, Leverkusen, Germany
PROF. DR. PETER GUMBSCH, Fraunhofer Institute for Mechanics of Materials (IWM), Freiburg, Germany
DIPL.-ING. MARTIN HÄGELE, Fraunhofer Institute for Manufacturing Engineering and Automation (IPA), Stuttgart, Germany
DR. MED. URSULA HAHN, Medical Valley Bayern e. V., Erlangen, Germany
DIPL. INF. VOLKER HAHN, Fraunhofer Institute for Computer Graphics Research (IGD), Darmstadt, Germany
PROF. DR.-ING. HOLGER HANSELKA, Fraunhofer Institute for Structural Durability and System Reliability (LBF), Darmstadt, Germany
DR. GUNTHER HASSE, VDI Technologiezentrum GmbH, Division EINS - Nanotechnology, The Association of German Engineers, Düsseldorf, Germany
DR. CHRISTOPHER HEBLING, Fraunhofer Institute for Solar Energy Systems (ISE), Freiburg, Germany
PROF. DR.-ING. MARKUS HECHT, Institute of Land and Sea Transportation, Technical University Berlin, Germany
DR.-ING. HELMUT HECK, Research Institute for Technology and Disability (FTB) of the Evangelische Stiftung Volmarstein, Wetter/Ruhr, Germany
PROF. DR. FRANK HEIDMANN, Potsdam University of Applied Sciences, Interface Design Program, Potsdam, Germany
DR. LARS HEINZE, VDI/VDE Innovation + Technik GmbH, Berlin, Germany
PROF. DR. RER. NAT. ANGELIKA HEINZEL, Institute for Energy and Environmental Protection Technologies, University of Duisburg-Essen, Germany
DR.-ING. MICHAEL HEIZMANN, Fraunhofer Institute for Information and Data Processing (IITB), Karlsruhe, Germany
PROF. ROLF HENKE, Institute of Aeronautics and Astronautics, RWTH Aachen University, Aachen, Germany
PROF. DR.-ING. FRANK HENNING, Fraunhofer Institute for Chemical Technology (ICT), Pfinztal, Germany
DR. RER. NAT. KATHRIN HESSE, Fraunhofer Institute for Material Flow and Logistics (IML), Dortmund, Germany
DR. ANDREAS HOFFKNECHT, VDI Technologiezentrum GmbH, Future Technologies Consulting, The Association of German Engineers, Düsseldorf, Germany
DR. DIRK HOLTMANNSPÖTTER, VDI Technologiezentrum GmbH, Future Technologies Consulting, The Association of German Engineers, Düsseldorf, Germany
DR. PATRICK HOYER, Fraunhofer-Gesellschaft, Munich, Germany
DR. HOLGER HUNDERTMARK, Max Planck Institute for the Science of Light, Erlangen, Germany

LUTZ-GÜNTER JOHN, VDI/VDE Innovation + Technik GmbH, Berlin, Germany
PROF. DR.-ING. MARTIN KALTSCHMITT, Hamburg University of Technology and German Biomass Research Centre, Leipzig, Germany
DIPL.-BIOL. THOMAS KASTLER, Institute for Geoinformatics and Remote Sensing (IGF), University of Osnabrück, Germany
PROF. DR.-ING. ALFONS KATHER, Institute of Energy Systems, Hamburg University of Technology, Germany
DR. RER. NAT. ROMAN J. KERNCHEN, Fraunhofer-Institute for Technological Trend Analysis (INT), Euskirchen, Germany
DR. RAOUL KLINGNER, Fraunhofer-Gesellschaft, Munich, Germany
DIPL.-ING. VOLKER KLOSOWSKI, TÜV NORD AG, Hannover, Germany
DIPL.-PHYS. JÜRGEN KOHLHOFF, Fraunhofer Institute for Technological Trend Analysis (INT), Euskirchen, Germany
DR. SIMONE KONDRUWEIT, Fraunhofer Institute for Surface Engineering and Thin Films (IST), Braunschweig, Germany
DIPL.-ING. ANDREAS KÖNIG, Leipzig Institute for Materials Research and Testing (MFPA Leipzig GmbH), Leipzig, Germany
PROF. DR. RER. NAT. KARSTEN KÖNIG, Faculty of Mechatronics and Physics, Saarland University, Saarbruecken, Germany
DR. UWE KORTE, Financial Services, BearingPoint, Düsseldorf, Germany
DIPL.-ING. SVEN KOWNATZKI, Institute of Energy Systems, Hamburg University of Technology, Germany
DIPL.-ING. PETRA KRALICEK, Empa – Swiss Federal Laboratories for Materials Testing and Research, St. Gallen, Switzerland
PROF. DR.-ING. MARTIN KRANERT, Institute for Sanitary Engineering, Water Quality and Solid Waste Management, Stuttgart University, Germany
DR. THILO KRANZ, German Aerospace Centre, Bonn, Germany
DR. OLIVER KRAUSS, VDI Technologiezentrum GmbH, Future Technologies Consulting, The Association of German Engineers, Düsseldorf, Germany
MENG JENS KUBACKI, Fraunhofer Institute for Manufacturing Engineering and Automation (IPA), Stuttgart, Germany
ANDREAS LA QUIANTE, Cysco Systems GmbH, Hamburg, Germany
DR. VOLKER LANGE, Fraunhofer Institute for Material Flow and Logistics (IML), Dortmund, Germany
PROF. DR.-ING. JÜRGEN LEOHOLD, Volkswagen AG, Research Base, Wolfsburg, Germany
PROF. DR.-ING. LEO LORENZ, Infineon Technologies AG, Munich, Germany
MSC EEIT MARCO LUETHI, ETH Zurich, Swiss Federal Institute of Technology Zurich, ETH Zurich, Switzerland
PROF. DR. TIM LÜTH, Dept. of Micro Technology and Medical Device Technology (MIMED), Technical University of Munich, Germany
DR. WOLFGANG LUTHER, VDI Technologiezentrum GmbH, Future Technologies Consulting, The Association of German Engineers, Düsseldorf, Germany
DR.-ING. ANTON MAUDER, Infineon Technologies AG, Munich, Germany
DR. BENJAMIN MERKT, E.ON Netz GmbH, Bayreuth, Germany
DR. WOLFGANG METT, German Aerospace Center (DLR), Cologne, Germany
PROF. DR. RER. NAT. ALEXANDER MICHAELIS, Fraunhofer Institute for Ceramic Technologies and Systems (IKTS), Dresden, Germany
DR. HELMUT MOTHES, Bayer Technology Services GmbH, Leverkusen, Germany
DR.-ING. WOLFGANG MÜLLER-WITTIG, Fraunhofer Institute for Computer Graphics Research (IGD), Darmstadt, Germany
WOLFGANG NÄTZKER, Fraunhofer-Institute for Technological Trend Analysis (INT), Euskirchen, Germany
DR. ULRIK NEUPERT, Fraunhofer Institute for Technological Trend Analysis (INT), Euskirchen, Germany
DIPL.-BETRW. ALEXANDER NOUAK, Fraunhofer Institute for Computer Graphics Research (IGD), Darmstadt, Germany
DR. RER. NAT. JÜRGEN NUFFER, Fraunhofer Institute for Structural Durability and System Reliability (LBF), Darmstadt, Germany

PROF. DR. MARKUS NÜTTGENS, Institute of Information Systems (IWI), University of Hamburg, Germany
PROF. DR.-ING. ANDREAS OSTENDORF, Ruhr-University Bochum, Germany
DR. GERHARD PAASS, Fraunhofer Institute for Intelligent Analysis and Information Systems (IAIS), Sankt Augustin, Germany
DR.-ING. STEFAN PALZER, Dept. Food Science & Technology, Nestlé Research Center, Lausanne, Switzerland
DIPL.-ING. KLAUS PISTOL, Leipzig Institute for Materials Research and Testing (MFPA Leipzig GmbH), Leipzig, Germany
PROF. DR. JÜRGEN PLATE, University of Applied Sciences, Munich, Germany
DIPL.-GEOÖKOL. CHRISTIANE PLOETZ, VDI Technologiezentrum GmbH, Future Technologies Consulting, The Association of German Engineers, Düsseldorf, Germany
DR. RER. NAT. JÖRN PROBST, Fraunhofer Institute for Silicate Research (ISC), Würzburg, Germany
PROF. DR. ALFRED PÜHLER, Centre for Biotechnology, University of Bielefeld, Germany
DR. GÜNTER REUSCHER, VDI Technologiezentrum GmbH, Future Technologies Consulting, The Association of German Engineers, Düsseldorf, Germany
PROF. DR.-ING. KARL ROLL, Production and Materials Technology, Daimler AG, Sindelfingen, Germany
DIPL.-ING. MATTHIAS ROSE, Fraunhofer Institute for Integrated Circuits (IIS), Erlangen, Germany
DR. KLAUS RUHLIG, Fraunhofer-Institute for Technological Trend Analysis (INT), Euskirchen, Germany
DR. ANDREAS SCHAFFRATH, TÜV NORD SysTec GmbH & Co. KG, Hannover, Germany
DIPL.-ING. SILKE SCHEERER, Institute of Concrete Structures, Technical University Dresden, Germany
DR. STEFAN SCHILLBERG, Fraunhofer Institute for Molecular Biology and Applied Ecology (IME), Aachen, Germany
ANSGAR SCHMIDT, IBM Research & Development, Böblingen, Germany
PROF. DR.-ING. KLAUS GERHARD SCHMIDT, Institute of Energy- and Environmental Technology e.V., University of Duisburg-Essen, Germany
DIPL. INF. DANIEL SCHNEIDER, Fraunhofer Institute for Intelligent Analysis and Information Systems (IAIS), Sankt Augustin, Germany
DR.-ING. HEIDE SCHUSTER, WSGreenTechnologies, Stuttgart, Germany
PROF. DR.-ING. KLAUS SEDLBAUER, Fraunhofer Institute for Building Physics (IBP), Stuttgart/Holzkirchen, Germany
DR. ULRICH SEIFERT, Fraunhofer Institute for Environmental, Safety and Energy Technology (IUSE), Oberhausen, Germany
DR. ULRICH SIMMROSS, Federal Criminal Police Office, Wiesbaden, Germany
PROF. DR.-ING. WERNER SOBEK, German Sustainable Building Council (DGNB) / Werner Sobek Stuttgart GmbH & Co. KG, Germany
PROF. DR. PETER SPACEK, ETH Zurich, Swiss Federal Institute of Technology Zurich, ETH Zurich, Switzerland
DR. GERHARD SPEKOWIUS, Philips Research Asia, Shanghai, China
DR. DIETER STEEGMÜLLER, Production and Materials Technology, Daimler AG, Sindelfingen, Germany
DIPL. KFM. KLAUS-PETER STIEFEL, Fraunhofer Institute for Industrial Engineering (IAO), Stuttgart, Germany
PROF. DR. ULRICH STOTTMEISTER, UFZ Helmholtz Centre for Environmental Research, Leipzig, Germany
PROF. DR. MED. RICHARD STRAUSS, University Hospital Erlangen, Dept. of Medicine I, Friedrich-Alexander University Erlangen-Nürnberg, Germany
DR. STEFANIE TERSTEGGE, Institute of Reconstructive Neurobiology, Life & Brain Centre, University of Bonn, Germany
DR. CHRISTOPH THIM, Bosch Siemens Home Appliances Corporation (BHS), Munich, Germany
DR.-ING. DANIELA THRÄN, German Biomass Research Centre, Leipzig, Germany
DR.-ING. THOMAS WALTER TROMM, Forschungszentrum Karlsruhe GmbH, NUKLEAR Program, Karlsruhe, Germany
PROF. DR. WALTER TRÖSCH, Fraunhofer Institute for Interfacial Engineering and Biotechnology (IGB), Stuttgart, Germany
DR.-ING. MATTHIAS UNBESCHEIDEN, Fraunhofer Institute for Computer Graphics Research (IGD), Darmstadt, Germany

Dr.-Ing. Jürgen Vasters, Federal Institute for Geosciences and Natural Resources (BGR), Hannover, Germany

Dr.-Ing. Rainer Völkl, Dept. of Applied Sciences, University of Bayreuth, Germany

Dr. Patrick Voss-de Haan, Federal Criminal Police Office, Wiesbaden, Germany

Dr. Volker Wagner, VDI Technologiezentrum GmbH, Future Technologies Consulting, The Association of German Engineers, Düsseldorf, Germany

Dr. Max Walter, Institute of Computer Science, Technical University of Munich, Germany

Prof. Dr. rer. nat. Alexander Wanner, Karlsruhe Institute of Technology (KIT), Karlsruhe, Germany

Prof. Dr.-Ing. Martin-Christoph Wanner, Fraunhofer Application Centre for Large Structures in Production Engineering, Rostock, Germany

Dr. Julia Warneboldt, Braunschweig, Germany

Dr.-Ing. Kay André Weidenmann, Section Hybrid Materials and Lightweight Structures, University of Karlsruhe, Germany

Prof. Dr. Ulrich Weidmann, Swiss Federal Institute of Technology Zurich, ETH Zurich, Switzerland

Dr. Birgit Weimert, Fraunhofer Institute for Technological Trend Analysis (INT), Euskirchen, Germany

Prof. Dr. Stefan Weinzierl, Audio Communication Group, Technische Universität Berlin, Germany

Dr. Horst Wenck, Research & Development, Beiersdorf AG, Hamburg, Germany

Dr. Georg Wendl, Institute of Agricultural Engineering and Animal Husbandry & Bavarian State Research Center for Agriculture, Freising, Germany

Dr. Thomas Wendler, Philips Research Europe, Hamburg, Germany

Prof. Dr.-Ing. Engelbert Westkämper, Fraunhofer Institute for Manufacturing Engineering and Automation (IPA), Stuttgart, Germany

Dr.-Ing. Reiner Wichert, Fraunhofer Institute for Computer Graphics Research (IGD), Darmstadt, Germany

Prof. Dr. Klaus-Peter Wittern, Research & Development, Beiersdorf AG, Hamburg, Germany

Prof. Dr. Volker Wittwer, Fraunhofer Institute for Solar Energy Systems (ISE), Freiburg, Germany

Prof. Dr. Stefan Wrobel, Fraunhofer Institute for Intelligent Analysis and Information Systems (IAIS), Sankt Augustin, Germany

Bernhard Wybranski, VDI/VDE Innovation + Technik GmbH, Berlin, Germany

Dipl. Holzwirt Tobias Zimmermann, Egger Group, Wismar, Germany

Dr. Dr. Axel Zweck, VDI Technologiezentrum GmbH, Future Technologies Consulting, The Association of German Engineers, Düsseldorf, Germany

Contents

The Tower of Babel (Pieter Brueghel senior, 1563)

The construction of the Tower of Babel is an allegory depicting the human trauma of not being able to communicate with someone because they speak a different language.

Today, this dilemma also exists in communications between experts from different scientific fields.

The Technology Guide seeks to make the language of experts easier to grasp, so that scientists from different disciplines can understand and inspire each other, and so that non-professionals, too, can join in on discussions of technical issues.

Technologies and the future

Automation, worldwide networking and globalisation are the buzzwords of our times. Social processes in all areas are becoming more intricate and less transparent, as most individuals in modern industrial societies would agree. By stepping out of nature into the increasingly anthropogenic environment of our culture, humankind has taken control of its own social development, and in the longer term probably even of its evolutionary development. In times when individuals find it difficult to comprehend the full scale of the developments that are happening in business, science and society, the positive and negative aspects of mastering this challenge are becoming increasingly obvious. Is the growing complexity of modern society truly inevitable? To put it succinctly: yes. Whichever area of society we look at, development always implies greater differentiation. A society in which the village chief is responsible for settling disputes is less complex than a society that has engendered specialised professions and institutions for this purpose in the form of judges and attorneys. We regard the security, conveniences and justice that are made possible by this growing complexity as achievements of our culture, and we no longer want to live without them. With regard to scientific and technical development, the situation is much the same. Today, technical innovations also tend to be derivatives of existing applications: They induce new markets, give rise to new job profiles, and create new lifestyles and social trends. This reciprocal action and interaction at the same time shows why it is now too narrow a concept to imagine innovation processes as being simple and linear. Innovations, defined for our present purposes as novel inventions that have gained widespread market acceptance, are created through close interaction between different players, social groups, and evolving technical possibilities. For the individual, it is not important to understand each differentiation in detail; what matters more is to master the complexity by knowing where to find which knowledge and what information, and by learning to apply them in the right context.

► **Interdisciplinary convergence.** It is remarkable that the fundamental scientific disciplines of physics, chemistry and biology are becoming increasingly dependent on mutual support and insights. This applies not only to the questions they set out to answer, but also to the methods they use. The blurring of traditional boundaries becomes particularly evident when it comes to transforming research findings into products and technical applications. This is clearest when we look at the sizes of the structures dealt with by scientists in each of these fields. In the last 50 years, as a result of advancing miniaturisation, the size of structures in applied physics has shrunk from the centimetre scale in electrical engineering, through electronics and microelectronics, to less than 100 nm in nanoelectronics. The scale of the structural features being investigated in biology, too, has diminished at a similar rate. From classical biology through cell biology and molecular biology, the biological sciences have now arrived at the stage of functional molecule design using the same tiny structures as in physics. At the same time, this opens up new avenues for functionalisation. Natural or modified biological systems of a size and structure that were hitherto customary in physics or chem-

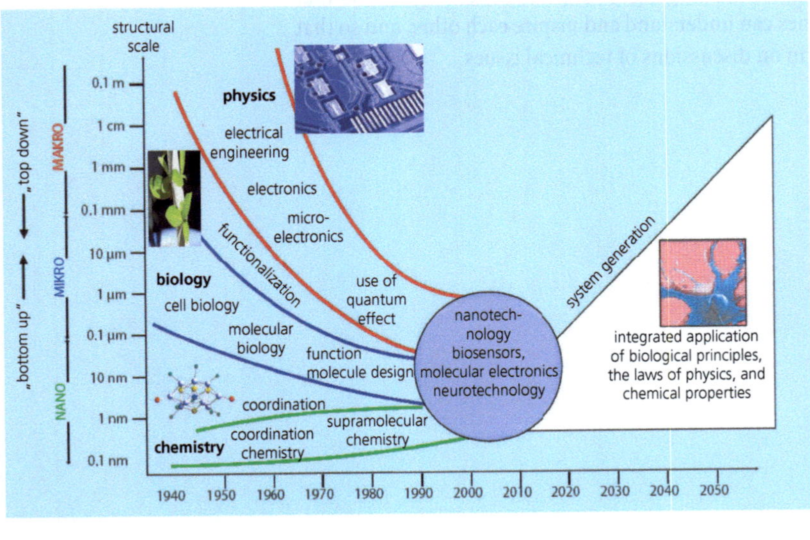

Development of structural scales studied in the disciplines of physics, chemistry and biology from 1940 to the present day. The orders of magnitude in the fields of work and study associated with these disciplines are converging. This will permit the integrated application of biological principles, physical laws and chemical properties in future. Source: VDI Technologiezentrum GmbH istry can now be used specifically for technical applications in biotechnology plants. The reverse tendency can be observed in the field of chemistry. Starting from traditional chemistry – with smaller molecules in inorganic and organic chemistry, at least originally – scientists have graduated via coordination chemistry and supramolecular chemistry to increasingly complex nanoscale structures. When we look at the sizes of structures, therefore, it becomes evident that the fields of work and study associated with the three fundamental scientific disciplines of physics, chemistry and biology are rapidly converging. The laws of physics, chemical properties and biological principles will be more closely interrelated than we could ever have imagined in the future.

Today's truly profound interdisciplinary understanding of modern materials sciences has paved the way for "tailor-made" new materials. Superalloys, nanomaterials, electrically conductive plastics and light-emitting polymers are examples of materials in which highly sophisticated technology is associated with significant value-enhancing potential. The interdisciplinary approach is also reflected in the discussion on converging technologies that is currently taking place on a more basic scientific level. Alongside nano-, bio- and information technology, the cognitive sciences have emerged as a crucial fourth element of the convergence process. Consequently, it will become less and less possible to assign future products and above all their production processes to any specific discipline. This becomes evident as soon as we consider individual fields of technology such as electronics – one need only think of areas such as mechatronics (e.g. antilock braking systems) or polymer electronics (e.g. conductive plastics for flat-screen monitors). Other examples include the convergence of disciplinary concepts towards biology, as we can easily see when we consider nano-biotechnology, neurotechnology or individual products such as biochips or drug delivery systems.

Hardly surprisingly, the concept of "interdisciplinary science" is also undergoing a semantic change. Whereas in the past it mainly signified the necessity of cooperation between different disciplines in order to gain new insights, today it is becoming an elementary requirement for translating the findings of basic and applied research into new products. This has made it a decisive factor in securing future markets. This development, in turn, has far-reaching consequences for our educational system, for the organisation of scientific work in industrial enterprises, and for publicly funded research. In the medium term, our understanding of interdisciplinary science will even affect the attitudes that we, as a society or as individuals, adopt towards future technologies and new products.

Long-term tendencies in technological development can be found in other areas as well. The best known example of this kind is Moore's law. Moore postulated in 1965 that the memory capacity and processing speed of semiconductor chips would double every 18 months. Like a self-fulfilling prophecy, his forecast became a principle in the semiconductor industry that is described today as a "law". Even if this prophecy is predestined to confront the known and anticipated basic physical limits one day, it is still likely to remain valid for at least the next decade – not forgetting that, time after time, technological barriers make it necessary to explore new avenues. A current example is the introduction of multicore processors, which is necessary because their increasing clock rate causes processors to dissipate too much power. However, developments of this kind and the technologies derived from them – not to mention their constant optimisation and evolution – are not an end in themselves. The true objective is to implement future products, and with them the envisioned relief, assistance and opportunities for large numbers of people, while minimising undesired side effects such as the depletion of resources or environmental degradation. In order to face up to the global challenges, scientific insights must increasingly be used in such a way as to yield significant improvements in efficiency and thus promote sustainability.

▶ **Sustainability through technology.** Looking at how their efficiency has developed from 1700 to today, we can clearly see the tremendous progress made in steam engines, lamps and light emitting diodes (LEDs). The efficiency of steam engines in the year 1712 was only around 1% as compared to around 40% for steam turbines in 1955. An even more rapid increase in efficiency can be observed for lamps and LEDs. While the efficiency of the first LEDs that entered the market in 1960s was lower than 1 lumen/watt, the efficiency has risen to more than 90 lumen/watt for today´s LED´s. Research and development activities are pushing for a more efficient use of resources and rendering technology affordable for growing numbers of consumers.

Such tendencies are not only evident in areas of technology that have evolved slowly over a long period of time. Comparable statements can also be made concerning the anticipated future cost of generating energy from solar cells, for instance. According to a recent estimate, these costs are expected to drop by almost a whole order of magnitude over the period from 1990 to 2040. Fuel cell system costs are also expected to fall in a similar way over the next 20 years. The same thing

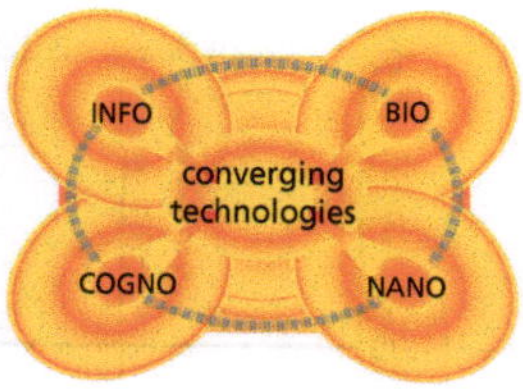

"Converging technologies" as a blending of different disciplines and their respective models: Particularly significant in this context is cognitive science, the findings of which open up new dimensions for technological applications. Source: VDI Technologiezentrum GmbH

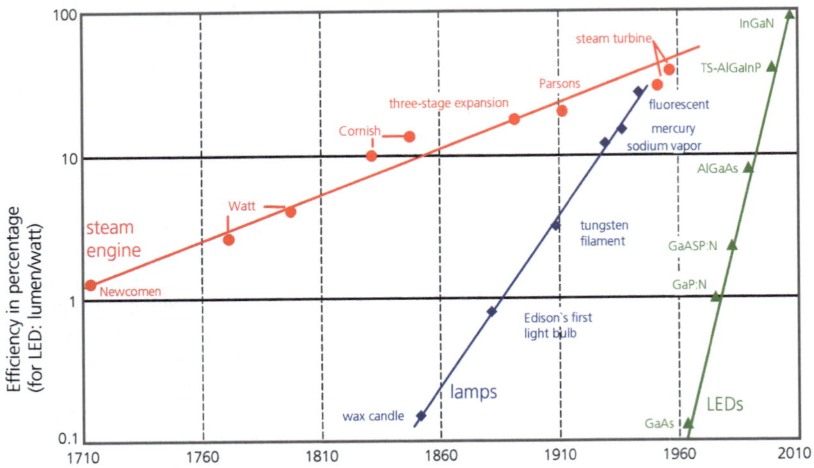

Development of the efficiency of steam engines, lamps and LEDs from 1710 to 2010: For steam engines the logarithmic chart shows an increase in efficiency from about 1% in the year 1712 to 40% for the steam turbine in the year 1955. A greater increase in efficiency can be observed for lamps: Traditional candles and the first Edison light bulb (around 1880) had an efficiency of less than 1%. The introduction of other types of filament (e.g. tungsten in about 1907) enabled light bulb efficiency to be increased to several percent. Around 1940, fluorescent lamps reached an efficiency of about 30%. A more recent example of an even greater increase in efficiency is the development of light-emitting diodes (coloured LEDs in this case). In 1962, the conversion efficiency of the first LEDs was inferior to that of wax candles. Ten years later their efficiency (lumen per watt) had increased roughly tenfold. This trend is ongoing. Source: VDI Technologiezentrum GmbH modified after C. Marchetti

can be said of electronics, where we need only think of the drastically reduced power consumption of TV sets in the past 70 years. The progress achieved from the first cathode ray tubes through to the arrival of transistors and integrated circuits, and culminating in the next generation of appliances illustrates the notion that if the 20th century is known as that of the electron, then the 21st century could be termed the century of the photon (optoelectronics, projection, LCD/plasma monitors).

Technological progress has brought about a substantial increase in efficiency, thus promoting sustainability. Sustainability, in this context, is understood in its widest sense as the coalescence of environmental, social and economic progress. In past centuries, technological progress unleashed the economic potential of the world's industrialised nations. The associated spread and availability of engineered industrial products to ever-wider strata of the population has been a major instrument of social change. Today's media, mobility and medicine are the best illustrations of how extensively our lives are ruled by technical developments. Until the mid-20th century, the ecological aspects of technological development tended to take a back seat. Looking at it this way, one might provocatively argue that technological progress over the last few centuries already unfolded and demonstrated its potential for social and economic sustainability. In this century, our primary concern is to safeguard these achievements in the face of numerous global environmental challenges by raising efficiency through lower resource consumption, minimised environmental impact, and more sustainable distribution. The majority of the technologies described in this Technology Guide illustrate the fact that technological development can do a great deal to help master the global challenges, including our homemade problems.

▶ **Invisible technologies.** Only a 100 years ago, smoking chimneys and gigantic overburden dumps were commonly accepted as visible symbols of the industrial age – despite their unpleasant side effects. Not until the water, soil and air, became palpably affected by pollution and the hazards to human health and the environment were becoming evident, were technologies devised for removing the damage from the immediate vicinity. Pollutants were distributed or diluted until they appeared to be harmless, at least in a local context. It is only during the last 30 years that the problem has been attacked at the roots. First of all, the emission of undesired pollutants was prevented with the aid of filters and treatment processes – a strategy that, though successful, was also expensive. The decisive step, and the one that really mattered in terms of the future of our industrial societies, was to start integrating anti-pollution measures into production processes. Avoidance now took top priority, followed by recycling and – only if this was not possible – disposal. New methods for conserving resources and reducing environmental impact constitute challenges to the economy, but they also provide a means of enhancing efficiency and stimulating innovation, as is shown by Germany's position at the cutting edge of environmental technology.

The application of new technologies will enable us to cut down "side effects" such as energy consumption or space requirements in future. Undesired side effects can be more rapidly identified, tackled and modified so that they only occur to a much lesser extent. Not only in this respect will technologies and their application be less conspicuous in future. The discussion about "pervasive" (Xerox Parc 1988) or "ubiquitous" (IBM 1999) computing gives us a foretaste of the extent to which future technology development will integrate itself even more smartly in our artificial and natural everyday worlds. Information and communication technology experts often also speak of "invisible" (University of Washington) or "hidden" (Toshiba 1999) computing. These terms make it clear that the very technologies that have disappeared from view will in fact have a major impact on our future. They are integrated in our everyday lives to such an extent as to become an indistinguishable part of them. We are growing accustomed to these technologies and losing our awareness of their very existence and of what they do to help us. We are learning to take them for granted. After shopping in the supermarket of the future, for instance, we can expect to check out and pay without having to line up the products on a conveyer belt in front of the cashier. The products will no longer have (only) a barcode, but will have electronic labels (RFIDs) that transmit the price information on demand without physical contact.

In our daily dealings with networked computers, too, we already make use of numerous automated services such as updates or security routines. The normal user is hardly even conscious of them, or able to comprehend them in any detail. On the one hand, this liberates us from having to manipulate everyday technology – still a very cumbersome task in many ways. In a world marked by sensory overload and a plethora of information, this gives us more room to think – to a certain extent, deliberately. On the other hand, this development entails the risk of losing our ability to perceive and understand technical and systemic interrelationships and thus also the supporting fabric of our modern-day life. We find it even more difficult in the case of applications such as self-organisation, artificial intelligence or autonomous robotics. In these areas of technology, due to the complexity of the processes involved, it is rarely possible to correlate the outcome with specific internal sequences of actions.

The progress made in these areas of technology essentially determines the manner in which we will master global challenges such as the growing world population, dwindling resources, sustainable economic development and other challenges. At the same time, technological developments are being influenced to a greater extent by socioeconomic decisions and social pressure. Businesses that focus solely on the development of technologies and products derived from them may find themselves among tomorrow's losers. Unforeseen, sudden changes in the trends followed by consumer markets may rapidly call for other technologies or products than those that are currently available. It is therefore becoming increasingly important for businesses to monitor long-term social and socioeconomic trends and to include them in the planning and development of their strategies and products.

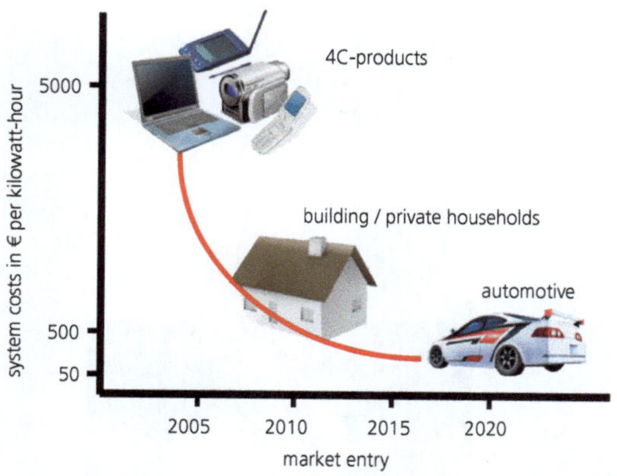

Fuel cells can be regarded as sustainable because they are more efficient and produce lower emissions than internal combustion engines, for example. In the future, they will be even more cost-effective and reliable. The diagram shows the anticipated development of system costs for generating electricity from fuel cells in euros per kilowatt-hour (€/kWh) over a period of 20 years. The system costs for applications in 4C products (camcorder, cell phones, computer, cordless tools) were at around 5000 €/kWh in 2005. Those costs will drop to a value of 500 €/kWh for applications in buildings by 2010. For 2020, an even lower value of 50 €/kWh is expected for automotive applications. Source: Technologiezentrum GmbH modified after WZBU 2007

In this context, the question arises as to whether technology is capable of solving all these challenges. The prospect of a successful technological development must not be euphorically put forward and proclaimed as "the universal problem solution". Global warming and dwindling resources make it clear that greater efficiency and technological innovations are significant factors in finding a solution. However, we must also change our habits, albeit with no restrictions in the quality of life if possible. This is a social challenge, especially for pluralistic-democratic systems, and increasingly calls for a sense of personal responsibility on the part of the individuals living in these societies. Today's heightened awareness of health issues illustrates how such a change of attitude can take place in our societies and force us to adopt greater personal responsibility. Altough preventive medicine is driven by advanced technologies (e.g. imaging), we will not reap their full benefit until a lifestyle that values keeping fit and eating a healthy diet is seen as a sign of affluence in our society and we make our own active contribution to our personal health.

DR. DR. AXEL ZWECK
VDI Technologiezentrum, ZTC, Düsseldorf

MATERIALS AND COMPONENTS

The development of civilisation is shaped by materials and their use. Whether wood, stone or ceramic – for centuries people have used the materials provided by nature. Whole epochs of human history have been defined according to the material dominant at the time, such as the Bronze Age or the Iron Age. In these periods, knowledge was acquired about the pertinent material and how to process it. With the discovery of ore smelting, mankind was no longer restricted to using materials as found in their natural state.

It is now customary to classify materials in three main groups:

- *metals* (e.g. steel)
- non-metallic inorganic materials (e.g. *ceramics*)
- organic materials (e.g. plastic)

There are also materials in the transitions between these categories: *semiconductors*, for example, are positioned between metals and non-metallic inorganic materials. The enormous variety of modern-day materials results from the vast array of different combinations that are possible using composites. These include fibre reinforced ceramics for high-temperature applications as well as multi-layer films for food packaging.

50 years ago materials research hardly existed as a discipline in its own right. Researchers had to seek out materials from the standard range available on the market and adapt them as far as possible for new purposes. Nowadays material scientists, chemists, physicists and even biologists create their own tailor-made materials. Today's knowledge of materials has grown exponentially over the past three decades. Around two thirds of all innovations depend on material-related aspects of development. In the western technology-based countries more than 70% of gross national product derives directly or indirectly from the development of new materials. The main contributors are energy generation, the automotive industry, mechanical engineering, the electrical and electronics industry and chemical production.

Development times for new materials are long. Progressing from the laboratory sample to a finished product can take a decade. Once a new material has been developed and tested, its properties, in particular its amenability to quality-assured and cost-efficient production, have to be verified. It also has to be borne in mind that the way a material is produced can have an influence on its structure and properties. Conversely, to achieve specific properties (e.g. hardness) specific processing methods are required.

Over the past decade nanotechnology has been the main driving force behind innovations in materials science generated from basic research, even though it is not an isolated discipline. The advance into these tiniest dimensions has proved useful in improving the resolution and understanding of material structures. The techniques developed now go beyond observation and facilitate specific processing. With small particle structures, completely new properties can be generated. It is now even possible to manipulate single atoms.

Lightweight materials dominate demand as far as the market is concerned, since the need to improve energy efficiency calls for materials that provide the same properties but at lower weight. The transport sector is the dominant player here.

▶ **The topics**. Around 80% of chemical elements are metals. Around the world they are the most commonly produced material. Although metals are among the oldest materials, they still require constant research: the sheer number of different alloys, containing various metals and other constituents, extends into the millions of possible combinations.

The Bronze Age was preceded by the Ceramic Age. The firing of clay is one of man's oldest cultural techniques. Today, distinction is made between utility ceramics and high-performance ceramics. The latter perform load-bearing functions such as in implants and technical functions such as in electrodes.

Synthetically produced polymers are relatively new. What started in 1870 with the invention of celluloid now encompasses an immense array of different organic materials. Products range from floor coverings and plastic bags to fishing lines (thermoplastics), from kitchen spoons and power sockets (thermosets) to rubber boots (elastomers). *Polymers* are long-chain molecules built up by multiple repetition of basic units with at least 1,000 atoms. Most polymers are made from mineral oil. In order to conserve this energy resource, sustainable materials are being increasingly used. They are mainly plant-based raw materials and biowaste, which are processed either to generate energy or to manufacture textiles, plastics and other chemical primary materials. The most important sustainable raw material is wood. Annual wood production exceeds that of steel, aluminium or concrete. Wood is relatively easy to process. In recent years technology has focused on how to process the large volumes available more efficiently.

In addition to the pure materials, *composite materials* have become established in a range of applications. A composite consists of two or more materials and possesses properties other than those of its individual components. Fibre composite materials are a dominant group. Polymers, metals or ceramics are used as the matrix and are interlaced with particles or fibres of other materials. The properties of these other components are therefore determined by their volume characteristics. The surface properties of materials are increasingly relevant. Tailor-made surfaces are used nowadays to attain important functions such as the "wettability" of liquids, barriers against gases, reflection of light and electrical conductivity.

Nanomaterials are being increasingly used both in the volume as well as in the surface layer. Although nanomaterials have become established in a wide range of applications, such as scratch-resistant car paints (surface) and sun creams (volume), expectations remain high for new applications. In *microelectronics* and *photonics*, major breakthroughs are anticipated from nanostructuring techniques and nanomaterials e.g. carbon nanotubes. Cheaper manufacturing methods will have to be developed, however, if widespread use is to be achieved.

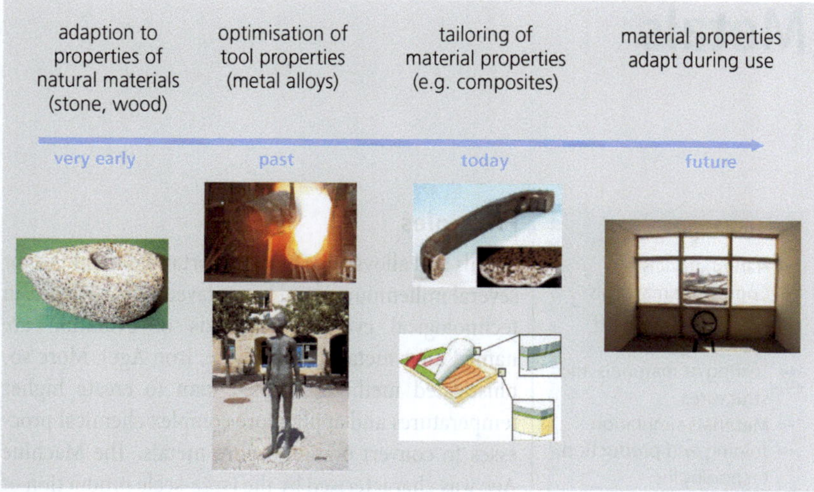

Material epochs: While early man merely processed materials as found in their natural form, the next technological stage of human history was characterised by new combinations of materials, e.g. steel as an alloy of iron and carbon or bronze as an alloy of tin and copper. Today, components or materials are specifically designed to match required applications and functions, e.g. as metal foams or multi-layered films. In the future materials will be endowed with the ability to change according to environmental conditions, e.g. windows will darken automatically depending on the amount of incoming light

Smart materials are acquiring a prominent yet not always very clearly defined place in scientific-technical discussions. They incorporate a component system in which materials change their shape or behaviour according to specific environmental conditions and thus appear to react intelligently. To this end, the sensor and actuator materials have to be combined with an electronic control device to create a composite unit. But it will be some time before an airplane can adapt its wings like a bird to match wind conditions.

With the increase in computer power, *material simulation* has also become established as a research field in its own right. Numerical simulation is used in industrial practice to develop new products in a shorter time and at lower cost, to design more-efficient production processes and to ensure the required product quality.

Self-organising systems are an area of research that is still in its infancy. Simple building blocks arrange themselves to create units of higher complexity which have different properties and a higher information content than the individual components. As this is the way all natural structures are formed, nature is once again guiding the way. ■

Metals

Related topics

- Nanomaterials
- Composite materials
- Surface and coating technologies
- Testing of materials and structures
- Materials simulation
- Joining and production technologies
- Building materials

Principles

Metals and alloys have been important to mankind for several millenniums, and have played a decisive role in technological evolution. Periods of evolution are named after metals (Bronze Age, Iron Age). More sophisticated methods enabled man to create higher temperatures and apply more complex chemical processes to convert ores into pure metals. The Machine Age was characterised by the large-scale production of train tracks and steam boilers of consistently high quality.

A metal is defined by its properties, such as good electrical and thermal conductivity, luster, strength, and ductility. Metals are grouped in the lower left part of the periodic table, and account for about 70% of all elements. In the table the bonding of the outer shell electrons becomes weaker from right to left and from top to bottom. The boundary between metals and non-metals is not clearly defined. Some elements, such as C, Si, Ge and others, exhibit metallic properties in certain circumstances.

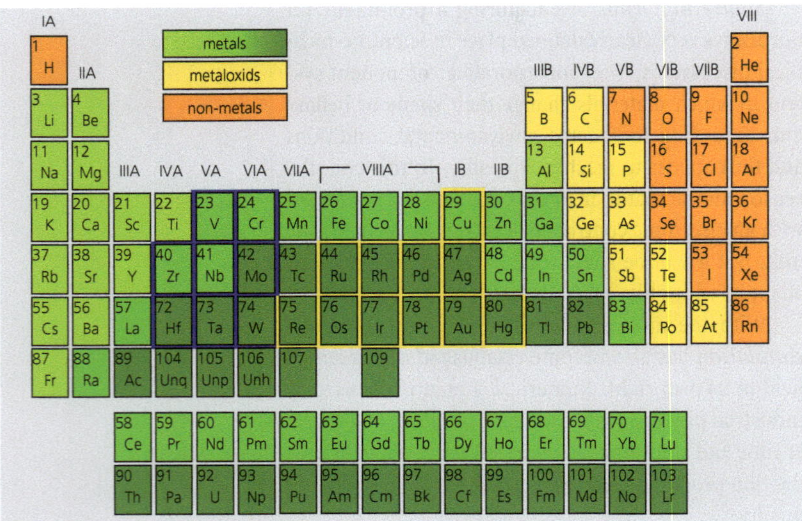

Periodic table of elements: Metals (green) are located at the lower left-hand side of the periodic system. Light metals (pale green), e.g. magnesium (Mg) and aluminium (Al), have densities of less than 5 g/cm³. Heavy metals are also classified by density: less than 10 g/cm³ (green) and more than 10 g/cm³ (dark green). Iron (Fe 26) is very important due to its low production costs and balanced properties. The blue-framed elements are metals with a high melting point (refractory elements, e.g. Cr: chromium). Yellow-framed elements are precious metals and platinum group metals – heavy metals which display virtually no oxidation.

► **Metallic bonding.** The weak bonding forces of the outer shell electrons cause an electron cloud to form when several similar metal atoms approach one another. The electron is then no longer attached to one special atom and can float around freely. This is the reason for the good electrical conductivity. The atoms can be imagined as hard spheres, just like billiard balls. These balls try to occupy the least possible volume. As a result, a large number of metals solidify in a crystal structure with the highest possible density of hard spheres, the latter occupying 74% of the volume.

► **Metals ordered by properties.** The properties of metals vary enormously. They can be classified into light metals (Mg, Be, Al, Ti – arranged by increasing density), low melting heavier metals (Sn, Pb, Zn), and heavier metals with a high melting point (Ag, Au, Ni, Co, Fe, Pd, Pt, Zr, Cr, V – arranged by increasing melting point). The refractory elements (Nb, Mo, Ta, Os, Re, W) are found at the upper end of the metals with a high melting point.

► **Alloys.** Pure metals are only used in very specific applications (e.g. for maximum electrical conductivity). For all other applications, a mixture of elements (an alloy) delivers much better performance. The easiest way to produce an alloy is to mix together specific quantities of pure metals in a crucible and heat everything to above the melting point of the mixture. The melt should not react either with the crucible or with the surrounding atmosphere. To prevent this from happening, melting often has to take place in a vacuum or inert gas and very specific crucibles.

In most cases the solid is a heterogeneous mixture, developed from several different phases. Homogeneous alloys are the exception. The different phases involve volumes of different composition and structure which strongly determine the behaviour of the alloy. This phase separation can be compared to the limited solubility of sugar in water if you cool down a cup of very sweet tea.

Assuming that there are 70 technical elements, approximately 2,500 binary phase systems have to be taken into account. If the concentration is varied in steps of 5% (100% Ni + 0% Cu, 95% Ni + 5% Cu, and so on), it is possible to distinguish 46,000 different alloys. Just recently, all technically relevant binary alloys have been determined.

The number of ternary alloys with three different elements is considerably larger, with 55,000 ternary phase systems and some 9 million different alloys (again in approximate steps of 5%). They have not yet all been classified. The number of alloys increases exponentially with the number of components used. Advanced alloys have four to nine different components. These multi-component alloys can only be developed along quite small paths, creating islands within this multi-dimensional space. Huge scope therefore still exists for discovering new alloys with properties we have only dreamt of until now.

The huge variety of possible mixtures is also reflected in the wide range of properties. One part of the periodic table shows the functional alloys for electrical and thermal conductivity (Cu and precious metals), magnetic properties (Fe and intermetallic phases), or structural properties such as strength, ductility, high temperature strength, fatigue testing (alternating tension/compression), or a combination of these properties.

▶ **Plastic deformation of metals.** An outstanding property of metals is their ability to deform to relatively large plastic strains (up to 100%) under great stress (several hundred MPa). Deformation can be subdivided into an elastic part (instantaneous and reversible) and a plastic part (at higher temperatures, time-dependent and not reversible).

Consider the deformation of a paper clip. Up to a certain force (deflection), the wire bounces back into its original position. Further increasing the force will at some point cause non-reversible plastic deformation, and the wire only slightly bounces back. This is a big advantage of metals and alloys in comparison to ceramics. Plastic deformation in metals and alloys can be explained by metallic bonding. The electron gas forms a strong but non-directional metallic bond. This allows defects to occur, and these defects may then move. In the case of plastic deformation, one-dimensional crystal defects, called dislocations, multiply and able to move by the force applied. A *dislocation* is characterised by its dislocation line, which separates the part of the metal that is already deformed from the part not yet deformed. Each dislocation represents an incredibly small plastic deformation on the atomistic scale of 10^{-10} m. However, the tiny plastic deformations of billions and billions of dislocations, which become longer and more numerous when an external force is applied, add up to a macroscopic deformation. When a paper clip is bent, the plastic deformation increases the length of all the dislocation lines in the paper clip by approximately 1,000 km.

In order to increase the strength of an alloy, obstacles to dislocation are introduced. The obstacle sizes and separations vary from 10^{-9} m (1 nm) up to 10^{-3} m. The volume fraction of obstacles can be adjusted by changing the composition of the alloy.

▶ **Manufacturing techniques.** Manufacturing techniques for alloys can be roughly divided into two main routes: the *melting route* and the *powder metal route* (comparable to ceramic sintering).

In powder metallurgy, a homogeneous mixture is created by mixing different powders. Compacting and sometimes shaping is carried out at room temperature.

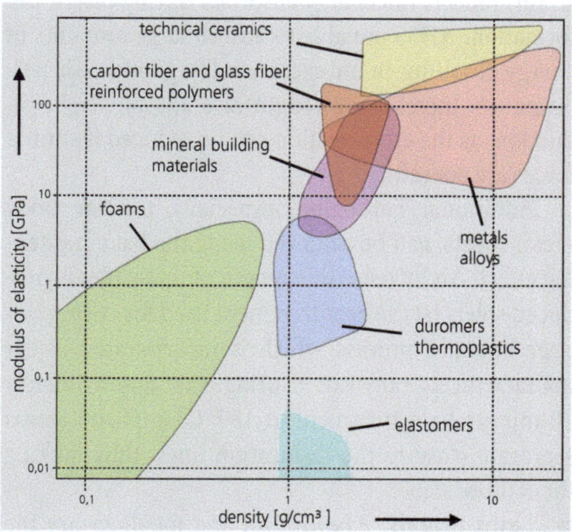

▲ Materials with different behaviour: Metals (red area) are materials with high elastic and tensile strength (vertical axis), but with the disadvantage of quite high density (horizontal axis). In comparison to technical ceramics (yellow area) metals can be strained to a great extent before failing. Metals and alloys are therefore the most commonly used structural material. Conventional polymers (duromers and thermoplastics) combine low density with a low price, as they can be produced in high volumes and have moderate stiffness and strength. Carbon fibre and glass fibre reinforced polymers (CFK and GFK) add stiffness and strength to conventional polymers but raise the price. Both polymeric and metallic foams have the advantage of much lower density than bulk materials. However, their low stiffness and strength confine their use to less demanding applications.

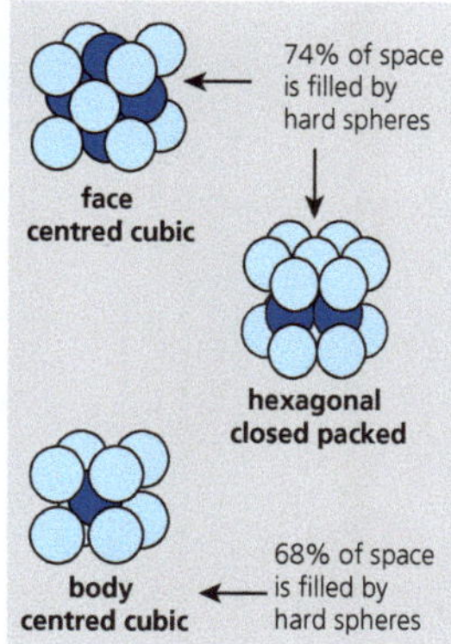

▶ Model of atoms in a metal: Most metals solidify in either face-centered cubic (fcc) or hexagonal close-packed (hcp) crystalline structures; hard spheres fill 74% of the space. Examples of fcc-metals are Al, Ni, Pt, Au and Ag; all of these have low strength but high ductility (the degree of plastic deformation before fracture occurs). Mg and Ti are metals with an hcp-structure, and are known to be difficult to deform at room temperature. Sometimes, the crystalline structure is body-centered cubic, occupying 68% of the space. Iron and tungsten exhibit this crystalline structure, with mechanical properties between those of fcc- and hcp-structured metals.

Sintering at temperatures below the lowest melting temperature reduces pore density, and connections form between the powder particles. Complex parts can be produced fairly cheaply without requiring finishing procedures. The pores cannot be eliminated, resulting in reduced strength compared to a 100% dense bulk material. The number of parts produced using the metallurgy route is quite small, but is increasing.

In the melting route, the parts are either shaped by mechanical shaping (milling and forging) or by direct casting in a mold. Milling of metal sheets (car body) and forging (crankshaft or connecting rod) can be carried out either at room temperature or at elevated temperatures. Only rarely, for small part numbers, is the final part machined out of a solid block.

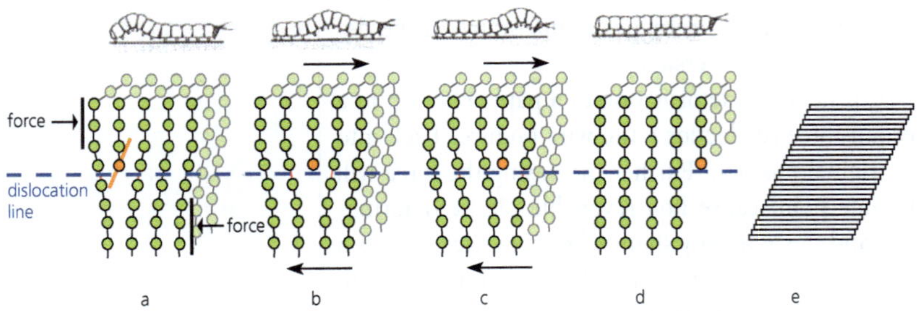

Deformation of metals: The plastic deformation of metals on an atomistic scale is similar to the movement of a caterpillar. A caterpillar needs less strength if it just makes a hump and moves the hump along its body than if it moves its whole body at once. a: In one-dimensional crystalline defects of metals, dislocations occur where atoms along the dislocation line do not have atomic bonds to some of their neighbouring atoms. b,c: When a force is applied not all atomic bonds along the force line are broken at once; only the atoms closest to the dislocation line move a short distance until new bonds are established. d: This "atomic hump" moves forward as long as the force is applied, until it reaches the surface of the metal part, where it leaves an atomic sized step behind. e: Billions of dislocations move simultaneously through the metal, which together result in a macroscopic deformation of the metal part.

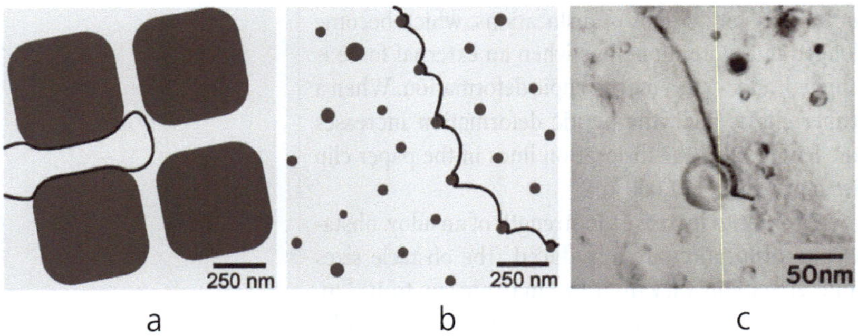

Influence of particle size and shape on the mobility of dislocations. Left: Diagram of the microstructure of a nickel-based superalloy. Cuboidal hard particles (dark grey shading) are embedded in a soft matrix (light grey shading). A dislocation (black line from left) cannot penetrate the hard particles and is forced to move through narrow matrix channels, giving rise to exceptionally high resistance to plastic deformation even at very high temperatures. Centre: Small hard particles (dark grey shading) evenly distributed in a soft matrix (light grey shading) exert forces of attraction on dislocations, preventing them from moving. Right: High resolution transmission electron micrograph of a dislocation pinned on small oxide particles. Source: E. Arzt, MPI for Metals Research, Stuttgart

Applications

► **Steel and cast iron.** Iron alloys make up 90% of the construction material used nowadays. Their use ranges from *building materials* (construction steels) to medical applications (stainless steels). Steel has a carbon content of less than 2 wt.%, while cast iron has a higher carbon content, 4 wt.% at most in technical applications. Steel has been used in large quantities since about 1850. World-wide production in 2008 was approximately $1.4 \cdot 10^{12}$ kg of crude steel. This amount would fill a cube with a border length of 560 m. Production in 2008 is approximately 50% more than it was five years earlier.

Despite its more than 150-year history, steel is a very innovative material. This is reflected in over 200 new steel standards per year and a total of more than 2,000 standardised steel qualities. The most recent developments are advanced high strength steels (AHSS). The product force (or stress) times the length (or strain) reflects the energy absorbed during plastic deformation. AHSS are able to absorb large amounts of energy resulting in an exceptionally good crash performance. Increasing strength also allows weight reduction, as the cross-section can be reduced if similar forces are operating.

Additional hardening, especially for car body metal sheets, can be achieved using the bake-hardening effect. An increase in strength of up to 10% is produced solely by the heat treatment used for curing lacquer layers, a process which is incorporated in car manufacturing anyway. During the approximately 20-minute heat treatment at 180 °C, impurity atoms segregate close to the dislocation lines, thus making them more stable.

► **Light metals.** *Aluminium* and its alloys are the most commonly used light metals. The food packaging industry accounts for approximately 40% (thin aluminium coatings on milk product storage devices, cans, aluminium foil). The *automotive* industry uses a share for parts such as the engine and gear box and, increasingly, for complex car body parts. All of these are manufactured by *pressure die casting*, a method in which the aluminium melt is squeezed into the mould at high pressure (up to 1,000 bar).

High strength aluminium alloys are usually strengthened by *precipitation hardening*. During heat treatment very fine particles (in the nm range) precipitate out of a super-saturated solid solution matrix.

These particles act as obstacles to dislocation movement. These aluminium alloys are classified not only by their composition but also by their heat treatment status. Heat treatment of high-strength aluminium alloys can be incorporated in the lacquer curing process, in a similar way to bake-hardening steels.

GLARE (glass-fibre reinforced aluminium) is an aluminium-glass-polymer composite with an outstanding weight-to-strength ratio and reduced crack propagation. GLARE was specially developed for aircraft bodies, and consists of multiple layers of aluminium and a glass-fibre reinforced polymer. GLARE is used in the forward and upper areas of the new Airbus A380 body.

The use of magnesium alloys is steadily increasing, but at a comparably low level, e. g. crank cases for car engines. These have been produced in series since 2004, using a die-casting process with an aluminium insert surrounded by magnesium. This produces the highest power-to-weight performance.

Titanium alloys play a significant role in the aerospace industry. In an aircraft turbine engine, about two third of the compressor air foils and the large fan, primarily responsible for propulsion are made from a titanium alloy (90% Ti, 6% Al and 4% V by weight). This alloy is an excellent compromise as regards to important material properties such as strength, ductility, fracture toughness, corrosion and oxidation. In-service temperatures can be as high as 550 °C. In an aircraft turbine, Ti alloys are also used as discs, bladed discs or bladed rigs.

A modern aircraft body also contains several tons of titanium threaded rivets. Intermetallic phases, such as titanium aluminides (TiAl 50:50) or nickel aluminides (NiAl 50:50) are under close investigation because their intrinsic properties, such as density and oxidation resistance, are exceptionally good. General Electric has qualified a titanium aluminide in the compressor section of its aircraft engines.

▸ **Copper alloys.** Copper has a good price-to-conductivity ratio. Copper with impurities of less than 1% is used as electrical wire. Small additions of cadmium, silver and aluminium oxide increase strength and do not significantly decrease electrical conductivity.

Brass (copper with up to 50% wt.% of zinc) possesses exceptionally good corrosion resistance and is often used in sea water environments. It has excellent failsafe running properties without requiring lubrication. Brass is therefore often used as a material for bearings.

▸ **Solder alloys.** Enormous efforts are currently being made to substitute lead with non-toxic materials. The classic solder alloy is made of lead and tin (63% and 37% by weight respectively – PbSn37). The use of lead in the automotive industry will be restricted by law during the next few years. The next generation of solder alloys will be much more expensive, as small amounts of silver seem to be necessary. These alloys will also have a slightly higher melting point, resulting in greater temperature stress on the semiconductor devices

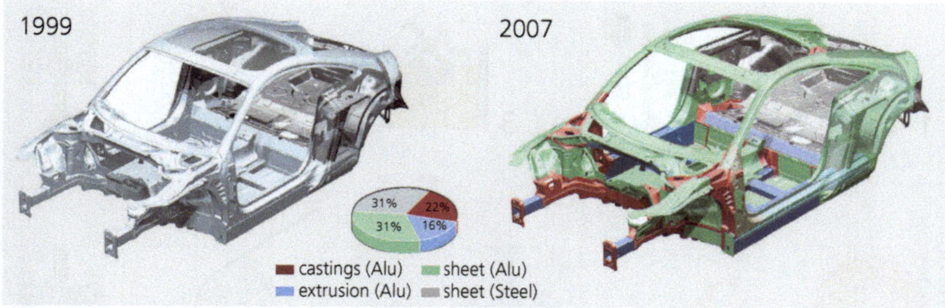

▣ Different steel grades and aluminium alloys used in car bodies over an 8-year period. More and more advanced high strength steels with increased yield strength are incorporated in car bodies, to reduce weight while at the same time increasing stiffness and improving crash performance. Left: The first generation Audi TT-Coupe (1999) made of 100% steel sheet material. Right: The second generation (2007) as a mixture of steel sheets with aluminium in cast, extruded and sheet material condition. Source: Audi AG

▣ Magnesium-aluminium compound crank case of the BMW 6-cylinder in-line engine for regular fuel. The cylinder liners with its water cooling channels are cast first. An aluminium alloy is used in order to withstand both the high mechanical and thermal loads produced by combustion, and the corrosion attack by the cooling water. In a second step the liners are recast with a magnesium alloy to save weight in this less severely stressed part of the engine. This achieves a weight reduction of 24% in comparison to a conventional aluminium crank case. Source: Fischersworring-Bunk, BMW

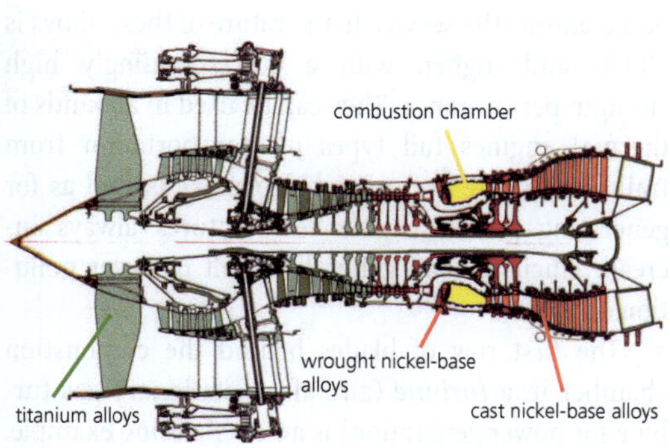

▣ Modern aircraft engine: The airstream flows from left to right. The large titanium fan mainly responsible for propulsion is followed by an air compression section, the combustion chamber and the turbine. The turbine drives the fan by a shaft that runs through the engine. Different materials are shown in different colours (green: Ti alloys, red: Ni alloys). Source: Esslinger, MTU Aero Engines

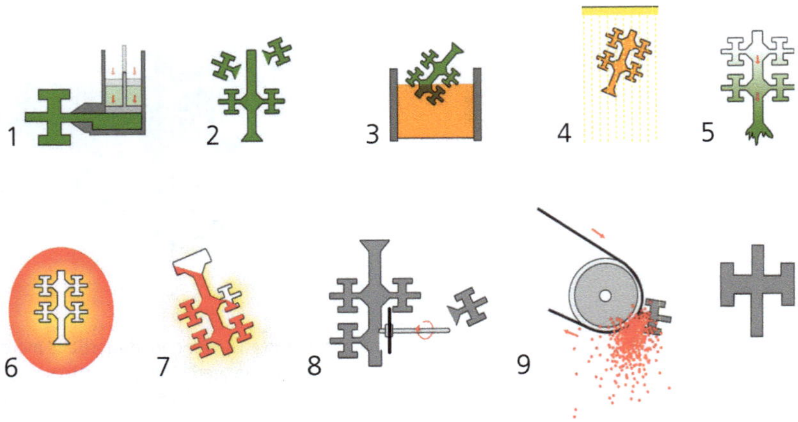

The lost-wax precision casting process. 1: Injection moulding of the positive wax model. 2: Assembling individual models in a cluster to save metal casting time. 3: Inserting into a ceramic slurry. 4: Irrigating with sand (repeating the slurry and sand steps (steps 3 and 4) several times with different consistencies until a thick ceramic mould has been built up). 5: De-waxing at approximately 180 °C. 6: Sintering the ceramic mould. 7: Pouring the liquid metal into the pre-heated mould. 8: Disassembling the cluster. 9: Sand blasting, chemical etching, final inspection of the part.

during the soldering process. The temperature window for the soldering process will be smaller in future.

▶ **High-temperature alloys.** Several high-temperature alloys are available commercially. The spectrum ranges from iron, cobalt and nickel to precious-metal based alloys. The service temperature of these alloys is 700 °C and higher, with a correspondingly high strength performance. They can be used in all kinds of thermal engines (all types of transportation from trains and ships to cars and airplanes) as well as for generating power. Higher temperatures always increase efficiency and generally result in lower pollution as well.

The first row of blades behind the combustion chamber in a *turbine* (aircraft or stationary gas turbine for power generation) is an outstanding example. These blades are exposed to harsh conditions:

- high temperatures (gas temperature up to 1500 °C, material temperature up to 1150 °C)
- solely due to centrifugal forces, the stress of rotating blades is in the range of 100 MPa (1,000 kg attached to an area of 1 cm^2)
- Aggressive burning gases impacting the blade at ultrasonic speeds.
- *thermo-mechanical fatigue* loading due to start and stop cycles. Temperature gradients between the hot outer surface and the cooled inner regions produce local stress
- unbalanced masses can result in high frequency loading of the material

Nickel-based superalloys (roughly 75 wt.% of Ni as the base, with additions of Al, Cr, Co, Mo, Ti, Ta, W and Re) are the only materials able to withstand these complex loading schemes. Directionally solidified or even single-crystal parts have been used for 30 years. In the directionally solidified state, the grain boundaries run parallel to the blade axis. Single-crystal blades do not exhibit grain boundaries at all, preventing grain boundary sliding and, as an important additional effect, reducing thermo-mechanical stress.

These blades are produced by the lost-wax precision casting process. This method of production has been known for at least 2,000 years – yet it is still the only way to manufacture such hollow complex geometries with precise inner contours.

▶ **Exotic alloy systems.** Metallic glass usually consists of zirconium-based alloys, and features a high cooling rate from the liquid to the solid state. This allows the non-ordered arrangement of atoms in the liquid state to be frozen into the solid state. The wide strain range, in which elastically reversible deformation can take place, is one of the advantages of metallic glass. Although the modulus of elasticity is low, the wide elastic strain range permits the storage and release of more elastic energy in metallic glass than in any other material. Most of the bulk production of metallic glass went into golf driver clubs to increase the range of the drive. Since 2008, however, the United States Golf Association has banned clubs made of bulk metallic glass in tournaments, as the ball speeds are too high and tee shots are too long.

Shape-memory alloys (e.g. Ni-Ti alloys) can resume their original geometry thanks to a memory effect. In these alloys, the usual dislocation mechanism is not activated, but deformation occurs by instantaneous and reversible phase transformations. As the transformations are reversible, a slight temperature increase causes the material to return to its original shape. These alloys are used in temperature-dependent controlling mechanisms and in medical applications. A high degree of pseudo-elasticity, again based on these reversible phase transitions, is another feature. These materials make it possible to achieve elastic strains that are approximately 10 times higher than for conventional metals. This characteristic is used in stents which fold down to an extremely small size during insertion and expand in the blood vessels in the final position, in dental correction braces, and in spectacle frames which are hard to deform.

Intermetallic phases are alloys of a fairly precise composition approximating stoichiometry, e.g. Ni_3Al. As a large amount of metallic bonding is involved, the composition may vary considerably. In the intermetallic phase Ni_3Al it varies from 73–75 atomic % Ni, in NiAl from 43–68 atomic % Ni. Intermetallic phases often have very exceptional properties (e.g. $Nd_2Fe_{14}B$, which is one of the strongest permanent magnetic materials) but are very brittle. This is because chemical ordering within the intermetallic phases does not allow dislocations to move. Large intermetallic phase parts are frequently manufactured by the powder metallurgy method. In permanent magnets, the powder is compressed within a magnetic field in order to align the crystals orientation in a magnetic favourable direction.

Trends

▶ **High-temperature alloys.** The high-temperature properties of nickel-based superalloys seem to be limited by the melting point of nickel at 1455 °C. However, these materials can withstand considerable stresses at temperatures up to 1150 °C (over 80% of the melting temperature). Much research is being carried out to improve the temperature capabilities of these alloys. Academic researchers are talking about the 4. and 5. generation of single-crystal nickel-based superalloys. Despite tremendous efforts, there is no other material in sight today that might be able to replace the nickel superalloys within the next 10–20 years. This is because these alloys have very well balanced properties, which include ductility at all temperature levels, combined with high strength at high temperatures. These alloys also possess fairly good oxidation and fatigue resistance. Special technical *ceramics*, alloys based on platinum group metals (Pt, Rh, Ir), and other high melting point systems such as Mo-Si are currently under intensive investigation.

Another strategy for optimising the development of high-temperature alloys is to reduce density and production costs by keeping the high-temperature strength at that level. This allows the overall weight of the system to be significantly reduced by decreasing the size of other load bearing parts. This is of specific interest for moving or rotating parts as the rotating speed can be considerably increased by using lower-weight rotating parts.

These two optimisation strategies seem to point in opposite directions, but both result in a comparable technological improvement. Though at first sight apparently different, both optimisation routes can be found in practically all alloy developments based on a specific application.

Prospects

- Cost is an extremely important overall material factor. Steels in particular can be produced in huge quantities using low-cost production, forming and molding processes. Steel will continue to make up a large part of car bodies in the near to medium term, as well as – of course – being used as a construction material. Particularly in view of the increasing cost and shortages of raw materials, the use of metals is becoming more widespread due to cost considerations and very efficient recycling possibilities.
- The demand for light metals and high-temperature metals to increase efficiency and reduce ecological impact will go on increasing as long as man is technologically minded. This applies to all kinds of transportation systems from bicycles to airplanes and energy production.
- Researchers are currently striving to develop lead-free solder alloys.
- Rare earth and precious metals are very expensive and only available in limited quantities, but are essential and often not replaceable in many high-tech applications (e. g. energy-efficient lamps).

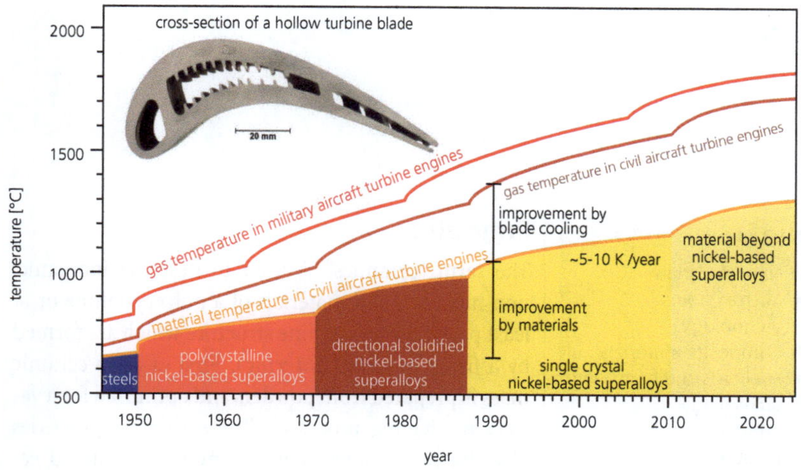

Increase in combustion gas temperature of aircraft turbine engines over several decades – the efficiency of a turbine is closely related to the gas temperature. After the Second World War, high-temperature iron alloys with an upper limit of about 700 °C were state-of-the-art materials. Nickel-based superalloys replaced iron alloys in the 1950s and are still the material of choice today. Leaps in technology meant that hollow turbine blades could be cast in increasingly sophisticated designs, improving cooling efficiency. Specifically designed nickel-based superalloys and adapted casting technologies enabled the change to be made from polycrystalline, to directionally solidified, to single-crystal turbine blades. The gas temperature was increased by omitting the detrimental interfaces between crystallites in single crystals. Due to the strong safety standards in civilian aircraft engines, nickel-based superalloys will not be replaced by ceramics in future. But insulation by a thick ceramic layer is and will be of increasing importance.

PROF. DR. UWE GLATZEL
DR. RAINER VÖLKL
Universitiy of Bayreuth

Internet

- www.worldsteel.org
- www.matweb.com
- www.tms.org
- www.mrs.org

Ceramics

Related topics

- Nanomaterials
- Microsystems technology
- Composite materials
- Fuel cells and hydrogen technology
- Implants and prostheses
- Automobiles

Principles

The term "ceramics" is used to describe inorganic, non-metallic solid materials of a polycrystalline or at least partly polycrystalline structure which are formed by a firing process (*sintering*). First of all, a ceramic mass is prepared from powders and optimised for a variety of shaping processes. These shaping processes give the components their geometry. Unsintered ceramic bodies are known as "green bodies". The subsequent sintering process causes the ceramic body to become compacted and take on its finished form, which may then require further machining.

The material properties, shape and size of the ceramic product are intrinsically linked to the way in which the respective process steps are carried out. The primary challenge is that of specifically influencing the structure by the choice of starting substances (powders, additives) and of selecting the parameters for the preparation, shaping and sintering processes.

Ceramics can be broadly categorised as traditional ceramics and advanced or technical ceramics. Advanced ceramics, in turn, are divided into structural and functional ceramics.

The technical progress made in traditional ceramics today (ornamental ceramics, tableware, bricks, tiles, sanitary ware, etc.) is taking place in the areas of production engineering and surface finishing. In the case of advanced ceramics, progress focuses on the synthesis of new materials and/or optimisation of existing materials, along with the new and continued development of the associated process engineering methods with respect to cost-cutting and quality assurance. Furthermore, there is an ever-greater focus on system-oriented solutions that call for adaptation of the engineering design, including the appropriate joining technologies, to the specifics of ceramic materials and the manufacturing process. The targeted combination of structural and functional ceramic technology permits direct integration of additional functions in ceramic components. Purely passive components can thus be transformed into active, "intelligent" components, making it possible to manufacture innovative products with distinct added value.

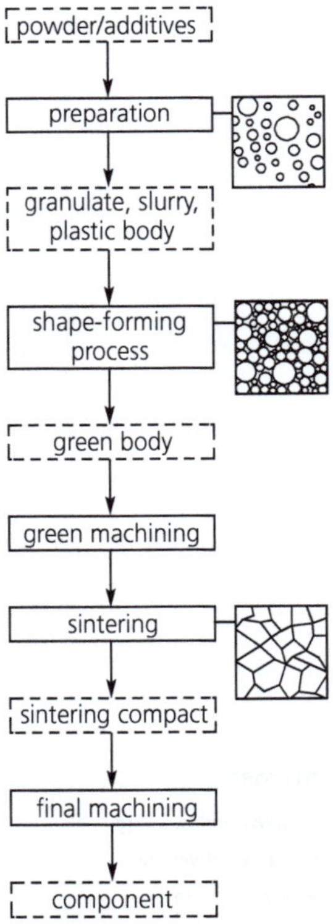

Process steps in the ceramic manufacturing process: Ceramics are shaped from powdery starting products at room temperature, but do not acquire their typical material properties until they are sintered at high temperatures.

	ceramics	metals
hardness	▲	▼
thermal resistance	▲	▼
thermal expansion	▼	▲
ductility	▼	▲
corrosion resistance	▲	▼
wear resistance	▲▼	▼
electrical conductivity	▲▼	▲
density	▼	▲
thermal conductivity	▲▼	▲

▲ tendency to high values ▼ tendency to low values

A comparison of the properties of ceramics and metals: Advanced ceramics are distinguished by their high strength and hardness as well as excellent wear, corrosion and temperature resistance. They can be employed for both heat-insulating and heat-conducting purposes, and the same applies to their electrical conductivity.

▶ **Structural ceramics.** Structural ceramics are remarkable for their outstanding mechanical and thermal properties. They are typically divided into two classes of materials: oxides and non-oxides. Among the most important oxide ceramic materials are aluminium oxide and zirconium dioxide, while the most important non-oxide materials are the nitrides and carbides of silicon, boron and titanium. Non-crystalline (amorphous) substances based on silicon, boron, nitrogen and carbon, which can be obtained from the early stages of polymers without a sintering process, are also gaining in significance. Structural ceramics are particularly suited to applications involving:

- mechanical stress
- corrosive stress
- thermal stress
- wear and tear
- biocompatibility

Because of their generally low density combined with a high fracture strength and great hardness, structural ceramics are ideal lightweight construction materials.

Applications

► **Ceramic implants.** For more than 30 years, structural ceramics have been prized as *implant* materials by surgeons and patients alike due to their biocompatibility, extreme hardness and long-term stability. The implanting of ceramic hip joints, for example, has become a routine procedure. Traditional hip joint prostheses consist of a metal joint head which moves around in a socket made of plastic or also of metal. This motion gives rise to exogenous abrasion, which causes "particle disease" and may, in the long run, necessitate the removal of damaged bones and the replacement of the implant.

Ceramic hip replacements are a good alternative, as this material has the lowest abrasion rate (on average 1 µm per year), and is most readily accepted by the body. Ceramic prostheses are thus intended for active, younger patients and for those who are allergic to metal. What is more, ceramics cannot be corroded by aggressive bodily fluids and exhibit high stability in absorbing extreme forces.

► **Ceramic diesel particulate filters.** *Diesel engines* are currently among the most efficient mobile combustion engines used in cars, trucks, off-road vehicles and stationary machines. However, the relatively high nitrogen oxide and particulate emissions from diesel engines are a problem. The use of highly efficient diesel particulate filters can drastically reduce particulate emissions. Porous ceramics are used as the filter material, as they are capable of withstanding the high thermal stresses occurring during the recurrent thermal regeneration, i.e. the burn-off of soot particles. Ceramics with a thickness of only 200–350 µm and a very uniform pore width of 10–15 µm are needed in order to precipitate 99.9 % of the nano soot particles from the exhaust. To keep the exhaust back pressure as low as possible, filter surfaces measuring several square meters are produced in cylindrical mould elements with a volume of just a few litres by forming the ceramic bodies as extruded honeycombs with a high number (200–300 cpsi = cells per square inch) of alternately sealed, thin parallel channels. The alternate sealing of the channels forces the exhaust gas through the porous ceramic walls, and the particles are precipitated on the surface of the inflow channels where they are removed by catalytic thermal regeneration.

In the production process, carefully selected ceramic powders and additives are kneaded to form a plastic mass, which is pressed through a honeycomb die, and the resulting honeycomb strand is cut to length. Once this has dried, the channels are alternately plugged at the front ends, before the ceramic filter is sintered and machine-finished, as required. Silicon carbide filters, which are used in nearly all cars because of their unusually high robustness, are joined together from small segmented honeycombs with a type of elastic cement that is stable at high temperatures. Cordierite and aluminium titanate honeycombs for application in heavy-duty vehicles can be processed as one monolith.

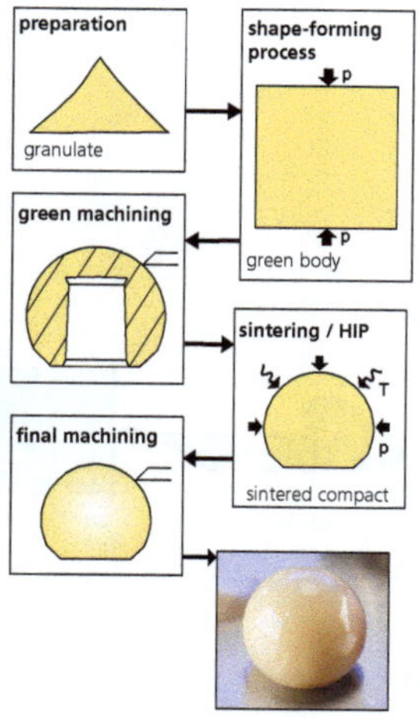

Process chain for the manufacture of a ceramic ball head for hip joint implants: The production of a ceramic ball head requires as many as 60 successive manufacturing and quality assurance steps. High-purity and sub-microfine starting powders are ground in a watery suspension, homogenised and then sprayed to form a granulate suitable for compression. The compression-moulded blanks are given their provisional outline by machining while in the "green state". A multi-stage thermal treatment process at high temperatures and under pressure gives them their sintered fine-grained microstructure, which is the precondition for a high and reproducible standard of mechanical properties. In a sequence of further steps, the now extremely hard ceramic body is precision-ground and polished to an accuracy of within 1 µm and polished to an accuracy of within 3 nm. Source: IKTS/CeramTec AG

► **Functional ceramics.** Functional ceramics is the term used for inorganic non-metallic materials that have a key physical function such as:

- electronic and/or ionic conductivity
- dielectric, piezoelectric or ferroelectric behaviour
- magnetic function
- electrochemical properties

Their range of applications includes numerous electronic components such as insulators, resistors, inductances, capacitors, overvoltage protectors, frequency filters, and sensor and actuator components.

► **Ceramic multilayer microsystems engineering.** Ceramic multilayer systems (LTCC, HTCC [Low/High Temperature Co-fired Ceramics]) are of great importance in 3D structural design and packaging technology for electronic components.

Recent developments are tending to expand the range of functions performed by ceramic multilayer systems. A transition from mere electronic packaging to integrated microsystems is now taking place. The capability to form 3D structures, which results from the screen printing/multilayer technology employed, leads to an extension of functions to include microfluidics, micromechatronics and sensors.

The variety of mechanical functional elements that can be produced is extensive due to the many possible ways of combining differently structured individual tapes. Membranes, cavities, channels, chambers, beams and springs can all be created in this manner. It is possible to build up complete *PEM (polymer elec-*

trolyte membrane) fuel cell systems in which not only the electrochemically active components (PEM stack), but also the entire system periphery (channels, valves, sensors, pumps, wiring) are fabricated in multilayer ceramics technology.

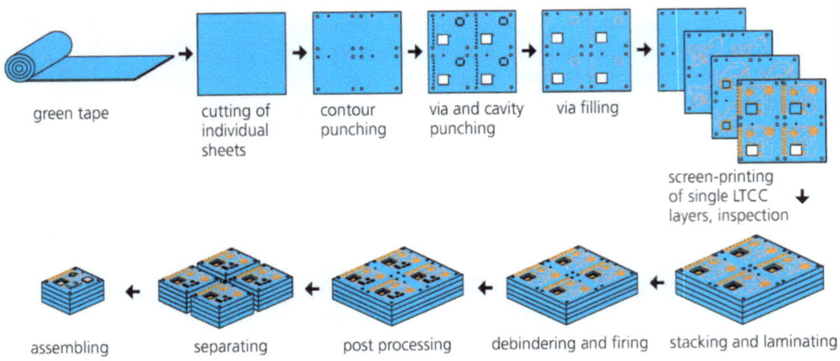

Process chain for the production of an LTCC (Low Temperature Co-fired Ceramic) multilayer system: The process chain begins with tape-casting to produce the ceramic tapes. The still unfired tapes are subsequently structured by micro processing (perforation, laser, stamping). Once the holes for continuous bonding (vias) have been punched and the vias have been filled, the conductor lines and other functional elements that will be concealed (buried) later are created by screen printing. The individual layers are then aligned, stacked, and bonded at a raised temperature (70–80 °C) and high pressure (200 bar) to form a multi-layer system. The next step after laminating is firing, which may or may not involve sintering shrinkage. This is followed in certain cases by further screen printing and firing of outer functional elements, separating the substrates by sawing or dicing (scoring, breaking), and assembly. Source: TU Dresden/IKTS

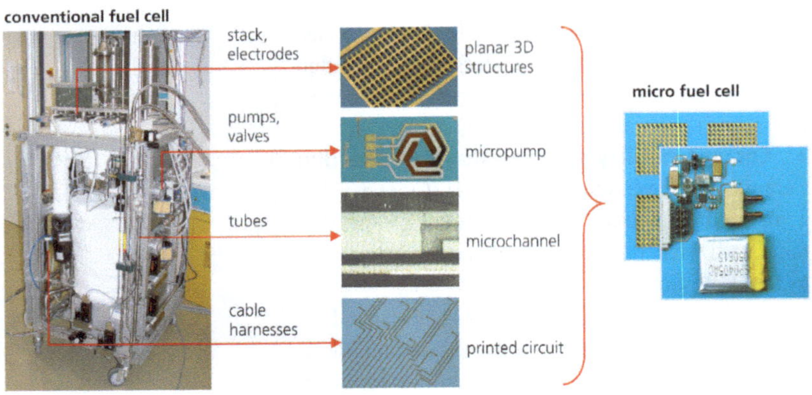

Multi-layer ceramics as an integration platform: Nearly all the process engineering components of a fuel cell system can be modelled by their micro-scale equivalents in LTCC ceramics. This enables complete, fully functional micro fuel cells to be constructed on the LTCC platform. Monolithic ceramic integration also increases reliability, as an LTCC system comprises far fewer components than a conventional system.

Diagram of the internal structure of an SOFC stack: Each repeating unit comprises a ceramic membrane electrode assembly (MEA) that is joined to a metallic interconnector with glass ceramic tapes. Several electron-conducting ceramic or metallic layers provide the electric contact between the MEA and the outside world. The metallic components are sealed off from one another with electrically insulating glass ceramic elements.

The production of pressure sensors in multilayer ceramic technology is another example.

In comparison with the manufacture of conventional ceramic pressure sensors, multilayer pressure sensors have several advantages that are associated with the type of multilayer technology applied. They include highly variable geometry requiring minimal processing effort, high integration density, a quasi-monolithic structure and the ability to integrate electronic circuits.

Trends

▶ **Ceramic high-temperature fuel cells.** Many hydrocarbons – from fossil or regenerative sources – can be very efficiently converted into electricity in ceramic high-temperature fuel cells (SOFC, solid oxide fuel cells) at very high operating temperatures of up to 850 °C. The central element of an SOFC is the stack, which is usually made up of planar individual cells. Known as membrane electrode assemblies (MEAs), these are built up using the classic methods of functional ceramics: a ceramic mass (slurry) is first prepared from the ceramic powder, then tape-cast to create a film. This ceramic layer is doped to make it stable and capable of conducting oxygen ions. It thus forms the membrane that separates the two combustion gases from one another. In the subsequent steps, this membrane is screen-printed with ceramic electrodes. After sintering, the ceramic MEA which is responsible for the electrochemical function of the fuel cell is ready for installation. In order to achieve the desired power density (kW-range) and operating voltage, several MEAs have to be switched in series or stacked to form

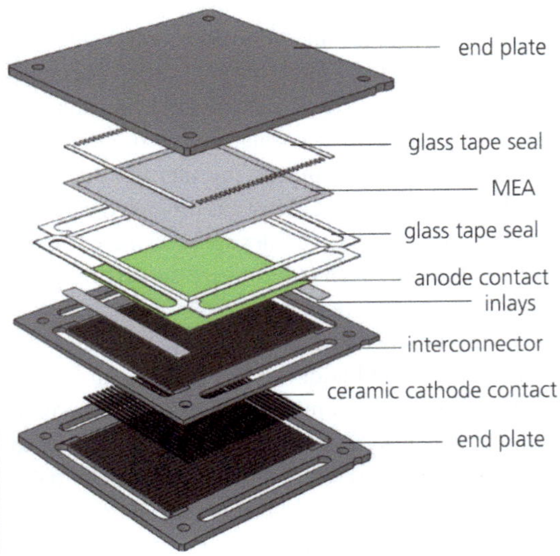

an SOFC-stack. This is done by placing the MEAs on a carrier (*interconnector*) which is usually made of metal. The mechanical link between the ceramic MEA and the metal interconnector is established by means of specially adapted, electrically insulating and gas-tight glass materials, which in turn are applied in the form of glass pastes or as a prestructured, stamped-out glass seal. SOFC-MEAs are provided with electric contacts via ceramic and metallic contact layers. SOFCs also contain large numbers of ceramic structural elements: for instance, the SOFC stack requires temperature-stable mechanical support and auxiliary elements to offset the ductile behaviour of many metallic construction materials at 850 °C, e. g. gaskets and spacers.

In addition to the electrochemically active SOFC stack, the periphery of the fuel cell system requires many other high-temperature components (reformers, afterburners, gas filters), which in turn necessitate ceramic materials. The reformers and burners, for instance, use foam ceramics and honeycomb grids made of silicon carbide, cordierite or aluminium oxide. Quite often, functional ceramic coatings are also applied (e. g. as a catalyst). Many metallic SOFC components are protected by ceramic barrier layers against chemical attacks by water vapour, hydrogen and CO in the combustion gas.

▶ **Multi-component powder injection moulding.** Components made from different types of ceramics or metal-ceramic hybrids are increasingly in demand not only in the medical field, for surgical instruments, but also for automobile parts. To produce material combinations of this kind, the multi-component injection moulding process used in plastics processing is being adapted for ceramic and metallic powders. In future, this process will enable the series production of complex multifunctional parts with highly diverse geometries. It will also reduce the number of production steps and simplify system integration. To manufacture such *composite material components*, the sintering and shrinkage behaviour of the materials used must be mutually compatible. In addition, the thermophysical properties must be adapted to ensure reliable operation in real-life conditions. Potential applications for material composites of this kind include precision forceps with an electrically conductive tip for use in minimally invasive surgery, and brake blocks for high-speed trains. But multifunctional components are also in growing demand for applications in automotive engineering and are currently being developed for manufacture using multi-component injection moulding. Ceramic spark plugs, ceramic cogwheels for fuel pumps and ceramic valve seats are a few examples.

There is now a trend towards new manufacturing concepts for the series production of complex material composite parts with thin functional layers.

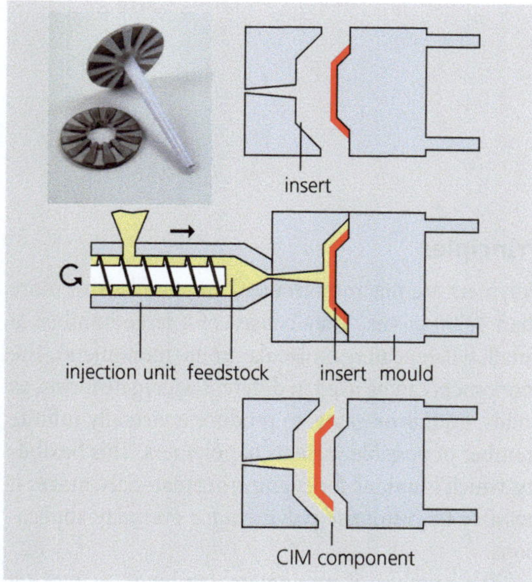

◼ Multi-component injection moulding. Right: The in-mould labeling of green tapes, a special form of multi-component injection moulding, involves inserting separately prefabricated pieces of tape (made of unsintered ceramics or powdered metal) into the mould and injecting the respective partner material behind or around them using a conventional injection moulding machine (centre). The inserts can be manufactured economically in large quantities by tape-casting with subsequent separation (below), while the injection molding process itself becomes considerably easier. Top left: Demonstrator thread guide green (with sprue) and sintered (front), component diameter 20 mm, produced by in-mould labelling of green tapes, in which a zirconium oxide ceramic material was injected behind a tape of powdered steel. Design: Rauschert Heinersdorf-Pressig GmbH

Prospects

Because of their unrivalled properties, advanced ceramics may be the solution to numerous key technological challenges of the future. A particularly crucial need can be identified in energy and environmental technology and in medical engineering, where technical breakthroughs can be achieved in the following areas thanks to advanced ceramics:

- Fuel cell technology for efficient, decentralised energy conversion with combined power and heat generation, and for mobile hybrid systems
- High-temperature materials for engines, block-type thermal power stations and turbines
- Membrane technologies for gas and liquid filtration (e. g. diesel particulate filters or wastewater treatment filters)
- Permeation membranes for gas separation, e. g. for more efficient combustion power stations
- New and improved ceramic materials for the storage of energy, e. g. in batteries and supercapacitors
- Large-scale integrated sensors and actuators for energy and environmental technology (e. g. monitoring pollutant emissions and setting optimum efficiency points)

PROF. DR. ALEXANDER MICHAELIS
Fraunhofer Institute for Ceramic Technologies and Systems IKTS, Dresden

Internet

- http://matse1.mse.uiuc.edu/ceramics/ceramics.html
- www.ceramics.org/aboutus/about_ceramics/index.aspx

Polymers

Related topics

- Process technologies
- Polymer electronics
- Composite materials
- Renewable materials
- Industrial biotechnology
- Food technology

Principles

Polymers are macromolecular compounds with more than 1,000 atoms. They consist of a large number of small, self-repeating molecular units (monomers). The monomers can be used in different aggregate forms, as solids, liquids or gases, to produce a virtually infinite number of possible variants of polymers. This flexibility, which is unique for engineering materials, makes it possible to customise polymers for everyday applications.

The number of monomers combined to make a polymer is known as the degree of polymerisation and is typically 1,000 to 100,000. The properties of these macromolecules hardly change when some of these monomers are removed or added. Most polymers consist of only one or two types of monomer, which already creates different structural variants.

In their simplest form, monomers are strung together like threads or chains. The image of a chain aptly describes the structure of the macromolecule because it implies a strong but flexible combination of elements. A chain-shaped macromolecule has a natural tendency to assume a haphazardly rolled-up shape like a ball of string – unless prevented from doing so by, for instance, other macromolecules. While chemical binding forces make the individual monomers join together to a polymer, physical binding forces create these “balls of string”. Physical binding is easily released by application of heat, solvents or mechanical force. Chemical binding forces, however, are much more difficult to break down. This often requires destroying the macromolecules completely.

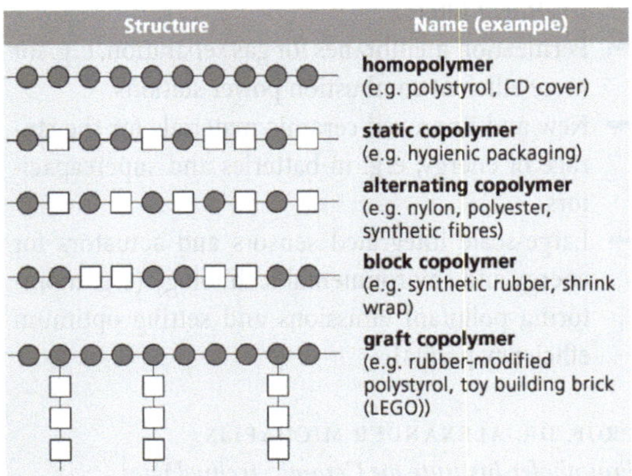

Structural variants of polymers: The circles and squares represent different monomers.

Polymers consist mainly of carbon-based organic compounds, sometimes in combination with groups containing oxygen, nitrogen or sulphur. However, they can also be inorganic in nature. Inorganic macromolecules that have industrial significance include silicons or polyphosphates. The organic molecules include natural polymers (biopolymers), chemically modified biopolymers or fully synthetic polymers. *Biopolymers* form the basis of all living organisms and no living creature can exist without natural macromolecular substances. Biopolymers include polypeptides (enzymes, hormones, silk, collagen), polysaccharides (cellulose, lignin, glycogen, chitin), polynucleotides (*DNA*, RNA) and polydienes (caoutchouc: natural rubber). Chemically modified biopolymers, also known as semisynthetic plastics, are obtained by chemical conversion of natural macromolecular substances, e.g. the cellulose derivative celluloid (nitrocellulose) and vulcanised fibre (parchment cellulose), artificial horn (casein plastic) or rubber (vulcanised caoutchouc).

The fully synthetic polymers are based on raw materials such as crude oil, natural gas and coal and, to a lesser extent, but with a slightly rising trend, on renewable resources.

Polymers can basically be divided into two classes: structural materials and functional polymers, also known as special effect and active substances. The structural materials include thermoplastics, thermosets, elastomers and foams. Functional polymers create special effects as active substances in their applications, such as colour transfer inhibitors in detergents, superabsorbers in diapers and additives for concrete.

Among the structural materials, a distinction is made between thermoplastics, duroplastics and elastomers.

▸ **Thermoplastics.** When heated, thermoplastics reversibly assume a plastic, i.e. deformable, state and retain their new shape after cooling. They consist of thread-shaped or only slightly branched chains of molecules.

- Standard plastics (mass plastics) comprise the thermoplasts polyvinyl chloride (PVC), polyethylene (PE), polypropylene (PP) and styrenic polymers (PS, SB) which are manufactured in large quantities. Typical products: shopping bags, drinking cups

- Engineering plastics have improved the mechanical, thermal and electrical properties of standard plastics. They can be exposed to complex stresses and can be used in many structurally demanding applications. They include the polyamides (PA), polycarbonate (PC), polyethylene terephthalate (PET), polybutylene terephthalate (PBT) and blends such as PBT/PC. Typical products: casings for computer monitors, vacuum cleaners, CDs, wheel caps
- High performance plastics are engineering plastics with outstanding properties, especially with regards to heat, although their processing is often more demanding. They include aramides, various polysulfones (PSU, PPSU, PAR, etc.), polyether ketones (PEK, PEEK), liquid crystal polymers (LCP) and fluoropolymers (PTFE). Typical product: microwave crockery

▸ **Thermosets.** Once moulded, these materials are no longer plastically deformable even when exposed to heat. They consist of macromolecules that are crosslinked in three dimensions. At normal temperatures they are hard or even brittle. Thermosets include phenol resins, polyester resins or epoxy resins. Typical product: electrical plugs.

▸ **Elastomers.** These (rubber-like) compounds are characterised by high elasticity over a wide range of temperatures. The molecular chains are cross-linked with each other. Typical examples are natural rubbers, polyurethanes and polyisobutylene. Typical product: tyres.

Most functional polymers contain structures with heteroatoms such as oxygen and nitrogen. As a result, they can carry positive and/or negative charges. This means they are often soluble in water.

▸ **Manufacture of polymers.** The chemical reactions leading from monomers to polymers are polymerisation, polycondensation and polyaddition. In the case of *polymerisation*, monomers that contain double bonds are converted by a chain reaction to long polymer chains. No re-arrangement or cleavage of molecular components takes place during this process, which proceeds in four phases:

- primary reaction: reaction that starts the chain reaction
- growth reaction: build-up of molecule chains
- chain transfer: branching of molecule chains
- chain termination: reactions that lead to the end of the chain reaction

Depending on the type of primary reaction, a distinction is made between ionic and radical polymerisation. The former is characterised by repeated linking of the monomers via positively charged ions (cationic polymerisation) or negatively charged ions (anionic polymerisation), and in the latter case by linking atoms and molecules by means of free, unpaired electrons (radicals).

In a *polycondensation* reaction, a low-molecular-weight compound is displaced from the monomers when they join together to form a macromolecule. The condensed compound is often water or an alcohol.

In *polyaddition*, no low-molecular-weight compounds are cleaved off and the monomers can join together directly. Polyadditions are behind many everyday applications. E.g. duroplastic epoxy resins are created through polyaddition of epoxy with hardening agents such as amines and acid hydrides. In the cross-linking reaction ("curing"), the epoxy ring is split open.

Applications

With an annual global production of more than 100 mio. t, structural materials play an important role in all areas of our daily life. The functional versatility and low weight of polymers make them ideal for producing many household goods.

▸ **Multilayer films.** Foods can be kept fresh for long periods when packaged in transparent, mechanically robust films with special barrier properties against oxygen, moisture and oils. The many different barrier properties are achieved by combining different plastics that are manufactured and processed in the form of thin (< 100 µm) and transparent multilayer films.

The market requirements of the film determine its chemical composition. In many cases, polyethylene (PE) or polypropylene (PP) are the basis of a multilayer film because of their high degree of flexibility, their

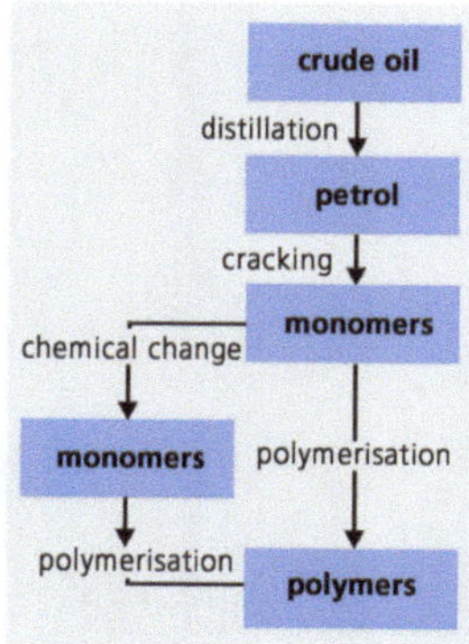

▪ Schematic process chain of plastic production

Behaviour	General structure	Appearance
thermoplastics	flexible linear chains	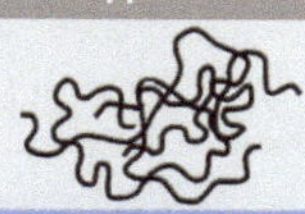
duroplastics	rigid three-dimensional network	 cross link
elastomer	linear chains with crosslinks	 cross link

▪ The commonest classes of polymers and their types of cross-linking barrier effect against water vapour and their sealability (i. e. bonding with other polymers by temperature and pressure). Thanks to these properties, the use of PE films as household cling film is a long-term success story. If the requirements are more exacting, however, a combination with polyamides is often chosen. PA has properties that PE and PP lack: it is tough and flexible at the same time, retains oxygen and carbon dioxide and keeps its shape well even at high temperatures. Depending on the application, barrier substances such as ethylene vinyl alcohol (EVOH), bonding agents, and adhesives that hold the different layers together, are also added.

One typical application for these composite materials is seen in the now highly popular dish-style packs used to present sausage and cheese slices in an attractive manner. The tough, thermoformed dish is made of a layer of polypropylene and polyamide joined by a bonding agent. The flexible, peelable and often re-sealable film cover consists of a combination of polyethylene and BOPA – a biaxially oriented polyamide – in which the molecular chains of the polymer are aligned in a highly organised manner, providing high tear resistance.

▶ **Manufacture**. Various methods allow to orient polymer molecules in a film. The orientation and stretch process can take place simultaneously or consecutively. All methods use an extruder to melt and transport the polyamide. The underlying principle is similar to that of a meat grinder. The extruder ends in a heated nozzle, whose specifications depend on the method used and the form of the film that is to be manufactured. A horizontally arranged nozzle produces a flat sheet, so-called cast film. A circular nozzle with a vertical haul-off unit "blows" a film directly into the form of a tube.

▶ **Superabsorbers**. These are crosslinked polymers made from acrylic acid that can absorb many times

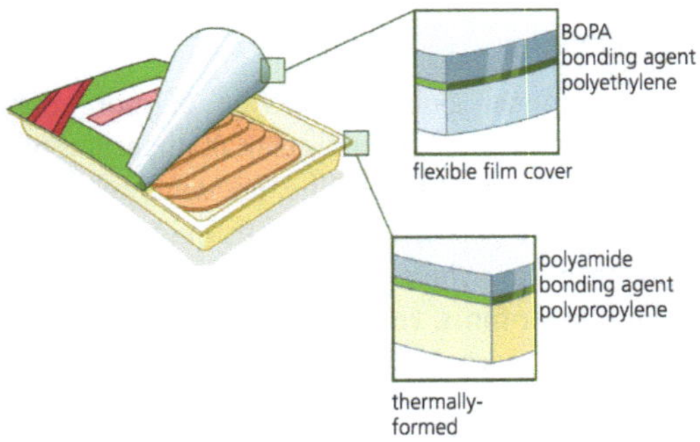

Multilayer films for food packaging: A multilayer film structure makes it possible to combine many different barrier properties. Top: Tear and perforation resistant thin biaxially oriented polyamide (BOPA) films with their longitudinally and transversely stretched, parallel oriented molecular chains. Bottom: Combining the good barrier properties of polyamides (PA) against oxygen with the good barrier properties of polyethylene (PE) or polypropylene (PP) against water vapour and moisture produces effective packaging systems even with low film thicknesses. Source: BASF

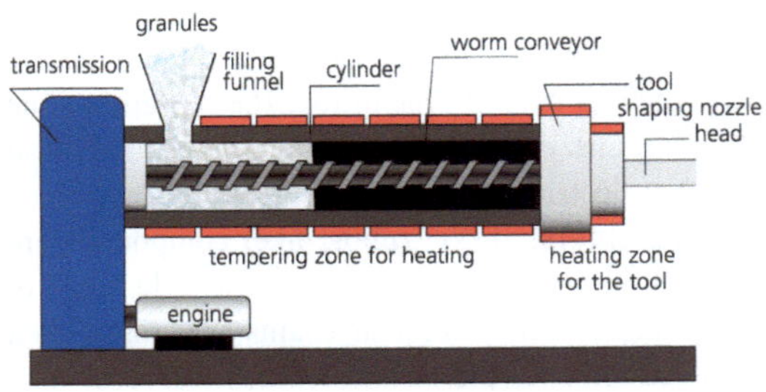

Extruder for processing polymers: The starting material (mainly polyethylene, polypropylene, polyvinyl chloride and polystyrene), usually available in the form of powder or granules, is poured into a feed hopper. The material is then heated and blended to produce a homogeneous mass. By means of a worm conveyer, the mass is pressed at the end of the extruder through a shaping nozzle head. Different products are obtained depending on the shape of the nozzle head.

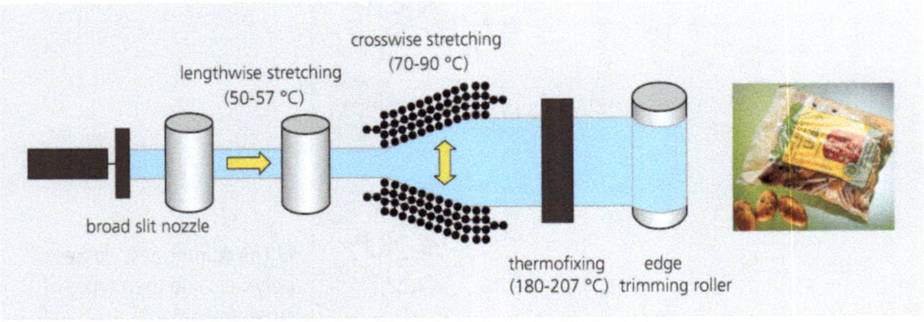

Manufacture of a step-by-step oriented cast film: In cast film extrusion, the melt exits through a broad slit nozzle. The film sheet obtained is then stretched lengthwise and crosswise to increase its strength. The improved mechanical properties allow the use of thinner films. The structure is then fixed by heating. Source: BASF their own weight in liquid. The granules, measuring only 0.1–0.8 mm in diameter, resemble crystals of table salt. The polymer granules do not dissolve in water, but rather absorb the liquid and swell. Superabsorbers can absorb 1,000 times their own weight in distilled water, 300 times their own weight in tap water or 50 times their own weight in body fluids such as urine. And this is how superabsorbers work: the individual acrylic acid molecules (monomers) are linked together by polymerisation to form a long chain. A downstream reaction with sodium hydroxide then produces negative charges on the chains that are neutralised by the positively charged sodium ions from the sodium hydroxide. On contact with water, the sodium ions dissolve in the water and the negatively charged strands of the polymer chain repel each other electrostatically. The polymer network swells and capillary forces draw up the water into the resulting cavities created between the molecules. In addition, an osmotic pressure develops that draws the water into the superabsorber.

Mass products of the hygiene industry account for 98% of the consumption of superabsorbers. The other, special applications are packaging, cable sheathing or auxiliary materials for agriculture in the form of geohumus used to increase the water storage capacity of soil. In industrialised countries, about 44 billion diapers are used annually. They consist of three layers. The first is a porous plastic film of polypropylene, which is in direct contact with the skin. Polypropylene is water repellent and does not store any moisture. Being porous, this film allows the urine to pass through it. Underneath this film we find a fabric made of cellulose incorporating the granular superabsorber. The *cellulose*, also known as fluff, absorbs the liquid and distributes it over the entire layer to allow as many absorber granules as possible to come into contact with the liquid and absorb it. The absorbent forces are so high that no liquid is released even if the baby slides along the floor with the diaper. The diaper's final layer is a water-impermeable polyethylene film.

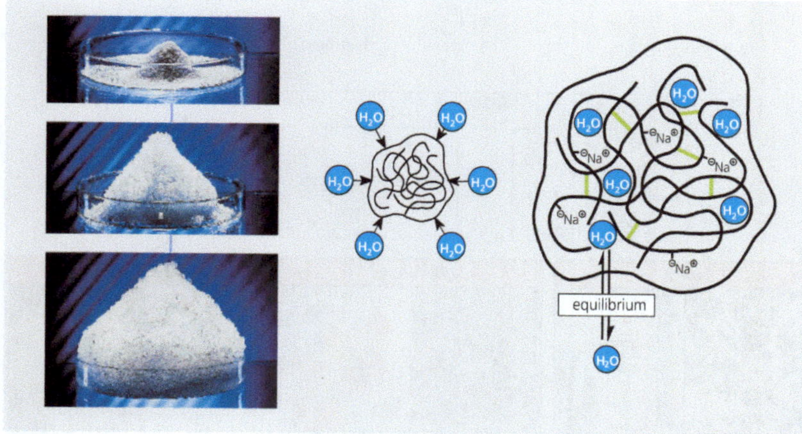

Superabsorbers: On contact with an aqueous liquid, they absorb up to 1000 times their own weight and convert it into a solid gel. Source: BASF

► **Additives for concrete.** *Concrete* is probably the single most important engineering material. About 7 billion m³ of concrete are used annually worldwide. A broad spectrum of characteristics and properties are required to cover the extensive range of applications for this material, e.g. in the transport sector (bridges) or as prefabricated concrete parts. These are achieved by adding small amounts of flow promoters such as

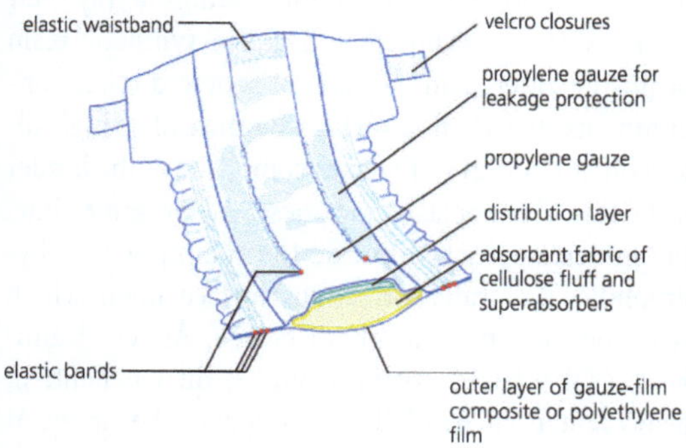

Structure of a baby's diaper with superabsorbers: It is both thinner and more absorbent than a conventional diaper. Even at elevated pressure, the absorbed liquid is not released. Source: BASF

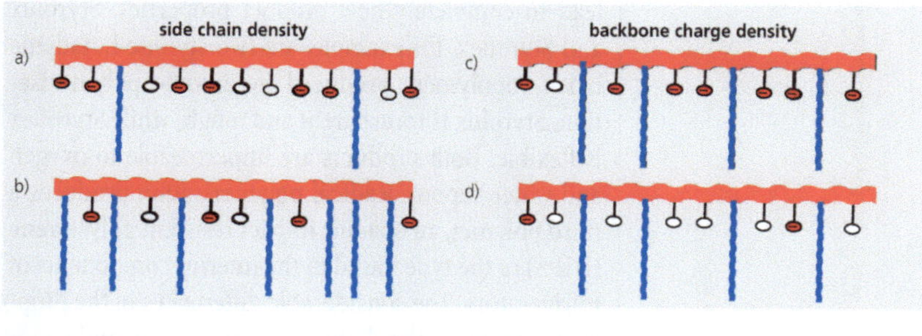

Structural variants of flow promoters for concrete: The comb-like architecture of flow promoters for concrete is tailor-made for the target application. Left: Increasing the number of side chains (a–b) leads to a bulky molecular structure that can stabilise the concrete for a longer time. Right: The same stabilisation effect is achieved when the charge density of the flow promoter is reduced (c–d). A stabilised concrete is used for ready-mix concrete. Flow promoters sketched in Figure a) or c) are used for precast concrete.

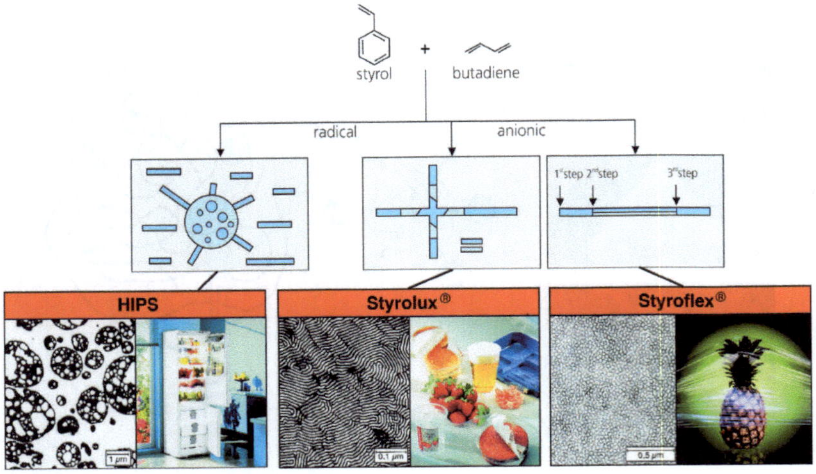

Product by process: Radical and anionic polymerisation lead to different supramolecular structures. The manufacturing process of the polymer made from styrene and butadiene determines whether the product is opaque or transparent. Source: BASF

those based on polycarboxylate ether. The desired properties of the concrete can be achieved by modifying the number of charges in the main chain and the number, length and distribution of the side chains.

Trends

Polymer development no longer involves the unspecific search for new polymers. Instead, polymers are tailor-made for specific applications. The control of structural hierarchies of nano- and microscale supramolecular structures are expected to generate new solutions.

▶ **Tailor-made property design**. Processing two defined components under different conditions can lead to completely new product properties. Styrolux and Styroflex, for example, are two styrene-butadiene block copolymers produced by anionic polymerisation. Styrolux is transparent and tough, while Styroflex is flexible. Both products are impermeable to oxygen and water vapour. Radical polymerisation produces a third polymer, an opaque, impact resistant polystyrene (HIPS) of the type found in the interior components of refrigerators. The considerable differences in the properties are due to the slightly differing micro- and nanoscale morphologies.

▶ **Nanoeffects**. Polymer foams such as expanded polystyrene (EPS, e.g. Styropor®) are widely used as insulating materials in construction. The insulating performance of the foam can be increased significantly by incorporating graphite in the sheets. Graphite scatters and absorbs most of the IR radiation (i.e. heat radiation) that conventional EPS allows to pass through. The particle size, shape and distribution of the graphite in the foam are key factors producing this effect. The insulating effect of a foam can even be increased still further if the diameter of the cavities is reduced to less than 100 nm. These tiny cavities now only have sufficient space for a single gas molecule and heat transport via groups of molecules is interrupted. Thus, the energy flux is further reduced, the insulation performance enhanced.

Polymers of one and the same type are subject to the same law: high flowability means reduced mechanical properties. Conversely, very tough types have much lower flowability. This long-established paradigm has recently been overthrown. Using a nanoparticulate additive, a polybutylene terephthalate has been created with unchanged mechanical properties but greatly improved flowability. This "high speed effect" hastens processing operations and reduces energy costs by 30–50%. It also enables the manufacture of highly intricate components.

Another approach to design new properties is the combination of different well-known materials. One example of the functional polymers is a novel binder (COL. 9) for facade coatings. It consists of a dispersion of organic plastic polymer particles in which nanoscale particles of silica, the basic constituent of glass and quartz, are embedded and evenly distributed. Thanks to this unique combination of elastic soft organic material and hard mineral, COL.9-based coatings combine the different advantages of conventional coating types. In contrast to brittle, mineral-based coatings, for example, the widely used synthetic resin-based dispersion paints are highly crack-resistant. But in summer, when dark house walls reach temperatures of 80 °C and more in the sun, these coatings betray their weakness: when exposed to heat the synthetic resin begins to soften, and particles of soot and other contaminants stick to their surface. Because of its high silica content, however, the nanocomposite of the binder lacks this thermoplastic tackiness. At the same time, the mineral particles provide the coating with a hydrophilic, i.e. water-attracting, surface upon which raindrops are immediately dispersed. As for cleanliness, COL.9-based coatings offer a dual benefit: in heavy rain, particles of dirt are simply washed away. At the same time, the thin film of water remaining when the rain has stopped dries extremely quickly. This prevents mould formation.

▶ **Phase change materials**. Phase change materials are used in building materials to effectively absorb day-time temperatures. These properties are brought

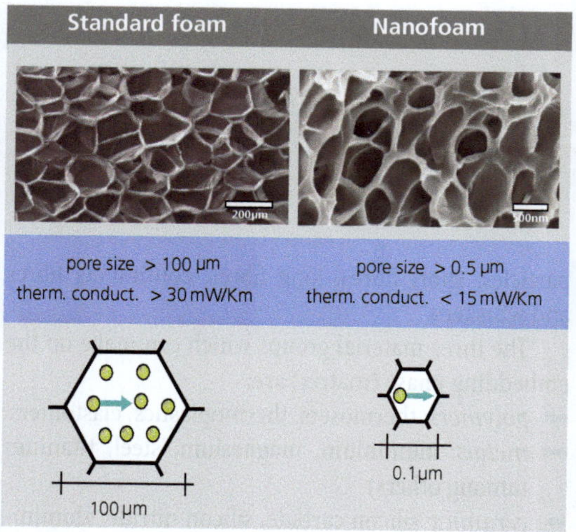

Functional principle of nanoporous foams: Standard foams have a pore size of 100 µm, which corresponds to 10 cells per millimetre. Nanofoams have a drastically reduced pore size of less than 0.5 µm, which corresponds to 2000 cells per millimetre. In nanofoams, the number molecules per cell is so low that the molecules do not "see" or transfer energy to each other. As a consequence, insulation efficiency is greatly improved. Source: BASF

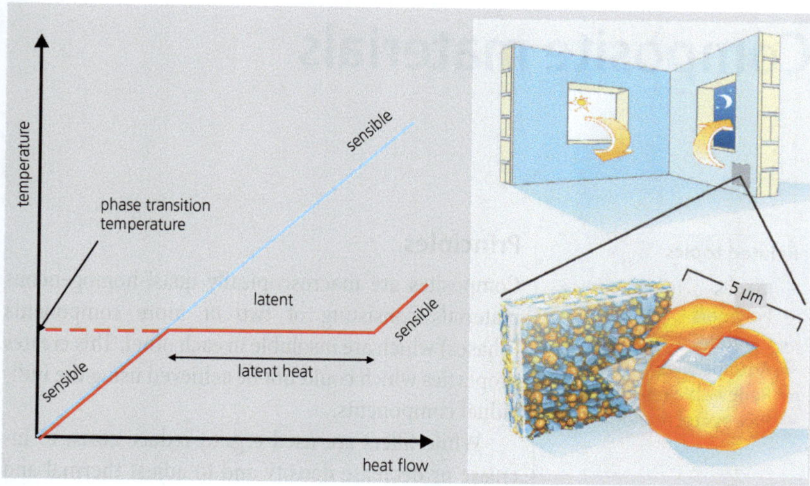

Latent heat is the amount of energy required by a substance when passing from one phase to the other – for example from ice to water or from water to steam. It is called latent, because it is not apparent in the form of temperature differences. This is why a melting ice cube removes large amounts of heat from the surrounding liquid and keeps a long drink cool for a prolonged period even in summer. Because melting 1 kg of ice at 0 °C requires a whopping 330 kJ – the same amount of energy could heat the cold water to about 80 °C. Right: The sun heats the room. At a temperature of 26 °C, the wax begins to melt and absorbs the excess heat. The wax is microencapsulated in a high-strength acrylic polymer shell and these capsules are embedded in a matrix of gypsum. Source: BASF

about by microencapsulated heat storers that use paraffin wax. This material acts as a thermal buffer: during the day, the wax heats up and liquefies, absorbing large amounts of energy. During the night, when the outside temperatures fall, the wax solidifies, releasing the energy. The amount of energy stored and the temperature at which the phase transitions occur depend greatly on the material. The length of the hydrocarbon molecule is the determining factor in paraffin waxes: paraffin with a chain length of 16 carbon atoms melts at about 20 °C, while chains with 18 atoms need 28 °C. Melting points between 6–60 °C can therefore be selected by choosing a different chain length. The challenge is to integrate the melting wax safely into building materials like wall plaster, mortar or gypsum boards. The process is called microencapsulation: tiny droplets of wax are enclosed in a virtually indestructible polymer capsule that can withstand even drilling and sawing.

Prospects

- Polymers will continue to stimulate future development trends in a number of industries, including automotive, packaging, construction and electrical sectors. They will increasingly supplant classical engineering materials, such as metals, as new properties (e. g. conductivity) are successfully introduced. Strong growth is particularly foreseen in medical and electronics applications, especially in connection with nanosized structures.
- Manufacturing plastics as bulk material will give way to custom designing of plastics for specific applications.
- A growing understanding of the effects of the supramolecular structures – and combining physical findings and phenomena with chemistry and biology – will open up a new range of applications for polymers in the years ahead.
- The combination of polymers and specific nanoscale structures will produce materials that play a key role in solving problems that must be faced in the years ahead, including resource conservation, reduction of energy consumption and generating power.

PROF. DR. FRANZ BRANDSTETTER
DR. JENS ASSMANN
BASF AG, Ludwigshafen

Internet

- www.e-polymers.org
- http://plc.cwru.edu/tutorial/enhanced/files/polymers/apps/apps.htm
- www.cem.msu.edu/~reusch/VirtualText/polymers.htm
- http://openlearn.open.ac.uk/mod/resource/view.php?id=196651&direct=1

Composite materials

Related topics

- Metals
- Polymers
- Ceramics
- Automobiles
- Renewable resources
- Process technologies

Principles

Composites are macroscopically quasi-homogeneous materials consisting of two or more components (phases) which are insoluble in each other. This creates properties which could not be achieved using the individual components.

While fillers are used e.g. to reduce costs, to increase or decrease density and to adjust thermal and electrical properties, reinforcing materials are used to improve the mechanical properties of a component. For a composite concept to be successful, the property to be optimised should differ by a factor of > 3 between the two phases, and each phase should constitute at least 10% of the weight.

Reinforcing components generally support loads, while embedding components provide the shape, ensure load application and transmission and also protect the reinforcing materials. Reinforcing components include fibres and whiskers (long monocrystals) as well as particles with an aspect ratio > 1. A distinction is made between composite materials reinforced with particles, short fibres, long fibres, continuous fibres and whiskers.

The three material groups which can make up the embedding phase (matrix) are:

- *polymers*: thermosets, thermoplastics, elastomers
- *metals*: aluminium, magnesium, steel, titanium (among others)
- *ceramics*: silicon carbide, silicon nitride, aluminium oxide, carbon (among others)

A distinction is consequently made between polymer compound materials (fibre plastic compounds), metal matrix compounds (MMCs) and ceramic matrix compounds (CMCs). Besides the performance of the matrix material and the reinforcing phase, the boundary layer between the reinforcing material and the matrix is essential for reinforcement. Depending on the combination of materials, adhesion is achieved with different levels of mechanical, chemical and physical bonding between the reinforcing components and the matrix. When combining the two components the aim is therefore to achieve complete wetting of the reinforcing components with the matrix.

The various reinforcing phases for composites are constructed in different ways. As a result, reinforcing fibres can be classified according to different criteria, for example according to their organic or inorganic structure, whether they are artificial or natural, or their high strength or rigidity. *Glass fibres* are generally produced by melt extrusion and stretching, and are subsequently coated with a protective sizing. *Carbon fibres* are produced by the thermal decomposition (solid-phase pyrolysis) of organic starting materials. Both types of fibres display homogeneous properties and generally have a diameter of 7–25 µm. The fibres are flexible and are bundled together into rovings – consisting of up to several thousand individual fibres – for efficient processing. The respective matrix materials are often reinforced by means of roving which have previously been processed to form textile semi-finished materials such as fabrics, non-crimp fabrics, fleece mats, etc.

▶ **Composites.** Composites can be classified as high-performance fibre compounds, which have continuous and aligned reinforcement, and as quasi-homogene-

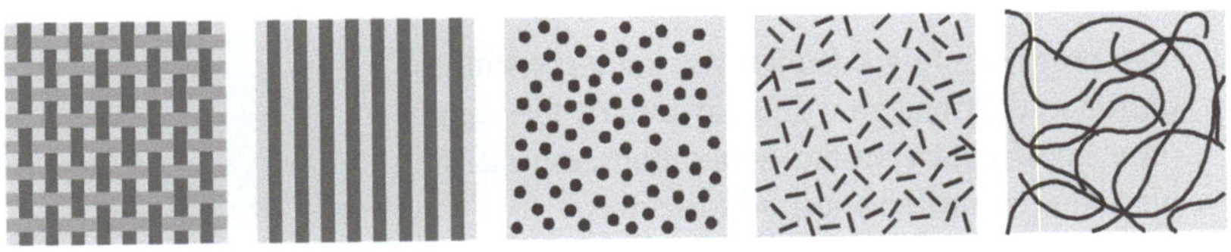

Schematic representation of composites reinforced with fabric, unidirectional long fibres, particles, short fibres (or whiskers) and long fibres

Material	Modulus of elasticity [GPa]	Density [g/cm³]	Specific modulus of elasticity [GPa cm³/g]
steel	210	7.85	26.8
aluminium	70	2.70	25.9
CFK	135	1.60	84.3

Specific modulus of elasticity of materials: The modulus of elasticity is a measurement of a material's rigidity. The specific modulus of elasticity refers to the density of the material. The advantage of fibre plastic compounds in terms of their high rigidity and low component weight is therefore clear.

ous fibre plastic compounds with short or long fibre reinforcement and a random fibre orientation. High-performance fibre composites comprise high-value materials such as carbon fibre reinforced polymers (CFRP), aramide fibre and also glass fibre reinforced polymers (GFRP), embedded in epoxy resins, phenol resins, polyetheretherketones, polysulfones and polyether imides. Before impregnation with the matrix material the reinforcement fibres are generally processed to form textile semi finished materials such as fabrics non-crips and mats.

The directional dependency (anisotropy) of the construction material's properties, along with suitable construction methods, means that these materials can be used to achieve a significant weight saving in the structure of components or systems.

It is therefore necessary to consider different material behaviour parallel to and perpendicular to the fibre direction and in the single fibre layer in textile structures in order to compensate for the different degrees of thermal expansion and prevent distortion of the component. Converting these materials into components by way of a suitable production process also represents a particular challenge. Especially where high-performance fibre composites are concerned, conventional production processes have so far failed to achieve adequate volumes for mass production, for example in automobile manufacturing.

The second group of short and long fibre reinforced plastics is most often used in semi-structural components. Unlike high-performance fibre compounds, quasi-homogeneous fibre plastic compounds can be mass-produced using processes such as injection moulding, extrusion and compression moulding. Thanks to the significantly higher productivity of these processes, cost-efficient starting materials can be selected. For this reason natural fibres are most frequently used for reinforcement, with a thermoplastic matrix of polypropylene or polyamide, or a thermosetting matrix of unsaturated polyester resin, vinyl ester resins and polyurethanes.

▸ **Metal matrix compounds (MMC).** These materials consist of a metal alloy and a reinforcing component. The combination of the two materials enables the mechanical and thermal properties in particular to be adjusted. The combination of metals with ceramics produces outstanding new physical properties such as high thermal conductivity combined with a low coefficient of thermal expansion (CTE). Aluminium alloys, for example, have high thermal expansion coefficients, which can be reduced appreciably by adding ceramic components with low CTEs. The density of this type of compound does not increase significantly, remaining under 3 g/cm³ (pure aluminium 2.7 g/cm³), while the lower thermal expansion coefficient leads to less distortion at high temperatures and consequently to a reduction of the internal stress. High thermal stability, hard wearing properties and corrosion resistance are particularly important mechanical properties. High thermal stability is a key factor in lattice dislocation stability at high temperatures, preventing deformation of the crystal lattice at high temperatures along with the creep effect.

Depending on the geometry and the distribution of reinforcing components, MMCs can be divided into different categories:

- Particle-reinforced compounds, which comprise reinforcing particles in a metal matrix: These particles generally consist of ceramic materials such as silicon carbide or aluminium oxide. The particle volume content can vary between 10–95%, although the proportion is around 50% in most applications. There is a high filler content in applications such as hard alloy cutting inserts.
- Fibre-reinforced MMCs which contain short or continuous fibres: while short fibres are randomly oriented, continuous fibre reinforced MMCs generally contain unidirectional continuous ceramic or metal fibres. The use of fibre preforms made of aluminium oxide or carbon fibres which are infiltrated by gas pressure infiltration or pressure casting is state of the art. The liquid metal wets the reinforcing structure under mechanical or gas pressure. Niche processes such as *compound ex-*

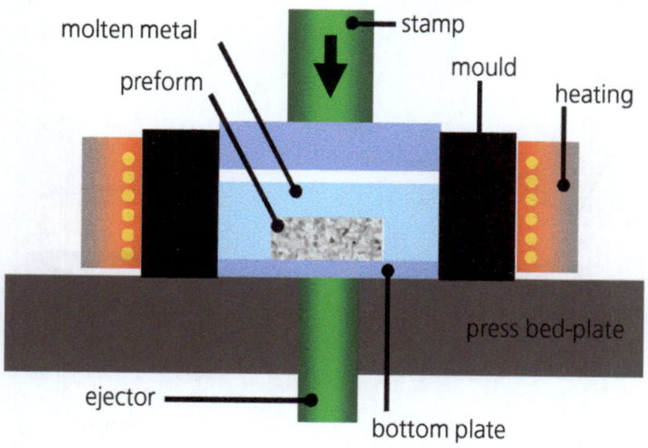

Squeeze casting: Instead of using a compressed gas, in squeeze casting the infiltration pressure is applied mechanically. The preform is placed in the reaction chamber, impregnated with liquid metal and subsequently infiltrated using ram pressure.

trusion of continuous fibre-reinforced profiles can also be applied.

The production of MMCs by *squeeze casting* is one of the most cost-efficient processes. The preheated preform is positioned in a mould, covered with molten metal and subsequently infiltrated under high mechanical pressure. Where gas pressure is applied, setting occurs much more slowly, allowing other mechanical properties to be achieved.

▶ **Ceramic matrix compounds (CMC).** CMC materials are high-temperature materials which can be used at higher temperatures than MMCs. In general, ceramics have high compression strength, high chemical and corrosion resistance and relatively low thermal expansion coefficients. Their exceptionally brittle behaviour at high temperatures, however, greatly restricts their use. This disadvantage can be overcome by applying fibre reinforcements which span possible fracture points and prevent or slow down crack propagation. The breaking elongation of the compound can be significantly increased. The type of load transmission and the avoidance of stress peaks make the interchange between fibre and matrix a key issue in improving the damage tolerance behaviour. Ceramic matrices are conventionally obtained from slip masses or by polymer pyrolysis. The *slip method* includes the infiltration of a fibre preform. In industrial ceramic production the components are finely ground in drum mills. After infiltration with a reinforcing component, the material is dispersed by the addition of water and/or alcohol with organic binding agents which are then mostly removed by filter pressing. The remaining filter cake is dried and may then be converted to a green compact, e. g. in a hot press. By contrast, it is also possible to obtain a ceramic material by pyrolysis of a polymeric starting material. As has already been described for the slip method, a fibre preform is infiltrated and subsequently transformed into a ceramic by heat treatment. As a result of the volume change in the conversion of the polymer into a ceramic, cracks form in the CMC material which can be infiltrated with metal alloys in a subsequent step. This process is state of the art in the manufacture of C/C-SiC brake discs.

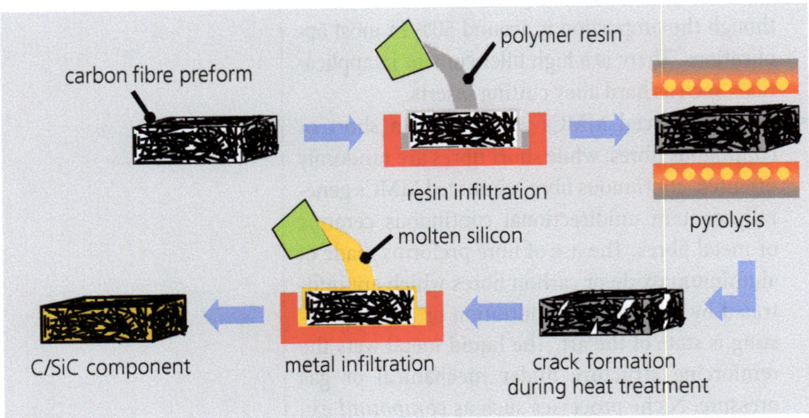

In the manufacture of C/SiC materials a carbon fibre based preform is first infiltrated with a polymer resin, which is then converted to a ceramic carbon matrix (pyrolysis). Thermal stresses lead to the formation of cracks as the material cools. The cracks are subsequently infiltrated with liquid silicon, which reacts with the matrix material to form silicon carbide (SiC).

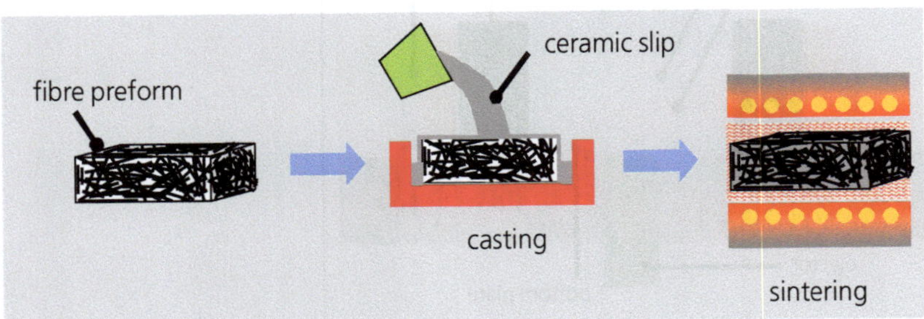

In the slip method a reinforcing preform (such as the fibre preform shown above) is infiltrated with ceramic slip and subsequently sintered using technical ceramics as a matrix material. One simple material system which can function without sintering is reinforced concrete, in which the matrix phase sets when exposed to air.

Applications

▶ **Composites.** Due to their specific property profiles, fibre plastic compounds exhibit significant advantages over homogeneous materials. These include good mechanical properties at relatively low weight (weight-specific properties) and also physical properties such as X-ray transparency (CFRP), electromagnetic screening (CFRP), adjustable thermal expansion within a defined temperature range, etc., and chemical properties such as corrosion resistance or stability on exposure to a range of media, depending on the selected matrix. This chemical stability has led to increasing application of FRP in the construction of chemical plants and in service pipes used in the construction industry.

▶ **Lightweight construction.** The main area of application for fibre composite materials, however, is lightweight construction with integrated functions. Fibre composites are most widely used in the aerospace industry, as well as increasingly in automobiles and rail transport.

The most common aim is to reduce the moving mass, which often leads to the application of these ma-

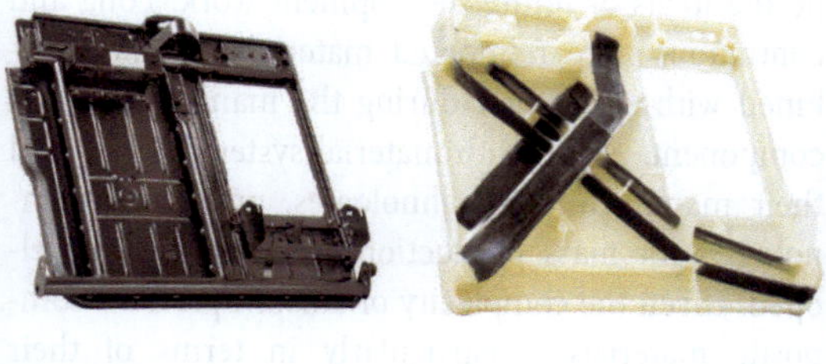

Backrest structure of a rear seatback. Left: The structure is the conventional steel component produced by joining together and subsequently painting various differently shaped semi-finished products. Right: The structure is the long fibre reinforced lightweight solution with local continuous fibre reinforcement in the load paths (black plastic matrix) and integrated functions, which is produced in a single manufacturing step. Source: Esoro/Weber

terials in energy plant construction, for example to increase the efficiency of wind plants or the speed and precision of automation technology.

In the transport sector, structural components are differentiated so that the anisotropy of the FRP can be exploited to achieve the necessary strength with the minimum weight and quantity of material. In road vehicles such components are incorporated in the primary structure of the chassis, crash boxes and absorbers, bumper crossbeams and shear field structures such as the vehicle roof. The aviation industry uses the highest proportion of fibre composites, including for the wing segments, aircraft floor and fuselage segments.

As well as the approx. 40 wt.% saving compared with metals, long fibre reinforced structural parts make it possible to completely integrate functional components, such as headrests, within a single processing step. This renders numerous secondary processing steps and joining processes unnecessary. The additional continuous fibre reinforcement positioned locally in the load paths means that crash loads can be withstood and operating loads absorbed. In addition the component can be mass-produced by extrusion in a commercially viable process.

A new development is the use of fibre-composite structures in safety parts such as car wheel rims. Lightweight construction has a great effect on the unsprung masses of an automobile. What's more, the high elongation at fracture compared with aluminium increases comfort and results in damage-tolerant behaviour, while press forming provides greater design freedom.

► **High-temperature materials.** The applications of metal matrix compounds (MMC) focus on the elevated temperature range and exploit the high temperature stability and wear resistance of the materials. Light metal wrought alloys based on aluminium, magnesium and titanium are generally used as matrices.

A significant increase in the efficiency of vehicle engines can be achieved by using MMCs in the pistons. MMC pistons, together with reinforced piston cavities, resist higher combustion pressures and temperatures, which means that these combustion parameters can be increased. *Diesel engines* in particular, as compression ignition systems become significantly more efficient. To match the pistons, reinforced cylinder walls are being mass-produced with the help of aluminium-infiltrated ceramic preforms. Their task is to improve the friction ratio between the cylinder wall and the pistons moving along it. In both cases, the metal matrix and the reinforcing ceramic phase form a three-dimensional interpenetrating network which creates exceptional properties.

In addition highly-filled composite materials with ceramic contents of 90% and more – referred to as hard metals – are state-of-the-art in metal machining.

► **Very high temperature materials.** At very high temperatures ceramic matrix compounds (CMC) can be used. An important area of application for CMCs is

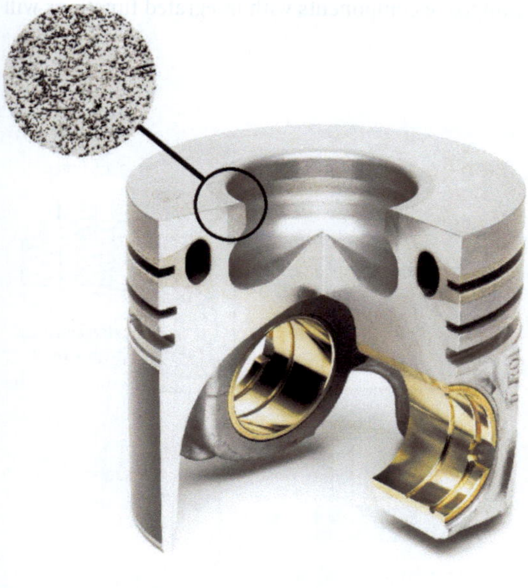

Diesel engine pistons with reinforced combustion chamber cavity: The ceramic reinforcing fibres can be seen in the micrograph as dark regions within the surrounding aluminium matrix. Commercially-used fibres consist mainly of aluminium oxide and have an average length of around 500 µm. Source: MAHLE, Germany

C/SiC sports car brake disc: The brake disc is screwed onto a steel carrier ring which connects it to the underbody.

in brake discs, particularly for high-performance cars and aircraft undercarriages. CMC brake discs are significantly lighter than conventional types and improve brake performance at higher temperatures. Here, too, unsprung masses in the underbody can be reduced, significantly improving the driving comfort. One disadvantage is that the brakes must be made of several joined parts to ensure connection to the underbody. What's more, the complex manufacturing technology and the low volumes mean that prices are very high and are not compensated by the longer life-span as compared with conventional brake discs.

CMCs are also used for the thermal insulation of reentry vehicles such as space shuttles, which are exposed to temperatures of around 1800 °C on reentering the earth's atmosphere. Applications also exist in the thermal insulation of combustion engines.

Trends

▶ **Composites.** The significant increase in energy costs around the globe, together with the need to find sustainable solutions to save resources and limit carbon dioxide emissions, means that lightweight construction, and particularly the use of composites, is becoming increasingly important. The further development of suitable processing and material technologies for the manufacture of high-performance fibre composite components with integrated functions will be the focus of future development work. Long and continuous fibre reinforced materials will be combined with each other during the manufacture of a component. Along with material systems, fibres and their manufacturing technologies, processing technologies for mass production are also being developed. Given the complexity of anisotropic fibre composite materials – particularly in terms of their changeable values over the operational temperature range – and the failure criteria dependent on the structure and processing, the methods, materials and processes must all be closely investigated. This includes the numeric simulation of anisotropic materials, the influence of processing on the composite and the durability of the fibre composite component.

▶ **Direct processes.** In the development of new processing technologies there is a clear trend towards direct processes. Starting from the original components, such as plastics, additives and glass fibres, the part is produced in a single processing step, eliminating the manufacture of semi-finished products and consequently the costly (intermediate) steps.

In the development of new high-performance composite components the trend is towards "smart materials". Special piezoceramic fibres can be integrated into this type of component. The fibres can then be electrically actuated, during which they expand or contract, functioning as actuators. This means that within certain limitations the local rigidity and/or

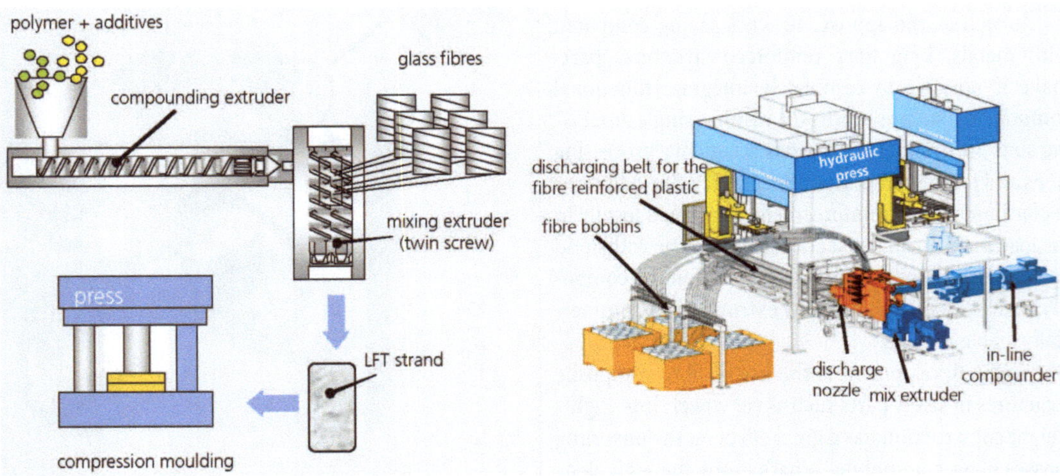

Direct process: One-step processes integrate material compounding and reinforcement in the moulding process, without cooling the intermediate product. The polymer is first melted, mixed with additives and then prepared for the integration of fibres. In a vertical mix extruder the reinforcing fibres are introduced into the hot melt and homogenised. The long fibre reinforced hot melt is then transferred directly into the press to form a component. These technologies save energy and permit the flexible combination of materials according to the product requirements.

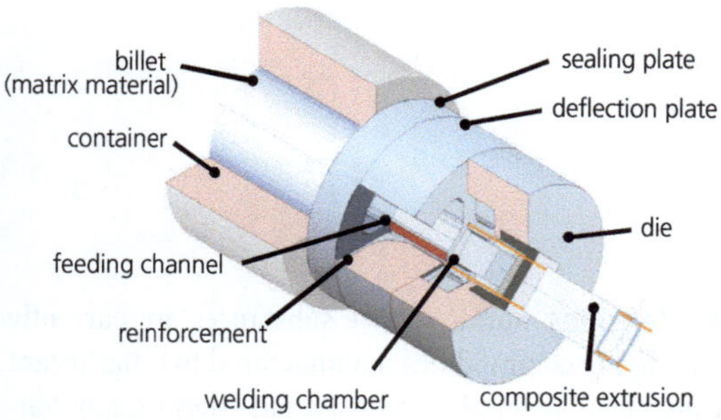

Composite rod extrusion: This process enables industrial-scale manufacture of unidirectional continuous fibre reinforced extrusion profiles. The process makes use of conventional rod extrusion with modified mould packages. Reinforcing material is added during extrusion and combined with the matrix material to form a composite extrusion profile.

shape of the component can be altered very quickly. Alternatively, *piezoceramic fibres* can be used as *sensors*, because when the component is distorted they produce electrical signals which can be measured and evaluated. Used in combination these sensors and actuators can change the properties of the component, for example to suppress or attenuate vibration. Damage detection, which is more complicated in fibre composite materials, is also a widening area of application, providing very early detection of fibre breakage or delamination which is not visible externally.

▶ **Metal matrix compounds (MMCs).** MMCs are already being used in drive technology, where the properties described above are an important requirement. However, more research is needed before they can be used as structural materials. For the efficient production of these materials, new, more cost-efficient processes are required than gas pressure and die casting, which are niche techniques. Exploitation of the outstanding physical properties of MMCs could open the door to further areas of application.

The widespread use of long fibre reinforced metals can only be achieved where cost-efficient processing technologies are available. Up to now they have chiefly been used in space technology – for example boron fibre reinforced aluminium – and in aviation, where higher material costs can be justified economically.

▶ **Ceramic matrix compounds (CMCs).** Research in the field of ceramic matrix compounds aims to further improve the fibre materials. A significant limitation to the temperature stability of these materials is the tendency of the fibres to oxidise in air at high temperatures, losing their rigidity and strength. If this problem were solved, application temperatures from approx. 1800 °C could be achieved. As an alternative to the conventional slip method, research is being conducted into processes for preform infiltration using gas-like components. Carbon fibre preforms are infiltrated by means of chemical gas phase deposition, whereby a chemical reaction causes the deposition of a solid component, such as silicon carbide, from the gas phase on the heated surface of the preform.

In principle CMC materials based purely on carbon could be used at temperatures of up to 2000 °C. However, beyond short-term exposure, application in oxidising atmospheres is not possible. For this reason, research is concentrated on improving oxidation protection systems. The low lateral strength, i. e. the low strength between parallel fibres of carbon fibre reinforced carbons, could be overcome by using three-dimensional fibre architecture concepts.

Prospects

- The proportion of composite material in industrial applications, pioneered by the aviation industry, has risen dramatically in the last few years. New aircraft concepts incorporate more than 50% composite material.
- Along with fibre plastic compounds, MMC materials are also being increasingly used in lightweight construction. Rod extrusion technology already makes it possible to use partially reinforced extrusion profiles in light framework structures. Despite these successes, further obstacles exist to the widespread use of composite materials, which represent challenges for the future and are already the focus of numerous research projects.
- The development of new production processes will permit a higher degree of automation, higher reproducibility and consequently the large-scale production of technically reliable composite components. Alongside the simulation of multiphase materials, the simulation of processing is also important in order to predict the properties of the material in the component.
- The broad application of hybrid material concepts inevitably requires suitable joining technologies.

PROF. DR. PETER ELSNER
PROF. DR. FRANK HENNING
Fraunhofer Institute for Chemical Technology ICT, Pfinztal

DR. KAY ANDRÉ WEIDENMANN
University of Karlsruhe

Internet

- www.ict.fraunhofer.de/EN/
- www.iwk1.uni-karlsruhe.de/index.php?id=96&L=1
- www.ikv.rwth-aachen.de/english/htm
- http://composite.about.com/
- www.science.org.au/nova/059/059key.htm

Renewable resources

Related topics

- Polymers
- Composite materials
- Industrial biotechnology
- Building materials
- Bioenergy
- Indoor climate

Principles

Since the 1990s, a new awareness of the finite nature of fossil resources and the need to diversify raw material resources has led to a sharp increase in research and development in the field of renewable resources in Europe and North America. There has consequently been a significant rise in the cultivation of plants that are not used by the food or feed industry but in energetic or industrial processes. In 2007, over 2% of the world's agricultural land was being used to grow renewable raw materials.

Wood, vegetable oils, cotton and other plant fibers, starch, sugar, and fuel crops are now the most important agricultural non-food commodities. Markets have also been established for cosmetic ingredients, herbal remedies and vegetable colors.

In principle, materials from renewable resources can be utilized in a cascading system: after their useful life, they can act as climate-neutral energy resources to produce power either through incineration or through fermentation in a *biogas plant*. This process basically emits only the same amount of carbon dioxide that the plants had previously assimilated while growing. From this point of view, heat and electricity thus generated from renewable resources do not pollute the earth's atmosphere.

Applications

Renewable raw materials have been used in *wood processing* and in the paper industry for a long time. Today, renewables are also moving into sectors that were previously dominated by petroleum-based chemicals.

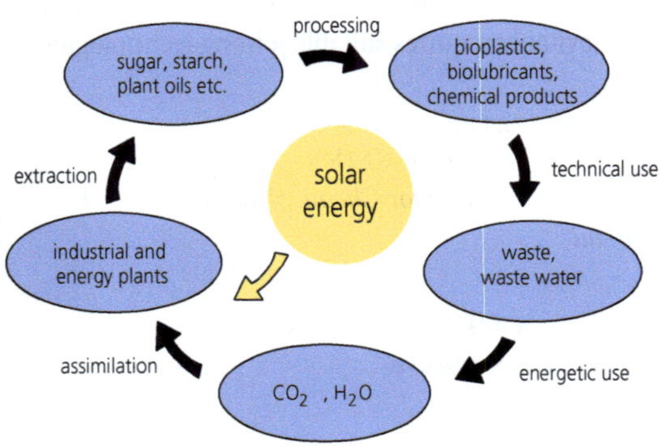

Cycle of renewable resources: Nature assimilates resources from water, CO_2 and solar energy. Various synthetic stages are used by humans, culminating in re-assimilation after decomposition or incineration of these products.

► **Tensides.** Surface-active substances are currently among the commodities manufactured to a significant extent from renewable raw materials. They usually consist of amphiphilic substances with both fat- and water-soluble components. Today, not only palm and coconut oils, but also sugar and amino acids, are processed on a large scale to produce surfactants. A special characteristic of sugar tensides such as alkyl polyglucosides and glucose fatty acid esters is their cutaneous tolerance. They are therefore used mainly in household cleaners or in cosmetic detergents and cleaning agents.

► **Biolubricants and biohydraulic oils.** The hydrocarbons contained in mineral oil lubricants and hydraulic fluids are eco-toxic, presenting significant environmental hazards when released. Rapidly biodegradable lubricants and hydraulic oils, on the other hand, cause little pollution if they enter the environment as totally expendable lubricants or due to leakage.

Lubricants are generally made from a mixture of base oils. Their final formulation is achieved by adding other substances (additives). The purpose of these additives, e.g. antioxidants, non-ferrous metal deactivators, wear protection products and corrosion inhibitors, is to improve the chemical and tribological properties of the lubricant.

Biohydraulic oils can be divided into two classes depending on requirements and the necessary technical performance. *HETG oils (hydraulic oil environmental triglycerides)* consist of refined rapeseed and/or other vegetable oils which are enriched by various additives to improve their performance and durability. HETG oils are cheap, easy to produce, excellent lubricants, and environmentally sustainable in the event of damage. They can be used at temperatures ranging from −20 to +70 °C, their only limitations being cold temperatures in severe winters or high loads in construction machinery. These limitations do not apply to HEES oils (hydraulic oil environmental synthetic esters). Also produced from vegetable oils, these fully synthetic, saturated products are completely esterified during the manufacturing process. Although the operating expenses are reflected in the price, this drawback is offset by a much longer service life.

Biolubricants are also used as engine oils, gear oils, lubricants, cooling lubricants, grinding oils, insulation oils and corrosion protection oils. In Europe, the sales volume of biolubricants is currently around 100,000 t/a. In the long run, their environmental and technical advantages will enable them to capture a 90% share of the total market for lubricants.

▶ **Building materials.** In addition to wood as one of the main construction materials, a wide range of construction products is available for today's environmentally conscious builders. Materials range from natural mineral fibrous insulating material made of plant fibres or sheep's wool, to modern floor coverings made of linseed oils, textiles, paints, transparent inks and plant-based varnishes.

The physiologically harmless properties of natural materials are a strong incentive to use them, as they cater to people's comfort, and to antiallergenic living. These natural fibre insulation products are capable of absorbing large amounts of water vapour, thus regulating indoor humidity. The insulating value of natural fibre insulation products is equivalent to that of synthetic insulation materials. They offer the additional special advantage of heat protection in summer.

Wood fibre insulation boards are used as interior insulation, for rafter und roof insulation, as flooring bodies and as plaster base plates.

Sheep's wool insulation materials are made partly from virgin wool and partly from recycled wool. To make them resistant to moth infestation and to minimise the risk of fire, these fibres are normally treated with boron compounds after washing and degreasing. Sheep's wool fleeces exhibit thermal insulation values similar to those of natural fibre insulation materials.

▶ **Bioplastics.** These products are made from carbohydrates such as starch, cellulose and sugar, and also from lignin and plant oils. They have properties similar to those of conventional plastics, and can therefore replace these in many fields of application. In addition, their service life can be adjusted to the specific application, varying from non-durable compostable food packaging to durable construction materials. As with all other plastics, different materials are produced depending on the commodity, the recipe and the process: transparent films, *textile fibres*, and stable thick-walled moulded paddings or foams, which exhibit either thermoplastic or duroplastic behaviour. Processing procedures for bioplastics are basically the same as for conventional plastics.

Three fundamental principles are currently in use for the manufacture of bioplastics. Numerous variants also exist, in which several of these procedures are combined:

- Physical processes: Natural polymers such as starch can be plastified in *extruders* with the aid of appropriate additives. To improve the characteristics of the material, the thermoplastic starch is then blended or compounded with other biodegradable polymers. Thermosetting packaging chips and moulded parts are manufactured in a similar way, but with the addition of water and subsequent foaming. This results in very lightweight products with an air content of about 90%.
- Chemical modifications: Reactive molecule components such as double bonds permit selective functionalisation. One example is the conversion of unsaturated fatty acids such as oleic acid or linoleic acid into plant oil polyols. These polyols can be transformed into bioplastics with the instruments of classical polymer chemistry. Not only resins, but also functional polymers such as polyurethanes or polyesters are produced in this way.
- *Microbiological processes*: Most enzymatic or fermentation processes utilise sugar and other carbohydrates as a source. Bacteria or enzymes break down the sugar in *bioreactors* and metabolise it to produce suitable monomers. This yields intermediates such as propane-1,3-diol, lactic acid or hydroxyl fatty acids, which generate the corresponding polyester. Transparent polylactic acid (PLA), for example, currently plays a prominent role in industrial biotechnology. Where its properties are concerned, it ranks with traditional mass plastics such as polyethylene or polypropylene. The procedures and facilities of the plastics industry are fully applied to PLA.

One notable use of bioplastics is for short-lived food grade packaging and garbage bags, as well as for cater-

Products from Renewables	Renewable Raw Materials used
building materials	wood, natural fibres, linseed oil etc.
surfactants, detergents and cleaning	fatty acids, fatty alcohols from plant oils, sugar derivates
lubricants and hydraulic oils	fatty acid esters from rape or sunflower oil
biopolymers for packaging	starch, sugar, plant oils or cellulose
construction materials	natural fibres, biopolymers from starch, sugar, plant oils or cellulose
polyurethanes and foam plastics	sugar, vegetable oils
paintings, coatings	linseed or tall oil, dye plants
adhesives	fatty acids
textiles	cotton, cellulose fibers such as rayon, modal or lyocell
chemical intermediates	sugar, starch and vegetable oils
cosmetics	natural fragrances and flavourings
pharmaceuticals	pharmaceutical plants

◘ Main areas of application for renewable resources

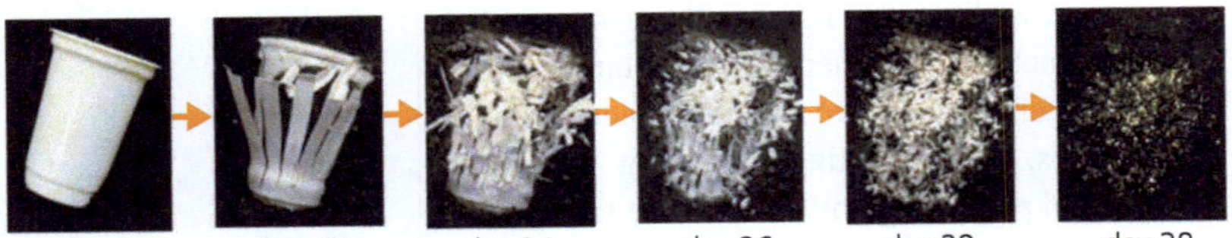

Biodegradation of bioplastics: In the composting plant, a cup is completely decomposed by microorganisms within about 5 weeks. Source: NatureWorks LLC

ing articles. The material decomposes with the aid of enzymes formed by microorganisms such as bacteria or fungi. However, bioplastics can be recovered by thermal recycling or by biochemical utilisation, for example in a *biogas plant*. Biodegradability is an advantage particularly in horticulture and landscaping. Biodegradable mulch films and binding twine need not be painstakingly removed, but instead naturally dissolve after a certain period of time.

Medical scientists appreciate the reabsorption of bioplastics. Fractures are stabilised with the aid of screws, nails or plates made from polylactic acid. The human organism metabolises these implants precisely when they have fulfilled their purpose, making it unnecessary for implants to be surgically removed.

Degradable sutures and substance depots have also been in successful use for a long time.

▶ **Natural fibre reinforced materials.** In the same way as glass or carbon reinforced materials, natural fibres also significantly upgrade the strength of polymeric materials. Composite materials with natural fibres are in demand for high-rigidity, low-weight compact mouldings, for instance in automotive engineering and body construction. Natural fibres are more flexible and lighter than the commonly used glass fibres. They are not quite so strong, but their share in the material can be increased because of their lower weight. The total weight and the material behaviour of glass and natural fibre reinforced materials are ultimately very similar.

Components formed from natural fibre reinforced materials: These are used especially in the automotive industry, for the interior linings of doors and luggage compartments, backlight shelf, transmission encapsulation and engine encapsulation. A modern car contains on average about 16 kg of natural fibres, mainly in the form of natural fibre reinforced materials. Source: Daimler AG

Bast fibres from flax and hemp are just as viable as the exotic fibres jute, kenaf, sisal and coconut, or regenerated cellulose. The polymer matrix is made primarily of mass plastics (polypropylene, etc.) and various resins. Due to the polar chemical structure of natural fibres, adhesives have to be added to link them to the matrices, which are usually hydrophobic.

Trends

▶ **Wood-plastic composites.** Composite materials made of plastic reinforced with wood, known as wood-plastic composites (WPC), represent a relatively new group of materials. WPCs contain wood fibres, usually recycled wood fibres and wood chips. The proportion of fibres may be as much as 90%. These are mixed with polymers and additives to permit thermoplastic processing. Because temperatures over 200 °C cause thermal damage to the wood components, particularly low-melting plastics such as polyethylene (PE), polypropylene (PP) and polyvinyl chloride (PVC), as well as starch and lignin-based bioplastics, are used as the matrix. Wood-plastic composites are manufactured by compounding the two components – fibres and plastic – either directly by extrusion processing to form continuous profiles, or by a granulation process. The most important sales market for wood-plastic composites are floor coverings for outdoor or garden applications such as verandas, terraces and outdoor stairs. The advantage of WPCs compared to solid wood products is their plasticity and their greatly superior weather resistance.

▶ **Outdoor applications.** The use of natural fibres in automotive engineering has so far been limited largely to interior applications. However, modern natural fibre components can meanwhile withstand exposure to extreme conditions, so even car bodies can be constructed from these materials. Natural fibre compounds have recently been utilised on the exterior of serial models for the first time, for example as vehicle underbody protection. The first prototype vehicles whose entire body consists of bioplastic-based natural fibre composite materials, or biocompounds, are currently undergoing tests.

Duromer composite plastics can be produced by a sheet moulding compound procedure, for instance. This involves processing plain semi-finished products from natural fibre fleece, impregnated with a polyester resin, in a hot tool to form the desired component.

Sheet moulding compound procedures produce large quantities of flat parts for outdoor applications with high mechanical requirements. Smaller quantities are produced by the low-pressure spray coating method for polyurethanes and the cold pressing method for materials such as polyesters. In this case, the plastic is sprayed onto a natural fibre fleece lying in a tool and subsequently pressed at room temperature. The achievable mechanical properties are significantly lower. Flat exterior parts with low mechanical requirements, such as engine encapsulations, are produced in this way.

Injection mouldings for natural fibre reinforced polymer granules are a promising method in the medium term. This method would significantly increase productivity and the component variability, as well as significantly minimising production costs.

▶ **Certification of biomass.** The rapidly growing global demand for biomass for non-food purposes, and the resulting expansion in trade flows, calls for steering instruments to avoid competition between non-food biomass on the one hand, and food and feed production coupled with the concerns of nature protection and landscape conservation on the other hand. Standards will be needed not only in the technical sense, but also to ensure compliance with environmental and social sustainability criteria.

To ensure that these standards are implemented in worldwide biomass production and trading, innovative certification methods have been developed. These provide verification for the sustainability of biomass production. The ongoing certification approaches draw upon four essential benchmarks:

- Certification must be effected independently of the use of biomass, and must take changes in land use into account. This prevents traditional production from shifting into new, ecologically sensitive areas while biomass for energy is produced on certified agricultural land ("leakage effects").
- Certification has been conceived as a meta-system. It builds on existing systems, e. g. agricultural or forestry systems such as the EC cross-compliance policies or other globally or locally accepted standards. This allows rapid implementation with low costs and broad acceptance.
- As bulk commodities, biomass and biofuels are traded globally. Tracing the commercial chain back to the source is only feasible to a limited extent and with great effort. The "book & claim" concept with tradable certificates comes off best in this situation. It distinguishes certificates from real flows of commodity and allows trading in virtual marketplaces. In this way, the book & claim concept establishes a direct connection between biomass producers and the final consumer without taking into account the diverse intermediate stages of processing and trading.
- To distinguish biomass production on the one hand from the greenhouse gas relevance of the product on the other hand, separate certificates are awarded for the sustainability of agricultural production and for the greenhouse gas effects of biofuels, for instance. The certification of crop cultivation approaches biomass at origin and rejects changes of land use and the loss of biodiversity. It helps to preserve land with a high carbon content and promotes compliance with social standards. Greenhouse gas (GHG) certificates, in contrast, describe the climate gas balance along the complete production and processing chain for bioenergy and products from renewable resources. The first step will define standard values (default values) by which to determine the GHG balance of the various products.

The first certificates for the verification of biomass sustainability are now being issued. In the long run it will be necessary to establish certification systems not only for the non-food sector, but also to regulate the later expansion of food and feed crops.

Prospects

- The depletion of oil reserves has aroused interest in renewable resources, not only as fuels for generating heat and power, but also increasingly for products in the chemical and manufacturing industries. They are capable of meeting sophisticated challenges and fulfilling highly demanding tasks. Renewable resources are becoming established today wherever they use chemical structures that nature has synthesised in plant materials, as well as offering technical advantages and high added value. However, they must be measured by typical commodity parameters such as availability, quality and pricing issues. In future, agricultural and forestry products will gradually expand their position as industrial raw materials and – depending on developments in the commodity markets – as sustainable resources even for mass products.
- There is sufficient agricultural potential to meet this need, provided that global agricultural production can be made significantly more efficient in the medium term and that losses along the entire production chain can be reduced to a minimum.

DR. TORSTEN GABRIEL
Fachagentur Nachwachsende Rohstoffe e. V. (FNR), Gülzow

WPC granules: The two main components of WPC are polyethylene and wood flour. Source: USDA

Garden terrace made of wood-plastic composites (WPC): Originating in North America, WPC technology is gaining acceptance mainly in garden design thanks to easy-care wood-like products. Source: Werzalit AG

Internet

- www.fnr.de
- www.european-bioplastics.org
- www.iscc-project.org

Wood processing

Related topics

- Renewable resources
- Laser
- Measuring techniques
- Bioenergy
- Sustainable building

Principles

Forest and timber products are an excellent example of an economic cycle in which a *renewable natural resource* (in this case wood) is used to make high-tech and innovative products and is ultimately employed as a CO_2-neutral energy source at the end of its life cycle. Processing the raw material of wood into high-tech products poses a variety of challenges. Its extraction from the forest has to be sustainable, i. e. the number of trees felled must not exceed the annual growth rate of the forest. Forests need to serve as recreational, natural landscapes for people and as climate regulators producing oxygen and consuming CO_2, while at the same time being an economic basis and resource for a very large and competitive industry.

Applications

▶ **Harvesting.** Forestry supplies industry with the renewable resource of wood in various forms. High-quality hardwoods such as beech are used as solid wood or veneer for furniture, while softwoods such as pine are used on an industrial scale as solid construction timber, for engineered wood products such as glue-lam beams, or for wood-based panels such as fibre boards. Moreover, softwood forms the basis for pulp and paper.

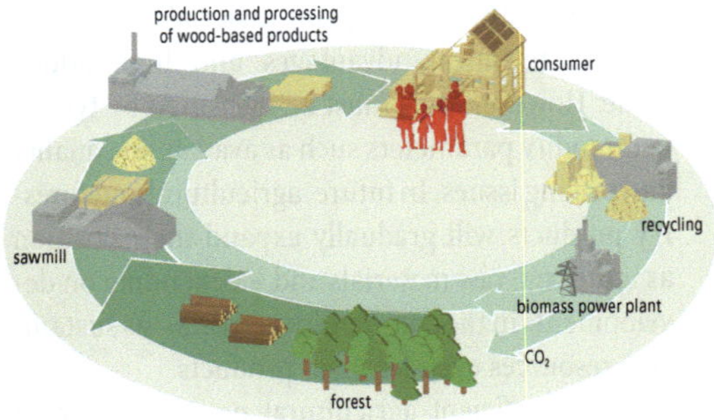

Eco-lifecycle of wood: Wood absorbs CO_2 while growing. Sustainable forestry practices involve harvesting only the annual increment of wood, which is then processed into lumber and further refined to make wood-based products such as panels for construction. At the end of their life time, the panels can be recycled, e. g. chipped and refined to be re-used as a resource for more wood-based products, or burned in biomass power plants to produce energy. The bound CO_2 in the wood is then released into the carbon cycle again in the same quantities as originally absorbed, which makes wood a CO_2-neutral resource. Wood acts as a "carbon sink" while being used, as the CO_2 is withdrawn from the carbon cycle for the duration of its use. Source: Egger GmbH

In large and relatively homogeneously grown pine stands, where profit margins for the raw material are relatively low, tree extraction has to be highly cost-effective while at the same time employing methods that avoid causing harm to younger trees and compacting the forest floor. Under these conditions, the generally labour-intensive felling and extraction of the trees is performed today by just one person operating what is known as a harvester.

A *harvester* is a tractor with a harvester head mounted on a crane arm. The harvester head embraces the stem with grippers and rollers and cuts the retained tree with an integrated chain saw. The cut tree is lifted up and turned without touching younger trees that must not be harmed. The rollers push the turned tree through the grippers to remove the branches (limbing) and immediately cut the stem into appropriate log assortments. During embracing and limbing, the harvested volume of the various log assortments is automatically calculated by measuring the length and diameter of each log segment.

▶ **Sawmills.** The logs are stored in the lumber yard to balance out times of intensive and less intensive logging so that a continuous supply of raw material can be provided in the quantities and quality required. In many cases, the stored logs are kept wet by sprinklers to prevent insect attack. Before being cut, the logs are scanned using a 3D *laser* to determine the best board dimensions and combinations to cut out in accordance with the order list so as to achieve the highest possible output. Each log is cut into log-specific beams, planks or slats. Cutting by circular saws or band saws produces a cutting gap and residual sawdust. In profile chipping, the wood is chipped off by a rotating knife, resulting in residual wood chips or strands of relatively homogeneous size. Both residual materials are used to produce wood-based panels such as particle boards or "oriented strand boards" (OSBs).

After cutting, the individual planks are subjected to a quality inspection. This can be done automatically or manually by a person, who marks faults with luminescent chalk. In the next step, a scanner identifies the marks and activates a compound saw which cuts out the fault. Automated quality inspections are carried out using laser scanners to identify faults such as knotholes, *imaging cameras* for detecting unwanted colour differences, *X-ray sensors* for establishing density profiles, and acoustic sensors for determining the stiffness of the planks. All the data is transferred to the compound saw, which cuts out the fault.

Planks that are cut because of faults can be automatically transferred to a parallel process line where the plank sections are automatically finger-jointed into planks of standard and ordered dimensions.

Mass customisation (every board is cut to order), quality and speed are the key factors in any competitive sawmill. Logs are driven through the cutting line and the cut boards inspected in-line at a speed of 150 m/min.

▶ **Wood-based panels.** Mass-produced wood-based panels such as medium-density fibre (MDF) board, particle board or oriented strand board (OSB) – coated or uncoated – are utilised in many everyday settings, e. g. as the substrate in laminate or wooden flooring (MDF), as the core material of virtually all home and office furniture (particle boards) or as the sheeting of wood-frame constructions (OSBs). As an alternative to solid wood, these wood-based products offer the consistent quality and engineered properties required in mass products, and serve as a raw material that can be further processed on an industrial scale. Moreover, wood-based panels fully utilise residual material of the sawing process, for instance sawdust, particles or chips/strands.

These boards are all manufactured in a continuous process that allows them to be mass-produced. A mat of particles, fibres or strands is formed and fed into a sometimes over 70-m long pressing line where it is slowly compressed at high temperatures to its final thickness and density. The only difference in the production technology to fabricate particle, MDF and OSB boards lies in the preparation of the particles, fibres or strands:

- Particle boards are manufactured from flakes, chips, shavings, sawdust and/or other lignocellulosic material in particle form (flax shives, hemp shives, bagasse fragments and the like) with the addition of an adhesive.
- Medium-density fibre (MDF) boards are manufactured from lignocellulosic fibers by a "dry process", i. e. with a fiber moisture content of less than 20% at the forming stage, and are essentially produced under heat and pressure with the addition of an adhesive.
- Oriented strand boards (OSBs) are engineered structural wood panels. Rather long strands of wood are bonded together with a synthetic resin adhesive. Sometimes in all three layers, but usually only in the outer layers of these panels, the strands are orientated in a particular direction.

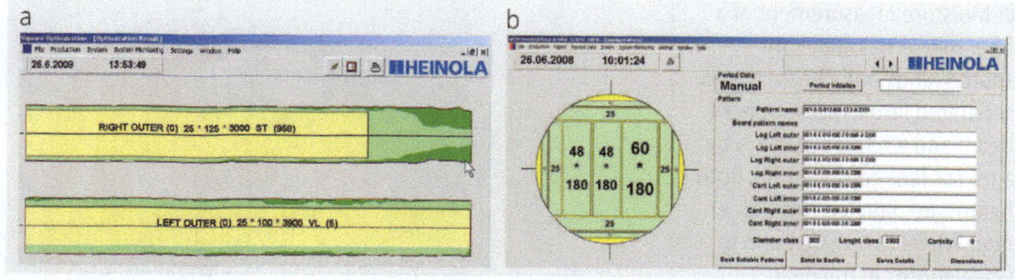

Log scanning: Before entering the cutting line, the logs are measured by a laser scanner to determine the optimal cutting programme for each one on the basis of the order list. The most suitable boards or planks, e. g. in terms of surface area (a) and thickness (b), are then cut out of each scanned log according to its dimensions and quality. In (a), the dark green area indicates unacceptable wood quality and limits the possible surface area of the boards. In the illustrated example, it was decided to produce a board measuring 25 x 125 x 3000 mm³ from the outer right part of the log; (b) shows the cross section of the log, again indicating the 25-mm thick board to be cut at the outer right edge. In a later step, the board will be trimmed to the desired 25x125 mm². Source: Heinola

▶ **Continuous hot press technology.** Today, nearly all mass-produced wood-based panels are produced using continuous hot press technology. After the raw material (i. e. the particles, fibres or strands) has been prepared and mixed with the adhesive, it enters the pressing process. This begins with the mat-forming system. The formed mats can be homogeneous single-layer mats for MDF boards, or three-layer mats for particle boards or OSBs, where the core and face layers differ, e. g. in particle size, to create a sheeting effect, or in orientation, to enhance the specific strength of the board in a preferred direction. In the case of MDF boards, the raw material (forest thinnings, sawmill byproducts) is reduced to fibres by steam and between grinding discs, which rotate at great speed. An adhesive, usually urea formaldehyde, and wax emulsion are applied to the fibres inside the inlet pipe (blow line) leading to the drying tube. The fibre/adhesive mix is dried in the drying tube. The dry fibres are then formed into a fibre mat on caul plates. The dry mat is pre-pressed to consolidate it and then cut and formed to press sizes. Following this pre-compaction, it is possible to pre-heat the resulting mat in order to raise its temperature before it enters the actual hot press. The

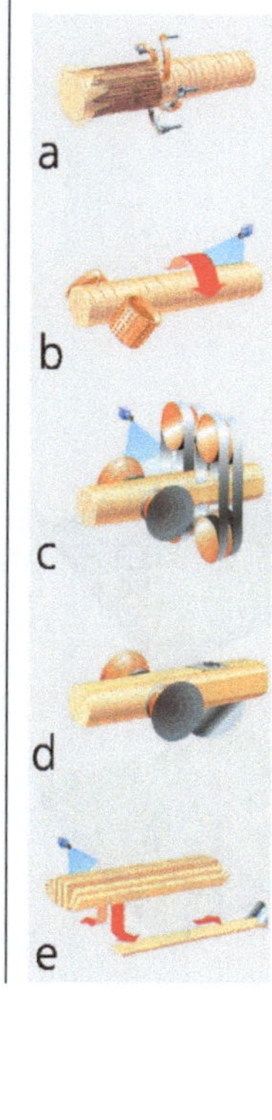

Typical cutting programme of a sawmill step by step: The log is continuously transported through each step at speeds of up to 150 m/min. First, the log is reduced and de-barked by rotating de-barking knifes (a). This is followed by log and cant positioning. As the logs are not symmetrical columns but are naturally grown with some twists and turns, they are rotated into the optimal cutting position according to their laser-scanned volumes (b). Then they are canted by rotating plates – this is known as profile chipping – and cut into planks, for example, by band saws (c). Alternatively, or in addition, they can be profiled and cut with circular saws (d). Finally, the boards are sorted and, if necessary, edged (e). Source: Soederhamn and Eriksson

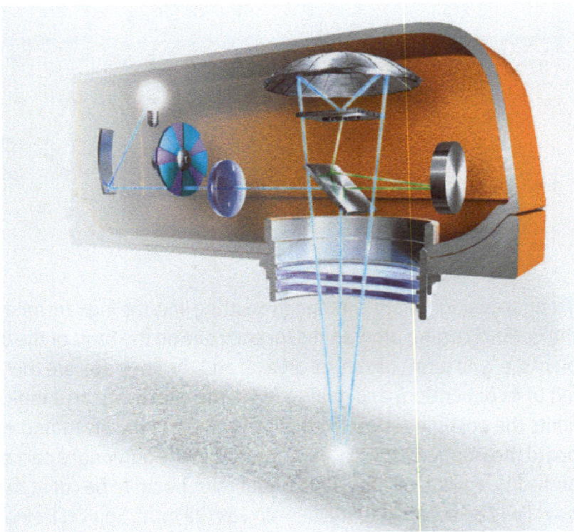

Moisture measurement of a fiber mat: A halogen light source sends out a beam that is separated into a measurement beam and a reference beam by a mirror-lens combination. Both beams are directed through a filter wheel to reduce them to wavelengths in the near-infrared (NIR) range, and are then directed onto the material. The intensity of the reflected beam varies as a function of the moisture content of the particles or fibres, and its difference with respect to the reference beam is monitored by the measurement head. Source: Grecon

goal is to heat up the entire mat quickly and evenly. The desired preheating temperature is set precisely by adjusting the amount of steam inserted into the mixture. This results in a more economical production process – especially for boards with greater thicknesses – and improved board properties as well as higher capacities, as the speed of the mat can be increased.

The mat now enters and runs through the press at a speed of up to 2 m/s, resulting in a continuous sheet of MDF with a thickness tolerance of 0.1 mm. Pressure and heat is transferred from the press plate to the press band via a carpet of steel rods. The pressure and temperature distribution in the transversal and longitudinal direction of the press can be varied in order to optimally control the moisture content and density distribution in the compressed mat or in the board, respectively. In the finishing area at the exit of the press, the sheet of MDF is trimmed and cut to produce the final board.

In continuous presses, the mats and boards are subjected to very high forward speeds (2 m/s), temperatures (240 °C) and pressure (110 bar). This pushes the process and process engineering to its physical and technological limits, e. g. with regard to the applied steam pressure and resulting plastic behaviour of particles within the mat and the board. This in turn makes challenging demands on process control and reliability. The press line is therefore equipped with various sensors to monitor the process parameters and adjust them in-line.

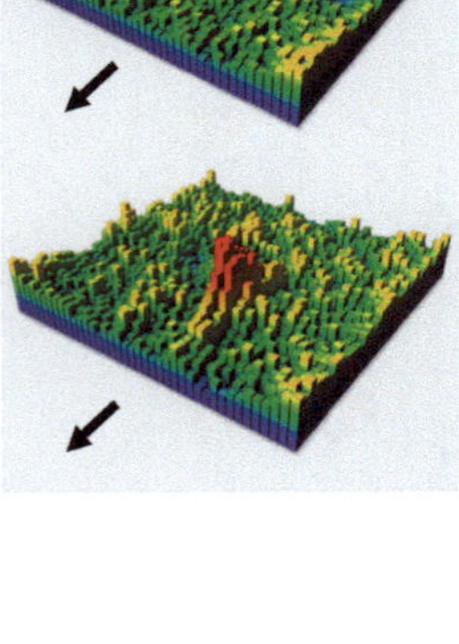

Overlapping X-ray scanning of the material mat before entering the press detects density peaks caused by such phenomena as small aggregates or lumps of glue and fibre. These areas of the mat have to be removed as they could harm the steel bands if too much material is compressed to the desired board thickness. The colours of the measurement results indicate acceptable density variations (blue and yellow), which are sometimes even desired in order to engineer certain mechanical board properties. If the variation exceeds the tolerance range (red), the material agglomeration is removed. Source: Grecon

To ensure a high product quality, it is necessary to continuously monitor the moisture content of the particles and fibres after leaving the dryer, before and after the application of the glue in the blow line – the pipe in which the particles are dried and mixed with droplets of glue in a continuous air flow – and before entering the press. The contact-free online measuring system applied in most cases is based on optics. Light in the *near-infrared (NIR)* range is absorbed by the material's moisture. The higher the moisture content of the material, the lower the reflection. The difference in applied and reflected NIR light is translated into the material's moisture content.

In addition, the mat is also completely scanned by overlapping *X-ray* areas as it moves along at 2 m/s before entering the press, in order to detect density variations. The X-rays penetrate the mat to varying degrees, depending on the amount of material present. On the basis of this penetration, a highly sensitive camera detector below the mat continuously calculates a precise density profile. The mat is also scanned for metal parts by magnet coils, and its height and weight distribution per unit area are continuously measured. If incorrect heights or densities – such as lumps of glue and fiber – or metal parts are detected, the sections are taken out of the production process before entering the press to prevent irreversible harm to the steel bands. The forming band can be withdrawn slightly from the steel band against the production direction. The mat section in question then drops through the resulting gap and out of the process.

After a finished board has exited the press, its density profile is also monitored online by X-ray. This makes it possible to adapt density-related mechanical properties during the pressing process. The online density profile measurement also reduces the time required to ramp up the press after a change in the product being produced (e. g. a shift from medium-density MDF boards used for furniture to lower-density MDF boards used as insulating material in construction), as the press set-up can be optimised on the basis of real-time feedback from the exit area. After density monitoring behind the press exit, a cut-off saw cuts the continuous board into separate boards of standard dimensions. The still very hot boards are then placed in special cooling devices to gently cure them before they are further handled, e. g. sanded or coated.

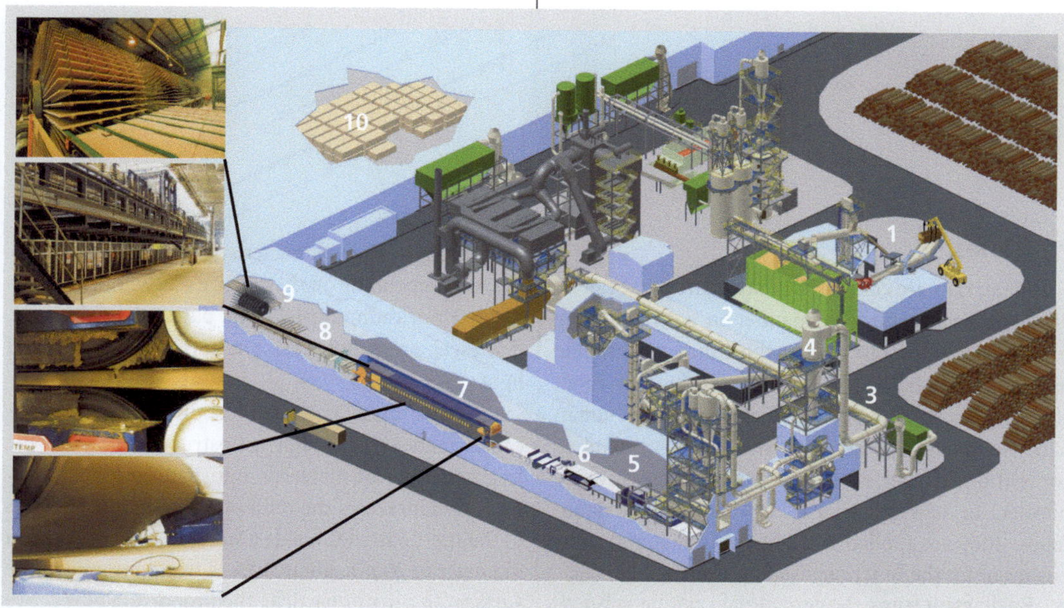

Medium – density fibre board process: The first step in the MDF process is wood handling, which includes de-barking, chipping, chip and bark handling systems and the separation of stones and other contaminations (1). Next the chips are washed steamed, pre-heated and defibrated. In the pre-heater, the chips are heated to a temperature of 160 °C, which makes the fibres soft and easier to separate. The soft chips are then transported into the refiner (2), where they are ground into fibers under a steam pressure of up to 8 bar. The fibre-steam mixture flows out of the refiner into the blow line, where it is covered with resin by the resin system, followed by the drying tube (3), where the moisture content of the fibres is reduced to around 10%. The fibre-drying stage involves one or two dryer cyclones and a 2-sifter to remove contaminations (4). The forming stage (5) forms the fibres into a mat, which enters the pre-press (6) before it goes into the hot press. There the fibre mat is compressed to its final thickness and the resin is cured due to the temperature increase (7). A modern continuous press can be more than 70 m long and 8–12 m wide. The last stage is handling, in which the MDF boards are cut into the correct dimensions (8), cooled down (9), sanded and stacked for delivery (10). Source: Siempelkamp

Lightweight honeycomb panel produced in a "frameless" continuous production line. Source: Egger

Trends

▶ **Lightweight panels.** The weight of panels and the optimisation of resources are important issues in the manufacture of modern wood-based products. The honeycomb panel, a traditional and natural lightweight building element, has been rediscovered and is produced as a frameless lightweight panel today. As the term "frameless" implies, the panel is produced without a frame on a newly developed continuous production line. The surface-layer boards are made of MDF, particle boards, OSBs, or other thin materials. The surfaces can be unfinished, sanded, laminated on one or both sides, or coated. The thickness of the surface-layer boards can range 3–10 mm.

The core materials used for the panels include honeycomb structures manufactured out of paper, corrugated core structures likewise made of paper, or lattice structures made of cardboard or MDF. The latter two are newer developments available as finished core material. Because of the high ratio between volume and weight during transport and storage (the material cannot be compressed), the traditional honeycomb structure made of paper is still the preferred core material. These honeycomb structures come pre-compressed. Their expansion takes place on honeycomb stretching and drying equipment.

Prospects

- The role of wood as a CO_2-neutral resource that can be burned in a biomass power plant at the end of its life cycle makes it a highly sought-after material in today's times of increasing awareness with regard to climate change and the scarcity of natural resources.
- The renewable resource of wood forms the basis for many and varied products. People often underestimate just how much wood is actually used and what technological properties it has to offer. It can be used for high-performance wood-based panels, as paper for printing, or even as a lightweight composite in airplanes, for example.

DR. RAOUL KLINGNER
Fraunhofer Gesellschaft, Munich

TOBIAS ZIMMERMANN
Egger Group, Wismar

Internet

- www.europanels.org
- www.se-saws.com
- www.siempelkamp.de/Plants-for-wood-based-products.572.0.html

Nanomaterials

Related topics

- Self-organisation
- Nanomedicine
- Intelligent materials
- Surface and coating technologies
- Polymers
- Process technologies

Principles

Nanomaterials are composed of structures that can be produced in a controlled manner in a size ranging from 1–100 nm in one, two or three dimensions. One nanometre is a billionth of a metre (10^{-9} m), or about 50,000 times smaller than the diameter of one human hair. Nanomaterials are nanostructured variants of conventional materials (e.g. metal and metal oxide powder) or new material classes like carbon nanotubes. They are applied as additives in material composites or for the functionalisation of surfaces.

▶ **Basic structures.** Nanomaterials can be subdivided into different basic geometric structures. Zero-dimensional basic structures are smaller than 100 nm in all three dimensions. This applies to almost point-shaped objects such as nanoparticles or nanoclusters. One-dimensional basic structures include tubes, fibres or wires with nanoscale diameters, while two-dimensional basic structures are represented by layers or films with a thickness smaller than 100 nm.

▶ **Material properties.** Nanomaterials differ from coarser structured materials through their ability to demonstrate highly altered characteristics relating to electrical conductivity, magnetism, fluorescence, hardness, strength, or other properties. Unlike macroscopic solids, electrons in a nanocluster can only adopt specific "quantitised" energy states, which are influenced by the number of atoms that are free to interact. This may strongly alter characteristic fluorescent properties, e.g. depending on the particle size within the cluster. A 2 nm cadmium telluride particle, for example, fluoresces green light, a 5 nm particle fluoresces red light.

In most cases, chemical reactivity can be significantly increased through nanostructuring, since the fragmentation of materials into nanoscale substructures also increases the ratio of reactive surface atoms to the inert particles inside a solid. In a particle with a diameter of 20 nm, about 10% of the atoms are on the surface. In a 1-nm particle, the proportion of reactive surface atoms is 99%.

▶ **Production process.** Nanomaterials can be produced by breaking up larger fractions of matter into nanostructured elements ("top-down") through mechanical crushing in ball mills, for example. Or they can be manufactured by relying on the specific structure of the smallest components of matter in which *self-organising processes* cause atoms and molecules to merge spontaneously to form larger nanostructured clusters of matter by way of chemical interaction ("bottom-up"). Such self-organising processes are the basis for life processes in biology, but they are also increasingly being researched for technical applications. Technically relevant processes for the production of nanomaterials include gas-phase synthesis (used to produce carbon nanotubes, for example) and processes in the liquid phase, such as the *sol-gel method.* In the latter case, the finely distributed nanoparticles (sol) in a solution are produced by a chemical reaction from chain-forming molecular source materials (such as metal alcoholates). These nanoparticles can either be turned into a porous material during the drying process, or into a solid gel, from which ceramic layers or moulds can be produced through heat treatment.

▶ **Carbon nanotubes.** CNT are cylindrical formations of carbon atoms arranged in regular comb-like hexagons along the cylinder's axis. CNT can also be regarded as unrolled graphite layers with one open and one closed end, having a tube with a diameter ranging between 1–100 nm and a length that varies from several hundred nanometres to a few millimetres. Due to the strong bonding forces between the carbon atoms and the regular arrangement along the cylinder axis, CNT are mechanically very robust. Theoretically, if the stability of one single tube could be expanded to macro-

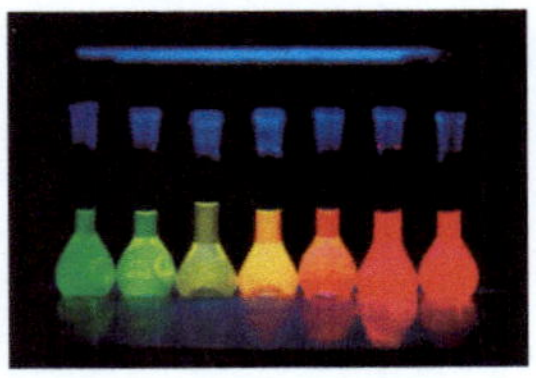

◀ Fluorescent properties depend on particle size. The pistons contain solutions with cadmium telluride particles of different sizes. Despite identical chemical particle compositions, the fluorescence changes colour from green to red as the particle diameter is varied from 2–5 nm. Source: University of Hamburg

Particle diameter [nm]	Total number of atoms	Proportion of surface atoms [%]
20	250,000	10
10	30,000	20
5	4,000	40
2	250	80
1	30	99

▲ Ratio of volume to surface: The smaller a particle, the larger the proportion of atoms on the particle surface. The ratio of surface atoms to total number of atoms varies particularly strongly in particles between 1 nm and 20 nm in size.

scopic dimensions, the tensile strength of CNT would be up to 20 times greater than that of steel. In addition, CNT demonstrate unique electrical properties based on the special arrangement of the carbon atoms and the percentage of movable electrons inside the molecular frame. Depending on the arrangement of the hexagon edges along the cylinder axis, CNT conduct electric currents significantly better than copper, or exhibit semiconducting properties.

► **Composite materials.** Composite materials contain nanoscale fillers which enhance the material's properties. One of the most common examples are polymers. These are filled with mineral, ceramic or metal nanoparticles, or even carbon nanotubes, to enhance their mechanical properties, electrical conductivity, gas impermeability or even the non-combustible properties of the polymer. The gas impermeability and tensile strength of polymers can be significantly improved by adding layer-like clay minerals (silicates). The thickness of the individual nanoclay layers stacked on top of each other is approximately 1 nm, the diameter of the nanotubes ranges between 50 to over 1000 nm. To produce polymer composites – e.g. through melt *polymerisation* – the gap between the silicate layers is first enlarged using organic molecules that accumulate between the layers (*intercalation*). This enables polymer chains to permeate the gaps, resulting in the separation and distribution of the individual silicate layers in the polymer matrix during the course of polymerisation (exfoliation). The parallel arrangement of the nanolayers in the polymer creates a barrier effect against diffusing gases. This is achieved by the fact that the gas molecules must move around the layers when diffusing through the polymer, which increases the diffusion path and reduces the diffusion rate. The increased gas impermeability of the polymer is of benefit in applications like fuel tanks or for plastic wraps. It also inhibits the combustibility of the polymers, since oxidising gases are retained in the event of fire. Due to nanostructuring, the property-enhancing effects are already achieved at lower weights than with conventional fillers, facilitating the technical processing required to achieve the desired polymer properties.

Applications

► **Catalysis.** Compared to coarser-structured materials, nanomaterials have two essential advantages in catalysis: on the one hand, the catalyst material, which is often expensive, is better utilised due to the high proportion of catalytically effective surface atoms. On the other hand, the radically modified electronic properties of nanomaterials enable substances to be used as catalysts that are otherwise catalytically inactive, e.g. gold, which is inert in its naturally dense structure, but is catalytically very active as a nanoparticle variant. Heterogeneous catalysis, in which nanoparticles of the catalyst materials are fixed on a solid substrate, is of particular importance. Here, nanoparticles are produced from suitable basic substances, either in the gas or liquid phase, and separated onto the substrate. It is possible to define and control particle size, e.g. through use of nanoporous substrates such as zeolites, aerogels and dendrimers. The size of the catalyst particles is in this case determined by the pore size of the substrate. Nanostructured catalysts are used in the production of chemicals, in fuel cells and in automotive catalysts.

► **Effect pigments.** *Interference pigments* provide varying colour effects depending on the viewing angle

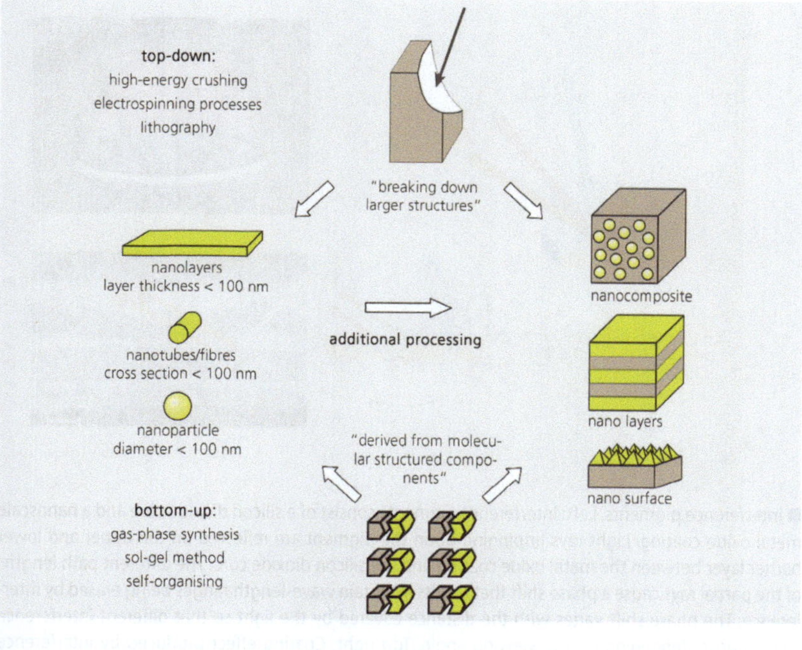

Principal approaches and processes for the production of nanomaterials: Nanomaterials and structures, including nano films, tubes, fibres and particles (left), and nano-structured surfaces, layer systems and composites (right), can be produced taking a "top-down" or a "bottom-up" approach. In "top-down" methods, nano-scale structures are extracted from larger solids, for example, by mechanical processes such as high-energy milling or by optical lithography processes such as those used in microelectronics. In the "bottom-up" approach, nanomaterials are produced from individual molecular or atomic units in self-organisation processes. When producing nanomaterials for technical applications, it is often necessary to combine different nano structuring methods with traditional material processing techniques.

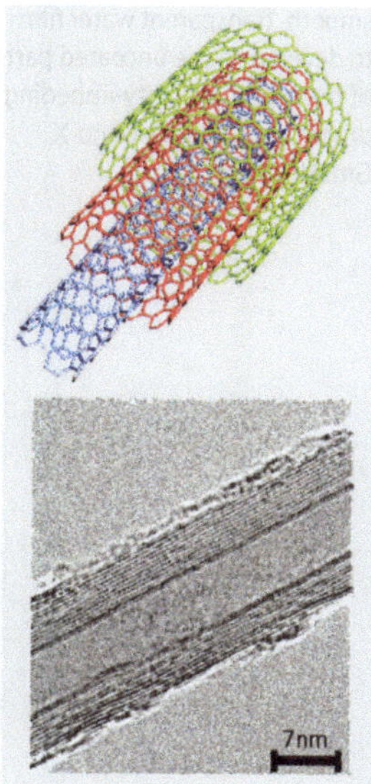

Carbon nanotubes (CNT). Top: CNT consist of carbon atoms arranged in regular hexagons forming a tube-like structure with a diameter measured on the nanoscale. In the multi-walled CNT shown here, several carbon layers are nested on top of each other. Bottom: The electron-microscopic picture of the cross-section of a CNT shows the individual carbon layers as dark lines. Source: Bayer Technology Services

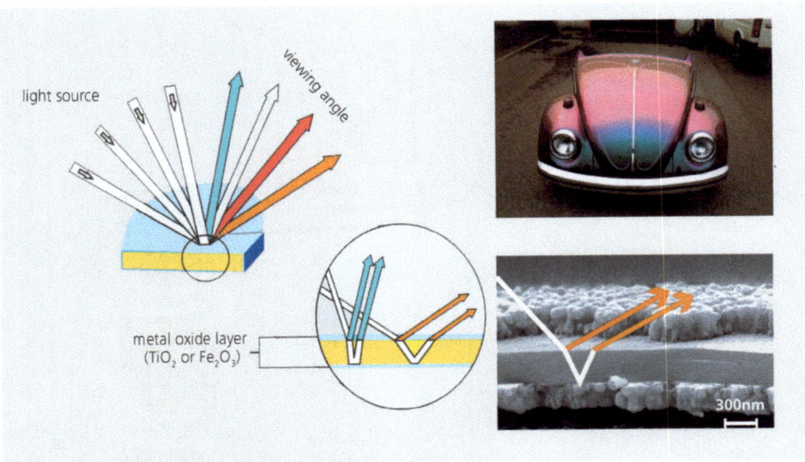

Interference pigments. Left: Interference pigments consist of a silicon dioxide core and a nanoscale metal oxide coating. Light rays impinging upon the pigment are reflected on the upper and lower barrier layer between the metal oxide coating and the silicon dioxide core. The different path lengths of the partial rays cause a phase shift that results in certain wave-length ranges being erased by interference. The phase shift varies with the distance covered by the light so that different interference colours arise depending on the viewing angle. Top right: Coating effect produced by interference pigments. Bottom right: The electron-microscopic photo shows the nanoscale titan oxide layer on the mica plate. Source: Dupont Deutschland GmbH (top right); Merck KGaA (bottom right)

Demisting effect produced by super-hydrophilic surfaces: The bottom left-hand side of the mirror is coated with nanoscale titanium oxide that causes a smooth, transparent water film to develop. In the uncoated part of the mirror, visibility-impeding drops occur. Source: Nano-X GmbH

of the observer. Interference pigments consist of a natural silicate with a very smooth surface and a plate-like structure –the mica core. This core is coated with one or more nanoscaled metal-oxide layers. Nanopigments can be produced by gas-phase separation: gaseous inorganic or metal-organic compounds are decomposed in an oxidising atmosphere at elevated temperatures and separated as uniform crystalline metal oxide layers on the mica plates. The colour effect is due to a combination of reflection, deflection and interference effects of the light rays meeting the pigments. At first, a part of the light is reflected on the surface of the metal oxide layer. The rest of the radiation penetrates the metal oxide layer through deflection of light and is only reflected after encountering the lower part of the metal-oxide barrier layer. The path difference of the reflected partial rays results in a phase shift, which causes certain wavelength ranges to be erased or enhanced by *interference* when the partial rays are recombined. Depending on the viewing angle, the path difference of the reflected partial rays varies, and thus the interfering wavelength ranges, resulting in changes in the perceived colour. Layer thickness in the nanoscale range is required to obtain the interference in the wave-length range of visible light.

► **Electroconductive polymers.** These polymers are used to prevent plastic casings from becoming electrostatically charged, such as those used for electronic devices, suction pipes, conveyor belts and fuel pipes. Antistatic properties are achieved by adding a conductive filler that permits the conductivity of the polymer to be adjusted according to requirements. Preference is given to a low filler content to avoid any impairment of the polymer's properties such as good processability. Carbon nanotubes show a significantly higher degree of electrical conductivity than conventional fillers like carbon black, so the desired conductivity can already be attained at lower concentrations. The polymer composite is produced by melting and mixing a polymer granulate and carbon nanotubes in an extruder. At just 1% CNT-content an interconnected network of carbon nanotubes (percolation network) forms in the polymer, thus making the composite electrically conductive.

► **Functional surfaces.** Nanoscale coatings allow the surface properties of materials to be specifically modified, such as their ability to adsorb or repel water and fats, depending on the coating materials used and methods applied. Using nanomaterials enables the creation of a wide range of hydrophilic (hygroscopic), hydrophobic (water-repellent) or oleophobic (fat-repellent) surface properties, which in turn have high market relevance as dirt-repellent or self-cleaning surfaces in endconsumer products. When a super-hydrophilic surface is affected by humidity, a smooth film of water develops on the surface preventing drop formation and thus a steaming-up of window panes and mirrors. The technical realisation of super-hydrophilic surfaces is made possible by applying layers of photocatalytic titanium dioxide. These are biocidically active, so that they also kill algae, mosses, microbes and germs adhering to the surface. The use of nanoscale titanium dioxide powder enhances the effectiveness of the layers and, due to the smallness of the particles, helps prevent light dispersion. This produces the kind of transparency required for application on window panes and other transparent surfaces.

Trends

► **Industrial production of CNT.** The material properties of isolated nanostructures under laboratory conditions cannot simply be transferred to technical products. Industrial scale production, however, is a prerequisite to the broad technical application of nanomaterials such as carbon nanotubes. A common production method for CNT involves their chemical vapour deposition on metal catalysts using carbon-rich gases. The free carbon atoms arising from the thermal decomposition of reaction gases grow layer by layer on

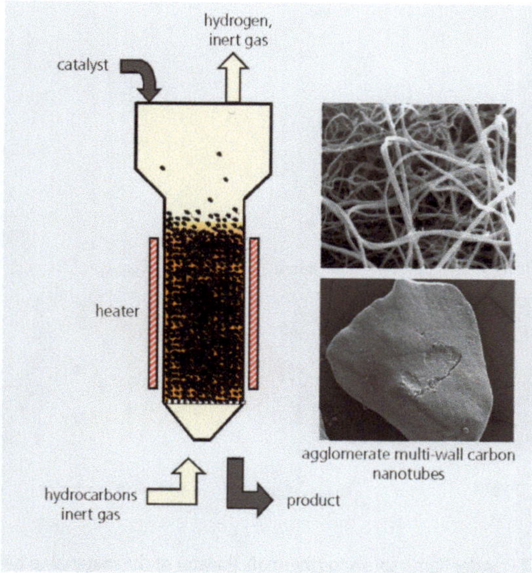

Continuous production process of multi-wall carbon nanotubes: In a heated reactor, gaseous hydrocarbons are catalytically decomposed in an atmosphere of inert gas. Carbon nanotubes form on the catalyst particles, are deposited as agglomerates on the bottom of the reactor, and are continuously discharged as product. The CNT are then cleaned, untangled and, if needed, chemically modified. Source: Bayer Technology Services

catalytic crystals to form nanotubes. Due to technical process improvements, such as the optimisation of the catalyst, undesired by-products such as carbon black can largely be avoided. Furthermore, the introduction of continuous reaction methods has paved the way towards industrial scale production of CNTs.

▶ **Switchable adhesives.** In conventional adhesive curing, heat has to be applied externally to the adhesive via the adherend. Adhesives filled with nanoscale magnetic particles (10 nm iron oxide particles), however, can be activated from outside by microwave radiation. Nanoparticles convert electromagnetic microwave radiation into heat, which heats the surrounding adhesive matrix and triggers the polymerisation reaction. The outcome is a significant reduction in curing time in comparison to conventional adhesive technology, since the heat required for curing is directly produced in the adhesive, therefore obviating the need to heat all components. Such adhesives can also support new recycling techniques through the thermal decomposition of the adhesive layer by the application of additional heat energy, enabling the non-destructive separation of components.

▶ **Risk research in nanomaterials.** Like all chemical compounds, nanomaterials do harbour risk potentials for human beings and the environment. Recently, the results of scientific research on the human- and eco-toxic hazards of nanomaterials, especially of nanoparticles, were the subject of an intense public and media debate. Attention focused in particular on free particles, which can enter the environment or the human body by various paths. Their effects are barely known. Respiratory absorption of nanoparticles in particular is regarded as a possible hazard to human health. Respirable, barely soluble particles (not only nanoparticles, but also particulate matter) carry the general risk of triggering inflammations in the lung tissue by overstressing the body's defence mechanism. Consecutive reactions could indirectly provoke cancer. Due to their small size also nanoparticles could enter the vascular system via the lungs and reach other organs where they could act as a trigger for diseases. By the same token, when discussing the risks of nanoparticles, it is important to remember that they are not only produced by targeted industrial production, but are also generated in large quantities by natural processes (such as erosion, volcanic eruptions, forest fires) and as by-products in technical combustion processes (diesel engines, welding, soldering).

Prospects

- Economically speaking, the most important applications for nanomaterials are currently nanostructured fillers (e.g. carbon black, carbon nanotubes or metal oxide nanopowder) used for optimising the properties of preliminary products like membranes, adhesives, substrates, and other products in the fields of chemistry, pharmaceuticals and electronics.
- Nanomaterials are increasingly being applied to develop new or improved properties for products used in daily life (cosmetics, sports equipment, textiles, household detergents, etc.). Nanomaterials provide increased mechanical strength for tennis rackets and hockey sticks, for example, self-cleaning and antibacterial properties for textiles, and even special optical features for cosmetics.
- In the medium term, gains can be expected in the high-tech applications of nanomaterials, such as materials with switchable properties, or functional nanoparticles, which can be used, for example, in medical diagnostics, therapy, high-efficiency solar cells or new electronic components.

DR. WOLFGANG LUTHER
VDI Techologiezentrum GmbH, Düsseldorf

Internet
- www.nanoforum.org
- www.nanowerk.com
- www.nanotechprojekt.org
- www.safenano.org
- www.nanoroad.net
- www.azonano.com

Surface and coating technologies

Related topics
- Metals
- Polymers
- Ceramics
- Nanomaterials
- Self-organisation
- Optical technologies
- Automobiles

The surface of an object is what defines our perception of it. Surfaces are not generally perfect, but their utility value and aesthetic appeal can be enhanced by suitable treatment. Coating techniques can be used to improve the way materials and products look and perform. Important examples include wear-resistant coatings on tools or machine parts and thin films on panes of glass that optimise transmission or reflection in selected parts of the electromagnetic spectrum. Many innovative products would not even exist without the special properties provided by thin films. Prominent examples are computer hard discs, optical data storage media like CDs, DVDs and Blu-ray, flat displays and thin-film solar cells.

Tailored surfaces enable the properties of surfaces to be adjusted, such as:
- mechanical (wear, friction)
- chemical (corrosion, permeation, temperature insulation, biocompatibility, wettability)
- electrical (conductivity)
- optical (transmission, reflection, absorption, colour)

Surface engineering is the activity of modifying, structuring or coating materials and components. A distinction can be made between "thick" and "thin" film technology. We can use the term thick film technology to refer to film thicknesses above 10 µm and thin film technology for film thicknesses between 0.1 nm and 10 µm, but in reality the demarcation line is not so clear-cut.

Principles

There are many different basic processes and related modifications for applying surface coatings. The main ones are painting, electroplating, thermal spraying, physical vapour deposition (PVD), chemical vapour deposition (CVD) and sol-gel coating.

► **Thermal spraying**. The coating material initially is a powder, a rod or a wire. After melting in an energetic heat source, small drops are transported to the substrate by a gas stream. The particle speed is between 100–1,000 m/s, depending on the process. The microstructure of the coating results from the solidification of the drops. In order to achieve good adhesion, the substrate surface has to be mechanically roughened (sand blasting) prior to deposition.

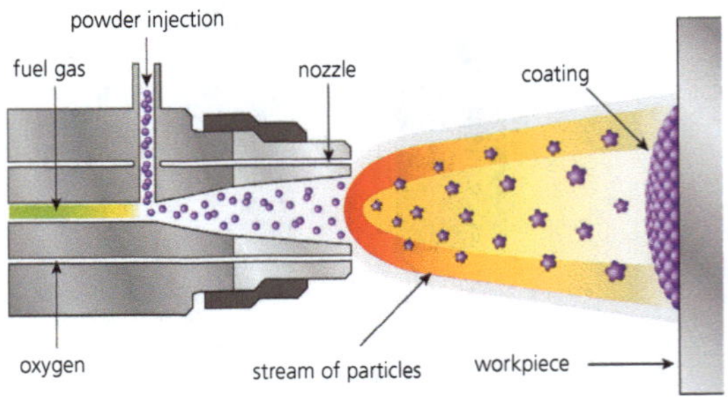

Powder flame spraying (thermal): Powder of the material to be deposited is heated using the chemical energy from the combustion of fuel gas in oxygen. The particles are melted in the flame and small drops are accelerated towards the workpiece where they form a localised coating which may be extended over a certain defined area if either the spray gun or the substrate is moved. Source: Fraunhofer IST

A wide range of metals, alloys and ceramics can be deposited. Thermal spraying is quite simple to handle, structured coatings and gradients are possible, and the temperature load on the substrate is low. The deposition rates are in the range of several µm/s and thus about two orders of magnitude higher than with PVD and CVD processes. The main areas of application are mechanical engineering, aerospace and automotive technology, as well as the medical sector, textile manufacture, paper production and the printing industry.

An important application of thermal spraying is the deposition of thermal barrier coatings on turbine blades for the aircraft industry and power generation. While the efficiency of the turbines increases with the temperature of the working gas, the temperature loading on the base material (a nickel alloy) has to be kept low. A ceramic coating based on zirconium dioxide protects the blades from temperature-induced corrosion.

► **Physical vapour deposition (PVD)**. The basic PVD processes are evaporation and sputtering. They are performed in a vacuum, because single particles have to be transported over a distance of several millimetres from the coating source to the substrate without significant energy loss. The coating material is solid, thin films are deposited by solid-liquid-vapour-

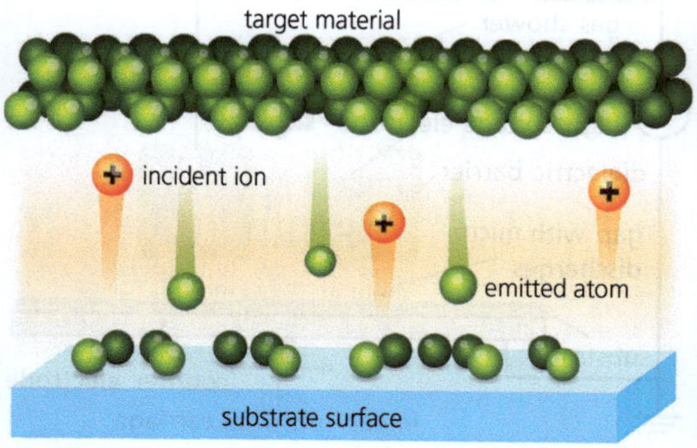

Physical vapour deposition: During a sputter process, positively charged ions created in a plasma are accelerated towards a negatively biased "target" consisting of the material to be deposited as a thin film. The ions transfer their energy to target atoms, some of these leave the surface and form a thin dense film on an opposite substrate. Source: Fraunhofer IST

solid phase transition in the case of evaporation and solid-vapour-solid transition in case of sputtering.

Evaporation is the oldest PVD process. Materials with a low melting point (this holds for many metals) may be evaporated by resistive heating, for materials with a higher melting point (e. g. metal oxides) electron beam evaporation is commonly used. Evaporation is quite a fast process with deposition rates in the range of several µm/s. Film quality and adhesion to the substrate may suffer from the low energy of evaporated particles. More sophisticated evaporation processes therefore make use of an additional plasma. The evaporated particles crossing the plasma zone are activated and ionised and consequently can form a much denser film. Plasma activated electron beam evaporation is nowadays the preferred process for producing antireflective coatings on lenses for optical instruments and eyeglasses.

Sputtering has been likened to a game of billiards with ions and atoms. Positive ions (in industrial processes mainly argon is used as process gas) are generated in a plasma which is ignited between a cathode (sputter target) and an anode (substrate). They are accelerated in the electric field, hit the target surface and transfer their energy to the target atoms, which are ejected and form a thin film on the substrate. If a reactive gas (e. g. oxygen or nitrogen) is added to the argon, metal oxides, nitrides or other compound films can be deposited. Compared with other deposition techniques, sputtering is a rather slow process (deposition rates in the nm/s range), but as the energy of sputtered atoms is around ten times higher than that of evaporated particles, dense and smooth thin films of high quality can be obtained. Sputter cathodes coat surfaces such as large panes of glass (3 m × 6 m) with highly uniform layers. Because the temperature loading on the substrate is very low, even temperature-sensitive substrates such as polymer web or polycarbonate discs (e. g. CD or DVD substrates) can be processed. With modern pulse magnetron sputter techniques, high-performance optical films can be deposited on glass and transparent plastics. New developments using very high-energy pulses aim to achieve even better film qualities.

► **Chemical vapour deposition (CVD).** In CVD processes thin films are formed directly from the gas phase. The source material, or 'precursor', is a vapour. Liquid material may be used but has to be transferred to the vapour phase. There are vacuum-based CVD processes and processes that take place under atmospheric pressure. The energy supplied to the precursor gas in the CVD reactor is used to break bonds, the film is formed from the fragments. Depending on the method of energy supply, a distinction is made between thermal CVD and plasma-assisted CVD (PACVD). Thermal CVD requires high process temperatures (up to 1000 °C) and is therefore limited to heat-resistant substrates, while PACVD is a low-temperature process and thus suitable even for plastics.

The deposition rates of CVD processes are in the range of several tenths of a nm/s to several nm/s. A typical application of PACVD is the deposition of diamond-like carbon (DLC). Such amorphous carbon coatings exhibit excellent wear and friction properties. They are therefore widely used in mechanical and automotive engineering. Acetylene (C_2H_2) and methane

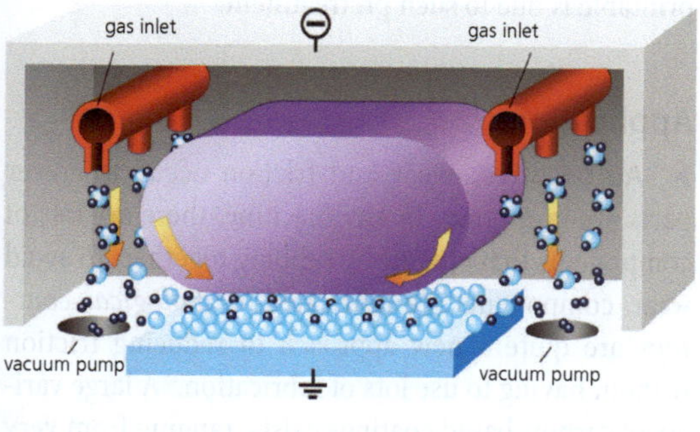

Plasma Assisted Chemical Vapour Deposition (PACVD): Molecules of a precursor are cracked in a plasma discharge. The fragments form a film. In this example, the particles represent CH_4 (methane) molecules (one light-blue with four dark-blue atoms), which are pumped into the chamber. The related coating is diamond-like carbon (DLC) containing a certain amount of hydrogen. Remaining hydrogen molecules (dark-blue) are pumped away. Source: Fraunhofer IST

Coating ceramic face seals by means of hot-wire CVD: An array of tungsten wires is spread out in a vacuum chamber and heated to temperatures above 2000 °C. In this example polycrystalline diamond is deposited using a precursor gas containing carbon. The coating is extremely wear-resistant and has a low friction coefficient.

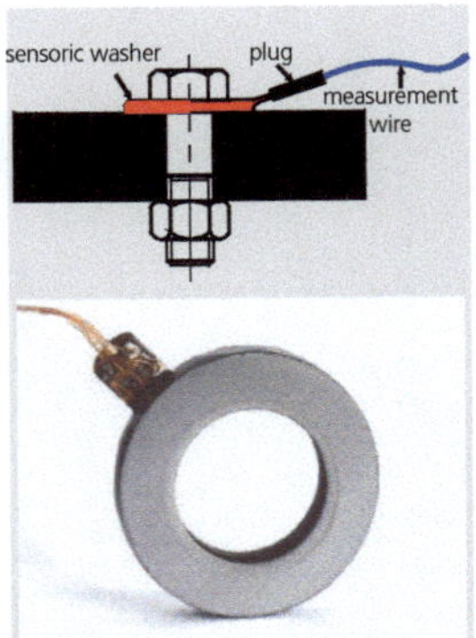

Intelligent washers based on sensoric DLC films: The films change their electrical resistivity under the influence of an external force.

(CH_4) are suitable precursor gases. At deposition rates of 1–2 μm/h and a required thickness of several micrometres the deposition process takes a few hours.

A further important PACVD process is the deposition of amorphous silicon from silane (SiH_4), a gas that contains silicon. *Thin-film transistors* that control the individual pixels in an active matrix liquid crystal display or absorber films in Si-based thin-film solar cells consist of amorphous *silicon*. Thus PACVD is a key process for making innovative products.

Hot-wire chemical vapour deposition (HWCVD) is a vacuum-based thermal CVD process using an array of tungsten wires heated to temperatures between 2,000–3,000 °C. Coatings of large two-dimensional plates (size up to 500 mm × 1000 mm) and substrates with a cylindrical geometry can be produced in different reactors. HWCVD has been used for the deposition of polycrystalline diamond on large surface areas. The feasibility of manufacturing amorphous silicon, microcrystalline silicon and silicon nitride has also been demonstrated. HWCVD may be an important step on the way to low-cost solar cell production.

▶ **Atmospheric-pressure plasmas.** Dielectric barrier discharges – also referred to as coronas – at atmospheric pressure can be used to modify, clean and even coat surfaces without need for expensive vacuum equipment. In a simple version, two rod-shaped electrodes powered by a high voltage face a grounded plane substrate at a distance of a few millimetres. Stochastic discharges occur within the gap. To avoid arc discharges, the electrodes are covered with a ceramic insulator (dielectric barrier). Such atmospheric-pressure plasmas are frequently used to treat plastic films, for examples, to modify their surface energy for better wettability. The fact that plastic shopping bags are printable is due to such pretreatment.

Applications

▶ **Automotive.** Wear and friction occur wherever parts are in motion. In any machine, there are lots of components that operate by rubbing together. To avoid wear, components are lubricated. *Tribological coatings* are quite a new approach to reducing friction without having to use lots of lubrication. A large variety of carbon-based coatings exists, ranging from very soft graphite to diamond as the hardest. The outstanding features of *DLC (diamond-like carbon) thin films* are very low friction coefficients and high hardness (low wear). DLC is about ten times harder than steel and its coefficient of friction is about five times lower. The use of DLC on engine components such as bearings, gears, camshafts, valves and pistons reduces fuel

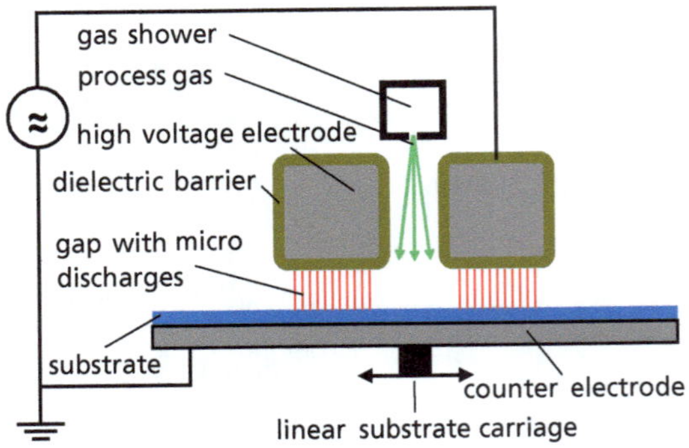

Atmospheric-pressure plasma process: Micro discharges are ignited in a gap between a grounded substrate and two rod-shaped electrodes covered with dielectric barriers. The process gas is introduced through a shower head located between the electrodes. Different process gases can be used to modify or functionalise the surface (e. g. hydroxy, amino or carboxy groups). PACVD processes can also be carried out by inserting special monomers to obtain e. g. hydrophobic or anti-sticking properties. Source: Fraunhofer IST

consumption and CO_2 emission. By reducing wear it also lengthens the component service life.

Besides their tribological properties amorphous carbon coatings can also be equipped with sensor functionalities. The DLC film changes its electrical resistivity with temperature and also under the influence of an external force. This behaviour provides the basis

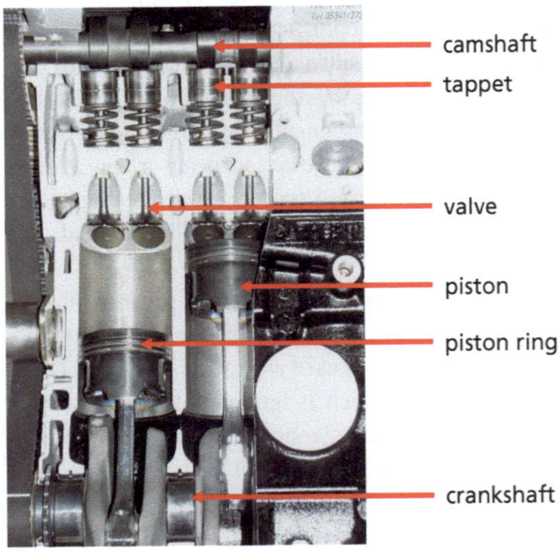

Diamond-like carbon (DLC) on engine components: The DLC coating reduces the friction of camshafts, crankshafts, tappets and pistons, for example, and protects the components against wear. This improves energy efficiency and fuel economy, lengthens service life and service intervals, increases loading capacity and reduces lubricant consumption. About 1% of fuel is saved just by coating the tappets.

for sensors that can be used to detect load, pressure, temperature and wear on plane and curved substrates.

▸ **Optical storage discs.** *CD*, *DVD* and *Blu-ray* Disc would not exist without thin films. The simplest medium is the pre-recorded CD, which only needs a 55-nm *aluminium* film as a mirror for the pick-up laser. The layer is deposited by sputtering, which takes around 1 second to perform. A recordable medium (CD-R) also needs this aluminium reflector, but in addition a recording film consisting of an organic dye is necessary to host the information. During the writing ("burning") process, the dye is locally heated after absorption of the laser radiation and a bubble is formed which marks a digital information unit. Writing and reading processes take place at different laser power levels.

A DVD consists of two polycarbonate discs (thickness 0.6 mm each) bonded together by a glue (thickness 50 μm). A DVD contains two information planes. The reflector film on the side of the incident laser beam has to be semi-transparent so that the beam can also be focused in the second plane.

▸ **Optical filters.** Optical coatings are based on the interference of light waves when passing through sequences of thin films with different indices of refraction. The main materials for such film stacks are MgF_2, SiO_2, Al_2O_3, Ta_2O_5 and TiO_2. They can be deposited by plasma-assisted evaporation, PACVD or sputtering. In general, the film thickness uniformities have to be kept within tolerances of 1%. The cold-light reflector is a mass product known from home lighting. The emission spectrum of a lamp consists of a highly desired visible and a highly undesired near infrared part. The coating on the reflector reflects only the 'cold' visible part, the 'hot' infrared part is transmitted to the back. Typical cold-light mirrors use alternating films of SiO_2 and TiO_2 with a total number of 20–30 films. Economic mass production is possible by means of modern pulsed PACVD processes. Rugate filters are made by depositing alternating layers of low and high refractive material. Their characteristic is a sine-shaped profile of the refractive index. Compared with conventional filters, sputtered rugate filters exhibit improved optical performance and very good heat resistance.

▸ **Coatings on glass.** Thanks to its excellent properties, the importance of glass as a construction material is steadily increasing. Some aspects of its use for architectural purposes are, however, problematic. Its low reflection in the far infrared part of the electromagnetic spectrum (room temperature radiation) causes undesired losses of the thermal energy needed to heat buildings in colder climates. On the other hand, its high transmission in the near infrared (solar radiation) increases the energy necessary for cooling buildings in hot climates. To cool a building by 1 Kelvin takes twice the energy needed to heat a building by 1 Kelvin. Coatings in the nanometre range help to save substantial amounts of energy and reduce CO_2 emissions. A low-emissivity coating based on a 10-nm silver layer that

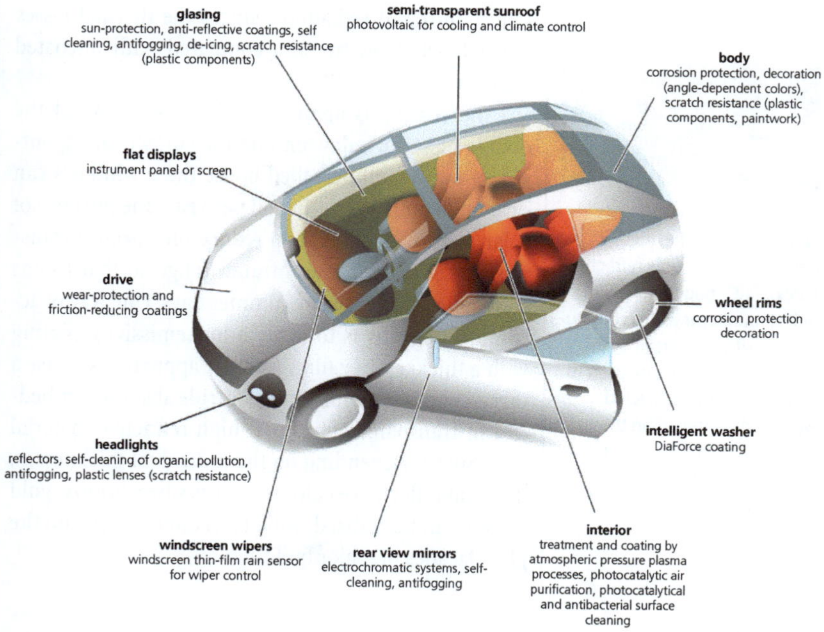

Surface engineering for automobiles

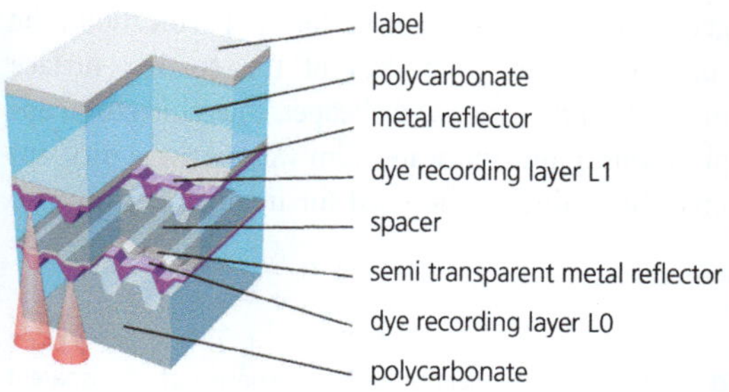

Recordable double layer DVD: The DVD contains two recording and two metal reflector layers. The reflector close to the incident laser beam has to be transparent so that the laser can also detect information from the second recording film. A recordable double layer DVD consists of several thin films sandwiched between two polycarbonate substrates. The label is printed on the surface opposite the incident laser beam. Depending on the laser focusing, information is written either in dye recording layer L0 or in dye recording layer L1. One of the metal reflectors has to be semi-transparent so that the laser can also write in or read information from the upper recording layer (L1). The spacer is a transparent glue bonding both polycarbonate discs together.

Cold-light reflector for home lighting: A multilayer stack of SiO_2 (low refractive index) and TiO_2 (high refractive index) reflects only the visible (cold) part of the lamp spectrum, thus avoiding an undesirable heat load.

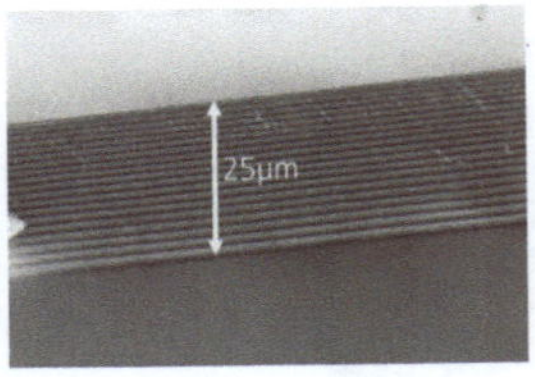

◼ Rugate filter: The filter consists of a SiO_2 (low refractive index)/Ta_2O_5 (high refractive index) multilayer with a continuous transition between both materials. The filters are used to reflect or transmit selected wavelength intervals in the ultraviolet, visible or infrared part of the electromagnetic spectrum.

reflects infrared radiation reduces the thermal losses through windows by 60% compared with uncoated windows.

The coating is applied on the inner pane of the double-glased window and on the surface facing outside. On a typical detached house these windows can save up to 500 l of heating oil per year. The purpose of solar control coatings is to reduce the spectral transmission of glass in the infrared range so that rooms stay cooler during sunny summer days. One way of accomplishing this is to apply a low-emissivity coating with a thicker silver film. Another approach is to use a three-layer stack with a metal nitride absorber embedded in transparent films of a high refractive material (e. g. SnO_2). Depending on the individual thicknesses, different reflectance colours such as silver, bronze, gold or blue can be realized, which can often be seen on the glass facades of major buildings.

◼ Low-emissivity coating and its location in a double-glased window. The coating reflects room temperature radiation, reducing losses of thermal energy by 60%.

In the architectural glass coating industry, films are sputtered on panes of glass of up to 6 m × 3.21 m, with a thickness uniformity of better than 2%. Productivity is very high, 18 m² of a low-emissivity coating can be manufactured in less than one minute.

▶ **Thin films in solar cells**. The absorbers in thin-film solar cells are made of several hundred nm-thick films of amorphous (a-Si) or microcrystalline (µc-Si) silicon, CdTe or Cu(In, Ga)Se_2 (CIGS). The *silicon* films are deposited by PACVD, CdTe is evaporated, and for the deposition of CIGS sputter processes are now often used. All thin-film solar cells need a metallic back contact that is also sputtered or evaporated and a transparent and conductive front contact facing incoming solar radiation. The transparent conductor is made of either In_2O_3-SnO_2 or ZnO:Al. Because indium resources are limited and it is used extensively as a transparent conductor in displays and photovoltaics, the price has increased sharply over the past few years. ZnO:Al is a promising alternative for many applications.

Trends

▶ **Photocatalysis**. Photocatalytically active coatings permit the decomposition of almost any kind of organic material by means of activation with UV light. In addition, such surfaces are completely wetted by water (superhydrophilicity). Photocatalytically active films consist of anatas, a crystal modification of TiO_2. The film thickness is crucial for the effect. It should be in the range of at least 300 nm. One of the important benefits of photocatalytically active surfaces is their antibacterial properties, which can destroy many kinds of germ. They can also be used for air purification. The superhydrophilic properties of the coating surface make cleaning easier and cheaper. In architectural applications a uniform water film washes away dirt, appreciably reducing the need for manual cleaning. To

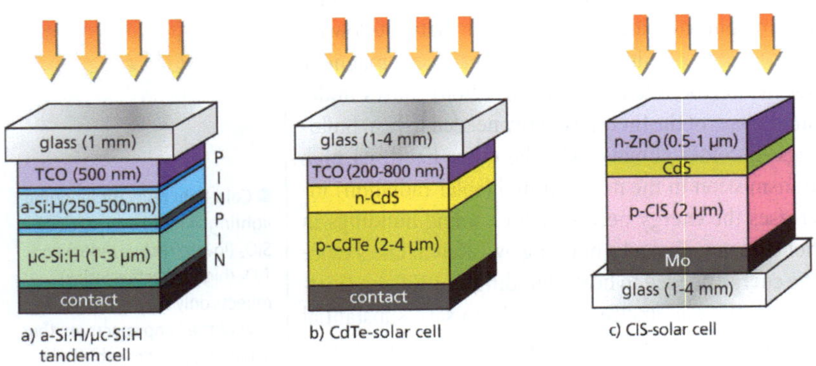

◼ Types of thin-film solar cell: Each type consists of a transparent conductive film (TCO or n-ZnO) on the sunlight incidence side, a metallic contact on the opposite side and a semiconductor film between the two electrodes. The hot solar radiation induces a potential difference between electrons and holes. This appears as voltage on the electrodes contacting the separated p and n doped regions (the "i" relating to the Si cell stands for intrinsic). The three different types of solar cell differ in their configuration (a, b: superstrate, c: substrate), the materials used, their band gaps and, therefore, in their efficiencies. To increase the efficiency of the Si cell (left), a combination of amorphous hydrogenated silicon (a-Si:H) and microcrystalline hydrogenated silicon (µc-Si:H) is used (tandem cell).

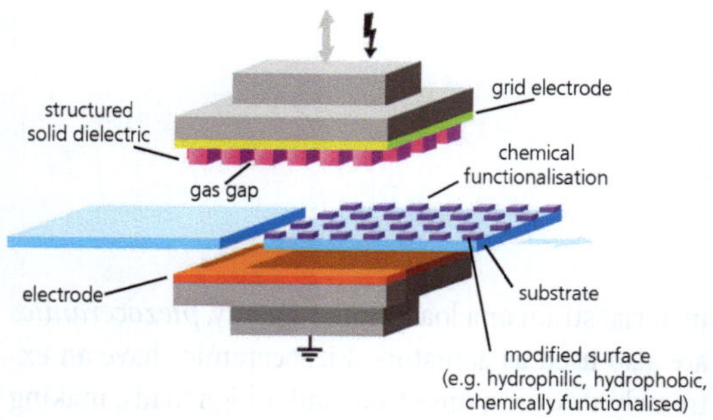

Localised surface modification at atmospheric pressure: Plasma printing enables the treatment of many insulating substrates, e.g. polymers or glass, at exactly defined locations. An appropriately patterned dielectric mask is used to form cavities in which a non-thermal atmospheric-pressure discharge is generated. The substrate is brought into direct contact with the discharge, between the patterned dielectric and the ground electrode. Depending on the process gases, the precursors and the discharge conditions selected, a multitude of functional groups, such as amino, epoxy etc., can be "stamped" onto the substrate.

improve the photocatalytical effect in natural light and to even make it available for indoor applications, the photocatalysis has to be shifted towards visible radiation. This can be accomplished by suitable doping with additional metals.

▶ **Microplasmas in microsystems.** At atmospheric pressure plasmas can be generated in cavities extending only a few tens of micrometre. This creates new possibilities for the localised surface modification of very small structures in microsystems technology.

The plasma printing process permits a localised surface modification or a structured coating deposition and/or functionalisation on large areas. The structures are generated by plasmas that are ignited in the cavities of suitably structured electrodes. As part of the process the surfaces delimiting the cavities can be activated, cleaned or coated. Furthermore, several chemically reactive groups, such as amino, epoxy, carboxy or hydroxy groups, can be locally generated on the surface, for examples, in order to couple biomolecules in biosensor and biochip systems. In addition, by chemisorption of palladium chloride to amino groups, plastic surfaces can be prepared for area-selective electroless metallisation with subsequent galvanic reinforcement. This opens up a wide range of possible applications, for instance, the low-cost production of biomedical and bioanalytical disposables.

Microfluidic disposables are usually produced by polymer injection moulding, a process that is able to generate complex structures with very high precision and resolution cost-efficiently. Microfluidic devices are presently undergoing intensive research and development worldwide as they represent new tools for the manipulation, reaction and analysis of small volumes of liquids – generally aqueous solutions. Because the surface properties of the polymers used are frequently not suited to the application, some kind of surface modification is often necessary. Microplasmas also enable the interior of microfluidic components to be treated and coated, even if they are already sealed, thus facilitating fast, highly efficient and above all inexpensive surface modification and coating deposition of disposables.

Low-temperature substrate *bonding* is decisively important for microsystem packaging. For many applications, however, thermal curing is problematic. The aim is to reduce the necessary temperatures considerably without weakening the bond strength. It has already been demonstrated that the annealing temperature can be lowered to below 100 °C for silicon. Furthermore, antistick surfaces can, if required, be produced by local etching. These results are currently being transferred to other silicon-based materials and to polymers.

Prospects

- Vacuum and plasma deposition processes make it possible to produce excellent thin-film coatings, but they are often expensive and too slow for economic mass production. Continuous efforts are needed to speed up these processes to achieve higher deposition rates and make them more cost-effective.
- One of the most important areas of research for future thin-film technology is the improvement of existing and the creation of new coatings to combat climate change. Thin-film solar cells in particular will play a key role in generating and saving energy.
- Technical surfaces with adjustable properties, so-called smart surfaces, are still in their infancy. In many respects, they still represent no more than a vision of the future. Research and development will focus on subjects such as adjustable adhesion, switchable optical properties and integration of various functions in clothing.

PROF. DR. GÜNTER BRÄUER
DR. SIMONE KONDRUWEIT
Fraunhofer Institute for Surface Engineering and Thin Films IST, Braunschweig

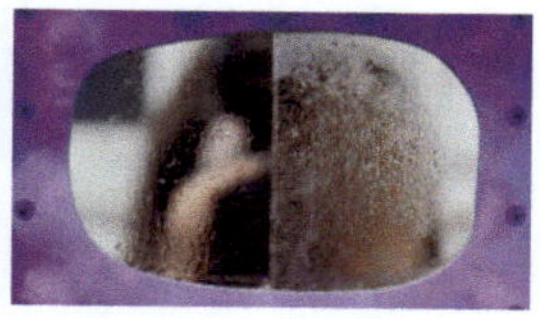

Half-coated rear view mirror. Left: Superhydrophilic surface. Illumination with UV light reduces the water contact angle of the TiO_2-coated surface. Water spreads across the surface, providing a better view through the mirror and improving driving safety.

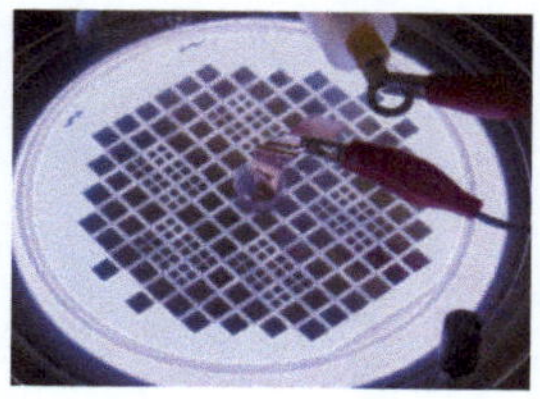

Treatment of an inert surface by plasma printing. Local hydrophilisation is achieved by generating, e.g. hydroxyl groups.

Internet

- www.ist.fraunhofer.de
- www.svc.org
- www.surfaceengineering.org/csg.php

Intelligent materials

Related topics

- Composite materials
- Ceramics
- Nanomaterials
- Sensor technology
- Bionics
- Microsystems technology

Principles

Intelligent materials are capable of responding to stimuli or changes in the environment and adapting their function accordingly. Piezoelectric materials, magneto- and electrostrictive materials, shape memory materials and functional fluids are all used to influence the mechanical properties of systems. The physical effect is not the same: sensors transform mechanical energy into measurable electrical parameters; whilst actuators transform other energy forms into mechanical energy.

These materials are not "smart" or "intelligent" in themselves, as they operate solely in the sensing or actuating mode. Only one effect can be used at a time. In order to make them "smart", the sensing and actuating materials are coupled together with an electronic control unit to form a composite material. The term "smart/intelligent", however, can also be applied to base materials and their transducing properties.

▶ **Piezoelectric materials.** These materials are currently the most widely used intelligent materials. When a load is applied, the direct *piezoelectric effect* generates a measurable electrical charge on the surface. Piezoceramics are therefore used as sensors for measuring mechanical parameters such as force, strain and acceleration. When an electrical field is applied externally, the indirect piezoelectric effect produces a material strain or a load. Consequently, *piezoceramics* are also used as actuators. Piezoceramics have an extremely short response time under high loads, making them suitable for controlling vibrations within the audible range. However, piezoceramics produce a relatively small amount of strain (0.1– 0.2%). This generally corresponds to a displacement of just a few micrometres. Further disadvantages are the high electrical voltage required (a few hundred volts, depending on the type of construction) and the relatively low energy density.

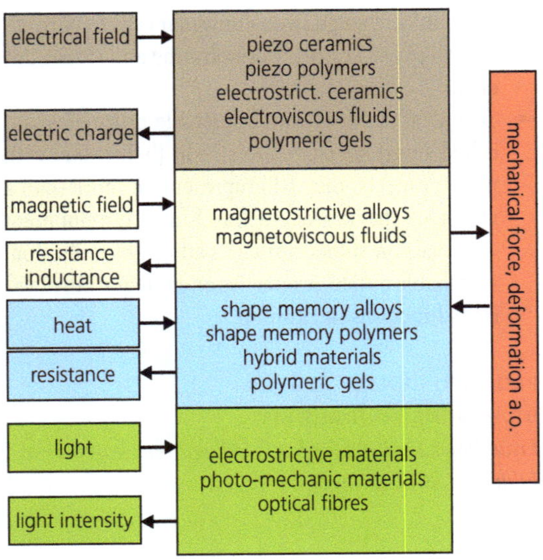

▶ The actuating and sensing effects of intelligent materials: An electrical field applied to the piezoceramics is converted into a mechanical deformation (actuating effect). Conversely, a piezoceramic material can transform a mechanical deformation into an electrical voltage (sensing effect). Source: Fraunhofer LBF

A significant advantage of piezo materials is that they are manufactured as polycrystalline materials and can be mass-produced. This has led to a wide variety of products and designs, examples being stacked actuators and bending transducers. The most significant advance has been the integration of piezoceramic fibers or foil modules into passive structures.

▶ **Shape memory materials.** Shape memory alloys (SMA) are metal alloys which can change their shape through microstructural transformation induced by temperature or magnetic fields. Thermally activated shape memory alloys are the type used almost exclusively at present.

The shape memory effect is based on a thermoelastic, martensitic, microstructural transformation in the alloy. From the high-temperature austenitic phase, a folding process known as twinning produces the face-centred cubic lattice of the low-temperature martensitic phase. The resulting martensite can easily be deformed, as the crystal's folds are extremely flexible. The structure can therefore "unfold" when the material is mechanically strained, enabling a maximum reversible deformation level of approximately 8% . This deformation is called pseudo-plasticity and is permanent for as long as the material remains at a low temperature. However, if the deformed martensite is heated beyond its transition temperature, it recovers its austenitic crystal structure, and hence its original shape. The resulting strain is technically exploitable as an actuating effect.

Apart from generating high strains, SMAs work extremely well at load levels that are high in relation to the volume of material used. The main disadvantage is their "sluggish" response to thermal stimulation. This affects the extent to which they can be used to reduce vibration. It is also difficult to implement the heating mechanism and to integrate the overall system into a structure. Unlike piezoceramics, there is no "inverse" effect, so SMAs can only be used as actuators.

▶ **Electro- and magnetostrictive materials.** *Electrostriction* is similar to the inverse piezo effect audit occurs in all substances. Here too, an external electrical field causes mechanical deformation. Some of the electrons' shells are displaced in relation to the core of the atom. A few specific materials, such as polycrystalline lead magnesium niobate (PMN), can be strained by approximately 0.2%, which is technically exploitable. Electrostrictors compete with piezoceramics in a few applications, as their basic characteristics – such as frequency range, strain and force generation – are comparable. In terms of hysteresis and intrinsic heating, they are even superior to piezoceramics, which favours their use in higher frequency ranges. However, their significantly higher specific electrical capacity and resultant higher power consumption cancels out this advantage. A significant problem is the fact that the physical effect depends strongly on temperature, which restricts their use if no additional provision is made for stabilising temperature. They cannot be used as sensors as they have no inverse effect.

Magnetostriction is the change in geometric dimensions under the influence of a magnetic field. All magnetic materials have magnetostrictive effects, but only the "giant" magnetostrictive effect is of technical interest. Magnetostrictive alloys such as Terfenol-D can be strained by approximately 0.2%. The disadvantages of magnetostriction are the weight and size of the magnetic coils and the activating magnetic field required. The magnetic field can cause electronic components to malfunction, for instance in the automotive industry. In the past, these restrictive factors have favoured the development of piezoceramics over other transducing materials.

▶ **Functional fluids.** The viscosity of electrorheological (ERF) and magnetorheological (MRF) fluids changes under the influence of electrical and magnetic fields respectively: if the field is strong enough, the fluid may even solidify completely. The fluid contains specific particles, which align themselves into long chains along the field lines. This effect can be applied in three different active modes:

- In shear mode, the fluid is located between two parallel plates, which are moved in relation to one another. The stronger the field applied between the plates, the greater the force required to move them.
- In flow mode, the fluids act as a hydraulic fluid. The applied field can adjust the pressure drop in the gap.
- In squeeze mode, changing the distance between two parallel surfaces enclosing the fluid will produce a squeezing flow. The resulting compressive cushion can be varied according to the field strength.

These fluids are mainly used in absorbers, valves or *clutches* such as those used for fitness training and physiotherapy equipment. When using this equipment, the trainee or athlete works against an ERF-controlled attenuator. This has the advantage that isodynamic training is possible. In other words, the resistance of the absorber can be adjusted depending on the velocity of movement. This attenuation profile is very similar to natural loads and is therefore gentle on the muscles and joints. The ERF characteristics can be controlled precisely and have low response times. A further advantage is that ERF technology does not require moving mechanical valve components, which eliminates wear and loud noise.

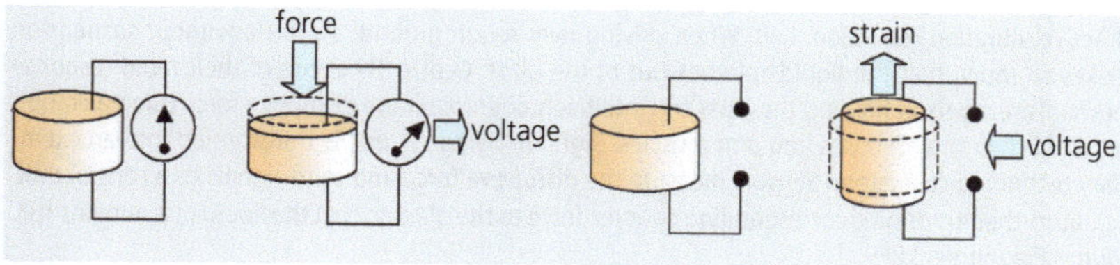

Direct and indirect piezo effects. Left: Direct piezo effect – the force applied deforms the piezoelectric body. This mechanical deformation changes the relative positions of the ions in the piezo's crystal lattice. The electrical charge of the ions causes an electrical displacement, which generates a measurable electrical charge on the crystal's surface (sensor effect). Right: Indirect piezo effect – an externally applied electrical voltage exerts a force on the ions. This force deforms the crystal's lattice structure. This change is detected macroscopically. The deformation is small: 0.1– 0.2% of the component's dimensions (actuator effect). Source: Fraunhofer LBF

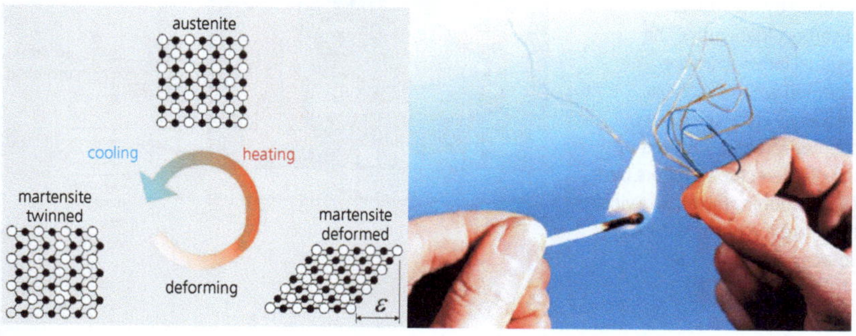

Thermal shape memory effects. Left: In its martensitic phase at low temperatures (below left), the material can be easily deformed and strained to a high degree (up to 8%). Temperature causes the material to "remember" its original shape: When the temperature rises, the material becomes austenitic; when the temperature drops, it returns to its original martensitic phase. Right: A bent paper clip made from SMA: The length of wire returns to its original shape after heating and cooling. Source: Fraunhofer LBF, Siemens

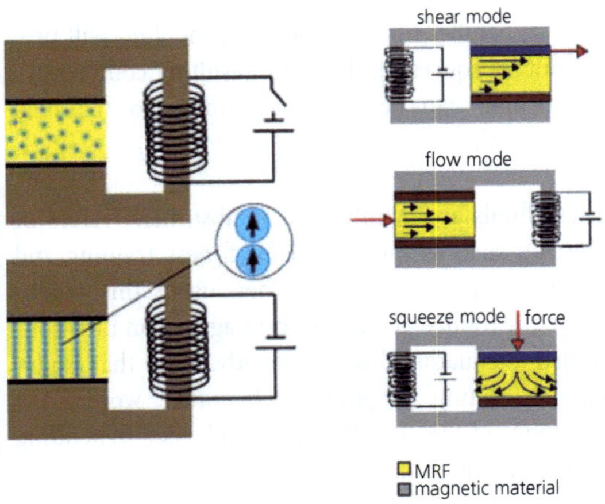

Magnetorheological fluid (MRF). Left: When a magnetic field is applied, the magnetic particles (blue) form chains in the fluid (yellow), increasing the fluid's viscosity. Right: In shear mode, the sliding force between the plates is adjusted. This mode is used in clutches, for instance. In flow mode, the pressure drop is controlled; this mode is used in hydraulic units and absorbers. Similarly, squeeze mode is used to adjust the attenuation characteristics which vary the pressure between the plates. Source: Fraunhofer ISC

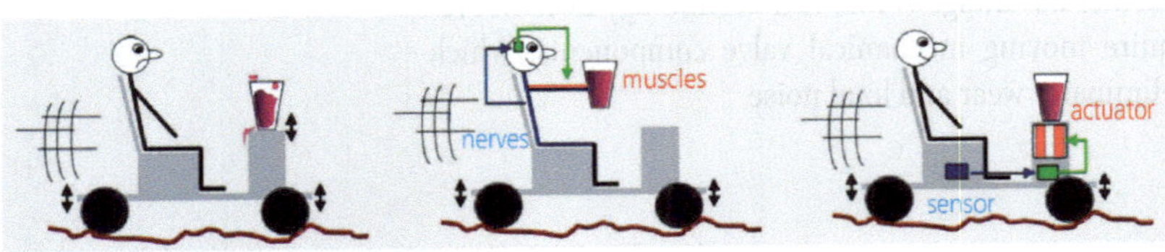

Active vibration reduction. Left: When driving over rough ground, a vehicle without suspension shakes so much that the liquid splashes out of the glass. Centre: By virtue of their rapid response mechanism, a person holding the glass will intuitively counteract the disruptive force using a combination of their brain, senses and arm muscles. Right: A "living system" is transformed into an exemplary technological system. Sensors measure the disruptive force and send signals to a control unit. Actuators then transmit a corresponding counter-force to the glass so that the liquid in it remains still. Source: Fraunhofer LBF

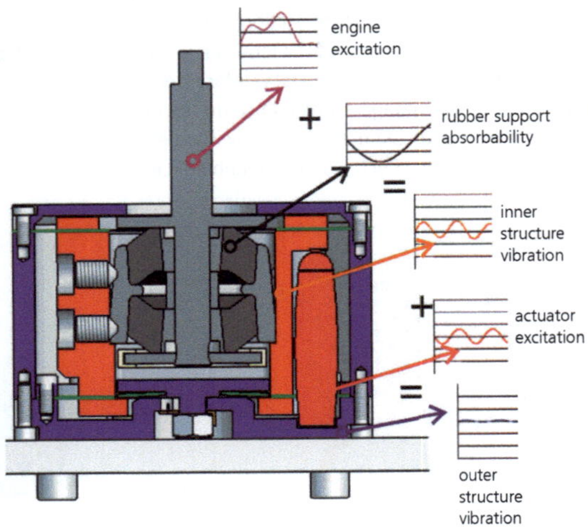

Active vibration reduction of a ship's diesel engine: The mounting points of the diesel engine are fixed to the upper rod of the active interface (grey). The low-frequency range of the engine excitation is damped by the rubber absorbers (dark grey). The residual inner-structure vibration with its higher frequency is canceled out by counter-forces introduced by the piezoelectric actuators (red). This decouples the engine vibration from the body of the ship. Source: Fraunhofer LBF

Applications

▶ **Active vibration reduction.** Looking to the future, active vibration reduction is seen as a key application for intelligent materials. Engineering structures subjected to external excitation can be stabilised, reducing the noise and vibration that impair comfort. This will be particularly important in the transportation industry.

Recently, the basic principle was used successfully to damp the engine vibrations on a *ship*. The ship's *diesel engine* was placed on active interfaces consisting of piezoelectric actuators, which mechanically decoupled the engine from the ship's chassis.

The theory of *active vibration reduction* can also be applied to audible vibrations. The equipment used in magnetic resonance imaging (MRI) may generate noise in excess of 120 dB, causing extreme discomfort to patients. To reduce noise, a prototype scanner tube was fitted with piezoceramic fibre modules. Actuators, sensors and a controller counteract the acoustically radiating surface and reduce the amount of noise emitted.

▶ **Glass facades.** Intelligent materials in the form of adaptive glass can autonomously regulate the amount of light entering a building. In contrast to solar control glass, adaptive glass enables the energy and light transmittance of glass panes to adjust to the actual requirements. Most systems also permit a view outside even when the glass is dimmed. They are distinguished according to the manner in which the optical switching takes place:

- *Electrochromic glazings* have transparent, electrically conducting, oxidic coatings that change their optical properties by assimilating or releasing charge carriers. If a voltage is applied, a charge carrier transfer takes place and the layer changes its sunlight transmittance. Some metal oxides exhibit color changes of this kind: tungsten oxide, for instance, turns blue.
- The transmittance of *gasochromic glasings* is changed by a supply of gas. In this case, too, the adaptive coating is made of tungsten oxide, which is applied as a transparent film to the inside of the outer pane. The glass is coloured by introducing gaseous hydrogen, which is provided by a catalytic layer. The gas permeates the porous tungsten oxide layer. The decolourisation process is initiated by oxygen flowing over the active coating. The gases are fed in via the space between the panes. Gasochromic systems need a gas supply appliance.
- *Thermotropic glasings* react by dimming or reducing their light transmittance when a certain temperature is exceeded. A thermotropic component such as n-paraffin, which is integrated in the matrix, then changes from a crystalline to a liquid state, causing the refractive index of the glass to change and a greater amount of light to be reflected.

Trends

▶ **Bionic concepts.** Given that the actuator, sensor and controller of a mechatronic system are similar to the muscles, nerves and brain of a living being, smart materials can be used to implement bionic concepts. This principle can be applied to adaptive form control, for instance, so that a system's external shape can be continuously adapted to the external circumstances, e. g. gusts.

A study of bird flight revealed that the wing profile of a Skua gull bulges at a few points in gliding flight. These "pockets" enable the bird to glide with less energy, and investigations into *aircraft wings* have revealed similar phenomena. Artificially thickening the wing's upper surface, by introducing bulges, can have a positive influence on flow conditions and thus save energy. As with the Skua gull, however, the location and size of the thickening must be variable so that it can form intelligently along the wing's upper surface. This is achieved by equipping the wing's upper surface with sensors that constantly measure the flow conditions. These data then enable the actuators distributed along the wing to "bulge" the profile accordingly at the right points. The actuators are based on shape memory alloys and are embedded in the wing. The flow processes on a bird's wing and on fixed-wing aircraft are not governed by the same physical laws, due to the different flight velocities, but the basic principle of minimising energy using adaptive shape control is common to both. This example shows how we can "learn from nature" by implementing intelligent materials in real products.

▶ **Intelligent packaging.** New developments in packaging with active additional functions are based on intelligent materials with sensory/actuating functions. Smart packaging can monitor and display the condition of packaged food during transportation and storage, usually by a coloured wrapper. Indicator elements integrated in the packaging show by a change of colour how fresh the food still is, or whether the recommended storage temperature has been exceeded. Time/temperature indicators, for instance, do this by diffusing coloured wax with various melting points on a paper matrix.

For easily perishable foods and for foods packaged in a protective gas atmosphere, oxygen indicators show whether the oxygen concentration in the packaging is too high. One such oxygen indicator is the redox dye methylene blue, which is incorporated in a polymer matrix, e. g. a hybrid polymer. After application and hardening, the indicator layer is blue. It loses its blue colour and is activated by exposure to UV light. If the packaging is damaged, allowing oxygen to penetrate, the indicator reacts with the oxygen and the reoxidation turns it blue again. This indicates to the customer that the contents may be spoilt.

Freshness indicators use antigen-antibody-reactions (*immunoassays*) to reveal the presence of spoilage organisms. In addition to smart functions, scientists are developing packaging solutions with active additional functions. These contain reactive substances which autonomously adapt the conditions inside the packaging to help maintain the quality of the contents. The most advanced developments are the absorption of residual oxygen from the ullage, and the antimicrobial coating of packaging materials.

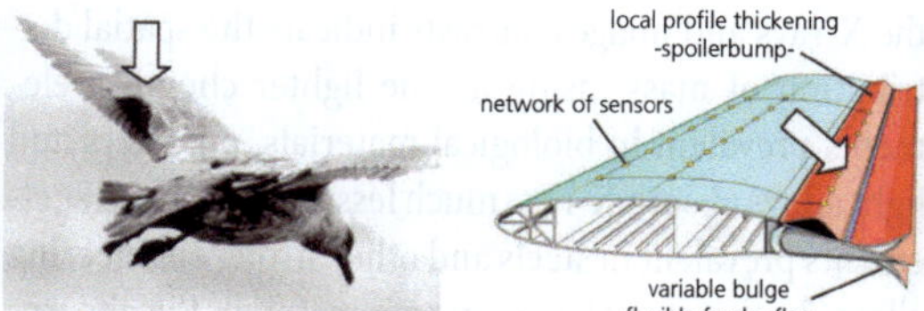

Adaptive wing. Left: Gliding flight of a Skua gull. Local thickening of the profile saves energy in flight (arrow). Right: Cross section of an adaptive aircraft wing. Adaptive local profile thickening ("spoiler bulges") is tested. A network of sensors on the wing's upper surface is required to measure the "pressure environment". Source: Ingo Rechenberg, TU Berlin, DLR

Prospects

- Intelligent materials can enhance product performance. By improving the mechanical properties of a system through adaptive assimilation to actual environmental conditions, future systems can respond independently to vibration, noise and other external influences and counteract them. Consequently, future products will save energy and require less material although they will have the same or even better properties than a passive system.
- Many ideas develop from observing nature and lend themselves to being translated into technological systems will, in the future, find their way into technical products that use intelligent materials. The concept of intelligent materials with sensing, actuating and control engineering functions will soon become a reality.

PROF. DR. HOLGER HANSELKA
DR. JÜRGEN NUFFER
Fraunhofer Institute for Structural Durability and System Reliability LBF, Darmstadt

Chromogenic materials. Above: Transparent yellow-red-violet switching (an example of a dye-doped hydrogel). Below: A demonstration window (100 cm x 50 cm) manufactured using a translucent gel. Left: At 20 °C. Right: At a switching temperature above 30 °C. Source: Fraunhofer IAP

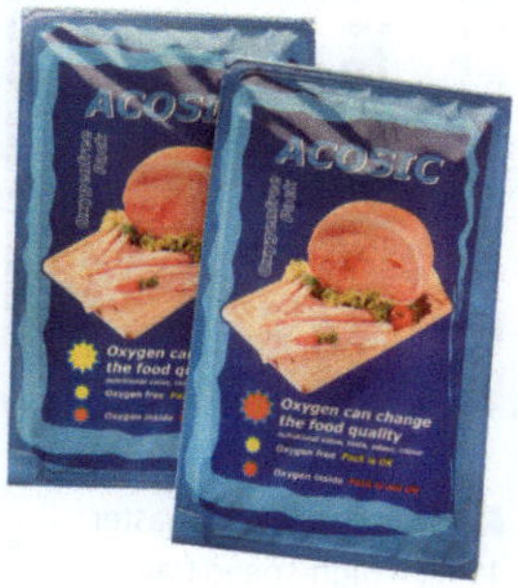

Packaging with an oxygen indicator: If oxygen penetrates the packaging, for instance through a defect in the heat-sealed joint, the indicator dye incorporated in the cover foil oxidises and changes colour from yellow to red (demonstration example). Source: Fraunhofer IVV

Internet

- www.lbf.fraunhofer.de
- www.fva.fraunhofer.de
- www.dlr.de/fa
- www.inmar.info

Testing of materials and structures

Related topics

- Materials simulation
- Optical technologies
- Laser
- Measuring techniques
- Sensor systems
- Joining and production technologies
- Metals

The Eschede rail disaster 1998: A high-speed train derailed as a result of an unnoticed fatigue crack in a wheel rim. Since then the inspection of wheel rims has been significantly improved and intensified.

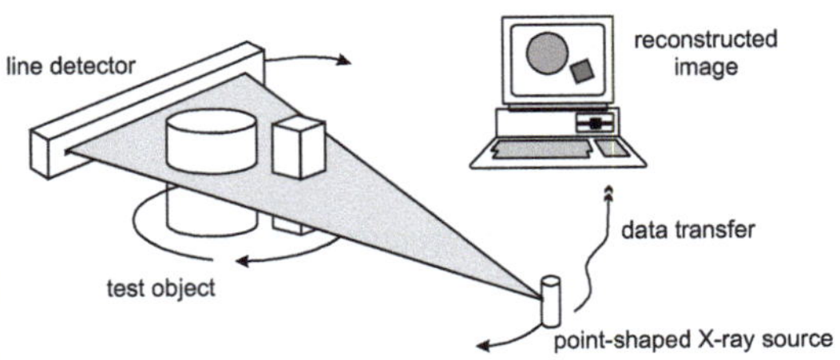

The principle of computed tomography (CT): X-ray radiographs of the object are taken from different perspectives and slices of the interior are computed using dedicated reconstruction algorithms. Source Fraunhofer ICT

Principles

Disastrous accidents like the high-speed train crash in Eschede (Germany) in 1998 are a clear reminder that reliable inspections of key structural components are vital for the safe operation of complex systems. Material and structural testing is essential in all phases of the product life cycle, from the design stage to production and operation.

There are two main types of material testing: destructive and nondestructive techniques, although the border between them is somewhat blurred. Destructive testing techniques indicate the conditions limiting how a material or component can be used. Destructive tests therefore usually entail applying a load until the test item fails, as manifested in irreversible deformation or fracture. This class of testing can be subdivided into static tests – such as tensile, compression and bending tests – and dynamic tests – such as high-cycle fatigue tests, notch impact tests and crash tests. All of these tests follow the same principles: a standardised specimen is mounted on the testing rig and subjected to a predefined deformation process, while the resisting forces are measured – or the loading history is predefined and the deformation response is monitored, e.g. by strain or deflection measurements. Testing a structural component can be a much more demanding task and may require elaborate testing devices and procedures as well as advanced evaluation methods, especially if complex service conditions are to be realistically emulated.

While destructive techniques are restricted to random testing, nondestructive techniques are in principle suitable for 100% testing, providing an essential and viable strategy for quality control. Nondestructive testing is becoming more and more important as the need for monitoring, testing and maintaining durable products, such as automobiles, and structures, such as buildings, bridges and power stations, is increasing constantly. The data required for decisions on safety issues must be collected without damaging the item or structure being tested. Ideally its operation should not be interrupted or compromised by the testing procedure.

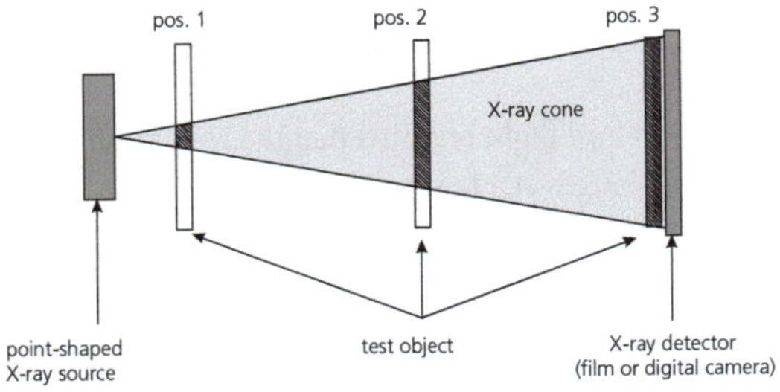

The principle of radiography: A projection of the test object is positioned between an X-ray source and a detector. The view of vision and the magnification of the image are controlled by geometric parameters such as the specimen-object-detector distances and the size and resolution of the detector.

Nondestructive testing is a diverse and dynamic discipline in which various kinds of waves are used to penetrate solid materials. These include electromagnetic waves, sound waves (typically in the ultrasonic regime) and thermal waves. The waves interact with matter in manifold ways, providing a wide range of information about the interiors of the specimens or components under test.

► **X-ray techniques.** Radiography is a versatile imaging method that enables analysis of internal defects like pores and cracks in the bulk of an object being tested. A projection of the test object is obtained by placing it between an X-ray source and a sensitive film or two-dimensional detector. The test object absorbs part of the X-rays and image contrasts indicate the spatial distribution of mass inside it. The lighter chemical elements prevalent in biological materials, polymers and light alloys absorb X-rays much less than the heavier elements prevalent in steels and other strong engineering alloys. In the case of a composite material, the absorption contrast between the reinforcement and the matrix may be large and thus radiographic techniques can be used to image the reinforcement architecture in general and to detect failures in the reinforcements in particular. The more homogeneous the base material, the easier it is to detect small anomalies and faint defects. The resolution of radiography depends on a number of factors, especially the size of the test object, the X-ray source, the film or detector and the view of vision. Under ideal conditions, resolutions in the micrometre range can be achieved. Computed tomography is a rapidly advancing extension of radiography. This technique permits the three-dimensional reconstruction of the test object from a number of radiographs taken from different perspectives.

As the wavelengths of X-rays are in the order of atom-to-atom distances, X-rays are diffracted by the crystallographic lattice. This effect is widely used to analyse the crystallographic structure of materials. X-rays are directed at the surface of the test object and the angular distribution of diffracted intensity is analysed by means of a movable point detector or by a stationary line or area detector. Such diffraction measurements can be conducted with very high precision, making it possible to resolve minute changes in atomic distance and to compute residual stresses from them.

Residual stress analysis is of enormous technological importance as residual stresses may significantly affect the performance and service life of a component – beneficially or detrimentally. X-ray diffraction-based stress analysis methods have thus experienced an enormous increase in use over recent years. Small test objects weighing up to about 5 kg are analysed by means of standard stationary *X-ray diffractometers* equipped with a sample stage on which the test object can be displaced and tilted. Heavier or immobile test objects call for more complex test setups where both the X-ray tube and the detector are moved around the stationary test item. Usually the residual stresses at the surface are of major interest because this is where fatigue cracks and other deterioration processes typically start. Lattice diffraction techniques can also be used to study stresses in the interior of materials and structural components, such as the load partitioning between the reinforcement and matrix in a composite material or hybrid structural component. This, however, requires special types of X-ray or neutron radiation only available at large-scale research facilities operated by national laboratories. Neutron reactors and high-energy synchrotron radiation facilities equipped with dedicated strain scanning equipment have been established at several national laboratories around the world, enabling the residual stress states to be analysed with unprecedented precision and also inside bulky components.

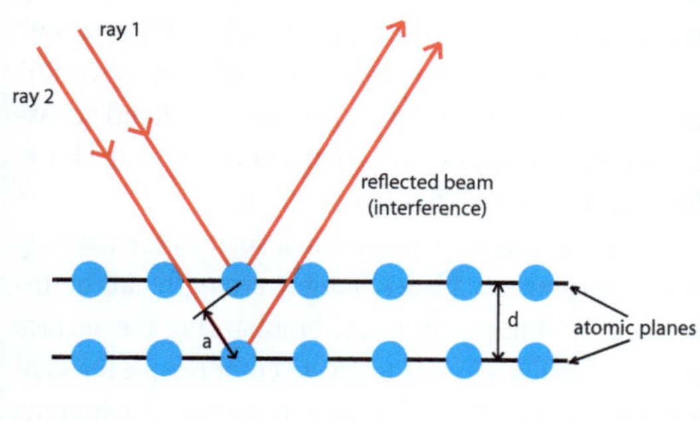

Diffraction measurement: Because crystal structures contain planes of atoms, each plane will reflect incident X-rays differently in such a way as to produce interference of the reflected beams. If two monochromatic X-ray beams (of a specific wavelength) strike a crystal structure at an incoming angle, ray 1 will reflect off the top atomic plane while ray 2 will reflect off the second atomic plane. However, because ray 2 has to cross deeper into the atomic plane, it travels a distance farther than ray 1. If the distance (2a) is equal to the integral number of wavelength of the two waves, they will be in phase, thus constructively interfere, when they both exit the crystal. If the distance d is changed as a result of elastic deformation, constructive interference is observed at a slightly different angle. This effect is used for X-ray-based stress analysis.

► **Ultrasonic testing.** In ultrasonic testing, a short pulse is transmitted into the specimen and picked up by a second transducer positioned at the same position (pulse-echo method) or at some other carefully selected location. A defect in its path will deviate, weaken and partly reflect the pulse. The location and size of the defect can thus be deduced from the arrival times and attenuations of the received signals. The resolution of ultrasonic techniques is largely governed by the wavelength of the ultrasound in the material being inspected: the shorter the wavelength (i. e. the higher the ultrasonic frequency), the higher the achievable resolution. However, since the attenuation of ultrasound caused by scattering and internal friction increases sharply with frequency, the optimum frequency for a given task is often a compromise between two conflicting objectives: spatial resolution versus signal-to-noise ratio. Various ultrasonic scanning techniques have been developed over recent years, which in combination with computer-based data processing and visualisation can provide a three-dimensional picture of the defect population. The transducers are usually coupled to the test object by means of a liquid medium. Because this is not always feasible, considerable effort is being focused on the development of air-coupled ultrasonic devices and other contactless techniques including laser-optical systems.

► **Acoustic emission.** In *acoustic emission analysis*, ultrasonic pulses are generated by the test object itself. Like seismic activities in the earth's crust, acoustic emission is the spontaneous release of localised strain energy in a stressed material. This energy release can, for example, be due to the nucleation of a new microcrack or the gradual propagation of an existing one. Acoustic emission tests are extremely useful for monitoring damage evolution under service conditions and

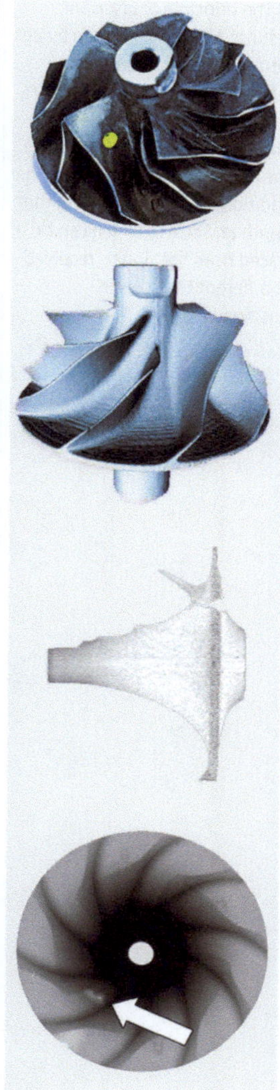

High-resolution CT of a turbine rotor showing a casting flaw. From top to bottom: Optical view, 3D volume reconstruction, 2D slice through 3D data set and single radiograph. The casting flaw is indicated. Source: Viscom AG

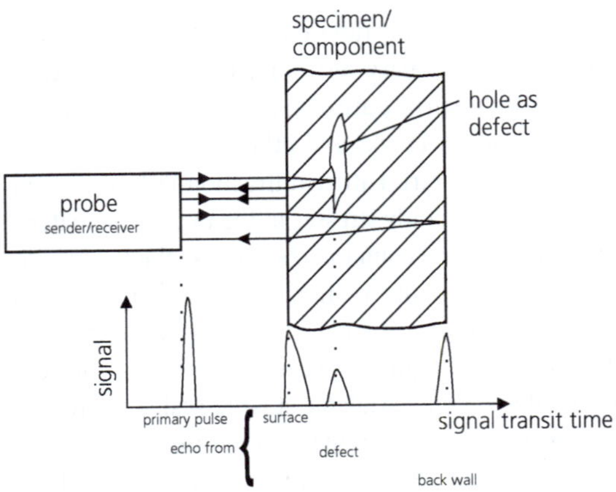

▸ The principle of ultrasonic testing: A pulse generated by an ultrasonic transceiver is reflected off the front surface and the back wall of the test object. If a defect is present in the interior of the object, additional echo is observed, the depth and size of which can be traced from the signals received at different transceiver positions.

forecasting catastrophic failure. In recent years, acoustic emission analysis has advanced increasingly as a quantitative tool, in particular in bulky construction materials such as rock and *concrete*, where tests can be performed on samples that are so large in all three dimensions that the wave propagation from the acoustic emission sources to the sensors is not affected by the boundaries of the sample. Under these circumstances, acoustic emission analysis can be used to determine not only the locations of newly formed cracks but also their size and orientation as well as the fracture modes by which they are formed. In studies on engineering materials, such as *metals*, *ceramics*, *polymers* and *composites*, feasible laboratory samples are typically much smaller and of restricted shape (e.g. plates or rods). In such media, elastic waves are guided by the boundaries and propagate in modes that do not exist in bulk media and that can exhibit substantial wave velocity dispersion. Acoustic emission signals are thus appreciably altered on their way from source to sensor. In recent years, it has become clear that acoustic emission signal analysis can be put on a broader physical basis if these wave propagation effects are taken into account. This deeper insight is also relevant for monitoring the condition and degradation of thin-walled constructions such as pressure vessels.

▸ Thermomechanical fatigue test on a high-strength nickel alloy for gas turbine applications: The rod-like specimen is mounted on a hydraulic test rig. While the specimen is subjected to periodically varying mechanical loads it is also exposed to varying temperatures by means of infrared radiation heating and gas-jet cooling. Integrated computer control of all the test parameters makes it possible to simulate the complex thermomechanical loading conditions that prevail in a turbine engine. Such laboratory tests are crucial for optimum design and reliable service life predictions. Source: Karlsruhe Institute of Technology

Applications

▸ **Turbine blades and discs.** Jet engines play a key role in the aircraft industry, and materials play a key role in jet engines. Components like the turbine blades become extremely hot on the outside and must therefore be cooled from the inside to prevent them from softening or even melting. The temperature gradients in these hollow structures produce thermal stresses that are altered every time the jet engine is started and stopped. The turbine blade material (typically a nickel-, cobalt- or iron-based superalloy) is thus subject to *thermomechanical fatigue*. To increase the service life of turbine blades, a lot of effort goes into optimising their design, improving the cooling concepts and developing better materials. It would be uneconomical to build and test a whole engine for each gradual change in the design or material. Rapid progress is achieved by first predicting the thermomechanical loading scenarios by numerical simulation and then imposing these scenarios on standard laboratory samples by means of a computer-controlled testing setup. This enables temperature and stress to be varied as required. An important strategy for increasing the service life of turbine components is to coat them with a ceramic layer acting as a thermal barrier. Acoustic emission analysis is a viable method for testing the integrity of thermal barrier coatings. The release of strain energy associated with cracking and spalling produces acoustic emission signals well within the amplitude and frequency ranges of state-of-the-art acoustic emission measurement systems – if the sensors can be placed in the vicinity of the test object and interfering noises from other sources can be excluded. Up to now this has only been possible under laboratory conditions, but great efforts are being made to extend acoustic emission techniques to applications in the field, i.e. for monitoring turbines in operation.

In some turbine components, such as the rotating discs carrying the blades, fatigue strength can be increased by shot peening, i.e. bombarding the surface with hard balls in order to create compressive residual stresses in the material near the surface. Such compressive residual stresses are desirable as they inhibit the nucleation of fatigue cracks. However, the fact that desired residual stresses may relax significantly during the service life of a component is a general problem. Applying and optimising this strategy requires a method for quantifying these residual stresses reliably and nondestructively. This task is accomplished by means of the X-ray or neutron diffraction techniques.

▸ **Reinforced-concrete structures.** Buildings are subject to gradual deterioration under normal circumstances and to sudden damage under exceptional circumstances (e.g. earthquakes). Reinforced concrete is an important and universal building material exhibiting very complex behaviour. Research activities are being conducted around the world to gain better insight into the deterioration and damage of this material and structures made from it. In recent years, laboratory experiments have shown that ultrasound inspection and other nondestructive testing techniques can provide a wealth of information about the condition of a reinforced-concrete specimen and about the evolution of internal damage under an applied load. The enormous potential of these tools for efficient condition management of structures in service has been recognised. In recent years, wireless sensor networks have been developed that can be used to monitor measurements taken in situ. Factors of key interest are the structure's own frequencies, humidity and temperature outside and inside the structure, unusual stress and strain and the detection of cracks and other deteriorations using acoustic emissions. Embedded sensor systems register the structural condition and a wireless network transmits the sensor data to a common base station where further analysis is performed. For example, a cross-check of acoustic emission activity with increasing strain or with a sudden or abnormal increase in the ambient or internal temperature can provide insight into the condition of the structure. Such sensor data correlations also reduce the amount of data that need to be transmitted and considered further.

Trends

The two major trends prevalent in nearly all branches of materials testing are in situ testing and computerisation. In situ testing means that nondestructive tests are performed while the component is in service or loaded as if in operation. Much better insight into the condition of an object can be obtained, for example, if a tensile test is carried out under high-resolution observation of the object rather than by just examining a sample under the microscope. This requires miniaturisation of testing devices so that they can be fitted into the chamber of an *electron microscope*, or the development of electron microscopes that are mobile and can be attached to universal testing devices. The intelligent combination of existing testing techniques is a vital source of innovation, as is the development of new and improved techniques.

Computerisation is a megatrend in materials testing, extending from the automation of complex testing procedures to the use of expert systems for evaluating test results. Ultimately, however, the education and experience of the humans involved in testing is the crucial factor for the quality of competitive products.

Residual stress measurements are analysed by means of X-ray diffractometry: If the test object is large and heavy like this turbine disc, it is held in place while the X-ray source and the detector are moved around it, both pointing to the desired measurement location. In the test setup these devices are held by two industrial robots and aligned by a laser-optical system. Source: MTU Aero Engines, Munich

Prospects

- Computer-aided and computer-controlled testing is a growing technology. It is impacting routine testing and permitting the experimental simulation of realistic loading scenarios. The progress in computer-controlled testing procedures and devices will enable us to simulate the most complex mechanical and thermal loading scenarios in laboratory experiments, facilitating reliable life-cycle assessment and service-life prediction.
- The rapid development of materials and processes by means of computer simulation will reduce the number of costly and time-consuming tests that have to be conducted in the laboratory or in the field, making the design process for structural components faster and more efficient.
- Technological systems are becoming more complex and approaching their technical limits as customer expectations in terms of reliability, economical efficiency and ecological compatibility continue to rise. Testing methods that ensure the quality of such systems will be integrated into their design, permitting sophisticated health monitoring over their entire service life.

PROF. DR. ALEXANDER WANNER
Karlsruhe Institute of Technology (KIT)

Acoustic emission measurements being taken on a reinforced-concrete structure under mechanical load: The acoustic signals emitted by cracking and other processes of deterioration inside the structure are picked up by several sensors and then transmitted wirelessly to a remote data storage and processing unit. The potential of this approach is currently being examined for the continuous or regular structural health monitoring of bridges. Source: Materials Testing Institute, University of Stuttgart

Internet

- www.iwk1.uni-karlsruhe.de
- www.bam.de/index en.htm
- www.cnde.iastate.edu
- www.sustainablebridges.net
- www.mpa.uni-stuttgart.de

Materials simulation

Related topics

- Metals
- Casting and metal forming
- Ceramics
- Automobiles

Principles

Materials simulation enables us to look into the future of materials or components, for instance through reliability predictions or life cycle analysis. It allows components to be tested while they still only exist in the heads of their developers or at most on blueprints. And simulation makes it possible to impose load scenarios which ought not to occur at all in practice or which would be technically too difficult to imitate in a test field.

The basis of materials simulation is the depiction of real or yet-to-be-developed materials or components as a computer model. Their structure, function and behavior are cast in models and described by mathematical formulae. The materials response to external stresses or other loading is then simulated. The quality of the computerized predictions is determined by the practical relevance of the model and by the parameters that are fed into it, which often have to be determined through elaborate and time-consuming experiments.

▶ **Simulation methods**. The "tools of the trade" on which simulation is based are the classical theories of physics and continuum mechanics, which are usually formulated as partial differential equations. The method most commonly used in practice is the *finite element (FE) method.* In FE modelling a component is broken down into a finite number of small elements to which the materials properties are assigned. The requirement that each element be in equilibrium with its neighbours results in a system of equations which is solved numerically. The state of equilibrium thereby produced can be seen as the response of the material or component to the external load.

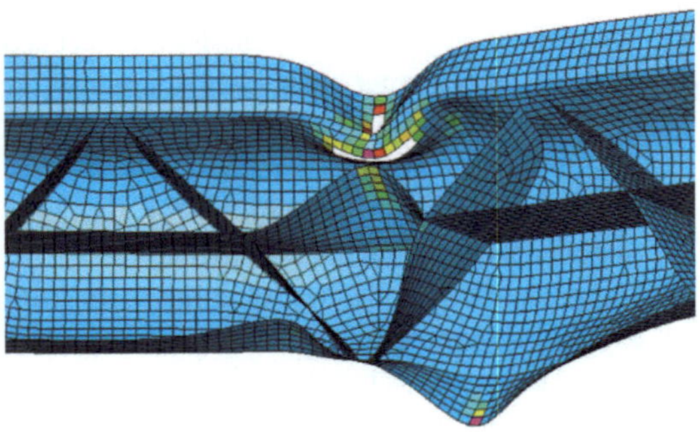

Simulation of the deformation of a car body structure: The element lengths in crash models for critical regions of a structural component are usually of the order of a few millimetres. With increasing computing power the minimum of the element size in crash simulation decreases to better reflect details of the geometry. At the same time more detailed and more accurate materials models are employed. Source: Fraunhofer IWM

Components are frequently designed in such a way that they do not change their shape significantly or permanently in operation. In this case it is sufficient to describe the material as elastic (elastic means that the material resumes its original shape when relieved of the load), which only calls for a comparatively small numerical effort. Greater computing power is required for problems involving major, permanent deformations of the kind that occur when reshaping metal car body parts or in a crash. The material is then usually described as elastic-plastic. More elaborate materials models are needed if, for instance, the internal material structure is strongly altered for example in or near weldments.

In contrast to the simulation of solid continuous materials, simulations of powders or liquids (such as for filling processes or flow simulations) make use of particle dynamic methods. In the discrete element method (DEM) or in the smooth particle hydrodynamics (SPH), the material to be simulated is put together from individual elements that may each have different shapes and properties. When filling a forming tool with ceramic powder, for instance, different grain shapes of the individual powder particles can be modeled. Their position or speed can be calculated after a certain time from the physical laws that govern the particles and the way in which they interact, e. g. by friction.

On the molecular level, forces of repulsion or attraction between atoms or molecules take effect. In this case one employs molecular dynamics simulations (MD) which are used to follow the time evolution of all the atoms involved. This can be most useful to study the spreading of a liquid or the details of the propagation of a crack. Quantum mechanical methods such as the density functional theory are employed to simulate chemical reactions. Their aim is to describe the interplay of atoms and electrons in search of an energetically balanced state. They can reliably predict the energy cost or gain in breaking and forming chemical bonds and give access to electronic or optical properties of materials.

The accuracy of materials simulation is determined by the complexity of the material model and the available computing power. This often means working with limitations and simplifications. A certain amount of experience is therefore required in the selection of the appropriate model to ensure that the simulation results are meaningful for practical application.

Applications

► **Virtual exhaust laboratory**. In a combustion engine, the exhaust manifold collects the hot exhaust fumes from the cylinder block and passes them on to the catalytic converter. Depending on the operational status of the engine, the exhaust manifold has to withstand temperatures from below 0 °C in winter when shut down, and up to 800 °C at high engine speeds. The combination of these temperature transitions with mechanical load cycles result in *thermomechanical fatigue* of the material and ultimately, after a certain number of cycles, crack formation and component failure. Numerous time-consuming component and field tests are necessary when designing and optimising such high temperature components and no exhaust manifold can be built in series until it can be guaranteed to have a sufficiently long service life.

Materials simulation enables an increasing proportion of the component development to be carried out on a computer. This presupposes a simulation model with which the complex behavior of the material at various temperatures can be described. The model has to describe the cyclic plasticity, rate dependency, creep and relaxation of the material in mathematical terms, and ultimately to predict the locations and lifetimes when micro-cracks will occur. The materials simulation is compounded by the fact that the mounting conditions of the exhaust manifold are extremely complex due to the threaded connections, the gaskets and the inhomogeneous temperature distribution depending on the geometry of the component.

The appropriate calculations are based on the FE method. Each element is assigned thermo-mechanical properties of the material from which the exhaust manifold is to be made. These include the elastic behavior, yield strength, deformation behavior and the coefficient of thermal expansion – all as function of temperature. To ensure that the simulations accurately reflect reality, the specific properties must be determined in complex experiments. Special fatigue tests are carried out at various temperatures to obtain these parameters and to determine damage parameters to be incorporated in the materials simulation. The computer model then undergoes the same load cycle that is experienced by the real-life component. The components are virtually heated up and cooled down again in a "virtual high-temperature laboratory". After a few hours of computing time, the critical points that limit the lifetime of the component can be pinpointed. The mechanical stresses and the associated damage suffered by each structural element become visible.

Although the new manifold still undergoes practical testing after its development in the virtual exhaust laboratory, there is no longer any need for expensive preliminary testing.

► **Powder technology**. *Ceramic* components are known for their high wear- and temperature-resistance, strength and hardness. Typical applications for ceramic components include heating elements, slide bearings, nozzles and ball bearings. In the manufacturing process, loose powder is filled into a die and pressed to form what is known as a green body. The green bodies are then sintered at high temperatures, causing the powder particles to fuse into dense semi-finished products or finished parts. The technological challenge is to fill the die evenly with powder, as different densities in different areas can not be balanced during pressing. Depending on the shape of the component, these density variations are amplified to a greater or lesser extent during pressing. Cracks can form after powder pressing, particularly during ejection of the green body out of the die. During *sintering*, the areas with different densities shrink differently, with the result that the parts warp – which is why it is usually desirable to achieve as homogeneous a density distribution as possible.

Distortions and crack formation generate high costs in modifications to machine tools. Materials sim-

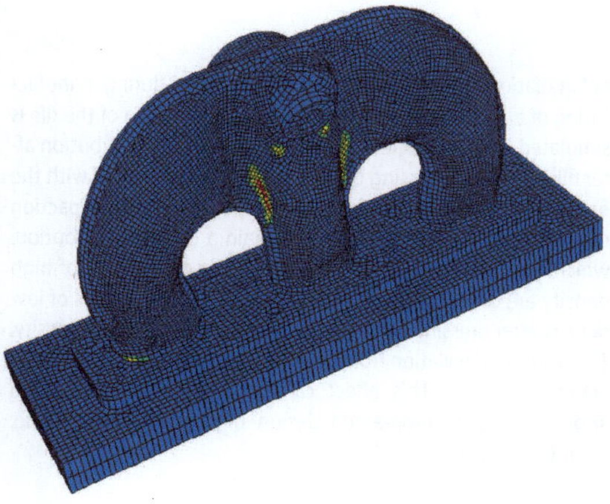

Finite-element simulation of an exhaust manifold which is subject of a virtual "high temperature component test". In red, the critical locations of the exhaust manifold where microcracks will appear. Source: Fraunhofer IWM ulation allows the filling process to be planned, tool shapes and pressing plans can be optimised before the tools are made, and suggestions for improving the finished parts can be submitted.

▶ **Filling simulation.** A widespread means of filling dies with powder is by using "feed shoes". These travel across the mould one or more times, discharging the powder they contain. In order to be able to dispense with elaborate trial-and-error loops to achieve uniform density distribution, the powder grains are numerically described using the DEM. Redistribution within the bulk goods is calculated by this method. The quality of simulation depends on a good representation of the grain shapes and the friction between the powder grains and on imitating the discharge behaviour from a moving feed shoe.

▶ **Simulation of powder pressing.** Knowledge of powder density distribution is essential to the realistic prediction of the distortion due to sintering or the formation of cracks, both during pressing and during subsequent sintering. To this end, the mechanical behaviour of the powder during compression – that is, under pressure and shear stress – must be described. One important parameter is the coefficient of friction between the powder and the tool, which must be measured and incorporated in the simulation. The density distribution can be calculated on the basis of plastic deformation and the consolidation of powder.

▶ **Simulation of sintering.** The *sintering* process takes place in three stages, during which the porosity and volume of the green body decreases significantly. In the first stage, the particles rearrange themselves relative to their initial position so as to better fill the space. Sinter necks form between adjacent particles. In the second stage, the porosity drops noticeably. In the third stage the strength of the sintered bodies is established by surface diffusion between the powder particles. Materials simulation models all three stages. Parameters such as the development of pores between the powder grains and the formation of contact surfaces must be described mathematically. The result is a description of how the component geometry develops during the sintering process.

◀ Simulation of three different processing steps during manufacturing of a aluminium-oxide seal disc. Top: The filling of the die is simulated with a particle code, giving the density distribution after filling. Centre: Pressing of the green part is simulated with the finite element method using a plasticity law for the compaction of granular materials. The result is again a density distribution, which is represented here through a colour code. Areas of high density are coloured red. Bottom: The sintered disc. Areas of low density after pressing shrink more than those with higher density. The result is a deviation from the desired geometry. The seal disc warps as shown. This effect can be eliminated through an appropriate simulation-aided design of the punches. Source: Fraunhofer IWM

Trends

▶ **Virtual process chain.** A car body sheet can have completely different material properties depending on how it was milled, pressed or bent. These varying properties are of particular interest in crash simulation, as plasticity is affected not only by the way the starting material is manufactured, be it by casting or by milling, but also by all the subsequent production steps such as machining, pressing and welding. Today's simulation models only factor this in to a limited extent. Conventional finite element modelling assumes that there are for example numerous randomly distributed individual crystallites in each finite element. However, the plastic deformation of a car body element is affected by the grain structure and crystallographic texture of the polycrystalline metal. This in turn depends on the deformation behaviour of individual micrometre sized crystallites.

The challenge is to trace the changes in the individual crystals throughout the various stages of manufacture by computation.

For simulations to produce an accurate materials-specific prediction, the most important mechanisms on the atomic level, the crystal level and the macroscopic (material) level must each be integrated in what is known as a *multiscale material model.* However, the computing power required in order to simulate a complete car body part on a crystal mechanics level, let alone the atomic level, becomes excessively large. It is therefore particularly important to filter out the most important information from each level without detracting from the predictive accuracy of the simulation.

▶ **Friction and wear (tribosimulation).** One of the main goals when developing new mechanical systems is to reduce energy losses and thus to conserve fossil fuel and cut emissions. A significant amount of the power of a vehicle engine is consumed by friction and converted into heat and wear. This is aided by countless tribocontacts, i. e. the contact made by two friction parts and any lubricant between them. In technological terms, tribocontacts are among the most demanding elements of a technical design. The complexity of friction phenomena is due to its multiscale properties, mechanical roughness being just as important as chemical consistency. The lubricant itself is a complex

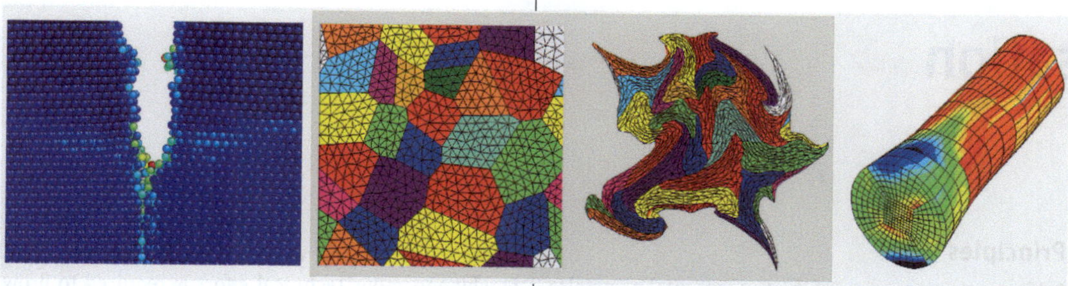

Multiscale material modelling when drawing out tungsten wire. Left: Crack along a grain boundary in tungsten at the atomic scale. Centre: Grain structure in tungsten before and after drawing at the microstructural level. The colours illustrate the different alignment of the grains. Right: Stress distribution in a polycrystalline tungsten wire during drawing (material level). Tension occurs during wire-drawing, and the specific conditions at the grain boundaries can cause longitudinal cracks in the tungsten wire. Source: Fraunhofer IWM

material with additives. Furthermore, new phases are created during the friction process. This is why the simulation of friction and wear processes has to cover all scales from the elastohydrodynamics of the lubricating gap to the quantum mechanical description of the bonding electrons:

Taking the example of *diamond-like carbon layers* that are used for wear protection, simulation methods can be illustrated on various scales:

- Quantum chemical simulations are used to calculate the fundamental properties of the coating material, based on bonding energies. It is also possible to simulate possible reactions between the basic lubricant, any additives, oxygen, and the surfaces involved.
- Molecular dynamic simulations give information on layer wettability or on friction values. A possible use of this method is to describe the early stages of abrasion or the dynamics of a lubricant film in microscopic terms.
- Mesoscopic simulations make it possible to depict the interrelationship between layer topography (surface structure) and friction value, and to simulate how topography affects layer parameters.
- Continuum mechanical simulations make it possible to simulate macroscopic crack formation. This method is used to create models to determine the service life of layered systems and calculate their load-carrying capacity.

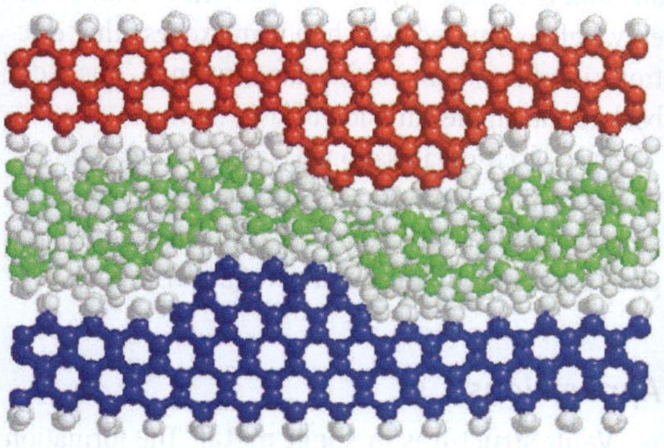

Simulation of tribocontacts at the atomic level: Two nanoscale bodies (diamond asperities, red and blue) are in relative motion lubricated by a simple base oil (hexadecane, green and white). Such simulations provide deep insights into the origin of friction, wear and running-in phenomena.

At quantum chemical level, it is economically feasible to simulate model systems with a few hundred atoms. However, millions of atoms would be necessary to describe even a single asperity contact in the sub-micrometer range. This is where multiscale material simulation becomes necessary. It involves mathematically combining the most important information from the atomic level, the micro-structural level and the material level to produce models that can be used for realistic calculations that will yield accurate predictions.

Prospects

- Materials simulation helps to model load scenarios and processes for existing and yet-to-be-developed materials and components on the computer, and to develop and improve these materials to suit their intended use.
- The atomic, molecular or crystalline microstructure of the material can be incorporated in simulations that will yield accurate predictions, depending on the requirements to be met.
- The present and future challenges are to comprehend more soundly the response of materials to the manifold loads and environments they are subject to. Better materials models, more accurate numerical methods and linking of different simulation methods are required.

PROF. DR. PETER GUMBSCH
THOMAS GOETZ
Fraunhofer Institute for Mechanics of Materials IWM, Freiburg

Internet

- www.simbau.de
- www.iwm.fraunhofer.de

Self-organisation

Related topics

- Metals
- Polymers
- Sensor systems
- Laser
- Nanomaterials
- Microsystems technology

Hive-like vault structure illustrated by a washing machine tub: A hexagonal vault structure was developed here. This is an evolution from the square structure, because it achieves torsion stiffness in all directions. Source: Miele

Principles

Self-organisation is one of the fundamental principles of structure formation and growth in nature. Individual building blocks join to form ordered, functioning units. This enables the development of complex systems and the assembly of galactic clusters, special types of landscapes, crystalline structures and living cells. Some influential factors, such as bonding forces between system components and special environmental conditions also play a decisive role in that process.

The principles of self-organisation are envisioned particularly in nanotechnology for the implementation of technological products and generations. The conventional manufacturing processes for the production of highly-ordered structures are reaching their limits in the lower nanometre range. The so-called *bottom-up manufacturing method* is a possible solution. This means building up the desired structures directly building block by building block – or atom by atom, if necessary. However, in order to apply this technique to produce macroscopic building blocks at reasonable cost, self-organisation mechanisms are used for building up the elementary building blocks and ordering them in periodic multiplicity. Control is not carried out by direct manipulation of the individual building blocks, but rather by adjusting the boundary conditions. This appears to enable spontaneous ordering.

At first glance, this phenomenon appears to contradict the Second Law of Thermodynamics, which states that physical systems will, as a rule, strive for a state of maximum entropy (maximum disorder). In order to recreate an ordered state, energy must be added. In fact, naturally existing ordered systems, such as crystals, often emerge under extreme conditions. That is why enormous technical effort is required to grow artificial crystals, like silicon monocrystals, that are sufficiently ordered and free of defects. All living organisms are also ordered systems, and they, too, need to put in a great deal of effort (breath, nutrition intake, etc.) to prevent the order from collapsing.

In fact, then, self-organisation does not actually violate the Second Law of Thermodynamics. Ultimately, when order is created in one system, a greater amount of disorder (heat) is created elsewhere. Self-organisation appears in open, non-linear systems under energy input. These systems are not in a thermodynamic equilibrium. For the sake of completeness, one should mention that breaking symmetry by so-called bifurcation, as often happens in non-linear systems, is another significant feature of self-organisation. This means that for the system, two more or less stable states exist as of a certain point (bifurcation point). The switch to one of those states often happens suddenly.

One technically relevant example, which could not exist without self-organisation, is the *laser*. When energy is input, as of a certain point quite distant from the thermodynamic equilibrium (laser threshold), the system abruptly switches into a state of emitting coherent light, i. e. light waves that oscillate in the same direction, at the same frequency and with the same phase. In a gas laser, for example, an electron around an atom is excited. When it drops back to a lower energy, light is emitted, which in turn excites other electrons in other atoms in the same manner. This feedback results in the emission of laser light.

Even if self-organisation plays a role in many areas, the following essay will only discuss the phenomena and mechanisms that address self-organisation as applied specifically to the build-up of material objects.

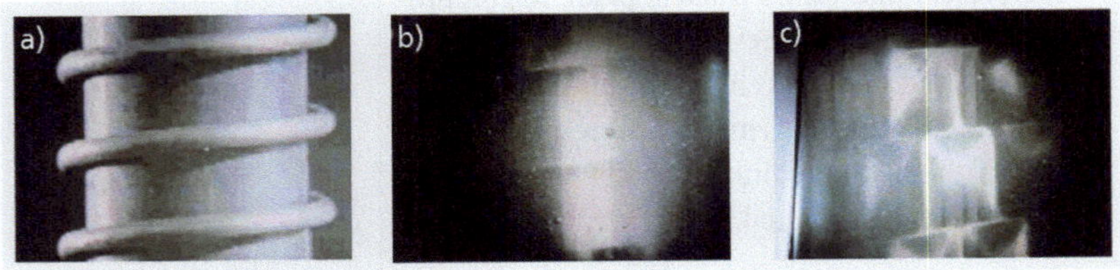

The principle of vault structuring: (a) The support structure consists of a cylinder around which a spiral rail has been affixed. (b) A piece of sheet metal is bent to the shape of a cylinder around the structure. (c) When pressure is applied from outside, a square vaulted structure forms on the surface in a self-organising manner and according to the minimum energy principle. Depending on the supporting structure, square, hexagonal or crest-shaped 3D structures appear that are up to eight times more rigid than the original material. Source: Dr. Mirtsch GmbH

Applications

▶ **Vault structures in sheet metal.** The formation of so-called vault structures in materials technology is an illustration of self-organisation. In the vault structuring process, multi-dimensional structures arise from the material by self-organisation according to the minimal energy principle, just like the shells of turtles or the filigree wings of insects. As happens in nature, this phenomenon arises, for instance, when sheet metal is bent to form a cylinder, which is then subjected to external pressure. The cylinder is supported by an inside spiral guide rail. Symmetrical surfaces take shape on the surface of the metal, without having been completely predefined. The advantage of structured sheet metal is that it has a far greater level of torsion stiffness than unstructured sheet metal. This enables manufacturing of the respective components with thinner walls but the same rigidity, which implies greater efficiency in material usage and energy. This sheet metal is used for the panelling between the rear seat and the boot of cars, for example, or for the tub of a washing machine.

▶ **Atomic force microscope.** The vault structuring is an example of a self-organisation phenomenon at macroscopic level (in centimetres). For far smaller structures (micrometres to nanometres), high-resolution analytical methods are an important condition for the technical evolution of self-organisation at this order of magnitude. For example, thanks to atomic force microscopes (AFM), individual molecules and atoms can be made visible on surfaces. The central sensor element in the atomic force microscope consists of a very fine needle with a diameter of no more than 100 nm. The needle is affixed to the end of a cantilever and is guided across the sample surface. The displacement of the spring hanger as it tracks the topographical structure of the surface is measured with a laser beam. Scanning electron microscopy (SEM) and transmission electron microscopy (TEM) are other significant analytical methods using imaging.

Trends

In the medium and long terms, controlled self-organisation processes are expected to enable product innovations and improvements as well as superior process engineering. Harnessing self-organisation for technical applications promises to generate simple and more efficient structuring processes and new and more sustainable process technologies and manufacturing methods at lower production costs. Additionally, it will allow structural dimensions that would not be accessible using conventional methods.

▶ **Electronics.** For the manufacturing of the smallest electronic components nowadays top-down processes are used. This involves using a macroscopic *silicon wafer* to produce micro- and nanostructures of the electronic components by removal, deposition or implantation of materials. *Lithographical processes* of this type are becoming more and more expensive with the advance of miniaturisation and will eventually reach their limits. So a bottom-up approach is gaining favour for structures measured in nanoscale sizes, in other words, the build-up of the entire system by linking its nanoscale components. To achieve this, a variety of ideas taken from self-organisation are being pursued.

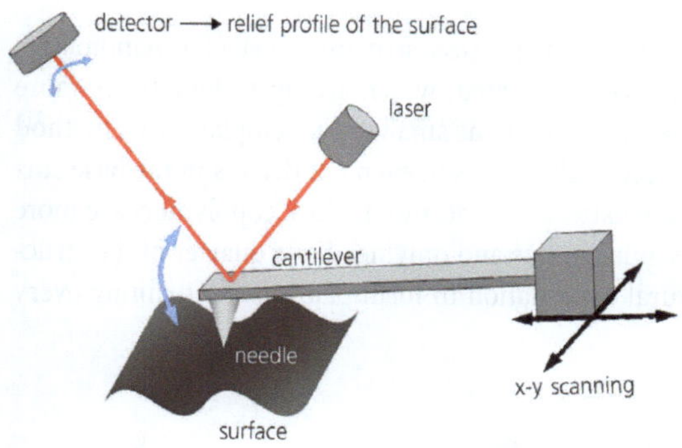

◂ Atomic force microscope: A needle is moved along a surface. This displaces a cantilever, which, in turn, projects a laser beam to different positions of a detector. Scanning over a surface generates a relief profile of that surface up to atomic resolution. Source: VDI Technologiezentrum GmbH

▶ **Block copolymers.** One possible approach to produce nanoelectronic structures is to combine conventional lithographical techniques with self-organising block copolymers. These consist of two or more polymer building blocks that are linked together. Block copolymers exemplify systems with special structures and interface active properties that enable the self-organised build-up of specific complexes. Given adjustable ambient variables, such as temperature and pH values, they will polymerise into a variety of structures. These can serve as templates for structuring and selective coatings of and with nanoparticles and nanowires.

This allows unheard of control in the manufacturing of nanoscale structures, thus providing a way to continue reducing the size of electronic components beyond the ability of conventional methods. This concept is very promising wherever relatively simple and cost-effective ordered structures are required, for example in *semiconductor memories*, waveguides, filters, high-density *magnetic memories*, or *biochips*. Applying this principle a method was developed by which a self-organised block copolymer film is used as a mask for an etching process. This allows to build “nanocrystal flash” memories or *DRAM* capacitors. The minimal size of such component structures is ultimately determined by the molecular weight of the polymers used, so the technology offers potential for future scaling. The advantage over conventional methods is the clear size distribution and more regular ordering, which, by the same token, requires less effort.

Another method is currently being explored using block copolymers: a dot pattern is drawn onto a carrier material by an *electron beam*. Block copolymers are then brought onto the surface (which may be chemically treated). Their long molecular chains spontaneously align themselves with the given structures. Coded information in the molecules determines the specifications, such as size and interval of the alignment. In the process, structures just a few nanometres in size are created, which are up to four times more dense and twice as small as the template. This method additionally allows to even out defects in the base surface pattern. Furthermore, block copolymers are more regular in size and only need one quarter of the structural information to form patterns. So defining every fourth point on the template is sufficient. The technique based on self-organisation is suitable, for instance, to store more data on a hard disc and to continue minimising electronic components.

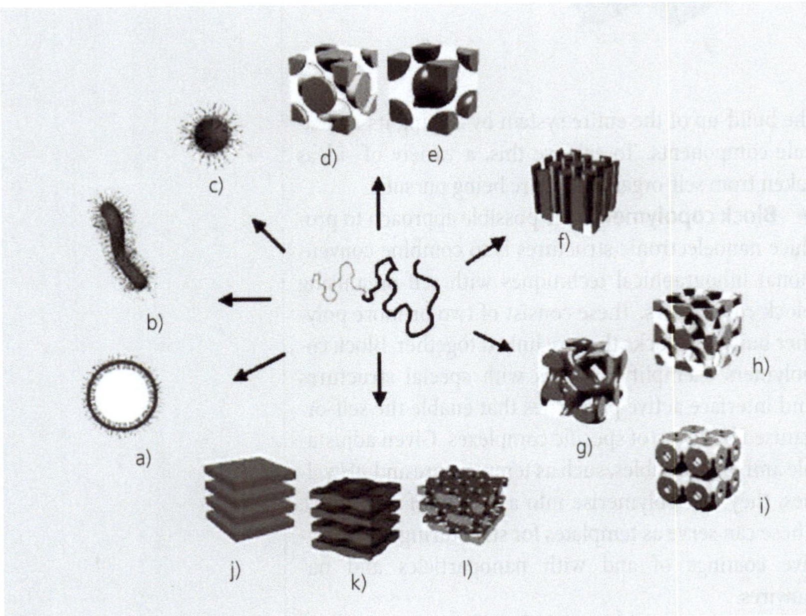

Self-organised nanostructures of block copolymers: The possibilities are (a) vesicles, (b,c) various micelle forms, (d,e,f) spherical structures, (g,h,i) porous or foam-like forms and (j,k,l) lamellae. The diversity of possible structures shows the sheer flexibility of the applications. Source: S. Förster, University of Hamburg

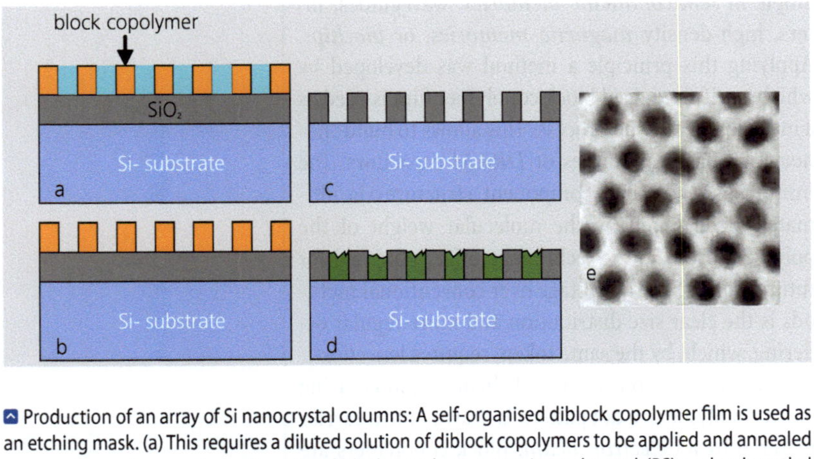

Production of an array of Si nanocrystal columns: A self-organised diblock copolymer film is used as an etching mask. (a) This requires a diluted solution of diblock copolymers to be applied and annealed on a surface using a spin coating. The diblock copolymer contains polystyrol (PS) and polymethyl methacrylate (PMMA) and creates a densely packed hexagonal structure of PMMA cylinders in the PS matrix. (b) The PMMA is etched away leaving the PS matrix. (c) The PS structure is transferred to the underlying oxide layer by an etching process. (d) A deposit of amorphous silicon ultimately creates nanocrystals. (e) The microscope image shows the finished nanocrystals (dark) with a diameter of about 20 nm. Source: IBM

► **Biomolecules.** Another vision is to use organic molecules and self-organisation to produce complex nanoelectronic switches and memories. Biomolecules with specific bonding abilities, like DNA, can be considered the bottom-up tools of the future for the build-up of tiny, complex structures. The process begins with the self-organisation of two complementary DNA strands to form a double helix. DNA, in its entirety, is composed of many building blocks, but essentially of four different ones, the so-called nucleotides. These nucleotides are attached to a skeleton of sugar and phosphate parts. There are always two nucleotides that have a particularly high affinity to one another, and so they tend to bond together (lock and key principle). This is the mechanism that allows the combination of two individual DNA strands to form a DNA double helix. The DNA strands can be used for technical applications by connecting them with molecular or nanoscale building blocks. The self-organisation process thus allows the construction of complex units for electronic applications.

► **Viruses.** Viruses can also be deployed to build up self-organising structures and objects. A technology has already been developed to construct magnetic or semiconducting nanowires using the bonding and self-organising traits of viruses. These viruses, so-called bacteriophages, are harmless for humans or animals, but they do attack bacteria. Bacteriophages are about 2,000 nm long and have a diameter of between 20–50 nm. Their outer shell consists of a single protein. It can be modified to specifically bond a particular type of conductive nanoparticle (for example, metallic gold or a semiconductor like zinc sulphide). When heat is applied, the metal particles melt and the phage vaporises. The result is a wire of nanoscale dimension. By further modifying the proteins at the "head" or the "foot" of the phage, one could devise specific self-assembled "wiring" on predefined areas of the surface of an electronic component. The length of the phage, too, is modifiable (within limits).

► **Biocybernetics.** Last but not least, the self-organising potential of neurons and neuronal networks are the subject of many research projects. The aim here is to control their growth in technical environments, to establish neurotechnological interfaces and in particular to electromagnetically address the neuronal grid. The complex nanoelectronic circuits and memories, which are built up by the self-organisation of biomolecules and nanomaterials, should enable the development of *neurochips*. Research in this direction is still very basic and aims towards developing biocybernetic systems. In the future, though, it could permit the implementation of fully functional neuroprostheses.

The decoding and application of biofunctional, self-organising mechanisms will continue to advance medical treatment and diagnosis, implantology and particularly tissue engineering (the regeneration of tissue by in vitro breeding of tissue taken from living cells).

▶ **Biomembranes and biosensors.** With so-called surface layer protein (S-layers), functionalised biomolecular membranes can be made for the development of biosensors. S-layers are two-dimensional and crystalline surface structures in the cell shells of many bacteria. They consist of regularly ordered membrane proteins. The average thickness of this type of layer is 5–10 nm. The membrane proteins are able to self-organise. They form different periodic orders on a scale of 3–30 nm consisting of subunits, which, owing to their periodicity, have very regular pores with diameters of 2–8 nm. By adding the appropriate reagents, S-layer proteins can be easily isolated and recrystallised on various surfaces to form diverse new patterns. This recrystallisation can be easily controlled through the surrounding conditions (for example, temperature, pH value, ion concentration) and particularly via the substrate surface, which paves the way to nanobiotechnical applications.

One great potential of S-layers as a membrane-like framework structure lies in the targeted immobilisation of functional bio-molecules on the surface or in the pores. This allows the implementation of membranes with a sensor/actuator effect that are permeable to solvents but do retain dissolved substances or react with them. These membranes or sensors can be used in areas where a high degree of separation at the nanometre level is important, for example production and product cleaning processes in biotechnology or pharmaceuticals.

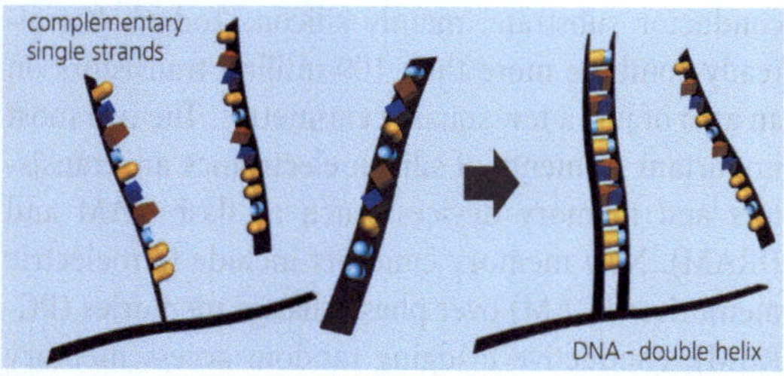

Self-organisation as exemplified by the combination of two complementary DNA strands: A DNA strand consists of a specific sequence of four different molecular components. Chemical bonding occurs because each building block has a complementary one on the opposite strand. For two DNA strands to combine to form a double helix, they must contain complementary molecule pairs. Source: VDI Technologiezentrum GmbH

▶ **Photonic crystals.** Self-organisation processes also play an important role in the manufacturing of optical components based on nanocrystals. Self-organisation phenomena are hereby used that can be influenced through the deposition rate, the lattice mismatch, temperature, electrical fields, solvents, concentration ratios, and so on. Photonic crystals (such as naturally occurring opals) are just one example of this. They are self-organised structures in which nanoscale material segments are embedded in an orderly fashion in a matrix with a different composition. Because of the periodically modulated refractive index, photonic band gaps are generated, in other words, the spread of light at wavelengths corresponding to the size of the grid constants is suppressed. Photonic crystals can be manufactured synthetically from polymer or metal oxide colloids by way of self-organisation in a bottom-up process.

The options for applications are diverse and range from novel components in information and communication technology, such as optical or data storage chips, *optical fibres, diodes,* colour *displays* and *effect pigments.* Controlled self-organisation processes not only lead to special structural formations in this case, but also to the creation of (quantum) physical parameters (such as emissions characteristics) that are of great significance for future applications.

Prospects

- Thanks to self-organisation, regular atomic and molecular structures can be synthesised on the basis of physico-chemical interactions and recognition mechanisms and integrated in functional units.
- In addition to nano(bio)materials and block copolymers, research and development is focusing on self-organisation in hybrid structuring processes.
- The technical use of self-organisation is still in its infancy, but it is no longer just a vision. The use of self-organisation phenomena for technical applications can contribute to cost-efficiency and ultimately to more sustainable production. Furthermore, it enables the implementation of structures that would hardly be accessible otherwise.
- The next commercial applications of self-organisation can be expected wherever they can be integrated in conventional manufacturing processes.

DR. HEINZ EICKENBUSCH
DR. GUNTHER HASSE
VDI Technologiezentrum GmbH, Düsseldorf

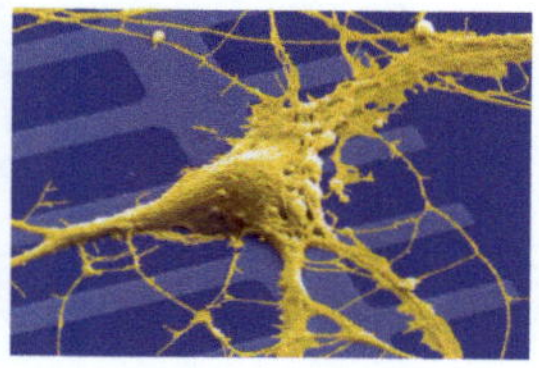

Combining biological and electronic systems: A nerve cell from a rat's brain on a silicon chip. Neurochips of this type are used to examine neuronal networks and should contribute in the future to the development of new drugs. Source: Fromherz, Max Planck Institute for Biochemistry

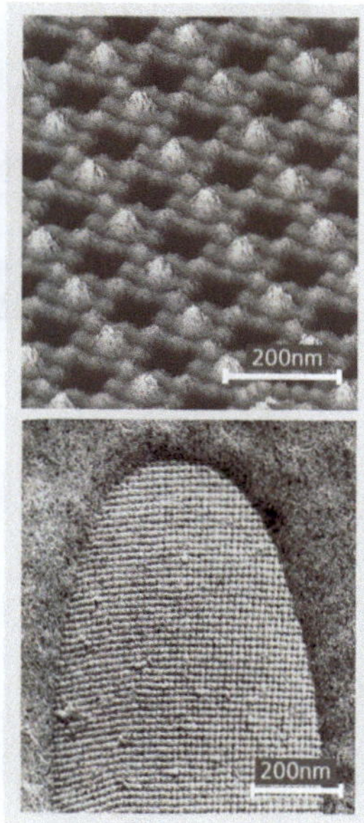

Periodic ordering of proteins through self-organisation. Top: In the cell wall of a bacteria. Bottom: A protein layer from the lab shows great symmetry. Source: Pum, University for Natural Resources, Vienna

Internet

- www.chemie.uni-hamburg.de/pc/sfoerster/forschung.html
- www.woelbstruktur.de/
- www.calresco.org/sos/sosfaq.htm
- www.cmol.nbi.dk/research.php

2 ELECTRONICS AND PHOTONICS

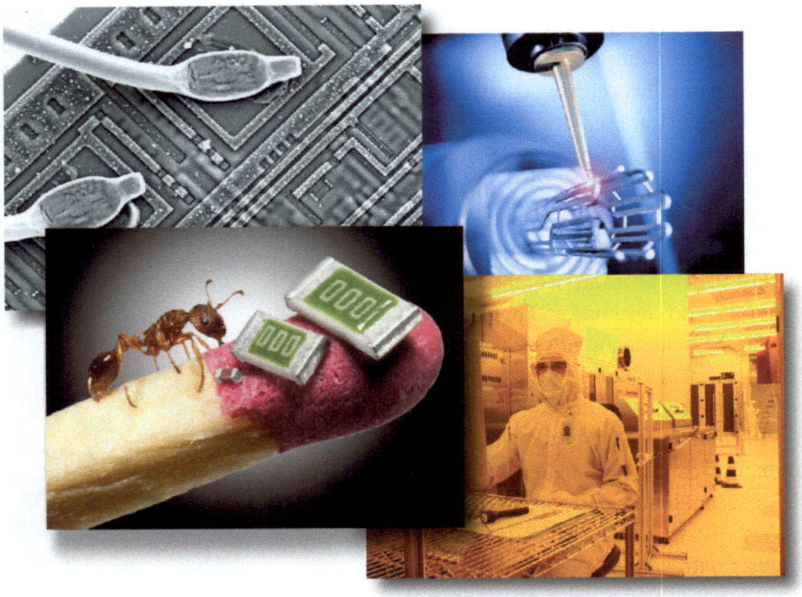

The diverse industries and lifestyles of society today would not be possible without the discovery and utilisation of electrons and photons. If we are able to exploit the technical potential of these two elementary particles, it is because modern science – in particular quantum mechanics – has done a great deal to explain their nature and their significance for solid-state physics.

In recent decades, electronics has contributed greatly to rapid technical progress in information and communication technologies, and will continue to be the enabler of many innovations and developments. World-wide sales in the electronics industry amount to about 800 billion euros, which makes it the leading manufacturing industry, even overtaking the automotive industry. *Semiconductor electronics*, which is largely based on silicon, has a market share of over 98.5 %. About 22 % of the industry's sales is reinvested in research and development, which is pursued intensely in this sector. Due to the favourable properties of silicon, the scalability of the electronics processing technology and advances in *photolithography*, there is still room for progress in reducing the size of components, as well as increasing the clock speed and number of components on *integrated circuits*. The greatest innovation potential is expected to be in highly integrated electronics and in energy-saving electronics. Research and development efforts are also going into introducing new materials, developing new component architectures, improving design, and developing new lithographic processes. The latter is closely related to the development of optical technologies.

The 20th century is often referred to as the century of electrons. The 21st century is expected to become the century of photons. Photons – light particles – can do far more than just illuminate rooms: the technical use of light – from lasers in the processing industry to scanners at check-out counters – has become part of our everyday life and is often connected with electronics. Light is a medium with unique properties and characteristics. It can be focused to a spot with a diameter of about one nanometre. *Lasers* can engrave words on a single hair. Of the many different laser types, there are high-power lasers capable of cutting steel, while in ophthalmology, lasers are used to cut thin flaps in the cornea of an eye. Photonics is the enabling technology for a number of industries and products, ranging from information and communication, biotechnology and medical technologies, to microelectronics, environmental technologies, industrial production and sensor systems.

► **The topics.** *Semiconductor* devices are the core of all modern electronic products. They consist mainly of integrated circuits (ICs), which combine the basic elements of electronic circuits – such as transistors, diodes, capacitors, resistors and inductors – on a semiconductor substrate, mainly silicon. Today's ICs already combine more than 100 million transistors on an area of just a few square centimetres. The two most important elements of silicon electronics are transistors and memory devices (such as flash RAM and DRAM). New memory concepts include ferroelectric memories (FRAM) over phase change memories (PC-RAM), conductive bridging random access memory (CB-RAM), and magnetic random access memories (MRAM). The latter is based on the tunnelling magneto-resistance (TRM) effect. Exploitation of this effect, and of the giant magneto-resistance (GMR) effect, gave rise to the new technological field of *magneto- electronics*. Its main areas of application are computer and automotive engineering. Computer hard discs have been fitted with GMR sensors, for example, which enable data to be saved on very small areas of magnetic hard discs, thus greatly increasing storage capacity. In automotive engineering, next-generation GMR sensors are used in passive and active safety systems such as ABS, EPS and airbags, to provide greater protection to drivers and passengers. They are also used to help optimise engine management and adaptation to ambient conditions, in support of lower fuel consumption and emissions.

Among the promising alternative materials in electronics are polymers. As a new technology platform, *polymer electronics* is paving the way to thin, flexible, large-area electronic devices that can be produced in very high volumes at low cost thanks to printing. A number of devices and applications with huge potential have already been created, for example RFID, thin and flexible displays and smart objects.

Advances in both silicon electronics and compound semiconductors are also contributing to further developments in power electronics. This particular field is concerned with switching, controlling and conversion of electrical energy. The idea behind the application of electronic switches is to control the energy flow between source and load at low losses with great precision. *Power electronic components* are already deployed in a variety of areas, notably energy transmission lines, engine control systems (such as variable speed drive), hybrid electric drive trains, and even mobile phones. Advanced power electronics converters can generate energy savings of over 50 %. One trend in this field is to miniaturise and enhance the reliability of power electronics systems, essentially through system integration.

System integration is also an issue in *microsystems*. The main aim of microsystems is to integrate devices based on different technologies into miniaturised and reliable systems. Microsystems are already in evidence in automotive engineering, medical applications, mechanical engineering and information technology. The devices are made up of electronic and mechanical as well as fluid and optical components. One example from optical components: an array of microlenses that concentrate light in the front of solar cells.

The term "*optical technologies*" describes all technologies that are used to generate, transmit, amplify, manipulate, shape, measure and harness light. They affect all areas of our daily life. The most obvious technical use of light is lighting. Biophotonics and femtonics are the trend in this field. For most of these applications the enabling device is a laser. Laser light

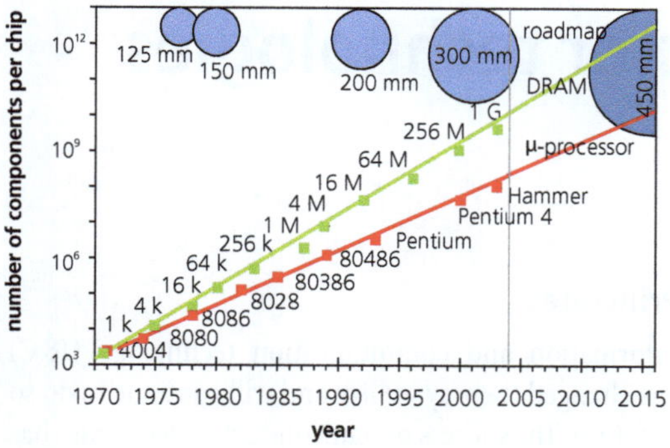

Development of silicon electronics: The graph shows the number of components per chip in microprocessors and DRAMs since 1970 and the continuing increase forecast in the International Technology Roadmap for Semiconductors (ITRS). The progression in size of the silicon wafers used in production is also shown. Source: VDI Technologiezentrum GmbH

differs from other light sources by its very narrow spectral bandwidth, as well as its high temporal and spatial coherence and directionality. High-power laser systems are used in material processing such as welding, cutting or drilling. Other fields of application are medical technologies, measuring technology, as well as information and telecommunications. *Optics and information technology* deals with the processing and generation, transfer, visualisation and storage of information. Optical data storage systems include CDs, DVDs and Blu-Rays. Systems for the visualisation of information would comprise, for example, plasma or liquid crystal displays as well as projectors. Holographic projection systems are a trend in this particular field. The technological progress and growing use of lasers in material processing and information and communication, for example, has led to a reduction in size and cost of laser systems over the past few decades.

Measuring techniques have profited from this development. Laser spectroscopy, laser radar, and gas spectroscopy are examples of applied measuring technologies. These techniques are particularly suitable for use in industrial automation and contribute to quality assurance, safety and efficiency in industrial production and process control. The enabling devices for measuring techniques in general are *sensors*, which measure changes in physical properties, such as humidity and temperature, or which detect events such as movements. The measured variable is then converted into electrical signals that can be further processed. In cars, for example, sensors are indispensable to the reliability and safety of the systems employed. ■

Semiconductor technologies

Related topics

- Microsystems technology
- Sensor systems
- Magneto-electronics
- Power electronics
- Computer architectures
- Optical technologies
- Solar energy

Principles

Information and communication technology (I&C) has changed the way we live and will continue to do so in future. The success of semiconductor electronics has played a key role in this development.

The properties of semiconductors lie between those of insulators and conductive metals.

This behaviour can be illustrated by means of the *energy band model*. Electrons in the valence band are integrated in the bonds between the lattice atoms of the solid body. If all of these bonds are occupied the electrons in the valence band cannot move and therefore do not contribute to the electrical conductivity of the material. In the conduction band, by contrast, the electrons are free to move and can therefore contribute to the conduction of the electric current. The relatively small band gap between the valence band and the conduction band in semiconductors can be overcome by electrons if there is an input of energy. This renders the semiconductor conductive.

In principle a distinction can be drawn between element semiconductors, here silicon is the most important representative for electronic applications, and compound semiconductors, which are composed of two or more chemical elements.

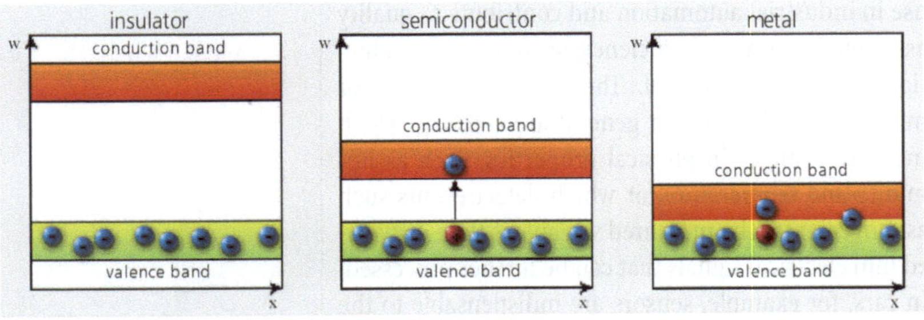

Electronic properties of semiconductors compared with insulators and metals in the band model. Left: In an insulator, the conduction band lies energetically above the fully occupied valence band and there are no electrons in it. Right: In metals, however, the valence band is not fully occupied and in most cases the two bands overlap, so that some of the electrons are always in the conduction band. Centre: Semiconductors too have a band gap between the valence and conduction bands, but this is relatively small. At low temperatures semiconductors, like insulators, are not electrically conductive. Through an input of energy, e.g. in the form of an increase in temperature, however, valence electrons can overcome the band gap. The electrons can then move freely in the conduction band (n-conduction). The remaining "holes" can move freely in the valence band and likewise contribute to the conductivity (p-conduction). Source: VDI Technologiezentrum GmbH

Compound semiconductors are formed by the combination of elements from the main groups III and V, II and IV or elements from group IV of the periodic table. The technically interesting compounds are in particular the *III-V semiconductors*, of which gallium arsenide (GaAs) is the most well-known representative. Compound semiconductors with a particularly wide band gap, such as gallium nitride (GaN) or silicon carbide (SiC), are referred to as wide-band-gap semiconductors. The wide band gap has the effect that components made of these materials display greater temperature resistance and possess a higher disruptive field strength than components made of silicon.

A decisive factor for the use of semiconductors in electronics is that their conductivity can be specifically controlled by the introduction of foreign atoms, referred to as doping atoms. In the case of silicon, atoms are selected which either have one external electron less than silicon (e.g. boron, gallium, indium) and thus create an electron shortage (acceptors) or exhibit one external electron more (e.g. arsenic, phosphorus, antimony) and thus provide an electron surplus (donors). A p-semiconductor is therefore doped with acceptors and an n-semiconductor with donors.

A division into direct and indirect semiconductors is particularly important for optoelectronic applications. In the case of direct semiconductors, such as GaAs or GaN, the energy required for the transition of an electron from the valence band to the conduction band can be provided directly by the absorption of a photon with the corresponding wavelength. Conversely, the recombination of a conduction electron in these semiconductors with a hole in the valence band entails the emission of a photon with the corresponding wavelength. By contrast, for quantum mechanical reasons, in indirect semiconductors, e.g. silicon, such transitions additionally entail coupling to a lattice oscillation in the semiconductor crystal and are therefore much less probable than in direct semiconductors. Indirect semiconductors are therefore not suitable as a material for light-emitting diodes or optical sensors.

▶ **Basic elements of silicon electronics.** The central component of semiconductor electronics is the *integrated circuit (IC)*, which combines the basic elements of electronic circuits – such as *transistors, diodes*, capacitors, resistors and inductors – on one semiconductor substrate. The two most important elements of silicon electronics are transistors and memory devices. Transistors can switch electric currents. For logic applications *MOSFETs (Metal Oxide Semiconductor Field Effect Transistor)* are used, in which the control electrode, the gate, is separated from the channel by an insulating layer. The characteristic dielectric constant of the insulating layer determines the capacitive coupling of the gate to the channel. The layer is therefore also referred to as the gate dielectric. Unless a voltage is applied to the gate there are no free charge carriers in the channel between the electrodes source and the drain. The application of a voltage to the gate creates an electric field which penetrates into the channel through the insulator and leads to an enrichment of the charge carriers there. As a result a conductive channel forms between the source and the drain – the transistor is switched to current flow.

▶ **Semiconductor memories.** A fundamental distinction is made between flash, DRAM and SRAM semiconductor memories. The structure of the *flash* cell is similar to that of a *field effect transistor FET* which has been supplemented by a floating gate (FG) of conductive polysilicon in the gate dielectric. Charges on the FG control the transistor characteristic and in this way represent the stored information. The flash memory is quite slow but is not volatile, i.e. the information is retained even if there is no power supply. It is therefore used in memory cards for cameras, mobile phones, handheld devices, MP3 players, etc.

Flash is not used as computer memory. Read/write access is too slow for this purpose. What's more, flash components can only withstand about 10^6 read/write cycles, which is inadequate for computer memory applications.

A *DRAM cell (Dynamic Random Access Memory)* consists of a transistor and a capacitor. The information is represented by charges on a capacitor. As the insulation of the capacitor is not perfect, charge is constantly dissipated. The stored information is therefore lost over the course of time, i.e. the memory is volatile. For this reason the memory has to be regularly refreshed. DRAM is mainly used as fast computer memory.

The *Static Random Access Memory (SRAM)* does not need any refresh cycles, but it loses its information as soon as the voltage supply is interrupted and is therefore volatile. An SRAM memory cell typically consists of six transistors which are connected to two coupled inverters. The SRAM cell is therefore relatively large but, because of its speed, is used as a buffer storage area (cache memory) close to the processor.

▶ **Manufacturing process.** For the manufacture of integrated circuits ultra-pure and almost defect-free *silicon* monocrystals are required. As lattice structure and impurities significantly influence the conduction properties of the substrate, the arrangement of the semiconductor atoms must continue as regularly as possible across the entire material. Furthermore, the requirements in respect of purity and absence of defects rise with increasing miniaturization of components such as transistors and memories. The smaller the structures, the more crystal structure irregularities become a factor and jeopardise the fault-free functioning of the component.

In the manufacturing process quartz sand, which consists to a large extent of SiO_2, is reduced to raw silicon. This still contains about 1% of interstitial atoms, including carbon, iron, aluminium, boron and phosphor. To further reduce these impurities the solid raw silicon is converted with hydrochloric acid into liquid trichlorosilane and distilled. In reaction with ultra-pure hydrogen solid silicon is obtained which is further purified in a zone cleaning process. For this purpose, the silicon is heated and melted using a high-frequency coil beginning at the top end in a zone of a few millimetres width. The area of the liquid silicon is now conducted downwards. The impurities concentrate in the liquid phase and migrate down to the bottom of the material.

At the end of the purification process the solidified silicon contains less than 1 ppb of impurities, which is less than one defect atom per billion silicon atoms. The silicon is, however, in polycrystalline form, i.e. with irregular distortions in the crystal structure of the bulk material.

Two methods have become established for producing the monocrystals:

In the *Czochralski process* the silicon is firstly melted and brought into contact with a monocrystalline seed crystal. This initiates a crystal growth process

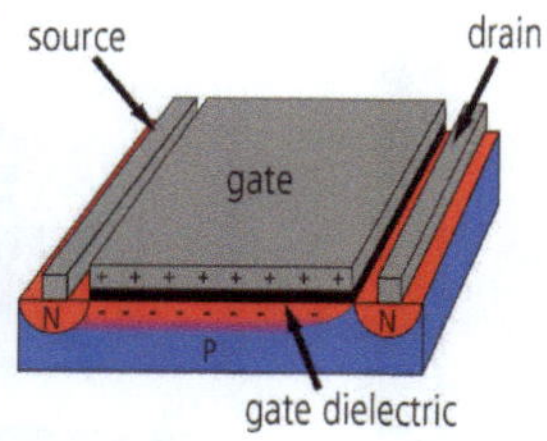

▸ Principle of a field effect transistor as exemplified by an n-channel MOSFET: The application of a positive voltage at the gate creates an electric field which penetrates through the gate dielectric into the channel and leads to an enrichment of negative charge carriers there. As a result, a conductive channel forms between the source and the drain – the transistor is switched to current flow. Source: VDI Technologiezentrum GmbH

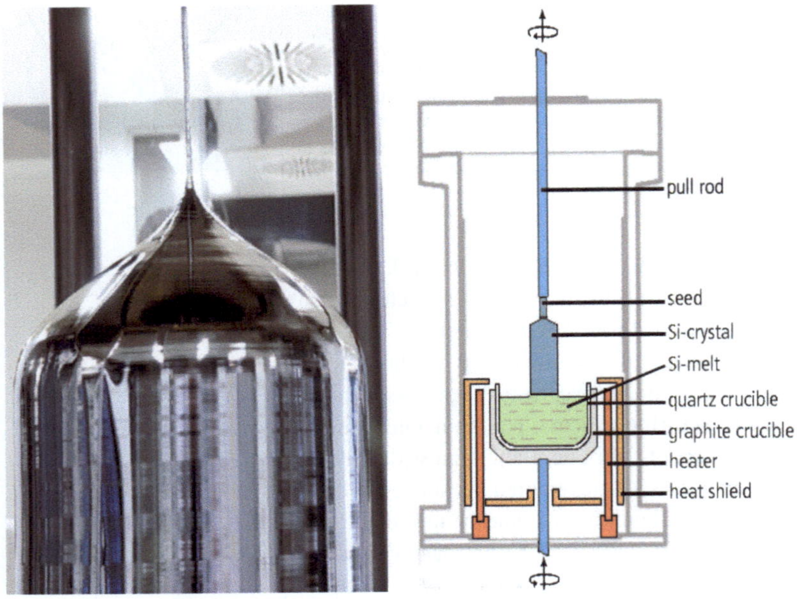

Crystal drawing unit incorporating the Czochralski process. Left: Silicon monocrystals are produced in crystal drawing units in a cleanroom environment, reaching a diameter of 300 mm. These monocrystal cylinders have a length of about 170 cm and weigh approx. 315 kg. From such a crystal up to 1,500 wafers can be made. Right: The crystal drawing unit comprises a graphite crucible and a quartz crucible supported in it containing the silicon melt. The melt temperature is controlled by means of a heater. The crucible area is thermally shielded from the outside by a heat-resistant enclosure. A small, ultra-pure Si monocrystal (seed) is fastened to a rotating rod and brought into contact with the melt. The melt crystallises in the area of contact and continues the lattice provided by the seed crystal. While rotating constantly the growing Si crystal is steadily drawn upwards at a speed of 3–20 cm/h. Source: Siltronic AG, VDI Technologiezentrum GmbH

in which crystallisation follows the given atomic lattice of the seed crystal. A rod of monocrystalline silicon of up to 300 mm thickness forms. Up to 1,500 wafers can be cut of this so-called "ingot".

In the zone melting method the polycrystalline silicon rod is melted locally using an induction coil in a described above for zone cleaning. At the top end of the silicon rod a small monocrystal is brought into contact with the topmost melting zone. The monocrystal provides the lattice structure. During local melting the crystallisation process continues successively over the entire rod.

ICs are made on thin wafers sawn from the crystal. In the production of ICs a distinction is made between the front-end and the back-end process. The front-end process includes all process steps through which the *wafer* passes entirely. The transistors, memories and their interconnects are produced on the wafers by means of *photolithography* in several process cycles. Firstly the wafer is coated with a *photoresist*. The desired structures, which are imprinted in a larger scale on photographic masks, are projected onto the wafer by means of *laser* light. In a chemical process the photoresist is removed from the areas exposed (or unexposed) to the light. The wafer parts beneath the remaining photoresist are protected from the next process step. This can be a chemical etching process or a doping step. Finally, the photoresist is completely removed. After a further step, in which an oxide or a metallising coating may be applied, the next process cycle begins. Modern integrated circuits need more than 20 such cycles. After completion of the front-end process the back-end process starts. The wafer is sawn into individual ICs, which are tested, packaged and bonded.

Compound semiconductors are made in epitaxy processes, i.e. by means of unified crystal growth. *Metal-organic vapor-phase epitaxy (MOVPE)* is mainly used because this method is particularly versatile and comparatively low-cost. In this production process the elements needed to grow the semiconductor are present in metal-organic or metal hydride precursors which can be brought into the vapour phase easily. These starting substances are mixed with a carrier gas and led to a host crystal substrate in a laminar gas flow, in which a complex chemical reaction leads to the formation of the desired semiconductor as well as gaseous by-products. The growing semiconductor continues the crystal structure of the host crystal, if the crystal structures do not exhibit any major differences. The entire process takes place in a reactor at moderate pressures (20–1,000 hPa) and high temperatures (650–1,100 °C).

In the MOVPE process it is possible to apply various semiconductor compounds in layers consecutively on the substrate. This means that electronic components can be produced by systematic arrangement of semiconductor materials with differing band gaps. At the junctions between two different materials internal electric fields arise which have an impact on the charge carriers and influence their transport through the structure. By means of this process, referred to as band gap engineering, heterostructure field effect transistors, for example, can be made which function in the same way as the MOSFETs in silicon technology, but which permit distinctly faster switching times because undoped semiconductor materials of particularly high conductivity can be used for the channel between the source and the drain. It has to be borne in mind, however, that the various compound semiconductors cannot all be combined with each other – their lattice structures must match each other to such an extent that no major defects occur at the transitions.

Applications

With a market share of over 98.5% semiconductor electronics is largely based on silicon. The rest is mainly accounted for by compound semiconductors, which are used in particular in high-frequency power and high-temperature electronics.

Another important area of application for compound semiconductor types that permit direct band transitions is optoelectronics, especially in the form of light-emitting diodes, but also as laser diodes (for *CDs*, *DVDs*, *Blu-Ray* discs and optical data communication) and optical sensors. In particular for the white *LEDs* based on wide-band-gap compound semiconductors, new lucrative mass markets are evolving in special lighting (e. g. on automobiles), for display backlighting, and further down the line in general illumination.

In 2004 the semiconductor industry for the first time produced more transistors than the grains of rice harvested worldwide. Semiconductor electronics is installed in electronic products mainly in the computer industry (approx. 40%), the communication industry (approx. 20%) and in entertainment electronics and other industry (in each case approx. 10%). Internationally, the automobile industry accounts for approx. 5% of the electronics market.

The proportion of semiconductor electronics in value added by manufacturing has risen steadily over recent decades and in some applications such as PDAs, mobile phones and digital cameras exceeds 30%. Semiconductor electronics is one of the main drivers of technological progress even though, being integrated in electronic components which are in turn installed in end products, the end consumer is unaware of this fact.

Trends

► **Silicon electronics.** The favourable properties of silicon, the scalability of *CMOS* technology and the progress made in photolithography have made it possible to steadily reduce the size of components. In 1965 Gordon E. Moore formulated the theory that the power of an integrated circuit and the number of components on a chip would double every 18–24 months – with the price per chip staying the same. At that time an integrated circuit had about 60 components. Today's processors have more than 150 million transistors and a clock speed of over 3 GHz.

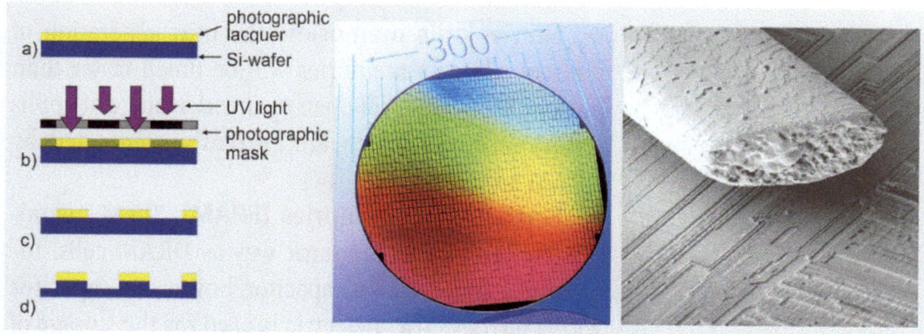

Production of integrated circuits. Left: Representation of the various photolithographic process steps. Firstly, the Si-wafer is coated with a photoresist (a). The desired structures are projected onto the wafer through a photographic mask (b). By means of a chemical development process the photoresist is removed at the unexposed places (c). The parts of the wafer under the remaining photographic lacquer are protected from the subsequent etching process (d). Centre: 300 mm wafer with memory chips. Right: Chip surface with fine conduction structures. To provide a size comparison a hair (diameter about 100 µm) is pictured. Source: VDI Technologiezentrum GmbH, ZTC; Infineon Technologies AG; Micronas

► **High dielectric constants.** For decades silicon dioxide was used as the gate dielectric, i. e. as the insulator between the transistor channel and *gate*. As miniaturisation progresses the thickness of the gate dielectric also has to be reduced in order to keep the gate capacity constant despite the smaller surface. If this is too thin, charge carriers can tunnel quantum mechanically through the layer, and the tunnel probability increases exponentially as the thickness decreases. The consequence is that the leakage current becomes unacceptably high. An alternative is provided by materials with a higher dielectric constant which, while providing the same capacity, permit a thicker layer, exemplified by various hafnium alloys, such as HfO_2/SiN. A general problem connected with the introduction of new materials is their compatibility with the production process, in which temperatures of approx. 1000 °C are necessary, for example.

► **FinFET.** With advancing *transistor* miniaturization the length of the channel shrinks. Correspondingly, the proportion of the channel influenced by the source and drain voltages increases compared with the proportion that can be switched by the gate voltage. In order to achieve a further reduction in structure sizes these "*short-channel effects*" must be minimised to ensure controlled activation of the transistors. A solution is provided by new transistor architectures with two gates or of FinFET design in which the channel is vertically structured and enclosed by the gate.

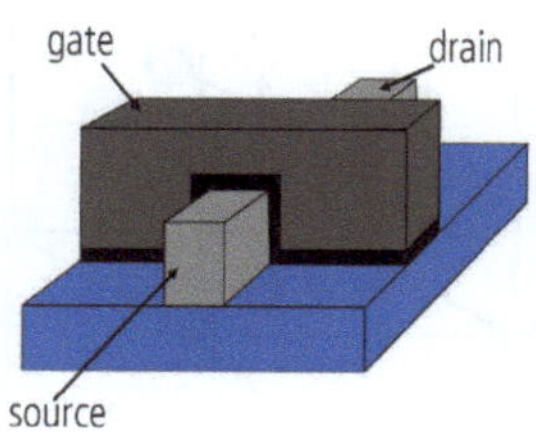

Principle of a FinFET: To reduce "short-channel effects" the channel between source and drain is three-dimensionally structured and enclosed by the gate. This increases the channel length on the same surface area. Source: VDI Technologiezentrum GmbH

► **Non-volatile memories.** The next generation of non-volatile data memories will be much faster than present flash. This will open up a wider range of applications in areas which are currently dominated by DRAM and SRAM.

► **Ferroelectric memories (FRAM).** These memories are built up in the same way as DRAM cells. Instead of a conventional capacitor, however, a capacitor with a ferroelectric dielectric is used for the storage of information. Ferroelectric memory cells have a domain structure in which spontaneous electrical polarisation occurs. By applying an electric field the polarisations of the various domains can be equidirectionally aligned. Even without an external field the material retains a specific polarization. This remanent polarization determines the property of the component and represents the stored information which is retained even without any supply voltage.

► **Magnetoresistive memories (MRAM).** These memories utilise the *Tunnelling-Magneto-Resistance effect (TMR)*, which describes the spin- and magnetisation-dependent change in a tunnel resistor. The basic element of the MRAM consists of two ferromagnetic electrodes which are separated by a non-conductive barrier of (sub-)nm thickness. Depending on the relative magnetisation of the two electrodes (parallel or anti-parallel magnetisation) upon voltage application a "large" (parallel magnetization) or a "small" (anti-parallel magnetisation) tunnelling current flows through the layer stack, representing the digital states "1" and "0".

► **Phase change memories (PC-RAM).** These memory cells are based on a a resistive phase change element. PC-RAMs use the reversible phase change properties of specific materials (e. g. chalcogens). They enable switching between an amorphous and a crystalline state according to the temperature applied. The electrical resistance of the two phases differs by up to four orders of magnitude and encodes the stored information.

► **Conductive bridging random access memory.** The CB-RAM uses a solid-state electrolyte with dissolved metal ions (e. g. Cu or Ag). Upon voltage application a path consisting of these metal ions is formed in the material, which lowers the resistance of the element. By applying an inverse voltage this can be reversed. CB-RAM components exhibit low power consumption and high switching speed. However, the memory cells are still considerably larger than, for instance, DRAM or flash components.

► **Chip design.** Today's processors already combine more than 100 million transistors on an area of just a few square centimetres, interconnected by several kilometres of wire. One of the challenges is to connect the transistors as effectively as possible. Whereas the circuitry layout used to be produced by hand, this work is based on automated processes nowadays. The problem is, however, that in recent years design productivity has not been able to keep pace with the increasing complexity of electronic circuits. Considerable efforts are therefore being made to improve electronic design automation (EDA).

► **New lithographic processes.** Present methods use laser light with a wavelength of 193 nm in order to realise structural sizes well below 100 nm. For the current 45-nm chip generation immersion lithography is deployed, maintaining the wavelength of 193 nm. In this process a water layer is introduced between the lens of the lithographic device and the wafer, which gives better resolution and depth of field. The next advance in miniaturisation will probably still be based on this technology. For later chip generations intensive work is being conducted worldwide on a lithographic process using *extreme ultraviolet light (EUV)* with a wavelength of 13 nm.

Domain structure in a ferroelectric memory: Application of an electric field causes the equidirectional alignment of the domains and thus a polarisation of the material, which is partially retained when the electric field is no longer applied (remanent polarisation; P_r). By means of a reverse electric field the polarisation can be "switched over". Left: The schematic shows the direction of polarisation and represents the "1" state. Right: The direction of polarisation is now "switched over". This represents the "0" state. Source: VDI Technologiezentrum GmbH

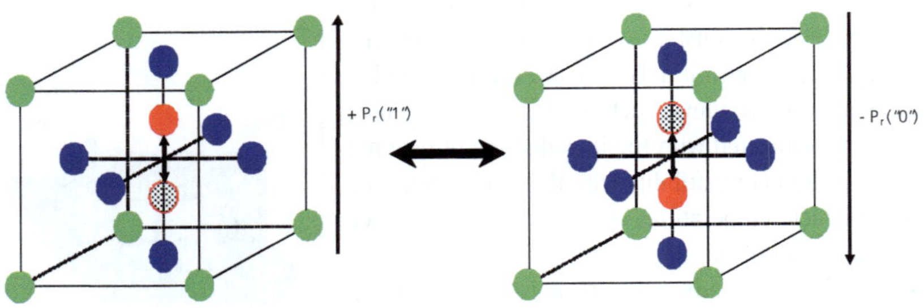

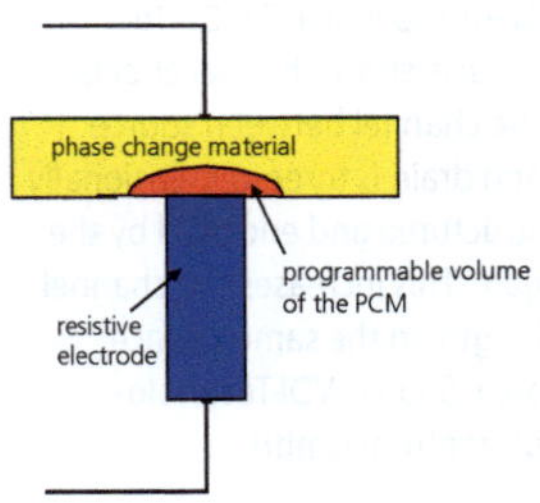

PC-RAM element: A reversible phase change material is arranged between the electrodes. A certain area of volume at the contact point to the heating electrode can be amorphised or crystallised by means of specific voltage pulses. In the crystalline state the material exhibits a low electrical resistance and in the amorphous state a high electrical resistance. To read out the information a voltage is applied across the memory element producing a current flow, the amount of which depends on the resistance. Source: VDI Technologiezentrum GmbH

► **Compound semiconductors.** The production capacities for compound semiconductors are being massively expanded at present. This expansion is being driven mainly by the strong increase in demand for white *LEDs*. At the same time, the power electronics market, which is important for compound semiconductors, is also growing faster than the electronics market as a whole. A major step can be made towards larger quantities by increasing the wafer size, and compound semiconductors hold considerable development potential in this respect compared with silicon. While 150 mm wafers are the standard today for GaAs, wafers made of SiC are smaller, measuring 75–100 mm. In the case of GaN, as needed for white LEDs, pure bulk substrates are difficult to realise and so GaN layers are produced by means of the MOVPE process on substrates made of a different material, in particular SiC and sapphire. Intensive efforts are being made to find alternative substrate materials whose lattice structure matches well with GaN and which are low-cost. At the same time research is being conducted into alternative growth processes for producing pure GaN wafers.

► **Alternatives to silicon electronics.** Even if it can be assumed that in the medium to long term silicon electronics will hold on to its dominant position, intensive work is being carried out on alternative electronics concepts.

The various approaches pursued differ fundamentally in their strategy. For instance, research is being conducted on alternative components such as single-electron transistors or resonant tunnel diodes which in theory can also be produced in silicon. In other approaches the aim is to introduce new functionalities exploiting physical effects. Other technologies, such as *polymer electronics*, are attempting in principle to transfer the proven silicon concepts to new material systems. Finally, completely new types of information processing are being pursued with the quantum computer, neural networks and DNA computing.

Candidates which are particularly interesting for the long term are *spintronics* and molecular electronics. Whereas conventional electronics uses charges and their energy, spin electronics (or in short spintronics) exploits the spin orientation of the charge carriers and spin coupling in addition to or instead of the charge. *Molecular electronics* focuses on components in which the information is processed or stored in single molecules, nanotubes, nanowires or clusters. In recent years a number of demonstrators have been built, but molecular electronics is more of a long-term option. Apart from numerous other aspects, to achieve technical utilisation of molecular electronics completely new production methods based on *self-organisation* will have to be used and fault tolerance concepts developed because self-organised structures can never be perfectly realised.

Memory type	Cell size	Read speed	Write speed	Numbers of cycles
DRAM	~70 nm	10-50 ns	10-50 ns	unlimited
SRAM	~400 nm	2-60 ns	2-60 ns	unlimited
Flash	~50 nm	60-100 ns	~10 µs	~10^6
FRAM	~130 nm	20 ns	20 ns	>10^{15}
MRAM	~90 nm	~10 ns	~10 ns	>10^{15}
PC-RAM	~90 nm	~20 ns	~20 ns	~10^8

Comparison of the properties of conventional semiconductor memories (DRAM, SRAM, Flash) and promising new, non-volatile memories (FRAM, MRAM, PC-RAM), showing in each case cell size, access speed and the mean number of write/read cycles of present memory components. Source: VDI Technologiezentrum GmbH

Prospects

- The semiconductor industry is the only industry whose products have experienced an extreme rise in performance while at the same time undergoing a great drop in cost.
- Silicon electronics is the key technology for information and communication technology. It will maintain a dominant market position in future.
- Its success is rooted in ongoing miniaturization, which is made possible by the favourable material properties of silicon.
- Great efforts will be needed on the R&D front if this is to continue in future, including the introduction of new materials, the development of new component architectures, improvements in design and the development of new lithographic processes.
- In many applications, where their superior properties come to the fore, compound semiconductors have gained a mass market. The key factor in maintaining this success in the future will be the further reduction of production costs.

DR. LEIF BRAND
DR. ANDREAS HOFFKNECHT
DR. OLIVER KRAUSS
VDI Technologiezentrum GmbH, Düsseldorf

Internet

- www.semiconductor-glossary.com
- www.st.com/stonline/books/pdf/docs/5038.pdf
- www.necel.com/v_factory/en/index.html

Microsystems technology

Related topics

- Semiconductor technologies
- Sensor systems
- Polymer electronics
- Magneto-electronics
- Surface and coating technologies
- Minimally invasive medicine
- Self-organisation

Principles

Microsystems technology (MST) encompasses technical systems consisting of functional components between 1–100 µm in size. The spectrum of functional elements integrated into microsystems ranges from electronic to mechanical, as well as optical and fluid components.

Because of their small size, microsystems save on material and energy. MST combines various fundamental technologies including mechanics, optics and fluidics, as well as new fields of technology, such as *polymer electronics* and innovative materials derived from *nanotechnologies*. MST is an enabling technology which can be applied to many areas like the automotive, medicine, machinery and information technology sectors. Essentially, MST integrates devices based on different technologies into miniaturised and reliable systems. To make use of a new *sensor* element, for example, more than the pure sensor element is required. In most cases, the sensor element has to be integrated with electronics, an energy supply, communication devices and a media supply in extremely miniaturised and sturdy housing like a heart pacemaker or an ABS sensor.

Just like biological cell systems, microsystems are more than the sum of their parts. Every living system consists of a number of specific microsystems, its cells. Despite its small size, each somatic cell represents a highly productive system. It transforms energy, produces body components and molecular cues, generates and conducts electric signals, communicates with neighbours and distant partners, and is even able to multiply or repair itself. Depending on its specific task inside the body, each cell is specialised in one operation or another. A complex macrosystem is enclosed within suitable packaging technologies, and new capacities are created by intelligent integration into compact systems.

▶ **Micromechanics.** An important raw material for MST is silicon, a semiconducting material, also commonly used in microelectronics, which can be used to build three-dimensional (3D) mechanical structures. Technologies already used in the semiconductor industry are applied: *lithographic techniques* to structure layers of protective lacquer (which have been previously applied to the silicon's surface) and etching processes that utilise the material's specific crystal structure. The silicon crystal layer has an optimum etching direction and can be used to create a width to depth ratio of up to 1:50 for the 3D structures. Dry and wet etching methods are combined industrially today. Using etching methods to resist protected areas, in order to create 3D structures, requires several processing steps, including layer by layer photoresist application, lithography and etching. Several sacrificial layers may also be included. This is why techniques based on semiconductor technologies are usually expensive and are best applied in mass production.

For other materials, such as metals, the *LIGA technique* is an important production method that is used in MST. It contains different steps that follow each other during the production process: lithography, galvanisation and moulding. To complete this process, a plane substrate such as a silicon wafer is deposited with a seed layer, which is then covered in a visible light- or X-ray sensitive photoresist (or Polymethyl methacrylate, PMMA).

The *photoresist* is exposed through a mask and removed by developing. This creates a negative mould of the intended metal structure. The metal is precipitated to the substrate during galvanisation only in those places where the photoresist was removed. After the removal of both substrate and seed layers, the galvanised metal structure is transformed into a mould insert, which can then be inserted into a moulding tool and used to produce plastic parts via injection moulding or hot stamping, for example. The LIGA technique is used for many applications, including the produc-

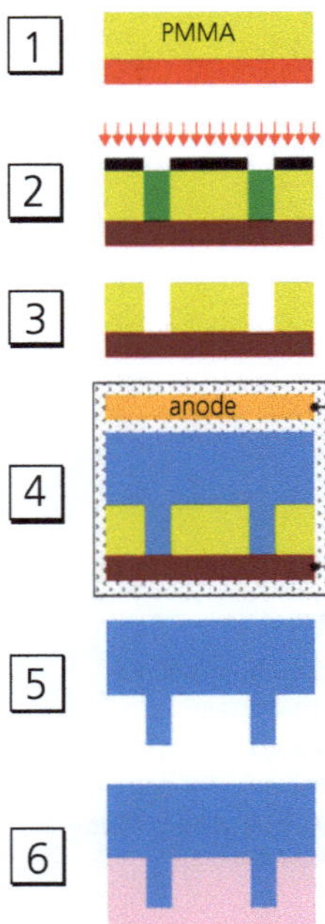

LIGA technique for mechanical microstructuring: The figures show the six steps required to produce a metal form for injection moulding. The form must be highly accurate, especially for mechanical 3D structures. The LIGA process can be used to make multiple copies of this form, depending on the size of the item.
(1) Application of an X-ray sensitive plastic coating (PMMA) – that can be up to 1 mm thick – to a base plate with a conducting surface (red)
(2) Lithographic deep structuring via parallel x-raying
(3) Removal of the exposed areas using a suitable developer
(4) Galvanic deposition of the metal (blue) inside the structure, up to 5 mm above the surface of the photoresist
(5) Metal structure after removal of photoresist
(6) Reproducing this metal structure via plastic moulding (pink)

tion of pinions for miniature drive systems, as well as injectors and optical components, such as *Fresnel lenses*, refractive optics and optical waveguide interlinks.

► **Micromachining.** For financial reasons, technologies applied in traditional precision mechanics are increasingly being considered for MST as well. Micromachining (microlathing, milling and drilling), for example, is becoming ever more applicable as tools are miniaturised and the precision of the machines and the machining processes is improved. Diamond and hard metal tools with a diameter of 50 µm are already commercially available and machine tools used in machining processes are now accurate to within a few micrometres. For ultra precision machining, tools are often covered in monocrystalline diamond. These tools create rugosities on surfaces of only a few nanometres.

► **Casting.** High-temperature tool steel is an excellent raw material for producing robust casting tools for micro hot stamping and micro injection moulding. Moulding techniques ensure low-priced production of 3D components with complex and fine structures.

The raw material (in the form of pellets or a paste) for injection moulding is forced into a heated tool under high pressure in order to create a work piece, which will be the original mould of an initial master work piece.

Both techniques make the mass production of plastic and ceramic parts with microstructures possible.

► **Micro-optics.** MST can be used to perform optical functions, in addition to purely mechanical functions, when combined with materials with the right characteristics. In miniaturised optics, lenses, mirrors and optical gratings are made from optical materials using micromachining techniques – lithography in combination with etching processes, for example. The micromirror array is an optical component that is already used for light deflection in video projectors.

Conventional optical and special glasses are the materials that are normally used. For mirrors and IR applications, silicon is especially useful because of its high refractive index. One of the benefits of micro-optical elements is that optical arrays (lenses, prisms, mirrors, filters, polarisers) can be produced, something that is hardly feasible on a macroscopic level.

These types of arrays can be produced inexpensively using single-chip fabrication techniques, as can the lens arrays with several hundred lenses used in *LCDs*. With microsystems technologies, light-conducting elements can also be created as planar systems on top of substrates, such as silicon, *photonic crystals* or polymers.

The application and subsequent structuring of light-conducting layers through lithography or laser beams is one way of creating planar optical wave guides. Complete microsystems with optical (or other) functions can then be equipped with separately pro-

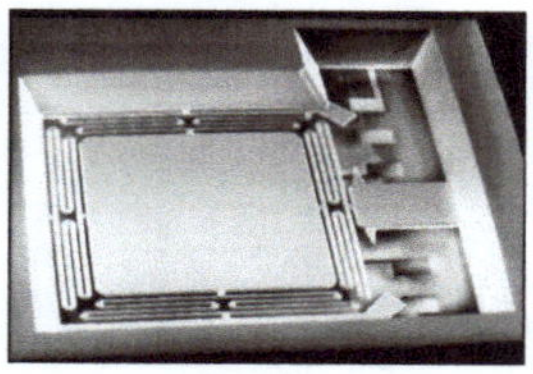

▶ Micromirrors: Mirror created by silicon micromechanics, with a 3x3 mm² surface area for a 2D scanner, movable in two directions. Another important application for micromirrors like these is digital light processing in video projectors. The mirror is driven by electrostatic attraction, through the impression of a high electric voltage between the mirror and the isolated electrodes below it (not visible). Source: Fraunhofer IZM

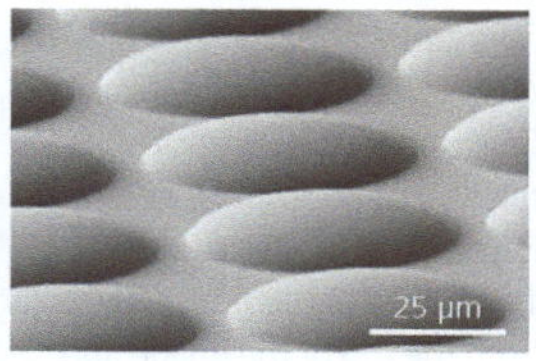

▲ Microlenses: This array was created on top of a borosilicate glass substrate via lithography and a wet chemical etching technique. The lens array is used, for example, as a light concentrator in front of III-V solar cells. Source: Fraunhofer ISIT

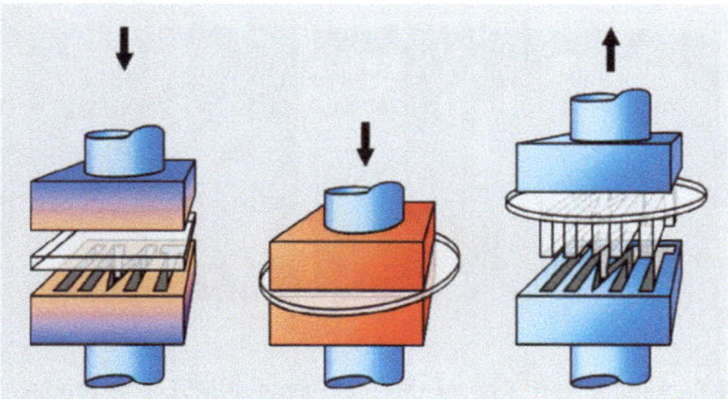

▲ Hot stamping of polymer plastics: The stamping forms, usually made of metal, are heated up before the stamping process, and cooled down before demoulding. Excess material is pressed out between the stamps and the surface tension forms a round plate. Structures accurate to within micrometres can be created. The image is not realistic: stamping tools are in fact much bigger than the structures that they help to create. Source: Forschungszentrum Karlsruhe

▲ Injection moulding. Left: Mould insert for injection moulding, created by bronze micromilling. Right: Microfluidic "Bio-Disc", a miniaturised disposable device used in bio analytics, which is made from plastic using the designated casting tool (left), and which can be used for biotechnological analysis. Disposable devices like these will complement standard procedures in the future, in pharmaceutical screening and testing, for example. The advantages of these disposable devices are the higher throughput created by parallel reaction, the smaller consumption of substances and the faster reaction times. Sources: i-sys Automationstechnik GmbH, Ernst Reiner GmbH

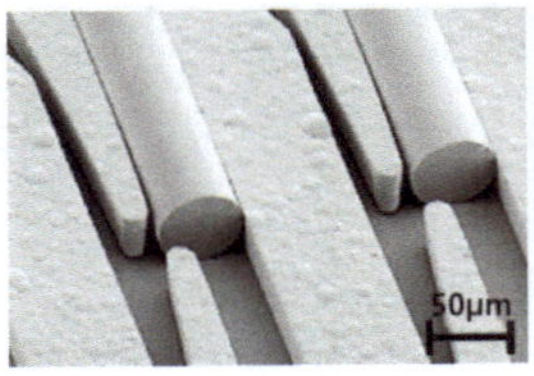

Optomechanic microsystem: Micro plastic component, which is created using moulding techniques, and is used as a fibre coupling system for high precision bonding of optical fibres. Channels for enclosing the fibres and springs for attaching them frontally are visible. The fibres are solely inserted from the top of the image and their counterparts are missing for reasons of clarity. Source: Institut für Mikrotechnik Mainz GmbH

duced optical microcomponents and microfibres, as well as micromechanical or optoelectronic components (like laser or photo diodes).

▶ **Microfluidics.** In MST systems, it is possible to control tiny amounts of fluids (liquids or gas). Functional components, such as channels, mixers, separators, heat exchangers, pumps and valves, which are less than 1 mm in size, can be produced using microtechnologies. Various materials can be used. For many microfluidic applications, all that is needed are polymers made of readily available materials, which are cost-effectively structured by hot embossing. These materials are used in bioanalytics, medical diagnostics and pharmaceutical research. Traditional chemistry can also be miniaturised through microfluidics in order to increase conversion efficiency dramatically. Microreactors can be used for production purposes; they can also be used to study the kinetics of reactions.

▶ **Packaging.** Packaging is central to producing microsystems. Packaging integrates the different components – electronic, optical, mechanical and fluidic elements, for example – into one system. Interfaces therefore need to be created between the overall system and the macro environment. A microsystem that monitors oil pressure and temperature inside an engine, for example, needs an interface connecting it to the oil sensors, as well as an electronic interface that transmits information to a control unit.

▶ **Wire bonding.** Electrically and mechanically integrating sensors or actuators into substrates calls for different techniques such as soldering, bonding or wire bonding. Wire bonding allows different components to be installed on top of the substrate and then interconnected using thin aluminium or gold wire.

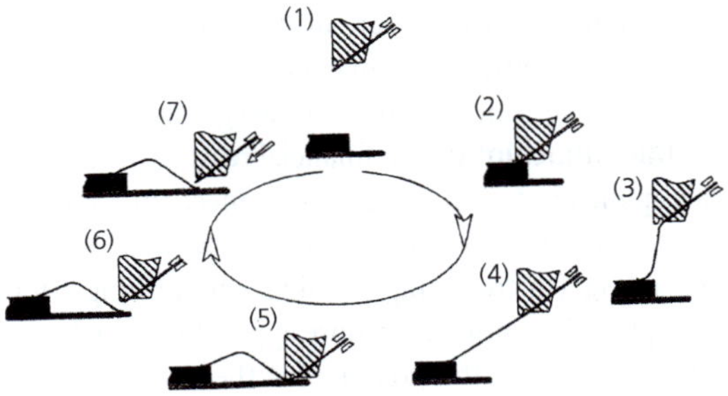

Supersonic bonding of wires: (1) A bonding wire is inserted into the borehole of a tool, which is at an angle to the root area of the tool. (2) After the tool is lowered, the wire is pushed onto the metallic conductor pad and fused by friction in the supersonic spectrum (e. g. 125 kHz). (3, 4) The tool is pulled towards the contact surface to be joined. The wire therefore winds off the spool. (5) The second joint on the contact surface is also created via supersonic fusion. (6) A clamp behind the bonding tool is then enclosed around the wire and moved so that the wire rips at the contact point without damaging the contact itself. (7) The clamp pushes the bonding wire through the borehole again so that its end lies beneath the root area. The bonding process can now be continued.

▶ **Flip chip technique.** The semiconductor components have small metal bumps (alternatively made of electrically conductive adhesive) that can be soldered or bonded onto the corresponding contacts on top of the substrate when these parts are flipped over. The advantage of this technique is that the complete surface of each chip can be used to ensure the contact and that

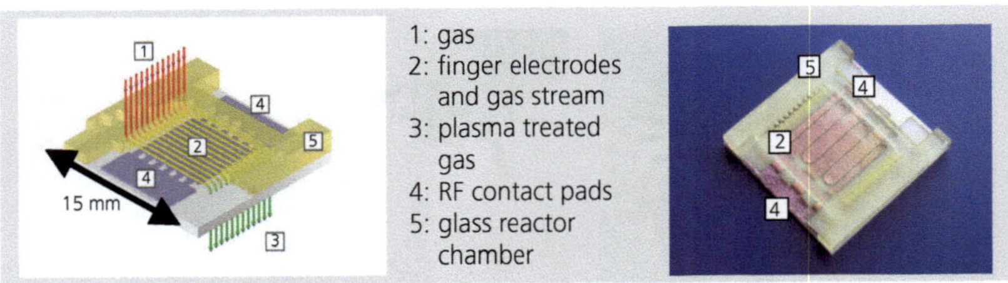

Microreactor for the electrical generation of plasma at atmospheric pressure using microstructured electrodes. Left: The microreactor consists of a glass reactor chamber and finger electrodes, which are connected to RF contact pads providing a radio frequency power supply. The size of the base is 15 x 15 mm^2. The gas flow is as follows: the waste gas (red lines) enters the reactor chamber through inlet channels on its left-hand side, it flows (yellow lines) across the finger electrodes to the outlet channels on its right-hand side where the plasma-treated gas (green lines) leaves the chamber. The electrodes ignite the gas discharge and this generates plasma. This non-thermal generated plasma might be used to create hard thin film coatings, for example. Unlike traditional coating processes, work at higher pressures is possible. In the long term, the use of microreactors like these will enhance many chemical and physical production processes by providing higher yield under more moderate conditions. Right: The photograph of the microreactor shows the transparent reactor chamber, the finger electrodes with the RF contact pads and the inlet and outlet channels. Source: L. Baars-Hibbe, PCI; IMT, TU Braunschweig

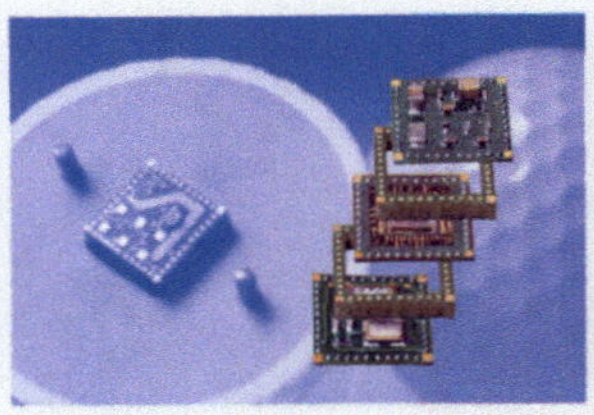

Packaging. Left: One of the main aims of MST is to produce more and more miniature devices and still ensure high levels of reliability. One approach is to stack the necessary electronic devices in three dimensions to save space and reduce wiring length. This packaging technique allows several circuit boards of equal size to be connected via frameworks and bonding, creating 3D packages. This technique might lead to new applications, such as highly miniaturised acceleration sensors with wireless data communication. Right: The picture shows one possible application of highly miniaturised microsystems: an E-Grain with an integrated acceleration sensor was incorporated in a golf ball and equipped with a process and a wireless communication interface. The ball may be used, therefore, to determine the performance of golfers by measuring the power of their stroke. Source: Fraunhofer IZM

Wire bonding. Left: A chip-to-substrate connection via small bond bindings. Right: A chip with a high density of connectors. The main purpose of the wires is to conduct electrical signals. In some cases, heat transfer is another important function that the wires perform. Source: Fraunhofer IZM

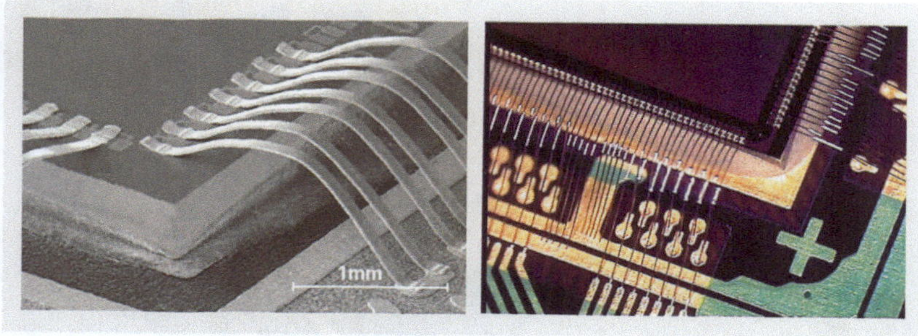

all the connectors can be connected simultaneously. Additional wires are not necessary.

Organic and *ceramic* substrates are often used to build microsystems. Organic substrates are less expensive, but they also have less favourable mechanical and thermal capacities. For reasons of miniaturisation, 3D integrational techniques are used increasingly often. The "stack" method is often applied. On the one hand, the casing needs to protect microsystems against extreme environmental conditions, such as high temperatures or aggressive media. On the other hand, the casing must also allow controlled access so that parameters such as temperature or acceleration can be measured.

► **Systems integration**. Systems integration technologies are used to integrate different parts of a microsystem. This often leads to conflict between the quest for miniaturisation and the costs of systems integration. The best approach is to build up everything using one technology, e.g. a *complementary metal-oxide-semiconductor (CMOS)*. But, in reality, this would lead to high specific costs per item for small volumes, especially at the beginning of a product's life cycle, and alternative integration methodologies are therefore required. A good case in point is the introduction of a new automotive sensor in small volumes in the high-price segment which starts off using a hybrid integration solution. At a later phase of the life cycle – when production volumes are higher – monolithic integration (with high development costs) might be more feasible.

Two ways of resolving this conflict have therefore emerged: *monolithic integration*, where all functions are performed using only one semiconductor; and *hybrid integration*, where single components are integrated on top of a different substrate. The costs of monolithic systems integration are rising because systems are becoming more complex and hybrid systems are becoming increasingly important. Improved technologies in the field of monolithic integration do not lead to the replacement of hybrid systems; they enhance their functionality. Consequently, the overall body of a system is always hybrid. Conductor boards used as substrates for (monolithically integrated) components in the completed system can be flexible or inflexible, and can carry up to twenty electric layers.

Applications

The main areas of application for MST are the automotive, medicine and machinery sectors. In the future, microsystems will play an increasingly important role in applications such as consumer electronics and biotechnology.

► **Micropumps**. In many areas of application, such as life sciences, analytics, diagnostics, the environment, high-end electronic chip cooling and fuel delivery cells, the active pumping of fluids and air is essential. Especially for precise mobile devices with low energy consumption, small dimensions, low weight, low noise and long lifetimes that do not require short service intervals, micropumps are more effective than conventional pumps. Micropumps can now be produced in plastic moulded devices with integrated inlet and outlet valves. They are able to pump liquids or gases precisely and reliably in quantities ranging from microlitres to millilitres. Using micropumps as air or vacuum pumps means aggressive media and fluids with functional particles can be pumped. The advantage is that the media is not in contact with the pump and the pump can be used permanently. Another option is to produce pumps or cartridges made of high performance plastic that can withstand the aggressive media. In mass produced quantities – ranging from hundreds of thousands to millions – micropumps can also be disposable.

For medical and environmental applications, mobile pump devices are particularly interesting for diagnostic and analytical tasks, such as vacuum pumping for painless sampling of cell fluid through human skin. Especially in the home care sector, patients want flexible, easy to use devices with noise levels that do not cause disturbance. Micropumps are effective for these applications, which include artificial nutrition, diabetes care and peritoneal dialysis.

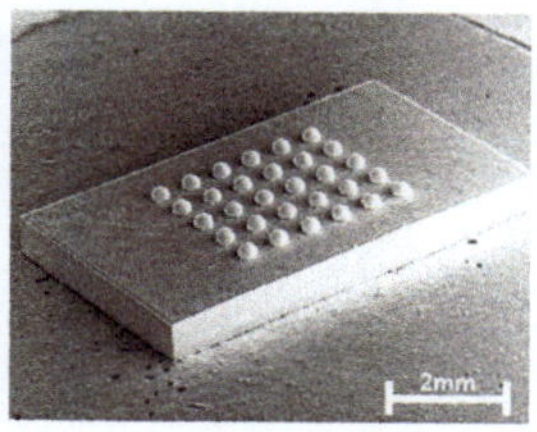

Flip chip technique: Balls of solder with a diameter of 300 µm are applied to a circuit. They are used to ensure the connection with the substrate. The circuit is flipped over and set down so that the solder bumps connect with the metallic contacts on top of the substrate. When heated, the solder melts and creates contacts between the circuit and the substrate. Sources: Atmel Germany GmbH

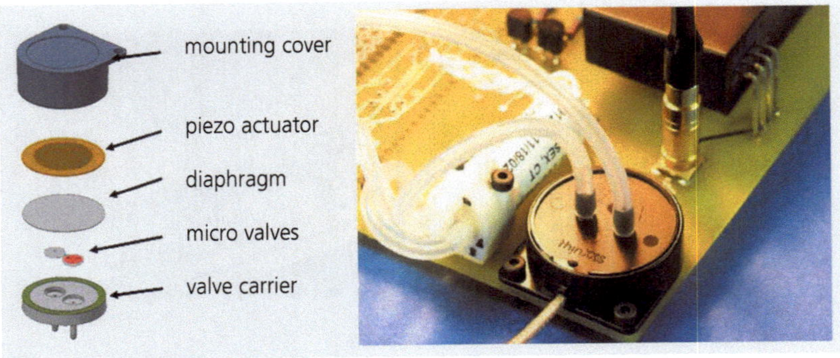

Micropump: The pump itself consists of a valve carrier with tubing ports on the connection side and passive check valves on the pump chamber side, which is covered with a plastic diaphragm. A piezo actuator is bonded to the membrane, including the electronic connections for the pump driver. For mounting and protection purposes, different covers are available. The pump platform can be easily customised to meet the customer's needs. Standard dimensions are a diameter of 23 mm and a height (without protective cover) of 3 mm (8 mm with tubing ports; not needed with on-chip solutions or when in-line ports are used). The overall weight of such a device is 3 g. PCB micropump drivers for mobile applications and battery use are available, allowing the driver to be directly integrated in the pump or in the application's main circuit. Left: Exploded drawing of a plastic micro diaphragm pump. Right: Pump integrated on a printed circuit board. Source: thinXXS Microtechnology AG

▸ **Demand-actuated maintenance.** MST-based solutions can prolong the life span and availability of machine tools. Cutting tools can now be equipped with structured sensory coatings that emit a signal in case the tool wears out. For this purpose, coating and laser machining processes have been applied to implement meander-shaped, electrically-conducting sensory layers for abrasion detection inside the wear protection coatings of reversible plates in their clearance areas. When the flank-wear land width has increased after a certain period of time to such a degree that the work piece is touching some part of the sensory structure (causing a short to ground), or has simply been interrupted, a distinct electrical signal is emitted. With its preset distance to the cutting edge and its progressive destruction caused by the cutting process, the sensor's microstructure provides information on the actual width of flank-wear land and therefore the total degree of wear to the tool. This results in maintenance that is more user-friendly and timely, as there is no exceptionally long downtimes; and tools no longer need to be replaced at fixed intervals as a precautionary measure, but only if they are really worn out.

Trends

▸ **Energy autonomous microsystems.** In the future, energy autonomous microsystems will play an important role in the field of diagnostic and condition monitoring. These mobile and easily upgradable microsystems are capable of generating all the energy they require for operation as well as data transfer from their environment (*energy harvesting*). Energy can be generated from sources such as thermoelectrical or *piezoelectrical effects*. One possible application is sensors for monitoring tyre pressure in automobiles. At the moment, these sensors are equipped with batteries that need to be regularly replaced. In the future, these sensors will generate their own energy from the vibrations caused by the movement of the car or from thermal changes. This stimulation will be directly transformed into electrical energy via piezo elements, for example, which function in a similar way to how a cigarette lighter produces sparks.

▸ **Micro-nano integration.** As miniaturisation continues to advance, it is predicted that, in the near future, microsystems components will become so small that conventional assembly methods, such as wire bonding, will no longer be possible. New assembly methods are therefore being developed for the transport, alignment and joining of nano-scale devices, and *self-organisation* principles are being applied for this purpose. The advantages of these methods are the great quantity of components that can be assembled simultaneously. These methods also signal the end of the road for very expensive high-precision assembly machines. Fluidic methods (fluidic self assembly) – which exploit gravity effects – are used, in particular, for very small devices like *RFID* transponder chips. Many alternative physical effects, such as electrostatic, dielectrophoretic and magnetic forces are currently being researched. *Dielectrophoresis (DEP)* applies a force to an uncharged particle in a non-uniform elec-

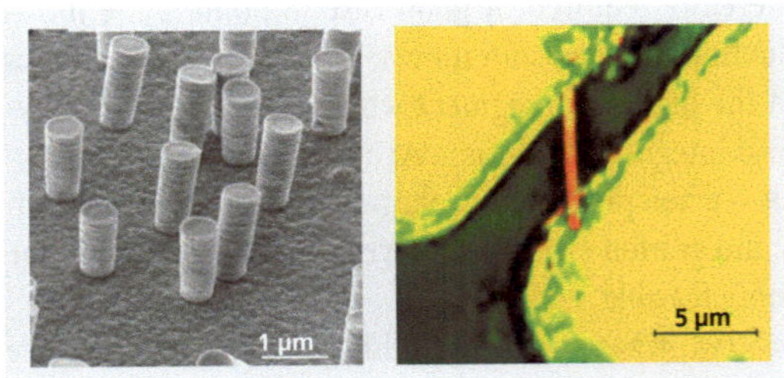

Dielectrophoresis for micro-nano integration. Left: Free standing gold wire, produced by galvanic plating of a nanoporous polymer template. The picture shows the metal filling of a single pore in the polymer template, i. e. a wire. Right: Nanowire (Gold, 500 nm diameter) assembled by dielectrophoresis. Two electrodes (yellow) have been driven with an AC signal (4 MHz, 7 V). A single gold wire (red), dispersed in an aqueous glycerol solution, has been trapped by dielectrophoretic forces, attracting the wire into the electrode gap. After careful removal of the solution, the wire remains in place. Source: Fraunhofer IZM

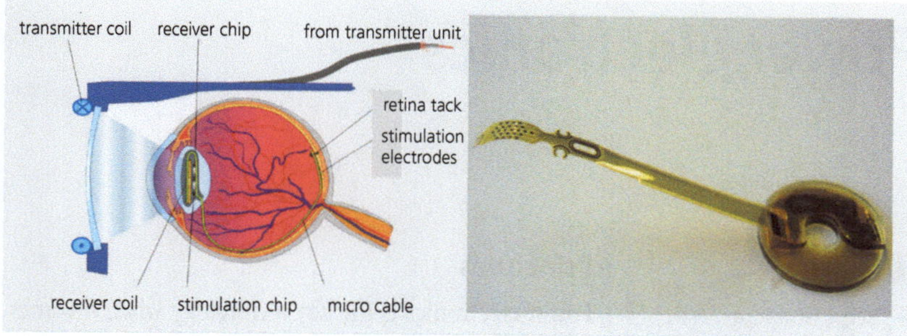

Retina implant system. Left: Energy and data are transmitted by a coil to the electronic part of an implant that is located in an artificial lens. According to the data received, stimulation pulses are generated that reach the stimulation electrodes via a microcable. Right: Artificial lens containing the electronic components. On the left-hand side, a thin polyimide foil is visible supporting the stimulation electrodes. A microcable electrically connects the electronic components with the stimulation electrodes. Source: Wilfried Mokwa, RWTH Aachen

tric field. The strength of the force relates to the particle's specific exhibited dielectrophoretic activity and can be used for the adjustment of particles.

▶ **Medical technology**. Instruments for *minimally invasive surgery* are the most common microsystems used in the field of medical technology. Various electrical, optical, mechanical, sensory and other functions are integrated in a small space. These tiny tools are applied in operations on very sensitive areas of the body, such as the brain. It is already possible to insert every type of diagnosis and operation instrument into the brain through one single working channel – the "trocar". Trocars are inserted into natural orifices or very small body incisions and only therefore cause extremely minimal traumata.

Other microsystems are used in the field of health surveillance. Small sensors are placed into or onto the bodies of patients in order to continuously measure blood pressure, intraocular pressure or glucose levels, and to transfer this data to the physician via *telemetric systems*. In the future, microsystems will increasingly be used in the medical field to help disabled people. *Implants* will help blind people with degenerated photoreceptors – e.g. Retinitis Pigmentosa patients – to recover their vision. If their optic nerve and retinal ganglion cells are still working, visual sensations may be restored by electrical stimulation of the retinal ganglion cells using an implantable microelectrode array placed onto the retina inside the eye. A *retina implant* system like this records visual images recorded by a camera integrated in a pair of glasses. The images are transformed by a signal processor into the corresponding data required to stimulate the retinal ganglion cells. This data, together with the energy required, is transmitted via an inductive link to the receiver unit of the intraocular implant. The integrated circuitry of this unit decodes the signals and transfers the data to stimulation circuitry that selects stimulation electrodes placed onto the retina and generates current pulses to these electrodes. By doing this, action potentials in retinal ganglion cells are evoked, which reach the visual cortex via the optic nerve causing a visual sensation. Microsystems like these were implanted into six legally blind patients with Retinitis Pigmentosa for a period of four weeks. All patients reported visual sensations such as dots, arcs, or lines of different colours and intensities.

Prospects

- MST is an enabling technology which has a groundbreaking and integrating influence on many areas of application like the automotive, medicine, machinery and information technology sectors. Essentially, MST integrates devices based on different technologies into miniaturised and reliable systems.
- The main areas of application for MST are the automotive, medicine and machinery sectors. In the future, microsystems will be increasingly used in applications such as consumer electronics and biotechnology.
- Smart systems integration: microsystems are currently integrated in many areas of application such as the automotive, medical applications, machinery and information technology sectors. In the future, microsystems will become more advanced and develop into smart systems. Future smart systems will possess their own "senses", and incorporate wireless networking and self-diagnosis. They will be energy autonomous, and capable of communicating and making decisions. Smart systems will be able to interact with each other and react sensitively to their environment. Generally speaking, smart systems will need cognitive qualities, integrated in highly miniaturised systems.

DR. LARS HEINZE
BERNHARD WYBRANSKI
LUTZ-GÜNTER JOHN
DR. GABI FERNHOLZ
VDI/VDE Innovation und Technik GmbH, Berlin

Internet

- www.mstnews.de
- http://cordis.europa.eu/fp7/ict/micro-nanosystems/home_en.html
- www.smart-systems-integration.org/public
- www.bmbf.de/en/5701.php

Power electronics

Related topics
- Semiconductor technologies
- Sensor systems
- Electricity transport
- Energy storage
- Automobiles

Principles

Power electronics is the key technology when it comes to accurately controlling the flow of electrical energy to match a source with the requirements of a load. It is only when an electrical load is supplied with the precise amount of electrical energy needed to meet the current requirement – e. g. in brightness control of lighting, rotation control of an electric motor, acceleration/braking control of trains – that minimum energy consumption and optimum performance of the load can be achieved. In order to minimise power loss when controlling the flow of electrical energy, it is crucial to select the appropriate power *semiconductor* devices to give the proper frequency and switching performance. At high switching frequencies, like those applied in power supplies for PCs (typically higher than 100 kHz), fast electronically-triggered switches such as power *MOSFETs* (metal-oxide-semiconductor field-effect transistor) are required. By contrast, electronically-controlled switches that have low conduction losses, such as IGBTs (insulated-gate bipolar transistor), are needed to control the rotation speed of electric motors (which have typical switching frequencies lower than 10 kHz).

However, in addition to controlling electrical energy flow, power electronics also ensures the reliability and stability of the entire *power supply* infrastructure of any given country, from electricity generation, transmission and distribution to a huge variety of applications in industry, transport systems and home and office appliances. Power electronics is not limited solely to high-power scenarios; as a cross-functional technology, it is used in all manner of systems, from extremely high gigawatt (e. g. in power transmission lines) to the very low milliwatt range (e. g. as needed to operate a mobile phone).

Power electronics technology enables the effective use, distribution and generation of electrical energy. Many areas of application could potentially benefit from the use of power electronic technology: home and office appliances, heating, ventilation and air conditioning, information technology, communication and computer electronics, factory automation and motor control, traction drives in trains and hybrid electrical vehicles, conventional and renewable energy generation, to name but a few. Advanced power electronic converters can cut energy losses by more than 50% in converting from one grid system's mains or battery to a flexible new power supply system.

Power electronics involves switching, controlling and converting electrical energy using power semiconductor devices and includes the associated measurement, control and regulation components. The challenge is to accurately control the energy flow between the source and the load using electronic switches while ensuring low losses.

A power electronic control unit's main role is to form a bridge between an energy source with fixed electrical data (voltage amplitude, frequency and number of phases) and a variable power supply system that has the voltage amplitude, frequency and number of phases required by the load.

Coupling an alternating current (AC) system and a direct current (DC) system involves four main functions:

- Rectification: Transforming AC into DC, where energy flows from AC to DC; used every time an electronic device is connected to the mains supply (e. g. computer, TV, etc.)
- DC/AC inversion: Transforming DC into AC, where energy flows from DC to AC; used every time a DC source (e. g. a battery) is used to supply a load (e. g. in the variable speed control of a hybrid electrical vehicle)
- DC/DC conversion: Transforming from DC with a given voltage amplitude and polarity to DC with a different voltage amplitude and – in certain cases –

◘ The power electronic system interconnects the energy source and the electrical load: The DC voltage polarity and current flow can be adjusted by applying a power electronic control unit. As is the case for AC, the voltage amplitude, frequency and number of phases (single-phase or multi-phase system, beneficial for the motor drive) can be regulated and adjusted to the desired value. Source: University of Erlangen-Nuremberg

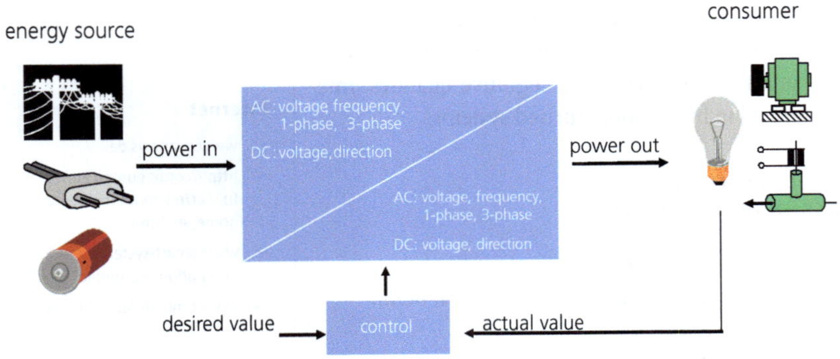

perhaps even with reverse polarity; used in most mobile devices (e.g. mobile phone, PDA, etc.)
- AC/AC conversion: Transforming from AC with a given voltage amplitude, frequency and number of phases to AC with a different voltage amplitude, frequency and – in many cases – even a different number of phases (used for variable speed control of motor drives, light dimmers, etc.)

These four basic electrical energy conversion functions are carried out by various types of power electronic controllers, known simply as power converters. In DC/DC conversion and AC/AC conversion, the flow of electrical energy is predefined. Generally speaking, AC converters and DC converters can reverse the direction of energy flow. All these basic power conversion functions come into play when an AC network is coupled with a DC network, and one or more of them are used when two similar networks are connected.

▶ **Basic layout of power electronics.** Power electronic systems consist of a control unit that detects the requirements of the load and a power switch that controls the amount of electrical power transmitted to the load. Depending on the application, the control units may vary from a simple voltage measurement in a PC power supply to a complex feedback loop that determines the actual power flow in a vehicle's hybrid power train.

Nowadays, most control units take the form of integrated circuits (ICs), or an electric circuit comprised of ICs, various sensor elements and a power switch drive stage. For lower power ratings, the control unit and power switch are often integrated within the same semiconductor chip. For higher power ratings up to several gigawatts, the control unit and power switch are kept separate for security reasons.

These days, power switches tend to take the form of power semiconductors because of their tremendous advantages over mechanical switches. Power semiconductors offer much higher switching frequencies and, with no need for a mechanical contact, are wear and arc-free. They require no maintenance and offer a much longer service life. Semiconductor switches also need less space and lower control power. Although mechanical switches continue to be used as circuit breakers in distribution grids and in conventional light switches in homes and offices, here too they are gradually being replaced by semiconductor solutions where additional features like automation or remote control are required.

Today's key devices are mainly based on silicon technology, although compound semiconductors are also used in niche applications. Power converters aim to transmit energy with a high degree of efficiency in the on-state (conduction status) and to stop energy flow with no delay or losses in the off-state. During the

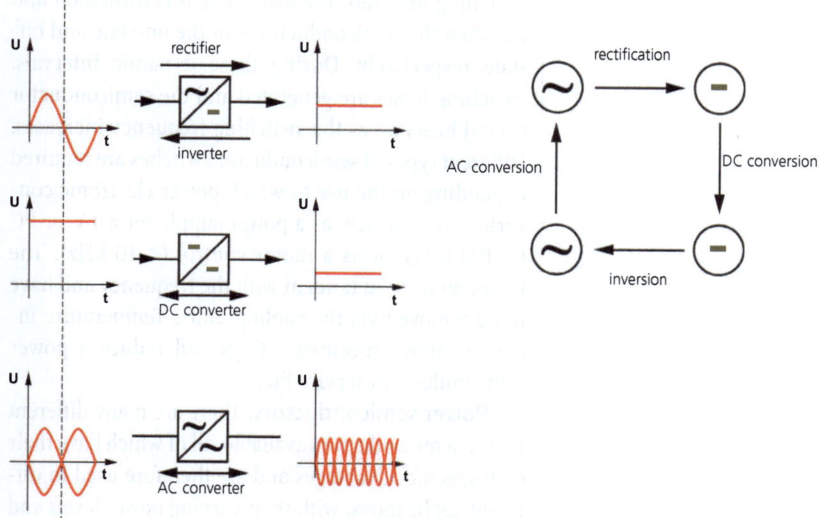

Basic function of a power electronic control unit. Left: Types of power electronic control units. Top: A rectifier transforms AC into DC and an inverter transforms DC into AC. Centre: A DC/DC converter transforms DC with a given voltage into DC with a different voltage. Bottom: An AC/AC converter transforms AC with a given frequency into AC with a different frequency. Right: Types of energy converters. Rectification, DC/DC conversion, inversion, AC/AC conversion. Source: Siemens AG

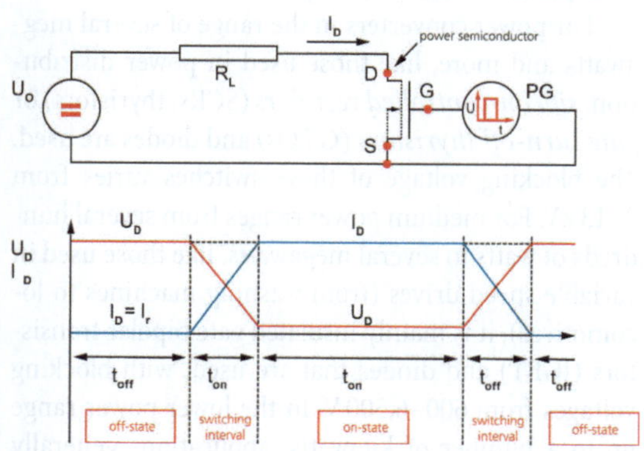

Basic circuit diagram of a power electronic controller. Top: Power semiconductor switch for controlling the flow of electrical energy. The power semiconductor device will be triggered directly by a microcontroller (μC) from a pulse generator (PG) for state-of-the-art devices. The three power switch terminals are called the source (S), drain (D) and gate (G). The gate controls the current flow (I_D), which is dominated by the resistor R_L. Bottom: Switching performance. During the switching intervals (t_{on} and t_{off}), the current flow increases when the semiconductor is in the on-state and is cut off when the semiconductor is in blocking mode. In blocking mode, there is a voltage across the drain U_D (red), while the current flow I_D (blue) is almost zero. In the on-state, the voltage across the drain is very low (a few volts), while there is a drain current I_D. The switching interval is the transition time between the on-state and the off-state (or vice versa). Source: Infineon Technologies switching intervals, the load current is turned on and off when the semiconductor is in the on-state and off-state respectively. During these dynamic intervals, switching losses are generated and the semiconductor crystal heats up as the switching frequency increases. Different types of semiconductor switches are required depending on the use to which power electronic converters are put, e.g. as a power supply for a TV or PC (> 100 kHz), or as a motor control (< 10 kHz). The losses increase in tandem with the frequency and have to be removed via the cooling units. Temperature increases above specified ratings will reduce a power semiconductor's service life.

▶ **Power semiconductors.** There are many different power semiconductors available, all of which have their own specific advantages and are therefore used in different applications, with their varying power levels and required blocking voltages. Depending on their switching behaviour and conductivity characteristics in the on-state (unipolar or bipolar), which are reflected in the semiconductor cell structure, power semiconductor components can essentially be divided into two types: voltage-controlled devices (e.g. power metal-oxide-semiconductor field-effect transistors (power MOSFETs) and current-controlled devices (e.g. bipolar transistors). These days, a power semiconductor is made up of several thousand to several million cells, all connected in parallel to carry the required load current.

For power converters in the range of several megawatts and more, like those used in power distribution, *silicon-controlled rectifiers* (SCRs, thyristors) or *gate turn-off thyristors (GTOs)* and diodes are used. The blocking voltage of these switches varies from 2–13 kV. For medium power ranges from several hundreds of watts to several megawatts, like those used in variable speed drives (from washing machines to locomotives), it is mainly insulated gate bipolar transistors (IGBT) and diodes that are used, with blocking voltages from 600–6,500 V. In the lower power range up to a number of kilowatts, applications generally tend to use power MOSFETs with blocking voltages from 20–1,000 V and nominal currents from several hundred mA to 100 A.

Power MOSFETs use a field effect by applying or removing a voltage to the gate in order to turn the load current on or off. Because they use the same control principle as the MOSFETs in ICs, power MOSFETs have very low static control power losses. Not only do power MOSFETs have much larger chip areas than MOSFETs in ICs, they also have different structures. In ICs, the current flow is always very shallow at the front surface of the semiconductor chip. In power MOSFETs, where high voltages and high electric currents have to be controlled, the current flows vertically

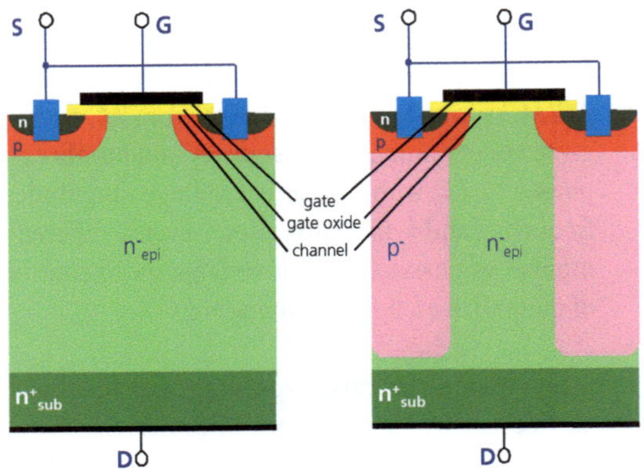

Power MOSFET. Left: Cross-section of a conventional vertical n-channel power MOSFET, in which the load current flows from the contacted n source on the front surface of the chip to the contacted n+ drain on the rear surface, and is controlled by the gate. In the on-state, a positive voltage is applied at the gate and a conductive inversion (i.e. electron) channel beneath the gate electrode is established in region p. In the off-state, the gate voltage is set to zero or even a small negative voltage. In the on-state, the device resistance is dominated by the resistance of the n- epi layer (deposited epitaxially on the substrate), which also serves as a voltage-sustaining layer. Consequently, to achieve a required blocking voltage, n-(donor) doping – and hence conductivity – must be fairly low. Right: Cross-section of a modern super junction vertical power MOSFET, which has essentially the same structures on the front side of the chip. Here, the voltage-sustaining layer consists of donor (n-) and acceptor (p-) doping, situated in two individual regions. In blocking mode, the difference between the donor and acceptor doping determines the blocking voltage. Provided the difference between the donor and the acceptor is small enough, the donor doping can be increased, and thus the on-state resistance reduced. Source: Infineon Technologies AG

through the semiconductor chip. The rear side of the chip serves as an electric contact and as a transition area for the device's power losses. Nowadays, both current and heat are drained via a large metal area in the device packaging.

Power MOSFETs are the most commonly-used switches in power electronics, given that these devices and circuits are used in almost every modern electronic device at home and in the office, from the power supplies for electronic entertainment systems and information technology appliances to modern lighting systems in the form of compact fluorescent or high intensity discharge lamps. Because they are so widespread, these devices are particularly important when it comes to energy-saving efforts.

▶ **Power electronic converter.** Power electronic converters are needed in all applications where the targeted energy flow must be controlled; in other words, where the operating frequency, number of phases and voltage amplitude must be changed with a high overall degree of efficiency (power semiconductor plus circuit topology) in order to optimise system behaviour.

In order to cover all fields of application, we need power converters ranging from a few watts (e. g. mobile phone battery charger) through several megawatts (e. g. the *traction drive* of a locomotive) and right up to gigawatts (as used in power transmission lines, e. g. HVDC (High Voltage DC) systems). In the lower power range – take the example of a mobile phone battery charger – a power converter is connected to the mains (230 V, 50 Hz or 115 V, 60 Hz), operating at around 100 kHz and generating perhaps 5 V DC at the output. By contrast, in the high power range (e. g. for the traction drive of fast trains like the TGV or ICE), using a power converter, the input voltage supply (25 kV, 50 Hz or 15 kV, 16.7 Hz) can be transformed by an inverter into a 3-phase AC output voltage (3,000–3,500 V with the desired, variable frequency). An AC converter regulates the rotation speed and torque of the motor drive in a locomotive (frequency of the AC voltage and current supplied by the AC converter respectively), and both the driving dynamics and the energy supply (likewise energy saving) depend solely on the power electronic controller.

There are numerous switching topologies for power electronic control units; these essentially take the form of one of two basic configurations:

- AC voltage source rectifier with a fixed AC voltage, from which an adjustable DC voltage is generated by modulation
- DC voltage source inverter with a fixed DC voltage, from which an AC voltage with a variable frequency, number of phases and amplitude is generated by modulation

Both variants are used in electric motor power supplies. Depending on the load conditions (single motor drive or multi-motor drive), one of the two will be the preferred option. In many fields of application – depending on the mains voltage source, as discussed above – direct entries (AC/AC converter) are needed to create from the mains a new power supply that has the variable frequency, number of phases and amplitude needed to operate different loads (e. g. variable speed-controlled motors) or to interface between different power supplies (e. g. between 50 Hz, 3-phase mains and 16.7 Hz mains used in some European train system power supplies).

Application

A power electronic control unit basically consists of passive and active power devices including smart cooling systems, energy storage components (e. g. capacitors and inductors), overall control electronics and EMI (electromagnetic interference) filters to eliminate high frequency oscillations. The efficiency and service life of power electronic control units are determined by the selection and division of their components, circuit topology and layout, and how the system is integrated. In the future, the primary driver will be integration of the electrical system and the mechanical actuator, or any other load (e. g. integration of the control electronics inside the socket of an energy saving lamp or motor body). Every day we routinely use a variety of electrical systems that incorporate typical ap-

power electronic converter	power range	switching frequency range	application examples
light-triggered thyristor	top end of power range: >1 GW	10 - 100 Hz	high voltage DC lines
GTO	high power range: 10 MW	100 - 500 Hz	high power supplies large motor control
IGTB	middle to high power range: 1 - 10 MW	1 - 5 kHz	traction drive for trains at several MW, motor control in industry hybrid electric vehicles (HEVs)
power MOSFET	low end power range: 10 kW	100 kHz and higher	power supply

Power range of typical power electronic converters as a function of the switching frequency of power semiconductor devices: The higher the converter's power capability, the lower the switching frequency.

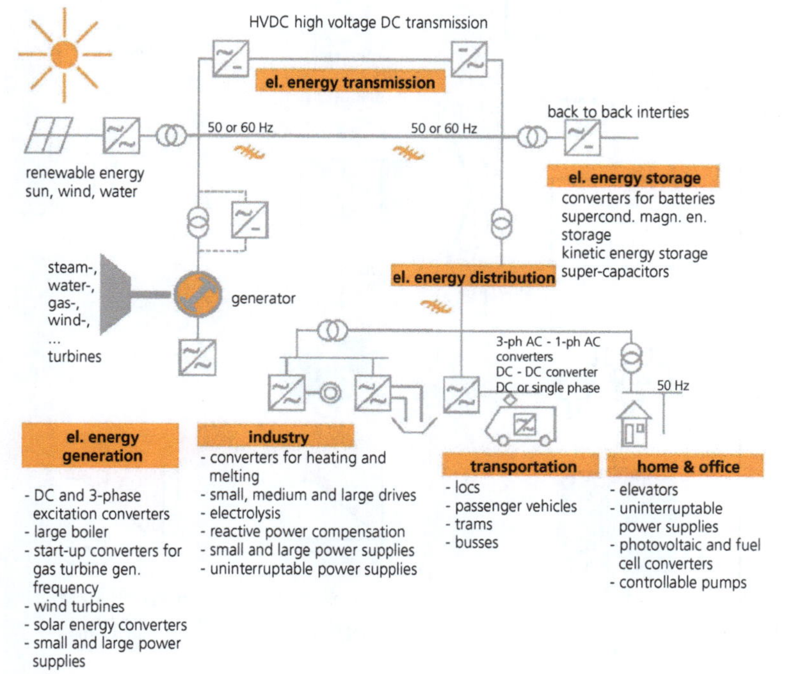

The entire chain of factors that influence a power electronic control unit: The chain extends from energy generation, transmission, storage and distribution to the end consumer in industry, transport systems and the home/office arena. Here, the most frequently-used conversion is AC/AC power conversion with an AC power grid that has a fixed frequency, number of phases and voltage amplitude. Source: IPEC 2000

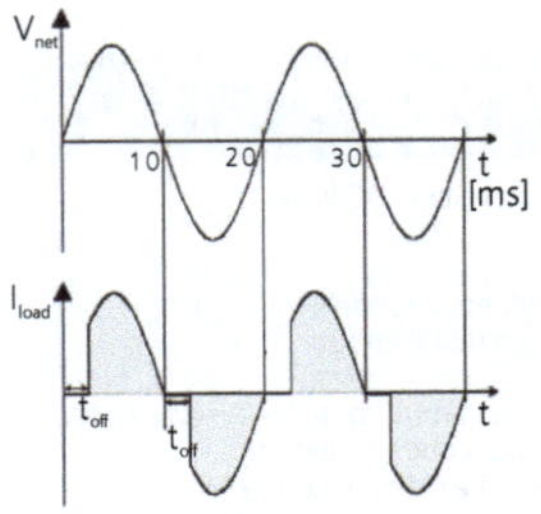

Leading-edge phase principle: Dimmers based on this principle are used, among other things, to adjust the brightness of lamps. The top curve shows the sinus waveform of the net voltage (V_{net}) while the bottom curve represents the current flow (I_{load}) to the lamp. The dimmer switches the current off at the start of each sinus half wave. After a period of time (t_{off}) that can be defined by the user, the dimmer switches the current back on and energy is again transferred to the lamp. When the voltage sinus wave reaches the zero point, the current is switched off again. This process is repeated every sinus half wave, thus every 10 ms (100 times per second) in a 50 Hz energy grid. The shaded area below the current curve corresponds to the brightness of the lamp. By varying the switch off-time, the brightness can be adjusted continuously.

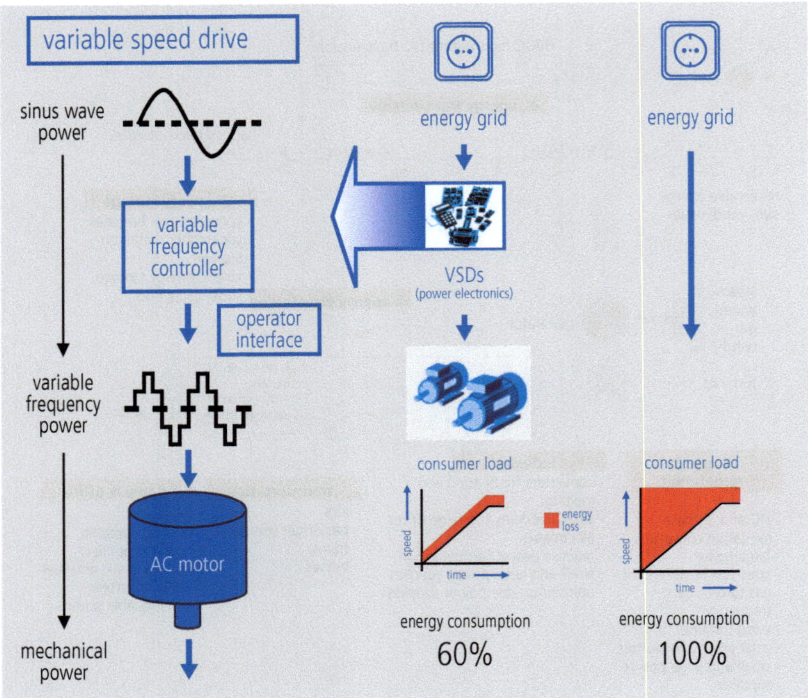

Electric drives with variable speed control. Right: The alternating current (AC) energy flow is traced from the energy grid to the consumer load (e.g. motor), as is its corresponding energy consumption. The right-hand side of the picture shows a direct flow from the energy grid to the consumer load. In this case, the energy loss corresponds to the red area above the line when the speed of the motor is plotted against time. Here, the energy consumption is 100%. An example of this would be a typical shunt, which uses different ohmic resistors connected in series with the motor via mechanical power switches. The appropriate resistor is chosen depending on the desired speed. This method of control wastes energy, reduces the lifetime and produces unacceptable noise during switching. Furthermore, only very coarse speed control is possible. The left-hand side of the picture shows an energy flow which is controlled by a variable speed drive (VSD), thus optimising the system. In this case, the energy loss is represented by the small red area above the speed-time-line. The energy consumption is about 60% of that in the uncontrolled example. Left: the VSD system comprises a control unit (variable frequency controller) with an operator interface and an AC motor. The speed of the motor can be adjusted by altering the frequency. The basic idea of a VSD system is to convert power from AC (sinus wave) to variable frequency to mechanical power – for example, as in the rotation of a fan. The variable-frequency controller is made up of power semiconductor conversion devices. Generally speaking, its function is to convert AC input power (sinus wave) to DC intermediate power using a rectifier. The DC intermediate power is then converted to quasi-sinusoidal AC power using an inverter switching circuit, thereby giving the AC power a variable frequency and voltage. Source: ZVEI 2006

plications of power electronic control. And now that governments around the world have adopted policies to protect the environment and save energy, power electronics will be paramount when it comes to meeting emerging requirements.

► **Dimmers.** A power electronic control unit (dimmer) can, for example, adjust the brightness of a lamp according to the level of daylight and/or working environment requirements. Not only can these electronic components save up to 25% electrical energy, they also extend the lifetime of a bulb. Without brightness control, the lamp would be operating at its limits, shortening its life span and consuming the maximum amount of energy. Dimmers use two fundamental modes of operation:

- The leading-edge phase principle, for pure resistive and inductive loads such as lamps and low voltage halogen lamps
- The trailing-edge phase principle, for loads demonstrating capacitive input behaviour such as electronic transformers. Universal dimmers detect the nature of the connected load and select the appropriate dimmer mode.

► **Motor control.** More than half of all electricity used for home applications or factory automation is consumed by electrical motors, e.g. in elevators, refrigerators, air conditioners, washing machines and factories. The majority of these motors do not have any electronic control. These simple electrical motors are either "fully on" or "fully off", which is like driving with the accelerator flat to the floor and then taking your foot off it over and over again. Everyone agrees this is not a good way to drive – nor is it efficient. By running all these electrical motors using variable speed control (in other words, by simply operating them at the current requirement at any given time), it is possible to cut power consumption by half. A *variable-speed drive (VSD)* is a system for controlling the rotational speed of an alternating current (AC) electric motor, and generally comprises a controller with an operator interface and an AC motor. The speed can be adjusted by altering the frequency of the electrical power supplied to the motor.

► **Hybrid (or fully) electric drives.** The transport sector, which accounts for around 27% of end-use energy, relies heavily upon petroleum-based liquid fuels. And this percentage is likely to increase further.

While in motion, energy is also wasted in breaking, idling, and stop-start traffic. Hybrid electric vehicles (HEVs) work on the principle of recapturing this wasted energy using an electric motor, storing the energy in the battery for later use, or stopping the engine fully during idling.

However, the benefits provided by HEVs are heavily dependent on the type of HEV system used. One process that is common to all HEV systems is that of power conversion, storage and later use of energy. And it is power electronics that makes this process possible.

A high voltage (HV) DC line runs from the battery to the sub-system and the various components of the HEV system. A DC/AC power converter supplies the electric (propulsion) motor that provides power to the wheels, and an AC/DC converter draws power from the generator (connected to the power train) to recharge the batteries.

Trends

The three most important trends in power electronics can be summarised as follows :

- Further development of existing power semiconductor devices to achieve higher cell density and smaller semiconductor volume, with the aim of reducing loss, increasing ruggedness and reliability, broadening the power range and ensuring functional integration, likewise development of devices based on new materials such as silicon carbide (SiC), gallium nitride, diamond, etc, which offer better electrical or thermal features than silicon.
- New technologies for chip contacting and overall system interfacing. On the one hand, a silicon chip needs a high voltage isolation capability, but on the other, it must also provide excellent heat transfer. Attempts are being made to develop new materials that match the thermal expansion coefficient of the overall system.
- Increased system integration density for complex power electronic controllers, circuit topologies demonstrating a high degree of efficiency, higher temperature ratings, improved EMI performance and zero-defect design for the overall power converter: IPMs (integrated power modules), where power switches – and at least parts of the control unit and auxiliary electronics – are integrated into a single package and facilitate the converter design, represent one step towards increased system density.

Prospects

- Technological changes in power electronics are currently being driven by the push towards rational use of electrical energy, miniaturisation of power electronic converters, improved efficiency and reliability even at higher operating temperatures, and in-system communication.
- Energy-saving outcomes achieved by the application of variable speed motor drives in the electrical system, further development of new lamp sources incorporating electronic control, improvements in the efficiency of power supplies in the consumer, information, communications and IT fields, and ultra-low-loss power semiconductor devices.
- Miniaturisation and improved reliability of power electronic systems are achieved, essentially, by the method of system integration. In the lower power range this means using system-on-chip integration, in the middle power range multi-chip integration on substrate layers, and in the high power range, power electronic building blocks (PEBBs).
- Future developments in power electronics will be driven by new areas of application in the fields of mobility (e.g. HEVs), mobile devices and alternative energy supplies, likewise by applications in the highest-power fields of energy transmission and storage.

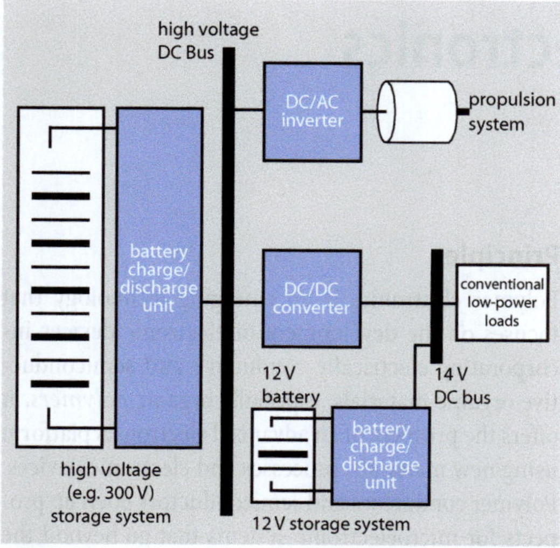

Hybrid electric vehicles: Microelectronics/power electronics is key to converting and controlling the different voltage states in hybrid electric vehicles (HEVs). The traction battery charge (boxes on the left – storage system with charge and discharge unit) is controlled by a sophisticated energy management system that supplies the high voltage direct current (DC) bus. The propulsion system – e.g. an asynchronous electric motor – requires multiple phase alternating voltage (AC). A DC/AC inverter therefore transforms the DC from the high voltage DC bus into AC. The conventional loads – e.g. lighting or actuators to open windows – are still connected to a low potential voltage DC bus, buffered by a standard 12 V battery, and a DC/DC converter transforms the high DC voltage (e.g. 300 V) from the DC bus into the required low DC voltage (e.g. 14 V) for these low-power loads.

PROF. DR. LEO LORENZ
DR. ANTON MAUDER
Infineon Technologies AG

Internet

- www.ecpe.org
- www.cpes.vt.edu
- www.pels.org
- http://services.eng.uts.edu.au/~venkat/pe_html/contents.htm
- http://ece-www.colorado.edu/~pwrelect/book/slides/slidedir.html

Polymer electronics

Related topics

- Polymers
- Semiconductor technologies
- Microsystems technology
- Logistics
- Solar energy
- Optics and information technology
- Sensor systems

Principles

Polymer electronics is an emerging technology that focuses on the development of electronic devices incorporating electrically conductive and semiconductive organic materials, especially organic *polymers*. It offers the prospect of an advanced electronics platform using new materials, processes and electronic devices. Polymer conductors and semiconductors open up projects for microelectronic systems that go beyond the scope of conventional electronics based on silicon as the *semiconductor*.

► **Properties of polymers.** Plastic materials are organic polymers, which means that they consist of large molecules with long repeating chains of smaller organic units. Depending on the structure of the materials, they are electrically insulating, conducting or semiconducting. The insulating properties of organic materials, especially polymers, have been used for a long time in electronic applications, for example as wire insulating coatings or as insulating housings.

Special polymers, such as conductive polyacetylene, consist of many identical single units, or monomers, combined in a chain. Typically, these organic polymers are based on chains of carbon atoms combined with hydrogen atoms. Conjugated chains comprise alternating single and double bonds between the carbon atoms, which result in delocalised electron states; in this case the polymer is semiconductive. It should be mentioned that these properties are not limited to polymers, and smaller molecules like pentacene or benzene also exhibit this behaviour. These semiconductive organic materials can be transformed within special devices like transistors to either a conductive or an insulating state. This is similar to what happens in conventional semiconductors such as silicon.

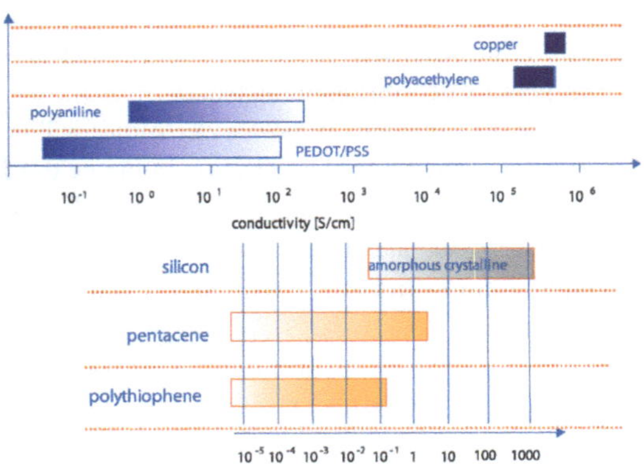

Semiconductors and conductors. Top: The electrical conductivity (in S/cm) of typical polymers is shown compared with conventional conductors like copper. The conductivity of organic polymers is much lower than that of copper. Polyacetylene is an exception, but this material is not stable in air. Bottom: The charge carrier mobility (in Vs/cm^2) of typical organic semiconductors like pentacene and polythiophene is shown compared with the conventional semiconductor silicon. Here it can be seen that the organic materials have lower mobilities, but the values of amorphous silicon and even polycrystalline silicon can be matched. Source: PolyIC

In order to achieve electrical conductivity, the polymer must be doped, which means that electrons must either be removed (by oxidation) or additional electrons must be added (by reduction). These holes or additional electrons now move along the conjugated chain within the polymer, rendering the material electrically conductive. Electrons can be removed or added by means of chemical doping, for example using acids.

In addition to conductive and semiconductive properties, some polymers also display electroluminescent properties and emit light when electric power is applied. The inverse effect also exists, whereby incoming light is converted into electric power. Other classes of organic materials exist that have special sensing properties or which can store information in a memory. As a result, polymer electronics based on electrically active organic polymer materials creates a new platform that makes it possible to produce high volumes of thin and flexible electronic devices covering the same full range of applications as conventional silicon-based electronics technology.

► **Production process.** The main advantage of polymer electronics compared with conventional electronics is the simplicity with which the polymer electronic devices can be produced. While conventional electronics demands high-class clean rooms as well as complex vacuum and high-temperature processes, the production processes in polymer electronics are significantly simpler. There are many different ways of producing polymer electronics, including vacuum processes, but the most economical process is printing.

Polymers can be dissolved in special solvents and used as a kind of *electronic ink* in different printing processes. This makes it possible to produce electronics in continuous printing processes on flexible substrates at low cost, as in standard printing methods, e.g. newspaper and book printing. A wide range of *printing processes* can be used, including gravure, flexo and offset printing, as well as many special methods and subgroups. In the case of gravure printing the printed structure is engraved on a steel cylinder, while in flexo printing the pattern is made in a raised structure of plastic material. In offset printing the print pattern is made by hydrophobic structures on top of a flat cylinder.

Digital printing is a completely different type of printing. The print pattern is applied directly onto the substrate. One example is inkjet printing, where single drops of ink are deposited one after the other onto the substrate. Consequently, no printing plate is needed.

The electronics printing process imposes high demands on the attainable resolution. Printing methods used today typically have resolutions in the range of about 100 µm. For polymer electronic devices like transistors resolutions well below 30 µm are necessary, the smaller the better. Electronic devices do not have the optical demands of a picture comprising many small coloured areas, but instead require long fine lines that are connected to each other. These lines are clearly separated from each other, but must also be as close together as possible. If one long line is not connected along the path, no electrical current can flow through the circuit, but two separate lines in contact at only one point will cause a short circuit. Both possibilities must be avoided to ensure that the electronics will work. Another aspect is that multiple layers, featuring different electronic materials, must be made with well defined interfaces and high overlay accuracy. In contrast to conventional printing, the print pattern and printed layers in polymer electronics need to have special electronic properties. While printing resolution is decisive, the electronic ink itself must also exhibit the right functionality (e.g. conductive or semiconductive) after printing – a completely new challenge for the well established printing technologies. What's more, additional electrical testing is necessary to provide quality control after printing, and this also represents a new aspect of printing technology.

▶ **Devices**. The simplest electronic devices are the passive elements such as electrical conductive coatings, resistors, conductive lines and capacitors. In these applications only conductive and insulating materials need to be printed. Some of these devices are already established, including printed resistors, printed circuit boards and printed capacitors. For active devices, however, semiconductive materials are also necessary, and this is a completely new field for polymer electronics. Active devices include transistors, diodes, solar cells and light-emitting diodes. They are usually made of several layers of polymer films, with different electronic properties. The film thicknesses and pattern sizes vary from several nanometres to micrometres. Complete devices, either as single elements like solar cells or as complex circuits, can attain surface areas significantly larger than is possible with conventional electronics.

▶ **Organic transistors (OFET)**. The basis of every logic circuit is a transistor, which is the element that controls electrical currents by turning voltages on and off. There are many different possible organic transistor setups. In polymer electronics the field effect transistor, also called the organic field effect transistor – OFET – is the most firmly established.

OFETs are built up from several layers of different thin films and are therefore often referred to as *organic thin film transistors* or OTFTs. On a flexible plastic substrate, e.g. polyester, a first fine patterned layer of conductive material is applied. This forms the bottom electrodes (source and drain) of the transistor. On top of this a thin film of semiconducting material, such as polythiophene, is applied. The next layer is an insulating or dielectric film which separates the electrode and semiconductive films from the top layer. This is again a conductive film that serves as top (*gate*) electrode. By applying voltage to the three electrodes – source, drain and gate – it is possible to switch and control the electrical current between the source and drain electrodes.

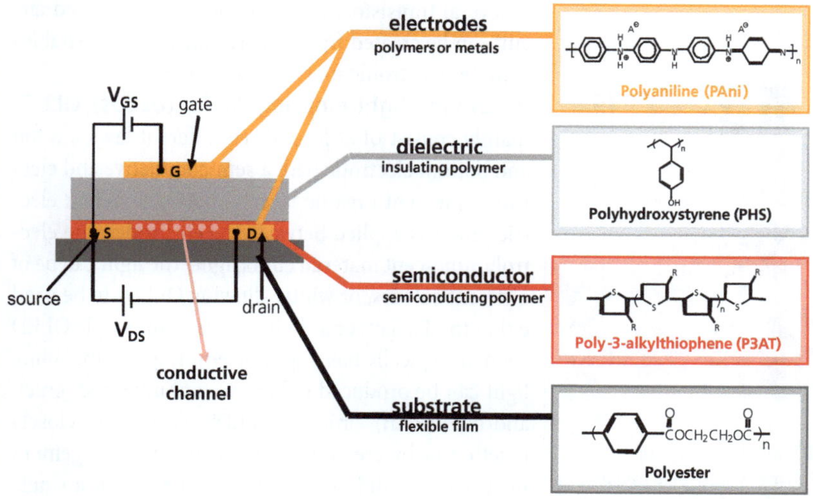

Organic field effect transistor (OFET): OFETs feature a multilayer system made of thin semiconducting, conducting and insulating layers, based on different materials. The single layer thicknesses typically range from a few nanometres to micrometres. By applying a voltage to the gate electrode, an electric field is created which passes through the dielectric layer to the semiconductor, where it leads to the accumulation of charge carriers, and a conductive channel is created. When a different voltage is applied to the source and drain electrodes, current flows between the electrodes. The size of the current can be controlled by the voltages applied to the electrodes. On the right side possible material examples are shown. Source: PolyIC

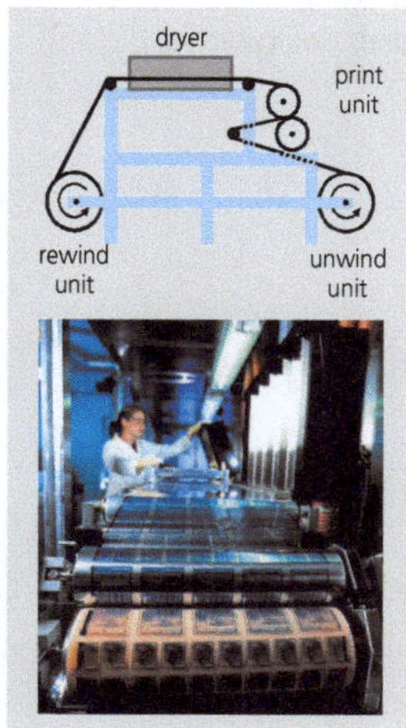

Roll-to-roll printing process. Top: For high volume printing roll-to-roll processes are used. The flexible substrate material is unwound from one roll onto a printing unit, where the pattern is printed on the substrate. After passing through a dryer, the substrate is rewound onto another roll. This process permits fast and continuous printing of polymer electronics on flexible substrates. In actual practice several printing units are mounted one after the other on the machine. Bottom: Roll-to-roll printing machine used for the development of printed electronics based on polymer materials. Source: PolyIC

Printed RFID tag. Top: Polymer-based radio frequency identification (RFID) tags used as electronic product codes (EPC™) on low-cost high-volume consumer goods. The aim is to replace the optical barcode to improve both the logistics process and customer satisfaction. Bottom: Printed transponder chips for RFID tags based on polymer logical circuits. Such chips provide the basis for the use of printed RFID on high-volume consumer goods. Source: PolyIC

If several transistors are combined, an integrated circuit can be created from plastic materials that enables complex electronic functions to be performed.

▶ **Organic light-emitting diodes (OLEDs).** OLEDs mainly consist of at least three different layers, a top and bottom electrode with a semiconductive and electroluminescent organic layer in between. When electric power is applied between the electrodes, the electroluminescent material emits light. The light can be of different colours, or white, allowing OLEDs to be used either for displays or as light sources. One single OLED element typically has a specific emitting colour. White light can be produced either by combining red, green and blue (RGB) emitting OLEDS placed very closely together or by creating a more complex arrangement incorporating different emitting materials in a single device. This will open up completely new applications such as large-area, thin, flexible lamps and displays.

▶ **Organic photovoltaics (OPV).** Organic photovoltaic cells convert incoming light into electric power. The arrangement is similar to that of OLED devices, but the semiconducting layer is optimised for high solar efficiency. The level of efficiency is lower than that of conventional solar cells, but the advantage of OPVs is that they can be produced in high volumes at low cost and with large surface areas.

Applications

▶ **Radio frequency identification (RFID).** RFID tags based on polymer electronics enable RFID to be used in high-volume, cost-sensitive areas, especially in consumer products. Initially the main applications will probably be electronic brand protection, ticketing and anti-counterfeiting. In later development stages they will be introduced in automation and logistics. In a final stage the aim is to provide a low-cost electronic product code (EPC™)on supermarket packaging to replace the optical barcode used today.

▶ **Thin and flexible displays.** Display technologies are of great market interest. Polymer electronics will open up new classes of thin and flexible displays. OLEDs will be used to make self-emitting displays that are even thinner and more brilliant than today's LCD screens. The first OLED displays are already being introduced on the market. *Polymer-based thin-film transistor (OTFT)* arrays as the backplanes for thin and flexible displays are also developing rapidly. Together with electrophoretic display elements, they are paving the way for applications such as electronic newspapers and rollable cellphone displays. OLED displays and OTFT displays can also be combined, and this is under development. In all of these displays, a matrix of electrode lines and columns is placed underneath the display layer, producing single pixels which build the image in the same way as today's TV screens.

▶ **Large-area lighting.** OLEDs can be produced as large surface areas and a wide range of colours can be displayed. White light for *lighting* can also be pro-

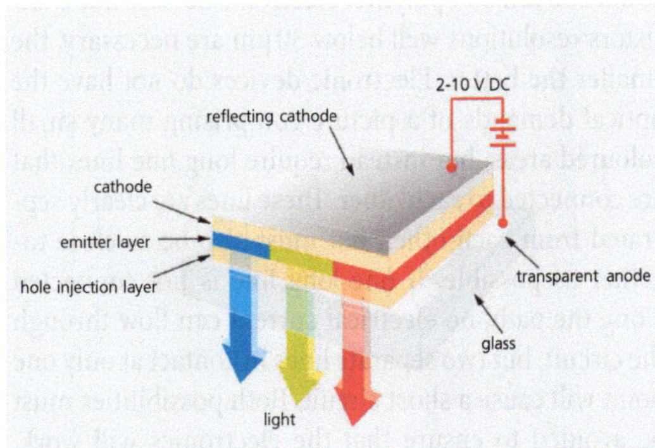

OLED schematic: An organic light-emitting diode consists of a layer of organic electroluminescent semiconductor material placed between two electrode layers. The upper electrode, referred to as the cathode, serves as the electron injector, the lower electrode serves as the hole injector. When the electron and the hole enter the semiconducting layer, light is emitted, which can exit through the transparent anode layer and the glass substrate at the bottom of the device. Depending on the semiconducting material, the emitted light has different colours. Materials can be combined to create white light. The display elements can also be structured in a matrix-like arrangement, producing OLED displays with high resolution and brilliance.

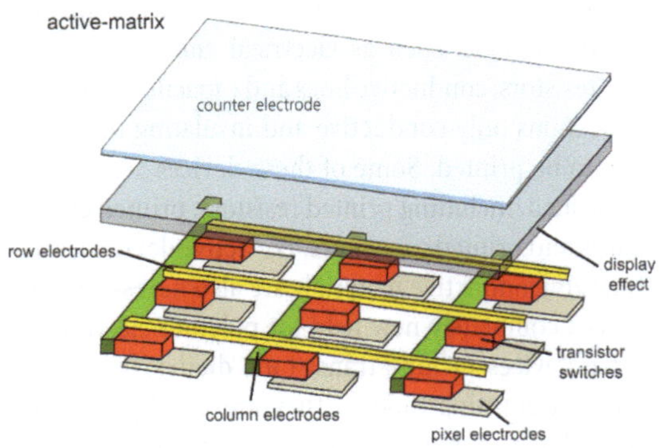

Active-matrix display: For the creation of images or videos on displays it is important to have a number of single pixels. These pixels are addressed by a matrix of electrodes placed in rows and columns. To achieve a good-quality display requires an active matrix in which each pixel electrode is driven by a transistor. In the first layer the electrodes are placed with a transistor switch at each crossing. A layer with the display effect is placed on top of this, followed by an optical clear counter electrode. As a result, each single pixel can be addressed separately in order to create a high-quality image on the display. Source: Philips duced. OLED lamps will permit completely new lighting applications, outstripping existing systems based on relatively small conventional lamps. Light sources could in the future take the form of a wall covering or a very thin, flexible sheet.

▶ **Sensors.** Conductive organic materials are often sensitive to factors such as temperature, humidity and pressure, but they can also be made sensitive to specific chemicals and parameters. This makes it possible to create large-area thin and flexible sensors, e. g. for position-sensitive pressure detection. Inexpensive single-use medical sensors can also be produced for the analysis of fluids e. g. to test blood sugar levels or to check for alcohol and drugs. In a temperature sensor, for example, the conductivity of a printed resistor depends on the temperature and the change in conductivity can be measured by simple electronic means.

▶ **Memories.** One of the main drivers of modern technology is the availability of large memory capacities. Memory systems are based on conventional technology with which data is stored either optically, magnetically or on silicon chips in separate devices. Polymer-based memories will make it possible to place memory chips on flexible substrates which can be directly integrated in consumer products and packaging. Like the displays described above, polymer memories are arranged in an electrode matrix structure. Instead of an OLED, a memory element is installed that changes conductivity or the dielectric behaviour according to the status "1" or "0".

▶ **Smart objects.** The materials and processes used in polymer electronics technology enable different electronic devices to be combined in a way that was not possible with conventional electronics. Devices such as a *solar cell* or a *battery*, a logic circuit, a *sensor*, a *memory* and a *display* can be integrated on one flexible substrate in a single production process. Such combinations are just starting to be realised today but they are destined to enter many new areas of application, for example as intelligent packaging, single-use sensors and new electronic games.

Trends

Polymer electronics is a new technology platform which will open up new applications where there is a demand for thin, flexible, lightweight and low-cost electronics, and where the functionalities of the electronic devices can be limited. It will not replace conventional electronics. Polymer electronics is still a young technology that is not yet fully established on the market.

The electronic parameters of organic polymer materials are lower than those of conventional inorganic materials like silicon and copper. The parameters referred to are the conductivity of conductive materials, the charge carrier mobility of semiconductors and the operating life time in general. Work in this field will focus on the further optimisation of the basic materials and processes with new solutions for electronic quality control methods, but also in the finding of new applications, starting in niche areas before entering high-volume markets.

One trend is the development of a *CMOS*-like electronic technology for polymer electronics, as is standard for conventional electronic systems. Not only organic molecules or polymers are of interest in this context, but also new nanomaterials, as well as inorganic printable formulations and other materials such as *carbon nanotubes*, nanosilicon and zinc-oxide. Along with the materials, the processes too still need to be optimised, as polymer electronics demands very high quality from processes which are not yet ready for this technology. The optimisation of circuit design is also essential to create new products for the market.

For many applications, especially products used in displays and *photovoltaics*, good barrier layers and subtrates are essential. A lot of work is being carried out to optimise the barrier effects against oxygen and water, which are the most critical factors. The challenge here is to produce thin, flexible and clear barrier films at low cost.

Another trend is the development of thin and flexible displays for new types of e-readers or *e-papers*. They will be bendable or rollable, which makes them interesting for use in electronic newspapers, mobile phones and organisers.

Prospects

Polymer electronics

- is a new technology platform that is paving the way for innovative thin, flexible, large-area electronic devices which can be produced in very high volumes at low cost.
- is based on conductive and semiconductive organic materials. These materials can be of many different classes, including organic molecules and polymers, nanomaterials (organic and inorganic), carbon nanotubes, and many more.
- can be produced in high volumes on flexible substrates, e. g. by printing.
- has huge potential in many areas of application, e. g. RFID, displays, lighting, solar cells, sensors, memories and smart objects.

DR. WOLFGANG CLEMENS
PolyIC GmbH and Co. KG, Fürth

▲ Flexible displays: Flexible displays are possible thanks to new developments in polymer electronics, either as OLED displays or as polymer-based transistor backplanes for other display principles.

▲ OLED as a light source: OLEDs make it possible to create new lighting concepts with very thin and large-area lamps. By combining different materials and structures white light and other well defined colours can be produced. Source: Fraunhofer

Internet

- www.oe-a.org
- www.polyic.com
- www.izm-m.fhg.de/en
- www.itri.org.tw/eng/Research
- http://organics.eecs.berkeley.edu

Magneto-electronics

Related topics

- Sensor systems
- Microsystems technology
- Semiconductor technologies
- Computer architectures
- Automobiles

Principles

Magneto-electronics is a new technology used to store, display and process information based on changes brought about by magnetic fields in the electrical properties of a material or material system. The magnetisation alignment can be influenced by external factors. It can be converted into electrical signals by means of quantum-mechanical effects.

▶ **Giant magnetoresistance (GMR).** A basic setup of a GMR system contains, for instance, two magnetic iron layers separated by a non-magnetic chrome layer just a few atoms thick.

Depending on the thickness of the chrome layers, which influences the magnetic coupling of the two iron layers, the magnetisations of these two layers are either parallel or anti-parallel. The magnetisations' alignment can be influenced even by weak magnetic fields like those produced by information saved on computer hard discs. It was recently discovered that the sandwich's electrical resistance changes dramatically when the iron layers' relative magnetisation switches from parallel to anti-parallel. This effect is accordingly termed giant magneto resistance (GMR). A large change brought about by a small magnetic field in an easily measurable electrical resistance is an ideal basis for developing the highly sensitive magnetic-field sensors that are now standard features in hard disc read heads. The first types of these sensors have also been employed in automotive and automation technology.

A material's electrical conductivity is related to the number of available charge carriers (for example, electrons) in the material and their mobility. If the electrical charge carriers hit obstacles or are deflected (scattered), the material's electrical resistance increases. Two types of charge carrier play a role in magnetic materials: spin-up and spin-down. Like mass, charge and energy, spin is a fundamental attribute of electrons. This attribute is quantised, i. e. it can only assume two different quantum states: spin-up or spin-down. Electron spin also gives a material its magnetic property. Every electron has a magnetic moment of a fixed value and two possible directions. In a ferromagnetic material, the number of free-moving, magnetically coupled spin-up electrons exceeds the number of spin-down electrons (or vice versa). If these numbers are identical, the material is not ferromagnetic. It is therefore a phenomenon of quantum physics on a macroscopic scale, which makes a magnet cling to a memo board. The magnetic coupling of the quantum spin states of individual electrons results in the ferromagnetism underlying the familiar behavior of a memo-board magnet. Magnetism is one of the few everyday phenomena that provide an insight into the workings of quantum physics. In general, spin does not play a role in transporting electric current in cables and wires because electrons with both spin directions contribute equally to the flow of current. In contrast, spin can become very important in narrow geometries, for example in very thin layers or inside an atom – particularly in the presence of magnetic fields or magnetic materials.

The thinner the material conducting an electric current, the lower the ability of charge carriers to scatter and thereby switch the spin. The GMR sandwich's layers, which are at most only a few nanometres thick, not only fulfil this condition, but also ensure anti-par-

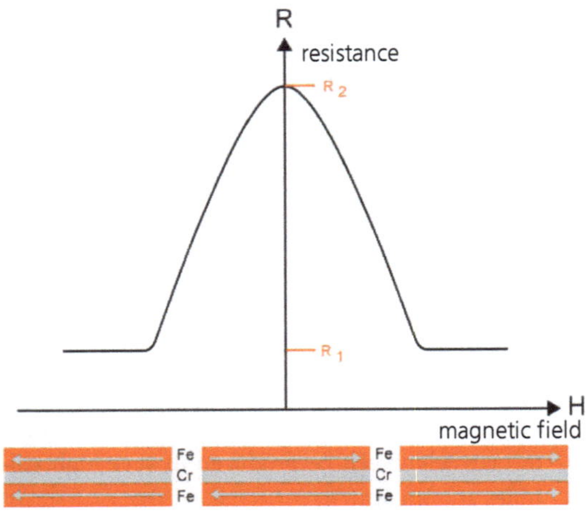

GMR effect: The parallel magnetic vectors (left-hand and right-hand extremities of the curve above) imposed by a large, external magnetic field H result in a low electrical resistance R_1. In the case of a small, external magnetic field, the layers' magnetic vectors remain anti-parallel, resulting in a high electrical resistance R_2. The difference between the resistances in the anti-parallel and parallel modes ($\Delta R = R_2 - R_1$) with respect to the basic resistance ($\Delta R/R_1$) is termed GMR ratio. The higher this value, the more efficiently changes in magnetic fields can be converted into easily measurable changes in resistance. By optimising materials and their layers, it is possible to achieve GMR ratios of several 10% at room temperature. These values are very large compared with those of previous technologies such as AMR (anisotropic magneto-resistance, 3–5%). Source: VDI Technologiezentrum GmbH

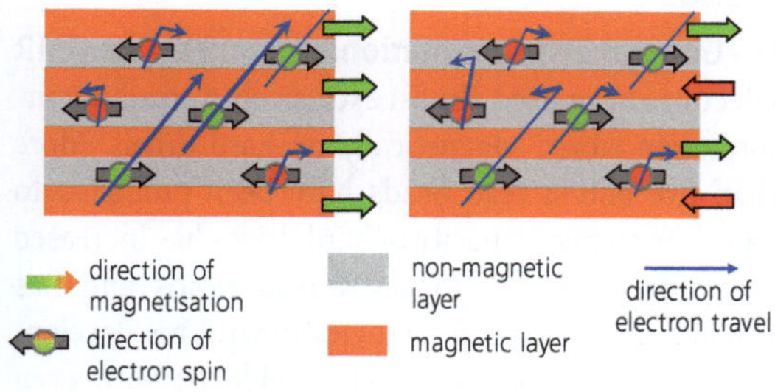

GMR effect. Left: Parallel alignment. When all magnetic layers are magnetised parallel to each other, the associated charge carriers (green) are hardly scattered at the layer boundaries; instead they are able to drift through the stacked layers without experiencing any additional resistance. The charge carriers not associated with the magnetisation (red) are scattered at the interfaces and experience an increased resistance. Right: Anti-parallel alignment. When the magnetic layers are aligned anti-parallel to each other, both types of charge carrier are scattered and experience an increased resistance. Consequently, the electrical resistance during parallel alignment is lower than that during anti-parallel alignment. Source: VDI Technologiezentrum GmbH

allel alignment of the individual layers on removal of an external magnetic field. If a spin-up electron strikes a layer containing a predominance of spin-down electrons, scattering at this interface will occur with high probability and raise the electrical resistance. If the charge carrier and the struck layer have identical magnetisation vectors, however, there will hardly be any scattering. The GMR effect utilises changes in electrical resistance dependent on layer magnetisation.

▶ **Tunnelling Magneto Resistance (TMR).** Replacing a GMR element's non-magnetic, metallic intermediate layer with a thin insulating metal oxide such as Al_2O_3 or MgO results in a tunnelling magneto-resistance element (TMR). The quantum-mechanical tunnel effect enables weak currents to flow through the thin, insulating, intermediate layer in the presence of an appropriate external voltage. Contradicting the laws of classical physics, this surmounting of the insulating, intermediate layer is achieved by the electrons' small yet viable, quantum-mechanical probability of being present on the other side of the barrier. Quantum mechanics describes and explains phenomena occurring on the scale of very small particles and lengths of a few nanometres and is not observable as part of everyday life. If an attempt is made to precisely measure the location of such a small particle, it turns out that its speed can no longer be precisely measured. Conversely, if an attempt is made to precisely measure the particle's speed, its location remains largely uncertain. Quantum mechanics also suggests that physical variables such as energy and magnetisation only assume discrete, quantised values. To nevertheless permit particle behaviour to be characterised, use is made of a wave function similar to that representing light waves. The wave function is a measure of the probability, for instance, with which a particle is present at a certain location or travelling at a certain speed. The term "tunnel effect" is used to imply that the particles, or electrons in our case, being unable to cross the barrier in the conventional sense, tunnel through it instead.

The tunnel current depends on the spin of the tunnelling charge carriers, i. e. on the relative magnetisation of the two magnetic layers. Parallel magnetisation leads to a higher current than anti-parallel magnetisation. Like GMR, a TMR element is therefore able to convert changes in an external magnetic field into easily measurable changes in resistance, and in doing so acts as a magnetic-field sensor. TMR elements are much more sensitive than GMR elements. Modern, sophisticated TMR magnetic-field sensors achieve a TMR effect ($\Delta R/R$) of up to 500% at room temperature. Their disadvantage compared with GMR sensors is that they are much more complicated to manufacture, especially in terms of reproducing the barriers, which are often thinner than 1 nm.

Applications

▶ **Computer engineering.** Computer hard discs have been fitted with GMR sensors since 1997. It is the sensitivity of a GMR sensor's magnetic field that made it possible to save data on very small areas of magnetic hard discs and greatly increase their storage capacity as a result. Today, as many as 28 Gbit can be saved on an area 1 cm^2. The even more sensitive TMR sensors, which are recently introduced on the market, raise the capacity of magnetic data storage elements still further. Their storage densities can reach up to 80 Gbit/cm^2.

▶ **Automobile sensors.** GMR sensors are already being employed by the automobile industry for registering mechanical variables such as speed, angle and rotational frequency. These sensors are installed on automobiles to measure, for instance, the steering wheel's angle for the *ESP* and driving dynamics systems, the wheel speed for the *ABS* system, and the exact position, rotational speed and rotational direction of shafts such as crankshafts in combustion engines.

GMR steering angle sensors have been used for ESP systems since 2006. ESP systems compare the steering angle – i. e. the driver's intended curve radius, which is measured by a GMR sensor – to the actual movement of the vehicle. In the event of deviations, for example, caused by skidding of the vehicle, the ESP system reestablishes a stable driving condition by influencing the braking and the drive systems. When used in active safety systems such as ESP, the GMR

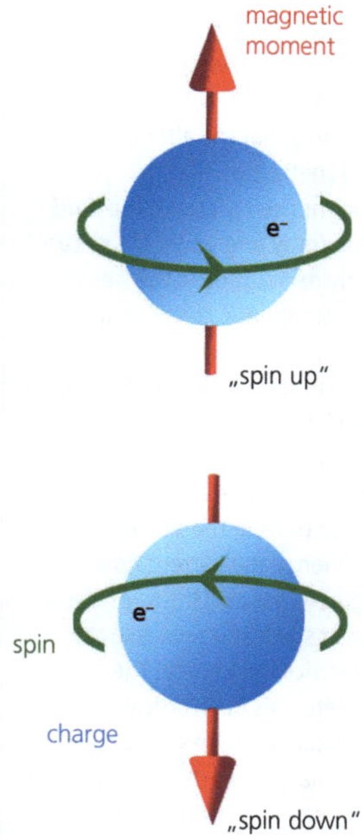

Electron spin: Spin is caused by intrinsic angular momentum. Spin-up charge carriers rotate in the opposite direction to spin-down charge carriers. This rotation is associated with a magnetic moment in addition to an electric charge. Source: VDI Technologiezentrum GmbH

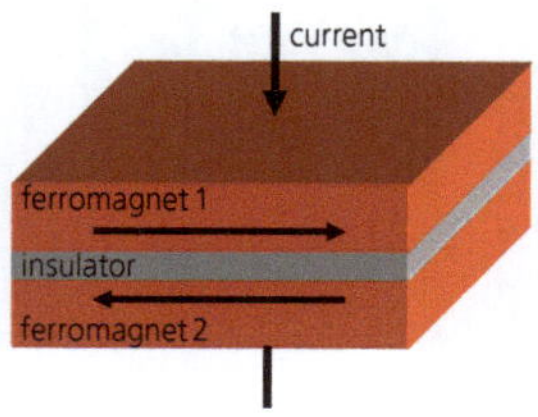

TMR element: The element consists of two ferromagnetic layers separated by an insulating, intermediate layer. If both ferromagnets are magnetised in parallel, a higher current flows through the stacked layers than in the case of anti-parallel magnetisation. This makes the TMR element a sensitive detector of magnetic fields. Source: VDI Technologiezentrum GmbH

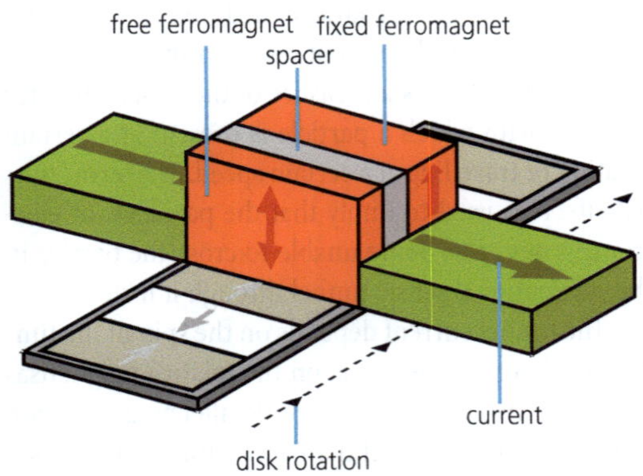

GMR read head: The non- re-magnetisable base layer (fixed ferromagnet, wide red layer) acts as a reference layer and ensures that the alternating magnetisation of the magnetised areas on the hard disc (gray and white areas) can only re-magnetise the free ferromagnetic layer (narrow red layer) of the sensor. The parallel or anti-parallel magnetisation in the layer package (red, gray, red) and the GMR effect produce different conductive conditions in the sensor. These influence the current flow (green) passing through the layer stack and thus also the electrical resistance. The differences in electrical resistance can easily be translated into the digital signal "0" or "1". Source: Forschungszentrum Jülich

sensor's high angle resolution (0.1°) and large measurement range of up to +/−780° help to increase the safety of the vehicle. Due to the magnetic measurement principle employed, these sensors do not require a standby electricity supply but can deliver the current steering angle immediately after turning on the ignition.

Monolithically integrated GMR sensors are developed by combining the GMR-related process modules with an integrated cicuit (IC) technology that provides the signal conditioning (iGMR technology). In contrast to discrete solutions with a GMR sensor connected to a signal processing IC by wire bonding or by a package-to-package system, iGMR technology exhibits its minimised capacitive/inductive parasitics as well as minimised sensitivity to disturbing signals, thereby enhancing the sensor performance.

Other mechanical variables such as pressure can also be measured via magneto-strictive GMR elements, for instance through changes in magnetization caused by mechanical stress. Just like an external magnetic field, mechanical tension produced in a thin magnetic membrane, for example by the air pressure inside a car tire, can change the membrane's magnetization. In this way, the GMR effect can be used to measure the tire pressure.

30% of all sensors in an automobile operate on a magnetic principle. Due to their improved sensitivity and meanwhile also their cost-effectiveness, GMR sensors are gradually replacing AMR (anisotropic magneto resistance) and *Hall sensors*, which have so far dominated in the measurement of magnetic fields and of variables derived from them.

TMR sensors, which exhibit greater sensitivity not only to measurement signals but also manufacturing environments, have yet to prove themselves in automotive applications. These sensors must meet the challenge of withstanding external influences such as temperature fluctuations, voltage and current peaks and vibrations, while remaining cost-effective.

The tera era: Terabyte hard disc with IrMnCr GMR read sensor. The storage density is about 28 Gbit/cm². The International Data Corporation (IDC) estimates that the volume of data worldwide in 2007 was 2.81 exabytes (EB), which is 2.81 million terabytes, and this figure is expected to increase tenfold every five years. A high storage density is therefore very important. Source: Hitachi Global Storage Technologies

Trends

► **General instrumentation.** Since 1997, the GMR effect has been used almost exclusively to read out information stored magnetically on hard drives. More than five billion read heads have been produced to date. The storage capacity of hard drives has increased substantially thanks to GMR read heads and has reached more than 1.5 terabytes today. This development is continuing and will provide us with even smaller electronic devices, which are able to deal with even larger amounts of data in even shorter periods of time with lower energy demand.

► **Next-generation automotive sensors.** The increasing use of GMR sensors in automotive engineering ensures greater protection through passive and active safety systems such as ABS, airbags and ESP. Such sensors also help to optimise motor control and adaptation to ambient conditions, thus lowering fuel consumption and exhaust gas emissions.

In a start/stop system, the combustion engine is stopped when the vehicle is standing (e.g. at a traffic light). The engine is re-started shortly before the vehicle starts moving again. This type of engine emits up to 10% less CO_2 than its conventional counterparts. Hy-

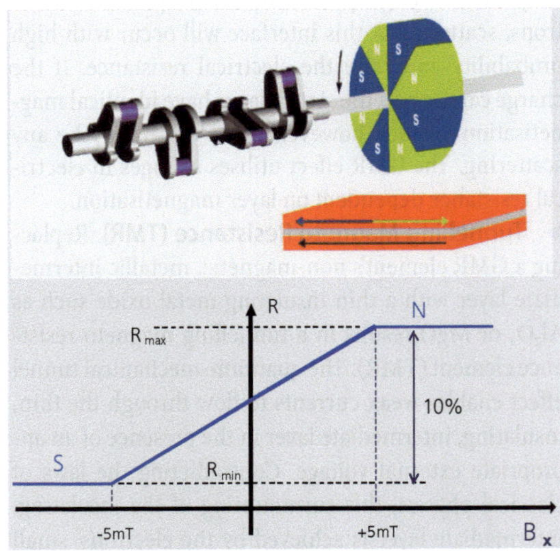

Principle of a GMR rotational speed sensor. Top: As a pole wheel affixed to the main rotational axle (in this example of a crank shaft) rotates, the differently magnetised sectors (S and N) on the pole wheel constantly re-align the upper magnetic layer. The GMR sensor converts this re-magnetisation into a constantly changing electrical resistance. Bottom: When the pole wheel rotates, the characteristic line of the GMR sensor is sampled from S to N and then from N to S, and so on. In this example, the characteristic line is cycled 8 times for each rotation of the axle. The resistance values R_{min} and R_{max} corresponding to the magnetisations S and N differ by 10%. The frequency of the changes in resistance is a measure of the rotational speed of the crankshaft. Source: VDI Technologiezentrum GmbH, Infineon Technologies AG brid vehicles, which have a combined system of a combustion engine and an electric motor, can cut CO_2 emissions by up to 30%. These drive concepts require highly precise and reliable information on the engine's condition, e.g. the position, rotational speed and rotational direction of the crankshaft, the camshaft or the throttle valve. Future GMR sensors with optimised sensitivity, angle resolution, repetition accuracy and noise performance will be able to fulfil this requirement profile.

▶ **TMR memory.** TMR elements also provide a basis for non-volatile magnetic data memory or *MRAM (magnetic random access memory)*. Despite a number of problems still to be solved, MRAM is a promising candidate for the universal memory of the future. This type of memory should be as fast and economical as *DRAM* while being able to save data like a *flash* memory does, i.e. without a need for continuous power. Bits are read out by measuring the electrical resistance of TMR cells. According to the original concept, information is written by re-magnetising a magnetic layer via the magnetic field generated by a neighboring conductive track. However, this cell concept has significant disadvantages in terms of energy consumption and miniaturisability.

This situation can be remedied by current-induced switching of MRAM cells. In this case, the spin-polarised tunnel current from one ferromagnetic electrode is used to re-magnetise the other electrode. Accordingly, the tunnel current can be set to a high value for writing a bit, and to a low value for reading a bit non-destructively. This concept appears to work for highly miniaturized MRAM cells and is being keenly researched worldwide. First demonstrator memory chips based on this advanced, current-induced switching scheme, known as spin-transfer MRAMs, have already been presented.

▶ **TMR logic modules.** In addition to non-volatile data storage, TMR technology also permits the configuration of fast, freely programmable, non-volatile logic circuits. The advantages of these modules over currently available *FPGAs (field-programmable gate arrays)*, also freely-programmable and non-volatile, are speed and durability. Existing FPGAs can be programmed just a few million times. The low programming speed does not permit hardware functions to be re-configured while a software programme is being executed. The speed forecast for TMR logic modules would permit this. Moreover, the life cycle of a TMR element is 1000 to 10,000 times longer.

Prospects

- Having represented the state of the art in hard disc read heads since 1997, GMR sensors are now acquiring an ever wider spectrum of applications. In addition to measuring magnetic fields (in hard disc read heads), GMR technology is now used to measure a variety of other variables such as angles, rotation rates, pressures and currents.
- TMR technology in the form of products known as MRAM and TMR hard disc read heads is either already available or will soon be introduced to the market. Its range of applications is also expected to broaden, as in the case of GMR.

DR. JOCHEN DRESSEN
VDI Technologiezentrum GmbH, Düsseldorf

DR. DANIEL E. BÜRGLER
PROF. DR. PETER GRÜNBERG
Forschungszentrum Jülich

GMR steering angle sensors. Left: Top view on GMR angle sensor chip for 360° measuring range. The white, shining, meander-shaped structures denote the single GMR resistances. To measure accurately and reliably despite the presence of interfering magnetic fields, two GMR sensor areas are connected in a bridge circuit. The two sensor areas are turned by 90° in order to extend the unambiguous measuring range from 180° to 360°. The length of the meanders serves to adapt the resistance to the read-out electronics. Active circuitry is located beneath the GMR structures. Centre: Packaged iGMR angle sensor. Right: High-resolution GMR steering angle sensor for use in ESP systems to increase driving safety. The circular hollow in the module encompasses the vehicle's steering column. In 2008, as many as 2.4 million GMR steering angle sensors were produced that were predominantly put to use in ESP systems. Other areas of application include adaptive bend lighting and rear-axle steering. Source: Infineon Technologies AG; Robert Bosch GmbH

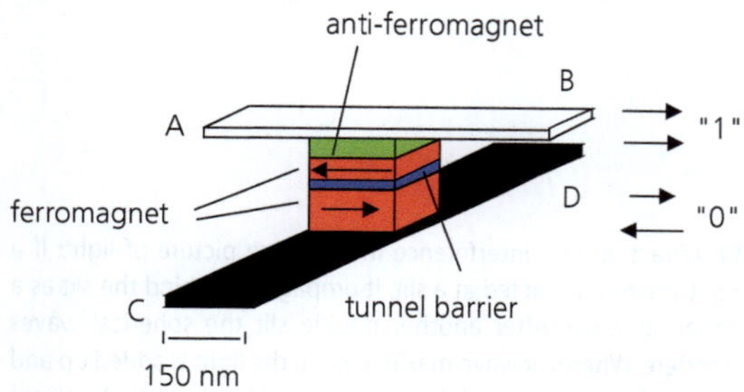

Memory cell of an MRAM: An MRAM saves information based on the relative alignment of the magnetic TMR layers (red). Values of "0" and "1" are assigned to these layers' parallel and anti-parallel alignments, resulting in associated values for the TMR element's changing resistance. The layers' alignment is retained even on disconnection of the power supply, so that the information in these memory cells is non-volatile. Bits are read by a current from A to D, and written by the total magnetic fields generated by sub-critical currents from A to B plus C to D. Source: VDI Technologiezentrum GmbH

Internet

- www.fz-juelich.de/iff
- http://www.research.ibm.com/research/gmr.html
- http://www.magnet.fsu.edu/education/tutorials/magnetacademy/gmr

Optical technologies

Related topics

- Optics and information technologies
- Laser
- Sensor systems
- Measuring techniques
- Minimally invasive medicine
- Solar energy
- Semiconductor technologies
- Self-organisation

Principles

Optical technologies encompass all technologies for the generation, transmission, amplification, manipulation, shaping, measurement, and utilisation of light.

In physics, visible light is identified as electromagnetic waves with wavelengths roughly in the range of 380–780 nm. Optical technologies also make use of the neighbouring areas of the electromagnetic spectrum, i. e. infrared and ultraviolet light up to the X-ray range. Light can be regarded as a form of electromagnetic waves or as a stream of particles (photons), which are governed by the laws of quantum mechanics. Photons are therefore also referred to as light quanta.

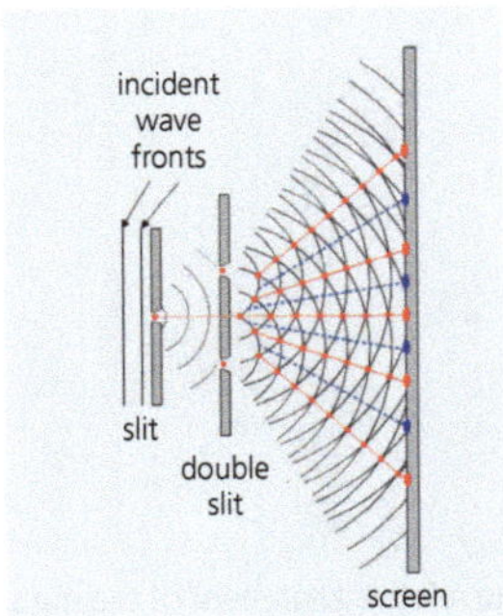

Diffraction and interference in the wave picture of light: If a light wave is diffracted at a slit, it propagates behind the slit as a spherical wave. After another double slit the spherical waves interfere. Wherever wave maxima meet, the light is added up and the respective spot on the screen appears bright (red). In places where the two waves annihilate each other, the screen appears dark (blue). Source: VDI Technologiezentrum GmbH

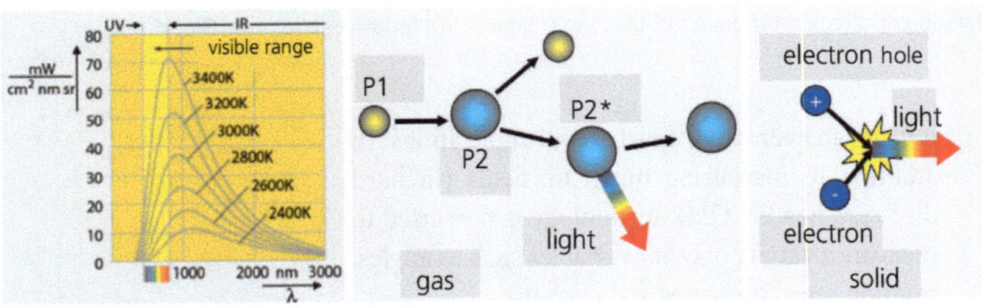

Three basic principles for the generation of light. Left: Radiation emission from a hot solid. The higher the temperature, the closer the maximum of the radiation emission moves to the visible part of the electromagnetic spectrum. Centre: Collision excitation of atoms, ions and molecules during gas discharge in fluorescent tubes. Particle P1 collides with particle P2. The collision lifts P2 into a state of higher energy (P2*). Upon de-excitation into the initial state, P2* releases its energy in the form of light. Right: Recombination of charge carriers in solids (LED). If an electron and an electron hole (positive charge carriers) meet, energy is released in the form of light. Sources: Osram GmbH, VDI Technologiezentrum GmbH

In principle, light or light quanta are generated when bound charge carriers in atoms, molecules or solids make a transition into a lower energy state and the released energy is emitted in the form of photons. Free charge carriers – such as in plasma – are stimulated to emit photons by external acceleration, deflection or deceleration and also through the transition into a bound state. The unhindered transmission of light takes place in a straight line. The wave nature of light is often neglected and its transmission described by ray optics, which allows the phenomena of light reflection and refraction to be interpreted. Wave optics is based on the model of light as an electromagnetic wave, and permits the interpretation of diffraction, interference and polarisation of light.

These basic phenomena in the manipulation of light are used in various optical elements:

- Mirrors are based on reflection.
- Lenses and prisms rely on refraction at interfaces.
- *Diffraction* can be employed in diffraction gratings. However, it limits the attainable spatial resolution of most optical systems.
- *Polarisation* and interference are of use in filters for the manipulation of light.
- Absorption is usually undesired but can be applied in apertures and collimators to shape light.

Light is measured on the basis of its interaction with charge carriers. The effect of the light on these charge carriers can be measured by electric or electronic means, such as CCD (charge-coupled device) or CMOS (complementary metal oxide semiconductor) image sensors in digital cameras.

Applications

The following basic properties of light are crucial for its technical application:

- Light is not affected by electric or magnetic fields and thus also not by other light. It spreads at almost the same speed (300,000 km/s) in a vacuum and in air, and does not need a "support medium".
- Light is suited for transmitting information over long distances.
- Light is suited for transmitting energy, which is used in production technology, for example.

▶ **Lighting.** The most obvious form of technical use of light is lighting. Three basic principles are employed today for generating light in lamps:

- Emission of radiation from a hot solid in thermal equilibrium (incandescent and halogen lamps)
- Collision excitation of atoms, ions and molecules during gas discharge in gases (fluorescent lamps)
- Recombination of positive and negative charge carriers in solids (*light emitting diodes – LEDs*)

Traditional incandescent and *halogen lamps* use the radiation emission of a hot solid. Due to its electrical resistance, the current-carrying filament heats up to the extent that the emitted thermal radiation stretches into the visible spectrum. The proportion of visible light increases with the temperature of the filament, which means that high temperatures (2,600–3,400 K) are preferred. This increases the thermal burden on the filament, which literally begins to evaporate, actually resulting in a gradual blackening of the glass bulb.

Halogen lamps use a chemical cycle process by which the evaporated material of the filament (usually tungsten) is transported back to the filament. Both incandescent and halogen lamps provide a light that is pleasant for humans; however, they have a poor light efficiency of only 3–5%. The remaining energy is emitted as heat radiation.

Fluorescent lamps are more efficient. They use the effect of gas discharge to generate light. Common fluorescent tubes, for example, contain mercury at a low pressure. Due to the applied voltage and the low pressure, free electrons reach energies of several electron volts before they collide with mercury atoms, thereby stimulating the atoms to emit ultraviolet radiation. The ultraviolet radiation is converted into visible light by a luminescent material on the inner side of the bulb. Compact fluorescent lamps ("energy saving lamps") work according to the same principle. High-pressure lamps, e. g. car headlights, also use the effect of gas discharge for light generation. Due to the high gas pressure, the motion of the free charge carriers in these lamps leads to frequent collisions with the gas atoms or molecules, causing these to heat up strongly. This induces an arc discharge and generates a light arc. The heat flow in the lamp leads to wall temperatures in the range of approximately 1,000–1,300 K. At these high temperatures, certain metals and metal halides evaporate in the lamp body, emitting a pleasant light in the light arc.

▶ **Optical tweezers.** Optical technologies enable the observation of biological objects in microscopy; they even permit their gentle manipulation. With "optical tweezers", individual cells from cell cultures or individual bacteria from a bacterial mixture can be selected and isolated under the microscope without damaging them. To achieve this, a laser beam is reflected into an optical microscope in such a way that it is focused on the object plane. The objects being manipulated need to be transparent at the wavelength used. If an object is in the focus of the laser, any positional deviation causes

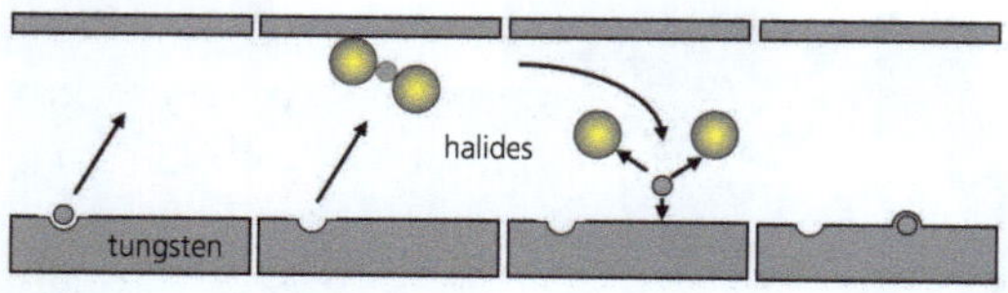

◘ Halogen cycle: During the operation of halogen lamps, tungsten atoms evaporate from the filament due to the high temperature. Near the wall of the glass bulb, where temperatures are below 1400 °C, the tungsten atoms form stable halogen compounds (halides) with gas atoms such as bromine. When these compounds end up back in the vicinity of the filament due to thermal flow, they disintegrate as a result of the higher temperatures. Some of the tungsten atoms are then re-adsorbed by the filament. This halogen cycle prevents the evaporated tungsten from precipitating on the inner side of the bulb and blackening it. Source: VDI Technologiezentrum GmbH

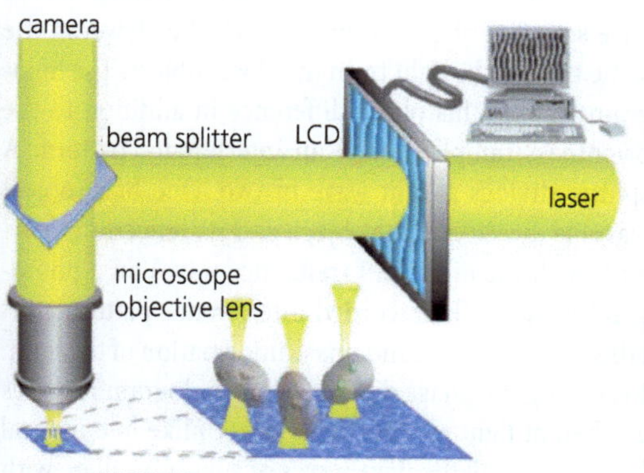

◘ Optical tweezers: They use the forces that light transmits to small objects. These forces move objects such as single cells towards the focus of the laser beam. The cells can be kept in the focus and can be moved with it. Source: Institute for Technical Optics, University of Stuttgart

◀ Measuring set-up for a car body: Vibrations are excited by an electro-dynamic shaker and the vibrational response is measured by scanning the car body with lasers and detectors mounted on an industrial robot. The resulting 3D map of vibration velocities can be transformed and filtered into amplitudes for single frequencies. The pseudo-colour picture of the car body displays the vibration amplitudes for a single frequency where red means low and blue means high amplitudes. The results can be used to reduce interfering vibrations by changing the car body, or to optimise vibration simulations in finite element models. Source: Polytec GmbH

a momentum transfer by refraction to draw the object back into the focus. These forces can be enough to actively move small objects

► **Optical technologies in production.** Laser technology has revolutionised some of the traditional manufacturing processes such as cutting, drilling and welding, but also the precision manufacturing of microsystems. In the field of electronics, optical lithography processes are the critical technologies at the beginning of the value chain in the production of chips. Optical manufacturing processes also enable components to be further miniaturised both in and outside the electronics domain.

Optical methods are particularly suitable for use in industrial automation and quality assurance in production – because of their speed and because they scan the respective work pieces without contact and thus without causing wear. They allow surfaces to be examined, the geometric shapes of work pieces to be measured, and parts to be automatically recognised. *Rapid prototyping* is becoming increasingly important in the area of macro manufacturing.

► **Laser Doppler vibrometers (LDVs).** The vibrational behaviour of a structure is an important characteristic with regard to functional mechanisms, resonances, durability, comfort, acoustics and many other aspects. When investigating vibrating systems in nature and technology, it is necessary to use sensitive and flexible measurement techniques that do not disturb the specimen or structure. Laser *Doppler* vibrometers (LDVs) are particularly well suited for measuring vibrations in cases where alternative methods either reach their limits or simply cannot be applied.

The measurement is based on the optical Doppler effect, which states that light waves experience a change in their vibration frequency after scattering at a moving surface. The frequency shift is directly proportional to the instantaneous value of the vibration velocity and, despite its incredibly small relative value of less than 10^{-8}, can be very accurately determined by interferometric methods.

A single laser vibrometer measures the projection of the sampling point's velocity vector along the vibrometer's optic axis. By using three independent vibrometers co-aligned to the same sampling point but at different interrogation angles, the complete 3D velocity vector at that point can be determined. This principle can be extended to full-field measurements by using three scanning vibrometers, where vibration measurements are made simultaneously from three different directions at each respective sampling point of a measurement grid on the surface. By further combining this technique with *robotics*, the complete measurement of large structures can be fully automated.

Examples of applications in which LDVs play a key role include the development of current high-capacity hard disc drive technology, the advancement of quiet brake systems for passenger cars, and the experimental testing of *microelectromechanical systems (MEMS)*.

► **Holography.** Holography makes use of the wave properties of light to create 3D images of an object. Common photography only records the colour and brightness of the light, while information about the distance between the camera and the recorded objects is lost. When you look at the photo, your brain therefore judges the distances according to its viewing habits.

In contrast, a hologram additionally contains spatial information. This is made possible by the following effect: if one object is further away from the viewer than another, the light coming from that object has to have set off earlier in order to reach the viewer at the same time as the light from the closer object. The hologram records this phase difference in addition to the brightness (amplitude) in an interference pattern. A splitter divides a light wave in two. One half (object wave) is directed at the object, and its reflection interferes with the other half (reference wave) on a photographic plate. The recorded interference pattern contains the brightness and phase information of the light. Recording the phase difference in a hologram requires a coherent light source (i.e. laser). Unlike in a natural light source, all the light waves of a laser oscillate with the same phase (which means it is coherent). Therefore, the different travelling times of the reflected light waves create measurable phase differences. Coherent light with the same wavelength is necessary to view the hologram. The hologram is illuminated with this light and creates a virtual 3D picture for the viewer.

► **Holograms.** Holograms are being used more and more in everyday products, e.g. on bank notes or smart cards. These "rainbow holograms" are also visible in natural light, but only at certain angles. They are stamped in plastics and used as protection against copying and forgery, as their production demands high technical precision. More advanced systems rely on the machine-readability of holograms. The German passport, for example, has invisible holographic information stored in the picture.

Trends

► **White LED.** New light sources, especially inorganic white-light LEDs, have the future potential to generate light with much greater efficiency than conventional incandescent or fluorescent lamps. These optoelectronic components are therefore becoming interesting for the domain of general lighting, too. Inorganic LEDs use the recombination of negative and positive charge carriers (electrons and holes) for light generation. The active part of these light sources consists of semiconductor layer structures. Through the targeted incorporation of foreign atoms, the number of the charge carriers available for current conduction can be adjusted. If the device is of the right design, then the energy released by the recombination of electrons and holes is radiated in the form of a light particle (photon). Its energy, i.e. the light's wavelength or colour, depends on the given material combination and can be adjusted specifically in the production of the LED structure. Furthermore, if a direct current in the correct polarity is applied, it continuously supplies charge carriers, and a continuous generation of light is stimulated. The light generated by an LED is always monochromatic, i.e. red, green, blue, ultraviolet or infrared. If one wishes to generate white light with a single LED, one needs to employ a so-called luminescent converter. This is a luminescent material which, for example, absorbs the originally pure blue LED light and converts it into yellow light. If the housing of the LED is constructed in such a way that only part of the blue light is converted into yellow, the result is a device that simultaneously emits blue and yellow. This combination then appears to the human eye as white light.

In addition to their higher efficiency, LEDs also have longer life spans. Up to now, they have been used primarily in electronic equipment as well as in vehicle and traffic control technology. If LEDs are to be used for general lighting, they will have to feature a greater light flux than those currently available. An alternative approach towards the usage of LEDs for general lighting is based on organic materials (polymers) and devices that are correspondingly called *organic LEDs* (OLEDs).

► **Biophotonics.** New biophotonic methods of observation could make it possible in the long run to track individual proteins within a living cell and to observe what function they have in that cell, without influencing their behaviour.

Optical illusion created by a two-dimensional photograph: All three pairs are of the same size but, because of their arrangement in the picture, the brain perceives them as being different sizes. In contrast, a hologram additionally contains spatial information. Source: A. Winter, GNU FDL

Holograms in everyday life. Top: Silver strip of a 20 euro banknote. Bottom: Holograms on a cash card. Source: PixelQuelle

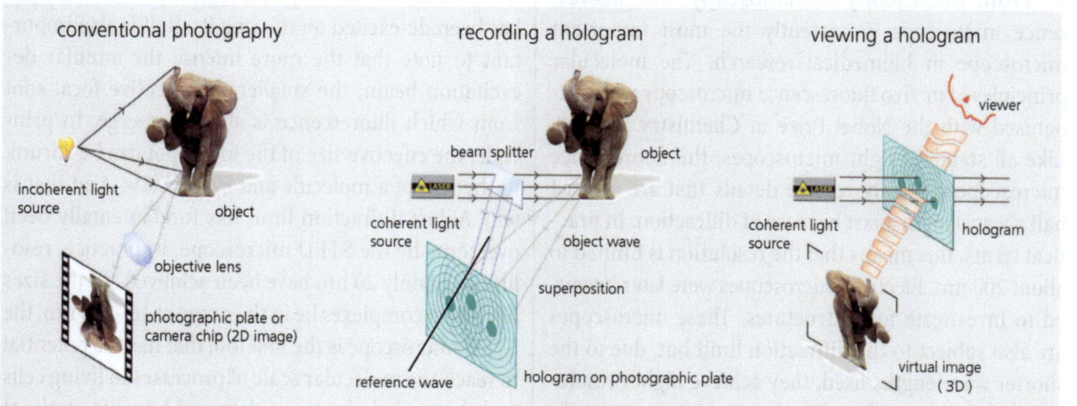

Holography. Left: Conventional photography. Only the brightness of those light rays that directly reach the camera from the object is recorded. Centre: A splitter divides a light wave in half. One half (object wave) is directed at the object, and its reflection interferes with the other half (reference wave) on a photographic plate. The recorded interference pattern contains the brightness and phase information of the light. Right: Viewing a hologram. Coherent light with the same wavelength is necessary to view the hologram. The hologram is illuminated with this light and creates a virtual 3D picture for the viewer. Source: Karl Marx

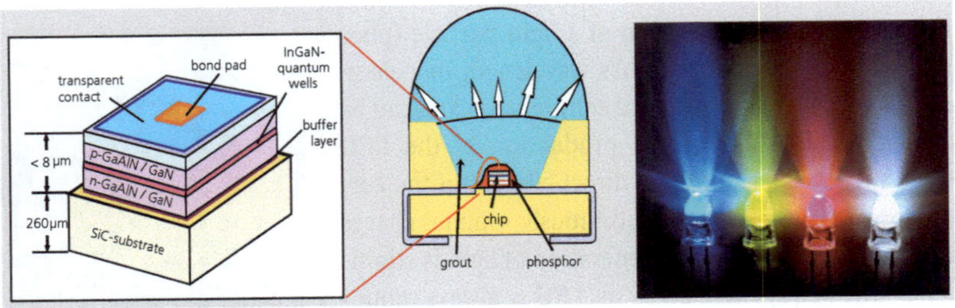

Configuration of an LED chip and chip housing: Left: On a semiconductor substrate approximately 260 µm thick lies a combination of thin, differently doped layers (n- and p-doped). The layer system is completed by a transparent contact layer. Upon recombination of electrons and electron holes, light of a defined wavelength is emitted. The LED chip is connected to the power supply by a wire attached to a bond pad. Centre: The chip is mounted on a heat sink and covered with an appropriate luminescent material ("phosphor"). Normally, blue LEDs and yellow phosphors are used. The mixing of blue LED light with the yellow light of the luminescent material creates white light. The chip is furthermore encapsulated with an epoxy to protect it against environmental influences. The light is transmitted upwards. Right: LEDs of different colours in operation. Source: Osram GmbH, Fraunhofer IAF

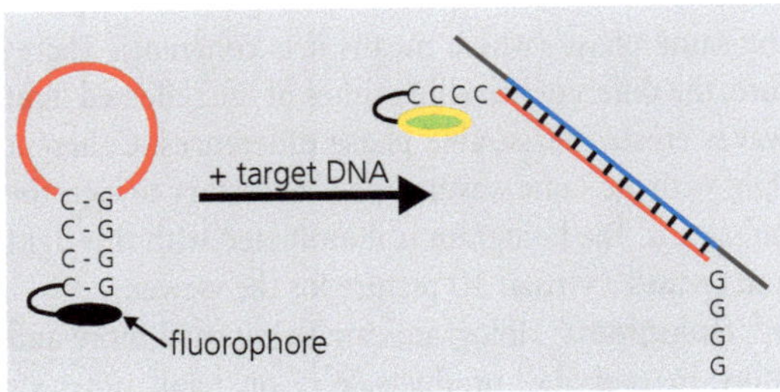

Smart probes: A short needle-shaped single strand of DNA (left side) is attached with a fluorophore (dye molecule) at its tip. The fluorophore is quenched by the guanine moieties (G) in the stem. If the binding sequence in the needle head is hybridised with the target DNA (blue), the needle head opens and binds to the target DNA sequence. This separates the fluorophore from G resulting in an increase of fluorescence intensity that can be detected. In this way, it is possible to quickly provide evidence for the presence of a specific target DNA, for instance for resistance testing. Source: Universität Bielefeld

"Smart Probes", for instance, serve to evidence the presence of a specific target *DNA sequence*. They are short, needle-shaped single strands of DNA (oligonucleotides), to which a dye molecule is attached. The dye is at the tip of the needle near several guanine moieties, which initially cancel its fluorescent activity. The needle head of the Smart Probe is specifically geared to the target DNA and opens in its presence. This distances the dye molecule from the guanine moieties, resulting in a drastic increase in fluorescence. This signal can be used to detect the target DNA. In this way, it is possible to quickly verify whether an infection with an antibiotic-resistant pathogen has occurred, for example. Resistance testing with conventional microbiological techniques can take several weeks.

► **From microscopy to nanoscopy.** The fluorescence microscope is currently the most important microscope in biomedical research. The molecular principles of in vivo fluorescence microscopy were recognised with the Nobel Prize in Chemistry in 2008. Like all standard light microscopes, the fluorescence microscope can only resolve details that are at least half a wavelength apart because of diffraction. In practical terms, this means that the resolution is limited to about 200 nm. Electron microscopes were later invented to investigate finer structures. These microscopes are also subject to the diffraction limit but, due to the shorter wavelengths used, they achieve higher resolutions. However, neither electron microscopes nor other types of high-resolution microscopes (< 200 nm) can image living cells, since they are limited to imaging surfaces or even require a vacuum.

In creating the *STED (stimulated emission depletion) microscope*, a variant of the fluorescence microscope, for the first time a way was discovered to overcome the Abbe diffraction limit. The key to this achievement is to make use of the fact that light can not only excite fluorescent molecules but also de-excite them. In practice, a focused (blue) beam is employed for fluorescence excitation. Due to the diffraction, this excitation beam has an extended focal spot. Fluorescent molecules within this illuminated spot are excited. A "doughnut"-shaped light spot (i. e. a light spot with a hole in the middle) of a longer wavelength (e. g. yellow), which is capable of de-exciting the fluorescent molecules (emission depletion), is then superimposed on the excitation spot. As a result, the fluorescent light (for instance green) only comes from the hole of the ring or doughnut, as fluorescence has been de-excited on the ring itself. It is also important to note that the more intense the annular de-excitation beam, the smaller the effective focal spot from which fluorescence is able to emerge. In principle, the effective size of the focal spot can be shrunk to the size of a molecule and even below. And that is why Abbe's diffraction limit has fundamentally been overcome by the STED microscope. In practice, resolutions of only 20 nm have been achieved. As the sizes of protein complexes lie in the range of 10–200 nm, the STED microscope is the first tool that has the potential to reach the molecular scale of processes in living cells and thus to help better understand how life works at this level.

► **Femtonics.** Femtonics deals with phenomena that take place within a femtosecond, i. e. a millionth of a billionth of a second. In nature, these are typically oscillations of molecules or the circulation of electrons around an atomic nucleus. This time scale can be technically measured by lasers, which send out femtosecond light pulses.

The light of femtosecond lasers has properties that are predestined for many technical applications. In materials processing, the interaction time is shorter than the time it takes for any heat to be transported. Therefore, the work piece stays absolutely cold. No tensions or cracks appear in the material and no melting occurs during the treatment of metals.

With femtosecond lasers, it is possible to cut explosives such as detonators for airbags without triggering a detonation. Furthermore, they are suitable for any applications requiring utmost precision, e.g. drilling holes in injection nozzles or structuring (honing) cylinder walls for motor vehicles. The lasers are also being tested for use in medical applications, and have been proven technically feasible for refractive cornea surgery for the correction of visual defects, and for painless, minimally invasive caries therapy. Femtosecond lasers are eminently suitable for data transfer. The shorter the pulses are in the time domain, the broader they are spectrally. That means they contain multiple "colours" at the same time. Each colour acts as a data channel, so it is possible to transfer data through numerous channels simultaneously with every single pulse.

▶ **Photonic crystals.** Photonic crystals are the semiconductor materials of optics. Current research on photonic crystals aims to open up the possibility of designing optical properties as accurately as electrical properties in semiconductors. First of all, you need a material – i.e. a photonic crystal – in which light of certain wavelengths does not propagate. This behaviour can be described by a wavelength band gap in analogy to a band gap in an electric semiconductor. If, in a photonic crystal, light propagation is impossible for a given wavelength band in all three spatial directions (and regardless of the polarisation of light), we speak of a complete band gap. The second step is to intentionally introduce defects into the photonic crystal, thereby making it possible to control the propagation of light inside it; this is analogous to doping the semiconductor, enabling precise control over its charge carrier density and thus its conductivity.

▶ **Photonic crystal fibres.** In a conventional optical fibre, the light is kept inside by total internal reflection. Imagine a circular, two-dimensional, photonic crystal structure with an infinitely extended third dimension – this is what the fibre is like. Due to the structure of two-dimensional photonic crystals, the light cannot propagate perpendicularly to the direction of the fibre. Now, if you leave a "hole" (defect) at the centre of the fibre, then the light moves exclusively through this "hole" in the fibre's direction. The result is a group of optical fibres with a wide range of special optical properties, which strongly depend on the specific design details. The materials being considered are glasses and polymers. Photonic crystal fibres are already commercially available. Potential applications include fibre lasers and amplifiers.

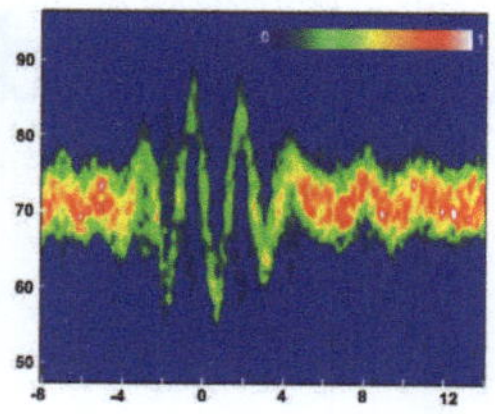

▣ Electric field of a 5-fs pulse at 760 nm; recorded with an attosecond streak camera. The x-axis shows the time delay (ns) and the y-axis shows the kinetic energy (eV). The colour distribution illustrates the number of electrons in arbitrary units. Source: MPI for Quantum Optics

Prospects

Optical technologies are key technologies, which are often at the beginning of the value chain and are crucial as enablers of certain applications:

- Advanced light sources based on LEDs and OLEDs will make lighting more efficient and environmentally friendly. At the same time, the diversity and flexibility of light sources will increase significantly.
- Optical measurement technology will increasingly contribute to improving process quality in manufacturing.
- The improved understanding of processes in living organisms as enabled by optical technologies can become the starting point for a revolution in medicine.
- Holography exploits interference effects to make use of the phase information of light. The applications range from forgery protection on banknotes and documents to imaging processes in medicine and further to material testing and new data storage systems.
- Femtosecond lasers enable materials to be machined with the utmost precision. Because of the short interaction time, no tension or cracking occurs. The lasers are also being tested for use in medical engineering applications, for treatments that are particularly gentle to the treated tissue.

DR. DIRK HOLTMANNSPÖTTER
DR. GÜNTER REUSCHER
VDI Technologiezentrum GmbH, Düsseldorf

Internet

- www.photonics21.org
- www.spie.org
- www.optics.org
- www.nanoscopy.de
- www.bmbf.de/en/3591.php

Optics and information technology

Related topics
- Optical technologies
- Laser
- Sensor systems
- Digital infotainment

Principles

Information and communication is the backbone of modern society. The underlying technology can be organised in four major tasks: processing and generation of information, transfer of information, visualisation of information and storage of information.

Although this is not always obvious, optical technologies have a deep impact on all aspects of information and communication, and even act as a driving force in these technologies. Integrated circuits, on which information processing is based, are produced by optical lithography, and the next generation of chips will depend on improved performance in lithography. The Internet, the foremost example of information transfer, is based on optical fibres and on lasers that feed those fibres. All kinds of displays and projection systems help to visualise information, and there is even a close relationship between optical technologies and printing with laser printers. Finally, most of today's data storage is optically based: CD, DVD and Blu-ray all use laser light and optics to record and read information.

Applications

▶ **Digital camera**. Conventional photo and film cameras are increasingly being replaced by digital systems. Compared to analogue reproduction based on a photosensitive emulsion, images are captured today as pixels using optoelectronic detector arrays, such as *charged couple devices (CCDs)* or *complementary metal oxide semiconductor (CMOS)* detectors. Both sensor types, CCD and CMOS, are based on the inner photoelectric effect, whereby electrons are emitted from a semiconductor after the absorption of light. The number of electrons generated is proportional to the energy absorbed, i. e. the brightness of incident light. At this point the picture is coded as an electron distribution in the sensor.

CCD and CMOS technology vary in how the stored electron distribution is read out. CCDs use a pixel-by-pixel transfer: each line is read by consecutively transferring the electrons from one pixel to the next. Subsequent to the read-out, there is only one amplifying system for all electrons, so that the camera is only then ready to take the next picture when all electrons have been transferred to the amplifier. This concept results in long duty cycles. Additionally, care must be taken that the sensor is not subjected to further exposure during read-out. Several techniques are used to increase speed and avoid unintentional exposure of the CCDs:
- Full frame CCDs make use of a mechanical shutter.
- Frame-transfer CCDs: in this case the whole image is transferred to a second sensor array which is kept in darkness. The first sensor is ready for exposure while the second sensor is performing the read-out.
- Interline transfer CCDs: the image is transferred to shaded intermediate lines located between the pixel rows.
- Frame interline transfer CCDs: a combination of frame transfer and interline transfer CCDs.

The read-out of a CMOS camera uses a different technique. Each pixel has its own amplifying electronics, which means that an 8 million pixel CMOS sensor makes use of 8 million amplifiers, resulting in short duty cycles. CMOS detector arrays are extremely flat and compact in size. Their low power requirements

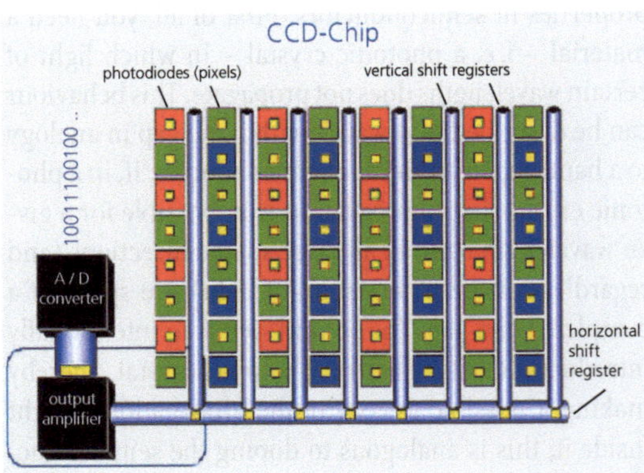

CCD colour detector: The read-out of the sensor is done by successively transferring the stored electrons pixel by pixel. There is only one amplifying and converter system for all electrons at the end of the read-out unit. In order to record a full colour image, a red, green or blue colour filter is placed in front of each detector pixel. The "Bayer-pattern" is the most common colour filter arrangement. To obtain a full-colour pixel, one red, two green and one blue pixels are combined. The twofold presence of green pixels corresponds to colour perception by the human eye.

also make them the preferred choice in mobile phone applications.

Regardless of whether CCD or CMOS technology is employed, the cameras must be modified to record colour images by placing red, green and blue colour filters in front of the detector pixels. While this decreases the camera's spatial resolution, it adds colour capability. The most commonly used colour filter matrix is the so called "Bayer-pattern", where one red, two green and one blue pixel are combined to create one full-colour pixel. The dissimilar representation of the three colours mimics the perception of the human eye, which is most sensitive to green.

► **Optical data storage.** *Compact discs (CDs), digital versatile discs (DVDs)* and *Blu-ray discs* are the most common digital optical storage systems. They have replaced most of the conventional analogue storage systems, such as records, audio and video tapes or films. These discs are most commonly used to store music, movies and electronic data.

Since these optical storage systems code information digitally, all information is coded in the two states represented by "one" and "zero". The translation into an optical read-out is done as follows: the disc has pits and lands, whereby the pits are approximately one quarter of the wavelength lower than the lands. A laser beam reflected from the land results in a bright detector signal and the reflection from a pit results in a dark detector signal, due to destructive interference. When the disc is rotating, a constant signal – bright or dark – is interpreted as a "zero" and a change in signal is interpreted as a "one".

The capacity of optical discs depends on data density, i.e. it is inverse proportional to the area required to store one bit. For example: a reduction of bit size by a factor of two results in an increase of capacity by a factor of four. This leads to storage densities and capacities that are proportional to the square of the wavelength of the laser and inversely proportional to the square of the numerical aperture (NA) of the lens used for read-out. Here the NA is defined as the fraction of the lens's radius to the distance between the lens and the disc. Progress in optical discs has been achieved by reducing laser wavelength from 780 nm in CD systems to 650 nm in DVD systems and 405 nm in Blu-ray discs. At the same time the NA has been increased from 0.45 to 0.6 and 0.85 respectively.

With the exception of the wavelength used and differences in NA, the optical setup is basically the same for all systems. The light of a diode laser is collimated and directed to the disc via a beamsplitter and an objective lens, of which the latter determines the system NA.

The light is reflected from the disc, passes through the same objective lens in the opposite direction and is directed to a focusing lens via the beamsplitter. The lens focuses the reflected beam which is then registered with a segmented detector system. This approach allows the simultaneous control of focus and track position in addition to data read-out.

A further increase in storage capacity can be achieved using double layer discs, writing information in different layers in different depths. Additionally there is the possibility to store information on each side of such a disc ending up with up to four data layers per disc.

Today different disc formats are commonly in use:

- CD/DVD-ROM is read-only data storage.
- CD/DVD-R allows one-time recording of individual data. Here the bits are coded by thermally induced modifications to the data layer's surface.
- CD/DVD-RW allows data writing, deleting and rewriting.

► **Optical fibres.** The major advantage of optical fibres is their low loss of signal when compared to copper cables. As a rule of thumb, light is attenuated by 0.15 dB, which is about 3.4%, when it has travelled through 1 km of fibre. This feature makes optical fibres well suited for applications in data communication and for energy transported in the form of laser light.

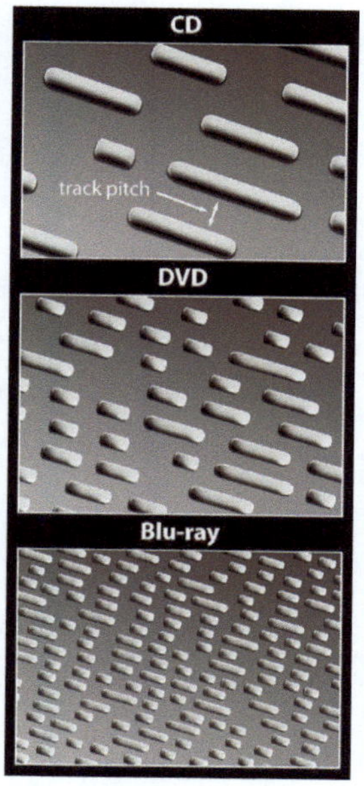

Comparison of the pits and track pitch in CD, DVD and Blu-ray systems: The size of the pits depends on the read/write wavelength used. CD: 780 nm; DVD: 650 nm; Blu-ray: 405 nm. The data capacity of the three systems corresponds to the track pitch. CD: 1.6 µm; DVD: 0.74 µm; Blu-ray: 0.32 µm Source: S. Ostermeier

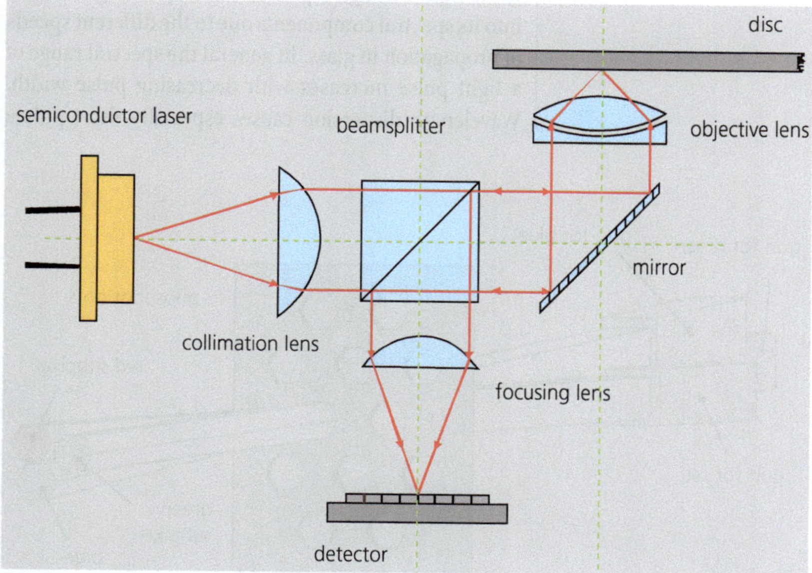

Sketch of a pick-up system for CD/DVD players: The light of a semiconductor laser is collimated by a lens. The collimated beam is guided through a beamsplitter and mirror into the objective lens. The objective lens focuses the light into a small spot on the disc. The reflected light is guided back the same way but is focused on a detector array. Source: Fraunhofer IOF, Cologne University of Applied Sciences

An optical fibre has a cylindrical, coaxial shape and consists of low-loss materials of high purity, such as silica glass, with varied and well-tuned refractive indices. The core always has a refractive index which is higher than that of the cladding. This leads to total internal reflection of light at the core-cladding interface. Detailed analysis shows that propagation of light is only possible at certain angles of reflection. These different propagations are said to be in different modes. Each mode has a different speed of propagation. The number of modes possible is in turn a function of the core's diameter. Fibres that can carry a single mode only have diameters of typically 8–10 µm. Fibres with larger core diameters are multimode fibres.

In *multimode fibres* the sequence of short and dense signals can become mixed up due to the different propagation speeds of the various modes. This undesired effect is called modal dispersion and does not exist in single-mode fibres, which is why these are used for high-speed data transfer. Multimode fibres are used, for example, to transport energy from a laser source to a welding station. Due to their larger core diameter multimode fibres are able to carry higher powers.

Short light pulses are needed to transport data at high speed through a fibre. These pulses are composed of light from a certain spectral range. Here a further aspect has to be taken into account: wavelength dispersion. This familiar effect can be witnessed by passing white light through a prism, whereby the light is split into its spectral components due to the different speeds of propagation in glass. In general the spectral range of a light pulse increases with decreasing pulse width. Wavelength dispersion causes especially short pulses to be stretched, since the integral wavelengths travel at different speeds. For a dense sequence of pulses this may – as in modal dispersion – cause the signal sequence to become mixed up, or in other words: data corruption. Fibres for high speed data transfer are made of special glass types, which show little wavelength dispersion. In addition, lasers are used whose wavelength matches the minimal wavelength dispersion of the glass, typically at about 1.32 µm.

The fibres discussed so far are so-called step-index fibres. Core and cladding have a homogenous refractive index, each with an abrupt change at the interface. A more complex approach is the *GRIN technology*, where the core shows a GRaded INdex profile. The core's refractive index decreases continuously from the centre to outside, following a well designed relationship. Since light travels faster in materials with a lower refractive index, the additional path length of higher modes can be compensated by a lower refractive index at the outer region of the core. Skillful index design results in a compensation of most of the modal dispersion. GRIN technology is expensive since during the growing of the pre-forms of the fibres the concentration of the dopant is continuously changed leading to an index variation.

► **Cathode ray tubes.** Over the past 100 years, the most common display was based on the cathode ray tube (CRT), invented by Ferdinand Braun in 1897. In an evacuated glass bulb electrons from a heated electron gun are accelerated by high voltage, focused to a beam by a Wehnelt cylinder, and directed to the phosphor-coated inner surface of the bulb, which acts as a screen. The electrons transfer their energy to the phosphor, resulting in a bright fluorescence. Magnetic scan coils guide the electron beam along the screen in horizontal lines, whereby rapid modulation of the intensity of the electron beam results in modulation of spatial image brightness, i.e. the image contents.

In colour CRTs, three electron guns are present. For the generation of a true-colour pixel, the three beams are directed through one hole in the so-called shadow mask. A short distance behind this mask the three beams separate again, albeit in close proximity to each other. Here, the beams impinge on the phosphor-coated screen, which is divided into red, green and blue fluorescent sub-pixels. To render true colour, beam intensity is modulated independently for all three beams.

The advantages of CRTs include high resolution, very high contrast, fast response, high colour saturation, colour rendering independent from viewing angle, and low price. Its disadvantages are size – especially depth – high weight, the emission of soft X-rays, and the system's sensitivity to stray electromagnetic fields.

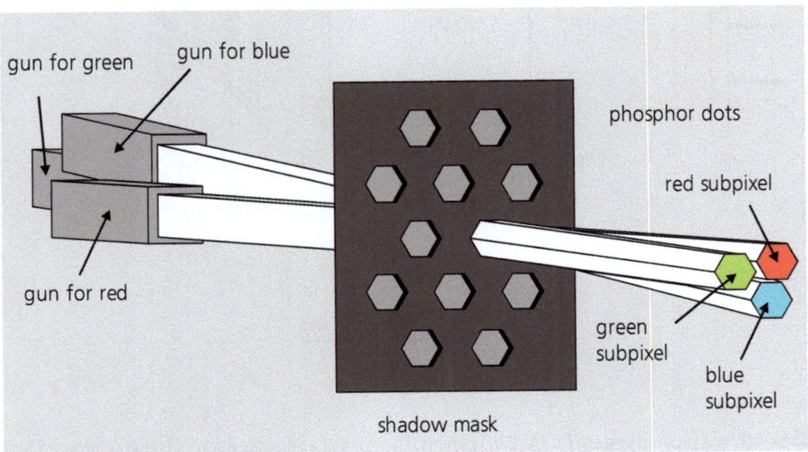

Three electron beams from different guns are directed through a shadow mask. Three subpixels with fixed geometry are formed. Source: Cologne University of Applied Sciences

▶ **Plasma displays.** Flat screens have become popular over the last few years. With an overall thickness of typically less than 10 cm, plasma displays are representative of this class of displays. Whereas a colour CRT consists of three electron guns in one bulb, a plasma screen consists of many individual small cells, equal to three times the number of full colour pixels displayed. In each cell a long and sharp electrode is used to initiate a plasma discharge in an atmosphere of noble gas. The glowing plasma emits UV radiation, which is in turn absorbed by the phosphors on the covering glass plate of the cells, forming the actual screen. The respective intensity of the emission of red, green and blue phosphors is regulated by varying the cell's current flow.

The advantages of plasma displays: large screen sizes at reasonable cost, high resolution, very high contrast (especially compared to LCD systems), flat construction and high luminance. Major disadvantages include the high power consumption required for the discharges and the potential for image burn-in.

▶ **Liquid crystal displays.** Similar to plasma displays, liquid crystal displays (LCDs) consist of pixels that are addressed individually and render colour by division of pixels into red, green and blue subpixels. Each pixel, and also every subpixel, contains liquid crystals embedded between two glass plates. These plates have additional polarisation functionality and are called polarisers and analysers.

Both glass plates are covered with polarisers so that they both act as polarisation filters. What this means can best be understood in the model of light as a transverse electromagnetic wave, noting that a transverse wave propagates in the direction perpendicular to the oscillation. In this model, light is seen as a wave with electric and magnetic field components, oscillating in phase perpendicular to each other. The wave propagates in the direction which is perpendicular to both field components. Light is referred to as being linearly polarised when the field components oscillate in fixed planes. The oscillation plane of the electric field component is then called the polarisation plane of the light. A linear *polarisation filter* can turn unpolarised light into linearly polarised light. Two linear polarisation filters are said to be "crossed" when all light is blocked by the combination of both polarisation planes being perpendicular to each other in a so-called "normally white" cell of an LCD. The liquid crystals between the glass plates influence the polarisation plane of the traversing light depending on the applied voltage, so that the cell can be switched continuously from dark to bright.

In most cases fluorescent lamps are used as backlights for LCD panels; the first systems with light-emitting diodes are, however, finding their way onto the market.

LCDs outperform CRTs with respect to power consumption and have the additional advantage of no X-ray emission. On the other hand, LCDs can suffer from poor viewing angle and slow response time, which causes jitter in rapidly changing images.

▶ **Projection.** In contrast to CRTs, plasma displays and LCDs, optical projection systems create the image on an external surface. Projection systems consist of three parts: the light engine for illumination, a display matrix for information inscription and projection optics for distant image formation.

The two most common display matrix systems are *liquid crystal (LC)* systems and *digital mirror devices (DMDs)*. Less expensive one-chip projection systems display true colour images by sequential projection of red, green, and blue sub-images. A colour wheel consisting of three sections (red, green, blue) rotates in front of the DMD. When the red section is in front of the DMD, the DMD displays a red sub-image. In the same way the green and blue sub-images are displayed.

One-chip systems are used to produce compact and cheap projectors. In the more complex three-chip version, the white light of a lamp is split up into 3 colour channels by a dichroic beamsplitter system. Each sepa-

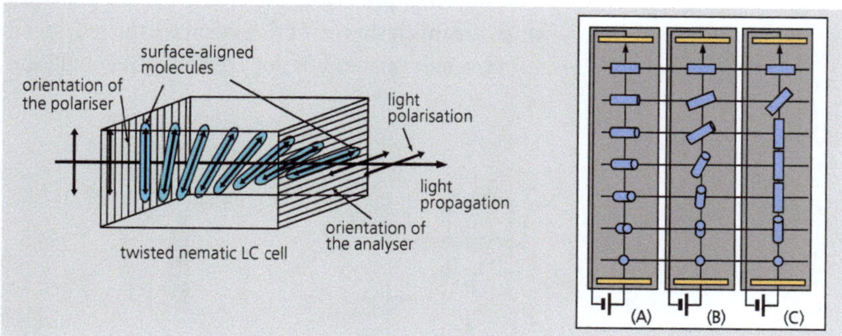

Principle of a "normally white" liquid crystal cell. Left: Liquid crystals (LC) are aligned between two transparent electrodes covered with glass plates. The glass plates in turn are covered with crossed polarisers. Normally (in an empty cell) no light can pass, since the incoming light is linearly polarised in one direction by the first polariser and blocked by the second polariser (called an analyser) which is rotated 90°. Now, the LC molecules are aligned along a virtual line rotated 90°. The incoming linear polarised light follows the twisted alignment of the molecules. The light direction is rotated by 90° and therefore the light can pass the analyser. Right: The cell appears bright when no voltage is applied to the cell (A) as is the case here: light can pass unhindered through the cell. If voltage is applied to the cell, the LC molecules align into the direction of the electric field. As a result, the virtual line is no longer twisted but linear. Since the light follows the direction of the LC molecules, in this case the polarization of the light is not rotated and the light is blocked by the analyser (C). To generate different grey levels, less voltage has to be applied to the cell (B) since this influences the alignment of the LC molecules and thereby the quantity of light that can pass the cell. Source: HOLOEYE Photonics AG

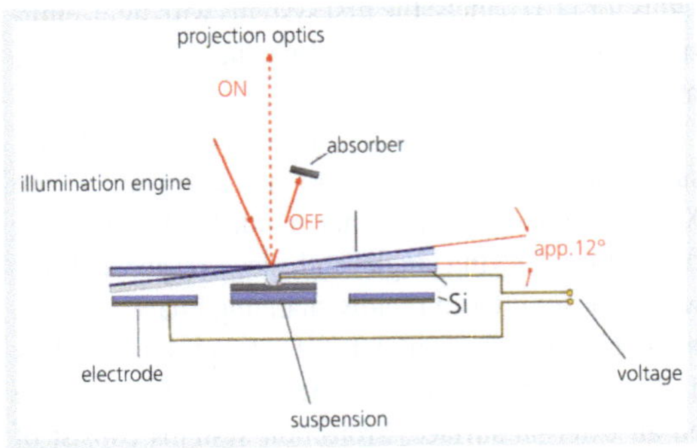

Digital mirror devices: A DMD consists of aluminium micro-mirror elements. Underneath each mirror a microelectronic circuit acts as a driver to actuate tilt in the mirror. In one position the light is reflected onto an absorber, causing a dark pixel, and in the other position light is reflected in the direction of the projection lens, causing a bright pixel. The switching speed of a DMD is very high (up to 10 kHz) and allows different brightness levels to be created by rapid switching between the on and off state. Source: Fraunhofer IOF

rate colour illuminates its own chip and is combined again with the other colours after modulation by the displays. In both systems, one- and three-chip, a magnified copy of the chips, i. e. the actual image, is created on a distant surface by projection optics. Image quality is better in three-chip systems but being more complex it is also more costly. In one-chip projectors, DMDs are preferred as the information display matrix.

Three-chip projectors preferably use LC matrices, which are much slower. At slower speed, the red, green and blue sub-images are not created sequentially on one chip, but simultaneously on three chips. Brightness modulation is caused by a continuous polarization rotation as opposed to the extremely fast switching employed in DMD systems.

Both DMD and LC chips are fabricated by micro lithography. This leads to small size and high integration. Pixel sizes of 10 µm are standard, so that a HDTV 1920×1200 pixel LCD device has a size of only 19.2×12.0 mm².

A different approach is offered by laser-based projection systems. Three lasers of different colours (r,g,b) are used as light sources. Two scanning mirrors for horizontal and vertical deflection virtually expand the point-like laser beams to a two dimensional area. High-speed intensity modulation of the laser beams renders brightness modulation and colour composition. Since laser beams are collimated, the image is sharp at any distance and no projection optics is required. This, in turn, allows projection on non-flat surfaces. Further advantages of laser projectors are extremely high contrast and a much larger colour gamut compared to other systems.

In projection technology image quality does not depend on the projection system only, but also on the projection screen. In conventional projection screens, observers not only see the reflected light of the image, but also ambient light. This results in low contrast and image degradation. Holographic projection screens (HoloPro) do not have this problem because they are transparent and none of the ambient light is reflected into the observer's eye. HoloPros consist of photographic film containing phase modulating structures, i. e. structures that only change the phase of incident

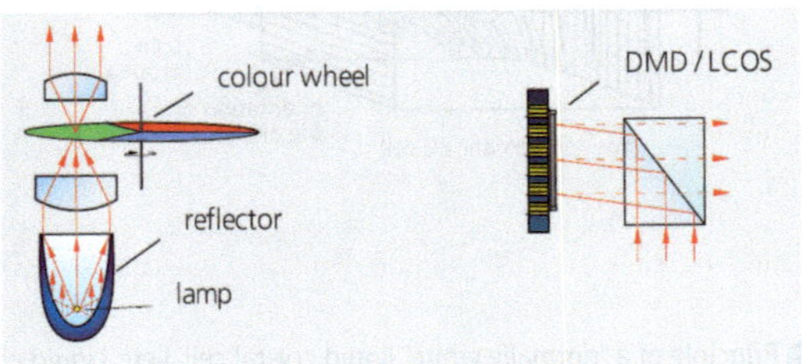

Digital projection system. One-chip-version: A true colour image is rendered by sequential projection of red, green and blue sub-images. White light from a lamp is separated into the colours red, green and blue by a rotating colour wheel. To obtain a white image point the display matrix stays in the on-state as long as all the three colours illuminate the chip (red + green + blue = white).

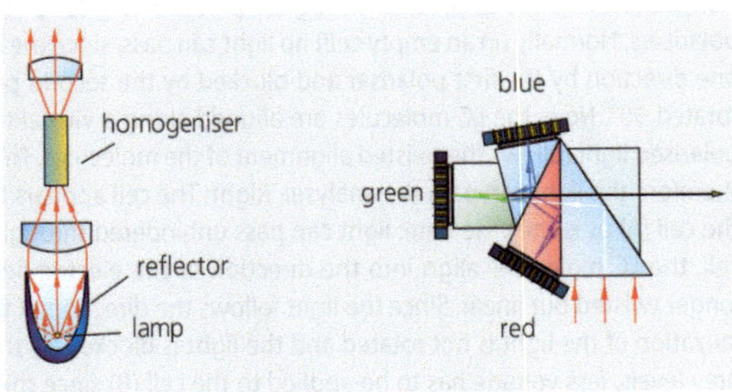

Three-chip-version: Light from a lamp is homogenised and collimated. Dichroic beamsplitters separate the light into the three colour channels red, green and blue. Each color simultaneously illuminates one chip. To get a true colour image, the monochromatic colour frames of the three chips are superimposed by the beamsplitter and guided to the projection optics. Source: Fraunhofer IOF light without affecting the intensity, creating a transparent appearance. The holographically recorded phase structures are basically thick diffraction gratings, which are effective for one certain angle of incident only. At this angle the projector will shine onto the HoloPro, redirecting the light to the observer's eye. Light from all other directions will pass the screen unhindered. Advantages of HoloPros are their suitability for daylight projection, and inconspicuousness: when the projector is turned off it has the appearance of a simple glass panel.

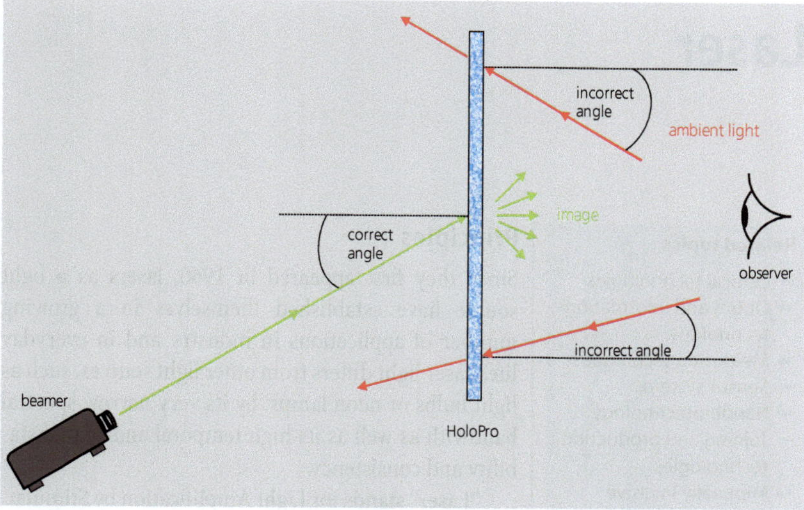

Holographic projection: A transparent phase hologram acts as projection screen. Basically consisting of thick diffraction gratings, only light from a certain angle is diffracted. Here the light of a beamer is redirected to an observer. Light from all other directions is not affected. Ambient light does not reduce image contrast, which makes holographic projection screens suitable for daylight applications. Source: Cologne University of Applied Sciences

Trends

▶ **Holographic data storage.** An emerging and innovative method of optical data storage uses *holography*. Here, data is stored in photosensitive materials by writing a hologram: an object beam, which contains the information, i. e. the text of a whole page, is made to interfere with a reference beam in the material. The interference pattern, i. e. the hologram, is stored in the material either by local phase or local transparency modifications. Shining the reference beam onto the recorded hologram reconstructs the object beam, i. e. the information stored.

High data densities can be achieved since hologram recording is not restricted to the surface of the recording material but can make use of the whole volume of the material. Writing different holograms into the same volume is called multiplexing. To make the different mulitplexed holograms distinct, they are either coded by different wavelengths or by different angles of the reference beam.

Most of the progress in holographic data storage currently arises from the development of photopolymers that, compared to conventional holographic materials, do not require chemical development after recording.

▶ **Holographic projection.** An alternative laser projection system that has neither need for projection optics nor for a scanner system is based on holography. Here an LC display is illuminated with a widened laser beam. The laser beam is diffracted by the display, i. e. the pixels themselves and the specific contents of the pixels. If the far field hologram of the desired image is written into the display, the image will appear at a large distance from the display without the need of any image forming optics. Closer projections are possible too, but require a more complex calculation of the hologram. Additional care has to be taken to remove higher diffraction orders from the image, which show up as undesired artefacts. This is usually done by applying the so-called iterative Fourier transform algorithm (IFTA). The major obstacle for this upcoming technology is the computer power required. Time will put this problem into perspective and the development of more elaborate and effective computational algorithms is under way.

Prospects

- Optical lithography will be pushed to the limits. Private libraries, the size of a credit card, are in sight.
- Diffractive optics will become part of everyday optical design. Together with advances in laser miniaturization this will be, for example, the enabler for lensless projectors integrated into mobile phones.
- New holographic materials that have no need for chemical development will enable many new applications: Holographic data storage with densities far beyond Blu-ray will become common. Holograms will become printable in modified desktop laser printers, opening up mass markets for 3D.

PROF. DR. STEFAN ALTMEYER
JOHANNES FRANK
Cologne University of Applied Sciences

Internet

- www.arcelect.com/fibercable.htm
- www.osta.org
- www.bell-labs.com/org/physicalsciences/projects/hdhds/1.html
- http://solutions.3m.com/wps/portal/3M/en_US/Vikuiti1/BrandProducts/secondary/optics101
- www.jimhayes.com/lennielw/jargon.html
- www.thefoa.org/user

Laser

Related topics

- Optical technologies
- Optics and information technology
- Measuring techniques
- Sensor systems
- Nanobiotechnology
- Joining and production technologies
- Minimally invasive medicine

Principles

Since they first appeared in 1960, lasers as a light source have established themselves in a growing number of applications in industry and in everyday life. Laser light differs from other light sources, such as light bulbs or neon lamps, by its very narrow spectral bandwith as well as its high temporal and spatial stability and consistency.

"Laser" stands for Light Amplification by Stimulated Emission of Radiation, whereby radiation in this context refers to a stream of light particles (photons). Light amplification depends on the existence of a laser-active medium – a material in which a discrete amount of energy can be deposited (absorption) and subsequently released (emission) at an atomic level. Emission can be spontaneous, being either induced by the medium itself, or stimulated by other photons, as is the case for lasers. The newly generated photons have the same properties as those already existing, in terms of colour of light or wavelength.

The wavelength of the generated photons can be modified according to the condition of the laser-active medium, which can be either a gas, a liquid or a crystal to which specific atoms have been added. This means that each laser can typically only emit photons within a small defined wavelength range. Since the wavelength not only represents the colour of the light particles but also their energy, photons of different wavelengths also possess different amounts of energy. Thus, light particles with a wavelength in the blue spectral range possess greater energy than photons with a wavelength in the red spectral range.

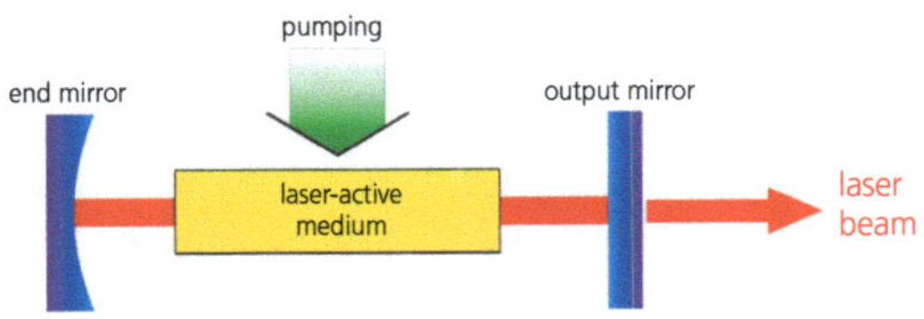

Laser setup: The mirrors serve to generate a light field from identical photons by backcoupling already generated photons. To maintain the laser beam, energy must be deposed continuously in the laser-active medium by "pumping".

Category wavelength	Application
gas laser (ultraviolet, visible, infrared)	material processing medical technology measurement technology
excimer laser (ultraviolet)	semiconductor photolithography LASIK eye surgery
solid-state laser (visible, infrared)	material processing medical technology measurement technology optical telecommunications
diode laser (visible, infrared)	material processing medical technology measurement technology optical telecommunications everyday applications
dye laser (ultraviolet to near infrared)	laser spectroscopy

Categorisation and characterisation of some laser types

Aside from the energetically loaded laser-active medium – the crystal, liquid or gas – it is necessary to feed back the generated photons into the medium using a *resonator*, so that a resonant field of laser light comprising a large amount of identical photons builds up by further stimulated emission. Typically, the resonator consists of two oppositely positioned mirrors for each wavelength. One of the resonator mirrors is highly reflective to laser light. The other mirror allows part of the laser light to exit the resonator. Energy must be deposited continuously in the laser-active medium, in a process known as "pumping". This ensures that sufficient stimulated emission is generated on a continuous basis to maintain constant laser operation. The pumping of the laser-active medium can be accomplished in several ways, such as generating an electrical field in the laser-active medium or by irradiating the medium with light.

Lasers are categorised in different ways, i.e. according to their operating mode (continuous-wave or pulsed) or the type of the laser-active medium (solid state laser, gas laser, diode laser, etc.).

Continuous-wave lasers are systems that emit laser light continuously. These systems are most easily char-

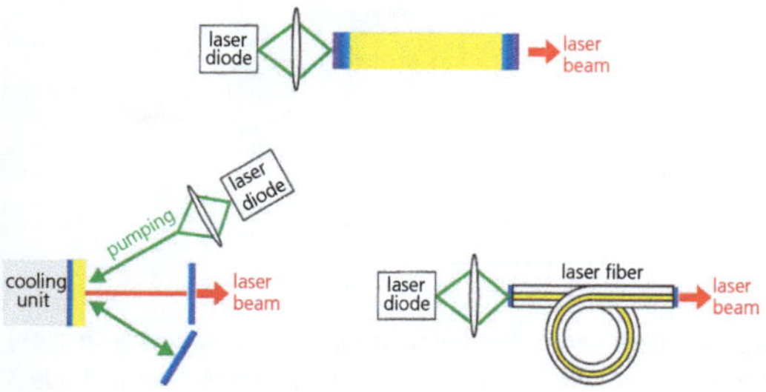

Realisation of solid state lasers on the basis of different geometries of the laser active medium: rod (top), disc (bottom left), fibre (bottom right).

acterised by being monochromatic. By contrast, the spectral bandwidth of emissions from a pulsed laser system grows steadily, the shorter the laser pulses become. The pulse duration of these light flashes may be in the range of femtoseconds (1 fs = 10^{-15} s). For instance, the emission spectrum of a laser emitting in the visible range, with a pulse duration of 10 fs, spans the entire spectral range visible to the human eye. Using these laser pulses, extremely high light field intensities, or photon densities, can be achieved for short durations; these may be higher than the average light intensity of sunlight. Such high photon densities are therefore particularly well suited for use in material processing, since very high levels of light energy can be applied over a short period of time.

► **Gas lasers.** These laser types consist of a gas discharge tube in which the laser-active elements are elevated to an energetically higher state by colliding with each other, with electrons or with other added elements. The energy thus stored can then be released by stimulated emission of photons and the necessary backcoupling. The gas laser family is typically represented by argon (Ar) ion lasers, helium neon (HeNe) lasers, carbon dioxide (CO_2) lasers, as well as excimer lasers (e. g. excited inert gases). The wavelengths provided by the different gas laser systems span from infrared light to the ultraviolet range. Gas lasers can be realised as continuous-wave as well as pulsed laser systems.

► **Solid state lasers.** The laser-active medium of solid state lasers can be of different designs such as rods, discs, or fibers. The laser-active medium is based on a host crystal in which laser-active atoms are embedded. The most important host crystals of solid state lasers are yttrium aluminium garnet, silicate glass, or sapphire, whereas neodymium, erbium, ytterbium, or titanium atoms are typical laser-active atoms in the host crystals. Solid state lasers are usually compact systems, which are not only used in the continuous-wave mode, but also specifically for the generation of extremely short pulses in the femtosecond mode of operation. Solid state laser systems are primarily pumped by optical means, particularly Xe-flashlamps which release a short intensive flash when discharged through a high-voltage capacitor, or through laser radiation from a diode laser.

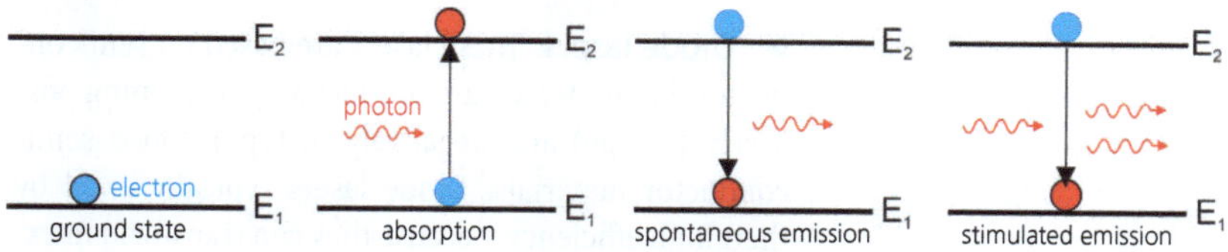

Laser light emission: To store energy in an atomic system an electron bound to the atom must be transferred from the ground state (E_1) to an energetically higher state (E_2). This is done by absorption of a light particle through the atomic system. While the electron decays back to ground state, the stored energy is released again by the spontaneous emission of a photon, i. e. by the atomic system itself, or by the stimulated emission of a photon, i. e. as a result of the irradiation by another photon.

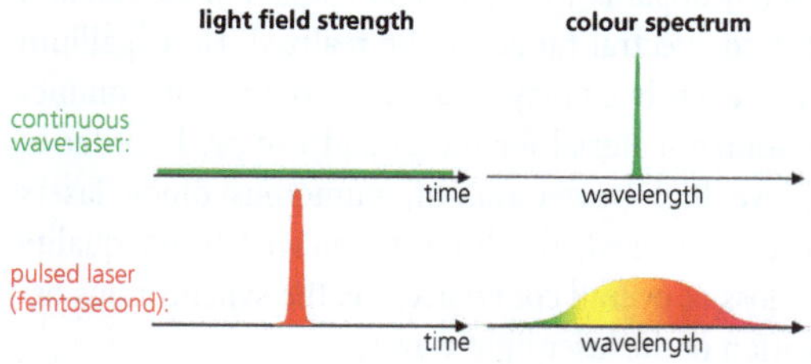

Light field intensity and spectral width of continuous wave and pulsed lasers: The average power of continuous wave and pulsed lasers is equal. The peak power of pulsed lasers distinguishes itself from that of continuous wave lasers by the fact that the total amount of photons is concentrated in a short period of time (left). At the same time, a broad colour spectrum is required to generate short pulses (right).

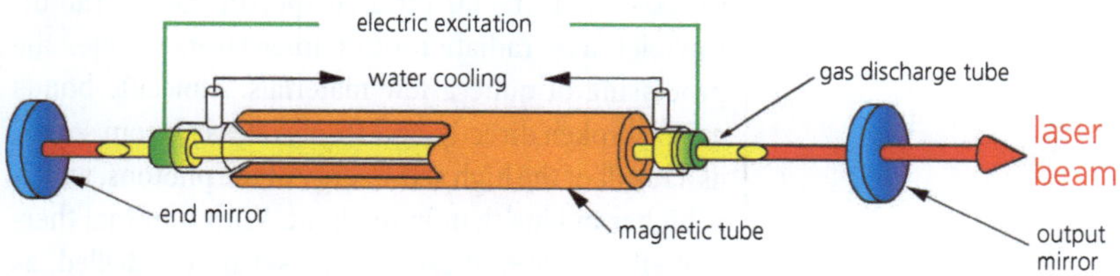

Gas laser setup: Energy supply takes place by gas discharge between two electrodes (green). The outcoupling mirror is partially transparent. A magnetic field focuses the discharge onto the central axis. Water cooling is required to dissipate the heat generated by the discharge.

▶ **Diode lasers.** These lasers are based on semiconductor diodes, which are produced by combining positively (p-type) and negatively (n-type) doped semiconductor materials. Diode lasers typically excel by their high efficiency, because they can transform in excess of 50% of the employed electrical energy into optical laser energy. The laser light is generated only in the boundary layer of the forward biased semiconductor diode. This area is merely a few micrometres in size and the laser mirror can be fixed directly to the semiconductor diode, making it possible to manufacture very compact diode lasers in the millimetre or submillimetre range. Depending on the choice of *semiconductor* material and the incorporation of foreign atoms (co-dopants), various light colours in the red and infrared spectral range can be realized. Here, gallium arsenite has been playing a major role as semiconductor source material for the p- and n-type. In order to achieve high-power output, numerous diode lasers must be stacked, resulting in reduced beam quality and loss of overall coherence, i. e. the synchronous oscillation of the laser light waves.

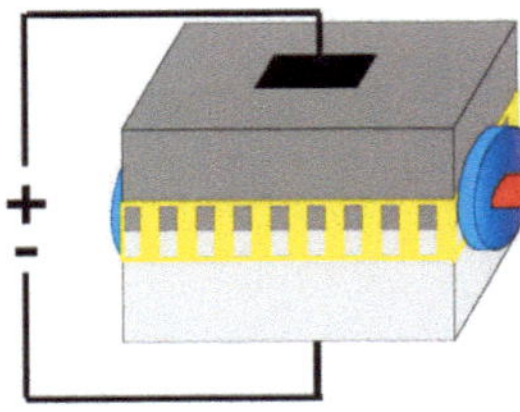

Diode laser setup: Positively (p-type) and negatively (n-type) doped semiconductor materials are combined. When the diode is electrically contacted in forward bias, a laser light field is generated in the junction layer (laser active zone) of both semiconductor materials.

Applications

▶ **Material processing.** All laser systems with an appropriately high output power are suitable for use in macroscopic as well as microscopic material processing. Amongst gas lasers, particular use is made of excimer lasers, which emit in the ultraviolet range, and CO_2 lasers, in the far-infrared spectral range. The ultraviolet laser radiation of excimer lasers enables the processing of quite a few materials. Molecule bonds can be broken directly, and atoms released from solids, as a result of the high self-energy of the photons, which is higher in blue than in red light. With minimal thermal effect, material can thus be separated, drilled, as well as ablated. Furthermore, the very short wavelength enables high focusability, which allows for the application of laser radiation in very small volumes. This makes it possible to modify or structure material in the micrometre range.

Micromaterial processing using excimer lasers: Preparation of minimal structures in polymers (left), ceramics (centre), or transparent materials such as glass (right). Source: Laser Zentrum Hannover e.V.

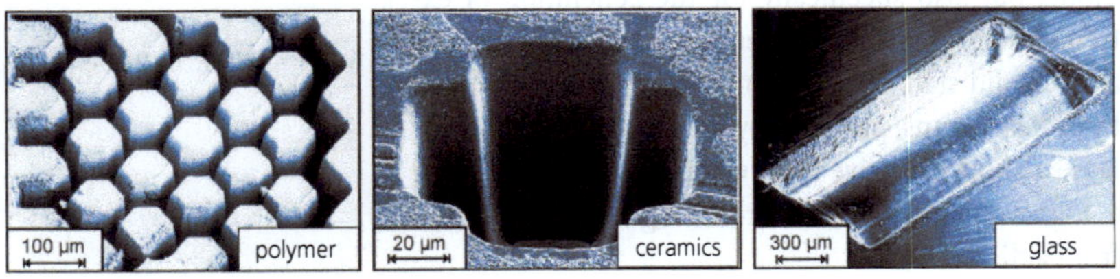

CO_2 lasers are among the most powerful laser systems available. With output powers of up to some ten kilowatts, they are ideally suited to material processing. Material is not only separated, drilled or ablated using laser radiation, different materials are also welded together. Here, in contrast to excimer lasers, material processing using CO_2 lasers is caused more by thermal impact than through direct manipulation of the material, as with ultraviolet laser radiation.

In recent years, solid state lasers have also increasingly been used in material processing. Solid state lasers with output powers of some kilowatts, based on host materials doped with neodymium or ytterbium, have thus become an alternative to CO_2 gas lasers in material processing.

Solid state lasers that emit extremely short pulses in the femtosecond range are also used in microprocessing of materials. In contrast to excimer lasers, the light energy is deposited in the material by a *multiphoton absorption process.* Hence, material processing only takes place at a well defined photon density, so that the intensity of this multiphoton process can be controlled via the pulse duration as well as the total power of the light field, i. e. the total amount of photons. This control capability enables drilling, cutting, ablating and structuring in the range of a few micrometres. The use of such lasers is not limited to industrial materials such as steel or aluminium, but allows for precise processing of transparent media, such as glass or very hard materials such as diamond, in the micrometre range. As a result of the selective and controllable multiphoton process, there is no thermal impact on, or damage to, the process surroundings.

Excimer lasers, which emit radiation in the deep ultraviolet spectrum, are largely used in microlithography. *Lithography* is the process primarily used in

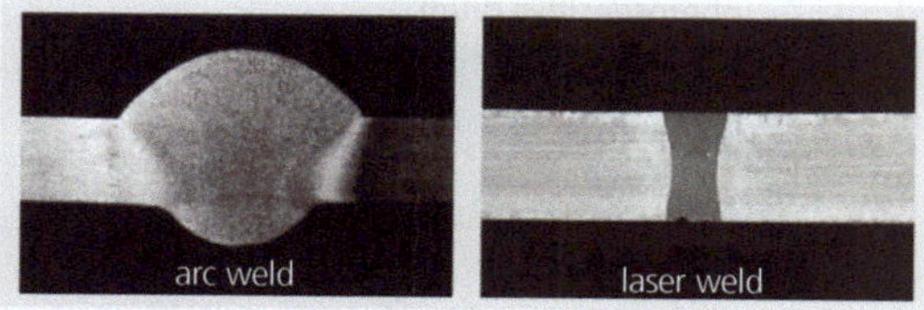

Laser welding: During conventional arc welding, an electric arc is used as heat source. The high temperature of the electric arc melts the materials to be joined at the weld, leaving a weld seam in the form of a bead (left). Laser beam welding allows the welding process to be better controlled. The thermal affect of the weld seam surroundings is minimal and a very flat weld seam can be produced (right). Source: Laser Zentrum Hannover e.V.

the microelectronics industry to manufacture integrated circuits. The process is based on the polymerisation of a resist by UV radiation and subsequent development and etching of the semiconductor substrate. As the minimum feature size that can be achieved by projecting a mask onto the wafer surface is almost proportional to the illumination wavelength, the trend is clearly towards shorter wavelengths in the UV range, enabling the production of chips of higher densities. Typical lasers are ArF-excimer lasers at a wavelength of 193 nm (or KrF-excimer lasers at 248 nm), operating at several hundred watts of output power to satisfy throughput requirements.

▶ **Medical technology**. In medical technology, different laser types are used. CO_2 lasers find their applications particularly in surgery and dermatology, because their laser radiation is readily absorbed by water in tissue. A very high energy load can be used in certain applications, such as the laser scalpel. Solid state lasers based on erbium or holmium are also used, since their emission wavelengths in the infrared are also readily absorbed by water in biological tissue, allowing an efficient energy deposition in tissue. In ophthalmology, particularly during *LASIK (Laser Assisted In Situ Keratomileusis)* operations to correct ametropia, excimer lasers as well as solid-state-based femtosecond lasers are used today.

Another important field is the use of lasers in dentistry. Lasers are well suited to processing hard dental tissue in preparing cavities for caries therapy. Due to their excellent absorption in water and hydroxyapatite, Er:YAG lasers, working in the infrared range at a wavelength of 2.94 µm, are nowadays used to structure dental enamel and cement, as well as bone tissue. Compared to conventional methods of dental care, operations take a lot less time and the surfaces require no further treatment. However, the rise in temperature during laser treatment can cause additional pain when using Er:YAG lasers. Alternatively, ultrashort laser pulses can be used to remove the material without any heating of the surrounding material. Lasers can also be used in endodontic therapy, i.e. treating the pulp of a tooth with the aim of elimination of infection and protection of the decontaminated tooth from future microbial invasion, as well as the cutting or removal of soft oral tissue. In parodontology, lasers have in some cases proved superior to avoid bleeding during dental work.

Epilation is a procedure to remove body hair, whereby the hair is completely removed including its root. Laser epilation, in which the skin is irradiated with laser pulses, is quite well established. The laser light is absorbed by the root of the hair and converted into heat. The local heat deposition leads to coagulation of the root. This procedure, however, only works well on dark hair. The lack of melanin in white or blond hair leads to unsatisfactory results. Melanin is a red, brown or black pigment which is responsible for skin and hair colour and acts as the absorber for laser light. Flashlamp pumped, pulsed lasers (< 50 msec), and even high-energy flashlamps on their own, are typically used as laser systems for epilation.

Independent of their type, laser systems in general

Cavity in dental enamel prepared by femtosecond lasers for caries treatment. There are no cracks and damages in the surrounding of the cavity which makes this process very comfortable for both the doctor and the patient. Source: Laser Zentrum Hannover e.V.

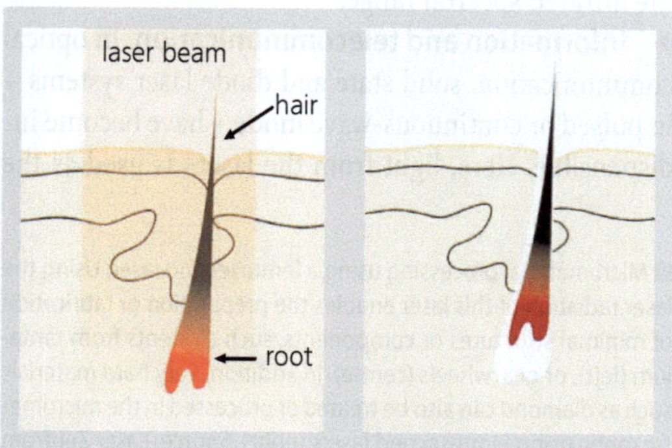

Laser epilation: This procedure removes body hair by irritating the skin with laser pulses. Left: The laser beam penetrates the skin deep enough to reach the root of the hair. The root absorbs the laser light and converts it into heat. The pigment melanin is responsible for the absorption. Right: The local heat deposition destroys the root and the hair falls out. Melanin which occurs only in dark hair is needed for this procedure. Therefore, laser epilation works best on dark hair. Since hair grows in three phases and laser can only affect the currently actively growing hair several sessions are needed to remove all the hair of a certain skin area.

Material processing using a CO_2 laser: Different separated materials and a CO_2 laser in use. Source: Laser Zentrum Hannover e.V.

are being increasingly tested and used for therapeutic purposes, e.g. in cancer treatment, etc. The narrow spectral bandwidth of the lasers used allows for a selective manipulation, activation and treatment of the tissue. Laser radiation is locally applied to the tissue via fibre-based endoscopes.

▶ **Metrology.** Spectrally broad short-pulse lasers are used in metrology, as well as single frequency (single colour) lasers whose wavelength can additionally be modified (tuning). The spectrally multicoloured light flashes that results from the generation of ultrashort laser pulses make these systems suitable for optical coherence tomography. In this case, laser light is divided into a reference signal and a measurement signal. Multidimensional minimal structures, such as different layers or cracks, can be analysed in the micrometre range in transparent materials and organic tissue through alterations to the measurement signal, when compared with the reference signal.

Single wavelength lasers are essential to *optical spectroscopy* and analysis technology. Elements, gases and other compounds absorb light differently depending on their wavelength, and can therefore be detected selectively in complex conglomerates such as air or exhaust gases.

Diode lasers, which can have a very narrow spectral range, are also used for such spectroscopic analyses. They allow a very sensitive and selective verification of single elements and compounds when combined with a tunable emission wavelength. Changes in the concentration of carbon monoxide or dioxide in combustion engines can be detected using a diode laser in the infrared spectral range.

▶ **Information and telecommunication.** In optical communication, solid state and diode laser systems – in pulsed or continuous-wave mode – have become indispensible. Here, light from the lasers is used as the information carrier. The actual information is "modulated" onto the light field, by varying the power of the information carrier, for example, and can then be transmitted via optical transmission paths such as fibres. The information transmission rates achievable by means of optical laser light are by some orders of magnitude higher than the rates of communication technology based on radiofrequencies. Such optical communication networks now span the entire globe and allow data transfer from one place to another within a fraction of a second.

One of the most prominent applications is *laser printing*. In order to perform the task of producing a printed version of a matrix image (a representation of a picture in form of a binary matrix) laser printers make use of a xerographic method. In this method, the matrix image is fed bit by bit to a laser diode which subsequently transfers the image to a revolving *photosensitive roll* via moveable transfer optics and an appropriate scanning algorithm. The surface of the photosensitive drum is negatively charged prior to exposure to laser radiation and it remains in this state as long as it is not illuminated. The essential idea behind the laser is thus to use a collimated light beam to discharge the roll's surface in a targeted way, i.e. scanning the surface with the laser that is turned on and off according to the matrix image's current bit value. Thus the resulting pattern on the drum is a negative image of the picture to be printed. In the next step the roll is exposed to toner particles which stick only to the charged points on the drum. Now the picture can

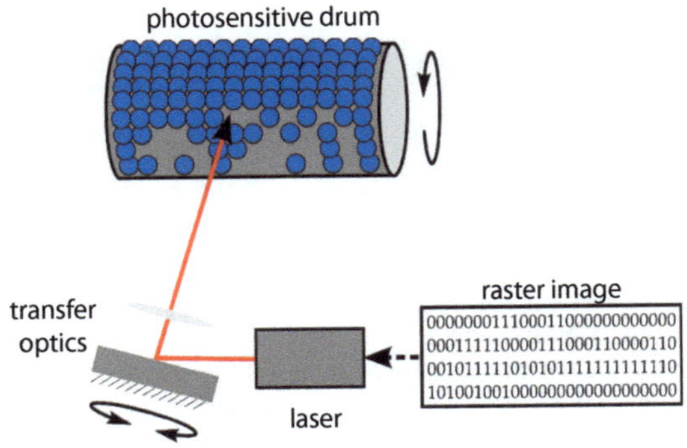

Principle of a laser printer: The printed text will first be digitalised into a matrix image and a laser transfers the information bit by bit to a photosensitive drum. The negative charges are removed by laser radiation so that the toner particles during subsequent printing only stick to the remaining negative charges. After the selective transfer of the toner particles to the paper it is heated to fixate the toner. The drum is then newly charged to prepare the next sheet.

Micromaterial processing using a femtosecond laser: Using the laser radiation of this laser enables the preparation or fabrication of minimal structures or components, such as stents from tantalum (left), or gearwheels (centre). In addition, very hard materials such as diamond can also be treated or processed in the micrometre range using femtosecond lasers (right). Source: Laser Zentrum Hannover e.V.

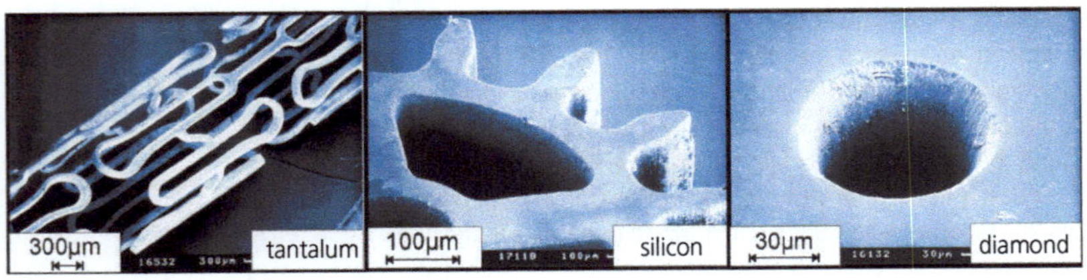

finally be transferred to a sheet of paper by rolling the photosensitive cylinder over the sheet and thereafter imprinting the toner on the paper by applying heat.

Trends

► **"Blue" lasers.** The development of laser systems, particularly of compact and efficient solid-state and diode lasers – with emission wavelengths in the blue and ultraviolet spectral range – is of specific importance to microtechnology and micromaterial processing. Structures of decreasingly small size, such as for electronic components (computer chips, etc.), can be produced due to improved focusability of laser radiation, which in turn are the result of laser beam sources of diminishingly short wavelengths.

Moreover, compact lasers, emitting in the blue and ultraviolet, are the basis for writing and reading systems in the field of data storage and holography. Since laser radiation can be focused onto smaller sizes, more information units per volume can be written, stored and read. This allows to double the storage capacity of a CD, for instance, by replacing the systems that have operated in the near-infrared with laser light sources emitting in the blue spectral range.

► **Higher ouput powers.** To increase the output power, modified or new geometries of the laser-active medium are being investigated, particularly for solid state materials, since some configurations, especially in the kilowatt range, have become limited with regard to capacity and feasibility. In this context, interest is focused particularly on combining the advantages of different designs, in order to reduce limits imposed by heating of the laser material, for instance, or in order to obtain a favourable ratio of input energy to extracted energy. Thus, the combination of fibre and rod design allows the thermal limits to be reduced by means of the fibre characteristics, while the rod shape also reduces limitations on power.

Laser systems with a higher ouput power are particularly suited for more efficient material processing. These high power laser systems can thus be used for welding or drilling of material of greater hardness and thickness.

► **Shorter pulse durations.** The further development of very short light pulses offers a further increase in precision in micromaterial processing as well as in medical technology. Decreasingly short pulses allow the light energy of the pulse to be used in a more targeted way, so that the thermal impact to surroundings can be reduced.

In addition, shorter light pulse durations enable a further increase of transmission rates in the field of optical telecommunication. The broadened wavelength range inherent to shortened pulse durations makes it possible to use further carrier signals from the emission spectrum of the laser to transport information with laser light. This is simultaneously supported by increasing the pulse repetition rate, or light flashes per time unit, since more information incidents per second can be transmitted in this way.

► **Miniaturisation and profitability.** Extending the use of lasers in everyday life (e.g. industry, science, private households) depends on further miniaturisation and improved cost effectiveness. Although some applications already exist in everyday life, lasers have yet to establish themselves on a broad scale. Compared with conventional devices or procedures, many of today's laser systems are still too big and/or not efficient enough in terms of investment cost and operation. Profitability can be improved, for instance, by increasing the pulse repetition rate. The latter increases the efficiency of processes and analyses in (micro)material processing, in medical technology or in metrology, because more processing or query light pulses per time unit are provided. Thus, femtosecond lasers with high pulse repetition rates could replace mechanical drills in dental treatment in the future.

Prospects

- High power laser systems are used in the field of material processing such as welding, cutting or drilling. Laser systems emitting extremely short laser pulses can be used as tools for highest precision processing of different materials in the micrometre range.
- Lasers represent light sources which allow a very precise detection and analysis of elements, molecules or other compounds in air or gas mixtures.
- In medical technology, laser beam sources are suitable for use not only as sensitive devices in diagnostics and therapeutics, but also as precise surgical instruments.
- In optical communication, laser light is used as a carrier of information and thus allows for considerably higher transmission rates compared with radiofrequency technology.

PROF. DR. ANDREAS OSTENDORF
Ruhr-University Bochum

DR. HOLGER HUNDERTMARK
Max Planck Institute for the Science of Light, Erlangen

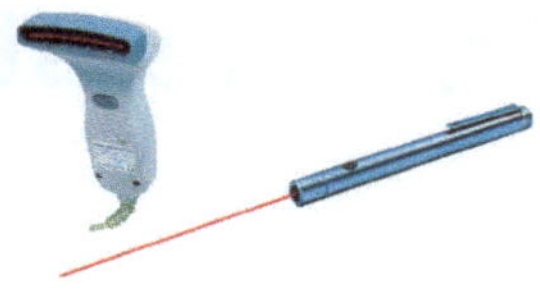

Lasers in everyday life: Scanning unit to read bar code data at cash registers (left). Laser pointer to be used as light pointer during lectures and presentations (right). Source: Laser Zentrum Hannover e.V.

Internet

- www.laserinstitute.org
- www.eli-online.org
- www.bell-labs.com/about/history/laser
- http://science.howstuffworks.com/laser1.htm
- www.opticsforteens.org

Sensor systems

Related topics

- Measuring techniques
- Image analysis and interpretation
- Ambient intelligence
- Intelligent materials
- Microsystems technology
- Automobiles

Principles

Sensors are devices that measure changes in physical quantities such as humidity and temperature or detect events such as movements. If the sensors are integrated in a feedback system, the measured variables can be used to automatically monitor or control processes.

Sensors convert measured variables into electrical signals. Nowadays, many sensors are mounted on an integrated circuit (chip) that converts the signal into digital form at the point of measurement. Some sensors have a simple two-state output (yes/no) indicating the state of their environment, e. g. smoke detectors. Other sensor output signals provide a direct readout of a range of values, e. g. the temperature inside a reactor. In some cases the sensor might be linked to an actuator, enabling the system to provide an "automatic" response. An example would be a system that measures the temperature and humidity in a room, and uses this information to automatically adapt the setting of the air conditioner.

The choice of sensors and their position relative to one another in a system is crucial to obtaining accurate measurements and assuring the correct functioning of the system. Since sensors generally have to be placed in close physical proximity to the phenomena they are meant to measure, they often have to withstand extreme conditions, e. g. temperatures of up to 1700 °C in power generating plants or corrosive environments such as molten glass or flue gases in furnaces. Thermocouples in glass furnaces are enclosed in a platinum capsule and electrically insulated with ceramics to protect the wires from being corroded by the molten glass.

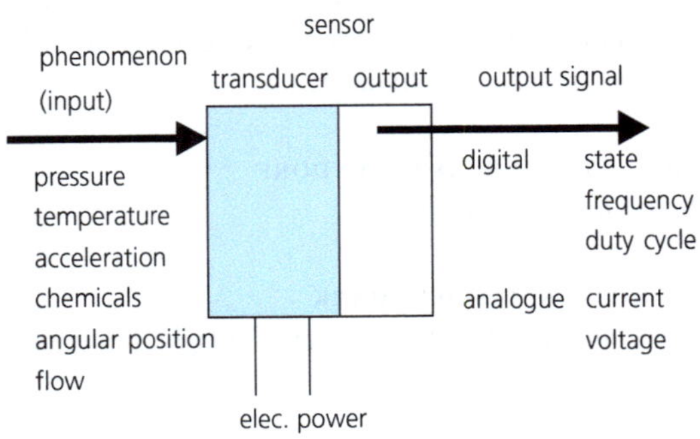

Sensor technology: The most important element of a sensor is its "contact" with the measured variable. The physical quantity or event provokes a change in the sensor, e. g. a change in the conductivity of the sensor material, which the sensor outputs as an electrical signal. This signal can be analogue or digital. In the first case, its amplitude, frequency or pulse duration is proportional to the measured quantity. In the second case, it provides simple binary information (yes/no). An electronic circuit converts the electrical signal accordingly.

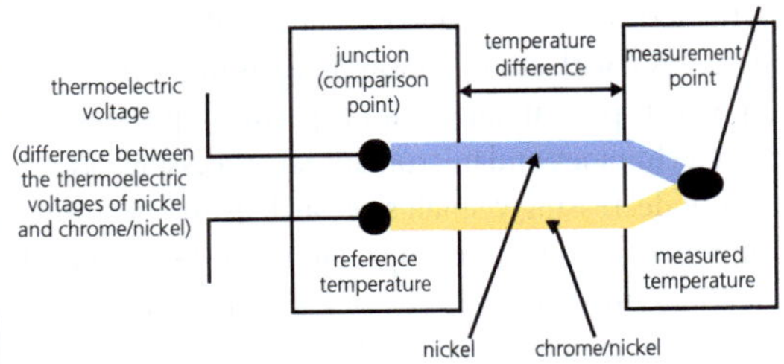

Temperature measurement using a thermocouple: The difference in temperature between the two electrodes, which are made of different metals, creates a voltage that is proportional to the temperature difference.

Sensors are classified, on the one hand, according to their function, e. g. temperature or pressure sensors, and, on the other hand, according to the principle on which they operate, e. g. chemical or optoelectronic sensors (which can in turn perform a variety of functions). In many cases, different principles can be applied to measure one and the same functional variable. The choice of the best sensor for a particular application thus depends on such criteria as sensitivity, reliability, suitability for a specific environment, and price. Temperature, for instance, can be measured either using a thermocouple (in direct contact with the measured object) or by means of a light beam (non-contact method).

► **Thermocouples.** When two different metals are joined, a thermoelectric voltage is set up at the junction between them if a temperature gradient exists along the metals (*Seebeck effect*). The potential divergence arises from the difference in energy of the electrons in the two materials, which is balanced out by the difference in electric potential. As a function of temperature this difference is in the order of microvolts per degree. One of the junctions between the two different metals – e. g. nickel and a chrome-nickel alloy – is situated at the measurement point, while the other is used as a reference. The thermoelectric voltage is measured at the reference point, and the temperature at the measurement point is deduced from this (accuracy approx. 0.1 °C).

► **Semiconductor gas sensors.** Semiconductor gas sensors are the standard electrical device for measuring gases. These metal-oxide sensors measure electrical conductivity to detect the presence of gases. The detection process works by chemical sorption. Molecules of the detected gas accumulate on the semiconductor surfaces of the sensor and form a chemical bond with the material. This changes the electrical resistance of the surface, enabling the presence of the gas to be detected via the change in current when a voltage is applied. The adsorption process is reversible, i. e. the surface becomes less conductive when the concentration of the gas diminishes. The use of gas sensors is particularly advantageous in applications requiring wide-area or permanent monitoring of hazardous substances. They are relatively cheap in comparison with electrochemical sensors, which are eventually "used up" by the chemical reactions. One common application is the control of air dampers for the ventilation of car interiors, where these sensors measure the air's carbon dioxide content.

Applications

Most modern premium makes of car contain something like 150 sensors. They provide input for such tasks as controlling the engine (pressure sensors for direct fuel injection), regulating ride comfort (information on pitch, roll and yaw for *ABS* and *EPS* systems), and activating safety-relevant systems such as airbags.

► **Lambda probe.** An engine consumes less fuel and produces lower emissions of toxic exhaust gases if the oxygen in the combustion chamber can be set to the optimum concentration that will ensure complete combustion of the fuel. For this reason, a lambda probe monitors the level of oxygen in the exhaust stream. At temperatures around 300 °C, doped zirconium dioxide starts to conduct enough oxygen ions for sensor applications. This conductor is in contact with both the exhaust gases and the outside air (used as reference). Electrodes made of platinum are attached at both points of contact, allowing the oxygen to be either incorporated in the zirconium dioxide (from the outside air, by taking up electrons in the form of negatively charged oxygen ions) or released from it (into the exhaust gas, by releasing electrons). On each side, the number of electrons required to feed this process corresponds to the number of oxygen molecules present. The differing concentrations of oxygen promote an exchange between the two areas through the ion conductor. The corresponding exchange of electrons takes place through a separate conductor. The electrical current thus generated serves as a measure of the concentration of oxygen in the exhaust gas. This signal enables the fuel intake to be finely controlled.

► **Pressure measurement.** Future emission-control laws in the United States will reduce permitted emission levels to no more than 10% of those allowed today. The pressure in the combustion chamber of a diesel engine is a vital parameter for controlling the ignition timing, and this in turn has a decisive effect on the reduction of nitrogen oxides and particulates. The sensor is incorporated directly in the glow plug. The glow plug descends into the interior of the chamber and transmits pressure data to the sensor. The sensor consists of a film of silicon in which the electrical resistance varies as a function of pressure (*piezoresistive effect*). A slight shift in the position of the atoms in the crystalline structure of the silicon modifies the elec-

▸ Operating principle of a semiconductor gas sensor: The oxygen (O_2) in the air separates into negatively charged ions (O^-) on the hot surface of the sensor. If a molecule of carbon monoxide (CO) lands on the sensor surface, it reacts by picking up two oxygen ions (O^-) to form carbon dioxide (CO_2). In the course of this reaction, the oxygen ions transfer a free electron to the sensor material. This causes the electrical resistance of the sensor film to drop. The change in resistance is proportional to the concentration of carbon monoxide in the surrounding air.

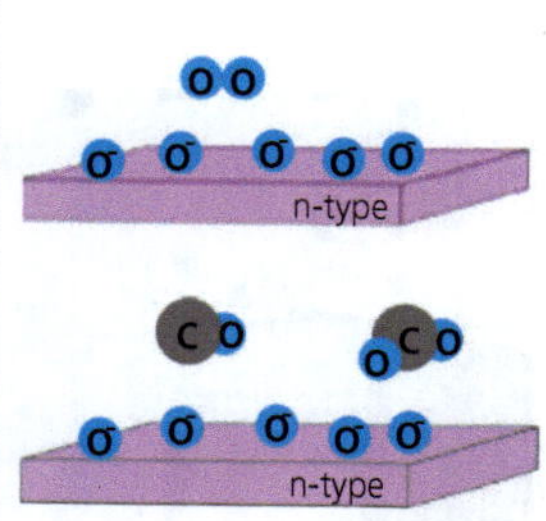

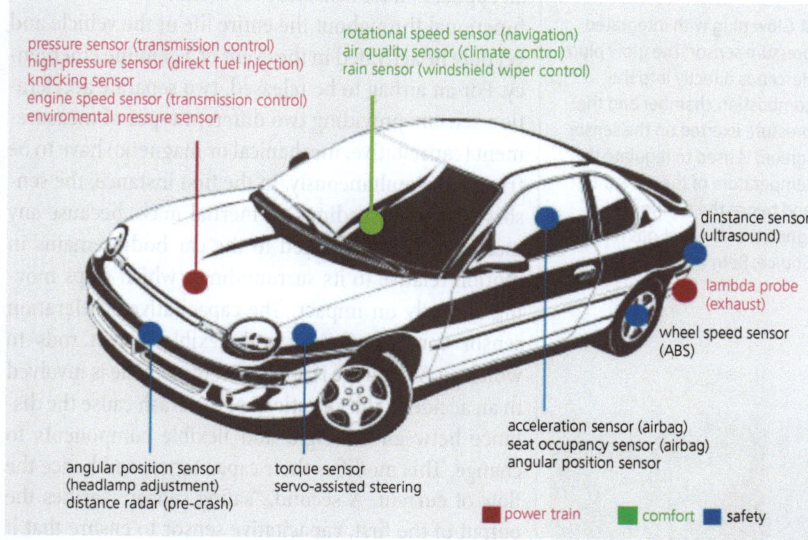

▴ Sensors in an automobile: A modern family saloon contains over 150 different sensors. The first VW Beetles built in 1954 did not contain a single sensor until the first fuel gauge was developed. Source: adapted from Bosch

▸ Lambda probe: The sensor measures the residual oxygen concentration in the exhaust gas stream relative to the oxygen concentration in the surrounding air. Oxygen ions are transported through the zirconium dioxide as a result of the differing concentrations. This means that electrons must be taken up from the outside air and released into the exhaust gas. The generated electrical current is measured.

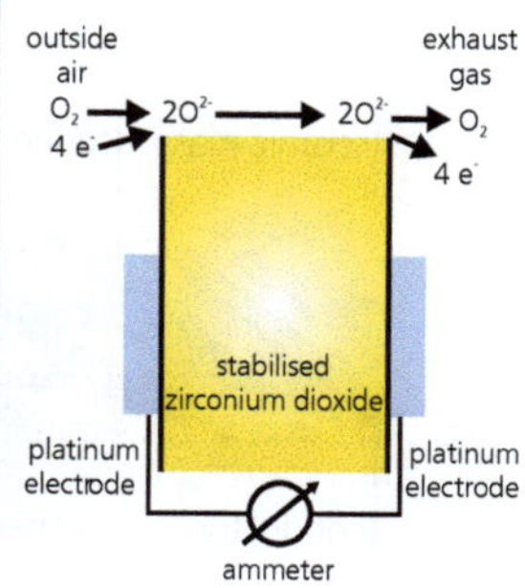

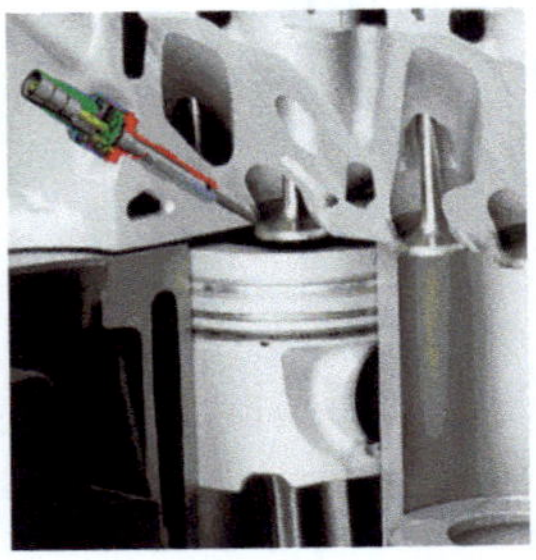

Glow plug with integrated pressure sensor: The glow plug descends directly into the combustion chamber and the pressure exerted on the sensor (green) is used to regulate the temperature of the glow plug and hence the timing of the ignition of the fuel-gas mixture. Source: Bern AG

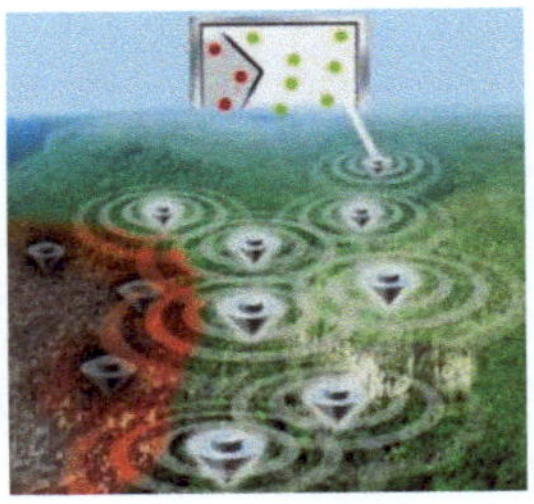

Sensors working together in networks. Thermal sensors spread out over wide areas of woodland measuring changes in temperature and communicating among themselves by means of miniature antennas. Critical events affecting the entire network, such as a widespread, significant rise in temperature or the failure of several sensors (which might indicate a fire), are reported to a control centre via a data link.

trons' mobility (density of states) and thereby increases the conductivity of the silicon film. Such sensors can withstand pressures of up to 200 bar in continuous operation, and temperatures of up to 140 °C. Changes in pressure are detected within 1 ms and are forwarded to the control unit. The ignition of the fuel-gas mixture is controlled by fine adjustments to the temperature at the level of the glow plug.

▶ **Airbags**. Airbags represent a special challenge as an application for sensors, because they have to remain functional throughout the entire life of the vehicle and yet only be activated in the event of a genuine emergency. For an airbag to be released, two separate acceleration sensors providing two different types of measurement (capacitative, mechanical or magnetic) have to be triggered simultaneously. In the first instance, the sensors base their readings on inertial mass, because any mass not firmly attached to the car body remains in motion relative to its surroundings, which stops moving abruptly on impact. The capacitative acceleration sensor consists of rigid and flexible silicon rods to which an AC voltage is applied. If the vehicle is involved in an accident, the vibrations of the crash cause the distance between the rigid and flexible components to change. This modifies their capacitance, and hence the flow of current. A second, "safing sensor" verifies the output of the first, capacitative sensor to ensure that it has not been influenced by other factors such as electromagnetic interference (electrosmog). This prevents the airbag from being inflated in situations where there has not actually been a crash. The safing sensor might be a magnetic switch consisting of a ring magnet mounted on a spring. The impact of the accident compresses the spring, moving the magnet with it and bringing two magnetised metal plates into contact (reed relay). Another type of mechanical safing sensor consists of a roller that is torn out of its anchoring in the event of a sudden impact, triggering a contact switch. The airbag's inflation mode is adapted to the anticipated force with which the person is likely to be projected towards it. Sensors in the seat measure the occupant's weight and pass this information to the control unit. Even the position of the driver and passengers is taken into account in the choice of inflation mode. Continuously operating ultrasound sensors determine whether the front-seat passenger is leaning forward, in which case this airbag needs to be activated sooner. These sensors are sited at various points around the seating area (dashboard, ceiling). They send out ultrasound pulses and analyse the returning echo to determine how far away the person is. In extreme cases (out-of-position), the system might even decide not to activate an airbag because, if the driver is leaning forward over the steering wheel (containing the airbag), the initial velocity of the expanding airbag could cause serious injuries. All of these data have to be evaluated within 20 ms for the driver's airbag, and within 5 ms for the side airbags – the maximum time span allowed before the airbags are deployed.

Trends

▶ **Sensor fusion**. Following nature's example, and on the same principle as human senses, information captured by different types of sensor is linked together to obtain a fuller picture, e.g. combined evaluation of images and sound. A camera can provide rapid image analysis over a wide area, while a supplementary acoustic analysis of the scene can locate sources of sound. Sensor fusion in this case results in an "acoustic camera", which can be compared to a multidimensional image. The technique employed to locate sounds is comparable to human directional hearing: the brain registers the different times it takes for a signal to reach the left and right ear respectively – similarly, a multitude of separate microphones measure the differences in propagation time. In this way, it is possible to establish a link between a sound and its source. This type of technique is employed to analyse noise pollution, for instance.

▶ **Sensor networks**. There is a strong trend towards the miniaturisation of measuring systems. It is becoming increasingly common to integrate the electronics for signal preprocessing into the sensor device itself, at the point of measurement. In future it will be possible to gather certain types of monitoring data requiring wide-area coverage, e.g. air quality, from intercommunicating wireless sensors linked in a decentralised ra-

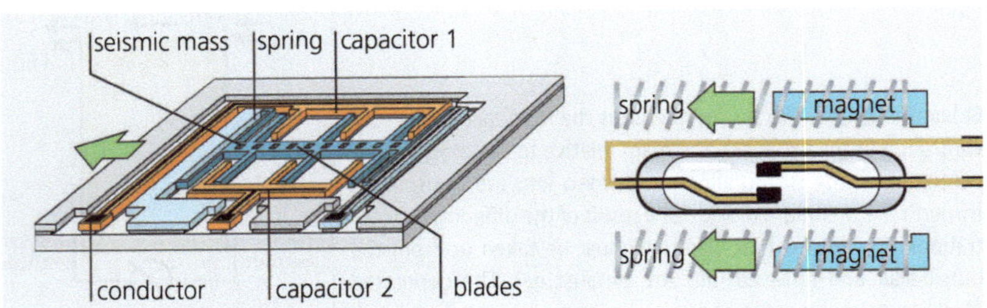

Airbag sensors: Two separate acceleration sensors have to trigger the command for the airbag to be inflated. Left: The silicon blades of the capacitative sensor carry a live electrical charge. In the event of a collision, the vibration brings them into contact, causing a current to flow. This type of sensor has a specific "sensing direction", i.e. a single direction in which it can measure acceleration. To provide measurements in 3 dimensions, several sensors have to be combined in an array. Right: In a magnetic switch, the force of the impact causes the magnet to shift to the left, causing the two live wires to come into contact, generating an electrical pulse.

dio network (*e-grains*). In addition to the sensor itself, each miniaturised device contains the integrated power supply that it needs for wireless communication. Certain sensor devices will even be able to tap the energy they need from the environment (energy harvesting).

▶ **Cellular nonlinear networks.** The human eye is more than just an excellent optical sensor. Over the course of evolution, human beings have developed a system of visual perception that relies on picking out distinctive shapes and outlines, and registering changes, thus enabling situations to be compared, interpreted and defined in many different ways. By contrast, when a digital camera analyses an image, it treats every single pixel as if it had the same significance, which is the limiting factor in computer-based image analysis. This is one reason why attempts are being made to accelerate computer analysis by applying human-type algorithms in so-called cellular nonlinear networks (CNNs). Instead of being digitised and then processed, the data are analysed directly in their analogue form on the chip, which results in high processing speeds. The data are not digitised until after they have been analysed, i. e. using a significantly reduced volume of data. A frequent problem in industrial image analysis is that of determining whether a component lies inside or outside a defined geometry. Metal sheets have to be clamped in a precise position before being welded. Sensors are used to verify that they are indeed in the correct position. In image analysis, this involves assigning each pixel that lies within the boundary zone of the geometry to a point either inside or outside the geometry. When using CNN, the output signal of each pixel depends not only on its own signal but also on the signals of adjacent pixels. In this way, the individual pixels have a "sense" of how the analysis is progressing in their immediate environment and in a specific direction – the analysis is integrated in the measurement and thus gains in resolution. For instance, contrasts are amplified to the point where a digital in/out decision can be made. It is possible to set different parameter values for the strength and range of interactions in the pixel "neighbourhoods", to suit the specific requirements of the task at hand. This concept is of special interest to applications where the input data arrive at a very rapid frequency, e. g. in high-speed moving systems (automotive, aerospace, production engineering). In such cases, image recording and processing speeds of several thousand frames per second can be achieved.

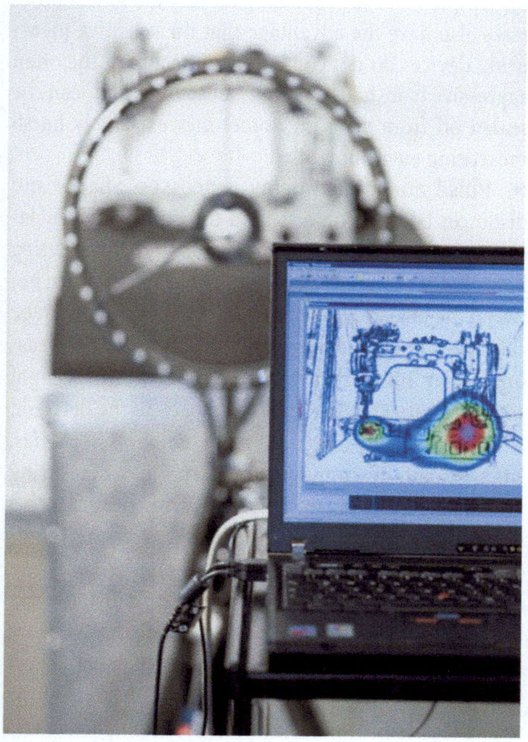

Acoustic camera: In addition to the visual image, the camera records acoustic data. A ring of microphones (visible in the background) identifies the origin of sounds by analysing the differences in propagation time. The visualised acoustic data are superimposed on the photographic image and displayed by the computer. In the example illustrated here, the result reveals that the noise produced by the sewing machine is not caused by the needle but by the electric motor. Source: Ansgar Pudenz

Prospects

- The design concepts for new devices linking sensors with the corresponding signal analysis functions are based on the trend towards even greater miniaturisation of microelectronic components, the networking of sensors, and high-speed signal processing. Applications aim to satisfy diverse needs for reliability and safety (industry, automotive, environment), and to simplify everyday life.
- In future it will be possible to process complex tasks much faster thanks to the integration of a variety of sensor sources into programmable analysis units.
- By networking low-cost miniaturised sensors, it will become possible to set up decentralised monitoring systems capable of analysing large quantities of data from a multitude of sources. This can provide benefits in many domains, ranging from production engineering and the control of flows of merchandise to environmental analysis and assisted-living projects.

DR. PATRICK HOYER
Fraunhofer Gesellschaft, Munich

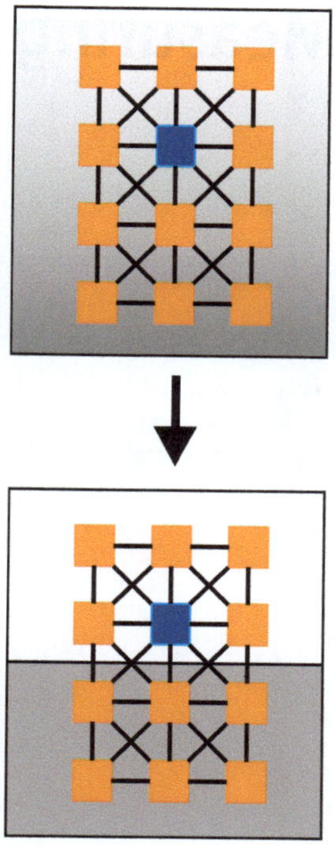

Cellular nonlinear networks: Each measurement point interrogates its neighbours before deciding on its own output value. In case of doubt, this enables the blue measurement point to opt for the value representing the bright area, in preference to the dark area, since its decision depends on the values measured by its neighbours, within the limits defined by the parameter settings.

Internet

- www.sensedu.com
- www.sensorportal.com/HTML/Sensor.htm
- www.automation.com/portals/sensors-instrument

Measuring techniques

Related topics

- Sensor systems
- Laser
- Image analysis and interpretation
- Process technologies

Principles

Since time immemorial, mankind has used various technical aids to measure physical parameters such as weight, temperature or time. In the case of direct measurement, the result can be read off the measuring instrument itself. Indirect measurements, on the other hand, deliver results in roundabout ways because a measurement device cannot be used directly. Modern measuring techniques need to be able to measure precisely and quickly even in complex measuring environments. Nowadays, it is possible to measure a number of different parameters in running processes with the utmost precision, and to use these measurement data for process control. Metrology thus plays an important role in a variety of application areas – from industrial production and process control through biotechnology, to safety, automotive, construction and medical engineering. The factors measured include not only geometric parameters such as distances, shapes or structures down to the micrometre range, but also the chemical composition of gases, liquids and solids.

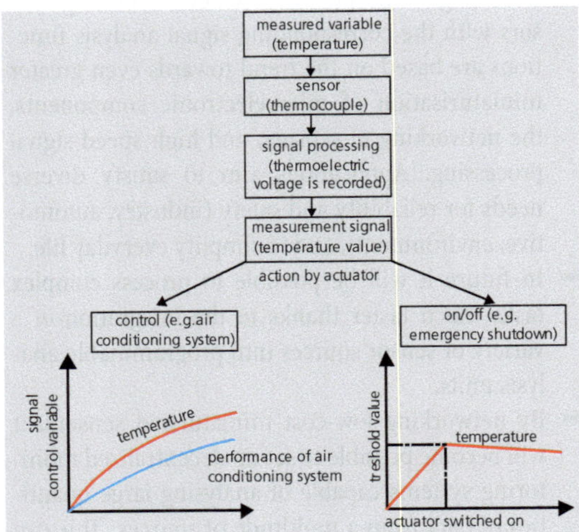

Principle of measurement technology: A sensor converts the variable being measured into an electrical signal. This signal is then used either for discrete switching (off/on) in the event that a boundary value is exceeded or undershot (right) or for continuous processing, i.e. for process control (left). The actuator reacts to the impulse or measurement variable and actively influences a process.

One important aspect of metrology today is signal processing. Increasingly large volumes of data have to be analysed and processed as an essential prerequisite for controlling even highly dynamic processes on the basis of measured data.

The various measuring techniques include mechanical, magnetic, electrical, electro-chemical, and – more and more frequently – optical processes. The rapidly growing importance of optical techniques over the past few years can be attributed mainly to the increasingly efficient lasers now appearing on the market. They are capable of contactless, non-destructive, high-precision measurements, which enables them to be integrated in production processes and to perform measurements on sensitive materials such as biological samples without damaging them. Contactless processes also have the advantage that the sensitive measuring devices do not come into contact with the often aggressive substances being measured and can be sealed off from what are sometimes extremely harsh measuring environments.

▸ **Pulse and phase delay processes.** Shapes and distances between objects tend to be measured by laser nowadays. Distances are measured by emitting light that is then reflected by the object. The space between the measuring head and the target object can be calculated from the speed of the light and the measured delay time taken for it to travel from the light source (emitter) to the object and back to the detector. The light sources employed are usually *laser diodes*. High-resolution scanner systems with high measuring speeds require finely focused measuring beams with a high beam quality. Fast and sensitive avalanche photodiodes (APDs) serve as the detectors. The two most important laser processes for measuring distance are the pulse and phase delay methods. Long distances are measured using pulse delay systems, which emit short laser pulses with a high output power. This technique makes it possible to measure distances of up to several hundred metres, with measurement inaccuracies in the centimetre range. In the phase delay process, the laser is modulated in such a way that the brightness of the laser beam constantly fluctuates sinusoidally. If the beam is reflected by an object, the phasing of the reflected beam compared to that of the emitted beam serves as the gage for measuring the distance. However, this gage alone is not conclusive, as different distances regularly produce the same phasing. A second, longer, modulation wavelength is therefore used in addition to the first one in order to obtain an unambiguous signal. This enables accurate, highly conclusive measurements in the millimetre range. In devices designed for everyday use, the laser output is limited to 1 mW to protect the user's eyes. Such devices are therefore generally able to measure distances of 30–50 m.

3D laser scanners are used to check what is known as the clearance profile of railway lines, for example. This profile frames the outer vehicle dimensions with defined safety margins. When transporting special, oversized loads, the measured clearance profile is used to determine in advance whether the vehicle and its goods can actually navigate the route. A scanning system mounted on the measuring train projects the laser beam into the route's clearance profile via a rapidly rotating mirror. The surrounding objects reflect the laser beams, and a detector picks up the light thrown back. If an object protrudes into the route profile, the light travels a shorter distance. From the altered light delay, the measuring system can deduce how far away the object is. The position of any object point can be calculated on the basis of these distance data and the rotation angle of the mirror. The initial result is a 2D cross-sectional profile. Once the measurement train starts moving, the forward motion adds the third dimension. The faster the vehicle is moving, the greater the spacing between the helical scan lines becomes. High-performance systems record up to 600 profiles per second, each with 3,300 measurement points. This means that even objects just a few millimetres in size can be detected at travelling speeds of up to 100 km/h.

Applications

▶ **Laser light-section technique.** A multitude of different techniques for measuring three-dimensional surfaces and structures have become established in everyday practice. The most suitable method for a given measuring task depends primarily on the size of the object being measured and the level of accuracy required. Optical techniques deliver fast measurement results in easy-to-process electronic form. The light-section technique, with resolutions down to the micrometre range and a measuring range of approximately one metre, lies between the alternative processes of high-resolution interferometry and delay measurement.

The light-section technique is based on *triangulation*: if one side and the two adjacent interior angles of a triangle are known, it is possible to calculate the distance between the starting points and the opposite corner. In the light-section process, the height profile of an object is measured with a narrow laser beam that is projected onto the object's surface at a defined angle. The light source generally employed is a semiconductor-based diode laser. A surface camera (*CCD* or *CMOS* matrix) positioned above the object registers the deformation of this laser beam on the object's surface. The beam's deviation from its original home position is the gage for determining the height of the object. The measuring range and resolution depend on the angle between the camera axis and the laser beam. The greater the triangulation angle, the higher the resolution and the lower the height measuring range, and vice versa. In any given case, the angle is selected according to the dispersion characteristics of the surface and the desired measuring accuracy.

In industrial production monitoring, the light-section technique is used to measure the surface profiles of components or tools. In the manufacture of screws,

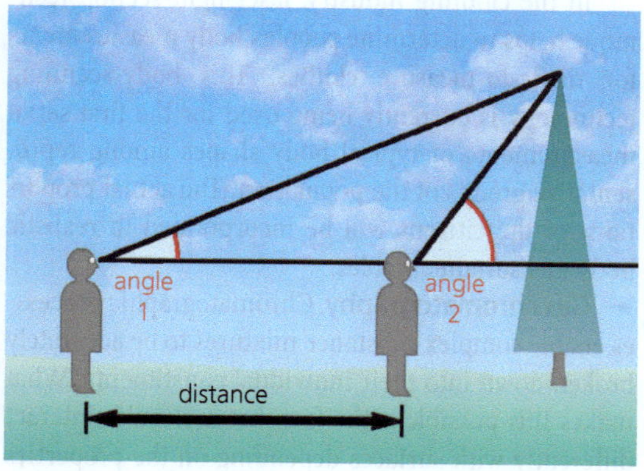

◂ Triangulation for measuring the surface of the earth: An object is viewed from two different distances, resulting in two different angles from which it is possible to calculate the distance between the two points of observation.

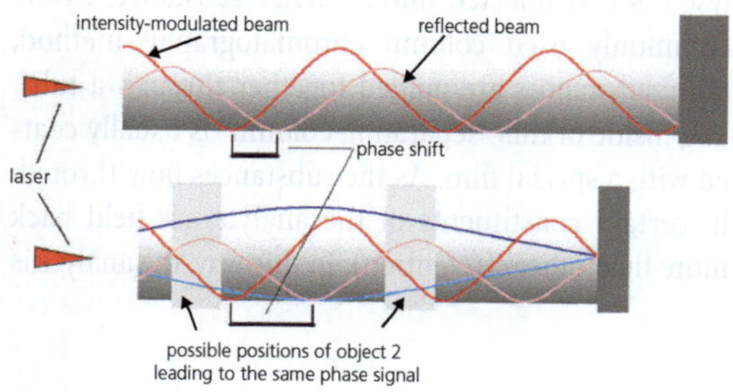

▴ Phase delay process with modulated laser light. Top: The distance is derived from the measured phase difference, the modulation frequency and the speed of light. High modulation frequencies are required in order to measure accurately. Bottom: Because the phase difference repeats itself periodically with increasing distance (e.g. every 1.5 m at a modulation frequency of 100 MHz), additional modulation is carried out at one or more lower frequencies in order to reduce the level of ambiguity. Source: Fraunhofer IPM

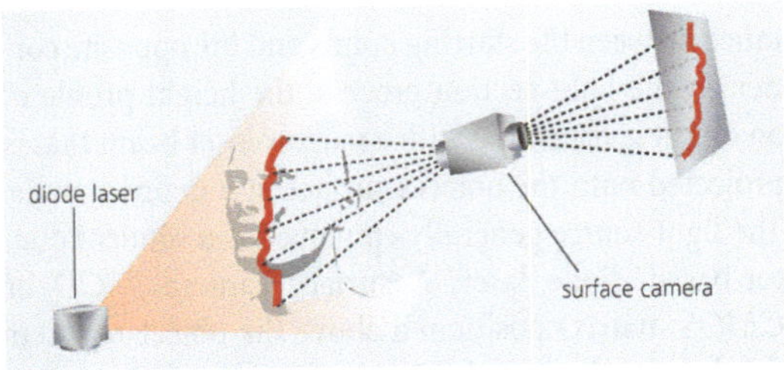

The light-section technique is a triangulation-based process: A narrow laser beam is projected at a certain angle onto the surface of an object and recorded by a surface camera. When the laser beam hits the object, it becomes deformed. The resulting deviation from its original home position reveals the outline and dimensions of the object.

for instance, thread cutting is increasingly giving way to chipless processes such as cold forming, with two thread-cutting rollers pressing the desired profile out of the solid material into the blank. Pressure and friction causes these tools to wear out, and even the slightest abrasion on a scale of just a few micrometres impairs the quality of the screws. The surfaces of the roller tools are therefore regularly measured by laser beams during the production process. High-resolution light-section technology helps to determine exactly when the tools need to be replaced.

In the clothing industry, laser light-section technique helps to determine people's body measurements for made-to-measure clothes. This body-scanning technology is currently being used for the first serial measurements of typical body shapes among representative groups of the population. The actual proportions of the citizens will be incorporated in realistic body measurement tables.

► **Gas chromatography**. Chromatographic processes enable complex substance mixtures to be accurately broken down into their individual constituents. What makes this possible is the fact that molecules interact differently with surfaces depending on the properties of the substance involved. The substance being analysed is first injected into a carrier substance. In the commonly used column chromatography method, both substances are guided together through a tube. The inside of this "separating column" is usually coated with a special film. As the substances flow through it, certain constituents of the analyte are held back more than others (retention). In this way, the analyte is separated into its various constituents. The separate fractions finally reach a detector at different times and are individually registered. With a careful combination of a carrier gas, a separating medium and a detector, together with defined measurement parameters (pressure and temperature, etc.), the constituents of the substance cannot only be separated but also clearly identified on the basis of their retention times.

Gas chromatography (GC) is one of the traditional laboratory techniques. Today, it also plays a prominent role in process analysis technology. The chemical and petrochemical industries, in particular, traditionally use GC techniques to monitor processes. The pharmaceutical industry, too, is increasingly employing the automated process measurement technology to enhance production efficiency. Although chromatography is inherently a delayed process that does not permit real-time online monitoring, its increasingly short analysis times now enable virtually continuous measurements. GC is used, for instance, in such processes as the desulphurisation of gases. The combustion of natural gas, which contains sulphur compounds, produces environmentally damaging SO_2. By feeding in a targeted supply of air, the hydrogen sulphide can be converted into elemental sulphur. A process gas chromatograph measures the concentration of sulphur compounds, thus helping to control the process.

► **Laser spectroscopy**. Spectroscopic techniques can be used to analyse solids, liquids or gases. Light of a specific wavelength is reflected, scattered or emitted by the sample under analysis. This happens in a specific way depending on the chemical substance, thus enabling qualitative and quantitative conclusions to be drawn about the compound. Optical spectroscopy, particularly infrared spectroscopy, is a standard method of analysing gas. If light of a defined wavelength passes a volume containing a certain gas, the gas molecules absorb specific parts of the light. The light that then hits the waiting detector shows spectral lines that are as characteristic of each chemical element as a fingerprint is for human beings. Absorption spectroscopy is able to detect even the smallest concentrations of gas. The sensitivity of the spectrometer depends essen-

Clearance profile scanner: Laser-based distance measurement systems are accurate to within a few millimetres, even at high travelling speeds. The laser beam is deflected by a rotating mirror and helically scans the route profile as the vehicle moves forward. The distance between the measuring head and the scanned points is calculated using the phase delay technique. Source: Fraunhofer IPM

Gas chromatography separates mixed substances into their individual constituents. What makes this possible is the fact that different molecules interact differently with different surfaces. If a mixed substance is injected into a carrier gas and fed through a coated tube, its various constituents stick to the tube's surface to varying degrees and are effectively held back. In this way, they are separated and can be identified and quantified by detectors, e.g. flame ionization detectors or thermal conductivity detectors.

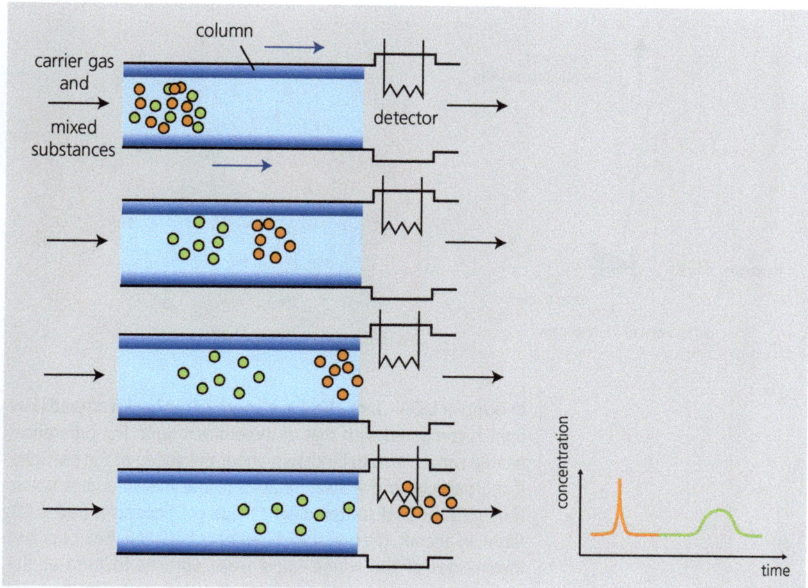

tially on the distance traveled by the measuring beam through the gas mixture, among other factors. Laser spectrometers can be used in situ, i.e. directly on site, and can continuously monitor the development of gas concentrations.

► **Gas spectroscopy.** *Optical spectroscopy* is a contactless process that does not necessarily require sampling. This makes it eminently suitable for industrial process control. In industrial combustion processes, for example, an important factor is the concentration of oxygen. By varying oxygen concentrations and providing a targeted supply of the gas, it is possible to optimise efficiency and reduce the amount of environmentally damaging exhaust gases produced. *Lambda probes*, which are mainly used to measure the residual oxygen content in exhaust fumes, operate with chemical sensors and are therefore not very suitable for use in aggressive, explosion-prone process environments. A better solution is to use contactless, optical oxygen sensors, which guide the laser beam directly through the sample compartment. They measure more accurately and without any noteworthy cross-sensitivity, i.e. without interfering effects caused by other measurement variables, due to their narrowband absorption lines.

During the manufacture of steel in an electric arc furnace, for example, part of the energy introduced is lost in the exhaust gas stream as incompletely combusted carbon monoxide (CO). By afterburning CO to CO_2 using oxygen, additional heat can be generated and used for the meltdown process. To this end, the concentrations of the relevant furnace emissions, CO, CO_2 and O_2, are measured by IR laser spectroscopy. The results are then used to actively control the oxygen supply, so that the energy potential of the afterburning process is exploited as efficiently as possible.

► **Laser radar (LIDAR).** The LIDAR (Light Detection and Ranging) technique is used in environment sensor technology to measure aerosols (clouds), the chemical composition of the atmosphere at different altitudes, or wind speeds. As in the pulse delay method, this involves measuring the offset between the time of emitting a laser pulse and the time it is backscattered from

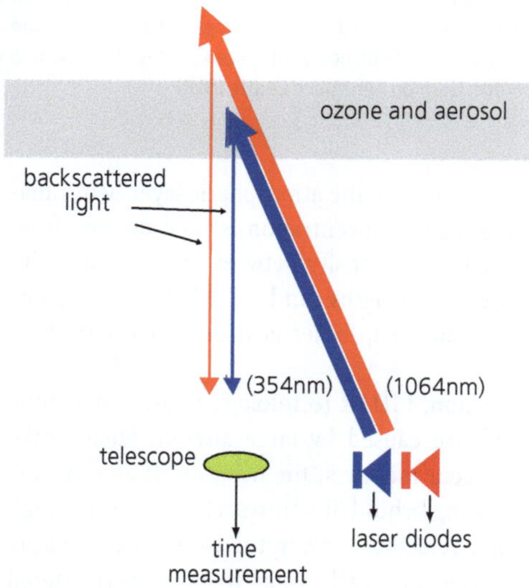

LIDAR measurement of aerosols in the atmosphere: A telescope collects the backscattered laser light and converts the light delay between the emitted laser pulse and the backscattered signal into a distance value. The varying intensities of the signals at different wavelengths enable conclusions to be drawn about the concentration of gases (e.g. ozone) in different atmospheric layers.

liquid droplets or particles in the atmosphere. In order to measure the concentration of ozone at different altitudes in the atmosphere, two laser beams of differing wavelengths (1064 and 354 nm) are emitted simultaneously. The 354-nm UV light is specifically absorbed and weakened by the ozone, which means that the backscattered light of the two laser beams is of differing intensity. The signal delay is evaluated to deter-

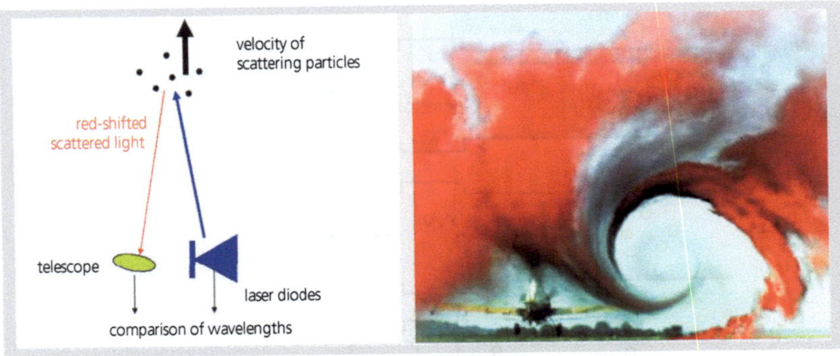

▲ Doppler LIDAR. Left: The wavelength of the backscattered laser light is compared with that of the emitted light. The differences enable conclusions to be drawn about the speed of the particles. If the particles move away relative to the measurement beam, they shift the backscattered light to lower frequencies (red shift). Right: In aircraft, the pressure differences between the upper and lower sides of the wings cause wake vortices to form at the wingtips. These vortices can be measured using the Doppler LIDAR technique and made visible by image post-processing (in this case, the frequency shifts of the Doppler LIDAR process are not directly visible). Airplanes drag these vortices along behind them. Source: German Aerospace Centre (DLR)

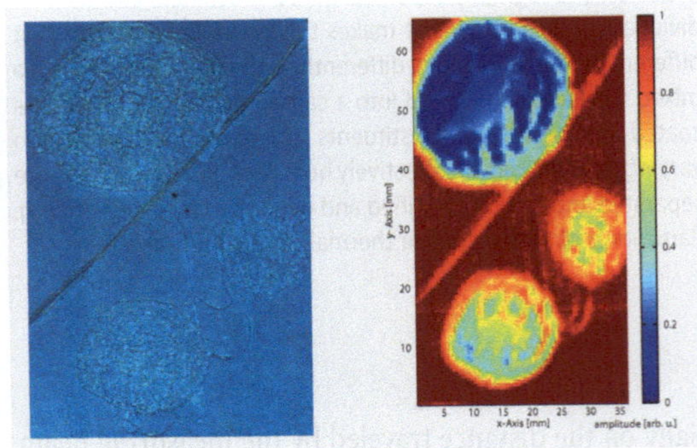

▲ Imaging terahertz (THz) processes are suitable for non-destructive materials testing: They can be used to measure geometric shapes and film thicknesses down to the micrometre range, concealed material inclusions or inhomogeneities. Left: A plastic sample made of two bonded parts with visible flow of adhesive material (round spots) and a scratch (diagonal) was analysed using THz imaging techniques. Right: The THz image clearly shows the scratch and the spread of adhesive material. The process was performed with transparent plastic to increase visibility, but it works equally well with opaque material. Source: Fraunhofer IPM

mine the distance of the atmospheric layer being analysed. The ozone concentration is then derived from the difference in intensity between the two laser signals. Other wavelengths can be used in the same way to quantify and map other environmentally relevant gases.

In aviation, LIDAR technology is used to analyse air turbulence caused by large aircraft. Such "wake vortices" occur mainly at the wingtips. If another aircraft following behind flies into such a vortex, the high cross-wind velocities may, in the worst case scenario, actually cause it to crash. Wake vortices are registered by measuring the backscatter of a laser beam from particles floating in the air. A Doppler LIDAR system additionally helps to determine the speed and travelling direction of the particles and thus also of the surrounding air. The *Doppler effect* is based on the principle that particles moving relative to the beam direction lead to an altered frequency of the backscattered light. A comparable, familiar effect is that of the siren on passing emergency vehicles: In this case, the frequency of the sound changes, depending on whether the vehicle is moving towards or away from the listener. The signal sounds higher as it approaches, and lower as it fades into the distance.

Recording the wavelength of reflected light is admittedly difficult due to the low signal intensities. However, because the wavelength of the emitted laser signal is precisely known, it can be used as a reference measurement for the backscattered signal. To this end, the output signal and the backscattered signal are superimposed. The waves influence one another in that they cancel each other out in the event of simultaneously occurring wave troughs, and reinforce each other during simultaneous wave crests (interference). This results in a regular pattern (beat), which enables accurate conclusions to be drawn about the wavelength of the reflected signal.

Trends

► **Terahertz measurement technique**. Broad ranges of the electromagnetic spectrum are being used for measurement purposes today. One final frequency still virtually unexploited until very recently is terahertz (THz) radiation. The past few years have seen the development of low-cost, easy-to-handle THz emitters and detectors that have now opened the door to the use of THz radiation in such fields as laboratory analysis, process measurement, quality assurance and security technology.

THz radiation unites several of the advantages of its bordering spectral ranges. Because it has a greater wavelength than visible or IR light, it can penetrate more deeply into any material being analysed. At the same time, it offers a better spatial resolution than the longer-waved microwaves in measurement applications. The following basic principles apply:

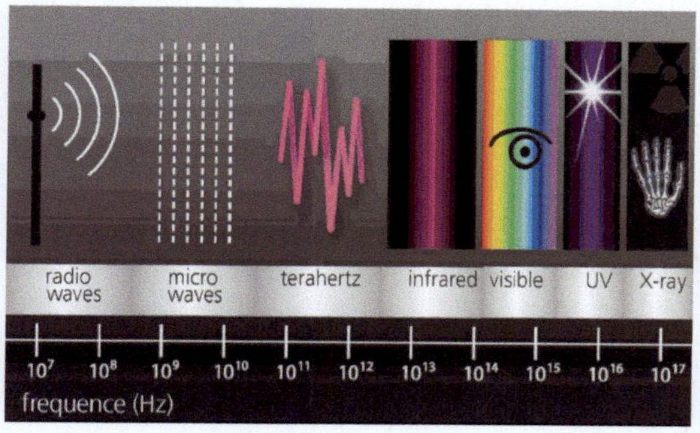

Electromagnetic spectrum: The terahertz range, with frequencies of 0.1–10 THz, lies between the infrared and microwave ranges. Its bordering frequency ranges are used in high-frequency electronics and optics applications. From a technological point of view, THz radiation lies on the border between optics and electronics, and offers several of the advantages of its neighbouring spectral ranges.

- High wavelength, i. e. low frequency: greater penetration depth, reduced energy
- Low wavelength, i. e. high frequency: better resolution, increased energy

Unlike X-ray or UV radiation, THz radiation does not trigger any changes in the chemical structure of the substances being analysed. It is therefore safe for humans. THz radiation can easily penetrate materials such as cardboard, wood, and various types of plastic. This makes it possible, for example, to test drugs for their quality and authenticity spectroscopically through their packaging. The quality of surfaces or composite materials, too, can be tested very accurately and non-destructively by means of THz spectroscopy or imaging. Water and metal, however, almost completely absorb THz radiation. This characteristic fundamentally rules out certain applications, including most medical scenarios and analyses in closed metal containers. However, it can also be useful, for example for highly sensitive measurements of residual moisture, or for detecting metal inclusions.

Prospects

- Measurement technology today contributes significantly towards quality, safety and efficiency in industrial production and process control.

Radiation band	Typical applications
X-ray	• Materials analysis • Medical examinations
UV	• Measuring of aromatic hydrocarbons that occur in the pharmaceutical and chemical industries (e.g. benzene, toluene)
Infrared	• Near infrared (NIR): Measuring of moisture, oxygen, proteins and oils, e.g. for food or refinery products • Medium infrared (MIR): Particularly suitable for analysing gas due to its high sensitivity
Terahertz	• Examination of large molecules, e.g. analysis of crystallisation processes • Materials analysis (delamination tests) • Analysis of substances through a variety of packaging materials (safety technology, quality inspection)
Microwaves	• Analysis of molecule properties, e.g. bond lengths, conformational structures

Use of the various spectral ranges for spectroscopy and imaging: The strongly differing wavelengths determine the characteristics of each radiation band of the electromagnetic spectrum. At one end of the spectrum, there are the radio waves with wavelengths of over 100 km, in stark contrast to the extremely shortwave X-ray radiation with wavelengths of just a few nanometers. In measurement technology, it is essential to achieve the best possible ratio between penetration depth and resolution. For certain measurement tasks, it is also important to ensure that the measurement radiation employed does not alter the sample material. High-energy radiation such as X-ray or extreme UV light has an ionising effect, which means it changes the chemical structure of the samples being analysed.

- Measurements integrated in the production process (inline measuring techniques), including realtime image analysis, enable processes to be controlled on the basis of measurement data.
- New measuring techniques and the linking of measurement data from different sources (information fusion) will open up more applications for measuring technology in the future, even beyond the industrial sector. New areas might include security technology, medical engineering or automotive engineering.

DR. ANNETTE BRAUN
Fraunhofer Institute for Physical Measurement Techniques IPM, Freiburg

Internet

- www.ama-sensorik.de
- www.euramet.org
- www.npl.co.uk

3

INFORMATION AND COMMUNICATION

Information and communication technologies (ICT) have become a major driving force of innovation worldwide. Modern electronics and microsystems have entered our everyday life, from the Internet to business software, from satellite navigation to telemedicine.

ICT forms the technological basis for the information- and knowledge-based society, as well as for the steady stream of new multimedia and services offered in business (*E-business*, *E-commerce*), public administration (*E-government*), health care and private life.

The ICT sector, which encompasses electronics, including micro- and nanoelectronics, communication technology, telecommunications and IT services, generates on average 10% of gross domestic product in the OECD countries, and the trend is strongly upwards. As the key technology of an increasingly knowledge-oriented business environment, ICT additionally act as a growth accelerator for many other sectors, including mechanical engineering, automobile manufacture, automation, training and development, the services industry, medical technology, power engineering and logistics.

The intermeshing of technology is of central importance, involving production on the one hand (computers, electronic components, telecommunications, radio and TV equipment and I&C services on the other (software development, data processing and telecommunication services).

As big as the industry is, its life cycles are short. The time it takes before memory chips, video screens, mobile phones and software are superseded by better successor products reaching the market, is becoming shorter and shorter.

Microelectronics remains one of the most important technologies for the information and communication industry. The computer power and storage capacity of microprocessors have risen a thousand-fold over the past 15 years. Today, they are found in many everyday things such as automobiles, cell phones and domestic appliances, as well as in industrial plant and equipment. Mobile terminals are developing into multipurpose systems. Many systems, even miniaturised ones like sensors, can be equipped at low cost with their own computer power and storage capacity, so that they can process and communicate data themselves.

In just about every area of life, society is increasingly dependent on an efficient information and communication infrastructure. The boundaries between telecommunications and information technology will disappear, as they will between mobile, fixed-line and intra-company communication. Services and applications are moving increasingly into the foreground. At any location, at any time and from any mobile or stationary terminal, users can access the applications they require to complete tasks (always-on services). Networks and IT infrastructures will become invisible to the user. This vision will become reality though the Internet as the dominant network protocol.

More than 90% of processors are not installed in PCs, but work away in the background as embedded systems e. g. in anti-lock braking systems, machine controls, telephone equipment and medical devices. Embedded systems are an example of overarching interdisciplinary research and development. From the various applications for embedded systems, interfaces are created to a number of other scientific disciplines. Electronic systems (hardware components), communication technology (networking platforms) and microsystems (sensors, actuators) deserve particular mention.

A paradigm shift has occurred in the development of *processor cores (CPUs)*. While processor clock speeds have risen constantly in the recent past, no further increase can viably be achieved above 4 GHz. The solution to the problem is provided by processors with more than one computational core, the multi-core CPUs, in which each core runs at comparatively low clock speeds. The downside is that this creates new challenges for the software needed to parallelise the programmes.

▶ **The topics.** Today's communication landscape is characterised by an increasing heterogeneity, particularly of access networks, as well as by the coalescence of fixed-line and mobile networks. Despite this heterogeneity and complexity, next-generation networks will have to allow an efficient, integrated, flexible and reliable overall system to evolve (convergence, virtualisation). What's more, networks should provide mechanisms which permit increasingly mobile users to use broadband multimedia services independently of the access network, securely and with the guarantee of privacy. Multimedia telephony, conferencing, interactive games, distributed services and IP telephony will additionally broaden the services already well established on the Internet.

Key focal points in the further development of *Internet technologies* include self-organising mechanisms for controlling dynamic network topologies and improved procedures which guarantee quality of service (individual service, web 2.0, Internet of Things). Worldwide communications and thus also globalisation are mainly driven by the Internet as a global network and by the continuous development of Internet technologies. Before Internet technologies were standardised, data communications were characterised by isolated solutions. At the end of 2008, more than 1.5 billion people were using the Internet.

Computer architecture creates the technical foundation for devices and systems to communicate with each other – by advancing the design of computers, particularly how they are organised, as well as their external and internal structures. Current areas of research include parallel computer architectures, configurable processors and distributed embedded systems.

An increasing part of system functionality is already implemented in *software*. Up to 80% of the R&D expenditure incurred in realising new communication systems is attributable to software development. Software engineers are particularly focusing on higher efficiency, quality and reliability. New mathematical models, data structures and algorithms further the exponential growth of computer power. Put together with improved hardware and increasingly accurate simulations, they achieve high precision, visualisation quality and speed. Work is being conducted increasingly on the modularisation of software architectures, in order to develop service-oriented systems (information management, service-oriented architecture SOA, open-source software and Java).

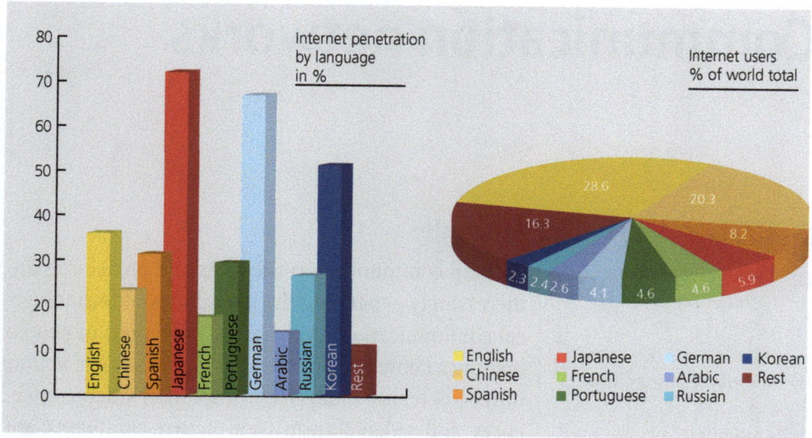

Internet usage. Left: Internet penetration by language. There is clearly a higher percentage of Internet penetration in small language areas – such as the German-speaking regions with just under 100 million inhabitants or the Japanese-speaking zones with approximately 128 million people – than in very large language areas such as the Chinese-speaking parts, which have a population of about 1.3 billion. Right: Tallying the number of speakers of the world´s languages is a complex task, particularly with the push in many countries to teach English in their public schools. In India and China a lot of users choose English as their Internet language. Source: internetworldstat

Despite the increase in computer speed, storage capacity and the logical design of computers, devices and robots with intelligence matching that of humans are still a long way off. *Artificial intelligence* is a subcategory of computer science which is somewhat difficult to define precisely. The interpretation of collected data in the context of the environment and the derivation of actions required, for instance, by an autonomous robot, presents a broad field with enormous research potential.

Many applications depend on computer-aided monitoring and assessment of the visual environment. The key is provided by automatic *image evaluation and interpretation*. New algorithms have the potential to remove all the manual effort involved in searching through or monitoring camera images, e.g. when looking for a particular subject in a photo collection or monitoring the movements of a suspicious person. ■

Communication networks

Related topics

- Optics and information technologies
- Laser
- Internet technologies
- Microsystems technology
- Ambient intelligence
- Information security

Principles

Modern communication networks and procedures are increasingly expanding the boundaries of technological communication. Holographic depictions of people holding a conversation, once the stuff of science fiction novels, is reality today. High-resolution video conferences and video transmission across continents are now possible in real time.

Communication networks are the backbone of modern information societies and today's globalised economy. Their importance has grown to an extent that bears comparison with other technological milestones, such as the introduction of the railway, the car and electrical power. The World Wide Web and the networks of globalised corporations and large service providers exemplify what is understood by communication networks.

▶ **Transmission technology.** The networks are based on physical media, via which the data are transmitted. These media can be:

- *Glass fibres* (OWG = Optical Wave Guide)
- Copper media (cables with 2, 4 or 8 wires; coaxial cables)
- *Wireless transmission*

Data can be transmitted using a range of methods that make the best possible use of the properties of the media employed. Different methods are necessary because the requirements to be met in terms of price, speed or distances to be bridged vary from case to case. If customers request a connection to the Internet via a telephone line, a *Digital Subscriber Line (DSL)* is often used via an existing copper telephone cable to bridge the distance to the next local exchange point. If the distance is short, a DSL option such as VDSL with a higher transmission rate can be used to bring TV, data and phone services ("triple play" offers) into the private home. As a rule, the greater the distance to be bridged, the smaller the available bandwidth.

▶ **IP networks.** In order to create a network, the individual transmission sections must be linked together to form a whole. This takes place at network switching nodes, known as routers and switches, which connect the different types of media. These must also "know" where to send the information to, so that the recipient can be reached. This means that they must have a globally valid addressing scheme. This addressing scheme is known as *"Internet Protocol" (IP)*, a procedure which has become standardised worldwide since the 1980s, and which is used independently of the underlying media and access methods employed.

All information to be transmitted is uniformly broken down into IP packets and transferred via the platform. Previously independent forms of communication such as voice and video are increasingly being connected to one another, creating mixed forms with their own, often new, requirements regarding the platform. A sound film, for example, will be transmitted via the IP platform in two separate streams (sound and image), which can each take different routes in the network. The platform must be able to synchronise both streams, at the latest by the time they reach the recipient. The slides of a presentation, shown online to participants in different places around the world, are saved at several points in the network, so that all the participants can scroll through the slides at almost the same time

In addition to data transport, the IP communication network must also control and manage an unlimited number of communication partners working in parallel or as a group. Thousands of participants in modern corporate communication networks, and millions of people on the Internet, can communicate with one another at the same time. Working at extremely high speeds, the equipment involved must be able to locate the respective recipients, control and regulate the traffic flows, and in some cases even monitor their compliance with certain restrictions.

▶ **Structure of a communication network.** A modern communication network consists of a core network and a range of access networks. This hierarchical

	Data transfer	Video transfer	Video transmission	Memory access	Communication services
Transport	IP(=Internet Protcol)				
Access technology	DSL, TV cable	structured in-house cabling	GSM (2G)	UMTS (3G)	WLAN, WiMax
Medium	copper cable	Glass fibre			Radio waves

Layers of a communication network: The network is based on the three media copper, glass and radio. The access technologies determine the different methods of data coding employed and regulate access of several participants to the media. IP acts as a uniform interface and abstract representation of the individual access technologies, in order to connect applications and services. A video transmission can then make its way from a terminal unit, connected by a copper cable, via a glass fibre core network to a radio-based mobile phone, all connections being uniformly linked through the universal Internet Protocol.

structure has proved to be a sound solution, allowing rapid modifications to be made to parts of the network, ideally without serious repercussions on the whole structure. Core networks are basically responsible for the rapid transport of data through the interlinked nodes and must be able to handle millions of data streams. The access networks lead the traffic via different connection alternatives, operating at a smaller bandwidth, to the core network, and thereafter transport the data away again. Besides their transport and connection function, the switching nodes between the access and core network (border or edge nodes) in most cases assume other tasks that include address forwarding, access security (firewall), traffic shaping and controlling access rights.

▶ **Glass fibre core networks**. The core network (core/backbone) is based on glass fibre cables, and in exceptional cases on copper, microwave or satellite connections. The common bandwidths on the connection paths are between 1–100 gigabit/s.

Lasers of different performance and quality are used for large distances, mainly with monomode glass fibre. These fibres are expensive to produce, but they are capable of transporting a monochrome light beam almost without distortion over large distances in a narrow 9 µm core. It is also possible to send parallel light waves with slightly different wavelengths through a monomode glass fibre, without interference between the individual light waves. This technology, known as Wave Division Multiplexing (WDM), multiplies a fibre's transport capacity.

▶ **Wireless connection technologies**. The wide scale use of mobile terminal units – laptops, PDAs, wireless tablet PCs, cell phones and others – is based on wireless radio network connection technologies. A clear trend towards FMC (Fixed Mobile Convergence) is underway to make all services and applications available to mobile devices, such as PDAs, that are already available to stationary equipment, such as the desktop PC. This applies to telephony, video telephony, email, phone and video conferences, television, as well as the ability to transfer and work on documents, and to collaborate with others through joint access to a single document.

There are two technologies available for the mobile connection of terminal equipment: WLAN and cellular radio networks.

▶ **WLAN/Wi-Fi (Wireless LAN/Wireless Fidelity)**. WLAN/Wi-Fi is available today at bandwidths between 1–300 Mbit per second. This proven technology is finding increasingly widespread use in preferred hot spots such as airports, railway stations, department stores, cafes and restaurants, but also inner cities, hospitals or university campuses – i. e. places where many potential users congregate. This technology works in the unregulated 2.4 GHz band or 5.0 GHz band. Furthermore, the regulated 4.8 GHz band is available for secure undisturbed communication. Spurious radiation sources such as video cameras or microwaves can easily disturb transmission in the unregulated bandwidths. Range and throughput depend strongly on micro-geographic conditions, such as the relative position of walls, cupboards, buildings, but also on the number of people in a room, etc. The range (distance between a station and an Access Point) extends from 50–200 m.

Two steps are necessary before user data can be exchanged via the Access Point. Firstly, each station must be connected to the Access Point. In a second step, the station is authenticated and security keys are agreed as the basis for encrypting the transmission. Various procedures, differing in expense and level of security, are available for encryption and authentication. In the simplest case, e. g. at public hot spots, or using the default settings of newly purchased equipment, there is no encryption at all. It would otherwise be impossible for users to establish a connection to the Access Point. In this case, users must themselves take care that encryption is configured after connecting to the AP, e. g. by setting up an encrypted tunnel into the company network.

In the simplest form of WLAN encryption, all stations are configured with the same SSID and an identical security WEP (Wired Equivalent Privacy) key. This

	Bandwiths
Voice over IP (VoIP)	30-100 kbit
Interactive Gaming	128 kbit - 6 Mbit
Internet Applications	500 kbit - 1.5 Mbit
Broadcast TV	3 Mbit - 5 Mbit
Voice in Demand (VoD)	3 Mbit- 6 Mbit
HDTV (MPEG-4)	6 Mbit - 7 Mbit

◘ Individual bandwidth requirements of applications: The access bandwidth and the total bandwidth required in the core network (e. g. 6–7 Mbit/s for HDTV) are based on the bandwidths of the expected applications.

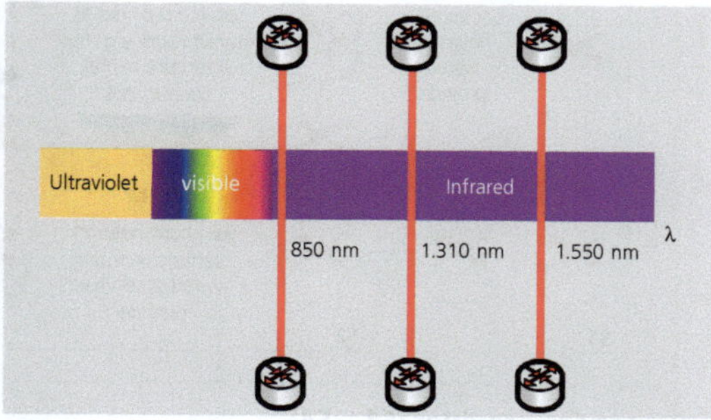

◘ Wavelength spectrums used: Data is transmitted via glass fibres in the non-visible spectrum at 850, 1310 and 1550 nm. The attenuation of the glass fibre cable is at its lowest at these wave lengths ("transmission window"). Glass fibre cables can bridge large distances with high bandwidths (> 1 terabit/s). From approximately 80 km upwards, additional regenerators must be fitted to refresh the signal between the connected units.

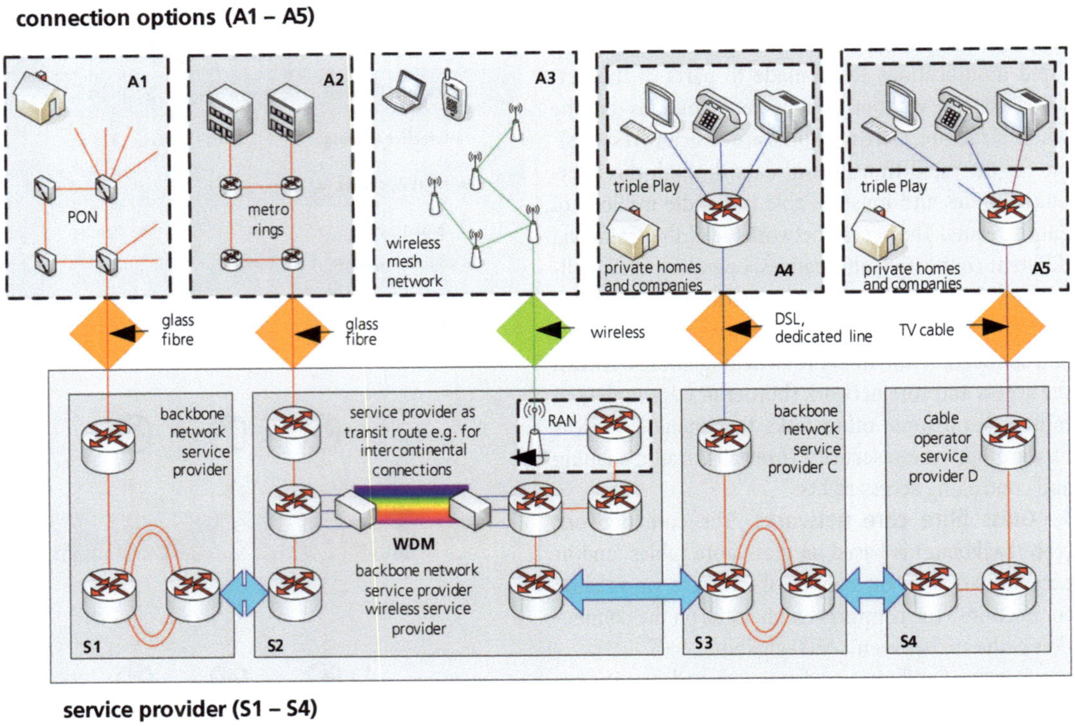

Connection of access networks to backbone networks: Five connection options (A1-A5) are shown as an example of connecting companies and households with service provider networks (S1-S4). Households and companies (A1) are linked at high speed to a service provider (S1) via a Passive Optical Network (PON) based on glass fibre. This service provider (S1) uses a dual ring structure between its core components to maximise availability. An actively controlled glass fibre network based on metro rings is primarily used by companies (A2) and offers a high throughput and extended services. Glass fibre connections are denoted by red links. A wireless connection (green link) for stationary or mobile end-users is based on WiFi (WLAN) or WiMAX technology (A3). The connection in the service provider (S2) area to the IP/MPLS-based packet network is provided by the Radio Access Network (RAN). As a third alternative (A4), small companies and households can be connected via telephone copper media using DSL technology. Copper connections, shown as blue lines, make up approximately 90% of household connections and are also used for connections over short distances. The service providers operate joint transfer points (blue arrows), providing end customers with an entire network, without demarcations. A further copper cable alternative (A5) is connection via cable (TV cable), most commonly used in the private sphere. Individual intermediate links or rings are designed with Wave Division Multiplexing (WDM) (S2) when several connections are brought together via a glass fibre.

key is static, however, and is very easily decrypted when data packets are intercepted, and so provides only a certain degree of security for private users. Within cities and for commercial use, each station should be individually secured by a dynamic encryption procedure with a higher level of security and an individual query for user name and password. In such cases more modern procedures are used, involving the WPA (Wi-Fi Protected Access) or WPA2 encryption method, for example, with a long key that is repeatedly and automatically exchanged at certain intervals. A WLAN transmission, secured using this modern method, is interception-proof.

► **Cellular radio technology.** The use of mobile or cell phones has grown at an incredible pace over the past 30 years, much in line with the development of other network technologies. Until the 1980s, analogue A, B and C networks or networks belonging to the first generation (1G) were predominant in mobile phones. Portable C network phones weighed several kilograms. The second generation (2G) of mobile phones entered the market around 1990, based on the cellular GSM technology. In Germany these were the D and E networks in the frequency range of 900 or 1800 MHz. A cell denotes the radio area within a radius of several kilometres of a transmitting mast. A transmitting/receiving area must have a sufficient number of intersecting cells to ensure connectivity at any point. Ideally, a mobile phone conversation can continue without interruption even as the caller moves from one cell to another ("handover/handoff") while driving in a car.

Connection-orientated GSM technology was extended by high-speed packet technology for data transmission through GPRS (Europe) and EDGE (USA) (2.5G). UMTS is the third generation of mobile networks (3G) and requires a new infrastructure, as the physical wireless and coding procedures have changed compared to GSM. HSPA (3.5G) has developed UMTS (384 kBit per second) further, to data transmission rates of over 3 Mbit per second. LTE (Long Term Evolution) and UWB (Ultra Wide Band) denote work being carried out on technologies of the fourth generation (4G). An important element here is the integration of various technologies, e.g. the seamless transfer or uninterrupted "roaming" of an IPT phone conversation between a WLAN network (in the office) and a GSM or UMTS network outside the building.

▶ **WiMAX**. Satellite communication and WiMax are alternatives to backbone technologies in areas that are inaccessible or thinly populated, where so-called backhaul connections replace backbone networks because it would not be viable to establish a glass fibre infrastructure. The WiMax standard has recently been extended to support mobile end-user equipment (mesh network or mobile WiMAX). In contrast to the older wireless standards such as GSM, IP is being used from the start as the universal transport protocol for WiMax.

▶ **Other wireless transmission technologies**. Bluetooth and Infrared act as wireless point-to-point connections between laptops, mobile phones, GPS receivers, car radios, headphones and other equipment over short distances (several metres). DECT technology is widespread for cordless phones. Here, several phones are connected to a base station at a range of up to 50 m. An identical numerical identifier is used to allocate the phones to the base station.

Applications

▶ **IP telephony**. IP telephony or VoIP (Voice over IP) have meanwhile become a common method of communication. Even if the terms are not identical, they all refer to calls that are made via an IP network, such as the worldwide Internet. The colloquial term "skyping", meaning to phone via the Internet (named after the Internet telephony provider Skype) shows the extent to which this technology is already rooted in everyday life.

In traditional analogue phones, voice signals are converted according to their frequency and volume and transmitted electromagnetically. Digitalisation through *ISDN (Integrated Services Digital Network)* or similar standards introduced the next step: signals were converted prior to transmission in the form of a bit stream, and reconverted by the receiver. Digitalisation resulted in the new systems being more flexible and scalable, meaning they could be expanded to include many more participants, and ultimately proved to be less costly. Nevertheless, it was nearly a decade before ISDN caught on in Europe. IPT is the next stage in the development of the telephony system. Rather than being transmitted as a bit stream, a number of bits that make up the digitalised voice stream are packed into IP packets. After adding a destination IP address, these are sent via the IP network (company network, network of a service provider or Internet). At the receiving end, the bits are removed from the IP packets and combined to form a bit stream that is converted and sent as analogue signals to the receiver's phone.

Certain requirements must be fulfilled by a network to enable voice transmission. If transmission between sender to recipient exceeds 150 milliseconds, this generates echo effects (you can hear yourself) that can only be suppressed by complex technology. The voice contents of the IP packets must arrive at the receiver's end at the same interval at which they were sent or gaps will occur in the voice stream, which is no longer synchronised. This can only be achieved by giving voice transmissions priority in IP networks. The worldwide Internet is still not configured in its entirety for this purpose, leading to diminished quality when making a phone call over the Internet: high loads on certain parts of the network often mean that IPT packets have to wait.

The migration from digital to IP telephony is now fully underway. As all the phone networks cannot be converted at once, both systems will exist over the next 20 years. There are transition technologies, which connect digital and IPT environments via gateways, so that participants in one environment can communicate effortlessly with those in another. This applies both to company phone networks and to phone networks of service providers, who are step by step replacing their traditional telephone networks, the PSTNs (Public Switched Telephone Network), with IPT networks. Three other characteristics of IPT are worth mentioning:

Range	Abbreviation	Technology
< 100 m	WPAN	Bluetooth, infrarot, Dect
< 300 m	WLAN	WiFi (=WLAN)
< 5 km	WMAN	WiMAX
< 30 km	WWAN	GSM(2G), GPRS/HSCSD (2,5G),UMTS(3G),HSPA(3,5G)
any	GWAN	Satellite

Range and terms used for radio technologies: The classifications above are dependent on the specific range of the respective radio technology. 1st: W=Wireless, G=Global, 2nd: P=Personal, L=Local, M=Metro, W=Wide, 3rd: AN=Area Network

DECT	Digital Enhanced Cordless Telecommunication
Wi-Fi	Wireless Fidelity (synonymous of WLAN)
WiMAX	Worldwide Interoperability for Microwave Access
GSM	Global System for Mobile Communications
GPRS	General Packet Radio Service
EDGE	Enhanced Data Rates for GSM Evolution
HSCSD	High Speed Circuit Switched Data
UMTS	Universal Mobile Telecommunications System
HSPA	High Speed Packet Access

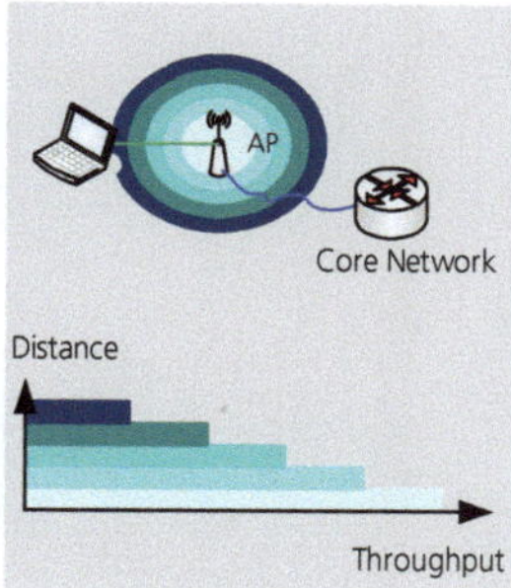

Communication with an Access Point: A station (laptop) communicates (green link) with an Access Point (AP). This establishes a connection in the core network (blue link). Both the station and the AP are in the same domain, using the same SSID (Service Set Identifier), although this does not provide protection against unwanted users. The greater the distance of the station from the AP, the lower the transfer rate. Several stations belonging to the same SSID share the available bandwidth.

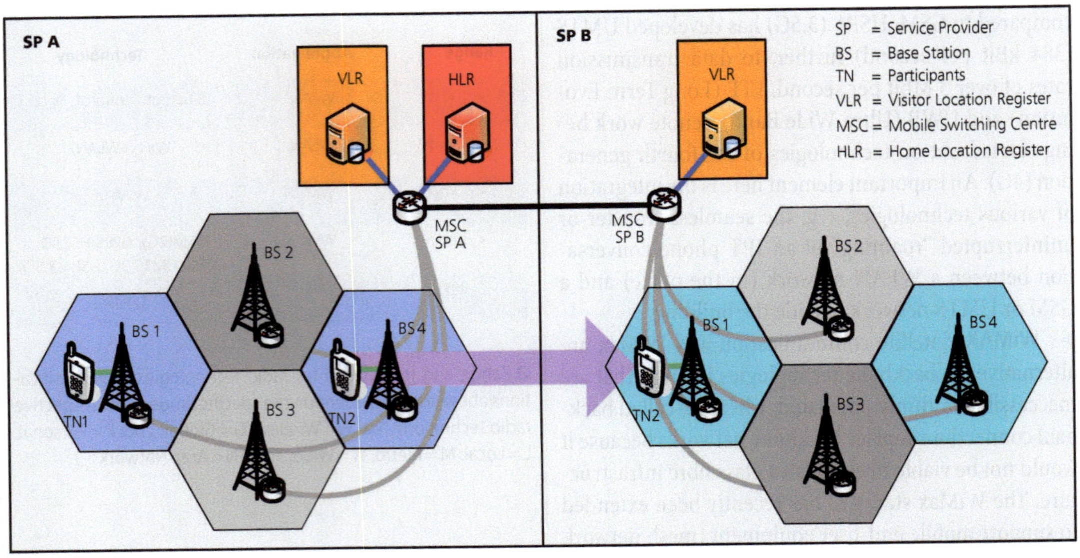

Connection of mobile phones: How does a mobile phone in country "A" find a mobile phone in country "B"? Subscriber TN 1 and subscriber TN 2 are both customers of SP "A", e.g. in Germany. When they switch on their mobile phones, they will be registered at the nearest Base Station (BS) in their service provider's service area and entered in the "Home Location Register" (HLR) of SP "A". If TN 1 now calls TN 2, the system queries the HLR of SP "A" to find out which base station is closest to TN 2 and then sets up the call. If TN 2 moves to the service area of SP "B", e.g. in France, the mobile phone of TN2 logs into the system of SP "B", via the next base station. SP "B" recognises that TN 2 is not one of its customers, but one belonging to SP "A". SP "B" enters TN 2 in its "Visitor Location Register" (VLR) and at the same time notifies SP "A" that TN 2 can be reached via SP "B". SP "A" enters this information in its HLR. If TN 1 now calls TN2 from the network of SP "A", the system determines via the information in its HLR that TN2 is in the area of SP "B". The network of SP "A" now routes the call into the area of SP "B", where TN 2 is temporarily registered, and SP "B" puts the call through. The precondition is that SP "A" and SP "B" have signed a roaming agreement, and that both TN 1 and TN 2 have enabled the "roaming" feature on their mobile phones, which carries an extra charge.

- The new IPT networks are compatible with each other as they are all based on the new SIP (Session Initiation protocol) standard for IPT.
- Voice can be compressed for transmission via IPT without any loss in quality. Less bandwidth per phone call is therefore required than was the case with ISDN, where a voice channel takes up 64 kbps. Voice can, however also be transmitted at 64 kbps (as with ISDN) or even at > 200 kbps, which is close to HiFi quality.
- In IPT, voice can be encrypted without additional equipment and transmitted without fear of interception. This degree of security was only achieved in the traditional telephone environment with expensive additional hardware.

Trends

Two important trends currently dominate the world of communication networks – convergence and virtualisation. Although these trends seem contrary, they fit together very well.

In the traditional world of communication a separate physical network existed for voice, TV, radio, telex, data and video, in addition to special networks for the police, intelligence services or foreign services, for every type of communication. The requirements placed on each of these systems were too different to be met by a single physical network.

These parallel networks still exist today. It is now possible, thanks to new technologies and functions, to incorporate the individual services in a single physical network. This "new" network is an NGN, a "Next Generation Network" based on IP, which is high performance, secure, redundant, multifunctional and suited therefore to all named requirements. Networks of service providers and companies are increasingly built on this principle; existing heterogeneous networks are being converted into a uniform NGN. There are two reasons for this trend. It is cheaper to construct and run a uniform physical infrastructure than several separate networks. The second reason is that a uniform NGN

Requirements of individual services: Each application requires certain grades of service, for example: bandwidth, upper limit for dropped packet, maximum delay, and limitation of variances in the transit time of the individual data packets.

	Voice	Video	Data Best Effort	Data Critical Data
Bandwidth	low to moderate	moderate to high	moderate to high	low to moderate
Random Drop Sensitivity	low	low	high	moderate to high
Delay Sensitivity	high	high	low	moderate to high
Jitter Sensitivity	high	high	low	moderate to high

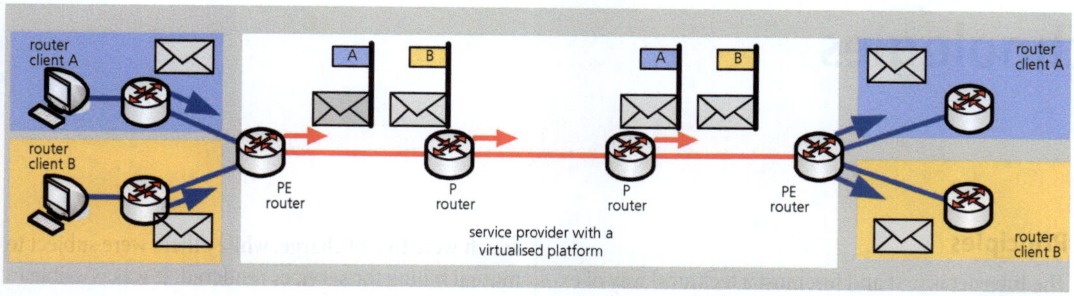

Multi-Protocol Label Switching: Two clients communicate separately, each via a MPLS-VPN. Service providers, larger companies and public institutions use Multi-Protocol Label Switching (MPLS) as NGN technology. MPLS ensures that the physical infrastructure is virtualised through a clear separation between individual clients. Each customer's packets receive a label (here, "A" and "B"), representing the customer, when they enter the NGN. This information is removed again when leaving the NGN. The NGN guarantees that there is no amalgamation and that each customer only receives his or her packets. The NGN service provider can make better use of the one physical platform and can provision clients and services efficiently.

can support modern business processes in a much better way.

▶ **Convergence.** The communication network provides the basis for transmitting many applications and services simultaneously, and on the same physical network. This is referred to as a multi-service network or a converged network. The network must guarantee the essential high grade of service for the formerly independent data streams of certain applications, such as the short transmission times required for a video conference. This grade of service is described as *Quality of Service (QoS)* and can be assigned individually per application. The base mechanism for QoS operates in such a way that the IP packets of a video conference can be identified by a corresponding flag (a numerical value in the header of the IP packet) and assigned to the category "Video traffic" when entering an IP network. All subsequent network nodes recognise this flag and process these IP packets with priority, which puts them ahead of other IP packets with lower priority.

A converged platform allows the end user to make use of all services and applications at any time and in any location. The operator of the platform can offer services efficiently and in line with market requirements.

▶ **Virtualisation.** Virtualisation is a complementary trend. Virtualisation means that a suitable physical infrastructure, such as a large multi-functional network, can be divided into logical networks. These logical – or virtual – networks then behave as if they were physically completely separated. Each user group can have different services such as data, voice and video transmission made available for them. This multi-tenant capability works independently of location.

Prospects

- Major advances in hardware technology in wired and wireless networks have resulted in leaps in the available bandwidth and in the processing speed of network components. This hardware is the basis of a new generation of IP networks, which simultaneously bring many services to many clients. Convergence and virtualisation now enable platforms to be used for all forms of communication. Fixed Mobile Convergence is an important trend – it aims to make all services and applications available, independent of the access technology.
- Modern communication technology now has a life-cycle of 3–10 years. The speed of technological revolution has certainly been the greatest in mobile phones, but it is also high in other areas, such as in WLAN and Ethernet. The available bandwidth increased 50-fold (11 → 540 Mbps) in WLANs between 1999 and 2009, and 100-fold (100 Mbps → 10 Gbps) in Ethernet between 1995 and 2006. Most technologies are fortunately backwards compatible, i.e. equipment of a newer standard remains compatible with previous standards.
- Participation in worldwide communication is now indispensable for all industrial nations, but also for less developed countries, particularly in education. Communication and the exchange of information between all participants, peer-to-peer networks and the "human Internet" have now become everyday usage under the term Web 2.0. Investment in Internet technology and in technologies such as WiMAX and UMTS, which can provide thinly populated sections of country with high bandwidth Internet access, are fundamentally important, as is the low cost provision of Internet access and inexpensive terminal equipment.

HEINZ DEININGER
ANDREAS LA QUIANTE
Cisco Systems GmbH, Stuttgart

Internet

- www.ietf.org
- www.wimaxforum.org/home
- www.wi-fi.org
- www.itu.int/net/home/index.aspx
- www.nem-initiative.org
- www.emobility.eu.org

Internet technologies

Related topics

- Communication networks
- Human-computer cooperation
- Business communication
- Information and knowledge management
- Virtual worlds

Principles

The Internet is first and foremost a technical way of exchanging all kinds of information with numerous partners throughout the world. The term "Internet" per se simply refers to the connection of two local networks to form a larger network (inter- networking). All computers in a network are able to communicate with all computers in another network to which it is connected. By linking further networks to this first pair, an even larger network is created. Today, the term "Internet" has long become synonymous with a system of networks that extends across the world. The only thing these groups of computers have in common is the protocol that they use – that is, the agreement as to how data are to be exchanged. Known as *TCP/IP (Transmission Control Protocol/Internet Protocol)*, the protocol used for the global network of computers enables data interchange through a wide variety of media including copper wires, fiber-optic cables, telephone lines, radio, satellites, and other technologies. For many private users, a fast ADSL connection (ADSL = Asymmetric Digital Subscriber Line) with a transmission rate of several megabits per second is already the accepted standard.

However, no single entity is responsible for the entire Internet. Instead, the operators of the individual sub-networks each bear the responsibility for their own network and its connection with neighboring networks. In the early days, these were mainly universities and research institutes (the ARPA network was originally a research network) along with a few companies that had participated in the development. Each of these providers also voluntarily supplied services for the whole network (e. g. forwarding of e-mails), some of which were free of charge, while others were subject to mutual billing for services rendered. It was possible in this way to reach any computer on the Internet from any other computer connected to the network. In the meantime, commercial Internet service providers play the chief role in the network, and it is they who constantly expand and enhance its capacities. Yet even the Internet cannot function entirely without rules. Appropriate organisations (which are usually state-supported) are at working to improve the protocols and further develop the whole network.

ARPA (the Advanced Research Projects Agency) was founded in response to the surprise head start that the Soviet Union made into space. It was designed to be a think-tank and a center of synergy between state institutions, universities and industry (NASA was founded shortly afterwards). ARPA was supported by funds from the U.S. Department of Defense (DoD), which is why the TCP/IP family of protocols was often termed "DoD protocols". The ARPANET, the predecessor of today's Internet, came into existence through an ARPA research project. Communications were only intended to take place between a transmitter and a receiver. The network between them was regarded as insecure. All responsibility for the correct transmission of the data was placed upon the two end points of the communication: the transmitter and the receiver. At the same time, every computer in the network had to be able to communicate with any other linked-up computer. These were the requirements on which the Internet protocol specification TCP/IP was built. Another milestone in the development of the Internet was the foundation of the NSFNET by the National Science Foundation (NSF) at the end of the 1980s, which gave American universities access to five newly founded supercomputer centers.

The Internet as we know it today is a network of over 500 million computers. Over the last two years, the number of connected computers has doubled roughly every 9 months. According to EITO (European Information Technology Observatory), 1.23 billion people were using the Internet by early 2008. At the beginning of 2008, slightly more than half of the European Union's 500 million citizens were making regular use of the Internet, while 40% do not use the

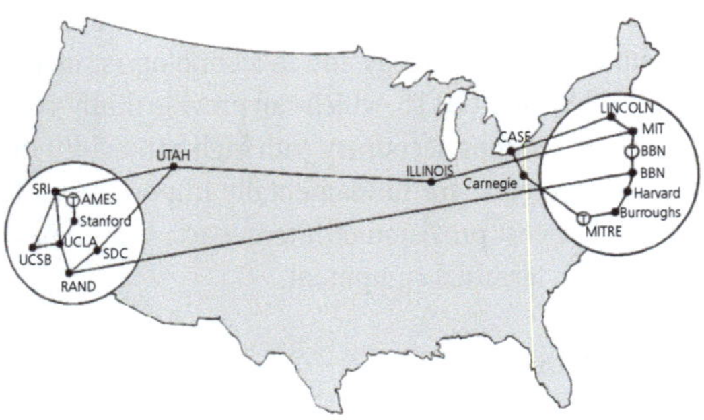

ARPANET: In 1971, the east and west coasts of the USA were linked by the ARPANET, a predecessor of the Internet. Which already had the typical mesh structure whose task was to ensure constant availability of the computers; and serious thought was already being given to the routing processes. The map clearly shows the universities and enterprises that were actively involved in the development.

Internet at all. The computers connected to the Internet are usually integrated in local area networks (LANs). LANs that belong together in organisational terms tend to be grouped in regional networks, which in turn have at least one interregional access point (WAN, Wide Area Network). The Internet thus presents a homogeneous outward image, although technically speaking it rests on a heterogeneous conglomeration of networks.

Applications

Unfortunately, the "Internet" – as a transport medium per se – is often put on a par with a few of its applications. "Internet" and "World Wide Web" are perceived as identical. This does not do justice to the medium, however: just as not only heavy-duty vehicles but a wide variety of vehicles travel along our roads, there are many applications besides the WWW that make use of the Internet. These include electronic post (e-mail), file transfer (FTP), terminal access to remote computers (Telnet, ssh), discussion forums (news), online discussions (chat), time synchronisation, Internet telephony (VOIP, Voice Over IP), and much more. Furthermore, new applications and protocols may be developed and introduced at any time – one need only think of the file exchange programmes for which new protocols were created just a few years ago.

► **Sending information.** Correspondence on the Internet works in a similar way to correspondence by post. There is just one important principle: data are not transmitted in a continuous stream of bits and bytes, but – as in local networks – are split up into small units of information ("packets" or "frames"). Not only are packets from different senders transported in the same network, but if an error occurs it is not necessary to resend all of the information. The information packet contains the addresses of the sender and the recipient, and is consigned to the network. When handling a letter, the local post office decides how the letter should be sent to the recipient on the basis of their address. Addresses contain elements that characterise the receiving post office (zip code, city) and elements that determine the recipient in the catchment area of this post office. This way of transmitting information has significant parallels with packet-transmitted communication on the Internet. The individual sub-networks of the Internet are connected by devices known as "routers". That assume the function of exchange offices and of the local post offices that deliver the mail. At present, Internet addresses take the form of 32-digit binary numbers – ideal for computers, but not for the average citizen. The numbers are therefore split up into four groups of 8 bits (or 4 bytes) each, charted as decimals, and separated by a period – for example 141.84.101.2, 129.187.10.25 or 127.0.0.1

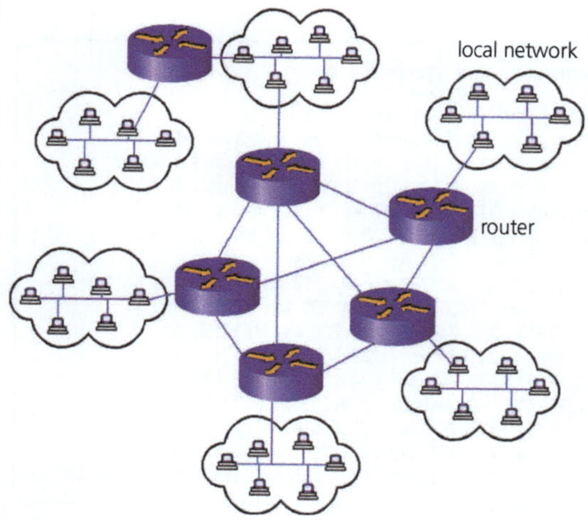

Networks on the Internet: All local networks on the Internet (shown as clouds) are interlinked via routers. Each router "knows" its neighbours and forwards the data packets to the desired destination. A data packet will usually travel through several networks and routers. The routers also have information on the "cost" of a connection, and select the shortest possible route. The concept of "cost" refers not only to the actual routing costs, but also the data rate or the number of other routers on the way.

► **Internet Protocol.** The Internet Protocol IP bears the main responsibility for ensuring that data take the correct path through the Internet. If a data packet has been correctly placed in an "IP envelope", it can be safely consigned to the network. However, the Internet also needs to be able to handle errors (loss or mutilation of packets). Nor is it the task of the Internet Protocol to handle data quantities that are larger than the typical maximum size of 1500 bytes, or to decide which application the data belong to. This is the task of the transmission control protocol TCP. Inside each IP packet is a data set that has been generated by the transmission protocol – a letter within a letter, so to speak. This "inner envelope" holds the information that deals with the problems mentioned above: the data packets are numbered consecutively and contain checking information to enable error recognition, repeat requests for defective or lost IP packets, and so on. The TCP data set contains the payload data to be transmitted. This is where it is decided which service will take care of the data (file transmission, WWW, e-mail, or even printing via the network or via network drives). TCP thus ensures the correction of errors, the correct sequencing of the packets delivered, and the associa-

```
traceroute  www.starbucks.com

 1  141.39.244.254 (141.39.244.254)  0.415 ms  0.347 ms  0.363 ms
 2  vl-3013.csr2-kbl.lrz-muenchen.de (129.187.0.161)  0.739 ms  0.723 ms  0.862 ms
 3  vl-3003.csr1-kbl.lrz-muenchen.de (129.187.0.137)  0.491 ms  0.479 ms  0.365 ms
 4  vl-3002.csr1-2wr.lrz-muenchen.de (129.187.0.133)  0.742 ms  0.726 ms  0.740 ms
 5  xr-gar1-te1-3-147.x-win.dfn.de (188.1.37.89)  0.741 ms  0.730 ms  0.738 ms
 6  zr-fra1-te0-7-0-1.x-win.dfn.de (188.1.145.53)  8.238 ms  8.348 ms  8.233 ms
 7  64.213.78.237 (64.213.78.237)  8.236 ms  8.228 ms  8.360 ms
 8  64.212.107.98 (64.212.107.98)  93.805 ms  93.792 ms  93.793 ms
 9  tbr1.n54ny.ip.att.net (12.122.86.102)  174.375 ms  174.104 ms  174.115 ms
10  cr1.n54ny.ip.att.net (12.122.16.161)  173.749 ms  174.228 ms  173.866 ms
11  cr1.cgcil.ip.att.net (12.122.1.190)  173.874 ms  173.606 ms  173.742 ms
12  cr1.st6wa.ip.att.net (12.122.31.162)  173.748 ms  174.111 ms  174.109 ms
13  tbr1.st6wa.ip.att.net (12.122.23.162)  174.498 ms  174.599 ms  173.990 ms
14  gbr2.st6wa.ip.att.net (12.127.6.178)  173.747 ms  173.605 ms  173.616 ms
15  12-122-254-74.attens.net (12.122.254.74)  173.872 ms  173.980 ms  173.867 ms
16  mdf1-bi8k-2-eth-1-2.seal.attens.net (12.129.1.218)  174.246 ms  174.223 ms  174.236 ms
```

Route of a data packet: The route taken by a data packet from the Munich University of Applied Sciences to Seattle can be traced as follows: Each line indicates the name, the IP address and the data throughput (minimum, maximum, average). As some of the routers have meaningful names, you can even deduce the path taken by the data packet. You can follow the route from Munich via Frankfurt (line 6: "fra"), New York (lines 9/10: "n54ny"), Chicago (line 11: "cgcil") all the way to Seattle. The website www.hostip.info/ helps with the search, specifying the location of almost any router. The next data packet may take a completely different route, depending on available network capacity.

tion with a certain application. This latter activity takes place via "ports", some of which are firmly allocated to particular services (e. g. port no. 80 for WWW or port no. 25 for e-mail).

► **Routers.** The network incorporates active components known as routers, whose task it is to direct the IP data packets of an e-mail from sender to receiver. Just as a switch distributes the data throughout the local network, the router forwards the packets at IP level. The transmitting computer sends the IP packets to the nearest router (if we send a message from home, that will normally be the connection to our provider). A router usually has several connections that lead to various other networks. The router decides on the basis of an internal table which of its "colleagues" it will forward the data to, and that second router does the same. The networks of the various providers, too, are all interconnected via routers. The packet thus travels from router to router until it reaches its destination. Each router normally also has a "default" entry in its table which enables it to forward even packets whose destination is not known to that router. In addition, neighboring routers communicate with one another by means of special protocols.

You can follow the route of a data packet on your own home computer. The "traceroute" command ("tracert" in Windows) supplies a list of the routers via which the data packet is traveling. It is by no means necessary for all data packets belonging to a particular file or e-mail to take the same route, as the TCP protocol will ensure that they arrive at the destination computer in the correct sequence. Route tracing can therefore never be more than just a snapshot. A traceroute from Munich University of Applied Sciences to a coffee shop in Seattle, for instance, passes through 16 stations.

► **Address nomenclature.** Because numerical IP addresses are not particularly easy to remember, it was not long before meaningful and easy-to-remember names were being called for. As a result, the Domain Name System (DNS) was devised to regulate the structure of computer names. It assigns a numerical IP address to each (worldwide unique) name. The names in the DNS are hierarchically structured. The entire Internet is divided into "domains", which are in turn structured into sub-domains. The structuring process continues in the sub-domains. This hierarchy is reflected in the name. The appropriate sections of the domain are separated by a period, e. g. "www.ee.hm.edu". The top-level domain ("edu" in our example) is at the far right. There were initially six top-level domains, and others were added later. Outside the USA, the top-level domain is abbreviated by the country code (e. g. "de" for Germany, "fr" for France, "uk" for the United Kingdom, and so on). Below the top-level domain come the domain names that have been agreed upon, together with the further structure of the namespace, within the respective organisations. At the far left stands the name of the computer. In order to establish a connection between two computers, the domain name of a computer must always be converted into the associated IP address. For certain services such as e-mail, a user name is also required. As everyone knows, this is placed in front of the computer name. The user name and the domain name are separated by the "at" symbol "@", e. g. "president@whitehouse.gov".

Trends

► **Individual services.** The whole Internet based on TCP/IP is simply a means of transport, in the same way that roads are the means of transport for vehicles. Whether you travel by motorbike, car or truck depends on the task to be performed. The individual services that can be used on the Internet today have developed gradually over time. In some of these services the client and the server remain in dialogue for a long time (e. g. for remote maintenance), while in others the connection is shut down again as soon as the requested data have been transmitted (e. g. on the WWW). First of all, the basis for text-oriented services such as e-mail

	Generic top-level domains
Generic	.biz .com .info .name .net .org .pro
Sponsored	.aero .asia .cat .coop .edu .gov .int .jobs .mil .mobi .museum .tel .travel
Infrastructure	.arpa
Deleted/retired	.nato
Reserved	.example .invalid .localhost .test
Pseudo	.bitnet .csnet .local .root .uucp .onion/.exit

The generic top-level domains currently in existence: The first were "com" (commercial), "org" (non-commercial organization), "net" (networking), "edu" (educational), "gov" (government) and "mil" (military). "info" and "biz" are welcome alternatives when the desired name under "com", "net" and "org" has already been assigned. Domains designated as "reserved" can be used for local networks.

and file transfer was created. Later came the graphically oriented World Wide Web (WWW), which is capable of integrating virtually all other services. The WWW was invented at CERN (the European Organization for Nuclear Research in Geneva) based on a technology known as hypertext. Hypertext presents the information in such a way that certain keywords in a text can be expanded to reveal further information (in a similar way to an encyclopedia). In the WWW, however, not only text but also images, sounds and animations are transmitted – all in the form of text with embedded formatting and reference information couched in a language called "*Hyper Text Markup Language*" (HTML).

Nowadays, the network is also used for telephony and other purposes. It was not long before goods were being offered on the WWW – the mail-order business had discovered a further communication route in addition to surface mail, telephone and fax. At the same time came search engines, which roam the Web and index millions of WWW sites. This index can then be searched for keywords. The best-known search engine is undoubtedly Google. Whilst people initially tried to make their domain name as meaningful as possible, today the page rank in the search engines is almost the only thing that matters. The search engine operators refuse to reveal how this ranking is performed. Besides the actual content of the web pages, a deciding factor seems to be how many other web pages point to a certain site.

The WWW is undergoing the same rapid change as all the peripheral technology (a cell phone is the only thing needed today in order to use e-mail and the WWW; computers are no longer necessary). Further Web applications such as auctions and job platforms were then added. The higher data rates made it possible to download music, photos and videos (e. g. Flickr, Youtube).

► **Web 2.0.** There is currently a transition of the user's role from a (passive) media consumer to an active web author who is a member of "communities" devoted to all kinds of topics with people who share similar interests. Not only can users explore sources of information via the browser, but the latest news can also be retrieved automatically (e. g. via "RSS feeds"). On the Internet, local limitations have been completely eliminated and replaced by topic-oriented groups. The publisher Tim O'Reilly coined the term "Web 2.0" for this second Internet boom. Wikipedia is a classic example. Prior to that there were encyclopedia providers such as Britannica Online, whose content was managed by a publishing house. With Wikipedia, the community is at the same time the provider of information and the controlling body (which does not always work as smoothly as it should). This entails other new developments: instead of a strict distribution of roles between information providers on the one hand and mere information consumers on the other, the roles have now been reversed. Users upload their own articles to the network (user-generated content), keep journals on the Web (weblogs, blogs), and even move private data to the public network. Nor is it customary any longer to use the various services separately; instead, the web content of a variety of services is seamlessly connected with new services (mashups) via open programming interfaces. In Web 2.0 the pages are often generated dynamically rather than being loaded statically. They change according to user input, for example. A role system enables users to be organised in groups with different rights, so that the content can only be altered or viewed by certain participants.

Users often have the opportunity to create a "personal profile" in which they place themselves in a certain category. Such a profile usually comprises the name, an alias (nickname), the date of birth, e-mail address, and gender. In addition, users are often asked to state voluntary information such as interests and hobbies. This not only brings together participants who share the same interests, but also more accurately customises sales offers and advertising from website operators. A user's profile can be accessed by other users, sometimes even publicly. Services such as "MySpace" enable social networks to be built up, and functions such as Instant Messaging permit spontaneous online communications.

XPort: To cater for the "Internet of things", Lantronix has developed a module called "XPort" which is only slightly larger than a network connector. Despite its small size, it incorporates a microcontroller that acts as an Internet server for the connected device. Circuits of this kind can provide today's devices with an Internet or USB connection without the need for elaborate new developments.

```
<script>

function verify(form)

  {if (form.elements[0].value.length < 5) }
  {alert(„Please enter at least 5 characters");return
  false; }

  elsif (form.elements[1].selectedIndex == 0)
  {alert(„Please select a field"); return false; }
  else {return true; }
```

Programming: Example of a simple form with a text field and an option menu. A few lines of programming code enable the input to be initially validated in the browser: Is the input in the text field at least five characters long, and has an option been selected? The programming code is embedded in the HTML document. If the input is incorrect (e.g. the input field is empty), a message window is displayed. This means that the document does not have to be transmitted from the browser to the server more than once. Nevertheless, final verification on the server remains indispensable.

```
<form action="auswert.cgi" onsubmit="return auswertung(this)">
<input type="text" size=20>
<select>
<option>Bitte eine Option au
<option>1. Auswahl
<option>2. Auswahl
<option>3. Auswahl
<option>4. Auswahl
</select>
<input type="submit" value="Abschicken">
</form>
```

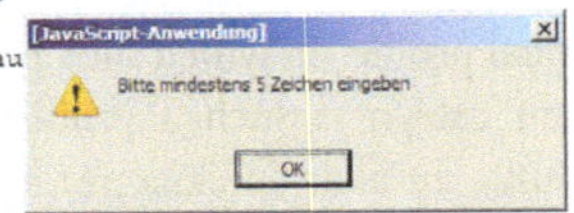

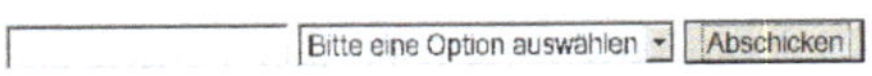

JavaScript helps to evaluate input forms: The error check is carried out in the local browser (in this case there is no input in the lower left-hand field) and enables the user to rectify the error. Not until the input is plausible are the data transmitted to the server. The illustration shows a section of the HTML source code and the form mentioned in the text. Because the input field has been left empty, an error message is issued.

A further feature of Web 2.0 are its intuitively operable pages and enhanced comfort. One way in which this is made possible is by using enhanced stylistic tools (Cascading Style Sheets, CSS), which permit the appearance of a page to be individually adapted, as well as extending the interactive capabilities of websites. This has been possible for some time with a programming language called JavaScript, whose programmes run on the local browser and which has meanwhile been so enhanced that a *JavaScript* program can now also retrieve data from the WWW server. By incorporating AJAX functionality (AJAX = Asynchronous JavaScript and XML), parts of a website can be replaced or expanded, instead of having to rebuild the entire page as in the past. The pages thus become more dynamic and user-friendly (e.g. by displaying auxiliary information when filling out a form).

► **Internet of things.** The latest development could be described as the "Internet of things". More and more devices have their own Internet connection; something that started in the office environment with printers, copiers and other computer-related devices is now becoming increasingly widespread. Access control systems, vending machines and beverage dispensers, facility management components and many other things already contain small integrated Internet servers which permit the retrieval of data and the control of the device. This is another reason why it is necessary to migrate to IPv6 as soon as possible, as every "thing" also needs its own IP address. With continuing developments in miniaturisation and declining costs, it is becoming not only technologically possible but also economically feasible to make everyday objects smarter, and to connect the world of people with the world of things. This is accomplished by embedding short-range mobile transceivers into a wide array of additional gadgets and everyday items, enabling new forms of communication between people and things, and between things themselves (machine-to-machine communication, M2M). A new dimension will be added to the world of information technologies: from anytime, any place connectivity for anyone, we will now have the same connectivity for anything. The concept, often referred to as ubiquitous computing, is not new. What is new are four key technologies, all at advanced stages, that will enable ubiquitous network connectivity: *RFID* (radio frequency identification) tags, *sensors*, *embedded intelligence* and *nanotechnology*. In this new world, for instance, clothes containing embedded chips will communicate with sensor-equipped washing machines about colours and their suitable washing temperatures and bags will

remind their owners that they have forgotten something.
An example for a smart doorknob could be the following: When you approach the door and you are carrying groceries, it opens and lets you in. This doorknob is so smart, it can let the dog out but it will not let six dogs come back in. It will take FedEx packages and automatically sign for you when you are not there. If you are standing by the door, and a phone call comes in, the doorknob can tell you that you have got a phone call from your son that you should take.

▶ **New Internet applications.** There will always be new applications on the Internet – the simulation system "*Second Life*", in which you can immerse yourself in a completely new, different kind of existence, is a case in point. While the aim of such applications may be open to dispute, there is undoubtedly a great deal of benefit in systems that permit direct booking of rail tickets or "electronic postage stamps", for example. The documents generated online can be printed out at home, and the rail ticket collector or the mail distribution facility verifies the printout by means of its two-dimensional barcode. Two relatively new applications are Facebook and Twitter, which also show that the range of Internet applications is still growing.

Facebook is a free social networking website operated by Facebook, Inc. Users can join networks organized by city, workplace, school, and region to connect and interact with other people. People can also add friends and send them messages, and update their personal profile to notify friends about themselves. The website's name refers to the paper facebooks depicting members of a campus community that some US colleges give to incoming students, faculty, and staff as a way to get to know other people on campus. The website currently has more than 120 million active users worldwide. Facebook has a number of features for users to interact with. They include the Wall, a space on every user's profile page that allows friends to post messages for the user to see; Pokes, which allows users to enter into contact with someone despite not being a member of their network or circle of friends (a notification tells the person that they have been 'poked'); Photos, where users can upload albums and photos; and Status, which allows users to inform their friends of their whereabouts and actions.

▶ **Twitter.** *Twitter* is a free social networking and micro- blogging service that allows its users to send and read other users' updates (known as tweets), which are text-based posts of up to 140 characters in length. Twitter allows short messages to be sent around the globe, much like the twittering of a great flock of birds. Updates are displayed on the user's profile page and delivered to other users who have signed up to receive them. The sender can restrict delivery to those in his or her circle of friends (delivery to everyone being the default). Users can receive updates via the Twitter website, SMS, RSS, e-mail or through an application. Several third parties offer an update posting and receiving service via e-mail. At the last estimate, Twitter had over four million accounts.

In May 2008, the earthquake in Chengdu, China buried tens of thousands of victims. Twitter users were the first to provide on-the-scene accounts. Suddenly, Twitter's triviality was no longer its most notable feature. People on the ground in Chengdu were able to report what was happening long before traditional media could even get close. It was a very important tool – an example of how useful it can be.

Prospects

- The separation of local and central data management is disappearing, as even users with average technical skills are using data storage facilities on the Internet (e.g. for photos). Local applications access network-based applications, and conversely, search engines can access local data. The separation between local and network-based applications is thus being eradicated. Programs automatically update themselves via the Internet and load modules when required; and more and more applications now use an Internet browser as their user interface.
- The dwindling number of yet unassigned IP addresses is another major problem. The division of available addresses makes routing increasingly complicated: the IP address is divided into two logical components, the network address and the computer address. To forward the data, the routers only look at the network component. The finer the division becomes, due to the need to economise, the faster the routing tables grow and the more complex the routing becomes. This is why scientists have been working for more than 10 years to implement the next-generation protocol IPv6, which will permanently solve the problem with 128-bit addresses. It will then be possible to assign a provider enough addresses to last the next 100 years, and then all routers for this provider will require only a single entry in the table.

PROF. DR. JÜRGEN PLATE
University of Applied Sciences, München

Internet

- www.isoc.org/internet/history/brief.shtml
- www.internetworldstats.com
- www.nsf.gov/about/history/nsf0050/internet/internet.htm
- www.searchandgo.com/articles/internet/net-explained-1.php
- http://computer.howstuffworks.com/internet-technology-channel.htm
- www.future-internet.eu

Computer architecture

Related topics
- Semiconductor technologies
- Software
- Internet technologies

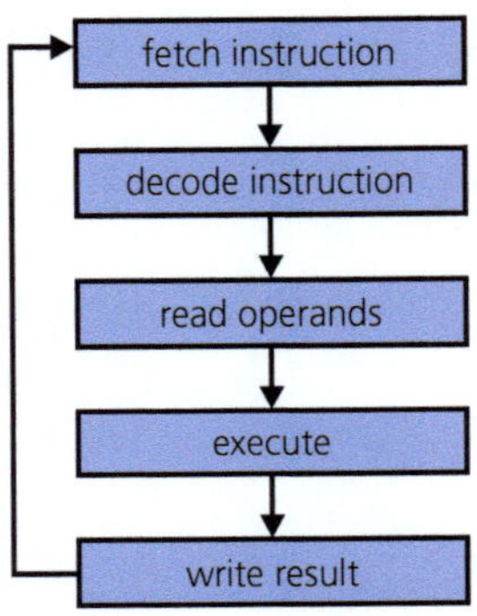

Instruction cycle of a von Neumann machine.

Principles

Computer architecture is the science of designing computers from lower-level components. A computer is an information-processing machine with a structure that is independent of the application in which it is used. A computer can process any information in any way possible, provided it is suitably programmed and its memory is large enough. However, as modern computers can only process binary information, both data and programmes have to be coded in this form. All computers – ranging from the control unit of a vending machine to the world's largest supercomputer – are based on the same principle.

▶ **The universal computing machine.** The origins of modern computers date back to the 1940s. The same basic principle, known as the von Neumann architecture, can still be found in all modern computers. According to this architecture, a computer comprises the following units:
- The *arithmetic logic unit (ALU)* performs basic arithmetic operations (addition, multiplication, etc.) and logical functions (bitwise operators: AND, OR, etc.).
- The control unit reads instructions from the memory, decodes these instructions and sends appropriate control signals to the ALU, the memory, and the input/output (I/O) units.
- The memory consists of consecutively numbered cells of equal size (usually 8 bit). These numbers are called addresses.
- The input/output unit controls all peripherals including the monitor, keyboard, and mouse as well as background storage (e. g. hard discs) and network interfaces.
- A *bidirectional universal switch (BUS)* provides the interconnection between these components.

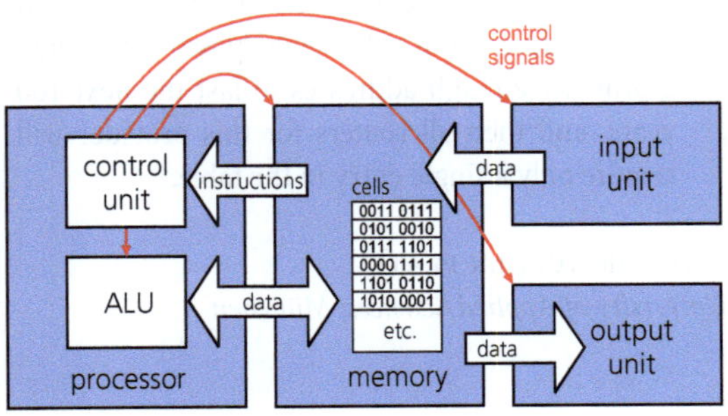

Basic units of a computer: The memory contains instructions which the control unit processes sequentially. The control unit decodes these instructions and sends corresponding control signals to the arithmetic logic unit, the memory, and the I/O units.

The memory stores data and programmes in binary form. Programmes are a set of instructions stored sequentially in the memory. Instructions are classified into arithmetic/logic instructions, move/load/store instructions and jumps. An instruction starts with an operation code (opcode), i. e. a binary number defining the type of instruction. A special register in the control unit, called the instruction pointer, stores the address of the memory cell containing the current instruction. Each time an instruction is executed, the register moves up to the next instruction. The execution of a single instruction is divided into the following phases:
- Fetch instruction: The instruction is copied from the memory into the control unit.
- Decode: The control unit reads the instruction's opcode, and depending on what this is, sends control signals to the other units in the subsequent phases. The type of instruction determines exactly what happens in these phases.
- Read operands (optional): The input parameters (e. g. the summands) are copied from the memory into the ALU.
- Execute (optional): The ALU performs an operation (e. g. addition). In the case of a jump instruction, the instruction pointer is set to the target address.
- Write result (optional): The result of the operation (e. g. the sum) is written to the register or a memory cell.

The von *Neumann architecture* is a universal architecture. In theory, as far as we are able to ascertain with current knowledge, a machine based on the von Neumann principle is capable of solving any problem that is possible to solve. There are no known problems that require a computer based on a more sophisticated design (e. g. quantum computers). The only advantage of such designs is their potentially higher performance.

▶ **Non-functional properties.** Although all computers are based on the same basic principle, they differ widely in terms of performance, dependability, security, physical dimensions, and costs. The following non-functional properties can be used to categorise computers:
- performance (storage and memory capacity, response time, throughput)

- dependability (availability, reliability)
- physical dimensions (size, weight, power consumption)
- security (confidentiality, availability, integrity)
- costs (acquisition cost, operating costs, end-of-life costs)

A machine's potential area of application depends on these properties. There is no single machine that is suited to all applications. For example, a high performance computer has a high memory capacity and high performance, but is too heavy for use as an on-board satellite controller; it is not sufficiently dependable for the financial sector; and is too expensive for use as a web server.

▶ **Microprocessors.** Nowadays, nearly all computers use a microprocessor as the *central processing unit (CPU)*. A CPU combines the ALU and the control unit on a single digital integrated circuit (chip). In most cases, this processor also includes parts of the memory in the form of caches. There is a tendency, particularly in embedded systems, to integrate all units of a computer (system on a chip) or even several computers (network on a chip) on a single chip.

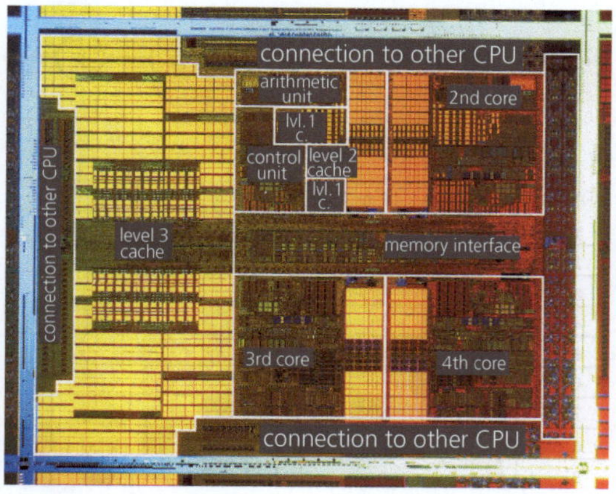

Quad-core multiprocessor: Modern microprocessors are integrated circuits consisting of billions of transistors. This processor has 4 cores, each containing an arithmetic logic unit and a control unit. Parts of the memory are also placed on the chip in the form of caches. Each core has its own level-1 and level-2 cache, whereas all cores share the level-3 cache. A memory interface connects the processor to its memory. It can also communicate directly with three other CPUs. Source: AMD

Property	Examples	Comments
performance	million instructions per second (MIPS)	the performance of the same computer may vary according to the area of application
	storage capacity	benchmark suites measure the performance according to a clearly specified set of applications
	response time throughput	
dependability	availability	this property is difficult to verify, as many systems have to be observed over a long period of time in order to obtain statistically relevant results
	reliability	
power consumption	watt	systems using a lot of electrical power are heavier and larger, as they need more powerful batteries and better cooling
security	confidentiality availability integrity	see article on IT security
costs	acquisition cost	energy costs and expenses due to planned and unplanned downtime are often neglected but have a major influence on the total cost of ownership (TCO)

Non-functional properties of computers.

Applications

▶ **Workstations and PCs.** These types of computers have an operand size of 32 or 64 bits. The operand size is the number of bits which can be processed by the ALU in a single operation. A computer with a lower operand size of 16 bits needs two steps to perform a 32-bit operation. Increasing the size to 128 bits or more is only beneficial if the application uses similarly large-sized primitive data types. However, this is not the case with most applications. Modern CPUs have special machine instructions that make optimum use of the performance of ALUs with a large operand size (e.g. 32 bits), in applications operating mainly with small operands (e.g. 8 bits). When carrying out an instruction like this, the ALU performs the same operation in parallel, on different parts of the operand (e.g. 4 parts with 8 bits each). This technique, called *single instruction multiple data (SIMD)* is often used to increase the throughput of multimedia applications.

The number of operations that can be executed per unit of time depends on the CPU's clock speed. This has increased exponentially in the last few decades and is now in the order of several gigahertz. Thus, a modern computer can execute several thousand million add operations per second. However, increasing the clock speed leads to a corresponding linear increase in power consumption. Quite apart from the ecological and cost aspects, this has a negative impact on battery life. Increased power consumption also leads to problems of heat dissipation, which cannot be solved even by significantly improving cooling performance. Consequently, future microprocessors will operate at only slightly increased clock speeds. Their performance can only be enhanced by executing multiple machine instructions concurrently. Pipelining and parallel programme execution are therefore important techniques for ensuring this performance/clock speed compromise.

▶ **Pipelining.** This technique is similar to the organisation of an assembly line: the execution of each machine instruction is divided into several phases, which are executed concurrently by different parts of the control unit or ALU. Typical phases are: fetch instruction, decode, execute, and write result to register. Several machine instructions can be executed in an overlapping way. This does not reduce the time it takes to

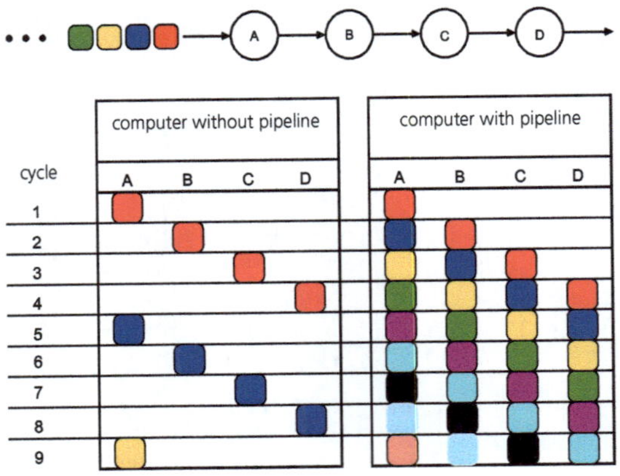

Pipelining: An instruction stream executed without pipelining (left) and with a four-deep pipeline (right). Each instruction is represented by a coloured box. Without pipelining: Each instruction must be fully executed before the next one can begin. Only one of the four components is therefore used in each cycle. With pipelining: The processor can work concurrently on up to four instructions. The time needed to execute a single instruction is the same in both cases.

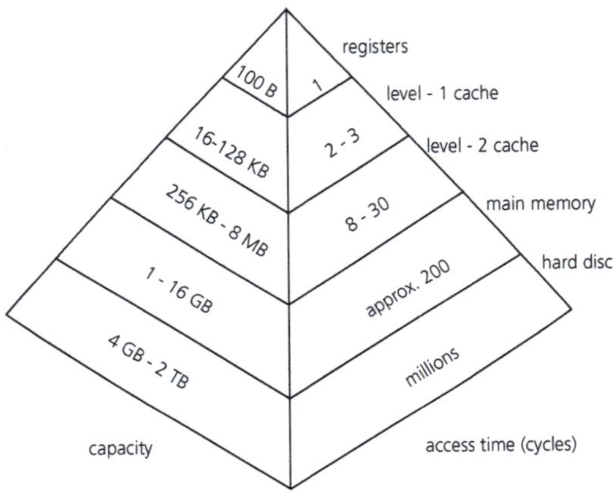

Memory hierarchy: Memory capacity is measured in bytes, each byte consisting of 8 bits. 1024 (i.e. 2^{10}) bytes are referred to as a kilobyte (KB); 1024 × 1024 bytes as a megabyte (MB); 1024 MB as a gigabyte (GB) and so on. Fast registers can only store a few hundred bytes of data. Typical level-1 cache sizes range from 16 KB to 128 KB. Level-2 caches have a capacity between 256 KB and 8 MB. A computer's main memory can store several gigabytes of data and a typical hard drive can store several hundred gigabytes.

execute each instruction, but it does increase the throughput of instructions through the processor as instructions can be executed concurrently. If each machine instruction can be divided into four phases, for instance, the processor can work on four instructions in a single clock cycle. In practice, however, the achievable time savings are lower, due to conflicts arising from jump instructions and data dependencies, which require a delay in the pipeline.

► **Caches.** Today, the main memory in a computer is typically made up of transistors and capacitors (so-called *dynamic random access memory, DRAM*). The access speed for this type of memory is about 100 times slower than that of access to internal CPU memory which is built from much faster, but also more expensive, *static RAM (SRAM)* components. As the execution of each machine instruction requires at least one memory access, a processor would be 100 times slower if it did not have cache memories. Caches are small but quick SRAM memories used to store copies of the most recently accessed memory cells. In practice, all programmes access memory cells many times consecutively. Only the first access has to wait for the slow memory, whereas the fast cache is used for subsequent read operations. Moreover, programmes often need to access neighbouring memory cells successively. Therefore, every time a memory cell is copied into the cache, all neighbouring memory cells are also transferred, increasing the probability that the content required by the next memory access will be found in the cache.

Trends

► **Parallel computers.** In addition to pipelining, modern systems can execute multiple processes, such as running programmes concurrently. Even if a system has just one control unit, it can switch between the different processes about 100 times per second, creating the illusion that many processes are taking place almost simultaneously. The computing power available for each process decreases as a function of the number of processes being executed. This can be avoided by using multiple processors, a technique commonly used in servers and high-performance computers. A single programme running on a dedicated machine only benefits from multiple processors if a programmer has explicitly divided it into independent subtasks. In most cases, this is a difficult and laborious exercise.

► **Multicore processors.** In the past decades, the number of *transistors* on a chip has been doubled every 18 months, a trend that is likely to continue to come. Until recently, the additional transistors have been used to increase operand sizes, length and pipeline depths, and to exploit fine grain parallelism at instruction level by integrating multiple ALUs on one processor. All these enhancements were transparent to the programmers: no rewriting of old programmes was necessary to benefit from the novel features of the new architectures. However, operand sizes, pipeline depth and instruction-level parallelism seem to have meanwhile reached their maximum limits, and hence other techniques must be used to benefit from the large number of transistors that can be placed on a chip. Besides increasing cache sizes, this includes placing multiple CPUs on a single chip resulting in so-called multicore systems. A multicore system is a parallel computer on a chip. Currently, dual- and quad-core systems are available, and larger systems are under development. A multicore CPU comprises multiple control units, each possessing multiple ALUs. Usually, each core has dedicated level-1 caches, whereas the higher-level caches might be shared by the cores. As in all parallel computers, application programmers have to divide the application into independent parts in order to enable a single application to benefit from the multiple cores.

► **Supercomputers.** Many scientific and engineering applications require an especially high level of performance. The accuracy of a weather forecast, for instance, depends on the amount of data processed – the greater the quantity of sensor data included in the computations, the more reliable the forecast will be. This, and many other applications, therefore have an almost infinite demand for computing power.

► **Parallelisation.** A parallel programme consists of several processes, each working on a part of the overall problem. In the best case scenario, the time needed to solve a given problem can be reduced by the factor N, where N is the number of processors working on the problem. For instance, if a machine with four dual core processors is set to work on a problem which would take eight minutes to compute on a single core machine, the fastest execution time we can expect is one minute. In practice, the runtime on the parallel machine will be longer, and may even exceed the eight minutes of the sequential machine for the following reasons:

- It is impossible to fully parallelise an algorithm. Some of the initialisation procedures, at least, have to be executed sequentially.
- If the workload cannot be evenly distributed over the eight nodes, some nodes will have a share of the workload which takes longer than one minute to compute. The slowest process determines the execution time of the overall program.
- In most applications it is impossible to divide the problem into completely independent subtasks. Instead, intermediate data has to be exchanged, which requires some processes to wait for slower nodes.
- It also takes time to transfer the intermediate data. In parallel programmes, data are exchanged frequently, but only in small amounts. The overall runtime is therefore very sensitive to data access latency.
- The most effective sequential algorithms are usually the most difficult to parallelise. An alternative algorithm is therefore often used for parallelisation, to solve the same problem. As this is not the most effective algorithm, it adds an overhead to the parallel computation.

Parallel programmes can exchange data either by using shared memory, or by sending messages over a network. A shared memory can be used to send messages, while some computers have special hardware, which can automatically transfer memory between nodes. This is similar to the way in which a globally distributed shared memory works. Both of these programming models can be applied on most computers.

► **Shared memory.** In a shared memory environment, variables of different processes can be mapped to the same physical memory cells. Thus, processes can

Component	Access Time	Comparison with familiar speeds
register	< 1 ns	supersonic jet (2000 km/h)
cache	< 5 ns	fast train (400 km/h)
main memory	approx. 60 ns	bicycle (33 km/h)
remote memory	< 1 μs	slug (2 m/h)
message over Infiniband or Myrinet	< 10 μs	(20 cm/h)
message in a local network	approx. 100 μs	bamboo growth (2 cm/h)
access to hard disc	approx. 10 ms	plant growth (5 mm/day)
message over Internet (DSL)	50-500 ms	growth of human hair (10 mm/month)
access to tape archive	6 s	continental drift (1 cm/year)

Access times for memories and networks: In modern computers, different types of memories are arranged hierarchically. The upper levels are closely connected to the control unit and ALU, and can be accessed in a few clock cycles. The lower levels are slower, but have a much higher capacity. Similarly, sending messages over a network depends on the distance between the nodes: local area networks are fast; it takes longer to access the worldwide web. A comparison is shown with familiar speed dates.

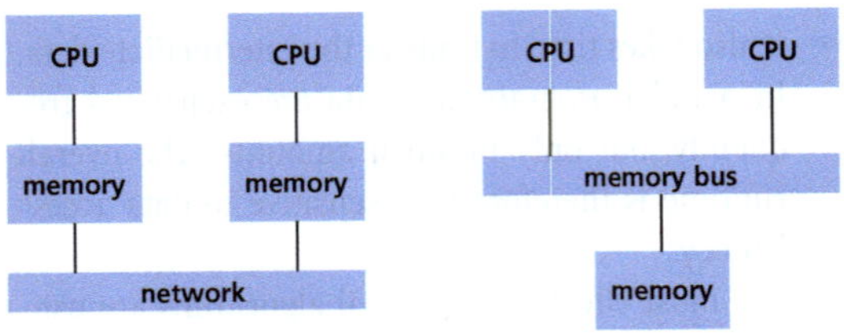

Parallelisation and shared memory: Parallel computers either communicate by sending messages over a network (left), or by using a shared memory (right). Sending messages requires many more modifications to the original sequential programme, as all communication has to be explicitly specified. In shared memory programmes, concurrent access to the same memory cells can result in errors and must be avoided.

communicate by simply writing and reading these variables. Although based on a very simple principle, programming shared memory systems can be a challenging task for the following reasons:

- Access conflicts can arise if several processes want to access the same set of variables concurrently. Therefore, access to shared variables must be restricted by synchronisation constructs, i. e. special functions which must be called before a shared variable can be accessed. If the variable is currently being accessed by someone else, the process is delayed. After the access, a release function must be called which automatically grants access to the next waiting process.
- The use of synchronisation constructs bears the risk of creating a so-called deadlock. For example, if processes A and B both want to access variables I and J, a situation might occur in which process A occupies I and is waiting to access J, whereas B occupies J and is waiting for I. As neither of the accessed variables is released, both processes have to wait infinitely.
- In many systems, accessing remote memories takes much longer than accessing local memory because in the former case a message has to be sent over the network. However, the majority of high-level languages make no distinction between different types of memory access. From the programmer's point of view, the same programming constructs are used for both local and remote memory access.

Supercomputer: The Roadrunner is currently the most powerful computer in the world consisting of 6,480 AMD Opteron dual-core processors and 12,960 IBM PowerXCell CPUs. It is about 100,000 times faster than a high-end PC, performing more than a million billion operations per second. It consumes 2.3 MW of electrical energy (cooling included), and has 300 racks (3 are shown in the picture). It has a footprint of approximately 500m². Source: IBM

As a result, it is usually a difficult task to find performance bottlenecks in shared memories without specific programming tools.

▶ **Message passing.** As an alternative paradigm, messages can be used as a means of communication between processes in a parallel programme. This approach requires the programmers to explicitly use "send" and "receive" operations every time communication takes place. In addition to basic point-to-point communication, most modern message-passing environments also offer efficient implementations of "gather" and "scatter" operations, which can be used to collect data from multiple nodes or distribute data to multiple nodes, respectively. The message-passing paradigm requires a lot more modifications to the original sequential programme, as all communication has to be explicitly specified. However, such explicit communication can be more easily analysed to track down performance bottlenecks.

▶ **Servers.** These computers allocate resources to multiple users via a computer network. The Internet is an important application environment for servers: for instance, WWW servers host web pages, and e-mail servers store and forward electronic mail. In local area networks, file servers provide shared background storage for workstations. Servers must provide high performance at low operating costs, so that they can offer their services to as many users as possible as cheaply as possible. As servers do not communicate directly with users, they do not require expensive I/O components like monitors or graphic accelerators. Instead, the emphasis is placed on fast network access and background storage. Apart from offering high performance, servers must be extremely dependable as a server malfunction affects a large number of users. Server availability is therefore increased by using components with a low failure rate, and by being able to quickly replace a component that has failed during runtime. Loss or corruption of data might have devastating consequences in some areas such as the financial sector. The systems used in these sectors have redundant components designed to tolerate faults and which prevent system failure.

▸ **Embedded systems.** The majority of all computing devices produced today are microcontrollers in embedded systems. These computers process information delivered by the system's sensors, and trigger actuators. For example, a microprocessor in a vending machine counts the number and denomination of coins inserted, compares the sum with the price of the chosen product, and activates the sensors to dispatch the product and return the change. If a new product is introduced or if prices change, only the machine's software has to be updated. Hardware does not need to be replaced. Embedded systems are ubiquitous: their applications range from domestic appliances to telephone switching centres, and from industrial automation systems to fly-by-wire controls in aircraft.

In most embedded systems, the focus is on lightweight, low power consumption, high dependability and cheap unit costs. Some applications, such as credit cards, also require high levels of security. These systems are therefore highly specialised to enable fast encryption and decryption of messages.

▸ **Reconfigurable hardware.** Reconfigurable architectures diverge substantially from the traditional von Neumann concept. Instead of having a fixed, universal structure, which can be programmed to solve a specific problem, a reconfigurable machine has a variable structure which can be adapted to the problem. Thus, a programme is no longer necessary, because the system can simply be reconfigured, depending on the given application. These systems are implemented using *field programmable gate arrays (FPGA)*, consisting of "logic blocks" and a programmable on-chip interconnect. Each logic block can be configured to perform one of the basic functions including AND, OR, and NOR, or to serve as simple mathematical operators. Other blocks are used as memories to store intermediate results. By reconfiguring the interconnections, blocks can be combined to create any type of digital circuit. FPGA-based circuits are slower and consume more power than their traditional counterparts. Furthermore, FPGAs can only be used with comparatively small circuits. However, they have the advantage that the digital circuit defining a specific function can be reconfigured at any time, even while the system is in operation. This is advantageous for the following reasons:

- The circuit can be optimised for use in a specific application. In many cases, therefore, a specialised FPGA will be much faster than a programme running on a general-purpose processor, even if it has a much lower clock speed.
- An FPGA does not need to be configured in the factory. The use of FPGA-based systems is therefore much more advantageous for low-volume products.
- This also means that they can be brought to market in a much shorter time. Developers often use FPGA-based systems as prototypes for future high-volume products.
- Design bugs can be fixed or updates can be made using FPGA-based systems, even if the system is already installed.
- Under certain conditions, reconfigured systems can tolerate permanent failures of FPGA blocks.

A *hardware description language* (HDL) defines an FPGA's functionality. As these languages are fairly low-level, attempts are being made to generate them automatically from high-level specifications resembling traditional programming languages. To combine the advantages of both approaches, reconfigurable architectures are often combined with a traditional microprocessor on a single chip.

Prospects

- Although all processors are based on the same design principle, they differ widely in terms of performance, dependability, power consumption, and costs.
- Future multicore processors will offer much higher levels of performance than today's computers. However, applications have to be parallelised in order to benefit from multiple cores.
- As general-purpose multiprocessors are optimised for the most common applications, they do not provide the best levels of performance for niche applications. Reconfigurable architectures will therefore become more common in these applications.
- There is a tendency to centralise computing power, memory and storage in the form of large clusters of servers. Users access these resources via the Internet and may use lightweight mobile devices with low power consumption.
- Dependability is becoming more important as resources are becoming more centralised. This contrasts with hardware components, which are becoming more and more prone to error. Fault-tolerant technology is therefore becoming increasingly important.

DR. MAX WALTER
Technical University of Munich

Internet

- www.top500.org
- www.scientific-computing.com
- www.mdronline.com
- pages.cs.wisc.edu/~arch/www/
- www.cs.iastate.edu/~prabhu/Tutorial/title.html

Software

Related topics

- Computer architecture
- Internet technologies
- Ambient intelligence
- Human-computer cooperation
- Information and knowledge management

Principles

Software comprises computer programmes and the data that the programmes use and produce. It is the software that turns a computer into a machine that executes the desired functions. Be it a video game, a control programme for an airbag in a car, a word processor to write this article, or a bank transaction system – all these applications have been made possible by specially developed software.

Ever since computer systems were first established, progress in microelectronics has led to increasingly powerful and at the same time cost-efficient hardware – a development that does not look set to end in the foreseeable future. This progress has enabled software to become more and more convenient and versatile in its use, with the result that practically every aspect of our life is now affected by information technology and thus by software. Parallel to this progress, software has become very complex in its logical structure as well as in its development. In today's information technology, the costs for software normally exceed those for hardware by a considerable margin.

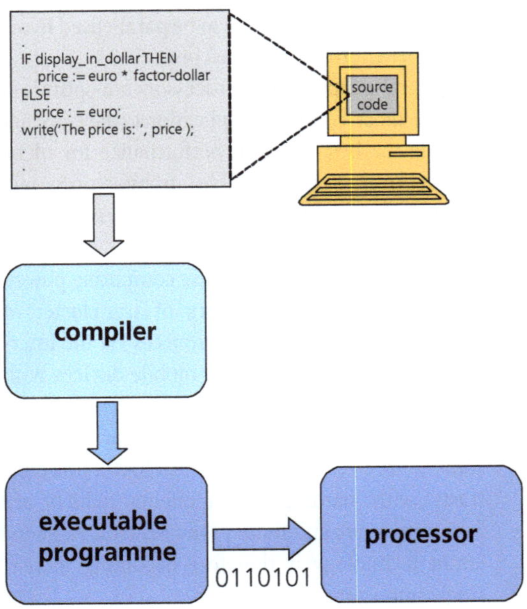

Source code, compiler and machine code: a compiler generates an executable programme (machine code) from the source code written in a high-level programming language. Only the machine code is directly executable by the computer's processor. Source: BearingPoint

▶ **Source code.** In practice, most software programmes are written in a high-level programming language that has a certain degree of similarity with the human language. Such representations are known as the programmes' "source code". The source code describes in detail what the computer should do on a level that can be understood by humans. However, the computer itself (respectively its processor) is only able to execute comparably rudimentary commands represented by sequences of bits, referred to as the "machine code". For this reason, a compiler has to translate the source code into an executable programme that consists of machine code directly executable by the processor. The compiler is nothing more than a special executable programme itself that provides the link between source code and machine code.

▶ **Software architecture.** In general, software systems are enormously complex. In order to reduce their complexity, they are decomposed into separate components and the relationships between them at an early stage during the design phase. The structure of this decomposition is called software architecture. It describes the system on a high level without concentrating on details.

A common example is the three-tier architecture, in which the architectural components of the structure are represented by three tiers: a presentation tier, a logic tier and a data tier. These tiers can each potentially run on a different computer, but it is also possible for them to be implemented as logically separated components on a single machine.

The presentation tier is the interface to the user. The corresponding software provides for displays and input fields such that the user is able to comfortably interact with the system. A well-known example of this tier is the incorporation of input/display functionalities in a browser.

The logic tier contains the application functionality of the system (often called the business logic in the case of applications used in a commercial context) and enables communication with the surrounding tiers by moving data between them.

The data tier stores information on a data carrier (e. g. hard disc) and retrieves it. The tier itself consists of a highly sophisticated system of programmes (called a database management system) that effectively organises data management processes to guarantee fast access to the information required. The data management system is normally neutral and independent of a given application, which means that different applications can use the same database management system. In practice, most database systems hold data in the logical format of tables.

► **Software engineering**. Professional software development comprises far more than simply writing programmes. Creating a complex software system of high quality with potentially millions of lines of source code involves many individuals with different tasks and roles. Moreover, in today's world of distributed and outsourced tasks, these people often work at different locations or companies. The coordination of all these tasks and roles calls for a systematic approach including project management and quality assurance. The development of professional software (and its operation and maintenance) is, in effect, an engineering discipline comprising standards and methods similar to those used in the development of traditional products. Various tasks are regulated in the software development process.

A requirement analysis results in system specifications as seen from a user's perspective.

For instance, a system that stores and manages the data of a company's business partners needs an input mask for entering names and addresses. The input field for the town is relatively simple, but even here, a number of questions have to be clarified in order to accurately specify the requirements, such as: How many digits may a field of the input mask have? Is the input of special characters like "@,#,~, .." allowed? If so, which ones are allowed, and does this hold for the first character, too? Is the input of numbers allowed? If so, does this also hold if the complete input string consists of numbers? What kind of orthography corrections should the system execute automatically? How should the system react to different input errors? Even this comparably trivial example results in a heap of work for an expert who has to analyse and formulate the requirements, so performing a requirement analysis for a complex system is obviously an even tougher task.

The system design comprises the technological aspects needed to turn the requirements into a working system. Whilst the requirement analysis essentially deals with the question of what the system should do, the system design basically answers the question of how this is realised. Relevant points include the software architecture, the description of the programmes to be developed, and the specification of software interfaces to other systems that may interact with the system being built.

During the development phase, the software developers actually programme the system on the basis of the system design using a concrete programming language.

Before the system is delivered for use, it has to undergo an extensive test phase to verify whether all the

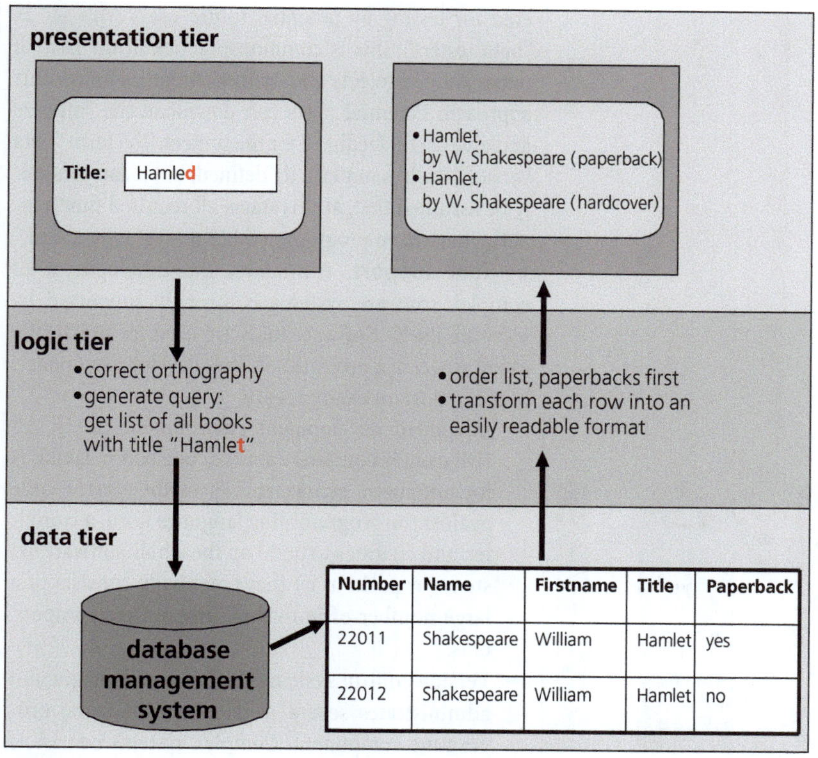

The tiers of the three-tier architecture: Illustrated on the basis of a bookshop application that is able, amongst other things, to display information on books belonging to a given title. The software corresponding to each tier fulfils different tasks. The presentation tier provides for displays and input fields. Users in the bookshop can enter the title of a book, upon which this tier will present them with the result of their query. The logic tier contains the application functionality of the system. The input from the presentation tier is processed (e. g. orthography corrections are automatically carried out) and transformed into a query that is manageable for the data tier. The response to the query carried out in the data tier is in turn processed and transformed into the format in which the answer needs to be displayed in the presentation (client) tier. In our example, the answer from the data tier would be a list of books. In this case, the processing step would involve ordering the list into the required sequence (e. g. paperback issues first) and transforming each row of the list into easily readable text that is handed over to the presentation tier for display. Source: BearingPoint requirements specified during the requirement analysis have been fulfilled. In addition, any errors detected during the test are corrected. Organising and performing systematic software tests is such a complex task that it can be considered as a discipline in its own right. People involved in organising the test phase of a software project often need to be qualified as certified test managers.

Traditionally, the testers are a direct part of the project staff. In addition, software that has already reached an acceptable standard (beta version) during the course of the development process is often delivered for testing by potential future users who act as "beta testers". This is common practice in the case of open-source projects and many companies follow this approach. Potential users can download the software and give useful feedback for the project. The term "beta version" itself is not exactly defined, but it can generally be assumed that, at this stage, all required functionalities have been programmed but not yet fully tested.

► **Tool support**. Nowadays, the development of complex software systems is strongly supported by software itself. Software tools are used for practically every aspect of a project's software development phase. Some important examples are:

- Integrated development environment (IDE): An IDE usually contains a source code editor, facilities for automatic syntax checks of the source code against the programming language used, a compiler, and facilities to build up the whole software system (or parts of it) that potentially consists of a large number of individual programme components.
- Version control system: This system manages and administrates several versions of the same programme component. Complex systems consist of thousands of programme components, each of which exists in several versions (e.g. a version a programmer works on, a version that has already been delivered for a test, and a version of a delivered release that is already in use). Often, the version control system is a further part of the IDE.
- Test tools that help to manage and carry out testing activities.

The use of the first two examples is standard practice. In recent years, test tools have become increasingly relevant, too. They support most aspects of the test management process. Testing activities are usually structured according to test cases. Each test case in turn reflects a certain requirement (or a part of it) from the viewpoint of the test. The test management software stores each case in the form of a script in a database system. For each test, it displays the test script(s) to be dealt with. The test script chronologically describes the activities to be carried out in order to check if the given requirement has been fulfilled. Example: there is a requirement that dictates that the field for the town in an input mask for addresses must not accept the character "#" (or other special signs) as the first character and that a corresponding error message has to pop up if such signs are used. The corresponding section in the test script could be: Enter "#Testtown" in the input field for the town and press the OK button. Check if the error message "invalid first character #" pops up. If the requirement is found to be fulfilled, the tester marks the test case as successful, and this information is stored. In this way, the status of each test case is recorded such that the progress concerning the quality of the software can be easily monitored and evaluated – even if thousands of requirements are relevant. It is also possible to automate the approach by means of what are known as capture and replay tools. These record the manual input necessary to carry out a test case, and afterwards it is able to repeat the execution of the test case automatically. This is very useful since tests have to be repeated many times: for instance, after correcting defects detected in the course of a test cycle, the tester has to check not only that defects have been corrected, but that those requirements not affected by the defects are still fulfilled, too.

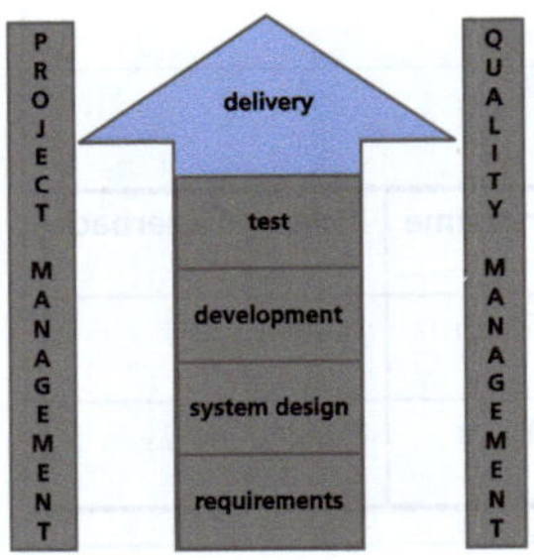

Software development process: The requirement analysis answers the question as to what the system should do. The system design deals with the question of how the requirements will be fulfilled from a technical point of view. During the development phase, the software developers actually programme the system on the basis of the system design. Before the software is delivered, it has to be tested thoroughly. In addition, all activities are steered and controlled by an effective project and quality management system. Source: BearingPoint

Applications

► **Information management**. Information technology has revolutionised administrative tasks. Among the various industries, the financial service sector has been particularly affected. Here, software systems support the complete value chain of a company. In an insurance company, for example, this will include the process in the following important fields of business:

- Policy management for handling all the policies the company has sold
- Claims management for handling claims against the company
- Commission management for handling commissions for the agents who sell the policies
- Administration and execution of monetary collections and disbursements

Software applications not only support these tasks, but are now also the tools that actually enable them. Moreover, even the products the company sells (i.e. the policies) essentially exist as virtual entities in a software system responsible for the policy administration. In practice, each of these application areas has its own software application. Typical mechanisms for managing the information and keeping the data up-to-date include:

- Manual input followed by automatic processing. Example: An employee of the company enters the key data of a new policy a customer has bought, such as the kind of contract (e.g. motor insurance) and the insured risks. Afterwards, the system automatically calculates the premium and stores all the information.
- The applications are interconnected. Example: The user of the system responsible for policy management has a look at some data related to a particular policy stored in the policy system. The policy system receives information from the collection system as to whether the premium for a policy has been paid, and displays this additional relevant information to the user.
- Receiving or sending data from/to external systems. Example: A policyholder's car has been damaged due to an accident. The customer enters the data relevant for compensation via a web browser from his home. This information is transferred to the claims system via the Internet for further processing.
- Automatically triggered processes. Example: The policy system automatically checks once a month for which policies the premium is due and provides the collection system with the necessary information in order to collect the money.

▶ **Service-oriented architecture (SOA).** In large companies, in particular, it is common for business-enabling software to consist of quite a large number of individual software applications. Moreover, each of these applications is often realised by means of a different technology. There are three examples of business applications in an insurance company:

- The "business partner" system is realised by means of modern Java technology and handles the names and addresses of all persons who interact with the company: individuals who own an insurance policy with the company, persons who have claims against the company, and many more.
- The "policy" system runs on a legacy system based on an old-fashioned technology that dates back several decades. It manages the insurance policies sold by the company. It holds information for each policy, for instance concerning the kind of contract (e.g. motor insurance or casualty insurance), the duration of the contract, the insured risks and the money to be paid in the event of a claim.

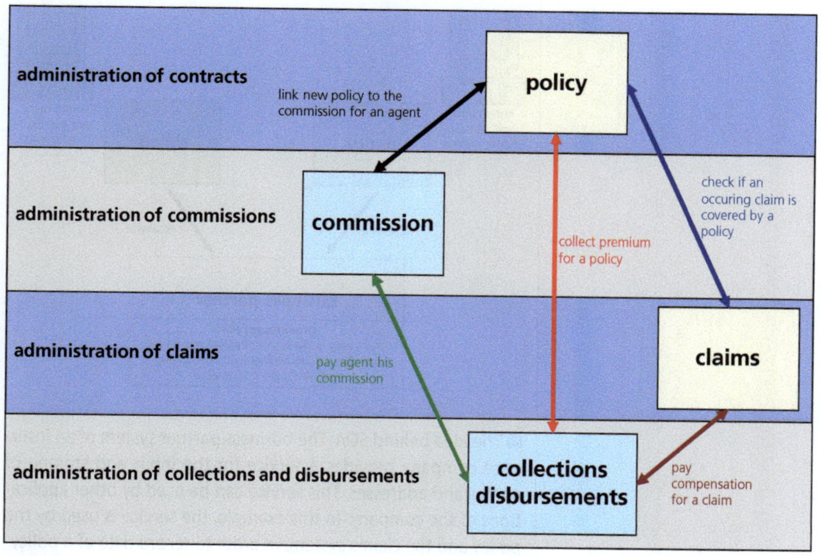

Example of information management: Some relevant tasks of the value chain of an insurance company and their corresponding software applications. The tasks in each of these application areas are executed or supported by their own logical tier or (in practice) even their own software application (represented by a box). The applications have to be interconnected (arrows), since a task executed by a certain system triggers activities belonging to another system. The text aligned to each arrow represents a relevant example for each case. Source: BearingPoint

- Finally, the "claims" system is again realised by means of Java technology. This system manages information concerning claims against the company such as: objects and events connected to a claim (e.g. damaged car due to a traffic accident), information concerning the legal validity of a claim (e.g. checks as to whether the claim is covered by an existing valid insurance policy) and the money that has to be paid to the claim holder.

The three systems outlined above do not operate independently, but have to exchange information. If a user of the policy system enters the data for a new policy, he also needs to enter the name and address of the policyholder. These data, however, are managed by the business partner system, and need to be automatically linked to the concrete policy the user is working on. Similarly, if a user of the claims system enters information concerning a concrete claim against the company, he has to enter the name and address of the person who is claiming compensation. Again, the name and address are managed by the business partner system and an automatic link has to be established between these data and the entered claim.

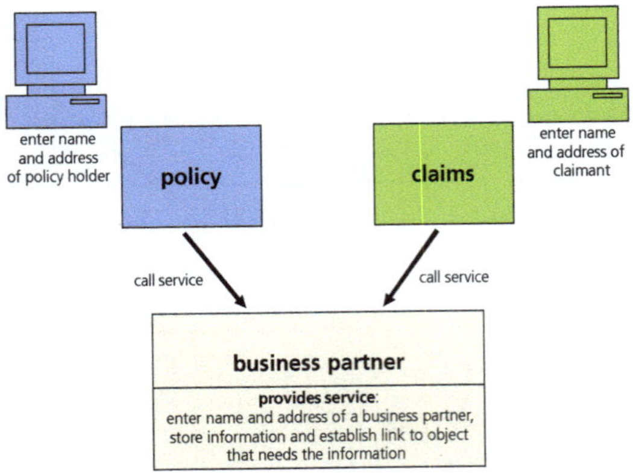

The idea behind SOA: The business partner system of an insurance company provides a service for the input and storage of names and addresses. This service can be used by other applications of the company. In this example, the service is used by the policy and the claims system, in order to record data of a policyholder and a claimant, respectively. Exploiting this concept, further systems (within or outside the company) can use the service, too. Source: BearingPoint

One possible solution would be to develop two specific software packages, each of which fulfils the above requirements and takes the various technologies involved into account: one package would enable users of the policy system to enter the name and address of the policy owner in the business partner system, and would establish the necessary link between these data and the policy. Similarly, the second package would enable analogous actions between the claims and the business partner system.

A more intelligent solution is based on the following approach: The business partner system provides a "service" that allows users of any calling system to enter the name and address of any business partner and to establish a link to the object (e.g. a policy or a claim) for which business partner information is required. This service is not restricted to the calling system nor to its specific software technology and can thus potentially also be called by further applications (even outside the company) that need the described functionalities – for instance by a system that handles the information about the agents working for the company and the commissions they receive.

Now, if several software applications communicate in a meaningful manner by various services as described above, we talk about a *"service-oriented architecture" (SOA)*. Establishing an SOA essentially poses two challenges: a logical and a technological one. The latter means that we need a "bridge" technology in order to connect applications of different software technologies. Various solutions of this type exist – a relatively common approach is to use the SOAP protocol specification. From the logical point of view, we have to design each service in such a way that various calling systems can potentially use it. Moreover, all services involved have to "play" together so that the whole system of services makes sense. This is called "orchestration" of the services, in analogy to an orchestra where the playing together of a large number of musical instruments results in a meaningful performance. Such a concept offers enormous flexibility in view of potential future requirements due to market changes. For instance, it reduces the effort involved if and when (old) software applications have to be exchanged or business processes have to be restructured.

Trends

▶ **Open-source software.** It usually takes a large number of skilled people to develop a software application. This is the main reason why software is so expensive – even when sold as a mass product. Software companies and vendors therefore have a natural interest in protecting their products, both legally and technologically (in the latter case, e.g. by applying technologies that prevent their software from being copied illegally). This often results in quite complex software licenses that regulate the use of the product as well as the corresponding source code. For many years, it was normal for customers of a software product to not even be able to look at the source code of the product they had purchased.

During the last decade, open-source software came up as an alternative approach, with the central idea that:

- the source code is freely available for anyone
- the software product may be copied, used and distributed by anyone at any time
- the source code may be changed and the changed version may be distributed freely

The open software approach has some specific advantages over conventional commercial software: the effort of developing the software can be shared by a large number of involved parties (e.g. individuals or companies) that eventually want to use it. This can indeed be cheaper than purchasing a commercial product. Moreover, users who want functionalities of a programme to be changed or customised to their own needs no longer depend on companies that own the exclusive rights to the software. In principle, they can even carry out the changes themselves or contract a third party without running into legal problems.

▶ **Java.** In recent years, Java has emerged as an important and popular software technology for modern software applications. Developed by Sun Microsystems, its essential components are the Java programming language and the "Java Virtual Machine" (Java-VM). The main reasons behind Java's popularity are:

- The Java programming language strongly supports object-oriented programming: on a logical level, the application is built up of objects that interact with one another – just like in the real world. For instance, a car in a video game is represented by logical attributes such as colour, brand, model, and maximum speed. In addition, there are algorithms (known as methods) that describe its behavior, e. g. how it moves. These properties are logically attributed just to cars, and not to any other objects.
- Java provides a wide range of what are known as application programming interfaces. This means that, for a large number of functionalities (e. g. programmes for graphical representations aimed at displaying or moving geometrical figures on the monitor), the programmer does not need to write the complete source code on his own. Instead, these functionalities are already provided by Java, and the programmer can incorporate them just by writing a limited number of commands following rules described in the interface specification.
- With Java, the vision of writing platform-independent software has been widely realized. The same Java program can in principle run on a PC, a powerful workstation or on a PDA, regardless of which operating system is installed.

The latter is a powerful property. It is enabled by a concept in which the compiler of a Java programme does not directly convert the source code into the machine code of the computer on which the programmes is intended to run. Instead, the source code is converted into a bytecode. This code is a compact sequence of bytes that represent commands for the Java Virtual Machine. The virtual machine is in turn software that plays an analogous role to the bytecode as the processor for the machine code. The virtual machine interprets the bytecode commands and executes them on the computer on which the programme is running. Exploiting this concept, a compiled Java programme (in the form of bytecode) can run on any computer on which a Java-VM is installed. Meanwhile, Java-VMs are available for most computers and operating systems, such as mainframes, Windows or MAC PCs, mobile phones/PDAs or even embedded systems. One advantage of this is that programmers can easily develop and test an application for a PDA on a PC.

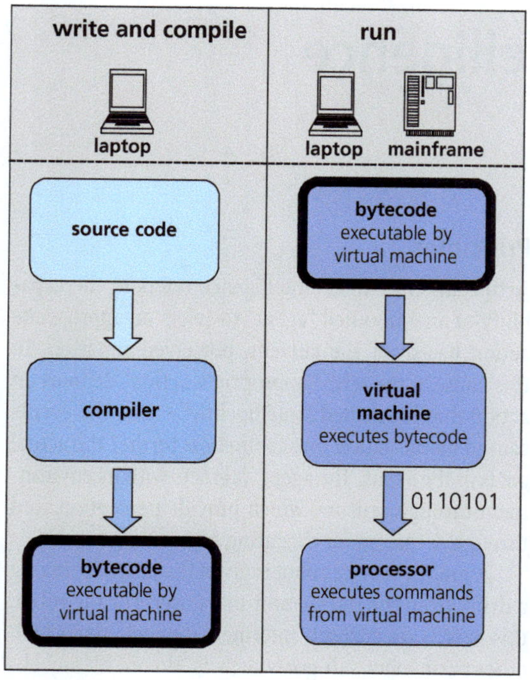

Concept of a virtual machine: The source code is compiled into a bytecode that is executable by the virtual machine. For instance, the source code can be written and compiled on a laptop (left). The compiled (bytecode) programme can run on any platform equipped with the virtual machine (right), e. g. on the laptop on which it has been compiled or on a powerful mainframe. Source: BearingPoint

Prospects

- Ever since the dawn of information technology, software has become more and more versatile and complex. This is an ongoing development.
- Software architecture helps reduce the complexity in the early design phase of an application.
- Software has to be developed following principles from traditional engineering. This is called software engineering.
- Software has revolutionised administrative tasks. Software applications have become the tools for carrying them out. Service-oriented architectures help to enable and maintain flexibility with regard to future requirements.
- Open-source software is becoming an increasingly relevant alternative to traditional software license models.

DR. UWE KORTE
BearingPoint, Düsseldorf

Internet

- www.softeng.co.nr
- www.istqb.org
- www.whatissoa.com
- www.opensource.org
- www.java.sun.com
- www.nessi-europe.eu/Nessi

Artificial intelligence

Related topics

- Robotics
- Human-computer cooperation
- Ambient intelligence
- Image evaluation and interpretation
- Sensor systems

An agent is embedded in its environment: The agent uses sensors to perceive and actuators to perform actions that result from an internal decision-making process. Actions projected into the environment may have causes. The rules or laws that govern transitions in the environment will entail certain effects. These effects may be perceived by the agent again, and the loop closes.

Principles

Artificial, "embodied" intelligence refers to the capability of an embodied "agent" to select an appropriate action based on the current, perceived situation. In this context, the term "appropriate action" denotes an action that is selected "intelligently" – i. e., the events caused by the action must somehow further the actual goals of the agent. The agent interacts with its environment through sensors, which provide perception, and through actuators for executing actions.

A mobile service robot is given the task of bringing a drink from the refrigerator to its user. For the robot, this task is its current, intermediate goal, along with other permanent sub-goals such as "do not hit any obstacles" or "keep your energy level above a certain threshold." Consequently, if the robot is confronted by a closed door, then we expect it to decide to open the door based on some form of internal processing. The robot has to detect the state "door closed" via its onboard sensors, to generate the actual decision to open it, and send an opening command to the actuators, which in turn open the door. The state "door open" / "door closed" can be perceived by the robot and changed by the robot. A robot such as this, which adapts its actions to different situations, can be seen as possessing a certain degree of "intelligence."

It is difficult to measure intelligence based on a single metric criterion, since this term covers many attributes. It is a very general mental capability that, among other things, involves the ability to reason, plan, solve problems, think abstractly, comprehend complex ideas, learn quickly and learn from experience.

IQ tests do exist, but they only cover certain aspects of cognitive skills, such as combinational thinking or the ability to concentrate. Humans' ability to use tools and play musical instruments, or other skills such as creativity and emotional or social intelligence can hardly be covered with typical IQ tests. For artificially intelligent computer programmes there exists the so-called Turing test. A human judge engages in a natural language conversation via keyboard and screen with one human and one machine each of which tries to appear human like. If the judge cannot distinguish between the machine and the human, then the machine has passed the test.

▶ **Game intelligence.** One key aspect of intelligence is the ability to make logical combinations, or "reasoning." This can be described as the capacity to solve a problem, for example, or win a game that requires intelligently-chosen actions. The winning situation is the goal that has to be achieved by the agent. The agent has to select the most beneficial actions with respect to its goals.

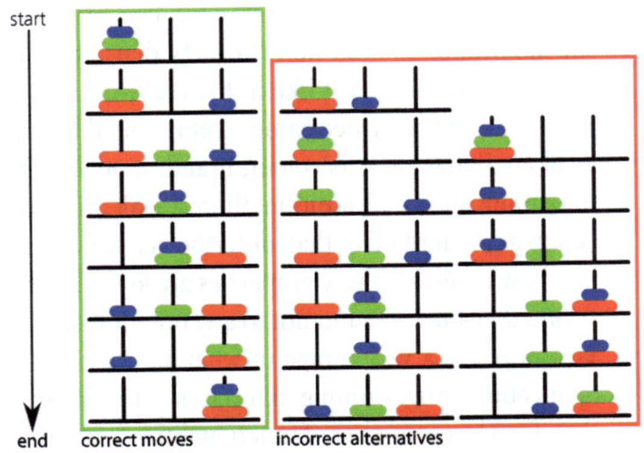

Tower of Hanoi: The figure shows the shortest possible solution for a three disc variant of the "Tower of Hanoi" game in the left (green) column. The boxes on the right contain other alternatives for each move that would also have to be considered in an exhaustive search. Endless loops are also possible, since earlier states can be reached again in the subsequent search.

In the "Tower of Hanoi" game, the goal is to move a stack of discs from the left pin to the right pin, using the centre pin as a temporary storage area. A constraining role is that a larger disc may never be placed on a smaller one. Since the number of pins and discs is limited, the possible number of disc configurations is also limited. A possible move can be modelled as a pair of two configurations: the preceding one and the resulting one. In this respect, all possible "histories" of the game can be thought of as one sequence of move pairs. In some games (or general systems), the set of all possible histories is also limited. In the case of the Tower of Hanoi game, the list is infinitely long since loops can occur, past configurations can be constructed again and again, and so forth. In some cases (such as the game Solitary) the situation is different: figures are placed on top of other figures to remove them, until only one ultimately remains. According to the rules, figures must be removed one at a time, and therefore some states cannot be repeated.

Since the number of possible actions (or moves) is

limited each time, a computer can simulate runs what may or may not lead to the desired goal configuration – the winning situation. There are two fundamentally different approaches. In a breadth-first search, all the subsequent configurations are listed and the step is repeated for all of them, and so on. In a depth-first search, one move is tried, then only one more, and so on. The moves that were already made at a given step must be stored. The Tower of Hanoi game can be solved using a breadth-first search, because in its simplest version, the depth search may get caught in loops.

Applications

► **Computer games.** There are currently many computer games in which you can play against automated opponents. A classic computer game that has been thoroughly studied is chess. Today, intelligent chess programs easily beat average chess players and often win against grandmasters. In order to benefit from the full theoretical capacity of a fast and exhaustive search, imagine a computer that performs a computational step almost instantaneously and possesses almost unlimited storage space. In this case, a simple breadth-first search of all possibilities would also be possible. This would result in the perfect player. Once the complete tree has been computed, the program would only choose those moves that lead to the final winning situation for the computer player. A hypotheses is, that two perfect player will always play to a draw.

Obviously, good human chess players were not beaten by chess programmers for decades because the assumptions stated above are not quite true. When just

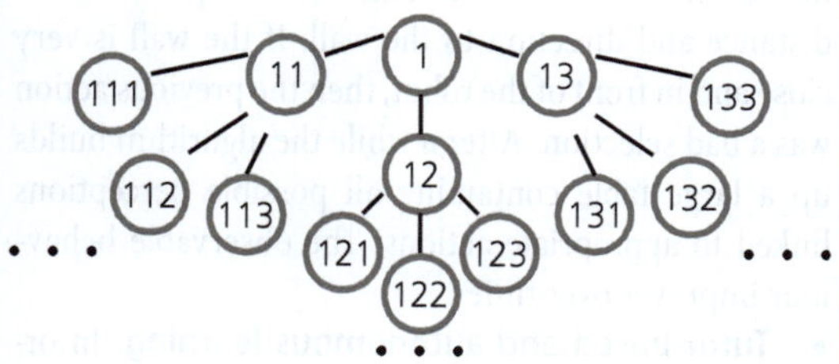

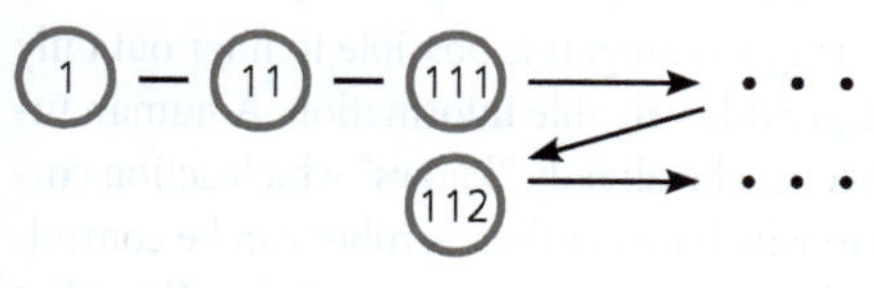

Different exhaustive search strategies: A breadth-first search (top) searches for a sequence of actions leading to the desired goal by listing all the possible next moves and then all the possible next, next moves and so forth. A depth-first search (bottom) tries only one selected move at a time until some end state is met (either a desired one or an unsuccessful one), and then returns to the last branch containing paths that have not yet been tried.

one possibility is added to the next level of search, a whole sub-tree spans of quickly growing in size. The state-space complexity in a chess game is estimated at around 10^{50} possibilities of positions that can be reached from the initial start setting. For comparison, the state-space complexity of the Tower of Hanoi with 3 discs is 27 (i. e. 3^3). There are other games that far exceed the complexity of chess, such as the Asian game Go, which has an estimated state-space complexity of around 10^{171}.There is still no computer programme available that can beat human Go masters.

The phenomenon of quickly-growing possibility trees is sometimes referred to as combinatorial explosion, and it makes it very hard to produce good moves in a short period of computing time. The human brain uses some "tricks" in order to prune the number of possibilities that need to be considered at length. For instance, pruning can be performed by selecting a "general" strategy with some loosely defined goals and/or milestones. A simple strategy may be (along with other strategies and moves in a chess game) to move a pawn towards the opponents' side of the board in order to exchange the piece for a queen.

Another way of selecting one good move out of the vast range of possibilities is by using the heuristics approach – or "rules of thumb" – that has often worked in the past. These rules are based on either individual or collective experiences. Humans are masters in the acquisition, usage and further distribution of rules that typically lead to success.

► **"Reflexes" for cleaning robots.** In the domain of real-world robotics, it is hard to translate the environment and rules that govern changes over time into a set of precise and usable computation rules, as is the case in the game examples above. Furthermore, immediate actions are sometimes required in real environments, thus ruling out the possibility of surfing through large state trees or calculating long plans. In order to build control systems for mobile robots, concepts such as "behaviour-based" or "reactive" controls exist. Parallel "behaviour modules" follow different goals simultaneously. Each behaviour module follows an individual goal that is important to the agent and suggests direct actions to the motor control parts. This method is described below using a fictional control system that could be of practical use, such as controlling a cleaning robot.

There could be one basic behaviour module, called "move", constantly sending control signals to two motors (left wheel motor and right wheel motor) that make the robot move forward. But the robot could immediately run into an obstacle or fall down (if there are stairs, for example). A second behavior module called "back up" could be introduced that mediates between

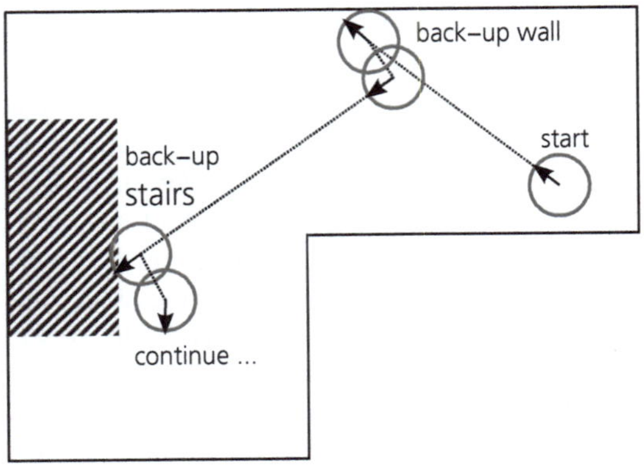

Possible movements of the "random wanderer": A simple control scheme with only two behaviour modules to cover a floor. Beginning at the starting position, the "move" behaviour causes a straight movement until the "back-up" behaviour interrupts (top) since a wall is detected. "Back-up" causes a short backwards motion and then turns in place at a random angle. After the next straight movement, stairs are detected (left) and "back-up" is activated again.

two sensor systems (inputs) and the two motors. This system is described below.

One sensor is a so-called bumper – a front part that moves slightly in relation to the chassis. Switches are mounted in between that emit a signal if the robot hits an obstacle. A second sensor system contains four IR distance sensors and looks downward to detect "lost ground." Both sensors trigger the "back up" behavior, which in turn causes the robot to make a short backwards movement and rotate randomly. The rotational movement is performed by setting the two wheel motors at the same speed in opposite rotation directions for a random period of time with preset limits. The "back up" module has priority over the "move" module. The robot could be called a "random wanderer" since it wanders around freely, continuously and randomly (due to the random turning), instantly triggering back-ups if there is a rigid obstacle in front of it or a steep slope. This control system could already be used on a cleaning robot that randomly moves around a floor.

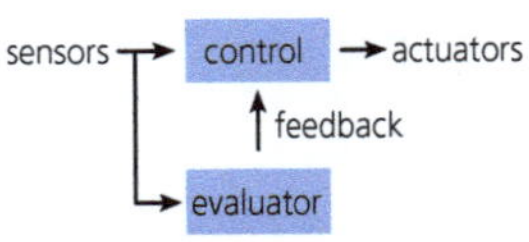

Setting up a reinforcement learning control system: The basic control within the agent receives additional feedback from an "evaluator" module. The evaluator lets a certain time interval elapse after an action is performed and then computes a feedback signal (good/bad) based on the perceived effects caused by the selected action.

A better version of the robot could include, in addition to the other sensor systems, distance sensors mounted around the chassis in order to detect obstacles further away. An "avoid" module could then be added that reduces the number of "back up" impacts that occur. It would read the distance sensors located at the front and triggers a slight curve to the left if something detected is on the right hand side and vice versa. This module has higher priority than "move", but lower priority than "back up." The "avoid" module can initiate slight or stronger curve movements by setting the two wheel motors to different speeds (with the same rotational direction) in order to prevent bomping into obstacles ahead of time.

Trends

▶ **Adaptive systems.** The term adaptive means that the control rules are not static, but that they can change over time. The system can "learn" how to achieve its objectives more efficiently or how to achieve them at all. As an example, one could extend the mobile robot control system by making the "avoid" module adaptive. It could learn how to set the internal parameters so that the robot moves as close as possible along walls without hitting into them. Such learning can be programmed by updating the utility value – a number that represents the goodness of being close to the wall and the badness of being too close to the wall – of certain actions in response to certain perceptions. To do so, the possible perceptions and actions can be converted into discrete values so that they can be numbered. For instance, if the distance sensor system consists of eight single distance sensors then one can compute an 8-dimensional state vector by using thresholds for "far", "near", and "very near." The overall number of possible perceptions is then 3 to the power of 8 (=6561).The commands can also be converted into discrete actions, e.g. "straight slow," "straight fast," "smooth curve left," "smooth curve right," "sharp curve left" and "sharp curve right." This represents 6 possible actions. The so-called q-learning algorithm then tries different actions in response to the different perceptions. It has an internal evaluation rule that computes a value for the next perception, which is the result of the current action. In the "avoid" module, this value could depend on the distance and direction to the wall. If the wall is very close and in front of the robot, then the previous action was a bad selection. After a while the algorithm builds up a large table containing all possible perceptions linked to appropriate actions. The observable behaviour improves over time

▶ **Tutor-based and autonomous learning.** In order to limit the number of samples that must be considered during learning, it is possible to filter out only those that provide valuable information. A human tutor is someone who already "knows" which action corresponds to which perception. A robot can be controlled externally (with a joystick, for example). The robot can record the externally-provided motor commands and monitor the perceptions ahead of them. If the tutor is a "good" tutor then all pairs he presents generate the correct result. This is of course a very simplified description. More complex learning approaches in modern artificial intelligence and robotics are currently being investigated. For instance, instead of using a joystick as an input device for providing good tutor samples, force-torque sensors or current measuring units can be used for robot manipulator arms. The robot arm can then be switched to a mode in which it has no physical resistance if moved by the tutor instead of by the robot itself. However, the robot records the complete trajectory of the arm movement internally. Together with the previous information provided by the sensor systems, even complex tasks can be learned.

The abstraction process has proven capable of distinguishing between the need to replicate a certain trajectory and the goal of hitting the nail. In the execution (or test) phase of the learning system, different nail positions not included in the training set were also hit. The robot needs about five demonstrations to learn that it is important to hit the nail, and that the position of the nail does not matter. An external stereo-vision system is connected to the robot in order to obtain sufficient context data (e.g. nail position, arm position, etc). The robot arm is also switched to a resistance-free mode in order to record the internal joint angle readings.

► **"Conscious" agents**. Consciousness in this context does not refer to our phenomenological experiences of the self and the associated feelings, but rather in a more technical context to explain certain underlying principles that could provide artificially intelligent agents with capabilities that fall into the scope of consciousness when referring to humans. Attempts were initially made to implement elements of consciousness theories in actual machines. The two main approaches that have been investigated and implemented on artificial systems are global workspace theory and simulation hypothesis.

Global workspace theory is a model that attempts to explain the mechanisms involved when certain items of thought become "conscious" and others not. The central concept is a working memory that contains internal speeches or images, which are selected by an attention selection scheme that can be thought of as a "spotlight." Different potential thoughts compete to enter the arena. The spotlight itself is controlled by other processes such as goal-directed reasoning. The selected content of the working memory is globally available to passive unconscious processes such as long-term memory or long-term planning. Parts of this theory have already been implemented as software systems, including robot control systems.

The simulation hypothesis is based on the idea that interaction between the agent and the environment can be produced or simulated internally, i.e. represented

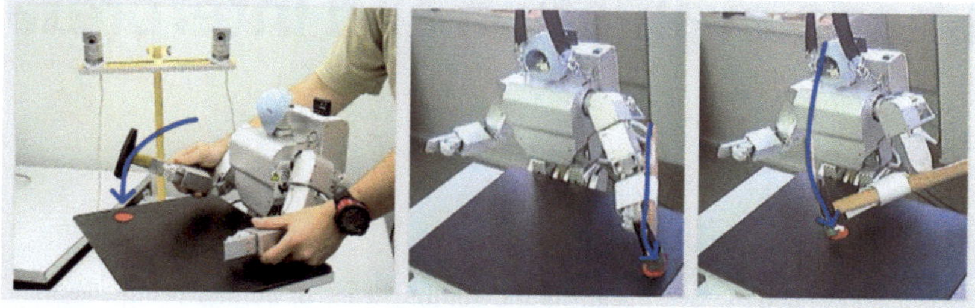

Robot and its tutor: A humanoid robot torso is trained to hammer a nail. The abstraction process must "learn" that the position of the nail varies but the effect of the hammering must be the same in all cases. Source: A. Billiard, Robotics and Autonomous Systems, Elsevier

for perceptual systems and motor systems as an internal "movie". These simulations can be used for planning future actions, to learn about opportunities or to generate new ideas. The ability to simulate is also related to the capacity to predict how the environment reacts to certain actions. This can be used to improve control systems, as they can act ahead of time. It is assumed that powerful internal simulation systems also require for a model of the agent itself. One theory describes the activation of these self-models as a basic of concious experience. Aspects of internal self-simulation have already been implemented as software systems as well.

Prospects

- Artificial embodied intelligence attempts to understand how agents that are coupled with the environment can produce intelligent behaviour.
- In limited systems such as games, it is possible to determine the correct subsequent actions through search strategies that "try" different alternatives.
- In analogy to biological control, multiple behaviours can be modelled by robot control systems, which can produce complex behaviour based on a set of simple control rules.
- Current research focuses on how learning through interaction with the environment and with a tutor can enhance robot control systems in order to provide more sophisticated cognitive abilities.
- Aspects of consciousness theory are also being investigated and implemented on artificial systems in order to provide more sophisticated cognitive abilities.

JENS KUBACKI
Fraunhofer Institute for Manufacturing Engineering and Automation IPA, Stuttgart

Internet

- http://ai-depot.com/LogicGames/Go.html
- http://chess.verhelst.org/1997/03/10/search/
- www-formal.stanford.edu/jmc/whatisai/
- http://library.thinkquest.org/2705
- http://plato.stanford.edu/entries/logic-ai

Image evaluation and interpretation

Related topics

- Human-computer cooperation
- Ambient intelligence
- Sensor systems
- Optics and information technologies
- Access control and surveillance

Principles

Sight is the guiding sense of human beings. Human eyes produce images which are processed and evaluated by the retina, the optic nerve, and the brain. This process is crucial in order for humans to be able to manage their environment. The human visual cortex can do extraordinary things in real time if we consider, for example, the ability to easily grasp and interpret even highly variable scenes with a large number of disturbing influences.

The problem involved in the technical imitation of visual cognitive abilities is that human beings are not explicitly aware of their subconscious evaluation mechanisms and are not able to transfer them directly into machine-suited algorithms. In the case of cognitive abilities – a perfect example being mutual recognition by people who have not seen each other for a long time – human beings are still superior to any technical system. If, however, the interest lies on absolute values such as dimensions, angles, and colours of industrially produced parts or on continuous and reliable image processing performance, the odds are clearly on the side of the machine.

Technically speaking, images are signals in an at least two-dimensional plane. If the image comprises three spatial coordinates, they are referred to as three-dimensional (3D) images. Such images may be obtained from tomographic reconstruction, for instance. If the plane comprises two spatial coordinates and one time coordinate, image sequences are obtained, e.g. video data. 3D image sequences are the most universal case.

Images can be divided into three value categories:

- Scalar: a value is assigned to each image point, e.g. gray level images.
- Vectorial: a vector is assigned to each point, e.g. colour/multi-spectral images or gradient vector images.
- Tensorial: a matrix is assigned to each point, e.g. images of stress fields.

With regard to location and time, an arbitrary number of compound tuples of values can in any case be assigned which can then be interpreted as generalised images. For example, optical images, radar images, and depth maps of a scene can be combined to form a multi-channel image of superior information content and expressiveness.

A typical image acquisition system consists of *CCD* (Charge Coupled Device) or *CMOS* (Complementary Metal Oxide Semiconductor) cameras which gain images in the form of a 2D illumination distribution of the scene. Since a real scene manifests itself in a wealth of different physical features such as colour, temperature, velocity, electrical conductivity as functions over time and space, it is obvious that images taken by a camera can only contain a fraction of the scene information.

Image evaluation and image interpretation endeavor to draw conclusions from images on the underlying scene, while usually a specific problem has to be solved.

However, the aim is not always a "fully automatic machine". There are tasks which require the union of a human being and a machine into a more efficient com-

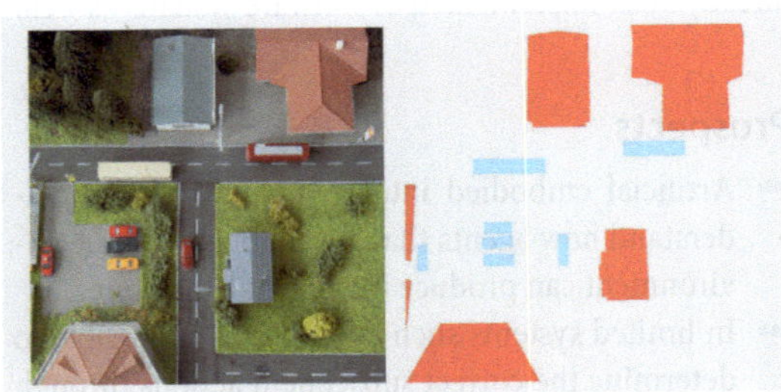

Example of segmentation. Left: Aerial view of an urban scenery model. Right: Segmentation of the scenery into vehicles (blue) and buildings (red). Source: Fraunhofer IITB

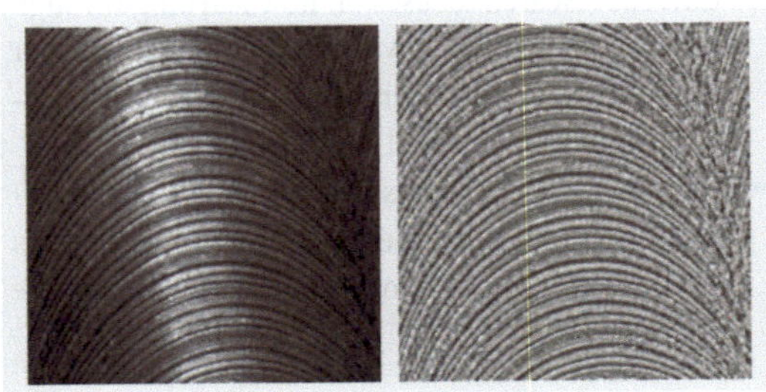

Example for preprocessing. Left: Milled surface with irregular illumination. Right: Image of the homogenisation. Source: Fraunhofer IITB

pound system. Such interactive image evaluation systems are applied to difficult problems, which require the precision and speed of a machine as well as the cognitive abilities, expertise, and intuition of human beings, or in cases where a human being has to take the final decision, as when evaluating medical images.

Automated image evaluation and interpretation typically passes the following stages:

- Automatic calibration: quantitative determination of the geometric and radiometric features of optical mapping during the recording of images; it is usually not repeated for each image acquisition and comprises factors such as the calibration of the camera with respect to the brightness of the object or scene to be observed.
- Preprocessing: suppression of disturbances, image improvement, image restoration; e.g. the homogenisation of an image, if, due to scene features, a homogeneous illumination is not possible, e.g. with regard to complexly shaped or highly reflective objects.
- Feature extraction: determination of characteristic features; locally or for the image as a whole, e.g. extraction of edges which are then used for geometric measurement.
- Segmentation: decomposition of the image into regions significant for solving a specific problem, e.g. the selection of image regions covered by vehicles in order to enable a traffic analysis.
- Object recognition: detection and classification of objects, e.g. the recognition of vehicles and determination of their current activity.
- Scene analysis: determination of relations between objects, recognition of situations, comprehension of the scene, e.g. for traffic analysis by determining the density and mutual relation of the vehicles.
- Decision: Triggering an action; e.g. separating a product which has been identified as being defective or releasing a status message that indicates a traffic jam.

However, not every phase needs to be carried out as an explicit processing step. Along the chain from the image towards a decision, irrelevant information is removed and information useful for the specific problem is worked out. The information becomes increasingly abstract and more compact and might even result in a binary decision (defective or not defective). Along this chain, helpful additional information in the form of mathematical or physical models as CAD data, for instance, as well as information from other sensors, is also included in order to increase the performance of the system.

Applications

▶ **Automated visual inspection.** Especially in automated factories, image analysis is applied for automatically inspecting the quality of the products as well as for monitoring and controlling the production processes. Examples include completeness checks during the assembly of computer boards, and inspections for technical and aesthetical defects in terms of the shape, dimensional accuracy or colour of products. In order to achieve systems with high performance and reliability, the illumination and the optical setup must be adapted to the problem to be solved. This means optimising not only image analysis but above all image acquisition. Multisensors can extend the observation to non-optical features, for instance to identify conductive or magnetic particles in automated bulk sorting systems.

▶ **Visual serving.** Image analysis is integrated in mobile systems, where it plays the role of a sensor for closed-loop control of the motion and position of the mobile system. Cars, for example, are equipped with automatic parking systems which measure the dimensions of a parking space using laser rangefinders or vision systems, controlling the steering angle and possibly the speed of the car accordingly in order to perform a safe and convenient parking manoeuvre. Spacecraft are able to perform automatic docking manoeuvres by means of vision-based navigation. In robotics, image analysing systems are increasingly being used to determine the position of the robot with respect to its surroundings (simultaneous localisation and mapping, SLAM). The combination of image-based and other

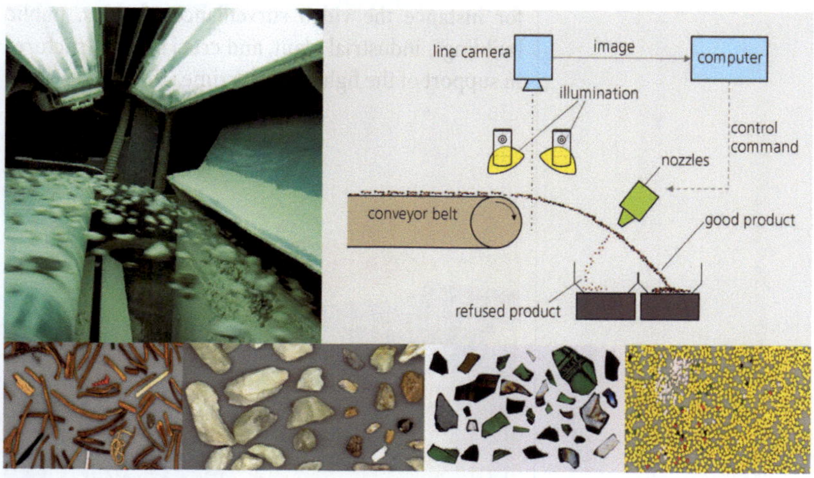

Real-time image analysis for the automatic sorting of bulk material: The bulk material being dropped from a conveyer belt is recorded by a line camera and the real-time image is inspected for criteria such as the size, shape, or colour of the particles. The well-timed triggering of pneumatic nozzles enables particles classified as defective to be blown out. This system is used for purposes such as the removal of contaminants from tobacco, the separation of ores, the sorting of coloured glass and plastic granulate material (bottom line from left to right). Source: Fraunhofer IITB

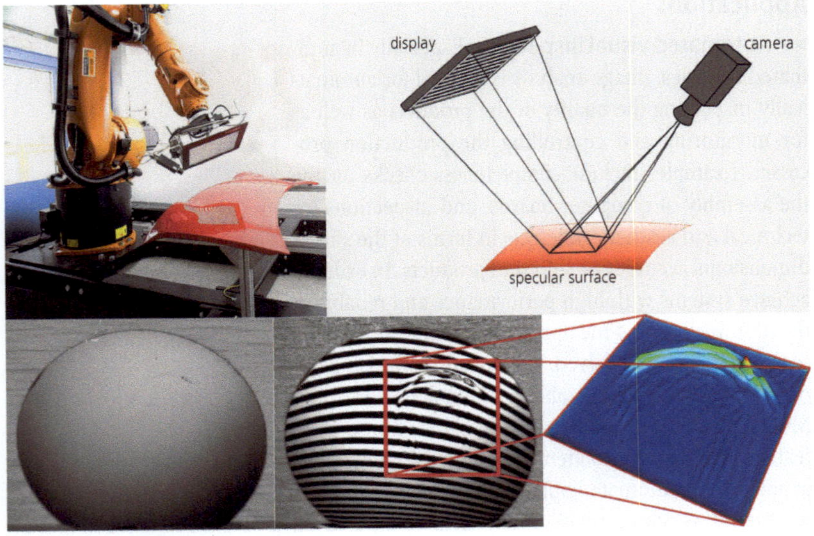

Deflectometric inspection of a specular surface for visual quality control. Top line: The surface is observed by a camera that looks at the specular reflection of a pattern shown on a display. The display and the camera combine to form a compact sensor head. When mounted on an industrial robot, a powerful inspection system for large specular parts such as car bodies is obtained. Bottom line: Deflectometry makes is possible to identify defects that would be hard to detect by any other inspection method. The images show a circular groove on a plastic lens which is hardly visible in daylight (left). The reflections of a stripe pattern in the lens reveal the circular groove and can be assessed in real time (centre). A deflectometric image evaluation shows the structure of the circular groove in detail (right). Source: Fraunhofer IITB

sensory information produces safe and highly dependable sensor systems.

► **Video surveillance.** The demand for greater security, which has increased especially over the past few years, has nurtured the market for video surveillance – for instance the video surveillance of cities, public buildings, industrial plant, and crucial infrastructures in support of the fight against crime. The reliable automatic identification of persons, critical situations and suspicious objects in video sequences based on characteristic behavioural patterns is a topic of current research and development.

► **Image-based land survey (photogrammetry).** Images of the earth's surface taken from *satellites* and airplanes are used for military and civil reconnaissance purposes. Thanks to multi-channel images covering the electromagnetic spectrum from ultraviolet across the visible spectrum to deep infrared, areas can be classified in detail on the basis of spectral reflection features that vary for different materials and geometries. Thus, for instance, the different reflections of fallow land and vegetation within the visible and the infrared spectrum can be used for classifying land utilisation, plant cover, or the expected yield of agricultural land. In addition, radar imaging techniques such as *Synthetic Aperture Radar (SAR)* supply relief data independently of the weather and can automatically detect moving objects and measure their speed based on the frequency shift of the reflected waves (*Doppler effect*). Professional image evaluation systems use robust correlation and transformation techniques to reference the diverse data with regard to existing map material (*georeferencing*). They subsequently merge information from different sources – e. g. images taken within the visible spectrum, infrared images, SAR images, images taken at different times – to form a superior global image which then serves as a basis for automatic segmentation, detection, and classification.

Trends

► **Mathematical methods.** The increasing availability of computing power makes it possible to utilise

Fused SAR (Synthetic Aperture Radar) image sequence of hill country taken from an airplane: The images in the sequence (grey) have been exactly aligned to form an image carpet, georeferenced with respect to a digital terrain model which shows the terrain relief in pseudo colours (coloured area). The result shows the SAR images at the correct location in the digital terrain model and thus facilitates the evaluation of the SAR image sequence. Source: Fraunhofer IITB

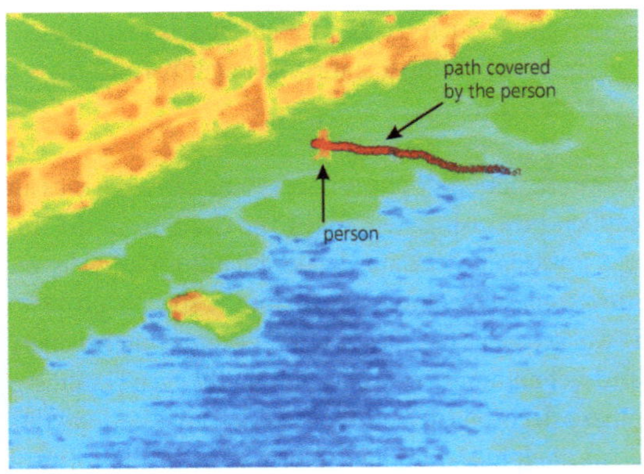

Automatic tracking of a person on a car park: An infrared image sequence enables the person to be detected and tracked on the basis of temperature, size, and motion features. Source: Fraunhofer IITB

advanced mathematical methods for image processing. Physical actions in scenes, as well as their mapping to image data, are being described by more and more realistic mathematical models that are matched with observed images. As an example, the classification of vehicles can be performed on the basis of models using parameters such as height, length or shape, which are estimated by image evaluation. Such a model-based procedure makes it possible to include prior knowledge effectively, and regularly produces better results.

An example for the inclusion of physical models is the analysis of medical ultrasound flow images, which enable conclusions to be drawn about the velocity field of the blood flow. Physics teaches that the flow field is subject to certain regularities. Thus, the observed image data must be consistent with those laws so that physics provides a further information source which can be used for validating and enhancing the image data as well as for optimising the image acquisition.

Models that are used for image analysis may also originate from computer graphics. For example, if the visual appearance of a surface of interest is expressed by a certain reflection model and acquisition constellation, computer graphics can produce a synthetic image of that surface that can be compared with a recorded image of the real surface.

With regard to tomographic problems, the raw data result from spatially integrating acquisition methods. To put it more simply, conclusions must be drawn from 2D projections on 3D quantities. This generally results in ill-posed problems which require careful mathematical treatment. While many applications for tomographs already exist in medical technology, there is a considerable demand for inexpensive systems with satisfactory 3D resolution in the area of material testing, not only for identifying even the smallest cracks in materials, but also for inspecting the contents of closed containers to detect dangerous objects and substances, for instance. Besides having a greater resolution, these tomographs must quickly produce useful results even from sparse raw data.

The introduction of such mathematical methods is promoted by the trend towards multicore hardware architectures. While only a few years ago, all processing stages from image acquisition to the final decision had to be either processed subsequently by a single-processor computer or split into several parallel tasks executed on computer networks, the multiprocessor and multicore architectures of today's computers make it easier to parallelise the complete processing sequence.

▶ **Adaptiveness**. Typically, many procedural parameters of image acquisition and evaluation, such as exposure time and filter parameters, have to be adjusted in order to receive a useful result. Any changes to conditions such as the ambient light or the object colour significantly impair the performance of automated systems. Adaptive approaches try to increase the robustness of systems by avoiding free tuning parameters. For that purpose, the values of the parameters are derived from current image data. A simple example is that of adapting to varying ambient light by automatically adjusting the exposure time based on the measured image intensity.

The concept of learning is closely connected to adaptiveness. Learning strategies aim to draw more extensive conclusions based on experience collected with an operating system. In this way, long-term structural modifications can be initiated and integrated into an image acquisition and evaluation system, for exam-

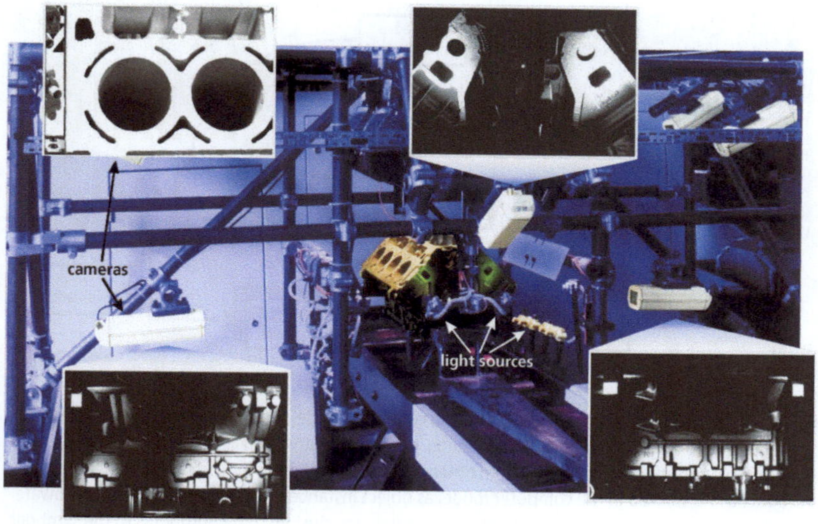

Fully automated visual quality control: 7 cameras and 25 controllable light sources inspect a cast engine block for dimensional accuracy and casting defects. The light sources are arranged and switched so that relevant object regions (e. g. cooling channels or ribs) in each camera image are illuminated with high contrast. The specific camera images (small images) are analysed in real time in accordance with the production cycle. Parts found to have defects are sorted out of the production process. Source: inspectomation GmbH

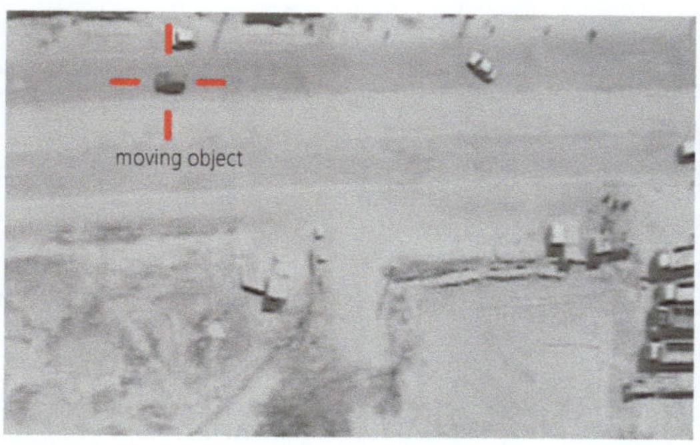

Detection of moving objects: A moving car that has been recorded in a video sequence from a fast flying aircraft is detected and marked. The detection is based on features that are characteristic for a moving car, such as its proper dynamics in the motion-compensated image sequence, or its dimensions and shape. Source: Fraunhofer IITB

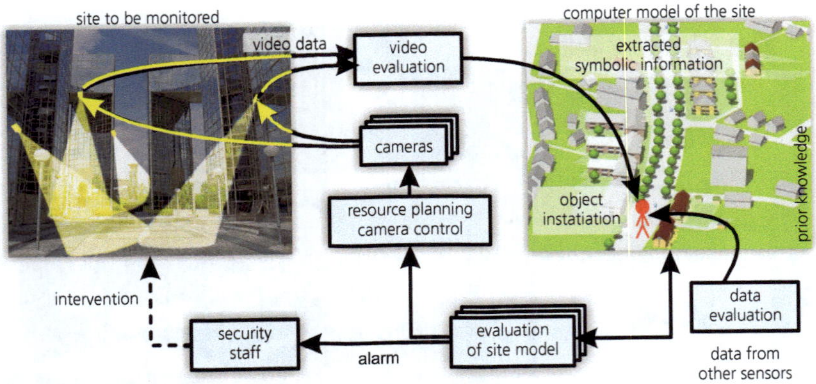

Monitoring of sites. The site is represented in a computer model containing all relevant information. Cameras collect visual information on the scene, and this is automatically evaluated to extract symbolic information about the site. Data from other sensors are evaluated similarly. Persons or objects are introduced and tracked in the computer model as object instances, thus providing situational awareness. The video data is thus not used to draw immediate conclusions, but contributes to the symbolic computer model of the site, which acts as a common memory for all sensors. Evaluating this symbolic representation makes it possible to infer actions such as acquisition commands to control the cameras. If the evaluation of the computer model reveals a critical situation, security staff can be alerted. Source: Fraunhofer IITB

ple by incorporating additional filtering steps in established processing schemes.

▶ **Camera arrays**. An alternative to single high-performance cameras are spatially arranged arrays of simpler cameras which, by employing different perspectives and potentially different optical characteristics, provide more and more diversified information about scenes. In particular, the multiple parallaxes of all cameras looking at the same objects can be used for the robust 3D reconstruction of scenes. In this context, multiple parallaxes are the different apparent displacements of close objects against their background that are caused by the different camera positions. The networked cameras represent a sensor array, which is embedded in a common coordinate system by means of geometric transformation, and can be used to generate 3D models from 2D images.

Virtual camera arrays emerge from different frames of an image sequence (e.g. a video sequence) from a moving camera. These image sequences can be fused to form mosaics, which are a generalisation of panoramic images, or can be analysed in order to extract a 3D scene description (stereo from motion). Eventually, it will be possible to combine images from different sources, e.g. fixed cameras, cameras on helicopters and satellites, maps etc., to produce a coherent patchwork. This involves various challenges: the mutual calibration of all image sources, the alignment of different types of image data, and the fusion of all images to deliver a consistent final result.

Surveillance cameras in cities, which can easily number more than a thousand, are a good example of a large camera array. Integrating these into a common system of coordinates, i.e. determining the relative camera positions, is a basic prerequisite for applications such as the automatic tracking of a suspicious person over several camera views.

▶ **Image series**. Besides the geometrical constellation of the illumination, the scene, and the camera during image acquisition, there are several optical degrees of freedom that include the spectrum, directional distribution and intensity of the incident light, polarisation, coherence, and so on. The parameters space of image acquisition is hence multi-dimensional, which consequently makes it difficult to determine the optimum values for the parameters of the task in hand. However, even the optimum single image is often not sufficient to provide a satisfying solution to a problem: Take *computer tomography*, where many 2D single images are necessary in order to reconstruct the 3D image e.g. of a brain tumour. One remedy is to

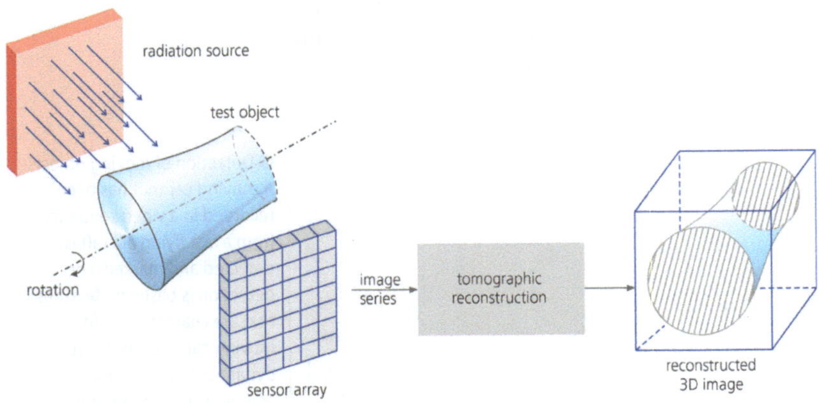

Generation of 3D images by tomographic reconstruction: The object to be inspected is rotated and subsequently projected onto a sensor array from different directions. An image series is recorded, on the basis of which the object is mathematically reconstructed. Source: Fraunhofer IITB

specifically vary the geometrical and optical parameters of an image series. Simple examples are series of images that have been acquired with illumination from various directions in order to determine the surface shape and reflection properties simultaneously, or series of images with varying focus settings for generating synthetic images with an enhanced depth of focus in microscopy. In the latter case, the different focus settings ensure that all relevant sections of a 3D object are covered in focus at least once in the image series. In a kind of patchwork, the focused regions of the single images are combined, thus obtaining a resultant image which is consistently in focus.

The challenges lie in the planning of such image series, in their explorative acquisition by means of closed-loop control for image acquisition (active vision), and in their exhaustive exploitation and interpretation, for instance by fusing them to produce meaningful resultant images. Multi-channel images, such as multi-spectral images, are special cases of such image series in which all single images coincide. They play an increasingly important role in image analysis for material classification on the basis of absorption or emission spectra, e. g. for sorting recyclable waste or verifying substances in the pharmaceutical industry.

▶ **Image retrieval.** The number of images stored worldwide is enormous, and is steadily growing. In order to maintain efficient access to those images, search engines for evaluating image content are under development:

- Inquiries can be made by a textual description, for example "front view Mercedes". This requires metadata which, ideally, are automatically extracted from the images offline. As an example, the result of an automatic object recognition of vehicles can be processed to create metadata.
- Alternatively, inquiries can be made with the help of sample images such as the picture of a sunset. For that purpose, significant features of the structure, shape, colors, textures, etc. are required. These can be extracted from the images during the image search and make it possible to measure the similarity between images by means of suitable metrics. In the sunset example, the search for further images could use features such as the shape and brightness of the sun, a darker surrounding with red colouration, and a horizontal colour edge characterising the horizon.

Such techniques become increasingly important in connection with data mining and for the semantic web.

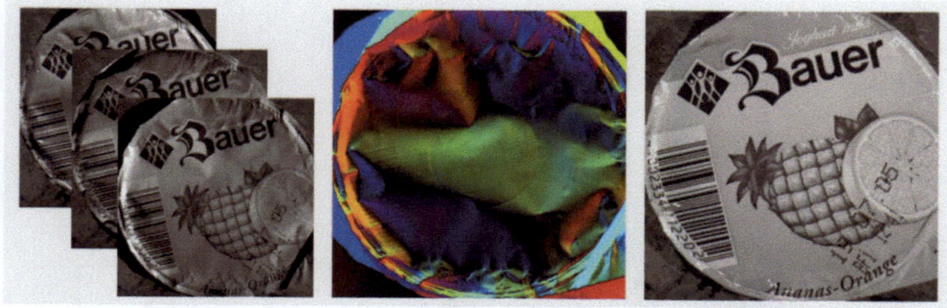

Combined acquisition of the 3D structure and reflectance of technical objects. Left: Image series of a printed deformed aluminium foil with illumination from varying directions. Centre: The first result of image fusion is a pseudo-colour image of the surface gradient (gradient vector image) permitting inspection of the seal, for example. Right: In the second result of image fusion, the printed image can be inspected without interference from the surface structure. Source: Fraunhofer IITB

Prospects

- Increase in evaluation performance: The constant growth of available computing power as well as more comprehensive physical and mathematical models make it possible to build image evaluation systems with a lower false alarm rate and fewer false conclusions. This means less uncertainty at a higher speed.
- More flexible evaluation: Adaptiveness and learning ability make image evaluation systems suitable for highly variable scenes by automatically adjusting the processing stages and parameters.
- Integral view: The inclusion and if necessary systematic manipulation of the geometrical and optical constellation during image acquisition facilitates and enhances subsequent image evaluation. Instead of solitary image evaluation components, there is an increasing demand for more complete systems, in which image evaluation solves particular subtasks in a superordinate system such as the visual perception of a robot.
- Semantic aspects: Automatic image interpretation requires formalised, machine-readable extraction and description of the image contents and structure. Automated inference processes can make use of such a representation, draw conclusions, and in consequence interpret images automatically.

PROF. DR. JÜRGEN BEYERER
DR. MICHAEL HEIZMANN
Fraunhofer Institute for Information and Data Processing IITB, Karlsruhe

Internet

- www.bruce.cs.cf.ac.uk
- www.vdma.com/visionfinder
- www.iapr.org
- vision.fraunhofer.de/en/0/index.html
- www.spie.org

4 LIFE SCIENCES AND BIOTECHNOLOGY

Biotechnology uses findings from biochemistry, microbiology and process technology to produce certain materials. This is done using *biocatalysts*, or so-called enzymes. In the process, catalytically active proteins are employed in a variety of ways:

- As a catalyst to convert substances (e. g. polylactide from starch or the decomposition of toxic contamination in soils)
- Directly as a product from industrial production (e. g. detergent enzymes or food additives)

As biotechnology comprises a very broad spectrum of applications, colour-coding is used to differentiate the various fields of application, these being denoted by red, green, white, grey, blue and yellow.

The past three decades has seen a rebirth of biotechnology, although biotechnical applications have existed for millennia: humans have used them, at times unknowingly, to manufacture foodstuffs or clothing. They were in fact a crucial foundation of human development. But it took the discovery of the structure and function of genetic material, *DNA*, to herald a new era in the life sciences. The sharp growth in the knowledge of how life functions has produced a great number of today's biotechnological applications – in the manufacture of drugs, in new diagnostic and therapeutic concepts, in the production of fine chemicals, or in methods to purify wastewater. No one doubts biotech's potential as an ecologically advantageous and economically promising technology, in many fields.

Biological methods are increasingly complementing the methods used in mechanics and electronics; in the commercial sphere a synergetic effect can be seen between established techniques and biological processes. This observation is exemplified by processes used to synthesise chemical substances, which are increasingly being replaced by processes from industrial biotechnology.

Biotechnology is destined to become more present in our everyday lives than ever before. New cellular therapy methods will be available in the medical field. It will be possible to manufacture drugs inexpensively using plants and animals acting as *bioreactors*. Medicine will be more individualised, and will include new methods in medical diagnostics, where *genechips* will find growing use.

However, some of biotechnology's applications continue to be controversial in society. These include interference with the human genome, especially when germline therapy is involved, where changes can be inheritable. As with *stem cell research*, stringent ethical standards must be put in place. An ethical framework must be set up that irrevocably establishes a connection between what can be done, what must be done and what is permitted. *Genetically-modified food plants* should be part of public debate on this issue. However, there is a demand to secure nourishment for a growing world population, and the products are meanwhile establishing themselves: In 2007, genetically-modified seeds were used for 64 % of the world's soy-bean crop.

▶ **The topics**. White – or industrial – biotechnology is still relatively young in comparison to red (pharmaceutical) and green (agricultural) biotechnology. It has two fundamental goals:

- Substitution of fossil resources with new renewable raw materials
- Replacement of conventional industrial processes with biological alternatives, which would increase efficiency in the resources employed

White biotechnology uses nature's tools for industrial production. This explains why the chemical industry, above all, is the current driver of industrial biotechnology. This development has been further accelerated by greater global competition and the resulting increase in energy and commodity prices, as well as by overall efforts to design more sustainable industrial processes.

Applying plant and *agribiotechnology* as a basis for new plant breeds represents an important resource economically for industrialised nations. While non-European industrialised nations have now come to accept the use of transgenic plants and animals, the majority of Europeans still have reservations about cultivating transgenic plants for food and animal feed and using transgenic animals in the foodstuff sector. EU legislation also reflects this attitude. According to the ISAAA (International Service for the Acquisition of Agribiotech Applications), the surface area used for the cultivation of genetically-modified crops grew sharply in 2007, by 12% to 114.3 million hectares. The leaders in the cultivation of genetically-modified crops are the USA, Argentina, Canada, China, Brazil, India and South Africa.

Biotechnological applications within the framework of stem cell technology and gene therapy have a great deal of potential. Stem cells are body cells that can grow into various cell types and tissues.

Over 200 different types of cells originate from the stem cells of an embryo. Research in the field may allow scientists to generate new tissue or organs in the future. In the field of gene therapy, new possibilities are arising to combat hereditary diseases or genetic defects. Gene therapy processes have not yet made their mark on a wide scale, but they could give rise to entirely new markets in the future: predispositions to certain illnesses might be identified, for example, allowing early therapy with customised medication.

Systems biology focuses on a quantitative understanding of dynamic life processes by creating computer-assisted models of them. It tries to understand and represent the complex and dynamic processes of a cell or an organ, for example, as it adapts to the environment, how it ages, or responds immunologically. A large amount of data has been gained from the various levels of the life process regarding individual cell components and functions (genome, proteome, metabolome), and it must all be put into an overall context. These complex processes can only be ascertained with fast computers and then transferred to the appropriate models. This has engendered an entirely new discipline known as bioinformatics. In the final analysis, this should complete the step from qualitatively descriptive biology to quantitative and theory-based biology. Nevertheless, much research effort will be needed before all biological processes within the human being have been completely clarified. There is one last riddle of the human being in particular, the brain, which is still a great mystery.

Biotechnological applications are differentiated by colour. The separate technical and scientific disciplines differ among themselves, but they do have something in common: Findings from biochemistry, microbiology and process technology can be used in the handling of microorganisms or their component parts. Source: Fraunhofer

Bionics focuses on the systematic transfer of solutions from nature to the technical world. Nature's evolutionary processes, which go back millions of years, have produced robust and optimised solutions that can serve as models when seeking to resolve technological issues today. And transferability is not limited to constructions or geometrical forms, but also to new joining methods or special communication techniques (e.g. underwater). ■

Industrial biotechnology

Related topics
- Renewable materials
- Bioenergy
- Polymers
- Food technology
- Process technologies
- Pharmaceutical research

Principles

Industrial or white biotechnology is the term used to describe the production of fine chemicals, active pharmaceutical agents, new materials and fuels from *renewable raw materials* (biomass) with the help of biocatalysts. The biocatalysts used in these production processes are either intact microorganisms or isolated enzymes.

Substances showing a high degree of functionality are defined as fine chemical products. The main distinguishing chemical features of these products are that they usually have several reaction centres and tend to create enantiomers – i.e. mirror-inverted molecular structures. Thus, typical synthesis routes for these substances consist of several reaction stages and often require the application of protective group chemistry entailing great expense, such as the use of expensive noble-metal or heavy-metal catalysts as well as of extreme reaction conditions. *Biocatalysis*, in contrast, allows the synthesis of complex compounds under far less extreme conditions, in particular as far as reaction parameters, such as pressure, temperature and pH value, are concerned.

The source materials used to manufacture products of high value, such as vitamins, antibiotics, amino acids or carbohydrates, include corn, grain, sugar beet, rape, soya and wood. These renewable raw materials generally contain complex carbohydrates (starch, cellulose and hemicellulose), proteins and fat/oil and may be converted into products of high value with the help of microorganisms and/or enzymes. The complete decomposition of high-molecular biomass (*cellulose, wood*) requires the use of additional physicochemical methods, such as heat treatment, extraction using acid solutions and supercritical gases.

Such substances are classified as bulk chemical products if the annual production volume exceeds 100,000 t. Even though biotechnologically manufactured products with a large market volume are mainly to be found in the food and animal feed processing industries, the large-scale synthesis of solvents such as bioethanol and ethyl lactate by conversion from glucose is becoming increasingly interesting from an economical point of view. Even basic chemicals such as acetone or butanol, which originally used to be synthesised using biotechnological processes before industry switched to the cheaper petrochemical route, may once again be produced using biotechnological processes in the future. This will be the case when petrochemical feedstocks become too expensive or new processes become available that completely eliminate the generation of unwanted co-products and/or byproducts.

The decisive factors likely to trigger a shift to biotechnological production are:
- Reduced consumption of raw materials and energy
- Simplified production processes: Replacement of multistage chemical synthesis by biotechnological processes (fermentation and/or enzymatic synthesis)
- Optimisation of product processing and purification as compared to chemical synthesis
- Elimination and/or reduction of byproducts and waste products

Product	Annual production [t/a]	Application
Acids		
Citric acid	1,000,000	food, detergents
Acetic acid	190,000	food
Gluconic acid	100,000	food, textiles
Amino acids		
L-glutamate	1,500,000	flavour enhancer
L-lysine	700,000	feedstuff
L-cysteine	500	pharmaceuticals, food
Solvents		
bioethanol	18,500,000	solvent, fuel
Antibiotics		
Penicillin	45,000	pharmaceuticals, feed additive
Biopolymers		
Polylactide	140,000	packaging
Xanthan gum	40,000	food
Dextran	2,600	blood substitute
Vitamins		
Ascorbic acid (vit. C)	80,000	pharmaceuticals, food
Riboflavin (B_2)	30,000	pharmaceuticals
Carbohydrates		
Glucose	20,000,000	liquid sugar
High fructose corn syrup	8,000,000	food, drinks
Cyclodextrines	5,000	cosmetics, pharmaceuticals, food

Selected products of industrial biotechnology. Source: DECHEMA 2004

▸ **Biotransformation with whole cell systems.** In the past, only whole microorganisms were used as "microreactors", e.g. Corynebacterium glutamicum in the synthesis of amino acids. The rapidly growing knowledge in the field of molecular biology now enables isolated enzymes to be used directly in catalytic processes (non-cellular systems). This is complemented by new *immobilisation* procedures for binding the isolated enzyme to a carrier substance. Immobilisation ensures that the isolated enzyme remains in the reactor and that the end-product is free of dissolved enzymes. Biotransformation with whole cell systems enables sugar or more complex carbohydrates (starch, cellulose) to be transformed into products of high value (alcohol, acetic acid, lactic acid, hydrogen and methane) without having to make use of the heavy-metal catalysts or aggressive solvents commonly employed in conventional chemical production processes.

▶ **Reactors.** To obtain an optimum yield from microbial metabolic processes, it is necessary to develop efficient techniques that stimulate the growth of the microorganisms. The most frequently applied approach involves the use of fumigated stirred-tank reactors. On account of its high stirring speed (up to 3,000 rpm) and its efficient gassing, this type of reactor benefits from a good mass transfer coefficient, i. e. it promotes intensive contact between the microorganisms and the nutrients they feed on. By selecting the appropriate type of gas (air, oxygen or hydrogen), both aerobic and anaerobic microorganisms may be cultured. The cell yield varies from species to species and lies between 100 mg (many archaeal species) and 200 g (E. coli) per reactor. The basic requirement for the successful development of an optimal bioprocess is the availability of suitably adapted online measuring systems, and the appropriate process modelling and regulation systems.

▶ **Membrane filtration processes.** Classic static filtration (dead-end-filtration), as used when filtering coffee, is applied in biotechnology for sterilisation purposes. These filters are perforated by pores measuring no more than 0.2 µm in diameter, and are thus impermeable for bacteria (size: 5 x 1 µm²). Membrane-based cross-flow filtration techniques play a major role in separation processes employed in biotechnological methods of production. These dynamic processes are used to concentrate and separate the product and allow high yields to be obtained from the processed fermentation product. These processes involve passing the solution to be separated over a membrane that divides it into two flows – the permeate and the residue retained. The permeate is the substance that passes through the membrane, leaving behind a concentrated residue of components of the solution containing substances of high molecular weight, such as proteins, resulting from the loss of water through the membrane.

Artificial membranes used in separation processes are made of polymers (polyethersulfone, polyacrylonitrile, cellulose acetate or thin layers of silicone on a polymer substrate) and are mainly produced by casting thin films. A more recent alternative is ceramic membranes produced by extrusion.

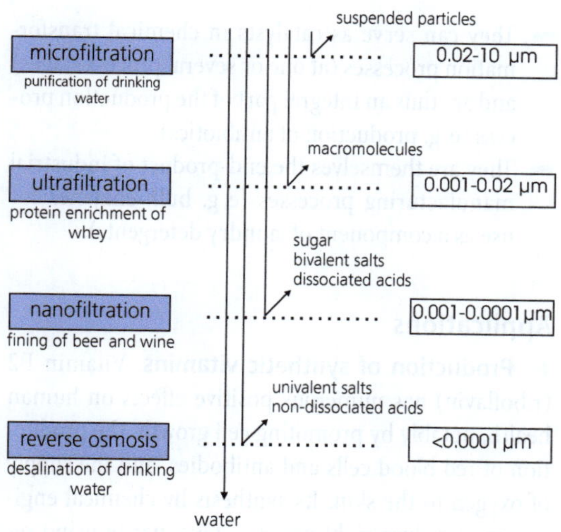

◼ Membrane filtration processes with the respective pore sizes and applications.

The membrane filtration processes are based on the following principles of separation:

- Transport through the pores of the membrane (filter mechanism, *ultrafiltration*, filtration); the moving force is the difference in pressure
- Difference in solubility and diffusion (*gas separation*, dialysis, *pervaporation*); the moving force is the difference in concentration (chemical potential)
- Difference in charge (*electrodialysis*)

▶ **Enzymes.** Enzymes (biocatalysts) are called the "motor" of industrial biotechnology. They are catalytically active proteins existing in all living organisms and responsible for the functioning of the most elementary processes of life. Enzymes are capable of carrying out complex biochemical reactions without themselves being affected or modified. They enable (bio)chemical transformations even in non-cellular systems, i. e. their activity is not restricted to living cells.

Biocatalysts work with greater precision than chemical catalysts, as they have a higher degree of selectivity, that is, they only transform certain source materials into defined products. This includes enantioselectivity, that is the ability to transform only such substances which are present in a certain spatial conformation. Many chemical molecules exist in two spatial arrangements, each of which is the mirror image of the other (enantiomers). Only biocatalysts are able to "recognise" these spatial differences and transform the correct enantiomers. One typical biotechnological product is the artificial sweetener aspartame, which tastes bitter in one conformation and sweet in the other.

Enzymes may come into play at various stages of industrial production processes:

- They can be used in the production of chemical feedstocks (e. g. acrylamide).

- They can serve as catalysts in chemical transformation processes (at one or several process stages) and are thus an integral part of the production process (e.g. production of antibiotics).
- They are themselves the end-product of industrial manufacturing processes (e.g. bulk enzymes for use as a component of laundry detergents).

Applications

▶ **Production of synthetic vitamins.** Vitamin B2 (riboflavin) has numerous positive effects on human health, notably by promoting cell growth, the production of red blood cells and antibodies, and the supply of oxygen to the skin. Its synthesis by chemical engineering involves eight process steps, partly using renewable raw materials but also involving the use of certain environmentally relevant chemicals. By contrast, the alternative biotechnological manufacturing process consists of a single fermentation stage that only requires a small quantity of artificial chemicals with a low environmental impact in addition to the renewable raw materials. The use of fungi as the main ingredient in the production process has enabled CO_2 emissions to be reduced by 30%, the consumption of resources by 60% and the output of waste by 95%.

▶ **Production of basic chemicals.** One example of the potential of enzymatic production methods in the manufacture of basic chemicals is acrylic amide, which serves as source material for the production of a broad range of chemical derivatives, both in monomeric form and in the form of water-soluble polymers. The main advantages of the biotechnological production process, in which the source material acrylonitrile is transformed into acrylic amide by hydration with the help of an enzyme catalyst, are the high degree of selectivity and the mild and environment-friendly reaction conditions. The equivalent chemical production process using copper as a catalyst, on the other hand, not only requires the removal of excess acrylonitrile and the used copper catalyst from the synthesis cycle but also generates waste in the form of byproducts and polymerisation products on account of the high reaction temperature needed of 100 °C.

Another focus of development work on biocatalytic production processes is the biotechnological production of monomer and polymer substances for the plastics and polymer industry. Here, two trends of growing economic significance are the substitution of petrochemical processes in the production of feedstock chemicals for the manufacture of plastics and the development of new biodegradable polymer products based on polylactide. In this process first starch is transformed into glucose by enzymatic hydrolysis. Then the glucose is fermented with the help of bacteria, producing lactic acid. Polylactic acid is obtained by ionic polymerisation of lactide, a ring compound consisting of two molecules of lactic acid. Ring-opening polymerisation takes place at temperatures of 140–180 °C and as a reaction to the addition of catalytic tin compounds. Thus, plastics of high molecular mass and stability will be produced.

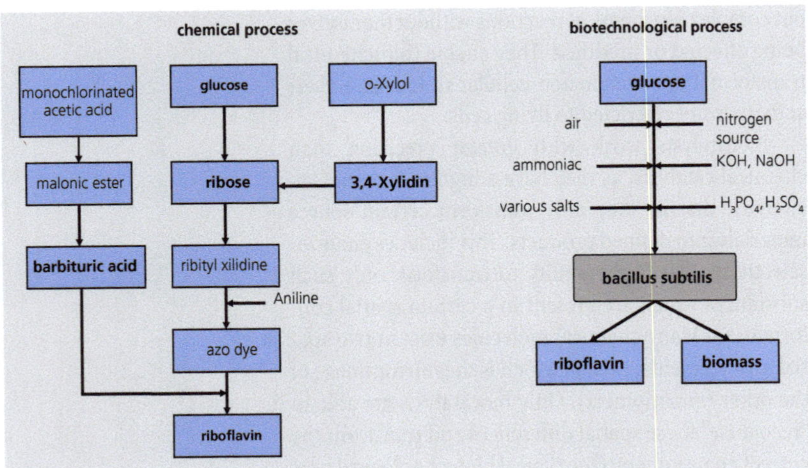

Comparison between the chemical and biotechnological production processes for vitamin B2. Left: The chemical synthesis process consists of multiple stages requiring the use of diverse environmentally relevant substances and resource-intensive processing conditions. Right: The biotechnological manufacturing process consists of a single fermentation stage that requires no additional process materials and only moderate reaction conditions. Source: Hoppenheidt

Trends

▶ **Biorefinery.** Biomass is often referred to as the "all-purpose weapon" among renewable energy sources. On the one hand this is true on account of the fact that biomass may be used for the generation of electricity, heat and fuel. On the other hand, and in addition to its significance for the food and feed market, biomass itself has also been used as a raw material traditionally – and increasingly in "new" markets as well. From the point of view of systems engineering, options which perfectly combine both a material and energetic use of a substance, thus maximising economic and ecological benefits, are especially promising. This is the case, for example, in the concept of a biorefinery using biomass to produce both raw materials to be used, for example, in the chemical and pharmaceutical industries, and fuels (e.g. bioethanol) – all of that, if possible, on the basis of organic substances which are available on the market at favourable and stable prices and for which there are only reduced potential alternative uses. The concept of such a future biorefinery implies the manufacture of diverse main products (i.e. bulk and platform chemicals) and a great number of additional products for specific small markets (e.g. fine and specialty chemicals) from biomass – comparable to the classical refinery processes.

▶ **Biocatalysis under extreme conditions.** Numerous industrial manufacturing processes require special biocatalysts which are characterised both by a high specificity and a great stability under unconventional conditions. Traditionally used enzymes soon reach the limits of their capacity and are destroyed. Enzymes generated from extremophile microorganisms, that is microorganisms living under extreme environmental conditions (e.g. in the arctic region at 0–5 °C, in hot springs at 70–130 °C, in salt lakes with a salt content of 20–30 % or at pH values between 0–1 and/or 9–12), offer a solution to that problem. The cell components of these microorganisms are optimally adjusted to their extreme environmental conditions and their properties make them interesting for a large number of biotechnological applications. Fields of application for enzymes generated from extremophile microorganisms are to be found, for example, in the detergent industry, as well as in the food, textile, paper and pharmaceutical industries. The production processes in those industries partly require temperatures above 60 °C or pH values below 4 or above 10. In addition, the enzymes generated from extremophile microorganisms are characterised by a high stability with regard to organic solvents, chelating agents, detergents and denaturing reagents, which are used in a great variety of industrial procedures and products.

▶ **Custom-designed enzymes.** In addition to the enzymes harvested in their natural state from natural resources, molecular biologists are developing increasingly targeted methods of modifying enzymes to meet the specific requirements of certain reactions or reaction conditions. In the first instance, this involves identifying new wild type microorganisms. In this context, the previously mentioned extremophiles are of special interest as examples of microorganisms that have adapted to the hostile conditions of their natural habitat. Normally, the enzymes produced by microorganisms capable of surviving under such extreme environmental conditions are only available in very small amounts and their natural host organisms often have difficulty growing under usual conditions of culture. Consequently, the genes for these often "extremely stable" enzymes (extremozymes) are isolated and transformed into appropriate production strains. The same purpose is served by databases containing the genetic sequences of e.g. soil bacteria, enzyme collections and strain collections, which can be searched using a high throughput screening method.

Basically, there are two molecular biological/genetic engineering approaches for modifying known enzymes (*protein engineering*). So-called "rational design" requires a detailed knowledge of the function-structure relationships of the enzyme and its related gene. First, the effects of specific modifications to genes and/or amino acids, e.g. in order to optimise a specific function, are visualised using computer *simulations*. On the basis of these simulations, a small number of specific variants are selected for experimental production and subsequent biochemical testing to detect any increase in activity, etc. The so-called "directed evolution" approach pursues a different strategy which, unlike "rational design", does not require detailed knowledge about function-structure relationships. Instead it imitates the natural processes of evolution. For millions of years, the living organisms on this planet have evolved as a result of naturally occurring, accidental modifications of genes and hence the proteins for which they code, combined with a process of natural selection that has ensured that only the best-adapted organisms survive ("survival of the fittest"). In the laboratory, huge numbers of mutations (10^3–10^6) are generated using various mutagenesis methods, and then the most suitable variants are identified using increasingly effective/faster screening methods. The short reproduction cycles of microorganisms allow successive generations to evolve at a very high rate, providing almost perfect experimental conditions for the molecular biologists. Robot-assisted screening systems allow more than 50,000 enzyme variants to be tested per day.

Prospects

- Using the modern technologies that have emerged in recent years (e.g. genomics, metagenomics, proteomics, metabolomics, directed evolution), it will soon be possible to employ custom-designed biocatalysts in the production of a broad range of chemicals.
- The concept of the biorefinery is analogous to today's petroleum refineries, which produce fuels and multiple products from petroleum. Industrial biorefineries have been identified as the most promising route for the creation of a new domestic bio-based industry. By producing multiple products, a biorefinery can take advantage of the differences in biomass components and intermediates and maximise the value derived from the biomass feedstock.

PROF. DR. GARABED ANTRANIKIAN
Hamburg University of Technology

Internet

- www.biobasics.gc.ca/english/View.asp?x=790
- www.suschem.org/
- www.europabio.org/positions/DSM-WB.pdf
- www.biocatalysis2021.net

Plant biotechnology

Related topics
- Industrial biotechnology
- Food technologies
- Systems biology

Principles

Modern techniques in biotechnology have rapidly expanded the horizons of plant breeding and crop improvement. Conventional plant breeding exploits mutagenesis and crossing within a species (or between closely related species) to produce new crop varieties or lines with desirable properties. In combination with improved cultivation methods, it has been possible to create crops with better yields, improved nutritional quality and resistance to stress or pathogens. In classical crossing procedures, the parental genes are randomly assorted. Since plants contain tens of thousands of genes, the alleles are extensively mixed during crossing, and selection procedures are required to identify progenies containing the combinations of alleles most likely to provide the desired properties. In contrast, gene technology enables the direct introduction of single genes resulting in a defined novel crop trait. In this process it is also possible to use genes from diverse species, such as bacteria, dissimilar plants and mammals, making the process even more target-orientated since it becomes possible to introduce traits that could never be obtained by conventional breeding.

To generate a plant with novel properties, a foreign gene (transgene), which has been isolated from another organism, must be stably integrated into the genome of a plant cell. As a result of this new information, the genetically modified plant produces a novel, recombinant protein, e. g. a pharmaceutical protein. Different transformation techniques can be used to transfer the transgene containing the novel genetic information into the chromosomes of the plant cells. Common to all procedures is that the transgene is integrated into a circular *DNA strand*, the plant expression vector. This vector contains all elements required for the transfer into the plant genome and the expression of the transgene, e. g. regulatory elements driving transgene expression. In addition, the expression vector carries a marker gene, often a gene conferring resistance to an antibiotic. The use of a marker gene facilitates the selection of genetically modified plant material since only cells carrying the marker gene and thus also the transgene can grow into intact plants on antibiotic-containing selection media.

Gene transfer into dicotyledonous plants such as tobacco, potatoes or canola is predominantly achieved through Agrobacterium-mediated transformation, exploiting the soil bacterium Agrobacterium tumefaciens. This channels the expression vector into the plant cells, then transfers the transgene and marker gene expression cassettes into the cell nucleus and from there into the plant genome.

Monocotyledonous plants, including maize, rice

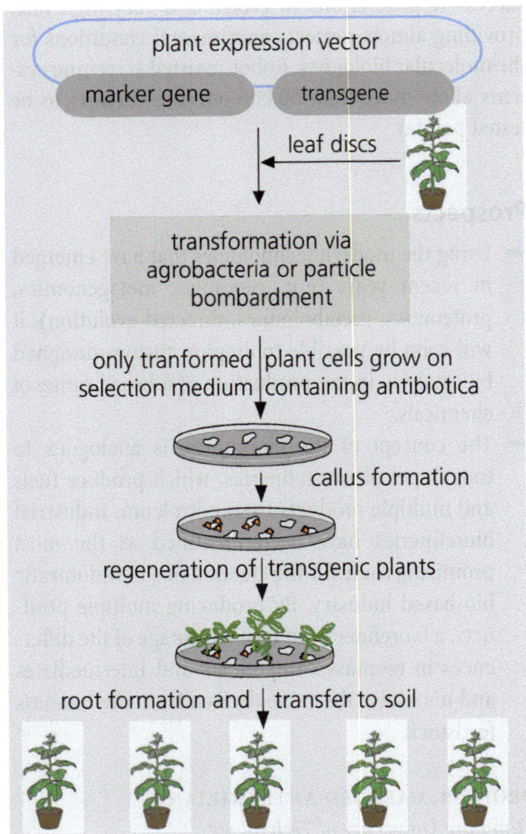

Generation of transgenic plants: Transgenic plants are most commonly generated by particle bombardment or Agrobacterium tumefaciens-mediated transformation. Both procedures enable the integration of the target gene (transgene) into the plant genome. In addition to the transgene, a marker gene is transferred which may confer resistance to an antibiotic or herbicide. Upon transformation only a tiny number of cells, which have successfully acquired the marker gene, and therefore the transgene as well, will survive on a selection medium containing the antibiotic or herbicide. Transformed cells form a callus containing undifferentiated cells. This is followed by the formation of roots and intact plants, which is induced by the addition of hormones to the culture medium.

and wheat are transformed by bombardment with DNA-coated gold particles (biolistics). In this process, the gold particles act as carriers for the plant expression vectors and enable the perforation of the plant cell wall, allowing the transport of the transgene into the nucleus where it integrates into the plant genome.

In principle, all important crop plants can be transformed by one of the above procedures. The integrated genetic information is bequeathed to the offspring so that all subsequent generations contain the novel gene's and consequently produce the recombinant protein.

► **Biosafety.** Despite the enormous potential and its advantages, green biotechnology has encountered resistance and the release of transgenic plants has been prevented in many areas. Critics fear the uncontrolled spread of foreign DNA to soil bacteria or via pollen to related wild species as well as the production of unwanted toxic compounds and unpredictable allergic reactions following consumption. It is known that the transgene can be transferred to wild species; this has been comprehensively documented, e. g. in the case of canola. However, extensive investigations have demonstrated that the cultivation of transgenic plants is safe, increasing yields and reducing pesticide use. Despite detailed analysis, the formation of toxic or allergenic substances has not been documented in genetically modified commercial crops. Interestingly, changes in protein composition are significantly higher when plants are treated with conventional breeding techniques such as irradiation, demonstrating that genetic engineering is a targeted approach to introducing novel traits into crop plants.

Furthermore, the use of marker genes conferring resistance to antibiotics and therefore facilitating the selection of transgenic plants has raised some concerns since it is not possible to rule out transfer of the resistance gene to bacteria. This additional acquired resistance would make those microorganisms difficult to control and could therefore harm animals and humans. However, the antibiotics used for selection are no longer used for medical treatment and all plants naturally teem with antibiotic-resistant bacteria with no evidence of any ill effects on animals. Alternative non-antibiotic selection markers have been developed and it is even possible now to remove the marker gene from the plant genome after transformation.

Applications

In 1996, the first genetically modified seeds were planted in the United States for commercial use. In the 2007 growing season, genetically modified crops were grown on 114 million hectar (ha) worldwide, an increase of 12 million ha over the previous year. The country with the greatest area of genetically modified crops is the United States, followed by Argentina, Brazil, Canada and India. The most important transgenic crops are canola, cotton, maize and soybean. In the EU, only one genetically modified maize event is cultivated, all in all on about 110,000 ha.

So far genetic engineering has been used predominantly to create crops with resistance to herbicides or insects. These properties are known as "input traits" and contribute to the reduction of costs reflecting the increased yield and reduced application of pesticides.

► **Herbicide resistance.** Herbicide-resistant crops are the most widely grown transgenic plants in the world. More than 70% of the genetically modified plants grown today are herbicide-resistant. Herbicide-tolerant maize, canola and soybeans contain an additional bacterial gene that facilitates the degradation of the active ingredient in the target herbicide, rendering it harmless. If farmers use herbicide-resistant crops, "non-selective" herbicides can be used to remove all weeds in a single, quick application. This means less spraying, less traffic on the field, and lower operating costs. In contrast, conventional agricultural systems can only use "selective" herbicides that can differentiate between plants that are crops and plants that are weeds. Such herbicides do not harm the crop, but are not effective at removing all types of weeds.

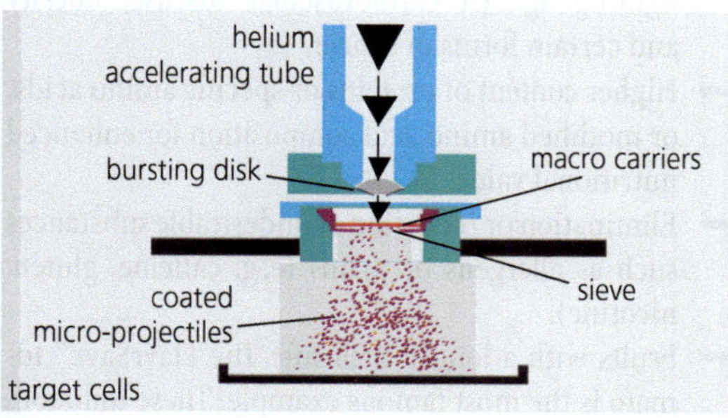

Gene gun: A suitable method of gene transfer is to bombard the leaves with tiny gold pellets to which the plasmid adheres. In a vacuum chamber, DNA-coated particles are catapulted through the walls of the plant cells by particle guns without destroying the cell walls. Helium flows into the accelerating tube until a certain pressure is reached and the bursting disc gives way. The resulting shock wave spreads out in the chamber below and accelerates a synthetic macro carrier, to which the micro carriers have been attached. While the macro carriers are held back by a sieve, the DNA-coated micro-projectiles pass through it and smash into the target cells. Source: Fraunhofer

Insect resistance: Cotton capsules damaged by insects (left and center) and genetically engineered Bt cotton producing the Bt toxin. Source: United States Department of Agriculture

Plant	Cultivation	Trait	Research and development
Canola	5.5 mio. ha (20%) in Canada, USA, Argentina	herbicide resistance	product quality, pathogen and drought resistance
Cotton	15 mio. ha (43%) in USA, China, India, Argentina, Australia, Mexico, Colombia	insect resistance (Bt toxin), herbicide tolerance	product quality (fibres, colour), cold, drought and heat resistance
Maize	35.2 mio. ha (24%) in EU, USA, Argentina, South Africa, Philippines and others	insect resistance (Bt toxin), herbicide tolerance	oil quality, drought and salt resistance, energy plant
Soybean	58.6 mio. ha (64%) in USA, Argentina, Brazil, Mexico, Paraguay, Uruguay, Canada, South Africa	herbicide resistance, fatty acid composition	product quality, pathogen, drought and salt resistance

The four most important genetically modified crops. Worldwide cultivation of genetically modified crops and their contribution to the total harvest in 2007. Source: gmo-compass

▶ **Insect resistance.** The next most common commercial trait in transgenic plants is resistance to insects, mainly used in cotton and maize. This approach exploits a gene from the soil bacterium Bacillus thuringiensis (Bt), which encodes a Bt-toxin. The plant-derived Bt-toxin kills certain insects such as the European corn borer, which can destroy up to 20% of a maize crop. Bt compounds have been used in biological plant protection for more than 50 years and have been approved for organic farming.

The use of Bt-plants has led to a significant reduction in the use of chemical insecticides and to an increase in yield. Critics claim that in some cases the use of insect resistant crops can harm beneficial insects and other non-target organisms. Extensive ecological impact assessments have been carried out to address these issues. In the field, no significant adverse effects on non-target wildlife or long term effects of higher Bt concentrations in soil have yet been observed.

Trends

As the knowledge of plant genomes continues to accumulate (at the current time, the genome sequences of Arabidopsis thaliana, poplar, rice, a moss and representatives of the green and red algae are available) and as we continue to unravel plant metabolic pathways, it is becoming possible to generate plants with improved resistance (input traits), improved quality (output traits) and novel applications (added value).

▶ **Pathogen resistance.** To date, most interest has been focused on virus resistance, but genetic engineering has also been used to confer resistance to fungi, bacteria and nematodes. The most common way of creating virus resistance is the integration of viral DNA into the plant genome. The plant recognises the foreign nucleic acid and degrades it through a process called virus-induced gene silencing. If the virus tries to infect the plants, the defense machinery has already been in progress and shuts down viral replication. There are already genetically modified virus-resistant plants on the market, e.g. papaya and squash.

Fungal resistance has been achieved by introducing genes from other plants or bacteria encoding antifungal proteins, such as chitinase or glucanase. These *enzymes* degrade major components of the fungal cell wall. As well as combating yield losses, preventing fungal infection prevents the build-up of mycotoxic compounds produced by some pathogenic fungi.

▶ **Stress resistance.** Researchers are using biotechnology to make plants better suited to environmental challenges such as drought, salinity or extreme temperatures. For example, tomatoes have been genetically engineered to withstand salty water by accumulating salt in the leaves. Other plants have been engineered with increased tolerance to drought, oxidative stress, poor soil quality, water logging, heat/solarisation and freezing.

▶ **Altered composition.** Researchers are increasingly developing transgenic crops that offer enhanced product quality or valuable new traits aimed at satisfying industrial and consumer demands. The most famous example is "golden rice" enriched with β-carotene and other carotenoids, which are precursors of vitamin A. Normal rice contains negligible amounts of β-carotene, so people that subsist on a rice-based diet often suffer from vitamin A deficiency, which leads to blindness. Through gene technology, two bacterial genes and one daffodil gene have been transferred to rice, resulting in significantly higher levels of β-carotene and other carotenoids. This has been further improved so that only one bacterial gene and one maize gene are required. A typical portion thus now provides the recommended daily allowance of vitamin A for a child.

Consumers may benefit from the next generation of transgenic plants, which could produce basic materials for industry or active ingredients, thus providing food and feed with improved nutritional value and constitutional effects, or food that reduces the risk of certain diseases. Research activities include:

- Modification of oil content and composition (e.g. polyunsaturated fatty acids such as linoleic acid or laureic acid) for maize, canola and other oil crops. These modified crops could be important in fighting against cardiovascular disease, obesity and certain forms of cancer.
- Higher content of proteins or specific amino acids, or modified amino acid composition for enhanced nutritional value.
- Elimination or reduction of undesirable substances such as allergens or toxins (e.g. caffeine, gluten, nicotine).
- Fruits with a longer shelf-life: The FlavrSavr® tomato is the most famous example. These tomatoes

were the first genetically modified fruit sold in the US and were sold as tomato purée in the UK. Apples, raspberries and melons with delayed ripening characteristics have also been developed.
- The transgenic "amflora" potato contains an increased level of amylopectin (an increase from 75–98%). This modified starch can be used in the manufacture of paper, textiles and adhesives.

▶ **Production of valuable compounds.** Plants are capable of producing an enormous number of useful substances. This spectrum can be improved by manipulating plant metabolic pathways to promote the production of specific metabolic compounds, or even expanded through the integration of novel genetic information, e. g. to produce pharmaceutical or industrial proteins (this application is known as molecular farming). Conventional systems for the production of pharmaceutical proteins include bacteria and mammalian cells. However, plants facilitate economical production on a large scale. Major plant production platforms include tobacco, maize and various others, and key products include antibodies, vaccine subunits, hormones, cytokines, blood products and enzymes. In the case of intact plants, proteins are extracted by mechanical disruption of the plant tissue followed by clarification of the plant extract through filtration and purification of the target protein via chromatographic steps. Alternatively, moss or plant suspension cells have been cultivated in bioreactors facilitating the isolation of the target protein from the culture medium. The most advanced projects in the field of plant-made pharmaceuticals are:
- A poultry vaccine based on a protein from the Newcastle disease paramyxovirus has been produced in tobacco suspension cells. Chickens immunized with the cell extract containing the vaccine showed resistance when challenged with the virus. The plant-derived vaccine was approved by the USDA in 2006.
- An antibody, which is used as an affinity reagent for the purification of a hepatitis B vaccine, has been produced in tobacco plants. The vaccine itself is produced in yeast, but in 2006 the entire production process was approved so the plant-derived antibody was evaluated with the same stringency as the pharmaceutical target protein.
- The most advanced pharmaceutical products intended for human use are gastric lipase indicated for cystic fibrosis (produced in maize), and an antibody against Streptococcus mutans adhesin protein indicated for the prevention of tooth decay (produced in tobacco), both of which have reached Phase II trials.

Golden rice: Through genetic engineering, the beta-carotene content has been increased in rice kernels. The intensity of the golden color correlates with the concentration of beta-carotene. Top: unmodified rice. Centre: Golden rice 1, Bottom: Golden rice 2. Source: I. Potrykus, ETH Zürich

- An Israeli company has produced the recombinant protein glucocerebrosidase in carrot cells, and has taken advantage of an abbreviated approval process that permits direct progression from Phase I to Phase III trials on the grounds that the drug, indicated for Gaucher's disease, falls within the scope of the Orphan Drug Act.

Moreover, plants have been exploited to produce valuable compounds for industrial applications, e. g. biodegradable plastics such as polyhydroxybutyrate (PHB) or spider silk proteins for the generation of flexible materials with a high mechanical load capacity, such as bulletproof vests.

Prospects

- Plants are an important contributor to our economic wealth. They serve as food for humans and feed for animals as well as commodities for the manufacture of clothes, paper, oils, dyes and pharmaceuticals. With the advances of the "green revolution" novel technologies such as pesticide treatment, irrigation, synthetic nitrogen fertilisers and improved crop varieties have been developed through conventional science-based breeding methods. Agricultural productivity has therefore improved significantly. However, today's challenges such as population growth and climate change require new strategies to satisfy industrial needs and consumer demands.
- Genome research and biotechnology have promoted the development of new technologies that provide the genetic information required to generate crops with high yields and improved quality. Nowadays, genetic modification is a major tool in plant biotechnology, supplementing conventional breeding and modern cultivation methods that are required for sustainable agriculture. A significant expansion of genetically modified commercial plants has taken place over the last decade and, with the development of novel traits, it is likely that this trend will continue in the future.

DR. STEFAN SCHILLBERG
Fraunhofer Institute for Molecular Biology and Applied Ecology IME, Aachen

Bioreactor: Production of pharmaceutical proteins in tobacco suspension cells under sterile and controlled conditions. The target protein is either secreted to the culture medium or retained within the cells. Source: Fraunhofer IME

Internet
- www.gmo-compass.org
- www.gmo-safety.eu
- greenbio.checkbiotech.org
- www.isaaa.org
- www.who.int/foodsafety/biotech/en
- www.epsoweb.org

Stem cell technology

Related topics

- Gene therapy
- Pharmaceutical research
- Systems biology

Principles

Stem cell is the term used to designate those cells of the body that are capable of dividing in their unspecialised form (self renewal) and yet still have the potential to develop into specialised cell types. The process by which stem cells develop into specialised cells devoted to a specific function is known as differentiation. In the course of this process, the immature, undifferentiated cells develop into specialised cells capable of performing one specific function in the adult organism. To reach this point, the stem cell has to pass through a series of differentiation stages. The morphology and function of differentiated cells is very different from that of their progenitor cells, and varies widely from one type of cell to another.

All organs and tissues in the human body originate from stem cells. Stem cells also play a major role in the natural capacity of many organs to regenerate. While there are many types of stem cells, most research in recent years has concentrated on adult (or tissue-specific) stem cells and on embryonic stem cells. The two categories exhibit many differences in terms of self renewal and their ability to develop into other cell types.

► **Adult stem cells.** Adult or somatic stem cells are found in numerous types of specialised tissue, such as bone marrow or skin. Their function in the human body is to maintain a stable state of the specialised tissues and to generate intermediate cells to replace cells that have been lost as a result of injury, allowing the tissue to regenerate to a greater or lesser extent. For example blood-forming haematopoietic stem cells help to generate sufficient numbers of red blood cells to meet the body's need for the renewal of 2.4 million such cells per second. White blood cells and platelets are similarly generated from stem cells.

► **Embryonic stem cells (ES cells).** Embryonic stem cells originate from cells of the inner mass of the blastocyst, the cluster of cells formed during the earliest stages of embryonic development up to the point of implantation of the embryo in the uterus (approximately 5 days after conception).

ES cells have the potential to develop into any cell type in the body. But their use in a research or medical context is sometimes contested on ethical grounds. Nevertheless, they offer many advantages over adult stem cells, including their unrestricted differentiation potential (pluripotency), their almost unlimited self renewal capacity, and their amenability to genetic modification. The technical processes available today allow ES cells to be developed into numerous types of mature tissue-specific cells, including neurons, heart muscle cells and insulin producing cells. Such methods are based on a combination of spontaneous and controlled differentiation processes, in which the cells are stimulated to differentiate in a targeted manner towards a specific function, through the addition of suitable factors. For body cells cultured "artificially" in the laboratory to be employable in a therapeutic context, they have to be extremely pure. If the cultured cell population contains immature ES cells, there is a risk that they might give rise to unwanted tissue growth or tumours – so-called teratomas – after transplantation. Furthermore, therapies involving the use of embryonic stem cells and derived differentiated cells carry the same intrinsic risk of rejection as any other organ transplant carried out using the methods in common practice today.

► **Nuclear transfer techniques.** The purpose of nuclear transfer techniques is to obtain pluripotent stem cells from the patient's own body. In theory, once the cell has differentiated in culture, it should be possible to carry out an autologous transplant (in which the transplant donor and recipient patient are identical). The process involves removing the nucleus from a donated non-fertilised oocyte and replacing it with the nucleus of a somatic cell (for instance a skin cell). By culturing this artificially created cell up to the blastocyst stage, it is possible to obtain embryonic stem cells

■ The early development stages of stem cells in humans: The first cells to develop by division after fertilisation are totipotent. By the time the fertilised egg becomes implanted in the uterus, these cells have reached the blastocyst stage. Embryonic stem cells are derived from the inner cell mass of the blastocyst. At this stage they are pluripotent and no longer totipotent. Source: ZUM

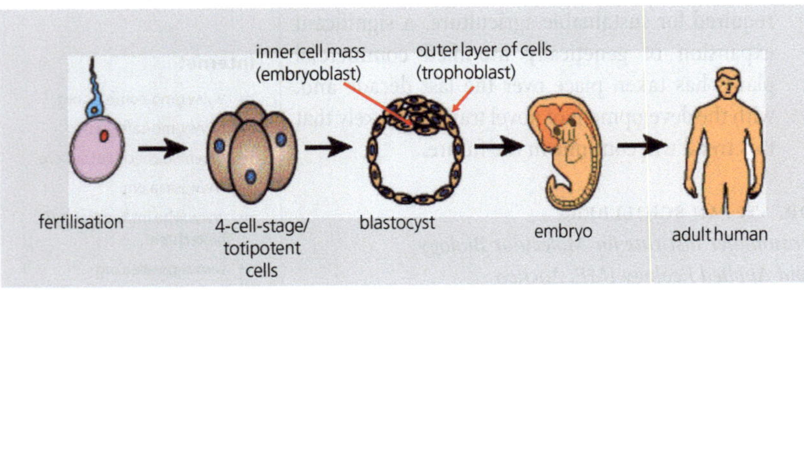

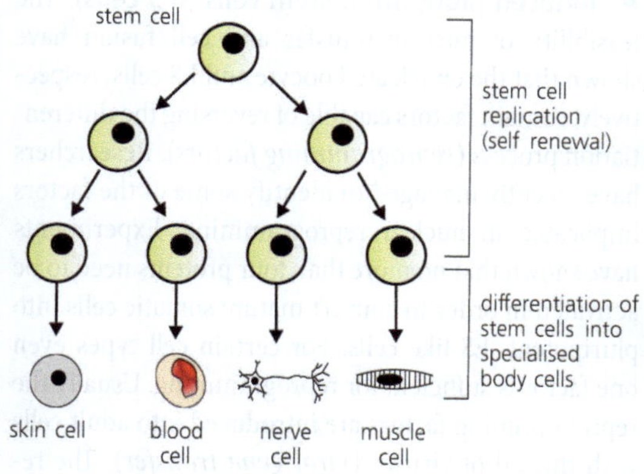

Cell differentiation: Stem cells are capable of developing into many types of specialised body cells, including blood cells, skin cells, nerve cells and muscle cells.

potency	example	differentiation capacity
totipotency	fertilised oocyte	totipotent cells can give rise to a whole organism
pluripotency	embryonic stem cells	pluripotent cells can generate virtually all cell types of the body, but are unable to form a functioning organism on their own
multipotency	blood stem cells neural stem cells	multipotent cells can give rise to a certain limited number of cell types (hematopoietic stem cells in the blood-forming system – the basis for all types of blood cell)
oligopotency	lymphoid stem cells	these cells have a limited capacity to yield specialized cells (hematopoietic stem cells are the progenitors of the lymphoid or myeloid stem cells out of which blood cells develop)
unipotency	skin cells	unipotent cells can only give rise to cells of the same type

The differentiation capacity of different types of cell: Potency is a measure of a specific cell type or tissue's ability to differentiate. The least committed of these is the fertilised oocyte, which has the capacity to develop into a whole organism. A skin cell, by contrast, can only divide to yield other skin cells – it is far more limited in terms of its differentiation potential. A cell's ability to differentiate decreases in gradual stages through the categories listed above; the transition between categories is overlapping, not absolute.

that are genetically almost identical to the donor somatic cells. Once these cells have differentiated into the required cell type – for instance heart cells – they could be used for transplantation. The advantage of this method is that it avoids the risk of organ rejection. The use of nuclear transfer is ethically and morally contested, owing to the fact that it involves the generation of a cloned precursor to the human embryo. Practical proof of the technical feasibility of nuclear transfer was demonstrated for the first time in 1997, with the birth of the cloned sheep Dolly. More recently, nuclear transfer as a method of producing pluripotent stem cells has lost much of its importance due to the progress accomplished in the generation of so-called induced pluripotent stem cells.

Applications

► **Blood-forming stem cells.** Nowadays, the use of haematopoietic (blood-forming) stem cells from bone marrow forms part of routine clinical practice. Cancer patients, for example, receive a therapy in which haematopoietic stem cells in their bone marrow are mobilised into the bloodstream by injecting growth factor G-CSF (granulocyte colony-stimulating factor), where they can be collected by taking blood samples. Reintroducing the stem cells obtained in this way permits the patient's blood system to reconstitute itself after chemotherapy, a treatment that not only destroys tumour cells but also other fast-growing cells such as haematopoietic stem cells. The same principle can be employed to collect haematopoietic stem cells from a compatible donor for transplantation to a leukemia patient whose own bone marrow cells have been eliminated by radiation. This method, known as peripheral blood stem cell collection, is employed today with over 75% of donors. The advantage of this technique is that, unlike bone marrow harvesting, it does not require general anaesthesia.

Trends

► **Patient- and disease-specific ES cells.** One of the latest trends in stem cell research relates to the generation of patient-specific or disease-specific ES cells. Disease-specific cell lines can be obtained from embryos in which a hereditary disease was identified during pre-implantation genetic diagnosis. An alternative method would be the nuclear transfer from somatic cells (e. g. skin cells) of a patient to enucleated oocytes. The great advantage of patient-specific ES cell lines is that, when they are transplanted, the patient's body does not recognise them as being foreign and therefore does not reject them. Disease-specific cell lines, on the

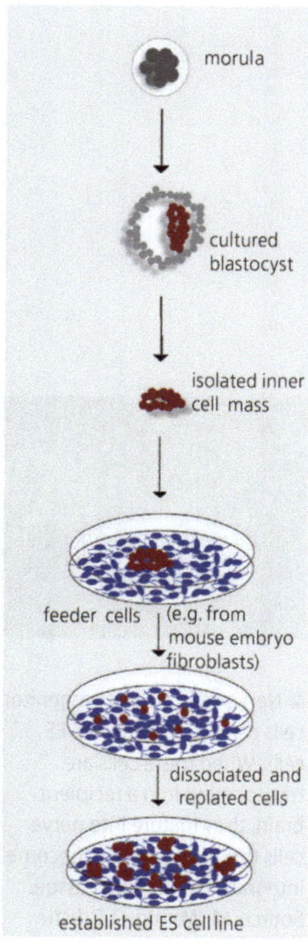

Harvesting ES cells: The first step in the process involves isolating the inner cell mass of the blastocyst. This is usually carried out by a technique known as immunosurgery, in which the outer layer of cells forming the blastocyst, the trophoblast, is destroyed by means of cell-specific antibodies. The remaining inner cell mass is then cultured in the presence of so-called feeder cells (usually mouse connective tissue cells). The feeder cells provide nutrients that support the growth of the inner cell mass and maintain the stem cells' ability to self renew. The result of culturing the inner cell mass is a new ES cell line. ES cells have an almost unlimited capacity to divide when grown in culture. Nevertheless, it is useful to be able to create new ES cell lines for certain applications. For example, by culturing disease-specific ES cell lines, researchers are able to investigate diseases at the cellular level.

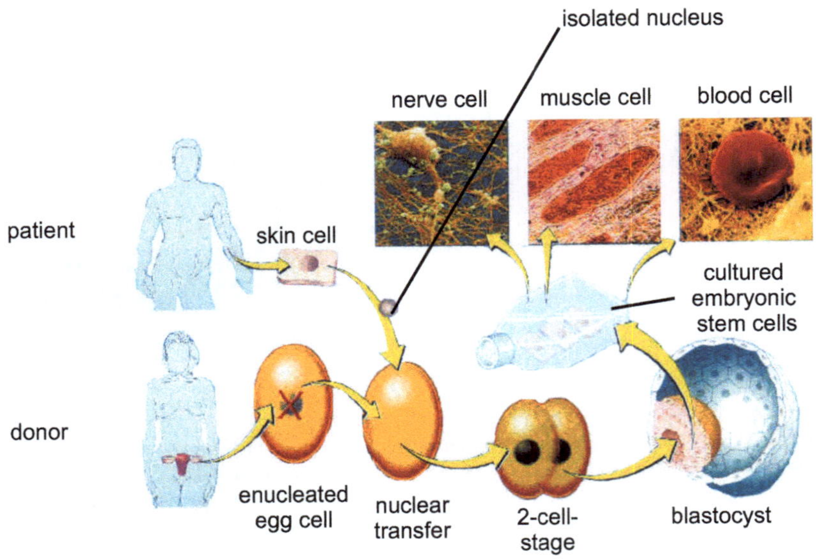

The nuclear transfer paradigm: The nucleus of a somatic cell biopsied from a patient is introduced into an enucleated non-fertilised oocyte. The resulting cell can be further cultured to develop into a blastocyst, from which ES cells can be generated and differentiated into tissue-specific cells. So far, this technology remains hypothetical and has not yet been translated to humans.

other hand, offer great potential for use in pharmaceutical research and development, because they allow human-specific pathogenic processes to be simulated at cellular level and can thus serve as more reliable disease models than the animal experiments usually employed for this purpose.

▶ **Harvesting pluripotent cells**. The two methods described above for harvesting pluripotent stem cell lines both involve destroying developing embryos at the blastocyst stage. Current research studies are therefore focusing on ways of bypassing this ethically controversial step. One possible solution is a technique known as "altered nuclear transfer" (ANT), in which a gene responsible for embryonic development in the nucleus of cells obtained from the patient is deliberately switched off. As a result, the blastocyst grown from the harvested cells after nuclear transfer is still capable of supplying ES cells but no longer has the capacity to develop into a viable embryo. For instance, the genome can be modified in such a way that the cells resulting from nuclear transfer lose their ability to implant themselves in the uterus. Another approach to avoid the blastocyst stage is to directly reprogram a mature somatic cell in order to artificially induce a state of pluripotency. There is experimental evidence that such reprogramming can be achieved by fusing ES cells with somatic cells (cell fusion). However, the resulting cells contain a fourfold set of chromosomes (tetraploidy), which prevents them from being utilised directly for medical applications.

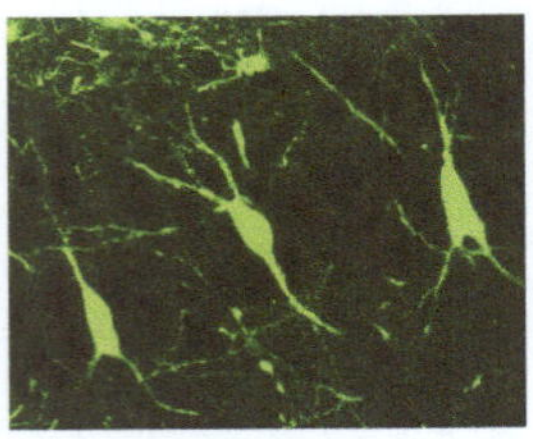

Nerve cells: Neural progenitor cells can be derived from ES cells. When these cells are transplanted into a recipient brain, they mature into nerve cells (bright green) and become integrated in the brain tissue. Source: M. Wernig, O. Brüstle

▶ **Induced pluripotent stem cells (iPS cells)**. The feasibility of nuclear transfer and cell fusion have shown that the enucleated oocyte and ES cells, respectively, contain factors capable of reversing the differentiation process (*reprogramming factors*). Researchers have recently managed to identify some of the factors implicated in nuclear reprogramming. Experiments have shown that no more than four proteins need to be activated in order to convert mature somatic cells into pluripotent, ES-like cells. For certain cell types even one factor is sufficient for reprogramming. Usually the reprogramming factors are introduced into adult cells with the aid of viruses (*viral gene transfer*). The resulting iPS cells are similar to ES cells in terms of their ability to self-renew and differentiate. IPS cells can be generated directly from patient cells, obtained, e. g. via a small skin biopsy. Since they carry the patients' own genome, their transplantation would be possible without any risk of rejection. Proof of principle of this method has already been demonstrated in animal experiments. Nevertheless, at the present time, there are limits to the therapeutic value of iPS cells in humans. Viral gene transfer and/or uncontrolled gene activation carry the risk of triggering the growth of tumours. Recent studies using mouse cells indicate that reprogramming can also be achieved with non-viral methods – albeit much less efficiently.

▶ **Cell therapy**. Research exploring methods to derive highly pure, functional tissue cells from pluripotent cells still has a long way to go. Experimental data show that it is possible to obtain functionally active heart muscle cells, insulin-producing cells and nerve cells, among others, from pluripotent cells. Tests in animal models have demonstrated the ability of nerve cells derived from ES and iPS cells to become functionally integrated in the recipient animal's brain and attenuate typical symptoms of Parkinson's disease. Similarly, in animal models, it is today possible to repair myelin defects, such as those that occur in connection with multiple sclerosis and other neurological diseases, with the help of ES cells. Yet, clinical applications still require extensive research and development, especially with respect to the generation of defined and purified tissue-specific donor cells and their functional delivery to the host tissue.

In addition to their potential use in cell replacement therapy for diseases such as diabetes, Parkinson's or myocardial infarction, progenitor cells derived from pluripotent cells offer interesting perspectives for cell-mediated gene transfer. The amenability of pluripotent cells to genetic modification enables the derivation of genetically engineered cells to produce factors with a therapeutic effect. Tissue-specific cells developed from the modified pluripotent cells could be transplanted in a diseased organ where they would serve as a vehicle for pharmacologically active substances. One such type of substance could be a neurotrophin, which promotes the survival of nerve cells.

► **Disease research and drug development.** Investigating the molecular pathomechanisms of various diseases at the cellular level is another area in which cell products derived from pluripotent cells are opening up new prospects. Such cellular disease models are useful for testing drug efficacy and toxicity directly in the relevant target cells. Previously inaccessible or poorly accessible cell populations such as nerve or heart muscle cells can be derived from pluripotent human cells with increasing efficiency and purity. This will ultimately lead to screening methods that more closely match real-life conditions, by replacing the current generic cell lines with the actual target cells affected by the individual disease. The ease with which pluripotent cells can be genetically modified will permit disease-associated genes to be introduced into these cells, enabling disease-specific assay systems to be developed for pharmacological screening. The still emerging technology of cell reprogramming will allow disease-specific cells to be generated directly from the patient's skin cells. Disease-specific cell models obtained in this way could one day be used to complement or substitute animal models in many areas of biomedicine.

Prospects

- Although adults stem cells are already being clinically used they are difficult to expand and differentiate preferentially into their tissue of origin.
- Pluripotent stem cells such as ES cells have almost unlimited growth potential, and unrestricted differentiation capacity. These characteristics make them highly suitable for a wide range of biomedical applications. In particular, the possibility of creating disease-specific pluripotent stem cell lines opens new perspectives for pharmacological screening and drug development. The amenability of pluripotent stem cells to genetic modification will permit them to be used as a vehicle for pharmacologically active substances.
- The use of reprogramming techniques could allow pluripotent cells to be derived directly from the patient and used to replace cells in another type of tissue in the same patient, thus avoiding adverse reactions of the immune system. Reprogrammed stem cell lines such as iPS cells can also be expected to gain in importance in disease research, as a means of studying pathogenic processes at the cellular level.

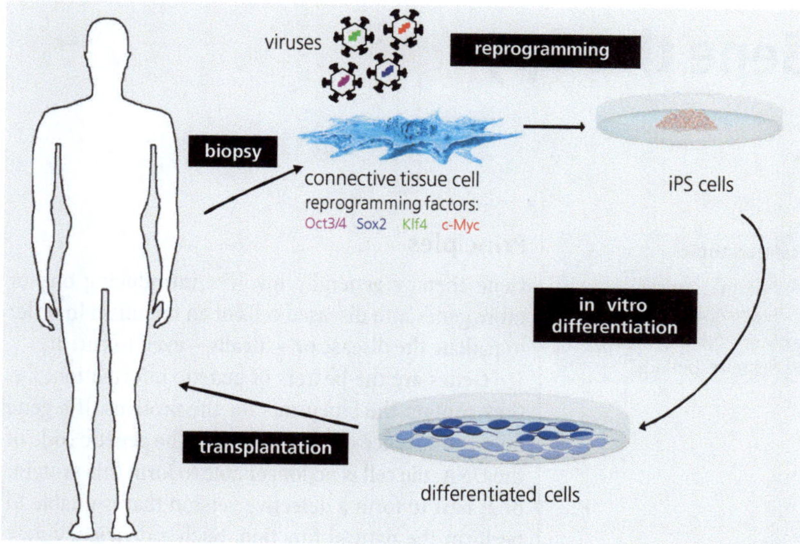

Overview of the production and use of patient-specific induced pluripotent stem cells (iPS cells): Pluripotency can be induced artificially. Cells are harvested, e.g. from the patient's skin by biopsy and cultured in vitro. The cells are subsequently reprogrammed. This generally involves the transfer of several reprogramming genes employing viral vectors. The viruses insert the protein-encoding genes in the cells, which develop into pluripotent cells. Cell reprogramming is a relatively lengthy process, and the underlying molecular mechanisms are still largely unknown. The reprogrammed iPS cells differ only insignificantly from ES cells with respect to their stem cell properties of pluripotency and self-renewal. The iPS technology could enable patient-specific pluripotent cells to be differentiated in vitro to create the desired populations for transplantation without the risk of rejection.

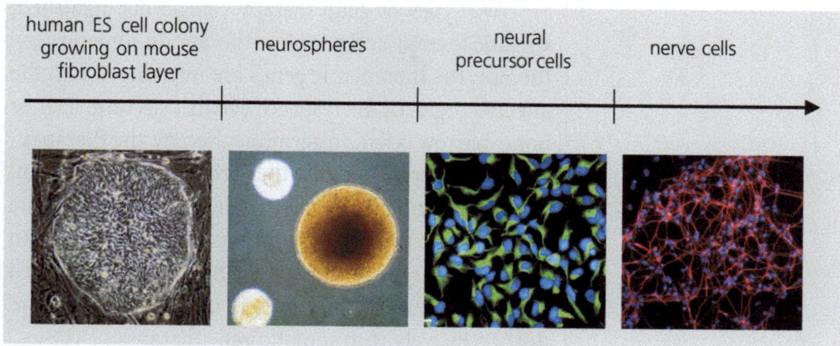

Human ES cells grow in the form of colonies: In culture, the differentiation of human ES cells can be directed via several intermediate stages to obtain nerve cells. The first differentiation stages can be multiplied in the form of aggregates (neurospheres). By breaking up these aggregates into single cells, it is possible to obtain neural progenitor cells capable of differentiating into nerve cells. In the images above, the neural progenitor cells and nerve cells are stained to show the characteristic protein components of these cells (neural precursor cells: nestin; nerve cells: β-III tubulin). Source: O. Brüstle

DR. FRANK EDENHOFER
DR. STEFANIE TERSTEGGE
PROF. DR. OLIVER BRÜSTLE
University of Bonn and Hertie Foundation, Bonn

Internet

- www.stemcells.uni-bonn.de
- www.stammzellen.nrw.de
- http://stemcells.nih.gov
- www.isscr.org
- www.eurostemcell.org

Gene therapy

Related topics

- Pharmaceutical research
- Stem cell technology
- Systems biology

Principles

Gene therapy generally involves introducing one or more genes into diseased cells of an organism in order to palliate the disease or – ideally – even to cure it.

Genes are the bearers of genetic information, i.e. they contain the blueprints for the proteins. If a gene fails, for instance due to an error in the genetic code of the DNA, the cell is no longer able to form this protein, or at best to form a defective version that is unable to perform the natural function. Such a deficiency may directly result in disorders such as heamophilia or Huntington's disease. Diseases caused by the failure of a single gene are known as monogenetic diseases. If more than one gene is responsible for the deficiency, the correct term is "multigenetic diseases". Gene therapy targets the origin of these diseases, the deficiency in the genetic material. Treatment involves introducing a faultless copy of the defective gene into the diseased cells, enabling them to synthesise the missing protein – at least to a certain degree.

Two types of gene therapy can be distinguished, somatic therapy and germline therapy. In somatic gene therapy, the therapeutic genes are introduced into the affected organ or the respective body cells (e.g. liver or bone marrow cells). In germline therapy, the therapeutic gene is introduced into the germ cells (female ovum or male sperm cell), thus passing down the genetic modification to later generations. In most European countries, germline therapy for humans is prohibited for ethical reasons. The present article will therefore deal exclusively with somatic gene therapy.

There are two basic types of treatment in gene therapy. In vivo treatment takes place directly in the diseased tissue of the organism. After being packaged in a carrier (often a modified virus), the gene is injected into the affected tissue and is absorbed by the diseased cells (e.g. the dystrophin gene is injected into the muscle of patients suffering from muscle weakness caused by damage to the dystrophin gene). In ex vivo (or in vitro) treatment using cell cultures, isolated cells are genetically modified outside the body in such a way that they have a therapeutic effect after (re-)implantation. For example, immune or tumour cells taken from certain cancer patients' own bodies are genetically modified to increase the body's immune defence against the tumour after re-implantation. If ex vivo gene therapy is performed with exogenous cells, immune suppression or encapsulation of the cells will be required.

▶ **Gene transfer systems.** The purpose of gene transfer systems is to carry the therapeutic genes into the body, or more precisely into the diseased target cells. These systems need to be as efficient as possible, i.e. to infect as many target cells as possible, while ideally excluding any risk to the patient's health. Both viral and non-viral gene shuttles are used to introduce therapeutic genes into the human body. Gene transfer with viral systems is termed "infection", that with non-viral systems is termed "transfection".

Viral gene shuttles have the advantage that the gene transfer is relatively efficient and the duration of gene expression can be set at anything from transient to stable by selecting the appropriate virus. Inherent drawbacks of viral gene transfer systems are the limited size of the therapeutic gene to be expressed, the possible immunogenicity or human pathogenicity, and their low selectivity relative to the target cells. Viruses basically consist only of their genome and a protein sheath. Their natural behaviour is to insert their genes into foreign cells. To do this, they dock onto the host cell and inject their gene material into the inside of the cell. This sets off a process in which they cause the cell's own synthesising mechanism to produce nothing but virus proteins for self-reproduction. Gene therapy makes use of this mechanism, employing viral gene shuttles to introduce therapeutic genes into diseased cells. To be useful in gene therapy, however, the viruses have to be modified so that they are no longer pathogenic but can still infect the cells. To this end, specific genes that are needed for virus reproduction and for building the virus components are removed from the viral genome and replaced by the therapeutic gene or

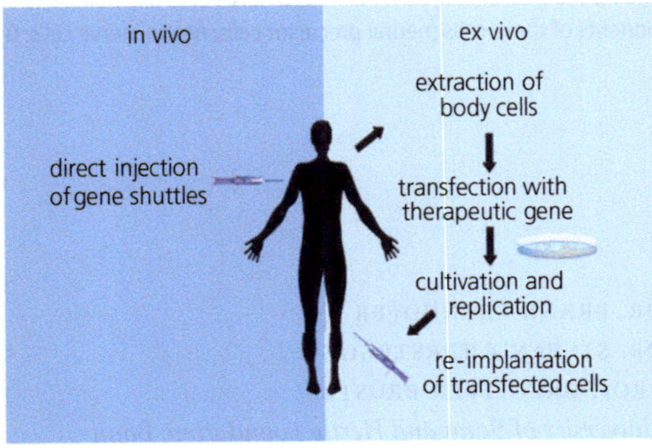

Diagram of the various forms of treatment used in gene therapy: For in vivo treatment, gene shuttles bearing therapeutic genes are injected directly into the organism. For ex vivo treatment, the body's own cells are isolated, transfected with the therapeutic gene, replicated in cultivation dishes, and re-implanted in the body. Source: VDI Technologiezentrum GmbH, ZTC genes. As gene therapy requires large quantities of viruses, these have to be replicated in special host cells (packaging cell lines). These cell cultures contain the deleted viral genes in their cellular genome, thus providing the proteins needed for the replication or packaging of the viruses.

► **Viruses as gene shuttles.** Essentially, four strains of viruses are utilised for clinical gene therapy studies: retroviruses, adenoviruses, adeno-associated viruses, and herpes-simplex viruses. They differ in terms of their gene transfer efficiency, their specificity relative to the target cells, their ability to incorporate themselves firmly in the genome of the target cell, their gene expression duration, the quantity of therapeutic proteins formed, and the maximum size of the gene sequence that can be incorporated into the virus to produce the therapeutic foreign protein.

► **Retroviruses.** Using retroviruses as gene shuttles eliminates the risk of an immune over-response. Retroviruses firmly insert their genome semi-accidentally into the genetic make-up of the host cells. The advantage of firm integration is that it normally has a lasting therapeutic effect. However, integration in the host genome also entails the risk that the expression of healthy genes (e.g. tumour suppressor genes) may be disrupted, or that pathogenic genes (e.g. oncogenes) may be activated. This phenomenon is called insertional mutagenesis. As retroviruses often integrate several copies into the host genome, the risk is correspondingly increased.

► **Non-viral vectors.** Compared to viral vectors, non-viral gene transfer systems are significantly less efficient and less cell-specific. Even the necessary transition of the therapeutic gene from the cell plasma to the cell nucleus, for instance, is a problem. This is particularly true of target cells that only have low division rates, as the nuclear membrane is dissolved during cell division. One advantage of using these systems is that the health risk for the patient is usually very low.

Non-viral gene shuttles include biochemical, chemical or physical methods. The basic principle of most biochemical or chemical systems is to shield or neutralise the negative charge of the DNA, so that the gene to be transferred can better penetrate the cell membrane, which is likewise negatively charged.

Biochemical gene transfer systems include cationic liposomes. In an aqueous phase, these liposomes form spherical structures with a positive charge on their surface. When the gene to be transferred is added, its negatively charged DNA bonds to the positively charged surface of the liposomes and can thus be absorbed into the cell together with the cationic liposomes by upturning the cell membrane. The transferred genes are subsequently released inside the cell.

Gene transfer systems			
Viral	**Non-viral**		
biological	**biochemical**	**chemical**	**physical**
retroviruses	cationic liposomes	polycationic peptides or polymers	microinjection
adenoviruses			jet injection
adeno-associated viruses	receptor targeting		particle bombardment
herpes viruses			electroporation

Viral and non-viral gene transfer systems are used to carry therapeutic genes into diseased cells. Source: VDI Technologiezentrum GmbH, ZTC

The overall process is known as lipofection. In receptor targeting, synthetic, virus-like molecular constructs are used for gene transfer. In this process, lipid molecules of the vector are aligned on special receptors on the surface of the target cell. Vector constructs can also be linked with magnetic nanoparticles, enabling them to be enriched in the target tissue with the aid of suitable magnetic fields.

In physical gene transfer systems, the therapeutic gene is introduced into the cell using physical methods. These include microinjection, jet injection, particle bombardment, and electroporation. In microinjection the gene is inserted directly into the nucleus of the target cell with the aid of a micro glass capillary. In the jet injection method, "pure DNA" is injected directly into the target cells in an air stream at high pressure. Particle bombardment involves loading small gold particles with the therapeutic genes and shooting them at the cells at high speed. While most of the particles puncture the cells, the genes are stripped off in the

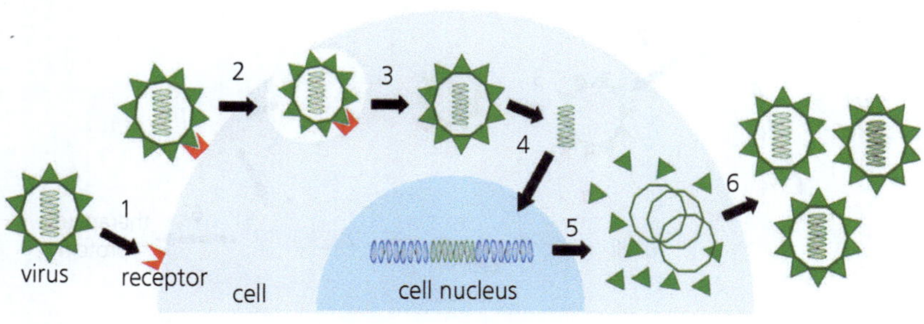

Infecting a cell with a retrovirus: The virus attaches itself to specific bonding sites (receptors) on the cell surface (1). The receptors enable the cells for infection to be accurately identified. At the bonding site, the cell membrane turns itself inside-out to enter the cell and forms a small bubble (vesicle) with the virus (2). Specific proteins help the virus in the cell to escape from the vesicle (3) and release its genes into the cell plasma. The viral genes migrate into the cell nucleus because a retrovirus integrates its genes in the genome of the host cell (4). In consequence, the primary function of the host cell is to produce new virus components or viruses (5), which are released and can in turn infect more cells (6). Source: VDI Technologiezentrum GmbH, ZTC cells and not infrequently reach the cell nucleus. In electroporation, short electrical impulses with a high field strength create short-lived pores in the cell membrane, enabling the DNA of the therapeutic gene to enter the cell.

Applications

Strictly speaking, no gene therapy applications have yet become established. 65% of the existing studies addressed cancer, 9% cardiovascular diseases, 8% infectious diseases and 8% monogenetic diseases.

► **Monogenetic diseases.** The failure of a gene occurring naturally in the cells is compensated by introducing a healthy, error-free copy of the gene into the cell. This copy then ensures that the correct protein can be synthesised again in the cell and, ideally, the disease pattern will be cancelled.

One example of this approach is the first gene therapy treatment on humans, which was conducted in the USA in 1990. Two young girls were born with the extremely serious hereditary disease known as ADA-SCID (Severe Combined Immune Deficiency). Certain cells of the immune system (T-lymphocytes) that are formed in the bone marrow lack the working gene to produce the adenosine-deaminase enzyme. The result of this ADA deficiency is that DNA replication is disrupted, damaging the B- and T-lymphocytes that are vital for immune defence. Children affected by it have virtually no protection against pathogens of any kind, and have to live from birth in a germ-free environment such as a plastic tent.

At the beginning of the somatic ex vivo gene therapy, the researchers extracted some of the girls' few remaining white blood corpuscles and enriched them in the laboratory with an already genetically modified retrovirus containing the "healthy" ADA gene. Having been adapted to carry its genetic material into the host cell as quickly as possible, the no longer replicable virus fulfilled its task as a gene shuttle and "vaccinated" the white blood corpuscles ex vivo with the healthy gene copy. These modified white blood corpuscles were subsequently placed in the girls' bloodstream by means of an infusion. The treatments were successful, but they had to be repeated regularly because the white blood corpuscles containing the new ADA gene had only a limited life span. The T-cell count increased, almost reaching normal levels again. Admittedly, while receiving gene therapy the children had continued to be treated with the drug PEG-ADA, a compound that contains the missing enzyme and can likewise cause an increase in T-cells.

These first attempts at treatment showed that, in principle, gene therapy is viable. Normal cell genes can be employed as therapeutic agents and restore the correct function of previously malfunctioning metabolic pathways. One of the two girls treated has since been able to live almost like a normal child of the same age. The second girl, however, produced antibodies against vector components and still has to be treated regularly with PEG-ADA.

► **AIDS.** A special strategy is used in the treatment of infectious diseases such as AIDS. The severity of HIV infections is due among other things to the fact that

Gene transfer system	Advantages	Disadvantages
retroviruses	• high transfer efficiency • only infect dividing cells (selectively in tumour treatment) • integration in genome, assuring long gene expression duration and inheritance by descendants • low immunogenicity	• only dividing cells can be infected • risk of insertional mutagenesis • small insertion size of therapeutic gene (8 kb maximum)
adenoviruses	• high transfer efficiency • capable of infecting even non-dividing cells • high gene expression rate	• high immunogenicity • may cause inflammations • small size of therapeutic gene (8 kb maximum) • no long-term expression
adeno-associated viruses (aav)	• no immune response • capable of infecting even non-dividing cells • partly specific integration in host genome	• small size of therapeutic genes (6 kb maximum) • partly unspecific integration in host genome
herpes viruses	• capable of infecting even non-dividing cells • larger therapeutic genes (up to 30 kb) possible	• high immunogenicity • low transfer efficiency • risk of cytotoxic infection • transient gene expression duration

Comparison between viral gene transfer systems. Source: VDI Technologiezentrum GmbH, ZTC

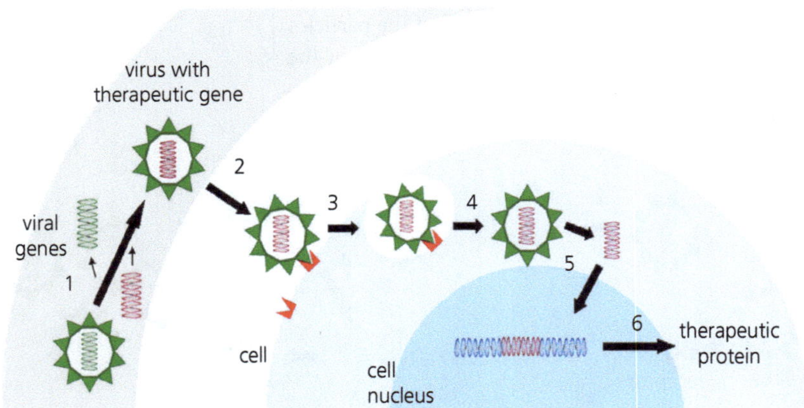

Gene therapy with a retrovirus: Before the retrovirus is used for gene therapy, viral genes are removed and the therapeutic gene is inserted in their place (1). The virus attaches itself to specific bonding sites (receptors) on the cell surface (2). The receptors enable the cells for infection to be accurately identified. At the bonding site, the cell membrane turns itself inside-out to enter the cell and forms a small bubble (vesicle) with the virus (3). The virus escapes from the vesicle (4) and releases the therapeutic gene into the cell plasma. The gene migrates into the cell nucleus and is integrated in the genome of the host cell (5). There it is read by the cell's inherent systems and the therapeutic protein is formed in the cell (6). Source: VDI Technologiezentrum GmbH, ZTC these retroviruses deliberately infect certain cells of the immune system, known as the T-helper cells. A decrease in the number of T-cells indicates the progress of the disease. T-helper cells were extracted from five patients with a chronic HIV infection who no longer responded to conventional therapy, and were genetically modified in such a way as to prevent replication of the HI virus. The gene shuttle used for this gene therapy treatment was an HI virus which had been rendered harmless, and was used as a "Trojan horse" to carry an antisense gene sequence. The treatment strategy in this case was based on the fact that the antisense gene sequence blocks the expression of a gene for a particular envelope protein (VRX496) of the HI virus, thus preventing it from replicating in the cell. The patients were given an infusion of ten billion modified T-helper cells from their own bodies, equivalent to 2–10% of the normal T-cell count. Contrary to the usual course of the disease, the T-cell count remained stable, and in two cases the number of HI viruses was even significantly reduced.

▶ **Tumour therapy.** The goal of tumour therapy is to specifically destroy tumour cells. To this end, scientists are devising gene transfer systems that carry a toxin gene which they deliberately insert into the tumour cells. When the gene is expressed in the tumour cells, they are systematically killed off. In another similar approach, the tumour is infected with a gene to produce a "suicide protein". In this case a non-toxic or only slightly toxic substance, the prodrug, is administered. The prodrug enters the tumour cells where it is converted into a toxic metabolite that kills the cancer cell. The toxic metabolites of the transfected cells are often transferred via intercellular links (gap junctions) to neighbouring tumour cells, which are then also destroyed (bystander effect).

Trends

▶ **Targeting techniques.** Different targeting techniques are being developed, for instance for viral and non-viral gene transfer systems, in order to increase specificity. What this basically involves is specifically adapting the gene transfer system to match the respective target cell. Body cells such as those of the liver possess tissue-specific surface molecules. The aim is to provide matching counterparts for the viruses or synthetic systems employed, so that they exclusively infect the desired cell type. If this proves successful, particularly in non-viral systems, it will minimise the risk of human pathogens. These systems can be further optimised by linking nuclear localisation sequences to the therapeutic gene, so that it is precisely targeted at the cell nucleus.

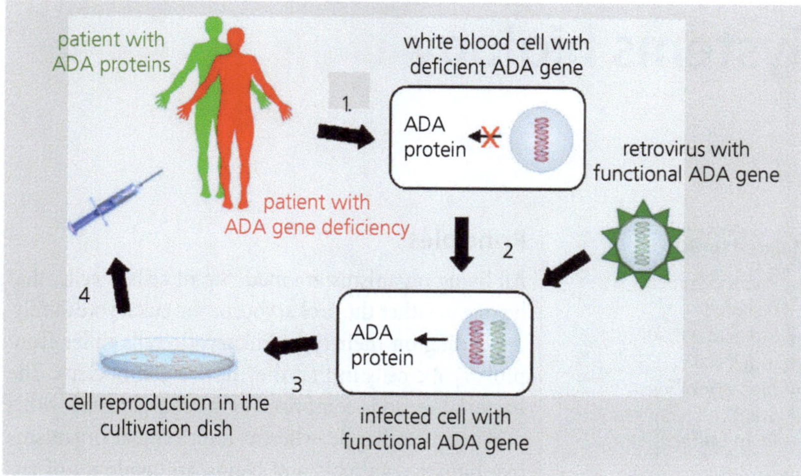

Somatic gene therapy (ex vivo) for ADA deficiency: This deficiency causes a disease of the immune system because the cells are no longer able to form the adenosine desaminase protein. Therapy consists in extracting white blood corpuscles (T-lymphocytes) from the patient that are no longer capable of producing the functional ADA protein (1). The therapeutic gene is introduced into the diseased cells with the aid of a retrovirus, restoring their ability to produce the ADA protein (2). The cells that have regained their normal function are reproduced in cultivation dishes (3) and re-implanted in the patient's bloodstream (4). Source: VDI Technologiezentrum GmbH, ZTC

To treat monogenetic diseases, attempts are now being made not only to introduce the healthy version of the failed gene into the cells, but to specifically replace the defective copy at the original place in the genome; a real genetic repair of defective body cells is performed. The method chosen to achieve this is by using zinc finger nucleases, a type of protein with finger-like protrusions that are held in shape at the base by a zinc atom. With these fingers they can literally reach out for DNA and systematically search for short DNA sequences identified by the target gene. It is hoped that the associated DNA-cleavage nucleases will enable the defective gene variant to be removed so that the healthy gene introduced can serve as a master copy for subsequent repair.

Prospects

- The fact that no gene therapy method has become established, despite the high number of clinical studies and the long time that clinical research has been taking place, proves how difficult it really is to perform genetic interventions in humans.
- The successes that have already been achieved, and the hope of being better able to treat diseases such as cancer or AIDS, provide motivation for constant improvement of the methodical approaches.

DR. MATTHIAS BRAUN
VDI Technologiezentrum GmbH, Düsseldorf

Internet

- www.genetherapynet.com/genetherapy.html
- www.ornl.gov/sci/techresources/Human_Genome/medicine/genetherapy.shtml
- http://learn.genetics.utah.edu/content/tech/genetherapy

Systems biology

Related topics

- Pharmaceutical research
- Molecular diagnosis
- Industrial biotechnology
- Stem cell technology
- Plant biotechnology

Principles

All living organisms are made up of cells – cells, that belong to either the prokaryote or the eukaryote family, depending on their type. Eukaryotic cells differ from prokaryotic cells in that they have a cell nucleus. The bacterial world is comprised of microorganisms with a prokaryotic cell type, whereas multicellular organisms like humans, animals and plants are made up of eukaryotic cells. If we want to understand the life processes of an organism, we have to analyse its constituent cells at the genome, transcriptome, proteome and metabolome levels. The genome comprises all the genes in a cell, while the transcriptome comprises all the transcripts (messenger RNAs), the proteome all the proteins, and the metabolome all the metabolites (metabolic products). The transcriptome, proteome and metabolome can all vary greatly depending on the condition of the cell. There are a number of inter-relationships between the genome, transcriptome, proteome and metabolome levels. The genome of an organism contains a large number of genes. *Gene transcription* creates transcripts, which in turn provide the blueprint for protein biosynthesis, a process known as translation. The majority of the synthesised proteins then act as biocatalysts and control metabolite biosynthesis. Cellular activities are highly regulated. Genes can be switched on or off, depending on the environmental conditions; the life span of transcripts and proteins is adapted according to requirements; and the biocatalytic activity of *enzymes* can also be modified. Thanks to genome and post-genome research, data sets that provide information on genes, transcripts, proteins and metabolites can now be produced at cellular level. However, it should be noted that these data sets merely provide snapshots of the condition of a cell under specific environmental conditions. Systems biology is currently seeking to create models that capture the interdependency of these data sets and describe it mathematically. This makes it possible to predict the behaviour of the analysed cells; the next stage is to conduct experiments to confirm the predictions obtained from modelling. However, systems biology is still a new discipline. In many of the areas discussed above, a lot more groundwork remains to be done before we can herald systems biology as a success.

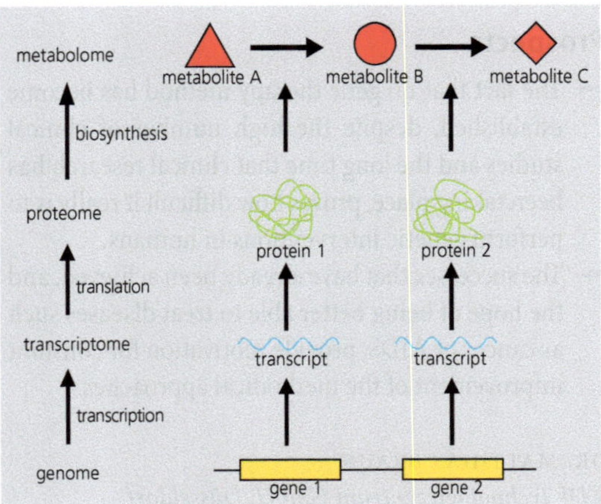

Interconnection of the genome, transcriptome, proteome and metabolome levels in a cell: Using genes, transcripts, proteins and metabolites, the diagram shows how the various levels are interconnected within a cell. Transcripts are created by gene transcription. Proteins are formed by the translation of transcripts. And finally, as biocatalysts, proteins are involved in metabolite biosynthesis. In this instance, two individual genes in a genome are shown, which produce two transcripts (mRNA) after transcription. These two transcripts are then converted by translation into two proteins, which are enzymatically active and synthesise metabolite C from metabolite A via one intermediate stage (metabolite B) in a metabolic pathway. Source: Bielefeld University

► **Genome research.** Once the structure of the hereditary molecule had been clarified in 1953, and the genetic code deciphered roughly a decade later, work on identifying the nucleotide sequence of DNA fragments started making progress. The first sequencing methods involving individual *DNA strands* were published in 1977, with Maxam and Gilbert describing the chemical degradation procedure and Sanger publishing the chain termination method. Sanger's chain termination method was the one that caught on, because it could be automated. Until recently, Sanger technology was used to determine the genome sequences of bacteria, yeasts, fungi, plants, animals and humans. Depending on the size of the genome, this often involved hugely complex projects, some of which were comparable to major industrial programmes. However, the latest technological breakthrough in sequencing has both radically shortened the time it takes to determine genome sequences and made the process cheaper. This breakthrough is based on a new analysis technique that can be massively parallelised. Sanger's chain termination method produces classes of individual strands that ultimately all have the same base although they vary in length; the position of their bases is then determined by measuring the length of the individual DNA strands using gel electrophoresis. This process could only be parallelised to a limited extent, though – automated sequencing machines were operated in parallel, perhaps involving up to only a hundred such gel electrophoreses in narrow capillaries. The new technology has now moved away from this cumbersome method of gel electrophoresis. Instead, it uses short DNA fragments that no longer have to be cloned (in contrast to the Sanger method), but simply amplified on a small bead or solid substrate by means of a *polymerase chain reaction (PCR)*. This process creates a large number of identical DNA fragments that are either bound to a small bead or clustered on a solid substrate. The actual sequencing is then done using DNA synthesis, in which various methods are employed to detect the individual replication stage on a small bead or in a cluster. This way of doing things means as many DNA fragments as desired can be sequenced simultaneously. These days, it is possible to process several million DNA sequencing reactions in parallel, so that each sequencing run produces several billion sequenced bases. Consequently, it is now possible to sequence absolutely any gene, no matter its size, without having to invest unacceptable amounts of time and effort in the process.

▶ **Transcriptome research**. Once an entire genome sequence has been obtained and all the genes in an organism have been identified, the next issue arises, namely its activity in relation to environmental conditions. This activity is determined by the number of gene transcripts. At the moment, the number of transcripts is identified using array technology, in which industrially manufactured DNA chips are widely used. This involves a massively parallel hybridisation procedure (i. e. a combination of suitable individual strands) that measures the activity status of all the genes in a cell. However, one shortcoming of this technology is that no quantitative statements can be made. But here, too, new procedures are emerging. Ultra-fast sequencing methods, in particular, are pointing the way ahead. Transcripts can be transcribed into DNA strands that can subsequently be sequenced. The transcribed genes can then be read from this sequence, and the frequency with which a specific transcript is sequenced in turn permits quantitative statements.

▶ **Proteome research**. So far, procedures are described involving all the genes and all the transcripts within a selected cell. The question now is whether all the proteins in a cell can also be recorded – this is proteome research. Proteins are first separated using protein gel electrophoresis. Then a mass spectrometer is used to identify individual protein spots. During this process, the protein to be identified is proteolytically cleaved, i. e. broken down into defined fragments using an enzyme, after which the fragment specimen can be identified in the mass spectrometer. If the genome for the organism to be analysed is available, the fragment specimen can then be used to determine the associated gene. This solves the basic problem – it is possible to clearly describe the proteins in a cell according to their correlation with genes. However, the matter of protein quantification continues to pose a huge problem that has not yet been resolved satisfactorily. Within a cell, proteins can also be modified by phosphorylation and glycosylation, and it still requires a great deal of time and effort to determine these modifications in detail. Ultimately, further knowledge is needed for all spheres of systems biology, e. g. information related to protein clusters, which can be studied using protein interactions.

▶ **Metabolome research**. This branch of research deals with the identification and quantification of metabolic products that are formed in a cell, i. e. metabolites. Here, too, the first step is to separate the metabolites. This can be done using gas chromatography, for example. They are then identified and quantified using mass spectrometry. Comparative substances are generally used for the identification process. Quantification is achieved by evaluating the peak areas that belong to the signals – e. g. for the Escherichia coli cell, an intestinal bacterium, a total of 800–1,000 metabolites is assumed. Cellular metabolic processes must be stopped abruptly at the precise moment the measurement is taken, otherwise significant distortions will occur between individual metabolite pools. It is necessary to record material flows in order to understand the metabolic process. One way to do this is to use specially-marked metabolites (e. g. C-marking).

▶ **Modelling**. Modelling lies at the very heart of systems biology. Any system to be modeled must be clearly defined in advance, since the complexity of the overarching "cell" system means it cannot yet be modeled in its entirety (and this will probably remain the case for the near future). It is more likely that a model will confine itself to partial aspects of a cell. Of course, the prerequisite is that the overall system can be broken down into individual modelable modules. Let's take the reconstruction of metabolic pathways as an example of modelling; if the annotated genome of an organism has already been recorded, its metabolic pathways can be reconstructed from the established gene functions. The metabolic pathways that are determined in this manner are validated, since the gene involved in each individual stage of biosynthesis is known. This type of metabolic network can then be recorded using a system of mathematical equations whose solutions allow predictions to be made regarding the behaviour of the system being modeled – for example, it is possible to make predictions about the growth behaviour of cells if the carbon source changes in the medium. Similarly, predictions can also be made regarding the viability of mutants if targeted mutations that disrupt the reconstructed metabolic network are induced. That said, it should never be forgotten that such predictions always require experimental verification.

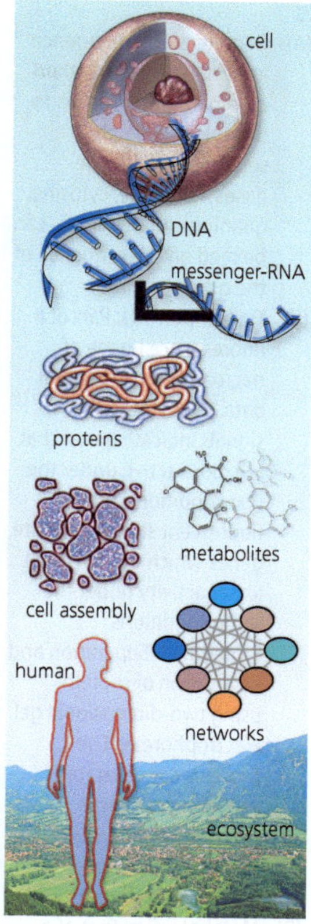

The various stages of systems biology: The individual cell has proved to be the smallest unit of life. Its construction and vital functions are stored as genetic information in the hereditary molecule DNA. This genetic information takes the form of genes, which are generally responsible for the creation of proteins. Along the path from gene to protein, however, we also find the transcript or messenger RNA. Proteins, which have an enzymatic function, are used to biosynthesise metabolites or metabolic products. The relationships between genes, transcripts, proteins and metabolites can be represented in the form of networks, which can be determined using molecular systems biology. Aggregation of individual cells leads to a cell assembly, and subsequently to multicellular organisms like people, animals and plants. Finally, all living organisms form an ecosystem with their surroundings; this is the highest level of complexity of systems biology.

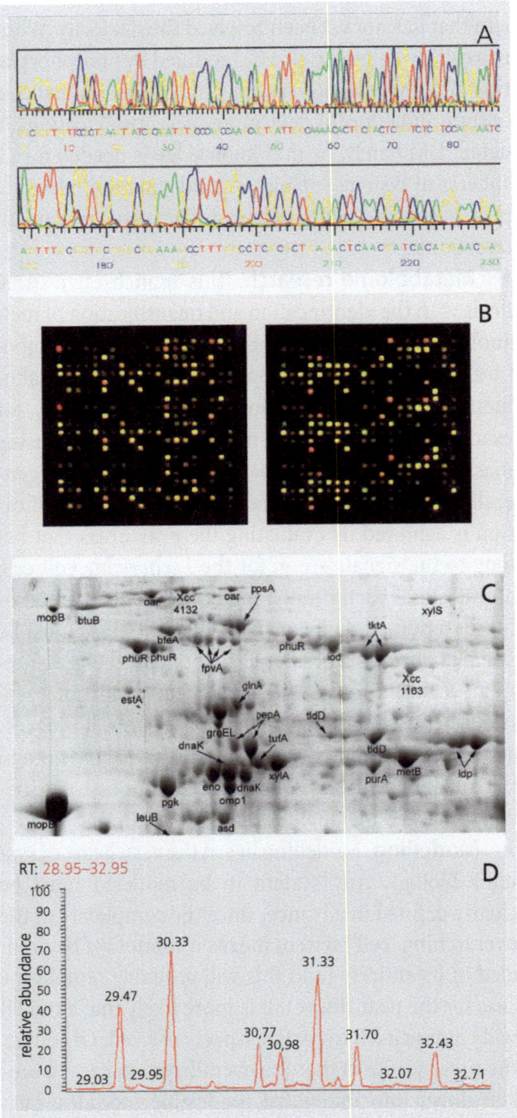

(A) Genomics: DNA sequence information as a standard chromatogram. Thanks to the colour code, the sequence of the four bases – adenine, cytosine, guanine and thymine – can be read off the sequence of the graph maxima.
(B) Transcriptomics: Part of a fluorescence image of microarray experimental data. Red fluorescent signals indicate genes that are more active under the test conditions; green fluorescent signals indicate genes which demonstrate lesser activity under the same conditions.
(C) Proteomics: Separation and verification of proteins using two-dimensional gel electrophoresis. The proteins are separated according to charge (horizontal) and mass (vertical). Each dot represents a single protein.
(D) Metabolomics: Part of a chromatogram of water-soluble metabolites. In conjunction with the retention time (RT) and the mass spectrum, individual peaks can be assigned to corresponding metabolites. Source: Bielefeld University

► **Bioinformatics.** The whole premise of systems biology is dependent on sophisticated bioinformatics. This refers to the administration and evaluation of genomic and post-genomic data, for which very powerful computers are generally required. The aim of bioinformatics is to extract information from the genome and post-genome data that will answer pertinent questions relating not only to fundamental research, but also to industrial biotechnology, agriculture, the environment and medicine. Another look at genome research can demonstrate the role played by bioinformatics: the data obtained from sequencing projects always relate to very brief sequence segments only. Using assembly programmes, bioinformatics will produce the complete genome sequence of an organism from a large number of these small sequence segments. Then the resultant genome sequence must be interpreted in detail. For example, the genes that occur in such a genome sequence must be found and – if possible – characterised functionally. Bioinformatics has developed gene identification and annotation programmes to do this. Annotation involves assigning a function to an identified gene, a procedure that is usually based on a sequence comparison with already known genes.

Applications

Organisms that are easy to cultivate and have already been intensively characterised are particularly well-suited for systems biology applications – the intestinal bacterium Escherichia coli, for example, or the hay bacillus Bacillus subtilis. Unicellular eukaryotes like yeast also play a major role as model organisms. However, the extent to which plant or animal cell cultures are suitable for systems biology analysis has not yet been established.

► **Recording cellular metabolism.** The term metabolism refers to the metabolic processes in a cell, which encompass both anabolising and catabolising reaction pathways. Catabolising reaction pathways are for extracting energy, while anabolising reaction pathways are for synthesising complex molecules, which ensure cell survival and reproduction. Using systems biology to study individual components of the metabolic process is a pioneering field of research. One such project is looking at the absorption of glucose and the subsequent initial steps of glucose processing in Escherichia coli. Both the glycolytic metabolic pathway and the enzymatic parameters of the biocatalysers involved are needed to be able to do the modelling and the mathematical description.

► **Observing gene regulation.** Once genome and transcriptome data have been gathered, another objective of systems biology is to observe gene regulation in bacteria. In an initial step, bioinformatics can be used to search the genome data for all the genes that may regulate transcription. Microarray experiments then enable identification of the regulation networks created by these regulators, thus clarifying the entire, complex regulation system for the cellular transcription processes within a cell.

► **Processing signal transduction.** Cells have to react to a multitude of external conditions: heat, cold, osmotic pressure and pH values all have to be noted, evaluated and – if necessary – responded to by producing a suitable reaction. This happens through so-called signal transduction, in which a receptor in the

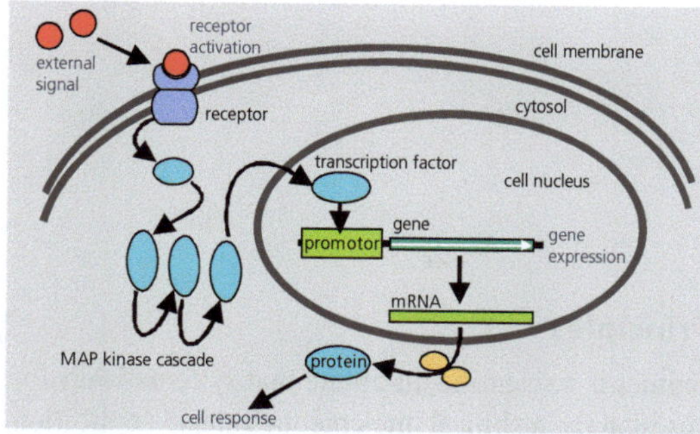

Signal transfer within a cell: The diagram shows a section of a cell. The cell membrane is adjacent to the outer edge of the cytosol. Within the cell itself, particular attention is drawn to the cell nucleus, which contains the genome. An external signal activates a receptor, which forwards the signal to the inside of the cell. The MAP kinase cascade then activates a transcription factor, which in turn triggers transcription of the affected gene through an interaction with a promoter. The synthesised mRNA is expressed from the cell nucleus and is responsible for synthesis of a protein, which represents the cell's response to the external signal. Source: Bielefeld University as per E. Klipp

membrane usually receives the external signal and sends it on within the cell by means of a signal transduction cascade. Phosphorylation reactions play a major role when a signal is forwarded in this way. The receptor protein is stored in the membrane in such a way that it exhibits an extra- or intracellular portion. The extracellular receptor portion then receives the signal, upon which the intracellular receptor portion phosphorylates itself by means of a conformational change. The proteins in the MAP kinase cascade then transfer the phosphate group in question to a transcription factor, which is activated as a result. This activated transcription factor in turn triggers a genetic programme, which produces the response to the original external signal. Such signal transduction pathways are particularly well-suited for use in systems biology applications: the membrane-bound receptor, the signal transduction cascade created by MAP kinases and the gene expression triggered by the transcription factors are all currently the subject of mathematical modelling.

▶ **Analysing the cell cycle.** In a eukaryotic cell, a cell cycle refers to the division of a cell into two daughter cells. This regulates the doubling of chromosomes, the division of the cell nucleus into two daughter nuclei and, ultimately, the division of the cell itself. The general cell cycle diagram for eukaryotic cells is always identical, from yeast to mammals. And since many details of the cell cycle of selected organisms like yeast are already known, it lends itself particularly well to systems biology analysis. The cell cycle of yeast has already been determined using systems biology; this in turn makes it possible to analyse the cell cycle of tumour cells in detail so we can understand which mutations of cell cycle proteins in genes lead to rampant growth.

Trends

▶ **Industrial biotechnology.** *Industrial biotechnology* uses the synthesis capability of cultivated cells to produce macromolecules and metabolic components. These can be therapeutic proteins, for example, or metabolic products such as vitamins, amino acids or antibiotics. Systems biology creates models of these synthesis processes in the cultivated cells and then converts the models into mathematical equations, thus enabling simulation of the production process. For example, modelling the material flow in the direction of the desired product means bottlenecks can be identified and instructions issued on how to achieve a higher rate of yield through targeted genetic modifications. For many decades, industrial strain development blindly followed the "mutation and selection process", with the result that both the site and the effect of the mutation remained unclear. Nowadays, systems biology is enabling rational strain development, i.e. it is using targeted mutations with predictable effects on the end product yield.

▶ **Development of novel organisms.** Systems biology is also concerned with combining genetic modules from different organisms in order to develop novel organisms that exhibit metabolic outcomes not yet identified in nature. For example, novel plants have already been created that emit light like glow-worms, synthesise biodegradable biopolymers or produce medically-effective antibodies. Particularly in the sphere of industrial biotechnology, we have seen the development of novel production strains that combine biosynthesis chains from different original strains. In this connection, systems biology will provide information regarding the extent to which the newly-introduced metabolic pathways are suited to the host organism and how to optimise their suitability.

Prospects

- Systems biology is a young science that seeks to gain an understanding of complex biological processes through mathematical modelling. It is the only way to keep control of the flood of data being produced by genome research in particular.
- Systems biology is an integrative science that needs not only biologists, chemists and physicists, but also engineers and mathematicians.
- The first successes chalked up by systems biology can be seen in the spheres of metabolism, gene regulation, signal transduction and the cell cycle.
- Future areas of application lie in industrial biotechnology, medicine, plant cultivation and the development of novel organisms.

PROF. DR. ALFRED PÜHLER
Bielefeld University

Internet

- http://microbialgenomics.energy.gov/MicrobialCellProject
- www.systemsbiology.org
- www.systembiologie.de
- www.nrcam.uchc.edu/login/login.html

Bionics

Related topics

- Microsystems technology
- Sensor systems
- Robotics
- Aircraft
- Textiles

Principles

Bionics is a scientific discipline that systematically focuses on the technical implementation and application of designs, processes and development principles found in biological systems. Bionics, also known as biomimetics, unites the fields of biology and technology and stands for a "symbiosis" of the two conceptual and working approaches. While basic biological research draws on modern technology and its methods and equipment, and to a certain extent also poses questions aiming at a deeper understanding of biological functions and systems, bionics comprises the actual transfer of biological findings to the technological domain. This is no direct transfer in the sense of copying, but rather an independent, creative research and development process – in other words, a nature-inspired process of "re-invention" usually involving several stages of abstraction and modification en route to application.

"Top-down approach" to developing a car tyre: The paws of big cats (the picture shows a cheetah) serve as a model, with their excellent ability to transmit power during acceleration, their tremendous grip in bends, and their method of splaying the contact surface when braking. These mechanisms were the inspiration for a non-slip, adaptable all-weather profile which, after being designed, was subjected to an artificial evolution, i.e. optimised specifically for cars and made fit for application using the evolution strategy. Source: Plant Biomechanics Group Freiburg, BIOKON, the Competence Network Biomimetics

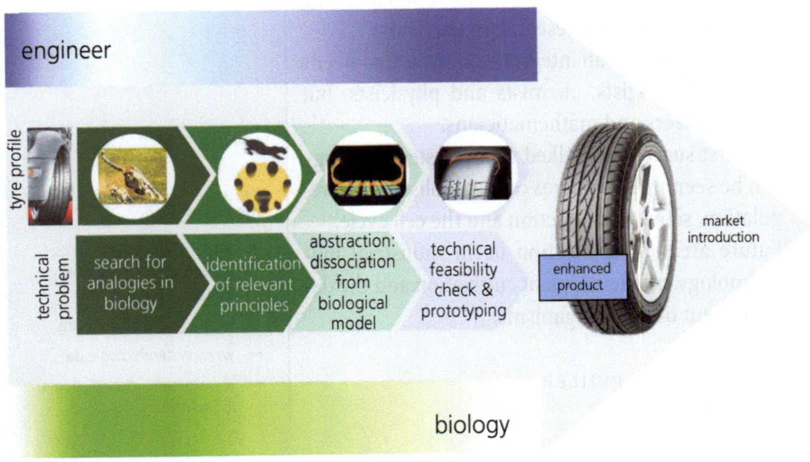

► **Working approaches to bionics.** There are two different approaches to bionic development. One is the "top-down approach", in which technical tasks stimulate a targeted search for solutions in nature that can help to develop novel products and tackle the problems involved. The selected natural models are used to analyse relevant functional principles from a technical point of view, verify their validity under altered conditions, and compare them with state-of-the-art technology. If they are found to have innovative potential, they are screened to determine the technical possibilities for developing a product with new (bionic) properties. The most decisive factor is functionality. The original form, i.e. the concrete manifestation, of the natural models is of virtually no importance to bionic implementation (bar a few exceptions).

The "bottom-up approach" sets out from phenomena and structural solutions found in the biosciences. These findings are used to develop principle solutions that are normally focused in a particular direction but can (multivalently) open out into different product lines depending on the area of application. In this case, a variety of material- and production-based considerations, design aspects and other factors are incorporated into the equation. Cost pressure, too, often makes modifications necessary.

Bionics is an iterative process, in which scientific and technological advances cross-link different disciplines and seemingly distant fields of application. Plant bionics, for example, is currently making surprising contributions to aircraft and automobile construction in terms of composite and gradient materials and growth algorithms. These approaches are complemented by findings from zoology, such as the mechanisms of adaptive bone remodelling.

Applications

► **Functional surfaces.** In fluid dynamics and other areas, it was believed for a long time that smooth surfaces offered the least resistance and were highly dirt-repellent, and that they were the ideal solution. However, such surfaces rarely occur in the natural, living environment. The discovery that the fine, longitudinal ribs on the scales of fast-swimming sharks actually reduce turbulent frictional resistance led to a paradigm shift in fluid dynamics research. The phenomenon was investigated and explained using surfaces with artificial rib structures. The resistance-lowering effect consists in the longitudinal ribs obstructing the lateral movements of the turbulent currents near the wall and thus slightly reducing the turbulent loss of momentum overall. In addition, the viscous underlayer in the rib valleys slightly thickens, causing the wall shear stress to be reduced. The surface roughness does not lead to any additional friction, as the grooves still have a hydraulically smooth effect due to their small size.

In experiments, razor-sharp longitudinal bands helped to reduce resistance by 10% compared to smooth surfaces. However, it was not possible to manufacture such filigree surface structures for practical applications. A compromise was reached in the form of ribs with triangular groove valleys, which can be stamped automatically into a plastic film. This bionic skin was applied to virtually the entire surface of the Airbus A320. Subsequent flight tests resulted in fuel savings of approximately 2% compared to an aircraft without a riblet film. Initial problems concerning the film's UV resistance and adhesive technology have meanwhile been solved. The film is also more resistant to dirt than a painted surface. As far as mounting and repairing are concerned, however, there is still a need for efficient, economical technologies for wide-scale application.

► **Lotus-Effect®**. During comparative studies on the systematics of plant surfaces, it was discovered that the leaves of lotuses possess astonishing self-cleaning properties. In general, it became evident that it is not the smooth surfaces but always the finely structured and extremely water-repellent (superhydrophobic) surfaces that remain the cleanest. Micro and nano structuring – usually in the form of multiple structuring on different scale levels – minimises the contact surface of the dirt particles and thus reduces their adhesive force, much like a bed of nails. The hydrophobic properties have the effect that water droplets do not run but actually contract into balls as a result of the surface tension, forming contact angles of almost 180° to the surface. This means that they barely stick to the substrate. In a similar way to droplets on a hot stove, they run across the surface like glass pearls, collecting all the dirt particles encountered along the way.

These particles now stick much better to the droplets than to the substrate, and a light drizzle is enough to fully clean the surface.

The same mechanism also works with other aqueous solutions, such as liquid glue or honey. While the microstructures of the plant surface structure themselves from wax deposited by the cells (formation of crystal-like molecular arrangements) and can regenerate if necessary, the bionic implementation of this principle called for suitable substances and technologies that could meet the various technical requirements and would have a stable, long-term effect. Scientists have now succeeded in transferring the lotus effect to a variety of materials such as facade paint, textiles, plastic surfaces, roof tiles, glass and, in test samples, also metal surfaces. For temporary applications, there is a spray that can be used to treat almost any desired object or material, for example rainwear or garden furniture. Similarly coated containers used to hold liquids can be completely emptied without leaving any residues, which means that they do not need to be specially cleaned and dried. Clean and dry surfaces also mean less corrosion. Buildings can be made more resistant to fungi and algae, and paint colours appear more deeply saturated thanks to the velvety matt lotus surface.

Initial feasibility studies show that textiles with superhydrophobic surfaces can stay dry even under water, just like the fishing spider. Next to the possibility of combining resistance-lowering riblet surfaces with the lotus effect for use on aircraft, it has been demonstrated in water that an artificial shark skin made of elastic material significantly delays biofouling, i. e. the growth of marine microorganisms on a surface. This bionic solution could help to avoid the use of toxic agents generally added to anti-fouling paints.

► **Optimisation processes**. In addition to concrete individual phenomena, bionics also deals with the uni-

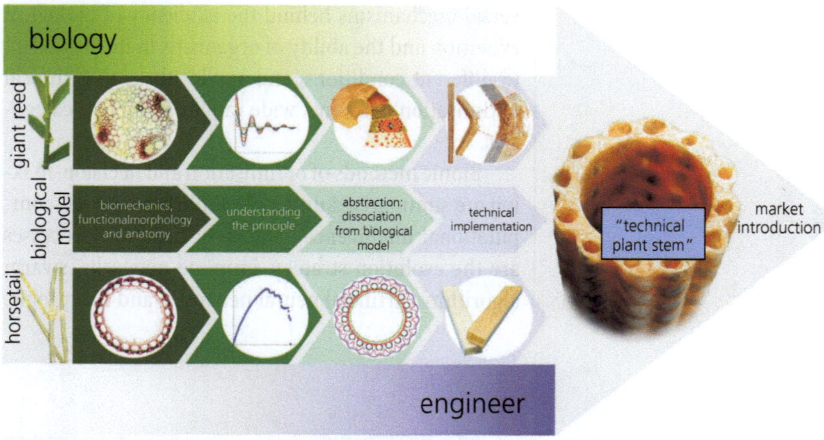

"Bottom-up approach": Transfer of biomechanical findings from botanical research to a novel technical product. Understanding the functional anatomy of plant stems, followed by abstraction, leads to a novel design for technical ducts of high specific strength. Source: Plant Biomechanics Group Freiburg, BIOKON, Competence Network Biomimetics

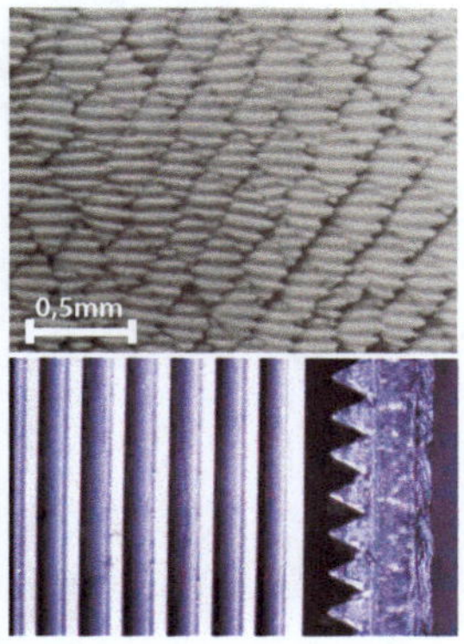

Artificial shark skin. Top: Shark scales with rib structure. Bottom: Plan view and cross section of a bionic shark skin (riblet film made by 3M) for reducing resistance on aircraft. Source: E. Reif, Munich Paleontological Society, Saarland University

versal mechanisms behind the efficiency of biological evolution and the ability of organisms to flexibly adapt to different conditions, and applies these to technical optimisation tasks in a wide range of problem categories.

Bionic methods of optimisation and decision-making are summarised under the umbrella term of "computational intelligence". The most common processes are the evolution strategy, immune networks, swarm algorithms, artificial neural networks, and fuzzy logic.

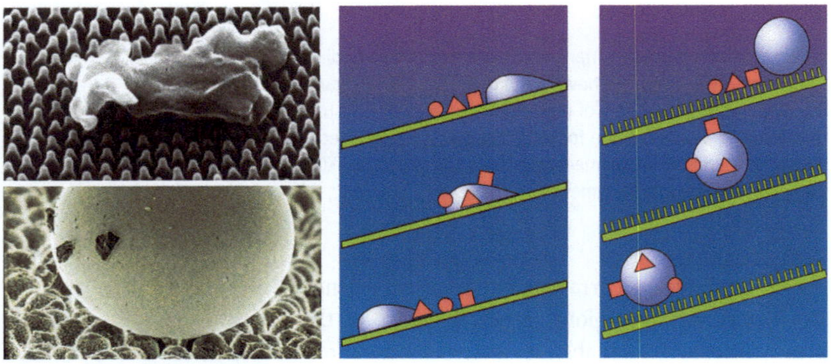

Lotus-Effect®: On the micro and nano structured, hydrophobic lotus surface, dirt particles perch on top, just like on a bed of nails (top left). Water droplets take on a spherical form, and the dirt particles stick better to the droplets than to the substrate (bottom left). While water droplets running off a smooth surface only carry the dirt particles a short distance (centre), those on a lotus surface clear them away completely without a trace (right). Source: W. Barthlott, Nees Institute, University of Bonn

▶ **Evolution strategy.** What biology and technology have in common is that they both start out with a basic model (e.g. a primeval insect or bird, or a functional model or prototype respectively) and try to make further improvements. The quality criteria involved can be restricted to specific individual parameters (e.g. weight, strength, production time or costs) or can cover a range of variables in the case of *multi-criterial optimisation*. The simplest form of evolution strategy implies that the starting configuration, or list of parameters, generates multiple descendants with randomly produced changes (mutations). The resulting technical designs or process cycles are implemented in simulations or in physical form in order to evaluate the quality of each descendant. The best one in each case is selected as the father or mother of the next generation, until the best possible overall quality has been reached.

Evolution strategies have proved to be very successful in numerous technical applications, particularly in the case of complex optimisation tasks in "interference-laden" development environments, e.g. parameter drift in the course of price developments or fuzzy quality specifications due to measurement errors. Such applications include optical systems, statics and mechanics (e.g. lattice constructions), support and fuselage structures in aircraft manufacture, flow-dynamic applications (e.g. tube elbows, multi-phase nozzles, wing profiles, multi-winglets and propellers), paints and varnishes, process optimisation of complex control systems, e.g. baking processes in the ceramics industry, engine control systems, optimisation of ma-

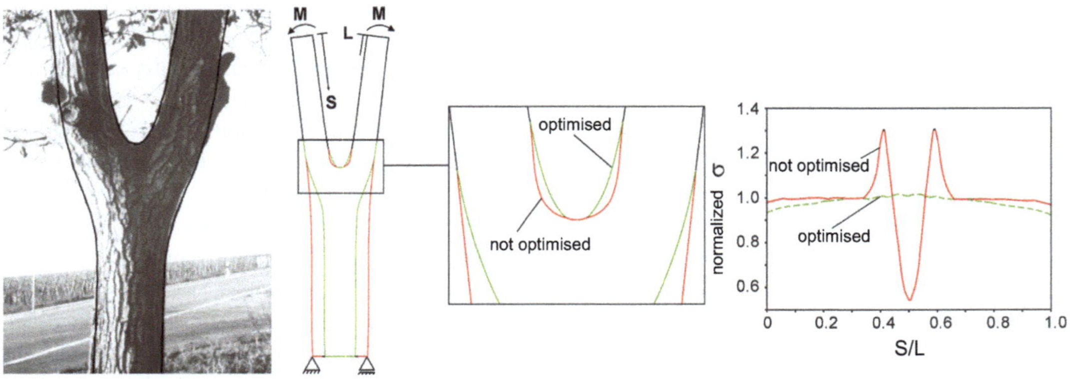

Shape adaptation: Trees grow according to the axiom of uniform surface tension, i.e. they offset tension peaks by growing extra material in affected places. In forked trunks, this results in a notch shape without any notch stress. By transferring this shape optimisation method to computer models – computer aided optimisation (CAO) and soft kill option (SKO) – it is possible to design stress-optimised lightweight components of any shape. Left: Forked trunk with optimised notch shape. Centre: Stress-optimised lightweight component. Notch stress is reduced by additional material in the notch, while material can be pared away at the base of the component without compromising strength. Right: Distribution of tangential stress along optimised and non-optimised notch contour from left to right (S increases to accommodate L). Source: C. Mattheck, Karlsruhe Research Centre chine systems, production processes and integrated production flows.

▶ **Adaptive component growth.** Trees and bones handle a variety of stress situations by means of cellular growth and flexible shape adaptation mechanisms. Thanks to an enhanced finite-element method, it is now possible to let technical components "grow" into their ideal shape on the computer. In living biological structures, new cells form in places subjected to high mechanical loads. Additional material keeps growing there until the tension is evenly distributed. Forked tree trunks, for example – unlike shape transitions commonly found in technical constructions – are free of notch stress, which would otherwise lead to breakage or cracking. Bones are additionally supported by scavenger cells, which remove superfluous material in unstressed regions. By implementing these basic principles in technical design and calculation programs – computer-aided optimisation (CAO) and soft kill option (SKO) – it is possible to create novel structures that can achieve optimum strength with a minimum amount of material and also last longer even if subjected to frequent load changes. These bionic processes have been extended to include complex, three-dimensional structures over the last few years, and are now in use in many industrial sectors.

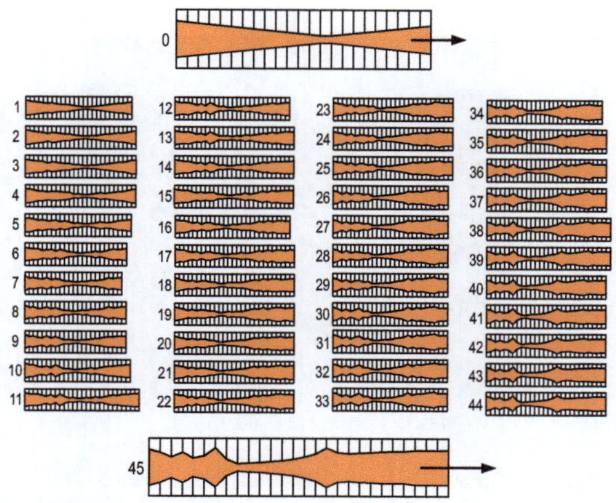

◼ Optimisation of a two-phase nozzle (hot steam jet) in a physical experiment using the evolution strategy. Top: Original configuration (classic nozzle shape). Centre: The best descendants of every generation (each generation is represented by a number). Surprisingly, during the course of the optimisation process, the nozzles acquired chambers which – as it later turned out – contain specially arranged ring vortices that improve the vaporisation process. Bottom: Optimised shape (efficiency improved by approximately 25%). Source: H.-P. Schwefel, I. Rechenberg, TU Berlin

◼ "Bionic Car" concept study: Bionics inside and out. The boxfish (top left) served as an example of efficient space utilisation and a streamlined shape (cw = 0.19). The car's highly resilient lightweight frame (top right) was optimised by the bionic methods of computer aided optimisation (CAO) and soft kill option (SKO). Source: Daimler AG

Trends

▶ **Glue-free adhesion.** On the micro and nano scale, bionics researchers have hit on a phenomenon that works in the exact opposite way to the lotus effect. Flies and geckos have adhesive pads made up of extremely fine hairs on their feet, which cling to surfaces in such a way that they can use the molecular forces of attraction (van der Waals forces). This enables them to walk across the ceiling and stick to smooth surfaces. Comparative functional examinations of flies, beetles, spiders and lizards show that the finer the hairs and the more densely packed they are, the greater their adhesive force. It is therefore logical that lizards should have the finest structures. Unlike the tiny micro and nano hairs found in biology, their technical counterparts are still tenths of a millimetre in size and therefore at the lower end of the performance scale. Polymer structures with synthetic micro-fibres currently stick about five times better than a smooth surface made of the same material. New technologies for producing finer structures will open the door to many possible applications.

▶ **Biomechatronics and robotics.** This discipline focuses on everything from intelligent micro and nano machines and micro actuators, miniature aircraft, a wide range of walking and swimming robots, bionic gripping systems and manipulators to artificial limbs, organ prostheses and special equipment, particularly for biomedical engineering.

The humanoid robot, for example, emulates the spectrum of mechanical functions of the human arm and hand on a 1:1 scale with artificial muscles and a total of 48 articulated axes (degrees of freedom). The pneumatic (fluidic) muscles act as powerful, ultra-light ac-

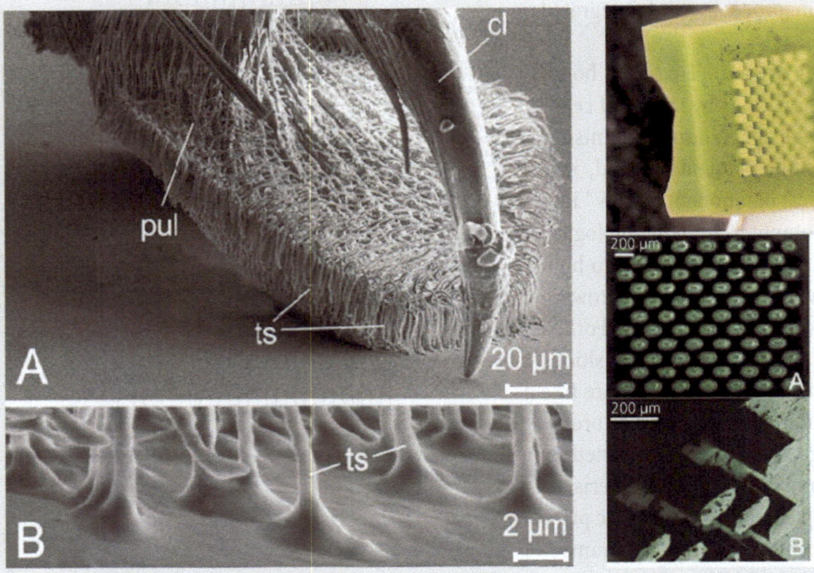

Glue-free adhesion. Left: SEM image of a fly's foot (cl – claw for hooking into rough-textured surfaces, pul – adhesive pad made of microscopically fine hairs (ts) for maximising the contact area on relatively smooth substrates). Right: Prototype of a bionic polymer surface with five-fold adhesive reinforcement. The dimensions of the individual hairs have not yet been optimised. Source: S. Gorb, Max Planck Institute for Metals Research

tuators and elastic energy reservoirs that enable flowing, flexible movements. Their tractive forces can be torquelessly channeled through several joints by means of artificial tendons (extremely tearproof Dyneema® cords) in order to provide leverage at the desired place. This keeps the mass of movable parts to a minimum. Intelligent mechanical coupling systems and an adjustable "muscle tonicity" enable the robot to perform humanoid motion sequences, ranging from relaxed dangling or swinging to force-controlled, high-precision movements. The movements of the arms and hands are controlled by computer or remote control using a data suit. This demonstrates how easy it will be for humans and humanoid robots to work together in the future.

► **Flow bionics.** The wings of modern large *aircraft* the size of the Airbus A380 generate immensely strong wake turbulences (wake vortices), which cause drag and can pose a threat to smaller aircraft. For safety reasons, it is necessary to wait a while after takeoff and landing until the vortices have died down before resuming operations. The problem could be alleviated by small winglets at the wing tips, just like those used by birds. Birds, however, spread whole winglet fans (their pinions), thus generating several small vortices instead of one large wake vortex. Optimisation experiments in a wind tunnel using the evolution strategy have confirmed that multi-winglets reduce edge resistance and the smaller wave vortices dissipate more quickly. In theory, the ideal solution would be an infinitely fine breakdown of the wake vortices to reduce the resulting drag. However, this is not possible in practice, as it would require an endless number of infinitely fine pinions, which at the same time would generate an infinitely strong frictional resistance. A practicable solution is to leave out the entire inner section of the winglet fan and instead to join the two outer winglets together to form a wing loop whose shape corresponds to the envelope of the winglet fan. By adapting the setting angles and the profile depth, the circulation distribution along the loop-shaped wing tip can now be adjusted in such a way that only a thin vortex layer is left behind in the wake. This layer cannot build up into a strong core vortex. As a result, less energy is left behind, too, and the resistance is reduced. This principle can also be applied to a propeller, in which case the winglets are not joined on each individual blade but are linked to their respective counterparts on neigh-

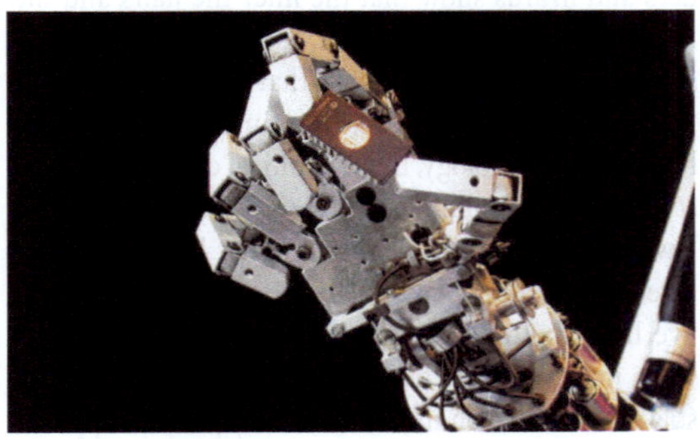

Hand of a humanoid robot with artificial muscles: Pneumatic muscles enable the robot to perform delicate movements. Source: R. Bannasch, I. Boblan, EvoLogics, TU Berlin and Festo

Bionic propeller. Top left: Splayed bird feathers (pinions) act like multi-winglets, reducing turbulence at the edge of the wing. If only the envelope of the winglet fan is adapted to form a wing loop, it is possible to achieve a similar flow-mechanical, vortex-reducing effect. Top right: When transferred to a propeller, this principle results in a completely new design. Bottom left: In the case of a normal propeller, strong core vortices form at the blade tips and lead to noise-producing pressure fluctuations during rotation. Bottom right: In the case of the bionic loop propeller, the wake field is enclosed by a uniformly thin vortex layer. The pressure field is more balanced and the noise emissions correspondingly lower. Source: R. Bannasch, EvoLogics GmbH

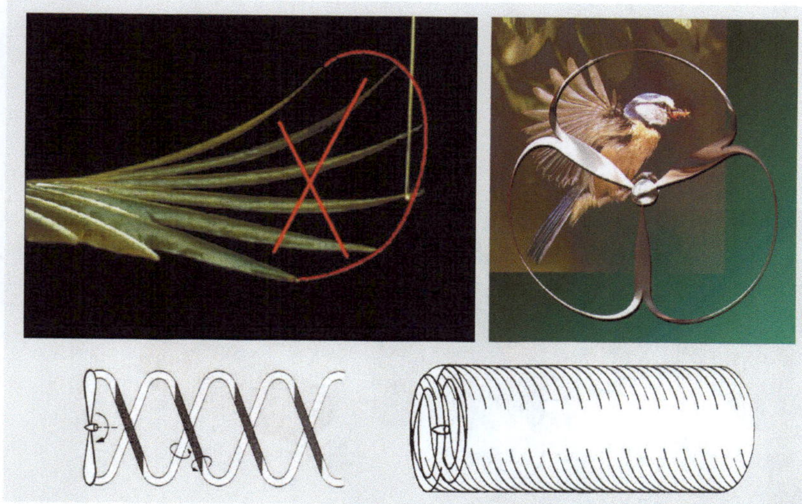

bouring blades. The result is a bionic loop propeller that generates a more balanced pressure field without the usual core vortices at its blade tips, and thus also operates much more quietly.

▶ **Communication and sensors.** In the sea, electromagnetic waves have a very limited range, while radar and radio waves are blocked completely. The only solution here is to work with acoustic signals, which are prone to disruption and sometimes fail altogether. Dolphins and whales have adapted to the situation: they chirp and sing across a broad frequency bandwidth. It was discovered that this continuous change of frequency not only serves to transmit information, but also to compensate for sources of interference such as echoes and noise. *Acoustic underwater modems* that "sing" in the ultrasonic range can reliably transmit digital data over long distances of up to several kilometres without harming the environment. Autonomous research and measurement submarines, oceanographic measuring stations and an innovative *tsunami early warning system* in the depths of the Indian Ocean have all been equipped with the new modems, and various applications for underwater systems in the offshore industry are currently in the pipeline. The dolphin-like carrier signals enable distances to be measured accurately and speeds to be determined, as well as supporting communications and spatial orientation. This means that multiple functions can be implemented in a single system. High-resolution measuring techniques for non-destructive materials testing and ultrasonic medical diagnosis are being tested in the laboratory.

However, dolphins are only one of the many natural role models on Earth. Bionics research covers a large number of different sensor principles and aims to adapt them for use by people and technical applications. Recent developments based on bionic principles include cochlea implants and artificial retinas, shake-proof camera systems, infrared sensors for detecting forest fires, artificial vibrissae (tactile hairs) and antennas for mobile robots, and many other types of sensor. Various hybrid techniques combining biological and electronic elements are also being tested in the laboratory.

Perspectives

- Bionics is a dynamic scientific discipline that has developed numerous new facets in recent years. The present boom can be attributed to the fact that we have now fully detached ourselves from the idea of trying to copy nature on a one-to-one basis. Biologists and engineers are working together to "crystallise out" the underlying functional principles, and to constructively transfer these to new technical applications.
- The main focus of bionic research and development is currently on areas such as micro and nano structures, self-organisation, intelligent materials, biomedical engineering, sensors and actuators, and robotics. In each case, different elements and principles of either botanical or zoological origin are combined and integrated in a variety of ways.
- Next to a steadily growing wealth of individual solutions taken from the living environment, scientists are also reaching a deeper understanding of the methods of biological optimisation and are turning them into universally applicable processes.

DR. RUDOLF BANNASCH
BIOKON e. V./EvoLogics GmbH, Berlin

Internet

- www.biokon.net/index.shtml.en
- www.kompetenznetz-biomimetik.de/index.php?lang=enwww.gmo-safety.eu
- www.extra.rdg.ac.uk/eng/BIONIS
- www.biomimicryinstitute.org

5 HEALTH AND NUTRITION

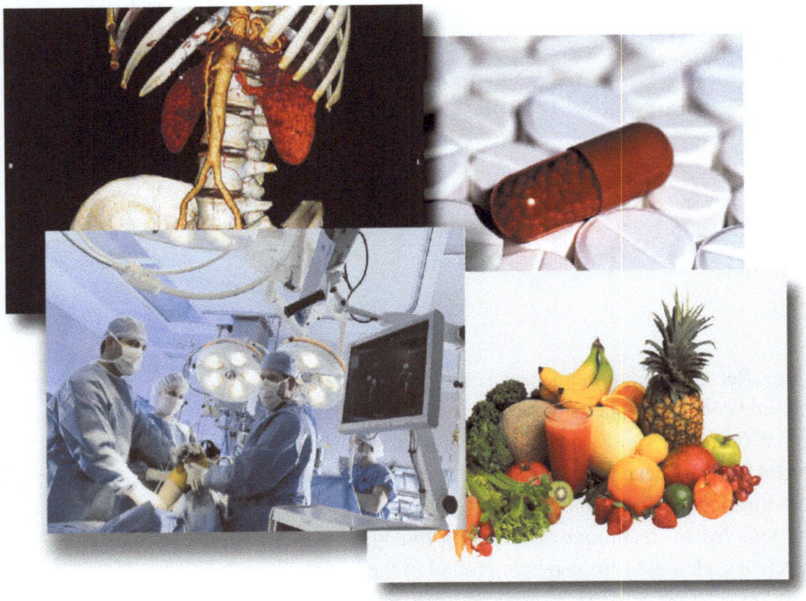

Good health is the result of complex interactions between ecological, behavioural, social, political and economic factors. The aims of research into health and nutrition are to live more healthily, longer and more actively, in keeping with the motto "Die young, but as late as possible!" The chances of ageing healthily are better than ever today. Advances in medicine mean that average life expectancy has increased continuously over the last 100 years, and is now over 80 years for women and 75 years for men in the Western industrialised world. In the last 20 years alone, life expectancy has increased by an average of 5 years.

Initially, hygiene, safer food and better working and living conditions created the pre-conditions for a healthy and longer life. Advances in medicine, such as new drugs, new methods of diagnosis and intensive care, have also contributed to this trend. It is not just a case of optimising the health care system from the viewpoint of offering the best technically feasible services. Aspects of cost also play a large role in the roadmaps of the future in medical and *pharmaceutical research.*

The highest costs by far are associated with expenses incurred in the final year of life – particularly for in-patient treatment and *intensive medical care.*

The health policy of this century will in this respect be determined by ethical issues. Will we be able to make every conceivable medical treatment available to every citizen, or will cost/benefit analyses determine medical treatment? Medical research is progressing, but will it remain affordable?

The term "*theranostics*", which combines the terms diagnosis and therapy, describes a new approach in medical care. It is principally based on the fact that each person is different. In the same way that each person looks different, each person varies in their predisposition to illnesses, and the medication dose required to achieve an effect will vary. In future, diagnoses at an early stage, and establishing a predisposition to illnesses (if necessary, by means of gene tests), will mean that illness can be treated before it "breaks out". The technological drivers of this are genome research combined with rapid *genome sequencing*, as well as compact analytical equipment (*lab-on-a-chip*): substances (markers) indicating an imminent illness can be identified at an early stage. Pharmacogenomics is the branch of pharmacology which deals with the influence of genetic variation on drug response in patients by correlating gene expression with a drug's efficacy or toxicity. By doing so, pharmacogenomics aims to develop rational means of optimising drug therapy, taking into account patients' genotype.

Human wellbeing is supported not only by health research but also, in particular, by the latest developments in medical engineering. One of the subdomains of this field is clinical engineering, which deals with medical devices used in hospitals. These include pacemakers, infusion pumps, heart-lung machines, dialysis machines and artificial organs, for example.

Devices for imaging diagnostics are among the most frequently used and most complex medical products encountered in hospitals. Imaging processes are being developed toward more flexibility, greater resolution, faster evaluation and fewer stressful procedures for patients.

► **The topics.** Advances in the areas of molecular biology, genome research, bioinformatics, screening and process technologies have revolutionised *drug research*. Nowadays, the starting point for this research is our knowledge about changes at a molecular level which are relevant to the illness. This results in the characterisation of pathogenic, endogenous factors (targets), which can then act as sites of action for future medication. High throughput screening is used to search for inhibitors of pathogenic factors. Millions of chemical substances are tested, with a daily throughput of 200,000 test substances. These target-oriented processes lead to the discovery of selective and highly effective agents. Precise knowledge of the molecular course of an illness also allows detection of the characteristic biomarkers, which, on the one hand, help to diagnose illness in the early stages and, on the other hand, help to predict how successful the effect of the medicinal agent will be.

Whereas nowadays the patient only goes to the doctor when experiencing symptoms, and only then can a diagnosis and treatment be initiated, it is expected that advances in molecular medicine will enable risks or current illnesses to be detected at a much earlier stage. These can then be specifically treated at a molecular level or preventive measures can be taken.

The use of information technology in medicine is resulting in fundamentally new applications, such as telemedicine and *assistive technologies* in patient and old-age care. Barrier-free mobility for maintaining independence is particularly important to old and disabled people. This includes building measures, user-friendly equipment and availability of information. The networking of information technology and the integrated processing of data also play a major role in the running of operation rooms and throughout clinics. Further developments in IT also mean that data extracted from imaging procedures can be evaluated much more quickly and clearly, supporting doctors in their work. It would now be impossible to imagine intensive care, which is concerned with diagnostics and the treatment of life-threatening conditions and illnesses, without information technology. The control units and monitors on an intensive care ward are powerful evidence of this.

Research in *molecular medicine* is resulting in new insights into the interaction mechanisms between biological systems and artificial materials on a molecular level. This, combined with findings about new materials, is leading to developments in the area of prostheses, implants and tissue engineering. In tissue engineering, the use of stem cells will also herald the start of a new generation of autogenous implants. Ultimately, it is hoped that this will result in the replacement of complex organs.

Minimally invasive and *non-invasive medicine* is another area of great potential for the future. Diagnoses and treatment can be carried out by means of minor, or even without incisions. This means that patients can have the operations as out-patients, under local anaesthetic and with less post-operative pain, and convalescence will be correspondingly quicker.

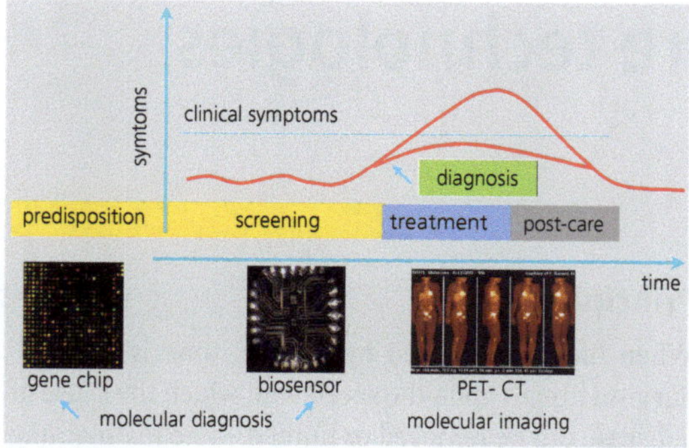

▲ Theranostics with molecular medicine: The combination of molecular diagnostics and molecular imaging will help to detect illnesses at an earlier stage. Treatment can then start as soon as possible, before symptoms have appeared. Molecular diagnosis using biochips provides evidence of illness-specific biomarkers or of a genetic predisposition to specific illnesses. Molecular imaging makes it possible to localise sites of illness (e.g. inflammations or cancer) in the body. Source: Philips Research Europe

Technology fields such as nanomedicine offer great advantages for pharmaceutics, *implant medicine* and *medical imaging*. Many promising agents cannot be considered as medicines, as they can only partly reach their sites of action despite their high pharmacological potential. It is expected that nanoscale agent carrier systems will protect agents from disintegration during transport to the target tissue, take them through biological barriers such as the blood-brain barrier and enable their controlled enrichment in the target tissue.

There is still much to be clarified about the interaction between nutrition and illnesses. However, the findings are becoming more and more clear – there is, for example, an obvious risk relationship between the mode of nutrition and the occurrence of circulatory illnesses. We can all choose the food that we eat, but not the way that it is prepared. Industrial *food technology* has a significant influence on the quality and safety of food. New foodstuffs, enriched with additional ingredients (e.g. vitamins, minerals, unsaturated fats) are being developed, which are expected to have a positive effect on health. The term for these foods, "nutraceuticals" (a combination of nutrition and pharmaceutical), suggests that the border between food and medication is becoming increasingly blurred. ■

Intensive care technologies

Related topics

- Medical imaging
- Medicine and information technologies
- Molecular diagnostics
- Minimally invasive medicine
- Nanomedicine
- Image evaluation and interpretation

Principles

When life is threatened by disease, time is survival. Survival hinges upon the speed with which the cause of a disease can be removed or limited while maintaining the patient's vital organ functions. *Intensive care units* (ICUs) have been established to help the patient survive by accelerating diagnostics, monitoring vital parameters continuously and initiating early therapy far beyond the means of standard hospital care. The faster and the more specifically any life-threatening situation is treated, the better the outcome for the patient. The outcome of intensive care today is evaluated by the method of "evidence-based medicine" sought through monitored, biostatistically scaled and synchronised clinical studies in hospitals around the world. To obtain an early diagnosis and start therapy, it is necessary to quickly and precisely integrate as much data and background information about the patient as possible. This is achieved by intensive personal communication with the patient – if possible – while gaining access to the patient's blood circulation and airways, and by obtaining organ images and vital blood parameters.

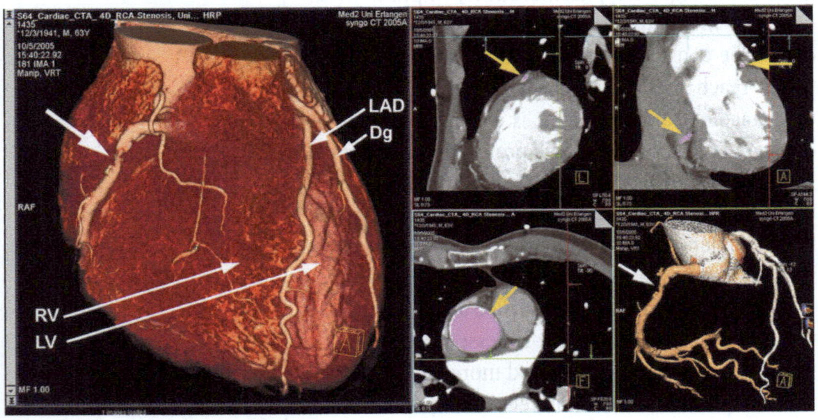

Visualisation of coronary stenosis: The image shows software-based visualisation tools for contrast-enhanced computed tomography (CT) of the heart to detect coronary stenoses non-invasively. Left: 3D reconstruction of the entire heart, which was automatically extracted from the chest volume. The surface-weighted volume rendering technique displays the heart from an anterior perspective. The left (LV) and right ventricle (RV) can be seen, as well as the coronary arteries on the surface of the heart (LAD = left anterior descending coronary artery, Dg = diagonal branch). The right coronary artery has a significant luminal narrowing (stenosis) in its proximal part (large arrow). Right: Extraction of an isolated coronary artery tree from the data set. The blood pool of the aortic root and coronary arteries is automatically identified, and cross-sectional images help to verify whether it has been detected accurately (top left and right, as well as lower left panel). The identified blood pool is highlighted in transparent pink (yellow arrows). A three-dimensional reconstruction of the identified blood pool (lower right) shows the isolated coronary arteries and once again the stenosis of the right coronary artery (white arrow). Source: Siemens AG, Healthcare Sector

► **Medical imaging.** Digitalisation, data compression and data archiving combined with non-invasive methods of visualisation by ultrasound, computer tomography, magnetic resonance tomography, conventional X-rays and endoscopy provide body images from the anatomical down to the sub-microscopic level. In the field of computer tomography (CT), investigations are performed in order to visualise the coronary arteries, for example, and detect coronary stenoses non-invasively. Because coronary arteries have very small dimensions (typically about 3 mm diameter) and move very rapidly, the CT scanner needs to have an extremely high spatial and temporal resolution to visualise them.

► **POCT (point-of-care testing).** This term has been coined to describe tests performed using new miniaturised, bedside laboratory equipment. POC blood or serum testing provides immediate results, of which the marker enzymes of myocardial infarction are the most important. If a coronary artery suddenly becomes clogged by a thrombus or a ruptured plaque due to underlying arteriosclerosis, the heart muscle area downstream of the blocked artery is no longer supplied with oxygenated blood. Its cells die and release proteins and enzymes in tiny amounts into the bloodstream. Their amount, for instance of troponin, myoglobin or creatine kinase, correlates very specifically with the size of the area at risk or the definite infarct. The substances can be detected in as little as 150 µl of venous blood drawn from the patient and placed into the test slot of a diagnostic analyser. In current POCT analysers, up to five different enzymes or proteins can be measured simultaneously at levels of down to 0.05 ng/ml. Diagnostic results are obtained within 12–15 minutes. In addition, the bar-coded wristbands of doctors and patients may be read by the analyser in this process to reduce potential errors regarding the drug, dose, time and patient identification. In this way, accurate and decisive diagnostic results can be obtained and validated at the bedside, reducing the previous "time to results" by up to 3 hours. In the case of acute chest pain indicating an acute coronary syndrome with looming infarction, this provides a new time frame in which to remove the thrombus or plaque from the coronary artery by catheterisation, restore arterial blood flow before the myocardium is fatally damaged and prevent the infarct.

Applications

► **Ventilation.** Oxygen is vital to all body tissues and is required continuously and in sufficient quantities. In the event of respiratory failure, the patient's gas exchange (i. e. oxygen uptake and carbon dioxide elimination) is inadequate and cannot meet the body's demands. Mechanical ventilation with or without an additional oxygen supply ensures adequate oxygenation (O_2 uptake) and ventilation (CO_2 elimination) during various forms of respiratory failure, for instance in connection with a severe trauma, pneumonia, or a pulmonary embolism. Mechanical ventilation is delivered by a face mask (non-invasive ventilation) or via plastic tubes (invasive ventilation) after intubation or tracheostomy. It is performed by positive-pressure ventilators which inflate the lung with a mixture of gases. Patients need to be unconscious or sedated to tolerate *invasive ventilation therapy*. The tube has a soft inflatable cuff which seals off the trachea so that all inhaled and exhaled gases flow through the tube. Microprocessor technology incorporated into the ventilator enables the degree of mechanical assistance to be adapted continuously to the patient's individual needs. Within predefined limits, ventilators can adjust the flow, pressure and time of each breath to diminish the risk of lung injury. The ventilator varies the inspiratory flow for each breath to achieve the target volume at the lowest possible peak pressure. This prevents "volutrauma" (ripping and tearing of the lung's intrapulmonal structures). Next-generation ventilators will be able to adapt themselves to data obtained from patients, responding to their changing needs ("closed-loop control ventilation"). Maintaining adequate tissue oxygenation in critically ill patients is of crucial importance, yet there are still no techniques with which tissue oxygenation can be directly measured.

► **Tissue oxygenation.** Haemoglobin molecules transport chemically bound oxygen to the capillaries, releasing it there before flowing back through the veins as de-oxygenated haemoglobin. Oxygenated haemoglobin levels therefore give the most direct approximation of peripheral oxygen supply. Oximetry, a non-invasive method to monitor oxygenated haemoglobin levels in the arterial blood, uses spectrophotometric measurements: the absorption of defined light waves through a medium is proportional to the concentration of the substance that absorbs the light and to the length of the path the light travels. Oxygenated haemoglobin and deoxygenated haemoglobin differ largely in their capacity to absorb light – at wavelengths of 660 nm and 940 nm respectively, the shorter length being the red region of the light spectrum (which is why arterial blood is redder than venous blood). Measurements with both wavelengths reveal how much oxygenated haemoglobin exists as a percentage of total haemoglobin in the blood. An oximeter usually takes the form of a small glove-like plastic sleeve that is placed over a fingertip or an earlobe, with light-emitting photo transmitters on one side and photoreceptors on the opposite side.

► **Pulse oximetry.** The advantage of this measurement method is that the photo detector amplifies only light of alternating intensity, so that only light absorbing haemoglobin in pulsating arteries is detected, while haemoglobin in non-pulsating veins is ignored, as is the background absorption of haemoglobin by surrounding connective tissue. Oxygen-saturated hae-

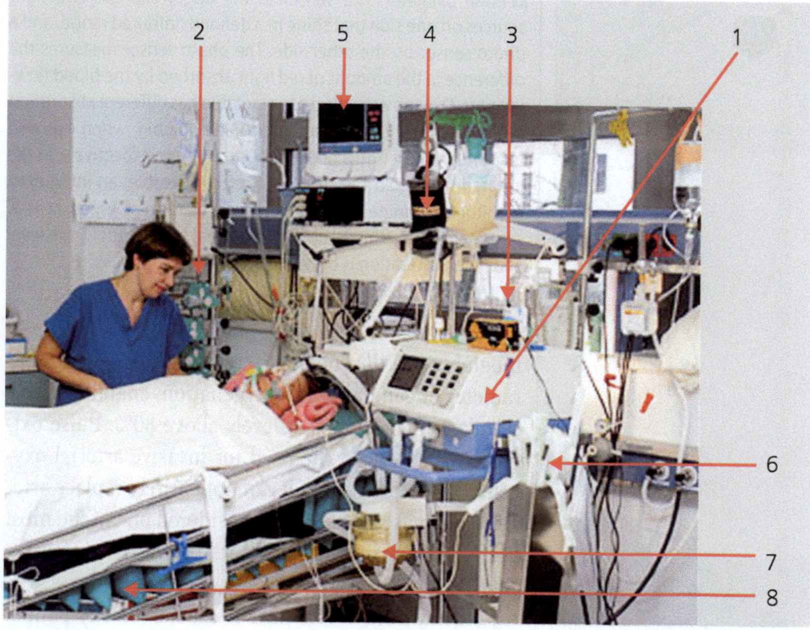

Intensive care bed: Intensive care patients are kept in beds with high-tech equipment that enables vital bodily functions to be maintained. Respirators (1) ensure a reliable gas exchange in the lungs by administering an oxygen-enriched gas mixture. (2) Electronically controlled precision pumps provide an accurate and continuous intravenous supply of highly effective drugs. Nutrition pumps continuously feed individually composed alimentary solutions into the patient's body through a plastic probe (3), which leads to the stomach or small intestine and can be sealed off from the oesophagus by a balloon so that no stomach contents can flow back. In the event of a major blood transfusion, the blood previously kept cool for storage purposes is heated to body temperature by blood heaters (4) prior to the transfusion in order to prevent the patient from cooling down. (5) Data concerning vital bodily functions such as blood pressure and heart rate are continuously recorded, saved and displayed on monitors. Acoustic and visual alarm functions immediately alert the doctors or nurses to any changes. (6) Soft suction catheters siphon off mucus and liquid from the upper air passages. (7) An active respiratory gas humidifier and heater supports the physiological aerolisation of the respiratory gases. (8) The blue air chambers of the special intensive care bed can be synchronised by wave mechanics to ensure even pressure distribution of the patient's body weight. This is an effective preventative measure against bedsores (decubitus ulcers). The bed can also quickly be tilted so that the patient's head is pointing either upwards or downwards if they are suffering from circulation problems. Source: Orthopaedic Hospital Vienna-Speising

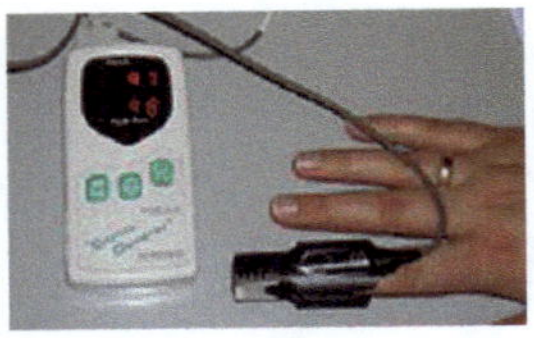

Pulse oximeter: The sensor in the finger clip has two light sources on one side that shine in a defined (infra)red range, and a photo sensor on the other side. The photo sensor measures the difference in the amount of red light absorbed by the blood flowing past. Oxygen-saturated haemoglobin has different absorption properties to oxygen-unsaturated haemoglobin, when exposed to light at wavelengths of 660 nm and 940 nm respectively, in the infrared range. On the basis of a reference table, an integrated monitor determines what percentage of the red blood cells is saturated. In healthy tissue with a normal blood supply, the level of oxygen saturation is between 96–100%.

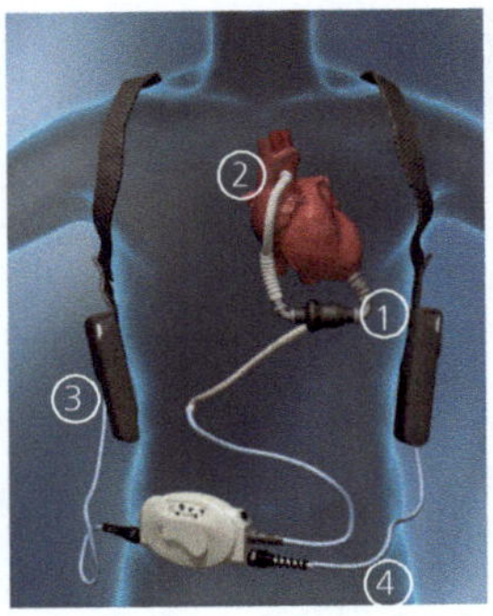

Ventricular assist device: The HeartMate II LVAS is a support system for the left heart. It works according to the axial pump principle, is designed to operate for long periods of time and has a pump output of up to 10 l/min. The inlet duct (1) is implanted in the left tip of the heart (apex), while the outlet duct (2) is inserted via a vascular prosthesis into the aorta. Weighing just 350 g, the implanted miniature machine pumps the blood continuously out of the left heart chamber and into the aorta with the help of a small propeller. The control electronics (3) are worn by the patient in a small case on a belt outside the body. The pairs of 12-V batteries required to power them can be carried in a shoulder belt. The electric cable for the power supply (4) runs from the abdomen through the skin of the stomach to the control unit. Source: Thoratec Corp., USA

moglobin can thus be monitored continuously as a parameter of general tissue oxygenation, enabling stable circulation conditions at levels above 80%. Pulse oximetry has reduced the need for invasive arterial oxygen measurements, but it can not entirely replace arterial blood gas samples. It is considered one of the most useful monitoring devices for patients in intensive care.

► **Fibrillation.** While time is crucial for any patient requiring intensive care, survival is essentially time-limited in the case of cardiac arrest. Cardiac arrest (and thus pump failure) arises when the heart stops beating completely (asystole), or when the rhythm is so fast (fibrillation) or irregular (tachyarrhythmia) that contraction of the heart muscle (myocardium) no longer correctly responds to normal nerval stimulation. In both cases, circulation stops or is minimal and no pulse can be felt. All organs become oxygen-depleted, the brain being irreversibly damaged within 3–6 minutes. An ECG can detect a fibrillation of the myocardium and effect immediate defibrillation. The myocardium consists of muscle cells that are closely linked through electrochemical and mechanical activity. The synchronised contraction of these cells is prompted by a coordinated wavefront of electrochemical activity (approx. 40 mV) stimulated by the hierarchical system of pacemaking and conducting cells of the heart. Each activation ("depolarisation") is followed by a short regenerative phase in which the normal membrane potential is restored.

► **External defibrillation.** When the ventricular myocardium has been injured as in an acute infarct or in shock, the activation sequence is lost, creating irregular activation waves throughout the heart – often at high frequencies. The external defibrillator overrides this fluctuating electrochemical chaos in a one-time depolarisation of a critical mass of heart muscle by delivering an electrical shock of 200–300 J through the skin above the heart (corresponding to approximately 750 V for 1–10 ms, of which approx. 10–20% actually reach the myocardium). The automatic external defibrillator normally comprises two paddles which are placed over the chest of the patient, and electrodes for

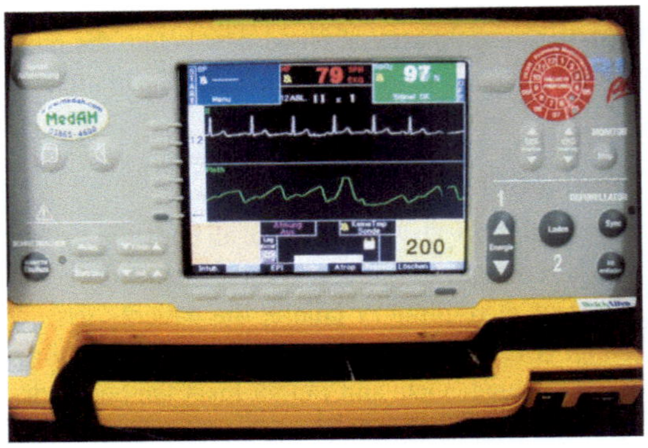

Defibrillator: The monitor shows the heart rate (red figure, top centre), the currently selected ECG lead (white progress curve, centre) and the oxygen saturation of the blood (green field at top right, green progress curve at bottom centre). The currently selected energy level for the defibrillator appears in the yellow field at the bottom right. The desired amount of energy (in joules) can be selected using the button at the bottom right next to the monitor. In the blue field at the top left, it is possible to display measurements taken by a blood pressure sleeve (not connected here). The function of an external temporary pacemaker can be displayed in the orange field at the bottom left.

the electrocardiogram, to analyse the heart's rhythm before applying the shock. The shock terminates fibrillation or tachyarrhythmia and allows the endogenous, spontaneous rhythm of the sinuatrial node to re-establish itself as the pacemaker. Biphasic defibrillation decreases the amount of applied energy required and has become the standard mode of application with success rates of over 90% for the first shock.

Trends

► **Implantable cardioverter defibrillator (ICD).** These small electronic devices are implanted underneath the pectoralis muscle covering the chest and they extend electronic leads through a major vein into the right ventricle or the right atrium of the heart. The electrodes become self-adherent to the endocardium, the inner lining of the heart. They contain sensors which pick up the *electrocardiogram (ECG)* from the endocardium, which is then analysed in the ICD's microprocessor. The multimodal processor can react both to a very slow endogenous pulse (bradycardia) and to an extremely fast and irregular pulse (tachyarrhythmia). Today's ICD microprocessors offer a great variety of stimulatory programs adaptable to patients' individual needs, as determined by the ECG. The lithium batteries used to power the ICDs last at least 5–10 years. In order to replace an ICD, it has to be extracted from its muscle pocket. The electrodes can be left in place and simply re-attached to the new device. However, the insertion sites of the leads may become fibrotic over time in a type of scar tissue, resulting in "undersensing" or "oversensing". Bradycardia is the classical pacemaking indication. If the patient's own pulse drops below 40/min, the ICD will send small pacemaking impulses of 2–5 V through the electrode to the right heart to generate extra heart beats. However, if the patient suffers from tachyarrhythmia, or has already survived an episode of fibrillation, an ICD may be implanted which is also programmed to automatically administer electric shocks to terminate the otherwise fatal arrythmia. These shocks are much more powerful than pacemaking impulses (10–40 V) – and can be painful – but they last only a few seconds. Current devices can sense ventricular tachyarrhythmias and terminate them without resorting to a painful shock. Diagnostic features in current devices now include a "second look" by rhythm analysis prior to the shock, so that energy delivery can be aborted if need be. Home-monitoring technologies are now integrated into the ICDs. All cardiovascular data and events registered by an ICD can be continuously transferred in real time from patients anywhere in the world to any doctor via an integrated antenna and the mobile communications network.

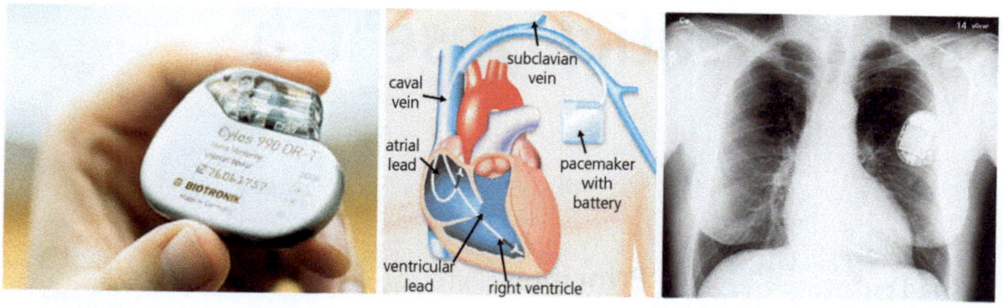

Implantable cardiac device (ICD). Left: Pacemaker that emulates the sinuatrial node – the natural pacemaker of the human heart. It stimulates the heart in accordance with the given level of physiological excitation, taking into account measurements of the patient's myocardial contractility, thus achieving improved heart rate responses and heart rate variability. Centre: The pacemaker is inserted to the left into a pocket of skin under local anaesthetic. The electrode probe pierces the cephalic vein, is pushed along with the flowing blood to the right atrium, or the top of the right heart chamber, and is anchored there. The pacemaker in this diagram has an additional atrial lead, which can both measure and stimulate electrical activity in the atrium. The tip of the electrode can be fixed actively (with a tiny helical screw) or passively. Right: Frontal X-ray image of a female thorax. The pacemaker and the electrode cable are impervious to X-rays, whereas the blood vessels and muscles are transparent. The bony thorax is visible, as are the heart's muscle mass as a silhouette and the cartilage-supported bronchial system. Source: Biotronik, tsp/Bartel

▶ **Mechanical ventricular assist device (VAD).** Such a device can be an option in the case of acute heart failure or as a temporary "bridge to recovery", especially if the patient is waiting for a heart transplant and conventional drug therapy fails. VADs are mechanical pumps that help the heart's left ventricle to pump blood to the aorta. They are implanted surgically. A new left ventricular assist system (LVAS) uses a continuous flow pump that constantly moves blood with an Archimedic screw, i. e. a spinning rotor. This allows the device to be slimmed down to a mere three inches in length and a weight of less than a pound. An electrical cable that powers the blood pump passes through the patient's skin to an external controller worn on the waist. The controller is either powered by batteries or connected to an electrical power outlet. The pump regulates the blood flow according to the patient's needs, and the controller's microprocessor monitors the pump's performance, sounding an alarm if it detects dangerous conditions or a possible malfunction. The system can operate on two external batteries, allowing the patient to move freely for up to 3 hours.

Prospects

- Technical advances in ICU diagnostics will be characterised by further miniaturisation, data integration and personalisation. New sensor technologies will replace ex vivo measurements and will measure such variables as glucose, electrolyte, pH or oxygen values directly in vivo either transcutaneously or intravascularly. Online in vivo measurements shorten the time to result, reduce the amount of blood drawn from the patient and avoid errors in sampling or labelling.
- Advanced patient data monitoring systems (PDMSs) will not only collect data from the patient electronically and establish an electronic patient record, but also process and analyse trends instead of single values (e. g. heart rate, serum electrolytes, glucose). Integrating physiological data by PDMS will help to reduce medication errors and adverse drug events.
- Future treatment will be more individualised, e. g. the requirement for mechanical ventilatory support will be measured on a breath-by-breath basis and adapted to the patient's needs. Real-time measurement of blood glucose will lead to the immediate adjustment of insulin dosing.
- There is a rapidly growing global shortage of highly trained doctors in the field of intensive care. This problem has been recognised by the health ministries in Europe, the United States and Indonesia and by the WHO but is estimated to require a multi-billion dollar investment which will only prove successful in the long term.

DR. URSULA HAHN
Medical Valley Bayern e. V., Erlangen

DR. R. STRAUSS
University Hospital Erlangen, Friedrich-Alexander University Erlangen-Nuremberg

Internet

- www.sccm.org
- www.esicm.org
- www.icu-usa.com
- www.aral.org.uk
- www.ccmtutorials.com

Pharmaceutical research

Related topics

- Stem cell technology
- Systems biology
- Gene therapy
- Industrial biotechnology

Principles

The development of new drugs is driven by the effective use and application of novel technologies, such as molecular biology and pharmacology, highly developed methods in chemical analysis and synthesis as well as powerful computational systems and robotic machines. Nowadays, the process is focused and based on the growing knowledge about molecular principles and processes as opposed to the phenomenological approach of former times. The molecular description of a disease enables the targeted development of a drug substance that constitutes the central ingredient of a marketed medicine or drug. All active drug substances are chemical compounds – in other words, molecules that interact with other (bio)molecules (e. g. enzymes and receptors) in the body to ensure the desired efficacy. The active drug substances are mixed with excipients and other ingredients in order to obtain the optimal application form, e. g. a tablet, an injection or a drinkable liquid. The development of a new drug takes 12–14 years from early research to market entry, with overall industry success rates of below 1%; the costs per commercialised drug prior to market entry are in the range of 900 million to more than 1 billion euros.

New-drug development can be described as having different phases, among which the research activities focus most of all on new and innovative technological methods.

▸ **Pharma research.** In most cases, a disease-relevant molecular target is needed to begin with, and is sought using biochemical methods and approaches in molecular biology. A target is a biochemical structure – normally a protein – that has the potential to significantly impact the disease process. Specific processes addressing disease-relevant mechanisms can be initiated once a drug has started binding at the desired target.

Not until a disease-relevant target has been identified does the drug-hunting process begin. One of the most common subsequent approaches is the *high-throughput screening (HTS)* process, in which several hundred thousand or more different chemical compounds – available at mostly internal and proprietary compound libraries at pharmaceutical companies – are screened in vitro against a target. The endpoint that determines a hit in most cases is the binding affinity. The screening hits identified via these methods are evaluated in terms of their developmental potential until a so-called lead compound is found. The chemical structure of this lead compound is then further optimised with regard to the desired profile using methods of medicinal and combinatorial chemistry in close coordination with biological methods. The most promising compound with the best profile is then picked for further development. The research process takes about 4–5 years on average.

▸ **Preclinical development.** Before clinical testing in humans can begin, the drug candidate is evaluated in animal studies with regards to efficacy and its behaviour in a living organism (absorption, distribution, metabolism, excretion: ADME). Potential toxicological effects are evaluated by means of an appropriate number of safety studies in animals.

▸ **Clinical development.** The drug candidate is tested and investigated for the first time in human beings during clinical phase I. For this purpose the drug candidate is appropriately formulated to support the desired application in humans, be it oral intake, intravenous administration or another option. The clinical phase I studies are mainly focused on tolerance and the ADME parameters in humans. There are normally less than 100 volunteers involved in these studies.

The first use of the drug candidate in patients who suffer from a targeted disease takes place during the clinical phase II study. The key parameter being investigated during this phase is whether the drug candidate shows the desired therapeutic effect. Additional outcomes encompass the optimal dosage as well as potential side effects. Most of these studies are run with a few hundred patients.

In the clinical phase III study, the drug candidate is tested on a higher number of patients – often more than 1,000 – and its efficiency and safety profile is thoroughly investigated. As such studies are usually longer-term, investigations are also carried out on side effects and potential interactions with other drugs.

Once the clinical phase III study has been successfully concluded, the pharmaceutical manufacturer or producer applies for regulatory approval from the responsible authorities. After regulatory approval, longer-term observations are normally undertaken in phase IV studies. These are run to observe potential rare side effects, which are often not discovered during the preceding phase III study due to the lower number of patients taking part.

Preclinical and clinical development takes about 8–9 years on average.

Applications

▶ **Target identification.** A fundamental aspect of pharmaceutical research and development involves the identification of disease-relevant targets. Knowledge gained from the human-genome gene sequencing project is often used in this respect. Disease-associated genes can be identified from publicly available gene databases in order to find novel targets for drug development. The gene sequences represent a blueprint for disease-relevant proteins. Among other reasons a "writing error" in a gene sequence can produce an incorrect or defective protein, which can then trigger a disease. The goal of these early research activities is to identify a specific active agent that positively modifies the disease-causing processes – either at the targeted protein or directly at the gene.

▶ **Proteins produced naturally in the body.** These proteins represent the preferential target class in drug research and development. Proteins are responsible for driving and controlling the essential processes in human cells – be it as a receptor, pump, transporter, channel, enzyme or messenger. Malfunctions in these processes can lead to the development of a disease or illness. Various methods can be applied to detect the targeted – in some cases defective or malfunctioning – protein, so that the protein patterns of healthy and ill patients can be compared. Research efforts focus on those proteins found only to reside in ill patients. The process is extremely complex due to the high number of different proteins and the corresponding messengers (mRNA, messenger ribonucleic acid) present in both healthy and diseased cells.

▶ **Ion channels.** Ion channels are transmembrane pores which allow the passage of ions (charged particles) into and out of a cell. Ions generally cannot move freely across cell membranes, but must enter or exit a cell through pores created by ion channels. Ion channels open and close, or gate, in response to particular stimuli, including ions, other cellular factors, changes in electrical voltage or drugs. There are hundreds of different ion channels which are distinguished according to their ion selectivity, gating mechanism, and sequence similarity. Ion channels can be voltage-gated, ligand-gated, pH-gated, or mechanically gated. These gating criteria, along with a combination of sequence similarity and ion selectivity, further subdivide ion channels into several subtypes.

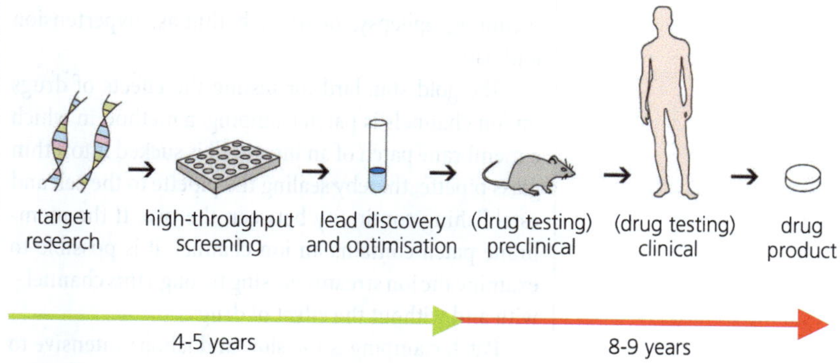

Drug development process: The research phase, in which the biological target is defined and the drug candidate is identified and optimised, takes 4–5 years on average. The preclinical (mouse, rat, dog) and clinical phase (human being), in which the drug candidate is tested with regard to efficacy and safety, takes another 8–9 years on average.

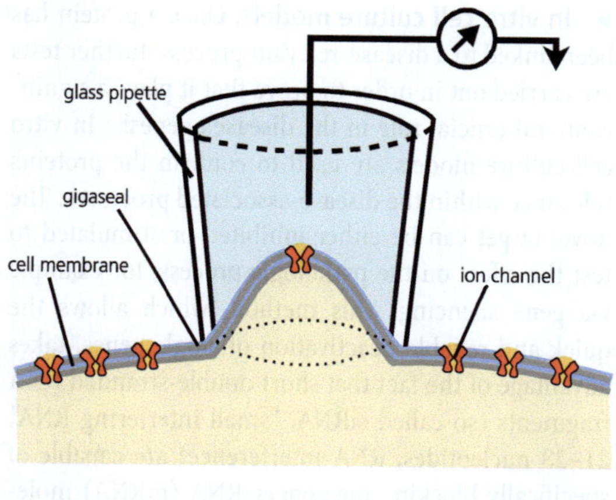

Patch clamp technique: A glass pipette with a very small opening is used to make tight contact with a tiny area, or patch, of a neuronal membrane. After applying a small amount of suction to the rear end of the pipette, it becomes so tightly sealed to the membrane that no ions can flow between the two. This means that all the ions released when an ion channel opens in this area must flow into the pipette. The resulting electrical current can be measured with an ultrasensitive electronic amplifier connected to the pipette. The patch clamp method allows experimental control of the membrane potential in order to characterise the voltage dependence of membrane currents. Source: Fraunhofer

Ion streams mediate information in the human body and play an important part in the regulation of the human metabolism. Small molecule compounds have been shown to both activate and inhibit ion channels. As a result, ion channels represent an important class of targets for pharmaceutical intervention in a broad range of disease areas. Drugs that target ion channels are used to treat a wide array of conditions, including epilepsy, heart arrhythmias, hypertension and pain.

The gold standard for testing the effects of drugs on ion channels is patch clamping, a method in which a membrane patch of an intact cell is sucked into a thin glass pipette, thereby sealing the pipette to the cell and establishing continuity between the two. If the membrane patch contains an ion channel, it is possible to examine the ion streams passing through this channel – with and without the effect of drugs.

Patch clamping is too slow and labour-intensive to be used in a primary screen. A much faster method for screening applications is to use voltage-sensitive dyes to measure changes in transmembrane electrical potential. Depolarisation across the cellular membrane is indicated by a rapid increase in the fluorescent signal within seconds. An appropriate membrane-potential-sensitive dye can be selected for the cell line and ion channel of interest.

▶ **In vitro cell culture models.** Once a protein has been linked to a disease-relevant process, further tests are carried out in order to prove that it plays a significant and crucial role in the disease's genesis. In vitro cell culture models are used to confirm the protein's relevance within the disease-associated processes. The novel target can be either inhibited or stimulated to test the effect on the pathologic process, for example via gene silencing. This method, which allows the quick and reliable deactivation of single genes, takes advantage of the fact that short double-stranded RNA fragments (so-called siRNA, "small interfering RNA", 21–28 nucleotides, RNA interference) are capable of specifically blocking messenger RNA (mRNA) molecules in the cells. If such synthetically derived inhibitory fragments are incorporated into the cell, the mRNA of the target gene is abolished ("gene knockdown"). The expression of the specific protein can be reduced or prevented – the observed changes compared to the "normal" situation allow deeper insight into the function of the targeted gene or protein, respectively. The search for and investigation of new drug candidates for the treatment of heart failure represents a good example in which an in vitro cell culture model is applied in a disease-relevant process: the hypertrophy (increased or immoderate growth) of heart muscle cells is a common symptom in this type of disease. In order to validate potential anti-hypertrophic targets, cells of a heart muscle cell line are treated with pro-hypertrophic substances, and the effect on cell growth is then determined. If these cells are initially treated with inhibitory RNA (RNAi) that is active against the target to be validated, only the expression of the target protein in the cell is blocked or at least reduced. Consequently, the effect of the pro-hypertrophic stimuli should be decreased or blocked completely if the corresponding gene indeed has a critical role in modulating hypertrophy. The development of a rare blood cancer called chronic myeloid leukemia (CML), for example, could be cleared up on a molecular level, and a tailor-made new drug – targeting critical mechanisms of the cancer cells – could be developed. It is known that, in patients with CML, a genetic error causes a defect in an enzyme controlling cell division. This then induces the uninhibited growth of white blood cells. The drug aims at and inhibits the defective enzyme and prevents further uncontrolled cell division. Healthy cells are not impacted.

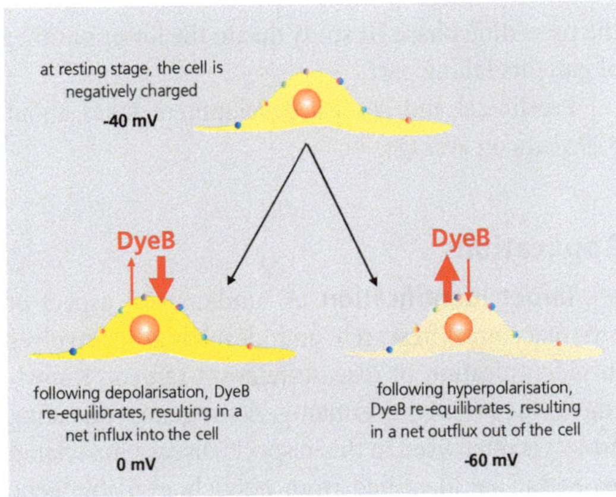

Fluorescent-based assay to measure cell membrane potential. Top: At resting stage, a cell is negatively charged. Following depolarisation, the negatively charged membrane potential FMP blue dye distributes itself passively across the cell membrane and moves into the cell. The more positive the membrane potential of a cell, the more dye will move into the cell, and the more fluorescence is recorded (left). In contrast, a negatively charged cell will force the dye out, resulting in a reduction of cellular fluorescence (right). Thus the intensity of the fluorescence corresponds directly to the cell membrane potential. Source: Bayer AG

▶ **High-throughput screening (HTS).** If a target – in other words, the disease-specific protein – is identified, it is incorporated into an in vitro test system known as an "assay". To simulate a situation comparable to the environment present in the human body, the DNA for the target is incorporated in the test cells in such a way that the target can, for instance, be expressed at the surface of the cells. Subsequently, chemical compounds are screened in a high-throughput format against the target to identify promising ligands with e. g. inhibitory effects – so-called "hits" – that can be further profiled or developed towards a lead structure. For this purpose, compound libraries containing as many as 2 or 3 million individual test compounds are tested against the target using sensitive biological test procedures. The goal of the screening effort is to identify those compounds which have the desired biological effect on the target. The trial-and-error-based screening process is supported by a variety of different robotic and automated systems. When a compound shows the desired activity, a signaling pathway incorporated in the cell starts a specific chemical reaction, which emits a light effect that can be measured with appropriate analytical photometers.

The HTS process is usually carried out to identify innovative small molecule hits and lead compounds. These compounds are supposed to have a molecular weight below 500 g/mol. For larger molecules (protein therapeutics or biologicals), other processes apply.

► **Lead structure identification**. Compounds that have been identified as promising hits following the high-throughput screening process are further profiled with regard to binding affinity and target selectivity. The most promising of these are further examined in more specific in vitro assays and tests and then finally, in most cases, in one or two animal models. Once the activity of the investigated compound has been established by means of significant and relevant in vivo data, the compound is classified as a lead structure or lead compound. This lead structure is then further optimised with regard to compound properties, ADME parameters, selectivity, etc. during the lead optimisation phase. The goal of the lead optimisation program is to identify and characterise the most promising active substance – the drug candidate – for further preclinical and clinical development towards a finally approved drug.

During the lead optimisation phase, derivatives of the lead structure are synthesised and tested e. g. with regard to their pharmacological and pharmacokinetic profile. A structure-activity relationship (SAR) of the compound class is developed as the optimisation phase progresses. This information about the SAR and its further development is critical for a successful lead optimisation campaign. Furthermore, the optimisation can often be supported by computational methods to design potential new molecules (computational chemistry). Another technology that is often used to expedite lead optimisation is parallel synthesis (combinatorial chemistry), whereby a high number of structurally closely related compounds can be produced quickly. The design of the produced compound libraries is often driven by the existing knowledge about the SAR at that stage of the lead optimisation process.

The novel drug candidate is the final result of lead optimisation. Prior to the start of further preclinical studies with the newly identified drug candidate, a high number of criteria have to be fulfilled and supported with relevant data sets, e. g. pharmacological parameters such as bioavailability, pharmacokinetic data such as half life (time taken for 50% of the applied drug candidate to be metabolised or excreted from the body), and non-prohibiting toxicological studies. The half life of the drug candidate in different species is often relevant for its future application form – e. g. once daily or twice daily. A number of further experiments are undertaken to elucidate potential drug-drug interactions or to further characterise the ADME parameters. For the latter, the binding profile to cytochrome P450 (CYP) enzymes is an important criterion. These CYP enzymes, located in the liver, play a crucial role in the metabolism of substances produced naturally in the body or those incorporated from external sources, e. g. drugs. Prior to its first application to human beings, the drug candidate needs to be extensively profiled in at least two toxicological animal studies.

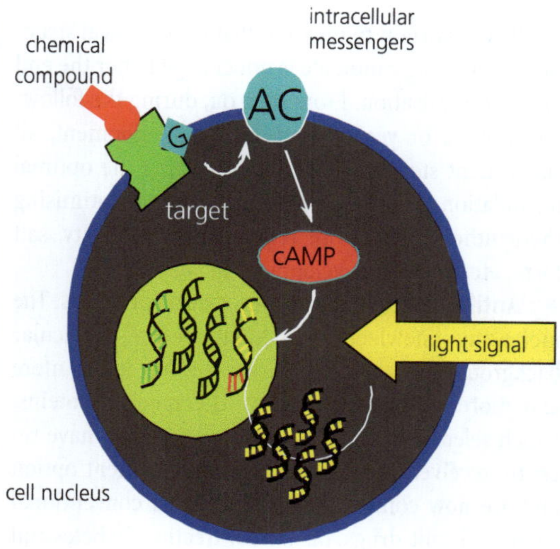

Tailor-made cells as test systems in high-throughput screening: If a chemical compound binds to the target, an intracellular signal cascade is initiated. The signal cascade activates a gene, which was built into the cell nucleus beforehand. The activated gene triggers the release of a fluorescent dye, which then causes a measurable light signal. The detection of the light signal via a spectrometer provides information on the binding process of the chemical compound to the target. Source: Bayer AG

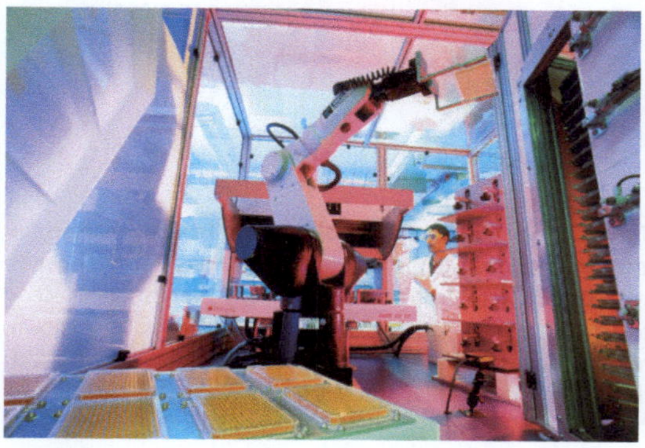

Robot system for high-throughput screening: This method enables 200,000 substances to be tested per day for their biological effect. Microtiter plates with 1,536 "wells" are used as multiple-reaction vessels. For the test, the robot brings the target into contact with the various different chemical compounds in the wells. Source: Bayer AG

It is important to mention that the chemical structure of the drug candidate is not changed after the end of lead optimisation. From then on, during the following months or years of preclinical development, all subsequent steps are focused on finding the optimal formulation for the drug (galenics) and on optimising the synthetic route with regard to costs, purity, salt form, etc. (process development).

► **Antibodies and other therapeutic proteins.** The increasing knowledge gained about the molecular background of a number of diseases has led to more and more options for therapy. Therapeutic proteins, which selectively bind to a targeted receptor, have recently received more attention as a treatment option and are now competing with the more conventional small molecule drugs. Insulin for treating diabetes and the coagulation factor VIII for the treatment of haemophilia are two examples of such therapeutic proteins. These proteins are usually manufactured using gene-based technologies. In both cases, the missing human protein in the body of the patient is replaced with the drug. Another important class of therapeutic proteins is monoclonal antibodies, which play a key role in immune response. They influence diseases by recognising specific molecules, or antigens, e. g. on the surface of disease-causing agents (pathogens). "Monoclonals" are specialised to recognise the specific target structure or symptom only. For instance, they have the potential to bind to specific surface molecules on tumour cells and to start a cascade that will destroy or inactivate the tumour, which allows their use in specific and targeted cancer therapy.

► **Natural products.** It is an option to screen potential active substances isolated from bacteria, fungi, plants or animals with regard to their activity against disease mechanisms – these compounds derived from natural sources are generally called natural products. Natural products can serve as lead structures that are further derivatised and optimised during a lead optimisation programme (semi synthesis). In many cases, the identified natural lead compound is fully re-synthesised in a "total synthesis", which allows flexible access to potentially desired derivatives. A total synthesis of sometimes very complex natural products is often time-consuming and expensive, but it has the advantage of not being dependent on the often very limited availability of the compounds from the natural source. Alternatively, a semi-synthesis using naturally produced intermediates can sometimes be performed. An example of a marketed drug arising from a natural source is represented by a specific taxane, a chemotherapeutic substance initially derived from the yew tree (taxus brevifolia).

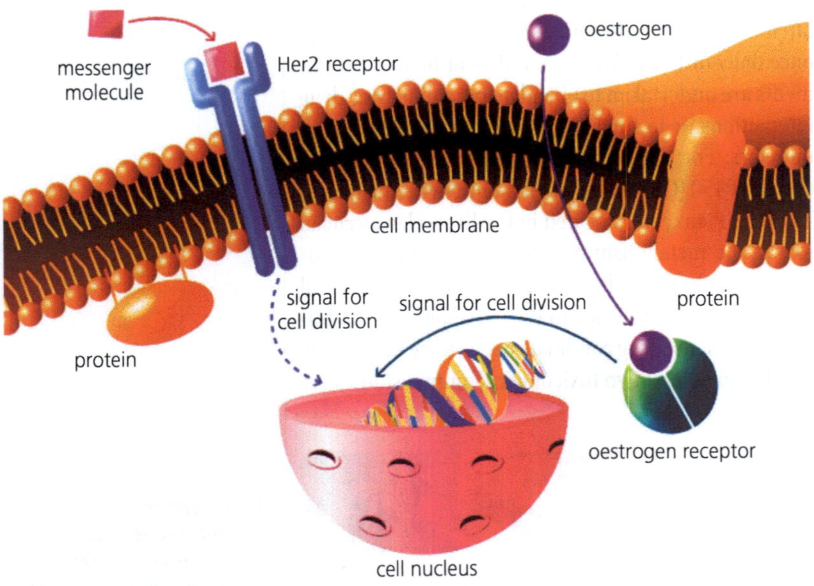

Biomarker for tumour diagnosis: The oestrogen and Her2 receptors represent two biomarkers for different breast cancers. Both receptors transmit signals to the cell nucleus and hereby activate genes that initiate cell division. While the Her2 receptor is located in the cell membrane, the oestrogen receptor is located within the cell leading to different drug interactions or signalling pathways. If it is known which receptor is significantly involved in tumour growth, the appropriate drug can be used to affect the desired signalling pathway and thereby the disease. Source: Bayer AG

Trends

► **Genomics.** It is the aim of genome research (genomics) to obtain a deeper insight into the gene-driven biological processes of the human body. The increasing knowledge gained about these processes and the key role that certain molecules play in disease genesis allows the development of optimally fitting active agents and drug candidates that are able to interact specifically with the targeted molecular structure. This increases the chance of discovering a drug with improved specificity combined with reduced side effects.

► **Pharmacogenomics.** A patient's individual gene pattern has a significant impact on the efficacy of the drug and on patient tolerance. For this reason, the effect that a drug has can be dependent on the patient's genetic pattern. Pharmacogenomics helps to better understand drug metabolism and enables predictions to be made on desired or undesired drug effects for individual patients. Such an analysis of genetic patterns provides information on the individualised efficacy of a drug and therefore generates the potential for personalised therapy, including the use of specific active agents for patients with a specific genome type (individualised or personalised medicine).

► **Biomarkers.** With the advances made in molecular diagnostics, biomarkers will increasingly become the focus of R&D efforts supporting the transfer of drug candidates from the research stage into the clinic, and finally into the patient. Biomarkers are molecules in the human body, such as specific proteins. Based on their involvement in molecular disease processes, these can act as indicators – be it for diagnosis or for treatment control. The group of potential biomarkers is heterogeneous and the characteristic biomarker profile can provide evidence for an optimal treatment strategy. Nowadays, breast cancer patients can already be specifically treated with the most promising, available drug, dependent on the individual diagnosis markers. A set of defined receptors – binding pockets on the surface of the cancer cells – serves as the biomarker in this case and allows the distinction and diagnosis of the tumours. The most promising treatment is selected on the basis of the diagnosed receptor type.

► **Tumour therapy.** There are many reasons for the degeneration of cells. It is therefore unlikely that the pharmaceutical R&D process will generate a single active agent that has the potential to block all potential pathways of tumour genesis. However, the combination of several active agents that address different pathways of tumour genesis could be a realistic option. These combinations could, for instance, inhibit the formation of new blood vessels in the tumour (angiogenesis) and at the same time block tumour cell division. An additional consequence could be programmed cell death (apoptosis) in the tumour cell. The use of drug combinations has the potential to provide long-term control of specific cancers and to upgrade cancer treatment to a status comparable to a chronic disease. This development has the chance to significantly improve the quality of life of affected patients.

The use of tumour-cell-specific antibodies as transporters of toxic molecules for cancer therapy is another growing technology. These antibodies bind specifically to surface molecules that exist exclusively on the surface of cancer cells, which in the ideal case ensures that the adhering toxic molecules are delivered to the targeted cancer cells only. Healthy human cells are not affected by this treatment, since the antibodies do not bind to the surface of those cells.

► **Galenics.** Technical progress in pharmaceutical R&D is not only focused on new active chemical entities – potential new drugs – but also on the development of new and innovative formulations and preparations of these new drugs. Such galenic technologies ensure that an identified active compound is administered in such a way – be it orally, intravenously or in another form – that the drug achieves the desired and best possible effect at the targeted location in the body. The wide variety of galenic technologies available influences which disease indications can be addressed with a drug. Tablets, i. v. solutions, creams, sprays, implants with a depot effect, and plasters (transdermal therapeutic systems) are some examples of potential formulations. The latter are used to ensure a regular and consistent release of the active substance through the skin into the human body. In some cases, this procedure can help to avoid frequent, daily oral administration of a drug.

More recent developments include a technology that allows the inhalation of anti-infectives in order to better treat infectious diseases in the respiratory tract or the lung. An inhalable insulin formulation is already on the market for the treatment of specific forms of diabetes.

New approaches using nanotechnology have recently been investigated to further optimise and broaden the scope of galenic technologies.

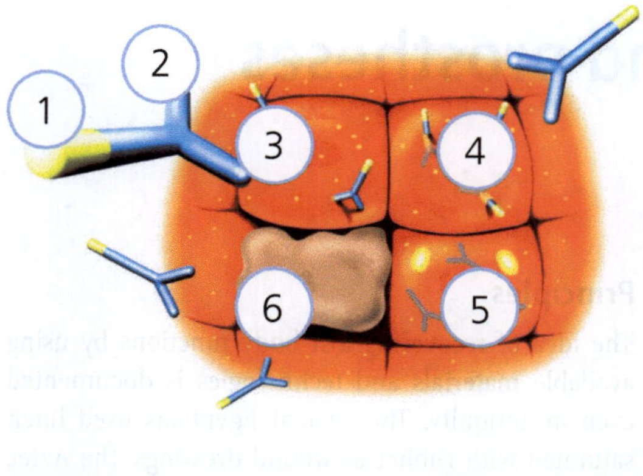

Therapeutic antibody: An inactive cancer drug (1, yellow) is transported to the tumour cell via an antibody (2, blue). The antibody binds to a surface molecule (3) that is specific to the cancer cell. The surface molecule initiates the transport of the anti-cancer agent carrying the antibody into the cell (4). In the environment of the inner cell, the link between antibody and active substance is cleaved. The active substance is activated (5) and the mechanism leading to tumour cell death is initiated.

Prospects

- The increasing knowledge gained about molecular processes (e. g. based on gene data) has delivered new insights into the molecular development of diseases and thus has allowed access to novel disease-relevant targets.
- The goal is to use this knowledge to develop novel drugs with disease-specific effects.
- High-throughput technologies and modern methods of medicinal and combinatorial chemistry allow the targeted and goal-oriented identification and further development of novel drug candidates.
- The identification and use of biomarkers has the potential to efficiently support the process from lead identification to clinical use in patients.

DR. BIRGIT FASSBENDER
DR. TIMO FLESSNER
Bayer Schering Pharma, Wuppertal

Internet

- www.pharmaprojects.com
- www.yalepharma.com
- www.the-infoshop.com/pharmaceutical.shtml
- www.leaddiscovery.co.uk/
- www.insightpharmareports.com
- www.sciencemag.org
- www.fda.gov/CDER/Handbook/develop.htm
- www.drugdevelopment-technology.com

Implants and prostheses

Related topics

- Metals
- Ceramics
- Polymers
- Stem cell technology
- Sensor systems
- Microsystems technology

Principles

The idea of recovering lost body functions by using available materials and technologies is documented even in antiquity. The ancient Egyptians used linen saturated with rubber as wound dressings, the Aztec civilisation used gold in dental fillings, Romans had urological catheters, while the first prosthetic hands, feet and legs appeared during the Renaissance.

Today, the development of implants and prostheses is driven by growing insights into interactions at a molecular level between biological systems and artificial functional materials. The aim is to meet given requirements – be they macroscopic (e.g. function, design) or microscopic (e.g. biocompatibility, bioactivity).

► **Prostheses.** (Exo-) prostheses are artificial extensions that replace missing body parts, either for cosmetic reasons, or to restore certain functions. Loss or restriction of body parts can be a consequence of trauma, tumour removal, or disease-related arterial stenoses. It can also be due to limbs missing or ill-developed at birth. In fact, the bulk of prostheses consist of prosthetic hands, arms and legs. Modern prosthetic arms and hands are usually made of polyethylene, foamed plastics and silicone, which make them appear more natural. Technology is very advanced in prosthetic arms and hands, replacing anything from a single phalanx to an entire arm. This includes active prostheses (i.e. driven by electrical motors and controlled via muscle potential measurement) and passive ones, which have to be brought in position with the sound hand.

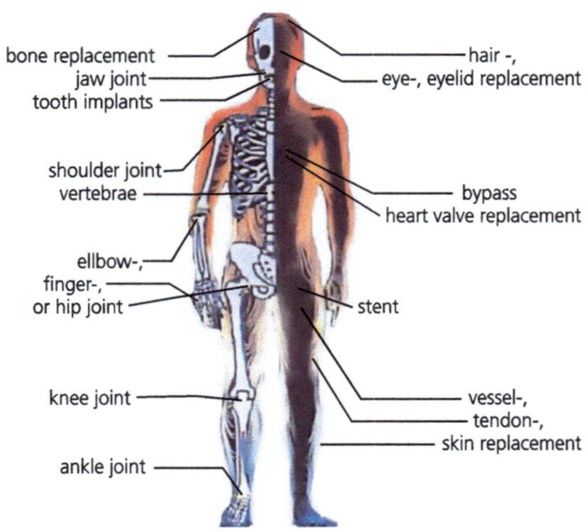

▸ The human body as storage room for spare parts: Examples for the application of implant materials. Source: H. Gärtner

Industrially made modular prosthetic legs consist of an interior steel pipe construction and a foam material lining that is used for aesthetic reasons. Prosthetic feet either have joints or flexible carbon springs (carbon-fibre reinforced).

► **Implants.** Implants include the following products:

- materials inserted into the human body by means of surgery
- materials replacing skin
- materials substituting the surface of the eye

Implants remain in the body either temporarily (less than 60 minutes), for short periods (less than 30 days) or for long periods of time (over 30 days). Most of the implants replace simple mechanical or physical functions of the human body. These functions might require substitution because of a singular tissue defect or as the result of a chronic disease, such as osteoporosis or cancer. These "classical" implants do not assume the function of an organ, nor do they replace entire organs. They are therefore not really an alternative to organ transplants.

Apart from being biocompatible and able to perform the required function, implant materials must be structurally compatible. In other words, they must provide the greatest transfer of loads and forces physiologically possible between the load-bearing implant and the surrounding tissue, such as bone. Modification of the implant surface with suitable biomolecules (e.g. peptides) helps accelerate the attachment growth of the body's own cell tissue.

In recent years, implanting has faced a new challenge in the growing risk of infections in hospitals brought about by strains of bacteria that are resistant to antibiotics, such as the methicillin-resistant strains (MRSAcycles). The most promising solution to the problem lies in implant coatings, which release defined doses of Ag^{+} or Cu^{2+} ions. These, in turn, are highly effective in denaturalising bacteria and preventing their cell division or proliferation; furthermore, the bacteria cannot develop any resistance.

► **Tissue engineering.** Tissue engineering, a subdiscipline of regenerative medicine, involves the in vitro culturing of tissue from the living cells of an organism in order to achieve a physiological substitute for traumatically or chronically damaged tissue, often in combination with extra-cellular components of biological or synthetic nature. In a second step, the regenerated materials or structures are implanted into the organism in question. There are three kinds of donor cells: xenogeneic (derived from a creature of a different kind), allogeneic (from an individual of the same species), and autogeneic (derived from the patient himself). The most obvious advantage of an autogeneic implant is that it is accepted by the patient's immune system, which will recognise the proteins on the surface of the cultivated cells to be its own.

Today, an estimated 25,000 Europeans are living with skin, cartilage or bone cells cultivated "in vitro". To obtain these cells, a tissue sample is harvested from the patient and individual cells are isolated. After a proliferation phase, these cells will grow into new tissue on so-called *scaffolds*. That tissue can then be reimplanted into the patient. The scaffolds are biocompatible and of different flexibility, they can be sterilised and structured. They may offer long-term stability or be biologically degradable, depending on the requirements of the intended application. These structures can be made up of spongy layers or watery and rubber-like gels. Scaffolds may be hard as cement or flexible as fleece. They may consist of woven or hollow fibres. Ideally, cell growth is further accelerated by means of controlled growth factor release. The players involved in tissue engineering hope that the disadvantages of present materials, such as poor processability into 3D structures (natural polymers), or negative physiological response to degradation products in vivo (polyester), will soon be overcome by more advanced materials.

There is a multitude of new ideas in clinical research, including new concepts based on the potential of adult stem cells. Adult stem cells are cells, which – unlike embryonic stem cells – continue to be present in organisms after birth. Their capacity for self-renewal, however, and their potential to develop into specialised cell types are considerably smaller.

Applications

Advanced materials and technologies used in prosthetics include:

- Titanium and carbon fibres, which provide prostheses with mechanical stability at minimum dead load
- *Magnetorheological fluids*, which change viscosity and have a damping effect when a magnetic field

Applications	Metals	Polymers	Ceramics
joint substitutes	titanium and Ti-alloys Co-Cr-alloys	polyethylene (UHMWPE)	aluminium oxide (Al_2O_3) zirconium oxide (ZrO_2)
bone cements		polymethylmethacrylate (PMMA)	calcium phosphates Bioglasses
dental medicine (fillings[1], crowns[2], bridges[2], dental implants[3])	amalgam (Hg-Ag-Sn)[1] Au-alloys[1,2] titanium[3]	dimethacrylate systems[1]	aluminium oxide[2] zirconium oxide[2] glass cerami c[2]
vascular surgery	stainless steel, Co-Cr-alloys	polyester polytetrafluorethylene	
tissue engineering, bioresorbable scaffolds		polyglycolid acid (PGA) polylactic acid (PLA)	

Some examples for application of basic non-biological implant materials.

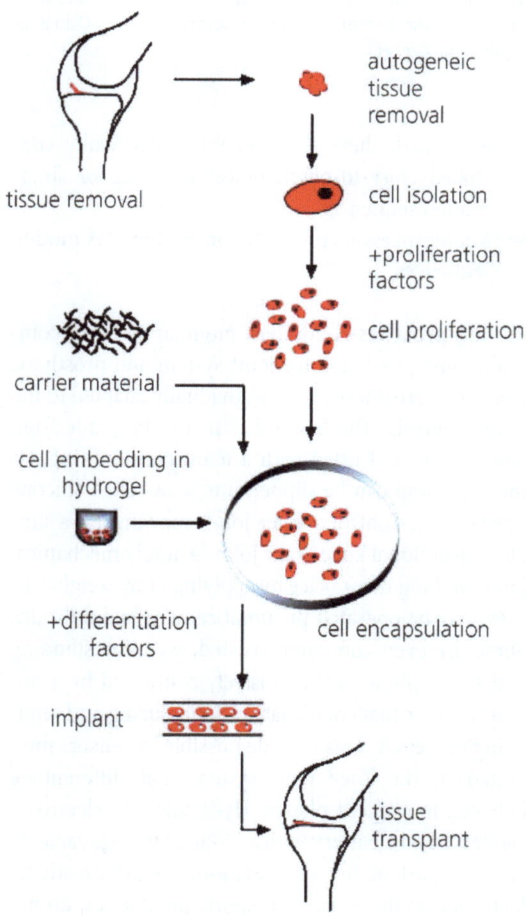

Tissue engineering: Healthy cartilage tissue is collected from an undamaged area of the joint. Cartilage cells are then isolated from the tissue and proliferated. The cartilage cells are then transferred onto a mechanically formable collagen carrier tissue and re-implanted into the patient. The carrier tissue will degrade within just a few weeks - leaving the cartilage cells to remain. Source: Charité - University Medicine Berlin, Germany

Cell scaffold: The bioresorbable dressings based on polyhydroxyethylester fibres are ideal for treating extensive or slow healing wounds. The inorganic material forms a stable, three-dimensional, guiding scaffold supporting the growth of new cells from the wound edge inward. The fleece slowly dissolves by hydrolytic degradation and is entirely resorbed by the body during the course of the healing process. Degradation times can be adjusted to specific applications by way of the materials' synthesis, and the manufacturing process can produce a variety of designs that can be used flexibly.

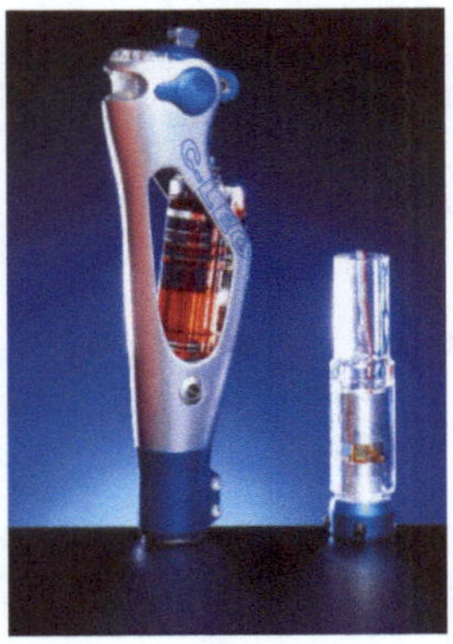

C-Leg: The hydraulic knee joint system (left) is used between a casing at the thigh stump and the tubing adapter for the prosthetic feet (right). The tubing adapter measures the bending moments in the ankle region via strain measuring tapes – simply put, it measures the actual motion sequence of the foot. It then sends a movement command to microprocessors in the knee joint to coordinate foot and knee movements. Source: Otto Bock Health Care GmbH

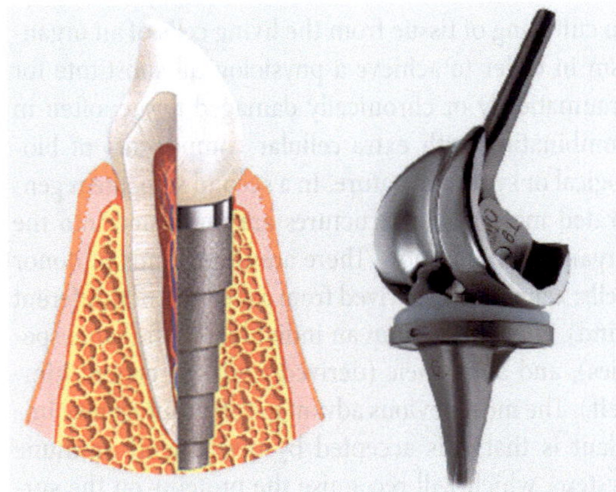

Standard implants. Left: Artificial tooth root made of pure titanium. Right: Prosthetic knee joint consisting of the femoral part (on top) and the shinbone part (below) with a light grey polymeric onlay (sliding surface) Source: Bundesverband Medizintechnologie e.V.

is applied; these fluids enable individually controlled close-to-physiological damping of single motion sequences
- Microprocessors precisely control complex motion sequences

▶ **Leg prostheses.** Modern modular systems, combining liner, socket, knee joint system and prosthetic feet, allow prostheses to be individually adapted to the wearer's needs. The femoral shaft can be made from hard plastic and fitted with a foam plastic lining. Or the leg stump can be slipped into a so-called silicone liner socket. Control of knee joints may involve a simple conventional knee brake joint (a purely mechanical joint; braking takes place by applying body weight), or they may be operated pneumatically or hydraulically. Some are even computer-assisted, whereby standing and swing phases are exclusively controlled by a microprocessor that coordinates all measuring and regulating processes. This is made possible by sensors integrated in the knee joint system that differentiate between movement phases. Hydraulic and electronic elements and a battery with a 35-hour storage capacity are also part of the unit. State-of-the-art prosthetic feet, such as those used for sports prosthetics, do not have ankle joints, but rather springs made of carbon, which are lightweight, extremely durable and elastic. They can be individually adapted to the wearer, who, given regular training, can even engage in high-performance sports activities.

▶ **Implants.** Different materials are used depending on application requirements. Metals such as titanium (alloys), stainless steel and cobalt-based alloys can be used for artificial joints and fixation elements such as plates, screws and nails. Pure titanium is used in dental implants, and medical grade stainless steel in (cardiac) stents. Ceramics, aluminium oxide (Al_2O_3) and zirconium oxide (ZrO_2) are used for dental prosthetics as materials for inlays, bridges or crowns, or in endoprosthetics such as artificial hip joint heads. Calcium phosphate cements and granulates as well as bioglasses are used as bone-replacement materials mainly in the craniomaxillofacial region. Among synthetic long-term stable polymers, ultra-high molecular weight polyethylene (UHMWPE) is the standard material for joint sockets or knee and finger implants. Polytetrafluorethylene (PTFE) is the standard material for use in vessels. Polyethylenterephthalat (PET) is also for vessels and for sutures as well, and, finally, aromatic polyamides (PA) are used for the manufacturing of artificial tendons and ligaments.

▶ **Hip joint endoprostheses.** These prostheses are typical examples of load-bearing orthopaedic implants, i. e. they must be able to transfer high static and dynamic loads (in this case multiple body weight) in the most physiologically accurate manner possible. In principle, they consist of three components. The hip prosthesis shaft is anchored in the femur, or thigh bone (materials: titanium, Ti-alloys), the hip joint head (Al_2O_3, ZrO_2) rides on the shaft and the acetabular cup is fastened in the pelvis. The acetabular cup consists usually of an outer metal shell (titanium) and a polymeric inlay, but it is also available in ceramics only (Al_2O_3). Two types of fixation used are:

- In uncemented prostheses, acetabular cup and prosthesis shaft are either screw-connected to the bone or the shaft is jammed into the bone. A roughened surface or a coating with porous metal structures or hydroxyapatite can promote bone attachment. In some types of prostheses, side rips or teethed flange parts enhance rotation stability.
- In cemented prostheses, the implant is fixed by means of polymethylmethacrylate bone cement (PMMA), which hardens fast and may contain antibiotics (such as gentamicin). Care must be taken to keep the open bone canal free of any tissue and blood remains and to avoid inserts in the cement filling as the prosthesis could otherwise become loose.

If indication allows, only one of the two components, cup or shaft, will be cemented while the other remains uncemented. These prostheses are referred to as hybrid prostheses. The choice of fixation technique depends mainly on the bone quality. If the spongiosa (the spongy bone substance protected by the outer hard corticalis) is still intact, as is the case in most young patients – meaning that it is flexible and elastic – cementing is not indispensible, though it is in the case of patients suffering from osteoporosis. Usually, however, no decision can be made until surgery is well underway and the bone quality can be adequately judged.

▶ **Stents.** Stenoses in coronary arteries are early stages of heart attacks. Approximately 280,000 people suffer heart attacks each year in Germany alone. It is one of the three most common causes of death in industrialised countries. Since 1986, stents – cylindrical meshes – have been implanted to keep arteries open. They are mounted on a balloon catheter and brought to the location of the stenosis and unfolded by inflating the balloon until it is firmly pressed against the artery walls. The stent remains there to support the structure. If lesions are long, several stents may have to be placed. Their position is monitored by X-ray. Once the stent is in place, the balloon is deflated and removed together with the catheter. Owing to its inelastic form, a stent will remain spread in order to keep the artery open.

Apart from meeting the mechanical requirements, such as flexibility, high plastic deformability and low elastic reset after dilation, stents must also offer variable mesh sizes and sufficient X-ray transparency. This is the reason why, at present, approximately 90% of all commercially available stents are made of stainless steel. Current alternative materials are tantalum and nickel-titanium alloys. In addition, there are stent types made of degradable, bioresorbable materials such as magnesium alloys.

In 25–30% of cases, the underlying disease (arteriosclerosis) or excessive cell growth on affected artery walls causes restenoses. Most of the current stent materials have been found to be thrombogenous, in other words, they promote the formation of thromboses, as subintimal vessel areas are exposed by the treatment, which, in turn, stimulates the attachment and agglomeration of blood platelet. In order to prevent tissue reactions of interior vessel walls, stents can be given a polymer coating, which releases cell growth inhibitors into the surrounding tissue via diffusion or degradation over a period of several weeks.

▶ **Bone cements.** At the current time, bone cements – materials for the replacement of bone hard substance – are mostly deployed in the field of endoprosthetics, where they are used for the anchorage of artificial joints. Emphasis here is on the mechanical characteristics of the cement. PMMA became the gold standard, even though it does have some disadvantages:

- The high polymerisation temperature can cause tissue damage.
- Monomers, which have not been chemically converted into polymer chains and are released again from the material, are toxic and therefore harmful to health.
- The polymerisation shrinkage may lead to a loosening of the implant.
- The missing osteoinductive effect (that is, promoting bone formation) can cause a degeneration of the bone substance in the long term.

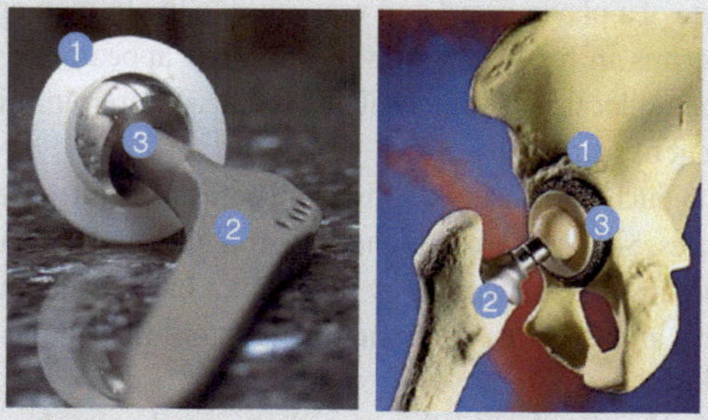

Hip joint prosthesis: Acetabular cup (1), femoral shaft (2), hip joint head (3) Source: Bundesverband Medizintechnologie e.V.

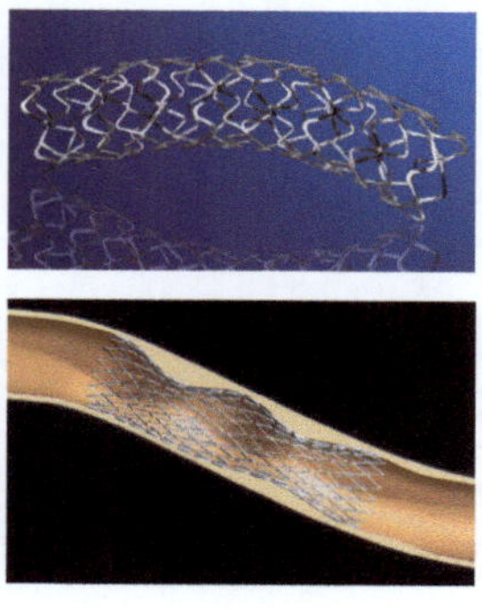

Stent. Top: A tubular metal mesh is mounted in folded state onto a catheter system. Bottom: After being brought into position by means of the catheter, the stent is unfolded by a controlled balloon inflation to match the blood vessel size. The stent remains permanently in the vessel. Source: Bundesverband Medizintechnologie e.V.

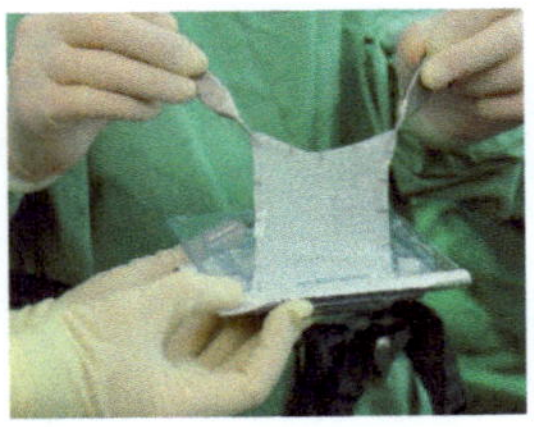

Engineered skin substitute: This skin is placed on a pretreated wound to grow together with the natural skin. Source: Bundesverband Medizintechnologie e.V.

Bone defects following tumour surgery, for example, have to be filled to remedy defects in the cranial area or to remodel ossicles. Biocompatibility is a must, as is low head effect of curing reactions, and bioactivity. The best inorganic bone cements for this purpose are those based on calcium phosphate or glass ionomer. Both alternatives are known for their osteoinductive potential. In the case of calcium phosphate (CaP) cements (hydroxyapatite), this is due to its close chemical relationship to the bone substance. Insufficient mechanical stability, however, limits the application range of this type of cement to non-load-bearing functions in the craniomaxillofacial area.

► **Skin replacement**. Worldwide, some 20 million patients are being treated for chronic wounds every year (venous ulcers, diabetic ulcers and pressure ulcers, for example), and another 6 million for severe burns or chemical burns. This number does not even reflect the increase in chronic wounds in an ageing society. When engineering substitute skin tissue, the precursor cells for epidermal ceratinocytes play a decisive part, as they are the cells that form the outer surface of our skin. Given conditions typical of organs, it takes approximately two weeks for an area as big as a stamp to develop into a three-dimensional piece of tissue the size of a playing card and very close to the human epidermis in texture. However, the outer appearance of the replacement skin available today is more reminiscent of scar tissue.

Such "artifical skin" is gained from in vitro cultures. Skin cells that still have their ability to divide are attached to the wound by means of a biological adhesive and they can continue to grow until the wound is closed.

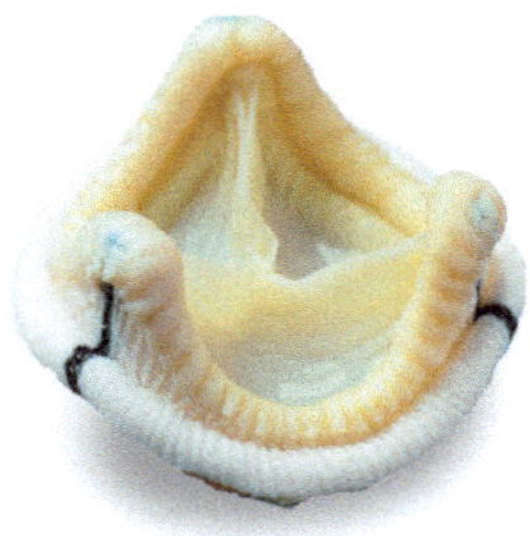

Heart valve: A polyester cuff with an attached biological heart valve (from a pig's heart) for inserting.
Source: Bundesverband Medizintechnologie e.V.

► **Heart valves**. While prostheses made of polymer or metal require a lifetime intake of clot inhibitors – because of the risk of clot forming – the disadvantages of heart valves of biological origin (pig or bovine) are less durability and the risk of an immune response. The starting point of tissue engineering in this case is the cultivation of "body-own" heart valves using autogenous cells. One basic approach is to use the collagen matrix of a pig's aorta valves as the basic structure. At first, all cells, cell components and all DNA molecules of animal origin are removed or destroyed. Some months after transplantation, these valves are seeded in vivo with fibroblasts (the connective tissue for building up cells) and endothelial cells (heart areas, blood and lymphatic vessels lining cells) to form purely autogenous tissue. The advantage of a biological heart valve is that it does not require continuous blood dilution by coagulate-restraining substances. It is, therefore, used in patients with increased risk of bleeding when treated with those substances. Nevertheless, the valve's average life span is limited to 10–12 years.

► **Cochlea implant**. Approximately 100,000 people worldwide have received cochlear implants, with recipients split almost evenly between children and adults. The main indication for cochlear implantation is severe-to-profound hearing loss that cannot be adequately treated with standard hearing aids. The present day multi-channel cochlear implants consist of two main components: the inner (implanted) section and the speech processor and headset.

The inner section is placed under the patient's skin behind the ear during the implant operation. The receiver-stimulator contains the circuits that send electrical pulses into the ear. An antenna receives radiofrequency signals from the external coil. The array with up to 22 electrodes is wound up to 25 mm through the cochlea and sends impulses to the brain through the auditory nerve system. The other parts of the implant system are worn externally. There are no plugs or wires connecting the internal and external components. The coil is held in place against the skin by a magnet and the microphone is worn behind the ear.

Trends

► **Neural prosthetics**. Neural prostheses are assistive devices that electrically stimulate peripheral nerves in order to restore functions lost due to neural damage. The main challenge here is to create a signal transducing and biocompatible interface between the nervous system and the technical system, in other words, a direct connection between nerve fibres and electrodes. One method uses 400-nm-thick platinum conducting lines enclosed in a polyimide capsule. At the contact points, the polyimide is etched off, which enables direct attachment of the nerve to the platinum. Another approach is to use a sieve electrode. Here, the nerves shoot through holes in a disc whereby the holes are surrounded by separately controllable electrodes. Another solution involves threadlike electrodes inserted into individual nerve bundles, or cuff electrodes, which are put around the nerve like a belt.

One example of applied neural prosthetics is the cyberhand. Nerves that are still intact in the patient's arm stump connect with a 10-µm-thick sandwich structure of conducting platinum lines enclosed in a polyimide capsule. The information can flow bi-directionally between the nerve system and the cyberhand. Electrical impulses of the motor nerve fibres are sent by telemetry to the prosthesis and then converted into control signals for six engines in total. In the other direction, technical sensors supply information about the position of the hand, the grasp strength and the risk of objects slipping. This information is also delivered via the electrode after having been "translated" for the human nervous system. As a result, a feeling for pressure and temperature is partly regained.

► **Regenerative medicine**. The targeted use of the human body's powers of self-healing substantially extends the field of medicine by therapeutic options described by the term "regenerative medicine". First steps toward the application of intelligent combinations of high-tech materials and cell cultures have already become reality in the fields of skin and bone replacement. In the long term, research will focus on entirely new therapeutic concepts, especially for life-threatening diseases and those widespread diseases that strongly impair the quality of life: organ lesions such as cardiac infarcts and/or organ failure, cancer, neurodegenerative illnesses, diabetes and osteoporosis. The regeneration and repair processes often begin in stem cells, which are, accordingly, a major subject of contemporary research. Numerous cell, tissue, or organ function processes and regeneration still have to be clarified in order to reach the ambitious final goal, which is the availability of individual stem cell banks that would enable the regeneration or breeding of whole organs or parts of the human body and thus making organ transplantations moot. Although this goal is still a long way away, nature does have numerous examples, for example, among amphibians. And this shows the enormous potential of this form of therapy.

Prospects

- The latest techniques combining material, steering and control are already used for arm, leg and hand prostheses. The implementation of bionic principles in technological developments and the realisation of a direct, bi-directional communication interface between biology (such as nerves) and technology to achieve a greater proximity to human anatomy and physiology is opening up a broad field for a new generation of prostheses.
- Today's classical implants are on a high technological level as far as design and function are concerned. However, there is room for further improvement in such parameters as structural compatibility, antibacterial release mechanisms, integration of drug delivery depots, and the friction-locked connection to the surrounding tissue by biologically effective cell surface modifications.
- The discipline of tissue engineering has already achieved impressive successes in the applied fields of skin, cartilage and bone replacement. New materials for three-dimensional scaffolds, the use of stem cells and new forms of therapy will lead to a new generation of autogenous implants in regenerative medicine, whose final goal is the substitution of complex organs.

DR. JÖRN PROBST
Fraunhofer-Institute for silicate research ISC, Würzburg

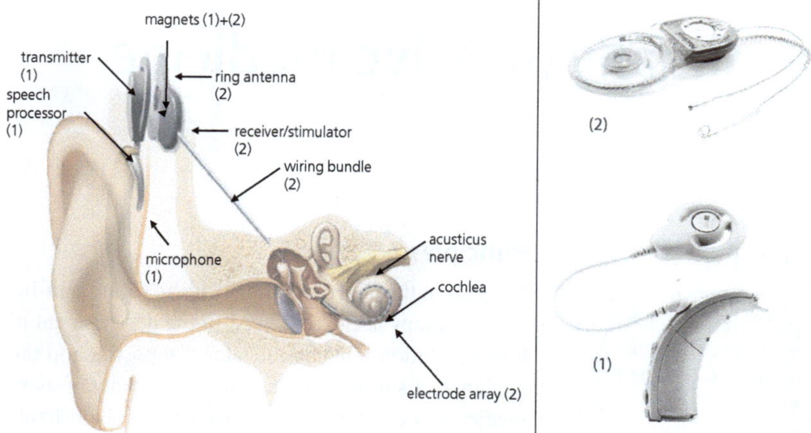

Cochlea implant: Outer (1) and inner (2) components of a cochlea implant and their positioning at and in the human ear.

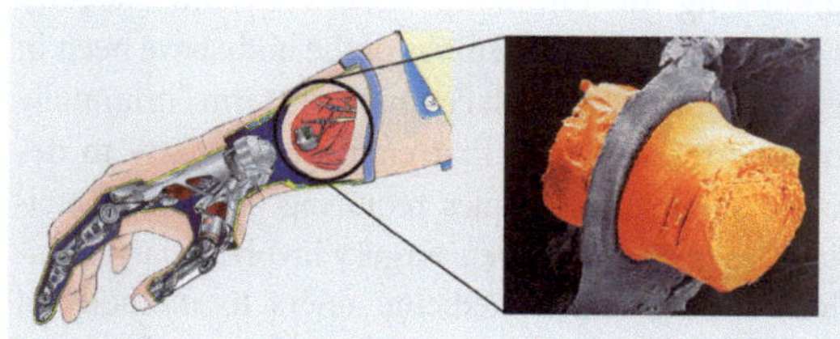

Cyberhand. Left: Inside the hand prosthesis, small engines respond to instructions from the brain. Sensors at the fingertips make palpation feelings possible. In contrast to standard hand prostheses, the cyberhand uses filament electrodes implanted into the lower arm and directly attached to the nerves. Right: The connection between a nerve fibre bundle (orange) and a sieve electrode (grey). Source: Fraunhofer IBMT

Internet

- www.biomat.net
- www.tissue-engineering.de
- www.lifesciences.fraunhofer.de

Minimally invasive medicine

Related topics

- Medicine and information technologies
- Image evaluation and interpretation
- Laser
- Sensor systems
- Microsystems technology
- Medical imaging

Principles

Progress in the field of minimally invasive and non-invasive therapy is closely linked to the development of high-resolution imaging tools for diagnostics and the guidance of surgical instruments. Minimally invasive medicine opens the way to low-cost, ambulant treatment under local anaesthesia with less post-surgery pain and improved cosmetic results. Microsystems engineering, nanotechnology, and laser technology, in particular, have contributed to the development of the required miniaturised high-tech medical devices.

Artificial light sources for the illumination and monitoring of intrabody tissues by introduction through the natural openings of the body have been in use since the nineteenth century. The term "minimally invasive therapy" was first employed in 1989 to describe surgical techniques requiring only small incisions, as opposed to open surgery involving large incisions. Non-invasive medicine refers to methods of examination and treatment without the use of surgical cutting instruments.

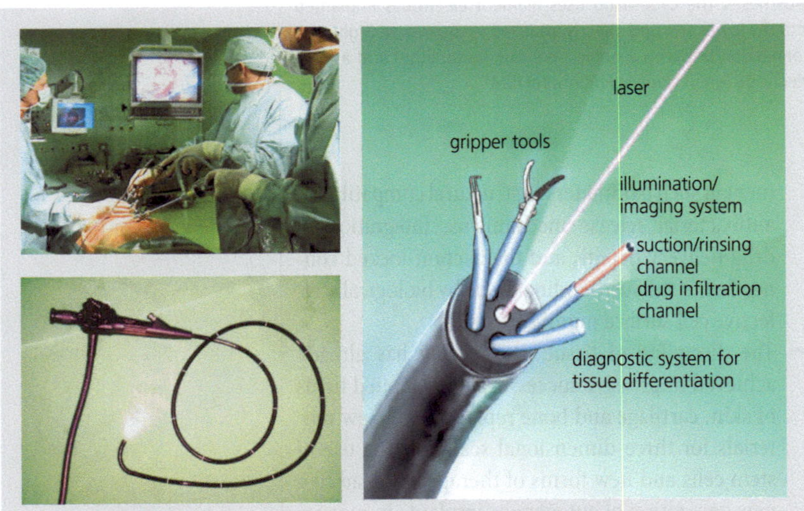

Endoscopy. Top left: Minimally invasive endoscopy through a small opening in the abdominal wall. The surgeons monitor the operating zone and the intrabody location of their instrumentation on the screen and perform the surgery by means of the CCD images. Bottom left: Flexible endoscope. Right: Multifunctional endoscope including a diagnostic system for tissue differentiation, an illumination system, a laser-guiding fibre, channels for drug infiltration and liquid removal and for the manipulation of gripper tools. Source: Karl Storz GmbH&Co.KG, Kalumet (GNU FDL)

Endoscopic surgery is employed widely in all fields of medicine, for instance for appendectomy and the removal of gallstones, laparoscopic examination of joints or the gastrointestinal tract (e.g. coloscopy), the harvesting of stem cells, and the treatment of herniated intervertebral discs. Endoscopic images are typically transmitted via a flexible bundle of *optical fibres* to a monitor. The endoscopic tool is manipulated by means of flexible cables integrated in the endoscope. Two working channels enable body fluids to be removed from the treatment area and permit the introduction of miniaturised surgical tools such as:

- Flexible cutting tools to obtain biopsies
- Flexible injection tools for the local administration of pharmaceutical agents such as chemotherapy drugs or interferon to stimulate immune response
- Flexible wire electrodes for electrocoagulation of blood vessels and tissue removal (ablation)
- Flexible optical fibres for optical coagulation and tissue ablation

The smallest mechanical surgical devices used in microendoscopy, which include micro forceps and high-frequency probes for thermal resection, have a diameter of 200–300 µm. Optical fibres may possess outer diameters of less than 100 µm, while an active core measuring as little as 5 µm can efficiently transmit laser light. Novel laser microendoscopes with a total diameter of 1.8 mm for brain imaging and intradermal imaging have been realised based on microstructured photonic crystal fibres (PCF), silver-doped *GRIN* lenses, and MEMS (*micro electromechanical systems*) scanning technology.

Non-invasive intrabody treatments have been performed with ultrasound, heated electrical wires, and pulsed *laser* light. Lasers are employed for example in ophthalmology for photothermal surgery on the retina and the photodynamic treatment of the macula without having to make surgical incisions in the outermost epithelium layer and without damaging other ocular structures such as the stroma, lens or eyeball. Optical techniques can also be employed to destroy blood vessels, hair roots, or tattoo pigments in the deeper layers of skin tissue without even touching the skin surface. The non-invasive laser treatment of inner organs can be realised by introducing miniaturised microscopes through the natural openings of the body (mouth, nose, urethra, rectum, nipple).

Meanwhile the biomedical use of *near infrared (NIR) femtosecond laser* technology enables cuts with a width as low as 100 nm, or one thousandth of the diameter of a human hair. This opens the way for nanoprocessing even below the fundamental limit of half the laser wavelength. This novel laser nanomedicine makes it theoretically possible to envisage ultraprecise nanosurgery within the human eye, within an embryo, and even within a single intratissue cell. A current application is the optical manipulation of human stem cells such as the targeted transfection by the laser-induced formation of a transient nanopore into the cellular membrane and the diffusion of foreign DNA before the membrane closes due to self-repair mechanisms. The use of femtosecond lasers for the optical knockout of unwanted cells in heterogeneous stem cell clusters has been demonstrated.

Nevertheless, minimally invasive and non-invasive medicine also has certain disadvantages. The surgeon has to rely exclusively on the information supplied by the imaging system, because he or she can no longer make use of the natural tactile feedback provided by his or her hands. Conventional low-cost light endoscopes used in conjunction with either the naked eye or a simple CCD camera as photodetectors can only provide 2D images. To obtain 3D information, it is necessary to use tomographic systems based on *X-rays*, ultrasound, *nuclear magnetic resonance (NMR)*, confocal reflection, optical coherence (OCT), or multiphoton effects. Of these, only multiphoton tomography is capable of providing a resolution in the submicron range.

Applications

► **Multiphoton tomography of human skin.** High-resolution multiphoton tomography based on NIR femtosecond laser excitation of endogenous biomolecules enables precise 3D images to be obtained of skin and other tissues by non-invasive optical sectioning, providing optical biopsies without any surgical procedure. A focused intense laser spot scans the skin in a particular tissue depth pixel by pixel and line by line (like a TV) using fast galvoscanners, and excites special biomolecules, causing them to luminesce. The intratissue excitation depth is varied by piezodriven focusing optics with submicron accuracy. 80 million pJ laser pulses are applied per second. The typical beam dwell time on one tissue spot (pixel) is in the order of 10 µs. The luminescence signal is obtained from fluorescent biomolecules such as the coenzymes NAD(P)H and flavins, the pigment melanin and the extracellular matrix (ECM) protein elastin. The fluorescence, which is the emission of light at a molecule-specific wavelength with a characteristic fluorescence lifetime, occurs from 50 ps (melanin) to 10 ns (porphyrins) after absorption of two NIR photons of an intense femtosecond laser pulse. A TCSPC (time correlated single pho-

diagnostics		therapy	
minimally invasive	non-invasive	minimally invasive	non-invasive
biopsy: removal of tissue for histopathology	camera pill for imaging the gastrointestinal tract	heart catheter, vascular dilatation, stents	photodynamic therapy of tumours and laser therapy of blood vessel anomaly
removal of blood e.g. for detection of blood sugar	multiphoton tomography of skin	LASIK eye surgery	laser therapy of the retina
injection of contrast agents for imaging	endoscopic imaging, e.g. ultrasound catheter	endoscopic microsurgery	laser lithotripsy, e.g. of kidney stones
intratissue sensors e.g. for measuring intracranial	fluorescence diagnostics of tumours	blood micropumps, e.g. in the case of heart attack	ionising radiation, e.g. for cancer therapy
pre-implantation diagnostics e.g. sampling of amniotic fluid	ultrasound and X-ray investigation	localised drug application by microsurgery microimplants	tattoo removal by laser
endoscopic imaging through keyhole incisions	positron tomography	drug infiltration	optical transfection, e.g. for gene therapy
	optical coherence tomography	harvesting of stem cells	heated electrical wires, e.g. in the treatment of prostata hyperblasia
	MRI		

Examples of minimally invasive and non-invasive medicine in diagnostics and therapy.

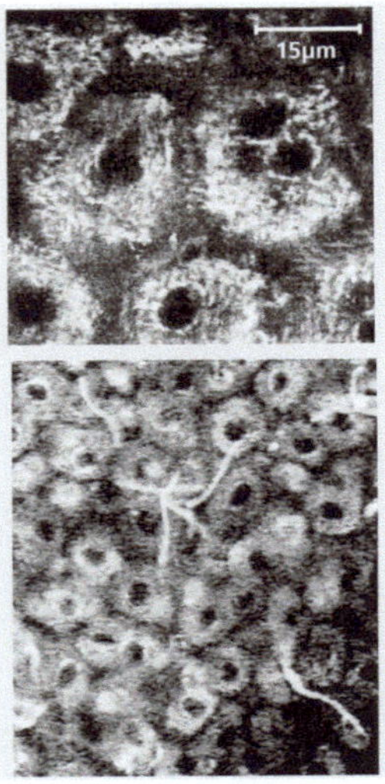

Top: Autofluorescence image of healthy intratissue skin cells using multiphoton tomography. Six single cells are depicted. Even single mitochondria (white organelles around the black nucleus) can be depicted due to the high submicron resolution. Bottom: Single cancer cells (melanocytes) of a patient with melanoma. The image shows the melanin fluorescence of these "spider" epidermal cells (dentritic cells).

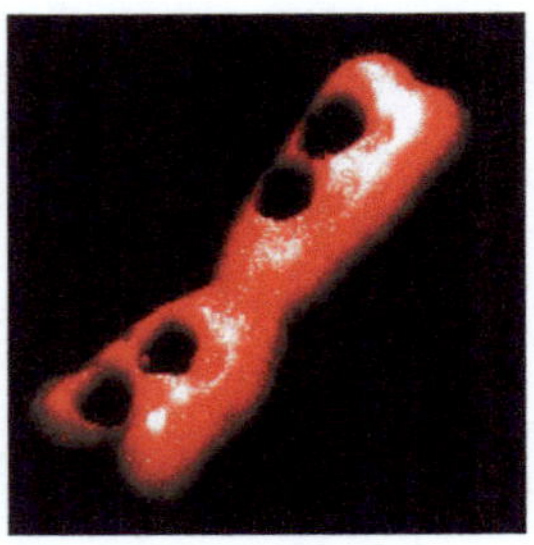

Nanosurgery with femtosecond lasers: Nanoholes with a diameter of less than 100 nm in a human chromosome are depicted.

ton counting) detector is used to detect the arrival time of the fluorescence photon and the fluorescence lifetime, respectively, and provides information on the concentration, the location and the type of the fluorescent biomolecule. The biomolecular structure and ECM protein collagen provide another type of luminescence signal, in which the incoming pulsed laser light is converted into an optical signal at half the laser wavelength by a process called second harmonic generation (SHG).

To date, SHG and fluorescence images produced by multiphoton tomography have been used in clinical centres in Western Europe, Japan and Australia to study the histopathology of melanoma and other tumours in vivo, to understand the effect of ultraviolet radiation on skin aging and keratosis, and to optimise therapeutic approaches for the treatment of chronic wounds and dermatosis. Researchers also use multiphoton tomography to address the hot topic of the biosafety of nanoparticles. They study in particular the accumulation of zinc oxide nanoparticles contained in commercial sunscreens in vital human skin as well as quantum dots and nanodiamonds.

► **Camera pill**. 90% of deaths due to cancer of the colon could be avoided by an early detection system. Current coloscopy techniques cause discomfort to the patient and do not allow examination of all regions of the entire colon (e.g. the small intestine). The use of a novel disposable miniaturised imaging tool for oral administration overcomes these limitations and enables the high resolution imaging of the entire intestinal tract as well as of the stomach. The images are transferred by telemetry to an external receiver carried on the belt. Typically, the "camera pill" takes 8–12 hours to pass through the body, sending 60,000 colour images (2 per second), before being excreted by the natural route. The high-tech camera pill is roughly the size of a normal antibiotic tablet. The device contains a light emitting diode (LED) to provide a white light flash, two microchip cameras, a telemetry chip, an antenna and a 1.5 V battery.

Trends

► **Heart catheter**. Minimally invasive examinations of the heart can be carried out by inserting a thin flexible plastic catheter into a venous or arterial blood vessel, e.g. in the groin, and moving it up into various regions of the heart. The pressure inside the ventricles and the surrounding vessels can be measured by mechanical sensors mounted, e.g. at the tip of the catheter. The physician can also inject X-ray contrast agents through the catheter. This allows images of heart and vessel structures to be displayed on the monitor (angiography). Hundreds of thousands of these investigations are performed annually in developed industrial countries. In addition to their use in diagnostics, heart catheters enable fast therapeutic methods such as widening of the coronary vessels by balloons (balloon dilatation) as well as the stabilisation of the vessel walls by means of coronary stents (tube implants). Novel implants based on nanoscale materials in combination with microsensors and microactuators are under development.

► **Refractive eye surgery**. Rapid technological progress is being made in the field of refractive eye surgery due to the application of novel ultrashort laser systems. More than half the human population has defective vision, requiring correction by means of spectacles or contact lenses. Alternatively, vision correction can now be performed using a laser to reshape the cornea. So far, millions of these *LASIK* procedures have been performed. Typically, nearsightedness (myopia) is treated using ultraviolet (UV) nanosecond excimer laser pulses to remove part of the corneal tissue beneath the epithelium (the stroma) with submicron precision, without damaging the outer layer of the cornea, or epithelium. Because UV lasers cannot penetrate the epithelium, a mechanical microdevice called a microkeratome blade is used to cut away a flap containing the nearly intact epithelium before the laser treatment. The horizontal 9-mm cut produces a flap of about 50–100 μm in thickness. The flap is folded back during laser ablation and replaced afterwards.

The advent of pulsed femtosecond lasers has enabled clinics to introduce a more precise optical tool to replace this minimally invasive mechanical device. So far, about one million patients have benefited from "blade-free" femtosecond LASIK procedures.

The use of femtosecond lasers as an optical flap-cutting tool is just a first step in efforts to optimise clinical laser processing. Researchers have set their sights on developing a non-invasive NIR femtosecond laser system for flap-less intra-ocular nanoprocessing of the cornea that avoids all risk of damage to the superficial epithelial layer. The proposed solution is based on a process called multiphoton-induced plasma formation, where an intense laser beam is tightly focused inside the stroma. Within the tiny focal volume only, tissue is vaporised into micron-scale plasma bubbles. Unwanted damage to out-of-focus regions such as the epithelial layer can be avoided by employing low-energy (nanojoule) laser pulses and focusing optics with a high numerical aperture.

► **Minimally invasive microtherapy**. The two major applications of microtherapy during outpatient minimally invasive surgery are targeted drug infiltration and microsurgery. Chemotherapy drugs, interferon, and even alcohol can be applied locally to destroy cancer cell clusters or to deactivate spinal nerve cells for analgesic therapy. Just a few drops of medication are required if they are applied directly to the target. This can be done by special microneedles under the guidance of imaging tools. Minimally invasive microsurgery is performed using miniaturised heated electric wires or fibre lasers. Typically the microinstrumentation is imaged by an open NMR scanner.

Prospects

- More widespread outpatient surgery based on minimally invasive or non-invasive techniques under local anaesthesia and the guidance of high-resolution imaging tools will reduce the operative zone, post-treatment pain, the duration of hospital stays, and working days lost.
- High-resolution, miniaturised imaging tools for the screening of specific areas of the body. The development of smart pill cameras will enable more site-specific gastrointestinal examinations coupled with highly targeted drug infiltration.
- Non-invasive ultrasound imaging is meanwhile the diagnostic method of choice, in preference to ionising X-ray analysis and other methods.
- Optical imaging tools are gaining in importance due to their superior resolution and ability to deliver optical biopsies on a subcellular scale. This opens up the prospect of obtaining online in vivo histopathological data without physically removing, slicing or staining tissue.
- We can expect the development of novel endoscopes based on high-resolution optical coherence tomography, ultrasound, magnetic resonance imaging and multiphoton tomography.
- Multiphoton tomography will be used in combination with other imaging techniques to detect single cancer cells, to monitor therapeutic effects and to support targeted administration of pharmaceutical and cosmetic drugs. The use of functional molecular imaging will enable the number of animal studies to be reduced.
- Active implants based on biosafe nanoscale materials, microsensors and telemetrically steered microactuators.
- Femtosecond light sources will become an important product in the multi-billion euro laser medicine market.
- Current medical femtosecond laser systems deliver optical biopsies with sub-cellular resolution. They are also employed as ultraprecise surgical tools in refractive eye surgery. With a precision in the sub-100-nm range, laser nanomedicine will finally be feasible.

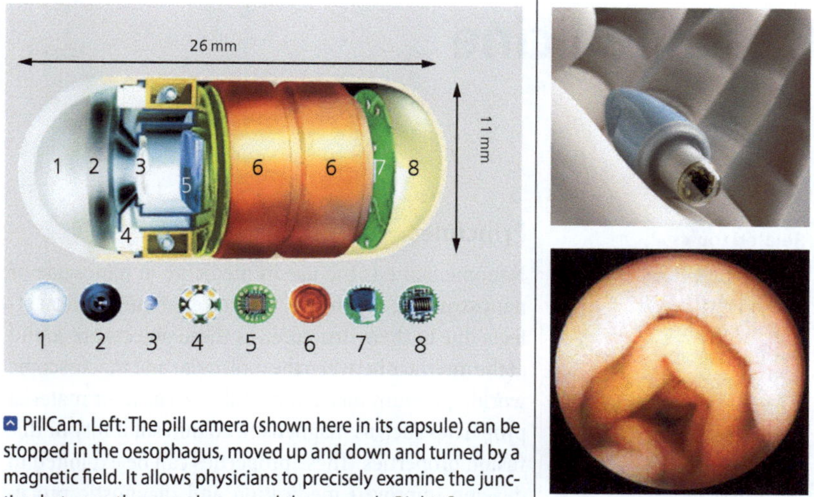

PillCam. Left: The pill camera (shown here in its capsule) can be stopped in the oesophagus, moved up and down and turned by a magnetic field. It allows physicians to precisely examine the junction between the oesophagus and the stomach. Right: Structure of the camera pill: (1) Optical window, (2) Lens holder, (3) Lens, (4) Illumination LEDs, (5) Camera, (6) Batteries, (7) Transmitter, (8) Antenna. Below: Image of a healthy small intestinal tract taken by the camera pill. The pill's transmitter operates at a frequency of 403 MHz and sends two colour images per second to the receiver on the belt. Source: Given Imaging Ltd., Fraunhofer IBMT

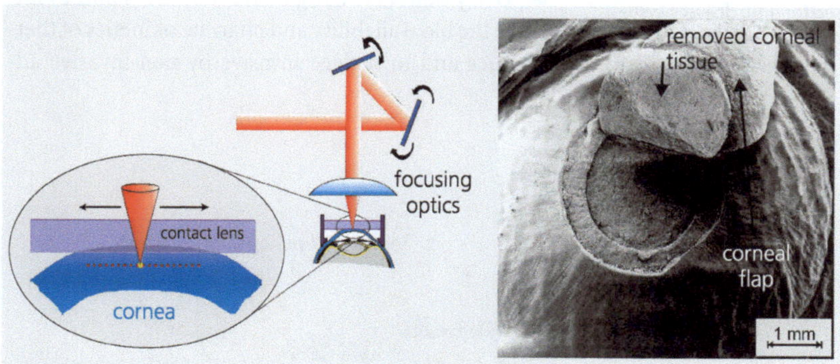

Use of a femtosecond laser in LASIK (Laser-Assisted In Situ Keratomileusis) surgery. Left: The cornea is cut only in the focus of the laser beam using multiphoton absorption. Right: Section of the retina cut away using a femtosecond laser. By removing the laser-treated portion of the lens, the cornea's refractive power and hence the eye's accommodation is modified after the flap has been folded back into place. Source: Laser Zentrum Hannover e.V.

PROF. DR. KARSTEN KÖNIG
Saarland University

Internet

- www.medtech-pharma.de/english/services/thematic-groups/minimally-invasive-medicine.aspx
- www.intralase.com
- www.americanheart.org
- www.ibmt.fraunhofer.de

Nanomedicine

Related topics
- Molecular diagnostics
- Implants and prostheses
- Gene therapy
- Pharmaceutical research

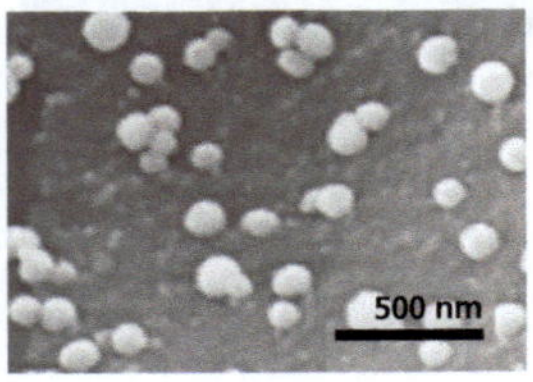

Electron microscopic image of albumin nanoparticles (albumin is a plasma protein) as used for drug delivery systems. Source: AG von Briesen, Fraunhofer IBMT

Principles

Nanomedicine is the use in medicine of nanoscale or nanostructured materials that have unique medical effects due to their structure. As these objects are found at the interface between the molecular and macroscopic world, quantum mechanics still governs their material properties such as magnetism, colour, solubility or diffusion properties. These properties can be exploited to develop improved medication and diagnostic procedures. Some effects of the interaction with cells and tissue are not restricted to objects with a scale of 1–100 nm – the technical definition of nanotechnology – but can also occur at significantly larger sizes. Therefore, the field of nanomedicine traditionally includes objects with a size of up to 1000 nm. The main areas of research in nanomedicine are drug delivery, diagnostics, and biomaterials/tissue engineering.

▶ **Drug delivery**. Drug delivery systems aim to improve the bioavailability and pharmacokinetics of therapeutics and to replace invasive by non-invasive administration. Nano drug delivery systems are a sub-class of advanced drug delivery systems with carriers smaller than one micrometre and mostly less than 200 nm. There is a great demand for nanoscale drug delivery systems, as they have the potential to:
- accumulate the drug in diseased tissue
- increase the solubility of drugs in aqueous media
- overcome biological barriers such as the blood brain barrier
- increase the chemical stability of the drug

Many negative side-effects of drugs are caused by their unspecific distribution within the patient's body. Therefore, the transport of the drug and its accumulation specifically in the diseased tissue has long been the subject of research by the pharmaceutical industry. Nanoscale drug delivery systems have shown the potential to reach this aim through an enhanced permeability and retention effect (EPR). The EPR effect is based mainly on differences between the vasculature in tumours and healthy organs or tissues. Blood vessels in tumours are leakier due to their accelerated growth. Furthermore, tumour cells are very often less densely packed than cells in healthy tissue. Thus, nanoparticles up to a diameter of 400–600 nm are able to diffuse out of the leaky tumour vessels and accumulate in the tumour tissues. Liposomes, which are closed vesicles formed from phospholipids, were the first nano drug carriers developed to accumulate drugs via the EPR effect. Drug molecules can either be entrapped in the aqueous core or intercalated into the lipid bilayer shell. Liposomes accumulated in the tumour are absorbed by the cells, where they degrade and release the drug.

Owing to their unique transport mechanism, nanoparticles are also of interest for the transport of drugs across biological barriers such as the blood brain barrier. The crossing of the blood brain barrier was first demonstrated in rats, using polymer nanoparticles coated with the surfactant Polysorbat 80. The crossing mechanism is not yet fully understood. However, the absorption of a certain blood protein on the particles seems to play a crucial role. Toxicity tests are currently underway to investigate whether this transport mechanism can be used to deliver hydrophilic drugs to the brain.

Research into this class of drug delivery system began in the early 1970s. However, scientists encoun-

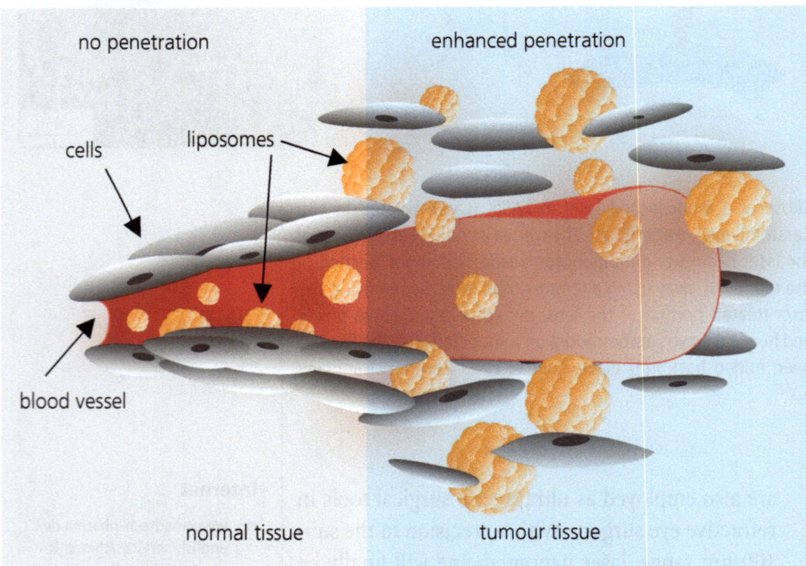

Passive targeting of tumours with liposomes. Left: In healthy tissue, liposomes cannot penetrate the vasculature. Right: The vasculature at a tumour site exhibits large openings through which liposomes or other nanoscale drug delivery systems can penetrate into the tumour tissue. This effect, in which the nano drug carriers accumulate in the tumour tissue, is known as the Enhanced Permeability and Retention Effect. Source: VDI Technologiezentrum GmbH

◻ Liposomes: Liposomes are nanoscale vesicles formed by the self-assembly of phospholipid molecules in an aqueous environment. They are the leading nanoscale drug delivery system, the first products having reached the market almost 20 years ago. Aqueous soluble drugs can be encapsulated within the central aqueous compartment or lipid soluble drugs within the bilayer membrane. Source: VDI Technologiezentrum GmbH, ZT

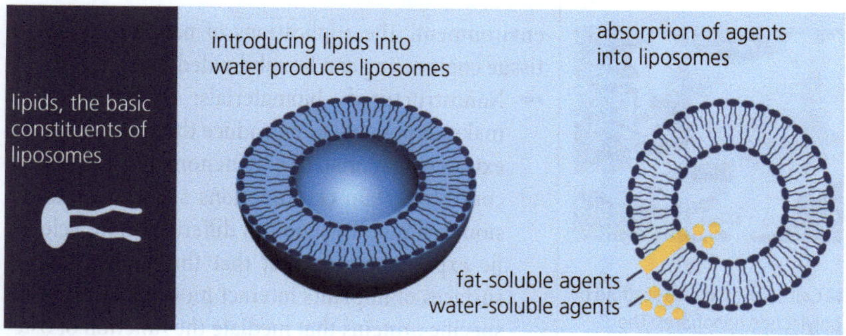

tered a variety of difficulties that have not yet been fully overcome, such as the toxicity of certain polymers and their degradation products, and the rapid clearance of particles from the blood by the liver and spleen. Due to these problems, the pharmaceutical industry has so far displayed little interest in this class of drug delivery systems, and only one anti-cancer drug on the market today is formulated with a biopolymeric nano drug delivery system.

Polymer therapeutics – a further class of drug delivery systems – is characterised by the use of single polymeric molecules that are attached to small molecule drugs or proteins to improve their therapeutic efficacy. Protein drugs often have a short plasma half-life and poor stability and often elicit an immune response. Attaching the chain-like molecule polyethylene glycol helps to protect the proteins from immune response and other clearance mechanisms. This reduces immunogenicity and prolongs the plasma half-life. One commercial example of this technology is pegylated interferon, which is used to treat hepatitis and whose antiviral activity is superior to that of the native protein.

► **Diagnostics.** *Nanotechnology* is applied to improve the selectivity and sensitivity of existing diagnostic methods and to develop new diagnostic techniques for diseases that are still difficult to diagnose.

Most nanotechnology research activities in the field of in vivo imaging have been focused on *Magnetic Resonance Imaging (MRI)*. MRI uses a magnetic field to align the nuclear magnetisation of hydrogen atoms in water in the body. Furthermore, a radiofrequency field is emitted. Under the influence of the magnetic field, hydrogen atoms absorb energy from the radiofrequency fields and produce a detectable signal. To improve the contrast between different types of tissue, contrast agents that accumulate in certain tissue types and alter the signal intensity are used. Superparamagnetic iron oxide particles with a diameter of 50–500 nm are one example of a nano imaging agent. Because of its minute size, this agent accumulates passively in the liver and the spleen, and can be used to image these organs. The contrast provided by superparamagnetic iron oxides is about two orders of magnitude stronger than that of molecular gadolinium compounds, the conventionally used imaging agent, so that the sensitivity of MRI is significantly increased. Ultra-small superparamagnetic iron oxides have an even smaller diameter of less than 50 nm, which delays their clearance by liver and spleen and allows the imaging of other organs. This type of contrast agent was developed for MR lymphography in the 1980s and has been successfully evaluated in various clinical trials for improved detection of lymph node metastases.

► **Quantum dots.** Nanotechnology for in vitro diagnostics includes the use of nanoparticles to improve existing diagnostic methods as well as the development of completely novel methods based on nanotechnological sensor concepts. Sensing systems based on quantum dots play an outstanding role in this area. *Quantum dots* are nanoparticles that consist of semiconductor materials such as cadmium telluride or lead selenide. They are characterised by high photostability, single-wavelength excitation, and size-tunable emissions. This combination of properties enables several different analytes to be measured at the same excitation wavelength with very high sensitivities. Imaging agents based on quantum dots have made it possible to visualise the HER2 protein in cells, a biomarker for a variety of solid tumours. Quantum dots have also been used in small animal imaging to visualise blood vessels, tumours or the reproduction of single cells.

► **Biomaterials and tissue engineering.** The ultimate goal of tissue engineering as a treatment concept is to replace or restore the anatomic structure and function of damaged tissue or organs by combining cells, biologically active molecules and biomaterials, and stimulating the mechanical forces of the tissue microenvironment. To date, most tissue engineering studies are focused on the investigation of macrolevel structures (supercellular structures > 100 μm and cellular structures > 1 μm) to build the essential gross morphology and generate real-size organ systems. However, to engineer the functional units of the tissue, subcellular nanostructures ultimately need to be constructed with exactly replicate the natural cellular

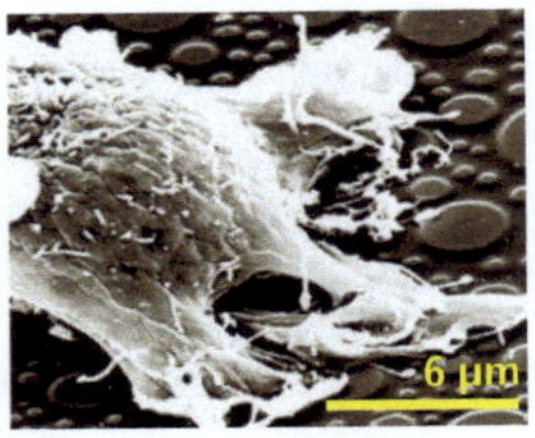

Cell cultivated on a scaffold of polylactide nanofibres: The nanofibres are produced in an electrospinning process that uses an electrical charge to draw nanoscale fibres from a liquid polymer solution. Source: Prof. Wendorff, University of Marburg

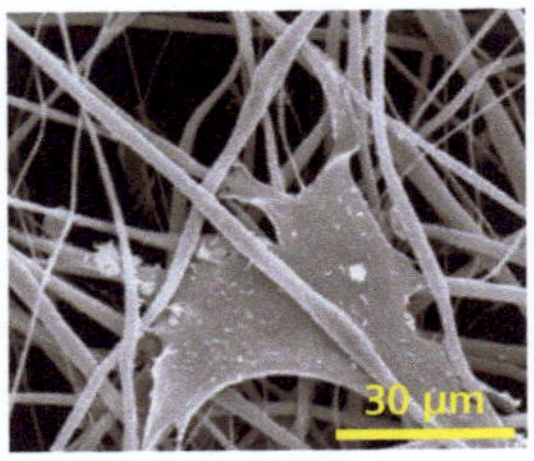

Cell cultured on a nanostructured plastic surface under investigation as an implant material: Cell cultures are grown on nanomaterials to investigate the effect of nanostructures on the adhesion, growth and reproduction of cells. Source: Dr Dalby, University of Glasgow

environment. The applications of nanotechnology in tissue engineering can be subdivided into:

- Nanostructured biomaterials: Nanotechnology makes it possible to reproduce the structure of the extra-cellular matrices at nanometre size, which controls certain cell functions such as cell adhesion, cell mobility and cell differentiation. Scientific experiments suggest that the nanostructured surfaces of implants interact more effectively with specific proteins that mediate the function of osteoblasts (bone cells).
- *Bioactive signalling molecules*: These molecules are naturally present in cells and act as signals within a complex system of communication that governs basic cellular activities and coordinates cell actions. Nanotechnology is used to develop extracellular matrix-like materials that include the specific proteins and peptides that act as cellular cues. Scientists hope to enable the development of bioactive materials capable of releasing signalling molecules at controlled rates. The signalling molecules would then activate certain cell functions, e.g. stimulate tissue repair at the site of the implant.
- *Cell-based therapies*: The major focus of cell-based therapies is the prevention or treatment of diseases by the administration of cells that have been altered ex vivo. Nanomaterials are used to leverage the self-healing potential of cells. Recently it was found that carbon nanotubes help adult stem cells to morph into neurons in brain-damaged rats. Furthermore, nanotechnology is used to develop polymer membranes to encapsulate transplanted cells. The nanopores of the membranes allow free diffusion of glucose, insulin and other essential nutrients while inhibiting the passage of larger entities such as antibodies. Inhibiting the contact between these immune molecules and the cell material can help prevent immune rejection.

Magnetic hyperthermia: Iron oxide nanoparticles in an aqueous dispersion. The patient procedure begins with the introduction of several millilitres of this magnetic fluid directly into the tumour, with the exact amount depending on the type and size of the tumour. Left: Following this minimally invasive procedure, the magnetic nanoparticles gradually distribute themselves within the tumour tissue. Centre: An alternating magnetic field is then applied to bring the nanoparticles into oscillation, without the need for any physical contact. This results in the generation of heat within the tumour which can be precisely controlled. Right: Once the target temperature (4 °C–70 °C) has been attained, the tumour cells become more susceptible for accompanying radiotherapy or chemotherapy, or in the case of higher temperature, are irreparably damaged by the heat. Source: MagForce Nanotechnologies AG

Applications

Liposomes for passive targeting of tumours. The first liposomal formulations of the anti-cancer drug doxorubicin were launched in the early 1990s. The liposomal formulations are designed as long-lived liposomes to achieve high accumulation in tumour tissue via the EPR effect. The cardiotoxic side-effects of the drug doxorubicin were significantly reduced in this way. However, owing to their high price, these first-generation liposomal drugs are only administered to risk patients who are suffering from cardiac diseases and are particularly susceptible to the cardiotoxic side-effects of doxorubicin. Another example of a liposomal formulated drug is Fungizone, which is administered to treat severe fungal infections. Again, a liposomal formulation was chosen to reduce side-effects, in this case the kidney toxicity of Fungizone.

▶ **Magnetic hyperthermia.** Nanoparticle based magnetic hyperthermia is a novel treatment in which iron nanoparticles are injected into the tumour and heated by a magnetic field. The glucose-coated nanoparticles are incorporated by tumour cells much faster than by healthy cells, by a mechanism that is not yet understood. The nanoparticles thus become concentrated in tumour cells. When a magnetic field is applied, the particles start to vibrate and to generate heat which in turn destroys the tumour cells. At elevated temperatures (41 °C–70 °C) the tumour cells become more susceptible to accompanying radiotherapy or chemotherapy, or – in the case of higher temperatures – they are irreparably destroyed.

This method makes it possible for the first time to focus the heat precisely on the tumour and thus prevent damage to healthy tissue. Furthermore, less accessible parts of the body such as the brain can be treated. This treatment concept is now undergoing phase II clinical trials for the treatment of brain and prostate cancer.

▶ **Formulation of poorly soluble drugs.** At present, about 40 % of the drugs in the development pipelines of pharmaceutical companies have poor solubility and therefore cannot be administered. The surface area can be greatly increased by reducing the size of the particles to some 10–100 nm. In addition, the increased curvature of nanoparticles results in a higher dissolution pressure. Together, these effects produce an increased dissolution rate. This technology might be applicable to 5–10 % of small drug molecules. One of the first drugs reformulated as nanocrystals was an immunosuppressant. Apart from the pharmaceutical benefit, the reformulation had the added effect of increasing the original compound's product life cycle.

▶ **Bone replacement materials.** Bone substitutes are applied to bone defects either as paste or as a pre-

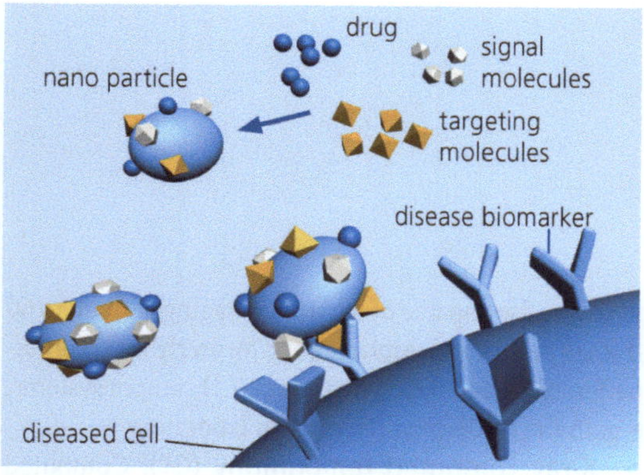

Multifunctional medical systems based on nanoparticles: They comprise 1) targeting molecules towards a cell surface marker of the diseased cell, 2) contrast agents for non-invasive imaging, and 3) drugs to treat the diseased tissue. The development of this kind of multifunctional drug delivery and imaging system is a long-term goal of nanomedicine. Source: VDI Technologiezentrum GmbH

cisely fitted bone graft. The aim is to develop bone substitutes that remodel to natural bone over the time of healing. Classic bone substitutes do not match the nanocrystalline structure of natural bone. This is seen as one of the reasons for the very slow degradation of bone substitutes in vivo. From the clinical point of view, the aim is to fully integrate a bone substitute into the dynamic human organism where bone is continuously dissolved and produced by remodelling. Scientists have therefore developed and investigated nanocrystalline bone substitutes with improved mechanical properties and a higher biofunctionality. One of the early nanomedical products on the market was a nanocrystalline hydroxyapatite bone substitute containing calcium phosphate nanoparticles. The substitute promotes three-dimensional regeneration of bone in the defect site into which it is implanted, and is resorbed as new bone material forms.

Trends

▶ **Multifunctional transport systems.** One of the strategies to reduce side-effects has been to use nanoparticles to accumulate the drug in tumours using the EPR effect side-effect. However, the goal is to carry and guide the drugs more precisely to the desired site of action in order to increase their efficacy. This involves chemical decoration to the surface of the drug carriers with molecules enabling them to be selectively attached to diseased cells and to deliver the drug. Attaching imaging agents to the drug delivery system makes it possible to create multifunctional transport systems that report the location of the disease by non-invasive signalling, deliver the drug, and follow up the progress of healing.

▶ **Sensitive imaging agents.** A further goal of nanomedicine is to develop more sensitive imaging agents by harnessing the specific physicochemical properties of nanomaterials that will enable diseases to be detected at the earliest state of development and treated even before the first symptoms appear. In the long run, this would promote a change from curative to preventive medicine.

▶ **Regenerative medicine.** The current strategy in regenerative medicine is to harvest stem cells from the patient, expand and manipulate them, and finally re-implant them into the patient. The trend followed by nanotechnology for tissue engineering is to develop materials that would be able to send signals to the stem cells already present in the diseased or damaged tissue niches, which would then trigger the regeneration process. Such a non-invasive method would open up the possibility to have cell free materials available off-the-shelf for use as and when required.

Prospects

Over the past five years, nanomedicine has seen a surge in research activities across the world. Nanomedicine has the potential to change the way we treat diseases through drug delivery, imaging and tissue engineering. Nanotechnology enables:

- Drug delivery systems that tackle key pharmaceutical challenges such as targeting the diseased tissue, insufficient solubility of drugs, and protecting against degradation
- Diagnostic methods that are more selective and sensitive than current methods
- Biomaterials that are able to control the regeneration of diseased tissue

In the near future, the goal will be to establish first- and second-generation nanomedicine products on the market. In the more distant future, research will focus on combining therapeutic and diagnostic functions to support a shift in medical care from treatment to prevention.

DR. VOLKER WAGNER
VDI Technologiezentrum GmbH, Düsseldorf

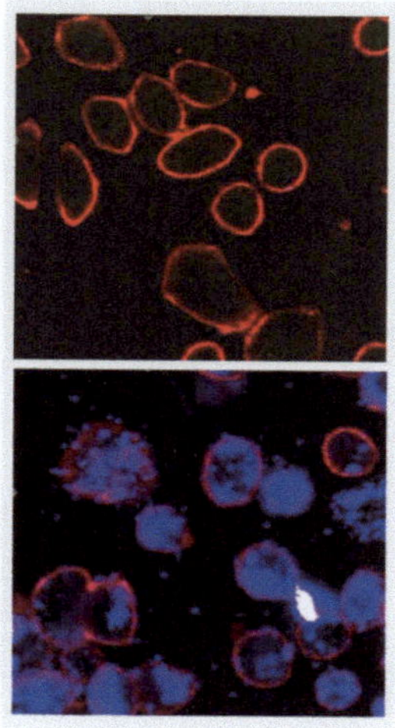

Accumulation of a blue tagged drug by a "normal" drug delivery system (top) and a drug delivery system that directly targets the surface structures of the diseased cells (bottom). Source: AG Langer, University of Frankfurt

Internet

- http://nano.cancer.gov
- www.nanomedjournal.com
- http://nanomed.uth.tmc.edu
- www.nanomedicinecenter.com

Medical imaging

Related topics

- Medical and information technologies
- Image evaluation and interpretation
- Minimally invasive medicine
- Optical technologies
- Pharmaceutical research

Principles

Medical imaging devices play a key role in today's healthcare technology. They have traditionally been focused on diagnostic purposes, but are today applied in many advanced therapeutic and monitoring procedures as well. While physicians are still the final decision makers, they are supported by a variety of imaging modalities and the corresponding image processing and viewing tools. The applied systems are selected on the basis of their sensitivity and specificity for the diseases in question.

▶ **Projection radiography**. This is the standard X-ray imaging technique. A limited part of the patient's body is irradiated by X-rays which are subsequently attenuated by the different body parts. As the body consists mainly of water, bones typically show a very good contrast, whereas the contrast exhibited by organs or muscles in X-ray images is very limited. This is why, for specific diagnostic tasks, iodine or barium contrast agents are injected or have to be swallowed by the patient. Due to their high atomic weight, these agents absorb X-rays better and thus enhance the image contrast of vascular structures or the gastrointestinal tract, for instance.

The projected image is captured by an X-ray sensitive two-dimensional detector system. While combinations of scintillating foils and film sheets or storage plates have been used in the past, direct digital systems – known as flat detector systems – are the trend today. These consist of an X-ray scintillator and an array of photo diodes, based for instance on amorphous *silicon thin film transistor technology* (a-SI TFT). These systems are also capable of dynamic imaging with up to 60 frames per second, replacing the formerly used combination of X-ray image intensifier/CCD camera systems, e. g. in the cardiac catheterisation laboratory.

▶ **X-ray computed tomography (CT)**. Computed tomography (cross-sectional imaging) is a scanning X-ray imaging procedure that builds a three-dimensional (3D) dataset of the patient's anatomy by merging individual X-ray projection images from different directions. A thin, fan-like X-ray beam passes through the patient and is attenuated according to the morphology of the body along its pathway. The projection image is registered by a line of discrete X-ray detector elements. Both the X-ray tube and the detectors are attached to a gantry rotating around the patient. Many projections need to be recorded in order to reconstruct 2D radiographic "slices" of the patient's body. Because the reconstruction algorithms are very time-consuming, they are implemented in modern CT systems by special hardware accelerators.

There are several scanning techniques including a single plane, an axial "step and shoot" acquisition, and spiral techniques in which the patient is continuously moved while the gantry is rotating. In recent years, CT systems in which the original one-dimensional detector line is replaced by an array of up to 256 lines, each consisting of up to 1024 individual detector elements (multi-slice CT), have become popular. These scanners provide a much faster acquisition time, which reduces motion artifacts and improves the workflow.

▶ **Magnetic resonance imaging (MRI)**. Magnetic resonance imaging is a tomographic imaging tech-

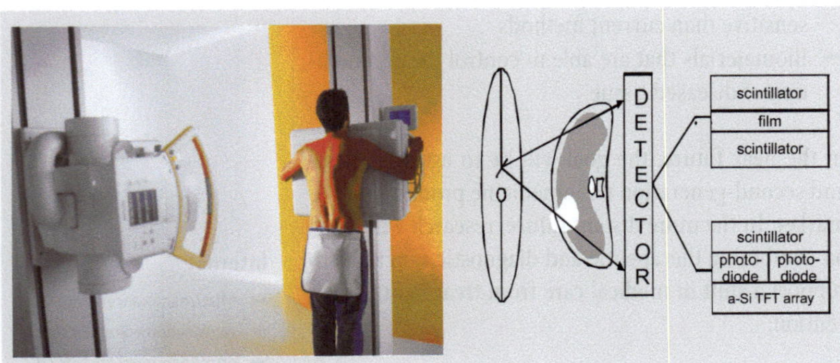

◘ Projection radiography system. Left: The system applied for lung imaging. Right: Schematic drawing of the set-up. The detection systems consist of a sandwich structure of X-ray converting scintillator screens and optical film or storage plates. There is a trend towards direct digital systems that combine a scintillating layer with an optical readout structure of photodiodes and an amorphous thin film transistor array. Source: Philips Healthcare nique that utilises the nuclear magnetic resonance (NMR) effect, primarily imaging the properties of the water protons in the human body. A static homogeneous magnetic field is applied by direct current (DC) coils arranged along the patient's body axis. This homogeneous magnetic field aligns the proton spins. For imaging purposes, the alignment is disordered by a short term radio frequency (RF) pulse. The subsequent relaxation of the protons generates another RF signal (NMR signal) which is captured by receiver coils arranged around the body region of interest. In a complex mathematical reconstruction procedure, it is possible to calculate high-resolution images from the NMR signal. MR images provide a good contrast between the different soft tissues of the body because the NMR signal differs for different body materials, e.g. muscles and fat. Small gradient fields are superimposed to define different imaging planes. For special purposes such as the imaging of vascular structures or tumours, specific contrast agents may be applied. Modern fast MRI systems even allow the imaging of moving organs or the beating heart. Besides the standard MR scanners, which look like a large tube, open MRI scanners which allow better access to the patient, e.g. for interventions, are available today.

► **Ultrasound imaging.** Ultrasound imaging, or sonography, is a widely applied medical imaging procedure. High-frequency sound waves are applied to the human body and their reflected echoes are detected to form an image. The reflective properties of human tissue depend on its density. Typically the shape, size, and structure of organs can be imaged or – one of the major applications – the growing fetus. In simple terms, the ultrasound system consists of a transducer probe, an image signal processing unit and a display. The transducer unit generates the sound waves and detects their echoes with a set of piezo-electric quartz crystals. As sound waves are strongly reflected at the intersection of media with very different impedance, e.g. air and water, a gel is commonly used to improve the contact between the transducer and the human skin. Transducers are available in several sizes and shapes to cater for the clinicians' various applications. Typical sound frequencies for medical imaging are 1–20 MHz depending on the application. Lower frequencies are used to image deeper tissue because they provide better penetration. Higher frequencies provide better resolution but have less penetration.

One important ultrasound technique is Doppler imaging. This technique makes use of the *Doppler ef-*

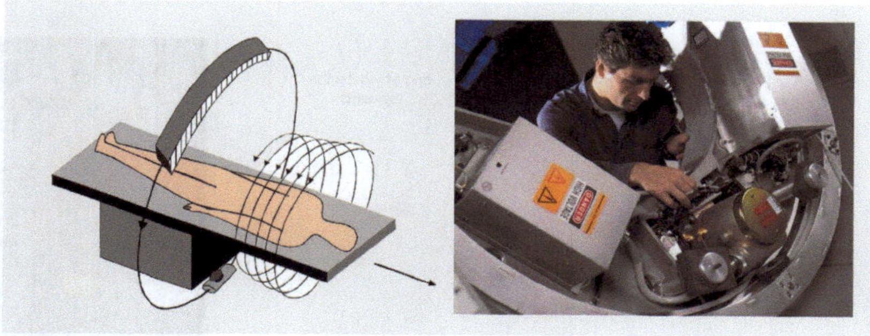

CT system. Left: Sketch of the CT principle. The X-ray tube (below table) and the detector array (above table) are mounted on a gantry rotating around the patient. Right: View of an open gantry during quality control. The X-ray tube (centre), high voltage electronics (up to 150 kV) and detector electronics yield a total weight of around 800–900 kg rotating around the patient at a speed of up to 3 rpm. Source: Philips Healthcare

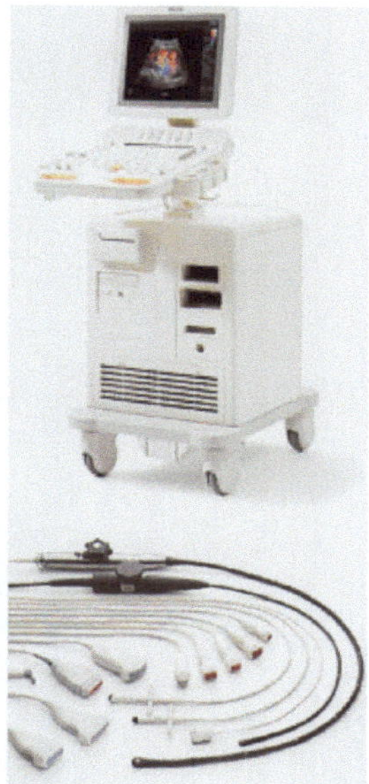

Ultrasound system. Top: A typical ultrasound system consisting of transducers, an imaging console and an integrated display. Bottom: A core component in ultrasound imaging is the transducer, which generates the sound waves and detects their echoes. Transducers are available in a large number of different shapes, sizes and frequency ranges to cater for the clinicians' various needs. Source: Philips Healthcare

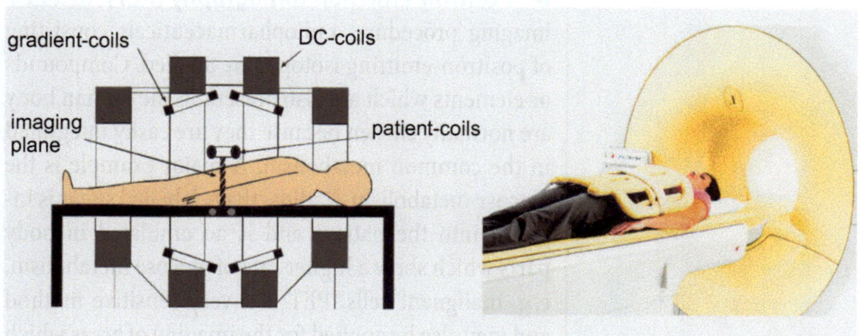

Magnetic resonance imaging. Left: A sketch of a typical MRI system consisting of several coil arrangements. DC coils generate a strong homogeneous field, typically 1–3 Tesla. This strong field aligns the magnetic moments of the nuclei in the body, mainly water protons. Because of the high currents required, the coils are commonly cooled down to become superconductive. Superimposed gradient fields define the imaging plane. An injected radiofrequency pulse generates the NMR signal, which is detected by receiver coils arranged along the region of interest on the patient's body. Right: Photo of a patient moving into the MRI scanner. The receiver coils are arranged on the patient's body. Source: Philips Healthcare

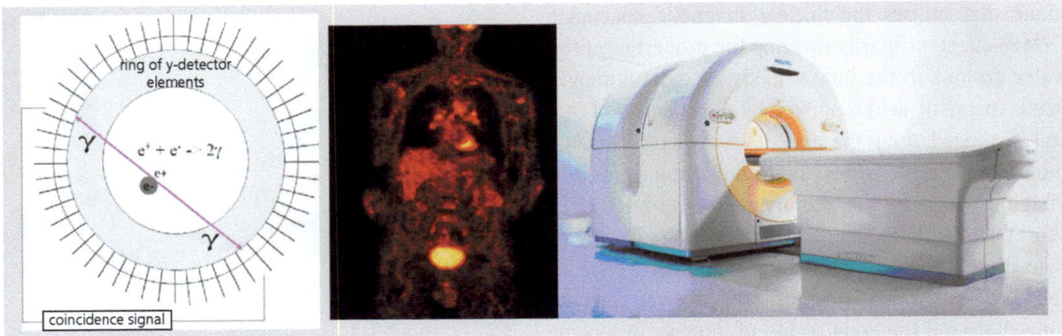

Positron emission tomography (PET): Biological active molecules, e.g. sugar, are labelled with a positron emitting radionuclides e.g. F-18. The compounds accumulate in parts of the body that have a high metabolism (e.g. heart, eyes, and tumours). Left: PET system principle. Within a few millimetres, the positron generated during the radioactive decay annihilates an electron emitting two γ-quanta in opposite directions. An array of γ-detectors arranged in a ring around the patient registers this event by a coincidence signal in two detector elements. The source of the signal can then be assumed along the virtual line connecting the two detector elements. Typically, some ten thousand coincidence signals need to be registered in order to reconstruct an image. Centre: Image generated by a whole-body PET scan with areas of increased metabolism shining brighter. Right: A modern combination of PET and CT. This PET/CT hybrid system is used for imaging in oncology. Source: Philips Healthcare

fect which occurs if the imaging objects are moving. The reflected sound waves show a higher frequency if the object is moving towards the transducer and a lower frequency if they are moving away. This effect makes it possible to image the flow of blood in the heart arteries in cardiology, for example. The velocity and direction of the blood flow can be displayed in different colours on the display (colour Doppler). In recent years, technologies for 3D imaging and even 4D (3D plus time) imaging have been developed finding specific applications in areas such as interventional cardiology and radiology.

► **Nuclear medicine**. In contrast to the imaging techniques discussed so far, nuclear medicine imaging applies no external radiation. A contrast agent which is enriched by unstable radionuclides is injected and accumulates in organs or diseased tissue. Subsequently, the γ-radiation emitted by the radioactive compounds is detected outside the patient's body and a radiological image is reconstructed.

► **Positron-emission tomography (PET)**. In a PET imaging procedure, radiopharmaceuticals consisting of positron-emitting isotopes are applied. Compounds or elements which are also present in the human body are normally chosen because they are easily integrated in the common metabolism. A major example is the glucose metabolism. Radioactively labelled sugar is injected into the patient and is accumulated in body parts which show a higher rate of glucose metabolism, e.g. malignant cells. PET is a very sensitive method and may also be applied for the imaging of areas which show a very low metabolism, e.g. parts of the heart muscle that are insufficiently supplied with blood after a myocardial infarction.

► **SPECT**. Single photon emission computed tomography is another tomographic imaging principle used in nuclear medicine. It generates images that allow body metabolism to be assessed. The radionuclides used emit γ-radiation which is detected by one or more γ-cameras rotating round the body. Compared to PET, SPECT uses radionuclides having a longer half-life, and has lower specific resolution and sensitivity. SPECT images are frequently used in cardiology, where the time period for measurement is synchronised with the heart beat (gated SPECT).

Applications

The imaging principles described above are applied in numerous areas of medical diagnosis and therapy. This article will focus on cardiology and oncology. Disease patterns in these medical disciplines are of particular interest to the healthcare system because they are frequent, life-threatening, and expensive for society.

One of the dominant procedures in cardiology is the diagnosis and image-based therapy of coronary artery disease. The main goals of the examination are the detection and assessment of narrowing in the arteries that supply the heart, their catheter-based repair, and the functional assessment of the heart muscle after insufficient blood supply (myocardial infarction, heart attack). The initial tests applied in cardiology are usually the electrocardiogram (*ECG*, which evaluates the electrical activity of the heart) and ultrasound imaging (echocardiography, which yields information about the cardiac anatomy, the dynamics of the heart muscle, and the condition of the valves). Colour Doppler ultrasound imaging can visualise the blood flow and its direction inside the vessels or heart chambers by superimposing colour onto the grey-scale image. Normal blood flow would be indicated by more or less homogeneous colour, while a stenosis would be indicated by complex flow patterns with typically increased speed and abrupt changes in the flow direction.

► **X-ray projection imaging.** For patients with suspected coronary artery disease, cardiac catheterisation is applied. In this interventional X-ray imaging procedure, a catheter is inserted into the arteries or chambers of the heart, an X-ray contrast agent is injected, and projection images of the arteries are made from two orthogonal imaging directions (coronary angiography). From these images, doctors can assess to what extent the coronary arteries are narrowed or blocked, and decide on the appropriate therapy. The projection images are stored as a video sequence.

Once coronary artery disease has been diagnosed, the same imaging setup is often used for catheter-based therapy. The narrowed artery can be treated by a group of interventions called angioplasty, procedures that reshape the inside of the arteries. In balloon dilation, for example, a balloon at the tip of the catheter is inflated in the artery, widening the vessel at the location where it is narrowed. A wired mesh tube (stent) may be inserted inside the artery to prevent it from reclosing. The correct positioning of the stent can be evaluated with the same X-ray imaging device.

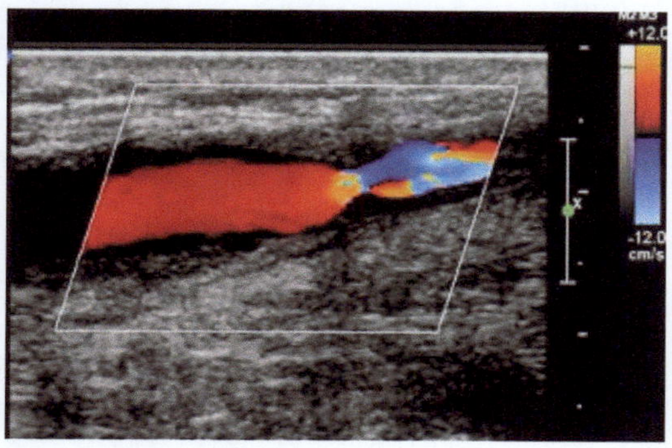

Colour Doppler image: Colour-coded visualisation of blood flow in a narrowed vessel, indicating the speed and direction of blood around a stenosis. The colour pattern indicates the direction and magnitude of the blood velocity. The colour fills the lumen of the vessel, and the blood flow is from left to right. The part on the left in the homogeneous red colour indicates a normal flow with a slight movement away from the probe. At the location of the stenosis, the flow pattern changes to all kinds of speeds and directions caused by the higher pressure and complex flow dynamics. Source: Philips Healthcare

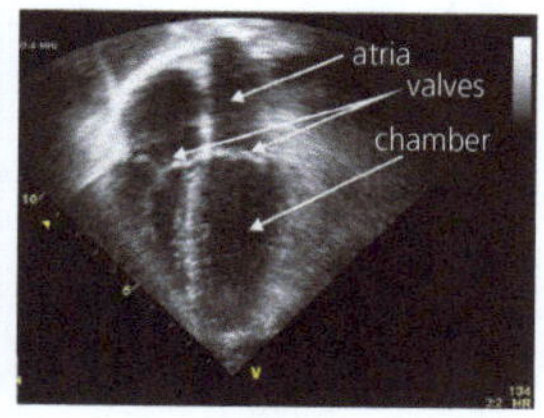

Ultrasound examination of the heart (echocardiography), showing heart chambers, atria and valves: The picture is part of a video sequence of the beating heart. Source: Philips Healthcare

► **Computed tomography angiography (CTA).** A more recent method for assessing the condition of coronary arteries is high-resolution CT examination of the heart. Coronary arteries can be visualised with high quality using slice images acquired in a CT scanner at high speed and resolution.

► **Magnetic resonance angiography (MRA).** Today, MRI is a vital option for cardiac imaging as well. Coronary arteries can be displayed in three dimensions (MRA, magnetic resonance angiography) with the aid of fast imaging sequences and the compensation of movement artifacts (synchronisation of heart beat and breathing).

Just as important as the assessment of the coronary artery system is the evaluation of the heart muscle after myocardial infarction to determine how much damage the heart attack has caused. The cardiac SPECT examination is the standard method for this purpose. In SPECT images, tissue with increased metabolism shines more brightly, as these are the locations in which SPECT-sensitive radio nuclides accumulate.

► **Combined imaging methods (PET/CT).** To diagnose pathological structures such as tumours, it is important to determine their morphology (shape and size), as is revealed by most imaging techniques such as X-ray, CT and MRI. It is also essential to acquire information about cell metabolism, which means that nuclear medicine scans such as PET are now standard imaging procedures in oncology. In PET, the increased metabolism of tumours can be visualised by the accumulation of radionuclides in the tissue. The combination and fusion of imaging techniques makes it possible to display the exact morphology and location plus the metabolic function in a single image. This has produced fully integrated devices such as PET/CT, SPECT/CT or PET/MR, which help to detect and classify tumours, for instance.

► **Image analysis and computer aided detection.** Progress in high-speed and high-resolution imaging has paved the way for new image processing algorithms

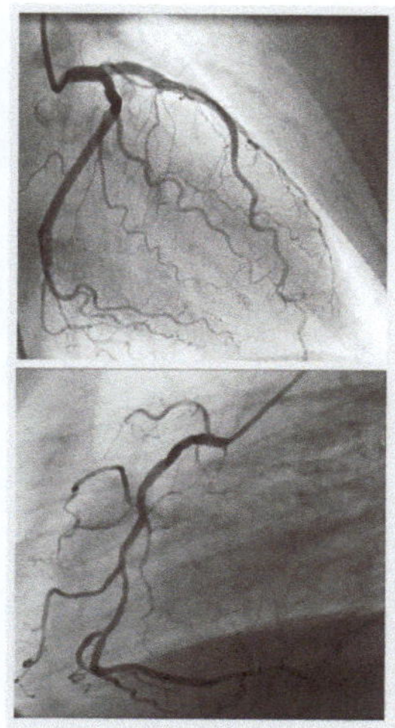

X-ray imaging during cardiac catheterisation: A catheter is inserted into the femoral artery in the leg and pushed up to the heart chamber and into the coronary arteries, through which a contrast agent is injected. The heart muscle itself is not imaged. Without a contrast agent, the vessels would hardly be visible. The two orthogonal X-ray projection planes (upper and lower picture) help the radiologist to develop a 3D impression of the scene for catheter placement and treatment. Source: Philips Healthcare

Imaging modality	Principle	Typical parameters	Technology trends
Projection radiography	2D registration of the projection of X-rays attenuated by the anatomy of the human body; dynamic imaging possible; subtraction angiography by application of contrast agents	resolution: 100 - 200 mm acquisition time: < 1 s (static) 60-90 frames / s (dynamic)	flat dynamic detector systems, rotational angiography for 3D imaging
X-ray computed tomography CT	scanning of radial X-ray projections and reconstruction of 2D slices	resolution: 500 µm - 1 mm acquisition time : < 30 s (whole body scan)	multi-slice detector systems, integration of photodiodes and amplifiers, e.g. in CMOS technology
Magnetic Resonance Imaging MRI	registration of nuclear magnetic resonance, mainly of water protons in the human body; tomographic reconstruction of 2D slices	resolution: 250 µm - 1 mm acquisition time : ~ minutes field strength: max 3 Tesla (higher field strengths for research applications)	open magnets; fast registration techniques and motion artifact compensation
Positron Emission Tomography PET	coincidence registration of two γ-quanta generated from the positron emission of radionuclides applied to the patient; reconstruction of 2D images	resolution: 4 - 10 mm acquisition time: < 1 hour	hybrid systems, a combination of PET and CT; time-of-flight PET registering the time-of flight time difference between the two γ-quanta; better localization of the annihilation, shorter acquisition times
Single Photon Emission Computed Tomography (SPECT)	detection of the density of γ-quanta emitted by radionuclides which are applied to the patient; tomographic reconstruction of 2D slices	resolution: 5 - 20 mm acquisition time: < 1 hour	hybrid systems, e.g. SPECT/CT
Ultrasound imaging	detection of echoes of ultrasound waves applied to the human body	resolution: 100 µm - 1 mm acquisition time: ~ seconds	real-time 4D (3D + T) imaging; miniaturized systems for intravascular imaging

Table: The most relevant imaging principles, their technical parameters and trends.

that support physicians in the detection of tumours by automatically marking suspicious structures in images. In detecting lung cancer, for instance, image processing methods have been optimised to search for lung nodules and differentiate them from other round structures such as vessels or bronchi. Particularly high-resolution CT images are the basis for techniques that automatically detect and classify tumours (CAD, Computer Aided Detection; CADx, Computer Aided Diagnosis).

In future, imaging procedures in oncology will be facilitated by much more sensitive contrast agents, which will be able to visualise metabolism and other processes in cells at molecular level (molecular imaging).

Trends

▶ **Resolution.** A crucial goal of technical development in most imaging procedures is to increase their temporal and specific resolution. There is certainly a trend towards 3D and 4D images, where 4D (3D plus time) means a real-time dynamic sequence of 3D data sets. This is relevant in areas such as cardiology, to display images of the beating heart.

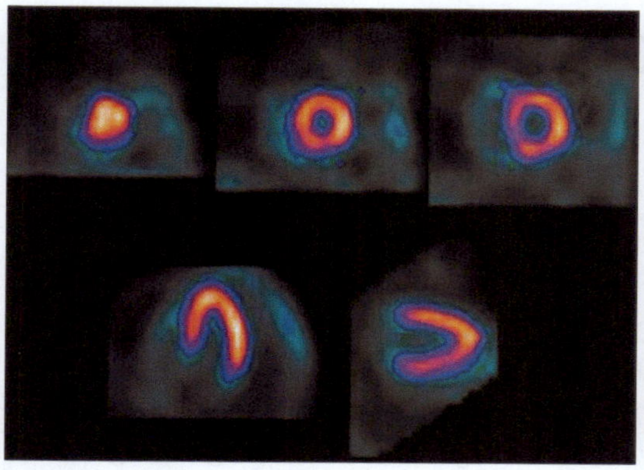

Cardiac SPECT images showing a cross section of the heart muscle: The perfusion of contrast agent into the muscle (red) indicates any damage that might have occurred during a heart attack. Source: Philips Healthcare

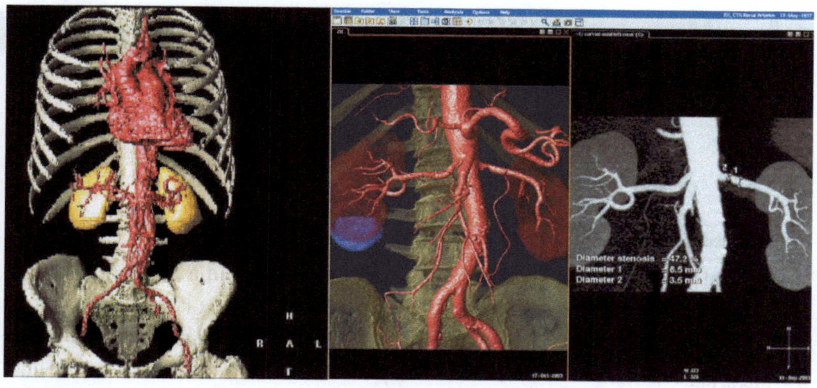

Tools for segmentation and quantitative image analysis: A 3D view of relevant body parts is generated on the basis of CT data sets. Left: Visualisation of bone, kidney, heart and vessels. Centre: Detailed 3D vessel system for blood supply of the kidney. Right: Quantitative measurement of the stenosis diameter in a kidney artery. Source: Philips Healthcare

▸ **Molecular imaging.** Many medical procedures rely on a thorough understanding of basic biological processes and events. Molecular imaging uses highly specific contrast agents for the non-invasive in vivo investigation of cellular molecular events involved in normal and pathologic processes. This principle is used in many applications for cancer, cardiovascular and neurological diseases. In comparison to conventional imaging, new agents act as biomarkers to probe particular cellular targets of interest. Various kinds of new imaging agents are under investigation for imaging methods such as MRI, SPECT, PET, ultrasound and optical imaging. Molecular biology and chemistry will play an important role in the future success of medical imaging.

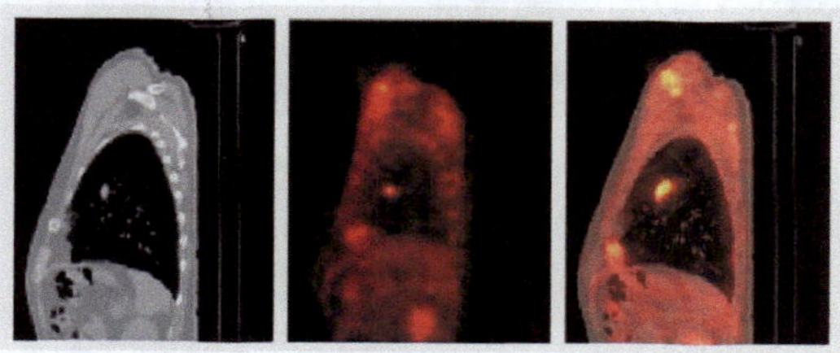

PET/CT: Lateral view of a lung image. Left: CT image showing the morphological information of the lung including the air-filled lung (dark area) and the ribs (white spots). White structures that might represent tumours become visible within the lung. Centre: Functional imaging with PET, showing regions with increased metabolism. Again, suspicious spots become visible. Right: A combination of PET and CT makes it possible to see morphological (CT) and metabolic information (PET) in a single image, facilitating the diagnosis of a tumour. Source: Philips Research Europe

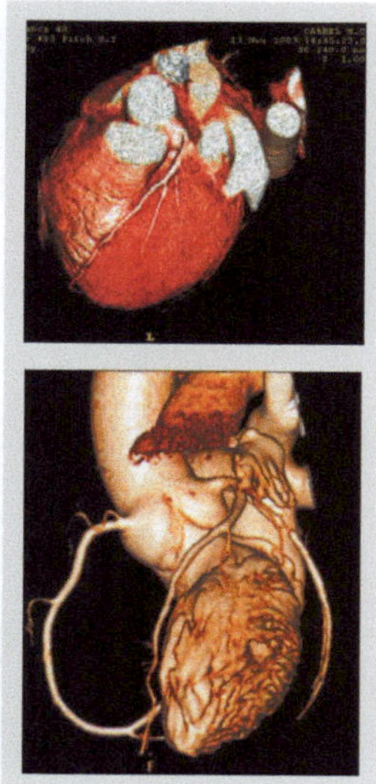

3D reconstruction of the heart and the coronary arteries. Top: Reconstruction from high-resolution CT slices (CTA). Bottom: Magnetic resonance coronary angiography using fast MRI imaging techniques and motion compensation. The 3D visualisation is generated by image processing algorithms applied to the slice images usually produced by CT and MRI. Source: Philips Healthcare

Prospects

- Medical imaging techniques have become routinely applied procedures in medical diagnosis and therapy.
- Although most of the imaging principles have been known for decades, recent technological progress has dramatically increased their usefulness in medical applications, resulting in more accurate diagnosis and more successful treatment.
- Increased computer power provides the technology for optimising medical imaging devices, and this development is justified by numerous new applications that deliver more accurate clinical parameters.
- For the future, we can expect to see closer integration of image-based diagnosis and therapy. Molecular imaging will provide considerably more specific diagnostic image content, yielding more precise information about body conditions at cell and molecular level.

DR. GERHARD SPEKOWIUS
Philips Research Asia, Shanghai

DR. THOMAS WENDLER
Philips Research Europe, Hamburg

Internet

- www.rsna.org
- www.medicalimaging.org

Medical and information technology

Related topics

- Medical imaging
- Virtual and augmented reality
- Minimally invasive medicine
- Human-computer cooperation
- Image evaluation and interpretation
- Robotics
- Implants and prostheses

Principles

Information technology has had a huge influence on many areas of medicine over the last 20 years. Large computer systems are now an established part of administrative systems in medical institutions – for hospital logistics, billing of services and storing patient data. In medical technology itself three-dimensional (3D) spatial and functional imaging (*X-ray, ultrasound, tomography*) has resulted in a changing awareness of the human body and in higher expectations of quality in diagnosis, therapy and rehabilitation. We not only expect a new knee joint to function reliably; it is also assumed today that both legs will be of equal length, the feet are aligned parallel to each other and the new knee can bear as much weight as the one it has replaced. Meeting such demands has only been possible since the beginning of the nineties and would not have been feasible without monitors, measurement technology and computer-controlled instruments in the operating theatre. We can only achieve optimal results if 3D image data is used to precisely plan the surgical procedure pre-operatively, and if these plans are executed precisely, to the millimetre.

▶ **Computer-assisted instrument navigation.** Instrument navigation allows the surgeon to guide and position the surgical instrument on or in the body of the patient with millimetre precision. The orientation of the instrument is based on a 3D image data set from

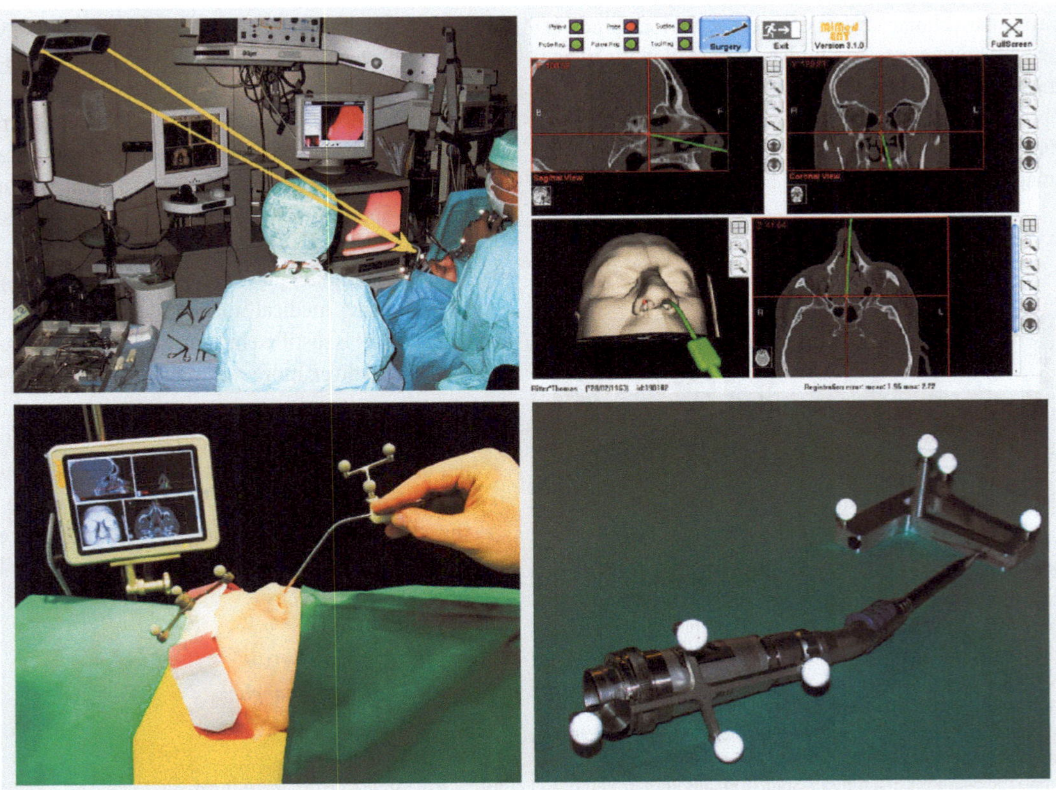

Computer-assisted instrument navigation. Top left: A stereo camera calculates the location (position and orientation) of the surgical instruments and the patient using light-reflecting markers attached to both the patient and the instruments. Bottom left: Although the instrument is inside the head, the surgeon can use the CT/MRI 3D images taken pre-operatively to navigate the instruments and define the position accurately. Top right: The monitor shows the precise position of the instruments in relation to the anatomical structures. Bottom right: The reflectors can be attached to suction cups, reamers (picture), saws, lasers, needles or probes. Source: T. Lüth computer tomography (CT) or magnetic resonance imaging (MRI), usually taken the day before the operation. Navigation has two advantages – the position of the instrument in relation to the patient can be defined precisely to 1–2 mm, so that even if the surgeon cannot directly see the instrument inside the body, the three-dimensional image data make it possible to follow the position of the instrument on the monitor. The process is based on a coordinate measuring system, which continuously measures the spatial position of the patient and instrument. Stereo camera systems, which emit 20 light flashes per second in the infrared range, have now become established in operating theatres. Stereo cameras can calculate the position of the light reflecting markers accurately to 0.2 mm. The image data set is registered when the reflectors have been attached to the patient and the instrument. It is then possible to select 3–4 reference points from the image data at which the instrument is brought into direct contact with the patient. The computer calculates the exact position of the patient from these four points and uses the attached markers to track changes in position.

The instrument and the patient image are only shown on the monitor when the data set has been registered. When the instrument is now moved inside the patient, the movement can be followed on the monitor. This navigation system has made it possible for surgeons to visually orientate the instrument in the patient's body without them being able to see the head of the instrument directly. They can easily identify the distance from sensitive structures and make allowances for them during surgery. The patient data should be recorded as close as possible to the operation itself, so that the images actually correspond to the current condition of the patient – if possible, on the day before surgery.

▶ **Computer-assisted planning.** Fluoroscopy and measurement of the body make it possible to accurately plan surgical procedures in advance using image data. These plans can include the following tasks:

- Precise targeting of a small anatomical structure with a needle, probe, laser or gamma ray
- Precise placing (position, angle) of a drilled hole
- Precise execution of an incision with a saw (bone), a knife (soft tissue) or a ray of hard gamma radiation (tumour treatment)
- Precise shaping of a cavity for inserting an implant (knee, hip, plastic surgery)

The advantage of planning the surgical intervention is that the postoperative result can be viewed on the monitor before the operation. Risks can be identified at an early stage and the best possible preparations can be made for the operation. Typical planning tasks are:

- Defining the safety areas, which should on no account be damaged during treatment e.g. nerves and organs
- Defining the areas of operation, where material will be treated, worked on or removed; drawing these areas precisely on the picture, such as incision lines, the final position for implants (e.g. screw implants, miniplates, prostheses); weight-bearing calculations

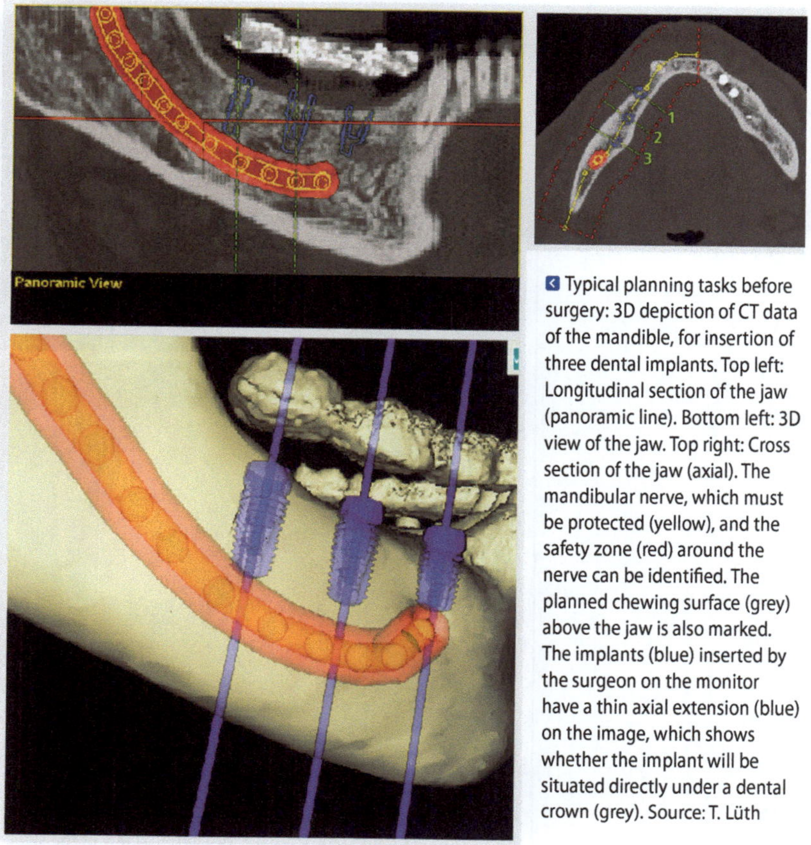

Typical planning tasks before surgery: 3D depiction of CT data of the mandible, for insertion of three dental implants. Top left: Longitudinal section of the jaw (panoramic line). Bottom left: 3D view of the jaw. Top right: Cross section of the jaw (axial). The mandibular nerve, which must be protected (yellow), and the safety zone (red) around the nerve can be identified. The planned chewing surface (grey) above the jaw is also marked. The implants (blue) inserted by the surgeon on the monitor have a thin axial extension (blue) on the image, which shows whether the implant will be situated directly under a dental crown (grey). Source: T. Lüth

Stress, distractions or stretched work capacities, such as occur during extended working hours in everyday clinical a practice, can mean that the operation does not proceed according to plan and may lead to complication risks, despite planning and instrument navigation. Systems were therefore developed to minimise complications during manual implementation.

▶ **Navigated control of instruments.** This technology developed originally from the practice of using robot systems in operations. It responds to the desire to work freely with the hands, while being able to execute

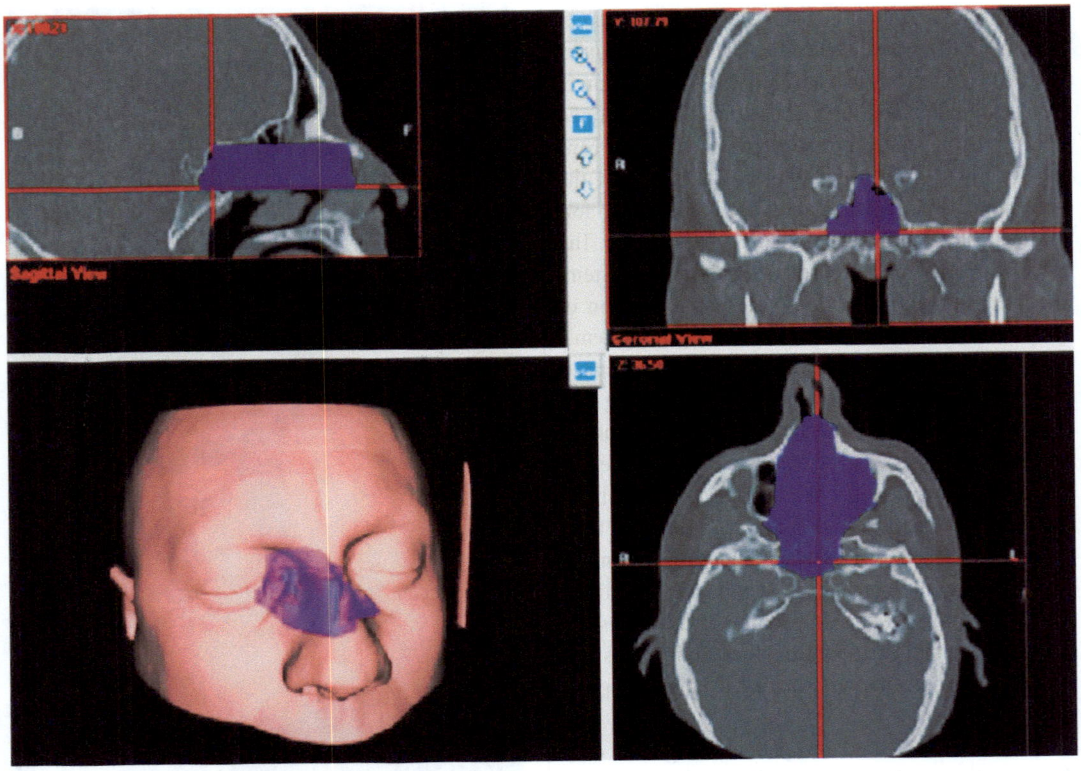

■ Typical planning tasks before surgery : 3D depiction of CT data of the upper facial bones. Top left: Longitudinal section (sagittal). Top right: Frontal section (coronal); bottom left: 3D view of the skull; bottom right: Cross section (axial). The surgeon has marked precisely the area to be excised (blue) in preparation for the removal of the paranasal sinuses. He or she also marks an outline (blue) in each exposure of the axial cross section of the CT image (bottom right). This preparatory work defines the areas of tissue which the surgeon must not leave when removing the polyps from the nose. Source: T. Lüth

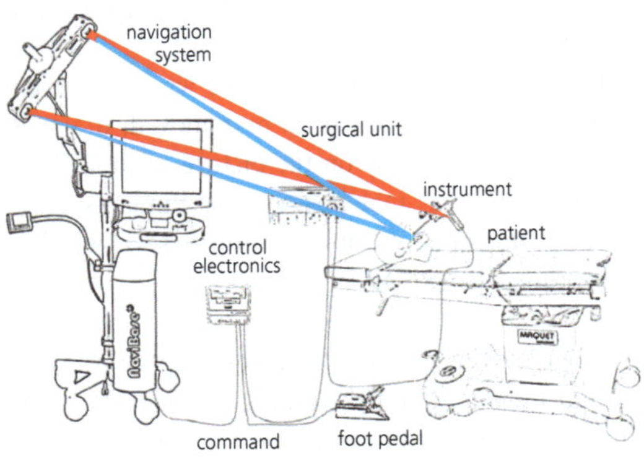

■ Navigated Control: The navigation system calculates the position of the instrument and patient. The control electronics check whether the instrument is moving out of the permitted working space or whether it can touch the border of a safety area. It does this 20 times per second. If there is a risk of the instrument deviating, electronic intervention is activated in the foot pedal control of the reamer within 100 ms. Source: T. Lüth

manual movements with the precision of a robot. The principle (navigated control) is based on switching off powered instruments (drills, reamers, saws) when they are positioned outside of a defined working space or at the edge of a safety area. Computer-assisted planning could, for example, be used to calculate the position of an artificial hip joint in the femoral neck. It is possible to plan the exact position and orientation of the endoprosthesis in the bone and therefore, also the exact shape and position of the tissue to be removed from the bone. Changes in shape could result in a lack of stability and changes in position could result in functional impairments or pain due to incorrect load bearing. Using instrument navigation the surgeon can see on the monitor where tissue still has to be removed. He or she controls the speed of the reamer with the foot pedal and tries to ream out the pre-defined cavity, accurate in shape and position, in the correct place on the thigh. The navigated performance control system supports the surgeon by preventing the reamer from removing any more tissue than necessary. As soon as the surgeon attempts to remove tissue from the wrong place, which would change the predefined shape, electronic intervention of the foot pedal control is activated, blocking the foot pedal. This process makes it possible to ream exact, precisely fitting shapes even in the dark. The surgeon now has a "three-dimensional electronic template".

► **Offline programmed robot systems.** In such systems, a robot, and not the surgeon, guides the instrument. Computer-assisted planning automatically creates a movement programme for the robot. The programme-controlled robot moves the surgical instrument, or radiotherapy device in the case of tumour surgery. These systems have been successful for many years in radiotherapy. Despite initial euphoria, however, robots have not become as generally accepted in emergency surgery and orthopaedics as navigation systems.

► **Telemanipulators.** In minimally invasive surgery, operations are carried out with instruments through small openings of the body. While the control handles of the instruments are outside the body, their effectors (forceps, scissors, suturing instruments) carry out the operation internally. All of the instruments are long and thin. The operator looks through an *endoscope* as he or she does not have a direct view of the operating area. The endoscope is a long and thin telescopic tube with two or three light conductors. One of the light conductors illuminates the operating area. The other light conductors show the surgical progress on the monitor via a camera. High-quality endoscopes project 3D stereo images externally. It is certainly not easy to manipulate the instruments inside the body using the handles on the outside. Suturing tasks, in particular, involving suture transfer and knotting, require much practice. The surgeon's hands become tired, they tremble, the instruments cannot be moved around freely and it is easy to become disoriented. Telemanipulation systems were developed to overcome these problems. When using these systems the surgeon sits at an operating console next to the operating table. He or she can see the operating area inside the patient, the organs and his or her instruments, three-dimensionally on a monitor, just as if they were inside the patient. Measuring or input devices, which track hand movements, are attached to their hands and fingers. These movements are now transferred to a robot, whose movements directly correspond to those of the surgeon. The doctor exerts only minimal force to manipulate the instruments and the robot does not replicate any trembling of the hands. Another advantage is that instruments can also be designed differently, if they do not have to be manipulated manually. Furthermore, instruments can execute large hand movements of the surgeon in very small movements. In theory, it is possible to use this technology over long distances in the form of telerobotics, although this idea remains a futuristic scenario.

► **Rapid prototyping.** Rapid prototyping and rapid manufacturing use 3D patient image data to produce patient-customised models and implants. 3D patient-specific models are produced which allow the surgeon to visualise the individual patient's anatomy and carry out test operations. The materials used are either plastics or silicones. In the case of maxillofacial surgery, for skull operations on children, the model is made of plastic. The surgeon surgically saws the model on the day before surgery as if he or she was carrying out the real operation and reseals the skull using small screws and metal splints. The model of the skull is in the theatre during the operation and the surgeon can proceed as previously planned. The first step of 3D printing for patient models involves spreading a layer of powdery

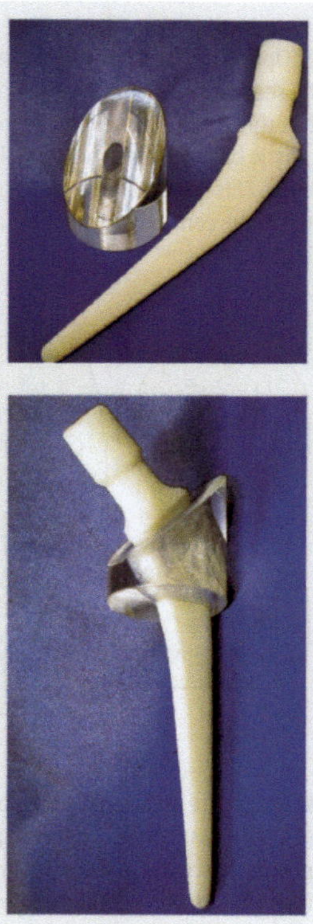

Plexiglas part with the cross section of a thigh bone and a hip prosthesis: The task consists of reaming a hole (cavity) in the plexiglas, so that the hip prosthesis is in exactly the right place and correctly aligned. There must be a perfect fit between the cavity and implant, even without a good view inside the patient's body. This process makes it possible to ream a cavity in the bone model which is precise in position and shape, with a surface deviation of less than 1 mm. The implant fits exactly into the cavity in the plexiglas. Source: T. Lüth

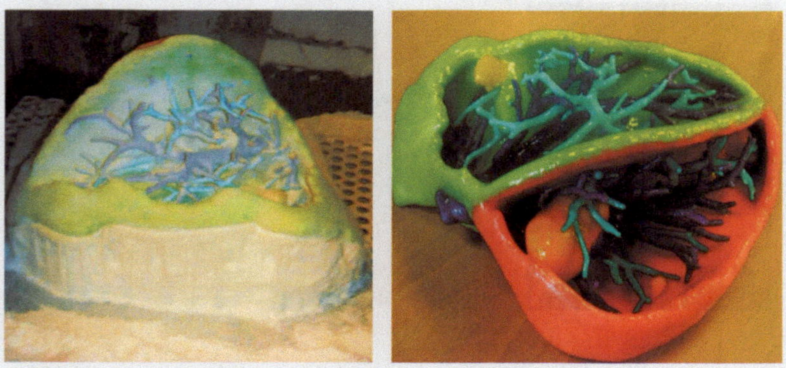

Rapid prototyping with 3D printing. Left: A patient's liver with exposed vascular tree of veins, bile ducts and tumour metastases, immediately after printing from plaster; the support structure under the object is clearly visible. Right: The finished rubber-like coloured liver model with the tumour metastases (orange) and the vessels (blue, turquoise). The parenchyma (liver tissue) has not been printed. Source: T. Lüth

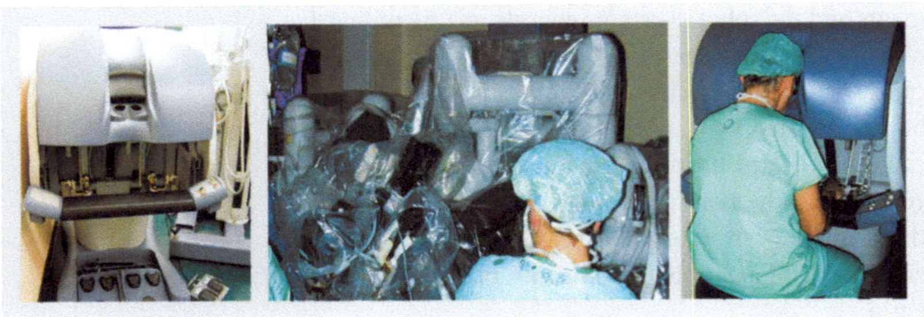

Operating console of a telemanipulator. Left: The foot switches and the "joysticks" for the hands can be seen, together with the two circular openings for viewing the stereo monitor. Centre: The robot itself has three arms – one for the endoscope and two for the instruments. The arms are sealed in special plastic bags to keep them sterile. The sterilised instruments are introduced via small incisions in the body. This is inflated with nitrogen gas, so that the surgeon can work in the cavity. Right: The surgeon sits at the console a few metres away from the patient and moves the instruments in the body. Source: T. Lüth

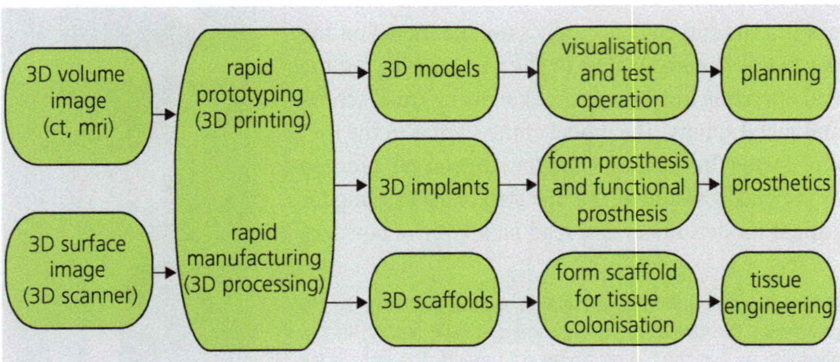

Procedure and applications of rapid prototyping in surgery: 3D image data from volume scanners (CT, MRI) or surface scanners (light incision method) is processed and used to control 3D printers or 3D processing centres . Visualisation models and dental, weight-bearing or joint prostheses can be created for each individual patient, as well as scaffold structures for tissue engineering.

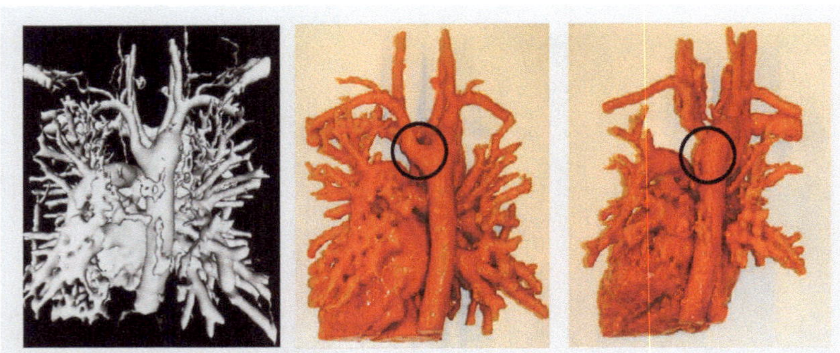

Rapid prototyping of a heart. Left: 3D computer model of the heart of a twelve-year-old child, produced by magnetic resonance imaging. Centre: A 3D rubber model was produced using 3D printing technology. The malformation of a main artery is clearly visible (circle). It is bilaterally branched instead of unilaterally, and envelops the esophagus. The operation can be planned before opening the ribcage. Right: Post-operative 3D rubber model, in which the main artery is posteriorly displaced and is easy to identify (circle). The left artery is now supplied by a different vessel. Being able to plan an operation on the basis of a model is a significant benefit, as there is not much time during the actual operation. Source: S. Daebritz, R. Sodian, LMU Munich

carrier material (e.g. plaster, starch, hydroxylapatite) on a lifting platform. The thickness of the layer is constant. A "special ink" consisting of adhesive/binders and colour is now sprayed on and into the uppermost layer of carrier material using a 2D printer head. The lifting platform is then lowered by 0.1 mm and another layer of powder is spread over it. The adhesive combines the individual layers, producing a 3D object. When the last layer has been printed any remaining powder is dusted off. The printed object then acquires either a very hard or a rubbery consistency after being dipped in special fluids. Prototyping methods are also used to produce patient-specific three-dimensional implants or prostheses as replacement material for bone or tissue. Biocompatible metals and ceramics are used for bone replacement and silicones for soft tissue. In plastic surgery, for example, where parts of the skull bone have been destroyed, replacement skull plates are made from titanium. Synthetic ears, eyes, noses and larger facial prostheses (epitheses) are made from silicone. 3D cameras are often used by dentists to measure the cavity for which an inlay is needed. This filling or crown is produced directly in the dentist's practice, shaped in three-dimensions, and ground to micrometre precision from a ceramic block of the correct colour. This process can also be used to produce patient-specific three-dimensional support structures with the same geometric shape as the bone or tissue it is replacing. Tissue cells are colonised on the scaffold structures, and grow into the scaffold, reproducing tissue and vascular structures (tissue engineering). Permanently biocompatible materials are used in this case, or bioabsorbable alternatives such as hydroxylapatite, which are broken down by the cells over time.

Applications

While the navigation procedure is well established in bone surgery, huge challenges still remain in its use in soft tissue surgery. Internal organs are soft, deformable and move constantly in pace with breathing patterns. Nevertheless, data processing also plays an essential role here in avoiding complications by supporting the preparation of surgical procedures, as in the case of liver surgery. The liver is a vital organ, with two vascular tree structures embedded in its tissue. Parts of the liver must be removed when there is a tumour. This is also the case for a living-donor liver transplantation, when the healthy liver of a donor is divided in such a way that both the donor and the recipient retains or receives a liver which is adequate in size. The liver can only regenerate itself and grow back, if the minimum size does not fall below approximately 20%. It is not so much the quantity of tissue, however, that is important, but whether the vascular trees inside the liver are retained. It is crucial to the success or failure of the surgical intervention that an accurate incision is made when removing tissue. The vascular tree structure means that even a slightly incorrect incision can have considerable effects on the success of the surgical intervention or the survival of the patient. This information must also be available in the operating theatre, so that any new planning can be adapted to other marginal conditions before and during the operation.

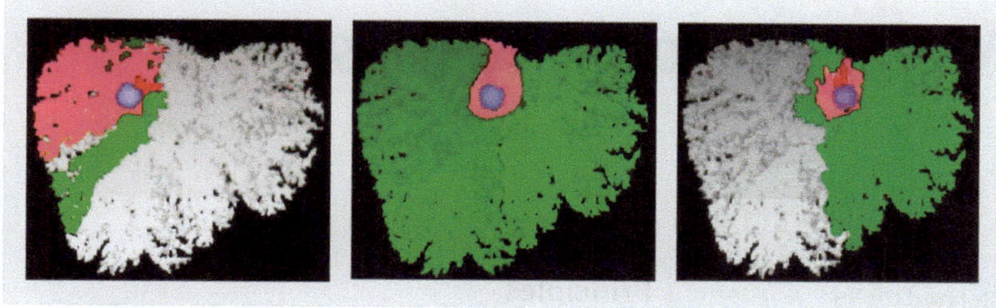

Removal of a liver tumour: The three images (left, centre, right) show the results of a tumour removal (purple point), with the tumour located in different positions, i. e. shifted approximately 3 cm further to the right in each case. In each image, the coloured areas of the liver are those no longer supplied by the vascular network after removing the tumour with a safety distance of 5 mm (dark red – small areas in the left and right images), 1 cm (pink) or 2 cm (green). The implications of the unintentional displacement of an incision are obvious. The middle image, for example, shows how an incision with a safety distance of 2 cm can destroy the whole liver. Source: Heinz-Otto Peitgen, Uni Bremen

Trends

Demand for visual information in the operating theatre will continue to grow. Operating theatres will increasingly make use of monitors and integrated equipment, accompanied by new technology based on monitors attached to ultrasound heads or instruments. There is also a new type of mobile image viewer, which makes it possible to see through the patient via a video camera, that provides superimposed image information. Such systems work in tandem with the navigation control system. Apart from calculating the angle of the monitor to the patient, they also work out what image information should be shown. The demand for small instrument guiding systems in robotics is increasing, as until now, robot systems have been large and cost-intensive. The need is for simple "servo" instrument guiding systems, which require only limited working space, and can be brought very close to the operating area. These small instrument guidance systems can also be used directly in imaging systems such as CT or MRI.

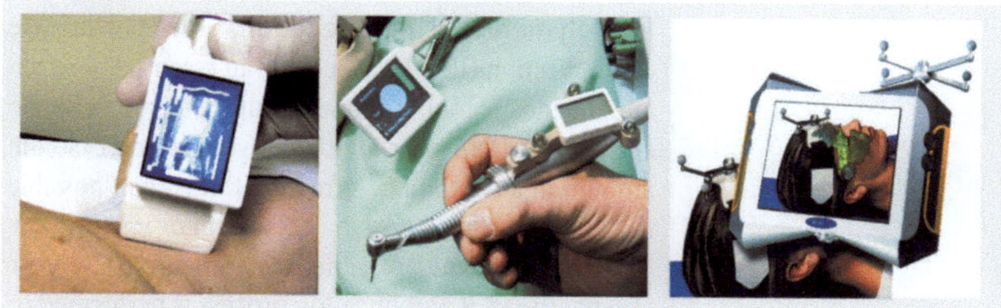

Integrated monitors. Left: Ultrasound device with integrated monitor. While image data are recorded, they are visible directly on the ultrasound head at the same time. Centre: Small monitors for instrument navigation, which can be attached directly to or in the operation area. The monitors can be much smaller, without any loss of information, as the distance to the eye is less. If a monitor's position in the area is known, it displays information relative to the situation, e. g. anatomical descriptions. Right: Navigated image viewer with reverse camera. The monitor makes it possible to carry out a review, with relevant 3D patient images superimposed in the exact position. It is almost like being able to see inside the patient. Source: T. Lüth

Prospects

- Mechatronics will increasingly be used in medicine due to the constant growth of computing power and microcontrollers, the continuing reduction in prices of flat screen monitors, and the wireless networking of computers.
- The linking of information technology and the integrated processing of data plays an important role in the building or renovation of operating theatres, clinics and surgeries. Instrument navigation or navigated performance control systems are now regarded as an integrated part of a modern operating theatre. In the past, technology tended to fuel developments in medical engineering. Communication between specialist disciplines has improved over the past few years and engineers have "learnt" to ask about real clinical requirements.
- Clinical studies, quality monitoring and assurance have now become the focus of attention. The clinical result is now the guideline for future developments, not technical progress. In the field of telemedicine and decentralised diagnostics, greater attention must be focused on the respective roles of the doctor and the technology involved. While it is not absolutely necessary that the devices in question be operated by doctors themselves, their assessment of the image data from a medical perspective is essential. Here too, it is foreseeable that technology will alter professional profiles and responsibilities.

PROF. DR. TIM LÜTH
Technical University of Munich

Internet

- www.mevis.de
- www.mimed.de
- www.iccas.de
- www.spl.harvard.edu

Molecular diagnostics

Related topics

- Gene therapy
- Image evaluation and interpretation
- Microsystems technology
- Forensic science
- Pharmaceutical research

Principles

Molecular diagnostics deals, among other things, with the detection of different pathogens such as bacteria and viruses and the detection of disease-relevant mutations of the human genome. Most techniques in molecular diagnostics are based on the identification of proteins or nucleic acids. Deoxyribonucleic acid (DNA) and ribonucleic acid (RNA) can be amplified and detected with diverse methods such as PCR, real-time PCR and DNA microarrays.

▶ **Polymerase chain reaction**. The polymerase chain reaction (PCR) has become an essential research and diagnostic tool. It allows scientists to copy a single segment of DNA over and over again to increase the amount of a certain DNA fragment. It is then possible to decide whether a specific bacterium or virus or any particular sequence of genetic material is present or absent. The PCR amplification process is performed in a PCR thermocycler. One cycle of this process can be subdivided into three major steps: denaturation, annealing and extension.

The PCR cycler repeats this cycle 30 times on average. Each newly synthesised fragment of DNA can act as a new template. It is therefore possible to amplify a single DNA fragment millions of times.

The technical challenges in developing PCR thermocyclers are:

- The desired temperature ranges have to be very precise simply because the PCR reaction is very sensitive. Therefore, minor temperature differences can lead to wrong results.
- Heating and cooling rates should be very fast in order to eliminate unspecific binding of the primers to the template.

▶ **Real-time PCR (RT-PCR)**. The real-time polymerase chain reaction (RT-PCR) follows the same principle as the PCR described above and passes through the same cycle steps. The special feature – and the advantage – of the RT-PCR is the possibility to closely monitor the amplification during the PCR process (in real time) using fluorescent dyes. The basis for monitoring the amplification process is the tendency of the fluorescent dyes to bind specifically to the amplified product. The greater the presence of amplified product, the greater the fluorescent intensity detected. The fluorescent markers are activated and the emitted light is filtered and monitored by a photodiode and graphically displayed by a computer. This enables the amplified DNA to be detected during the PCR process. When the fluorescent intensity exceeds a specific threshold after a number of cycles, the "threshold cycle" has been reached. On the basis of this threshold cycle, scientists can calculate the amount of DNA that was originally

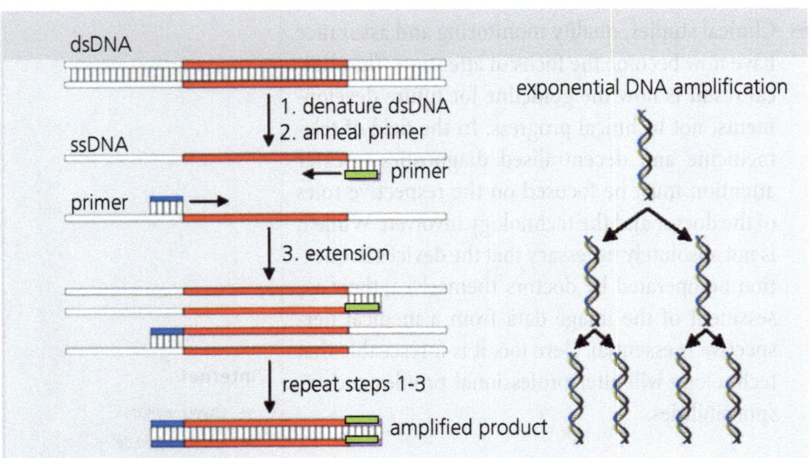

Polymerase chain reaction. Left: The PCR can be subdivided into three major steps during one cycle: The first step is the denaturation of the double strand DNA (dsDNA) to create single strands (ssDNA). These single strands act as templates for the subsequent annealing process. Second step: Annealing of two different short sequences of DNA (primers, blue and green) that bind to the DNA single strand (template). The primers have been designed by the scientist to bind in each case at a well-defined sequence of the DNA strand. The third step is the extension or elongation of the bound primers. The DNA polymerase, an enzyme that synthesises new DNA strands, makes a copy of the templates starting at the DNA primers. The cycle is typically repeated 25–40 times to build enough amplified DNA product for analysis. Right: All DNA double strands are denaturated by heating and serve as new templates. Therefore, the amplification process proceeds exponentially. One double strand produces two new double strands; these are separated into four single strands and thus produce four double strands, and so on. Source: Agilent Technologies present in the sample. With this information, two or more samples can be compared with respect to their starting DNA quantity.

The additional technical challenges involved are:

- Permanent excitation intensity of the fluorescent dyes
- Precise monitoring of the fluorescence signal via filters and lenses

▶ **Microarray.** Microarrays are glass or plastic slides with different molecules immobilised on their surface. Depending on the type of molecules, one distinguishes between DNA and protein microarrays (also called DNA chips or protein chips, respectively). DNA microarrays use DNA fragments (nucleotides) as part of the detection system whereas protein microarrays use small peptides for the detection of proteins.

▶ **DNA microarray.** DNA microarrays consist of thousands of microscopic spots of different small DNA fragments (probes) chemically immobilised on the solid surface. They are typically manufactured by a *photolithographic method.* A quartz wafer is coated with a light-sensitive chemical compound that prevents binding between the wafer and the nucleotides of DNA probes. Specific regions of the wafer surface are unprotected by illumination through a mask. Now nucleotides of the probes can bind to these unprotected spots. Subsequently, the next spot is unprotected and another probe with another sequence is attached to that region. This is repeated until all regions are occupied by probes with different sequences. The immobilised DNA fragments on the slide are able to stick to complementary DNA sequences (targets) in a test sample.

The target nucleic acids are labeled with fluorescent dyes first. Secondly, the prepared sample is hybridised to the microarray. This means that DNA strands from both the target and the probe, which contain complementary sequences, can create a double strand. Fluorescence-based detection shows the areas where target sequences have bound. This enables conclusions to be drawn about the composition of the sample.

If two differently marked samples are mixed and hybridised to the microarray, it is possible to reveal which nucleic acids can be found in both samples and which are specific to just one of them. This enables scientists to discover RNA and DNA patterns, for example, which are known in some cases to classify diseases.

▶ **Protein microarray.** For the manufacture of protein microarrays, small protein fragments (peptides) or antibodies are immobilised in a 2D addressable grid on a glass slide. To immobilise proteins on a slide, the glass surface has to be modified so that a cross-link to proteins is possible (e.g. via a cross-linking agent that reacts with primary amines). The attached antibodies on the slide can extract and retain proteins (targets) from a test sample: a lysate containing proteins is spread over the array, and some of these proteins are bound to the array. They can subsequently be detected, for example, via fluorescence or radioactive labelling. Knowledge of the specific position of an immobilised peptide or antibody on the slide, together with the detected signal, provides information on the protein composition of the sample. Protein microarrays could thus become a high-throughput method of profiling protein changes in diseases.

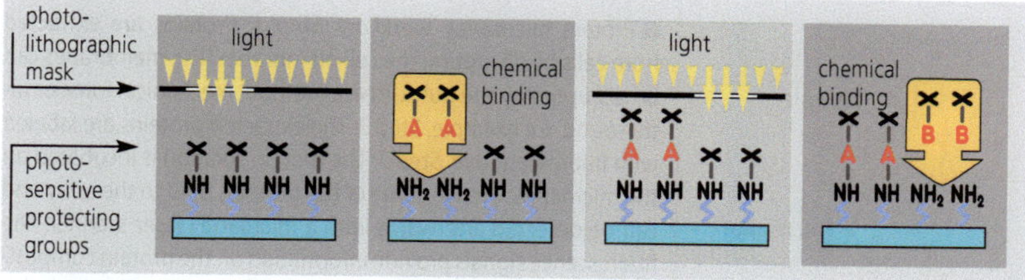

Manufacturing a DNA microarray: A quartz wafer is coated with a light-sensitive chemical compound (X) that prevents binding between the wafer and the nucleotides of DNA probes. Illumination through a photolithographic mask destroys the protecting groups (X) and enables nucleotides of the probes (A) to bind to the unprotected spots. The bound nucleotides from the probe are again protected with the light-sensitive chemical compound (X). A further illumination of a different region of the slide permits the attachment of another probe (B) with another sequence. This procedure is repeated until different probes occupy all regions. Source: Fraunhofer

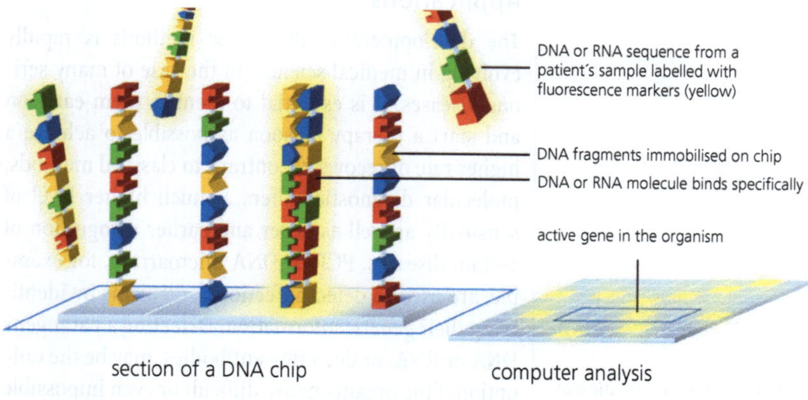

Detail of a DNA chip: On a thumbnail-sized DNA chip, a large number of different small DNA fragments are chemically immobilised. These fragments on the slide can stick to complementary DNA or RNA sequences from a patient's sample that is spread over the array and is labelled with fluorescence markers (yellow). The binding takes place according to the rules of base pairing: Guanine (G) pairs with cytosine (C) and adenine (A) with thymine (T) in DNA or uracil (U) in RNA, respectively. Nucleotides that do not match are removed by a series of washing steps. The labelling of the nucleic acids from the sample with fluorescent markers leads to a pattern of shining dots on the slide. A computer interprets this pattern and can identify activated genes related to a specific disease, for example. Source: Bayer AG

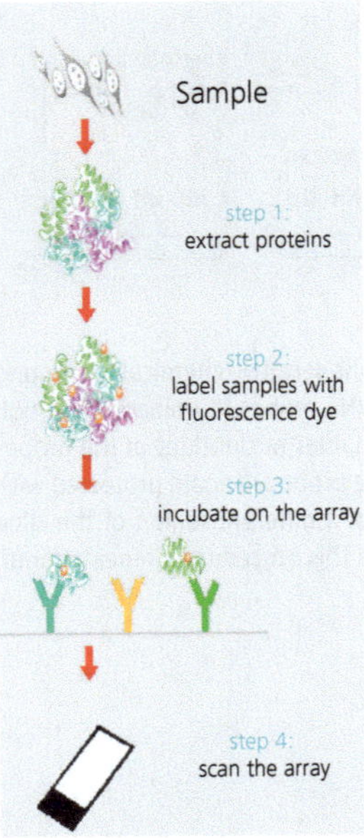

◀ Protein microarray workflow. Step 1: Proteins are extracted from cells by disrupting the cell membranes. The membranes can be disrupted by the quick-freeze method, mechanical force or ultra sound, for example. Step 2: The extracted proteins are labeled with fluorescent dye. Step 3: The protein solution is incubated on the microarray. Step 4: Some of the proteins bind to the array and can be detected afterwards with a microarray laser scanner. The fluorescence signals provide information on the protein composition of the sample. Source: Sigma-Aldrich

One difference between protein and DNA microarrays is that the range of protein concentrations in a biological sample is several orders of magnitude greater than that of nucleic acids. This means that detector systems for protein microarrays need to have a very broad range of detection. Another difference between proteins and nucleic acids is that proteins have a defined structure (secondary and tertiary structure). This structure is responsible for their biological function and their ability to bind other proteins. Different proteins will behave in different ways when they are exposed to the same surface. For this reason, the technical challenges in producing protein microarrays are:

- Obtaining suitable surface modification of the slides that allows the immobilised proteins to keep their native conformation and thus their functionality to bind proteins from the sample
- Retrieval of high-specificity antibodies that exclusively identify the protein of interest

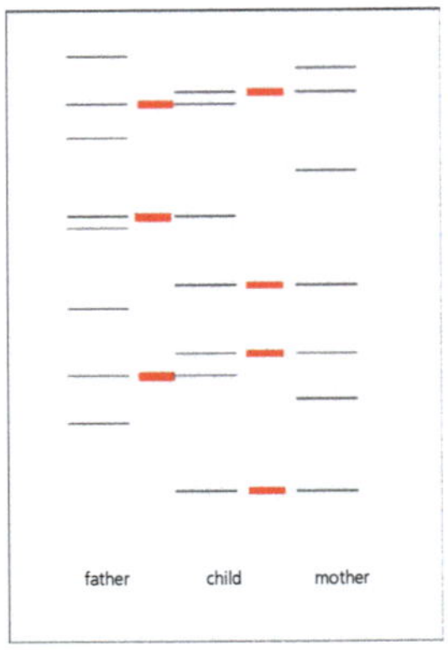

▲ Paternity test: Schematic illustration of the results of a gel electrophoresis test. The amplified DNA products from father, mother and child are separated by their size. The child has inherited parts of the genetic information of both parents (marked in red) and has its own individual DNA profile, a genetic fingerprint.

Applications

The development of diagnostic methods is rapidly evolving in medical science. In the case of many serious diseases, it is essential to identify them early on and start a therapy as soon as possible to achieve a higher rate of recovery. Contrary to classical methods, molecular diagnostics offers a much higher level of sensitivity as well as faster and earlier recognition of certain diseases. PCR or DNA microarrays, for example, are used to detect infectious organisms by identifying their genetic information. Detecting a pathogen's DNA or RNA, or detecting antibodies, may be the only option if the organisms are difficult or even impossible to culture. Molecular diagnostic tests (e.g. PCR or DNA microarray) for recognising infectious diseases have several advantages over conventional techniques based on culturing the organisms or detecting antibodies:

- The presence of viral or bacterial nucleic acids can be detected earlier than antibodies that are built up in the patient as a result of an infection.
- The use of PCR permits direct recognition of the pathogen itself, while the detection of antibodies with immunoassays provides indirect evidence only. Immunoassays are based on the specific binding of an antibody to its antigen. After binding, the antibody can be detected. There it has to be labeled. The immunoassay does not provide information on whether the organism is currently present in the patient or whether the antibodies are simply detectable due to earlier exposure to the pathogen. In order to know whether an infection is acute or not, it is necessary to detect at least one other antibody.
- Molecular diagnostic tests are more sensitive than protein immunoassays.

▶ **Genetic fingerprinting.** Genetic fingerprinting is a technique that enables individuals to be identified exclusively using samples of their DNA. This technique is used to verify family ties, to establish the paternity of a child or to identify lawbreakers in forensic medicine. The vast majority of human DNA is identical and is known as "encoding genetic information". These genes are actively translated into proteins. But not all of the human DNA is translated in this way. The part that is not translated is called "non-coding" DNA and differs strongly among individuals. This makes it possible to distinguish people's identity. DNA can be cut into well-defined fragments of different sizes by enzymes known as restriction endonucleases. These recognise specific patterns in the DNA (restriction sites) and cut the DNA sequence into as many pieces as patterns found. These pieces are amplified by PCR and can subsequently be separated by size via gel electrophoresis. DNA molecules are negatively charged because of their phosphate backbone. Therefore, they move in an electrical field to the positive charged pole (anode). The effect that smaller DNA fragments move faster than larger fragments in a gel matrix (e.g. agarose or acrylamid gel) is used to separate the DNA fragments by their size.

The pattern of DNA bands resulting from the gel electrophoresis provides a unique profile of an individual's DNA due to the distinctive restriction sites inside it. Because "non-coding" parts of the DNA are passed on from parents to their children, it is possible to establish whether people are related by comparing their DNA profiles.

Trends

The major trends in molecular diagnostics include the development of automated platforms resulting in faster, more flexible, more reproducible and more patient-oriented diagnostic applications. Chips can be designed and custom-built according to specific diagnostic purposes. The miniaturisation of analytical systems makes it possible to reduce costs and sample amounts.

▶ **Lab-on-a-chip.** Lab(oratory)-on-a-chip systems are universal diagnostic platforms that integrate one or several laboratory functions on a glass or polymer slide the size of a chip card. This technology has been steadily developing over the past few years, and can handle extremely small amounts (as little as picolitres) of reagents and substrates. The system is based on a network of micro-channels and wells that are created by chemical etching or abrasion. Liquids move through the channels to the wells by capillary force, electrical fields or vacuums/pressure. The analytical processes (physical, chemical or biological) take place in the wells. It is possible to mix, dilute or separate samples and to stain and detect nucleic acids or proteins on a single chip. The lab-on-a-chip technology also provides the opportunity to perform whole reactions such as PCR, to run antigen-antibody reactions (immunoassays) in order to detect bacteria or viruses, and to carry out gel electrophoresis for analysing and quantifying DNA, RNA or proteins.

The main advantages of lab-on-a-chip systems are:

- Very low sample volumes and therefore lower costs and less waste
- Faster analysis by massive parallelisation and short distances, and additionally by small volumes. As a result, shorter heating and cooling rates (e. g., for PCR) can be achieved.
- Improved reliability and reproducibility as a result of predefined assays and reduced manual work
- Greater functionality
- Automated data acquisition and analysis

The lab-on-a-chip system could become a key technology for performing molecular diagnostic tests with no, or at least little, laboratory support, for example in poorly equipped clinics, care centres or even in patients' homes.

Method	Application
DNA microarray	classification of human breast cancer using gene expression profiling; prognostic tests on risk of relapse
	detection of prostate carcinoma
	drug development
	identification of infectious agents
	prognostically useful gene-expression profiles in acute myeloid leukemia
Lab-on-a-chip	concentration and detection of influenza
RT-PCR	detection and identification of bacteria (e.g. chlamydia trachomatis)
	HIV and borrelia diagnostics
	determination of disease markers (e.g. tumour markers)
Protein chip	protein expression profiles of neuromuscular diseases
	protein expression profiles of diabetes
	development of vaccines

Examples for the various applications of molecular diagnostic methods

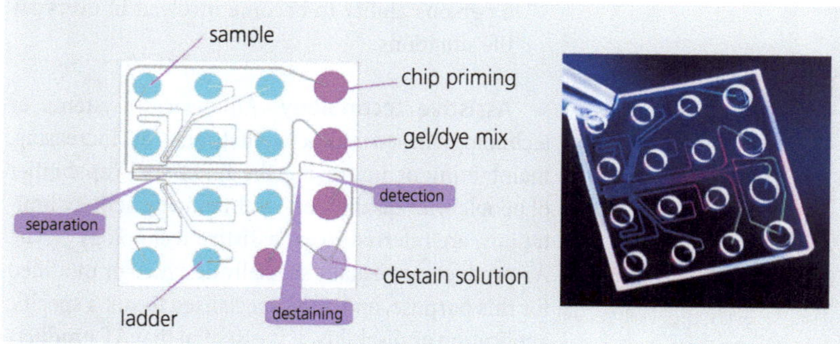

Protein chip. Left: Layout of chip. A sample containing proteins of different sizes is loaded into the sample well. The ladder well has to be filled with a ladder containing proteins of a known size. The ladder serves to determine the size of the different proteins from the sample. The proteins are separated by size via gel electrophoresis in the separation channels. The separated proteins are stained. The gel subsequently has to be destained, while the proteins remain stained and can therefore be detected. The protein chip described is used instead of a protein gel. The procedure on the chip is less time-consuming and less labor-intensive than conventional protein gel electrophoresis. Right: Photograph of a LabChip. Channels and wells are abraded or etched into the surface. Source: Agilent Technologies

Prospects

- Molecular diagnostics is a rapidly developing field of medical science.
- It features top quality tests based on the detection of nucleic acids or proteins and enables earlier diagnosis and better disease prophylaxis.
- Continuously improved methods and technologies will make it possible to recognise certain diseases faster, earlier and much more accurately. As a result, therapy could begin at an earlier stage of the disease and would probably lead to a higher rate of recovery.
- The development of all-purpose or patient-specific diagnostic platforms is a further step to becoming more flexible and more patient-oriented. Furthermore, these platforms will help to reduce costs and patient sample material.

DR. JULIA WARNEBOLDT, BRAUNSCHWEIG

Internet

- www.tecan.com/platform/apps/product/index.asp?MenuID=1865&ID=2352&Menu=1&Item=33.3
- www.bio.davidson.edu/Courses/genomics/chip/chipQ.html
- http://molecular.roche.com/roche_pcr/video_pcr_process.html
- www.chem.agilent.com/Scripts/Generic.ASP?lPage=6700&indcol=N&prodcol=N

Assistive technologies

Related topics

- Robotics
- Medical and information technologies
- Microsystems technology
- Sensor systems
- Internet technologies
- Ambient intelligence
- Human-computer cooperation
- Implants and protheses

Principles

A person's functional capability and disability are viewed as a complex interaction between the individual's health condition, other personal factors, and contextual factors concerning the environment.

► **Disability.** Dysfunctions can be found at the following levels:

- Impairments are problems concerning the physiological functions of body systems (physical, sensory, cognitive, or mental functions) or concerning anatomical parts of the body, for instance in the case of a significant deviation or loss.
- Activity limitations are difficulties in the execution of a task or action.
- Participation restrictions are problems concerning a person's ability to become involved in everyday life situations.

► **Assistive technology.** All devices, systems or technology-based services that aim at increasing, maintaining or improving the functional capabilities of people with disabilities, or at reducing activity limitations, are referred to as "assistive technology" (AT). AT products are usually specially designed or modified for this purpose, and often specialised to suit a specific application or disability. It is essential that AT products be adapted to the special needs of individual users, especially when it comes to the human-machine interface. It is also important to avoid a stigmatising design so that the products are more readily accepted by the users.

► **Design for all.** "Design for all" takes an opposite approach to AT: devices, software, environments and services, etc., are ideally designed in such a way that they can be used by all potential users; or that they can at least be adapted to users' individual needs. This ensures that nobody is discriminated against and that no one is unable to use a technology due to a disability. In cases where a universal design is not possible, the product has a technical interface that enables the user to operate it with the help of a personal AT device.

■ Wireless phone "designed for all": This device has a clear screen with a high contrast and clear lettering that can be read even by people with poor vision. Its volume can be widely adjusted to fit the needs of people who are hard of hearing. It has large control elements so that it can be operated by people with poor dexterity. Its functionality is reduced to the most essential functions and avoids cascading menus, so that it is not too complicated for people with reduced cognitive abilities.

■ Barrier-free access to a train: This is achieved by an extension board that automatically closes the gap between the train entrance border and the platform edge. The remaining small step is less than 3 cm high and is thus regarded as barrier-free for wheelchair users. This function is activated by the button below the standard door-opening button.

► **Barrier-free facilities.** Buildings, transport systems, technical devices, information and communication systems and other environments designed for everyday life are regarded as barrier-free if they can be accessed and used by people with disabilities more or less normally without any great difficulty and, most importantly, without the help of others.

Applications

► **Smart electric wheelchair.** Electric wheelchairs are classical AT devices that support mobility. Problems in wheelchair usage comprise:

- overcoming obstacles such as steps, gaps and stairs
- limited manoeuvrability in narrow or cluttered rooms
- stability, safety and speed outdoors
- transfer between bed/seat (car seat, chair)/bathtub/wheelchair
- comfortable seating
- transportation and operation of the wheelchair

A distinction is usually made between indoor and outdoor wheelchairs. While indoor wheelchairs need to be small and highly manoeuvrable, with the ability to rotate on the spot, outdoor wheelchairs need to be more robust, capable of overcoming steps and gaps, strong and stable for driving on steep or rough paths, fast, and powered by enduring batteries for long distances.

Typical electric wheelchairs have 4 or 6 wheels, 2 of them powered. The non-powered wheels are either steered directly or they are free castor wheels that can move in any direction (indirectly steered wheels).

There are different concepts for increasing manoeuvrability:

- Indirectly steered wheelchairs are usually able to rotate on the spot by simultaneously turning the 2 powered wheels in opposite directions. A new approach is to apply a variable wheelchair geometry, with a long distance between the axes to improve stability outdoors and a short distance between them to improve maneuverability indoors.
- While none of the common wheelchairs are able to move sideways, omni-directional drive concepts are attempting to achieve this by using either 4 powered and directly steered wheels or what are known as "mecanum" wheels.

▶ **Intelligent collision avoidance.** Severely disabled wheelchair users benefit from distance *sensors* (infrared, ultrasonic, laser or camera sensors) that help to avoid collisions, especially when driving backwards, or can prevent people from driving down steps. Distance sensors are also used for active steering support, e.g. automatic centrical alignment while driving through narrow doorways, or keeping a constant distance from walls while driving alongside them. Incremental sensors at the wheels enable frequent manoeuvres, such as passing through the bathroom door, to be recorded and automatically executed, including in reverse. Such semi-automatic control of the wheelchair can even be extended to a full-automatic mode where users simply indicate their destination, e.g. by a speech command, and the electric wheelchair navigates like a mobile robot using an environmental map for path generation and distance sensors to avoid collisions with dynamic obstacles. A wheelchair with such navigational intelligence can also be commanded in remote mode. For those who are not able to operate the standard input device, i.e. a joystick, alternative input devices exist, including single-button input, speech control, head movement and gaze detection systems.

▶ **Aiding mobility by virtual reality.** Some patients with Parkinson's disease sometimes experience a sudden blockade in their movement ("freezing") while walking, which can last for several seconds or minutes. This unsafe state of freezing can be overcome with the help of certain visual or acoustic cues. For this purpose, the patient wears special semi-transparent glasses with an integrated virtual reality screen. In the event of a sudden freezing attack, a portable computer is activated that displays a video on the screen showing virtual obstructions placed in the way. These obstructions can simply be pieces of paper lying on the floor, for example. The patients' reflex is to step over the obstacle, which stimulates their mental concentration and, as a result, releases them from the freezing state so that they can continue walking. This human reaction, known as "kinesia paradoxa", has not yet been fully understood.

▶ **Rehabilitation robots.** The purpose of a rehabilitation robot is not automation, but instead to activate and restore the user's functional abilities. Most rehabilitation robots are therefore operated in the manipulator mode, where the user is part of the control loop of the robotic system. The user steers the robot arm directly and observes the movements, which act as feedback for steering. Consequently, most manipulators do not need to apply external sensors. A *manipulator* arm usually has a two-finger gripper capable of gripping all kinds of objects or special tools, e.g. a rubber finger for turning pages. Users of manipulators normally also use an electric wheelchair. The robot arm is therefore often attached to the wheelchair to make it a mobile and versatile tool, mainly for pick-and-place tasks in an unstructured environment. Manipulator arms usually have 4–7 rotational or translational axes. Steering these 4–7 degrees of freedom requires very good spatial cognition. The standard input device is either a two-dimensional joystick (part of the wheelchair) or a special keyboard. The efficient and intuitive

Omni-directional wheelchair: Omni-directional driving concepts enable wheelchairs to move sideways – by using either 4 powered and directly steered wheels or mecanum wheels. The mecanum wheel can move in any direction. All 4 wheels are independently powered and controlled. As well as moving forward and backward like conventional wheels, they allow sideways movements if the front and rear axle wheels are spun in opposite directions.

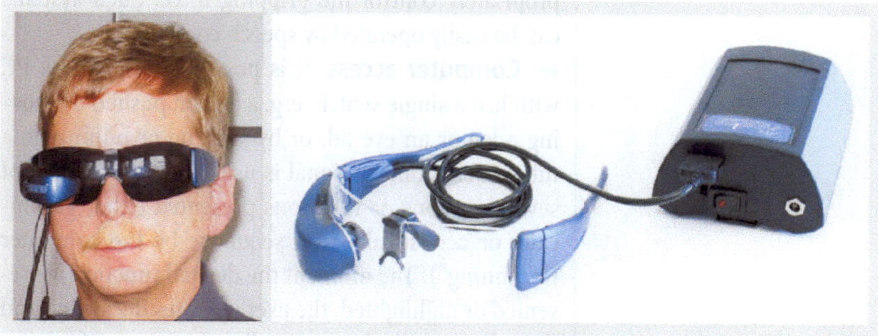

INDIGO system: This system helps patients with Parkinson's disease to overcome sudden blockades in their movement while walking. The degree of these mobility difficulties can vary from person to person, and can depend on the time of day and the stage of the disease, but are always accompanied by slow movement or complete freezing (phases called akinesia). The system works by displaying "moving visual cues" in the user's peripheral visual field. This is done through semi-transparent glasses, running MPEG video software on a dedicated portable computer. If these visual obstructions are placed in their way, e.g. a piece of paper on the floor, some people with PD undergo a dramatic release from the symptoms and can suddenly stand up straight, speak strongly and walk normally. This effect is called "kinesia paradoxa". Source: ParkAid, Italy

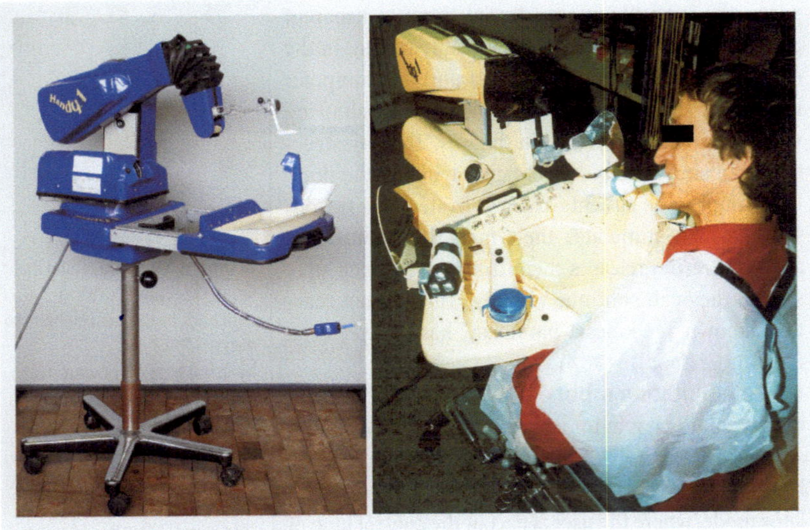

Rehabilitation robot: The robot consists of a simple 5-axis arm and a set of exchangeable, automatically detectable application trays, e.g. one for "meals" with a dish, spoon and cup (left) or one for "personal hygiene" with a water bowl, soap, sponge, electric shaver, electric tooth brush, water cup, and mirrors. The basic equipment includes an electric fan for users to dry their face. Each tray has a set of predefined robot functions which are sequentially displayed with LEDs and can be activated through a single switch. This allows even severely physically disabled people to control the robot.

operation of the robot arm's movements is an important research topic. In contrast to the general-purpose manipulator arm, there are also robots which perform semi-automatic tasks in a structured environment. In this case, the user can select predefined arm movements to be performed. More sophisticated robots apply a camera system for scene analysis, supporting the automatic control of fine motions for gripping an object. They also have force sensors in the gripper to appropriately control the gripping force. Such systems can be easily operated by speech commands.

▶ **Computer access.** It is possible to operate a PC with just a single switch, e.g. a button pushed by moving a leg or an eye lid, or by sucking or puffing on a tube. Such a binary signal is used to select one out of several options or functions presented either graphically or acoustically in a sequence by the computer ("scanning"). The moment the desired function is presented or highlighted, the user operates the switch and thus activates the corresponding function. Of course, the speed of the automatic scanning process needs to be adapted to the user's capabilities.

▶ **Text writing support.** In order to write a text, an onscreen keyboard is displayed on the computer screen together with the target word processing application. Scanning the alphabet character by character would be very time-consuming and tedious. Therefore, blocks of characters such as the row of a keyboard are scanned first; after the user has selected the appropriate block, the characters within it are scanned. Scanning can be optimised by placing the most frequent or most likely items first in the scanning sequence.

To save keystrokes during typing, user-defined abbreviations are automatically replaced by the corresponding words or text phrases. An intelligent user can define a system of mnemonic codes for several hundred text modules. A complementary approach is word prediction: as language contains a high degree of redundancy, a few written characters are often sufficient to predict which word is (probably) being typed. The prediction software offers a list of suggestions for completing the current word or the next words, based on the characters that have been typed in so far.

▶ **Output devices for blind users.** In standard computer applications most information is presented visually to the user, in the form of text or graphics. For blind computer users, there are two main alternatives, i.e. audible and tactile forms of presentation.

Plain text files can be read to the user by text-to-speech software. When textual and graphical information is mixed, as is usually the case in world wide web browsers, special screen reader software has to analyse the information displayed on the screen before reading it to the user. In order for this to work properly, the information – e.g. a web page – needs to be well structured and the graphics need to have a textual equivalent assigned to them in accordance with the Web Accessibility Guidelines (WAI). Technically, this has to be done by the information provider by "tagging" the content with additional structural information.

When editing one's own files or reading computer source code, for example, a tactile form of presentation is preferable to an auditive one. Braille displays make it possible to read the text line by line, character by character, or to move back and forth.

Efficient presentation of graphical information to blind users is still a research topic. Tactile graphics displays can give at least a "black/white" presentation of simple graphical information, e.g. the curve of a mathematical function.

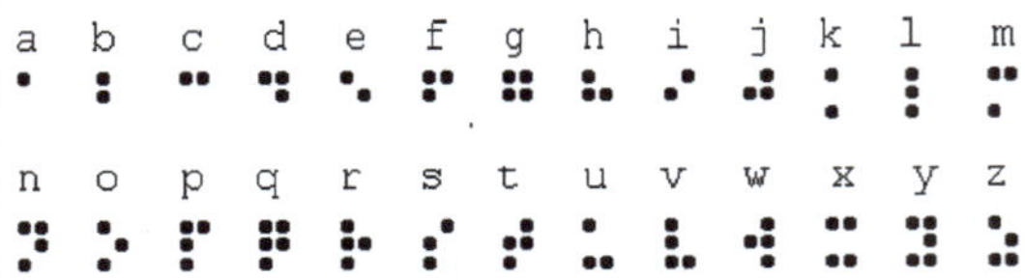

Braille code: To make a text "tactile" for blind readers, all characters need to be given a tactile form. The Braille coding system uses cells of 6 dot positions to code the standard character sets. In tactile form, the dots are embossed to be "readable" with the fingertips.

Trends

▶ **Public transport information system.** Although many improvements for disabled people have been achieved in public transport systems, not all facilities are barrier-free. However, all elements of the travelling chain, including buildings, vehicles, and information systems, need to be accessible for the traveller. Current routing algorithms in travel information systems find travel connections from any given starting point to any destination. These systems need to be enhanced to include detailed information on the accessibility of all relevant facilities in the regions covered. In addition, the person wishing to travel needs to define their individual user profile, indicating special requirements concerning accessibility. Then the routing algorithm searches for the best travel connection that matches the user's profile. If additional real-time information on train arrivals or on the availability of necessary elevators in a railway station building is available, then the information system can react to unexpected deviations from the travel schedule or other malfunctions, recalculate the planned journey, and inform the traveller en route.

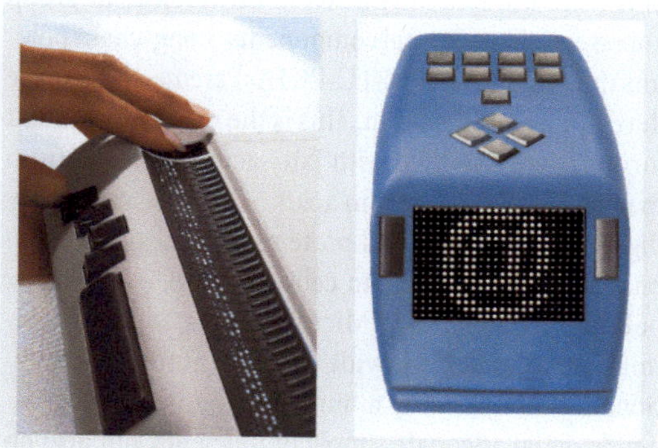

Braille display and tactile graphics display: A Braille display (left) is a tactile device that enables blind people to work with a computer. It consists of a row of Braille cells, where each dot is represented by a pin. The pins are electronically controlled to move up or down respectively, displaying the characters of the source text. Long Braille displays code 80 piezoelectrical characters to display a whole line in a typical word processing system. Right: Tactile graphics display for touch-reading graphical representations on the screen. The graphics are transformed into tactile images consisting of 24 × 16 pins. To this end, the graphical representation is prepared using image processing techniques. The tactile graphic dynamically adapts itself in real time to changes in the screen representation. Source: Handy Tech Electronic Gmbh

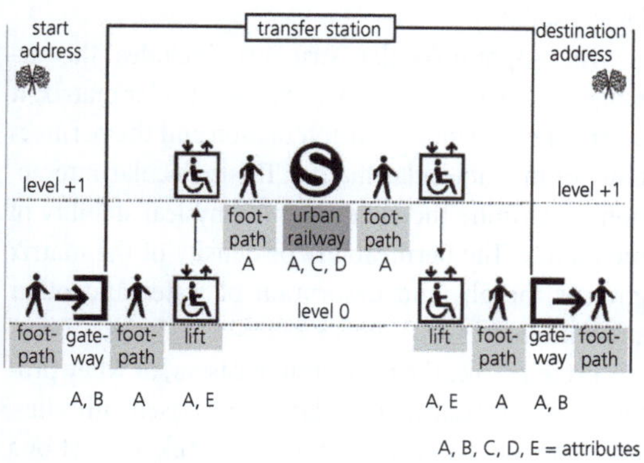

Barrier-free travelling: Routing barrier-free travel connections requires modelling the accessibility (features such as the steepness of a ramp, the width of a door, the availability of a blind guidance system, or free space for wheelchairs in a vehicle) of all facilities along the way. A travelling chain is only barrier-free if all the necessary elements are accessible. Source: IVU Traffic Technologies

▶ **Tele-eye system.** Blind or poor-sighted people, many of them elderly, face the problem of not being able to identify food packages; the printed list of ingredients is often difficult to read and understand. This is a particular disadvantage for people with special dietary requirements. A product information system linked to a database can identify a product via its bar code, which can be read with a laser scanner or just photographed with a mobile phone camera. Using the bar code, the corresponding product can be identified and displayed or read to the user. With the product identified, the list of ingredients can be checked against a user profile with dietary restrictions and a warning (visual or audible) can be issued. That way an elderly customer can be supported while shopping, and a blind person can identify consumer packages.

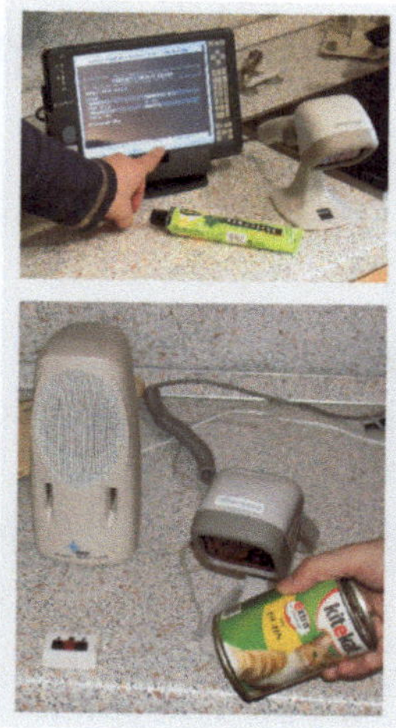

Tele-eye systems identify a consumer product via its bar code, provide product information, analyse the ingredients list and issue warnings in accordance with individual user profiles. They consist of a bar code scanner and either a touchscreen or a loudspeaker and 3-button operation element. The online database with product information and the knowledge database on ingredients are not visible to the user.

Prospects

- High potential for developments in assistive technologies lies in the information and communication technologies, especially in mobile applications, the miniaturisation of actuators and sensors, technology-supported social networking, and "assistive systems" that combine compatible AT devices.
- In view of the aging societies assistive technology is becoming increasingly important in compensating for disabilities.
- The application of "design for all" principles in all user-relevant products and services is extremely important in order to prevent the social exclusion of elderly people or people with disabilities, and should be compulsory for all relevant technical developments in the future.

DR. HELMUT HECK
Research Institute for Technology and Disability (FTB) of the Evangelische Stiftung Volmarstein

Internet

- www.tiresias.org/
- www.ftb-net.com/entwz.html
- www.trace.wisc.edu
- www.w3.org/WAI/
- www.assistivetechnologies.co.uk

Food technology

Related topics
- Process technologies
- Industrial biotechnology
- Renewable materials

Principles

► **Consumer needs.** Food technology encompasses all the know-how required to transform raw materials into semi-finished or finished food products. Over the past 50 years, consumers have gained greater access to information relating to the composition of food products. They are becoming increasingly aware of the implications of their diet to their well-being and the prevention of diseases. This has led to a steadily growing demand for products rich in fibres, whole grain, vitamins and minerals. During the second half of the 20th century, the living habits and the energy expenditure of the population in the industrialised world changed dramatically. Since diets remained more or less unchanged, the surplus energy consumed led to obesity, and its co-morbidities became a serious health concern in many parts of the world. This has resulted in a steady rise in the demand for nutritionally balanced products with a lower energy density. The beginning of the 21st century saw double digit growth rates in the market for products containing bio-actives that deliver specific health benefits (so-called functional food). Today's consumers also link nutrition and health with natural and organic foods, leading to a higher demand for such products in the developed world.

Approaching 7 billion people, the world's population is also growing rapidly. An increasing caloric consumption per capita is expected to cause serious food shortages in developing countries, and agricultural and food technology will have to meet this challenge.

► **Food composition.** Food products are made up of agricultural raw materials. They are composed of carbohydrates, proteins, lipids and small quantities of minerals, organic acids, vitamins and flavours. The digestible carbohydrates serve mainly as a source of energy. While lipids act as an energy store as well, some of the unsaturated fatty acids in this molecular group are essential nutritional components. Long-chain polyunsaturated fatty acids (LC-PUFAs) are needed for the development of the brain. This is the reason why infant formulas are enriched with fatty acids extracted from marine biomasses. Amino acids, which are derived by digestion or hydrolysis of proteins, are one of the main building blocks of human cells. They are essential for building muscle mass. Micronutrients, such as vitamins and minerals, provide different nutritional functionality. How healthy a human being is not only depends on an adequate supply of nutrients, but also on the microorganism flora of the human gastro intestinal system. It has been proven that certain strains of microorganism (probiotic bacteria) can defend the human gastro-intestinal system from infections by pathogenic bacteria. Probiotic bacteria or pre-biotic carbohydrates, which foster the selective growth of such bacteria species, are therefore sometimes added to food products.

► **Food structure.** The properties of foods are determined not only by their composition but also by their supramolecular, micro- and macrostructure. The supra-molecular structure and microstructure determines the texture of a food product, its digestibility, and the product's physical, chemical and microbiological stability.

The supramolecular structure includes the arrangement of molecules in a given molecular matrix. It determines the molecular interaction and the permeability of the molecular matrix. The molecular interactions determine the viscosity and physical stability of the matrix. The permeability or density of the matrix governs the ab- and desorption of water and other small molecules.

Proteins, like the milk protein casein, or whey proteins, might exist in micellar form. Casein micelles, which are naturally present in fresh milk, consist of a number of casein molecules that are cross-linked through calcium bridges. Whey proteins, lactoglobuline and lactalbumine, can form gel structures or microparticles. The molecular interactions in a micellar or microparticulated structure are reduced. Accordingly, even highly concentrated protein solutions ex-

A selection of supra-molecular food structures. From left to right: Protein micelles (e.g. in whey or milk), starch granule with crystalline and amorphous growth rings (e.g. in native flour), fat crystals (in a margarine product) and sucrose crystals (table sugar).

hibit a comparably low viscosity. By contrast, gelling the protein solution delivers viscoelastic textures. The gelling of caseins and whey proteins is what gives yoghurts and cheeses their semi-solid texture. Carbohydrates and lipids, for their part, exist either in an amorphous or crystalline state, depending on the velocity of drying or freezing. In the liquid-like amorphous structures, the molecules are randomly arranged, while in crystals they are ordered according to a recurring geometrical pattern.

Most food products are dispersed systems that consist of at least two different phases. There are emulsions (liquid droplets in a surrounding liquid phase, as in mayonnaise, salad dressings, butter, cream), foams (gas bubbles in a continuous liquid or solid phase, as in ice cream, mousse au chocolat, bread), powders (solid particles in a continuous gas phase, for example instant coffee, milk powder, infant formula, dehydrated soup), and suspensions (solid particles in a continuous liquid phase, like melted chocolate, chocolate drinks).

Taking a look at emulsions, one finds two basic types: water droplets in a surrounding oil or fat phase (for example, butter or margarine) and oil droplets in a surrounding aqueous water phase (for example, mayonnaise or milk). The texture of such emulsions strongly depends on the interaction between single droplets.

There are a number of food products that are considered foams. Ice cream is a typical example of a frozen foam, which typically contains 50 vol.% of air. Bread is another example: baker's yeast used in the dough produces carbon dioxide during leavening. Carbon dioxide bubbles are embedded in a protein/starch matrix, which is then solidified during the baking process.

Some food products are suspensions, in which particles are suspended in a liquid phase. Salad dressings and mustard are suspensions, as are chocolate drinks, which consist of an aqueous phase containing cocoa particles. In melted chocolate, mass cocoa particles are suspended in liquid cocoa butter.

▶ **Processes.** For manufacturing food products, a variety of transformation processes (called unit operations) is applied.

Since its beginning, food technology has sought to provide safe food products that have a longer shelf life, allowing them to be widely distributed and made available irrespective of harvesting seasons and the location of production. The aim is to eliminate potential pathogenic bacteria or spoilage organisms that might be present in raw materials, and to reduce the impact of other microorganisms to a level that ensures the product's quality until the end of its pre-defined shelf life. Losses of bioactives and vitamins, changes in colour, or unpleasant alterations to flavour, must also be avoided.

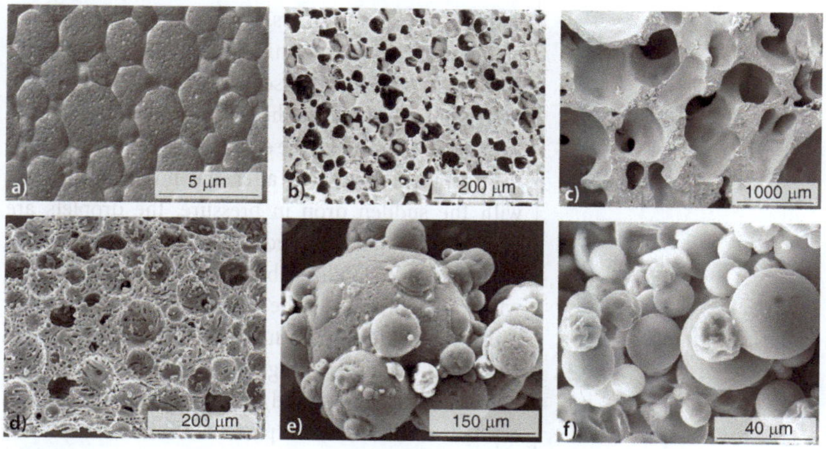

Microstructures of different food products.

a) Full-fat mayonnaise is an emulsion containing 90% oil. The thick smooth texture of the mayonnaise is the result of the interaction and friction between the tightly packed droplets.
b) Ice cream is a frozen foam that is stabilised by a network of milk-fat globules. Ice crystals are embedded into the freeze-concentrated solution filling the voids between the air bubbles.
c) Wheat bread is a highly porous semi-solid gluten/starch foam. During the baking process, this foam solidifies due to denaturation of the proteins (e.g. gluten) and the swelling of starch.
d) Freeze-dried coffee is a solid foam obtained by evaporation of water from a frozen coffee extract. These highly porous structures re-hydrate rapidly.
e) Milk powder particles are composed of a glassy lactose/protein matrix incorporating milk fat globules. During spray-drying, smaller particles adhere to larger ones forming the agglomerates shown here.
f) Encapsulated flavour particles produced by spraydrying: the glassy matrix that makes up the particles contains small oil droplets conveying the flavour components.

Certain processes have been applied by mankind for thousands of years. A number of these, such as fermentation processes, simply replicate processes that take place in nature. In food technology, the principle of these traditional processes is maintained, while the control of processing parameters and the equipment employed is improved in order to guarantee safety, consistent quality and a pleasant sensorial profile at affordable costs. Examples of such processes are fermentation of beer and proofing of raw dough.

Applications

▶ **Manufacturing ice cream.** Ice cream, a product invented 5,000 years ago by cooks serving the Chinese emperors. It is a frozen foam stabilised by a network made of milk-fat globules and milk proteins. The fat droplets are stabilised by specific food grade emulsifiers derived from vegetable fats and oils. Small ice crystals are embedded between the air bubbles in the continuous freeze-concentrated solution. To achieve this microstructure, a liquid ice cream mix is prepared using milk as its main ingredient. Fruit, sugars, flavours and food grade emulsifiers are then added to the liquid preparation before the mix is filtered and heated to 80 °C in order to pasteurise it. The liquid blend is then homogenised to reduce the mean droplet diameter to less than 2 μm. Due to the elongation of the droplets in the turbulent flow-field after the nozzle, combined with the sudden drop in pressure, the droplets are destabilized and ruptured. The smaller oil droplets generated are stabilized by the emulsifier molecules which accumulate at the oil/water interface. These milk fat globules form a network stabilising the foam generated during freezing. Once the liquid has been homogenised, it is stored at 4 °C to allow the fat to crystallise and to desorb proteins from the surface of the fat globules. The matured liquid mix is frozen rapidly at −6 °C in a scraped surface heat exchanger, a so-called freezer. In order to ensure that the ice crystals remain small, they are exposed to the high shear stress generated by knifes which rotate quickly and continuously inside the freezer. Ice crystals larger than 20 μm would produce a sandy texture. During the freezing process a large volume of air is pressed into the product, introduced as bubbles with a mean diameter of about 5–100 μm. These air bubbles, which are embedded in a network of agglomerated fat globules, are stabilised by a flexible shell built of whey proteins, casein micelles and emulsifiers. The void between the air bubbles is filled by a cryo-concentrated solution containing single ice crystals. The generated frozen foam is filled into dies or boxes and then further solidified at −40 °C. In order to avoid structural changes before consumption, the product has to be kept below −18 °C during storage and distribution. At higher temperatures, the ice crystals and air bubbles start to grow and eventually create an unpleasant sandy or sticky texture.

▶ **Manufacturing chocolate.** Chocolate is made from cocoa beans, which are known to have been cultivated as early as in the Maya and Aztec civilisations. The cocoa beans are harvested and fermented to reduce their bitter taste, whereby certain characteristic notes in their flavour evolve. The beans are then dried and roasted to generate the typical flavour profile. The next manufacturing step involves breaking the beans and then milling them to a paste. The obtained cocoa mass is mixed with milk powder, sucrose and flavours. This mix is kneaded and ground to a finer particle size in roller refiners. To avoid a sandy texture, the mean particle diameter has to be reduced to values below 30 μm. The mass is then subjected to several hours of permanent shear stress applied through rollers or stirrers at about 80 °C. This takes place in so-called conches (named after their mussel-like form). During conching, lecithin and cocoa butter are added to the mix. The solid particles are suspended in the continuous fat phase and sugar crystals and cocoa particles are coated with lecithin, while unpleasant volatiles like acetic acid simultaneously evaporate. The viscosity of the resulting smooth chocolate mass depends on its particle size distribution and the viscosity of the continuous fat phase. During subsequent tempering, the chocolate mass is cooled to about 30 °C to initiate crystallisation of cocoa butter. Seed crystals might be added to facilitate crystallisation in the desired β-modification. Crystallisation into the stable β-crystals reduces the formation of fat crystals on the surface of solid chocolate pieces (so-called "fat blooming") and provides the typical melting characteristics. The finished chocolate mass is filled into forms and further cooled down to solidify it entirely.

Process category	Unit operations	Applications
Size reduction	• grinding/milling • cutting	flour grinding cutting of mixed meat
Size enlargement	• agglomeration	agglomeration of coffee powder
Solidification Thickening	• gelling • drying • freezing	gelling of marmalade or fruit jelly drying of milk to milk powder freezing liquid ice cream foam
Structuration	• emulsification • extrusion • encapsulation	emulsification of mayonnaise extrusion of breakfast cereals encapsulation of flavours
Separation of components	thermal: • evaporation • distillation/rectification physical: • filtration • extraction with water/CO_2	 concentration of sweet and condensed milk distillation of liquor filtration of beer extraction of spices with supercritical CO_2
Changing molecular composition	thermal: • baking • cooking/blanching non thermal: • mixing/blending • fermentation	 baking of bread cooking of vegetables, meat and fish mixing of seasoning blends, bakery mixes fermentation of sauerkraut, yoghurt, sausages
Preservation	thermal processes: • pasteurisation (60 °C) • ultra high temperature (130 - 140 °C) • dehydration by roller- ,spray- or freeze-drying physical processes: • sterile-filtration • high pressure • UV light	 pasteurisation of fruit juice UHT treatment of milk dehydration of milk, infant formulas, beverages, vegetable powders, pastes, fruits freezing of meat and fish filtration of beer and water high-pressure treatment of oysters and ham UV light treatment of water

Unit operation for food processing

▶ **Bioactives and micronutrients.** In developing countries, food products are sometimes fortified with minerals and vitamins in order to compensate for local nutrient deficiencies and to fight malnutrition. In industrialised countries, consumers are offered so-called functional foods containing bioactive phyto-nutrients such as polyunsaturated fatty acids, performance boosting substances (e. g. caffeine) and antioxidants (e. g. catechins, flavones and anthocyanins).

Amongst the most important bioactives and micronutrients are essential polyunsaturated fatty acids, e. g. Arachidonic acid. These substances, which cannot be synthesised by the human body itself, are indispensable for the development of the human brain. Infant formulas are therefore enriched with such fatty acids, which are extracted from marine biomasses using solvent-free extraction processes. During these processes, *super-critical* CO_2 (180 bar and 30 °C), or food liquids such as vegetable oil, are used to extract the bioactive molecules.

It is often necessary, during their shelf life, to protect sensitive micronutrients and bioactives from oxidation and other degradation processes. There is also a need to ensure high bio-availability through the controlled release of the active components during digestion, and efficient resorption of the liberated substances into the blood stream. The pharmaceutical industry tends to blend the synthesised and purified active components with a small amount of powdered exipient, and to press the powder blend into tablets or fill it into capsules. The food industry, for its part, uses either food ingredients for encapsulation (starch, gelatine, alginate), or the food matrix itself to stabilise and deliver the active component. For *encapsulation* of lipophilic (fat- or oil-soluble) substances, the active components are often mixed into vegetable oil, which is then emulsified into a concentrated aqueous maltodextrine, dextrose syrup (both dextrose polymers) or gelatine solution. This emulsion is then spray dried or spray chilled to produce a free-flowing powder which can be dosed accurately. The generated particles are composed of a continuous carbohydrate or gelatine matrix, in which fine oil droplets containing the sensitive bioactive component are embedded. The produced microcapsules are occasionally coated with vegetable fats to protect them from moisture and oxygen. These microcapsules are used in the manufacturing of various liquid or solid food products, such as beverages or vitamin-enriched breakfast cereals. Structured emulsions can also be used to deliver bioactives in liquid food products.

▶ **Ultra-high temperature treatment (UHT).** Today, the most common way to preserve food is by applying thermal processes. The so-called ultra-high temperature treatment is frequently used in the preservation of milk. Fresh milk is clarified (removal of particles by centrifugation), standardised (adjustment of fat content by centrifugation) and homogenised (reduction of droplet diameter). The pre-treated milk is pre-heated to denaturate the milk proteins. The milk is then heated rapidly to a temperature of 140 °C by injecting steam or by indirect heat in a heat exchanger. This temperature is maintained for up to 4 s. Finally, the hot liquid is rapidly cooled down to avoid loss of micronutrients and browning reactions. Since, at the

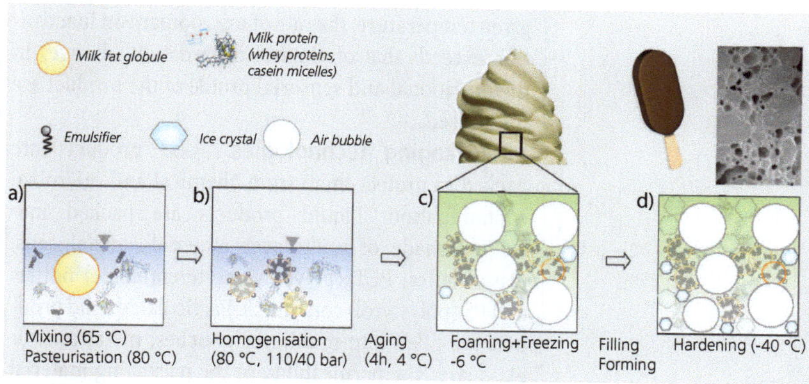

Stages in the production of ice cream and the evolvement of its microstructure

a) Preparation of the liquid mix and pasteurisation to inactivate micro-organisms
b) Homogenisation to reduce the oil/fat droplet size in order to stabilise the foam generated during freezing by a network of agglomerated milk fat globules
c) Freezing at –6 °C and applying high shear forces in a freezer. The volume is simultaneously doubled by pressing air bubbles into the freezing liquid.
d) Frozen foam is then hardened at –40 °C

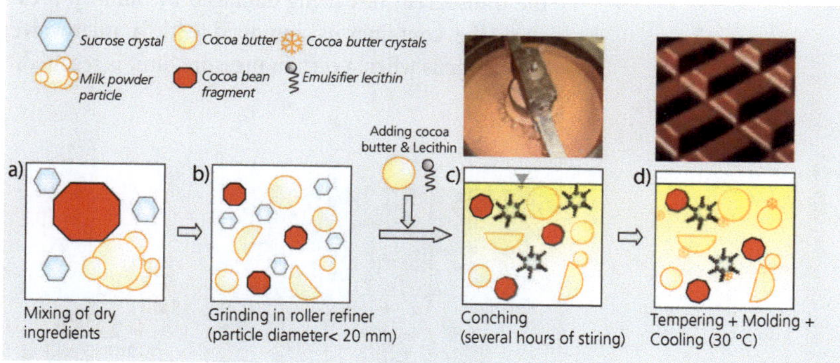

Unit operations and structural changes during chocolate production.

a) Fermentation, roasting and grinding of cocoa beans
b) Mixing of cocoa mass with milk and sugar and grinding to avoid sandiness
c) Addition of cocoa butter and lecithin, followed by conching to create a smooth texture
d) Tempering and cooling to crystallise cocoa butter

given temperature, the rate of microorganism inactivation exceeds that of vitamin degradation, changes in the nutritional and sensorial profile of the product are minimised.

▶ **Packaging technologies.** Food products are packed to protect them from chemical and microbial contamination. Liquid products are packed into briques made of multi-layer materials, metal cans, glass bottles, PET (polyethylene-terepthalate) bottles and PS (polystyrol) containers. Particulated solid products are filled into multilayer pouches, metal cans or glass jars. The permeability of the packaging material to light, oxygen and moisture is reduced by application of thin metal layers.

Aseptic filling technologies are becoming increasingly important, and smart packaging solutions now exist that release vapour or carbon dioxide liberated by the product. One example of such smart packaging is the sealed plastic box developed for chilled pizza dough. The dough contains living yeast which produces carbon dioxide during its shelf life. In order to avoid the transparent box being damaged by built-up pressure, the container is equipped with a microvalve which opens when a certain pressure limit is reached.

Trends

▶ **High-pressure treatment.** One of the aims of food preservation is to minimise thermal impact in order to maximise product quality. High pressure non-thermal preservation technology is already being used today in the pasteurisation of oysters, ham and marmalade. The packed food product is immersed in a liquid such as water or glycerol that for several minutes transmits hydrostatic pressures of over 6000 bar. The treatment is performed in discontinuous or semi-continuous plants. The application of pressure at moderate temperatures inactivates salmonella and other spoilage organisms while minimising the degradation of vitamins and aroma molecules. High pressure is not widely used today, but it might well become one of the future technologies used for preservation of semi-finished or heat sensitive premium food products.

▶ **Personalisation of food.** Products can be personalised in terms of nutritional profile, composition and sensorial properties. Personalisation of food in terms of sensorial properties can be achieved by offering a wide variety of single-servings that allow consumers a choice according to their actual preferences, such as the capsules inserted into domestic coffee machines

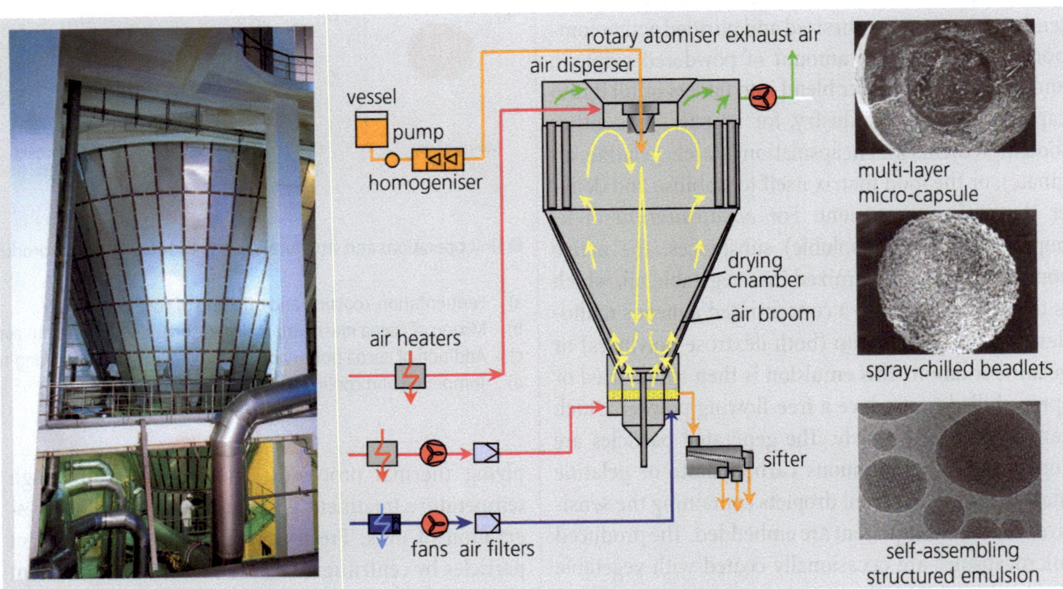

Principle of a spray dryer used for the encapsulation of vitamins, lipids and bioactive: Lipophilic bioactives or vitamins are dissolved in vegetable oil and the oil is then emulsified into a carbohydrate or gelatine solution. The emulsion is homogenised and dried in the tower. Right: Examples of microcapsules used for the delivery of flavours, vitamins, minerals and sensitive bioactives. Top: Multilayer capsule obtained by applying several coats on a core made of the bioactive component. Centre: Micro-beadlet produced by the so-called powder-catch process. This involves spray-chilling a gelatine/oil emulsion. The gelatine beadlets are coated with starch. Bottom: Structured emulsion droplet for solubilisation of amphiphilic or lipophilic bioactives. The self-assembling system is made of micelles that are made up of specific food grade surface-active ingredients. These micelles, which carry the active component, are dispersed in oil droplets which are surrounded by a continuous water phase. This structure is formed spontaneously at specific temperatures/concentrations. Source: Niro/DK, Glatt GmbH, DSM/NL for a single serving of espresso, available in many different varieties.

Products specifically adapted to the nutritional needs of infants and toddlers are another form of personalisation. Nutritional personalisation can also mean adapting a diet to the actual metabolic status of the human body or to a predisposition of an individual towards specific illnesses. Adaptation of the daily diet to the metabolism can only be implemented if the food industry and its partners are able to offer suitable in-home diagnostic tools. Such tools can be used to measure the concentration of defined *biomarkers* in the human body. The analytical results allow the micro- and macronutrient needs of individuals to be evaluated and their diets to be adjusted accordingly. As a complement to this service, the food industry may well offer special products tailor-made to specific situational requirements. *Genomics* present further possibilities in analysing the disposition of an individual towards certain illnesses. Armed with this information, people can adjust their daily diet through adapted food products that reduce the risk of the identified illness breaking out. This development will obviously lead to a greater variety of food products, and the food industry will have to develop and install more flexible manufacturing technologies, such as modular systems or micro-processing approaches, and new systems that will make it possible to cope with the increasing supply chain complexity.

Prospects

Tomorrow's food and agricultural technology has to solve some of mankind's most pressing problems:

- The growing world population: One major future challenge faced by agricultural and food technology will be the need to feed a growing world population through optimised use of available resources. Obviously, these efforts have to be complemented by a serious programme to control birth rates in the countries concerned.
- Global warming: The problem of global warming is causing growing concerns about the emission of greenhouse gases during the production of agricultural raw materials, and a responsible use of energy during processing. Agricultural techniques and manufacturing technologies have to be modified accordingly.

The food industry must respond to an increasing demand for natural food, which is ethically produced without generating a detrimental impact on climate.

- Consumer demand for natural and ethically produced food: Consumers in regions where food is readily available and affordable, and where food safety and convenience is taken for granted, are demanding healthy and natural food products that are produced organically and on ethically supportable terms (fair trade).
- Personalisation of food: Individualism becomes increasingly important in a world characterised by increasing anonymity. Furthermore, progress in analytical and medicinal technologies opens new opportunities to personalise food nutritionally. Food can be personalised by adapting it to the actual nutritional requirements of individuals which are analysed through specific biomarkers. However, such personalisation does mean increased complexity and possibly higher production costs (decreasing economies of scale).

DR. STEFAN PALZER
Nestlé, Lausanne

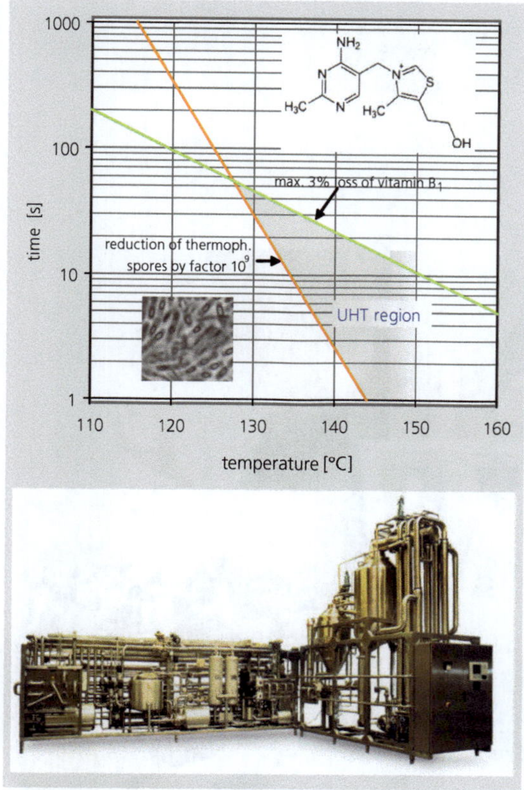

Ultra-high temperature (UHT) treatment. Bottom: A plant for UHT processing of liquid food materials. The liquid is heated indirectly with steam while passing through double-jacketed pipes. The liquid is exposed for some seconds to a temperature of 140 °C. Top: The short exposure to heat inactivates microorganisms, such as spores, and minimises the loss of vitamins like B12. Source: Niro/DK (Picture), deVries, Kessler

Internet

- www.sciencedirect.com/science/journal/09242244
- www.food-info.net/uk/index.htm

6 COMMUNICATION AND KNOWLEDGE

Information and knowledge are proving to be an essential resource within an information and knowledge-based society. Growth in knowledge-intensive services and the rapid development of global digital networks such as the Internet prove how important it is to organise and manage knowledge as a factor of production.

Knowledge is the key to social and economic development. The ways and means in which we access information, how we appropriate knowledge and how we pass it on, have been revolutionised by the rapid advance in *information and communication technologies* and the associated social change, which continues apace.

It is becoming increasingly difficult to handle knowledge through conventional forms of management and the use of ICT: supporting knowledge-intensive work processes through modern ICT becomes a question of negotiating the narrow path between control and self-organisation. The interaction between humans and computers must achieve a sensible combination of machine (syntactic) and human (semantic) information processing.

Fast, clearly structured and secure access to shared knowledge is of key importance in developing a knowledge-based society. Globally, great efforts are being made to continuously shorten innovation cycles and the time required to circulate new scientific findings. Knowledge technologies play an important role in this context, particularly the transformation of the World Wide Web into a semantically based and readily accessible knowledge network. Innovative knowledge technologies will make it possible for computers to interpret data content from the enormous quantity of electronic information available, and to present it to humans in a manner appropriate to their needs. These knowledge technologies link information logically, that is then stored and made available in different contexts. The generation and distribution of knowledge will be of major importance in the years ahead to the added value of products, and will provide an anchor in all aspects of how society perceives itself.

Modern ICT covers an extensive range of applications:

- Dealing with social challenges: Health, social integration, mobility, environment, governments
- For content, creativity and personal development: Media, learning, cultural resources
- In support of companies and industry: Companies, business processes, manufacturing
- To promote trust: Identity management, authentification, privacy, rights, protection

The development of technology within ICT applications ranges from initiatives to promote broadband and digitalise the media, to building networks for electronic business operations, to new *e-government* strategies for comprehensive on-line administration services, through to the introduction of electronic identity cards and electronic health insurance cards.

▸ **The topics.** The usability of computer technology is becoming a key quality feature within technology development. Successful terminal units must be:

- Useful for the tasks which have to be carried out
- Useable in terms of being intuitively understood
- Aesthetically pleasing and fun to use

A paradigm shift is underway in the interaction between technological systems and human beings – it is not the user that adapts to the machine, but the machine that needs to understand the human being. Human forms of communication such as language, gestures and emotions will be interpreted by technological systems. *Human-machine interaction* will become human-machine cooperation.

Customary uses of the media will increasingly change with the phasing out of analogue technology. Video, radio, television and music are all offering end-users new services, commonly categorised as *Digital Infotainment*. Information and entertainment are linked via a relevant human-computer interface. A man-machine interface system in motor vehicles, for example, makes it possible not only to listen to a radio programme, but also to receive written updated traffic information within seconds, as well as to display navigation system maps and use the telephone.

Virtual technologies can support industrial work processes by offering new solutions in the value-added chains of high-tech sectors, such as in the production of automobiles and aircraft. Complete products are now compiled digitally so that product features can be tested on these models. In *virtual reality* the user is completely integrated in the world of digital models and interacts with the virtual product.

The boom in computer games, where people play on their computers individually, has been followed by further development in virtual worlds. A *virtual world* is an interactive, simulated environment, to which many different users have access at the same time via an online Internet access. A virtual world comprises software, physical laws and the boundary conditions of the virtual world, and at least one avatar per visitor. Interactive games with players anywhere in the world, or commercial transactions such as shopping in a virtual shop, are equally possible.

In the future, with the wealth of information now available, we will need information that meets demands appropriate to the situation (*information and knowledge management*). Information becomes knowledge by indexation of the contents. Precise search machines can be developed with the aid of ontologies and linguistic analysis methods. Ontologies provide a vocabulary, which enables a content-related description of knowledge domains. Knowledge can therefore be automatically linked to new knowledge.

Unrestricted access to corporate information and keeping in touch with the right people in the company is becoming a high priority within *business communications* (Intelligent information assistants). The technological breadth of terminal units is correspondingly large: IP and fixed network telephone, video conference systems, cellphones, notebooks, PDAs, fax, etc. A unified communication concept involves combining telephone/video calls, web conferences, voice mail play-back and access to company-wide contact registries under a common interface.

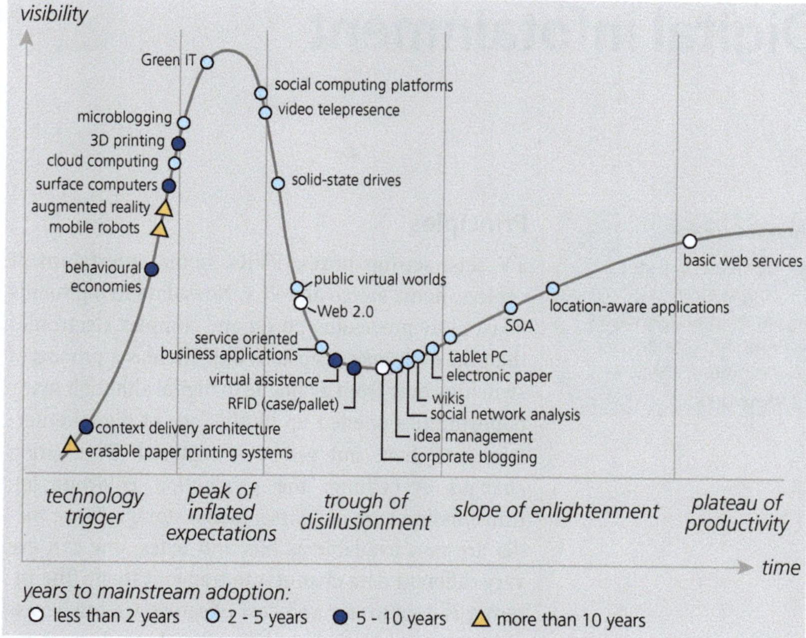

Hype cycle of IT technologies: Technology developments normally follow a distinct general pattern. The initial technological developments and successes of the proof of concept are followed by a phase of euphoria and enthusiasm. Many research teams take part in the further development and a series of application scenarios are publicly discussed. There are often setbacks, however, and a phase of disappointment then follows, as the essential preconditions for successful introduction to the market often do not exist. And then, slowly at first, after progressive development and adaptation, a new technology gains acceptance. It may then experience a rapid upturn after entry onto the market. In Gartner's Hype Cycle, emergent IT technologies are analysed in terms of their position on the technology development curve. The time axis does not have a scale as each new technology needs different periods of time to go through the whole development curve. "Virtual Assistants", for example, are already approaching the disillusion phase, but will need another 5-10 years before they are ready for the market. Developments such as "Green IT" have, on the other hand, only just reached the "Peak of Inflated Expectations", but will probably need less than 5 years to become established in the market. Source: Gartner Group 2008

Nowadays, the term "*e-services*" covers all internet-based services. Whereas e-commerce, offering goods for sale over the Internet, is now relatively well-established, other applications, such as e-government or e-learning, still have more potential for development.

Many of these stated trends are integrated in *Ambient Intelligence*. This digital environment unobtrusively and efficiently supports people in carrying out their tasks by giving everyday objects an "intelligence" and linking them to one another. Challenges here include the interaction in such systems, the guarantee of security and privacy, the management of dynamic, widely-distributed systems, and teaching such systems to act intelligently. In future, ambient intelligence will open up many new applications, in the household environment, in industrial automation, logistics and traffic technology, and in the health sector, amongst many others. ■

Digital infotainment

Related topics
- Human-computer cooperation
- Internet technologies
- Ambient intelligence
- Communication networks

Principles

TV sets, set-top boxes, PVRs, home entertainment centres, home stereo and PCs: Nowadays living rooms boast more processing power and complex electronics than the early space capsules did. There is a process of switching over from analogue to digital although many opportunities opened up by the "age of digital entertainment" have not yet been explored. Digitisation changes everything: the production environment, transmission and, of course, media storage. Since media are now available as bits and bytes, one can use very different data channels to transmit them. The Internet is easy to use and cost effective: we can access video, music and photos. Groundbreaking changes are being made in radio, films, TV and how literature and journals are viewed. These media will be made available to us in a completely new way.

▶ **Digital data**. Everything is going digital in the first place to cut costs. Moore's Law tells us that digital circuits are becoming more powerful, less expensive, more reliable etc., whilst getting smaller. Therefore digital processing is ultimately less expensive than its analogue counterparts. Digital data are easier to transmit and less susceptible to errors. But what exactly does "digital" mean? Using numbers and certain instances in time. An example from everyday life: a traditional instrument used to measure temperature does not show any numbers, but we can figure out the temperature by looking at the length of the mercury column with theoretically infinite possibilities. Of course we can't see that and the column would not be accurate enough anyway, but digital thermometers show us, for example, just half a centigrade as the basic setting. The continuum of values of the analogue measurement device gives way to a limited, discrete number of possibilities in the digital device. The other function of going digital is the sampling rate. We end up with a series of numbers. Every number can naturally be given in binary form, so we get a series of zeroes and ones.

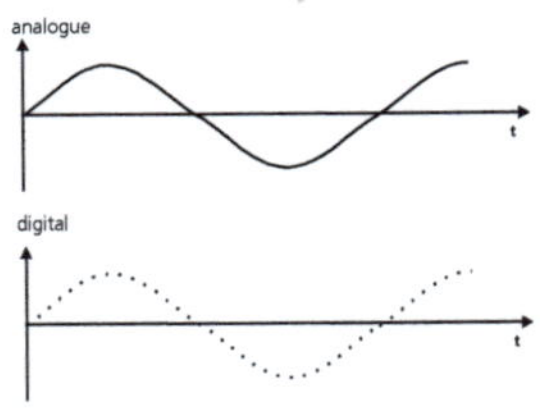

Analogue and digital signals: The analogue signal (above) theoretically has infinite possibilities, whilst the digital signal (below) has a finite number of possibilities and is measured at certain instances in time (sampled).

▶ **Audio compression**. We often talk about data rate in connection with media data. This is the number of bits or bytes we use to store or transmit the data. On *CD*, music data are stored 44,100 times per second (sampling rate) and quantized to 16 bit accuracy (65,536 possibilities for the signal amplitude). This gives a data rate of 44,100 times 16 bit times 2 (for the two channels of a stereo signal), which works out at about 1.411 Mbit/s. Digital music in MP3 format sometimes uses a data rate of just 128 kbit/s, which is about twelve times less. How does this work? The ears are very fine instruments, but we do not hear everything at all levels. We all know that very faint sounds are inaudible, but masking also plays a part: loud sounds mask fainter sounds. If we are at an airport talking to each other, the noise of an airplane starting or landing might make it impossible to talk. Scientists have produced formulae to describe this masking effect more accurately. This science is called psychoacoustics and we can best describe the effect of masking by looking at how the different components of a sound mask the sound at different frequencies. Modern audio compression algorithms (MP3, AAC, etc.) analyse the music to find frequency components. For example, in an MP3 we separate the music into 576 different frequency components (using a filter bank) and then analysed how important each frequency is for the overall sound. This depends on the actual music, so the sound has to be analysed about 40 times per second. More important parts are transmitted more accurately, whilst less important parts are transmitted less accurately or not at all. This is a very complicated process, which must not change the audio quality of the music at all, but the audio encoder does this very quickly. Audio compression algorithms have other features, such as Huffman coding of the frequency components (using shorter code words to transmit more probable digital numbers). The data rate can ultimately be reduced by a large factor without compromising the perceived quality of music at all.

▶ **Compression of video**. Standard resolution digital video signals, as used for TV, are transmitted at a data rate of about 250 Mbit/s. We can calculate this rate from 25 full pictures per second, each with about 400,000 pixels (720 by 576). Each pixel is composed of 8 bits of green, blue and red components respectively. This data rate is not readily available when distributing digital video over broadcast systems (terrestrial, cable or satellite) and even less so for Internet streaming. It is therefore even more important that video data are being reduced (compressed) than audio data. Some modern video coding methods can even achieve compression factors in excess of 100:1. They are the MPEG-

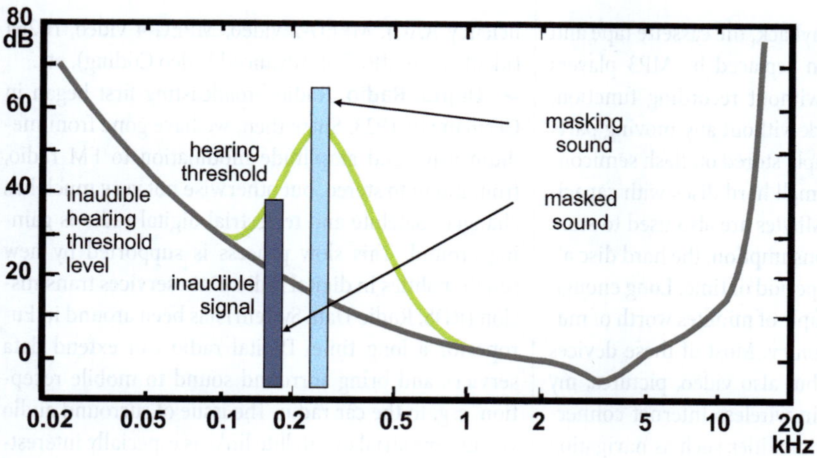

Sound masking: Audio coding algorithms like the MP3 use psychoacoustics. The grey line denotes the level below which humans cannot hear, from the lowest to the highest audible frequencies. No sound below this threshold can be heard (by human beings) at all. A louder sound (masker) modifies the hearing threshold (green line). Signals below the green line are inaudible (masked). Source: Fraunhofer

2 Video or the even more efficient and newer method of MPEG-4 Advanced Video Coding (AVC, also known as ITU-T H.264). The basic idea of these video coding systems is as follows: a video is essentially a sequence of pictures. First, we can code single pictures in a very similar way to how pictures in a digital camera are stored (JPEG picture coding). By doing this it is typically possible to reduce the amount of data by, for example, a factor of 10, but this is not enough. Therefore we exploit the fact that successive pictures in a film often have very similar content: the background may not change at all and other parts are very similar to preceding pictures (called video frames). So normally only the elements that have changed since the picture before are stored. If parts of the picture are similar to the preceding picture, but are in a different position (like a moving car), the change in position and the difference between this segment of the picture and the previous one (called a macroblock) are stored. This method is called "motion compensation". Therefore only the so-called "key frames" are transmitted from time to time and at other times information and motion vectors giving the differential image details are transmitted.

▶ **Surround sound.** The dream of high fidelity reproduction has traditionally included the ability to reproduce immersive sound. Stereo techniques have tried to give us the feeling of being in the middle of a concert or film. Modern surround sound technology requires at least 5 loudspeakers around the listener's area. The front loudspeaker directs the sound of the dialogue towards the front, even if the listener is not in exactly the ideal listening position. ("sweet spot"). The rear loudspeakers reproduce ambience or special sound effects. Quite often, there is an additional loudspeaker for the lower frequencies (e.g. below 120 Hz). This is used to reproduce loud sounds at low frequencies (LFE, Low Frequencies Effects channel). Modern main loudspeakers (front and back) are also often smaller and have a very limited ability to recreate sound at low frequencies. One "subwoofer" loudspeaker carries the low frequencies from all the channels including LFE. This technique (called "bass management") works because our ears are less able to tell which direction sound is coming from at low frequencies. Much higher data rates, up to 3.7 Mbit/s uncompressed, are needed to fully carry or store a 5.1 (5 channels + LFE) sound signal.

Applications

▶ **Homeserver and multimedia player devices.** In the age of digital media, new types of device come and others go. The classic home stereo system has given way to A/V-receivers, which control both audio and video from devices like the combined CD/DVD player or can receive analogue (FM) radio, digital satellite radio or Internet radio. Portable player devices (sometimes with record function) make all our media, i.e. music, pictures and films, portable. Modern TV sets may also have set-top box functionality and hard disc recording of the digital TV signals. Other set-top boxes are connected to the Internet. Some can deliver IPTV services to the TV set and some can stream programmes received from the satellite dish to every computer in the connected home. Modern PCs all have sound output and often they are even connected to active loudspeakers in a surround setup. Tube displays are long gone. In *LCD* technology, currently at the forefront of display technology, the lines are starting to blur between computer monitors (which can display signals from a set-top box) and TVs (which can display the output signal of a PC). High-end home cinema systems use projector systems.

In the field of audio playback, the cassette tape and MiniDisc (MD) have been replaced by MP3 players with, or, in most cases, without recording function. Many of these devices work without any moving parts and the music data are simply stored on flash semiconductor memory devices. Small hard discs with capacities of up to hundreds of Mbytes are also used to store music. To reduce power consumption, the hard disc always runs for only a short period of time. Long enough to just be able to store a couple of minutes worth of music in *semiconductor memory*. Most of these devices do not only store music, but also video, pictures, my calendar, etc. Some contain wireless Internet connectivity and additional functionalities such as navigation (*GPS*) or WWW access. Nowadays, mobile phones, called smartphones, have the very same features. These "do it all" devices can meet all our communication and entertainment requirements, wherever we go.

There are a number of new types of device in the home. Internet radio devices do the same as the Internet radio software on the computer, but they look more like a radio, disguising their ability to operate like a little computer with Internet connection (Ethernet or *WLAN*) and audio output. Other playback devices play video files back. They can be connected to the TV set, computer monitor or *LCD projector*. The media can remain on an internal hard disc or be accessed via Internet connection. Home servers store data which can be accessed by other devices to play back music or video. Recent standards, e.g. the certification guidelines by the DLNA (Digital Living Network Alliance, using the UPnP, Universal Plug and Play and WiFi standards) enable player devices (including some A/V-receivers and TV sets) to access music, video and pictures on many connected devices. These include PCs, some NAS (Network Attached Storage) devices or even set-top boxes. As different applications may use different data formats, most devices support a multitude of audio and video formats, including MP3, AAC, HeAAC (High efficiency AAC), MPEG-2 video, MPEG-4 video, H.264 (identical to MPEG-4 Advanced Video Coding), etc.

Living room: In future, TV sets will just display video media. The film on the screen might be stored on a portable multimedia device in the viewer's pocket or on the home server in a different room. The user will use a remote control, as we know it, or voice commands to access the media.

▶ **Digital Radio.** Radio broadcasting first began in Germany in 1923. Since then, we have gone from medium wave and amplitude modulation to FM radio, from mono to stereo, but otherwise not very much has changed. Satellite and terrestrial digital radio is gaining ground. This slow process is supported by new functionalities in digital radio. Data services transmission (RDS, Radio Data System) has been around in Europe for a long time. Digital radio can extend data services and bring surround sound to mobile reception, e.g. to the car radio. The issue of surround audio versus terrestrial or satellite links is especially interesting: standard surround sound formats need higher data rates. This is at odds with the digital radio requirements for the lowest possible bitrates. A lower bitrate means a higher number of entertainment or information channels, which is always the point of the exercise. MPEG has recently set the standard for MPEG surround. In this technology, the additional channels are not transmitted as separate data, but just as parameters describing the surround sound. These parameters account for only a small amount of the data rate and can be hidden in the "auxiliary data" fields of standard compressed audio formats. Older decoders still decode the standard two channel stereo signal of such a transmission, whilst new MPEG surround enabled devices detect the parameters and reconstruct the sound stage of the original signal. DAB surround uses this technology to broadcast surround over the European DAB digital radio system.

A number of digital broadcasting standards have been introduced in the last few years. They include the DVB (Digital Video Broadcasting) family of services in Europe and other parts of the world, the systems standardised by the ATSC in the US or ARIB in Japan and many more. Audio benefits from satellite services provided by XM radio, Sirius and WorldSpace, DAB in Europe and, recently, DRM (Digital Radio Mondial) for long-range medium and short wave services. DRM illustrates the advantages of digital radio very clearly: advanced digital modulation techniques (like OFDM, Orthogonal Frequency Division Multiplex) and forward error correction algorithms use the same bandwidth of radio waves to achieve much better sound fidelity and higher reliability during difficult wave propagation conditions.

Digital broadcasting services allow more than just audio and/or video to be broadcast: the DVB standard contains data fields for additional information about the current programme. In addition, the "data carousel" is used to send whole computer data files to either everybody (for example, containing update software for set-top boxes) or just to individual recipients (for example, containing keys for access to paid for, encrypted premium services). In DAB the UMIS service can turn the radio into a travel guide by combining position data (from a GPS receiver, for example) with information sent over DAB data channels. This service can be personalised; each user can set a personal profile to receive only certain kinds of data.

▶ **New TV standards.** The biggest change in cutting edge TV technology is the transition from standard to high definition broadcasting. The standard resolution of digital TV is 720 by 576 pixels. For HDTV broadcasting, the standardised resolutions are 1280 × 720 and 1920 × 1080 pixels. This produces much sharper pictures. HDTV on a large flat panel screen is more like cinema or a window on the world outside than a traditional TV image. To get the most from this technology, the whole media chain has to become high definition: cameras, production tools, video editing equipment, broadcasting standards and, lastly, TV monitors. Older movies can be scanned with HDTV resolution. Picture quality is not improved if lower resolution material is simply interpolated to HDTV. But it can be watched without block artefacts on the large screen display. As well as image resolution options, all HD material also has different frame rates. Standard definition TV benefited either from 50 Hz interlace (Europe and other countries) or 60 Hz interlace (USA and other countries), depending on the country. Interlace denotes the ability to display half the display lines in one picture and the other half in the next picture. This technique made it possible to watch TV in normal living room light conditions. In a cinema, where there is less ambient light, the frame rate is 24 Hz. For HDTV, the options are now usually 1280 × 720 pixel progressive scan (no interlace, called 720p) or 1920 × 1080 pixel interlace (1280i), both at a number of possible frame rates including 50 Hz and 60 Hz. An increasing number of displays (and consumer camcorders) are able to display (or record in the case of camcorders) 1920 × 1080 progressive scan. This is sometimes called "full HD".

Whilst HDTV has been introduced in the United States using MPEG-2 Video coding, most current DVB broadcasts just use SDTV MPEG-2 Video. The DVB-S2 standard (second generation DVB for satellite distribution) can use MPEG-4 AVC (a.k.a. H.264) to minimise the bitrates required. This codec permits much lower bit rates with similar video quality.

▶ **Mobile TV and IPTV.** DVB-H (Digital Video Broadcasting Handheld) technology is geared towards meeting the requirements of mobile receivers and therefore enables TV to be broadcast on mobile phones. To this end, image resolution is reduced to, for example, 320 × 240 pixels. The required low data rates of about 300 kBit/s are achieved by reducing the image resolution and using a compression method like MPEG-4 Advanced Video Coding (AVC). Delivery must use a low amount of power if at all possible so as not to drain the battery too quickly. The DVB-H data of a specific programme are therefore not transmitted contin-

Mixing desk: Multichannel systems are always used in audio production. Films are mixed from hundreds of separate audio tracks. Even pop music is mixed from many different microphones and computer synthesised music tracks. Nowadays the production chain involves the actual recording, electronic effects such as room simulation algorithms, storage of all the primary data on hard disc recording systems and, finally, the mix to the output format, 2 channel stereo or 5.1. To finish off, the signals are digitally processed using dynamic range processing (to make the signal sound louder) and/or audio coding to one of the compressed formats (MPEG Layer-2, MP3, Dolby Digital, AAC etc.) before they are broadcast.

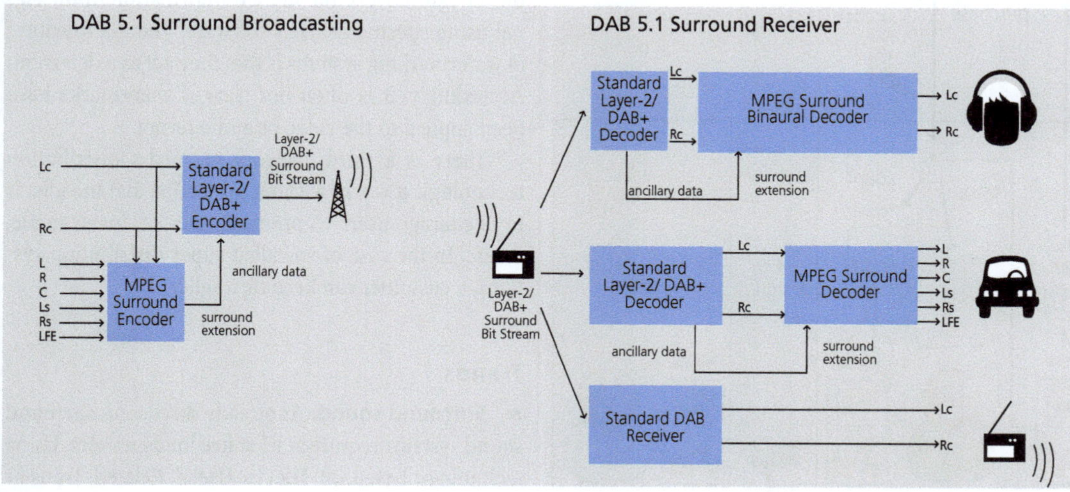

Working principle of DAB Surround: The transmitter is shown on the left. This is where the six channels for 5.1 surround sound are compressed and then transmitted as DAB Surround bit stream. MPEG Surround technology is used to compress the sound. Surround sound can accordingly be compressed at data rates which used to be standard for stereo sound. The working principle of DAB Surround receivers is shown on the right. DAB Surround users can experience the new surround sound on the move and at home: Portable DAB receivers play back surround sound over headphones. The car radio plays back high quality 5.1 surround sound over loudspeakers and all standard radios play back the same high quality DAB stereo signal.

ually, but only at set points in time. The receiver then receives all the data needed to watch a programme without any break in transmission until the next burst of data is received. The receiver can briefly switch off between data bursts to reduce power consumption.

Broadcasting TV images and videos over the Internet is also gaining in importance. By using compression methods like MPEG-4 Advanced Video Coding (AVC), video and audio data can be broadcast over the Internet in real time. The data rate for films is reduced to 0.5–1% of the original rate, but the quality of the image remains almost the same. A television connected to the Internet downloads a film from the video shop that a user has already chosen on a webpage.

▶ **Media distribution.** Many rights owners believe that legal distribution of media using Internet services must be governed by some form of copy protection. Over the last few years a number of so called DRM (Digital Rights Management) solutions have been developed and introduced onto primarily music distribution markets. DRM with copy protection wraps a secure envelope (for example, by using some way of scrambling data) around the actual media. Only users with the right to play the media can get unrestricted access to the video or audio media. Standard restrictions placed on copy protected media include the number of PCs which can play the media and the number of different playback devices (such as MP3 players) the media item can be copied to. In the case of subscription services, playback is limited to the time the user actually subscribes to the service.

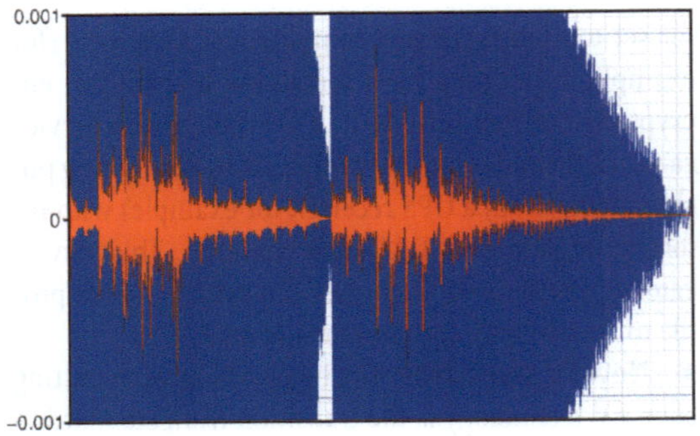

Watermarks: The information (red) is inserted into the music (blue) in such a way that it is inaudible. Horizontal axis: time, vertical axis: signal amplitude. Source: Fraunhofer IIS

No universal standard for copy protected distribution of music has been produced as yet. Different distribution services (e.g. Apple iTunes) only work with specific brands or types of hardware devices. This lack of interoperability has greatly compromised user acceptance and has led to a global move towards dispensing with protected media distribution or using less stringent methods.

Watermarking has gained acceptance as an unobtrusive method of copy protection: an invisible or inaudible signal containing data, for example, about the customer who has purchased the media is introduced. If unauthorised distribution is suspected, the rights owner can retrieve the hidden information in the signal using specific analysis software. The main feature of watermarking systems is that they act as a deterrent. Accordingly, it is often not clear if watermarks have been applied to the video or audio or not.

There is a third category of media distribution technology. It works without protection and the idea is to encourage users to promote sales of, for example, music. In the case of so called super distribution systems, a customer can become a seller.

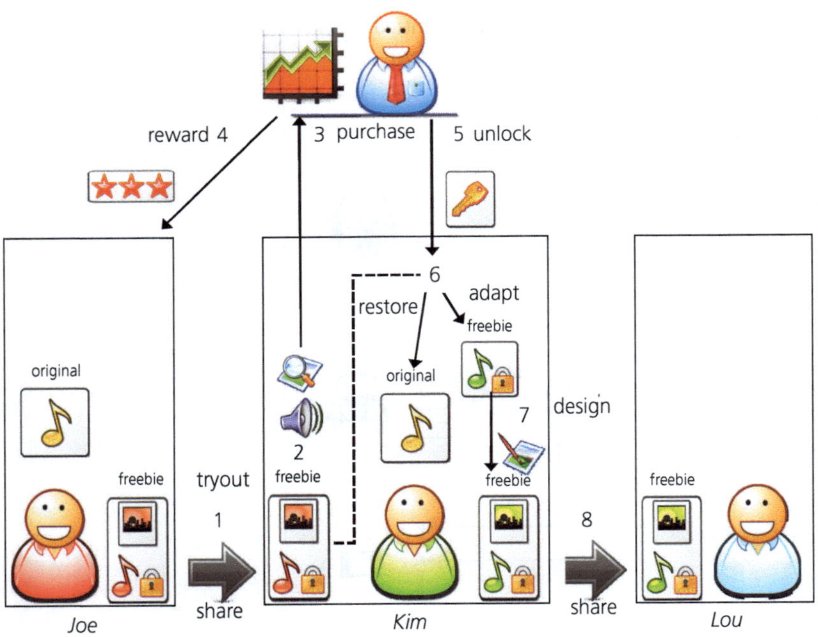

Freebies: "Freebies" are partly encrypted media files that lend themselves to super-distribution. If a buyer pays for and downloads a music file from an online portal he or she automatically receives a Freebie in addition to his or her original music file. The customer can use this Freebie to recommend the music he or she has bought to his or her friends – and is rewarded for every successful promotion with a credit, rebate or commission. Soure: Fraunhofer IDMT, Icons by IconEden.com, David Vignoni

Trends

▶ **Surround sound.** As already discussed, surround sound systems require at least five loudspeakers. Using technology based on HRTFs (Head Related Transfer Function) and room transfer functions, it is possible to receive surround sound on standard headphones. The systems modify the sound coming into our ears from a distance in the same way that sound waves are modified in a room until our ears pick them up.

Wave Field Synthesis (WFS) is another method

Wave field synthesis. Left: Whilst one loudspeaker is the source of just one wave front, a number of small loudspeakers recreate a complete wave field in a room by superpositioning all the sounds. A separate feed is calculated for each loudspeaker. Right: A cinema with more than one hundred loudspeakers, organised into loudspeaker panels mounted on the wall. Source: Fraunhofer IDMT

that can be used to reproduce sound even more faithfully. A number of small loudspeakers are arranged in a circle around the room. The input signal for this system is not a fixed signal for each loudspeaker, but a stored sound track for each sound source, e. g. musical instruments, a singer, somebody walking by, etc. The rendering computer knows the exact position of each loudspeaker. It uses the information about the virtual location of the sound objects to calculate a separate, optimised signal for each of the loudspeakers. Wave fronts are created by superpositioning all these sounds. The result is an immersive sound experience: listeners can really feel totally immersed in the sound and the sound (even if it or the listener is moving around) is much more realistic than the sound reproduced by any other sound reproduction system.

▸ **Home networking.** Electronic devices are increasingly using Internet technology to connect. Nowadays WLAN (WiFi), Ethernet or simply USB connectivity is a feature of devices such as TV sets, set-top boxes, Internet radio devices, media players (for mobile or stationary use), PVRs (Personal Video Recorders), NAS (Network Attached Storage), etc. Many devices can be controlled via their built in WWW server. These hardware devices are complemented by PC software that helps to organise the media data, prepare play lists, edit video and audio and retrieve media from the Internet. Now it is not so much a question of how to get access to media or how to store them, but of how to navigate through the seemingly endless options: home servers may contain years, worth of music or a month's worth of video. 20,000 photos are not thought to be very many. Ad hoc networks and better ways of searching and taking up recommendations will help to organise the "connected home". You will be able to see a film onscreen on your mobile phone and your latest music purchases will be downloaded to your car radio every time you drive your car into the garage.

MPEG Surround

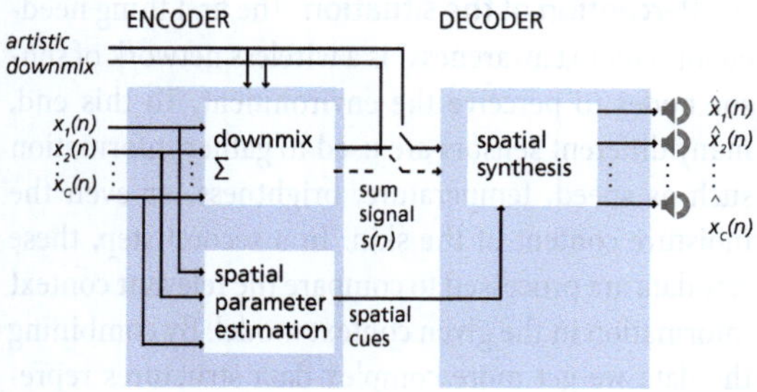

Working principle of MPEG Surround: The encoder receives the channels to be compressed and performs an automatic stereo or mono downmix. At the same time, surround sound parameters that specify the surround sound are set. These parameters are very compact. The parameters and the downmix are then transmitted together as a mono or stereo signal. Conventional receivers ignore the parameters and play the downmix. MPEG Surround receivers use the parameters to reconstruct the original multichannel signal from the downmix. A sound mixer can use his or her own downmix instead of the automatically generated one. After this process, MPEG Surround data is hardly any bigger than standard stereo data. MPEG Surround music is also played back on all standard devices in stereo, whilst new MPEG Surround devices reproduce the same high quality multichannel signal.

Prospects

- The digital media age is in its infancy. New additional services that make digital radio much more attractive are just one example.
- Progress made in compression technologies even enables TV to be broadcast on mobile devices and surround sound to be reproduced using current digital radio networks.
- Outmoded consumer electronics devices are giving way to multimedia servers and small, but very powerful media players to playback video or music. Media storage has been diversified: on the home server, only accessible from wide area networks or on the move with the PDA or smartphone. The age of physical storage of media is coming to a close; the main carrier for media distribution is networks, both local and wide area.
- All devices will be connected and have distributed functionalities. Every type of mobile device is incorporated into this system.

MATTHIAS ROSE
Fraunhofer Institute for Integrated Circuits IIS, Erlangen

PROF. DR. KARLHEINZ BRANDENBURG
Fraunhofer Institute for Digital Media Technology IDMT, Ilmenau

Internet

- www.chiariglione.org/mpeg
- www.dvb-h.org
- www.drm.org
- www.ensonido.de
- www.dmpf.org
- www.drm.org

Ambient intelligence

Related topics

- Human-computer cooperation
- Digital infotainment
- Assistive technologies
- Sensor systems
- Artificial intelligence
- Communication networks

Principles

Ambient Intelligence (AmI) is about sensitive, adaptive electronic environments that respond to the actions of persons and objects and cater for their needs. This approach includes the entire environment – including each single physical object – and associates it with human interaction. The option of extended and more intuitive interaction is expected to result in enhanced efficiency, increased creativity and greater personal well-being.

Ambient Assisted Living (AAL), which concentrates on supporting persons with special needs, is viewed as one of the most promising areas of AmI. We live in a society with a steadily increasing number of elderly and solitary people, who are quite distinct from today's youth and the community as a whole. This development has created a growing need for new means of orientation, support, and help for both young and old. We therefore need technical systems that will facilitate or take over certain everyday tasks – wherever required. The aim is to provide integral solutions that will encompass convenience functions and user support in the areas of independent living, home care, and residential care in nursing homes. The main goal is thus to prolong the time for which the elderly can continue to live in their home environment by increasing their autonomy and helping them to perform their daily activities. AAL seeks to address the needs of this ageing population by reducing innovation barriers with a view to lowering social costs in the future. A further focus is on the areas of rehabilitation and preventative care in order to reduce periods of illness.

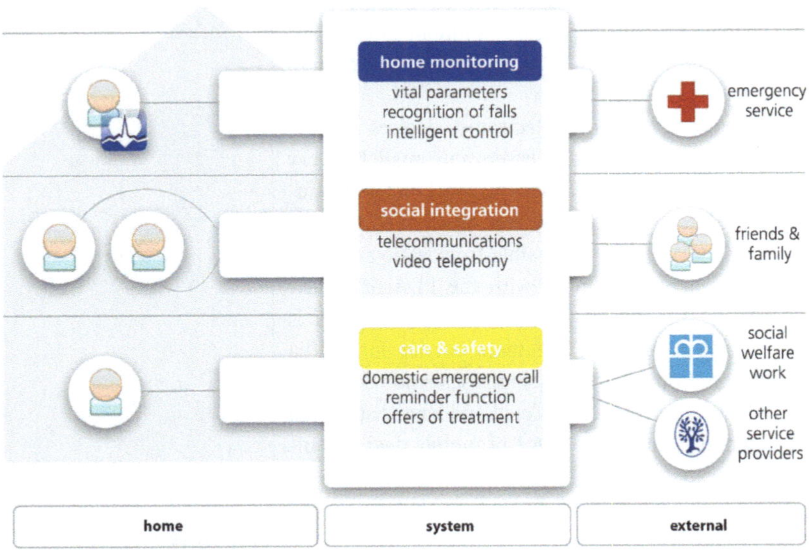

Areas of application for AAL technologies: AAL developments or "Assistance systems for a healthy and independent senior lifestyle" are to be expected primarily in the areas of home monitoring, social integration, and care & safety.

In a world supported by AmI and AAL, various devices embedded in the environment collectively use the distributed information and the intelligence inherent in this interconnected network. A range of information from sensors, such as lighting conditions, temperature or location, and also vital signs such as heart rate or blood pressure, is recorded by distributed devices in the environment. This cooperation between natural user interfaces and sensor interfaces covers all of the person's surroundings, resulting in a device environment that behaves intelligently; the term "Ambient Intelligence" has been coined to describe it. In this way, the environment is able to recognise the persons in it, to identify their individual needs, to learn from their behavior, and to act and react in their interest.

▶ **Perception of the situation.** The first thing needed for "context awareness" is a wireless network of sensor nodes to perceive the environment. To this end, many different sensors are used to gather information such as speed, temperature, brightness, or even the moisture content of the skin. In a second step, these raw data are processed to compare the relevant context information in the given context model. By combining the data we get more complex data structures representing the current state of the AmI environment. This step is also called sensor fusion. The computing involved in minor tasks, such as distinguishing between day and night or between loud and quiet, can be performed directly by the sensor node. If, however, more complex contextual information has to be classified, e.g. when several devices are needed in order to perform a specific action, the computer power is outsourced to high-performance nodes. To gain information on a higher semantic level, the received contextual information must be interpreted in a third step. Conclusions can then be drawn by combining various facts such as particular occurrences. This requires context models of the environment and specific personal behaviour patterns, e. g. what a certain person habitually does in a certain place at a certain time. The desired proactive reactions of the environment can be derived from this information. In the field of AAL, one of the most critical situations is when a person has fallen. A fall can be detected with great reliability by combining several factors, e. g. by integrating an acceleration sensor in a walking stick to show that the stick has fallen to the ground, or by including a signal from a camera-based analysis of the posture of a human body, indicating that the person is in a horizontal position.

It is becoming increasingly important to determine the position of a user, as the surrounding devices cannot be properly integrated unless it is possible to identify the position of a user and even the direction in which he or she is facing. Outdoors the position of the user can be identified by means of a satellite-based GPS device, for example. This method having proved its value in vehicles, more and more mobile end devices are being equipped with the respective receivers to facilitate navigation for pedestrians as well. Unfortunately, this technology can only identify the user's position or the way he or she is facing. Additional sensors such as an electronic compass or acceleration sensors are therefore used in order to directly identify or indirectly deduce the person's viewing direction. If a user enters a building, however, the direct visual connection to the GPS satellite is cut off and the positioning fails. In this case we have to use other positioning methods which usually provide a higher degree of accuracy. It is possible to mark a room electronically with active or passive beacons (position markers) and to identify it clearly by suitable detectors. Depending on the local information now available on the user's computer, the corresponding plans of the building must also be transferred so that the user's current position can be indicated.

▶ **Ubiquitous access.** To ensure ubiquitous access, digital media such as audio, video, images, and text information must be available at any time, in any location, and on any device. The access required comprises the retrieval, transfer, and reproduction of media which must be partly outsourced to more powerful (stationary) hardware due to the insufficient computing power of the existing end devices. Furthermore, digital media require either local storage media or storage media embedded in the broadband communication network. Local storage media can take the form of a *semiconductor memory* or an optical drive. Stationary servers with a high storage capacity are used as network memory devices. The network terminals are accessed by means of primarily wireless streaming of information from a storage medium to the mobile or stationary devices.

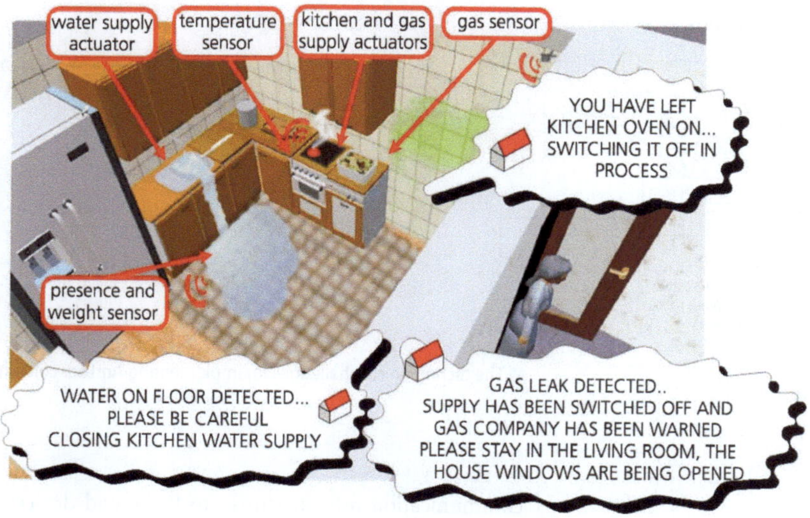

AAL technologies to increase independence and support everyday living.

Even today, server-based solutions synchronise data for the user between the mobile phone, PDA, laptop, and server. However, these solutions require a working Internet connection with sufficient speed. This means that the data cannot always be accessed and that data alignment may be slow or impossible. The problem can be solved in an approach where the data "move" with the user and are updated automatically and transferred if necessary. A physician making a home visit can thus receive the information needed to form a picture of the person's clinical status, and access it during the consultation. If, however, he or she suddenly has to leave on an emergency, all data necessary for this case are compiled and transferred. The home visit can then be taken over by another doctor. This does not entail any extra work for the first physician. He does not need to copy the data laboriously, as the system itself is able to recognise the change and automatically has the information transferred to the new location. If the data required are too complex to be processed on a locally available mobile end device, stationary computers are used and the end device is only used for presentation of the results.

▶ **Natural interaction.** Natural forms of interaction will have a crucial impact on the everyday life of the user. Concepts for the interactive control of objects in AmI environments will no longer be centrally implemented as on today's PCs. Instead, they will be implemented by networks of computing nodes that interpret the user's instructions and distribute them via existing

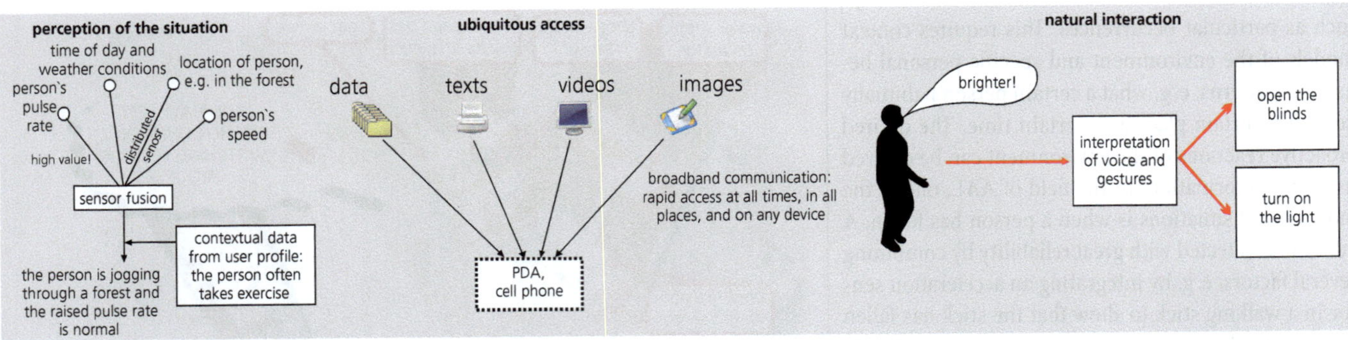

The technological challenges for implementing AmI are context awareness, ubiquitous access and natural interaction.

communication infrastructures to those end devices that can best implement the task.

Approaches involving multimodal interaction concepts such as speech and gesture recognition or computer vision require sophisticated algorithms that can only be implemented by stationary computers. If, in addition, an intelligent derivation from the existing information is required, the increased computing effort creates special peaks that can only be realised within a sufficiently short time with distributed computing nodes. Such interaction concepts might include interfaces for speech connection, 3D interactive video interfaces, or emotional interfaces for robots.

The options afforded by the new forms of interaction can be made clear if we take a look at the home environment. In addition to the latest concepts of central (tele-)control in which the functionality is programmed laboriously and the user also has to remember which features are activated by which buttons, interaction in the AmI environment is disconnected from the hardware. Instead of issuing commands to control the devices, the user defines objectives which are then interpreted and automatically implemented. If, for instance, the user acoustically defines the objective "brighter", the first thing that happens is that the system identifies which room of the house he or she is in. Then the system checks out which options to increase the brightness are available: Are there any shutters that can be opened? Which lamps are there? The state of the environment is also recorded for all actions, as it would not make sense to open the shutters at night. In addition, the preferences or further intentions of the user are considered. The system could thus choose indirect lighting for an evening in front of the television, and more direct lighting for a working situation or for reading a book.

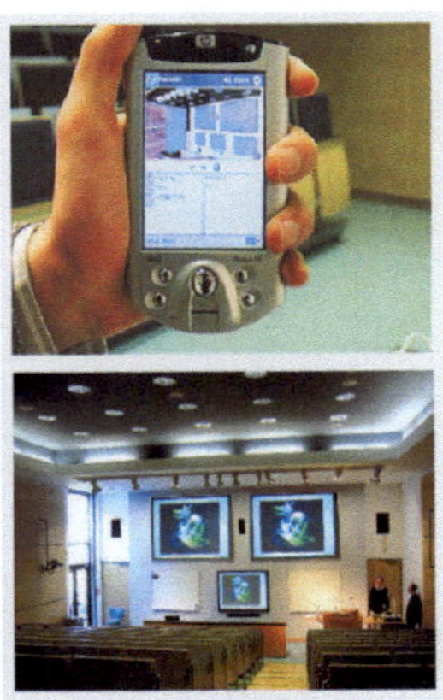

Digital assistance system. Top: On his or her PDA the user sees an automatically generated view of a conference room (bottom) with the corresponding media devices (e.g. the 5 monitors, 3 at the bottom and 2 at the top). In this way, the user can intuitively interact with the devices of the real environment. Source: Fraunhofer IGD

Applications

▶ **Intelligent operation assistance.** Today's multimedia environments, such as modern conference rooms, contain quite a number of intelligent devices including audio-visual equipment (e.g. big screen projections and surround sound), complex lighting infrastructures, HVAC equipment (heating, air handling and air conditioning systems), media storage devices, and personal devices such as notebooks, tablet PCs, PDAs, MP3 players, or smart phones. Users need to interact with all these different, heterogeneous devices. Intuitive handling, with the different systems aligning autonomously, would be a desirable solution.

A digital assistance system aims to provide a personal operation assistant that will help users to handle most of these different devices and systems in a straightforward manner. Someone on a business trip would thus have direct access to the technical infrastructure in his or her immediate environment, e.g. in the conference room, independent of their actual location.

The software components also support the user by allowing him or he to access distributed personal multimedia data such as texts, graphics, or video sequences. The system also "memorises" the particularities of the user and controls the devices, e.g. to set the volume or brightness the user normally prefers. In this way, he or she can intuitively handle the multimedia coordination in the conference room – be it the beamer, room illumination, microphone, or display – and compile or display their presentations or video clips without any problem.

The assistance system consists of four functional components which are realised by a dynamic and distributed composition of personal applications and the surrounding devices:

- The Personal Environment Controller recognises and dynamically controls the devices available in an environment. It also links recognised devices, their location, and their virtual, graphical representations. This means that the user gets a presentation of the actual room on his or her PDA, helping them to find their bearings in unfamiliar and complex environments.
- The Personal Media Management component provides the data relevant to the user's tasks on the right device at the right time. The system "knows" which files the user needs for a presentation and automatically to the locally available computers so that the user can use them for a presentation or can work on them.
- The Environment Monitoring and Personal Agenda components enable situation-driven assistance. The Personal Agenda knows all of the user's current and planned activities, and – by means of the Environment Monitoring function – can for instance turn up the heating in the meeting room in time to create a comfortable work environment before the meeting begins.

▶ **Intelligent home environment for the elderly.** The apartments of the future will be equipped with fall recognition sensors, motion detectors, intelligent lighting systems, etc. The intelligent environment recognises specific problems, analyses them, and if necessary alerts relatives or a service centre at an institution providing assistance for elderly people. Service centres offer comprehensive services for individual use 24 hours a day, ranging from sports and fitness programmes to domestic services and leisure activities. New technological developments enable the residents of the service centres and people in need of help in a domestic environment to be extensively supervised. Elderly people at home can thus be offered a high degree of security and autonomy, e.g. in the control of their body functions, and receive comprehensive medical care by telemedicine and telecare, e.g. to control their well-being. Telecommunication gives disabled or infirm persons the feeling of being accepted and integrated in society.

▶ **Monitoring bodily functions.** In order to measure vital data with as little discomfort as possible to the patient, sensor modules with skin-friendly dry electrodes can be integrated in tailor-made sensor shirts. In the senSAVE project, the electrode wires and the electronic circuit board are covered by a layer of fabric to prevent any unwanted contact with the patient's skin. All the electronic components except the rechargeable battery can remain in the shirt when it is laundered. *ECG* readings permit the automatic recognition of events. Until now, oxygen saturation has been measured on the finger or the ear lobe, which makes the device very uncomfortable to use however. Depending on the oxygen saturation of the blood, different amounts of light penetrate the respective part of the body. These data make it possible to monitor diseases of the lung. In future, however, this type of solution can be worn on the wrist as comfortably as a watch or a bracelet. The vital data are transmitted to a central receiving node, which deduces the pulse wave transit time from the ECG and pulse wave information and radios it to a server which, in turn, evaluates the information and – in an emergency, depending how urgent the situation is – establishes contact with a service provider or a physician. For the sake of acceptance, it is important for the data to be encrypted and signed so that only authorised persons can gain access to the collected information.

▶ **The intelligent living room.** Unlike conference rooms, which are typically planned and installed as an overall system, the technical infrastructure of private households is more heterogeneous and subject to

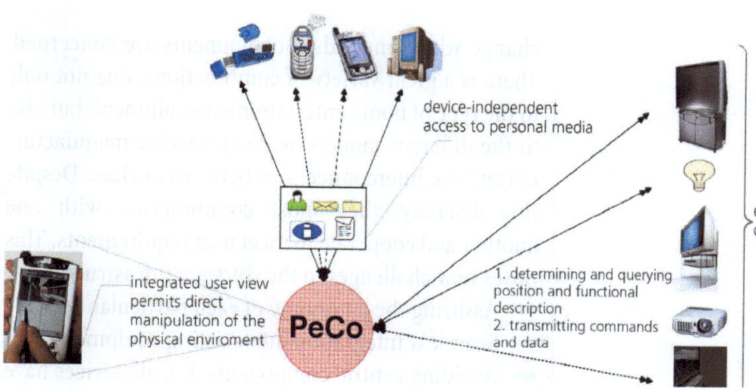

Personal Coordinator (PeCo): Users can view the present room situation in three dimensions by glancing at the display of their personal PDA. This 3D visualisation links the real environment with the virtual world, thus also providing interactive access to devices that cannot be seen with the naked eye. In this way, the user can see that the room is air-conditioned, for instance, and that the system can automatically adapt the temperature to his or her needs. Source: Fraunhofer IGD

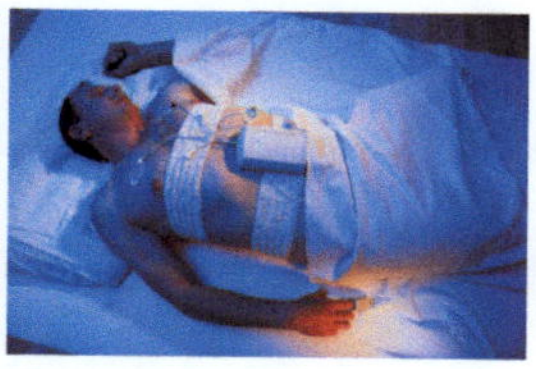

Monitoring bodily functions: A miniaturised system for sleep monitoring in the home. Source: Fraunhofer IIS

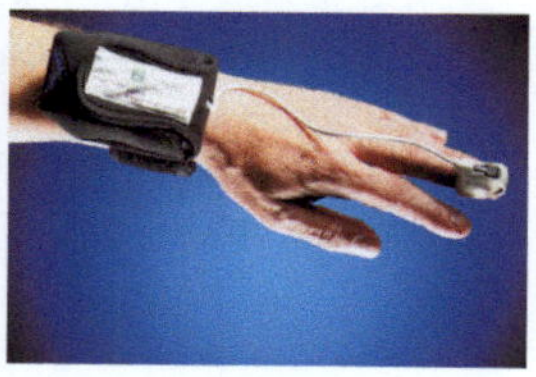

OxiSENS: Wireless pulsoximeter for measuring oxygen saturation on the finger. Source: Fraunhofer IIS

change where individual components are concerned. There is a great variety of combinations, due not only to the type of home entertainment equipment, but also to the different models made by specific manufacturers and the interconnections between devices. Despite this diversity, they must communicate with one another and cooperate to meet user requirements. This poses new challenges to the software infrastructure:

- Assuring the autonomy of each particular device
- Seamless integration into existing equipment
- Avoiding central components, i. e. all devices have the same status
- Dynamic integration of new devices and removability of devices, i. e. the connecting network must be capable of flexible adjustment
- Solution of conflicts in the case of rivalling devices, i. e. the devices must "come to an agreement"

Besides the technical networking of the devices, it is always important that the user is not overloaded with technical questions and decisions, and that the devices cooperate "reasonably" in the interest of the user.

Each device involved brings along components in a topology where they intercommunicate via shared channel structures. The news generated by the different components is published and distributed to the appropriate receiver components by the semantic channel strategies. A user's spoken command "TV night with crime thrillers", for instance, is first distributed via the event channel. The resulting objective, to "watch a movie", is distributed via the target channel. Finally news like "switch on the television" is distributed via the function channel before feedback is given to the user via the output channel. The physical device limits are thus dynamically overcome. The devices "see" the information input on other devices. The different components can apply for news or orders in their channels.

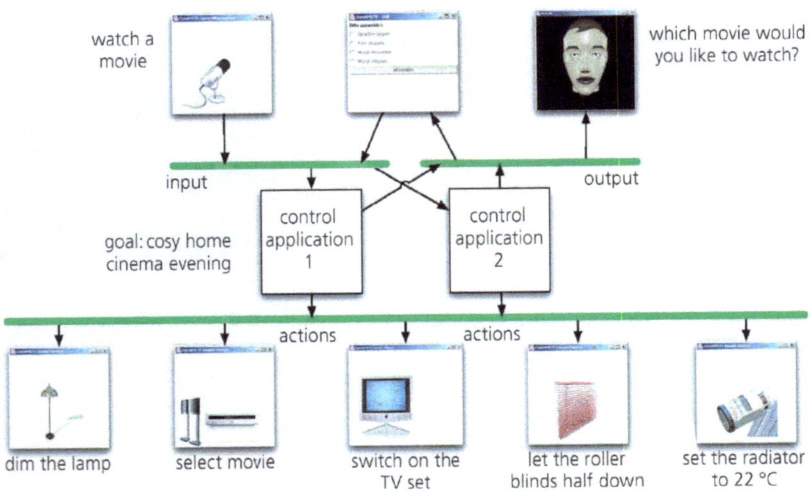

Components interact invisibly: The wish to "watch a movie" is implemented in the background, the actions are selected, and the corresponding devices (lamp, video and audio equipment, television, blinds, radiator) are controlled. If the devices "argue" as to how the task should be performed (e.g. the roller blinds and the lamp as to the optimum amount of lighting), these conflicts are resolved by the software. Source: Fraunhofer IGD

▶ **Components of an intelligent environment.** Input components must be able to capture a user's direct expressions (explicit interaction) such as speech or gestures, but also implicit interactions such as messages from movement sensors or pressure sensors. Combined with other sensor data such as a localisation system for persons and directional microphones, the system can detect who has expressed this wish. The user target of arranging a pleasant home cinema evening for Mr. X can thus be deduced from the implicit and explicit user expressions identified. However, as no specific movie request has yet been identified, the dialogue component can enquire about the type or title of the movie. The environment makes use of an output component for this purpose. Dialogue components can help to further specify a user goal via this output component, e. g. by speech or interactive display.

Finally, strategy components are able to generate exact function calls or sequences of function calls from an abstract semantic goal generated by the dialogue components. Thus, in the above scenario, a strategy component breaks down the recognised abstract goal "cosy home cinema evening" into the functional sequence "dim the lamp", "let the roller blinds half down", "set the radiator to 22 °C", "switch on the TV set", and "start the movie determined by the dialogue component". Actuators are ultimately responsible for the execution of this device functionality. These actuators are the real devices, such as the TV, lamp, roller blind, or DVD player.

We can imagine that not only "atomic" expressions of wishes – i. e. certain clearly defined words – are included, but also combinations and sequences of expressions.

These scenarios can only be realised if they are backed by infrastructures that are also able to avoid conflicts between devices or, alternatively, to solve them. Such conflicts may be caused by rivalling devices or components during the runtime of a device ensemble. Devices with a similar functionality may thus compete for orders or resources. If, for instance, a movie is to be recorded, and the device ensemble con-

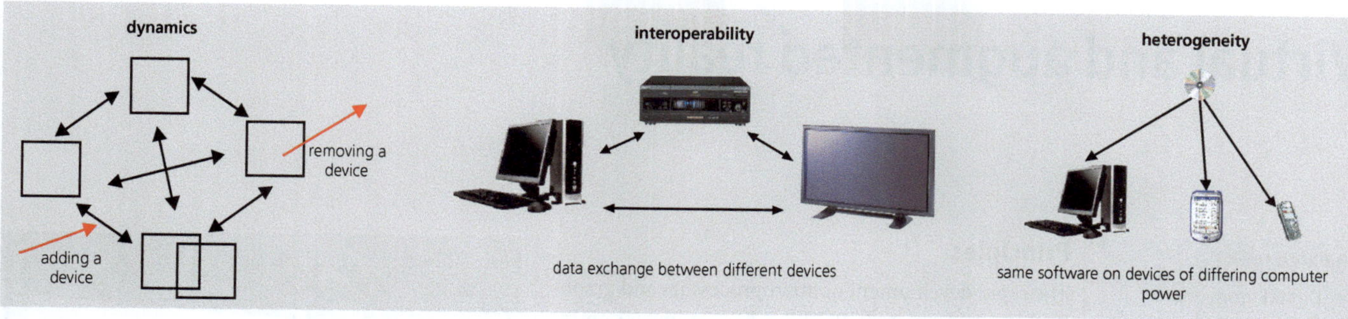

Conditions for a properly functioning AmI network. Left: It must be possible to add devices to a network and to remove them as required, without causing the network to break down (dynamics). Centre: Data exchange between completely different devices must be assured (interoperability). Right: Software has to be capable of running on devices of differing sizes (heterogeneity). Source: Fraunhofer IGD

tains different devices such as a video and hard disc recorder both seeking to perform that task, the conflict must be solved by appropriate mechanisms and strategies at the software structure level. In an ensemble, this is indispensable for transparency reasons. Such conflicts cannot be handled centrally (due to the demand for local organisation) or within the components themselves (due to the need for exchangeability).

Trends

AmI applications are developed in a very different way to the classic "single box" design. The concurrence of numerous components calls for special labs in which to develop and test new technologies and applications with a user-centred approach. With such a great number of objects involved in handling a problem, the interaction of the various components calls for new strategies. This is rendered more difficult by the fact that the devices have to "team up" to form ensembles. Three main points must be considered:

- Interoperability: This is the capability of independent heterogeneous units to cooperate and exchange data. The devices of more than one manufacturer have to cooperate – which eventually means that general standards must be defined, describing not only the interfaces but also the semantics of the data exchanged by the different devices.
- Heterogeneity: This term designates the ability to run software on different devices whose computing power differs substantially. In a device ensemble, the software infrastructure used must be capable of scaling the necessary computing operations to different units and of employing data exchange concepts that are compatible with different communication protocols.
- Dynamics: This is the capability of the network to adapt itself to changes. An environment must be able to respond flexibly to changes in the infrastructure, if for instance devices are added or removed.

Prospects

- Ambient Intelligence introduces a new paradigm for future electronic environments. The developments focus not only on the purely technical possibilities of device ensembles, but on the performance of the applied technology with regard to the assistance and added value for the user.
- Technical infrastructures consider the presence of persons, provide personalised interaction, and adapt to the person's behaviour. This paradigm can also be transferred to ambient assisted living, which is an area of increased priority due to the changing age structure.
- Devices cooperate and operations are performed according to users' needs, taking their wishes and preferences into account. It can already be predicted that technological advantages such as progressive miniaturisation, new lighting and display technologies, and advanced interaction concepts will allow "invisible" electronic devices to be integrated into the utility items in the user's everyday environment. In the future, the required technologies will autonomously form networks and communicate with each other. However, they will remain in the background, invisible to the user.

PROF. DR. EMILE AARTS
Philips Research, Eindhoven

DR. REINER WICHERT
Fraunhofer Institute for Computer Graphics Research IGD, Darmstadt

Internet

- www.research.philips.com/technologies
- www.igd.fraunhofer.de/igd-a1
- www.fraunhofer.de/EN/institutes/alliances/ambient_assisted_living.jsp
- www.aal-europe.eu
- www.eit.uni-kl.de

Virtual and augmented reality

Related topics

- Digital production
- Ambient intelligence
- Human-computer cooperation
- Image evaluation and interpretation
- Testing of materials and structures
- Automobiles

Principles

The rapid development of microprocessors and graphics processing units (GPUs) has had an impact on information and communication technologies (ICT) over recent years. 'Shaders' offer real-time visualisation of complex, computer-generated 3D models with photorealistic quality. Shader technology includes hardware and software modules which colour virtual 3D objects and model reflective properties. These developments have laid the foundations for mixed reality systems which enable both immersion into and real-time interaction with the environment. These environments are based on Milgram's mixed reality continuum where reality is a gradated spectrum ranging from real to virtual spaces. In this context, the term 'mixed reality' describes the fusion of the virtual and real worlds as follows:

- In virtual reality (VR), the user interacts with 3D digital objects in real time. Using stereoscopic projection systems and multimodal interaction devices, the user is completely immersed in a virtual world.
- The term augmented reality (AR) describes the real-time overlay of reality with digital information. A typical augmented reality system includes a mobile computing unit, a video camera – which captures the user's environment – and a head-mounted display for visualising the digital information superimposed onto reality.

Photorealistic rendering: Photorealistic presentations are vital to the product development process in the automotive industry. A key factor in design review is the capacity to reproduce the correct dimensions with respect to the actual environment. This allows designers, for example, to see their own hand beside the virtual car. Modifying the design of the life-sized models has to be performed in real time and with high levels of quality: For example, the presentation needs to be updated instantly when the viewing angle or colour is changed. Source: Fraunhofer IGD

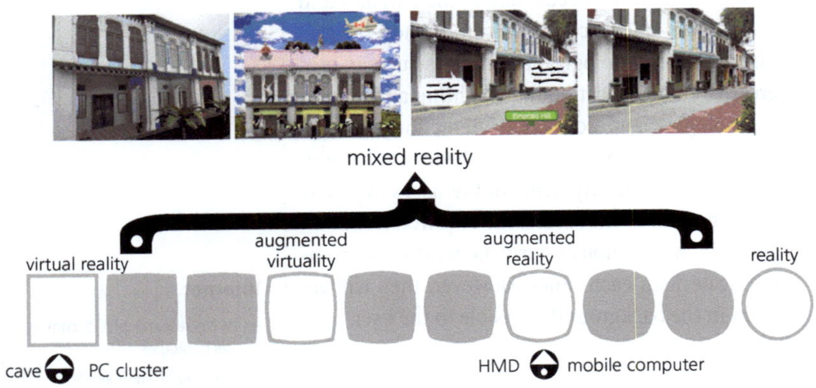

Mixed reality continuum: Mixed reality systems bring together VR and AR functionalities such as telepresence in VR applications or real-time simulations in AR scenarios. Hardware platforms range from mobile computing units in augmented reality to PC clusters in virtual reality. Display systems range from head-mounted displays to highly sophisticated immersive projection technology such as CAVE. Source: Fraunhofer IGD

Standard virtual reality applications are computer-generated 3D worlds in which the user navigates in real time (e.g. walkthroughs). These applications are used to assist the planning of architecture and to verify design studies. The automotive industry uses virtual reality to generate high quality and photorealistic virtual prototypes which can also be combined with simulations (e.g. airflow simulations in a virtual wind tunnel).

Telepresence is a specific kind of VR application where the behaviour of objects in the virtual world is calculated and based on real sensor data. For example, sensors are attached to a robot in an environment which is dangerous in reality. Users can control the robot safely in a virtual world performing critical and risky tasks. These users get the impression that they are at the real location but of course face no danger.

► **CAVE.** The Cave Automatic Virtual Environment

(CAVE) is an output medium for virtual reality. CAVE is a cube-shaped room where digital worlds are presented on each of the walls in real time using rear projection. A stereo technique also produces stereoscopic output. Two images are generated on each projection screen for the left and right eye respectively. They are separated using optical filters (e.g. for colour, polarisation, etc.). These optical filters are integrated in a pair of glasses worn by the user. One image therefore passes through the eyewear for the left eye but is blocked for the right eye, and vice versa. The result is 3D perception. Each projector is driven by a PC operating as an image generator. That results in an 11-node PC cluster is needed for a 5-sided CAVE. This PC cluster which distributes and synchronises the rendering algorithms.

A frame rate of 30 Hz is needed to provide seamless image generation that is easy on the eye. CAVE offers users the impression that they are entering a hologram. Users are immersed in the virtual world and feel part of this world. They are not represented by a virtual avatar – the 1 : 1 scale allows users to interact with the virtual environment directly with their own bodies. Users interact with and manipulate virtual objects with their own hands – this introduces an added dimension and a new quality of interaction. Users can move freely in the projection room (CAVE) exactly as they would in real life. This technique can be instantly and intuitively applied for each user.

▶ **Multi-touch table.** Multi-touch technology is one of the most dynamic areas in human-computer-interaction (HCI) research, and has laid the foundations for many of the new techniques and technologies creating new ways of working with computers. One can look at multi-touch interaction in two ways: as interaction using more than one or two fingers simultaneously; and as a means of working collaboratively in a multi-user scenario. The multi-touch technique has great potential because it allows people to seamlessly interact with whatever they are seeing simply by touching it. The technique feels very natural, and may lead to even more sophisticated interaction technology. New ways of generating immersion are therefore possible: for example, through tabletop interaction. Most people are comfortable working at tables because it is so familiar.

The goal is to improve immersion in VR by recreating the conditions people expect and are familiar with when working with each other at their desks. The multi-touch table's dimensions are very similar to an ordinary desk. The table therefore has a monolithic and streamlined design providing a sufficiently large display and enough space to allow people to work together. The display's size would typically be about $150 \times 90\ cm^2$.

▫ Virtual reality: Simulation of assembly processes using a 5-sided CAVE. The window opening mechanism (green-coloured area) has been inserted into the virtual door. This mechanical component was "taken" from the shelves (to the right) and integrated in the door using a data glove. If necessary, the user can grasp and turn the door in order to review it from a different perspective. Source: Fraunhofer IGD

The image(s) are created by a standard projector starting at a resolution of 1400 × 1050 pixels. The projection unit is embedded inside the table. A wide-angle optic and a mirror system create a clear and high-resolution image on the table's surface, although space on the table is limited. Screen width determines the table's final dimensions, which means that space is still an issue.

The tracking process detects illuminated fingertips in the greyscale picture of a video camera, and generates events if a finger touches, moves or disappears from the surface.

The finger tracking procedure is based on blob detection and blob tracking. Blob detection processes the recognition of bright spots in the image, which results in a set of 2D images showing fingertip positions. Blob tracking assigns a unique ID to each blob and tracks this blob from frame to frame. This is how finger movements are detected.

An optical touch-sensing method tracks the user's fingers using computer vision techniques. The multi-touch table's acrylic sheet is illuminated by infrared light. As the refractive index of acrylic is relative to air, light does not escape but is subject to total internal reflection. Whenever an object comes close enough to the surface, the total reflection is hindered; light therefore dissipates and illuminates the object. This would also illuminate a fingertip touching the surface and a

Multi-touch table: An entire plastic manufacturing process can be presented and explored by visitors using their fingers. Components can be selected and process animation can be activated by touch. Plant visualisation renders invisible processes visible, and informs the user in a very intuitive, definitive and visual way (the user can examine pipelines and components while they are in operation). Source: IGD, Coperion, Design and Systems

camera trained on the table's surface can now capture the resulting light blobs from fingers.

▶ **Augmented reality systems**. Augmented reality systems consist of a mobile computing unit (a PDA or smartphone, for example) connected to a head-mounted display (HMD) to which a miniaturised camera is attached. Optical-see-through or video-see-through systems are used as head-mounted displays. An optical-see-through system has semi-transparent displays presenting computer-generated information directly over the real world. In a video-see-through system, the video camera records live pictures that are shown on an LCD display together with overlaid virtual 3D objects.

The camera image is used as the basis for tracking. This means the viewing position is correctly registered and oriented with respect to the environment.

Augmented reality: The AR Chinese Language Learning Game is an interactive hands-on platform for teaching and learning Chinese. Actual Chinese characters interface with virtual 3D pictorial representations of their meanings. When the card is placed within view of the web camera, the AR application will display a model depicting the meaning of the word. The corresponding audio pronunciation of the Chinese character is then played. Source: IGD / CAMTech

The miniaturised camera therefore captures the real environment from the user´s perspective. The AR system processes these live images in real time extracting features for the identification of landmarks (specific characteristics of the real environment). Landmarks in a house, for example, are door or window frames. The AR system has to distinguish between dynamic and static objects since the door can be opened or closed. Landmarks are identified using point and edge detectors. This technique reconstructs the position of the camera in 3D space, based on the landmarks detected in 2D images. Epipolar geometry methods are applied.

Landmarks are either put in context with previous video images (frame-to-frame tracking) or correlated with a 3D digital model of the real environment. Tracking therefore correctly matches the virtual and real worlds.

Applications

▶ **Product development process (CAD/CAE)**. VR is normally applied in the early phases of a product's life cycle when no actual products are available. In contrast, AR applications require the existence of actual physical products. AR is therefore applied during the later stages of a product's life cycle.

The automotive industry adopted this technology early on, and has set the pace for applications using virtual and augmented reality. The industry uses this technology in various fields such as design review, assembly/disassembly processes, ergonomics and visibility studies, photorealistic visualisation, exploration of simulation data, or maintenance support. VR can present and evaluate phenomena before physical models are actually built. VR has had a fundamental impact on product development processes, and is used to manage the increasing variety and complexity of products in a more effective way since various layouts can be explored and evaluated. Design review sessions are a key element in the cooperation process. These sessions bring together product engineers, designers and clients to evaluate, amend and refine the 3D model using iterative processes during the various stages of the product development cycle.

The SketchAR system is used for 3D modelling in virtual and augmented reality. The SketchAR computer-aided styling system moves the product development design process into the realm of virtual reality, where new and user-friendly tools must be developed for designers and engineers. This requires real-time and effective visualisation methods, as well as interaction techniques that allow the user to manage virtual constructions in an intuitive way. The designer is able to develop conceptual sketches or a fundamental de-

Augmented reality: This technician is equipped with a mobile head-mounted display. He is guided through the assembly process by additional information superimposed onto his view (e.g. the direction of screws). A video camera (attached to the goggles) continuously tracks head movements, matching virtual information with the real environment. Source: Fraunhofer IGD

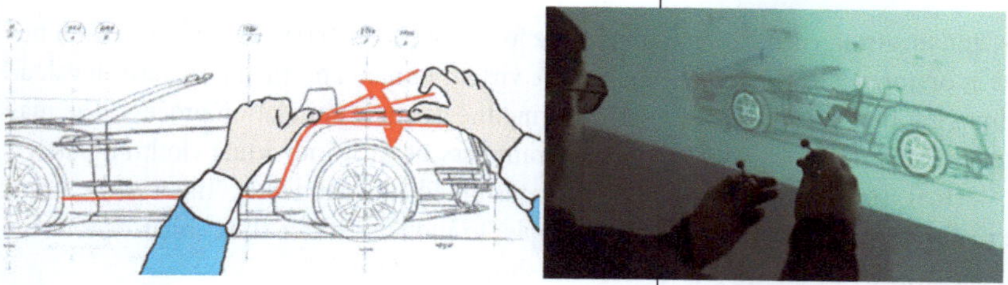

Virtual tape drawing: VR application (left), with sketch of work flow (right). The fixed position of the virtual tape is ensured by the non-dominant (left) hand while the dominant (right) hand controls the tangent of the polygonal line. The generated line, represented by a mathematically correct 3D curve, can be used as a CAD model for subsequent processes. This therefore reduces the manual and digital post-processing of actual, physical tape. Source: Fraunhofer IGD

sign in a semi-immersive environment using innovative interaction techniques.

In contrast to the majority of VR and AR solutions, which only work with tessellated models (graphic primitives such as triangles or quadrilateral meshes), SketchAR is based on a CAD kernel which directly operates at the models' topological and semantic level. In addition to traditional functions such as the generation of 3D base elements, the system focuses on the generation and manipulation of freeform surfaces. Mesh-based surfaces or connected surface sets can be easily modelled with the help of simple interaction metaphors.

Innovative pen-like interaction devices like the wireless Cyberstilo assists the designer, especially for freehand sketching in 3D space. Automobile design taping is also a popular technique for portraying a car's characteristic curves. Virtual taping simulates the procedure as a two-handed interaction technique in 3D space. Depending on the setup, it is possible to tape on virtual as well as real physical objects.

Augmented reality technologies integrate virtual and real models, leading to new opportunities for rapid product development: for example, augmented reality simulations. A prototype's various dynamic and static properties can be identified by overlaying specific virtual simulations onto the physical model: for example, computer-generated flow fields around a physical mock-up. The engineer is able to analyse and evaluate airflows around an existing prototype without needing to conduct expensive and time-consuming tests in a wind tunnel. AR technology can also help verify if a newly designed (virtual) component fits into an existing real environment. Interactive virtual simulations can therefore test actual physical characteristics.

► **Simulated reality.** Simulated reality combines advanced simulation and optimisation with visualisation and interaction – such as virtual reality. This helps engineers develop optimal solutions.

Crash simulations are just one of numerous and varied possibilities for research. This type of simulation is based on a description of the vehicle components and applies a continuum theory, using empirical material laws for deformation and fracture. A car's functional features are also reproduced using a wide variety of simulations. In crash-worthy simulations, it is possible to predict damage to the front of the car, the shape of the doors after the crash, and the impact on passengers and pedestrians. At the first stage of the vehicle development process, the maximum damage is defined. The simulation is then used to optimise these measures while keeping production costs, relating to weight, etc. low. The challenge is to find the optimum trade-off between all the various parameters.

Visualisation and interactive inspection of huge amounts of time-dependent crash test data still presents a challenge to computer graphics algorithms, in particular where the issue of the efficient storage and transmission of time-dependent data is concerned. GPU programming is used extensively to manage the huge data sets required for vehicle body simulations and allows the user to interactively explore the simulation results of different car body variants. Visual inspection of simulation results can be performed using a desktop VR workplace, or a large projection screen together with tracked input devices.

In the future, simulated reality will allow active safety components to be incorporated and optimised: for example, smart electro-ceramic coatings used as distance and impact sensors, or switches for safety features (airbags, fuel cut-offs, etc.). Simulated reality helps take things much further therefore than passive safety designs using components made of optimised material composites.

Simulated reality has an economic impact on every aspect of the automobile, reducing the costly process of determining characteristic data, which involves

Augmented reality: The tablet PC is used as a mobile "window" which captures the environment through a web camera. This brings the poster to life by superimposing the real environment as illustrated: the manga character symbolising the element "fire" is presented and animated as a 3D object. Source: IGD / CAMTech, TKG Comic Circle

Virtual flow simulation: The person developing the real object can visualise the car's virtual airflow trajectories. Using the Cyberstylo, this person is able to select any area of interest: for example, the front left-hand side of the car. This process is similar to specifying the location of smoke sources in a wind tunnel. A ten-minute introduction is all that is needed to introduce the user to use this system. Source: Fraunhofer IGD

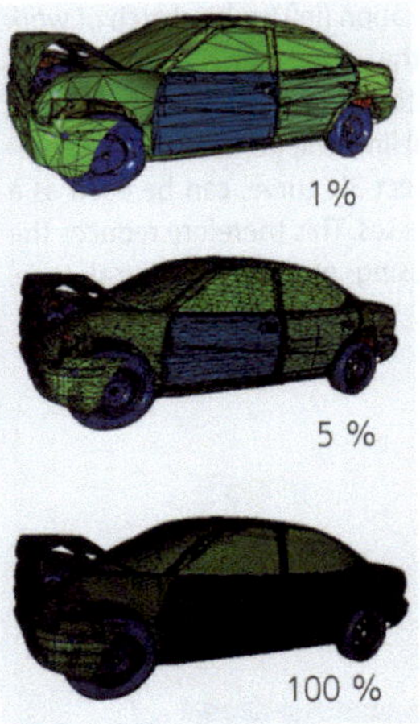

Simulated reality: Crash simulation sequence with different resolutions. Source: Fraunhofer IGD

many trial and error cycles, and ensuring correct system behaviour that can be monitored physically, as well as computational engineering that can be represented visually using every relevant scale.

▶ **Garment simulation.** Virtual Try-On uses new VR technologies to provide realistic, 3D real-time simulation and visualisation of garments. This allows real customers' virtual counterparts to try on the clothing, ensuring an individual shopping experience. Developments have been made in the field of contactless 3D body measurement (using a laser-scanner and light stripes for 3D surface reconstruction). Virtual selection, fitting and evaluation of customised clothing can now be performed.

Once a customer has chosen clothing from an electronic catalogue, either in a clothing store or from a home PC, he or she can virtually try on their selection and see how they look using their own 3D physical scan. In the future, clothing stores may offer body scans as a direct service: a camera scans the body, taking pictures and measuring the customer's body dimensions.

A virtual prototyping system for simulating fabric can also be used. This saves time and costs in the design and manufacturing process, because fewer garment prototypes need to be produced to compare and check the style and fitting.

Certain methods can be used for the real-time simulation of clothing. The garment's CAD data (the individual 2D cuts) can be positioned around a virtual human without user input. Based on these clothing patterns, suitable digital geometric representations can be calculated along with the necessary information about how the pieces should fit together, just like real life seams. The pieces are then placed around the person's body to produce two adjacent cloth patterns that need to be sewn together. With physically-based simulation, the pre-positioned garments are stitched together along the seam lines and the fitting of the garment is calculated. The triangulated representations of the garment patterns are used as the basis for a particle system, where each vertex of the triangle network identifies a mass point. Gravity, material properties, curvature forces in the material, as well as friction between the virtual human and the fabric, are all calculated during the garment simulation process. The material parameters of each individual clothing pattern can be determined independently. This can be used, for example, to model a stiffer shirt collar or to reinforce the shoulders of suits.

Based on the numerical solution for the equation of motions, the folds of the fabric and the fitting of the garment are determined in real time, depending on the material parameters.

During the simulation, collisions within the fabric itself and collisions between the fabric and the virtual human are taken into consideration and evaluated using distance field based methods. This technology is efficient and robust, since a distance field can rigorously split the space internally and externally. The clothes can then be visualised.

Methods exist for illustrating real-time simulated garments and virtual people, including shadows.

▶ **Architecture Design Review.** The Architecture Design Review application developed for Messe Frankfurt's new exhibition hall features scalable 2D architectural blueprints of several floors. Many architects can move and zoom through these plans simultaneously, and also navigate through a high-quality rendered 3D model of the building.

The key feature of this multi-touch application is the imitation and digital enhancement of an architect's own work environment. It recreates everything in a way that is immediately familiar to the architect: for example, the daily usage of paper plans. The application shows valid real-time visualisation of a whole building on the table. As a rule, blueprints are decoded using special symbols and a huge amount of information that normally only professionals can read and decipher. One can now see and control a high-resolution rendering of the blueprint that the architect is viewing. All plans are simply modelled as textured plane geometries. The plans can therefore be moved, rotated and scaled in the same way as the standard multi-touch applications just described. The user simply moves the plan around with a finger. Stretching two fingers apart scales the blueprints and rotates them by positioning an angle.

The truly innovative feature of this application is the navigation it offers through a 3D view of the building. A tile showing the 3D visualisation lies on the virtual table next to the blueprints. It is also modelled as a textured plane because it has to be transformed in the same way as the blueprints. But, whereas their textures are static, the texture displaying the architectural visualisation must be dynamically rendered frame by

Virtual Try-On. Left: A 3D body scan of the customer is generated using a commodity laser. The characteristic feature points which are decisive for the fitting of clothing are specified (green points). Centre: Virtual try-on of clothing on customer-specific 3D body scans using photorealistic illumination. Right: Comparison of simulation (left) with real clothing (right). Source: IGD

frame according to the movement of the virtual camera. Architects can now move, rotate, and scale the 3D visualisation just like any other plan. The view is controlled using simple gestures such as two fingers of the left hand for grabbing a plan just like one would hold a sheet of paper while writing on it. One finger of the right hand then points at the construction plan, and controls the camera's movement through the 3D model. A second finger of the right hand defines the direction to look at and controls the camera's orientation.

Trends

▶ **Mixed reality systems.** Consumers have more and more access to augmented reality systems allowing personalised and context-sensitive real-time visualisation on mobile computing devices such as smartphones. One example is a mobile AR system for tourists providing additional information about a historical site (e.g. illustrations). Virtual 3D reconstructions of the historical building, overlaying the real site of the historical area, are superimposed over the tourist's view. An AR tourist information system like this uses marker-less tracking methods to blend virtual objects seamlessly over the real environment.

The tourist uses a mobile computing unit, a semi-transparent display and a mini camera. The AR system analyses the video image and detects specific features of the historical site: for example, a temple in ancient Olympia. The tourist's position and viewing direction are also identified with relation to these detected landmarks. Finally, the position calculation allows the correct presentation of the computer-generated historical reconstruction on the semi-transparent display enhancing the tourist's view. The tourist can then take a virtual tour through ancient Olympia.

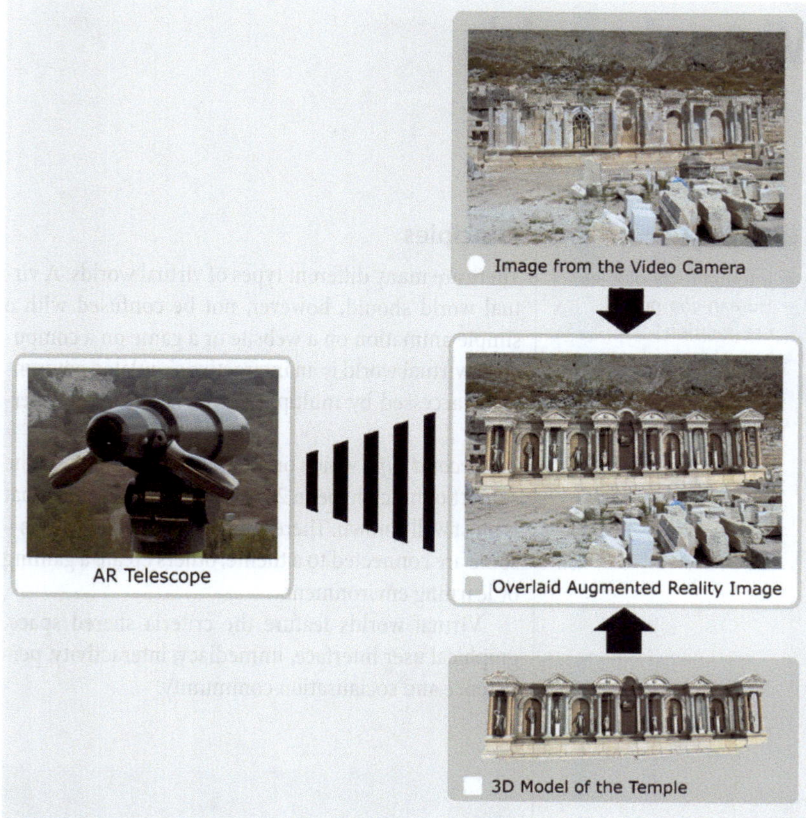

Augmented reality telescope: When a tourist looks at the above temple using a coin-operated telescope, additional digital information can be superimposed precisely over the view: for example, a 3D model of the temple exactly as it was in ancient times. Source: Fraunhofer IGD

Prospects

- Virtual and augmented reality technologies are opening up new possibilities in various areas of application such as medicine, cultural heritage, edutainment, or production, where CAD/CAE methods are seamlessly combined with mixed reality.
- As computing power continues to advance, more and more realistic simulations are possible. In particular, the parallel processing power of GPUs – which surpasses even Moore's Law – offers further cost-effective solutions for high-end graphics as well as computationally expensive non-graphic related problems (general programming GPU or GPGPU). The miniaturisation of computer power and components (e.g. in PDAs, mobile phones) will also make mobile augmented reality systems even more popular with users, and incredibly wide coverage and dissemination should occur in the very near future.
- New human computer interfaces will offer more intuitive interaction beyond the traditional keyboard and mouse. Subsequently, mixed reality will support work flows not only in the office but also in many other areas such as maintenance, assembly, and also for surgery. Further developments in mixed reality technology will also enable its use at work and even at home, as the technology itself recedes into the background.

PROF. DR. DIETER FELLNER
DR. WOLFGANG MÜLLER-WITTIG
DR. MATTHIAS UNBESCHEIDEN
Fraunhofer Institute for Computer Graphics Research IGD, Darmstadt

Internet

- www.web3D.org
- www.augmented.org/
- www.architecturemixedreality.com/Virtual_Reality/Virtual_Reality.htm

Virtual worlds

Related topics

- Internet technologies
- Human-computer cooperation
- Virtual and augmented reality

Principles

There are many different types of virtual worlds. A virtual world should, however, not be confused with a simple animation on a website or a game on a computer. A virtual world is an interactive simulated environment accessed by multiple users via an online interface.

Second Life, which originated in 2003 and was the subject of much hype in 2007, is one of the worlds that is most well-known. There are also many other worlds – some are connected to a theme, others create a gaming or learning environment.

Virtual worlds feature the criteria shared space, graphical user interface, immediacy, interactivity, persistence and socialisation community.

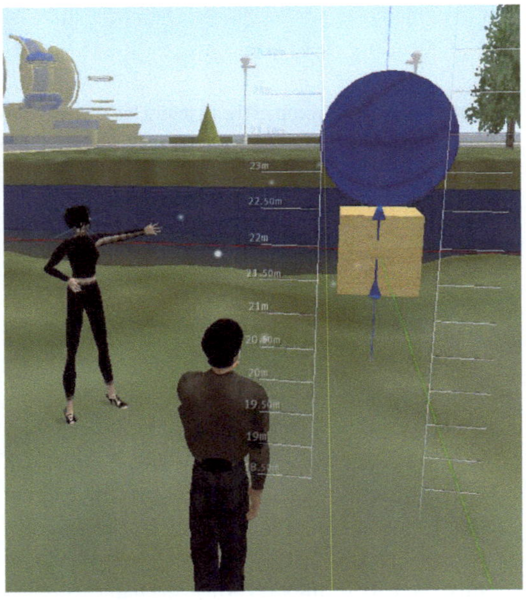

Changes: An important feature of virtual worlds is the possibility given to normal users to modify that world. Users can act together simultaneously. A system controls users' rights to ensure that objects cannot be modified, copied or moved without authorisation. In the picture the perspective is from the viewpoint of the lower avatar. He is working with the block of wood, while at the same time the left female avatar is working with the upper ball (shown by the small white points and her outstretched hand). The system shows the aids that can be used for working on the object (only on this particular user's screen). In this case it shows the position of the object in the room (21 m).

▶ **Shared Space.** As a virtual world exists online different users can visit it and experience it together, instead of just one computer user visiting it at a time. Different users are all in a virtual room at the same time, so that any changes to objects or their movements can be seen by other avatars. They can, for example, work or shop together. All the participants, however, are in the same room. This differs from an installed game or Flash animation, where the same room is generated for each user individually. Users in a virtual world can see one another and experience the same room together. This shared interest is important – it emphasises the social nature of virtual worlds.

▶ **Graphical User Interface.** Virtual worlds are represented graphically – in contrast to chat rooms (websites or programmes where different users write and read in real time, making it possible for them to have text-based discussions). This can be very simple 2D graphics or high-resolution '*CAVE*' rooms (special rooms where a beamer projects the image of a virtual room onto walls, ceilings and floors; special 3D glasses make the images three-dimensional and give the impression of standing in a virtual world). The sensation the users have of seeing themselves in a world in the form of avatars is the key factor - not the quality of the images.

▶ **Immediacy.** The actions of users or of their avatars can be seen by all the other avatars in their environment at the same time. This makes group activities (tours or discussions) possible.

▶ **Interactivity.** The world allows users to make changes, e.g. to build or create. The users can (but do not have to) generate the contents of the world. It is similar to constructing a model railway – the users work on objects and create a world. They are not just visitors.

▶ **Persistence.** The virtual world and its objects are persistent, i.e. they are not located on the computer of the user. The virtual world is accessible to everyone regardless of whether a user starts the program. The crucial point is not whether the virtual world server runs on the user's laptop or on a server, but that there is a server, which saves data independently of clients.

▶ **Socialisation/Community.** The avatars can see one another, move around amongst each other, forming and taking part in social communities.

The output of the server depends on the number of users and the complexity of the world. The complexity is calculated on the basis of:

- The number of objects (does a chair consist of 6 individual objects or 100, because it is represented in very precise detail)
- The complexity of the programmes integrated in the objects to make them interactive
- Their physical characteristics

Demands on the server can be divided into two areas:

- Tasks involving many transactions: All movements of avatars, all modifications to an object (by a programme or avatars) must be constantly saved on the server and distributed to all users logged in to this area. This very quickly produces several 100 to 1000 updates per second.
- Computationally intensive tasks: Collision calculations have to be made for almost all objects in a virtual world. Avatars moving through houses are not allowed to go through walls and tables, but they can instead climb a flight of stairs. It is possible to 'physically' control individual objects, by calculating them in a much more complex way. The mass of each object is calculated according to its material (wood, metal, glass) and its size. The forces (gravitation, wind, impact) affecting this mass are then calculated.

▶ **Hardware**. A virtual world consists of different servers, which undertake different tasks. Combining these tasks gives users the impression of a complete world. All of these servers must work together before users can move in a world and see anything:

- The user server manages all the information about the users who have registered in this world (name, password, home location of the avatars). The server sends an enquiry to the grid server about a position in the world the user would like to visit, e.g. which island he or she would like to stay on. Each island can run on a different server and only the grid server knows the technical details (Internet address) of the servers.
- The grid server checks the given co-ordinates and sends information to the user server about the simulation server to which the user would like to proceed.
- The simulation server animates the environment. This includes all the geometric objects in the world (chairs, tables, cars, houses: referred to as primitives, or prims), and the landscape. This includes a view of neighbouring islands, visible from the island represented by the simulation server. It also links up with other servers and receives visual information and updates when changes occur. Calculations are only made, however, on the current island. The 'physical engine' is an important part of the simulation server. It calculates the physical features of each object. It not only takes into account the fact that objects consist of a mass, and that a ball rolls down an inclined slope, but also that avatars cannot go through walls.

Complexity: Objects consist of individual primitives (spheres, cuboids, etc.). More complex objects, e.g. a chair, are generated from many of these objects. Left: The chair is kept simple, it just has a seat, a back and 4 legs. It does not require a lot of memory capacity and can be calculated quickly. Right: More complex chairs require more primitives and therefore more computing power, memory and bandwidth during data transfer. The aim in designing objects is to make them appear complex, even though their construction requires few primitive objects. Images are then applied to the primitive objects to make them appear complex. A bridge parapet, for example, does not have to be made from 500 individual objects, but from just 1 or 2 primitives. These are then overlaid by an image of a very complex bridge.

The asset server and the inventory server manage the primitive objects of the virtual world. Inventory is the term used for objects which avatars have in their own archive (for example, a hat that is not being worn at that moment). The asset server saves those objects that are available in more than one world (e.g. a three-dimensional company logo, installed on different islands). This means that not every simulation server has to save the corresponding data; it is managed at one central point.

Applications

▶ **Social networks**. Human beings are social creatures, who inevitably follow an impulse to form interest groups and swap ideas – and not only since the dawn of the Internet. Social networks such as MySpace (the largest global Internet platform for self-expression,

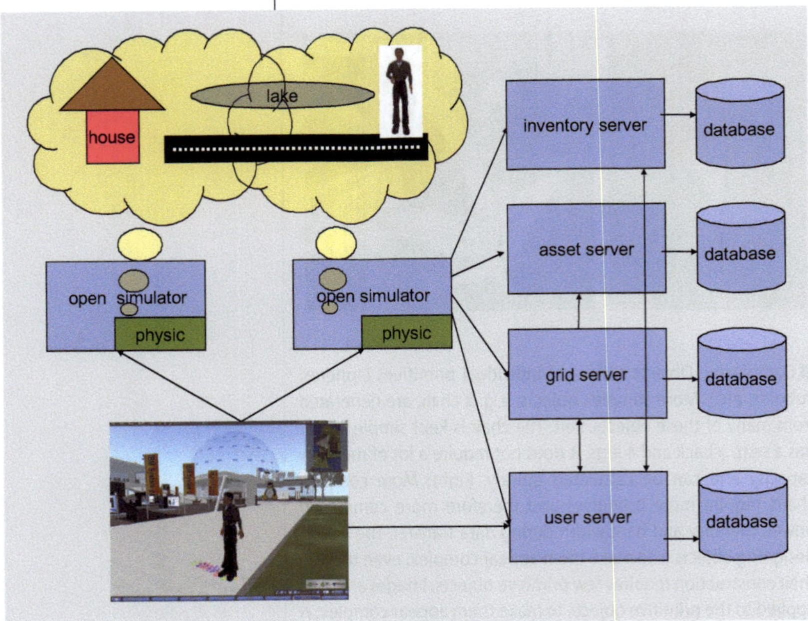

Interaction of the servers: Both the actual simulator and various administration servers manage the objects, avatars, etc. Bottom left: The view of the users. Their viewers connect to the open simulator, which calculates and simulates the part of the world where their avatars are standing at that exact moment. All scripts and physical calculations are produced here. However, these only give an abstract description of the world. The image is only then generated by the client, from descriptions of the world and its constituent textures. The simulated world is not calculated on one open simulator alone, but on many distributed simulators and computers. These computers can be located anywhere in the world. When avatars roam through the virtual world they will be passed from one simulator to the next. If, for example, avatars just move from a street into a house, their clients may change connection from a computer located in Europe to one located in Australia. A sophisticated protocol ensures that the users are not aware of this change. A viewer is therefore not connected to just one simulator, but to all the simulators within visibility of the avatars. Source: IBM

communication and forming social networks) are extremely popular. Users have compiled their own personal pages and shared them with others since the WWW began. Virtual worlds take this a step further. Users can generate their own 3D world and use a programme to design objects interactively. They see their "visitors" and have a "digital home".

▶ **E-business.** If furniture or clothing is going to be sold online, the buyer will want to see and try out the objects in 3D. These are products, which must individually suit the buyers or their home. An advantage of virtual worlds is that the clothing can be tried out on customised avatars.

Once the avatar has the same proportions as the user and the user's home has been modelled, the viewer can try out new furniture and see how it feels. What does the chair look like? Is the refrigerator in the best position in the kitchen? The communal character of virtual worlds also helps to simulate workflows, e.g. the motion sequences of two people when they are preparing a meal in a fully-equipped kitchen.

▶ **Avatars.** Avatars are an important element of a virtual world. They show the digital self of the users in the virtual world. Users see themselves in a virtual room. In reality, we have a sense of our body (without seeing our body). This makes it easy for us to assess, for example, how many people can fit comfortably into a room. We do not have this sense in the virtual world. We can only see a moving image. We are not able to assess what the scale of our environment is from the supposedly natural 'first person view' (through the eyes of avatars). It is only when we can see ourselves, including our hands and feet, that we can gain a sense of our body and judge the virtual room around us more naturally.

Another advantage of users having their own visible avatars is that they can control and hence trust them. Avatars only move and act if the users tell them to do so. They do not move a hand or leg unless users tell them to via the interface. The users increasingly trust their avatars and accept that they themselves are in the virtual room. This trust continues to grow. If users meet other avatars they know intuitively that there are real human beings behind these avatars. As the room is consistent in itself, that is the graphic quality of the avatars is comparable with their environments, there is no mismatch in the media experience, such as is the case in video conferences.

▶ **Generating avatars.** All avatars start off looking very similar - normal human bodies with T-shirts and trousers. Each part of the body can be changed – from the colour of the eyes to the length of the leg. The avatars consist of polygons (small triangular surfaces that create a 3D shape), which can show organic objects accurately.

Users can put different layers of clothing onto the avatars, and modify their bodies. The clothing itself is like a paper template. A painting programme is then used to paint an image or pattern (including a material or leather pattern) onto this. Parameters in the pro-

E-business: Shopping on the Internet will also mean that users shop together. This can be a spontaneous conversation while buying a book or installing a new kitchen together. Users can view simple computer-generated photos of the new kitchen on the Internet, and walk around the kitchen with their own avatars. They can try out the kitchen's functions and rearrange appliances and cupboards.

Avatars: These are the users' "ambassadors" in the virtual world. They are highly individual and can adopt different appearances. They can either be extremely similar to a human, or very symbolic, depending on the virtual world. Each individual area of the face or body can be modified and adapted to individual requirements, e.g. a large nose, blue eyes or a fat stomach. Left: Recently generated avatars are initially kept neutral. An editor, simply integrated into the viewer, can give avatars any appearance desired. No programming knowledge is required, but a slide control can adjust features such as the size of the nose. It is possible to make more complex modifications (such as the appearance of a dog), but these require an external editor. Right: In contrast to objects in the virtual world, avatars do not consist of primitives, but of so-called meshes. Tightly packed triangles form the surface of the avatars' bodies. The faces of the avatars have the finest resolution, making it possible to show facial expression.

gramme can be used to change the size and shape of the template. The "paper" can be made flexible, so that the clothing is as realistic as possible. The clothing then falls naturally around the body, blowing in the wind when the avatars move.

Animations are used to generate every movement of the avatars. Specific movements of the avatars are based on choreographic sequences. Users can adjust every angle of the body on a timeline, using programmes that are in some cases free of charge. Users must specify when movements start and finish and how long they last. The programme computes the smooth transition of these movements. It is possible to upload this animation into the virtual world and to use it (or sell it) in the same way as clothing or images. Users do not always, however, have to create animations, clothing and objects themselves. Several thousand of these digital assets are already available and can be acquired for free or for a small charge.

Only ready-made facial expressions can be applied to the hands and face area. These are combined with animation and sounds to create gestures. In contrast to a real human face which has 26 muscles, avatars have significantly fewer possibilities for facial expressions – these centre around the corner of the mouth, the eyebrows and the eyes.

Avatars, consisting of individual parts, are uploaded and saved onto a central server for the virtual world. This has the disadvantage that separate avatars must be created for each world. Future systems will make it possible to exchange avatars. Different avatars will be created for different occasions and used in one world, depending on the situation, e.g. the respectable business avatar and casual shopping avatar both have exactly the same measurements as their owner, who can now shop in virtual worlds.

► **Agents**. Avatars are often confused with agents. While users control their avatars in the virtual world like a puppet, agents consist of "intelligent" software that can perform tasks for users. *Software* (non player character) in the virtual world automatically controls the "agent avatars". The software tries to recognise the environment of the avatars and to interact within it. It works on the descriptions of the objects and not on their graphical 3D implementation. If the agents are, for example, sales assistants in a shop, they recognise other avatars as customers in the shop. They know the floor plan of the shop and all the products (size, number, price). The sales assistant agents move between the display items and the customer avatars and initiate talks about sales. Their interaction is based on natural voice techniques.

► **Generating objects**. All objects in the virtual world consist of primitives, either a box, cylinder, prism, sphere, torus, tube or ring. Sculpted prims are another option, in addition to these very simple object forms. The distance from every point on the surface of a sculpted prim to its middle point is defined. This makes objects such as a banana or an apple easy to model. On the other hand, a sundae with cocktail umbrellas cannot be modelled as it has different surfaces. An object like this would be generated from simpler individual objects (a scoop of ice-cream, goblet, plate, glass, spoon, etc.) and then combined so that they can be manipulated like an object.

A snowman, for example, is relatively easy to create, comprising 3 spheres to which a carrot nose, coal eyes, stone buttons and stick arms are added. If stones are used for eyes, they need a more complex object description. First of all, a programme has to generate a 3D description of the stone and upload it to the server. This object is generated offline on the computer and not on the server, in contrast to primitives. It is then sent to the server.

When clients create new objects, the server generates a new entry with a specific code number in the database. Different features can be allocated to these objects, in addition to the geometric dimensions. This includes, for example, whether the object is transparent, whether it exists physically or whether it is phantom (i. e. avatars can run through it as if they are running through fog). Texture and colour are other features of the objects. Textures are deposited in the server database in the form of images.

When users modify or generate objects on a client computer, the modifications are sent by the client to the server, which in turn updates its internal database. An update of the features of the objects is sent to every client within view of the objects. The clients reload the update if the texture changes. Clients only receive information about descriptions of the objects and then compute the view. This ensures that users can interact in the world without having to wait. They can move quickly and the images are smooth.

▶ **Interaction with objects.** Interaction with objects is another important feature of 3D worlds, in addition to modelling 3D objects. Small programmes (scripts) in the objects make this possible. If, for example, a snowman is to move his nose and greet an avatar approaching him, the action (the avatar approaching) must first of all be detected. A programme activates a scanner at regular intervals (e.g. every 5 seconds), which looks for avatars within a specific radius. This then activates a greeting programme, which has been deposited on the server. This programme has a list of avatars and other information.

These programmes use resources sparingly. A greeting programme, for example, does not run if there are no avatars nearby (in a virtual world a falling tree does not make a noise if there is no-one nearby to hear it.) Most programmes run on the server of the virtual world and modify the shapes of objects or generate new objects. The world is consistent for all viewers, since the programmes run only once at a central location.

Contact between objects does not have to be actively initiated. Suitable programmes, which are activated when the object is touched, can notify a central location and receive a signal when this happens. These programmes sleep if no contact takes place. There is no need for an active element, which activates itself at regular intervals. As soon as the signal is released, a programme is executed, which rotates the snowman's head by 90°. This gradually changes the angle of rotation. It is, however, computationally intensive and demanding on the network to rotate an object like this. An update must be sent every second to all clients in view, stating that the head has rotated. In this case, it is possible to send an update to all logged on clients, informing them that the snowman's head is rotating at a certain speed. The clients can then calculate the rotation themselves, making it much smoother (not jerky every second, but a smooth movement). The disadvantage is that clients do not all see exactly the same rotation. The world is therefore no longer consistent. Whether consistency is important can depend on the

Object generation: To generate an object, users of the virtual world can use simple tools to design their environment. Many worlds make it possible to generate and modify simple objects using an "In-world editor" and to fit them together into more complex objects, e.g. a snowman made of different spheres and cylinders. Objects generated like this look very simple and unrealistic, but users can generate them after just a few hours of practice. Other virtual worlds offer very good graphics, but are consequently not easy for users to modify. In the picture, the nose of the snowman has just been generated. On the left is the feature menu with the choice of colours and the texture of the nose cylinder. On the right is the choice of colours, any of which can be allocated to the object.

```
// When script runs first time
  state_entry()

  {
    llSay(0, "Hello, my name is Peter!");
  }

// When someone touched the object
  touch_start(integer total_number)

  {
     // Get the Avatar Name and say a sentence.
     llSay(0, llDetectedName(0) + " just touched me.");
     // Start to rotate the Object
     llTargetOmega(<0.0,0.0,0.5>,TWO_PI,1.0);
  }
```

Programme for rotating the nose when the object is touched: To make the virtual world more interactive, users can write simple, small programmes (scripts), which interact with objects or users. Clients, like the objects themselves, offer an environment suitable for developing these scripts. The scripts run on the server and are event-based. They are thus made available to all users at any time. Event-based means that they are not activated if specific events do not occur. These scripts are the only way in which the virtual world can become interactive and animated. In this example, the object says its name when it is generated for the first time. It then waits until an avatar touches it. It then finds out the avatar's name and sets the object in rotation. Green: General remarks; Black and Red: Programme lines (If ..., then ...)

situation – for example, whether a revolving roulette wheel should spin continuously.

Trends

► **Interactive television.** Many technical details are collected during sporting events these days. These are only made known to viewers, however, as statistics or numbers. Cameras record the images spectators see from various angles. At special sporting events, spectators can sometimes even select their preferred camera perspective themselves. However, they cannot experience the game, for example, from the eyes of the players or the ball. A realistic picture of the event can be computed in real time using the technical information already acquired and the technical possibilities offered by a games console. Since the world is created on the computer it can be viewed from any possible angle, e.g. from the viewpoint of the referee. It is also possible to pause the game at any point and observe various individual steps. These are not real images, certainly, but it is real data and therefore 'reality'. This technology is now used at tennis tournaments to check the decisions of the linesman (Hawk-Eye).

Television series, especially soap operas and science fiction films are, in addition to sporting events, an example of how television and virtual worlds are merging. Spectators can access the complete film set, which has already been generated on the computer - particularly for more expensive science fiction series. They can be a participant in a series, or they can become a part of the series by being in a game, which runs in the world of the series. They can even shoot their own episodes in the series.

Interactive television: Virtual worlds have the advantage, in contrast to television, that the angle of observation can be changed at will. The position of the ball, the speed and the angle of flight are now all recorded at tennis tournaments. These statistics can be used to reconstruct the tournament in the virtual world (Hawk-Eye at US Open). The positioned camera records the tennis tournament. It is also possible to see the reality reconstructed in the virtual world in real time, from any angle. The image shows the viewpoint of the ball boy. The viewer can pause the game and repeat a sequence at any point. In order to ascertain whether or not the ball was out, the observer can fly with the ball or take up the linesman's position. It is now possible to combine the graphics performance of modern game consoles with data from sports events recorded in real time. Spectators can now both observe and take part in the game – provided they do not have to influence real events, e.g. they can cycle on an exercise bike in real time during the Tour de France.

Perspectives

- Virtual worlds will become the norm within a few years. In the same way that users 'chat' today, meetings in the future will be held in 3D.
- There are still many technological possibilities, which have been known for some time, that have not made it in the mass market, because there is no application for them. Virtual worlds can use these developments to their advantage.
- Virtual worlds will not supersede the Internet (or the WWW), or real life. However, they will help communication on the global level, expanding essential areas of non-verbal communication and social integration. This will make global communication simpler, more efficient and more humane.

ANSGAR SCHMIDT
IBM, Research and Development, Böblingen

Internet

- http://opensimulator.org
- http://osgrid.org
- www.the-avastar.com/
- http://shifthappens.wikispaces.com
- www.virtualworldsreview.com/info/whatis.shtml
- http://sleducation.wikispaces.com/virtualworlds

Human-computer cooperation

Related topics

- Artificial intelligence
- Image evaluation and interpretation
- Ambient intelligence
- Digital infotainment
- Information and knowledge management

Principles

Whether it be DVD-players, automatic ticket machines, cell phones, PDAs, machines, computers or even airplane cockpits, an increasing number of products at home and at work have a user interface, in other words a place where humans and machines meet. Direct contact with interactive technology was previously limited to the workplace. Now, there is hardly a social group that does not come into contact directly or indirectly with computer or information systems. However, the continuous increase in the performance and functionality of information and communication systems has resulted in rising complexity in controlling such systems. The practicality of user interfaces has become more and more a key criterion for acceptance and success in the market.

Human-computer cooperation involves the analysis, design and evaluation of information and communication systems, whereby focus is on the human being and his or her individual and social needs. It offers methods and techniques that can ease both the tension between increasingly technology-oriented environment-related information overload, and the resulting excessive demands being made on the user, which can cause dissatisfaction. The range of possible input interfaces includes auditory (spoken language), visual (graphical user interfaces) and haptic (touch) modalities, as well as increasingly specialised combinations, so-called multimodal interfaces (for example, combined spoken and touchscreen input). Additional input modalities such as eye movement, gestures and facial expression will play an important role in many devices in the future, as will physiological parameters. On the output side, a distinction can be made between audio (sound and speech), visual and haptic modalities.

Communication channel between humans and systems: The bandwidth of communication between the human and the computer has been broadening continuously as different generations of user interfaces were developed. This broadening of the communication channel, however, has so far mainly happened on the output side, less so on the input side, where important modalities of human expression such as speech or gesture are still underused.

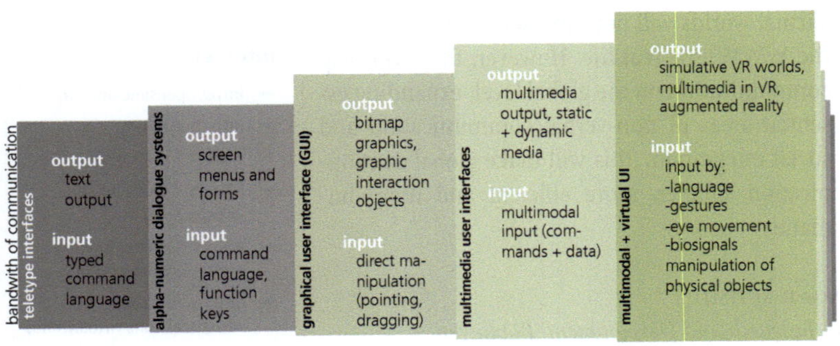

▶ **Speech recognition**. Speech recognition systems are generally classified as either discrete or continuous, or speaker-independent or speaker-dependent. In continuous speech, words follow each other without a break – the individual words are spoken without any audible pause between them. In contrast, discrete speech recognition requires there to be interrupted speech with artificial pauses between the words. Speaker-independent speech recognition enables the user to immediately begin working with the system without a training phase. Speaker-dependent speech recognition requires the user to be trained in pronunciation before beginning, so applications with frequently changing users are not possible.

Speech recognition is based essentially on a comparison of patterns. Acoustic signals are compared with previously learned reference patterns. A speech recognition system contains a large database of reference patterns (codebook), which serve as "templates" for the acoustically recorded words. In order to carry out speech recognition, digitalised audio data must first be divided into their constituent frequencies: every 10 ms, spectral parameters are taken from the Fourier-transformations and summarised as "feature vectors". The vectors are then compared with the prototypes from the relevant active word database. The words in the database are recorded as sequences of phonemes. Phonemes are the smallest differentiator of meaning in spoken language, but they do not themselves carry any meaning.

Like syllables, phonemes break down words into their smallest components. When two sounds occur in the same sound context, but belong to two different words, they are considered different phonemes. The words "fine" and "mine", for example, differ only in the phonemes "f" and "m". Each phoneme can be described by its frequency sequence in the time axis. A spoken phrase or word is recognised if the sequence of phonemes matches the phrase or word in the database.

Because a user does not always speak in the same way, these variations in the phonemes are compensated for with the help of a statistical program. As soon as a prototype is recognised, the system begins to search for the word with the help of a language model. This model includes the probability that a certain word is followed by a certain other word. It contains statistical and stochastic information derived from texts within the relevant context.

If an acoustic unit is associated with just one reference pattern, this is known as discrete word recognition. If one or more reference patterns are assigned to an acoustic unit it is called continuous speech recognition. This makes it possible to recognise continuously spoken phrases without explicit pauses after each word. A central method in the recognition of continuously spoken phrases is the use of so-called Hidden-Markov-Models (HHM) – stochastic models used to describe and analyse real signal sources and to extract the phonemes that fit the input signal best.

In addition to the actual performance of the speech recognition system, a well thought-out dialogue design is particularly important for the success of language portals, especially telephone-based ones such as travel schedules. Three design areas for *voice user interfaces* (VUI) can be distinguished that are in line with current standards such as Voice XML or SALT:

- Audio system outputs (prompts)
- Number of user expressions recognised by the system (grammar)
- Definition of the possible dialogue flows, for example, when the caller is not understood (call flow)

Simply put, the speech recognition system functions best when it has the smallest possible grammar. However, a small grammar limits the user to only a small number of expressions. Unless the user is explicitly presented with all the possible input alternatives before each input, exact knowledge of the input possibilities is required to control the application efficiently. This has consequences for the design of both prompts and grammar: the prompts have to be designed so that the functions are made clear to the callers. They must know what they can say and how they should speak. Clear explanations are often significantly longer and less effective than the corresponding model examples of audio system outputs. As for the grammar, it is important that the user expressions recognised by the system be as intuitive as possible, relate to the everyday language of the target group and be easily learned, remembered and spoken.

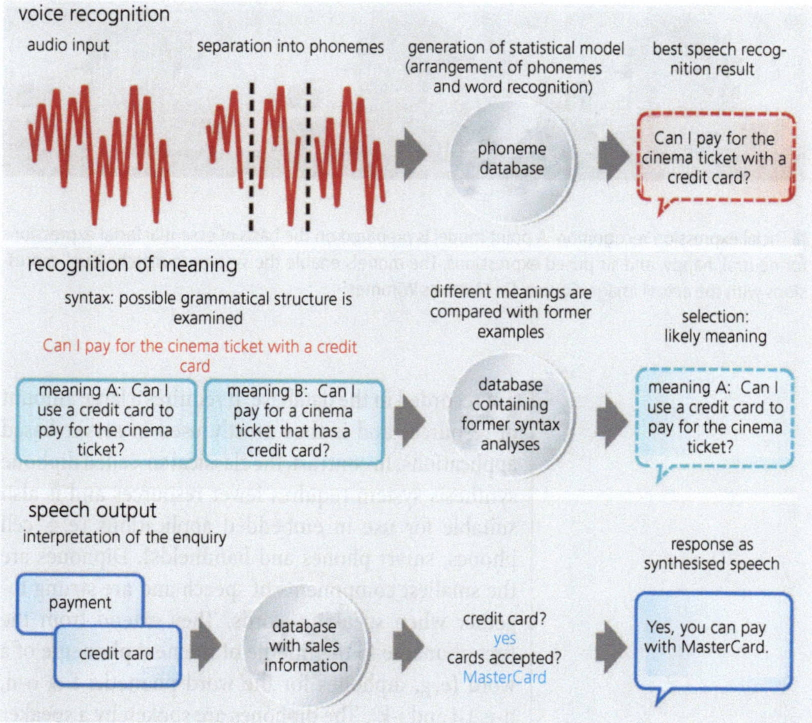

Semantic analysis. Above: The utterance is subdivided in phonemes. During the speech identification procedure they are arranged by a statistical model and the words are recognised. Middle: once the meaning of a sentence is recognised, its grammatical structure is compared with different examples in the database. Bottom: In the case of audio response, the enquiry must first be interpreted. The system can then use the database to form an answer through speech synthesis.

► **Speech synthesis**. Text-to-speech (TTS) systems enable a computer to produce spoken language. The goal of a text-to-speech system is to speak any text understandably and naturally. A text-to-speech system can be used if a display is not suitable as a human machine interface, for example while driving a car or with a telephone-based information system. The TTS system functions like an audio printer that does not understand the meaning of the reproduced text. The main challenge lies in the most exact possible reproduction of the text by the computer, based on a human voice that follows exact phonetic rules.

The newest generation of TTS software can reproduce any text in an understandable way with a naturally sounding human voice. The best fitting speech units are drawn from a large database and linked to each other flexibly (unit selection). These units are previously spoken by a speaker and recorded. The results sound natural (similar to the original speaker) but the system is inflexible with regard to situations that are

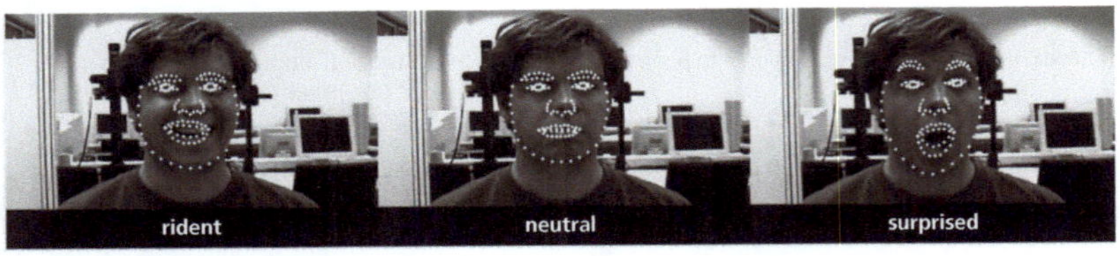

Facial expression recognition: A point model is prepared on the basis of essential facial expressions for neutral, happy, and surprised expressions. The models enable the system to match visual expressions with the actual image. Source: Dr. Matthias Wimmer

not recorded in the database. It requires a large amount of resources and is thus mostly used in server-based applications. In contrast, the classical so-called diphone synthesis system requires fewer resources and is also suitable for use in embedded applications (e.g. cell phones, smart phones and handhelds). Diphones are the smallest components of speech and are strung together when speaking words. They extend from the first phoneme to the middle of the next phoneme of a word (e.g. diphones for the word phonetic: f-o, o-n, n-e, t-i and i-k). The diphones are spoken by a speaker and recorded. In order to attain a natural speech melody in a text that is to be spoken, punctuation such as full stops, commas and question marks are also utilised.

► **Facial expression recognition**. Humans can recognise people's facial expressions very accurately and with little effort. The goal of computer-supported facial expression recognition is to make a user's complex facial expressions comprehensible to a computer. Knowledge of the user's emotional state will lead to more individualised human computer interaction. The basic principle of facial expression recognition is the video-based comparison of patterns between a recorded point model of central facial contours and the actual recorded facial pictures. There are six different fundamental emotional states: happiness, sadness, surprise, disgust, anger and fear. Current facial expression recognition systems work with three or four emotional states. They can recognise neutral, laughing and surprised facial expressions very reliably. And they can identify confirmation (nodding) and negation (shaking of the head). Facial expression recognition has a wide range of applications. It can be used in computer games, for recognising emotional states and signs of tiredness when driving a vehicle, and for communicating with service robots.

► **Handwriting recognition**. Online handwriting recognition represents an alternative input method particularly for compact end devices, for which a keyboard would be impractical. These comprise PDAs (personal digital assistants), for example, pocket PCs, Webpads or tablet PCs. While spoken language is the more effective input method for large amounts of text, handwriting is more suitable for jotting down notes or appointments. Handwriting recognition can also be an alternative to speech recognition when, for example, error-free speech recognition is not possible in an environment with loud background noise.

In contrast to off-line recognition (text recognition of scanned documents), on-line handwriting recognition does not make use of visual information; instead, the movements of the pen or stylus represent the writing in the form of a temporally-ordered sequence. The pre-processing for recognition is divided into phases of noise suppression and writing normalisation – the levelling of small irregularities with mathematical algorithms. This frequently corrects not only the writing speed, but also the slope, orientation and scale of the writing. The processing steps that follow are similar to those of speech recognition, particularly for continuous joined-up writing, which cannot be segmented into single letters. This can be done with a *support vector machine (SVM)*, for example.

In first-generation PDAs, a special letter alphabet had to be learnt and input was restricted to predetermined writing areas. Current software for handwriting recognition allows for more or less natural handwriting without a long learning phase and restricted writing areas. The ciphers, letters and joined-up writing are recognised with great accuracy. Additional functions increase performance by suggesting words after only a few recognised letters ("ger" > "Germany", for example). Furthermore, the device understands simple pen gestures like "cut and paste" or "undo". A particularly high hit rate can be achieved by holding the pen as horizontally as possible and by writing sufficiently large letters. However, the input with a pen cannot – and is indeed not intended to – compete with input using a standard size keyboard.

Applications

► **Location-Based Services (LBS)**. A location-based service provides the user with location-relevant information by means of a mobile appliance. When a request is made, it is initially passed on to a location service provider or geoportal together with the position of the appliance. The location service provider then processes the request - sometimes with the support of other service providers who supply geographic data or other services (driving directions, for instance). The answer from the location service provider is then delivered to the user. The most well known applications are navigation systems in cars and, increasingly, for pedestrians. These provide visual and spoken directions. In addition to supplying information on how to travel between two points, requests can also include the location of the nearest service facilities such as hotels, restaurants and shops. The following information can be communicated visually:

- position – actual location, points of interest (POI)
- landmarks and other anchor points in the real world
- lines of interest (LOI)
- areas of interest (AOI)
- objects/people
- search results – objects, distances, relations, etc.
- events and conditions

From the perspective of human-computer interaction, the challenge lies in providing the appropriate contextualised forms of visualisation and interaction for the relevant location. In the future, however, geoinformation will not be visualised solely using conventional maps. Aerial and satellite pictures or hybrid views, photos or videos of target objects as well as audio commentaries will be used for the visualisation and audification of navigation information. The destination of a journey might be displayed along with a current photo. Comparing the real-life representation with the user's actual view might enable a faster decision to be made on whether the chosen route is correct than would be possible using an abstract conventional map. In addition to the appropriate form of visualisation, a further challenge is to design the interface of the mobile unit to interact with the user as intuitively as possible. Current solutions are still closely tied to an eye and hand-based interaction pattern. The information appears on small displays and can be controlled by a pen or touch. The mobile user is therefore still very much tied to the system. New input and control technologies with haptic and voice interfaces are available in principle. However, they still need to be developed into marketable products. The same applies to new output technologies (foldable or rollable displays) and to the enhancement of the user's three-dimensional field of vision with digital information through augmented reality.

Trends

▶ **Voice User Interface (VUI).** Automatic speech recognition (ASR) is so effective, that a recognition rate of over 90% is possible even with speaker-independent systems. Telephone-based speech applications are already widely used. Even speech control of

◼ Location-Based Service: High-resolution maps on mobile phones facilitate detailed pedestrian navigation in an urban environment. Multi-sensor positioning puts pedestrians on the map using A-GPS (Assisted Global Positioning System). Landmarks include street names, buildings and public transport stops. "Breadcrumbs" mark the route taken, so that users can see where they're going and where they've been. Source: Nokia

appliances is now technically possible for televisions, for example, or digital video recorders and systems for heating, lighting or even drawing the blinds. But further research is necessary before a coherent multimodal control concept is available that meets with the acceptance of end-users. Speech interaction helps to simplify the complexity of the control function (e.g. for video recorders) and can in some cases even make independent control a possibility (say for user groups with limited physical mobility). The application of speech interfaces also makes sense when some manual activity would otherwise have to be interrupted, for example, when driving. Speech input in combination with gestures will play a particularly important role in the field of ambient intelligence.

▶ **Driver assistance systems.** Driver assistance and information systems, as well as communication and entertainment functions, are changing and extending the tasks facing the average driver, who is being confronted with a steadily increasing flow of information. The development of human-computer cooperation in future driving scenarios exemplifies how the newest sensor technologies can be combined with alternative multimodal forms of interaction (for example, speech control of navigation, air conditioning and entertainment systems). There are many helpful safety and comfort features, such as collision and lane departure warnings, night vision or parking assistance, but these must be carefully allocated to the various warning modalities (acoustic, optic, haptic) to avoid overloading the driver.

New ways of presenting exact information to the driver are showing a great deal of promise, for example *head-up displays* (HUD), which allow information to be blended directly into the driver's field of vision. Basically, head-up displays consist of a display, an optic module and a combiner. The display produces a picture, which is projected onto the combiner (usually the windscreen) with the help of a lens and a hollow mirror or using holographic technology. Multiple mirrors within a closed box deflect the light rays before the picture is projected onto the windscreen. The resulting picture appears approximately 2 m in front of the driv-

Head-up display: These displays present important information directly in the driver's viewing field. A unit in the dashboard projects the data onto the windscreen as a clear, easy-to-read image, adjusted for ambient light conditions and with a focal point just above the bonnet. This means that information can be read faster and the driver's attention is never distracted from the road ahead. Source: BMW AG

er over the hood of the car. The mirrors can also correct the distortions caused by the curve of the windscreen. *LED*s are currently being used as a source of light.

The use of speech for choosing selected functionalities and parameters (for instance, number selection, navigation requests or the choice of a radio station) further increases safety by reducing eye movement and keeping hands on the steering wheel.

► **Brain-Computer Interfaces.** This type of computer operation is based on changes in *brain waves* that precede the carrying out of an action by approximately half a second. Brain waves can indicate whether a person wants to move his right or left hand. Brain waves, which produce only a few millionths of a volt, can be measured using an *electroencephalogram (EEG)*. Software recognises the target signal within the diffuse amount of data and assigns it to the relevant thought, for instance a certain letter. By comparing the measured brain waves with those in the database, the computer can convert the user's thoughts. After about 20 min, the user is able to control a cursor with his or her thoughts. Brain waves can be converted into a large variety of commands depending on the computer programme used. They might be used to operate prostheses in the future, or bridge damaged connections between the central nervous system and muscles in paraplegics.

Multi-touch gestures denote a set of interaction techniques that allow computer users to control graphical applications with several fingers. These gestures include scrolling by making a circular motion, moving pictures or documents with a flip of the finger, and zooming in or out by making a pinching gesture, i.e. by touching and holding two fingers on the screen while moving them apart from or towards each other.

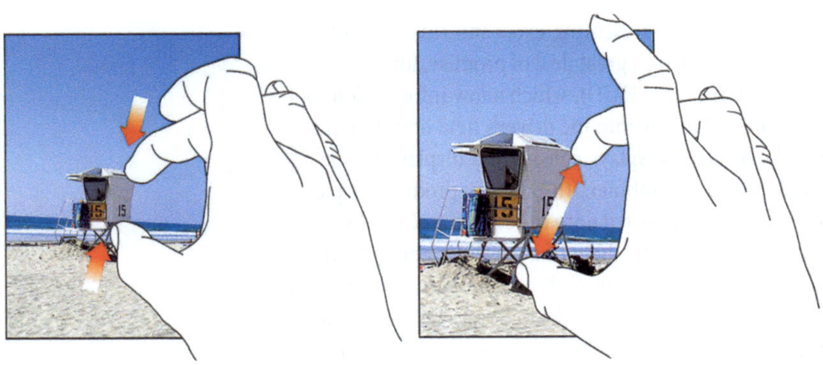

► **Multi-touch interfaces.** Multi-touch interfaces combine input and output modalities in a touch-sensitive user interface. They do not require a common input device such as a mouse, keyboard or stylus. Multi-touch interfaces are operated with fingers only and can process many inputs simultaneously. Thus, different fingers or different hands can carry out interactions at the same time. Objects like windows or photos can be directly manipulated on the screen. In order to zoom or move a window, for example, the user merely needs to perform the action with his or her finger on the screen. One does not need to go into a menu or input a command. This makes the operation of multi-touch devices far more intuitive than conventional computers.

The availability of SDKs (software development kits), for the Apple iPhone, for example, make it possible to create new multi-touch gestures. The contact and movement of the fingertips on the display are read in detail and interpreted as gestures. This opens up new possibilities for intuitive human computer cooperation, particularly for smart phones and other mobile devices that do not use a keyboard or mouse. In addition, multi-touch interaction with large wall displays will find increasing favour with teams working on projects.

► **Multimodal Input/Output Interfaces.** In multimodal interfaces the exchange of information between humans and machines takes place simultaneously over many channels of communication (speech, gesture, facial expression, etc.) as is the case in the interaction between humans. In principle, multimodality means sharing a broader range of information using a technical system – with fewer errors thanks to redundant information analysis. The technical challenge on the machine's input side is to produce sensors for collecting

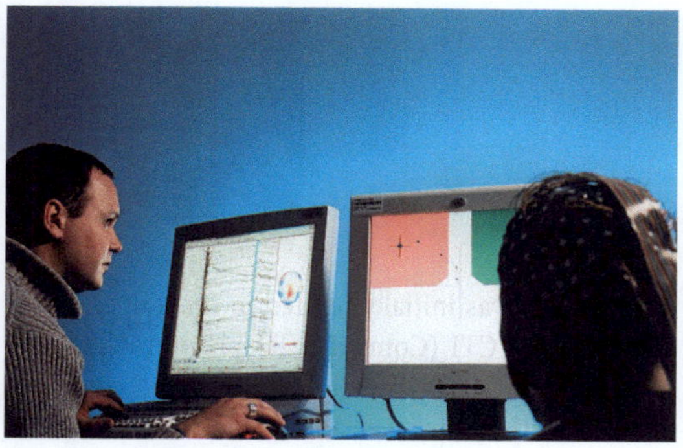

Brain-computer interface: These allow for a direct dialogue between man and machine. Cerebral electric activity is recorded via an electroencephalogram (EEG): electrodes, attached to the scalp, measure the electric signals of the brain. These signals are amplified and transmitted to the computer, which transforms them into device control commands.

and analysing the relevant multimodal actions of the user, including audio, visual and haptic inputs (recognising body language such as facial and emotional expressions, body, hand and eye movements, even speech recognition and lip reading).

Multimodal applications could be deployed in meeting rooms and cooperative environments, for example. Using hand gestures or laser pointers, users can select specific objects on a large display while using voice instructions to call up the content that should appear on the screen.

In addition to the combination of speech and facial expression (speech recognition and lip reading), multimodal systems are being developed that combine eye movement and speech recognition with more standard technologies such as keyboards or touch screens. This involves using movements not only for focussing on screen elements (scrolling, zooming, panning), but also for triggering actions on the screen. Registering eye movement is done by the cornea-reflex method, whereby infrared light reflexes are picked up by the cornea. Infrared light rays are used to illuminate the eye, while it is being filmed using a telephoto lens and a monochrome video camera. Cornea reflex and pupil coordinates are determined. The angle of vision can then be calculated and corrected for relative distance between reflex and pupil. The measurement accuracy of touch-free systems is less than 0.5°. Pilot applications have been carried out already in a medical context: radiologists can analyse X-ray and CT/MRT images more efficiently using eye movement and speech control. For example, focussing on an area of the lung automatically causes its image to be enlarged. The method can also determine which areas the eyes have already looked at and which still need to be covered.

Prospects

- Technologies and services can only be successful economically if focus is continuously kept on the needs of potential users. Whatever the product, the user experiences it through the interface. Intuitive control, high task relevance and positive emotional responsiveness are key acceptance factors.
- The goal has gone beyond merely designing the interplay between humans and technology as a simple interaction that follows a set of clearly defined rules. Multimodal human-computer cooperation with human beings and their natural perceptions and emotions will become the measure of all things:
  Computers will understand human speech much better than they do today; computers will be able to read handwriting and react to gestures and facial expressions; applications and devices will adapt to the context, preferences and location of the users (smart environments and ambient intelligence).
- The role of standard user interfaces will gradually become less important with the rise of an intelligent environment. However, in the short and medium term, the desktop metaphor with its keyboard and mouse control will continue to dominate the workplace. By the same token, the continuous advance of computing in everyday life and the enormous variety of mobile information systems will call for the creation of new interfaces suitable for everyday use. But the vision of invisible computing will remain just that, a vision, at least for the near future.
- Better interfaces can only be produced through interdisciplinary design and by focussing on the user. Interface design involves a step-by-step co-evaluation of users, user contexts and systems. The diversity of applications, contexts and technical possibilities are producing a steady stream of new opportunities for higher degrees of freedom. The human-computer interface will have to be designed more openly to allow users greater flexibility. The interface of the future will have to adapt to the human being and be available for use in ways other than originally intended.

PROF. DR. FRANK HEIDMANN
Potsdam University of Applied Sciences, Potsdam

Internet

- www.hcibib.org
- www.media.mit.edu
- www.cs.umd.edu/hcil
- www.hml.queensu.ca

Business communication

Related topics

- Communication networks
- Internet technologies
- Information and knowledge management
- Human-computer cooperation

Principles

Business communication is the term used to designate communication that is supported by information and communication technology, between employees in a company and in company networks. A great diversity of technologies is available, from telephone, fax, e-mail and messaging services to portal solutions, unified communications and collaboration systems.

Business communication in companies is based on server systems which store documents and business data as well as back-office applications like Lotus Notes or Microsoft Exchange. Telephony, which has been the traditional backbone of business communication for many decades, will increasingly and comprehensively be integrated in data traffic in the coming years owing to the shift towards Voice over IP. More sophisticated services, such as videoconferencing over the Web and the highly advanced telepresence systems, allow individuals to maintain eye contact over long distances while discussing matters and working on documents.

Applications

▶ **Voice over Internet Protocol.** For many years now, telephony has been undergoing fundamental change. This was initially limited to functional extensions such as CTI (Computer Telephone Integration) which allowed users to dial directly from a computer and to have the relevant customer information selected automatically from databases during an incoming call. UMS (Unified Messaging Services) drove integration further by adding fax services to e-mail boxes and allowing voice messages from the phone to be transferred to the e-mail box as well. The most fundamental development, however, was Voice over Internet Protocol (VoIP) which involves using the same network and the same transmission protocols for telephony as for conventional data traffic. This has produced enormous advantages for telephony: CTI and UMS are a lot easier to implement, the costs per call are considerably lower and the administration of the systems is far simpler because VoIP components are basically IT systems and can therefore be handled by the IT specialists already on the premises. It is believed that VoIP will entirely replace conventional telephone networks by 2020.

▶ **Business data.** In the field of business data, the technical foundation consists of server systems or clusters in which all company data are stored centrally. These types of systems must be considered part of business communications, even though they have nothing to do with "direct" communication between people. Nevertheless, these systems are used for sharing and processing information in the form of documents – often the most important data within the company's value chain and business processes. A wide-ranging development from plain file servers to integrated portal solutions has taken place over the last few years. These systems come in the shape of intranets or extranets and comprise document management systems, team and project calendars, task lists, wikis, blogs, discussion forums and other applications such as workflows. Workflows are largely automated, standardised process chains in which technology is used to convey a task to the next person in the chain. In an approval workflow, a document is processed and might then be routed automatically to a superior for approval. The approval task is then monitored to ensure timely completion. The integrated document management systems are intended, first and foremost, to simplify the administration and above all the retrieval of information. Records management functions are available for the administration of documents throughout their

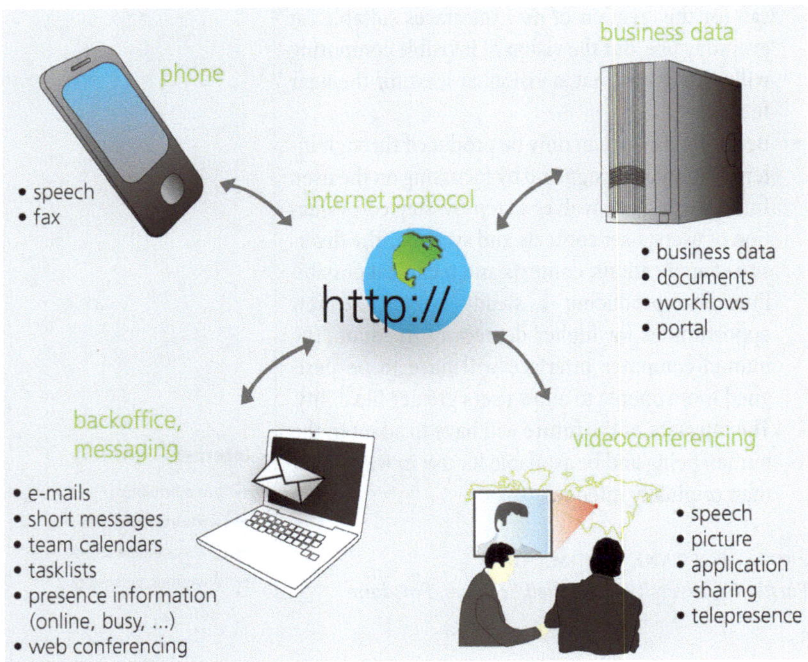

An overview of business communication: In the future, all communication will be conducted via Internet protocol. This will allow the main components of electronic corporate communications – phone, business data, back-office and videoconferencing – to be connected and combined far more easily.

life cycle. These functions ensure revision-safe archiving and storage, as well as deletion of out of date documents. You can configure the system to send a document to the archive after 3 years and have it deleted after 10. Furthermore, portal solutions make messaging functions available that inform the user either by e-mail or RSS (Really Simple Syndication) when a document has been processed by other members of the team.

▶ **In the field of business applications.** In business applications (e. g. SAP), the top priority is continuous integration of data and applications. The primary aim of integration is that an item of information need be generated only once in order to be available to all other users and applications. This is the only way to keep company information efficient, consistent and up to date. It is certainly advantageous for evaluations and management reports if the information can be read and processed directly from the business application, so that the aggregated data can be accessed on the company intranet in real time. This is termed "business intelligence functions".

▶ **Back office and messaging.** This is where one finds the groupware systems (e. g. Microsoft Exchange, Lotus Notes), which are now virtually taken for granted in companies. In addition to mailing and address management functions, they perform task and calendar management for defined teams of any kind. This fundamentally improves the efficiency of day-to-day office and administration processes like setting up appointments, reserving space for meetings or delegating and monitoring tasks.

Generally speaking, communication these days means that everyone is reachable at all times. This does have two significant disadvantages, however: first, people are forced to become familiar with a variety of terminals such as laptops, PDAs, cell phones, business phones, and so on, each with a different system and user interface. Some services, such as SMS (Short Message Service), are only available on certain terminals. Secondly, permanent availability means that they are exposed to constant interruptions which can make normal work almost impossible.

As a result, messaging solutions are now becoming increasingly widespread in business. The aim, in part, is to limit the e-mail flood of recent years. By publishing current presence information (e. g. online, busy, in a meeting, do not disturb, etc.), communication can be optimised to such an extent that the caller only reaches those team members who are available – or sends an e-mail to the unavailable person and expects to wait longer for an answer. This saves futile attempts at getting in touch and avoids interrupting a communication partner at an inopportune moment.

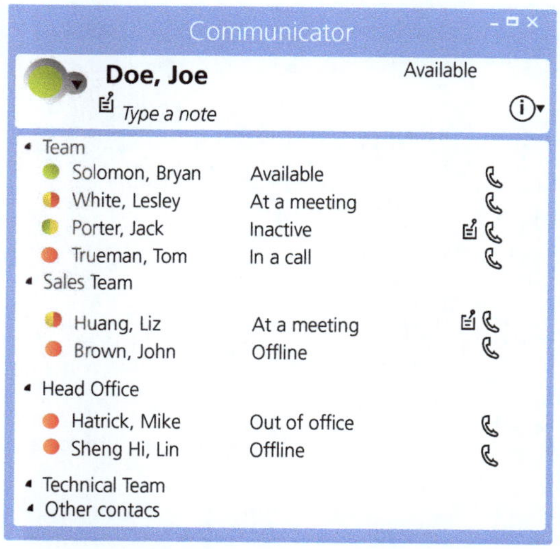

◼ Messenger: The presence information function of a messenger shows at a glance which team members are available at any given moment. At sign-on, the system automatically posts "available". The user can then change his or her status to "Busy" or "Do not disturb". Other presence information, like "At a meeting", is automatically retrieved from the personal calendar. If the user is not actively using the computer, the status automatically changes to "Inactive" after a while, and later to "Away".

▶ **Mobile work.** The mere online availability of documents via Intranet is not sufficient for mobile work, as not every place has access to the Internet – despite UMTS (3G wireless communication). Additional functions such as offline synchronisation of directories for mobile terminals (laptops, PDAs and smart phones) are therefore essential: once the mobile device is online again, all changes made to clients and servers since the last synchronisation are compared and the systems are updated.

Trends

▶ **Unified communications.** As has been shown above, the current presence information of colleagues can be easily traced in real time by messenger services. This information becomes even more interesting when it can be accessed on all electronic communication clients, in other words not only the messenger, but also the e-mail clients for mails, faxes, phone calls, voice mails, and so forth, and on the online portal. The current presence information of a colleague is retrieved from various linked sources. For instance, the appointment book of the groupware system reveals how long the colleague will be at a meeting, the VoIP telephone system displays the status of a phone call, and the messenger will show the colleague's own information as to whether he or she is busy or does not wish to be disturbed. Taken as a whole, the data clearly show whether a colleague can be contacted, and if so, when. Making contact is just as easy. A person can be called, mailed or contacted by a click of the mouse. All the contact information, such as the e-mail address, phone number, and so forth, is deposited in a central directory that is created at the same time as the user account.

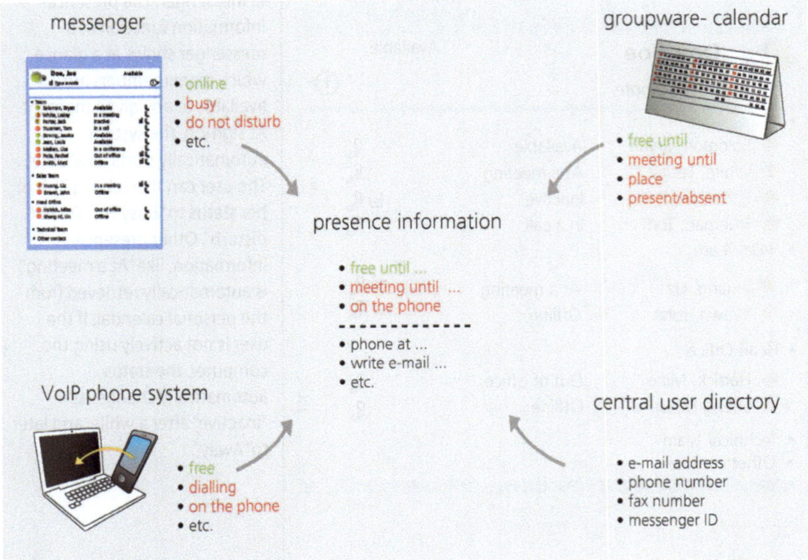

Presence information sources: In unified communication systems, the data for current presence information is drawn from a wide variety of sources. The messenger delivers the information provided by the user, e.g. "Do not disturb". The information about how long a contact person will be in a meeting is retrieved from the appointment calendar released for viewing, while the VoIP telephone system supplies the current phone status. The contact information – telephone number, e-mail address, etc. – is taken from a central user directory such as the Active Directory.

Nevertheless, this field of company communication is still in the early stages of development. As with so many new developments today, there are still many proprietary solutions that only function when the communication partners are using hardware and software made by the same manufacturer. It is essential that unified communications can be used across corporate borders, however, encompassing all functions and regardless of the devices or software employed. The solution providers (Cisco, Microsoft, IBM, Nortel, for example) will need to create the appropriate interfaces. A few promising ideas already exist, but these will have to be elaborated and optimised in the coming years.

The integration of all communication channels is

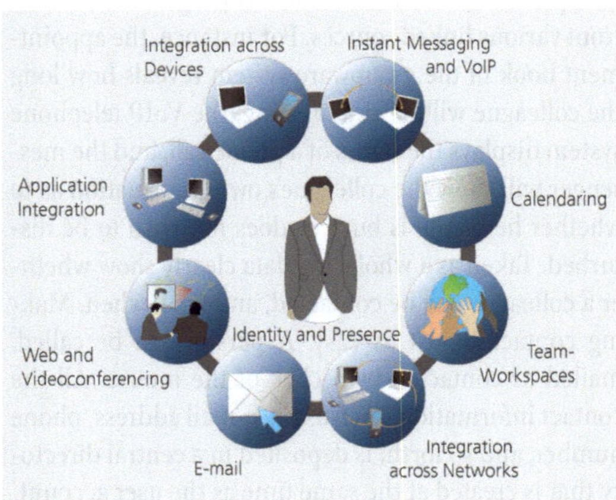

Unified communication and collaboration: By integrating all communication channels, presence and availability information is combined with contact information about the desired communication partner. This information is then made available in real time to all applications and on the online portal.

known as "unified communication and collaboration". It makes status and connection information available to all applications relevant to communications, i.e. the communication clients, the online portal ("Document last modified by …") and, naturally, the online address book. In addition, employees can indicate their fields of expertise in their personal profile. This offers the possibility of systematically searching out colleagues with specific abilities – which is a great advantage, especially in large enterprises. If a person with the required knowledge cannot be reached right away, the same system can be used to find someone else who is just as qualified to answer the question. Attempts are also being made to establish intelligent knowledge networks in which incoming e-mails from employees are analysed to determine the areas in which this colleague is an expert. Much development work is still needed before these semantic knowledge networks can function reliably.

▶ **Personalisation.** Classic Intranet portals are presented to users as a collection of websites whose content they cannot change in any way. This is not ideal, as employees in any given company have differing tasks and hence different interests when it comes to the content displayed. A top manager is likely to be most interested in the current sales figures and profit margins, and a sales employee will want to know about the sales opportunities in his or her region. There needs to be a way of aggregating these various interests on personalised web pages, or better yet, employees should have the possibility of assembling what they consider to be interesting information themselves. There are already a few promising approaches in this area, such as Microsoft's "Webparts", which enables users to compile their own individual web pages.

▶ **Electronic paper.** Electronic communication requires other technologies as well, not only for transmitting information, but also for input and output. The latest display technologies are one aspect of this. Mobile work above all requires flexible, power-saving, portable displays that come close to having the same resolution, display quality and feel of paper. Years of development work have already gone into e-paper and many promising approaches now exist, the most significant being electronic ink and polymer displays.

Simply put, electronic ink is a technology that involves placing electrically charged black and white particles between two electrodes. When a voltage is applied, the black or white particles adsorb on the transparent surface to produce a black-and-white image. This hardware can be incorporated in almost any carrier material (glass, plastic, cloth, paper). Tomorrow's displays will thus be able to assume all sorts of shapes which are very different from today's rigid and heavy monitors. Polymer displays, often referred to as organic "*OLEDs*" (organic light-emitting diodes), are based on special coloured polymer particles that begin to emit light when a low voltage is applied. This permits the creation of power-saving, flexible displays that can be viewed from as many angles as paper.

▶ **Intelligent information assistants.** These systems make sure that the users only receive the information that they actually need at a given time. Today, the set of tasks associated with these systems comes under the heading "information logistics", but most of the applications available are prototypes. So far, they have been based on relatively simple rules that ensure, for example, that an employee only receives very specific information while a meeting is taking place. For example, the system can block phone calls to people at a meeting, but let e-mails through. This is done by accessing several different directories and applications (appointment calendars, location systems) and deducing from them what information is currently needed. The problem with such systems is that there are frequent exceptions in the course of everyday work, and these cannot be depicted using such a simple set of rules. Here too, there is a need for self-learning, intelligent systems that can ascertain the situation as a whole, "comprehend" it, and draw the right conclusions.

Prospects

- High-quality applications that support mobile collaborative work and company communications are already available. Their technical implementation is becoming steadily easier, and the challenges are increasingly shifting to organisational issues.
- Voice over IP makes telephony a fixed element of information technology. This facilitates its integration with IT systems so that functions such as automatic dialling, call protocols, unified messaging and online meetings will very soon be a matter of course in companies.
- Unified communications connect the existing communication channels with the user's current presence information, thus allowing far simpler and more efficient communication. Depending on the presence status of the communication partner, the most suitable communication channel (telephone, e-mail, text message, and so on) can be selected by mouse click.
- Intelligent assistants with a wide variety of tasks will significantly determine the applications of the future – providing that computer performance continues to increase as rapidly as it has done so far. Today's computers are still too slow to accommodate such developments.

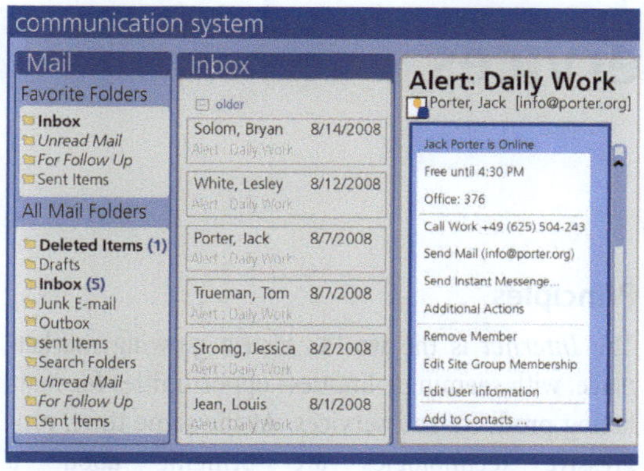

Status and connection information: Unified communications show at a glance whether and for how long a communication partner is available. His or her e-mail address and telephone number are also visible. A telephone conversation can now be initiated by a mouse click, for instance, thanks to computer-telephone integration (CTI).

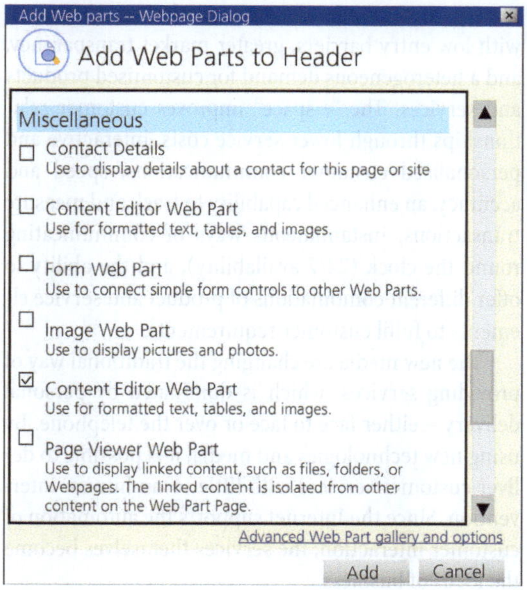

Webparts: Web pages that can be embedded at any point on another website; enables users to create web pages in the simplest way possible so that they will have a page displaying precisely the information that interests them most.

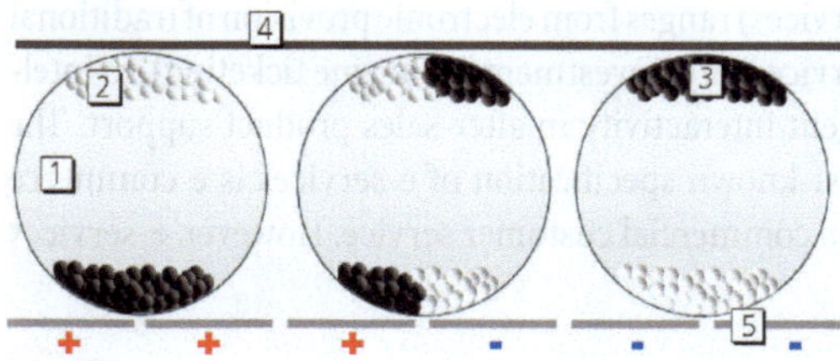

Electronic ink: Capsules filled with a clear liquid (1) and with a positively charged white pigment (2) and a negatively charged black pigment (3) are placed between two electrodes. The upper electrode (4) is transparent and can be made of various materials. When a voltage is applied to the lower electrode (5), the differently charged pigments re-organise themselves. Source: E Ink Corporation

DR. WILHELM BAUER
KLAUS-PETER STIEFEL
Fraunhofer Institute for Industrial Engineering IAO, Stuttgart

Internet

- http://communication.howstuffworks.com/instant-messaging.htm
- www.voip-info.org/wiki
- www.businesscommunicationblog.com/blog/?s=business+communication
- www.cisco.com/en/US/products/sw/voicesw/products_category_business_benefits.html
- www.nem-initiative.org

Electronic services

Related topics

- Internet technologies
- Information security
- Software
- Digital infotainment
- Business communication

Principles

The *Internet* is the world's fastest growing marketplace, with seemingly limitless opportunities for marketing products and services. At the same time, progressive technologies are bringing about a transformation from an industrial to an information society. This creates dynamic market environments with low entry barriers, greater market transparency, and a heterogeneous demand for customised products and services. The "e-space" improves customer relationships through lower service costs, interactive and personalised customer communications, speed and accuracy, an enhanced capability to track and measure transactions, instantaneous ways of communicating round the clock (24/7 availability), and the ability to offer different combinations of product and service elements to fulfil customer requirements.

The new media are changing the traditional way of providing services, which is dominated by personal delivery – either face to face or over the telephone. By using new technologies and media, it is possible to deliver customer services with little or no human intervention. Since the Internet supports the automation of customer interaction, the services themselves become the focus of business.

The scope of electronically delivered services (e-services) ranges from electronic provision of traditional services (e. g. investments or airline ticketing), to intelligent interactivity in after-sales product support. The best-known specification of e-services is e-commerce as a commercial customer service. However, e-services cover a broader spectrum of services. They also include non-commercial services (e. g. e-government) and the provision of pure information services.

From the technical point of view, high processing capacities, enhanced bandwidth and the ubiquitous availability of systems and networks act as innovation drivers for e-services. System architectures for e-services are increasingly characterised by modular design.

Tier models provide bases for modular architectures comprising self-contained services. A modular architecture allows modules to be delivered by different sources. This idea can be clarified by the following example, which describes a customer solvency check in an e-commerce transaction. The core business of e-tailers is to sell goods and services. At the end of a sales process, the customer generally has to pay and can often choose between different modes of payment (e. g. credit card, purchase on account, bank transfer, advance payment, PayPal). Besides the possibility of financial fraud, some of these payment modes constitute a greater or lesser risk that the e-tailer will not receive the money due to solvency problems on the part of the customer. Because an e-tailer does not normally possess the knowledge and data to accomplish solvency checks during a clearing process, this task is usually performed by a financial service provider. If the source of the solvency check is self-contained, this module can be linked to the e-tailer's software.

Every application in the e-space has to fulfil general security aspects. In order for the persons and organisations involved to be able to trust the application, the technological system architecture must fulfil the following requirements:

- Authentication: Individuals must be uniquely identified and their identity verified.
- Integrity: Only authorised individuals are able to modify systems, resources, files and information.
- Privacy and confidentiality: Access to information is only available to authorised recipients.
- Non-repudiation: An individual can not repudiate his or her actions.
- Availability: A system has to fulfil its purpose with a given degree of success.

Architectures for the provision of e-services must fulfil these requirements by supporting suitable security methods.

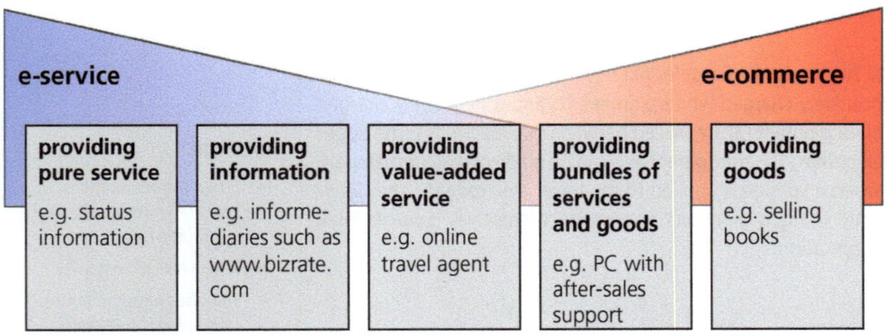

E-service versus e-commerce: While e-commerce is a widely known e-service, e-services also comprise non-commercial services such as e-government. Hence, there exists a broad spectrum of e-services from pure services (delivered free or as part of a service contract) to pure product sales via Internet technologies (e-tailing) with little or no service content. Source: Voss

Applications

▶ **E-commerce.** There are three main types of commercially oriented e-services:

- *B2C – Business-to-consumer*: Online businesses aim to reach retail consumers (e. g. e-tailers such as Amazon.com).
- *B2B – Business-to-business:* Businesses focus on dealing with other businesses (e. g. electronic, web based markets such as eSteel.com or Covisint.com).
- *C2C – Consumer-to-consumer*: Supported by electronic marketplaces, consumers are able to trade with one another (e. g. auction and trading platforms such as eBay).

Since customers and suppliers do not meet face to face, customer service applications in e-commerce are more critical than in conventional sales. By using Internet instruments as recommendation platforms or search engines, customers are able to evaluate and to compare benefits as well as checking out the pros and cons of competing merchants. Since there is no significant switching cost involved in changing one's provider, customers have no inhibitions about selecting a new provider with the click of a mouse. Merchants, on the other hand, find it difficult and expensive to acquire customers and promote customer loyalty.

E-commerce applications are tending to move towards self-service applications (shopping at Amazon.com). The e-tailer supports the customer with presales and after-sales services. Electronic services offered to customers to aid their online shopping process include search support, e-response to customer queries, orders and transactions, e-payment, e-transaction record management, e-assurance and trust, e-help and other online support in the B2C space.

A popular example of B2C e-commerce is Amazon.com. This e-tailer acts predominantly as a goods marketer in e-space. Amazon.com also sells digital goods such as e-books and offers a platform for other retailers. Moreover, the Amazon "marketplace" principle provides a basis for further B2C and C2C e-commerce activities which Amazon.com supports by providing technological platforms and additional services such as a payment service (Amazon payment).

Every time a user visits the Amazon.com page or transacts an order, the e-tailer collects user data. The use of data mining methods allows Amazon.com to make a personalised recommendation each time a recognised user visits Amazon.com (via personal login or via cookies on the user's computer). Once a user is recognised, Amazon.com provides a personal start page with personal recommendations. Moreover, users can customise the services themselves. As an example, a user can choose the menu item "improve your recommendations" to articulate explicit preferences. This method of customisation is called "personalisation by user". Amazon.com also offers recommendations for unrecognised users. Once a product has been viewed, along with detailed product descriptions and pictures, Amazon.com provides various aids to support the decision for a product:

- Frequently bought together: A product bundle is recommended.
- Customers who bought this item also bought: Recommendation for products bought by users with (probable) similar interests.
- Customers who viewed this product also viewed: Recommendation for products viewed by users with (probable) similar interests.

Collaborative filtering systems provide a basis for the personalised recommendations described above. Three groups of recommendation systems exist:

- Content-based filtering: These systems compare and analyse the contents and content-based patterns of documents (e. g. web pages, e-mails) for recommendations.
- Link-based collaborative filtering: These systems use graph-oriented algorithms to generate recommendations. Amazon.com uses an "item-to-item"-based collaborative filter. Rather than using explicit user preferences, this filter takes product similarity as a basis for recommendations.
- Collaborative filtering systems: This group of systems generates recommendations by relating the

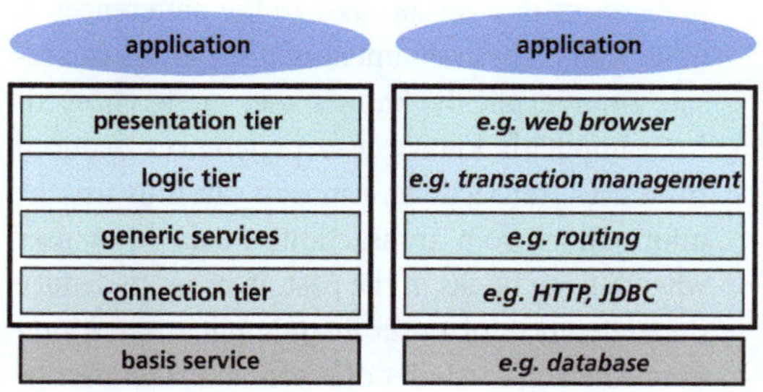

Typical tier architecture: E-services and mobile services can be provided by using modular service platforms. Left: In these flexible modular architectures, the modules are vertically related. The lower tiers are specifically for the technological components of processing, storage and connectivity. Each tier depends on the subjacent tier and contributes to an application on the highest tier. Moreover, the tier architecture allows the disclosure of individual modules. If modules are disclosed, they can be detached from the process and provided externally. Right: In the present case, tiers can be specified as follows: The presentation tier displays content and receives user input. The logic tier or business logic controls an application's functionality in terms of coordination, and comprises transaction management or security functionalities, for example. Generic services may comprise messaging services or routing services. The connection tier serves as an adaptor to attach different types of interfaces and protocols (e. g. JDBC – Java Database Connectivity and HTTP – Hypertext Transfer Protocol). The first-line tier comprises basic services such as databases. Source: M. Huber preferences of a certain user to the preferences of other users. The assignment of preferences can result from explicitly known user preferences or from implicitly known user preferences deduced from user transactions. Amazon can thus use the information from transactions made by a user when buying books in the past, in order to deduce the preferences of a user in an implicit way. Moreover, Amazon can tailor offers for a certain user by comparing that user's purchases with the transactions of other users who have similar preferences.

Doing business on the Web is riskier than doing business with local customers. The use of stolen credit card data, disputed charges, offshore shipping destinations, fraud and the lack of international laws governing global e-commerce problems are just some of the security problems for merchants in the e-environment. Customers take the risk of losing privacy, not getting what was paid for, or even paying but not receiving anything. From the technological point of view, there are three key points of vulnerability when dealing with e-commerce: the client, the server, and the communication pipeline. The concepts of public key encryption are routinely used to secure the communication channels. The most common way of securing channels is through the *Secure Sockets Layer (SSL)* of TCP/IP. The SSL protocol helps to secure negotiated client-server sessions.

To achieve the highest degree of security possible, IT security technologies can be used, but they alone cannot solve the fundamental problem. Organisational policies and procedures are required in order to ensure that the technologies are not subverted. Moreover, industrial standards and government laws are required to support the protected transfer of data, information and goods in commercial transactions.

► **E-government.** Public services are being supported to a growing extent by web-based applications. Moreover, the possibilities of Internet technologies make it possible not only to provide e-public services for businesses and citizens, but also to link applications between different governmental organisations. E-government services can be differentiated as follows:

- E-public services: E-services for citizens or businesses, to date at different levels of maturity depending on the respective council, e. g. online citizens' advice bureaus as information platforms or as partial substitutes for local bureaus offering transactions in digital form
- *E-democracy*, e-participation: E-services for voting
- *E-organization:* E-services to support organisational processes within and between governmental organisations

The key concept of e-government is to relieve the official channels and to accelerate processes in the public sector. E-government thus comprises not only the provision of information for citizens and business via web portals (front office), but also the virtualisation of governmental organisations (back office) and the integration of processes (front and back office).

A widely known e-public service application in both the government-to-citizen and the government-to-business area is e-taxes. Applications of this kind aim to set up integrated processes for tax declarations on the basis of ICT systems that will replace paper-based processes with numerous interfaces between governmental organisations and taxpayers.

Because e-government applications deal with confidential information about individuals and companies, a secure architecture is inalienable. Interoperability is an important aspect in the integrated delivery of e-government services.

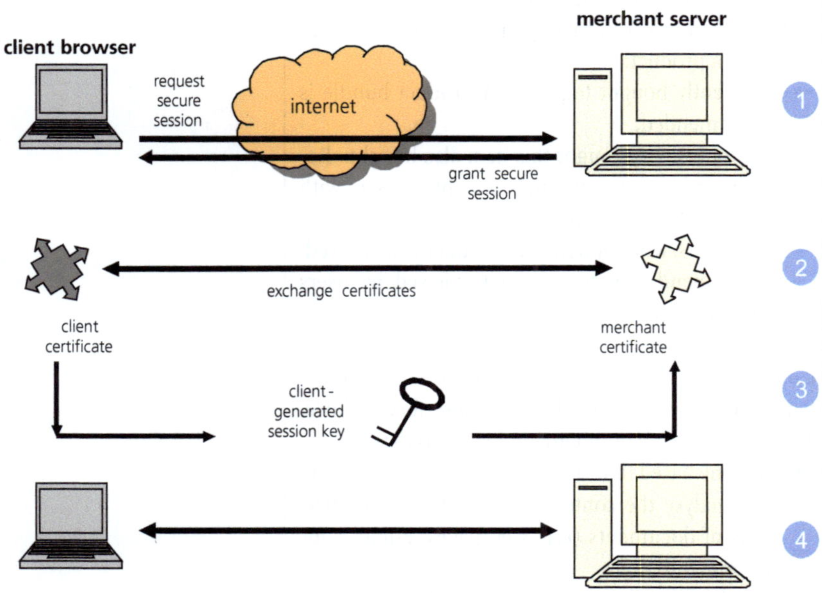

SSL encryption. 1: The customer's computer (client) and the merchant's server (server) negotiate the Session ID and encryption methods. 2: After an exchange of certificates, the identity of both parties is established. 3: The client then generates a session key, also known as a content encryption key. This is used to encrypt the information. The content encryption key itself is encrypted with the server public key and is sent to the server together with the encrypted information. By using its private key, the server is able to decrypt the content encryption key and thus also the information. This procedure is also known as a "digital envelope". 4: Finally, on the basis of the client-generated session key, the transmission of encrypted data for the established session is guaranteed. Source: K. Laudon, C. Traver

Trends

► **E-commerce.** Pertaining to trends in e-commerce, the *web 2.0* paradigm increasingly affects e-commerce services. Customers recommend products and services on specific platforms (e.g. www.epubliceye.com/). In the case of Amazon.com, customers can add "tags" to products. Tags are keywords that categorise a product according to a personal opinion. Besides the product description given by the retailer, tags help customers to search for products with certain characteristics. Moreover, tags help a customer to organise his or her interests by generating lists of similar products. Tags can be combined in "tag clouds" to describe the content of a web page.

Technological prospects, too, play a role in future e-commerce services. The concept of *software as a service (SaaS)* is increasingly expected to support e-commerce applications. According to this concept, applications are no longer locally installed and accordingly do not run on a retailer's infrastructure. Instead, the software is hosted by a service provider via Internet technology. The provider is then responsible for operation, maintenance and support of that application. This enables the retailer to reduce hardware costs. One problem associated with this solution is the interlinking of internal and external applications.

Moreover, new business models are appearing. Besides its core competence, e-tailing, Amazon.com also offers web services and SaaS. The service named "elastic core cloud" allows customers to arrange a virtual IT environment made up of operating systems, databases and applications which are all provided as services by Amazon.com.

Prospects

- As technology continues to progress, service providers will be able to establish new electronic means of service delivery. At present, the field of e-services comprises a broad range of electronically delivered services including both the delivery of information and the sale of physical products using Internet technologies as new distribution channels.

Tag clouds: The Amazon.com tag cloud shows the most popular tags given to the e-tailer's products by its customers, arranged alphabetically. To show the difference in importance of the tags, the more frequently used tags are larger, while the more recent tags are darker. By clicking on a tag, e.g. "biography", the customer sees a list of all products that have been characterised with this tag by other customers. Source: Amazon

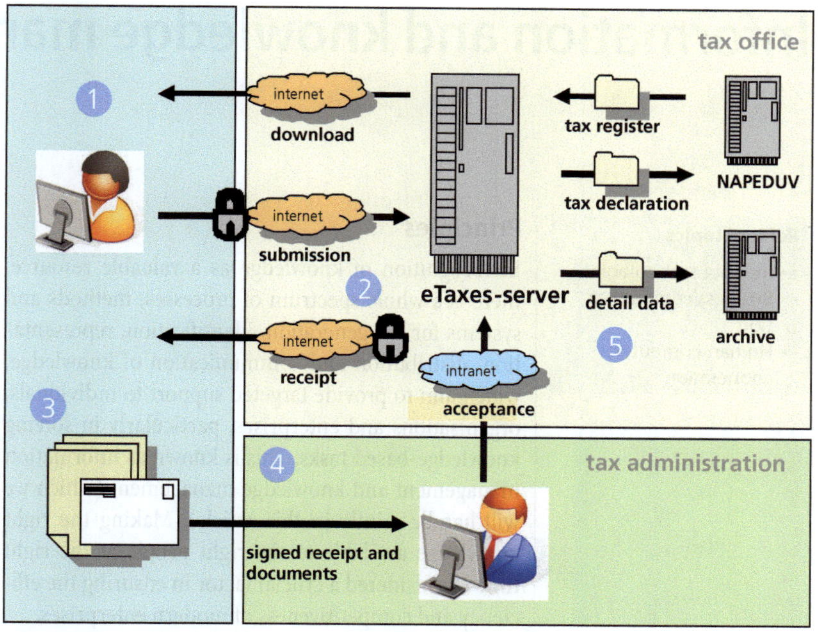

E-taxes process: The e-taxes process can be exemplified by the "eTaxes" concept of the Swiss canton of St. Gallen. 1: The taxpayer downloads the "eTaxes" software, including the tax declaration form, from the cantonal tax office web page. He or she also obtains documents and a registration number, and is sent a password by mail to log on to the "eTaxes" server. The "eTaxes" server also provides default data for taxpayers (e.g. name, date of birth, religion). Once the default data are communicated to the server, they are automatically inserted in the tax declaration. 2: After filling out the tax declaration form electronically, the taxpayer transmits the form to the "eTaxes" server electronically via Internet and receives a receipt. The preliminary tax declaration is stored on the server. 3: The taxpayer prints and signs the receipt and sends it by mail to the local tax administration together with any documents (e.g. invoices). 4: The local tax administration verifies receipt of the documents. After verification, the employee in the administration logs onto the "eTaxes" server and confirms the tax declaration form with the registration number. 5: The confirmed tax declaration is sent to the NAPEDUV application (Natürliche Personen EDV-unterstützte Veranlagungen, meaning "Computer-assisted tax assessments for individuals"). The tax declaration stays in NAPEDUV until the final assessment is generated. The procedures involved are stored in an archive. Source: Canton of St. Gallen

- E-commerce is a widely known e-service application, changing traditional customer interaction and the way in which information, services and products are provided to the customer.
- E-government uses the web-based e-paradigm to facilitate the provision of public services to citizens and businesses as well as advancing the integration of front-office and back-office processes within and between public organisations.
- The Web 2.0 paradigm is playing an increasing role in e-services. New forms of customer participation are being created. Moreover, further technological developments are enabling providers to develop new business models.

PROF. DR. MARKUS NÜTTGENS
NADINE BLINN
University of Hamburg

Internet

- www.ecominfocenter.com
- www.epractice.eu
- http://ec.europa.eu/egovernment_research
- www.egov4dev.org/mgovernment

Information and knowledge management

Related topics

- Internet technologies
- Business communication
- Human-computer cooperation

Principles

In recognition of knowledge as a valuable resource, there is a whole spectrum of processes, methods and systems for the generation, identification, representation, distribution and communication of knowledge, which aim to provide targeted support to individuals, organisations and enterprises, particularly in solving knowledge-based tasks. This is known as information management and knowledge management (which we will handle jointly in this article). Making the right knowledge available to the right people at the right time is considered a crucial factor in ensuring the efficiency and competitiveness of modern enterprises.

Traditionally, knowledge is considered to be bound to an individual where it may be implicit or explicit. Implicit knowledge is often subconscious and internalised, and the individuals may or may not be aware of what they know. It embraces informal skills and competences as well as mental models, convictions and perceptions, and arguably constitutes the largest and most important part of an individual's knowledge. Only a small part of knowledge is explicit, conscious and can readily be communicated to others.

More recently, knowledge is also deemed to implicitly reside in the structures and processes of any organisation, in addition to the collective knowledge of the participating individuals. Consequently, knowledge management in the broadest sense also refers to non-technical means of dealing with knowledge in an organisation. This could take the form of training or suitable communication structures, for example. Finally, knowledge about knowledge – i.e. knowing which knowledge exists, how to find it and how to use it – is referred to as meta-knowledge.

Increasingly, however, we tend to think of knowledge as a resource that can also be represented by a large variety of media. While these were primarily physical in the past, recent advances in information technology have meant that they are now predominantly created in electronic and digital form, consisting of numbers, text, speech, audio files, images or video of a structured or unstructured nature.

Knowledge management in the more technical sense of this article then refers to the methods and systems needed to effectively access these different digital media, to search within them and structure them comprehensibly, to decide which content is relevant for which people and tasks, and ideally to understand the media in a semantic way to be able to make inferences and draw conclusions. These methods and systems are described below. The process of turning human knowledge into physical or digital media, however, is not the subject of our considerations here.

▶ **From physical to digital media and knowledge.** In order for knowledge to be handled by information technology, it has to be captured in digital form. This includes the digitisation of existing physical media, primarily paper (text documents, drawings, descriptions of technical workflows, etc.), and also of communicative events such as presentations, conferences, meetings and telephone calls. This results in raw and unstructured digital representations stored on mass storage devices.

Raw digital data such as paper scans and audio or video recordings can be printed or played back, but have no internal structure that would enable searches or more advanced knowledge management. In the case of scanned text documents, the next basic step in this direction is *optical character recognition (OCR)*, which transforms the scanned text image into computer-readable text. New text documents today are usually created on computers from the outset, which of course obviates the need for OCR. As for drawings, comparable techniques are available only for very spe-

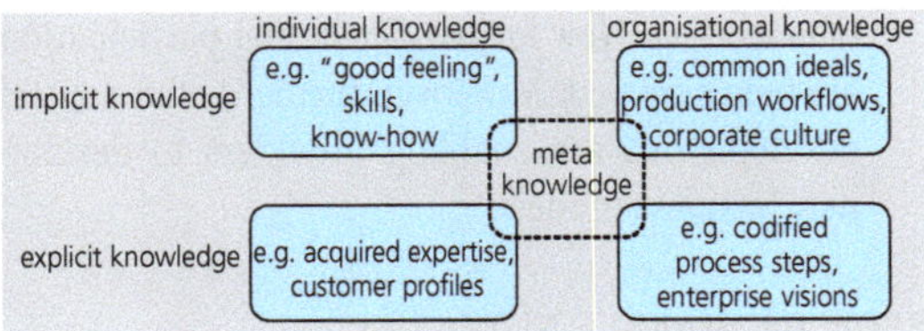

Forms of knowledge: Knowledge within an organisation exists in several different forms, as explicit or implicit knowledge of individuals or of larger teams/organisations, and as meta-knowledge.

cial cases like chemical formulas. Speech recordings can be automatically transcribed, to a limited extent, into computer-readable text, and content can automatically structured.

Once in a structured, computer-readable form, digital text can be searched effectively for user-specified words using search engines. For more advanced knowledge-oriented access, such as determining relevance, classifying, or drawing inferences, the digital documents have to be interpreted and linked. Continuous, to some extent media-specific, enrichment of the digital data with meaning (semantic annotation) generates information that permits statements in a specific context. By combining this information with further facts and background insight, it is ultimately possible to create knowledge that can help to solve problems.

To round it all off, it is necessary to enable other individuals and groups to acquire knowledge, too. This can be done by a number of efficient sharing and structuring systems that flexibly provide the relevant people with access during the problem-solving process.

Applications

► **Search engines.** Search engines permit the retrieval of documents in the World Wide Web (Internet search, e.g. Google or Yahoo) or in the document repository of an organisation (intranet search). After entering one or more search words ("terms"), the engines return a sorted list of documents containing the specified words. In addition, small parts of the documents are shown to indicate the context in which the words were found.

A dedicated web crawler (search program) collects web pages from the Internet, following links to further web pages. Indexed pages are revisited at regular time intervals in order to capture any changes. New web pages are included following explicit registration, or via links on known web pages. The indexer extracts all words from the collected web pages and documents and stores them in an "inverted" file. It stores each word together with references to the documents in which it is contained.

If the term "Berlin" is entered as a query, the search engine extracts the references to all documents containing "Berlin" from the inverted file. In addition, the documents are sorted according to their relevance, and the most relevant documents are presented to the user. Relevance is not only determined according to the number of occurrences of the search term in the document but also on the basis of the structure of links between web pages. A web page is ranked particularly highly if it has a link from another highly rated web page.

While search engines have been extremely successful, their results today are far from perfect, often containing irrelevant documents and missing others. A search engine cannot decide if "jaguar" refers to a car or a wild animal; this would require further annotation and/or semantic disambiguation. Furthermore, it would be desirable for the search to consider not only the exact search term itself, e.g. "computer", but also relevant semantically related terms, e.g. "mainframe", "PC", or "laptop". Such additional services would require further processing, structuring and semantic annotation, as described below.

► **Structured documents and annotation.** Most structured data on computers are databases consisting of attribute-value pairs in tabular form (e.g. name: Merkel; first name: Angela; etc.). Semi-structured contents such as web pages use XML (Extensible Markup Language) elements for partial annotation. The majority of content, however, is contained in unstructured, continuous text documents.

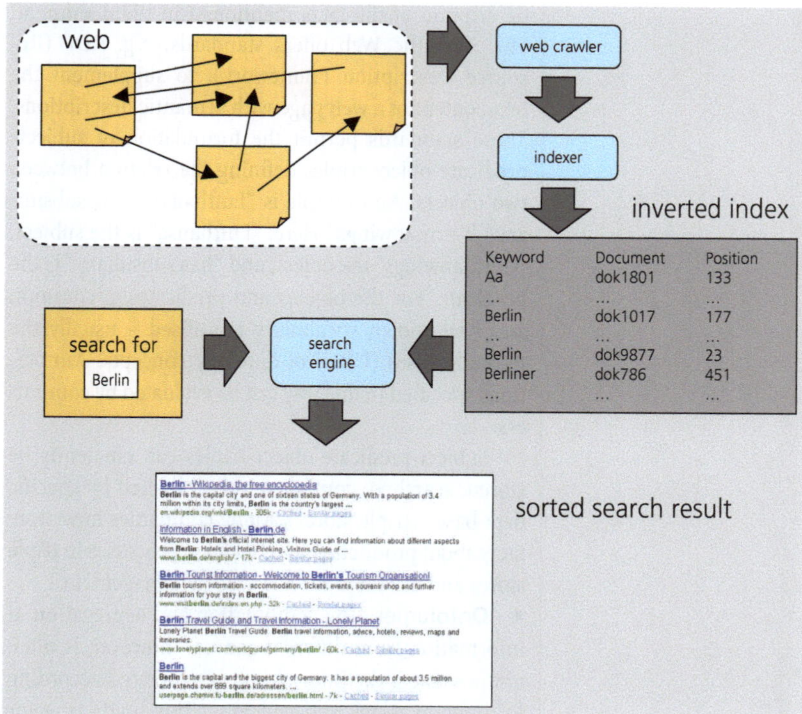

Full text search engine: The index is filled by a web crawler searching the web pages of the Internet. Each word in a document is stored along with its position in the text. For a query, e.g. "Berlin", the search engine compiles all documents containing the search term(s) and outputs a result list sorted by relevance.

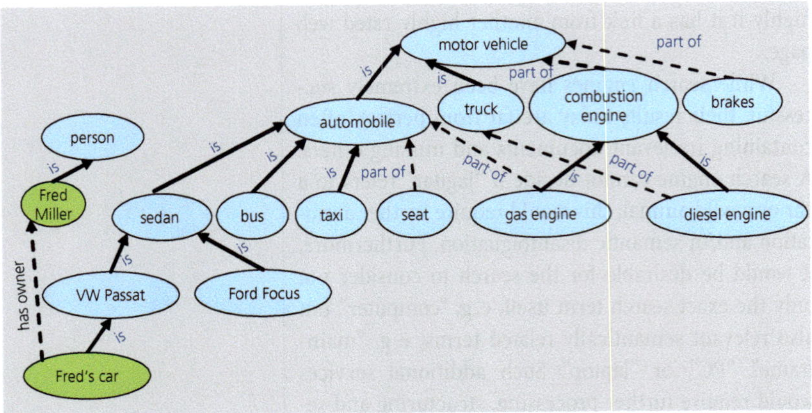

An ontology about the car domain: Concepts (blue) and instances (green) are linked by relations (arrows). By reasoning, new knowledge can be inferred, e. g. Fred's car has brakes.

The Semantic Web is an enhancement of the World Wide Web that allows authors to define the meaning of web contents by providing structured annotations. This can be high-level information about the document (such as the creator, title, description, etc.) or a description of the facts mentioned in the document. The Semantic Web offers standards, e. g. RDF (Resource Description Framework), to supplement the text content of a web page with semantic descriptions. These standards permit the formulation of subject-predicate-object triples defining the relation between two objects. An example is "Lufthansa" "has subsidiary" "Germanwings". Here, "Lufthansa" is the subject, "Germanwings" the object, and "has subsidiary" is the predicate. For the objects and predicates, a common and well-known vocabulary is utilised – usually the web addresses (URIs) of ontology concepts. Annotations specified in this way can be evaluated by computers.

Subject-predicate-object triples can efficiently be stored, searched, combined and evaluated by specific data bases (triple stores). Many companies now store facts about production and business processes in triple stores and use them for knowledge management.

► **Ontologies.** The combination and aggregation of information from different sources, however, is often not possible if the information is annotated according to different terms and concepts. This holds true for both attributes in databases or XML annotations and phrases in continuous text. One solution is to use an ontology with a common vocabulary for the domain of interest. Ontologies not only contain the specific vocabulary used to describe a certain reality, but also include relations between the concepts and a set of explicit assumptions regarding the intended meaning of the words.

The concepts of an ontology may be connected by taxonomic "is a" relations (a "bus" "is a" "automobile"). Terms expressing the same concepts are called synonyms. In addition, there may be instances of concepts, e. g. "Fred's car" is an instance of "VW Passat". Finally, there are relations between concepts or instances, e. g. "gas engine" is part of "automobile". This allows semantic restrictions to be formulated between concepts. Knowledge formally represented in this way can be processed by computers. For example, a machine can infer from the relations "gas engine is part of automobile" and "sedan is an automobile" that "gas engine is part of sedan".

Complex knowledge bases may contain millions of concepts, instances and relations. In this case, "inference mechanisms" are required to draw new conclusions using the available relations. Meanwhile, there are special data bases that permit knowledge to be effortlessly browsed and inferences to be made for given facts.

► **Semantic search.** A semantic search engine does not only retrieve documents containing specific character strings but also analyses the meaning of terms to arrive at highly relevant search results. Search queries are automatically parsed to detect the annotation of words and phrases, e. g. names of people and locations, or ontology concepts. In case of doubt, the user may select one of several alternatives. If, for instance, the query "bank (concept = `geology´), location = `Italy´" is entered, then documents are returned which contain the word "bank" in the sense of `geology´, i. e. river banks, and the name of a location in Italy. In contrast to customary search engines, no financial institutions are retrieved. Note that background knowledge in ontologies may be used during a search. For the above query, an inference engine may deduce that "Rome" is a part of "Italy" and include the corresponding documents in the result set.

► **Text classification.** As described above, search engines enable the retrieval of documents that contain particular search terms, or that have been annotated with particular ontological concepts. In many cases, however, the important question is how to identify documents of relevance to a particular novel task or query that cannot be expressed as a set of search terms or search concepts.

Text classification involves automatically assigning documents to classes on the basis of a class from predefined set of classes. An example is the classification of news stories into the classes "politics", "sports", "economy", etc. The assignment prescription (classification model) is automatically extracted from a set of training documents (training set) which are labeled with the target classes. Subsequently, the classification model can be used to automatically classify large document knowledge bases without human support. The accuracy of modern text classification systems rivals that of trained human professionals.

In recent years, theory based arguments have led to the development of text classification algorithms with impressive classification accuracy. Their starting point is the representation of a document by a word vector containing the frequency of words in a document or their information content (vector space model). As a training set often contains far more than 100,000 words, the word vectors of the different classes form sets in a high-dimensional vector space. The "support vector machine" then uses a separating plane between sets belonging to different classes in this vector space. There are generally many different separating planes. The support vector machine selects the plane that is furthest away from the sets of word vectors. The wide empty margin between the separating plane and the class sets ensures that new word vectors are classified with high reliability.

Sometimes the class sets are not separable by a hyperplane. In this case, the number of word vectors on the "wrong side" of the plane is reduced by penalty terms to yield a reliable separating plane. In addition, there are simple extensions of the support vector machine that allow the separating plane to be replaced by a curved (nonlinear) separating surface.

In order to determine the reliability of a classifier, it is applied to a new set of documents which have not been used during training (test set). The fraction of correctly classified documents can be determined and is often well above 80% for all classes.

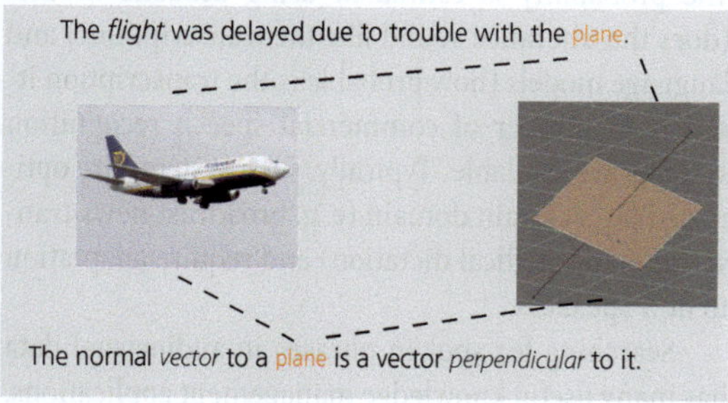

Disambiguation: In order to infer the meaning of a word (e.g. plane), its neighbouring words are analysed. If the word "plane" is close to the word "flight", the meaning "aircraft" is more probable, whereas the mathematical plane is supported by the words "vector" or "perpendicular".

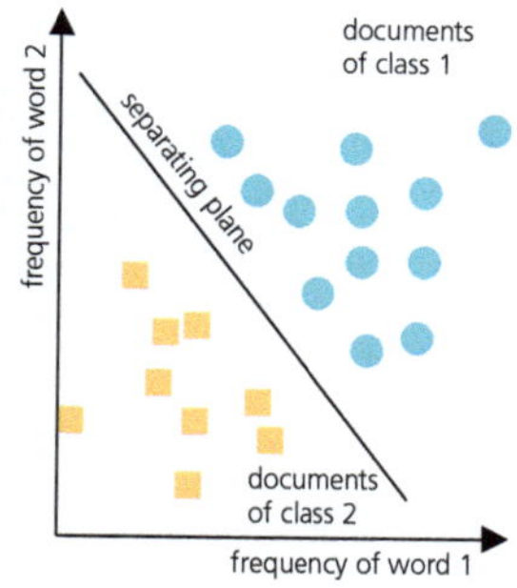

Support vector machine: The training documents are represented by vectors containing the frequency of words. The documents of each class form a set of vectors in a high-dimensional vector space. The support vector machine describes the separating plane between vector sets that has the largest distance to the sets.

Within a company, a small part of the knowledge base may be labeled manually to train a classifier. The definition of classes can be adapted to the requirements of the business processes. The classifier model can then be applied to all documents in the knowledge base. Documents belonging to specific classes and also containing certain keywords can be retrieved using a search engine.

▶ **Disambiguation techniques.** As pointed out above, terms with multiple meanings (such as "plane": aircraft, mathematical surface, tool, etc.) pose special challenges to search engines. Text mining techniques can be used to help disambiguate such terms on the basis of their context, i.e. map them to unique meaning.

A word is disambiguated by employing it together with its properties (e.g. part of speech, capitalisation, etc.) along with the sequence of neighbouring words and their respective properties. Quite simply, if the word "plane" is close to the word "flight", the meaning "aircraft" is more probable. As a variant of classifiers, such algorithms are trained using a training set with labeled entities. Depending on the domain and the quality of the training material, an accuracy of about 80–90% is achieved in many cases.

An alternative is to use clustering methods that group words according to the similarity of their neighbouring words. As a result, the word "bank" is automatically assigned to different clusters depending on whether it is used in the sense of a financial institution or the confinement of a river, etc. These topic models are able to separate terms belonging to different semantic areas without the need for any human-labeled training data.

In a similar way, such techniques can help to handle to cope with "named entities", which are proper names for people or objects that are constantly being newly created in all languages and are therefore difficult to handle. These include not only words but also phrases, e.g. "Fred Brown" or "Escherichia coli". Proper names are often not covered in fixed dictionaries, so have to be used to identify them learning algorithms.

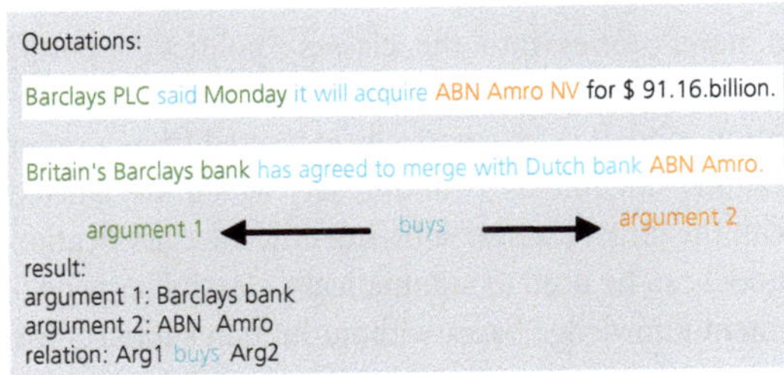

Relation extraction: In the case of a target relation, e.g. "company A buys company B", potential arguments (companies) have to be identified first of all. Then the existence of the relation is inferred by statistical and linguistic analysis.

Trends

▸ **Extraction of relations, opinion mining.** The annotation of text by simple semantic attributes of words and phrases (e.g. personal names, ontology concepts) has been developed over the last few years and can be utilised in knowledge management systems. The research focus is currently on the automatic extraction of complex semantic statements, especially relations between two entities or ontology concepts.

In the field of molecular biology, for example, many hundreds of thousands of research papers are published every year and made accessible in digital form on the Internet. However, due to the sheer number of papers available, even professionals are not able to track all developments. A first step towards automatic exploitation of this knowledge is to identify technical terms, e.g. names of proteins and genes, using the approaches described above.

A second step is to extract relations between terms, e.g. the effect of a chemical (e.g. a drug) on a protein. By combining linguistic analysis (e.g. parsing the syntactic structure of the sentence) with pattern- and statistics-based methods, it is possible to infer a number of useful relations. These methods have a high potential for improving knowledge management in enterprises, e.g. for retrieving patents or analysing insurance claims.

Sentiment analysis, or opinion mining, is of particularly high practical relevance. It aims at automatically identifying the subjective opinion of the author, as expressed in messages from customers, news stories, or web logs, for example. A frequent application is the analysis of customer e-mails in a customer service department. A typical example is "Titan 320 has an unreliable hard disc." In this case, a computer make (Titan 320) and a component (hard disc) are mentioned together with an opinion about their quality ("is unreliable").

▸ **Multimedia knowledge management.** The amount of available audiovisual data has increased drastically over the last few years. Examples include broadcaster archives, Internet video portals such as YouTube, or recordings from meetings or telephone conferences in companies. In order to make this information accessible, it is becoming increasingly important to employ automatic audiovisual analysis techniques.

The techniques described in this section can be used to generate a rich transcription of audiovisual documents, i.e. a transcription that contains both structural information, such as an acoustic segmentation or a speaker characterisation, and information about the actual spoken content. Such rich transcriptions can be used to describe and classify an audio document at a high level of abstraction.

▸ **Audio and speech.** In a preprocessing step, the audio signal can be divided into homogeneous acoustic segments. This will separate not only speech segments from silence, but also label speaker turns or changing acoustic backgrounds (for example if a vehicle drives past a reporter during an outside recording). A speech detection algorithm can separate speech segments. The speech part of an audio document carries a wide range of information such as the actual spoken content, but also information about the speaker.

Automatic speech recognition systems convert spoken words to machine-readable text. Modern speech recognition systems are based on a statistical approach, generating the most probable transcription. The probability is estimated using acoustic models (does the utterance sound like the transcription?) and language models (how probable is the transcription itself?). A number of commercial speech recognition systems are available. Typically, such systems are optimised for a certain domain (e.g. broadcast news transcription or medical dictation) and require adaptation to new speakers.

Searching for spoken phrases in audiovisual data has many useful knowledge management applications. A user of a company archive wants to know what reasons were discussed for introducing a certain production technique during a team meeting, and he may want to locate the corresponding video in the archive. This task is referred to as the *spoken term detection (STD)* problem. A number of reliable detection sys-

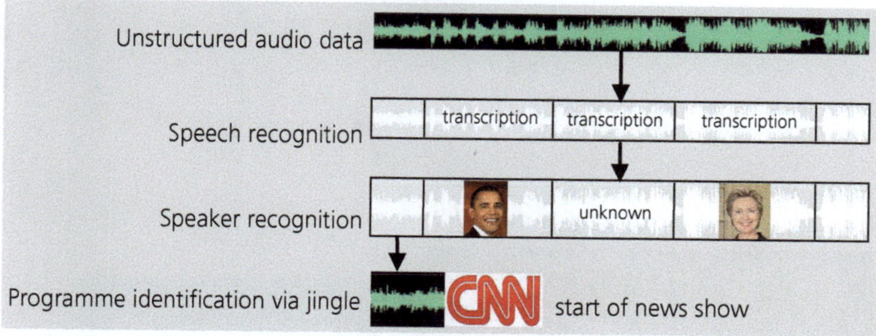

Structural analysis cascade for audio: A common workflow for extracting structural information from audio data. On the basis of the unstructured audio input (sound recordings), more and more complex features are derived in a sequential classification process. Apart from the results of the preceding steps, each extraction method typically requires various low-level features derived directly from the audio data. For example, gender detection is carried out only on speech segments, and additional estimated pitch features are used for the classification.

tems that use words, syllables or phonemes as indexing units were recently proposed.

Speaker characterisation algorithms can be used to estimate the gender or the age of the speaker. On the basis of this speaker information, different utterances made by the same speaker – for example in a discussion – can be labeled using automatic clustering techniques. If prior information about a speaker is available (i.e. if a model was trained using acoustic data from the speaker), speaker recognition can be applied to identify and label known speakers. Acoustic events – ranging from precisely known program jingles (signature tunes) to fuzzy audio concepts such as 'bomb explosion' – can be identified using classic pattern matching algorithms based on low-level acoustic features. Merging audio concept detection with visual concept detection leads to more robust cross-media detectors, which promise to overcome the limitations of mono-media detectors in complex acoustic and visual environments. The resulting structural information can be used to extract further details at a higher level of abstraction. Examples include interview detection, program type detection for TV and radio data or discourse analysis of meeting data.

► **Image management.** An image retrieval system is a computer system for finding images in a large database of digital images. Most current systems add some metadata, e.g. keywords or descriptions, to the images and retrieval is based on these metadata words. However, manual image annotation is time-consuming and expensive, so current research is concentrating on identifying the objects contained in the images of a collection automatically. Ideally, the user poses a query like "find pictures of a cask beer tap". This type of open-ended task is very difficult for computers to perform – pictures of different makes of beer taps may look very different. Current image retrieval systems therefore generally make use of lower-level features like texture, color and shape, although some systems take advantage of very common higher-level features such as faces or buildings. There is a very active research community in this field.

There is great potential in cross-fertilisation between audiovisual analysis and text mining. While the speech recognition algorithms could generate input for the text mining algorithms, text mining results (e.g. from named entity recognition) could be used to enhance the speech recognition process. However, the probabilistic nature of the recognition results poses a particular challenge with regard to text mining.

Prospects

- Knowledge management refers to the spectrum of processes, methods and systems available for generating, identifying, representing, distributing and communicating knowledge for the benefit of individuals, organisations and enterprises.
- In a technical sense, it refers to methods and systems needed to effectively access, search, structure, classify and semantically process digital media.
- Search engines are a powerful means of retrieval based on search words, but they currently lack semantic capabilities.
- The Semantic Web offers formalisms and ontologies for semantic annotation.
- Text mining approaches can be used to automatically classify documents on the basis of examples and automatically provide the necessary annotation for semantic searches.
- New trends include services such as topic and opinion detection and tracking, sentiment analysis, and relation extraction.
- Recent advances have enabled the structured handling of multimedia documents. While speech can be transcribed into text, there is now also basic support for images and video.

DR. GERHARD PAASS
DANIEL SCHNEIDER
PROF. DR. STEFAN WROBEL
Intelligent Analysis and Information Systems IAIS, Sankt Augustin

Internet

- www.kmworld.com
- www.dmoz.org/Reference/Knowledge_Management
- www.kmresource.com

7 MOBILITY AND TRANSPORT

Efficient transport systems and infrastructures are vital elements of a functioning modern economy. Mobility is essential for the well-being of the individual and of society in general as it facilitates social contact and cultural exchange as well as access to goods and services.

The objectives of research in the mobility sector are as follows:

- Transport safety, e. g. airbags and driver-assistance systems in automobiles
- Transport reliability, e. g. fewer breakdowns owing to excessive IT complexity in automobiles
- Traffic infrastructure reliability, e. g. no delays in rail or air transport
- Conservation of resources, e. g. lower energy consumption
- Comfort, e. g. cleaner public transport

Over the past 40 years, traffic volume in the OECD countries has more than doubled, with road and air transportation posting the highest growth rates. At the end of the 1990s, more than 600 million motor vehicles (75 % of which are passenger cars) were registered in the OECD, and the figure worldwide exceeded 900 million.

In GDP terms, the transport sector generates 10 % of wealth in the EU and more than 10 million jobs depend on it. The continuing growth in mobility is putting an excessive burden on the infrastructure, especially for travel by road and air. Although specific fuel consumption (l/km) is falling, more fuel is actually being consumed worldwide owing to the increasing use of transport.

In this context, EU transport policy seeks to reduce the negative effects of mobility by

- Better combined use of transport modes (co-modality)
- The advancement of green drive systems for more efficient energy use
- Wider introduction of urban transport concepts and infrastructure utilisation fees to reduce congestion, pollution and traffic accidents

A European transport network is regarded as being essential if the common single market is to function and the associated freedom of movement for people, goods and services is to be realised.

Over recent decades road traffic has increased sharply in the EU. In the past 30 years the distances traveled by vehicles have trebled. At the same time, the number of private cars has increased. More and more goods are transported over longer and longer distances by road. With the introduction of fees for road use, the environmental and health costs caused by a specific vehicle can be charged to the owner. Intra- and intermodal transport strategies are being deployed with the aim of achieving a significant improvement in traffic flow and safety. In this context, new concepts are also being developed for the collectivisation of individual traffic on the one hand and for the individualisation of local public transport on the other. The aim is to increase the average occupancy (at present 1.2 people) of a motor vehicle in commuter traffic. To encourage car-sharing, advanced I&C technologies are required to quickly synchronise supply and demand through control centres.

Railroad transportation is one of the more environmentally friendly means of transport. However, it has a negative impact on the environment, such as air pollution, noise and land consumption, but to a much lesser extent than for road transport. The biggest challenge for rail transportation is to overcome the different technical structures and inadequate connections between different countries and to facilitate cross-border freight and passenger traffic. Carrying capacity should be improved by modern control and signalling technologies and efficient route management e.g. to pave the way for successful intermodal strategies.

In air transport technical developments and statutory regulations mainly focus on air and noise pollution.

Shipping is one of the more environmentally friendly means of transport with its lower emission of air pollutants and noise. Technological advances are being made in ship propulsion and safety.

Intermodal transport is defined as a transport system in which at least two different modes are used in an integrated manner to form a complete transport chain from door to door. Modern I&C technologies have a key role to play here in optimising the entire system. Intermodal transport can be used both in the freight sector as well as in local public transport, especially in the major conurbations.

► **The topics.** The transport networks of the future will have to provide an optimal combination of all transport modes. The object will be to meet the demand for individual mobility and the transport of goods as economically, efficiently and ecologically as possible. *Traffic management* will seek to improve traffic flows in the existing transport infrastructure and improve the transport systems themselves. On the basis of modern digital technology, sensor systems will be used to observe traffic, communication networks to report events and I&C systems to direct traffic flows.

The key focus of development for the *automobile* will be to reduce fuel consumption. Hybrid systems combining electric power with the internal combustion engine are gaining prominence on the market. In terms of safety, great progress has already been made, as reflected in the steadily decreasing number of traffic fatalities in Germany, for example. Developments will continue and future driver-assistance systems will improve both safety (e.g. night-vision devices) and comfort (e.g. parking aids).

On the rail front, technology needs to be developed at European level in order to move towards standard power supply and signalling systems. For this mode of transport, too, issues of energy efficiency are in the forefront. Further development work will have to be done to deal with the problem of noise. And, of course, the reliability of the entire infrastructure, as reflected in the punctuality of rail traffic, requires further technological and organisational improvement.

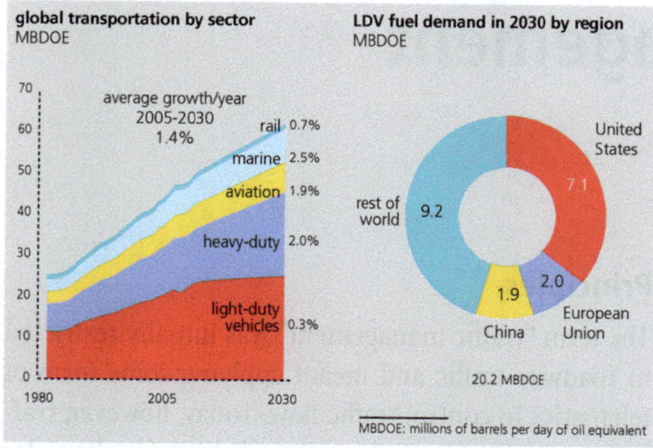

Global transportation. Left: The light-duty-vehicle subgroup – which includes personal cars, light trucks and sport utility vehicles – remains a key user of energy in transportation, though its rate of growth is projected to slow dramatically. In fact, through 2030 growth will be strong in all transportation sectors except for light-duty vehicles, in large part due to measures to increase fuel economy in private automobiles. From 1980 to 2005, light-duty vehicle energy use grew by 70%. From 2005 to 2030, however, growth is expected to be less than 10%, reflecting significant efficiency gains. Right: The United States will account for about one-third of global light-duty-vehicle (LDV) energy demand in 2030, using about 7 MBDOE, which is a decline of greater than 20% from the 9 MBDOE used in 2005. The European Union and China will each account for about 2 MBDOE, and the rest of the world will make up the balance. Source: ExxonMobil 2008

Key aspects of energy-efficient *ship* design and construction include developments in hull shape (upward thrust) and in propulsion (fuel-efficient operation) as well as the optimisation of production costs (group technology methods).

Development work in *aircraft* construction is mainly focused on energy-saving shape and functionality, lightweight construction with new composite and glass-fibre materials, modern electronics and further improvements in passenger comfort. As regards the size of passenger airplanes, the new superjumbo (licensed to carry more than 800 passengers) has taken us to the present limit; any further increase in size would have a severe impact on efficiency.

Space technology no longer attracts as much public attention as it did in the 1970s and 80s. Even if we leave manned space travel to one side, many key aspects of our modern technical world would not be possible without satellites, including telecommunications, earth observation and satellite navigation. ■

Traffic management

Related topics

- Automobiles
- Rail traffic
- Aircraft
- Ships
- Logistics
- Sensor systems

Principles

The term "traffic management" was initially restricted to roadway traffic and meant applying some form of telematics to control traffic flow. Today, however, traffic management is used as a generic term for all modes of transport to describe a wide variety of measures designed to optimise operations for users, operators and others affected by the transport mode. It includes measures on the planning, technical, organisational and legal levels.

The goal of traffic management is to develop and implement systems that enable all types of transport infrastructure to operate at maximum capacity while maintaining service quality, maximising safety, and minimising environmental impact and costs. Different modes of transport emphasise different qualities in their transport management systems: Safety is especially important for guided transport systems and aviation, while road transport often focuses on capacity and traffic flow.

Traffic management is a continuous process of cybernetic cycles. First of all, targets are quantified. Next, traffic management measures are implemented. Thirdly, results are compared with reality and then the cycle begins again. In cases where there are notable differences between targets and reality, additional measures are developed and implemented to achieve the targets.

There are two basic levels of traffic management:

- Strategic traffic management (long-term-oriented): initial state – analyse new demands – plan new or improved infrastructure – design – implementation – operation – adjustment – efficiency control.
- Tactical traffic management (short-term-oriented): existing infrastructure – operation – traffic observation – bottle neck/disturbance detection – intervention – efficiency control.

Strategic traffic management considers system operation in an off-line manner, i. e. its goal is to design new or improved transport systems appropriate for future operational needs to the maximum degree possible. In contrast, the task of tactical traffic management is to develop and implement short-term strategies needed to manage existing traffic demand in a way that ensures maximum capacity use.

Tactical traffic management is based upon the following elements:

- Sensors detect the location of vehicles and their movements.
- Communication networks transfer data gathered by the sensors to control centres.
- Central data systems gather and analyse the traffic data.
- Information networks provide additional data on events that could impact traffic flow (e. g. construction projects, special events, technical faults etc.).
- Control centres use the traffic and additional data to determine what strategies (often from a set of predetermined measures) to implement.
- Information systems transmit the instructions to vehicle drivers and other involved actors.
- Data storage and analysis systems evaluate past events and the effectiveness of strategies implemented.

In addition to these activities, traffic management for guided transport systems must also guarantee operational safety and help support efficient resource management (vehicles, infrastructure and staff).

Applications

▶ **Traffic management on trunk roads.** Roadway traffic management is most often used on motorway networks with high traffic volumes. Often these are located in metropolitan areas. Roadway traffic management systems consist of the following sub-systems:

- Motorway guidance systems to guide traffic along motorway sections. They communicate speed information, warn of congestion and incidents, and control lane use. Their principal goal is to assure the roadway section's operational availability and traffic safety.
- Network management systems to direct traffic to alternative routes when motorways are oversaturated, using variable direction signs and text-message signs.
- *Traffic control systems* include motorway access control devices and tunnel traffic control, amongst other measures. By consulting video surveillance information in tunnels, the operator can manually select preprocessed and standardised operational

states and according signals. Any warnings due to critical events, impaired visibility or high concentrations of exhaust emissions are issued automatically.
- Traffic information systems play a supporting role in traffic management. They communicate information on current traffic conditions and deviation recommendations to drivers.

An essential element of motorway traffic management systems are traffic analysis and forecasting models. These models use real time traffic data such as average speed, traffic volume, and occupancy rate measured by sensors to describe traffic conditions and develop appropriate strategies to control the traffic management devices and provide transport information (e.g. for display on the variable message signs).

Congestion on motorways often forms at heavily used entries. This has recently been counteracted by so-called *ramp metering* which is a part of motorway guidance systems. A traffic light at the entry ramp yields green for only one vehicle (2 seconds) followed by red for a longer period. By adjusting the duration of the red light, the entry flow can be restricted to as little as 200 vehicles per hour according to demand. By controlling the ramp inflow in this way, ramp metering prevents the entry of congestion-causing vehicle platoons and ensures that the given motorway capacity is not exceeded.

This system is only activated in the event of traffic disturbances and whenever a congestion risk persists. The recognition of traffic disturbances is based on the analysis of local traffic flow data aggregated in short time intervals (e.g. 30 seconds). The relevant data consists of traffic volume (vehicles per interval), detector occupancy (percentage of interval, during which the detector is occupied by passing vehicles), and speed (derived from the ration volume/occupancy). The control algorithm detects congestion risks whenever predefined thresholds regarding speed or occupancy or a combination of volume and occupancy have been surpassed or not reached over the course of a few time intervals.

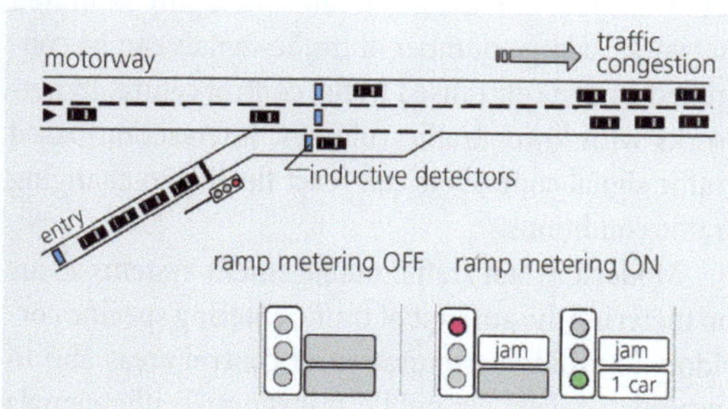

Principle of ramp metering: The system is activated in the event of traffic disturbances that could lead to a traffic jam. A red light causes vehicles to stop at the entry ramp. The light periodically yields green for 2 s only, allowing one single vehicle to enter the motorway. A control algorithm determines the duration of the red light in order to restrict the entry flow according to demand. Controlling the ramp inflow avoids exceeding the given motorway capacity and keeps traffic flowing. Source: ETH/IVT

Inside a motorway traffic control centre: the compilation of current traffic information issued by traffic control centres is largely based on police dispatches and on manual evaluation of video images derived from strategic motorway network monitoring. Such traffic information is usually routed to superior traffic information centres, which then supply it to the media, radio, and GPS devices by means of RDS-TMC (Radio Data System–Traffic Message Channel). Source: ETH/IVT

▶ **Railway traffic management.** Railway traffic management systems have a strictly hierarchical structure and must operate on several levels. The architecture and subsystems of railway traffic management systems are dominated by very stringent safety requirements. An important consideration is that railway system components have very long life cycles, which means that traffic systems are limited by the reduced functionality of older sub-systems.

A key factor in rail traffic management is the train protection system, which transfers movement authorities and speed limits from the interlocking to the train. In territories with line-side signals, the train control systems are used to detect errors by the train drivers (passing closed signals or speeding) and help to avoid accidents. In areas without line-side signals, the train control system provides the necessary data for cab signalling. Depending on the system, data transmitted from the track to the train are communicated either at intermittent points or continuously. The data transfer can be carried out through transponders, electro-magnetic systems or mechanical devices. The functionalities of different train control systems vary and not all of them are necessarily implemented. Possible functions include:
- Automatic warning for trains approaching a closed signal
- Automatic train stop in the event of a train passing a closed signal
- Supervision of the braking curve or speed profile and triggering of an emergency brake if the permissible maximum speed is exceeded

Today, a large number of interlockings at stations and junctions are operated by remote control. The size of these centralised traffic control areas is limited, since a small amount of manual control is still needed in case of disturbances. In Europe, traffic management centers supervise and control trains over networks of up to 5,000 km of track; in the USA, where train density is lower, traffic management centres control networks of up to 50,000 km of track.

At traffic management centres, dispatchers determine how to handle the traffic in the event of a delay, conflict or event. In this way, trains can be assigned with new routes, connections can be broken, or supplementary trains added to the schedule. However, dispatching measures are only communicated and not directly linked with the control or operation of interlockings. Greater delays require affected resources (staff, rolling stock) to be rescheduled and new rosters to be drawn up. Dispatchers are also in charge of alerting the responsible authority in the event of an accident, organising alternative transport (e.g. bus) and providing passengers with appropriate information. Advanced software tools support the dispatchers in all their tasks, including supervision and control of train operations.

▶ **Urban traffic management.** Most large metropolitan areas have urban traffic management systems. They include the following sub-systems:

- Traffic signal controls at intersections
- Parking space management aims to optimise the limited supply of urban parking space. Parking is managed by means of legal regulations concerning access, parking duration, and parking fees or by using planning regulations. Parking guidance systems are used to direct drivers to parking facilities and to distribute automobiles throughout the control area. These systems consist of directional signing showing the number of vacant spaces in parking facilities.
- Public transport system vehicle location and operational control. Most urban public transport systems use their own control systems to supervise and manage operation. Schedule deviations are displayed to dispatchers who initiate actions to address the problem. Man-machine interfaces in vehicles can show schedule deviations, helping drivers better maintain their schedules.

In contrast to motorways, urban traffic flow is influenced by the large number of intersections in small areas, by the interaction between roadways/intersections, and by the variety of transport modes that operate on urban networks (e.g. automobiles, trams, buses, bicycles, and pedestrians). The focus of urban traffic signal control is on optimising intersections.

Traffic signals at intersections are used to allocate temporary driving permission to individual transport modes and traffic streams in a repeating cycle. Traffic signals at several intersections can be coordinated and this coordination can be adjusted in real time or in a fixed manner (e.g. according to the time of day). Certain traffic signals can also be set to give priority to certain transport modes and streams. In highly saturated networks, a large number of traffic signals can be controlled from a centralised traffic control centre. In networks with lower traffic volumes, intersection-based traffic signal controllers can react flexibly to changing traffic conditions.

Modern urban traffic management systems focus on metering the amount of traffic entering specific corridors to avoid oversaturation of control areas and to minimise delays for public transport. Traffic signals can be programmed to give priority to public transport by either advancing or prolonging green phases when detectors identify public transport vehicles nearby. In such cases, the cycle time increases in a traffic-actuated manner. Parallel vehicle streams can instantly benefit from this priority, whereas the affected conflicting streams can receive compensatory green time during later cycles.

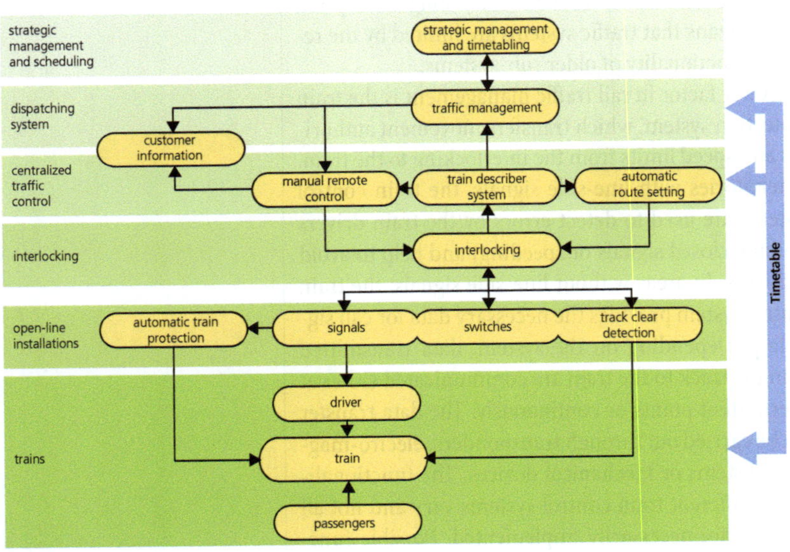

Railway traffic management systems: Railway traffic management consists of the following subsystems: Blocks are segments of rail line between stations used to ensure safe operations on open lines. Routes are segments of rail lines comparable to blocks but within station areas. Line-side signals at the beginning of each segment indicate to train drivers whether they are allowed to enter the section. Train protection systems assure that trains stop when signalled to do so by triggering an emergency brake if train drivers do not take appropriate measures in time. Track clear detection systems report on the occupation state of block or route sections. Interlockings are used to set and lock routes ensuring conflict-free movements for all trains. Remote centralised traffic control centers automatically operate advanced interlockings from large distances on the basis of data from the train describer system. Traffic management centres determine and implement dispatching measures used in the event of a delay, conflict or incident. Passenger information systems provide passengers with information on train operations via acoustic and visual displays at stations, in trains, or on mobile and electronic devices. Source: ETH/IVT

Trends

▶ **European Rail Traffic Management System.** The main traffic management trend in railway systems is to continue developing and implementing the European Rail Traffic Management System (ERTMS). The ERTMS is an integral part of the European Union's technical specifications for railway interoperability, which will enable barrier-free operation of trains throughout the European rail network. ERTMS consists of two parts: the European Train Control System (ETCS) and the harmonised radio system, the Global System for Mobile Communications-Rail (GSM-R).

Different levels were defined for ETCS to adapt the various specific conditions and requirements involved.

Depending on the level, different components are used on the track, at the control centre and on the train, and there are also variations in how the information is transferred. Level 0 describes the actual state, in which no ETCS components are used. Level 1 is based on fixed blocks and real-time information transmission using specific balises and (normally) track-side signals. Level 2 is also based on fixed blocks, but train control information is transferred continuously to the driver's cab by GSM-R. Both levels are in commercial operation, mainly in Europe, and more than 11,000 km are to be operated under ETCS worldwide by 2012. In the future, the Level 3 system will replace the discrete fixed blocks with a continuous block system, which will require an on-board train integrity system. Solutions for highly reliable train integrity systems are very complex and difficult to transfer to composed trains such as freight trains.

▶ **Railway dispatching systems.** A second major trend is the direct transmission of control measures from railway dispatching systems to the interlocking, which minimises the time and effort needed to implement dispatching actions. In the long term, dispatching systems will detect and solve conflicts automatically on the basis of advanced algorithms. Conflict-free schedules that are accurate to within seconds can then be transmitted to driver-machine interfaces (DMI) in the cab, communicating optimal driving instructions. Trains can be controlled very precisely using this integrated real-time rescheduling system. This will reduce the number of unnecessary stops, thereby lowering energy consumption and improving system capacity. In closed systems such as metros, more automated (driverless) systems will be operated.

▶ **Multi-modal control and information systems.** The main trend in urban traffic management systems is the continued development of integrated real-time multi-modal control and information systems. Pilot applications are already in operation, for example real-time information about an intermodal itinerary that provides both roadway congestion and railway delay information. New systems will be deployed that provide users with multi-modal travel information in real time; for example informing users about the availability of park and ride facilities in combination with alternative transport. Intermodal route planners are also under development for freight traffic. Foreseeably, these planners will be able to provide real-time monitoring information on the freight's position and load within the container terminal.

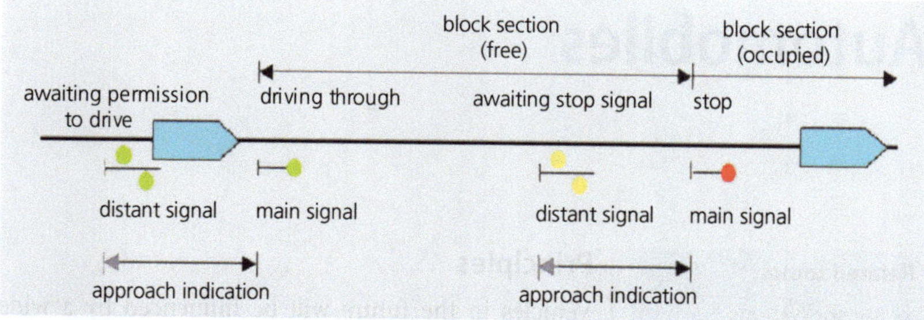

Principle of train separation in fixed block distances with one block signalling: Only one train can use a discrete fixed block section at any given time. Main signals in front of each block inform train drivers whether they are allowed to enter the next block section. Distant signals indicate the state of the main signal in advance and allow drivers to slow and stop safely in front of the next closing signal.

Prospects

- The main traffic management trend in private transport is the increasing development of in-vehicle driver support systems. These systems include new developments such as automated cruise control (ACC), lane keeping systems, object recognition systems, and anti-fatigue driver alarm systems, as well as improvements to well-established systems such as anti-lock braking systems (ABS) and electronic stability control (ESP).
- In the railway sector, improved information technology will be applied to increase long distance control of interlockings, and to improve the precision of train dispatching. A special issue will be to increase European railway network interoperability. Fully automated operations will increase, but only on underground railways and special guided transport systems. Finally, traffic management systems will become much more intermodal – providing comprehensive information and helping to control multimodal transport networks.

PROF. DR. ULRICH WEIDMANN
MARCO LUETHI
PROF. DR. PETER SPACEK
Swiss Federal Institute of Technology Zurich, ETH Zurich

Internet

- www.fhwa.dot.gov/tfhrc/safety/tms.htm
- www.highways.gov.uk/knowledge
- www.ertms.com
- www.uic.asso.fr
- www.ec.europa.eu/transport/rail/interoperability
- www.hitachi-rail.com/products/operation_and_management

Automobiles

Related topics

- Sensor systems
- Polymers
- Human-computer cooperation
- Bioenergy
- Fuel cells and hydrogen technology
- Energy storage
- Laser
- Joining and production technologies
- Optical technologies

Principles

Vehicles in the future will be influenced by a wide range of different customer demands on design, environment, dynamics, variability, comfort, safety, infotainment and cost effectiveness.

▶ **Reduction of fuel consumption.** The power train is the core component of an automobile. The principle behind it has remained almost unchanged since the motorcar was invented. The market is still dominated by internal-combustion piston engines, in which petrol or diesel is combusted. The pressure generated by the hot combustion gases produces a stroke movement of the piston which is converted into a rotary movement by the connecting rod and crankshaft. The power coming from the engine has to be transferred to the wheels; this is achieved by connecting the engine speed and torque via a gear system. The efficiency of internal-combustion engines in converting the mechanical energy generated is about 30%. The rest is lost as friction and heat. Great potential for reducing fuel consumption therefore continues to be offered by the consistent further development of internal-combustion engines to increase their efficiency.

Alternative fuels such as natural gas, liquid gas and biofuels have not yet reached widespread use, but are gaining in importance as oil reserves become depleted. Electrification of the power train is increasing with the use of hybrid systems, but the success of electric drives depends to a large extent on technical advances in energy storage.

Like the drive system, vehicle weight plays a decisive role in fuel consumption. As in the past, present-day vehicles are mainly made of cost-efficient sheet steel and high strength steel. Light metals such as aluminium and magnesium as well as fibre-composite materials are, however, being increasingly used in order to reduce vehicle weight.

▶ **Greater safety.** In the efforts to reduce weight it must be borne in mind that the car body has to meet the constantly rising requirements of passive safety. Passive safety embraces all the structural measures which serve the purpose of protecting the occupants of

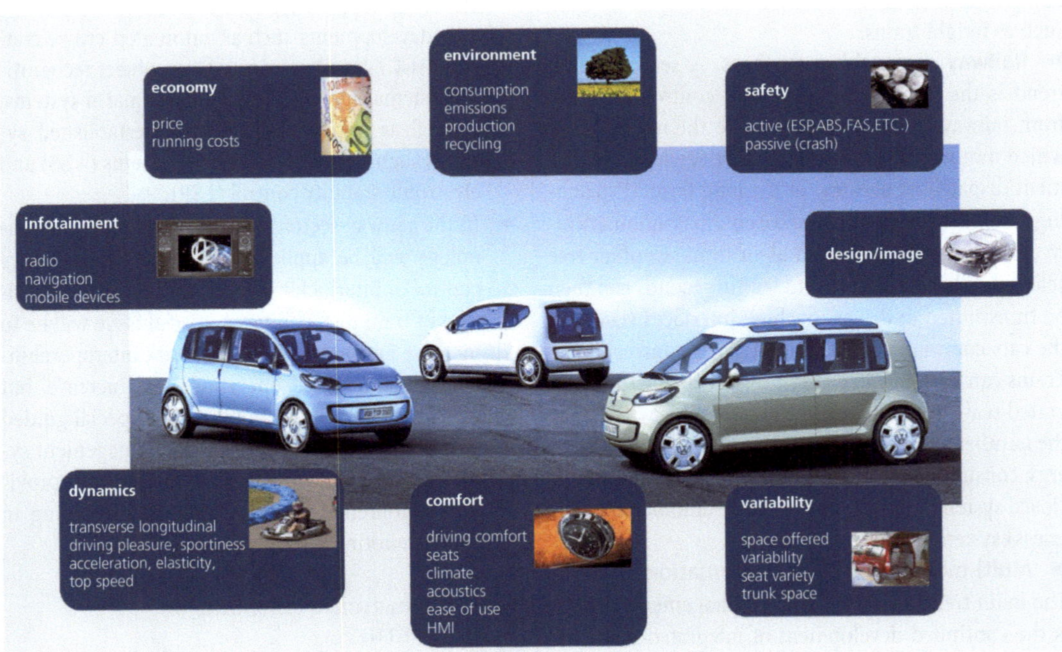

Future customer demands: Tomorrow's vehicle concepts are characterised by a wide range of different customer demands. The models exhibited by Volkswagen at motor shows in 2007 are presented here as examples of new vehicle concepts.

the vehicle from injury in the event of a collision. The most important passive safety features of modern automobiles include the rigid passenger cell and crumple zones at the front and rear of the vehicle, as well as safety belts and *airbags*. Active safety covers everything that can prevent accidents. The most important active safety features include systems which improve driving stability (suspension, curve stability, steering precision, brake performance), perceptual safety (lights, allround vision), operating safety (arrangement of switches and controls), as well as systems to reduce the burden on the driver (assistance systems, noise, climate control). Active safety features, together with elements of passive safety, help to provide optimal protection for all occupants of the vehicle.

Applications

▶ **TSI engines.** The efficiency of spark-ignition engines has been steadily improved in recent years. The introduction of innovative supercharger systems in connection with petrol direct injection and modern downsizing concepts has improved vehicle dynamics and at the same time reduced fuel consumption.

The term downsizing refers generally to a reduction in engine size, leading to a commensurate reduction in mass and friction. This results in lower consumption because such engines can be operated in higher efficiency ranges. Compared with a naturally aspirated engine of the same power, the operating points can be shifted to a higher load, which means that less work needs to be expended for expulsion and aspiration in the charge cycle (lower throttling losses).

The reduction in torque and power is overcompensated by means of one- or two-stage charging.

During charging, pre-compressed air is supplied to the engine. This enables the engine to combust more fuel at the same displacement, and engine performance is increased. A distinction is made between mechanical superchargers and exhaust turbochargers. Whereas a mechanical supercharger is driven, for example, by a belt directly from the crankshaft, an exhaust turbocharger is located in the exhaust train and draws its drive energy from the very hot exhaust gases discharging at high speed. This drive energy is conducted to a turbocompressor wheel which compresses the aspirated air.

In double supercharging the two systems are combined. While the mechanical supercharger provides the necessary air throughput up to engine speeds of 2400 rpm, the exhaust turbocharger takes over upwards of this engine speed. The challenge to be overcome with this combination lies in controlling the two superchargers.

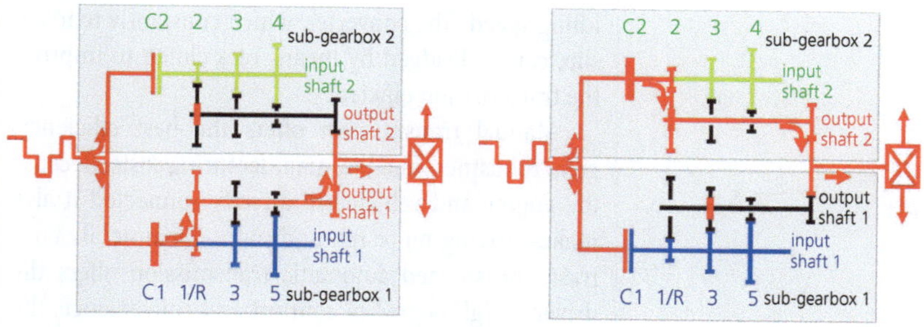

Dual-clutch gearbox: Mode of operation illustrated with reference to the six-gear DSG® from Volkswagen. If, for example, the first gear is synchronised, clutch C1 can be closed so that the power flows to the wheels through sub-gearbox 1 (left). Sub-gearbox 2 freewheels with it because clutch C2 is opened, so that the second gear is already synchronised. The actual shift from first to second gear is effected by closing clutch C2 while at the same time opening clutch C1 (right). The gear change takes place with no interruption of traction.

The TSI engine combines downsizing and double supercharging and therefore constitutes an engine which achieves extremely high performance with better fuel economy.

▶ **Dual-clutch gearbox.** The transmission systems in current automobiles continue to be dominated in Europe by the classic manual shift transmissions and in the USA and Japan by stepped automatic transmissions with hydrodynamic torque converter. Both transmission types possess specific advantages and disadvantages.

Manual shift transmissions in the classic sense comprise stepped gearboxes with 4–6 forward gears and one reverse gear. They normally take the form of spur gears, in which the teeth are constantly engaged. The actual gearshift action is carried out by synchronisations which firstly equalise the difference in rotational speed when the gear is changed and then produce a positive engagement. These transmissions feature a dry starting clutch which has to be actively operated by the driver during the gear change. This causes a brief separation of the power train and an interruption of traction.

Stepped automatic transmissions for automobile applications in the conventional sense are converter automatic transmissions. These are configured as planetary gears and various shafts are braked or coupled by means of brakes and clutches. This makes it possible to have various gears (4–8 forward gears and 1 reverse gear). In the automatic gear change one or several brakes or clutches are activated at the same time. The transitions can be ground to permit a gear change without any interruption of traction. A hydrodynamic torque converter serves as the starting element and is responsible for the characteristic forward creep at idling speed. The converter, which constantly tends to slip, can be bridged by means of a clutch to improve the transmission efficiency.

Manual transmission offers the best efficiency, great robustness and low manufacturing costs. Because the engine and vehicle are directly connected it also makes driving more fun and more dynamic. By contrast, the stepped automatic transmission offers the driver a high degree of comfort and convenience; the car moves off smoothly from a stationary position and the jerk-free automatic gearshifts do not interrupt traction.

In this context, it is logical to combine the advantages of both transmission systems in a new transmission generation, the dual-clutch gearbox. This transmission consists of two partial transmissions combined in one housing. The odd gears and the reverse gear are assigned to partial transmission 1, and the even gears are arranged in partial transmission 2. Each sub-gearbox has its own clutch. This enables the gears to be changed with no interruption of traction as two gears are synchronised and one clutch simultaneously closes while the other opens.

The dual-clutch gearbox can be designed either with wet-running or dry-running clutches. The term wet-running applies to transmissions with oil-cooled clutches, which are better at dissipating friction than transmissions with dry-running i. e. air-cooled clutches. Wet-running dual-clutch gearboxes are therefore mainly used for higher-performance classes of vehicle.

In dual-clutch gearboxes only one clutch is ever closed. In the other a differential speed is always maintained. This leads to greater friction losses in wet clutches than in dry clutches, owing to oil's higher viscosity – compared with air – in the gap between the friction surfaces. Dual-clutch gearboxes with dry friction contacts are used in vehicles of low to medium performance and offer particularly high efficiency.

► **Driver-assistance systems.** In recent years the use of driver-assistance systems in automobiles has steadily increased. Their purpose is to relieve the burden on the driver, produce a more comfortable driving environment and enhance the safety in critical situations. Driver-assistance systems incorporate various sensors which monitor the environment of the vehicle.

Automatic distance control represents a longitudinal guidance system based on speed control. The key component is a radar sensor (77 GHz) with a range of up to 200 m and a horizontal opening angle of 12°. From the signals supplied by the radar sensor the control unit calculates the distance and relative speed to the vehicles in front and, if the road has several lanes, their laterally offset position. If the automobile gets too close to a slower vehicle in front, or a vehicle has cut in the same lane, the automatic distance control slows down the pursuing vehicle by intervening in the motor management and the brake system (if necessary), and thus adapts the speed to the vehicle ahead, all the way to a complete stop (follow-to-stop function). This relieves the burden on the driver and at the same time increases safety.

The *lane-keeping assistant* senses if the vehicle strays from the lane it is travelling in and assists the driver with a corrective steering intervention. A correcting steering action takes place continuously and softly and can be overruled at any time by the driver with just a little force. The system does not of course relieve the driver from his or her responsibility to remain in full control of the vehicle.

A camera built into the inside rear view mirror detects the lane markings and evaluates the position of the vehicle. Beginning at speeds of 65 km/h the system is activated as soon as two clear lane markings to the right and left of the vehicle are recognised.

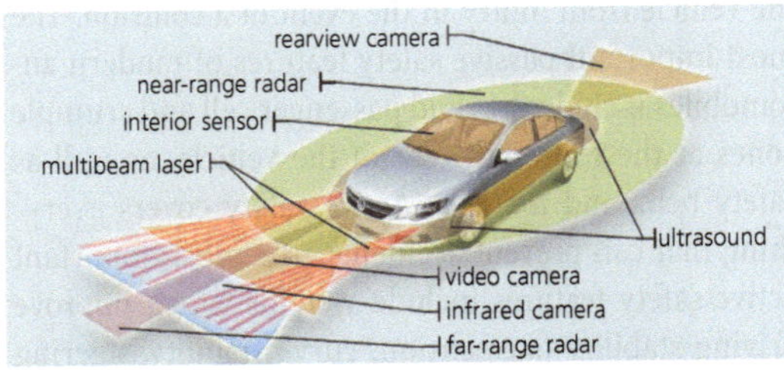

Driver-assistance systems: By combining a number of sensors and camera systems it is possible to monitor the vehicle's entire surroundings without interruption, through both longitudinal and lateral assistance systems.

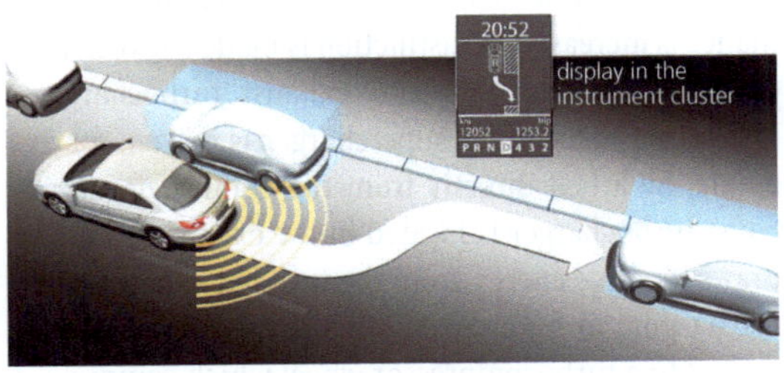

Park Assist: Ultrasonic sensors measure potential parking spaces at the roadside as the car travels past and checks whether they are big enough for the vehicle concerned. During the parking manoeuvre the vehicle automatically steers into the parking space on the ideal line in one single reverse movement. The driver just has to operate the accelerator and brake, but maintains control of the vehicle at all times. If the driver operates the steering wheel, or uses the brake to bring the vehicle to a stop, the steering assistant immediately switches off.

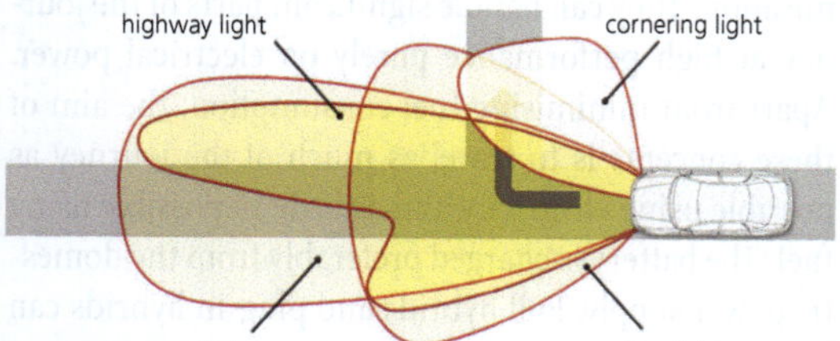

Adaptive light systems: They are able to adjust their illumination of the road dynamically to match different visibilities, driving conditions and traffic situations. Functions available today, such as curve light and masked permanent full beam, will be extended in future by the targeted illumination or suppression of individual objects.

The side assist makes the driver aware of any approaching vehicles before a manoeuvre to change lane. For this purpose, two 24-GHz radar sensors permanently detect the traffic on both sides and to the rear. An optical warning signal in the exterior mirror indicates critical situations, caused for example by the blind spot. The sensors have a range of approx. 50 m so that the driver can be warned even when vehicles are approaching at high speed.

► **Innovative vehicle lighting.** The demand for automobiles which consume less fuel, combined with various special design requirements, are driving the development of car lights in new directions. In addition to the established lighting technologies such as incandescent lamps and Xenon lights, LEDs are being increasingly used as the light source. In many vehicles functions such as parking, brake, indicator and day-driving light are already being performed by LED modules. The main advantages of LED lamps are their lower energy consumption, long service life, faster response time and new design potential. The first vehicles are now being fitted with full-LED headlights. The high light power required for low-beam and full-beam LED headlamps means that a large amount of heat is generated. Fans are used to remove the hot air and direct it past the front of the headlamp. The heat energy removed to the outside is used to demist or de-ice the glass cover.

The maximum connected load of the electrical consumers in the vehicle exterior lighting system can be reduced by up to 75% by replacing the conventional lights with LED lamps.

The present signal profile of vehicles is largely static, i. e. the lights operate independently of the visibility conditions and traffic situation. New systems are being developed, however, which are able to adapt dynamically to the illumination of the road in various driving circumstances. Xenon headlamps with integrated dynamic curve-driving light, for example, make it possible to improve the illumination on bends by up to 90%. From a speed of 15 km/h the headlamp cones move a maximum of 15° in the direction of steering wheel turn and provide a better view of the road ahead.

In combination with a cornering light, the visibility of objects in crossing and cornering situations can be further improved. For this purpose, an additional light in the headlamp is activated when a sufficiently high steering wheel angle is reached, or by operation of the indicator, widening the light cone by up to 35°.

It is also possible to adapt the illumination produced by the vehicle independently of speed and environment. In urban traffic, the headlamp cones are lowered and swivelled outwards to improve the visibility of objects in the near range. On highways and expressways the headlamp cones are swivelled towards each other and moved upwards to illuminate more strongly the road far ahead of the vehicle. In addition, the increasing use of environmental sensors in vehicles enables illumination to be adapted to the particular traffic situation. The full beam can for example be automatically switched off when vehicles are detected in front or approaching from the opposite direction. When there are no further vehicles in the scanned area the full beam automatically switches on again. This offers the advantage that the driver always has an optimal view and oncoming traffic is not dazzled.

Trends

► **Electrification of the power train.** Hybrid drives are drives with at least two different energy converters (e. g. internal-combustion engine and electric motor), as well as the associated different energy storage units (e. g. fuel tank and battery). Hybrid drives are differentiated according to their drive train topology (parallel hybrid, power-split hybrid and serial hybrid) and also according to their functional characteristics (microhybrid, mild hybrid, full hybrid and plug-in hybrid).

In parallel hybrid drives the outputs from both drive systems can be combined more or less to any desired extent. A transmission system transfers the drive power to the wheels. Output-branched hybrid drives separate the energy from the internal-combustion engine into a mechanical and an electrical path. In these concepts, electrical machines handle a large part of the transmission's function. In general torque/speed conversion takes place continuosly. In the serial hybrid, the entire drive output from the internal-combustion engine is converted into electrical energy by a directly coupled generator. As in an all-electric drive, the vehicle is driven by a further electrical machine which draws its energy from the battery and/or generator.

Full-LED headlight: The world's first full-LED headlight is offered as an option on the Audi R8. Altogether 54 LEDs handle all the front light functions: low-beam headlamp, full beam, day-driving light and indicators. The heat generated by the high light power is directed forwards by two small fans to the glass headlamp cover, de-icing it in winter.

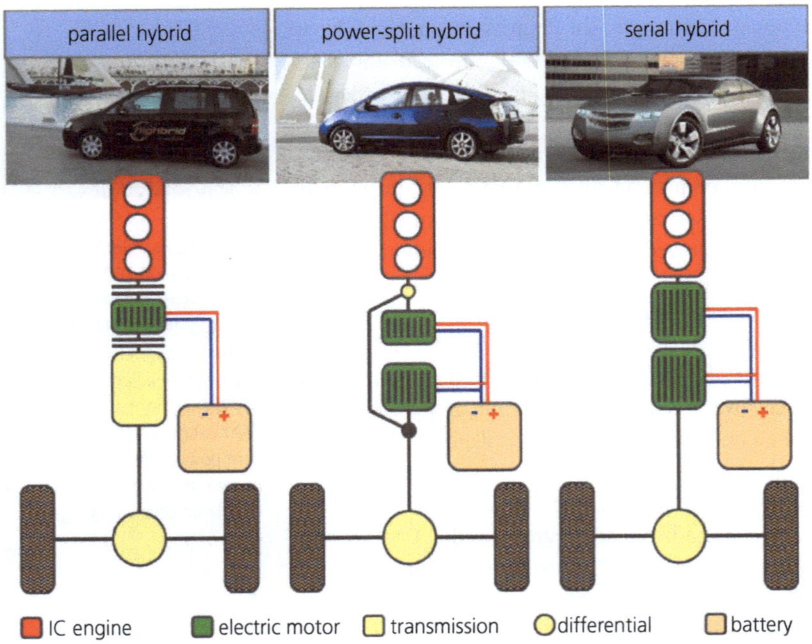

Hybrid concepts: Hybrid concepts are differentiated according to their topology. In the parallel hybrid both the electric motor and the internal-combustion engine act on the drive train, which means that they can be designed with a lower power rating. In the serial hybrid the internal-combustion engine is not connected mechanically with the drive axle and the electric motor has to provide the required torque and power. In the power-split hybrid, part of the power from the internal-combustion engine is transferred to the wheels mechanically and a further part electrically by a generator-motor combination.

The fuel economy of a hybrid drive depends on the efficiency of energy transmission, and in particular on how the hybrid works.

Whereas microhybrids perform the start-stop function with an enhanced starter motor – and in rare cases feature a very restricted energy recuperation system – mild hybrid drives can recover a significant amount of the braking energy. If necessary the internal-combustion engine is assisted by a boost function. Microhybrid and mild hybrid drives are typically designed as parallel hybrid drive types. Full hybrid drives can turn the internal-combustion engine on and off while the vehicle is in motion and thus perform the function of electric propulsion in weak load phases while also achieving even more efficient brake energy recovery. Generally speaking, the electrical components of full hybrid drives are designed in such a way that in typical motoring conditions a consumption minimum is reached which could not be further reduced even by more strongly dimensioned systems. Plug-in hybrids have electrical components that are more powerful than required for consumption minimisation. They can handle significant parts of the journey at high performance purely on electrical power. Apart from minimising fuel consumption, the aim of these concepts is to travel as much of the journey as possible using electricity and as little as possible using fuel. The battery is charged preferably from the domestic power supply. Full hybrids and plug-in hybrids can be produced in any of the three basic structures presented here, whereby the dimensioning of the electrical and IC drive components allow total flexibility in the transition between power sources, extending all the way to an all-electric vehicle.

▶ **Automatic driving**. Automatic driving can be regarded as a logical and consistent further development of driver-assistance systems. The major technical challenge to be overcome is twofold: the vehicle environment relevant for the driving task has to be reliably identified by suitable sensors, and on the basis of this information the correct decisions have to be made.

To achieve complete environmental scanning, different detection methods have to work in harmony: laser scanners scan the stretch of road to be negotiated directly in front of the vehicle and identify any unevenness or dangerous sections up to a distance of about 22 m. Cameras observe the road conditions up to a distance of about 50 m. Information about the road ahead and traffic over a longer distance is provided by radar systems. *GPS* location helps in planning the route to be taken.

All the different systems are operated by computers which merge the sensor data and analyse the infor-

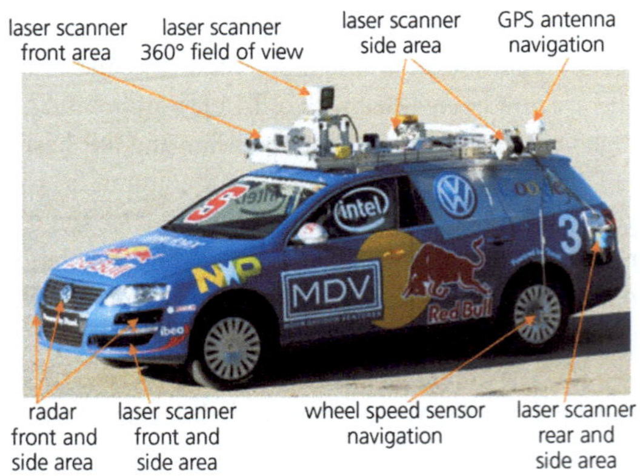

Automatic driving: The VW Passat Junior took second place in the DARPA Urban Challenge in 2007, a race in which the vehicles had to automatically perform driving tasks in an urban environment. The vehicle is fitted with a complex environmental sensor system: Laser scanners to identify the condition of the road ahead and lane markings, a 360° scanner to identify obstacles and other road users, and five 77-GHz radar sensors mainly for use at road intersections.

mation. By combining all the individual pieces of information they produce a virtually seamless picture of the environment. The software used is divided into five modules:

- An interface module receives the data from the environmental sensors and the vehicle.
- A perception module breaks down the environmental data into moving vehicles and stationary obstacles.
- The navigation module determines the route and how the vehicle should approach the journey.
- Control commands for the accelerator, brake and steering are given to the vehicle by a drive-by-wire module, causing it to start off, execute forward and sideways movements or to brake.
- A support module serves the purpose of monitoring the entire system and recording data.

▶ **Car-to-X communication.** Car-to-X communication stands for wireless exchange of information between a vehicle and a communication partner outside the vehicle system. The communication partners could be other vehicles (car-to-car), traffic signal systems or stationary Internet access points.

Car-to-X communication stands for wireless exchange of information between a vehicle and a communication partner outside the vehicle system. The communication partners could be other vehicles (car-to-car), traffic signal systems or stationary Internet access points.

WLAN and cellular mobile radio (for example UMTS) are established and affordable transmission technologies that provide a suitable technological basis for car-to-X communication. WLAN or cellular mobile radio alone each present disadvantages, which can be compensated by combining the two radio technologies. Stable car-to-car applications essentially require that car manufacturers use standardised data transfer protocols and that full infrastructure coverage is provided.

Traffic safety, traffic efficiency (information about the current traffic situation, route planning suggestions etc.) and infotainment (news and media on demand) each represent specific fields of applications.

Traffic safety information can inform the driver about hasards and could support the operation of the vehicle in order to avoid accidents. For example, a vehicle which has accidently stopped just around a bend, or just over the top of a hill, can immediately send a warning signal out to following cars. Environmental monitoring systems and onboard diagnosis modules receive, process and analyse such information. This data can be passed on to following vehicles and be displayed visually or acoustically onboard via navigation systems, or in the combined instrument panel, thus allowing drivers prepare and respond accordingly. Depending on the level of detail provided and the driver-assistance systems installed, the vehicle could also react automatically and avoid the critical situation.

Car-to-X communication: In the future, wireless radio technologies will enable an area-wide exchange of data between vehicles, traffic infrastructure, points of interest (PoI) and Internet access points. This would help to increase traffic safety as well as traffic efficiency while providing local context-oriented information.

To facilitate in-car Internet use, the operability and selection of the displayed content must be adapted to the situation in the vehicle. This might require preprocessing of web content as well as the use of specialised search methods. The navigation system can, for example, provide latest information about the current parking situation, nearby attractions, events, and even customers' ratings of restaurants in the vicinity. Bookings and reservations could be made by voice input.

Prospects

- The desire for individual mobility will remain strong, but efforts to reduce the burden on the environment and resources will play a dominant role.
- Sustainable mobility will require the interplay of different drive systems and alternative fuels, including the use of renewable energy in the form of biofuel, electricity and hydrogen.
- New vehicle concepts will be geared more strongly to the individual needs of customers, such as for example small electric town cars for one or two people.
- The increasing density of traffic will impose a heavier burden on the driver and both these aspects will have direct consequences on traffic safety. Vehicle and driver-assistance systems will be increasingly used, extending to partially automatic motoring in driving situations of reduced complexity.

PROF. DR. JÜRGEN LEOHOLD
DR. LARS GOTTWALDT
Volkswagen AGw

Internet

- www.autoalliance.org
- http://ec.europa.eu/information_society/activities/intelligentcar/technologies/index_en.htm
- www.hybridcars.com/shop-by-technology

Rail traffic

Related topics

- Traffic management
- Energy storage
- Composite materials
- Communication networks
- Power electronics

Principles

The market share of rail is low compared with other modes of transport, but it offers some great advantages which are prompting a renaissance in this means of transportation. Three key advantages of rail transport are outlined below.

▶ **High energy efficiency.** For a distance of 500 km the energy consumption is only 10 grammes of fuel per person/km. There are two (and in certain cases three) reasons for this low emission rate:

- Low friction between wheel and rail, which are both made of steel. The rolling resistance is much lower than for road vehicles (only about 20%).
- Low aerodynamic resistance because of the shape of the train. The cars between the first and last vehicle benefit from the slipstream effect like racing cyclists riding in single file.
- On electrified lines regenerative train braking can save energy by feeding the electricity generated back into the power system. On AC (alternating current) systems this is relatively simple to do, but on DC (direct current) lines special substations are needed to convert power from DC to AC.

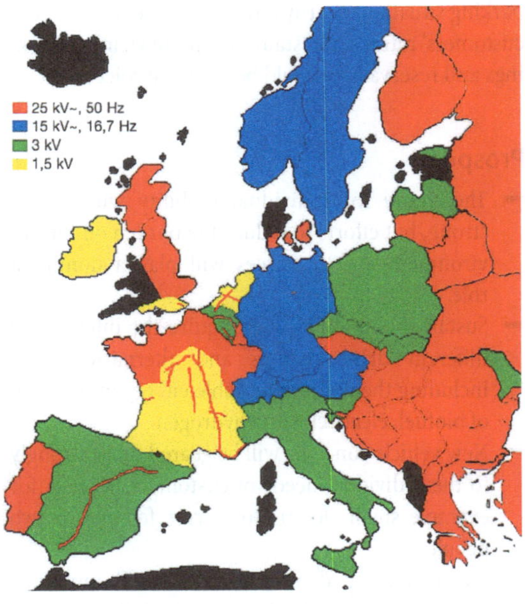

Distribution of the four different power systems: In Europe four different power systems are used on electrified railway lines: 1.5 kV DC, 3 kV DC, 15 kV 16.7 Hz AC and 25 kV 50 Hz AC. This stems from historical reasons and the systems not only vary from country to country but sometimes even within the same country. It would be far more expensive to harmonise the power systems than to use multi-voltage vehicles and so the standardization of vehicle types throughout Europe needs to be considered as an option.

▶ **Low CO_2 emissions.** Thanks to the high energy efficiency of rail transport, low CO_2 emissions are attained. The amount of CO_2 emitted by electrical systems depends on the source of the electricity and can be reduced by using green power (electricity generated by wind or water). Up to now, however, green energy has seldom been used because of its higher price.

▶ **High transportation capacity.** As more and more people live in cities, where space is a scarce commodity, mass transit is becoming essential for a healthy environment. A high-capacity mass transit system with two tracks and a 10-meter-wide traffic route including sidewalks for safety and maintenance can transport 80,000 people per hour in each direction.

In the freight sector, ore and coal trains weighing up to 35,000 t with a 40-t axle load are used on dedicated heavy-haul routes whereas on the European intermodal network the maximum train weight is 6,000 t. In general, on the European continent the maximum axle load is 22.5 t, while on some special routes and generally in Britain it is 25 t.

There are two main reasons for the relatively small market share of rail transport. The first is the high investment cost. Rail vehicles have a service life of about 40 years and the traffic route about 100 years, which means that the investment costs cover roughly half of the life-cycle costs (LCC).

The second reason is that railroads have a rather complex organisation with many interfaces.

Applications

▶ **Electric infrastructure.** For decades the transport policy of the European Union has been to increase the market share of trail traffic because of its environmental advantages. Nevertheless, the volume of rail freight is still comparatively low with a modal split of 10.5% internationally for EU 27 in 2006, compared with road transport's modal split of 45.6%. Because of the strong national roots of the former state railways, the market share achieved by rail in national traffic is generally higher than in international traffic. As the distances travelled in international traffic are, of course, longer than in national traffic, the aim is to increase the volume of traffic in the international transit corridors. The drawbacks are not only organisational but also technical.

The different power systems represent one major technical hurdle. The 1.5 kV DC system is not able to provide high power to vehicles at acceptable cost, so new long-distance railroad lines in 1.5 kV areas like the Netherlands or France south of Paris are in most cases electrified with 25 kV and 50 Hz. This is the newest and most efficient system, as public current can be used without having to be converted.

Different power systems can be used by multi-system locomotives and multiple units. Because of the enormous investment cost, older systems are upgraded to 25 kV and 50 Hz only in exceptional cases.

Public transport systems such as metros and streetcars generally use direct current, as they require less power per train and can therefore be run on a cheaper power system than with AC equipment. They do not need a transformer but the line equipment is more expensive. Because of the high operating frequencies and high vehicle numbers per line length, a DC system is cheaper, whereas for high-speed long-distance traffic the opposite applies.

▶ **European Train Control System ETCS.** Throughout Europe, the train control systems are even more diverse. They are needed to provide safe train control, for instance to prevent a train passing a warning signal by stopping it immediately.

Typically, every country has two or even three different train control systems – an older, a newer and a special system, e. g. for tilting trains. Most of these systems are not compatible with those in the neighbouring country. To equip locomotives and multiple units with all the systems required would not only be extremely costly, but the lack of space for the antennas under the vehicle floor would also be a problem.

For this reason the decision was made about 20 years ago to develop the European Train Control Systems ETCS, which was introduced 5 years ago with the help of the EU. There are two levels available now and a third level is in preparation.

Level 1 relies on the basic principle of railway operation that only one train may be in a specific line section, known as the block, at any time. This block is limited and secured by signals. As a train leaves one block and enters another, two blocks are occupied by one train. The signal is controlled by the switch tower where the safety-relevant mutual dependencies are organised and implemented. The vehicle and its brake are governed by a Eurobalise, which prevents signals being passed at danger. At level 2 there are no signals and the information is passed to the train driver's desk almost instantaneously. This improves line capacity because it enables moving blocks to be realised that cover the braking distance of each train. The train driver is informed about the track ahead well beyond the visible distance, encouraging the use of regenerative braking to save energy (see below). A safe, completely reliable radio link between switch tower and vehicle is essential, and this is provided by the GSM-R system (Global Systems for Mobile Communication-Rail).

At level 3, which is not available yet, no switch tower is needed. The trains organise themselves by communicating with each other. It is not likely that level 3 will be introduced as originally planned, however, be-

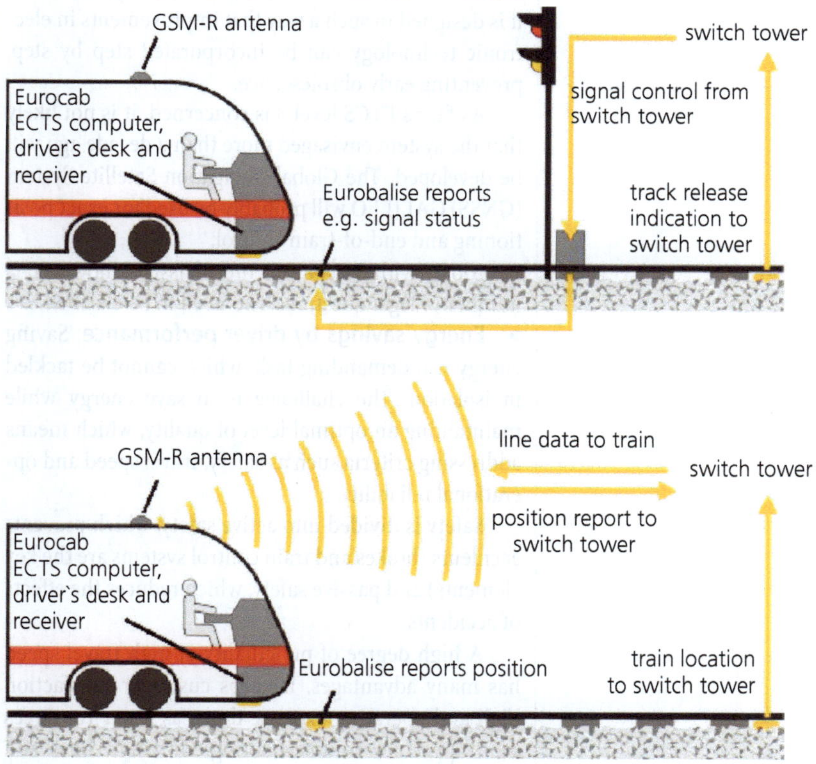

European Train Control System (ETCS). Top: Level 1 still has fixed signals at the end of each block length. A Eurobalise transmits the signal information to the vehicle and brings it to the required speed if the driver does not react. Bottom: At level 2 there are no signals. According to the speed and location more trains can travel along the line. A safe GSM-R radio control link transmits data, mainly the position of each train, to the switch tower. Instructions to the driver are also displayed onboard. The higher performance of the level 2 system means that it is much more expensive than level 1. In both cases, the responsibility for safety remains in the switch tower, which must not fail.

▶ The graph shows a typical speed distance diagram for a public transport system covering the distance between two stops. The two principles applied are shortest travel time and low energy consumption, along with a longer travel time and poor adhesion conditions caused by wet rails. Starting (solid line) always is to be performed with the maximum acceleration available. In poor adhesion conditions less acceleration is possible (dot and dash line). As soon as the power limit of the vehicle is reached, acceleration is reduced gradually (b) as traction force is the result of power divided by speed. If the travel time is intended to be as short as possible, the vehicle should accelerate up to the admissible top speed of the line and stay at that speed (c solid line). If there is some spare time the speed may be a little lower (c dashed line) than the admissible to speed and there may be a longish rollout phase without traction force (d). If there is no surplus time the rollout phase may even be reduced to zero. To save energy, soft braking with regenerative braking must start quite soon (e dashed line), whereas strong braking with all brake systems – mainly thermal braking as disc braking or electromagnetic rail braking – provides stronger deceleration and therefore the shortest travel time (e solid line). In poor adhesion conditions very strong braking is not possible and so the driver has to start braking a little bit sooner (e dot and dash line). If the time buffers are not needed for unexpected events or to compensate for poor adhesion conditions, they can be used to lengthen the travel time in order to save energy (dashed line).

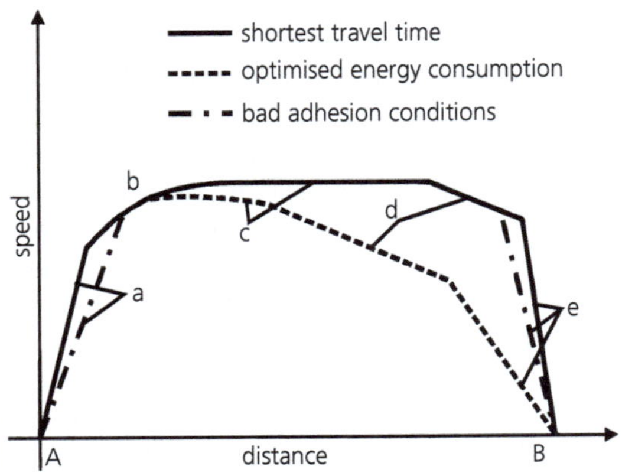

cause it fails to utilise the advantages of modern satellite navigation systems.

With its features ETCS will be more than just an additional European system on top of the existing national ones. It has an open standard, so that competing industries can offer the same functionality. Moreover, it is designed in such a way that improvements in electronic technology can be incorporated step by step, preventing early obsolescence.

As far as ETCS level 3 is concerned, it is not likely that the system envisaged more than a decade ago will be developed. The Global Navigation Satellite System (GNSS) GALILEO will probably be used for exact positioning and end-of-train control.

Along with various technical issues, however, a number of legal questions also need to be clarified.

▶ **Energy savings by driver performance.** Saving energy is a demanding task, which cannot be tackled in isolation. The challenge is to save energy while maintaining an optimal level of quality, which means addressing criteria such as safety, travel speed and operational reliability.

Safety is divided into active safety, which prevents accidents (brakes and train control systems are the key elements) and passive safety, which reduces the effects of accidents.

A high degree of punctuality at high travel speed has many advantages. It keeps customer satisfaction high and operating expense down as fewer resources are trapped for troubleshooting. On single lines and strongly crosslinked networks, high punctuality is essential.

To achieve operational reliability, it is not possible to construct timetables on the basis of shortest travel times. Some time buffers must be reserved for unforeseen disruptions caused, for instance, by poor adhesion coefficients, unexpected stops on the line or station delays. These reserves can be used for saving energy.

Trends

▶ **Rolling noise reduction.** The big environmental challenge for railroads is to reduce the noise they cause. The noise source differs according to type of vehicle and operating conditions. It may derive from power equipment such as the diesel engine or gearbox, auxiliary equipment, such as compressors for air-conditioning, or rolling noise and wheel squeal. Today all noise sources except for rolling noise, can be reduced significantly. Even with smooth wheel and rail surfaces as well as wheel damping, the best values achievable today still exceed the admissible noise emission values in critical situations of 10–15 dB.

One solution may be to insert a rubber layer in the head of the rail to reduce the vibrations in the rest of the wheel and the tie. A theoretical and experimental feasibility study has proven that this can reduce noise by as much as 12 dB, but huge efforts would be needed to develop a rubber-layered rail with the required safety level for practical application. Questions remain with regards to durability, failure behaviour, current transmission and weldability between rail segments

▶ **Increasing energy efficiency.** There is no straightforward way of reducing energy consumption. In each case the most-efficient priorities have to be set to meet the respective constraints. Several issues will therefore be addressed here and the most relevant aspects can be selected for the specific situation at hand.

Many ideas have not been developed further because of cost:

- Equipment installed to raise power efficiency increases weight and therefore operating costs as well as vehicle investment and maintenance expense. These additional costs would have to be offset by the future savings from reduced energy consumption.

- The biggest gain is achievable in the use of brake energy (commuter traffic, high-speed traffic and operation on steep lines).

There is a huge difference between electrical systems powered from an overhead wire or third rail, which can handle a very sharp burst of high power, and autonomous systems like diesel traction. For diesel, the big question is what can be done with the huge waste power that is briefly available. The problem known from the automotive sector is in principle the same but the quantities are much larger as brake power may exceed several megawatt. On hydraulic drive systems it can be transferred to hydro-gas accumulators.

On diesel-electric systems it is possible to charge batteries or capacitors. Several attempts have been made to use flywheel storage systems but they have failed to reach an adequate level of efficiency.

On electric locomotives, brake power is typically limited to values of around 150 kN, whereas the traction force reaches values of up to 300 kN for a four-axle locomotive. The reason for this difference is the potential derailment of a two-axle car behind the braking locomotive because of lateral forces that may occur in a change of tracks.

For the traction phase too, there is a big difference between electrical and diesel systems. The efficiency of the traction unit itself is around 80–90% for electrical systems (efficiencies in power generation and transmission vary between 30–90%) and is typically around 30% for diesel systems. Diesel is therefore much more the focus of attention. Energy from the exhaust fumes and the cooling systems may be used for optimisation, for instance using a steam cell.

Unfortunately, the difference between the energy available from full power and idling is very big and these are the two situations that mainly occur in operation, whereas partial load happens much less often than in the automotive sector.

The key issue for diesel engines is, of course, their emissions. Larger *diesel engines* in particular (typically 800 kW to 4.5 MW) still have a huge need for improvement. Railroad electrification and the use of green energy must therefore be increased in future.

The energy needed for cooling and heating of the passenger cars and the driver's cab will become more and more relevant.

The present practice of using fresh electric power for heating will be replaced and this holds the greatest potential for saving energy.

The use of heat pumps for cooling and heating can reduce energy consumption by up to 70%. The heat pumps designed for cooling, however, tend to suffer from icing when used for heating. They also need to operate at a high volume and are very heavy.

▶ **Reduction of train resistance.** Two areas are currently attracting particular interest – high-speed passenger and freight. The maximum speed for scheduled high-speed passenger services is 350 km/h (e. g. Madrid-Seville and Madrid-Barcelona), whereas the world speed record achieved in 2007 with a modified TGV-Duplex (Train à Grande Vitesse) (5 car bodies with 19.6 MW power) on the French LGV-Est (Ligne à Grande Vitesse) is 574.8 km/h.

The key question connected with introducing an operating speed of 400 km/h on wheel-rail systems is whether the energy costs are viable. The propulsion resistance of the pantographs, for instance, must be reduced. The present design incorporates single-arm pantographs with lots of struts and simple sheet metal plates for contact force adjustments as a function of speed. These induce considerable turbulence. New designs such as telescopic pantographs with active adjusted contact force cause much less turbulence and therefore far less aerodynamic drag. Aerodynamic noise from the turbulence is also reduced.

As the energy at 400 km/h is the same as a free fall from a height of 630 m the active safety measures such as wheel integrity surveillance (making sure that no dangerous wheel or axle cracks exist) and stability surveillance (making sure that no unacceptable oscillations occur in the wheel-rail contact that may lead to heavy wear or even to derailments) are essential. The windscreen has to be designed and tested to withstand the huge impact of a 1 kg aluminium ball traveling at a speed of 160 km/h, which means 560 km/h for a 400 km/h train.

As far as freight trains are concerned, running resistance on curvy tracks and air resistance of, for example, fast-running container trains are also topics for the future.

Running resistance on curvy tracks can be reduced significantly by radial steering bogies such as the so-called Leila. The radial steering is achieved by rolling radius differences between the outer and inner wheel in a curve. Because of the fixed installation of the wheels on the rotating axle shaft, the rotational frequency of both wheels is identical. But the shape of the wheel profile increases the rolling diameter the closer the contact point is to the wheel flange. As a result, the wheel on the outer side of the curve travels a longer distance than on the inner side and steering occurs. However, it is essential not only that the wheelset can

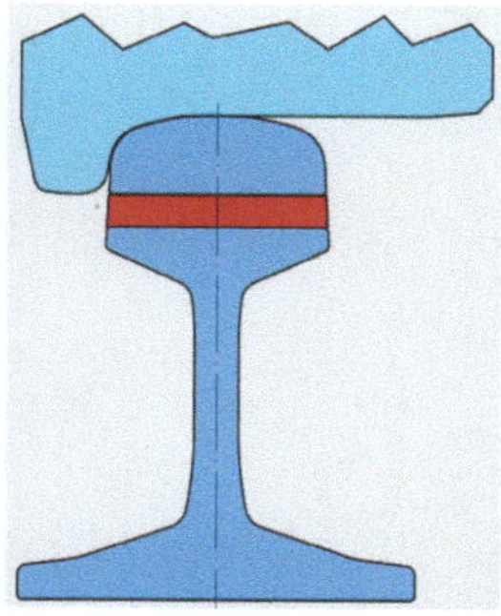

Quiet rail: A rubber layer in the head of the rail reduces rolling noise significantly. As the mass of the wheel is much bigger than the section of rail above the rubber layer the wheel dominates the displacement of the rail head. The lower part of the rail and the ties are insulated by the rubber layer. Compared with the conventional system, the vibrations of the wheel, of most of the rail and of the ties are reduced and less noise is emitted. The vibrations in the head of the rail are increased, but as its surface is relatively small it does not contribute significantly the amount of noise produced.

steer, but also that the bogie design enables the axle boxes to move back and forth.

Although the required displacements in the axle boxes measure just a few millimetres, it is not easy to manage this dynamically, i. e. to avoid unwanted oscillations or instability when running fast on a straight track. Leila bogies solve this problem by having crosslinks that couple the steering mode of the leading and the trailing axles to anti-phase. If instability occurs, the phase between the movement of the leading and the trailing axles is typically around 90°. Anti-phase through the crosslinks provides 180°. In this mode one axle supports the other to avoid instability but does not impede radial steering in curves, where anti-phase movement between the leading and trailing axles is ideal.

► **Streetcars.** Streetcars are enjoying a renaissance world-wide. Compared to buses, they are much more energy- and space-efficient. Studies show that in situations where streetcars are used under the same conditions as buses, the acceptance of streetcars is significantly higher, some 20% and more. Infrastructure costs are also a lot lower than for metro systems. Using automatic guidance, streetcar trains can be formed that are far longer than buses (up to 60 m with one articulated vehicle). This gives a much greater capacity than a bus system and employee productivity is higher too. On the negative side, catenaries can be a problem in picturesque old cities, as can electromagnetic interference in sensitive areas, for instance near research institutes.

Using power cups (capacitors) several kilometres with a few intermediate stops can be covered without an aerial contact line. This equipment can also be used to store brake energy in sections with a catenary, if the catenary is not able to take additional energy because it is already carrying too high a voltage.

► **Flexible freight push-pull train.** In passenger transportation major improvements in productivity (10–25%) were gained with the introduction of push-pull trains that have a locomotive permanently at one end of the train and a car with control cab at the other. The locomotive pushes in one travel direction and pulls in the other on the return journey.

Attempts to copy this in the freight sector have failed up to now as a car with a control cab is too expensive. Instead, the new concept of a flexible freight push-pull train provides for a detachable freight cab, which can be attached to any freight car fitted with an automatic coupler.

For 40 years all European freight cars have been equipped and adapted for central couplers. The cab allows freight trains to run at full speed (typically

Leila (light and low-noise) with crosslinks for radial steering. Top: View from below. Bottom: Overall view. The crosslink equalises the steering of the leading and the trailing axles and stabilises movement at high speed. Apart from improving energy efficiency, this bogie also reduces noise significantly thanks to a special low-noise wheel design, damping in the high-frequency range with rubber springs and hydraulic instead of friction damping. Reliability is also improved over conventional designs by reducing the number of friction surfaces. Source: Josef-Meyer Transportation

100 km/h and sometimes 120 km/h) cab forward with full train control features.

The standard diesel or electric locomotive is operated by safe remote control and application of the emergency brake is governed by the train pipe.

For operational reasons, the remote control connection must, of course, also be safe. To avoid any disruptions, the locomotive must be able to be started from the cab regardless of its position. With the huge ferromagnetic obstacles encountered on the railroad, radio shadows are not completely avoidable. Multiple antennas and multiple frequency bands for the control operation solve this problem by providing alternative systems that have different radio shadows.

The cab has its own energy source for control purposes and driver comfort. Present and future safety requirements are met, including measures to prevent accidents, such as a train control system and warning horns, as well as measures to reduce the impact if an accident occurs. The latter comprise anti-climbers to prevent vehicles from overriding, deformation elements and driver airbags. On conventional locomotives with end cabs the driver can take refuge in the machine compartment in the event of a collision. Here the same safety level must be reached with a rigid cab that provides survival space and airbags that soften the impact on the human body. With these measures even severe accidents can be survived with minor injuries.

The advantages of the system are that it shortens train reversing time significantly and reduces the infrastructure required.

These advantages really come to the fore on modern freight cars carrying containers, timber, powder, mineral oil, chemicals, automobiles, automobile parts, coils, ore, coal and pallets of any kind, which can be loaded and unloaded very quickly.

In a growing market where infrastructure construction is lagging behind because of cost and lack of space the reduced amount of infrastructure required for this system is a major advantage. No locomotive reversing tracks or additional switches are needed and the loading tracks can be short.

For long-distance travel train coupling/train sharing can be used. This increases line capacity and one driver can control several units with multiple control functions requiring just a little more functionality than for remotely controlling one locomotive from the cab. As train coupling and sharing takes only a short time it can generally be carried out in passenger stations when the track is not needed for passenger trains.

Prospects

- Over recent years the modal split has changed to the advantage of railroad systems. This process will continue but developments are still needed to improve safety, reliability and environmental compatibility.
- The question is not whether rail travel is better than another mode of transport, but whether the negative impact of traffic on the environment and the quality of life can be reduced. Railway systems pose a challenge in this respect because the technical solution alone is not a guarantee of its success. Many organisational interfaces also need to be addressed and improved.
- Passenger and goods traffic is paid for on the basis of the distance covered and expense increases mainly according to time, whether the trains are traveling or standing still. Enormous efforts are therefore being made to increase travel speed and vehicle operation and reduce standstill times.

Electric streetcar with power storage device. This streetcar features an inductive power pickup under the floor. On specially equipped lines power can be received inductively, so that catenary wires can be dispensed with in sensitive areas such as picturesque old cities. Because of the lower efficiency compared with the pantograph and overhead wire system this method of power pickup should only be used over short distances. An intermediate power storage device on the roof with power cups is also needed to equalise the fluctuating power requirements of the drive system. Source: Bombardier

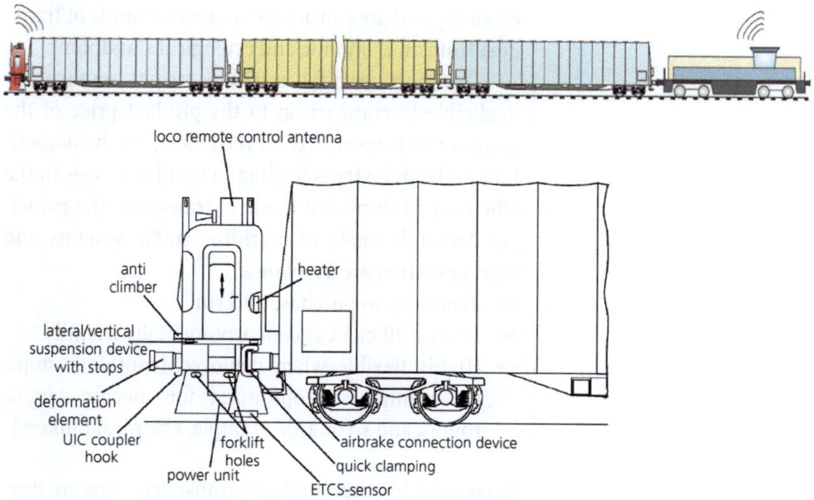

Freight push-pull train. Top: The whole arrangement. A cab is attached to an ordinary freight car with a bar inserted in the space for a central coupler, for which all European cars are equipped. The locomotive is operated by safe remote control backed up by the existing direct pneumatic emergency brake line. In the locomotive forward running direction the cab is shut down to the functionality of an end-of-train device. Bottom: The elements of the cab which are needed for full-speed mainline operation. UIC horns are fitted as a warning device, along with windscreen wipers for rainy conditions, ETCS sensors for train control purposes, as well as a heater and air conditioning for the cab. Lateral and vertical suspension devices provide vibration comfort. Stops limit deflection in case of an overload. The cab has its own auxiliary power unit. Remote control antennas are fitted for communication with the locomotive at the other end of the train and with switch towers. In the event of a crash, anti-climbers prevent vehicle overriding, and deformation elements reduce the impact on the cab. The cab can be placed in position by means of a container crane or forklift. By using quick-clamping and airbrake-connection devices it takes just a few minutes to attach the cab. If another train is attached in the case of train coupling/train sharing a UIC hook is used.

PROF. DR. MARKUS HECHT
Technical University Berlin

Internet

- www.vdv.de
- www.fav.de
- www.allianz-pro-schiene.de
- www.era.europa.eu
- www.unife.org
- www.railteam.eu

Ships

Related topics

- Joining and production technologies
- Logistics

Principles

The construction and operation of ships is one of the oldest large-scale technologies in human history. Today, ships form the backbone of the globalised world economy and are considered a superior mode of transport from an economic and ecological standpoint.

Transport costs for maritime freight are virtually negligible in comparison to the product price of the goods to be transported. Nevertheless, harsh competition in the industry is leading to rapid advances in the efficiency of ships as a mode of transport. The principal demands made of maritime traffic systems and their operation are as follows:

- Economic to purchase and run
- Energy-efficient and environmentally friendly
- Highly flexible when deployed (container ships, for example) or optimised for specific cargos, routes, and so on (for example, chemical tankers)

In contrast to other modes of transport, ships are usually unique, in that they are designed for very specific transport tasks.

Today, one distinguishes between merchant ships for carrying freight and passengers, special ships, such as fishing vessels, ice-breakers and even floating offshore structures as well as naval vessels for use above or under water.

Applications

▶ **Container ships.** These vessels carry normed containers. Container ships are dimensioned for specific routes with regards to sailing speed, the number of containers by size, type and weight, as well as the peripheral handling equipment for their loading and unloading. The external form of these vessels is a compromise arising from the requirements for maximum load and minimal flow resistance. Many container ships carry refrigerated containers, though those containers mostly have their own cooling unit. An alternative solution is to arrange for additional cold storage below deck.

Notable features of so-called "post-Panmax ships" – vessels that are unable to traverse the locks of the Panama Canal – are a load capacity of over 10,000 twenty-foot *containers* (TEU) and high speed. These workhorses of modern container traffic can only call at very few of the world's harbours.

The focus on a few harbours for container traffic has increased the significance of so-called container feeder ships that are used to convey the containers to many smaller harbours. These ships often have no hatch cover, which enables quick transhipping. They do have a crane for loading and unloading. If a combination carrier is used, the containers can also be unloaded over a trailer ship, ramps or a rear hatch. In other words, transfer to "roll-on, roll-off" (RO-RO) or "roll-on, load-off" (RO-LO) ships can be done smoothly.

On the whole, medium-sized container ships are the most logical choice for shorter distances. This is also confirmed by constraints during the passage (like the locks of the Panama Canal), the large number of harbours and inadequate port infrastructure that makes cranes on a ship a must. Fast ships are especially interesting for well-travelled and long routes, for example from Europe to the USA or Asia, or for military purposes. Container ships that can do up to 30 knots fall in this category. The hull of these fast container ships is hydrodynamically optimised, with a short middle section. Thanks to the shape of the hull, they can carry more containers. This shifts the centre of gravity upwards, however, which must then be compensated for by adding a considerable amount of ballast water. The high speed is guaranteed by a 12-cylinder diesel engine that generates 70,000 kW at 100 rpm. For reasons of stability and because of the transom stern, which is designed for high speed, the power plant and hence the deck house have to be built in the middle of the vessel, which implies a very long propeller shaft.

▶ **Ship engines.** Ships of the merchant marine are driven either by slow two-stroke diesel engines (60–250 rpm) or medium-speed four-stroke diesels (400–1200 rpm). Usually they consume cheap heavy fuel that is very viscous and contains impurities. It must be heated up before being injected into the engine. Both engine types are very economical, achieving fuel consumption of up to 170 g/kWh under full load. The combustion air is pumped into the engine using a turbocharger. An intercooler increases the air density and lowers the combustion temperature, which improves the engine's efficiency. At the present time, a single engine can develop up to 80,000 kW. With engines of that size, the diesel exhaust fumes are often guided through an economiser that produces steam for a turbine generator. This new hybrid machine concept also includes an exhaust turbine that uses about 10% of the off-gas stream to generate electrical energy. The electric current produced by the auxiliary and turbine generators can be used in turn to power two compact electrical booster engines on the main drive shaft. These direct-drive supplementary engines increase propulsion and act as flexible consumers of the surplus energy produced by the recycled off-gases. They also support the starting up of the main engine and can be used as an emergency plant in the event of main engine failure.

For fast ferries, cruise ships, yachts and navy ships, gas turbines are used for propulsion. Aeroplane turbines are adapted for operation in ships by channelling the thrust to a power turbine. *Gas turbines* have a low power-to-weight ratio, are very reliable and can be started up very quickly. But they also use a lot of fuel and require expensive oil distillates to run.

With slow-speed diesel engines, the propulsion power is conveyed directly to the propeller via a shaft. Deployment of medium- or high-speed engines requires a reduction gear unit in order to slow the propeller's rotation to an ideal hydrodynamic speed.

The conversion of engine power to propulsion is done by a propeller or jet propulsor. The fixed propeller is very widespread. The degree of efficiency of a propeller is calculated using the propeller's diameter and the derived pitch, whereby the size of a propeller's diameter does determine the degree of efficiency. If high manoeuvrability is required, then a variable pitch propeller is used. The pitch or angle of attack can be modified by turning the propeller blades. Azimuth thrusters can be deployed for even greater manoeuverability. These systems can attack in any angle to the longitudinal axis of the vessel. They are a combination of engine and rudder, like the rudder propeller or pod engine that are used very frequently in modern passenger ships. A cost-effective alternative to the pod engines are transverse thrusters, which draw in water and then propel it through a tunnel-like structure to port or starboard, thus allowing the ship to move sideways. The water-propulsion system is used for fast vessels. The water is propelled through a jet in the opposite direction of the ship's travel. The propulsion takes place owing to the repulsion, much like a rocket. The ship's course may be modified by rotating the movable jets.

A *rudder system* is needed to maintain the ship on a certain course or to change its direction. Essentially, a rudder works by generating a perpendicular force to the surface of the rudder from oblique currents. Rudder force is divided up into a longitudinal component against the direction of travel and a perpendicular component, which is logically perpendicular to the direction of travel. The perpendicular component creates a twisting moment by attacking an angle via the

Bow section of a fast containership built on keel blocks: In shipbuilding today, the individual sections are pre-fabricated and then assembled to form the ship's body. Thanks to a system of rails, the ship can be moved about the shipyard according to work requirements. Source: Fraunhofer AGP

lever of the stern. Modern rudders have a streamlined profile that produces greater rudder forces. Very effective rudders have been developed lately, such as the rotor rudder, which has a rotor on its leading edge that prevents the flow from separating.

Trends

▶ **Optimised ship operations.** Fuel consumption is especially important. Low required power can only be achieved by reducing resistance and speed as much as possible and similarly increasing the degree of efficiency of the propellers and of the interaction between the ship and its propellers. Since transport effectiveness increases proportionally to speed, the choice of the most economical speed depends on the current market situation – insofar as a ship can influence its operating speed at all. The ship's resistance and the degrees of efficiency are predetermined and optimised using *CFD (Computational Fluid Dynamics)* and verified in model tests. In terms of its key dimensions, there is still a great deal of leeway still in varying a ship's form so that wave resistance can be lowered. When it comes to friction resistance, large ships are more effective. Because the wetted surface area is lower in comparison to the weight. Large flow separation areas must be

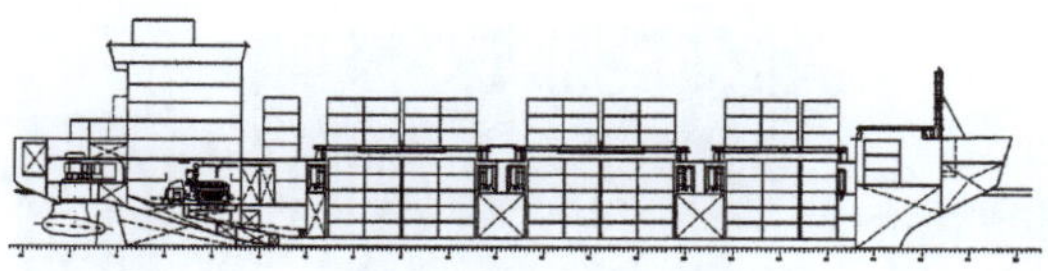

Ice-going container ships: The ice-going container ship Norilskiy Nickel can be operated in both directions (double-acting concept) thanks to the special design of the hull and power plant. If ice conditions are very bad, the ship can be used in reverse, which allows it to plough its way through ice up to 1.5 m thick. Once the vessel reaches waters with thinner ice, it can be turned 180°, thanks to its pod (rudderless) drive. These ships can navigate polar waters year-round without assistance. Source: WADAN Yards

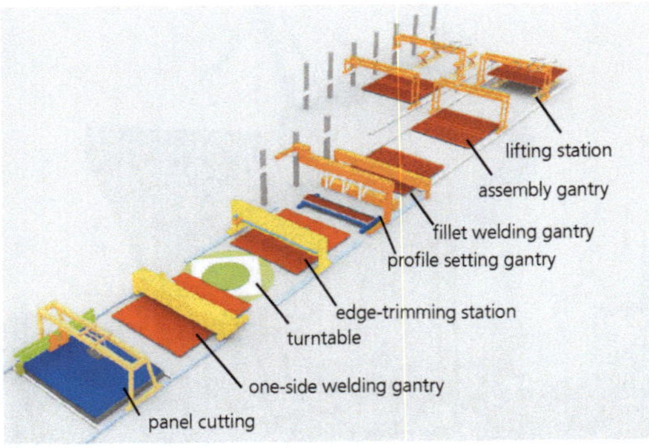

Automated manufacturing: The manufacturing of panels (stiffened panels) and sections is done on flow lines. After the panels are cut, they are welded together at a one-side submerged arc welding station to form panel fields. The panel field can be turned on a turntable, if needed, and cut to exact size, including openings, etc., at an edge-trimming station. At the profile setting gantry, profiles are taken from a comb pallet, pressed and tacked to the panel and then welded to the panel field with the fillet welding gantry. The stiffened panel is now finished. The completion of the section with frame girders is carried out during the next stations. In some shipyards, the stations are equipped with robots. Source: Fraunhofer AGP

avoided to keep pressure drag down as well. Moreover, additional resistance elements, propellers, for example, should be carefully designed with regards to flow. Efficiency is determined, too, by the shape and alignment of the propellers and rudders, the number of propellers and the flow conditions in the stern. Here, too, optimisation work is underway.

Appreciable savings can also be implemented in running a ship by taking appropriate measures. The outer hull and the propellers must be continuously monitored. Heavy growth of barnacles and other marine life on hulls and a deformed propeller can easily double the power required to move the ship. Trim can also be optimised for performance for each draft and speed. Modifying drafts and trim can vary performance by up to 10% for ships with a broad transom stern and a prominent bow bulb. A relatively new idea is a routing selection system that takes current and future weather data (wind and sea conditions) into account, as well as currents or possible shallow water effects. Plus, given sailing lists and harbour slots, the system can also compute the optimal route and speed for the current weather conditions and the local conditions expected during the journey. Such systems have to calculate higher consumption due to wind and sea-state resistance and in doing so must also consider such elements as drift caused by side winds and a turned rudder.

Dock assembly: Deviation for sections with dimensions over 10 m may not surpass 5 mm, so they fit together during dock assembly. This is especially complicated with three-dimensional curved sections, as exemplified in the picture by the assembly of a bow section.

► **Unconventional hull shapes.** A lot of the innovation in shipbuilding involves unconventional hull shapes. These are differentiated in their way of enhancing buoyancy.

Buoyancy generators by displacement include classic ships (displacers), fast monohull types, conventional catamarans, SWATH ships (Small Waterplane Area Twin Hull) and trimarans. The fast monohull types are displacers in which the hulls main displacement is performed by a low, underwater ship that is favourable to the flow. This hull shape reduces resistance at high speeds and increases engine efficiency. The torpedo-shaped underwater ship is very difficult to build, however. The advantage of a SWATH ship is its outstanding seaworthiness and the large deck surface. There are disadvantages, too, however: greater flow resistance, sensitivity to changes in the displacement and trim position and greater draft. A few years ago, a variant was developed called the High-Speed SWATH (HSS). This type of vessel combines a SWATH bow with a catamaran stern including water jet propulsion in order to converge the favourable seagoing features of a SWATH with the advantageous flow resistances of the catamaran. Catamarans are very successful wherever speed is required. The main advantages of catamarans over displacers are the higher speeds with the same power output, the large deck surface, less rolling, greater stability and maneuverability. The disadvantages are the rapid vertical and pitching movements. Lightweight construction (aluminium, FRP) are especially important for catamarans. Recently, so-called trimarans in sizes up to frigates were developed from catamarans.

The Hybrid Small Waterplane Area Single Hull vessels (HYSWASH), gliders and hydrofoils make use of buoyancy generation through hydrodynamic lift. Gliders are mostly sports boats that glide over the water when moving at high speeds. The HYSWASH type was developed by the US Navy, but it has not gained acceptance yet owing to high construction costs for the complicated hull technology. For hydrofoils, the situation is different. In contrast to conventional ships, they go a lot faster and offer passengers far more comfort. Even in very rough seas, the deployed foils prevent strong vertical motions, rolling and pitching. The main disadvantage is the very limited load capacity and the large draft. There are two types of *hydrofoil*. One has semi-submerged foils, i. e. they are vessels with a laterally angled foils that are partially submerged during the journey. Their lateral, longitudinal and vertical stability is permanently and independently controlled by the change in the submerged foil brought about when the vessel leaves the static equilibrium position. The construction of the semi-submerged ships is simple and robust, but the limited planing height is a disadvantage, as the ship is then at the mercy of the sea. The higher the boat lies during the journey, the wider the foils have to be. This presents physical limits, since the weight of the vessel increases in three dimensions, while that of the hydrofoil only does so in two dimensions. The depth of submergence and stability have to be controlled by additional systems that can modify the angle of incidence of the foil or foils. Fully submerged foils promise better seakeeping and a greater planing height, preventing the craft from hitting the crests of the waves. In spite of considerable advantages, such as low input power at very high speeds, high seakindliness and safety from mines, hydrofoils never really gained acceptance as military craft. The decisive factor is the vulnerability of the fully submerged vessels (without a stabiliser system with sensors, computers and effectors, it cannot be operated safely).

▶ **Hovercrafts.** Hovercrafts and Surface Effect Ships (SES) are in the aerostatic buoyancy corner. The principle of the hovercraft involves pressurising air under the craft's floor. The air cushion formed lifts up the craft and keeps it over water while another engine propels it forward. The greater the planing height, the larger the interval between the floor and the surface of the water. In turn, more air escapes, the pressure drops, the lifting force dwindles and the craft lands on the water's surface. To avoid loss of pressure, fixed sidewalls or a flexible apron is attached to the craft's side, which are entirely submerged when the vessel is at rest. The advantages of air cushion-borne vehicles is their enormous speed of up to 70 knots, their small draft and the almost non-existent underwater signature. The possibility of actually sailing on land within certain parameters makes them particularly interesting for military applications. The disadvantage of air-cushion vehicles is their sensitivity to side winds and the very slight change in trimming. Today, hovercrafts are used almost exclusively in the military sector. The SES are situated somewhere between SWATH types and hovercrafts. Calling them "surface effect ships" is in fact misleading, because the SES do not really make use of surface effects, rather they are a hybrid concept, a combination of air-cushion craft and SWATH. With the SES, the space between the two hulls is sealed off fore and aft with a flexible rubber apron and inflated with fans to create and maintain overpressure. Even slight pressure from the cushion will raise the craft so far out of the water that 80% of the cushion's total weight will be carried. The advantages of the SES are the high speed (up to 60 knots) at low power input, the small draft, good seakindliness and almost no underwater signature. The disadvantages are the rapid loss of speed in the face of heavy frontal swell, the loss of amphibian capacity as well as the enormous manufacturing and operating costs.

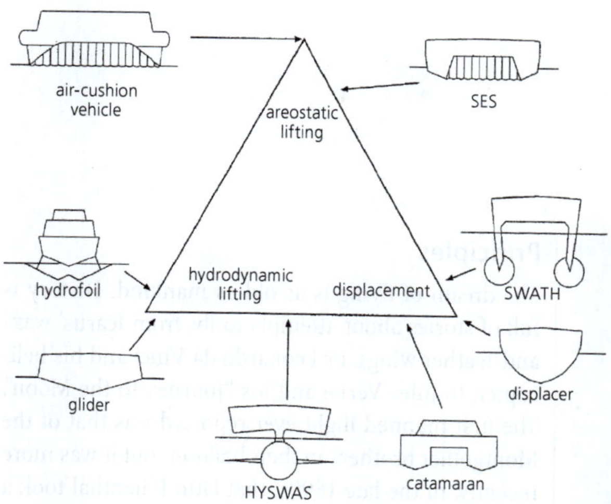

Morphological triangle. Left: For different hull shapes, the "way of generating lift" is the differentiating feature. Right (from top to bottom) SWATH: The high-speed SWATH Stena Explorer combines a SWATH bow with a catamaran stern in order to join the favourable seakeeping features of a SWAT with the advantageous flow resistances of the catamaran. The ship is 127 m long, can carry up to 375 passengers and travels at up to 40 knots. Hydrofoil: Fully submerged hydrofoils have only gained acceptance in niche segments so far. The fully submerged hydrofoil Jetfoil Terceira from Kawasaki, 50 knots, 350 passengers. Air cushion vehicle: Air-cushion vehicles are tend to be the used by the military only as a landing boat because of the high costs. LCAC 61 of the US Navy can transport 70 t of cargo up to 200 miles at 40 knots and with a water displacement of 180 t. Source: Fraunhofer, Ships of the world, Hynds, Fast Ferries

Prospects

- As a mode of transport, ships represent the backbone of the world's globalised economy. That is why it is absolutely necessary to continue developing the appropriate technologies, even considering the altered framework conditions such as the shortage of energy and a lack of properly trained personnel.
- The trend today is toward energy-efficient ships. This does not necessarily mean that tomorrow's ships will be especially slow. At any rate, the amount and diversity of goods to be transported will continue growing. Unconventional hull shapes represent enormous potential that can be leveraged thanks to modern technologies. The support of marine personnel will continue to grow. But topping the list of innovations in seafaring is security.

PROF. DR. MARTIN-CHRISTOPH WANNER
Fraunhofer Application Centre Large Structures in Production Engineering, Rostock

Internet

- www.vsm.de
- www.sname.org
- www.seehafen-verlag.de
- www.hro.ipa.fhg.de

Aircraft

Related topics

- Composite materials
- Metals
- Sensor systems
- Traffic management
- Weapons and military systems
- Bionics
- Logistics

Principles

The dream of flying is as old as mankind. History is full of stories about attempts to fly, from Icarus' wax-and-feather wings, or Leonardo da Vinci and his helicopter, to Jules Verne and his "Journey to the Moon". The first manned flight ever reported was that of the Montgolfier brothers in their balloon. But it was more recently, in the late 1890s, that Otto Lilienthal took a scientific approach to aerodynamics and flight mechanics. He studied the cambered wings of birds, and his "Lilienthal Polar Diagram" is still in use today. Shortly afterwards, the Wright brothers made their famous powered flight, ultimately based on Lilienthal's theories.

There is a wide variety of vehicles in the air today, from unmanned aerial vehicles (UAV) and transport aircraft to vehicles for space tourism. In the early decades of the last century, it was not at all clear which system would be best: rotary wings, i.e. helicopters, or fixed wings. Mainly due to maintenance problems and speed limitations, the helicopter lost that race and is nowadays used only for very specific tasks. But we can now see both principles realised in the same vehicle, e.g. tilt rotor aircraft such as the Bell-Boeing V-22 "Osprey". Recent predictions of the future of aeronautics include short-range application scenarios using such transformation vehicles.

In brief, the forces acting on an aircraft in flight are:

- Weight: In general, this comprises the payload, the fuel, the equipment and the primary structure. The latter amounts to some 30% of the total weight. A 10% weight reduction, e.g. by using advanced materials, thus cuts the maximum take-off weight to just 3%.
- Lift: As for a propeller or a helicopter rotor, the air accelerated by the propulsion system is directed downwards by the cambered wing, producing a force known as lift. The lift force must be as high as the weight in order for the aircraft to stay in the air. In today's configuration with wing and tailplane, the centre of gravity of the total aircraft is forward of the wing lift force, and the horizontal tailplane produces downward lift at the tail. The three vertical forces are thus evenly balanced. The distance between the point where the lift force acts and the centre of gravity is a measure of stability and controllability: the more stable an aircraft is, the more power is needed to control it by means of direction change. Conversely, the more unstable an aircraft is, the less power is needed to steer it. Military aircraft are unstable by design; the stability is then generated by the flight control system. In future this will apply to commercial aircraft too.
- Drag: The major drag components are friction drag, resulting from the flow around the aircraft surface, and lift-induced drag. Other forms of drag are interference drag between aircraft components, wave drag in the case of high-speed aircraft, and excrescence drag due to gaps, rivets, antennas, etc. Today's transport aircraft have a lift-to-drag ratio of around 20, and this figure rises to 70 for high-performance gliders.
- Thrust: This is produced by the propulsion system, and it must compensate for drag. Propeller engines, which have a short response time between power setting and thrust build-up, offer the best

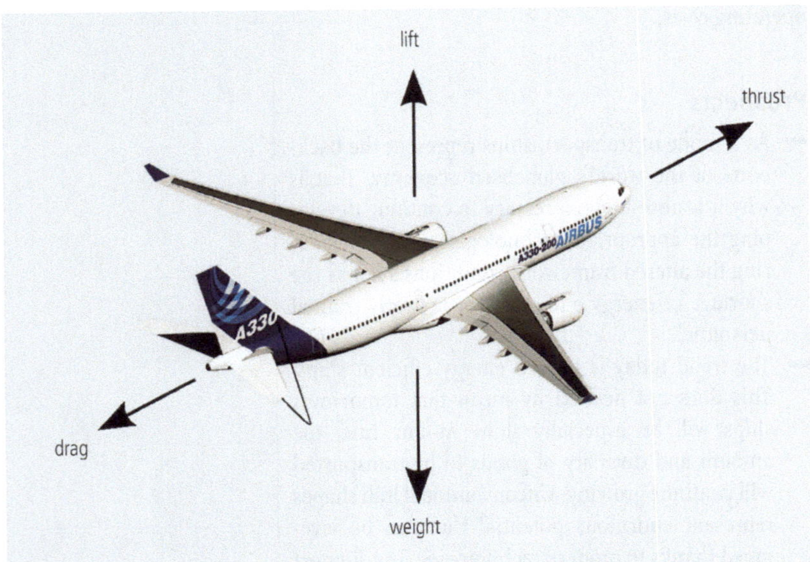

Balanced forces in flight: During stationary cruise flight, lift and weight must be equal, as must thrust and drag. In the case of engine failure in a twin engine aircraft, the remaining engine has to deliver the entire thrust needed. This produces yaw and increased drag, which again has to be handled by the remaining engine. This is an example of engine design requirements, most of which concern off-design conditions. Source: RWTH Aachen performance for low-speed aircraft. Jet engines, which are used for faster aircraft, are less dependent on flight velocity; they have a response time of up to 8 seconds between power setting and full thrust availability.

There is almost no physical lower limit to the size of an aircraft. Unmanned Aerial Vehicles with a size of an insect, even carrying a camera, exist today. At the other end of the scale, however, there are limits to the maximum size of an aircraft. The weight of a wing increases disproportionately to its size, so the engines have to produce more and more thrust simply to cope with the weight of the primary structure until no more payload can be carried. In addition, the ground operations of a super-large aircraft cause many problems at airports in terms of boarding, the area needed, and the number of gates. From today's viewpoint, the Airbus A380 has almost reached the maximum aircraft size.

Applications

▶ **Requirements and operation.** The shape and size of an aircraft, or in short its configuration, are determined by its transport task, i. e. what payload needs to be transported at what range in how much time, and sometimes also: at what cost? Below are a few examples:

- Military fighter airplanes: These are designed for air combat, so they need to be fast and agile but do not need long range capability or a heavy payload. Because of the speed requirements, they have highly swept wings, one or two powerful engines, and often an additional control surface at the front, known as a "canard wing", as on the Eurofighter Typhoon. This artificial instability increases manoeuvrability.
- Transport airplanes: These are designed to carry food or material by air to conflict or disaster areas. This mission calls for good low-speed performance, very good loading/unloading devices such as large back doors, plus long-range capabilities. If the transport needs to be fast, jet engines will be used, but many of these aircraft are propeller-driven because of the low-speed requirements.
- Passenger aircraft: After many configuration trials from the 1940s to the 1970s, the most common configuration today is with two to four engines, usually under the wings, plus a tailplane. Smaller airplanes and business jets have the engines at the back, while the large engines of bigger airplanes are under the wing to alleviate the load, for structural reasons and to facilitate maintenance.

The development process starts as soon as the top-level aircraft requirements have been defined, culminating in the certification phase which includes ground tests and flight testing. After entry into service, the aircraft is operated for many years, and its structural integrity must be regularly verified by various tests. Civil aircraft are visually inspected daily, and undergo a complete check down to sub-component level depending on the loads experienced and measured. Non-destructive testing is carried out in the same way as during certification, e. g. by ultrasonic measurements, computer tomography and magnetic resonance tomography. Parallel to these tests, load cycles are simulated on a prototype using a highly condensed process for the whole life cycle. This enables the aircraft life of some 60,000 flight hours to be tested within a few years, so that parts can be replaced before they experience fatigue or other problems.

▶ **Materials and structure.** The fuselage, wing and tailplane structure consists of the skin, stringers, and bulkheads or ribs. An integral tank in the wings holds most of the fuel. Sometimes, a central tank in the fuselage or an optional tank in the horizontal tailplane are used as well. To set up the cabin, insulation material is inserted in the fuselage structure and shaped with a kind of plastic wallpaper. All the cables lie within that insulation material; in the case of the Airbus A380, the total length of the cables adds up to 500 km.

Aircraft have been made of metal for the past 80 years or so; such structures have reached an optimum of sorts. The metals mainly used are aluminium because of its low weight and high stiffness, titanium for greater rigidity, and aluminium-lithium for even lower weight. However, metal is heavy due to its high density. In addition, since the early days of the Comet B4 aircraft, fatigue in metallic materials has become a severe problem in aeronautics: because the wing experiences many load cycles, e. g. in turbulence, structural cracks and crack propagation limit the service life of the aircraft structure.

Fibre-reinforced plastics have been in use since the 1970s. While these at first consisted of glass fibre, carbon fibre-reinforced plastic (CRFP) has meanwhile become more common. In addition, the Airbus A380

▲ Eurofighter: The additional control surface at the front, known as a canard wing, gives this fighter aircraft greater agility. Source: Refuelling Ltd.

▶ The smallest and largest passenger transport aircraft. Top: The CriCri is one of the smallest aircraft. With just a single pilot, it can fly some 400 km at 280 km/h, consuming 25 l of fuel. The wing area is 3.1 m², and the two engines together have a power output of 22 kW. Bottom: The largest passenger transport aircraft, the Airbus A380. It can carry some 550 passengers at 960 km/h over a distance of 15,000 km, using 310,000 l of kerosene. The wing surface area is 845 m², the span is almost 80 m. Source: AirVenture, Airbus

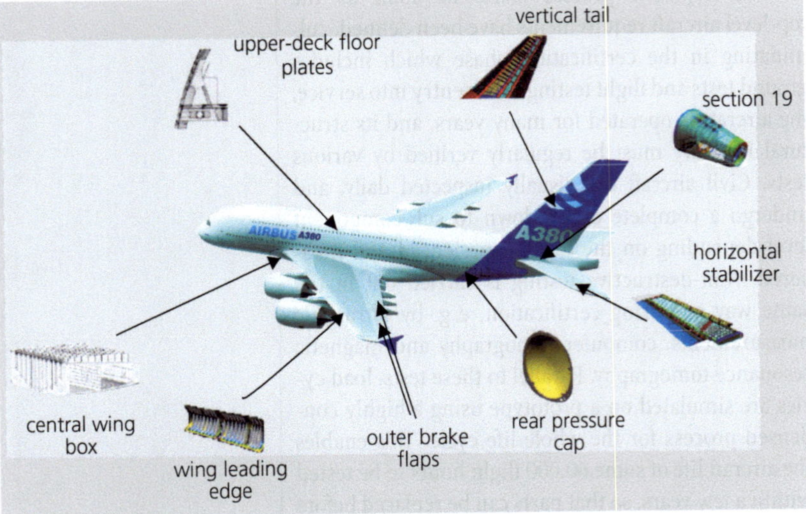

Fibre-reinforced plastics (FRP): In the Airbus A380 primary structure. Apart from these major components, many other parts such as water tanks, pipes for the air-conditioning system, window frames and floor plates are also made of fibre-reinforced plastics (FRP). Source: Airbus

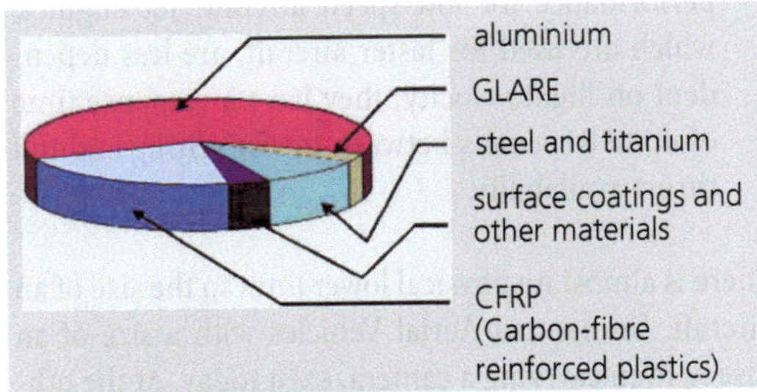

Material breakdown in the Airbus A380: CFRP fuselages have already been realised in smaller aircraft, but it is a slow process to replace aluminium in transport aircraft. Airbus started with the tailplane, and is gradually making the load-carrying structure out of FRP instead of metal. During these first steps, the general structure and arrangement of parts stays the same for FRP as for metal, using spars, stringers and ribs. Source: Airbus

uses *GLARE*, a hybrid material made up of aluminium and glass fibre in layers. The fibres in the layer can absorb the mainly longitudinal forces, while the resin matrix provides elasticity and allows the flow of forces. This calls for a very sophisticated arrangement of the fibres in the force direction, which is difficult to realise if that direction fluctuates. Another advantage of *fibre-reinforced plastics (FRP)* is that they are not limited by fatigue, so CFRP is frequently used not merely for weight reasons, but also because it reduces maintenance.

Often, FRP is applied in a sandwich structure, with some light material or even foam placed between two or more layers of FRP. The vertical tailplane of an Airbus aircraft is made of glass fibre for the nose boxes and carbon fibre for the main box.

One of the challenges when using FRP in aircraft is the range of operating conditions: a dark-painted outer skin of an aircraft on the ground may heat up to over 90 °C, while it cools down to around minus 70 °C in cruise flight. It has to withstand heavy loads on touchdown, plus bird strike at any speed, and will be operated even in hail or in ice clouds. Under all these conditions, the material must be shown to meet the whole range of requirements in terms of safety, reliability, stiffness, elasticity, ductility, etc. This is difficult for FRP as, so far, not as much experience is available as there is for metals.

Both fundamental materials, metal as well as FRP, call for specific joining techniques. To date, FRP design is similar to that of metal structures, mainly because the repair methods must also be similar. As airports usually do not possess an autoclave, damage to FRP structures is repaired by riveting sheets onto the outer skin. But the aim is to have specific adhesives available in the near future so that the material can be bonded rather than riveted. Friction steer welding has been introduced as a new joining technology for metals, and laser beam welding has already been in use in the production process for several years.

► **Flight control and fly-by-wire.** Flight control means deflecting control surfaces in order to influence the motion of the aircraft: rotations around the lateral axis by using the horizontal tailplane rudder, i.e. "pitch", around the vertical axis using the vertical tailplane rudder, i.e. "yaw", and around the longitudinal axis using the ailerons at the wing tip, i.e. "roll". These are the primary flight control elements, and produce a corresponding change in the direction of the flight path. Deflecting the high lift devices, i.e. the flaps and slats at the wing leading and trailing edge, is known as secondary flight control. This latter type of control can be supported during take-off and approach by the ailerons, which are then deflected symmetrically to act as flaps as well as being asymmetrically deflected to permit roll motion. The spoilers are also used as flight control surfaces, as they reduce lift if deflected on only one wing side, causing increased roll capability. If they are used symmetrically on both sides, they reduce lift and increase drag, thus reducing the forward speed and/or increasing the descent speed.

In the more distant past, control surfaces were manually deflected by the pilot, using bars and linkages. This system was later augmented by hydraulic and/or electrical servos, so that the pilot controlled the deflection merely by switching a hydraulic actuator. The Franco-British supersonic passenger aircraft Concorde, which made its maiden flight in the late

1960s, could not be operated with this type of direct control system. Transonic and supersonic phenomena, such as rudder reverse and the difficult trim situation, called for electronic flight controls, which were eventually introduced for that aircraft. This idea ultimately led to today's Electronic Flight Control System (EFCS), a system that was first implemented in the Airbus A320 which went into service in 1987.

The principle of the EFCS system, or fly-by-wire as it is often called, is very different from that of previous flight control systems. The pilot's command is no longer passed directly to a control surface, but is first entered into a flight control computer. Together with flight data delivered by sensors, such as the speed, attitude, altitude and temperature, and based on the flight control laws implemented, the computer checks whether or not this operation is permissible within a safe flight envelope. Only if the answer is positive does the computer send electrical signals to the actuators to deflect the control surface. What sounds like a simple change in philosophy is almost a revolution in aircraft control, even raising ethical questions on responsibilities, but it ultimately paves the way for many new technologies which could not be implemented without EFCS. Gust and manoeuver alleviation, overspeed and stall protection are now possible, increasing safety, reducing the pilot's workload, and even achieving a more efficient flight path in terms of noise and/or emissions.

Finally, EFCS enables pilots to fly different kinds of aircraft in a similar manner. This is called cross crew qualification (CCQ), and enables a pilot to fly a small Airbus A318 the same way as the huge Airbus A380. According to the manufacturer, CCQ reduces the direct operating costs, which include crew cost, by about 5%. This is more than many basic aerodynamic or structural technologies offer.

In standard operating conditions, the EFCS works according to "normal law", but in certain conditions it may be switched to "direct law", under which the pilot's commands are passed directly to the control surfaces.

The information from the flight control computer can be passed to the actuators electrically, which is called "fly-by-wire", or by fibre optics, which is then called "fly-by-light", but both methods rely on the same EFCS principle.

Trends

Aircraft design, development, production and operation involve many disciplines and areas of activity. Each of these exhibits future trends such as the development of Computational Aero Acoustics in flight physics, Fiber Bragg Gratings for deformation measurements in the field of structure, more advanced Integrated Modular Avionics in systems, etc. In order to implement the following three examples of technology trends, all major disciplines in aeronautics need to act in a concurrent, multidisciplinary way to increase overall aircraft performance.

Cockpit of an Airbus A340: The arrangement of instruments in the cockpits of large transport aircraft, even in different models, is very similar. However, the in-flight behaviour of different aircraft types can vary widely. Electronic flight control can overcome these differences, so that pilots need less familiarisation time when flying different aircraft types. Source: Airbus

► **Laminar technology.** The friction drag resulting from the airflow along a surface is a major element in overall drag; it accounts for about half of the total drag depending on the type of aircraft. Friction drag evolves within a thin layer of the airflow: the boundary layer. Directly at the surface, the flow speed must be zero; this is called the adherence condition. Within the boundary layer perpendicular to the surface, the flow accelerates up to the outer air speed. There are two known types of boundary layer: the flow starts with a laminar boundary layer, which is smooth and streamlined. After a certain length, depending on surface roughness, flow velocity, temperature and pressure, a transition takes place from laminar to turbulent. The turbulent boundary layer is chaotic, and the drag is of an order of magnitude higher than that of a laminar layer. Due to the conditions described above, the flow over almost all parts of a transport aircraft is turbulent. If one could make that boundary layer laminar over a certain part of the wetted surface, this would save a great deal of drag. Glider airplanes do this simply by shaping the airfoil of the wing in the right manner; this passive method is called Natural Laminar Flow. This is not sufficient for fast and large transport aircraft; the boundary layer must be "helped" to stay laminar around the nose region of the wing by active means, e.g. by boundary layer suction supported by profile shaping. The use of active means in conjunction with passive profile shaping is called "hybrid laminar flow control (HLFC)". The gross benefit of this technology is some 16% of the total drag if HLFC is applied to the wings, tailplane and nacelles. Subtracting the penalties due to the suction system and some surface treatment, the net benefit will be around 12%. This is by far the highest value that any single technology can offer. Boeing has carried out HLFC flight tests on a modified B757 wing; Airbus has done the same on a modified A320 vertical tailplane.

Because of the pressure due to fuel prices, it can be assumed that laminar flow control will be applied in the next-but-one generation of aircraft.

► **Adaptive wing.** Adaptivity is a basic principle in nature and designates the change of attributes of a body, usually its geometry, in order to cope with changing requirements, e. g. in different missions or in different phases of a single mission. Even today's aircraft are adaptive: flaps are deflected during the low-speed phases, changing the geometry of the wing in order to produce enough lift even under these conditions. But it is not yet possible for a solid aircraft to adapt in the same way as a bird by changing almost any parameter of its wing during each mission: camber, sweep, aspect ratio, twist, tip geometry.

Adaptation of a tern's wing when swooping for prey ("high speed") and when drinking ("low speed").

Nevertheless, modern aircraft requirements call for more adaptive capabilities, and new materials and systems make allowances for this. Features to be considered might include a change in camber to cope with different weights, called "variable camber", or slight but fast changes in the airfoil type to cope with gusts, or changes in the surface curvature to cope with certain high-speed phenomena, called Shock Boundary Layer Interaction.

In Europe, new control surfaces called Miniature Trailing Edge Devices (Mini-TEDs), which can be used for gust alleviation, roll control, rapid descent, etc. have been studied up to the point of flight tests on an Airbus A340. Almost two decades prior to this, NASA carried out the "Mission Adaptive Wing" programme using an F-111 flying test bed with variable profile geometry and variable sweep.

A320. Left: Interior of the A320 test aircraft with a suction system for the A320 HLF Fin Flight Test. Nine suction tubes coming from the fin on top run into a settling chamber, while the suction power on the right hand side is supplied by the compressor at the bottom. A bypass duct on the left-hand side ensures that the compressor is always running at nominal power in terms of mass flow. Right: Exterior of the A320 test aircraft with a laminar fin. Clearly visible are the suction nose boxes, the three rows of pressure tappings, and a traversable wake rake behind the fin. The fairing for the infrared cameras on the upper side of the horizontal tailplane, and the suction system exhaust above the penultimate window can also be seen. The whole setup was designed for an experimental flight test aiming to verify technical feasibility. For an aircraft in operation, it needs to be downsized and situated behind the bulkhead.

Given that adaptivity offers high potential in terms of fuel and noise reduction plus enhanced comfort, mainly in combination with the load control technologies described below. Since modern materials such as shape memory alloys or piezo-ceramics plus the EFCS afford the opportunity to introduce adaptive surfaces, more adaptive wings are likely to feature in the next generation of aircraft. These developments may culminate in a "morphing wing", which NASA has proposed for about 2050.

The technology targeted in the area of helicopters is almost the same. Individual blade control by adaptive technologies will make the present swash plate redundant, resulting in lower fuel consumption, less vibrations, less noise, less maintenance, less weight and, finally, lower cost. This objective, too, calls for a sophisticated EFCS.

► **Load control.** The lift load over an aircraft's wing varies along its span; this distribution has a major impact on the induced drag explained above. In aerodynamic terms, an elliptical load distribution would produce a minimum induced drag, but would increase the wing root bending moment, requiring reinforcement and thus resulting in additional weight which counterbalances the benefits of the decreased drag.

If the wing is exposed to turbulence or gusts, and for certain manoeuvers, it would be advantageous to control the wing loads, primarily to shift the resulting lift force inboard for a given period of time. This calls

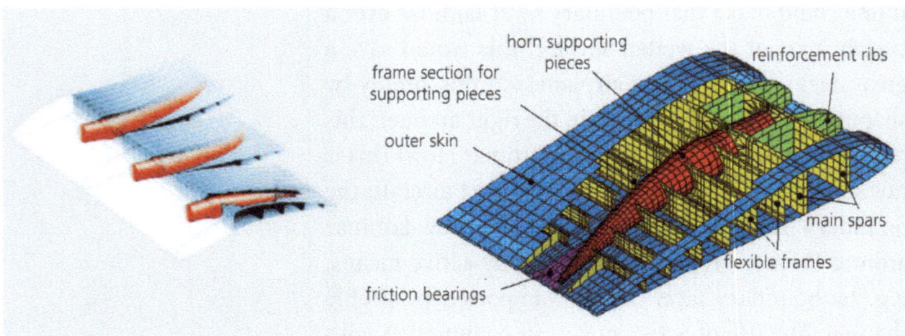

Horn concept: Torsion of the horns continuously changes the form of the flaps, and is comparable to different flap positions. Separately controlled horns would permit differential spanwise flap deflection, which would facilitate load control. Source: Daimler, EADS

for active load control, either in a spanwise direction, or in a chordwise direction, e.g. to control flow separation.

In addition, the wing is flexible and therefore bends and twists under load. This aero-elastic behaviour is different for metal and for FRP wings. Metal wings have certain self-alleviation characteristics due to their elasticity, while CFRP is too rigid and requires active load alleviation.

Today's aircraft already have manoeuver load alleviation systems. For certain manoeuvers, ailerons plus spoilers are used to shift the load inwards. This can be done in cases where the EFCS is expecting the manoeuver. But there is no way of knowing when gusts will hit the aircraft. Today's philosophy is to carry out some gust load alleviation by shifting the loads inwards on entering a gust field, but this has a negative effect on fuel consumption. It would therefore be desirable to have a turbulence sensor to measure the speed some 70 m in front of the aircraft, thus giving sufficient time to deflect the control surfaces if a gust is detected before it hits the wing.

Such a system could be used for two objectives. On the one hand, it could increase the comfort in turbulence, which also increases safety on board. In addition, and given that the wing structure is designed to withstand severe gusts, a reliable gust alleviation system could relieve the structural requirements and substantially reduce the weight of the wing.

Several turbulence *sensors* have been tested in the past; most recently, Airbus has carried out flight tests with a gust sensor based on an infrared *LIDAR system*. Given the high potential for weight reduction plus the increase in safety and comfort, a load control system with a turbulence sensor of this kind is likely to be introduced in the next-but-one generation of aircraft.

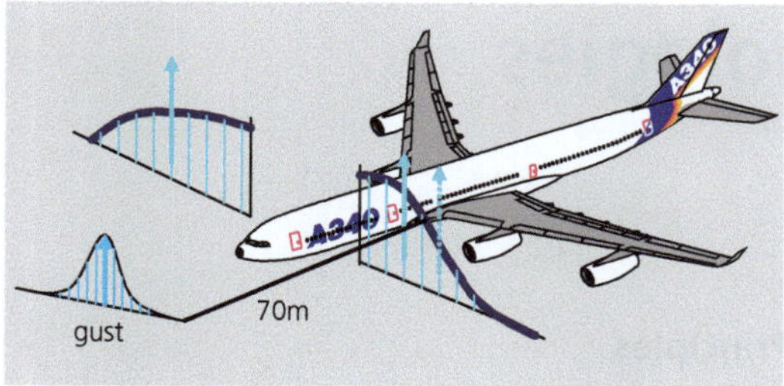

Load distribution: Cruise load distribution with corresponding wing root bending moment (left). Inboard shifted load distribution after control surface deflection, following an upwind gust as detected by a turbulence sensor (right). A reduced root bending moment alleviates the impact of the gust load and enhances comfort in turbulence. The turbulence sensor has to measure the gusts some 70 m in front of the aircraft, in order to have a minimum measuring and reaction time of 0.3 s (at a speed of about 850 km/h).

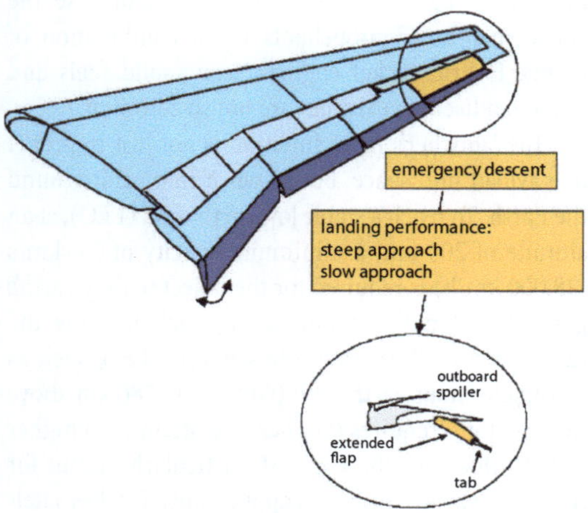

Wing with multifunctional control surfaces: Different target functions define the range of adaptivity. Multifunctional control surfaces are designed to permit load and roll control, glide path control during climb, approach and emergency descent, gust and manoeuvre load alleviation, and better overall flexibility under high loads and at speed and altitude. Source: Airbus

Prospects

Aerial vehicles can be expected to undergo drastic changes during the next few decades; the increasing pressure on the overall traffic system demands new solutions, not only in terms of emissions and noise, or product life cycle cost, but even more so in terms of operations in a saturated air traffic system (ATS). As a typical aircraft has a service life of some 50 years, so the Airbus A380 may still be in operation in 2050, there are two conceivable types of solution:

- Evolutionary developments will continuously improve existing aircraft and their derivatives, including new materials, active or passive flow control, and new electronic flight controls. Within this process, oil shortage will put high pressure on fuel-reduction technologies such as laminar flow.
- Revolutionary developments can also be expected, though they will not dominate air traffic for the next two decades because the present aircraft will be in operation for many more years to come. But ground handling will change, and the way will be paved for new configurations such as the Blended Wing Body, or aircraft with extremely short take-off and landing (STOL) characteristics.

PROF. DR. ROLF HENKE
RWTH Aachen University

Internet

- www.ceas.org
- www.ec.europa.eu/research/transport/news/article_2324_en.html
- www.adg.stanford.edu/aa241/AircraftDesign.html
- www.century-of-flight.net/index.htm
- www.hq.nasa.gov/office/pao/History/SP-468/contents.htm
- www.acare4europe.com

Space technologies

Related topics

- Communication networks
- Robotics
- Weapons and military systems
- Environmental monitoring

Principles

Space begins at an altitude of 100 km. Travelling beyond this frontier requires cutting-edge engineering and an enormous amount of energy – the prerequisites for any utilisation or exploration of space. The first space launch took place just over 50 years ago.

Space can be used as an asset to improve our daily lives and these days we almost take for granted the many applications provided by space infrastructures: the weather forecast and the transmission of TV programmes to our homes, satellite guided navigation and telecommunications services to even the most remote regions, as well as the extremely valuable capability to monitor climate change and disasters. All of these technologies rely on orbital infrastructures, and their influence will grow as society progresses towards a more and more globalised world.

► **Space transportation.** Operational use of in-orbit hardware requires space transportation systems to launch the hardware into space. Space transportation relies on rockets, and the most powerful propulsion systems currently available. Rocket engines generate their thrust through the chemical reaction (combustion) of a propellant (fuel) with an oxidiser material. In most cases, the oxidiser is oxygen. These fuel/oxidiser combinations can contain both liquid and solid components. In solid rocket engines, both propellant and oxidiser are pre-mixed and cast into a compound. Aluminium powder and ammonium perchlorate (NH_4ClO_4) are the main reaction components in this case. Kerosene or liquefied hydrogen are the most commonly used liquid fuels, burned in combination with liquid oxygen (LOX). Since hydrogen and oxygen have to be cooled to very low temperatures to stay liquid they cannot be stored for very long inside the launcher's tanks without evaporating. Commonly used, storable liquid rocket propellants are hydrazine and its derivates (e. g. UDMH – $C_2H_8N_2$ – fuel) and nitrogen tetroxide (N_2O_4 – oxidiser). Both components are either highly toxic or corrosive, which complicates their handling. Most space launch systems use the above mentioned propellants (or a combination of them). Hybrid rocket engines, using solid fuels and liquid oxidiser, do exist but are not so common.

The launcher's main function is not just to propel its payload into space, but to put it into orbit around the Earth. To reach a stable low earth orbit (LEO), at an altitude of 200 km, a minimum velocity of 7.8 km/s (28,000 km/h) is required for the object to defy earth's gravitational pull without falling back towards the earth's surface. Reaching orbits beyond LEO, such as geosynchronous earth orbit (GEO) – 35,786 km above the equator – requires the rocket to attain even higher initial velocities. The GEO ist particularly useful for satelite telecommunication applications. It takes satellites in GEO 24 hours to orbit the Earth – these satellites therefore have the same angular velocity as the Earth in rotation. To an observer on the earth's surface, the satellite remains constantly at the same position in the sky, and this ensures stable communications or

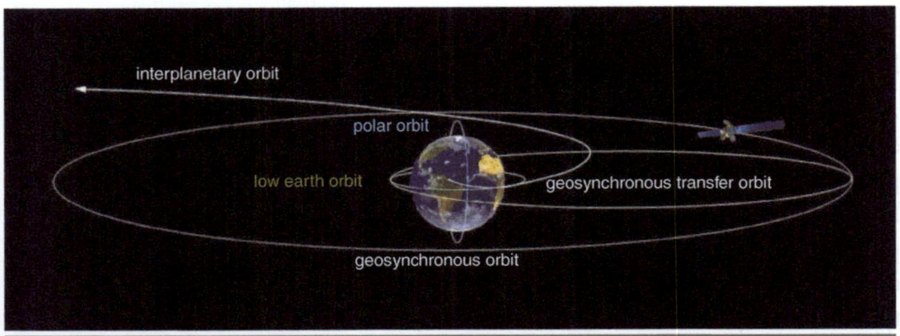

Orbit	Abrv.	Inclination	Altitude	Period	Typical payloads
low earth orbit	LEO	any	200 – 1000 km	78 – 95 min	space shuttle, ISS
sun synchronous orbit	SSO	95° – 100°	400 – 900 km, circular	100 min	earth observation satellites, telecommunications constellations, military surveillance
medium earth orbit	MEO	any	3000 – 36000 km	10 – 20 h	navigation satellites
geosynchronous transfer orbit	GTO	0° - 20°	500 – 36000 km	10 h	satellites aimed for GEO
geosynchronous orbit	GEO	0°	35786 km circular	24 h	telecommunikation satellites, weather satellites
interplanetary orbit	IPO	-	-	-	space probes

Earth orbits for satellites: most payloads are transported into low Earth orbits with inclinations ranging from equatorial to polar. Most modern telecommunications satellites are stationed in geosynchronous orbit, appearing at a fixed position in the sky. Before reaching geosynchronous orbit, satellites are commonly placed into transfer orbit. From there they reach the final orbit using their on-board propulsion systems. In order to reach other bodies within the solar system, space probes are put into interplanetary heliocentric orbits.

Ion propulsion. Left: The atoms of the propellant are first ionised, i.e. electrons are removed from the atomic shell. The now positively charged ions are accelerated and then ejected by applying a strong electric field. Excess electrons are emitted into the ion beam in order to avoid charging the engine. Right: Ion engine of the Deep Space 1 probe during testing.

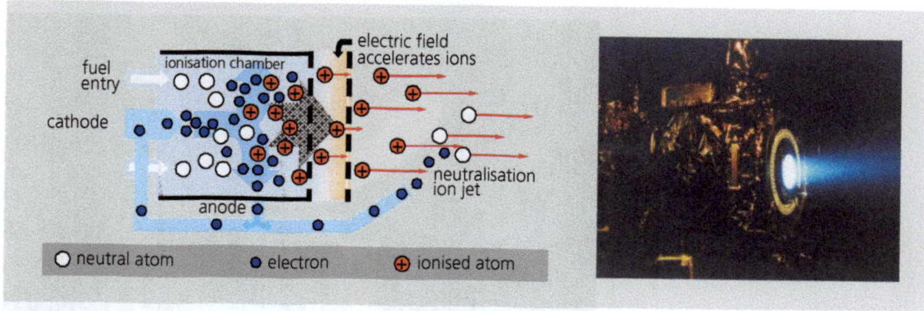

broadcast links with fixed satellite dishes on the ground.

▶ **Spacecraft.** Once a spacecraft has reached its intended orbit in space, it will be released from the launcher and continues on its own, requiring no further thrust to remain in orbit. All satellites have some sort of satellite bus, which is the main structure carrying the satellite's mechanical, electrical and propulsive sub-systems. The mission's payload, *solar arrays* (mostly deployable) and communication antennas are mounted on the bus. Depending on the satellite's mission, these buses ensure the spacecraft's overall integrity. They weigh between 1 kg (CubeSat) and approximately 7 t (cutting-edge telecommunications satellites). To adapt or change their orbit, most satellites have on-board *propulsion systems*, consisting of a set of thrusters, supplied by a propellant reservoir inside the satellite's bus. Modern telecommunications satellites also use electrical or ion propulsion modules for this purpose. The electrical energy required by the onboard instruments is usually generated by solar panels, placed either on the satellite's bus or on deployable structures. Space probes travelling to outer regions of the solar system – where the sunlight is not strong enough for solar cells to convert solar energy into electricity – need radioisotope thermoelectric generators, which convert energy from decaying radioactive isotopes. Individual control centres or a network of ground stations usually communicate with the satellites. Because the ground path of the satellite's orbit largely passes through foreign or inaccessible regions – as well as over the oceans – ground stations cannot provide permanent contact with the satellite. Therefore Satellites are usually equipped with onboard data storage and pre-processing systems, which provide permanent data acquisition, even if the downloading to ground stations is limited to certain visibility slots. For some types of predominantly manned space missions, however, permanent contact with ground stations is required. In this case, data relay satellites can be used. A good example is NASA's tracking and data relay system (TDRS), which currently consists of a constellation of nine satellites in geosynchronous orbit, providing inter-satellite data connections that can be relayed to ground stations worldwide. It is only through TDRS, for example, that live connections with the International Space Station (ISS) can be established.

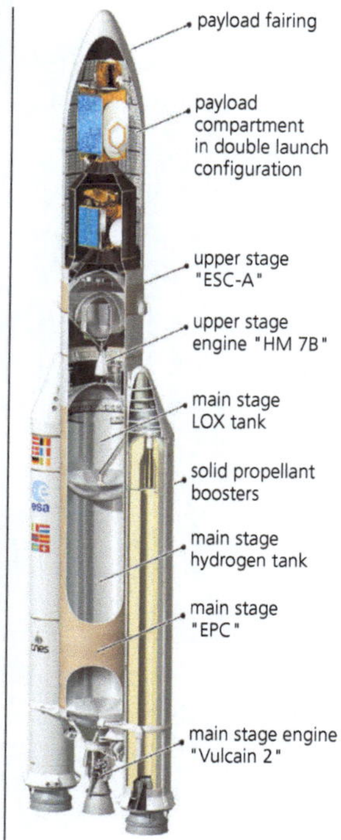

The Ariane rocket: The Ariane 5 launcher uses a combination of liquid and solid propulsion elements. Ariane is able to boost into orbit two telecommunications satellites with a combined mass of up to 9 t.

Applications

The utilisation of space comprises a wide variety of applications, the most prominent of which are telecommunications, earth observation and environmental monitoring, and satellite navigation. Space applications are also very important for manned space exploration and space science.

▶ **Telecommunications.** Two major concepts are used within telecommunications today: fixed satellite services (FSS), where the broadcast or data transmission is relayed between fixed ground stations; and mobile satellite services (MSS), which relay the transmission to portable terrestrial terminals. All FSS satellites are located in geosynchronous orbit, where they are mainly used for direct broadcast television services to small, private satellite dishes. FSS satellites use the C-band (3.4–4.2 GHz) and K_u-band (10.7-12.75 GHz) of the electromagnetic spectrum for broadcasting purposes – the K_u-band providing higher data rates. The uplink and downlink of data via the satellite's channels is carried out by transponders. The uplink signal is received, transformed and pre-processed, then relayed by the transponder. In the K_u-band, the transponder bandwidth is typically around 30 MHz, providing a transmission rate of 40 Mbps, which is equivalent to 16 digital TV channels. The emitting power of direct-to-home transponders is up to 240 watts per transponder. On a high-end satellite, the number of transponders is therefore generally limited to around 60. Solar arrays

Telecommunications satellite: Artist's impression of the Inmarsat XL latest generation telecommunications satellite, which is based on the European Alphabus platform. The satellite has a deployable antenna reflector with a diameter of 12 m, and its solar arrays will generate 12 kW of electrical power. The spacecraft's launch mass will be over 6,000 kg.

X-band radar image of the pyramids of Giza, taken by the German TerraSAR-X satellite: The image was taken with the satellite's High Resolution Spotlight Mode and has a full resolution of 1 m. ©DLR

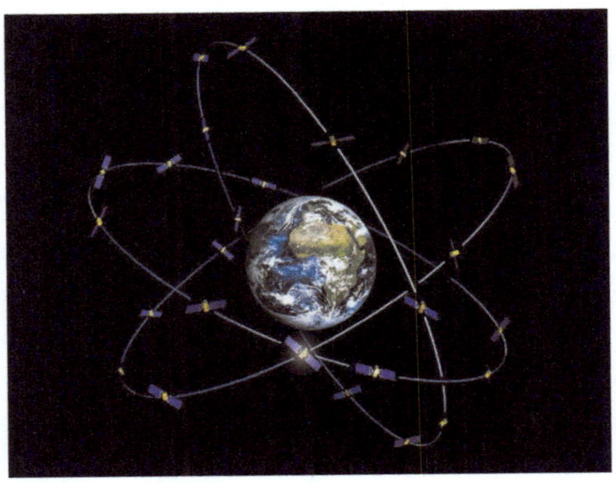

Satellite constellation: Once it is fully deployed, the Galileo constellation will consist of 30 satellites located in three orbital inclinations. This will ensure that, for most locations, at least four satellites are above the horizon at any given time.

provide the 14 kW of electrical power needed. With the emergence of high-definition television, satellites need even more capacity. With a current worldwide requirement for around 20 new satellites to be launched into GEO each year, the state of the telecommunications market is considered to be healthy and buoyant.

MSS satellites can be located in geosynchronous orbit, but constellations in low earth orbit or elliptical orbits are also used. Mobile satellite terminals such as handheld telephones or satellite radio receivers in cars cannot be equipped with fixed satellite dishes and therefore need a stronger satellite signal. The geosynchronous satellites dedicated to MSS (e. g. Thuraya, Inmarsat) have huge reflectors (12 x 16 m²) that ensure communications with individual satellite telephones on the ground. In order to provide effective global coverage, a fleet of at least three geosynchronous satellites is needed. A good example is the Inmarsat Broadband Global Area Network (BGAN) service, which uses a fleet of three modern Inmarsat 4 communications satellites. These satellites are currently in orbit over the Indian, Atlantic and Pacific Oceans and provide voice, IP data, ISDN, fax, and SMS services everywhere except the extreme polar regions. Satellites in low earth orbit require real satellite constellations to provide global coverage: the Iridium and Globalstar networks rely on 66 and 48 active satellites respectively. MSS communications mostly use lower L- and S-band frequencies (from 1-4 GHz) that still offer high capacities, but relatively low interference with terrestrial systems and low atmospheric absorption.

► **Earth observation.** Satellites in orbit around the Earth are ideally positioned to provide a global view of a wealth of topological, environmental and climatic phenomena on the ground and in the atmosphere. Most earth observation spacecraft are in low polar or sun-synchronous orbits, where satellites circle the Earth in about 90 minutes. While the Earth rotates below, the satellite covers its entire surface approximately every three days. The data can be used to improve our knowledge and to advance the fields of atmospheric and climate research, geophysics and cartography, agriculture, prospecting, oceanic and cryosphere research, disaster monitoring, and military intelligence. In recent years, *synthetic aperture radar (SAR)* technology, in particular, has been developed sufficiently for use on spaceborne platforms. SAR is an imaging radar technique that uses a moving radar to survey targets on the ground. The SAR instrument can measure the reflected signal's intensity and phase, and this can be used to generate high resolution radar images. Because the SAR imagers provide their own illumination (radar pulses), the system does not require solar illumination and can also work at night. Furthermore, SAR can also "see" through cloudy and dusty conditions, which is not possible with visible and infrared instruments. This is because radar wavelengths are much longer than those of visible or infrared light. The maximum resolution of civil systems, such as Germany's TerraSAR-X or Italy's COSMO-Skymed, is one meter. For military SAR systems, this information is usually classified, but sub-meter resolutions can be obtained.

► **Satellite navigation**. Modern *traffic management* relies on the operational availability of space-based navigation systems. Today, nearly all traffic systems make use of satellite navigation signals. A space-based navigation system consists of a global satellite constellation. The individual satellites emit a continuous timing signal, which is generated by extremely accurate onboard atomic clocks. The receiver calculates its three-dimensional location by *triangulation* of the received signal propagation time, which is measured from at least three satellites. Position accuracy depends on the synchronisation of the satellites' atomic clock and the receiver's internal clock. For more precise measurements, the signal of a fourth satellite is required to fine-tune the timing signal. In addition to the positioning measurement, analysis of the received signal's Doppler shift provides a precise measurement of the receiver's actual location.

The US *Global Positioning System (GPS)* consists of around 30 active satellites in medium earth orbit. They are distributed in such a way that, for any GPS receiver on the ground, at least four satellites are above the horizon, from which signals can be received. Positioning precision for authorised users is around 1 m. The available resolution for the general public is still within 10 m. The GPS timing signal is accurate to 100 ns and can be used to synchronise spatially separated processes.

Once it is fully deployed in orbit, the European *Galileo* navigation system will use the same concept. The Galileo system is currently under development and shall be operational by 2013. Galileo will provide a publicly available Open Service, with an accuracy that is comparable to GPS, as well as a subscription-based Commercial Service, which will be accurate to within less than 1 m. For safety and security applications, special signal modes will resist jamming interference. The Galileo and GPS signals will be compatible so that users will be able to receive signals from both systems: this will ensure more accurate services and wider geographical coverage.

► **Manned missions**. Space is a unique environment: this is what makes research and exploration programmes so interesting. However, the technical effort required for astronauts to work in space is enormous. A highly robust life support system is a must for manned spacecraft or laboratories. Although the structure of the spacecraft itself provides some degree of passive shielding from the harsh environment of outer space, there can be nothing passive about the life support system inside the spacecraft. The astronauts on board need fresh air and a controlled climate, as well as water and power supplies. The oxygen supply is one of the main issues. On board the *International Space Station (ISS)*, oxygen is generated electrolytically from water, and the station's large solar arrays provide the electricity required. The excess hydrogen from the electrolytic process is vented into space. Water is also precious, of course and must be recycled. All waste water on the station is collected, purified and later reused as drinking water by the astronauts. Despite its complexity, the life support system must be extremely robust, and able to operate without major disruptions for many years.

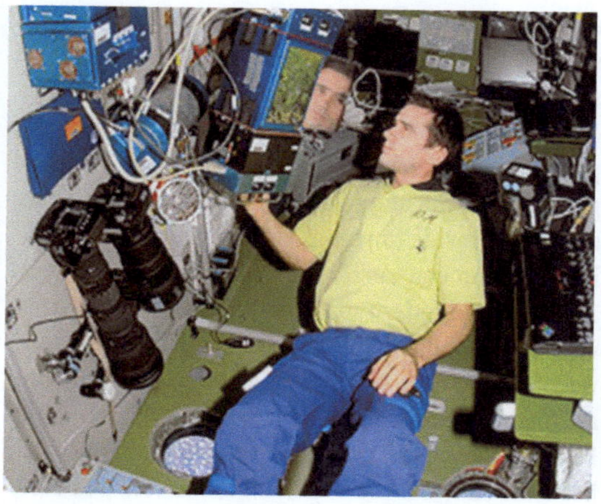

On board the ISS: As the ISS is an active laboratory, almost all of the room on board is reserved for hardware, and the stowage of equipment, tools and supplies. In the picture, Yuri Malenchenko checks the growth of pea plants. Through the windows below, clouds in the Earth's atmosphere are visible.

The ISS is the most complex space infrastructure currently in existence. In total, 40 space missions will be required to complete the ISS. Eventually, the 10 main modules and nodes will house 1200 m^3 of living and working space, and will weigh a total of 450 t. The wingspan of the solar array will be 80 m. The solar array will provide 110 kW of energy, supplying – among others – the 52 computers needed to control all of the systems on board the ISS.

Approximately 2,000 hours of extravehicular activity will be required to construct the ISS, and to install and connect all of its components. In order to work in space, astronauts use special and individually adapted space suits that protect them from the vacuum and radiation. Space suits provide all the astronauts' life support functions, including their air supply, thermal and pressure control, and micrometeorite shielding. Space suits also provide functions that are used in the working environment, such as communications, cameras and lighting. Modern space suits such as the US Extravehicular Mobility Unit (EMU) can be used for spacewalks lasting over eight hours. During that time, the astronaut travels more than five times around the Earth, spending half that time in bright sunlight, which heats the suit; and the other half in the cold and dark. In order to withstand these extreme conditions, the suit has several layers of different materials, ranging from the liquid cooling and ventilation garment, which is worn next to the skin, and is used to control the body's temperature, to the thermal micrometeoroid garment, which consists of seven layers of aluminized Mylar laminated with Dacron. The outermost layer is white Ortho-Fabric, made from Gore-Tex and Kevlar. A fully equipped EMU weighs 88 kg. In orbit, this weight is, of course, compensated by microgravity.

The end of any manned space mission is its re-entry into the earth's atmosphere. As the spacecraft enters the atmosphere at velocities that are just short of the orbital velocity, the spacecraft's kinetic energy increases significantly. Once inside the earth's atmospheric layers, the friction of the air molecules starts to transform the kinetic energy into heat, ionizing the local gas, which is turned into extremely hot and glowing plasma. During this phase, certain exposed areas, such as the wing edges or the Space Shuttle's nose cap, are heated to 1500 °C. In order to withstand these thermal stresses, all spacecraft need a heat shield that protects the inner structure during re-entry. The Space Shuttle's heat shield mostly consists of ceramic tiles, but carbon *composite material* is also used for the most exposed areas. The Russian Soyuz capsules use "ablative" heat shields that scorch and erode during atmospheric re-entry. This ablative shield is very effective, although it cannot be re-used.

Trends

All space hardware must be carefully engineered and qualified to withstand the extreme physical conditions in space (and on the journey into space). Once a technical system or a particular material has proven its ability to serve its purpose and operate faultlessly in space, manufacturers and users tend to be very conservative and stick with the proven solution. Nevertheless, the development of space hardware is extremely costly, and innovation is still one of its most important drivers.

► **Communication technologies.** Satellite communications is at the forefront of modern utilization of space. The need for high data rates is driving the development of technology, particularly in the field of telecommunications, where the commercialisation of the K_a-band is an ongoing trend. Because of its higher frequency compared with other microwave bands – and its smaller beam width – the K_a-band is very well suited to providing faster broadband services. Satellites are equipped with multi spot beam (18.3 to 20.2 GHz) transmitters that do not cover an entire region evenly but provide smaller footprints, allowing for considerable frequency reuse and much higher power. Many satellites with K_a-band payloads have been ordered in the last few years and feedback from their initial use in orbit has been positive. Use of the K_a-band provides the capability to increase data rates to 800 Mbit/s. It can also be used for non-commercial missions with higher bandwidth communication requirements, such as multi-spectral instruments for earth science applications.

To obtain even higher data rates, *optical data transmission* technologies and laser communication systems are required. Such systems are currently being tested and made ready for space applications. Laser communication systems work much like fibre optic links, except the beam is transmitted straight through free space or the atmosphere. Laser communication terminals offer several advantages for satellite applications: in addition to their high data rate, they are lightweight and efficient in electrical power consumption – two critical factors for orbital infrastructures. *Laser*

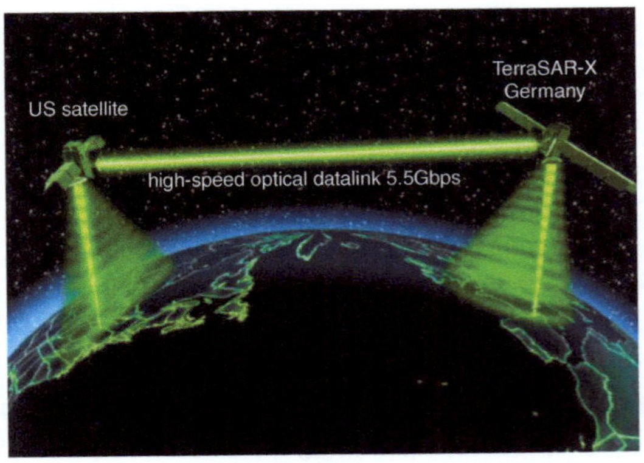

Communication technologies: High-speed optical inter-satellite laser communication is a field of technology where major progress has recently been made. Today, data rates of up to 5.5 Gbit/s have been demonstrated. Source: Fraunhofer HHI communication systems can also establish communication links with a data rate of up to 10 Gbit/s. In recent years, optical inter-satellite links, satellite-to-aircraft and satellite-to-ground links have been successfully created. In March 2008, a link was successfully established between the German TerraSAR-X and the US NFIRE satellites with a demonstrated data rate of 5.5 Gbit/s over a distance of up to 5,500 km. The error rate was better than 10^{-9}. Data relay will be the most beneficial, future application for laser communication systems operating between satellites in earth's orbit and beyond. The sensitivity of the laser beam to atmospheric cloud absorption, however, means the involvement of ground stations in the network will remain limited.

▶ **Robotics.** Robotic explorers have provided us with almost everything we currently know about the geological formations and soil compositions of other planets. Putting a space probe onto a planet's surface requires an extremely autonomous robotic system that can function without immediate control from operators back on Earth. Because of the long distances between Earth and most extraterrestrial bodies, the signal travel times can vary from minutes to hours. Operations cannot therefore be carried out in real time and need to be highly automated. The rovers on Mars are equipped with high-resolution stereo cameras that allow operational teams to generate a three-dimensional view of the area. This information is used to plan the rovers' operations. Once the signal for a specific action has been sent by the control station, it can sometimes take hours to see the end result. Most recently, a highly-sophisticated robotic laboratory landed on Mars on board the Phoenix spacecraft. Its largely autonomous scientific payload provided the first direct proof of ice on Mars' surface. The robotic laboratory did this by digging samples from the Martian soil and analysing them in the thermal and evolved gas analyser, which is an automated high-temperature furnace with a mass spectrometer.

More and more powerful robotic applications are being used on space missions each year. Robotics will be one of the key technologies in the future exploration of space – for manned and unmanned missions alike.

Robotics: Phoenix is the most advanced laboratory to be placed on another planet. Shortly after it landed in the polar regions of Mars, Phoenix directly detected ice by analysing the Martian soil.

The Orion spacecraft: Currently under development, Orion will succeed the Space Shuttle as the next generation manned space transportation system. NASA also plans to use Orion to return to the Moon.

Prospects

- In just over 50 years of space exploration, countless space applications have made their presence felt in our daily lives. In a more and more globalised world, the utilisation of space has become crucial. The fields in which space applications currently contribute the most are: telecommunications and broadband data relay, satellite navigation, earth observation with weather, climate and environmental monitoring, space science and medicine
- Human space exploration is another of the main factors driving space technology development. When NASA decided to discontinue the Space Shuttle programme in 2004, one design requirement for its successor – the Orion spacecraft – was its ability to reach beyond earth's orbit and to allow missions to the Moon and, maybe, even to Mars. It will take at least another decade to undertake these new exploration efforts, but one day Man will surely walk on Mars. In all probability, space exploration will be just as challenging in 50 years' time as it is today: space utilisation and exploration will always require the most advanced technologies.

DR. THILO KRANZ
German Aerospace Center, Bonn

Internet

- www.dlr.de
- www.esa.int
- www.nasa.gov
- www.russianspaceweb.com
- www.astronautix.com
- www.estp-space.eu
- www.isi-Initiative.org

8 ENERGY AND RESOURCES

Energy forms the very basis of our existence. Around 13 billion years ago the expanding universe developed from one point with unimaginable energy. And energy is the pre-condition for human life, without which it would not have come into existence, and without which it cannot exist. For our industrial society today, energy is particularly associated with the production of goods and foodstuffs, mobility and communication.

World energy requirements currently stand at 107,000 TWh/a. We consume around 13 billion litres of crude oil and approximately 14.7 million tonnes of hard coal and brown coal daily on this earth, most of this for the generation of electricity and heat, and for motor vehicles, airplanes and trains. Burning these fossil fuels causes approximately 25 billion tonnes of the greenhouse gas, carbon dioxide (CO_2), to be released each year into the earth's atmosphere. Apart from the finite availability of these fossil fuels, emissions are also contributing to the new global problem of climate change. The increase in greenhouse gas concentrations is causing the atmosphere to warm up – by 0.4 K in the last 15 years alone. This is causing changes in the climate, especially an increase in natural catastrophes such as strong storms and floods.

It is assumed that the steadily growing world population, and the growth in demand in developing countries as they strive for a better standard of living, will lead to a further increase in energy requirements of around 1.3 % annually by 2030. It is true that advances in energy efficiency will moderate this increase, but they cannot fully compensate for it. It is foreseeable that fossil fuels will still provide the significant proportion of future energy requirements. Despite the rates of increase expected in regenerative energies, it is expected that by 2030 these forms of energy will only represent a small proportion of total energy requirements – notwithstanding present growth rates of 9 % per annum reached by solar energy, largely thanks to government funding and support. This scenario means that CO_2 levels will continue to increase steadily, intensifying the greenhouse effect. Rigorous efforts must therefore be made to implement those strategies which reduce CO_2 output. These are:

- Energy savings through a change in attitude, not using energy, or additional measures to reduce energy consumption, such as renovation of buildings
- The development and use of more energy-efficient products
- Increasing the degree of efficiency when generating electricity
- The development of cost-efficient regenerative forms of energy

As the largest source of the earth's energy, the sun consistently supplies about 10,000 times the amount of energy needed worldwide. This should foster an understanding of the need to overcome the challenges associated with the use of this sustainable and emission-free form of energy.

► **The topics.** Even if the *fossil fuels, oil and gas*, are finite, they will still ensure an energy supply for the world for the foreseeable future. It is therefore worth using efficient technologies to discover new deposits and to extract the raw materials. Increasingly, this also includes areas which are difficult to tap, such as under the deep sea. However, in addition to these important energy providers, gas and oil, we must not disregard the fact that raw materials such as metals, from which our products originate, also have to be extracted from the earth. There are already shortages of some of the rarer raw materials, so that lower minimum amounts have to become viable for exploitation, whereby more efficient methods must compensate higher costs.

The following energy sources nowadays serve the generation of energy: *fossil energies* (coal, gas), *nuclear power* and regenerative sources. The latter subdivide into *biomass, wind, water power, geothermal energy* and *solar energy.* While traditional power stations are already optimising their efficiency and tests are currently being carried out for separating CO_2 and depositing it in underground storage areas, the security aspect is still at the forefront of the use of nuclear power. At the same time, the aspects of operational safety and the security of the ultimate storage of deposited radioactive waste must be taken into account. Nuclear fusion is still a vision or a future option for mankind - although it will not become reality for several decades, if ever.

Increasing energy extraction from biomass as a substitution for fossil fuel, mainly the extraction of ethanol or biodiesel, is currently a subject of debate, particularly in view of safeguarding global nutrition needs. Important in this context is the development of efficient processes to effectively produce alternative fuels from non-food plants.

We can use the energy of the sun by transforming it in different ways. Electrical energy can be generated directly by means of photovoltaics or in solar thermal power stations via heat generation, so that steam drives conventional turbines. Heat is also used directly in solar collectors. The availability of wind energy, similar to solar energy, is also very dependent on daily and seasonal patterns, and on the chosen site. The strong upswing in power from offshore units is particularly noteworthy: wind speeds are higher on the open sea, and noise disturbance and impairment of the landscape (caused by increasingly large rotors) are not a problem.

Apart from the conventional use of water power at appropriate sections of a river, or reservoirs, attempts are also being made to extract energy by making use of movements in the sea. Both the energy of individual waves, and the energy from sea and tidal currents are being considered.

A further large supply of energy is concealed in the earth itself. *Geothermal energy* partly originates from the heat remaining from the time when the earth came into existence, and partly from radioactive processes of decomposition. Geothermal energy has already been intensively used in appropriate high enthalpy storage locations close to volcanoes. Deep geothermal processes require thorough advance investigation of the geological pre-conditions and there are still some technical challenges to be overcome. It calls for drilling to several thousand metres, combined with the installation of a closed water circuit.

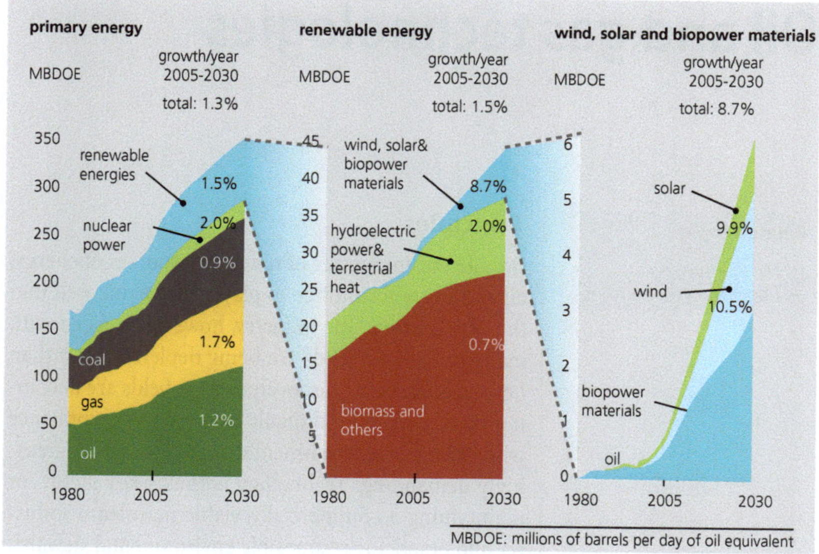

Global energy requirements by energy sources: It is predicted that global energy requirements will increase annually by 1.3% until 2030. Renewable energy sources will increase disproportionately, particularly wind and solar energy. However, as these currently only make up a small proportion of the primary energy supply, fossil energies will continue to dominate. Source: ExxonMobil

Energy storage devices must be developed to cover the transition between energy generation and energy use, so that regenerative (primarily wind and solar) energy can also be integrated into base load supply, Energy storage can take place in different forms, including chemical (battery), pneumatic (compressed air) and hydrogen alternatives. Hydrogen is under much discussion in this context because it is not a source of energy, but simply a carrier. In nature, only chemically bound hydrogen exists, and it can only be produced as a pure element when energy is employed. The fuel cell is the preferred system for the use of hydrogen. It enables "hydrogen energy" to be transformed directly into electrical energy without combustion. Applications of fuel cells are currently being tested on different scales.

Energy storage devices include *micro-energy storage* systems: although hardly relevant to levels of mass energy production, these systems ensure that appliances in the home or in communications can increasingly be operated on mobile networks. New developments are expected, extending beyond the traditional battery. Small individual consumers, in particular, should be able to extract their energy requirements directly from the environment (energy harvesting). ■

Oil and gas technologies

Related topics

- Fossil energy
- Measuring techniques

Principles

Despite the emergence of renewable sources of energy, oil and gas will continue to play a major role well into the 21st century. The industry, however, is faced with many challenges: fields are being depleted faster than new ones are being discovered; large fields are becoming more and more difficult to find; and compliance with stringent environmental regulations is increasingly demanding. Technology is thus a key factor for maintaining a commercially viable petroleum industry and practicing responsible environmental stewardship.

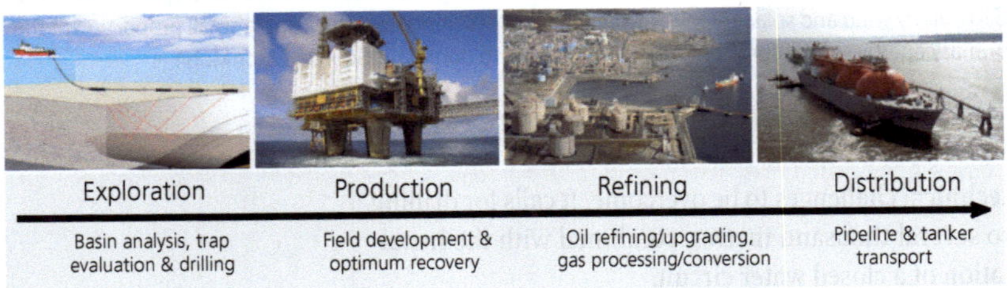

Simplified hydrocarbon value chain. Source: StatoilHydro ASA

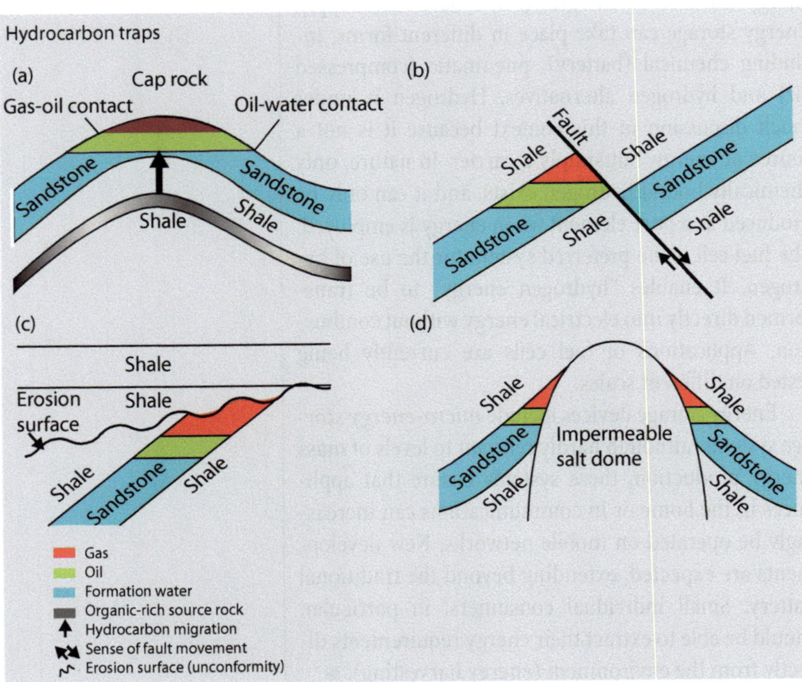

Common hydrocarbon traps. (a) An anticlinal trap where hydrocarbons have accumulated in an arched or domed bed of porous and permeable reservoir sandstone, sealed by impermeable cap rock made of shale. If gas is present, it will rise to the surface of the reservoir and will have successive underlying layers of oil and formation water due to density stratification. (Some formation water is also present in the gas and oil zones.) The arrow indicates the migration of hydrocarbons from an organic-rich source rock into the overlying reservoir. (b) A fault trap, where a bed of sandstone reservoir rock is juxtaposed against impermeable shale due to a sealing fault (a fault is a plane or zone of weakness along which rock masses are broken and displaced.) (c) An unconformity trap, where a tilted reservoir sandstone bed is sealed by impermeable shale sitting on top of an old erosion surface. (The flat beds overlying the erosion surface do not conform to the underlying tilted beds.) (d) A salt dome trap, where a reservoir sandstone bed abuts against a rising dome of impermeable low density salt, causing the host sediments to bend upwards against its flanks. Source: StatoilHydro ASA

Exploration. The entrapment of oil and gas in sedimentary basins (i.e. subsiding areas of the earth's crust that are filled with thick sequences of sedimentary rocks) depends on the timely and fortuitous combination of the following:

- A petroleum charge system: source rocks (mainly shales) that are rich in insoluble organic matter (kerogen) and capable of generating and expelling hydrocarbons
- Reservoir rocks: predominantly porous and permeable sandstones and limestones in which the hydrocarbons may accumulate
- Effective topseals: impermeable rocks (mainly shales) that impede the hydrocarbons from leaking out of the reservoirs

As the source rocks are buried and subjected to increasing temperature (and pressure), the kerogen is chemically broken down and converted into hydrocarbons. Optimum oil generation occurs at temperatures around 100–110 °C, followed by gas. The hydrocarbons then migrate (move) from the source rocks into overlying sediments, where some of them may be trapped. The explorer's job is to find such accumulations.

The subsurface (underground) geology of prospective basins is mainly reconstructed from 2D seismic

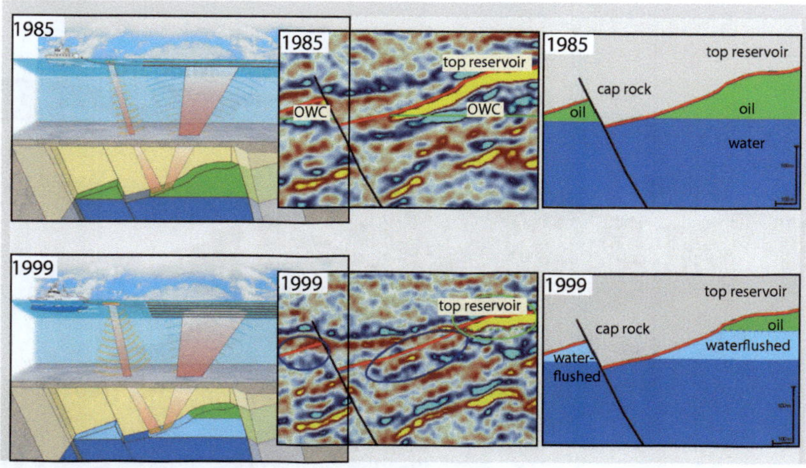

A well-known example of 4D seismic surveying from the North Sea StatoilHydro-operated Gullfaks field: The figures to the left illustrate the standard 3D seismic acquisition concept, in which a survey vessel tows a sound source and multiple streamers fitted with hydrophones. The hydrophones collect the sound waves as they are reflected back from the subsurface rocks. Because of the close spacing between the streamers and survey lines, computers are able to represent the information as 3D images for geological interpretation. The changes in the seismic data between the 1985 and 1999 surveys (centre) are interpreted in the right-hand figures. The oil/water contact (OWC) in the top reservoir to the right of the fault has risen, and the oil from the smaller reservoir to the left of the fault is fully depleted. Source: StatoilHydro ASA

images, assisted by detailed information from any nearby wells or outcrops displaying the same sequences. The images are produced from seismic data acquired by transmitting sound waves down through the underlying rocks, and recording the waves as they are reflected back to the surface from contrasting rock surfaces. The geological reconstructions are then used to identify potential traps and assess how likely they are to contain hydrocarbons - and if so, if there is a sufficient quantity to justify drilling. This requires the interpretation, modelling and simulation of the processes that led to their development over tens to hundreds of millions of years (structural, sedimentary, physical, chemical, biological and fluid dynamic processes).

Exploration seismic data sometimes hint at subsurface fluid content (gas/water), but electromagnetic sounding is far more effective for suggesting that oil and gas are present prior to offshore drilling. Nevertheless, the risk of drilling "dry holes" remains high (i. e. exploration wells that do not reveal commercial hydrocarbons). This is hardly surprising when one considers the uncertainties involved, and that most oil and gas accumulations are found some 2–4 km below the earth's surface. Crude oil may even be found at depths down to 6 km, while the depth of natural gas deposits may be greater or lesser (>9 km).

► **Production.** Appraisal and production technology is used to evaluate the commercial potential of a new field and the optimum means of maximising hydrocarbon recovery. A field normally contains a stack of several reservoirs separated by non-reservoir rocks (e. g. shales). Predictive computer models are used throughout a field's lifetime to simulate reservoir and field performance according to various development strategies.

Production wells convey hydrocarbons to the surface. Oil production may initially rely on natural reservoir energy because reservoir pressure is far greater than the pressure at the bottom of the wellbore. However, artificial lift systems (such as pumps) are often introduced as the differential pressure declines. Even so, much oil is left behind.

Greater recovery can be obtained by injecting external fluids through injection wells positioned in reservoir intervals which are in direct (fluid) contact with their production counterparts. The water-alternating-gas (WAG) procedure is one example where the higher-density water is injected to push the remaining oil from the lower parts of a reservoir towards production wells, and the lower-density gas is injected to push the remaining oil from the upper parts of a reservoir. Declining reservoir pressure is also boosted in this way.

Production is monitored using well data and geophysical survey techniques such as time-lapse and 4D *seismic methods*. In the 4D seismic method, the results of 3D seismic data acquired precisely over the same area are compared at different points in time (before a field starts production versus post-production stages, for example). Time is the 4th dimension. Reliable differences between repeated, high-resolution 3D seismic surveys can be attributed to changes in fluid and pressure fronts under suitable reservoir conditions. The use of 4D seismic surveys can therefore be invaluable for identifying drained areas, locating remaining pockets of oil, and helping to optimise the positioning of new production wells.

To ensure safe, uninterrupted and simultaneous transport of the produced gas, oil and water through flowlines and pipelines from reservoirs to processing

Modern Ramform Sovereign survey ship acquiring a swathe of 3D seismic data. The ship tows a multiple array of 20 streamers studded with hydrophones and multiple sound sources (air guns) located between the streamers and the ship's stern. The streamers are typically 6 km long and the spacing between the hydrophones is about 12.5 m. Source: Petroleum Geo-Services (PGS)

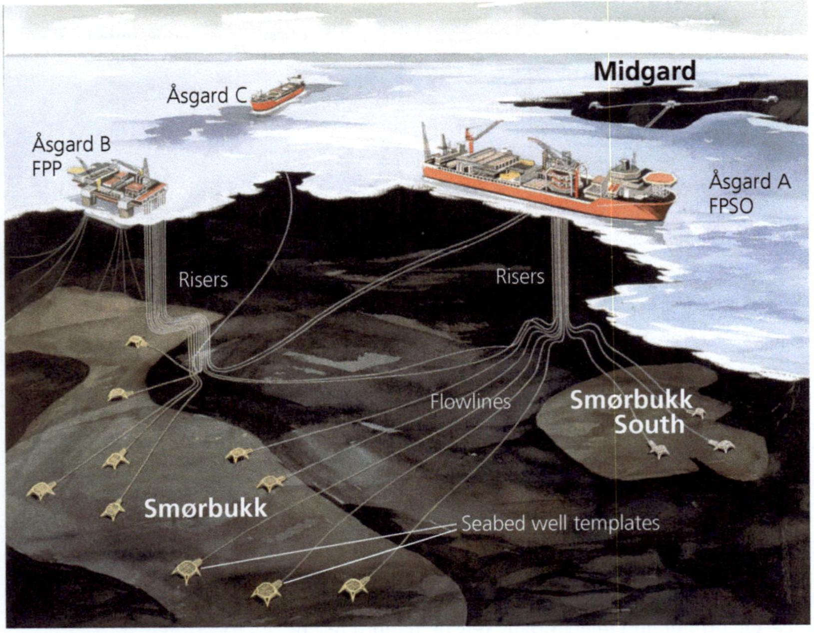

The complexity of subsea flow assurance at the StatoilHydro-operated Åsgard field: The field in the Norwegian Sea is located 200 km offshore and in water depths ranging from 240–300 m. The "field" actually consists of three separate fields developed as one: none of the three would have been economically viable if developed alone. Smørbukk contains gas condensate (gas vapour that condenses to light oil at the surface); Smørbukk South contains oil and condensate; Midgard contains gas. The multiple, stacked reservoirs from the Jurassic age (not shown) are buried as deeply as 4.85 km below the seabed. The production and injection wells have been drilled and completed subsea through well templates (manifolds). They are connected to two production facilities by hundreds of kilometres of flowlines/pipelines and risers, through which the hydrocarbon well streams and injection fluids flow. Åsgard A is a permanently-moored floating oil production, storage and offloading ship (FPSO), almost 280 m long and equipped with oil processing facilities; Åsgard B is a floating gas production platform (FPP) equipped with gas processing facilities. The permanently moored vessel in the distance (Åsgard C) is a condensate storage tanker which receives condensate from Åsgard B. Flowlines linking the Midgard accumulation to the Åsgard B riser system are up to 53 km long. The oil is temporarily stored at the field before being shipped to shore by tankers. The gas is piped to a processing plant on the Norwegian mainland at Kårstø. Source: StatoilHydro ASA

facilities, flow assurance and multiphase transport technology are necessary. Flow may be obstructed in two ways:

- Physical obstruction, by gas and liquid surges (slugs) and the deposition of small reservoir rock fragments that are dislodged during oil production and incorporated in the well stream (produced sand)
- Physico-chemical obstruction, by the formation of asphaltenes, wax, emulsions, gas hydrates and scale

The industry is capable of eliminating, controlling and inhibiting these phenomena (at least to some degree). For example, the addition of a thermodynamic "anti-freeze" inhibitor – usually mono-ethylene-glycol (MEG) – may effectively counteract gas hydrate precipitation.

▶ **Processing and distribution**. Some primary processing takes place at the fields, but the majority of the work is performed in natural gas processing plants and oil refineries. Processing involves the separation of raw well streams into their constituent parts (gas, oil and water), followed by fractionation and purification of the hydrocarbons so that they meet sales or delivery specifications.

Natural gas is purified until it consists of almost pure methane by removing unwanted components, such as carbon dioxide and hydrogen sulphide (e.g. using amine), water (e.g. using glycol dehydration) and liquid hydrocarbons. Once separated, crude oil is divided into fractions by distillation and further treated in a variety of large chemical processing units before the extracts are sold as fuels and lubricants. Some of the lighter distillation products form the basis for petrochemicals.

Oil is distributed to the market by road, by sea in tankers, and through pipelines. Gas is usually transported through pipelines. In areas lacking pipelines, the gas can be liquefied by cooling to about -163 °C at atmospheric pressure (*Liquefied Natural Gas – LNG*). This results in a 600-fold reduction in volume, allowing large quantities to be distributed in specially-designed LNG tankers. Liquefied and compressed natural gas can also be carried in road tankers over shorter distances.

Applications

▶ **Deep electromagnetic sounding**. Controlled-source electromagnetic (CSEM) sounding or "seabed logging" is a rapidly evolving offshore technique for revealing the presence of oil and gas prior to drilling. The principle is based on contrasting resistivities between reservoir rock containing hydrocarbons (high resistivity) and rock containing saline formation water (low resistivity). Resistivity is detected by measuring the subsurface response to the passage of ultra-low fre-

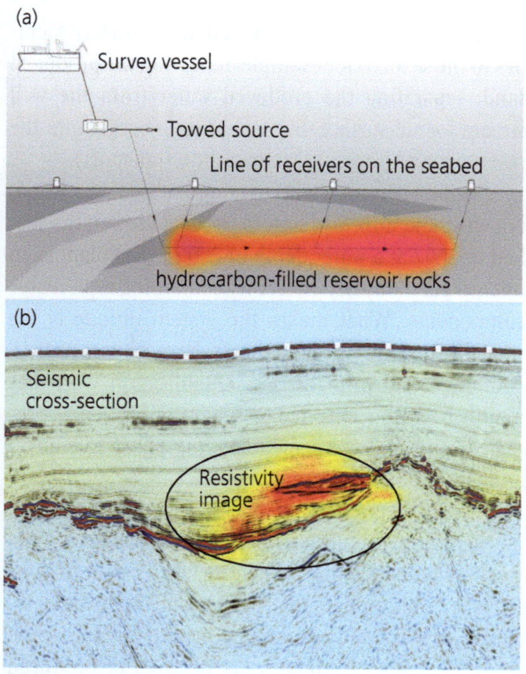

◼ Deep electromagnetic (EM) sounding: (a) The surveying concept. The downward EM fields pass laterally through the targeted reservoir rocks and return to the surface, where they are recorded by receivers on the seabed. Their strength is only slightly diminished when they pass laterally through hydrocarbon-filled reservoir rocks (high resistivity), but they are weakened (attenuated) if the reservoir rocks contain saline formation water. (b) A resistivity image superimposed on a seismic cross-section (the fuzzy red area represents higher resistivity; the fuzzy yellow area represents lower resistivity. Source: ElectroMagnetic GeoServices (EMGS)

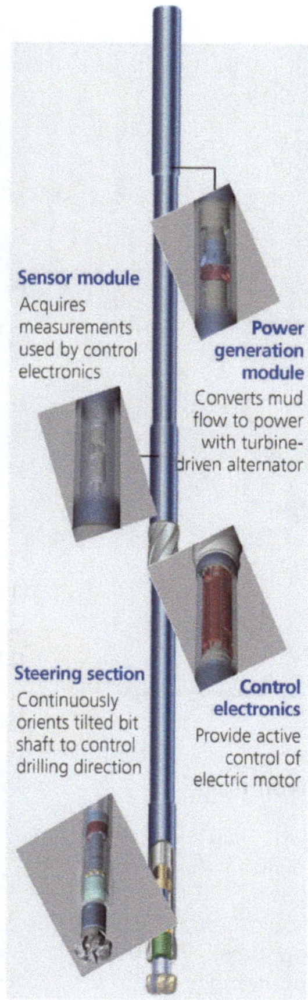

◼ Components of a rotary, steerable system for drilling extended-reach wells (PowerDriveXceed): The internal steering mechanism continuously orients the bit shaft to drill in the desired direction. Source: Schlumberger

quency, diffusive electromagnetic (EM) fields (subsurface resistivity is the resistance of the rocks and their contained fluids to the passage of an EM field).

The basic method uses a powerful, mobile electric source, which transmits frequencies between 0.01–10 Hz down through the sediment column. An array of EM receivers is deployed on the seabed over a targeted, but still unproven, hydrocarbon reservoir to measure the returning signals at the seabed as the source is towed over them. Each of the receivers contains an instrument package, detectors, a flotation system and a ballast weight. In general, the detectors consist of three-component electric and magnetic dipole devices, capable of recording vertical and horizontal components of the EM field. Once the survey has been run, the receivers can be retrieved by detaching the ballast weight.

The returning signals are recorded against time and analysed by computers to determine the overall electrical resistivity of the target. However, 2D and 3D images of resistive bodies can be converted from the time domain to the depth domain using a technique known as inversion. The images can also be superimposed on depth-migrated seismic sections to link the inferred presence of hydrocarbons to potential geological traps. The method is far from foolproof but is an important addition to the exploration tool kit.

▶ **Advanced production wells.** Extended reach wells (also known as horizontal wells if the well bore departure is approximately 80° or more from the vertical plane) are designed to maximise production by laterally penetrating reservoirs that are wider than they are thick, or by penetrating multiple targets in complex geological settings. Such wells are capable of draining the same amount as several wells with less deviation, and may extend up to 10 km or more from the *drilling* site. Their construction requires a combination of directional drilling and geosteering (well positioning) capabilities.

Directional drilling essentially involves pointing the drill bit in the direction that one wishes to drill. Over the years the procedure has evolved from early techniques such as deflecting the borehole with whipstocks (inclined wedges that force the drill bit to drill away from the well bore axis) to advanced rotary steering systems. The latest models have an internal steering mechanism which continuously orients the tilted bit in the desired direction (despite the rotation); a control system (an electronic sensor package that takes measurements to control the steering assembly); and a turbine-based power generation module supplying power for steering and control. Continuous rotation of the drill string (the drill pipe, bottomhole assembly and other tools) provides better borehole quality and faster penetration than previous techniques.

Geosteering is used to guide the path of the advancing borehole relative to the geological strata, often at extremely long distances and depths of several kilometres. Pre-studies quantify uncertainties about:

- the desired well path
- the subsurface geological model of the area
- the size and geometry of the target

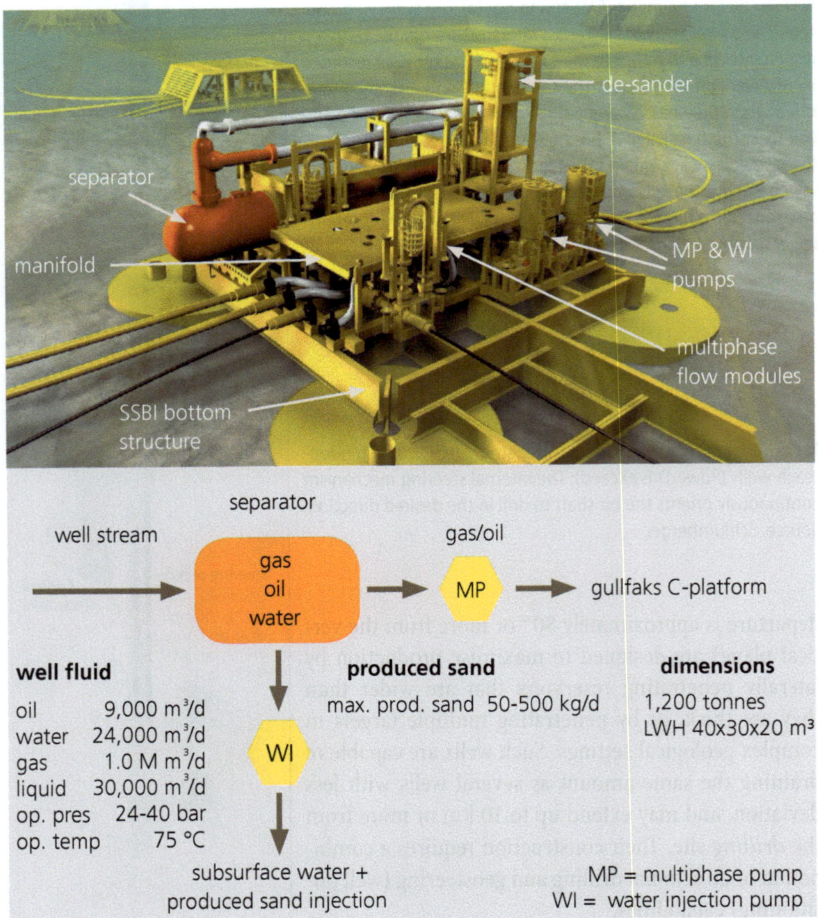

Subsea separation, boosting and injection system (SSBI): The system increases oil recovery from the Tordis satellite field, 2 km below the North Sea. The well stream enters the 180 t gravity-based separator, which is designed to remove up to 24,000 m³ of produced water and process 30,000 m³ of liquid per day. The 60 t de-sander unit removes the produced sand. The separated oil and gas is remixed, boosted by the multiphase pump (MP), and transported to the North Sea Gullfaks C platform. The untreated produced water is mixed with the produced sand and injected into an underlying, saline, sandstone aquifer. These measures strongly reduce pressure at the well head, thus increasing the pressure differential between the well head and the reservoir. This, in turn, increases oil recovery. Source: FMC Kongsberg/StatoilHydro ASA

The subsequent integration and visualisation of real-time, measurement-while-drilling data (measurements of rock properties taken at or near the drill bit) provide the basis for making directional adjustments as drilling proceeds.

► **Subsea processing**. Remotely-controlled subsea processing systems promise to play an important role in field development. The aim is to transfer some of the processing tasks normally carried out in surface facilities to the seabed (for example, removing the produced sand, separating the produced water from the well stream for subsurface disposal or re-injection into the reservoir, and boosting hydrocarbon transport).

The first commercially-operated subsea separation, boosting and injection system was installed on the seabed in 2007 at a water depth of 200 m. The plant is an integrated, modularised facility containing replaceable components. What makes the system unique is the way the marinised components have been adapted, combined and miniaturised. Qualification work has mainly been concentrated on separator efficiency, produced sand management (including pump tolerance) and electrical power supply.

The plant removes produced water from the well (subsea) and re-injects it back into a deeply buried, water-filled sandstone formation via a separate well (subsea). This reduces the backpressure at the field, resulting in increased production rates and hydrocarbon recovery. It also eliminates the disposal of produced water into the sea. The hydrocarbons are piped to a platform for further processing, storage and export.

Trends

► **Diver-free gas pipeline intervention**. Adding branch lines to major subsea gas pipelines is difficult, especially when they are located in water depths exceeding statutory diving limits: hence the emergence of diver-free, remote-controlled hot-tapping systems. The equipment is lowered by crane from a support vessel, guided into place by an *ROV (remotely operated vehicle)*, and operated via an electrical-hydraulic control system attached to the support vessel by an umbilical.

The fully automated system is able to cut a hole into a steel membrane in a pre-installed Tee-junction (installed on a pipe prior to pipe laying) while the mother pipeline continues to transport gas under full pressure. The initial system is qualified for use in water depths up to 1000 m and can handle the rotation of Tee-junctions up to 30° from the vertical, which may occur during pipe laying. The maximum operating pressure is 250 bar and the maximum diameter of the cut hole is 40 cm.

Further work has resulted in the remote installation of retrofit hot-tap Tee-junctions - once again with the mother pipeline transporting gas under full pressure. Benefits include the optimal positioning of hot-tap tie-in points and emergency intervention and repair if the pipe becomes blocked or damaged. The system is designed for a maximum water depth of

2000 m, although a remotely controlled hyperbaric welding technique (hyperbaric gas metal arc welding) may extend the depth capability to 2500 m.

▶ **Subsurface oil sand recovery**. Oil sands (or tar sands) are unconsolidated sediments impregnated with high-density, high-viscosity oil. They are commonly referred to as bitumen in Canada and extra heavy oil in Venezuela – the two countries which are thought to hold a considerable volume of the world's remaining producible oil resources. Their viscosity, however, makes production difficult and recovery is poor. Low surface and low reservoir temperatures in Canada make the bituminous oil exceptionally viscous and nearly impossible for subsurface production (as opposed to open cast mining) with the current cold production techniques. Thermal recovery techniques have thus come to the fore.

Steam-assisted gravity drainage (SAGD) has evolved over a number of years and employs twin horizontal wells, consisting of a steam-injection well located about 5 m above a parallel production well. The viscosity of the oil in the steam chamber is lowered, and the softened oil then flows downwards under the influence of gravity to the production well.

The *toe-to-heel air injection* (THAI) method, which is being field tested in Canada, uses a vertical air injection well in conjunction with a horizontal production well. The bitumen is ignited to produce heat, which mobilises and partly upgrades the oil ahead of the vertical fire front, driving the lighter fractions towards the production well from "toe" to "heel". This is thought to be more energy efficient and to have less of an environmental impact than SAGD, as it eliminates the need for an external energy source such as natural gas and reduces water usage.

The ultimate goal is to increase recovery using low-energy recovery methods and to mitigate carbon dioxide emissions.

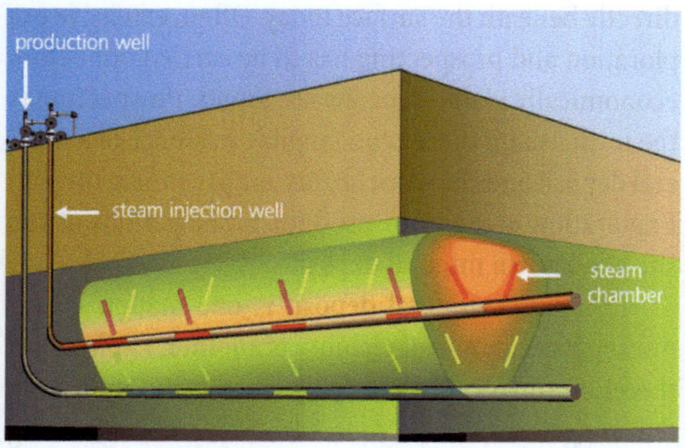

The steam-assisted gravity drainage (SAGD) method for improving bituminous heavy oil recovery. Condensing steam from the upper well forms a chamber within which the heavy oil is softened. The mobilised oil flows under the influence of gravity into the production well below and is pumped to the surface. Source: EnCana

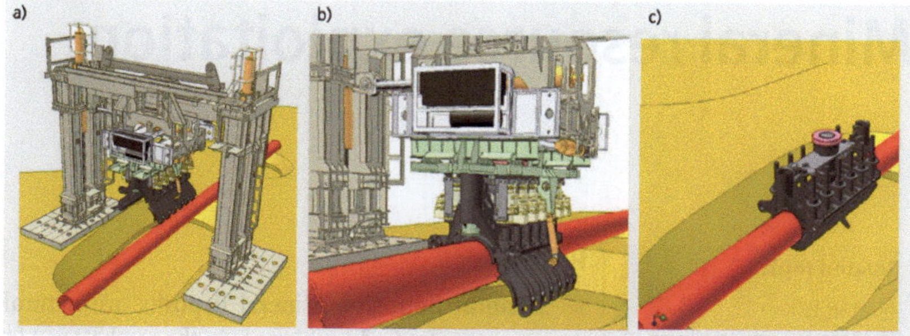

Principle of a diver-free subsea gas pipeline intervention: Retrofitting a Tee-junction to a mother well (red) while transporting gas under full pressure. (a) The H-frame containing a retrofit Tee being lowered onto the pipeline. (b) The Tee landing on the pipeline with the clamp still in the open position. (c) The retrofit Tee clamped to the pipeline and ready for welding the internal seal and then attaching a branch line. Source: Stolt Subsea 7JV / StatoilHydro ASA

Prospects

Major developments are anticipated:

- The merging of diverse, higher-resolution geophysical data to reduce exploration risk and improve production monitoring (seismic, electromagnetic, gravitational)
- More wide spread use of advanced well configurations; more fully-automated drilling operations; improved inflow control
- More comprehensive subsea processing solutions tailored to specific reservoirs and their development in more challenging environments (e.g. sub-ice and ultra-deep water)
- Gas pipeline intervention in increasingly deep water
- The emergence of new oil sand recovery processes, such as injecting vaporised solvents (propane/butane) – a cold production technique similar to SAGD; adding bacteria to generate methane; using catalysts and intense heat to upgrade the oil in situ

ANTONY T. BULLER
Formerly of StatoilHydro ASA

Internet

- www.StatoilHydro.com
- www.Geoexpro.com
- www.slb.com
- www.emgs.com
- www.gassco.no

Mineral resource exploitation

Related topics

- Metals
- Measuring techniques
- Image analysis and interpretation
- Process technologies
- Sensor systems

Principles

▶ **Mineral deposits.** Mineral deposits are natural occurrences of valuable elements in the earth's crust in a higher concentration than the average for those elements. They are found in rock series that have specific characteristics for each individual deposit type. Mineral deposits are unevenly distributed around the earth.

The elements of the periodic system are ubiquitous in the earth's crust, at times only in very low concentrations of grams or even milligrams per tonnes. Useful minerals or elements need to be enriched by natural concentration processes when tapping "economic mineral deposits", which are accumulations of minerals.

Minerals occur in different types of deposit. Basically, the deposit's geometry permits differentiation between seams, veins and massive deposits, which occur at varying depths and spatial positions within the earth's crust. Coal always occurs as a seam embedded in sedimentary strata. The thickness of the seam and its relative depth determines whether strip or underground mining will be used to exploit it, and whether to deploy continuous or discontinuous mining and transportation systems.

Metal	Abundance in the earth crust in % by weight	Minimum grade required for economic extraction in % by weight	Required factor of natural upgrading
Mercury	0.0000089	0.2	22,500
Lead	0.0013	4	3,100
Gold	0.00000035	0.0003	900
Uranium	0.00017	0.1	600
Zinc	0.0094	3	300
Copper	0.0063	0.3	50
Nickel	0.0089	0.3	35
Iron	5.80	30	5
Aluminium	8.30	30	4

Metals in the earth's crust and in mineral deposits: Metal deposits are unevenly distributed around the world. Aluminium, for example, constitutes 8.3% of the earth's crust. By contrast, precious metals such as gold occur over 20 million times less. If a mineral deposit is to be exploited economically, it must be upgraded by natural processes that will increase the concentration of valuable minerals and elements. According to the specific mineral value that requires upgrading, the factors that produce a mineral deposit are quite variable. Source: BGR

▶ **Mineral beneficiation.** Beneficiation is a process by which raw materials (concentrates) are produced from run-of-mine ore for subsequent metallurgical processes. Normally, valuable elements or minerals that are mingled with gangue minerals or country rock in the ore are concentrated by means of specific physical or mechanical separation processes. The purpose of the metallurgical processes performed after beneficiation is to separate the concentrated minerals into their elemental or metallic components. For example, in a copper mine exploiting an ore whose main component is chalcopyrite, containing 2% of copper on average, the beneficiation process could theoretically yield a 100% concentrate of the mineral chalcopyrite ($CuFeS_2$). Because this mineral contains the three elements copper, iron and sulphur, the stoichiometric fraction of copper in the concentrate is at most 36%.

Applications

▶ **Exploration.** Mineral deposits are geochemical anomalies in the earth's crust that are rarely discovered directly beneath the surface today. Often, extensive exploration and prospecting has to be carried out before economically viable deposits are found. Prospecting is the term used if there is no tangible evidence of a mineral deposit and the work begins as a greenfield project. Exploration is the term used for investigations at the site of a known mineral deposit.

Locating a mineral deposit is one matter. Determining its economical viability is another. Exploration therefore involves identifying proximity indicators that suggest a valuable concentration, and contrasting the deposit body with the surrounding rock. Proximity indicators are mineralogical or geochemical signatures that often appear in conjunction with known mineral deposits. The presence of certain metalliferous minerals in greater or lesser quantities may indicate their proximity to a deposit such as porphyry copper.

The Cu-Au-Mo core would be surrounded by zones with an increased concentration of zinc and lead in both lateral and vertical directions, phasing out into a wide "halo" prominently featuring the mobile elements As, Ag, Sb, Hg, Ti, Te and Mn. The presence of these elements and minerals in the lithosphere and their spatial distribution pattern can indicate the proximity of a mineral deposit.

The methods applied in prospecting and exploration are chosen according to the mineral resource being investigated and the type of deposit sought. Discovering mineral deposits buried under a rock cap, for example, first of all requires a combination of indirect methods borrowed from geology, geophysics and geochemistry. These methods define targets that must then be further investigated by direct methods. A near-surface deposit will necessitate prospecting trenches and pits. Anticipated deposits that are deeper underground will require drilling.

Indirect investigative methods always produce ambiguous results. A geochemical soil analysis, for instance, can produce the same result for a laterally distant mineral deposit hidden by only a thin cover, or for one that is in close vertical proximity but under a thick rock cap. Similarly, the physical or chemical behaviour of minerals near the surface can differ substantially from that of minerals deep underground. For example, copper, lead and zinc are elements that frequently occur in massive sulphide zones of volcanic origin. In atmospheric conditions, zinc is chemically much more mobile than lead. Zinc occurring near the surface can therefore be detected over more extensive areas and in higher concentrations than the other elements. Lead, in contrast, will be found in much higher concentrations only directly over the deposit, as it is relatively insoluble compared to copper and zinc. There are several ways in which the elements can migrate away from the deposit such as the mechanical transport of rock fragments, the motion of gas, or the diffusion of ions in fluids. The path from the origin of the deposit can be traced by empirical motion modelling, which plots the transport velocity as well as the possible transformation and absorption processes of minerals and elements under the ground.

▶ **Radio spectroscopy**. Different materials – rocks, minerals, soils and vegetation – have different absorption and reflection bands in the electromagnetic spectrum and can therefore be identified by comparing these spectra. Radio spectroscopy is a contactless investigative method that makes use of the radiance of sunlight or radar or acoustic sound waves that are emitted or absorbed depending on the ground's properties and conditions.

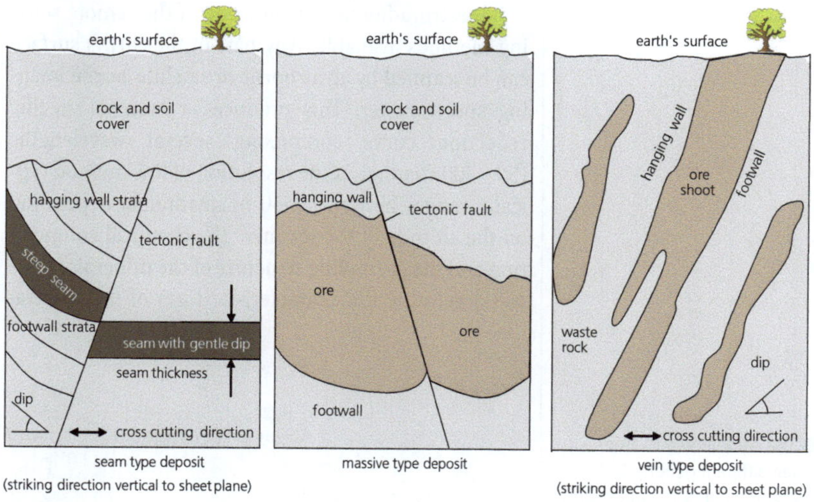

Deposit types. Left: Seam type deposits are mostly of sedimentary origin and have only a gentle dip, like the coal seams in the carboniferous strata of Northern Germany. Centre: Massive deposits are frequently formed by the emplacement and crystallisation of siliceous rock melts (magma) in the upper part of the mantle and crust of the earth. An example of this is the intrusive nickel-copper complex of Norilsk, Siberia. Right: Vein type deposits typically have a relatively low thickness and a two-dimensional spread. In most cases they dip steeply. Often, vein type deposits result from hydrothermal solutions filling the fissures and fractures in the earth's crust. Examples include the historically significant lead-zinc veins in the upper part of the Harz mountains, or many shear-zone hosted gold deposits such as the Eastern Goldfields in Western Australia. Source: BGR

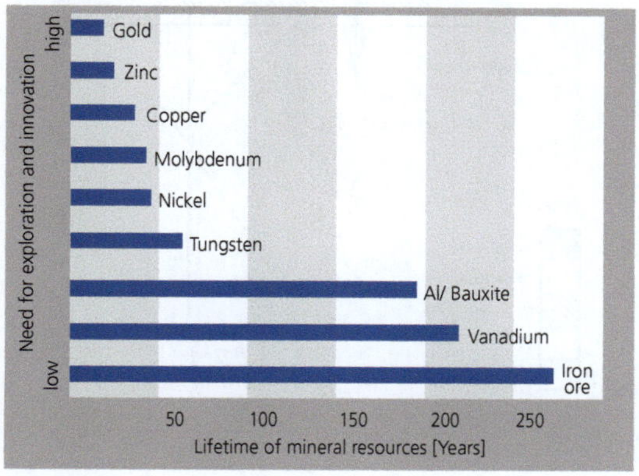

The static lifetime of mineral resources is defined as the quotient of actual mineral reserves and production rate. Because supply and demand of mineral resources are interlinked in a loop, in which exploration and innovation counteracts to supply gaps, the virtual lifetime for most metals has remained unchanged in the last decades and static lifetime is not equal to the ultimate point of depletion. For abundant metals such as iron, vanadium and aluminium, there is little demand for greenfield exploration. Due to the backlog of resources, investments in exploration activities for these metals would not be economically viable. Investors can earn faster returns with searches for underexplored metals such as gold, silver, zinc and indium, or in activities that promote the recycling and efficient use of metals such as tantalum that are in shorter supply. The need for exploration of these metals is considered to be high. Source: BGR

Spectroradiometry is just one of the remote sensing methods available. Any part of the earth's surface can be scanned by an airborne or satellite-borne imaging spectrometer. This produces a material-specific reflection curve comprising several wavelengths (50 – 300 channels). The resolution of this method typically ranges between 5–30 m quadrants, depending on the altitude of the scanner. The chemical composition and the crystalline structure of the minerals in the soil determine the characteristic form of the spectral curve and the position of the absorption bands within this curve. These can then be analysed on the ground using a field spectrometer, which provides a basis for calibrating the spectral data recorded from satellites or airplanes. This method identifies the position of rocks on the earth's surface and allows conclusions to be drawn as to the possible presence of underground mineral deposits thanks to proximity indicators. The same method can be applied to identify soil pollution caused by mining activities.

Several complementary geophysical methods are sometimes applied simultaneously to position a mineral deposit more accurately. They produce parallel cross-sections through the deposit that are combined to create a 3D model.

► **Electrical self-potential.** This property can be demonstrated by a passive method that is predominantly applied in the exploration of sulphide ore. Groundwater can act on massive sulphide ore bodies to produce a weak electrical charge. Sulphide and metal ions form a solution on the surface of the deposit, generating an electric voltage (the battery principle) due to the different oxidation state of the ores underground. Systematic measurements of the surface voltage may show a significant change when passing across areas of massive sulphide mineralisation.

► **Electromagnetic properties.** This method can be used to detect the different electromagnetic properties of sulphide and oxide mineralisation by applying a VLF (very low frequency) signal to the underlying strata. Certain geological structures, such as mineralised fault zones, will produce measurable secondary waves that can be recorded on the surface. The parameters measured are the phase difference, the amplitude ratio between incoming and outgoing signals, and the decay curve induced in the receiving coil by eddy currents.

► **Direct-current geoelectrical sounding.** This is an active potential method applied particularly when exploring for groundwater and near-surface mineral deposits. The parameters measured are changes in electrical resistivity and conductivity underground in the vertical plane. A current is passed underground via fixed electrodes located on the surface or in a drill hole, allowing the electrical resistivity to be measured as a function of the variable distance of the measuring electrodes. The standard measurement configuration formerly comprised only 4 electrodes, but today the computerised multi-electrode method with up to 150 electrodes is preferred. The latter facilitates the preparation of tomography-like resistivity images of the subterranean areas.

The system uses multi-core cables that contain the same number of individual wires as electrodes. These

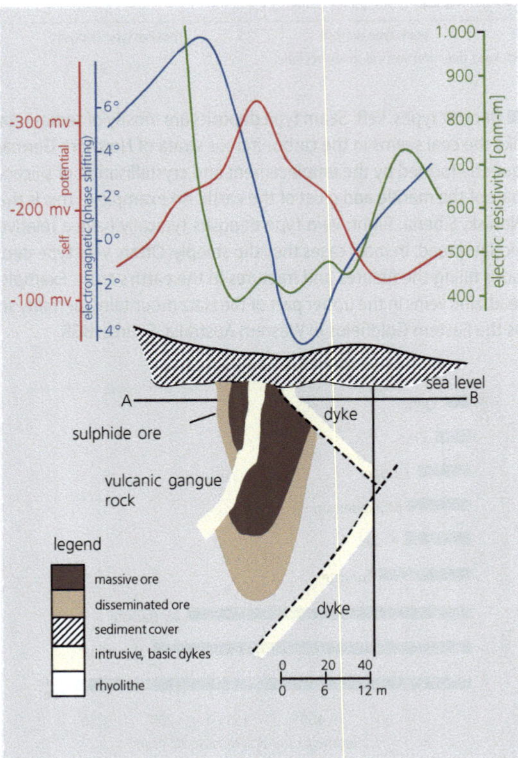

The combined data from three different geophysical measurements show anomalies in the proximity of a body of copper sulphide ore (bottom). Red: Increase of self-potential in the proximity of the sulphide ore. Green: Decrease in resistivity, as sulphide minerals exhibit lower specific resistivity than the surrounding rock minerals. Blue: Extreme phase shifts of secondary electromagnetic waves suggesting the presence of deposit interfaces. None of these methods by themselves represent conclusive evidence of the existence of a mineral deposit; only the combined conclusion from all three methods permits the deposit to be localised with a high degree of probability. Source: BGR

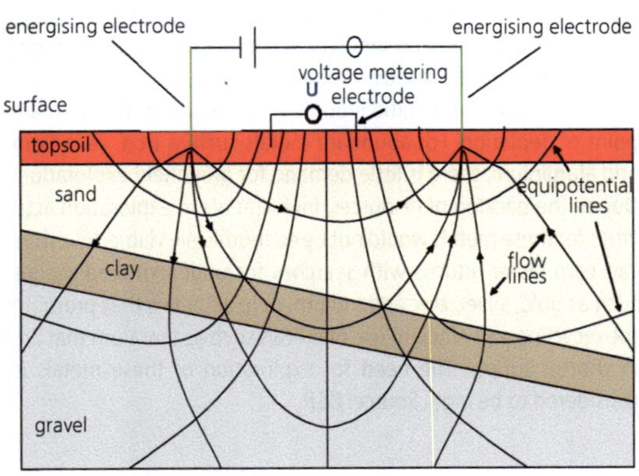

In direct-current geoelectrical sounding, a direct current is induced at two locations. The measured variable is the potential drop that occurs between the resulting equipotential lines. These depend on the characteristics of the underlying soil, rock and groundwater. The specific resistivity of the underground strata can then be calculated, providing information on the underground structures. Source: BGR are placed at varying intervals of 2–10 m. The measuring unit includes relays that automatically carry out a specific sequence of readings stored in its internal memory. The idea behind this set-up is to take readings for numerous combinations of transmission and reception pairs and thus to achieve some kind of mixed profiling/sounding range. The total length of cable is defined by the spacing of the electrodes multiplied by the number of electrodes: 240 m for 48 electrodes at 5-metre intervals. This determines the maximum depth of the geoelectrical investigation, which is approximately 20% of the total cable length. In other words, penetration of about 50 m can be achieved using 48 electrodes spaced at 5 m (total length: 240 m). Underground profiles based on the geoelectrical method provide information on the composition and stratigraphic sequence of the underground. However, in order to interpret the results properly, the profiles must be calibrated by a direct sampling method.

► **Extraction of mineral resources.** The extraction process can be subdivided into three operations: freeing the ore from the rock mass, loading it onto a means of transportation, and finally carrying it to the mineral processing plant. The established method of freeing ore from the solid rock mass is by blasting, in which explosive charges are inserted into blast holes and ignited. This liberates huge volumes of fragmented rock and ore. Clearly, blasting operations imply the discontinuous extraction of minerals. Continuous extraction methods, such as rock- or coal-ramming, shearing, cutting, and hydro-mechanical methods, are often applied in the case of low-strength ores. In industrialised countries, coal is extracted from underground or surface mines almost exclusively with cutting and shearing machines in continuous operation. The waste rock or soil cover of the deposit is removed either by blasting or by mechanical excavation and hauled to an internal or external waste dump.

► **Mineral processing.** Mineral processing is used to increase the share of marketable minerals in a concentrate or improve the technical properties of the raw material. In the case of concentration, the valuable minerals have to be separated from the waste rock or gangue. The first step is the comminution of the coarse run-of-mine ore to obtain the valuable minerals. Minerals are enriched by physical and chemical means that make use of the specific chemo-physical properties of minerals and waste rock, such as density, magnetic and electrical properties, the colour and spectral properties of the minerals, or the different wettability of the mineral surface. The latter is the principle on which mineral froth flotation is based. Most of the contemporary copper ores come from porphyry copper ore de-

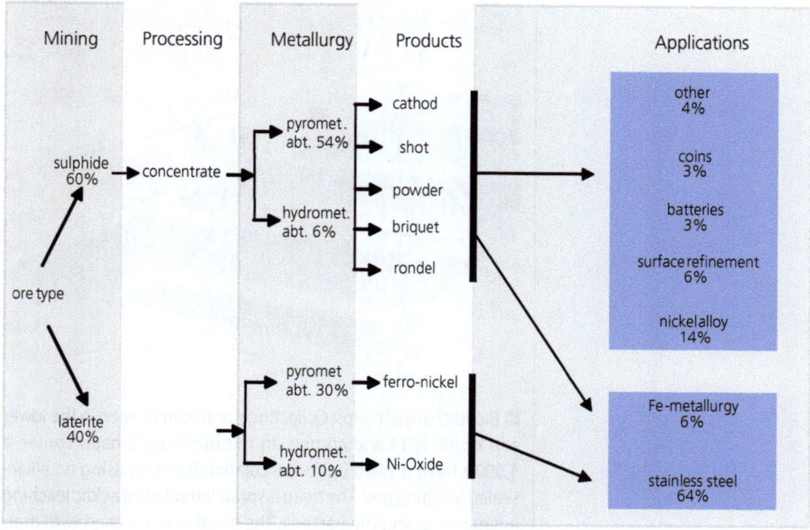

Production of nickel from different ore types. Only nickel sulphide ores are processed. The nickel sulphide minerals are concentrated to grades of 4–20 % Ni from a typical content of 0.3–8 %. This is usually achieved by mineral froth flotation and magnetic separation. The latter removes worthless pyrrhotite from the concentrate. Source: BGR

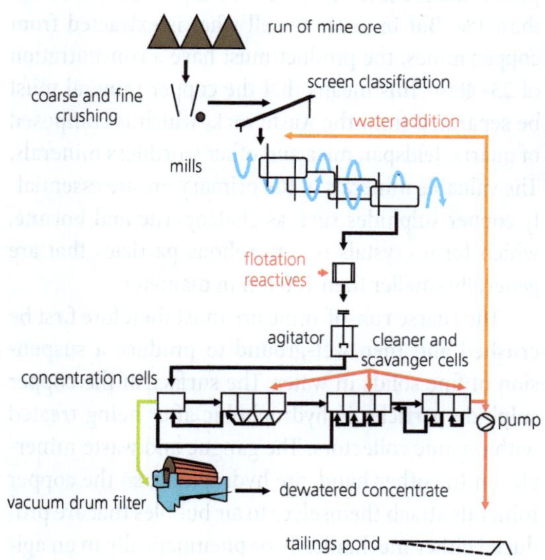

Flow sheet of copper mineral processing by froth flotation: Run-of-mine ore is comminuted by coarse and fine crushing, then ground in a wet mill to a diameter of less than 100 µm. As a result, the valuable copper minerals are exposed to a high degree. Organic reagents intensify the hydrophobic properties of the copper sulphide minerals, which attach themselves to air bubbles in the flotation cells. Subsequent flotation steps (concentration, cleaning and scavenging flotation) produce concentrates with high metal grades and tailings with low metal grades. The flotation process results in an upgrading ratio of 1 : 25–40 in the copper concentrate. The residual copper grade in the tailings is about 0.05 %. Source: BGR

A cutting wheel excavator strips waste material from a brown coal (lignite) seam. Once the seam is exposed, additional equipment is brought in to mine it. Source: BGR

Bioleaching of heaps: Collecting ponds can be seen in the lower and upper left-hand corners. In the upper right-hand corner, a 1,000-t heap is piled up for percolation leaching using an alkali-sealed leaching pad. The heap is sprinkled with the acidic leaching agent containing the bacteria. The solution is enriched with metals and then drains into the collection ponds by gravity. The blue colour of the leachate in the upper collection pond indicates the presence of copper sulphate. Source: BGR

posits that are known to have a copper content of less than 1%. But in order to sell what is extracted from copper mines, the product must have a concentration of 25–40%. This means that the copper mineral must be separated from the waste rock, which is composed of quartz, feldspar, mica and other worthless minerals. The valuable minerals in the primary ore are essentially copper sulphides such as chalcopyrite and bornite, which form crystals or amorphous particles that are generally smaller than 100 µm in diameter.

The coarse run-of-mine ore must therefore first be crushed and then wet-ground to produce a suspension of fine solids in water. The surface of the copper sulphide particles is hydrophobic after being treated with organic collectors. The gangue and waste minerals, on the other hand, are hydrophilic, so the copper minerals attach themselves to air bubbles that are produced either mechanically or pneumatically in an agitated flotation cell. The air bubbles lift the mineral grains to the surface of the pulp, forming a dense mineral froth that can be skimmed off. The flotation process requires several cleaning and scavenging steps in order to obtain a marketable copper grade in the concentrate. The wet concentrate has to be dewatered to about 10% residual moisture. This is done by means of thickening and vacuum filtration. Mineral flotation capacities in the range of 100,000 metric tons of ore per day and more are common in large-scale mining operations.

Trends

► **Bioleaching**. This mineral processing method uses microbial activities to facilitate the extraction of not readily soluble metal compounds from minerals. These metals could not be extracted by the conventional leaching methods that utilise acidic or basic solutions. Bioleaching allows mineral ores to be mined economically that would otherwise be unworkable because of complicated mineral compositions or low grades – copper, gold and uranium ores, for example.

Low-grade ore from open pit mining operations – typically a suitable candidate for bioleaching – is stacked on alkali-sealed leaching pads. The surface of the leaching heap is sprinkled with an acidic solution that includes microorganisms. The solution percolates through the heap, causing the minerals to dissolve with the help of the bacteria and concentrating the metal compounds in the leach agent. The leach agent is then collected in a tank and either returned to the heap or passed on to the hydrometallurgical extraction process, depending on the actual metal content. The leaching kinetics and the production rate depend on several factors: ore fragmentation and porosity, climatic conditions, and the species of bacteria that is used to support the leaching process. The duration of bacterial heap leaching operations varies from a few weeks to several months. The ore volumes treated in bioleaching operations range from several hundred metric tons to over 10,000 t of ore per heap.

Bioleaching utilises chemolithotrophic microorganisms that favour an acidic environment, e.g. acidithiobacillus, which produce their life energy by oxidising iron and sulphur compounds. These bacteria can be found in the acidic drainage of mining environments. Therefore, to cultivate leaching bacteria, one need only identify and activate them, and create optimal living conditions. They must be protected from sunlight and supplied with nutrients that are essential for their metabolism and the supply of oxygen.

Thanks to their enzymatic activities, leaching bacteria produce iron sulphates and sulphuric acids that help to break down the insoluble metal sulphides by oxidation and to dissolve the metal compounds. Supported by bacteria that act catalytically, this disintegration and dissolving process is up to 300 times faster than the regular chemical oxidation and dissolution process of metal sulphides.

► **Deep-sea mining**. The sharp increase in demand for metals combined with rising market prices have turned the mining industry's attention to the potential exploitation of ocean floor deposits. The interest in deep-sea mining extends to both base and precious metals – Co, Cu, Ni, Mn, Mo, Ti, Au and Pt – as well as a number of strategically important and innovative metals – Ga, Se, Te and In. There are two different types of deep-sea mineral deposits that are especially promising: manganese nodules and crusts, and seabed massive sulphides. The technology for the recovery of seabed ore is not yet ready for industrial operations. Nevertheless, two basic approaches for seabed mining have been defined:

- Remotely operated *underwater vehicles* – tethered underwater robots that are controlled by a person aboard a vessel.
- Dredging with a device that scrapes or vacuums the seabed. The excavation head is connected to a mother vessel that conveys the propulsion and stripping energy by a mechanical link.

Both technologies require a system to raise the ore to the surwface and adequate stocking and reclaiming equipment for periodical removal of the ore by freighters.

▶ **Sea floor massive sulphide deposits**. These sulphur-rich seabed ore bodies are produced in underwater volcanic regions of the world by "black smokers". These are formed when seawater seeps through the porous seabed, is heated, and re-emerges through vents carrying dissolved minerals. When the hot water hits the cold water on the seabed, the minerals precipitate, creating chimney-like towers called "black smokers". Over time, these towers collapse and accumulate to form ore deposits, some of which are rich in gold, silver, copper, lead and zinc.

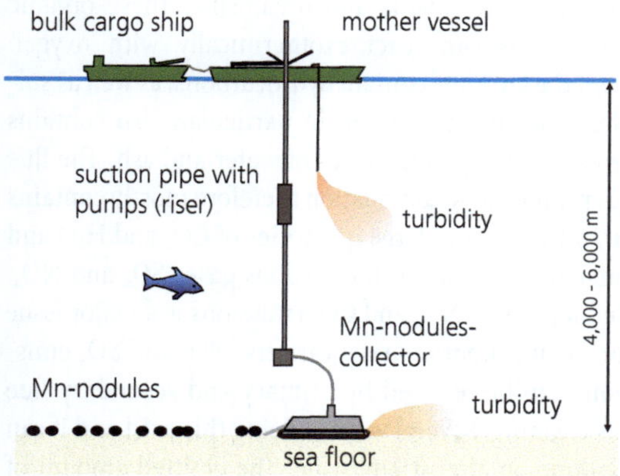

Concept for the extraction of manganese nodules: The nodule collector is controlled remotely from on board the mining ship. It advances along the seabed and picks up nodules. The nodules are pumped or lifted to the surface by an airlift system. The nodules are then removed from the transporting fluid and stored in the hopper of the mining vessel. A freighter periodically picks up the nodules and carries them to the on-shore processing plant. Source: BGR

This type of deposit usually contains a high concentration of base and precious metals – Ag 20–1200 grams per metric ton, Cu 5–15%, Zn 5–50%, Pb 3–23% – and is usually fairly lightweight, in the range of 0.25–18 million t per mound.

▶ **Manganese nodules and crusts.** Manganese nodules lie on the seabed and are often partly or completely covered by sediment. They vary greatly in abundance, in some cases touching one another and covering more than 70% of some areas. The total amount of polymetallic nodules on the sea floor is estimated at 500 billion tonnes. The copper content of this potential resource corresponds approximately to 40 times the present level of global copper mining production. Nodules can occur at any depth, but the highest concentrations have been found on vast abyssal plains in the ocean at depths of 4,000–6,000 m.

Prospects

- The current global consumption of mineral resources can be expected to double in the next 30 years. In that same period, extraction of raw materials will become more demanding because of increasing environmental and social demands on mining, a disproportionate surge in operating costs, and a steady decline in grades and good quality mineral deposits in mineral zones that are located close to demand or existing transport infrastructures
- In addition to minimising the ecological damage and environmental stress caused by mining activities, mining companies will have to put more effort into ensuring the social sustainability of their future mining projects.
- Exploration and technological efforts in the mining sector should be directed towards targeted exploration of mineral deposits in remote geographical locations or on the ocean floor and towards improving the utilisation of energy and labour in mineral extraction.
- In the field of mineral exploration engineering, the priorities should be on refining geophysical, geochemical and remote sensing methods. This would improve the discovery of anomalies that indicate the presence of deep-lying mineral deposits, and cut costs and speed up the investigation of extensive surface areas.

DR. JÜRGEN VASTERS
Federal Institute for Geosciences and Natural Resources (BGR), Hannover

Manganese and iron make up about 35% of the nodule. But the real value comes from the comparatively low grade nickel, copper and cobalt content. Source: BGR

Internet

- http://pubs.usgs.gov/info/seal2
- www.amebc.ca/primer2.htm
- www.jogmec.go.jp/english/activities/technology_metal/exploiting.html
- www.dmtcalaska.org/course_dev/explogeo/class01/notes01.html
- www.etpsmr.org

Fossil energy

Related topics
- Oil and gas technologies
- Carbon capture and storage
- Metals
- Renewable materials
- Bioenergy

Principles

Fossil fuels are very diverse, because they are formed in different ways and have different states of aggregation (solid, liquid and gaseous). Every fuel has specific characteristics, as well as advantages and disadvantages. Solid fuels are most difficult to handle, and the milling, transportation, storage and ash removal after combustion procedures are extremely intensive. Hard coal is located up to a few thousand metres beneath the earth's surface. Conventional mining only goes up to a few hundred metres, however. Hard coal consists mainly of carbon, as well as hydrogen, oxygen, sulphur, nitrogen, water and ash. Lignite (brown coal) is mostly located close to the earth's surface and can therefore be mined more easily. It consists of about 50% water and has a lower heating value than hard coal. Crude oil mainly consists of various hydrocarbons, as well as nitrogen, oxygen, sulphur and small amounts of metal. Depending on its source, crude oil has a specific chemical composition that influences its physical properties, such as its viscosity and colour. Oil-based fuels are easy to handle and almost ash free. Natural gas contains very little sulphur and is also ash free. No flue gas cleaning is required after combustion. Due to the fact that natural gas has a higher hydrogen content than coal, it emits less CO_2 related to the same amount of produced heat.

▶ **Availability**. Reserves and resources of non-renewable primary energy carriers are obviously limited, and the jury is out concerning the long term availability of fossil fuels. Hard coal is the primary energy carrier with the largest global reserves and resources and can meet demand for hundreds of years to come. Given the pending shortage of crude oil, and the increasing demand for energy from developing and emerging countries, it is assumed that use of hard coal as a primary energy carrier will increase. Reserves of lignite will also be available for hundreds of years to come. Because of its high water content, however, lignite's specific energy content is rather low and its transportation is limited to the area around the coal mine. No global market for lignite therefore exists. In all probability, crude oil can only meet an increasing global demand over the next few decades. Nobody can calculate or guess what will happen when maximum worldwide oil production is exceeded, because further development depends on so many factors, which are still completely unknown. Natural gas will also be available for many decades to come. Until the infrastructure for the transportation of LNG is fully developed, however, the gas must be transported via pipeline and, in the medium term, consumer countries will rely heavily on a few major supplier countries.

More than 80% of the worldwide commercial energy supply is based on fossil primary energy carriers. *Coal* accounts for 25.3% of primary energy; oil, 35.0% and natural gas, 20.7%. These energy sources are used to produce room and process heat, mechanical energy and electrical power.

Fossil primary energy carriers were generated from primeval plants and organisms. These organic compounds can react exothermically with oxygen from the air. They contain hydrocarbons, as well as sulphur and nitrogen. Coal, in particular, also contains remarkably high quantities of water and ash. The flue gas produced by combustion therefore mostly contains air, nitrogen, and large quantities of CO_2 and H_2O and smaller quantities of the noxious gases SO_x and NO_x. Reducing SO_x, NO_x and CO_2 emissions is a major issue for fossil primary energy carriers. NO_x and SO_x emissions can be reduced by primary and secondary flue gas cleaning. Several ways of doing this exist and it can be done on almost any scale. The emitted amount of CO_2 per kilowatt hour of useable energy is directly proportional, however, to the fuel's carbon content and the efficiency of the energy conversion process. Reducing CO_2 emissions from large power plants to zero is not yet state-of-the-art and will be connected with high investment and operational costs. But by this, power plants can play an effective role in environmental protection because they are large-scale and stationary sources of emissions.

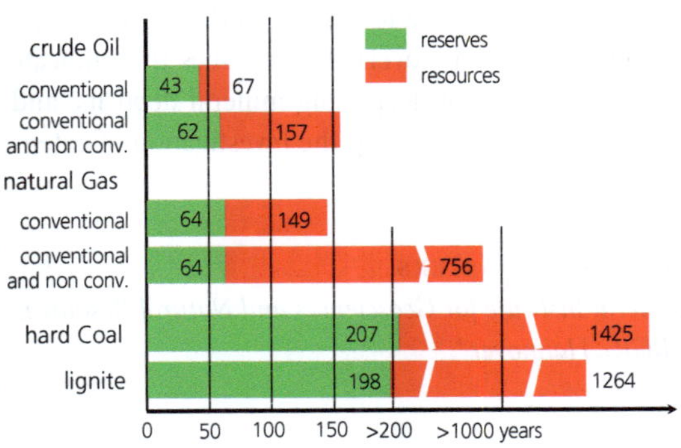

▶ Statical ranges of fossil fuels assuming future fuel consumption remains constant. Reserves are the amounts that can be produced economically using methods currently available. Resources include detected and assumed amounts. Non-conventional reserves and resources include production from tar sands, oil shale, gas hydrates, etc. Source: BGR

Application

▶ **Oil.** Petroleum products, such as light fuel oil, gasoline and kerosene, are predominantly used in the transportation and room heating/warm water production because they are much better suited to these sectors than any other energy carrier from a technical point of view. Natural gas is also used in these sectors, but to a much lesser extent. Houses with special thermal insulation and heating systems using condensing boiler technology are increasingly reducing the demand for crude oil and natural gas vehicles with more efficient engines, and hybrid power trains can also reduce the demand for oil.. Fuels made from *biomass* (synthetic biofuels) are also increasingly replacing fossil fuels. Crude oil, however, currently accounts for the highest proportion (40%) of yearly production of all non-renewable primary energy carriers. Because of its relatively high price, only few power plants are oil-fired.

▶ **Coal.** Hard coal and lignite are primarily used for power generation in steam power plants. The finely pulverised coal is burned in a steam generator. This vaporises and superheats pressurised water in hundreds of parallel tubes. Live steam is generated at approximately 600 °C and at more than 250 bar. The live steam flows into the turbine and expansion produces mechanical energy, which is converted into electrical energy in the generator. The steam is subsequently condensed to liquid water at a temperature slightly above ambient temperature (28 °C) and at very low absolute pressures – less than 0.04 bar. Powerful pumps route the water back to the steam generator.

The higher the turbine's temperature and pressure gradient – which uses the parameters of the live steam generated and the parameters in the condenser – the higher the power plant's efficiency. Significant gains in efficiency can be made by increasing the live steam parameters, fluidic optimisation of the turbines and decreasing the condensation level (23 °C at 0.029 bar) through improved cooling, as well as fluidic optimisation of the turbines.

After heat is transferred to the water/steam cycle, the flue gas is used to preheat the combustion air. This saves fuel. An electrical precipitator, a De-NOx plant (only for hard coal) and a flue gas desulphurisation unit clean the flue gas. The flue gas is then released into the atmosphere via the cooling tower or stack.

▶ **Natural gas.** The standard technology used to produce electrical power from natural gas is the combined gas and steam power plant, where the gas turbine process and the steam power plant process are combined.

Conventional natural gas-fired gas turbines drive a power generator. The hot flue gas from the gas turbine is used to produce steam in a *heat-recovery steam generator (HRSG)*. The steam is expanded in a conventional steam power plant cycle and produces about one-third of the electrical power of the entire power plant. The electrical efficiency of the combined process is very high. To further increase steam and power production, supplemental firing is possible. This, however, reduces the efficiency of the plant.

▶ **Combined heat and power.** Steam power plants release more than 50% of the supplied thermal fuel heat as waste heat into the environment via the condenser and the cooling tower or other similar devices. One way of decreasing these losses is through the combined generation of heat and power (CHP). CHP raises the back pressure and therefore the turbine's condenser temperature so that all the condensation energy can be used for district heating. CHP slightly reduces the electrical efficiency of the plant, but strongly increases its overall efficiency (power and heat). The primary en-

Power plant in Niederaussem, Germany: A modern lignite-fired 1,000 MW_{el} unit in the foreground and (up to) 40-year-old units in the background.

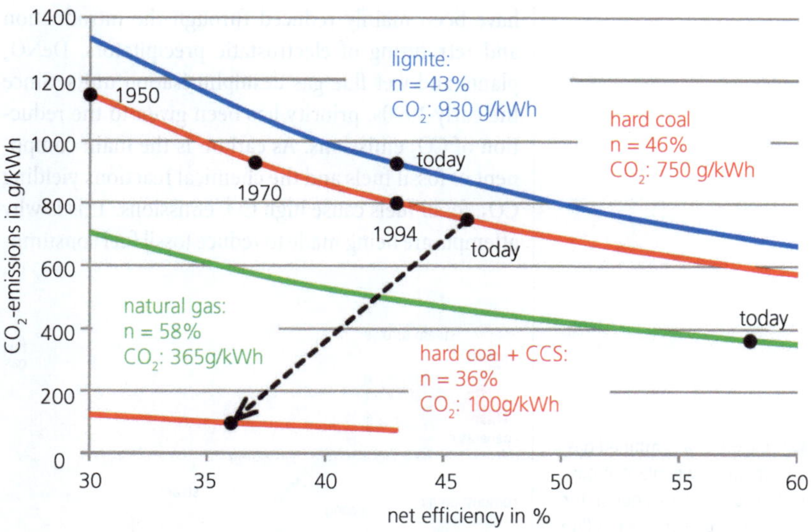

Specific CO_2 emissions versus net efficiency for different fossil fuels – with and without carbon dioxide capture storage (CCS). The higher the efficiency, the lower the specific CO_2 emissions. Adopting CCS at power plants results in lower CO_2 emissions as well as lower net efficiencies, i. e. higher specific fuel consumption.

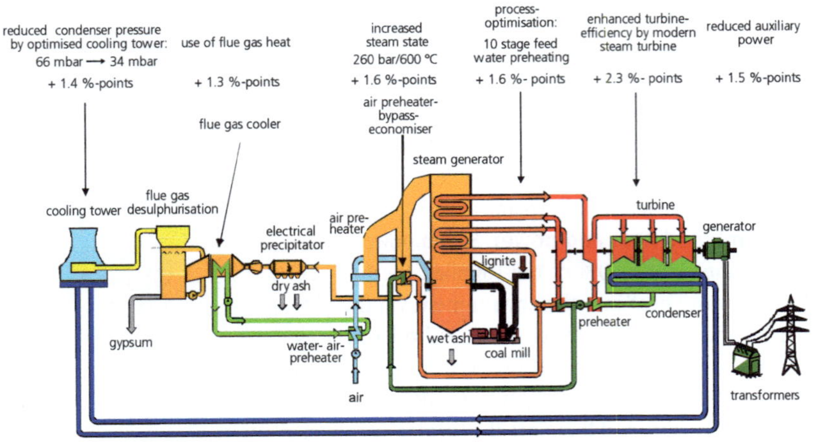

Principle of a modern lignite-fired power plant: After milling, lignite is burned in the steam generator: transfering heat to the water/steam cycle. After combustion, the flue gases are cleaned and enter the atmosphere via the cooling tower. Transferring heat from the flue gases in the flue gas cooler to the water/steam cycle reduces losses of efficiency. Feed water is routed to the steam generator where it is heated, evaporated and superheated. The greater the pressure and temperature of the steam leaving the steam generator, the higher the efficiency. The steam produced drives the steam turbines which is connected to the generator. The steam leaving the turbine is liquefied in the condenser and routed back to the steam generator. The lower the condenser pressure – and, therefore, the temperature – the higher the efficiency. The increase in efficiency is related to the lignite fired units erected in the early seventies. Source: RWE Energy AG

ergy contained in the fuel is utilised to a much greater degree and less waste heat is emitted into the environment.

Trends

Over the last few decades, dust, NO_x and SO_x emissions have been mainly reduced through the introduction and retrofitting of electrostatic precipitators, $DeNO_x$ plants and wet flue gas desulphurisation units. Since the early 1990s, priority has been given to the reduction of CO_2 emissions. As carbon is the main component of fossil fuels and the chemical reactions yielding CO_2, fossil fuels cause high CO_2 emissions. This is why attempts are being made to reduce fossil fuel consumption by increasing the use of renewable energies. However, the use of fossil fuels for power generation is increasing worldwide. Efforts to reduce CO_2 emissions caused by fossil fuels therefore have to be made.

CO_2 emissions from fossil fuels can be reduced in two ways:

- By increasing efficiency, i. e. using less fuel to produce the same amount of electrical power. Less reaction products are generated and therefore less CO_2.
- By separating CO_2 from the fuel or flue gas and storing it underground (*carbon capture and storage: CCS*)

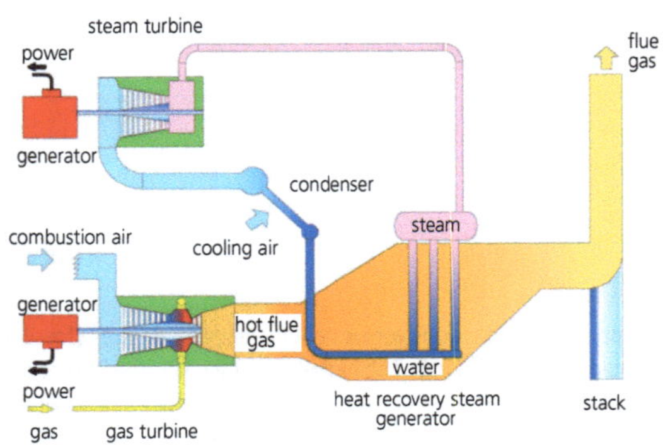

Principle of a combined gas and steam power plant. A gas turbine unit drives a generator to produce power. The hot flue gas from the gas turbine is used to produce steam, which drives a steam turbine to produce additional power. Source: Ingenieurbüro P. Lehmacher

▶ **Steam power plant.** Efficiency can be increased by improving individual aspects of the whole system. The specific CO_2 emissions of a modern 1,000 MW lignite-fired power plant, for example, are reduced by about 30% in comparison to older units from the early 1970s. Especially for lignite, which contains more than 50% water. Coal pre-drying offers a further potential to reduce the CO_2 emissions by about 10%. Extracted steam from the turbine is used for this purpose. This technology, which is currently undergoing intensive development, would raise the net efficiency of lignite-fired power plants to around 46% – approximately the same net efficiency as a modern hard coal-fired power plant.

R&D activities are currently focusing on raising the net efficiency of coal-fired steam power plants to more than 50%. One significant way of doing this is to increase the steam parameters at the turbine inlet to 350 bar and 700 °C (a modern plant currently has live steam parameters of 290 bar and 600 °C). To increase the live steam parameters, some parts of the steam generator and the turbines, as well as the live steam pipes, have to be replaced with expensive nickel-based alloys.

▶ **Gas and steam power plant.** The highest net efficiency achieved to date with a gas and steam power plant is about 58.4%. To attain this value, a gas turbine inlet temperature of 1,230 °C and a heat recovery steam generator (HRSG) with three pressure levels, one reheat and a condensate pre-heater. This stepwise steam heating transfers the energy from the hot flue gas to the condensate from the steam turbine in the most efficient way. A further increase of the gas turbine inlet temperature and minimalisation of the temperature differances between flue gas and water/steam cycle should lead to a net efficiency of about 63% within the next 20 years. This minimisation requires complex interconnections of the HRSG heating surfaces.

New materials need to be developed in order to increase the gas turbine inlet temperature. Cooling concepts for the turbine blades, which are in contact with the hot flue gas, must also be optimised.

▶ **CO_2 capture and storage.** Increasing the efficiency of fossil-fired power plants around the world can reduce the CO_2 emissions of these plants by about 35%. To reduce CO_2 emissions further, the CO_2 must be separated almost completely from the flue gas. Depending on the separation technology and power plant type, significant losses of net efficiency result, which means that to produce the same net power, the demand for fuel increases by around 20–25%. Investment costs are also high.

▶ **CO_2 capture by chemical washing.** In conventional coal-fired power plants, CO_2 can be captured through chemical washing. Based on today's technology, this method has the highest costs and efficiency losses. However, with steam power plants, an end-of-pipe solution such as scrubbing the CO_2 from the unpressurised flue gases using a solution of monoethanolamin (MEA) and water, for example, would be favourable. Extracted steam from the low pressure turbine is then used to regenerate the MEA solution. This and other power requirements – namely CO_2 compression and liquefaction – reduce the net efficiency by about 10–12% points.

▶ **Oxyfuel process.** This process is based on the conventional steam power plant process, but combustion takes place in an atmosphere of almost pure oxygen, delivered from an air separation unit, and re-circulated flue gas instead of air. The main components of the flue gas are therefore CO_2 and water vapor, because almost no nitrogen finds its way into the combustion process. The flue gas leaving the steam generator is dehumidified and has a CO_2 content of roughly 89%, which allows for energy efficient CO_2 capture. The remaining flue gas largely comprises excess oxygen, argon and small amounts of nitrogen, as well as oxides of sulphur and nitrogen. Net efficiency is reduced by about 9–11% points.

▶ **IGCC process with CO_2 capture.** While the IGCC (Integrated Gasification Combined Cycle) process is not the primary option for reducing CO_2 emissions by increasing efficiency, it looks more promising when combined with an additional CO_2 capture unit. CO_2 enrichment is necessary for effective separation to take place. This can be achieved with few technical modifications before combustion in the gas turbine takes place. Once the coal is gasified with oxygen, the CO contained in the fuel gas produced and additional steam are converted to hydrogen and CO_2 via a catalytic reaction (shift conversion) with steam. After this, CO_2 and other undesirable gas components are removed from the pressurised fuel gas by physical scrubbing (cold methanol is used as the reclaimable, circulating washing solution). The fuel gas consisting mainly of H_2 is then used in a gas and steam power plant. Net efficiency is 8–10% points lower compared to a modern IGCC or steam power plant without CO_2 separation.

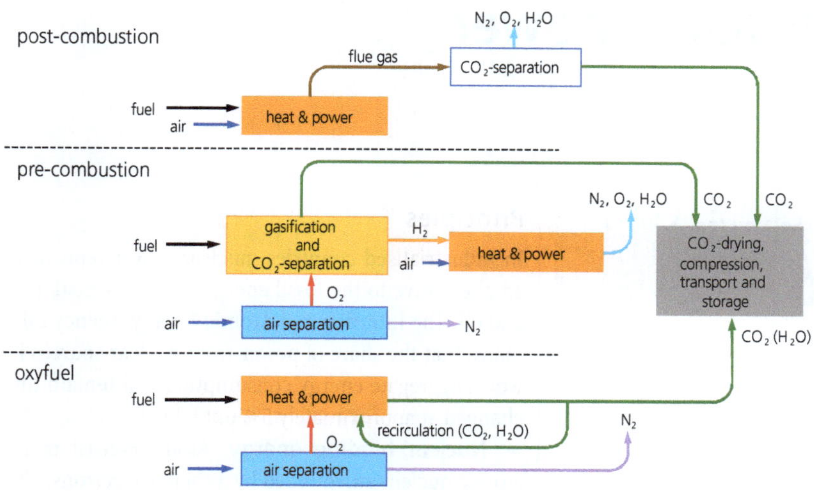

Different possibilities for CO_2 separation. Post-combustion: The fuel is burned with air in a conventional steam generator, and the CO_2 is separated after combustion in a flue gas scrubber. Pre-combustion: Coal is gasified, which produces a gas consisting mainly of CO_2 and H_2. The CO_2 is separated and the H_2 rich gas is burned. Oxyfuel: The fuel is burned in an atmosphere of O_2 and re-circulated flue gas, generating a flue gas that consists mainly of CO_2 and water. Source: VGB

The fuel gas generated by coal gasification can also be used to produce synthetic fuels. This is why the IGCC process with CO_2-capture is being developed. It is seen as a viable way of reducing our dependency on oil.

CO_2 separation technologies can result in separation levels of 90%. Before these technologies can be adopted on a wide scale, however, a secure storage for the separated CO_2 is required.

Prospects

Fossil fuels are the most important primary energy carriers. There are two points of departure for producing power from fossil fuels with low CO_2 emissions:

- Increasing the efficiency of coal-fired steam power plants and natural gas-fired gas and steam power plants
- Separation of CO_2 from steam power plants and gas steam power plants, either by chemical washing or an oxyfuel process, and from IGCC plants by physical washing

PROF. DR. ALFONS KATHER
SVEN KOWNATZKI
Hamburg University of Technology

Internet

- www.steinkohle-portal.de
- www.vgb.org
- www.cooretec.de
- www.iea.org
- www.worldcoal.org

Nuclear power

Related topics

- Plant safety
- Metals

Principles

In industrialised countries, nuclear power represents an alternative to the fossil energy resources, coal, gas, and oil. The International Atomic Energy Agency estimates that the share of power from nuclear sources in world aggregate energy consumption will remain unchanged at approximately 5% until 2020.

▶ **Nuclear binding energy.** Atoms consist of an atomic nucleus surrounded by a shell of electrons. The nuclear components (nucleons) are positively charged protons and uncharged neutrons. The formation of an atomic nucleus out of nucleons entails a mass loss (mass defect), as the combination of protons and neutrons into a nucleus causes a small part of their masses to be converted into energy and released. The average binding energy per nucleon is at its highest in nuclei with mass numbers of between 40–100, and decreases towards lighter and heavier nuclei. There are two basic possibilities for using nuclear binding energy:

- Very light nuclei are fused, giving rise to heavier nuclei whose nucleons are bound more strongly. This is associated with a mass defect and, thus, a release of energy. This principle applies to energy releases inside the sun and in fusion reactors.
- Heavy nuclei are split into medium-heavy nuclei. As the binding energy for each nucleon is higher in medium-heavy nuclei than in heavy ones, this, too, gives rise to a mass defect and, consequently, a release of energy. This is the principle on which energy generation in nuclear power plants is based.

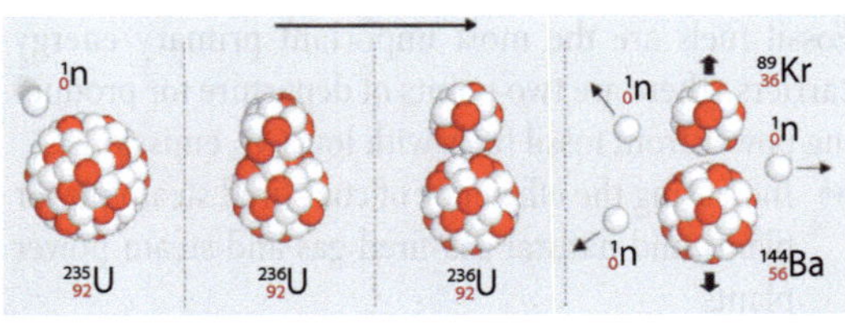

Splitting of a uranium-235 nucleus by a thermal neutron. From left to right: Attachment of the neutron produces a uranium-236 nucleus excited by the binding energy and the potential kinetic energy of the neutron. As the binding energy of the neutron exceeds the fission barrier of the uranium-236 nucleus, that nucleus disintegrates into two medium-heavy nuclei (in this case, e.g. krypton-89 and barium-144) and three neutrons, which are able to trigger further fissions. In the nuclear reactor, the reaction is controlled in such a way that the reaction rate remains constant in normal operation. Source: Informationskreis KernEnergie

▶ **Nuclear fission reactors.** Nuclear fission reactors are plants in which self-sustaining nuclear fission reactions take place in a controlled fashion. The purpose of civilian nuclear reactors currently in operation is electricity generation. These reactors are made up of five basic components:

- A sufficient mass of fissile material
- A material to slow down the neutrons (a moderator – with the exception of fast breeder reactors)
- Control rods for neutron capture
- Coolant to remove heat from the reactor
- Barriers for radiation protection and for the retention of radioactive substances

As a basic principle, any atomic nucleus can be split. Specific uranium and plutonium isotopes are particularly easy to split by means of neutrons. Splitting these atomic nuclei releases more energy than is consumed for this purpose. In the light water reactors generally used, the nuclei of uranium-235 are split by moderated slow neutrons. Most of the fission products arising in this process are radioactive. In addition to fission products, an average of 2 or 3 neutrons are released which, in turn, split uranium or plutonium isotopes. This enables a self-sustaining nuclear fission process (chain reaction).

The slower the neutron, the more probable its capture by a U-235 nucleus and the fission of that nucleus. That is why the fast neutrons produced by nuclear fission in a nuclear reactor need to be slowed down by means of a moderator. A *moderator* is a material such as graphite, heavy water or normal water. Collisions with these atomic nuclei slow down the neutrons, but still allow them to be available for the chain reaction.

In *light-water reactors*, the water serves as a coolant and moderator. If the quantity of nuclear fissions is increased, the power is increased and the water's steam content rises. This results in a dilution of the moderator and thus a reduction in nuclear fissions to prevent the chain reaction from continuing uncontrolled. The chain reaction is also regulated by means of the control rods of a nuclear reactor, which may be made out of cadmium, gadolinium or boron. Adding these substances to, or removing them from, the reactor core allows the reactor to be controlled.

In the steady-state mode, there is just enough neutron-absorbing material in the core to ensure that, on average, only one of the neutrons released per nuclear fission is available for further nuclear fissions. All the other neutrons are absorbed, for instance, by boron or cadmium or are lost to the chain reaction in other ways. The number of nuclear fissions per unit time remains constant, and a constant output of power is released as heat (critical state).

If the power of a reactor is to be reduced, its reactivity is temporarily decreased by adding neutron-absorbing materials, e. g. by inserting the control rods. So more neutrons are absorbed than would be necessary to maintain steady-state operation. A reactor in that state is referred to as being subcritical. The addition of sufficient amounts of neutron-absorbing material allows the reactor power to be reduced to zero and the reactor can then be shut down.

In order to raise the power of the nuclear reactor again, neutron-absorbing material is removed from the reactor core, for instance by withdrawing the control rods. Consenquently, more than one neutron per nuclear fission is made available for further fissions, and the number of fission processes, and thus the power of the reactor, increases (supercritical state).

The nuclear fuel cycle involves all production steps and processes involved in the supply and waste management of nuclear power plants.

► **Uranium mining**. Supplying the nuclear reactors with uranium is the starting point for using nuclear power. Uranium is a heavy metal that occurs in the earth's crust 100 times more frequently than gold or silver. Uranium deposits are found primarily in rocks, in the soil and in sea water in dissolved form. 2 t of uranium ore yield approximately 1 kg of uranium oxide (U308), also known as "yellowcake", which contains around 80% uranium.

► **Conversion, enrichment**. Uranium is present in its natural isotopic composition (approx. 99.3 % U-238 and 0.7% U-235) in yellowcake, the product of uranium ore processing. The proportion of fissile U-235 must be increased to between 3–5 % for use in a nuclear power plant. For this purpose, the uranium concentrate is converted by a series of chemical reactions into a gaseous state (conversion). In gaseous form the uranium is then present as uranium hexafluoride (235-UF_6 and 238-UF_6), which enables separation of the isotopes.

► **Fabrication of the fuel assemblies.** Uranium hexafluoride (UF_6) enriched with U-235 is converted into uranium dioxide (UO_2) by means of a wet chemical process to produce uranium fuel assemblies. The UO_2 then exists as a grey powder, which is pressed into pellets that are subsequently sintered (baked) at a temperature of 1,700 °C. The pellets are ground to a specific size and placed in cladding tubes made of zircaloy. The tubes are flooded with helium and sealed gas-tight at each end with weld-on caps.

► **Nuclear power plant.** The energy (heat) released during nuclear fission is used in nuclear power plants to generate electricity. The fuel assemblies are used in a reactor for up to seven years. During this time they are placed in different positions inside the reactor in order to achieve optimum burn-up. Spent fuel assemblies, which have a high level of specific activity and thus generate a great deal of heat, are stored inside the nuclear power plant in spent fuel pools filled with water for cooling. Following this initial cooling phase, spent fuel assemblies are deposited in interim storage facilities before being transported to the conditioning plant.

► **Waste treatment and conditioning**. There are two possible methods of waste treatment: reprocessing of plutonium and uranium or direct storage of the spent fuel assemblies. During reprocessing, plutonium and uranium are separated from the fission products by extraction and used again in fresh fuel elements. After reprocessing the highly active substances, heat-generating fissile product solutions are vitrified and packaged in stainless steel canisters. They are now ready for storage in a final waste repository. A conditioning facility can also dismantle and package spent fuel assemblies in such a way that the packages created are immediately ready for final disposal. A so-called multi-barrier system consisting of several independent areas can effectively prevent potential release of

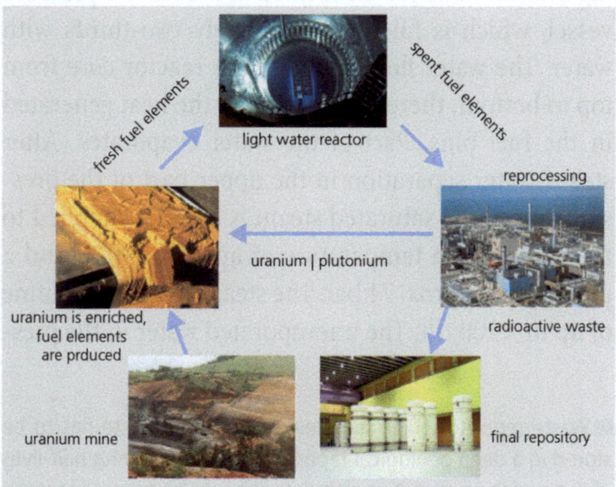

Fuel cycle: After being extracted from the mine, the uranium is converted and enriched before it is used in the reactor. The interim storage of used fuel elements removed from the reactor and their direct storage in a final waste repository are illustrated on the right-hand side. The used fuel elements can also be reprocessed by separating the emerging fission products and the fresh fuel.

pollutants from the repository. The innermost technical barrier is vitrified waste (HLW glass) and spent nuclear fuel, enclosed in a thick-walled container. The next geo-engineered barrier consists of materials minimising the void space remaining in the repository so as to increase mechanical stability, prevent solutions from penetrating, and chemically influence any solutions that may have penetrated. Moreover, these materials adsorb any radionuclides released, and enable the afterheat generated by the highly radioactive waste to be removed. The outermost barrier system is constituted by the geologic barrier consisting of the ambient host rock and the cap rock. In order to guarantee the principle suitability of the designated geological disposal site, all aspects of the system have to be analysed in a safety assessment. This includes the conditioning of the waste, e. g. the vitrification of the high-level radioactive waste, and the selection of a suitable cask. The safety concept for the mine includes the host rock and the effectiveness of the interaction between the geological and technical barriers.

Applications

With the exception of the Phénix fast breeder reactor in France, only light-water reactors (LWRs) – boiling-water and pressurised-water reactors – are operated in power mode for electricity generation. In principle, the reactors in nuclear power plants, submarines and aircraft carriers all produce the same basic fission reaction, and merely differ with regard to the type of coolant employed and the way it is managed.

► **Boiling water reactor.** The fuel elements containing uranium dioxide are located in the reactor pressure vessel, which is filled approximately two-thirds with water. The water flows through the reactor core from top to bottom, thereby transferring the heat generated in the fuel pins. Part of the water evaporates. After steam-water separation in the upper part of the pressure vessel, the saturated steam is directly supplied to the turbine at a temperature of approx. 290 °C and a pressure of approx. 71 bar. The steam reaches a volume of up to 7,200 t/h. The unevaporated water in the pressure vessel flows back down inside the annulus between the pressure vessel and the reactor core, mixing with the feedwater pumped back from the condenser. The pumps installed in the pressure vessel recirculate the coolant. Changing the speed of these pumps allows the circulating coolant volume to be modified and thus the reactor power to be controlled. The control rods containing the neutron-absorbing material are inserted into the reactor from below or withdrawn again either by electric motors (standard mode) or by hydraulic means (scram).

► **Pressurised water reactor.** The heat generated in the fuel elements is removed by the cooling water. To prevent boiling, the operating pressure in the main coolant system is set at 157 bar and controlled by a pressuriser. The coolant enters the reactor pressure vessel at a temperature of 291 °C and, after having been heated in the reactor core. It leaves at a temperature of 326 °C. The heated water transfers its heat in four steam generators to the water of a secondary system, which evaporates, supplying 2.1 t/s of saturated steam at a temperature of 285 °C and a pressure of 66 bar. A dual-circuit system of this type ensures that the radioactive substances occurring in the reactor coolant remain restricted to the main cooling system and cannot enter the turbine and the condenser. The steam generated is used to run a turbine. The control rods of the reactor use an alloy such as silver, indium and cadmium as the absorber substance. For fast control processes, such as fast reactor shutdown, or scram, 61 control rods can be inserted fully or partly into the reactor and withdrawn again. For slow or long-term control processes, boric acid is added to the reactor cooling water as a neutron absorber. The fuel elements are contained in a pressure vessel made of special steel (25 cm wall thickness) which, together with the primary system, is installed in a double-walled containment.

The European Pressurised Water Reactor (EPR) of the third generation currently under construction in Finland and France has been designed to cope even with very unlikely severe accidents involving core meltdown. This is achieved, among others, by safely collecting a core melt on the massive reactor foundation and cooling it passively. Core meltdown can occur when, in the event of a cooling failure, a fuel rod control defect or other accident sources, post-decay heat (the fission products of nuclear fission continue to decay after reactor shutdown and cannot be influenced by the control rods) heats up the fuel elements to such an extent that they begin to melt. If this condition persists for a sufficiently long time, the molten fuel may collect at the bottom of the reactor pressure vessel. This would raise temperatures in the fuel to more than 2800 °C.

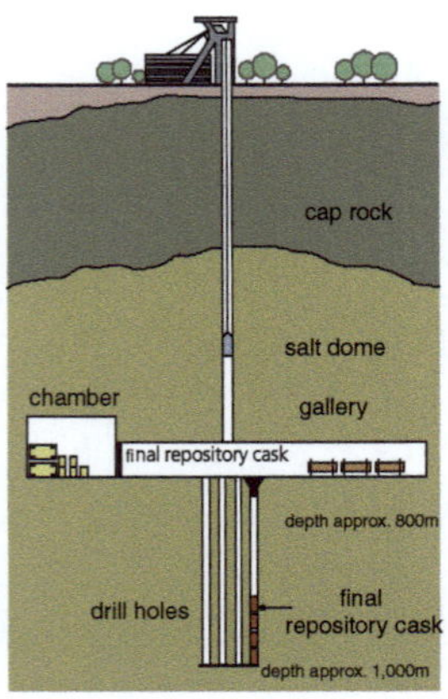

◘ Waste repository: The high-level radioactive waste has to be stored in a deep geological repository. Due to the long half-lives of some of the radioactive elements, it has to be ensured that the stored waste will not come into contact with the biosphere for very long periods of over 100,000 years. The most favorable host rock systems considered worldwide are clay, granite and salt. At present, 250,000 t of used fuel elements have to be disposed of worldwide, 40,000 t of which are in Europe. The required geological disposal system will therefore have a volume of several 100,000 m^3 and will operate for about 50 years.

► **Fast breeder reactor.** In boiling water and pressurised water reactors, the only naturally occurring uranium isotope that can be split is uranium-235, due to the slow neutrons. This uranium isotope makes up 0.7 % of natural uranium, and is enriched to up to 3.5 % in the fuel elements.

Uranium-238 cannot be used as a fissile material in these types of reactor. However, the nucleus of a uranium-238 atom can accommodate a neutron of medium or high velocity and subsequently transform into plutonium-239 over a number of steps. Pu-239 is split most effectively by fast neutrons. In fast breeder reactors, both processes are initiated deliberately: nuclear fission for energy generation, and conversion of part of the uranium-238 into fissile plutonium-239 (breeding process).

The concentration of fissile material in the compact core of a breeder reactor is much higher than that in a light water reactor. Breeder reactors allow the process to be controlled in such a way that more fissile plutonium-239 will be produced from uranium-238 than is consumed by nuclear fission. As fast neutrons are used both in nuclear fission and in plutonium breeding, this reactor line is also referred to as a "fast breeder". At the present state of the art, it is possible to make roughly 60 times better use of natural uranium in this way than in light-water reactors.

Water cannot be used as a coolant in breeder reactors, as it would slow down the fast neutrons too quickly and would not be able to effectively remove the heat produced in the core because of the high energy density. Consequently, liquid sodium is used as a coolant, which enters the reactor core at a temperature of 400 °C and is heated by 150 K. The melting point of sodium lies at 98 °C, its boiling point at 883 °C. In an intermediate heat exchanger, the liquid sodium of the primary system transfers its heat to the sodium of a secondary system. Another heat exchanger constitutes the steam generator in which the liquid sodium of the secondary system generates steam of approx. 500 °C (at approx. 180 bar), which is then used to drive a steam turbine. A three-cycle system is preferred for reasons of technical safety.

► **Thorium high-temperature reactor.** This type of reactor allows relatively high temperatures to be produced. While light-water reactors achieve coolant temperatures of up to 330 °C and fast breeders up to 550 °C, helium coolant in high-temperature reactors reaches 750 °C and above. This produces not only steam for driving turbines, but also process heat (e. g. for generating hydrogen).

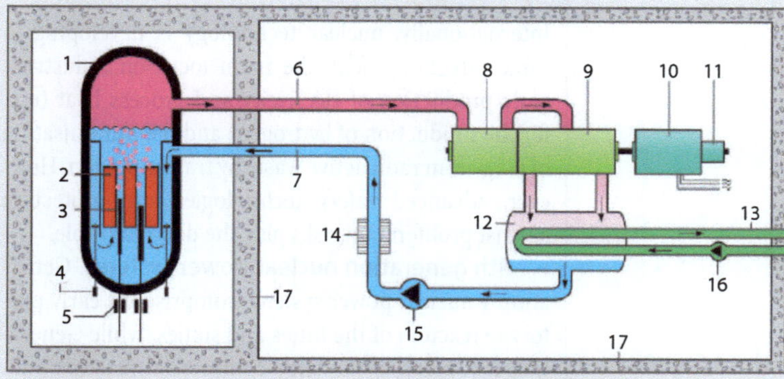

1 reactor pressure vessel
2 fuel elements
3 control rods
4 recirculation pumps
5 control rod drives
6 main steam
7 feedwater
8 turbine high-pressure section
9 turbine low-pressure section
10 generator
11 exciter
12 condenser
13 river water
14 preheater
15 feedwater pump
16 cooling water pump
17 concrete shield

Boiling water reactor: Cooling water flows through the reactor core (left), absorbs the heat generated in the fuel pins, and partly evaporates in the process. This saturated steam is fed straight to the turbine, then condensed and returned to the reactor. Source: Informationskreis KernEnergie

Nuclear power plant: A boiling water reactor (front left) and a pressurised water reactor (right) with the typical dome of the reactor containment and the cooling towers at the back. The reactor containment has a typical volume of about 50,000 m³, and its walls are more than 1 m thick. The cooling towers have a height of more than 100 m. This is because most existing light-water reactors generate over 1,000 MW_e, so the cooling towers have to be correspondingly large.

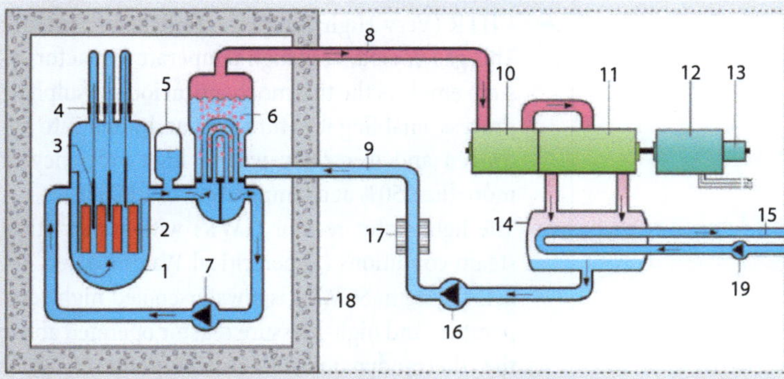

1 reactor pressure vessel
2 uranium fuel elements
3 control rods
4 control rod drives
5 pressuriser
6 steam generator
7 coolant pump
8 main steam
9 feedwater
10 turbine high-pressure section
11 turbine low-pressure section
12 generator
13 exciter
14 condenser
15 river water
16 feedwater pump
17 preheater
18 concrete shield
19 cooling water pump

Pressurised water reactor: Heated water under high pressure transfers its heat in steam generators to the water of a secondary system under lower pressure, thus causing the water to evaporate. In this way, the radioactive materials occurring in the reactor coolant are limited to the main cooling system and do not enter the turbine and the condenser. The steam drives a turbine (high-pressure section, low-pressure section), is condensed and returned to the secondary system. Source: Informationskreis KernEnergie

Trends

Internationally, nuclear technology is developing in three directions, with the main focus on industrial-scale production of electricity and process heat (e. g. for the production of hydrogen) and the minimisation of long-term radioactive waste by transmutation. However, advanced safety technologies and protection against proliferation risks play the dominant role.

▶ **4th generation nuclear power systems.** Generation 1 nuclear power systems comprise the early prototype reactors of the fifties and sixties, while Generation 2 consists of the large commercial reactors built since the seventies and still in operation. Generation 3 refers to advanced reactors currently on offer in new projects or already under construction, such as the EPR in Europe. The strategic objective of the "Generation-4 initiative" is to develop nuclear power systems that are able to supply energy as electricity and process heat, e. g. for the production of hydrogen, and to minimise long-term radioactive waste by transmutation. Generation 4 comprises the entire nuclear fuel cycle, from uranium extraction to waste management. Priority is given to criteria such as sustainability (minimisation of raw material consumption and waste), safety and reliability, resistance to proliferation, physical protection and economic performance.

Three nuclear power systems appear to be favorable and warrant further in-depth development. They are based to some extent on experiences made with nuclear power systems that have already been built or at least developed.

- VHTR (Very High Temperature Reactor System): This gas-cooled, very high temperature reactor system employs the thermochemical iodine-sulphate process, enabling it to turn heat and water into hydrogen and electricity with a high efficiency of more than 50% at a temperature of 1,000 °C.
- The light-water reactor (LWR) with supercritical steam conditions (Supercritical Water Cooled Reactor System, SCWR) is a water-cooled high-temperature and high-pressure reactor operated above the thermodynamically critical point of water (374 °C, 22.1 MPa).
- The technology of the Sodium-cooled Fast Reactor System (SFR) has already been demonstrated in principle. Research therefore is geared more towards improving the passive safety and economic performance of these plants.

Generally, all Generation 4 reactor systems still face numerous technological challenges, such as the development of new kinds of fuels and materials, e. g. high-temperature alloys and coatings, and sustaining core outlet temperatures of 850–1,000 °C or even higher. Some designs require new concepts, such as a high-power helium turbine, which makes even the first prototypes of such plants unlikely to appear earlier than 20 years from now.

▶ **Partitioning and transmutation.** In view of the long half-life and the hazard potential (radiotoxicity) of some radionuclides, it must be ensured over very long periods of time that no radioactive substances will be released from a repository into the biosphere. This is why alternatives to the final storage of long-lived radionuclides are being studied. The radionuclides are separated from the spent nuclear fuel by suitable processes (partitioning) and then converted into stable or short-lived isotopes by neutron reactions in special facilities (transmutation). This would allow the radiotoxicity of the waste destined for permanent storage to be reduced roughly to the level of natural uranium after several hundreds of years, thus minimizing the long-term hazard potential.

▶ **Fusion reactors.** The objective of fusion research around the world is to create an electricity-generating fusion power plant which, like the sun, produces energy from the fusion of atomic nuclei. For this purpose, a thin gaseous mixture of the deuterium and tritium hydrogen isotopes must be heated to 100 million °C. At these temperatures, any kind of matter will be transformed into a plasma state, i. e. the atomic structure is dissolved, and atomic nuclei and electrons are no longer bound to one another. A plasma thus consists only of charged particles. This property is used to confine the hot plasma in what is known as a "magnetic cage" without allowing it to contact the walls. External heating of the plasma to 100 million °C is achieved by three mutually supplementary methods:

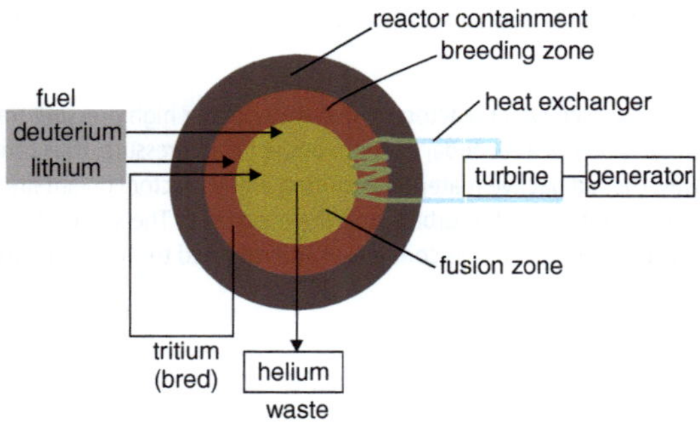

Fusion reaction: Differences in binding energy make it possible to gain energy from fusion. The mass defect or binding energy is the energy applied to separate a nucleus from the other nuclei of an element. A deuteron (deuterium nucleus) and a triton (nucleus of tritium, which is bred by the lithium reaction) fuse to form a highly excited compound nucleus, which immediately breaks down into a helium nucleus and a high-energy neutron. This neutron receives 80% of the released energy. The heat produced can be used in a classical type of power plant, such as a coal or nuclear power plant. Source: Forschungszentrum Karlsruhe GmbH

- An electric current induced in the plasma heats the plasma to approximately 15 million °C
- Injection of neutral particles: fast hydrogen atoms are injected into the plasma and transfer energy to the plasma particles.
- Heating by microwaves

The reaction products arising from the fusion of deuterium and tritium are a high-energy neutron and a helium nucleus (alpha particle). The energy of the helium nucleus is 100 times higher than that of the plasma particles. This "surplus energy" is transferred to the plasma particles by interaction. When the reaction rate is high enough, this "alpha heating" alone is sufficient to keep the plasma at operating temperature. This state is called ignition.

Safe confinement of the plasma requires a twisted (helical) magnetic field. In the case of the tokamak principle, the helical magnetic field is generated by superposition of three magnetic fields. This is the furthest-developed principle of all, but it has the drawback that a tokamak, at the present state of the art, can be operated in pulsed mode only. The stellarator principle, by contrast, allows the magnetic field to be operated in a steady-state mode because its helical magnetic field is produced solely by ring-shaped non-planar magnet coils.

Unlike the doubly-charged helium nucleus, the neutron arising from the fusion reaction is not enclosed in the "magnetic cage", but impinges on the 500 – 600 °C hot wall of the plasma vessel and is slowed down in the process. The heat generated in this step is extracted by a coolant (helium) and converted into electricity in a conventional circuit.

▶ **Experimental fusion plants**. A number of experimental fusion plants are in operation around the world to establish the basic physics and technical principles of a fusion power plant. In the JET (Joint European Torus) experimental facility, which employs the tokamak principle, 65 % of the power input was recovered once for a few seconds. The ITER experimental reactor currently being built in worldwide cooperation (construction began at Cadarache, southern France, in 2008) is to have a power of 500 MW and generate ten times more energy than is supplied externally to heat the plasma. The main objectives of ITER are to demonstrate a long-burning plasma typical of a reactor, and to test key technologies such as superconducting magnets, plasma heating, tritium breeding, energy extraction, and remote handling technologies.

In the light of findings arising from operation of the ITER experimental reactor, a decision is to be taken in around 2025 with regard to planning a demonstration power plant with a helium cooling system.

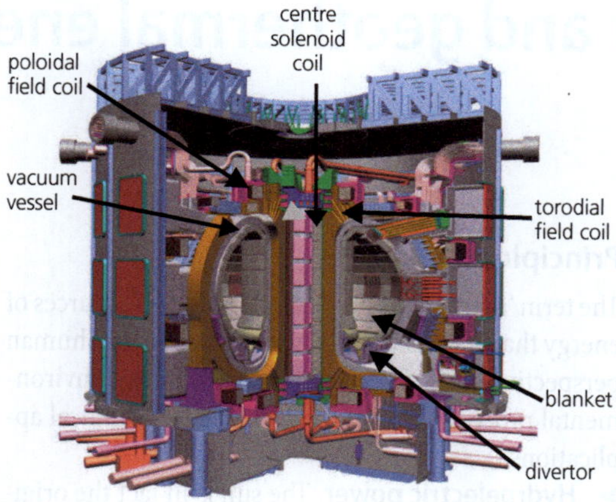

■ ITER experimental fusion plant: The toroidal and poloidal magnetic field coils serve to magnetically confine the plasma and have to be supported by various support structures. The double-walled vacuum vessel is lined by modular removable components, such as blanket modules, divertor cassettes and diagnostics sensors, as well as port plugs for limiters, heating antennas, diagnostics and test blanket modules. The total vessel mass is ~10,000 t. Components absorb most of the radiated heat from the plasma and protect the magnet coils from excessive nuclear radiation. The initial blanket acts solely as a neutron shield, and tritium breeding experiments are carried out on test blanket modules inserted and withdrawn at radial equatorial ports. The divertor design is made up of 54 cassettes. Large openings between the inner and outer divertor balance the heat loads in the inboard and outboard channels.

Such a plant could supply the power grid with electricity from a fusion reactor for the first time from the mid-thirties onwards. Commercial fusion power plants could contribute towards supplying commercial energy as of the middle of this century.

Prospects

- The share of energy from nuclear sources contributing to cumulated world energy consumption will continue to be around 5% until 2020. Electricity and energy generation in nuclear power plants represents a safe and economical way of producing energy in the form of electricity, hydrogen and heat, provided that a closed fuel cycle is used.
- There ist currently no alternative to final storage in deep geologic formations when it comes to the safe management of highly radioactive waste. The introduction of Generation 4 reactors and the strategy of partitioning and transmutation represent a possibility to alleviate the problems of high-level radioactive waste and its final storage.
- Fusion offers the prospect of a future energy supply characterised by practically unlimited fuel reserves and favourable safety qualities.

DR. THOMAS WALTER TROMM
WERNER BAHM
Forschungszentrum Karlsruhe GmbH

Internet

- www.kernenergie.de
- www.efda.org
- www.euratom.org
- www. euronuclear.org

Wind, water and geothermal energy

Related topics
- Energy storage
- Electricity transport
- Solar energy

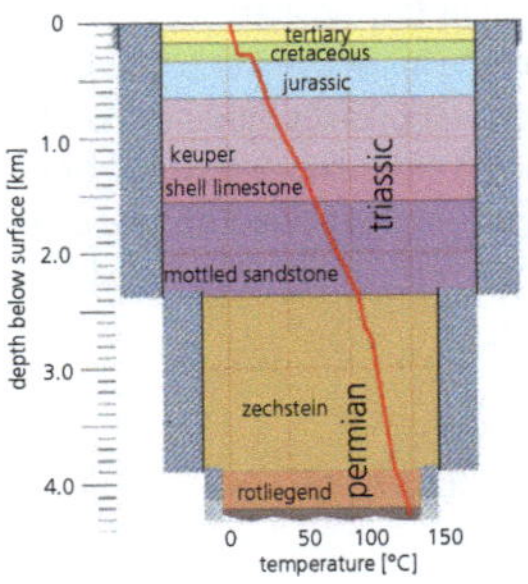

Temperature depth profile and geological structure of the subsoil at the drilling site in Groß Schönebeck, Germany. Source: GFZ Deutsches GeoForschungsZentrum

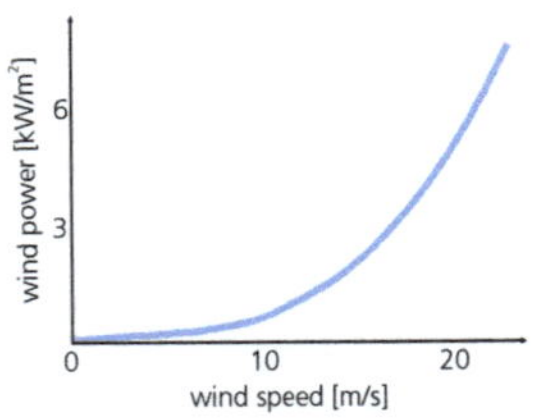

Exponential dependence of wind power on wind speed: The operating range of modern wind turbines extends from 3–5 m/s (light breeze) to around 25 m/s (gale).

Principles

The term 'renewable energies' refers to those sources of energy that are inexhaustible when seen from a human perspective. They are derived from ongoing environmental processes and made available for technical applications.

▶ **Hydroelectric power**. The sun is in fact the original source of most forms of hydroelectric power. It evaporates surface water by the energy of its radiation and keeps the global water cycle in constant motion. The kinetic and potential energy of water from this cycle is turned into electrical current in hydroelectric power plants. Only the energy obtained from tides is produced by gravitational effects – mainly from the Earth-Moon system.

Hydroelectric power makes up 17% of electrical power generation worldwide and thus has the highest share of all renewable energies. There are still large unexploited potentials all over the world, particularly in Asia, South America and Africa. However, the use of hydroelectric power does have adverse effects on nature and landscapes.

▶ **Geothermal energy**. Geothermal heat is the thermal energy stored in the earth. This energy originates mainly from the fused core inside the earth from where it rises to the surface. Rock and soil layers, as well as underground water reservoirs, are heated by the heat flow. Around 30% of this heat is derived from the original heat produced during the earth's formation approximately 4.5 billion years ago, and around 70% from the ongoing decay of natural radioactive isotopes. In addition, the earth's surface is warmed by solar radiation and thermal contact with the air.

There are very large amounts of thermal energy stored in the earth, but the interior of the earth generally only produces a small heat flow. To ensure the economic use of plants, therefore, geothermal energy exploitation must be especially designed to meet the risk of the local area of soil cooling too strongly. Geothermal energy can be used directly for cooling or producing heat for individual or industrial consumption. Indirect use involves converting heat into electricity. To optimise conversion efficiency, power plants that generate electric power and heating energy jointly have proven to be particularly economical.

On average, soil temperature increases by 3 K per 100 m. A distinction is made between near-surface (down to 400 m) and deep-seated geothermal energy (1,000–3,000 m), depending on the actual extraction area. Near-surface geothermal energy is typically used for heating and cooling owing to its low temperature. Deep-seated geothermal energy can often be utilised for power generation as well.

▶ **Wind energy**. Wind is generated by the sun's rays hitting the earth's surface. The atmosphere, land and water warm up to different temperatures, creating high- and low-pressure areas between which air currents are produced that balance the differences. The formation, strength and direction of the wind are also influenced by the rotation of the earth, the varying heat capacities of water and land, and elements of the landscape such as mountains.

Modern wind turbines start to rotate at wind speeds of 3–5 m/s. Building a wind power plant is profitable in areas with an annual average of wind speed of 4–5 m/s at an altitude of 10 m – the speed at a turbine hub height being considerably higher. Selecting particularly windy sites is important, because the kinetic energy of wind increases with the cube of its speed. High wind speeds are reached above oceans or in coastal areas, further inland the wind is slowed by friction with the earth's surface. But good wind conditions are also found in flatlands and at some higher altitudes.

The global, installed wind energy capacity reached about 94,000 MW as of December 2007, with an annual growth rate of 26%. It generates 200 TWh per year, which corresponds to 1.3% of global electricity consumption.

In some regions of the world, however, growth is reaching its limits, since the number of windy sites available is limited. The choice of sites is also limited because large-capacity wind turbines must be kept at a certain distance from residential or protected areas. Therefore, better use will have to be made of the potential of favourable sites by replacing technologically obsolete wind turbines by new plants with higher power and availability levels. Offshore areas will also be used increasingly.

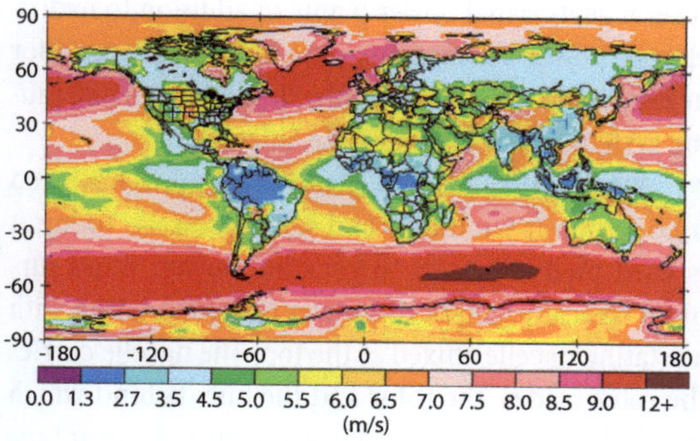

Average annual wind speeds at 50 m height (July 1983 – June 1993). Source: NASA

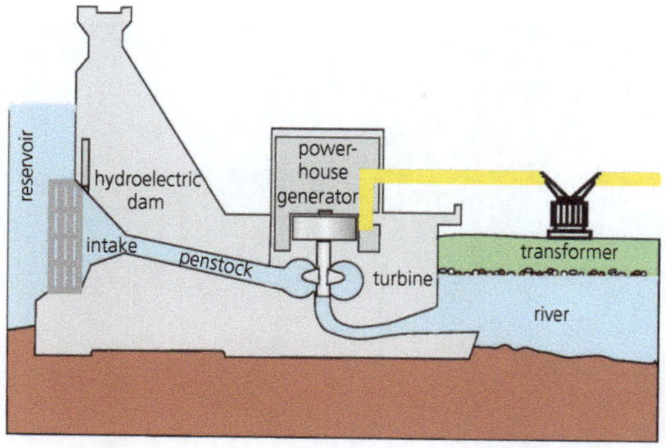

River power plant: Water is dammed up and then drives the generator via a turbine when it is drawn off into a downstream reservoir.

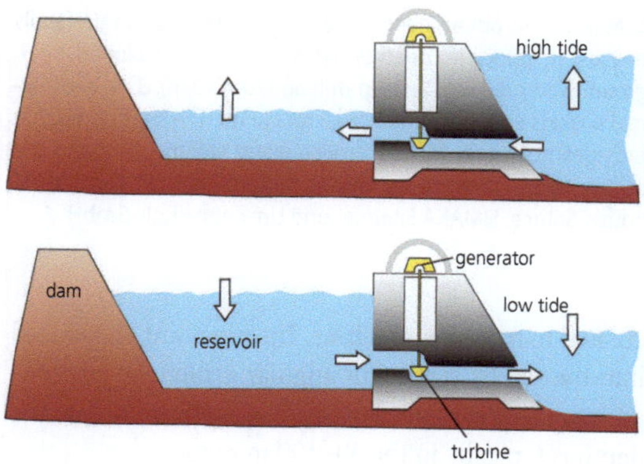

Tidal power station: Located on the Atlantic coast in the Rance River estuary near Saint-Malo/France, the power station can generate electrical power both at low and high tide. At high tide, the water is channelled through a turbine and into a gully used as a reservoir. When the water level in the reservoir is above sea level at low tide, the water is then let back into the sea driving turbines in the other direction.

Applications

► **Hydroelectric power plants.** In earlier times, the mechanical energy obtained from the kinetic and potential energy of water flows was used directly, for example in mills. Today, conversion to electrical energy predominates. This is mostly done at river power plants, where the flow of a dammed river or canal drives generators via a water turbine. This type of power plant is used to generate a base load. The 26 turbines of the Three Gorges Dam on the Yangtze River in central China are designed to provide a total of 18,200 MW and thus the largest plant capacity worldwide. However, the actual energy yield could be lower, since the maximum gross head of 113 m required for maximum output is in conflict with the other purpose of the dam, namely flood protection. The project has also sparked a controversy because of the geological hazards it has created and its ecological and social impact.

Hydroelectric power plants that are not operated continuously generate intermediate and peak load power. *Storage power stations*, for example, use the high head and storage capacity of reservoirs and mountain lakes for electrical power generation. In a pumped-storage power station, the storage reservoir is not filled from natural sources of water, but rather by water pumped up from the valley below. In this way, conventionally generated electrical power is temporarily stored as potential water energy in off-peak periods and can be turned back into electricity via a turbine in peak-load periods.

Dams have served as a source of water power for quite some time, but generating energy from the sea is still in its infancy. Tidal power, by which the ebb and flow of tides is used to drive turbines, has been exploited for some decades now. However, the tidal range must be more than 5 m if a tidal power station is to be operated economically. Such large differences in elevation are reached only in bays or estuaries, and that means that there are relatively few suitable sites, one being the French Atlantic coast.

► **Near-surface geothermal energy.** Near-surface geothermal heat (drawn from a depth of 400 m) is mostly used directly for heating or cooling. The prime application is natural cooling, whereby water at the temperature of shallow ground is used to cool buildings directly. The heat level can be exploited by means of geothermal probes, for example. These probes consist of bundles of pipes containing a carrier liquid cir-

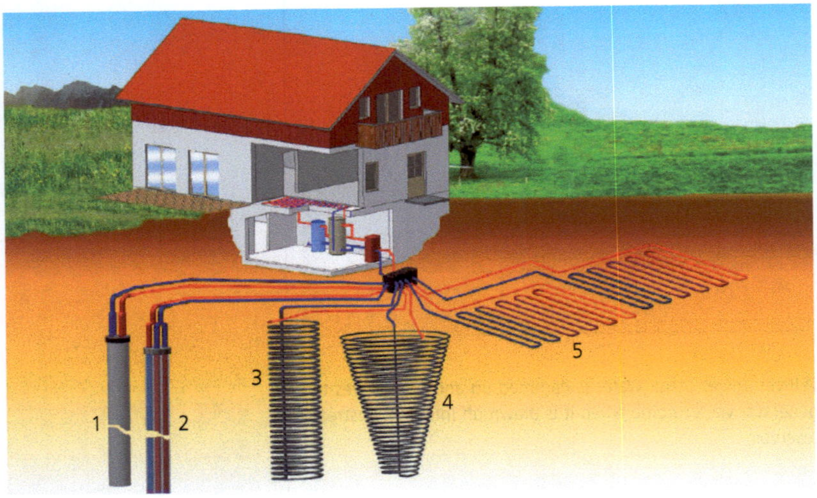

Methods to obtain geothermal heat for heating single-family houses: In principle, geothermal heat can be exploited by 1. groundwater drilling, 2. deep drilling (70–150 m), 3./4. geothermal baskets or 5. horizontal collectors (2 m). The heat pump that provides hot water for the service water storage tank and the buffer storage tank for the heating system is installed inside the house. Source: SIMAKA Energie- und Umwelttechnik GmbH

culated to transport the heat. The pipes are introduced into the soil by vertical or angular drilling or integrated into concrete foundations. For a private home, the depth is typically in the 50–100 m range.

For heating buildings using near-surface geothermal heat, the temperature must first be raised to a level suitable for heating with a heat pump that runs on electricity. These systems do require higher investments at the beginning, but they are competitive thanks to their low running costs and are therefore increasingly found in private homes as well.

Some geothermal anomalies result in water being heated to temperatures of several hundred degrees fairly close to the surface. They are particularly suitable for generating power. Iceland, for example, has a volcanic landscape. Its power supply is guaranteed by several geothermal power plants in addition to hydroelectric power. Furthermore, the hot water is used for heating buildings and footpaths through insulated piping systems, even over great distances.

▶ **Wind power plants.** Historically, wind has been used to power windmills and sailing vessels. Today, it is used predominantly to generate power. Wind turbines basically consist of a tower on a foundation with a rotating nacelle affixed at the top. The nacelle carries the rotor blades, the rotor and the hub, behind which one finds the generator and in some cases the gear box. Control and grid connection systems complete the installation. In today's wind power plants, three-bladed upwind *rotors* with a horizontal rotational axis have become the norm. In terms of rotation control they pose the least problems since the pressure load on the axis and tower remains relatively even during rotation. The rotor blades have an aerodynamic airfoil shape similar to that of aeroplane wings. The difference in the velocity of the air streaming around the sides of the blades causes the torque. Theoretically, the maximum percentage of energy that can be drawn from the wind in this way is 59%. This is about three times as much as the amount generated by the now rarely used drag-type rotors, which turn the drag of the rotor blades into rotation. Neither have wind turbines with a vertical rotational axis been very successful, as their profitability is low.

Modern generators in wind *turbines* are highly effective at a variety of speeds. Special annular generators with large diameters, whose speeds are within the range of the rotor speed, do not necessarily need a gear box. The frequency of the voltage generated fluctuates depending on the speed, so it has to be rectified, filtered and transformed to the right voltage and frequency using a converter. More modern turbines can be controlled by continuous adjustment of the rotor blade's angle of attack, which is another way of ensuring grid compatibility. This serves to reduce their feeding capacity and prevent overvoltage in the grid system. A storm control system prevents the turbine from

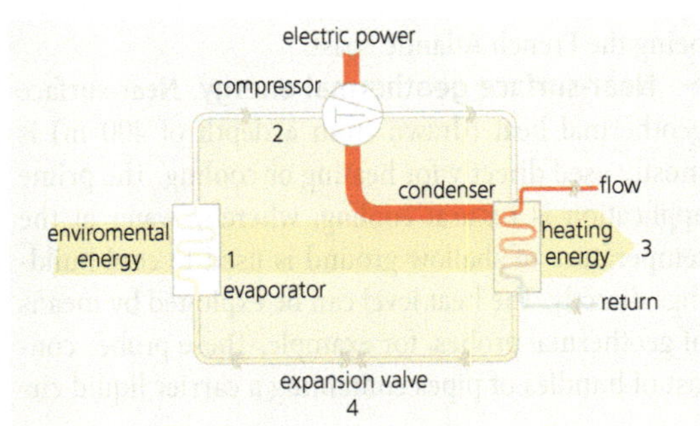

Heat pump: The thermal energy from the environment is transferred in the evaporator to the circulating refrigerant, which vaporises and expands (1). The gaseous refrigerant is then recompressed in the compressor, which raises the temperature (2). This heat can then be transferred to the building's heating system by means of a second heat exchanger (3). The refrigerant pressure is reduced again in the expansion valve (4). In the process, 75% of the energy obtained for heating is derived from geothermal energy, the rest is provided by the electrical energy required for running the system. Source: Stiebel Eltron GmbH stopping abruptly when the maximum permissible wind speed is exceeded.

Over the past decades, the technical trend has been towards larger and larger wind turbines. Currently, turbines are being built with a rated power of around 5-6 MW and total heights of over 180 m (120–130 m hub height, 110–130 m rotor diameter). Given a favourable location and thanks to the increase in technical availability, modern turbines can generate the energy needed for their manufacture within a few months.

Trends

The increasing cost of fossil fuels and the efforts at reducing greenhouse gas emissions are making it likely that renewable energies will make up a larger share of the power market in the future. Changes in the energy mix, however, are also impacting grid integration and its limits. The exploitation of wind energy over large areas through offshore wind farms, for instance, causes fluctuations in the energy supply. A higher percentage of the total power generation covered by these energies will create a need for measures for storing them, as well as flexible energy consumption patterned more closely on the available energy supply.

▶ **Reservoir-pressure waterpower machine.** Today's water turbines and generators are highly advanced technologically. Their effectiveness, availability and durability are quite satisfactory. In recent years, too, the cost and effort of running small hydropower stations, in particular, has been reduced. Among other things, the reservoir-pressure waterpower machine was developed. This new technology makes efficient use of even low dam heads and allows the water inhabitants pass through it unharmed. While still in the demonstration phase, the machines will permit further harnessing of the potential of water power.

▶ **Tidal farm.** Several concepts for harvesting energy from the sea are currently being tested. There are some prototypes of tidal farms, which are power plants producing electricity from underwater rotors activated by the natural flow of the tides, or strong, permanent ocean currents. Site factors for cost-efficient power production are a water depth of 20–30 m and a medium flow velocity generally in excess of 2.5 m/s. These conditions are found frequently off the British and French coasts.

▶ **Wave farms.** Different concepts exist for the generation of power from waves. The first wave farm has been operating since the year 2000 off the Scottish coast. Incoming waves press water into air chambers and each trough drains it again. This oscillating water

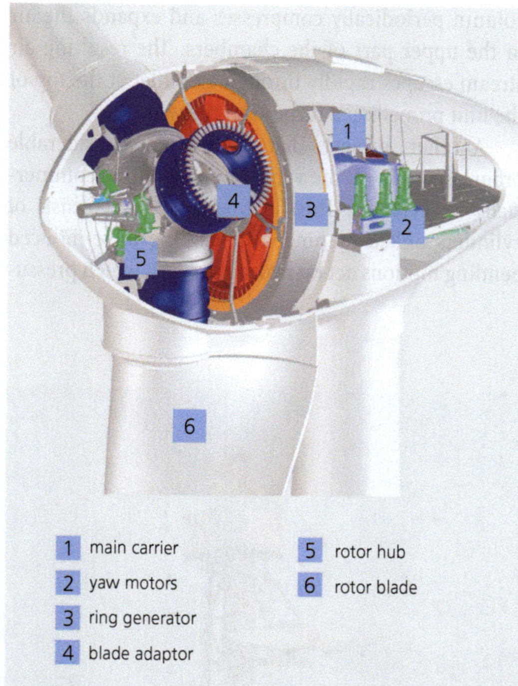

Nacelle of a 2 MW wind turbine: The blade adapter connecting the rotor blades with the hub permits active adjustment of the rotor blades to set an optimum angle of attack. In this model, the rotor hub and the ring generator are directly affixed to each other without a gear box. Output voltage and frequency vary according to speed and are converted for transfer to the power grid via a DC link and an AC converter. This permits high speed variability. Yaw control is provided by the horizontally rotating yaw motor. Source: Enercon GmbH

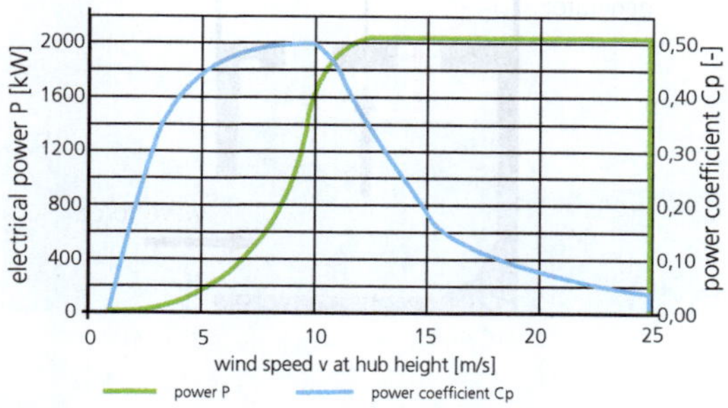

Power curve of a modern 2 MW wind turbine: Power and power coefficient as a function of the wind speed at hub height. The power coefficient is the quotient of the wind power used, divided by the wind power available. At a wind speed of roughly 2.5 m/s, the turbine automatically switches on. Under partial load, the blades are set to the angle generating the highest starting torque. Maximum power (rated power) of 2 MW is reached when the wind is blowing at 12 m/s. If the speed gets any higher, the power output is kept constant by moving the blades out of their optimum angle of attack, thus reducing the aerodynamic effect. In order to avoid damage caused by mechanical overload due to extremely high wind speeds, the turbine is switched off at 22–28 m/s (wind force 9–10).

column periodically compresses and expands the air in the upper part of the chambers. The resulting air stream escapes rapidly through an outlet at the top of the unit propelling a turbine on the way.

Another project that has made considerable progress and is on the verge of going online commercially is "Pelamis". This floating structure consists of cylinders linked by hinged joints. The wave-induced bending motions act on hydraulic rams, which pressurise oil that then drives hydraulic motors and the generators connected to them.

Tidal farm: First commercial prototype "Seaflow" tested off the coast of southern England with an 11-m rotor and a rated power of 300 kW. The rotor has been pulled above the water surface for maintenance purposes. Source: Marine Current Turbines TM Ltd.

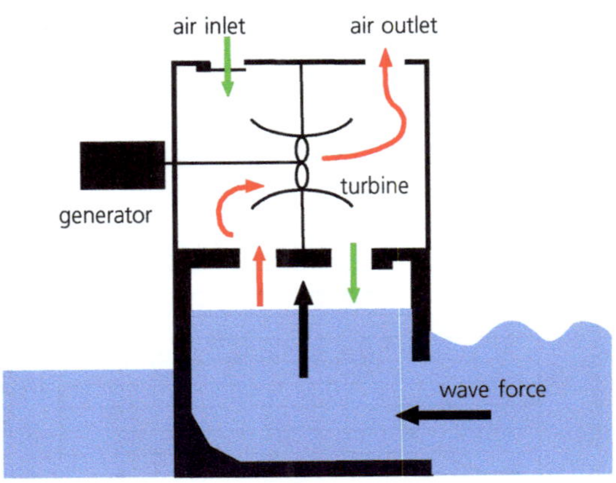

Wave farm off the coast of Scotland: Power is generated by a water column that surges under the force of incoming waves, thus causing a stream of air that propels a turbine.

The "Wave Dragon", which is still in the experimental phase, is based on a particularly simple principle. It is a stationary floating structure consisting of a water reservoir that is filled by waves coming over the top into V-shaped ramps. The water flows back into the sea through turbines that drive a generator.

▶ **Deep-seated geothermal energy**. In non-volcanic areas, deep drilling is usually necessary to reach higher temperature zones. Considerable progress still needs to be made in increasing the yield of the existing sources and in predicting their productivity. Underground water reservoirs with temperatures of 40–100 °C required for heating can often be made directly accessible from deep-seated geothermal resources at drilling depths of 1,000–3,000 m. In such hydrothermal systems, water is brought to the surface via a borehole. Some of the water's thermal energy is extracted with heat exchangers, after which the water is reinjected through a second borehole.

Another method of extracting heat from the earth is *petrothermal systems*. It involves making cracks in dry ground by hydraulic fracturing. Water is then allowed to circulate between at least two deep wells through which it enters and exits the system. Closed systems are also practicable in deep-seated geothermal energy generation. Geothermal probes can reach depths of 2,000–3,000 m. The fluid circulates in coaxial tubes – cold water flows downwards in a thinner internal tube, while the heated water rises up in the external tube. In theory, boreholes and shafts of closed-down mines could also be used to harness deep-seated geothermal energy.

Other pilot projects for generating power from deep-seated geothermal energy sources are also being conducted at the present time. Steam deposits with temperatures of more than 150 °C can propel turbines directly, for example. Moreover, new power station

Wave Dragon: Principle and prototype of the wave farm operating off the Danish coast. Ocean waves swell over the top of the Wave Dragon and are held in a reservoir above sea level. The sea water is subsequently let out through a number of turbines that generate electricity. Source: Wave Dragon ApS

technology is also enabling power generation at lower temperatures of around 80 °C. This is done using so-called organic Rankine cycles, which use a hydrocarbon as a working fluid. One outgrowth of this technology is the Kalina cycle, which is already deployed in some facilities. It uses a mixture of ammonia and water as a working fluid. The Kalina cycle requires much larger heat-exchanger surfaces. The facilities are more complex and costly, but they are a lot more effective as well.

► **Offshore wind farms.** Currently, the most important trend in wind energy production is the construction of offshore wind farms. The corrosive effects of the salty maritime air make considerable demands on corrosion prevention in turbines. High reliability and low maintenance are important efficiency factors, since maintenance or even the replacement of larger assemblies is much more complicated at sea than on land. The design of offshore wind farms must take high wind speeds into account, as well as the impact of ice and waves. Meteorological forecast systems are being developed that can estimate the power that will be fed into the grid in the hours or even days to come. Load management can then be used to better balance wind-induced fluctuations in the power output of wind farms.

Prospects

- The use of water power on land has a long tradition and is very advanced technologically. Many existing hydropower stations, however, are in great need of modernisation. The exploitation of ocean currents and tides, on the other hand, is still in its elementary stages.
- As for wind power, larger and more modern plants will do a better job at harnessing the potential of existing sites that enjoy high wind levels. New sites are likely to be mostly offshore, since both the acceptance and the energy potential are greatest there. However, more technological expertise is needed and grid integration must be optimised.
- In order to efficiently generate electricity from geothermal energy, combined heat and power is usually a must. Furthermore, the risks involved in accessing deep-seated sources must be mitigated. The use of near-surface geothermal energy for cooling and heating will increase significantly, even in the short term. Despite higher initial investments, near-surface exploitation is profitable because of low operating costs, so considerable growth can be expected. There is still room for reducing costs in this growing market as well. Given the almost inexhaustible energy resources stored in our planet, geothermal energy has the potential to provide a significant contribution to energy supplies in the long run, particularly if the technology is developed further and price ratios change.
- Generally, sources of renewable energy hold a great deal of potential. Further technological innovations are needed, though, to lower the additional costs spawned by new developments that are motivated by consideration for the climate and environment rather than by economics.

DR. ULRIK NEUPERT

Fraunhofer Institute for Technological Trend Analysis INT, Euskirchen

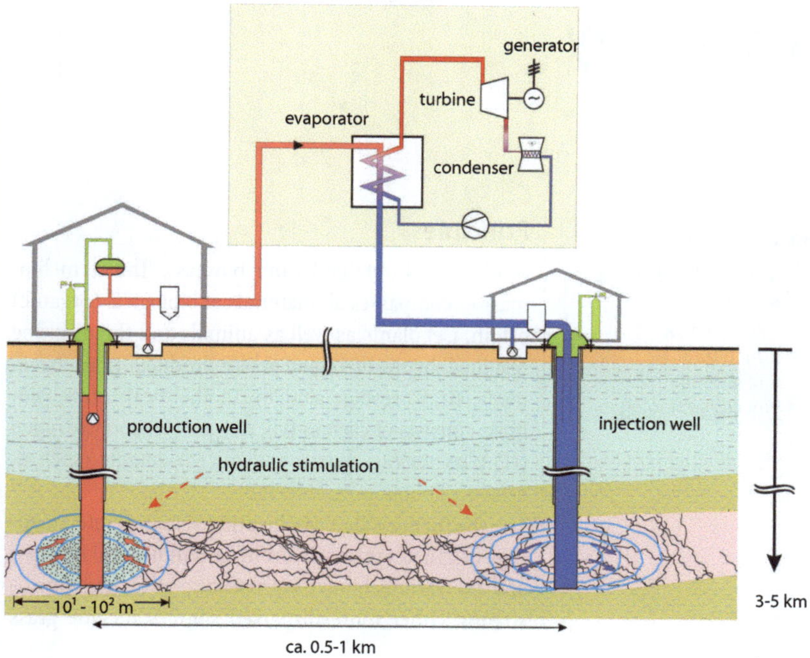

▲ Geothermal power generation requires two boreholes, a sustainable thermal water cycle and a surface power station. Water is extracted from the earth via the first borehole. Once it has been used at the power station, it is fed back into the reservoir through the second (reinjection) borehole. The thermal energy of the water is transferred by means of a heat exchanger to a working substance that boils at low temperatures, driving the power generator in a secondary cycle. Source: GFZ Deutsches GeoForschungsZentrum

▶ Offshore wind farm "Horns Rev" off the Danish coast in the North Sea, where 80 wind turbines (each 2 MW) generate electricity. Source: Vestas Deutschland GmbH

Internet

- www.wwindea.org
- http://iga.igg.cnr.it
- www.hydropower.org
- www.cleanenergydigest.com
- www.energy.gov/energysources
- www.erec.org/reh/heating-cooling.html

Bioenergy

Related topics

- Renewable resources
- Automobiles
- Fossil energy
- Agricultural engineering
- Process technologies

Principles

Bioenergy is obtained from biomass. The term biomass encompasses all materials of biological (organic) origin, i.e. plants as well as animals and the resulting residues, byproducts and waste products (e.g. excrement, straw, sewage sludge). In chemical terms, biomass consists of carbon (C), oxygen (O) and hydrogen (H) plus various macro and microelements (such as nitrogen, potassium, chlorine).

The biomass that can be utilised to generate bioenergy originates primarily from agriculture and forestry, and the various biomass-processing industries downstream. Other more dispersed sources include grass cuttings from roadside maintenance, organic residues from milk processing, and the organic fraction of household waste. Not all of this biomass is available for producing energy, however. Only the biomass left over after the demand for food and animal fodder has been satisfied, and the demand for biomass as a raw material for use as industrial feedstock (e.g. for timber derivatives, pulp and paper, bioplastics, cosmetics) has been met, can be used as an energy carrier. Certain plants can also be grown specifically as energy crops on land not required any longer for the production of food and/or animal fodder.

Seen on a global scale, biomass could potentially make a significant contribution towards meeting the given energy demand, from an energy-management point of view. Estimates range between 20% and over 100% of present levels of primary energy consumption, depending on the anticipated yield of various energy crops and the availability of additional arable land not needed for food and fodder production. (in comparison, the current share of biomass used to cover global demand for primary energy lies at around 10%.) It is foreseeable that Western industrialised countries will be able to increase the land area available for growing energy crops – and hence the biomass yield – as a result of enhanced methods for the production of improved plants for food and fodder. Such improved plants will allow increasingly higher yields to be obtained on less land in the future. By contrast, the potential volume of residues, byproducts and waste products is likely to remain constant.

Bioenergy is generated in a multistage supply chain that starts with the collection of residues, byproducts or waste or the cultivation of energy crops. This biomass then undergoes a variety of processing, storage and transportation steps and industrial conversion processes to produce the secondary energy carriers or biofuels, which can be used to meet the given demand for different forms of useful energy. There are numerous processing options for biomass, involving different conversion routes.

▶ **Thermo-chemical conversion.** Biomass can be converted into useful energy or into secondary energy carriers using processes based on heat. In such conversion processes, solid biofuels are mixed with an oxidising agent (e.g. air, water) below the stoichiometric

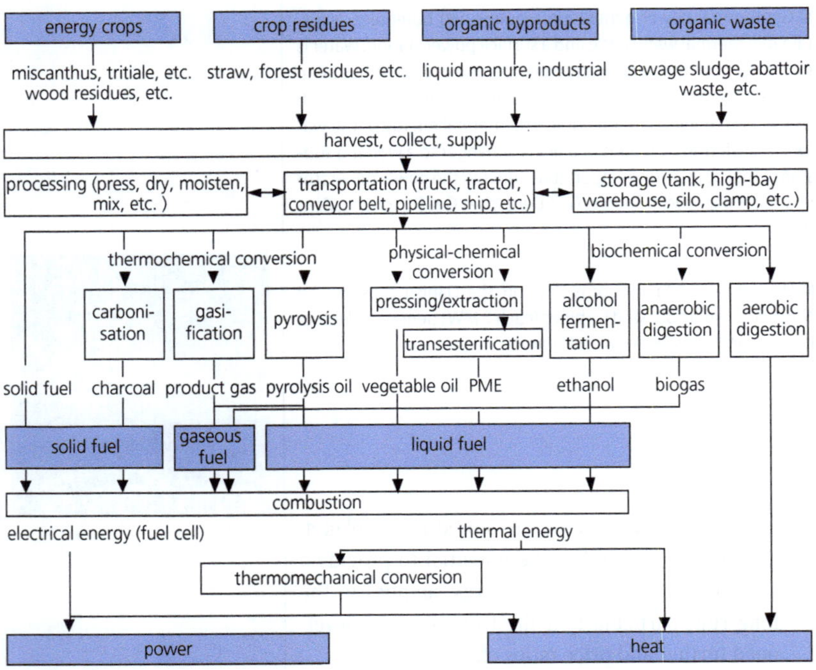

Range of options for producing energy from biomass (fields shaded in blue: energy sources, fields without shading: conversion processes). The various biomass fractions first have to be gathered up and delivered to a conversion plant. The main steps involved, where relevant, are harvesting, transportation, mechanical processing and/or storage. The biomass can then be converted into solid, liquid and/or gaseous secondary energy carriers for a broad variety of possible uses (so-called biofuels) using thermo-chemical, physical-chemical or bio-chemical conversion processes. Straw, for example, first has to be brought in from the field, pressed into bales, and transported to the processing site. There, if necessary after being placed in temporary storage, it can be used to produce a synthesis gas for subsequent conversion into methane suitable for use as an engine fuel in the transportation sector. Alternatively, the straw can be saccharified and then converted into bioethanol using a bio-chemical process. In this case too, the product is suitable for use as an engine fuel. Source: DBFZ 2007 concentration under defined conditions (pressure, temperature, etc.). This enables them to be converted into solid, liquid and/or gaseous secondary biofuels. All processes of this type are based on the same fundamental thermo-chemical reactions, but the composition of the resulting secondary biofuel varies depending on the selected process parameters and the type of biomass involved:

- In *gasification*, the solid biofuels are converted as completely as possible into an energy-rich producer gas or synthesis gas and thus into a gaseous energy carrier. After several cleaning processes, the gas can be burnt directly without any further chemical transformation, e.g. in an engine to provide heat and/or electricity, or undergo a subsequent synthesis process to be converted into a liquid or gaseous transportation fuel with defined properties (e.g. Fischer-Tropsch diesel, synthetic natural gas (biomethane), hydrogen).
- In *pyrolysis*, solid biomass is converted solely by the (short-duration) application of heat into solids (charcoal), liquids (bio crude oil), and gases. The aim of this conversion process is to obtain a maximum yield of liquid components. The byproducts, i.e. combustible gases and solids, can be used to provide e.g. the process energy. The resulting pyrolysis oil or bio crude oil is refined and can be utilised either as feedstock for the chemical industry and/or as liquid fuel for engines and/or turbines.
- In biomass *carbonisation*, the solid biofuel is converted through the application of heat on the basis of the same thermo-chemical reactions in such a way as to maximise the yield of solid carbon (i.e. charcoal), which can then either be used as an industrial feedstock (e.g. activated charcoal for filters) or as a source of energy (e.g. for leisure activities).

▶ **Physico-chemical conversion.** This method is used to produce liquid biofuels from biomass containing oils and fats (e.g. rape seed, sunflower seed). The oil is separated from the solid phase e.g. by mechanical pressing and/or extraction. The vegetable oil can be used as an energy source both in its straight form and, after conversion (i.e. transesterification with methanol), as FAME (fatty acid methyl ester or biodiesel). Both vegetable-oil-based liquid fuels are ready for market as far as the transportation sector is concerned, but only FAME has gained broader market significance so far.

▶ **Bio-chemical conversion.** In this method, the biomass is broken down by microorganisms, and hence its conversion is based on biological processes.

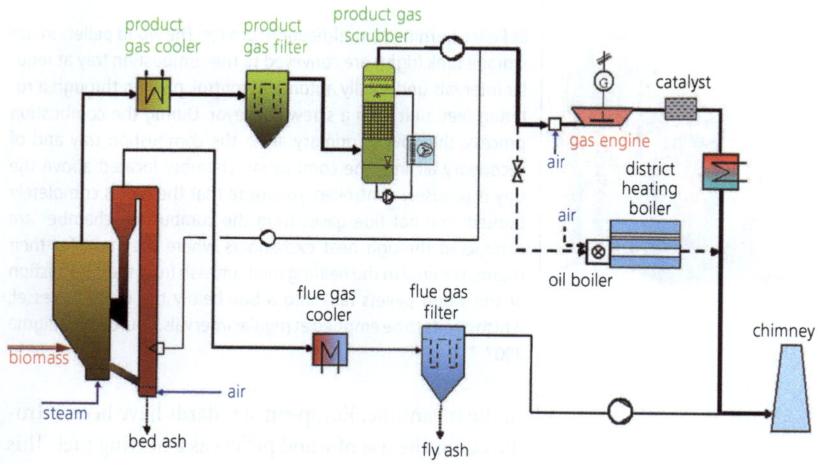

■ Thermo-chemical plant for the gasification of biomass and the generation of electricity from the resulting producer gas: The solid biomass is gasified with steam in a reactor to produce a medium calorific value (MCV) gas that, at this stage, is contaminated with tar and dust. The MCV gas is then cooled down. After this, the tar and dust are removed in a two-stage cleaning process to meet the prescribed specifications for its use as engine fuel. The energy required for the gasification process is obtained by burning the charcoal produced as a waste product of the gasification process itself. Finally, the filtered and scrubbed MCV gas is fed to a CHP unit (i.e. gas engine) to generate electricity and heat. Source: repotec Vienna 2006

- In alcohol fermentation, the sugar, starch and cellulose contained in the biomass is transformed into ethanol which, after distillation and dehydration, can be employed as pure alcohol or blended with conventional gasoline to fuel engines and/or turbines.
- In anaerobic *fermentation*, organic matter is converted into biogas (roughly 60% methane and 40% carbon dioxide) by microorganisms. This gas can be used in combined heat and power (CHP) generating stations to produce electricity and heat, or alternatively, after further processing, upgrading, and where necessary distribution through the existing natural gas pipelines, used to fuel vehicles that run on natural gas.
- Other bio-chemical conversion options currently under discussion include the production of bio-oil (known as algae or algal diesel) or hydrogen with the aid of microscopic algae.

Applications

▶ **Solid biofuels.** Solid biofuels (e.g. wood logs, wood chips, wood pellets) are the "original" biofuels most widely used throughout the world. Such solid biofuels are produced with a broad variety of processes characterised by widely differing degrees of mechanisation. For the provision of heat and/or electricity, such fuels can be used in fireplaces, furnaces, and other combustion devices with all thermal capacities demanded on the market. The biggest share of these solid biofuels in current use to cover the given energy demand is employed in private households for the provision of heat in small-scale heating devices. Especially the use of wood pellets has been gaining in popularity in recent years. In Central Europe this is particularly true for domestic heating systems operated on a very small scale (i.e. often below 15 kW thermal capacity).

Pellet-burning central-heating furnace: The wood pellets in the storage tank (right) are conveyed to the combustion tray at regular intervals under fully automatic control, passing through a rotating feed unit with a screw conveyor. During the combustion process, the flow of primary air to the combustion tray and of secondary air into the combustion chamber located above the tray is precisely controlled, to ensure that the fuel is completely burned. The hot flue gases from the combustion chamber are conducted through heat exchangers where they transfer their thermal energy to the heating medium. Ash from the combustion of the wood pellets falls into a box below the reaction vessel, which needs to be emptied at regular intervals. Source: paradigma 2007

In the meantime, European standards have been introduced for the use of wood pellets as a heating fuel. This combined with modern pellet burning technology allows highly efficient furnaces to be operated fully automatically and to be characterised by very low airborne emissions (especially emissions of particulate matter).

In terms of combustion efficiency, wood's low ash content and favourable chemical composition make it a highly suitable candidate for use as a solid biofuel. Compared to this, the combustion of straw and other herbaceous biomass fractions is more challenging due to their comparatively low ash softening point (the temperature at which the ash becomes fluid and therefore starts to bake onto the walls of the combustion chamber or the surface of the heat exchanger elements). Such non-woody solid biofuels also often contain a high proportion of undesirable trace elements (such as chlorine), which can have an elevated corrosive effect and/or lead to the release of highly toxic airborne pollutants. These unfavourable properties are the reason why solid biofuels derived from herbaceous plant material have not so far been used to any greater extent in small-scale as well as in large-scale combustion plants.

Power generation from wood in district heating or *combined heat and power (CHP)* plants has so far mostly been based on conventional steam processes and thus forms part of normal power-plant engineering practice. This type of power generation in plants with an output of 2 MW and above is market-mature and widely used within the global energy system. By contrast, methods of generating electricity or combined heat and power from solid biofuels in plants with a capacity in the kW range are still at the development stage. Such plants nevertheless offer the possibility of opening up new applications in the context of decentralised power generation (e. g. combined heat and power generating plants for public swimming pools or schools, for small district heating systems, or for large farming complexes).

In a district heating plant fueled by solid biofuels, the biomass is combusted for example on a grate. The fuel loader deposits the wood fuel on this grate through which primary air and in some plants additionally recycled flue gas is blown from below. This supply of air serves firstly to drive out the residual moisture in the fuel, and subsequently to assist with its pyrolytic decomposition and gasification as well as with its partial combustion. The combustion gases rising up from the moving bed are fully oxidised in the hot burning chamber located above the grate with the addition of secondary air. The hot flue gases are conducted through the steam generator and subsequently fed to the flue gas treatment unit where in most cases only dust is removed to fulfil the given emission regulations. The steam generated by the process is depressurised in a turbine and used to supply district heat.

► **Gaseous biofuels.** Gaseous fuels of biogenic origin can be used in gas-fired heating devices for heat provision and for the combined generation of power and heat in gas engines, turbines, and certain types of fuel cell, including those used to power road vehicles. After processing the gaseous fuel to obtain a quality comparable to natural gas, this biofuel can be fed into the natural gas distribution network. Consequently, the fuel is available for use not only at the site of the conversion plant but also at any other site connected to the existing natural gas supply infrastructure. Alternatively, the gas can be transformed into a liquid secondary energy source (e. g. methanol).

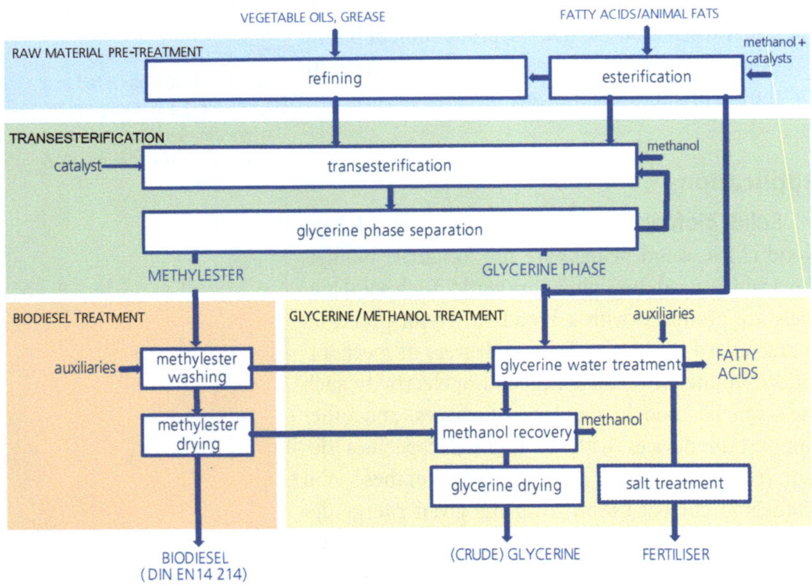

Production of biodiesel from colza oil: The vegetable oil is converted into fatty acid methyl ester (FAME) or biodiesel by transesterification, a process in which the trigliceride molecules in the vegetable oil are split open by a catalyst, and methanol is added to produce monoesters of fatty acid and glycerin. The resulting FAME is then purified and, if necessary, conditioned/winterised. It can then be used as a direct substitute for fossil diesel fuel. In parallel the glycerin is cleaned to be sold on the market.

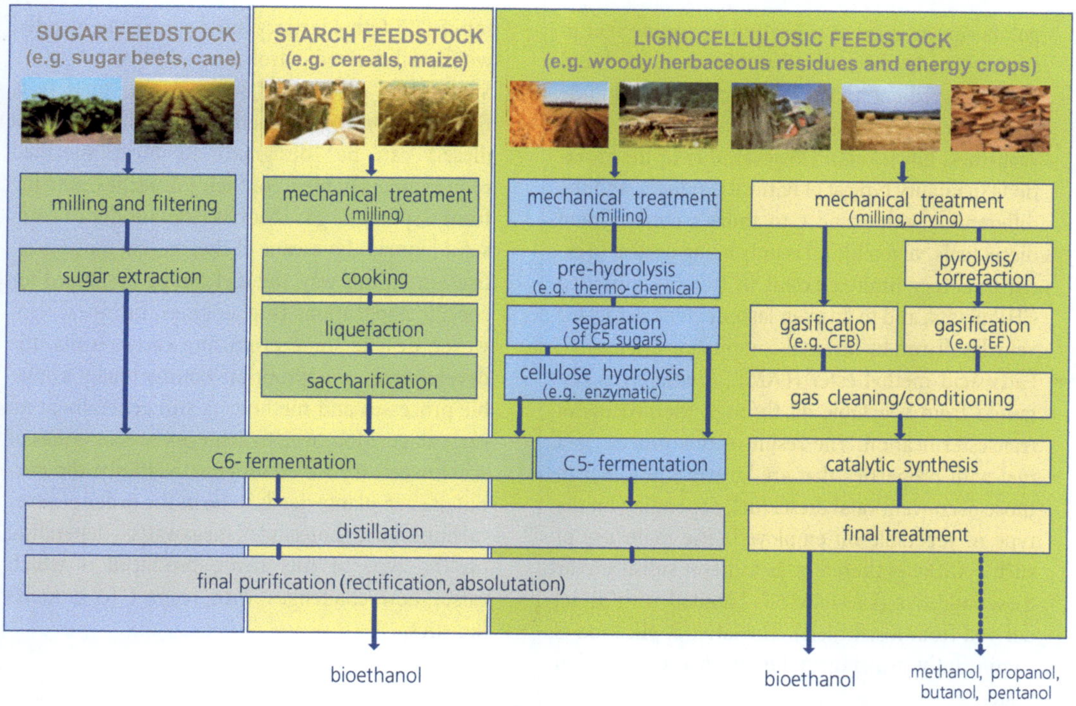

Possible ways of producing bioethanol from biomass: Basically all pathways to produce bioethanol based on bio-chemical conversion include the process steps of fermentation (i.e. converting sugar to ethanol), distillation (i.e. separating ethanol and stillage), and purification (i.e. to provide pure ethanol). If biomass containing sugar is used as a feedstock (e.g. sugar beet, sugar cane), an additional step is required to remove the sugar from the biomass to improve the efficiency of the bioethanol production process. If the ethanol production process is based on starch extracted from a cereal crop, additional liquefaction and saccharification stages are needed to allow the microbes to convert the biomass into bioethanol. Similar conditions apply to the use of celluloses as a feedstock. Here too, a technologically demanding saccharification stage is needed, e.g. based on hydrolysis. An alternative method involves converting the celluloses into synthesis gas using thermo-chemical gasification, and further processing the output by a catalytic synthesis into bioethanol. Source: DBFZ 2008

A number of very different techniques and processes are available to produce biogenic gaseous fuels.

- State-of-the-art bio-chemical biomass conversion plants for the production of biogas (i.e. biogas plants) typically have an electrical output ranging between tens or hundreds of kilowatts and several megawatts. They are usually designed to process organic residues, byproducts and waste (e.g. liquid manure, the organic fraction of household waste, sewage sludge) and/or energy crops (e.g. corn silage). In most cases, the resulting biogas is used to generate electricity in combined heat and power (CHP) plants (i.e. gas engines). But the produced biogas can also be upgraded to natural gas quality, fed into the natural gas distribution network and used, e.g., in the transportation sector as so-called "green" gas.
- Thermo-chemical plants for the production of producer gas from solid biofuels and the use of this gas for the generation of heat and/or electricity exist at present in the form of demonstration and/or pilot plants only. The advantage of such gasification plants compared to ordinary combustion units is their theoretically higher conversion efficiency to electrical energy. The downside is that such gasification-based conversion devices require a fairly complex technology that is only on the brink of commercialisation so far and thus characterised by relatively high risks. As an alternative and/or supplementary solution to the direct conversion of the product gas into heat and/or electrical energy, it is also possible to convert the producer gas to obtain a gaseous secondary energy carrier with defined properties. One example involves converting the gas into biomethane (synthetic natural gas or Bio-SNG) that meets natural gas quality specifications. This Bio-SNG can then be fed into the natural gas distribution network and used in the same way as natural gas, e.g. within the transportation sector.

▶ **Liquid biofuels.** Liquid biofuels are used as fuel for road vehicles or in (small-scale) CHP units. The energy sources in question are:

- Vegetable oil derived from e.g. rape seed or oil

palms can be used as a heating or engine fuel in addition to its more familiar uses as a food and fodder ingredient and as feedstock for a multitude of industrial products (e.g. cosmetics, paints and varnishes, lubricants). However, due to the variations in quality typical of natural products, and the differences with respect to today's conventional diesel fuels, vegetable oil is only being used at present to a very limited extent in suitably modified CHP plants, and to an even lesser extent as fuel for vehicles driven by internal combustion engines.

- Fatty acid methyl ester (FAME, *biodiesel*) is obtained from vegetable oil through the process of transesterification. The result of this process is a fuel with properties that are largely equivalent to those of conventional diesel fuel, depending on the type of vegetable oil employed. The main use of such FAMEs is therefore as a diesel substitute or diesel additive (i.e. FAME is blended with diesel derived from fossil fuels). The latter application is significantly growing in importance on a global scale.
- *Bioethanol* – in addition to its uses as a chemical feedstock and as a food (i.e. an ingredient in certain beverages) – is produced from biomass containing sugar, starch, or celluloses. The production of ethanol from sugar or starch is a centuries-old technique. This is not true for cellulose-based feedstock due to existing technological challenges relating particularly to the conversion of cellulose into sugar. Bioethanol can be used both as an additive blended with conventional petrol and in its pure form in suitably adapted engines. Both of these applications have already been commercialised (e.g. in Brazil). Alternatively, ethanol can be converted into ETBE (ethyl tertiary butyl ether) and blended with fossil-derived petrol as an additive.
- Synthesised gaseous or liquid engine fuels (e.g. Bio-SNG, hydrogen, methanol, Fischer-Tropsch diesel) can be "designed" to meet specified requirements in terms of their properties, starting from a synthesis gas that has been produced from solid biomass by means of thermo-chemical gasification. In this way, the fuel can be optimised for specific applications (e.g. engines, turbines). However, despite some promising early results, the development of appropriate commercially available processes and methods is still generally at an early stage. Moreover, to achieve the economies of scale necessary for commercial viability, the output of such plants needs to be in the order of several hundred to thousand of megawatts – depending on the type of fuel being produced – which raises new challenges with respect to biomass logistics.

Trends

▶ **Domestic heating furnaces.** There is a distinct trend towards high-efficiency, low-emission systems for generating heat from solid biofuels such as wood logs, wood chips and wood pellets. Emissions of particulate matter – an issue accorded a high priority on the political agenda these days – can be minimised by selecting the type of furnace best adapted to the solid biofuel in question. Additionally, the implementation of appropriate primary (i.e. related to the combustion unit) and secondary measures (i.e. flue gas treatment) might be necessary; for example, low-cost filters for removing particulate matter emissions from the flue gas of small-scale combustion units are already available on the market. It is also important that users should respect the prescribed operating conditions issued by the company producing the combustion device (i.e. by not misappropriating the heating furnace to burn household waste). The new types of condensing furnaces or boilers are likely to become more widespread in the years to come, because in addition to an improved overall efficiency they also feature significantly reduced pollution (i.e. emissions of particulate matter). It is also possible that systems designed for the combined generation of power and heat (e.g. based on the Stirling engine), especially for small-scale domestic applications, might one day become more common, although this is an area of power engineering that still needs intensive research.

▶ **Biogas production and utilisation.** In the industrialised nations, there is a tendency towards building

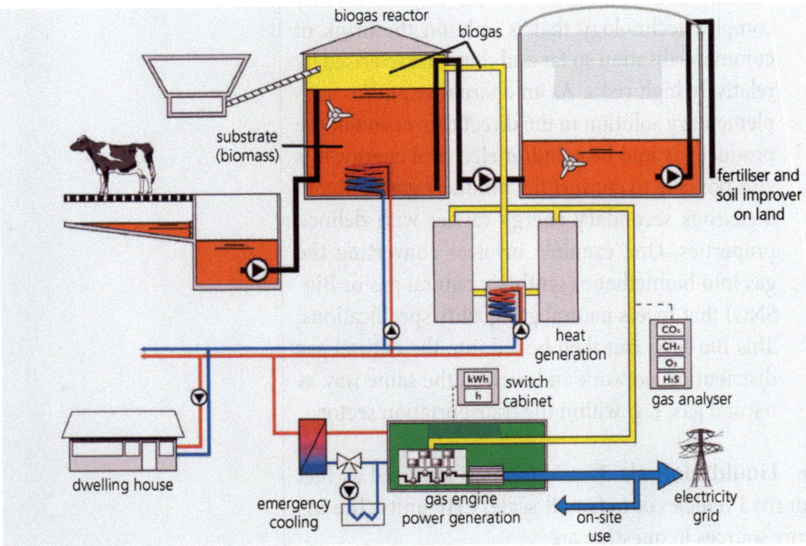

Biogas plant: The main components of the plant are the fermenters (top), in which microorganisms convert the substrate (brown) into biogas (yellow) in the absence of oxygen. The resulting biogas is cleaned and used in a combined heat and power (CHP) generating unit (i.e. gas engine, green). Most of the electricity is fed into the utility grid, while the provided heat is used locally. The fermented substrate is subsequently spread as fertiliser /soil improver on arable land. Source: Modified according to FNR 2006 higher-output biogas generation plants to process organic waste or energy crops. The main technological challenge, which has still not been satisfactorily resolved, is that of controlling and steering the *anaerobic digestion process* – and hence the production of biogas – in such a way that the output remains stable on a high level and predictable even when using a broad variety of input substrates. If this problem can be resolved, such plants will be able to ensure a reliable and highly efficient supply of high-quality biogas. One problem in this respect is that the bacteria responsible for the anaerobic digestion process react highly sensitively to changes in the properties of the substrate. There is therefore a need for a microbiology that is more tolerant of fluctuations in substrate quality, while at the same time breaking down the biomass to methane and carbon dioxide faster and more completely. For the input materials, operators are turning towards the use of waste streams – preferably monosubstrates – that in the past have only been used occasionally or not at all. There is also a growing trend towards the use of deliberately cultivated renewable resources, which are constantly being optimised by breeding varieties with improved biodegradability and significantly higher yields. Increasing quantities of refined biogas are also being fed into the natural gas distribution network, because this represents a very flexible and highly efficient means of making the gas available for use in numerous applications, including transportation. New methods of upgrading biogas to biomethane that meets natural gas quality standards will enter the market in the years to come. They include processes based on CO_2 removal with pressurised water or amines, or alternatively pressure swing absorption. Other methods (e.g. using *membrane* filters) are currently being tested. Although such methods have been in use for many years in the industrial processing of natural gas, there is still potential for improvement in plants operating at capacity ranges typical for biogas plants, which needs to be harnessed in the years to come.

In developing and newly industrialising countries, there is a trend towards the use of extremely simple biogas plants in which organic farming and household waste can be fermented to produce energy (i.e. biogas) and also provide valuable nitrogen fertiliser as a by-product.

▸ **Biofuels**. There is a need for industrial plants capable of converting biomass more efficiently into bioethanol. At the same time, there is growing political pressure to make more extensive use in the future of lignocellulosic biomass (i.e. biomass from plants or parts of plants that are not suitable for use as food). This explains the trend towards the development of integrated biorefineries. In such fully integrated biomass conversion systems, a wide range of different types of biomass are used to produce different types of fuel and other feedstock products (e.g. platform chemicals) with the least possible output of waste. This is true for processes based on bio-chemical biomass. In parallel, more advanced thermo-chemical processes are being developed that are designed to transform solid biomass into a synthesis gas (i.e. mixture of carbon monoxide and hydrogen) which can then be converted by various routes into gaseous (e.g. biomethane or Bio-SNG, hydrogen) or liquid fuel (e.g. Fischer-Tropsch diesel, DME). The latest biorefinery concepts aspire to a promising combination of bio-chemical and thermo-chemical processes.

Prospects

- Biomass can be converted into final or usable energy through a multiplicity of more or less complex supply chains – and hence using a wide variety of technical processes. This makes biomass a highly flexible renewable energy source.
- The use of solid biofuels for the provision of heat for cooking and heating purposes is the most important application from a global point of view. This explains why biomass currently contributes most to covering the given energy demand compared to all other renewable sources of energy.
- As energy prices continue to fluctuate widely at short notice, it can be expected that the demand for biomass-based energy solutions will increase across all markets. Consequently, there is an urgent need for new processes and techniques that will ensure high conversion efficiency, low energy provision cost, a significantly reduced environmental impact, and clearly defined specifications of the provided energy carrier to ensure that the biofuels or bioenergy fits easily into the given structures within the existing energy system.
- Given that the generation of energy from biomass will always have to compete with other forms of land use, dedicated to the production of food, fodder and industrial feedstock, it is vital that measures should also be taken to increase the productivity of agricultural and forestry land through the provision of improved seed material and better management schemes.

PROF. DR. MARTIN KALTSCHMITT
Hamburg University of Technology

DR. DANIELA THRÄN
German Biomass Research Centre, Leipzig

Internet

- www.ers.usda.gov/features/bioenergy
- www.repp.org/bioenergy/index.html
- www.ieabioenergy.com
- www.biofuelstp.eu
- www.nrel.gov

Solar energy

Related topics

- Semiconductor technologies
- Polymer electronics
- Energy storage
- Microenergy systems
- Sustainable building

Principles

Every hour, enough solar radiation reaches the earth to meet the entire annual energy demand of the world's population. The solar radiation at a given location varies according to the time of day and year, and prevailing weather conditions. Peak values for solar radiation on clear sunny days are around 1000 W/m². In temperate climatic regions, the annual total for solar energy is 1,000–1,200 kWh/m²; it can reach 2,500 kWh/m² in arid desert zones.

Solar radiation consists of two components. Direct (or beam) solar radiation is dependent on the position of the sun and can be concentrated with the aid of lenses or reflectors. It is primarily available when the total radiation value is high (clear sky). Diffuse solar radiation comes from the whole sky as scattered light and cannot be concentrated by optical devices. At medium latitudes, global or total radiation consists of 60% diffuse and 40% direct radiation. When the total radiation value is low (cloudy and foggy days), diffuse radiation predominates. For this reason, attempts are made to use both types of radiation.

Solar radiation can be converted into different forms of usable energy:

- Heat: With the help of solar collectors, the sun can make a major contribution towards producing low-temperature heat in buildings: for domestic hot water and space heating, as well as for cooling purposes via thermally driven chillers. If direct sunlight is concentrated, very high temperatures can be produced. This allows process heat to be generated by large, central heating systems for industrial purposes or for generating electricity in thermal power stations.
- Electricity: *Photovoltaic generators* convert direct and diffuse solar radiation directly into electricity. The energy output of these systems is essentially proportional to their surface area and depends on the global radiation value.
- Chemical energy: *Photocatalytic processes* in chemical reactors or reactions similar to natural photosynthesis convert direct and diffuse solar radiation into chemical energy. When irradiated with sunlight, carbon dioxide and water are converted into biomass. However, technological imitations of photosynthesis have not yet proved successful.

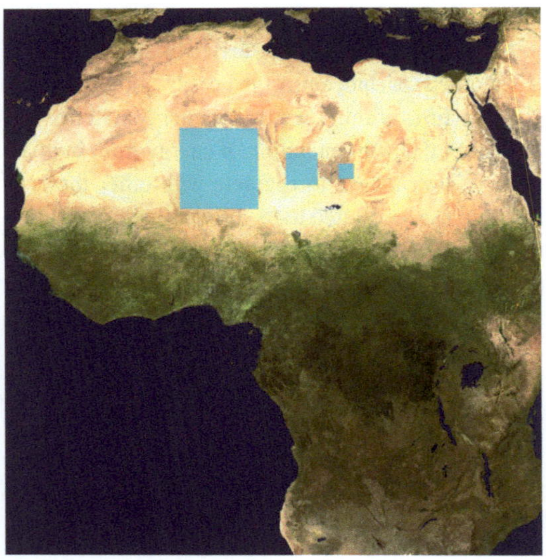

Solar radiation and energy requirements: The rectangles symbolise the surface area required – using solar cells with today's technology (efficiency of 12%) in the Sahara – to meet the energy demands of the whole world (large square with the length of approx. 910 km), Europe (medium-sized square) and Germany (small square).

▶ **Solar collectors.** Solar collectors absorb all solar radiation, both the direct and diffuse types, over the spectral range from ultraviolet through visible to infra-red radiation. Solar collectors then convert solar radiation into heat. The radiation directly heats a metal or polymer absorber, through which a heat transfer medium flows. This conveys the generated heat to the user or a storage unit. Optically selective absorbing materials, highly transparent covers and good thermal insulation in the collectors ensure that as much solar energy as possible is converted into useful heat. Temperatures of 50–150 °C can be reached. Water with an anti-freeze additive is most commonly used as the heat transfer fluid. In evacuated tubular collectors, the space between the outer glass tube and the absorber is evacuated. This reduces thermal losses caused by air conduction or convection (thermos principle) so that higher operating temperatures can be achieved.

► **Photovoltaics.** The internal *photoeffect* in semiconducting materials is the physical basis for solar cell operation. If radiation is absorbed in a semiconductor, electrons are excited from the ground state (valence band) into the conduction band. For the radiation to be absorbed, its energy must exceed a certain minimum level, which depends on the *semiconductor* material. Solar radiation with less energy (longer wavelength) is not absorbed by the semiconductor and thus cannot be converted into electricity. Positive charges - known as "holes" - are created in the valence band by the excitation of the electrons. By doping the semiconductor appropriately, zones can be produced which preferentially conduct negative or positive charges (electrons or holes respectively). These zones are called n-type or p-type. The photo-excited electrons tend to concentrate in the n-type layer; the electron holes in the p-type layer. This charge separation creates an electric potential, which can be tapped by applying metal contacts to the layers. The value of the voltage depends on the type of semiconductor material used, and is usually between 0.5–1 V. The usable current is proportional to the incident light intensity to a first approximation. When exposed to non-concentrated solar radiation, the maximum current density for silicon is 45 mA/cm².

Applications

► **Solar collectors.** Solar collectors are most commonly installed on individual houses to produce domestic hot water. However, as the number of houses with new windows and thermal insulation increases, and their heating energy consumption decreases, the use of solar energy for space heating is a growing trend. This has resulted in the installation of collector systems with a large surface area in combination with large thermal storage tanks.

► **Solar-thermal power stations.** Different types of optical systems - tracking lenses or reflectors - concentrate direct solar radiation onto an absorber. The energy heats a medium, which most commonly generates steam in a secondary circuit. The steam then drives turbines to generate electricity.

The three basic types of power station apply different principles to concentrate solar radiation onto a focal point or line. These power stations have different concentration factors, average operating temperatures and output power values. Parabolic trough power stations are already extremely reliable; solar tower and

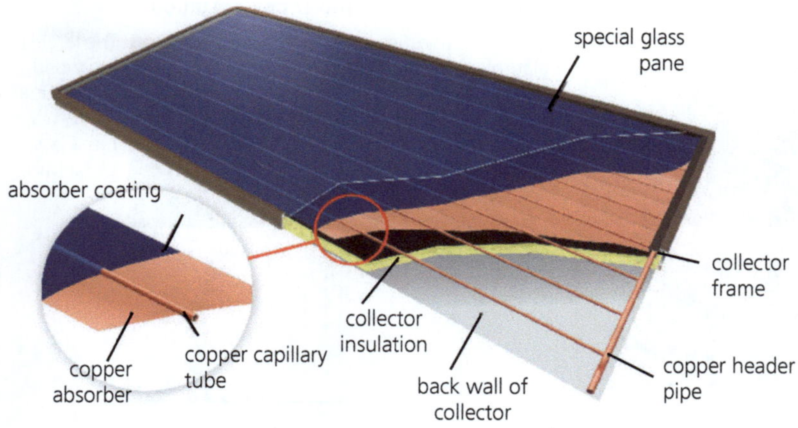

Solar collector: Solar radiation is transmitted through the transparent upper glass pane and is then absorbed by the dark, selective absorber coating, heating the underlying copper sheet. The radiative selectivity of the absorber and the presence of the cover prevent heat being radiated from the absorber (greenhouse effect). The medium circulating in the pipes, usually water, transports the heat out of the collector. Source: Fraunhofer ISE

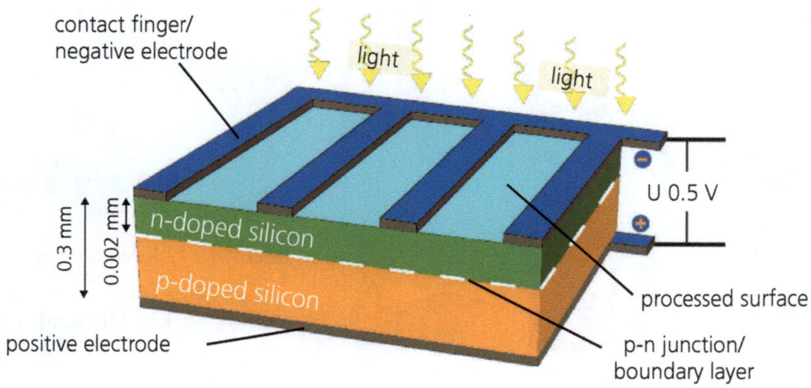

Photovoltaic cell: In the internal photoeffect, light excites an electron in a semiconductor material (e.g. silicon). The electron is raised from the valence band (ground state) into the conduction band. The missing electron in the valence band is described as a "hole" with a positive charge. An electric field arises at the interface between two differently doped semiconductor layers (p-n junction). This causes the electrons and holes to separate spatially. If a load is applied, the electric circuit is closed. In order to reduce light losses caused by reflection, the upper surface - usually a glass pane - is optically enhanced. As the electric circuit's negative electrode must be connected to the cell's upper surface, the contact fingers are kept as thin as possible to minimise shading losses. Source: BINE information service

parabolic trough power station

Kramer Junction, California.

The troughs are more than 100 m long.

Thermal oil flows through the absorber pipe, which is positioned along the focal line.

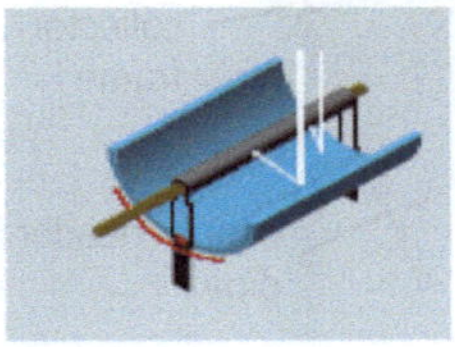

tower power station

The largest station in the world, "Solar One" in California, has a power output of 10 MW.

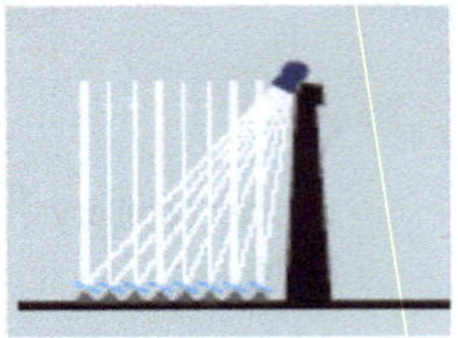

parabolic dish system

A paraboloidal reflector focuses the direct solar radiation onto a Stirling engine, which drives an electric generator.

The reflector tracks the sun continuously.

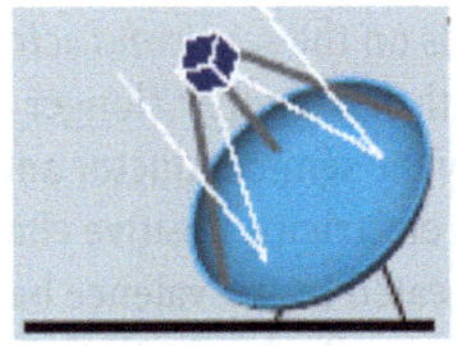

	parabolic trough power station	tower power station	parabolic dish system
efficiency value today	11%	8%	23%
concentration factor for solar radiation	40 - 80	200 - 1000	1000 - 2000
temperature of heat-transfer	< 400 °C	< 1000 °C	< 800 °C
power	10 - 400 MW	10 - 50 MW	5 - 50 MW

Solar-thermal power stations. Sources: Sandia National Laboratories, DLR, Nicole Krohn

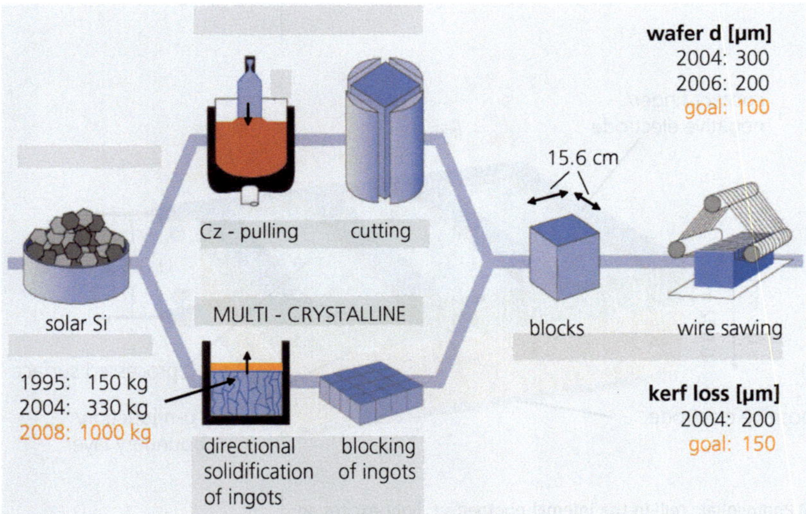

Production of silicon solar cells. Upper section: In the crucible pulling or Czochralski method solar silicon is melted in quartz crucibles. A seed crystal of high-purity monocrystalline silicon is introduced into the melt and slowly retracted under rotation (2–250 mm/h). High-purity monocrystalline silicon is formed on the seed crystal, leaving almost all impurities behind in the melt. Lower section: In the block crystallisation method, silicon granulate is melted in large scrucibles. It thereafter solidifies slowly into polycrystalline silicon. The wire sawing process to produce wafers is the same for both methods. The cylinder or square block is cut into wafers of 200 µm thickness, whereby 50 % of the material is lost as cutting waste.

parabolic dish power stations, however, are still in the testing phase.

In a parabolic trough power station, reflecting troughs concentrate solar beams onto an absorber pipe which is installed along the focal line. Thermal oil is heated, which drives a turbine via a secondary steam circuit. At present, further developments are being made in direct evaporation technology. The efficiency increases if the thermal oil is replaced by a water-steam mixture that can be heated to higher temperatures. In the future, improved, selective solar absorber coatings should produce temperatures in excess of 500 °C.

In a parabolic dish system, a paraboloidal reflector with a diameter of several metres focuses direct solar radiation onto a hot-air or *Stirling engine*, which drives an electric generator. The Stirling engine is a machine in which a gas is heated in a closed space as the working medium. The energy produced by the change in volume is transferred to a piston and is converted into mechanical energy. The gas then cools and is compressed again by heating: the process is therefore cyclical. The reflector continually tracks the sun. If the sun is obscured, the systems can be operated with fossil fuels. Current developments are therefore focusing on solar/chemical hybrid systems – some of which use biomass.

In solar towers, solar radiation is concentrated onto a receiver at the top of a tower by many reflectors (heliostats) on the ground. The system consists of an absorber (black wire matting, porous ceramics, pipe bundles) and a heat transfer medium, which can be air, steam, liquid sodium or molten salt. The efficiency value of solar towers would be dramatically improved if the air that is heated to very high temperatures can be used to drive a gas turbine in a primary circuit. The surplus energy can be used to generate steam which drives a steam turbine in a secondary circuit.

► **Crystalline silicon photovoltaics.** At present, the dominant semiconductor material for solar cell production is silicon. 90% of all solar cells currently consist of monocrystalline or multicrystalline silicon. Monocrystalline solar cells are made of 0.2 mm thin *wafers* that are cut from a single crystal created by drawing from a silicon melt. These cells are of high quality, but the high price of the wafers also makes the cells expensive. The crystal facets in multicrystalline solar cells are irregularly distributed, because the wafers are cut from a cast silicon block. They are, however, less expensive than monocrystalline cells. The efficiency value of industrially produced multicrystalline cells is about 14% – compared to 16% for monocrystalline cells. However, pilot production lines are already producing cells with an efficiency value exceeding 20%. For special applications, modules made of selected cells with efficiency values exceeding 20% are also available.

The wire sawing process to produce wafers is the same for both methods. The cylinder is cut into wafers of 200 µm thickness, whereby 50% of the material is lost as cutting waste.

First large-scale production plants for the selective coating of absorber plates used in solar thermal collectors. These plants typically use in-line putting technology and can coat up to one million square meters of copper or aluminium foil per year. Solar thermal will cover 50 % of the heating demand in Europe; in the long term the technologies will be used in almost every building, covering more than 50 % of the heating and cooling demand in refurbished buildings and 100 % in new buildings.

Grid-connected systems are currently the main application for photovoltaic systems. These systems are most common in Germany and Japan because of the legal regulations governing feed-in tariffs. The total global installed peak power value exceeds 9 GW.

Trends

► **Thin-film solar cells.** Many other semiconductor materials – in comparison to crystalline silicon – absorb light much more strongly. The best-known semiconductor materials in this category are amorphous silicon (a-Si), copper indium diselenide (CIS or, with gallium, CIGS) and cadmium telluride (CdTe). Many other candidate materials and combinations of different materials (hetero-junction solar cells and tandem cells) are also available.

Solar cells made of these materials can therefore be thinner - usually several µm thick - and can be manufactured on flexible substrates. A further advantage is that semi-transparent photovoltaic converters can be produced with the thin films. These are particularly suitable for architectural applications. Large-area coating technology can be applied to deposit thin-film solar cells onto inexpensive substrates such as window glass or metal foil. Single cells (voltages between 0.5–1.0 V) can be connected automatically in series during production. The first production lines are currently being set up for many of these new types of technology. It remains to be seen whether this type of solar cell can be manufactured to generate significant amounts of power at lower prices than crystalline silicon systems, and with sufficient long-term stability. The photovoltaic market is growing so rapidly at present that crystalline silicon is predicted to dominate the market for the next decade.

In *CIGS solar cells,* the light-absorbing layer consists of compounds of the elements copper (Cu), indium (In), gallium (Ga) and selenium (Se). The complete cell consists of a number of different types of thin films. Efficiency values exceeding 19% have been measured in the laboratory, compared with efficiency values between 10% and 13% in industrial production. Development work is continuing on the CIGS deposition process itself and the CIGS/buffer/front contact interfaces, as these have a decisive influence on solar cell performance and stability.

Attempts to reduce cell thickness using crystalline silicon are also being pursued. However, this approach requires solar radiation to be "trapped" using optical processes (multiple reflection) in a 5–30 µm thick crystalline structure (normal thickness is 250 µm). These cells can be prepared from paper-thin wafers or

Photovoltaic systems for different power ranges.

10 mW: Pocket calculator

100 W: Parking voucher dispenser

450 kW: Photovoltaic system on the terminal roof at Munich Airport, consisting of 2,856 polycrystalline silicon modules over approx. 4000 m². Source: Werner Hennies / Flughafen München GmbH.

5000 kW: One of the largest solar power stations in the world, Leipziger Land, consisting of 33,264 solar modules in a solar generator array covering 16 hectares. Source: GEOSOL.

Thin-film solar cell: This crystalline silicon cell demonstrated that wafers with a thickness of 30 µm are sufficient to produce solar cells with efficiency values exceeding 20%. Source: Fraunhofer ISE

from thin films that are deposited onto suitable, inexpensive substrates directly from the gas phase (epitaxy).

For *space applications*, where price is a less important factor, not only crystalline silicon modules are used but also modules made of so-called *III–V semiconductors* such as gallium arsenides (GaAs). These cells offer relatively high efficiency values (approx. 20 %), low specific mass and low sensitivity to cosmic radiation. Stacked solar cells significantly increase the efficiency of III–V solar cells. Each partial cell in the stack is optimised to convert a certain segment of the solar spectrum into electricity. Under concentrated sunlight, the triple solar cell produces efficiency values of almost 40%. This type of solar cell is developed for space applications and terrestrial power generation. It uses strongly concentrated solar radiation with concentration factors of 500 being typical. In sunny regions, it should be possible to generate photovoltaic electricity more economically with these cells than with crystalline silicon cells in large plants.

► **Concentrator modules.** In a similar way to thermal solar power plants, photovoltaic systems can also use concentrated solar radiation and, in principle, higher efficiency values are feasible. Lenses and reflectors are used as concentrators. One advantage of concentrating photovoltaic systems is that they can be scaled to the required dimensions. This means that small systems with high efficiency values can be constructed. The most common systems employ square Fresnel lenses with an edge length of 5–10 cm. These lenses concentrate the incident radiation by a factor of 500. The solar cell's active surface area is several square millimetres. The systems efficiency values for direct solar radiation exceed 26 %. The individual cell-lens systems are series-connected in sun-tracking modules with a typical surface area of several square metres.

► **Polymers and organic molecules.** Whereas current research in photovoltaics is concentrating on inorganic semiconductors, organic materials offer new alternatives in respect of photosensitivity and the photovoltaic effect. The main advantages compared to conventional silicon solar cells are:

- less expensive production methods and less material consumption
- flexible and light solar cells
- high levels of environmental friendliness
- coloured solar cells (architecture, design)

Polymer solar cells usually consist of an electron-donating and an electron-accepting material. The films can be produced by spin-coating or, for organic molecules, also by a vacuum process. The materials and technology are closely related to those used for organic light-emitting diodes. The rapid growth in that sector has also had an influence on these solar cells. With typical values in the range of 3–5%, the efficiency values for organic and polymer solar cells are still low in comparison to inorganic solar cells. Long-term stability cannot yet be guaranteed for these cells either.

► **Nanocrystalline dye-sensitised solar cells.** These cells consist of a very porous, electrically conductive substrate (e. g. n-type titanium dioxide), which is penetrated by an electrolyte or an ion-conducting polymer. The actual solar absorption process occurs in the dye molecules on the nanocrystalline structure's

	materials	laboratory cell efficiency value	industrial cell efficiency value
crystalline	monocrystalline silicon	25 %	15 - 17 %
	multicrystalline silicon	21 %	13 - 15 %
thin-film	amorphous silicon	12 %	5 - 7 %
	copper indium selenide (CIS)	17 - 19 %	10 - 13 % (pilot production)
	cadmium telluride (CdTe)	10 - 15 %	8 -9 % (pilot production)
	gallium arsenide (GaAs)	23 - 30 %	limited series production (space applications)
organic solar cells		3 - 5 %	(still under development)

Efficiency values of solar cells based on different types of semiconductor material.

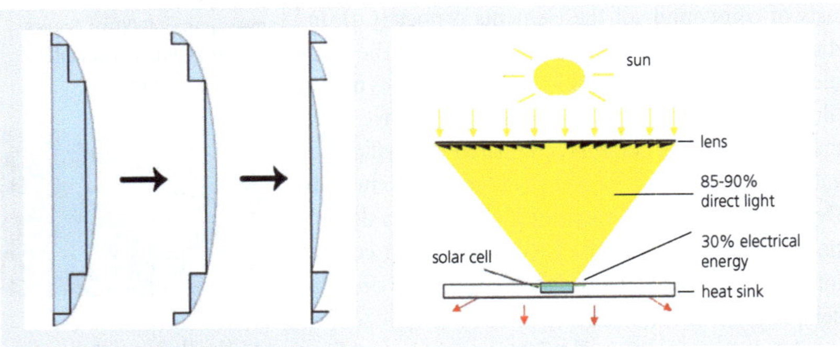

Concentrating sunlight. Left: Compared to a conventional lens, the volume of a Fresnel lens is much smaller, as it consists only of circular segments. Right: Direkt sunlight is concentrated by a factor of approximately 500 onto a small solar cell. The conversion efficiency is very high (36.7 %; 41.1 % in laboratory).

surface. Absorbed light excites the electrons in the dye, which are then transferred extremely rapidly to the semiconducting material and an electrode. This type of solar cell can be manufactured with simple technology (screen-printing) and does not require extremely pure semiconductor materials. It is also possible to choose from a wide variety of dyes and ion-conducting materials. Problems, which are currently being addressed, include long-term stability, which is presently inadequate, and the high cost of the dyes used.

▶ **Solar updraught tower power plant.** In contrast to power plants based on focused light, an updraught power plant makes use of the physical principles of the greenhouse and chimney effect. The solar irridation warms the air beneath a glass roof measuring up to 40 km². The expanded hot air is channelled by the funnel-shaped roof towards the turbines and tower, reaching temperatures of up to 70 °C (air collector). The air is driven through a tower measuring as high as 1,000 m. This air currect pulls the air mass under the collector roof after it (chimney effect). This creates a constant flow of air to drive the six turbines located at the base of the funnel, which together generate a total of 200 MW.

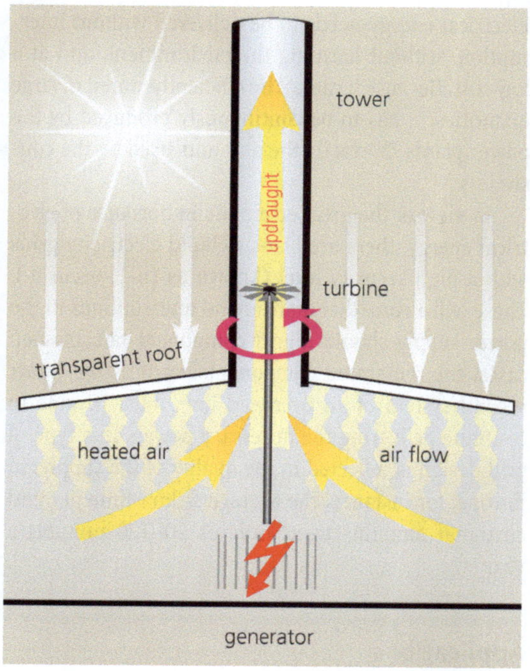

Solar updraught tower power plant: Air is heated up under a collector roof by the sun. The hot air rises through a chimney, driving turbines located inside the chimney.

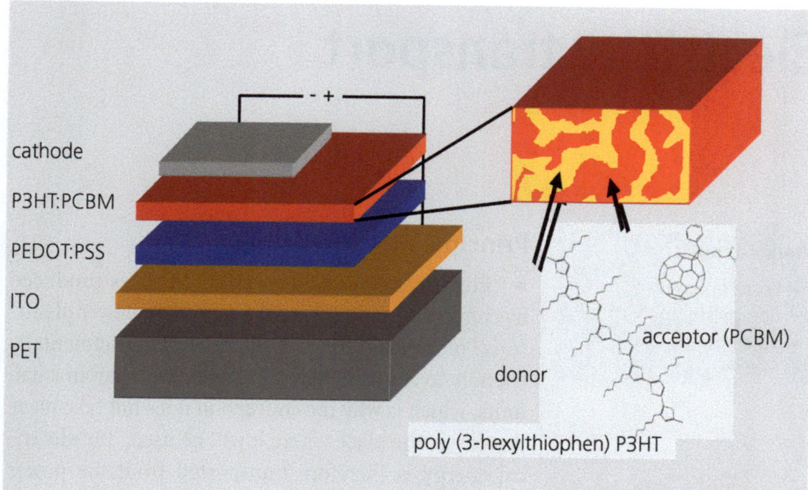

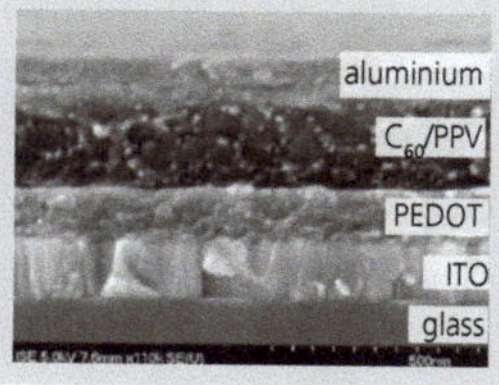

Polymer solar cell: Figure (above) and microscope photo of a cross section. The substrate is usually a glass or plastic film. The first layer is the transparent electrode (ITO – Indium Tin Oxide). This layer is followed by a p-conducting polymer (Pedot=Polyethylenedioxythiophene), which ensures a good electrical contact with the adjacent photoactive layer (C60/PPV=Fulleren/Polyparaphenylenevenylene). This red layer contains two components: an electron conducting material – in this case, a modified C60 molecule – and a p-type organic material (PPV). This layer absorbs the photons, electron-hole pairs are generated and are ultimately separated. The holes are transported to the lower electrode; whilst the electrons are transported to the upper metal electrode (Al). The whole system is less than 1 µm thick.

Prospects

- Two key research goals in thermal solar energy applications are the integration of systems into buildings, combining energy conservation, efficient energy conversion and solar energy usage, and the generation of process heat (and electricity). Use of solar energy, combined with other regenerative energy sources (wind, water, biomass, geothermal energy) can make a major contribution towards meeting our energy demands.
- Cost reduction is the primary goal of research into photovoltaics. This goal is being pursued by reducing consumption of silicon material (thinner wafers) and improving efficiency by applying new and less expensive production technology.

PROF. DR. VOLKER WITTWER
Fraunhofer Institute for Solar Energy Systems ISE, Freiburg

Internet

- www.eupvplatform.org/
- www.univie.ac.at/photovoltaik/pv5.htm
- www.iwr.de/solar
- http://esttp.org/cms/front_content.php

Electricity transport

Related topics

- Fossil energy
- Energy storage
- Solar energy
- Metals

Principles

▶ **Electrical energy.** Electrical energy is produced by converting primary energy sources in power plants. It is more efficient, reliable and environmentally friendly to convert energy in large units than in small units, which is why the conversion does not take place directly at the place where it will be used. The electrical energy is therefore transported from the power plant to the user via an electricity grid. Some of the energy is lost during transport. These losses increase in proportion to the ohmic resistance of the wire connections, but decrease quadratically in relation to the transport voltage. Long distances are therefore covered at high transport voltages, which are then transformed back to the required low voltages in the vicinity of the consumers. Nominal voltages of 220 kV and 380 kV are the standard levels in Europe. Although even higher voltages are technically feasible and are used in some countries such as Russia (500 kV, 750 kV), they are only economical when covering extremely long transmission distances due to the high cost of the equipment involved (insulation, pylons). *Long-distance energy transport* and distribution among consumers takes place at optimised, graduated voltage levels between 400 V and 380 kV, depending on the distances covered and the power ratings being transmitted.

In order to be able to transform the different voltage levels, power is transported as alternating current and voltage, with three phases interconnected to form a rotary current system (an interlinked three-phase alternating current system). Direct current could not be transformed directly. Three-phase systems also have the following advantages:

- They transmit triple the amount of energy using three cables, while direct or alternating current would require six cables to do this.
- They generate a magnetic rotary field, which makes it easy to set up three-phase motors.

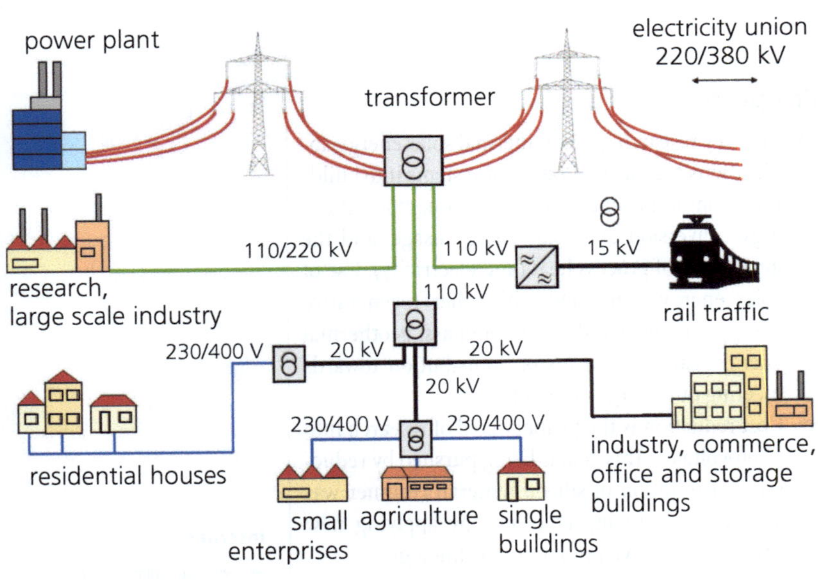

Transmission and distribution system for electrical energy: On leaving the power plant, the electricity flows through cables with a voltage of up to 380 kV to the main consumer centres, where it is delivered to the distribution grids at voltages of between 110,000–10,000 volts and finally to the consumers at 400 /230 volts. There are six voltage levels in total. Source: VDN

Electrical energy needs to be delivered without interruption, without harming the environment, and at a low cost. Because it cannot be efficiently stored in large quantities, it has to be continuously produced by the power plants in exactly the amounts used by the consumers.

In view of the great economic importance of electrical energy, there are plans to build electricity grids with a high security level (known as (n-1) security). These will ensure that there are no interruptions in the power supply should a grid component fail. In such cases, the remaining components, such as the intact second system of an overhead line, must be able to continue delivering the necessary power. This type of grid design is reflected in the quality of the supply: in Europe, for instance, the average failure time per end customer amounts to as little as 20–170 minutes a year.

Applications

▶ **Power system setup.** The different voltage levels of the electricity grid are interconnected by transformers. These connections between the voltage levels are switchable, as are the individual electric circuits within each voltage level. In order to be able to interrupt

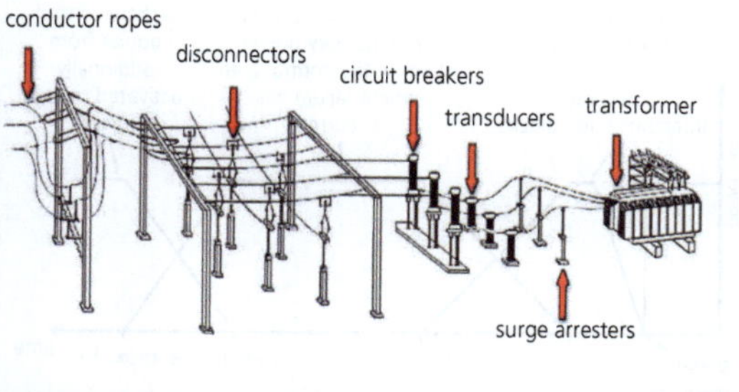

Setup of a substation: The electrical energy is transformed by power transformers into the distribution-grid voltages required to pass it on. In order to be able to interrupt the current flow, particularly in the event of a failure and for maintenance work, the substation is fitted with circuit breakers that can switch off both small and large currents, and disconnectors which subsequently ensure that the electric circuits remain separated. Surge arresters discharge temporary voltage peaks that can occur in cables in such events as a lightning strike. Transducers enable voltage and current flow to be measured for the purpose of grid control and billing. Source: Arbeitskreis Schulinformation Energie

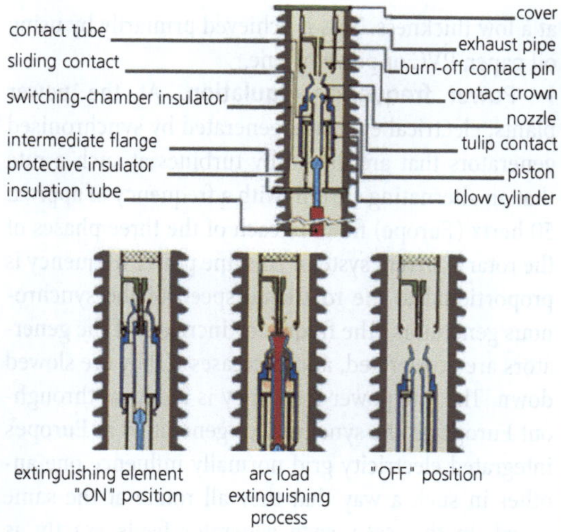

Circuit breaker: This component protects against the effects of an overload or a short-circuit by automatically opening itself and cutting off the affected section from the system. In the "ON" position, the conductor is connected. During the extinguishing process, the conductors are disconnected but the current still flows through the electric arc, which is cooled and extinguished by blowing in an extinguishing medium (e.g. SF_6). In the "OFF" position, the current flow is interrupted. Source: Arbeitskreis Schulinformation Energie

the flow of electricity, particularly in the event of a failure, the substations are equipped with circuit breakers and disconnectors. Connections between the transmission grids of different grid operators at home and abroad exist exclusively on the highest voltage level. The sub-grids are not interconnected on the lower voltage levels: firstly, in order to regionally limit the effects of a failure and, secondly, to limit the high currents that occur during short-circuits so that they can be safely switched off.

▶ **Transport through lines.** Energy transmission in the extra-high voltage grid (380 kV, 220 kV) takes place almost exclusively through overhead lines. In order to achieve both low electrical resistance and high mechanical strength, the conductor ropes are made of aluminium with a steel core. They are supported by steel lattice masts and carried by insulators made of porcelain, glass or plastic. Because transport losses cause the conductor ropes to heat up, the maximum transmission voltage is limited by the highest admissible temperature in the ropes and the resulting cable sag. E.g. in Germany's extra-high voltage grid, overhead lines bridge distances of over 100 km.

The transmission losses of an overhead line system are composed of current-dependent and voltage-dependent losses:

- Current-dependent losses are caused by the ohmic resistance of the conductor rope (current heat loss). They increase quadratically in relation to the current. Current-dependent losses can be reduced by lowering the ohmic resistance, for example by increasing the cross-section of the conductor rope.
- Voltage-dependent losses comprise dissipation and corona losses; they are caused by discharges in the air, which occur in the electric fields between conductor ropes, and between the conductor ropes and the ground. These voltage-dependent losses increase quadratically in relation to the transport voltage. *Corona losses*, which are caused by electric discharge in the strong electric fields near the conductor (glow), can be reduced by using what are known as "bundle conductors", in which two, three or four conductor ropes of a given phase run in parallel. This reduces the boundary field strengths on the conductor ropes.

The voltage-dependent losses in a 380-kV extra-high voltage overhead line amount to 2.5 kW/km, while the current-dependent losses lie at 88.5 kW/km at a current of 1,000 A.

Electrical grids also contain underground cables. These have an advantage over overhead lines in that they have less visual impact on landscapes and townscapes and are less exposed to the weather. The basic technical challenge in constructing such cables is that, while a thick insulation against the surrounding soil and the other conductors increases electric strength, such efficient electric insulation also hinders the dissipation of lost heat produced in the conductor. The solution is to use insulators that can achieve a high electric strength and are sufficiently heat-conductive even

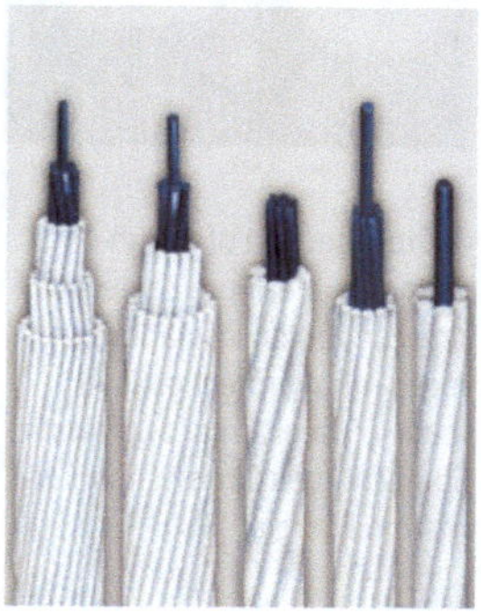

Conductor ropes in the extra-high transmission grid: The cables consist of the conductor (made of aluminium or copper), a relatively complex insulating layer, and an external mechanical protection layer. The insulation is made of various plastics or oil paper depending on the voltage level. Source: Arbeitskreis Schulinformation Energie

Insulators of a 380-kV transmission line: Made of porcelain or glass, the insulators carry the conductor ropes, which have a very high voltage compared to the earth potential of the masts. They ensure that this high voltage does not lead to a flashover (similar to lightning) between the conductors and the mast. Source: Arbeitskreis Schulinformation Energie

Cross-section of a high-voltage cable. From center to rim: conductor, insulation, mechanical protection layer made of steel armor and plastic. Source: Arbeitskreis Schulinformation Energie

at a low thickness. This is achieved primarily by using oil paper, PVC or polyethylene.

► **Power frequency regulation.** At the power plants, electrical energy is generated by synchronised generators that are driven by turbines at such a rate that an alternating current with a frequency of approx. 50 hertz (Europe) flows in each of the three phases of the rotary current system. This line power frequency is proportional to the rotational speed of the synchronous generators. The frequency increases if the generators are accelerated, and decreases if they are slowed down. The line power frequency is identical throughout Europe, as the synchronous generators in Europe's integrated electricity grid normally influence one another in such a way that they all rotate at the same speed. In this case, each generator feeds exactly as much electrical energy into the grid as is provided by the rotating turbine. The customers then consume this energy. To put it in simplified terms, the generators are accelerated by the turbines and slowed down by the consumers.

Any imbalance between the amount produced by the generators and that consumed by the users in the integrated grid has a direct effect on the line power frequency. If, for example, a power plant failure causes a power deficit in the grid, the generators are slowed down and the line power frequency decreases. In the event of a power surplus, which can occur when a large-scale consumer shuts off its power, the synchronous generators accelerate and the frequency climbs. This means that the line power frequency is a good indicator of the grid's power balance. Due to the mechanical and electrical characteristics of the generators and the consumers, the line power frequency has to be kept within a narrow window of around 50 hertz, i. e. the power balance must be as even as possible at all times. Otherwise, there is a risk of interruptions in the supply and damage to the generators and the consumers.

In the European integrated grid, the power balance is continuously stabilised by a sophisticated system of regulating mechanisms that employs various types of control power: primary, secondary and tertiary (minutes reserve). These types differ in terms of their activation and alteration speeds. Primary and secondary control power is accessed automatically from controllable power plants by the transport network operators and is almost constantly required in varying amounts and directions. Primary control power has to be provided in the required amounts within 30 seconds in each case, and secondary power within 5 minutes. The power plants' primary controllers respond exclusively to the line power frequency, providing additional power or reducing the plants' supply in proportion to the deviation from the rated frequency of 50 hertz. The secondary controllers likewise respond to the line power frequency (storage power stations, coal- and gas-fired steam power stations). In addition, however, they observe in which control area (the integrated network consists of numerous control areas) the power deficit or surplus has occurred. In this way, the secondary controller can take over from the primary controller and adjust the power balance by adapting the control area responsible for the imbalance. The secondary control power is then relieved by the minutes reserve so that it is available again as soon as possible. This transition is carried out manually by the grid management staff in agreement with the power plant operators or industrial consumers.

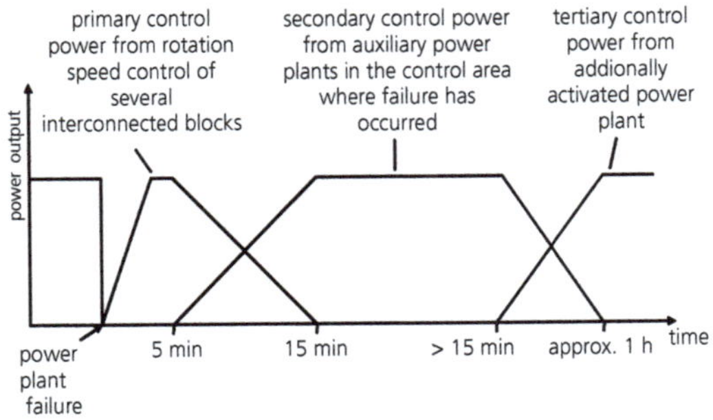

Timeline for activating control power after a power plant failure: The power shortage following a power plant failure is initially offset by power plants equipped with primary controllers, which respond directly to the drop in frequency caused by the power deficit. Within 5 minutes, the power-plant blocks connected to the secondary controllers increase their power supply in response to the control commands issued by the grid control station. Subsequently, in the tertiary control phase, the grid control station manually activates additional power plants to obtain energy that is then available for later failures.

Trends

► **Virtual power plants.** Europe is experiencing a continuing boom in decentralised production plants. These include both plants based on *renewable energy* sources, such as *wind-power* and *photovoltaic plants*, and small cogeneration plants. At present, the amount of electricity fed into the grid by these generators is not oriented towards customer demand, but instead depends on the availability of wind and solar radiation, or on the demand for heat in the case of combined heat and power plants. This fluctuating supply of electricity into the grid calls for efficient forecasting systems and the provision of additional balancing energy. Moreover, the supply from wind energy plants in particular calls for an expansion of the electricity grid, as plants in very windy regions are already supplying many times more energy than is actually consumed in those areas. In order to avoid having to back up each fluctuating production plant with an equivalent amount of power from calculable, conventional power plants in the future, it is necessary for wind-energy, photovoltaic and other fluctuating production plants to become involved in the balancing of power deficits and surpluses in the supply grid.

One solution would be to combine numerous different fluctuating producers to form virtual power plants, which are supplemented by reservoirs or conventional production units to create a calculable, controllable unit overall. A virtual power plant of this type could then participate in the energy market like any conventional power plant.

The decentralised production units can communicate with one another via a network of dedicated access lines and dial connections. The online regulation of biomass plants and block heat and power plants is preferably carried out using dedicated lines and ISDN connections. This enables single values for maximum and current electrical and thermal power to be transmitted in an ongoing exchange of data, for instance at one-minute intervals. Dial-up connections, on the other hand, are sufficient for the multiple-hour power schedules of photovoltaic plants or household fuel cells. The value series transmitted during a connection inform such decentralised production units about how much energy to provide over the next few hours. Additional connections can be established spontaneously if and when required, i. e. if unforeseen events make it necessary to recalculate. Dial-up connections can also be used for meteorological services to transmit weather forecasts.

▶ **Superconductivity**. Around the world, tests are being carried out on the use of superconductive cables for supplying electricity. A 250-kg superconductive cable that transmits electricity without resistance and with virtually no losses can replace a conventional underground cable weighing 18,000 kg.

Superconductors are any metals, metal compounds or other materials that lose all measurable electrical resistance on reaching a specified "transition temperature" and consequently also displace magnetic fields from their interior. The transition temperature is a material-specific parameter that differs for each superconductor. Some materials only become superconductive at very high pressures and low temperatures (e. g. germanium at 4.8–5.4 K and approx. 120 bar, or silicon at 7.9 K and approx. 120–130 kbar). If these materials were used to transport electricity, an extreme amount of cooling power would be required along the entire transport route in order to maintain the necessary low temperature levels. Scientists are therefore seeking solutions that offer realistic prospects of a technically viable and economical implementation. The highest transition temperature currently on record is 133 K ($HgBa_2Ca_2Cu_3O_8$).

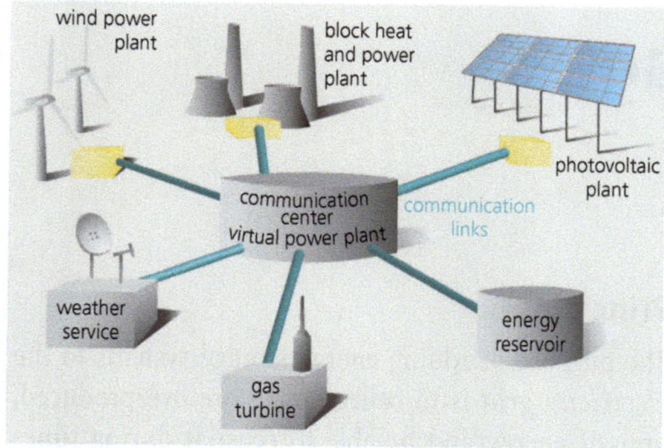

The virtual power plant: Several decentralised producers are combined to form a virtual power plant. The fluctuating power supply of the individual units is counterbalanced by conventional, calculable producers and reservoirs. Together, the various units form a calculable virtual power station that can be treated like a large-scale plant. This involves comparing a large number of target and actual values, controlling automation units, and obtaining forecasts on the sun, the wind, and consumer behaviour.

Prospects

- A reliable supply of electrical energy is ensured by an extensive transmission and distribution system that can cope with the failure of a power plant or a cable section without any resulting interruptions in the supply.
- Electrical energy supply grids are set up as three-phase rotary systems, as this makes it possible to dispense with return conductors and to produce magnetic rotary fields for three-phase motors.
- In electric power supply grids, the power generated must at all times correspond to the power consumed. The line power frequency is an indicator of the power balance in the grid. In the event of a power deficit, the frequency decreases, while a surplus makes it rise.

DR. BENJAMIN MERKT
MATTHIAS BOXBERGER
E.ON Netz GmbH, Bayreuth

Internet

- http://mrec.org/pubs/06%20MREC%20DeMarco%20PSERC.pdf
- www.raponline.org/pubs/electricitytransmission.pdf

Energy storage

Related topics

- Solar energy
- Wind energy
- Automobiles
- Full cells and hydrogen technology
- Microenergy technologies

Principles

The purpose of adding energy storage systems to the *electricity grid* is to collect and store overproduced, unused energy and be able to reuse it during times when it is actually needed. Such systems essentially balance the disparity between energy supply and energy demand. Between 2–7% of installed power plants worldwide are backed up by energy storage systems.

Electrical energy can be converted and stored by either mechanical or chemical means. Mechanical energy can be stored as potential energy by using a pumped hydro or compressed air storage method, and kinetic energy can be stored by a flywheel. Chemical energy can be stored in accumulators, for example lead, lithium or redox flow batteries. The advantage of these high-energy storage methods – excluding flywheels – is that they allow greater amounts of energy to be stored than other alternatives. Storage losses usually become significant during conversion processes, but are negligible within the storage system itself (exception: flywheels).

Alternatively, electricity can be stored directly in electrical form: either electrostatically, such as in supercaps (electrochemical double-layer capacitors, EDLCs), or electromagnetically, for instance in superconducting magnetic coils. These high-power storage mediums are not suitable for storing large quantities of energy but can dramatically improve short-term energy balancing and overall power quality in local electricity grids.

Applications

▶ **Lead-acid battery.** Lead-acid batteries are state-of-the-art technology. They are the most economical (in terms of price and efficiency) and most widespread batteries. However, due to the excessive maintenance they require and their short life spans, their innovative

Installed energy storage capacity: The installed energy storage capacity amounts to 46 GW in Europe. This is in absolute and relative terms, with 6.6% being the highest global amount. The most significant differences emerge between different power plant parks and as a result of regional restrictions on electricity grids.

	Installed storage capacity [GW]	Installed generation capacity [GW]	Relative ratio
North America	26	1200	2.2%
Europe	46	700	6.6%
East Asia	25	875	2.9%

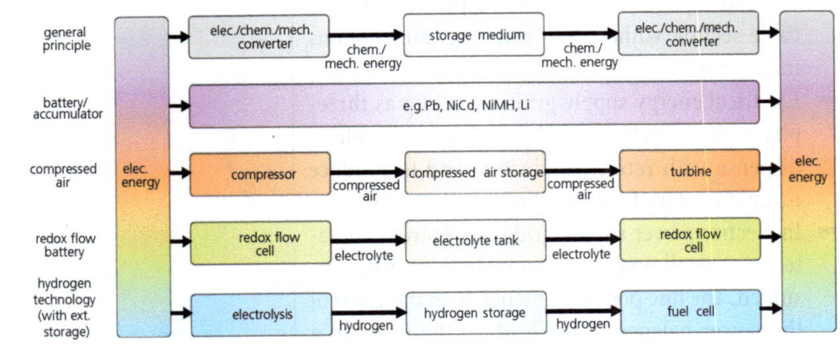

Electrical energy storage: Energy storage occurs in three steps. Initially, the electrical power is transformed into a form which can be stored. Then the storage medium (e.g. compressed air, hydrogen) retains the stored energy. Finally, when the energy is needed, the storage medium discharges the stored energy and transforms it back into electrical energy.

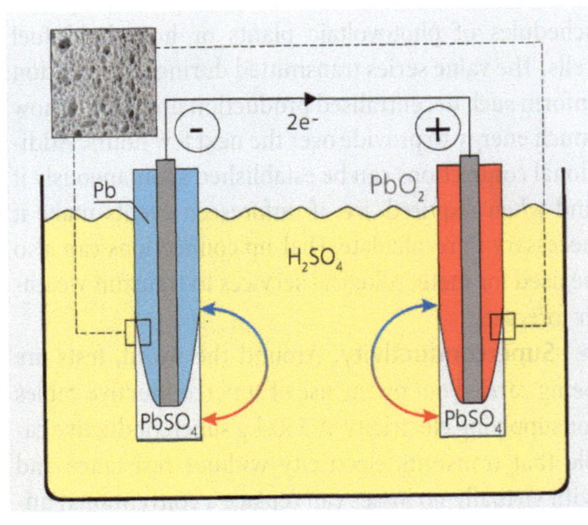

Lead-acid battery: The lead-acid battery is currently one of the most established storage systems. It is applied both in cars to start the motor and in self-sustaining energy supplies. Essentially, the battery consists of two electrodes which utilise diluted sulphuric acid as the electrolyte. During the discharging process, chemical reactions occur at the surface of the electrodes. The lead from the negative electrode and the lead-dioxide from the positive electrode are converted into lead(II)-sulfate. The charging process ensues inversely.

potential is too limited to really consider incorporating them into future energy systems. Over 125 MW-capacity lead-acid batteries have been installed worldwide, but some of these are already out of service.

▶ **Hydrogen**. The possibility of storing electrical energy in hydrogen form by *electrolysis* is an interesting topic, especially with regard to energy density. Energy stored in this state could potentially become fuel for cars in the future. However, while the volume and weight of the storage system are important for such mobile applications, they play less of a role in stationary ones. Hydrogen storage systems do not have a practicable future in this regard, due to their higher costs and their overall efficiency of approximately 26% (present) to 48% (goal), which is significantly inferior to current storage systems.

Grid-integrated energy storage systems have a multitude of applications and benefits, including smoothing peak demand loads, distributing load flow more effectively, storing bulk power, integrating renewable energy, ensuring an uninterruptible power supply (UPS), and balancing isolated grids.

Energy storage will play an increasingly important role in the future, and there will be a growing demand for reserve power that is able to regulate and balance the integration of fluctuating renewable energy sources. At present the main demand is for wind power, but the increasing integration of *photovoltaic power cells* into the grid could further add to this demand in the future. *Wind power* alone, without a storage mechanism, can only provide 5–7% of its useable output energy, which means that there is massive energy saving

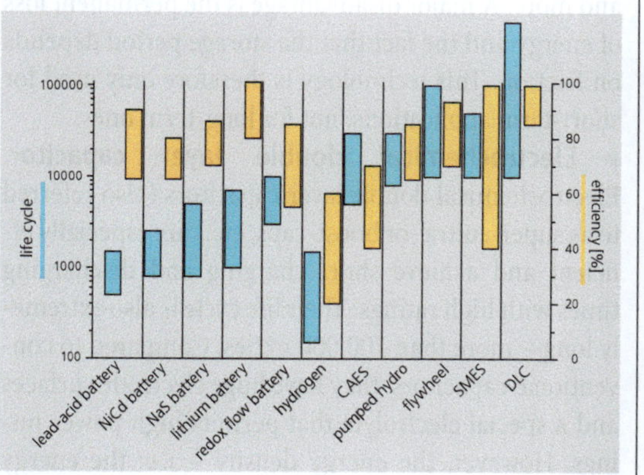

◘ Overall efficiency and lifetime of storage systems: Overall efficiency is the relationship between discharging and recharging energy. The balance boundary is the alternating current generator of the supply net needed in order to achieve an objective comparability. High efficiencies can be achieved by batteries such as lithium based batteries; however their lifetimes and number of cycles are limited. In contrast, mechanical storage systems, e.g. pumped hydro and compressed air, have significant advantages concerning both factors. Lastly, hydrogen fuel cells have particularly low efficiencies which are typically under 50%. Source: Fraunhofer IUSE

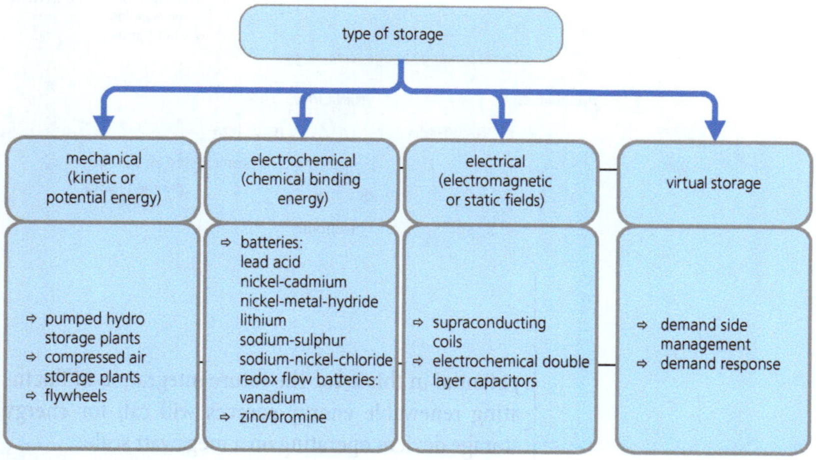

◘ Electrical energy storage technologies: Pumped hydro power plants are the most common storage systems and a typical example of storing electricity in mechanical form.

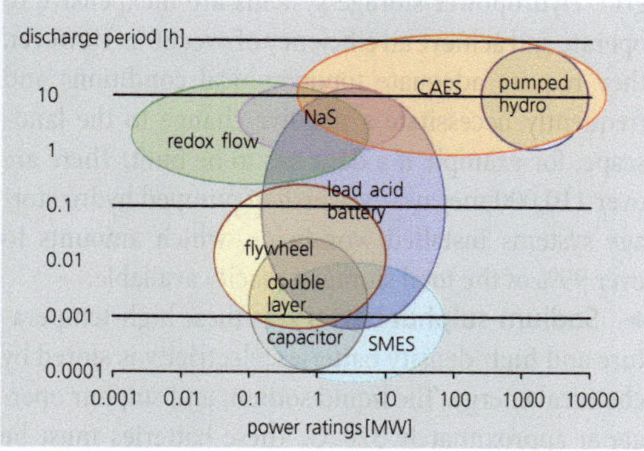

◘ Power ratings and discharge periods of typical storage systems: Pumped hydro and compressed air are currently the storage systems with the highest power ratings and discharge periods. Higher ratings are expected from redox flow, double layer capacitors, and SMES in the future. These new technological fields have only begun research recently. Double layer capacitor and SMES are usually only used for rapid and short term energy storage functions. Lead acid batteries are more versatile because they are applicable in both short and long term storage. In addition, their modular design enables the implementation of minor or major ratings. Source: Fraunhofer IUSE

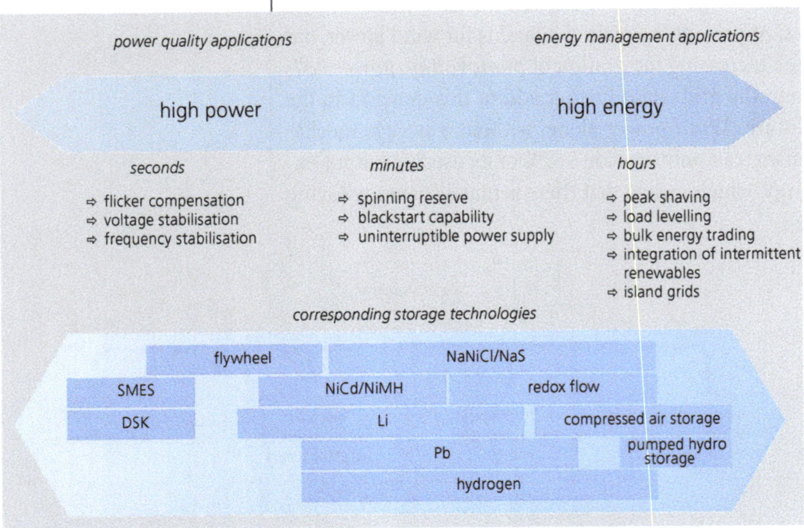

◂ Fields of application and implementation of typical storage systems: Systems with very short storage periods (seconds) are used for grid stabilisation. One example includes SMES technology which shows high ratings for low stored electricity. At the other end of the spectrum, storage systems exist with the ability to store significant amounts of energy, e.g. pumped hydro. They possess storage times of several hours or more and compensate for the fluctuating integration of energy, such as wind power, and are able to smooth peak demand loads. The intermediate spectrum operates in intervals of minutes ensuring uninterruptible power supply.

potential in this area. The future integration of fluctuating renewable energy sources will call for energy storage devices operating on a megawatt scale.

► **Pumped hydro storage.** During off-peak times, excess power is used to pump water out of a basin into a reservoir situated higher up. Then, when the demand for energy reaches its peak, the potential energy stored in the reservoir is released and converted back into useable energy. This is done using a system of turbines which harvest energy from the flow between the different elevations and integrate the energy back into the grid. Hydropower storage systems are inexpensive to operate and achieve an efficiency of over 80%. However, they require adequate topographical conditions and frequently necessitate a massive change to the landscape, for example if a dam has to be built. There are over 110,000 megawatts' worth of pumped hydro storage systems installed worldwide, which amounts to over 99% of the total storage capacity available.

► **Sodium-sulphur battery.** In these high-temperature and high-density batteries, electricity is stored by chemical energy. The liquid sodium and sulphur operate at approximately 320 °C. These batteries must be designed specifically to prevent heat loss and leakage. They are made up of 50-kW battery modules and are about the size of two tractor trailers for a capacity of 1.2 MW with 7.2 MWh. They have an efficiency of over 80% and are very durable, but are currently only established in the Japanese market (196 facilities with a total of 270 MW of installed capacity). Only a few demonstration projects in the United States are known to support grid stability at present.

► **Flywheel.** Flywheel systems store energy as kinetic energy by means of a rotating rotor. An electric motor accelerates the flywheel during the charging period and acts as a generator during the discharging period. For design reasons, this technology is made for high-power output and not for high-energy storage capacity. Traditional low-speed systems made of steel are only built for high power outputs, whereas modern high-speed composite rotors are marked by a greater energy-storage capacity and lower losses. A major advantage of this technology is its long lifetime with about 100,000 cycles, and its high efficiency of about 80% and more. A major disadvantage is the permanent loss of energy and the fact that the storage period depends on friction. This technology is therefore only used for short-term applications, not for long-term ones.

► **Electrochemical double layer capacitor.** Electrochemical double layer capacitors (also referred to as super-, ultra- or boost-caps, etc.) are especially efficient and achieve short charging and discharging times with high ratings. Their life cycle is also extremely long – more than 100,000 cycles. Compared to conventional capacitors, they have huge electrode surfaces and a special electrolyte that permits high power ratings. However, the energy density – i.e. the energy content per volume – is very low, so EDLCs are not suitable for major energy loads. They are therefore applied for rapid buffering in grids or in conventional battery systems.

► **Compressed air energy storage.** Compressed air energy storage (CAES) plants store energy by compressing air into subterranean caverns. During peak loads, the compressed air is directed to a conventional gas turbine with higher ratings because no energy is needed for compressing the combustion air. The caverns required for this technology comprise volumes of between 300,000 and 600,000 m^3 and are operated under a pressure of about 40–70 bar. Feasible plant sites are limited to special geologic formations containing subterranean caverns (e.g. empty salt domes), because only these formations provide low-cost storage volumes. At present, only two CAES plants exist at utility scale: one of them is operated in Huntorf (Germany, 1978) with a rated power of 321 MW over 2 hours, and the other one is located in Alabama/McIntosh (USA, 1990) with a rated power of 110 MW over 26 hours. The plant in Germany was built to deliver electricity during peak load hours. Its efficiency (42%) is low, because the thermal energy generated by compression is not used. Increased efficiencies are possible by installing an internal heat recovery system (recuperator). The CAES plant in the USA operates at an efficiency of 54%.

CAES plants offer promising future options, as they take advantage of well-known technologies (e.g. electrical machines, compressors, expanders) and high storage volumes leading to better power ratings. In the future, efficiencies will be raised to up to 70% in these adiabatic CAES plants by storing heat from compression and feeding it back into the process.

Decentralised plants of this type (micro-CAES with about 1–30 MW) are operated under higher pressure. The compressed air is stored in subsurface pressure vessels comparable to natural gas pipelines. In addition to the adiabatic construction not needing auxiliary energy, micro-CAES plants are independent of geological formations, adding yet another advantage.

▶ **Redox flow battery**. Contrary to conventional batteries in fixed units with traditional electrodes and electrolytes, redox flow batteries have separated parts for conversion (power conversion system – PCS) and storage (storage system). Each power conversion system contains two electrodes and a membrane in between to form a single cell. These cells are connected to form a stack which is comparable to the stacks in fuel cells. The storage of electrolytes takes place in external tanks, so that the power conversion system and the storage system are separated. This makes it possible to adapt them individually to the application. Conventional batteries lack this ability.

Redox flow batteries based on vanadium – which exists in different oxidation states (V^{2+}/V^{3+}, V^{4+}/V^{5+}) in diluted sulphuric acid – are currently preferred to other systems. Advantages of this system include: uncontaminated electrolytes due to the transferability through the membrane, an aqueous reaction which increases speed, and the potential of such systems to reach efficiencies of 70%.

At present, there are several demonstration plants that exhibit "peak shaving" practices of integrating wind power into energy grids. Scaling up this technology makes it viable to build plants on the megwatt scale. This power range is necessary for wind farms to provide reliable electricity or for large-scale consumers to shift peak loads. A major advantage is that it is very cheap to extend the capacity of redox flow storage systems through additional external tanks filled with vanadium in diluted sulphuric acid. But, at the mo-

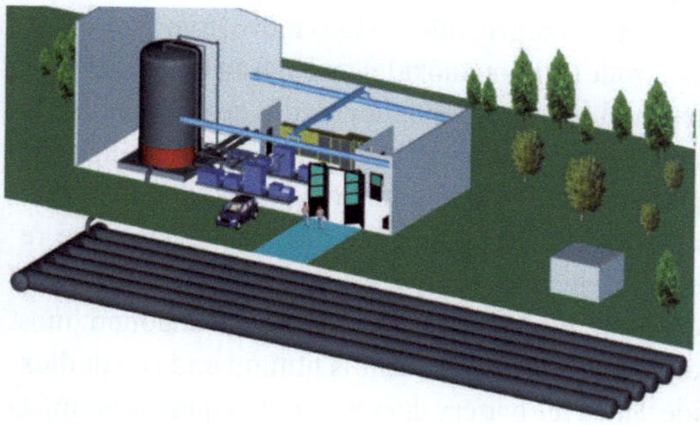

Micro-CAES system: In this concept the air pressure will be stored in subsurface steel tubes up to 200 bar. For 2 MW and 10 MWh storage, a tube with an area of 540 m^2 is needed. The heat produced from the compression process will be separated and stored in a high temperature heat storage. The compressed air will then be reheated before recirculation.

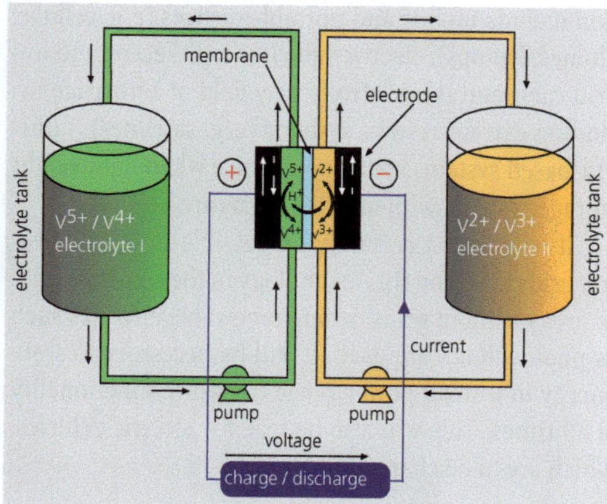

Outline of a vanadium redox-flow battery: The vanadium redox-flow battery is composed into a stack, in which chemical reactions take place in two electrolytes separated by a membrane. Charging and discharging are reversed chemical reactions either produced by electrical energy consumption or generation. The charged or discharged electrolytes are stored in external tanks; this is the major advantage, easily allowing a significant amount of energy to be stored.

ment, the initial investment costs for the power conversion unit (power electronics and stack) are very high. For this reason, the system has to be advanced with regard to efficiency and cost reduction.

▶ **Lithium ion battery**. There is a plethora of lithium-based systems with cathodes, anodes and electrolytes made of different materials. Most lithium batteries use intercalation electrodes. Intercalation means the inclusion of atoms in crystal lattices of multilayer crystals, e.g. the intercalation of lithium in graphite during the charging of lithium ion batteries. This type consists of two different intercalation electrodes. A separator provides the electrical isolation. If such a battery is charged, lithium ions move from the positive electrode (lithium dioxide) through an organic electrolyte and a porous separator to the negative intercalation electrode, where they adsorb. Discharging takes place the other way round. Both procedures are reversible, which means that lithium ion batteries are rechargeable. The positive electrode is produced out of different lithium compounds. The compound most commonly used at present is lithium and cobalt dioxide. Since the battery does not contain pure lithium, its typically high reactivity is inhibited, so there are no security risks in standard applications.

Lithium ion batteries cover a range of efficiency-oriented battery types all the way up to systems with high energy densities, with room for even greater potential. They show good operating performances concerning life span and efficiency, but need protection against overvoltage in every battery. Fields of application include mobile and portable devices (e.g. cellular phones, laptops), electric vehicles (e.g. electric and hybrid cars) and the relatively new field of aerospace including extraterrestrial systems (e.g. satellites). Lithium-based systems are a good choice when lightweight storage systems with long life spans are needed.

It will almost certainly be possible to install electric grids based on this technology in the coming years. In areas without grids, where renewable energies such as photovoltaics are used, it will be necessary to store energy in a small power range to ensure functionality at all times. This will also be true for electric vehicles, which are to be charged by the grid.

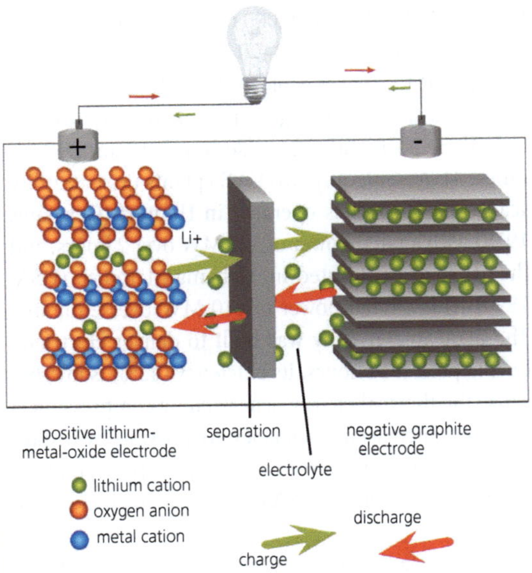

▲ Lithium ion battery. The charging works as follows: Lithium ions move from the positive electrode (lithium dioxide) to the negative intercalation-electrode and adsorb there. During the discharging process ions move backwards to the positive electrode out of a metal dioxide (e.g. cobalt dioxide). Lithium ion batteries have high efficiencies and high power ratings. One essential part of these batteries is the controlling device, which monitors the voltage in each cell and regulates it. In this way deep discharges and overcharges causing reductions of life span are avoided.

▲ Automotive applications: Batteries based on lithium polymer technology possess a wide range of sizes and forms. The figure shows one cell with a capacity of 40 ampere-hour (Ah) and a nominal voltage of 3.7 V, used in automotive applications.

▲ Small scale application of a lithium ion battery: The cell has a capacity of 15 mAh and 3.7 V. It is used in hearing aids and delivers electricity for one day.

Trends

▶ **Virtual energy storage**. Virtual energy storage systems compensate for off-peak and peak loads, just like real storage systems. In contrast to real storage systems, which are charged or discharged depending on supply and demand, virtual storage systems manage the compensation by automatically controlling the consumers. There are currently two different systems: "demand side management" and "demand response". "Demand side management" uses private or industrial consumers (e.g. private households with electric heaters, or ground-water pumps in coal-mining areas) which can be switched off for a short period. "Demand response" is a more challenging technology. During light load periods, more energy is purchased than actually required, e.g. the temperature in a cold store is decreased. During peak loads, the cold store can be switched off to unload the grid because the thermal energy remains stored. Virtual storage systems thus operate like real ones, shifting amounts of energy. In reality, the storage occurs via thermal energy, i.e. as cold in the case of the cold store, or as heat in the case of heat pumps with integrated storage. Managing different consumers by "demand side management" or "demand response" calls for smart grids with built-in intelligence. Developers are currently working on various IT technologies that allow numerous dispersed consumers to be managed in the grid by virtual energy storage.

These systems are of relevance to the industrial sector – where virtual storage is already in use for financial reasons – and to private households, where it is barely applied at present. Significant advantages include the system's low cost (control devices) and the possibility of decentralised integration. One disadvantage is that only a small part of the theoretical potential can be easily utilised because industrial processes sustain partial production loss if demand side management is used. In a private household, demand side management can cause a loss of comfort due to more determined private processes like using the washing machine.

▶ **The vehicle-to-grid concept.** In future, the energy stored in the batteries of electrical vehicles could offer a new storage alternative. This "vehicle-to-grid" concept requires vehicles with combustion engines to be replaced by vehicles with electric motors. The automotive industry is pursuing this scenario on a global scale via the intermediate step of hybrid vehicles. The intelligent use of large numbers of such decentralised storage units represents an interesting approach. There are three different concepts:

- Intelligent charging: A major purpose of electric vehicles will be to drastically reduce connection time to the grid. An intelligent control system will be required for charging in order to avoid new peak loads in the grid and to adapt the fluctuating integration of renewable energies. This concept is similar to "demand response" in that it includes the possibility of virtual energy storage.
- Grid stabilisation: In addition to intelligent charging, automotive storage will be able to integrate currently unstable cases into the grid. In this way, a more secure supply will be achieved despite the fluctuating source. Moreover, the estimated extra costs due to battery wear are moderate because the secondary load is small. However, there are still some problems to solve concerning automation, infrastructure and legal regulations.
- Periodic load compensation for the grid: This concept is based on the idea that electrical vehicles are typically idle for more than 20 hours per day and could be connected to the grid during this time. In doing so, the batteries of the vehicles serve as grid-integrated energy storage systems. This makes it possible to store and feed in a considerable amount of energy. The concept will be restricted by battery degradation, leading to reduced life spans and higher costs. These limitations are out of all proportion to the costs and storage services involved.

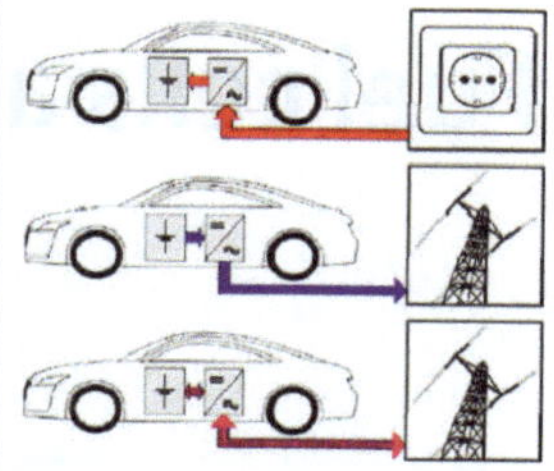

Vehicle to grid: There are three alternatives for integrating electric vehicles in the grid. Top: The first is a grid capable of controlling charging, to maintain the grid as consistently loaded as possible. Centre: The second alternative is the application of batteries to assure the operation of the grid with the possibility of energy reintegration to a small extent. Bottom: Thirdly, the periodic load compensation means electric vehicles are used to fully compensate the fluctuating integration of renewable energies. However, the reduced life span of the batteries is a serious hindrance where cost is concerned. Source: Fraunhofer IUSE

Prospects

The integration of fluctuating renewable energies – such as wind and photovoltaic power – generates severe load variations in the grid. To compensate for these variations and to assure reliability and steadiness, it is not only necessary to reinforce the grids but also to provide backup and reserve power which, in the medium term, calls for energy storage. Both real and virtual energy storage systems can charge energy in the event of excess wind power, and are also able to integrate power into the grid during periodic levelling. In the medium term, the following four systems have the best development potential:

- Compressed air energy storage: If their efficiency is increased to up to 70%, central large-scale plants will be able to compensate for the fluctuations generated by wind parks. These could be huge CAES systems with underground storage or more decentralised systems (micro-CAES) with steel pipes as storage vessels.
- Redox flow batteries can be used to standardise energy consumption, ensure integration, and compensate for the loads of major consumers. However, their costs need to be reduced and their life spans increased.
- Lithium-based batteries will probably become established in various decentralised applications, including electric and hybrid vehicles, but also in decentralised storage systems for remote areas with a minor demand for electricity.
- Virtual energy storage systems will be applied under economically viable conditions in the future. They are the best solution regarding costs, but have limited potential due to their difficulty of use. IT technology needs to be developed for controlling smart grids with dispersed virtual and real storage devices.

DR. CHRISTIAN DÖTSCH
Fraunhofer Institute for Environmental, Safety and Energy Technology IUSE, Oberhausen

Internet

- www.electricitystorage.org/technologies.htm
- http://mydocs.epri.com/docs/public/000000000001001834.pdf
- http://zebu.uoregon.edu/2001/ph162/l8.html
- http://ec.europa.eu/research/energy/nn/nn_rt/nn_rt_st/article_1154_en.htm

Fuel cells and hydrogen technology

Related topics

- Energy storage
- Microenergy technologies
- Automobiles
- Ceramics

Principles

The energy technology of the future will have to meet a dual challenge: to deliver a secure and sufficient supply of energy to a growing world population, despite limited energy resources, and to curtail emissions that have a negative impact on the environment. Electricity and hydrogen are clearly evolving into the most important energy vectors of the future. Fuel cells fit in with this trend, as they are able to convert hydrogen efficiently into electricity and heat in an electrochemical reaction, their only by-product being water. Fuel cells can continue to operate as long as fuels are available. These systems achieve a high electrical efficiency of about 60%, even in the low power range of several hundred watts, in contrast to conventional power plants, which need to be dimensioned in the MW range to achieve high electrical efficiency values.

Given that efficient energy conversion is the most important goal of energy technology, stationary fuel cell systems aim primarily to produce a combined heat and power supply with high electrical efficiency. Their greater efficiency makes it worthwhile to operate them even with natural gas or other fossil or biogenic fuels that are already widely available today, in which case hydrogen is generated inside the fuel cell systems.

As a fuel, hydrogen can provide emission-free propulsion for cars, boats and other mobile applications.

Fuel cells are also an option as a battery replacement in electronic devices, whenever it is easier to handle a fuel than to charge a battery.

▶ **Operating principle and types of fuel cells.** Electrochemical energy converters such as batteries and fuel cells have three main components: the anode, the cathode and the electrolyte. Hydrogen is oxidised at the fuel cell anode, and oxygen from the air is reduced at the cathode. The function of the electrolyte is to transport ions in order to close the electric circuit and safely separate the two gases. Electricity, heat and water are generated. A single cell delivers a voltage of about 0.5–1 V during operation, depending on the load. Practical voltage levels can be achieved by connecting numerous cells in series to form a fuel cell stack.

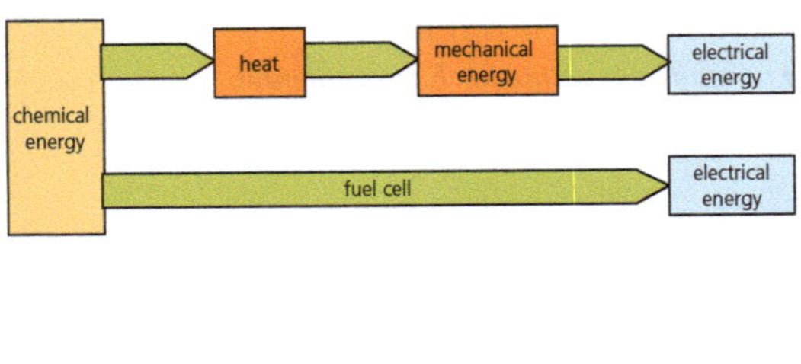

◘ Energy conversion in conventional power plants (top) and in fuel cells (bottom): Fuel cells convert chemical energy directly into electrical energy. In conventional power plants, the conversion takes place via intermediate thermal and mechanical steps.

Series connection is usually achieved by using a configuration of bipolar plates as cell frames. One side of the bipolar plate contains the channels for spatial hydrogen distribution, while the other side contains those for air distribution. The bipolar plates are made of an electrically conductive material, in most cases graphitic or metallic materials. A carbon cloth or piece of felt is placed between the bipolar plate with its gas channels and the electrodes as a gas diffusion layer. Its purpose is to mechanically protect the thin electrolyte membrane and to enable gas to reach the total electrode surface adjacent to the channel and below the ribs.

Besides electricity and water, the fuel cell reaction generates heat. In order to remove this heat from the fuel cell stack at operating temperatures, appropriate cooling measures must be taken. In lower-temperature fuel cells, for example, air or water cooling channels are incorporated in the bipolar plate to ensure homogeneous temperature distribution. For high operating temperatures, the problem could be solved by an excess air supply to the cathode or by integrating the heat-consuming generation of hydrogen from fuels such as natural gas. Fuel cells are thus able to provide consumers with electricity, heat, and even water, if required.

Various types of fuel cells have been developed, based on different combinations of materials, which are classified according to the type of electrolyte used. The operating temperature and all other materials are chosen and developed according to the ionic conductivity of this electrolyte.

▶ **Alkaline FC.** Historically, the first fuel cell was the alkaline fuel cell (AFC), which had 30% potassium hydroxide in water as the electrolyte. This fuel cell requires pure hydrogen and preferably oxygen as its fuel, since carbon dioxide (CO_2) causes the precipitation of potassium carbonate, which has to be removed.

▶ **Polymer electrolyte membrane FC.** The PEMFC is the type most commonly used at present, as the solid polymer electrolyte provides greater safety and a higher power density. The membrane materials used thus far typically require liquid water for ionic conduction, which limits their operating temperature to 90 °C.

Catalysts containing noble metals are used in the electrodes to ensure a fast electrochemical reaction at these relatively low temperatures. Besides pure hydrogen, reformates generated from fossil or *biogenic fuels* can also be used. Reformates typically contain CO_2 and traces of carbon monoxide (CO), the latter being a catalyst poison. The amount of CO must therefore be limited to 10–50 ppm, which requires sophisticated purification methods to be used for the reformate. Special anode catalyst materials have been developed to improve CO tolerance.

▶ **Direct methanol FC.** Not only hydrogen, but also methanol can be electrochemically converted in a PEMFC. This is then known as a direct methanol fuel cell (DMFC). The DMFC is attractive because methanol, being a liquid fuel, is easy to transport and handle. Due to the low operating temperature, the catalysts are sensitive to poisoning. Since CO is also formed during the electro-oxidation of methanol, special CO-tolerant catalysts are used for DMFCs. However, these have a much lower power density despite the typically high noble metal loading of the electrodes. In addition, the energy efficiency of DMFCs suffers from high electrode overpotential (voltage losses) and also from methanol losses by transfer (permeation) through the membrane.

▶ **Phosphoric acid FC.** The PAFC has proved reliable in field trials covering more than 50,000 operational hours, and has high availability. Phosphoric acid is a suitable electrolyte for operation up to 200 °C. It is combined with a reformer system for natural gas, and is typically used in stationary combined heat and power supply systems. PAFCs are usually operated with reformates. Due to the higher operating temperatures, the CO tolerance is improved to about 1% by volume. Nevertheless, a good reaction speed can only be achieved with the aid of noble metal catalysts.

▶ **Molten carbonate FC.** The MCFC is operated at about 650 °C, above the melting point of a mixture of lithium and potassium carbonate. The electrolyte is fixed in a porous ceramic matrix. At these relatively high temperatures, the noble-metal anode and cathode catalyst can be replaced by nickel and nickel oxide respectively. The molten salt is a corrosive liquid, requiring ongoing optimisation of the electrode materials in order to achieve longer cell lifetime and lower degradation. As ionic conductivity is obtained by carbonate ions which migrate from the cathode to the anode, reacting there with protons and thus forming steam and CO_2, the CO_2 has to be re-circulated to the cathode side of the cell. Oxygen ions need CO_2 as a "vehicle" to be able to migrate through the carbonate electrolyte and reach the hydrogen ions. This is why pure hydrogen is not suitable as a fuel for MCFCs, and reformate is typically used. CO poisoning of the catalysts no longer occurs at these high temperatures.

▶ **Solid electrolyte FC.** A solid ion-conducting ceramic material, yttrium-doped zirconium dioxide, also used in *lambda oxygen sensors* in automotive applications, is used as the electrolyte in solid electrolyte fuel cells (SOFC). Even with very thin electrolyte layers, it is not possible to achieve sufficient conductivity below 900 °C. High operating temperatures mean fast electrode kinetics. Again, nickel-based electro-catalysts can be applied on the cathode side, with $LaMnO_3$ being used as the anode material. Given that SOFC stacks also consist of a high number of cells, each cell having a three-layered electrode/electrolyte/electrode assembly, thermal stress – especially during the start and stop procedures – is an important factor to consider. The materials are chosen carefully according to their thermal expansion coefficient. Glass sol-

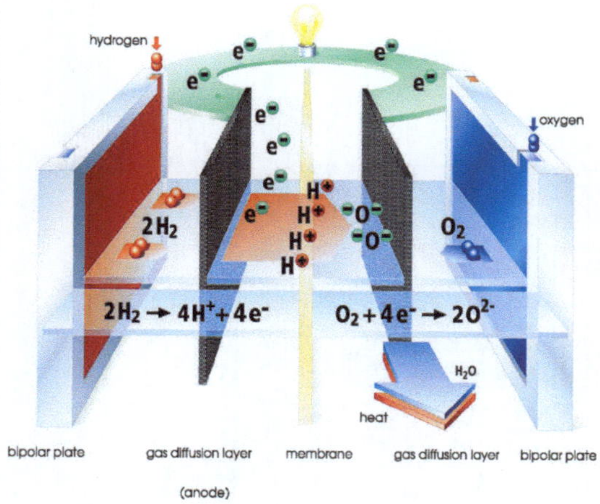

Operating principle of a fuel cell: Hydrogen molecules adsorb on the anode catalysts (left), and protons and electrons are formed by the resulting electrochemical reaction. The electrons flow through the external circuit, delivering electrical energy (lighting lamp). Finally, the electrons reduce oxygen to oxygen ions at the cathode (right). Hydrogen ions meanwhile migrate through the electrolyte membrane and react with these oxygen ions, forming water as a waste product.

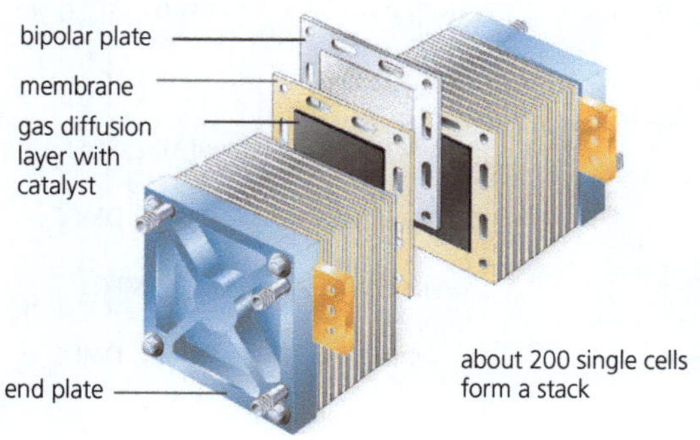

Fuel cell stack: As a single fuel cell only generates a maximum voltage of 1 V. Two hundred or more cells are connected in series to achieve the desired operating voltage.

ders have been applied thus far, as it is difficult to find sealing materials for these high operating temperatures. Different types of cell geometries have been developed, allowing thermal expansion with a minimum of mechanical stress.

The various types of fuel cells can be used for different applications. The highest power density and fast load changes can be realised with hydrogen-fueled PEMFC systems. This combination has also been developed for electric traction and back-up power. The DMFC is the best choice for long-term low-power applications. Methanol, being a liquid fuel, has a high energy density (4400 kWh/m^3) and is easy to handle; empty cartridges can also be refilled. Natural gas is usually available as a fuel for stationary applications, and moderate requirements with respect to start-up times and load changes permit fuel cells to be combined with a *reformer* system for the internal generation of hydrogen.

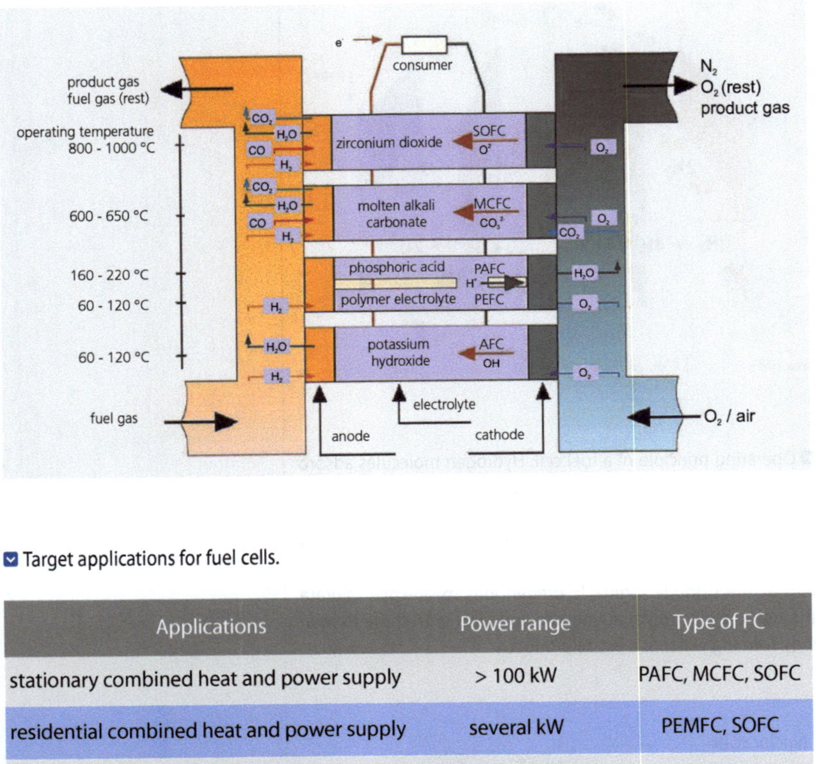

Operating principles and temperatures of the different types of fuel cells.

Target applications for fuel cells.

Applications	Power range	Type of FC
stationary combined heat and power supply	> 100 kW	PAFC, MCFC, SOFC
residential combined heat and power supply	several kW	PEMFC, SOFC
electric drives, cars and buses	10 kW – 200 kW	PEMFC
auxiliary power units, trucks	1 kW – 30 kW	PEMFC, SOFC
specialist vehicles	1 kW – 50 kW	PEMFC, DMFC
remote and back-up power	< 100 kW	PEMFC
portable generators, micro power	< 1 kW	PEMFC, DMFC

▶ **Hydrogen generation technologies.** Although hydrogen is the most abundant element in our world, it is not available in its elementary form. State-of-the-art technology is capable of generating hydrogen from sources such as hydrocarbons by reforming, or from water by electrolysis. Energy is required in order to split these chemical compounds. In the steam reforming process of natural gas, methane molecules are converted into hydrogen, carbon dioxide and carbon monoxide at temperatures of about 800 °C by reaction with steam in contact with special catalytically active materials. The subsequent shift reaction converts the carbon monoxide into carbon dioxide and hydrogen at temperatures of 200–400 °C with a surplus of steam. An alternative to steam reforming is exothermal partial oxidation, in which methane reacts with a sub-stoichiometric amount of air – less than required for complete oxidation – to form a hydrogen-rich gas mixture. The hydrogen content of the product gas is significantly lower, as it is diluted with nitrogen from the air and part of the methane is burnt in order to generate the necessary amount of heat.

These processes for generating hydrogen from natural gas (in which methane is the main component) are being developed to meet the special requirements of the different fuel cell types. The main difference is the tolerated CO concentration in the fuel; the PEMFC requiring the lowest level of less than 50 ppm. Other fuels, such as diesel, gasoline, methanol and ethanol, can also be used for the reforming process. Given that all of these hydrogen generation processes using fossil fuels emit the same amount of CO_2 as conventional combustion processes, the advantages of fuel cell technology lie in the high energy efficiency and low emissions of CO, NO_x, SO_2, hydrocarbons, and of noise.

The electrolysis of water, using electricity from regenerative sources, is a long-term hydrogen supply strategy. Electrolysers are already available in a broad power range, and conversion efficiencies higher than 75% have been achieved. Hydrogen and oxygen are generated in reverse process to the fuel cell reaction, typically using an alkaline electrolyte. The above-mentioned drawbacks for AFC have no effect on electrolysers. On the contrary, non noble metal electrocatalysts can be used.

Finally, hydrogen can be generated through the *gasification* of coal. Again, steam is used to produce a gas containing mainly carbon monoxide and hydrogen. Coal gas used to generate electricity via an SOFC, for example, is converted into water and CO_2. Condensing the water would thus make sequestration of CO_2 more possible compared to the scrubbing of flue gases for conventional power plants.

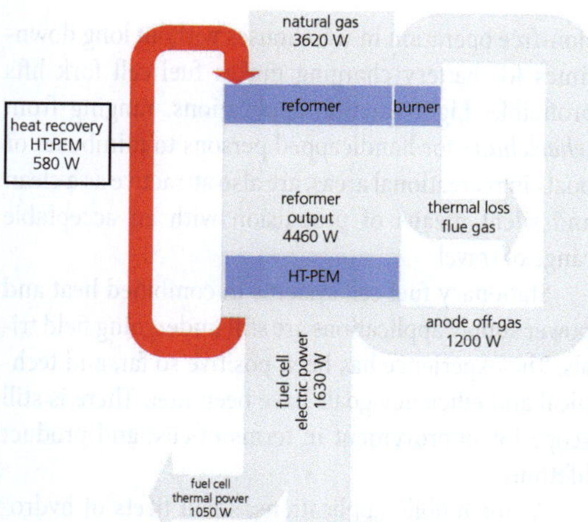

Steam reforming of natural gas: For a fuel cell system with a gas processing unit, natural gas is fed as fuel. In this example, the amount corresponds to a lower heating value of 3620 W. The steam reforming process requires heat, as it is operated at about 750 °C. This heat demand is covered in part by a burner, using natural gas and the anode off-gas of the fuel cell itself. Another part of the heat required to generate steam could come from the cooling circle of the fuel cell, which has a temperature level of 160 °C in the case of a high temperature membrane fuel cell. Thus, hydrogen with a lower heating value of 4460 W can be generated and is fed to the fuel cell. Inside the fuel cell, the hydrogen is converted to electricity (1630 W) and thermal power (1050 W), also delivering an anode off-gas with 1200 W due to the limited fuel utilisation of 80% and the high temperature required for the steam reforming process.

▶ **Transport and storage of hydrogen**. Despite state-of-the-art pipelines and pressure vessels, the transport and storage of gases is more difficult than that of gasoline. Hydrogen gas has an energy density of 3 kWh/m³, natural gas of about 10 kWh/m³, and (liquid) gasoline of 10,000 kWh/m³. The widespread use of hydrogen in future will only be possible if better transport and storage methods are developed. For mobile applications, lightweight pressure vessels made from carbon fibre materials and a gas-proof metallic liner have been developed for pressures up to 700 bar. Nevertheless, these tanks are larger and heavier than a gasoline tank with the same energy content. What is more, compression requires energy, and the energy needed to achieve a pressure of 700 bar amounts to over 10% of the energy content of the stored hydrogen.

Liquid hydrogen has an energy density of 2350 kWh/m³, but the boiling point of liquid hydrogen is extremely low at –253 °C. This requires special cryogenic tanks with extremely good insulating properties. One solution has been to use vacuum super-insulating technology. Nevertheless, it is impossible to prevent roughly 1% of the liquid hydrogen from evaporating every day. Although the hydrogen can be liquefied, this wastes energy because about 30% of the energy content of the stored hydrogen is consumed in the process.

A third possible option are *metal hydrides*. Several metal alloys are capable of forming a reversible bond with hydrogen, allowing canisters with nano-powders of these metals to be used as a storage system. Storage or adsorption takes place at elevated pressures (approximately 10 bar), and desorption can take place at close to ambient pressure. Due to the heat of the reaction, desorption produces a cooling effect which slows down the discharge process. This effect is beneficial for safety reasons but makes the hydride systems

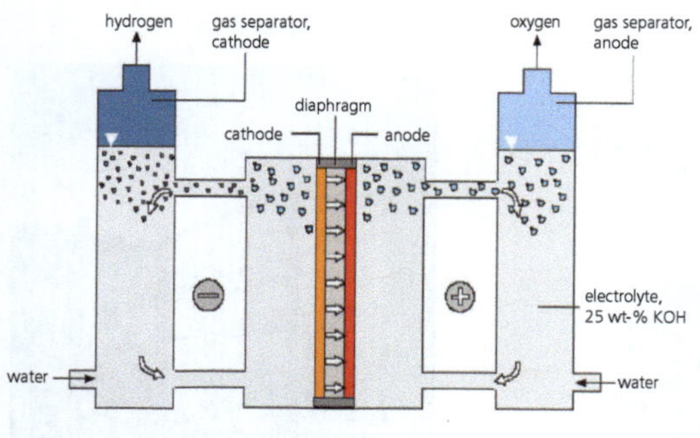

Alkaline electrolysis: In an electrolysis cell, water is split by applying a direct current voltage of more than 1.5 V to produce hydrogen at the cathode (negative pole) and oxygen at the anode (positive pole). The electric circuit is closed by hydroxyl ions migrating in the electrolyte from the cathode to the anode compartment. A separator prevents the product gases from mingling. Source: Fraunhofer INT

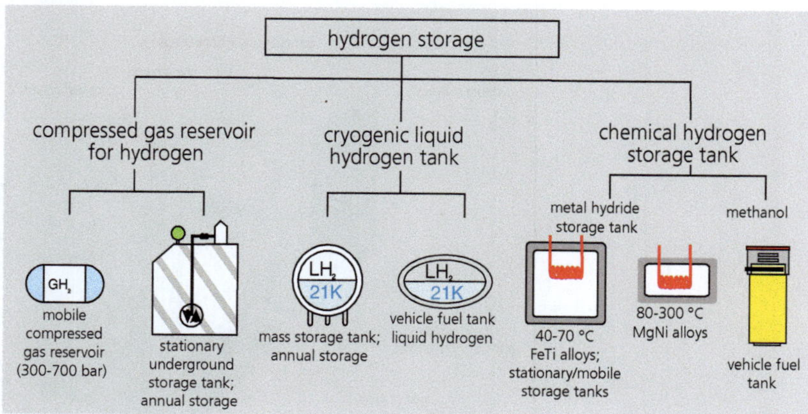

Storage options for hydrogen.

unattractive for electric tractions that require fast discharge properties. The energy density per volume is quite high (1,500-2,000 kWh/m^3), but gravimetric energy density is low due to the high weight of the metal powder.

Finally, hydrogen can be transported via pipelines. Several hundred kilometres of hydrogen pipelines already exist in Europe and the USA, connecting the companies that generate hydrogen with those that consume it. The technical feasibility of feeding hydrogen into the pipeline network as a partial substitute for natural gas has been investigated, and figures of 20% of hydrogen in natural gas have been published as an acceptable level. Seasonal storage in underground caverns is a state-of-the-art technique for natural gas, and could probably be realised for hydrogen as well.

Applications

The first fuel cell applications were built to supply power for space missions, using the hydrogen and oxygen on board the *spacecraft* as propellants for the rockets. In *manned spacecraft*, the fuel cell system also delivered drinking water. Another early application was for the *propulsion* of submarines during dive cruise, enabling greater diving ranges than a power supply using conventional batteries. Large hydride storage tanks for hydrogen and tanks for liquid oxygen were integrated in the submarines to provide a fuel supply.

DMFC systems in the power range of several tens of watts are in use on the military and leisure markets, and several companies manufacture hydrogen-fueled PEMFCs for ground-breaking remote and uninterrupted power supply applications.

Forklift trucks were another early market. Emission-free operation in warehouses without long downtimes for battery charging makes fuel cell fork lifts profitable. Light traction applications, ranging from *wheelchairs* for handicapped persons to minibuses or boats in recreational areas, are also attractive as a clean and silent means of propulsion with an acceptable range of travel.

Stationary fuel cell systems in combined heat and power supply applications are still undergoing field trials. The experience has been positive so far, and technical and efficiency goals have been met. There is still scope for improvement in terms of cost and product lifetime.

As for mobile applications, small fleets of hydrogen and fuel cell powered vehicles – buses and passenger cars – are on the roads. Their suitability for everyday use is being tested in pilot projects at sites where hydrogen fueling stations are available. The creation of a hydrogen infrastructure is seen as one of the next major challenges.

Trends

▶ **Lifetime enhancement.** All types of fuel cells still require significant lifetime enhancement and cost reductions if they are to compete with established technologies. Cost ceilings are the most difficult requirements to meet in automotive propulsion systems, as engines are manufactured in high-volume mass production processes at an extremely low cost. Trends in fuel cell development include reducing the content of expensive materials such as noble metal catalysts, and adapting the cells for mass production.

▶ **Larger-scale stationary applications.** In combined heat and power supply systems, the goal is to de-

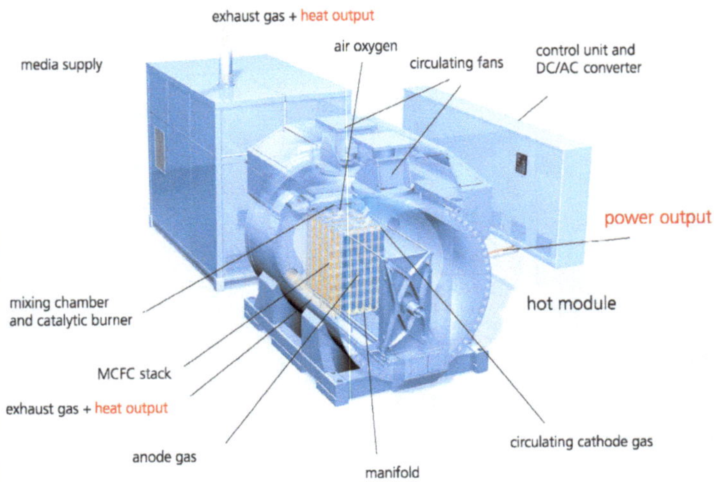

MCFC stationary fuel cell. Left: Schematic configuration. Right: System in a field trail. Source: MTU onsite energy velop systems that will deliver higher power. Flexibility in the type of fuel employed is also being investigated. Viable fuel options (besides natural gas) are biogas, sewage gas, landfill gas and finally gasified coal and *biomass*. SOFC and MCFC power plants consume hydrogen and carbon monoxide as their fuel, and can thus generate CO_2 without dilution by nitrogen as would be the case in a typical combustion process. The off-gas can be used for CO_2 sequestration.

▸ **Small-scale fuel cells.** The major goals for small-scale residential fuel cell systems are to simplify the balance of plant (BOP) and to improve lifetime and reliability. Lifetime goals of 40,000 or even 80,000 hours with the longest possible service intervals are under discussion. As reformate purification for stationary PEMFCs is a complex process, higher operating temperatures would be beneficial. So far, the polymer electrolyte material has limited the temperature to below 100 °C. High-temperature membranes are still being developed for PEMFCs, with various polymer materials now available for operation at 120 °C or even 160 °C. These membranes tolerate higher CO concentrations in the fuel and deliver cooling water at a higher temperature level. The latter is also an attractive option for fuel cells used for electric traction in automotive applications, as it is difficult to get rid of the heat at a temperature level of approximately 60 °C at high ambient temperatures. The gas process can be simplified by omitting the fine purification stage, and its efficiency can be improved by using the heat from the cooling circuit of the HT-PEMFC to vaporise the water for the steam reforming process. It is thus possible to increase the overall electrical efficiency of the fuel cell system by several percentage points.

Various types of hybrid systems for passenger cars are under discussion. One favourable combination is a high-power high-energy lithium battery as a power reserve and for regenerative braking, together with a fuel cell for long-distance driving. These electric drives could make the vision of clean and quiet urban transport come true.

The demand for electrical energy in trucks is steadily increasing, making it attractive to develop batteries with higher energy density and improved auxiliary power units for long-term power supply. Fuel cells are capable of continuously generating electricity even when the engine is off. Delivering power for trucks when stationary, avoiding idling of the engine, significantly reduces emissions.

As in spacecraft, fuel cells can also generate a supply of power and water for aircraft, provide clean power for ships laid up in harbours, and serve numerous other applications.

▸ **Portable applications.** Considerable effort has been undertaken to develop miniature fuel cells for a variety of portable applications such as laptop computers, mobile phones, or on-chip power supplies. The demands in terms of reliability and safety are very high, primarily due to the required passive operation in a consumer environment. Various concepts using either hydrogen or methanol as fuel are currently being pursued, the latter for a DMFC or reforming process. As methanol can be reformed to generate hydrogen at relatively low temperatures of about 300 °C, it can be combined favourably with a PEMFC using the new type of high-temperature electrolytes. Compact systems with a high power density can be produced by the thermal integration of heat-consuming reforming reaction with heat-releasing fuel cell operation.

For consumer applications, however, many detailed problems have to be considered, since operation and environmental conditions can vary widely. Examples include freezing, operational requirements that are completely independent of orientation, possible coverage of air breathing electrodes, and limitations with respect to outer surface temperatures. Nevertheless, small fuel cell systems are an attractive option to batteries as a power supply for electronic devices.

Fuel availability and infrastructure, together with power and energy demand, will ultimately decide which fuel and therefore which fuel cell system is implemented for all these new applications.

Prospects

- Fuel cells are efficient energy conversion devices and can help to save energy in combined heat and power supply systems with natural gas in the near future.
- The widespread use of fuel cells for electric traction depends on the hydrogen supply and infrastructure, and thus on political decisions regarding how to cope with rising oil prices and limited resources. In the future, all hybrid electric systems are likely to be the propulsion method of choice.
- In other applications, such as portable and micro fuel cells, the advantages for the consumer are the main argument for investing R&D efforts in fuel cell development.
- International energy policy will be the primary factor in determining how soon (renewable) electric energy and hydrogen become the main energy vectors.

PROF. DR. ANGELIKA HEINZEL
University of Duisburg-Essen

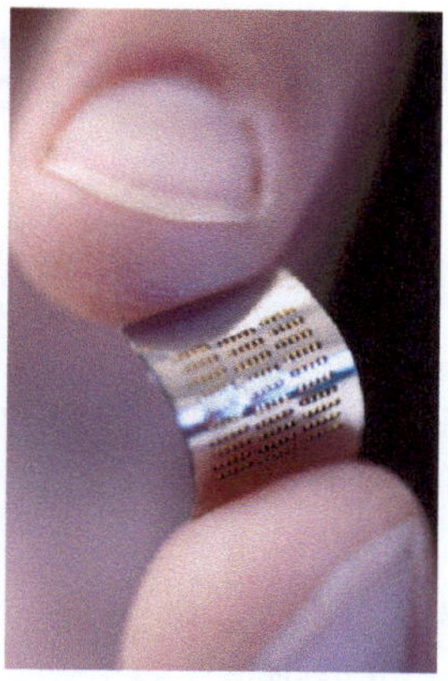

Miniature fuel cell: One of the smallest fuel cells as a prototype for powering a mobile phone.

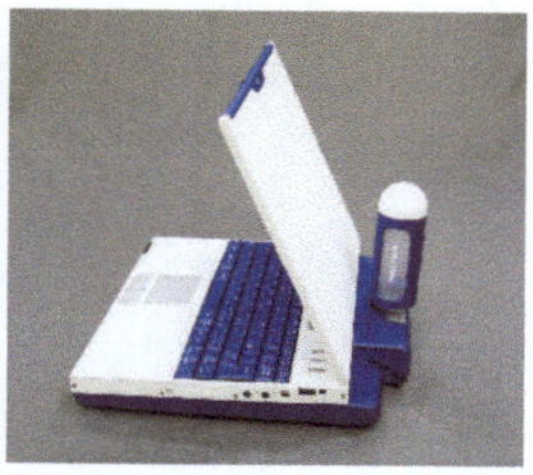

Prototype of a laptop computer with a DMFC as its power supply: The methanol cartridge is applied only during refill processes. Fuel cells are ideal for telecommunications systems, as they permit long, uninterrupted operation of electronic devices. Source: NEC Corporation

Internet

- www.fuelcells.org
- www1.eere.energy.gov/hydrogenandfuelcells
- www.fuelcelltoday.com/reference/faq

Microenergy technology

Related topics

- Solar energy
- Fuel cells and hydrogen technology
- Energy storage
- Microsystems technology

Principles

The term microenergy technology refers to the conversion, electrical conditioning, storage and transmission of energy on a small scale. Its goal is to improve the energy density and thus extend the operating time of battery systems, as this can be a limiting factor for many electronic applications. A fundamental distinction is made between technologies based on the conversion of conventional energy carriers with a relatively high energy density (such as hydrocarbons and alcohols) and those which passively transform energy from their local environment, such as light, heat and motion, into electrical power (and heat). The latter are grouped together under the term "*energy harvesting*", reflecting the passive way in which the energy is transformed in the absence of chemical energy carriers.

In addition to conversion technologies, key importance is also attached to the power management of the entire system. This includes the optimal tuning and adjustment of peripheral components, e.g. for the transport of reactants or in order to avoid parasitic leakage currents through needs-based power distribution. Additional factors include optimising the voltage level of electronic components using high efficient DC/DC converters, real-time *battery* monitoring and energy efficient operation of the entire system. Finally, wireless energy transfer to electron storage devices (inductive methods have already been implemented in items such as the electric toothbrush) offers further possibilities for sophisticated system architectures by means of microenergy technology.

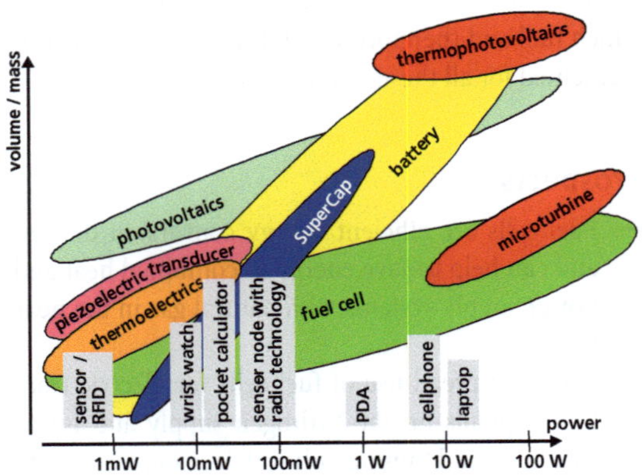

Power consumption of typical consumer applications and the conversion technologies suitable for them: While sensor networks or RFID tags, which have an energy consumption of just a few milliwatts, can ideally be supplied using energy-harvesting technologies (vibration transducers, thermoelectrics, photovoltaics), applications with higher energy consumption have to be supplied using fuel-operated transducers (fuel cells, microturbines, thermophotovoltaics). Fuel cells and microturbines have considerable advantages over batteries in terms of size and volume, particularly for high-power applications or when long operating times are required. Piezoelectric and thermoelectric transducers are limited to the output range below one watt, while photovoltaic systems and even fuel cells can be scaled up to the MW range as modular technologies. Source: Fraunhofer ISE

Conventional, macroscopic energy conversion systems are based on the initial generation of heat, which is first converted into mechanical energy using thermomechanical processes and then into electrical energy using mechanoelectric processes. The conversion efficiency increases at higher temperatures, but intermediate conversion places thermodynamic limits on efficiency. These limitations, however, do not apply to the direct conversion principles of electrochemical charge separation in fuel cells and *photovoltaic* charge separation in *semiconductor* solar cells. Here, conversion efficiencies of over 50% can be achieved, if hydrogen is used as an energy source for fuel cells, or monochromatic radiaton for the operation of photovoltaic cells.

As one moves from the energy conversion using conventional, macroscopic systems to microsystems involving nanoscale structure sizes further physical parameters and boundary effects gain in importance. These will give rise to new functionalities and design concepts:

- The extreme surface-to-volume ratio in microdevices helps maximise the energy conversion rate per volume unit and the energy flow per unit area through the device surface. The importance of phenomena induced by surface area also increases, as exemplified by the improvement of reaction kinetics in catalysis.
- Considerable changes take place in the ratio between the force within a material and its surface forces, such as Van de Waals, adhesion and electrostatic forces. This enables novel concepts to be de-

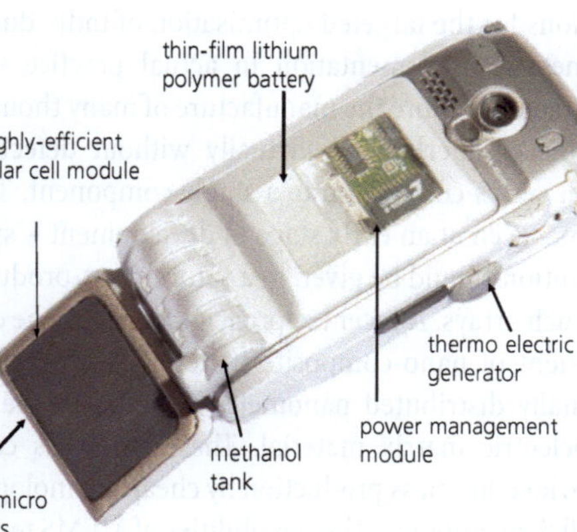

Vision of a mobile phone incorporating elements of micro-energy technology: Planar, air-breathing fuel cells convert stored methanol or ethanol into electrical power and charge a lithium-ion polymer battery via an energy management module. Similarly, a highly efficient Si solar cell module ensures unlimited stand-by operation through the conversion of ambient light. Side-mounted thermoelectrical elements can increase operation time by converting hand heat into electrical current.

veloped, such as highly dynamic systems involving tiny moving masses.

- Modification of surface areas makes it possible to maximise the thermal transfer to the environment. Furthermore, non-linear optical effects and resonance or tunnel effects within a device can be utilised in radiative exchange on a nanostructured surface.
- Non-continuous heat exchange occurs in gases if the typical mean free length of path between collisions of the gas molecules is of comparable order of magnitude to the length scale of the system.
- The radiation pressure of light is irrelevant at macroscopic level. On a micro- or nanoscale, however, piconewton forces can be used in the same way as light from a laser pointer to produce extremely small devices.

Microenergy conversion principles based on silicon technology permit the integration of moving components in order to perform physical and analytical functions in addition to their electrical functions (*microelectromechanical systems, MEMS*). Microfabrication of silicon-based structures is usually carried out by repeating sequences of *photolithography*, etching and deposition to produce the desired configuration of features. Miniaturised elements are also incorporated, such as traces (thin metal wires), vias (interlayer connections), microchannels, membranes, microreservoirs, microvalves, micropumps and sensors with a structural accuracy under 100 nm. These power MEMS can even be used to realise microdrive systems for pulse generation on a micronewton scale in the precise positioning of *satellites*, or *microreaction systems* for the mixing, measuring and processing of very small quantities of chemically active substances.

Applications

▶ **Photovoltaic generators.** Absorbed photons in solar cells, possessing an energy level above the material-specific band gap of the respective semiconductor, generate electron-hole pairs (charge carriers) which can be used in the form of electric power. The efficiency of a single p-n junction made of monocrystalline silicon is theoretically limited to about 28%, commercial systems for roof applications currently reach a module efficiency of 15–20%. As microsystems will not necessarily be exposed to direct solar radiation, the efficiency of the different cell types under low-intensity radiation is crucial. The average desk lighting level for example is 0.5 mW/cm². For simple applications such as calculators, the low-light efficiency of amorphous solar cells at 2–3% is sufficient to produce the few mW necessary. For applications with higher power requirements, such as mobile phones or small sensors, highly efficient silicon solar cells are required, which can achieve an indoor efficiency level of up to 20%. If monocrystalline *wafers* with a film thickness of under 80 μm are used in processing, the solar cell module will be as flexible as paper and easily mouldable into an ergonomic shape. This property

Individual solar cells are usually series-connected to match the charging voltage of a Li-Ion battery. Left: A shingle system can be used to optimise the output as the bus bar for each solar cell lies under that of the adjacent cell. Centre: Depending on geometric specifications, circular solar cells can also be produced which achieve a voltage level suitable for recharging lithium ion batteries, even under low-light conditions. Right: Thanks to their mechanical flexibility and low weight, organic solar cells can be integrated in wearable electronic systems. Source: Fraunhofer ISE

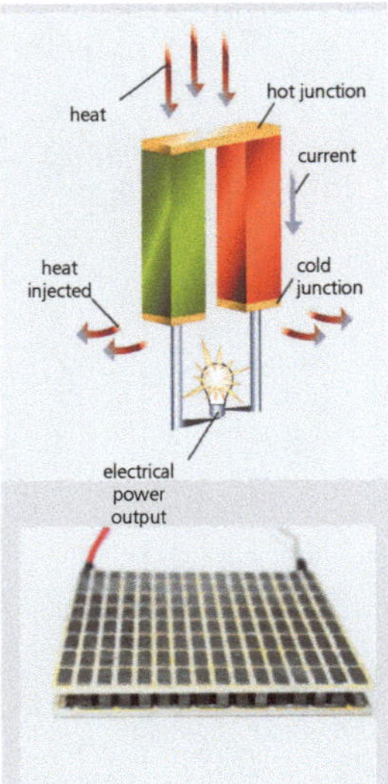

◂ Thermoelectric system. Top: A permanent heat source heats one side of a thermoelectric module, while the other end is being cooled. Through the permanent temperature gradient thermoelectric voltage is generated, which can charge batteries or directly operate an electrical device. The temperature difference may be just a few Kelvin, as in the example of a wrist watch, or several hundred Kelvin as in the thermoelectric use of hot exhaust emissions. As only a few millivolt of thermoelectric voltage is generated per branch, several hundred elements have to be electrically connected for the module to have a useful cut-off voltage. Bottom: A realised thermoelectric module. Source: Fraunhofer IPM

makes it possible to create applications such as "smart clothing" or car roofs with integrated solar modules.

▸ **Thermoelectric generators.** If a sustained thermal gradient is present in a system, thermoelectric elements can produce a thermoelectric voltage. The maximum attainable energy yield is physically limited by the applied temperature gradient. Ideally, in order to maintain a constant temperature gradient in the application, the electrical conductivity of the thermoelectric modules should be very high and their thermal conductivity as low as possible. These two parameters determine the thermoelectric efficiency of the device. High thermoelectric voltage of up to several hundred microvolt per Kelvin can be attained using appropriately doped semiconductor combinations, in which charge carriers are lifted into the conduction band in the warm region and diffuse to the cold region to balance the potential difference which arises from the gradient. Such a thermoelectric component consists of interconnected n- and p-type semiconductors and the usable voltage can be tapped from the ends, which are kept at different temperatures. With typical "*Seebeck coefficients*" of 400 μV per branch and 10 K of temperature difference the voltage rating is 4 mV, which means that several hundred elements are needed to generate sufficient usable voltages to charge a battery. As thermal conduction in semiconductors primarily takes place via crystal lattice vibrations (phonon vibrations), research into the suppression of these vibrations focuses to a large extent on the development of nanomaterials such as semiconductor-hetero-structures, on nanorods with a small rod diameter and also on crystalline superlattices, i. e. stacks of epitaxially grown, individual layers just a few nanometres thick. The basic principles rooted in scattering phonons on each of the many boundary surfaces, which reduces thermal conductivity. The highest levels of efficiency have been reached by applying this approach, and the superlattices are an excellent example of how the macroscopic volume properties of materials can be altered by means of nano-technological material modifications for the targeted optimisation of individual parameters. Implementation in actual practice would ultimately require the manufacture of many thousands of these superlattices, virtually without defect, and their series-connection to a single component. This is why – even at an early stage of development – special attention should be given to a future mass-production of such arrays. A recent approach focuses on the development of nano-composite materials with 3-dimensionally distributed nanometre particles in the thermoelectric matrix material. These materials can be provided for mass production by cheap technologies in sufficient amounts. The possibilities of MEMS technology open up significant development potential here, not only in the generation of thermoelectric power, but also in the efficient cooling of micro components using the reversed *Peltier effect*.

▸ **Micro fuel cells.** *Fuel cells* convert the chemically bonded energy of fuel directly and efficiently into electrical energy in an electrochemical process. The oxidisation of the fuel's hydrogen and the reduction of atmospheric oxygen occur in two partial reactions, spatially separated from one another, which are connected to the electrolytes via an electrically isolating ion conductor. The polymer electrolyte membrane (PEM) fuel cell is particularly suitable for use in microsystems thanks to its low operating temperature of 60–70 °C and its relatively simple design. Apart from hydrogen, it is also possible to use liquid alcohols like methanol or ethanol directly as fuel, but this would require a more complex catalyst mixture and system design. In micro fuel cells power densities of more than 250 mW/cm² are achieved with hydrogen and, owing to the reduced reaction kinetics, 50 mW/cm² in methanol-operated systems.

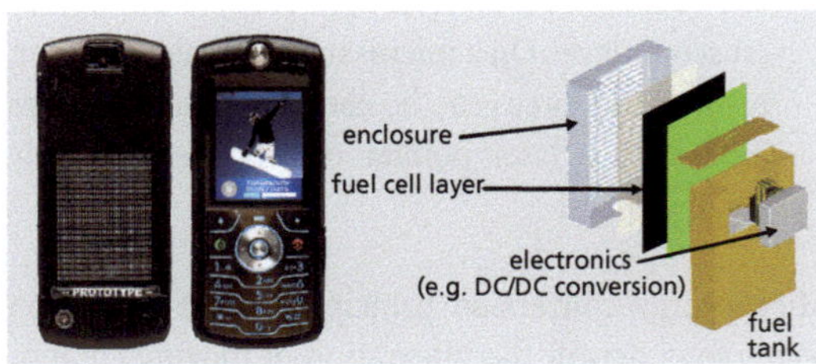

▴ Fully device-integrated fuel cell system in a cell-phone which exceeds the energy density of the conventional battery (left). The fuel cell system is based on series-connected, planar and air-breathing single cells and hydrogen as a fuel (right). The hydrogen is stored in a sophisticated, high storage capacity metal-hydride container under slight overpressure without any subsequent complex pressure reduction and security measures, respectively. Also, such passive air-breathing fuel cell systems can be highly attractive charging units for Li-Ion batteries in hybrid systems. Source: Angstrom Power

In addition to minimising electrochemical losses, the development objective for micro fuel cell systems is to increase the reliability and lifespan of the cells themselves as well as of the peripheral microelements in the system, such as fans (cooling, air supply for the electrochemical reactions), pumps (transport of liquids), valves (dosing media) and *sensors* (temperature, methanol concentration), and of the highly-efficient power electronics which control the entire system. Furthermore, a great deal of research is being conducted into system designs aimed at replacing these peripheral components with a highly passive system architecture. This would mean that some of these functionalities would be achieved through surface modifications of the individual fuel cell components, but would also remove some of the active control variables required for reliable operation at extreme temperatures or under dynamic loads. An important parameter for fuel cell operation is the relative humidity of hydrogen and air. On the one hand, the proton conductivity of the membrane increases with its water content, which is in balance with the gas humidity, while, on the other hand, condensed water leads to higher mass transport losses in porous gas diffusion media, or to the flooding of cell areas and their subsequent deactivation.

Trends

▶ **Vibration transducers.** Kinetic energy in the form of vibrations, random displacements or forces can be converted into electric power by piezoelectric, electromagnetic and electrostatic transducers. Suitable vibrations can be found in numerous applications such as common household appliances (refrigerators, washing machines), industrial equipment, manufacturing plants, road vehicles and structures such as buildings and bridges. Human movements are characterised by low-frequency high-amplitude displacements.

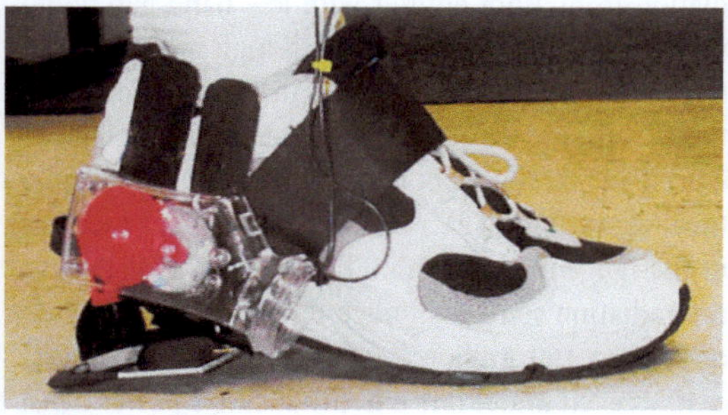

Studies have shown that a person weighing 68 kg produces around 67 W of energy in the heel of their shoe when walking. Although generating significant energy from the heel would interfere with the normal gait, it is clear that this presents a potential energy-harvesting opportunity, probably the best opportunity to parasitically extract significant energy from human activity. The picture shows a simple electromagnetic generator as a strap-on overshoe which produces 250 mW during a standard walk – enough to power a radio. Source: Joseph Paradiso, MIT Media Lab

Kinetic energy harvesting requires a transduction mechanism to generate electrical energy from motion. The generator requires a mechanical system that couples environmental displacements to the transduction mechanism. The design of the mechanical system should maximise coupling between the kinetic energy source and the transduction mechanism, whereby the characteristics of motion in the respective environment are decisive. Vibration energy is best suited to inertial generators, with the mechanical component attached to an inertial frame, which acts as the fixed reference. The inertial frame transmits the vibrations to a suspended inertial mass, producing a relative displacement between them. Systems of this type possess a resonant frequency which can be designed to match the characteristic frequency of the application environment. This approach magnifies the environmental vibration amplitude by the quality factor of the resonant system.

Piezoelectric generators employ active materials that generate a charge when mechanically stressed. Electromagnetic generators employ electromagnetic induction arising from the relative motion between a magnetic flux gradient and a conductor. *Electrostatic generators* utilise the relative movement between electrically isolated charged capacitor plates to generate energy. The work done against the electrostatic force between the plates provides the harvested energy.

Each electromagnetic carrier has specific damping coefficients which must be observed when designing the system. Typical excitation ranges in dedicated applications, such as household appliances, lie between 60–200 Hz, while the movements of people or moving buoys in water are examples of even lower-frequency excitations in the 1 Hz range. Maximum energy can be extracted when the excitation frequency matches the natural frequency of the system.

The following applies as a first approximation for vibration transducers:

- The electrical power is proportional to the oscillating mass of the system.
- The electrical power is proportional to the square of the accelerating amplitude.
- The electrical power is inversely proportional to the natural frequency of the generator.

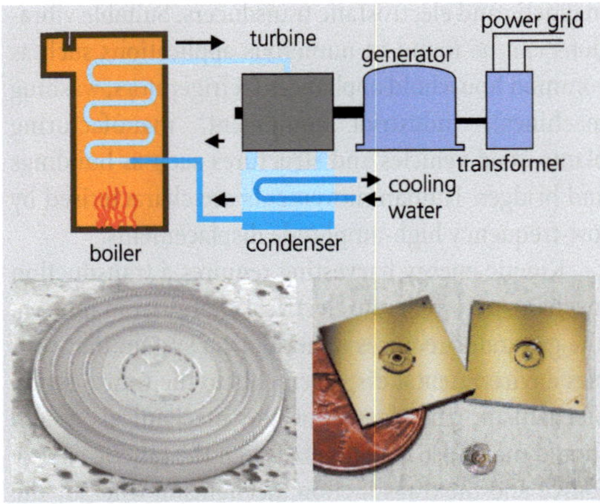

Microturbines. Top: Conventional steam turbines power an electromagnetic generator using thermomechanical processes. The heat sources used are typically fossil or nuclear energy carriers. The electrical output from a power plant can reach the GW range. Bottom left: Micro steam turbines are based on the same principle, but the mechanical requirements are disproportionately higher than in macroscopic systems. As the rotor blades have to attain a rotational speed of up to 1 million rpm, only materials with absolutely no defects can be used (right). In systems measuring just a few millimetre an electrical output of several tens of watts can be generated using high-density energy sources, such as propane or diesel or slightly lower power levels by scavenging waste heat from engines or industrial processes. As a result, microturbines are the transducer systems with the highest power density. Source: C. Lee, L. Fréchette, University of Sherbrooke

On this basis the transducer should be designed in such a way that the resonance frequency is at the lowest available fundamental frequency of the excitatory frequency range. Furthermore, the mass of the mechanical structure should be maximised within the given size constraints in order to maximise the electrical power output. Piezoelectric transducers have been shown to generate 200 µW of electrical power for a volume of around 1 cm^3 and an excitation frequency of 120 Hz (at 2.25m/s^2 acceleration). It should be mentioned that an alternating voltage is generated here, which first needs to be rectified in order to charge a battery. In general, the application vibration spectra should be carefully studied before designing the generator in order to correctly identify the frequency of operation given the design constraints on generator size and maximum permissible mass displacement.

Further fields of research in the low-energy area are the conversion of acoustic noise or the use of seismic movements. The electrical power achievable in these fields is in the range of just a few picowatts, this might be sufficient to supply very simple sensor networks.

► **Microturbines.** A microturbine is a conventional gas turbine on a kW scale. These systems have a volume of less than one cubic centimetre and a theoretically expected output of up to 10 W/cm^3, or a thermal conversion efficiency of 5–20%. Microturbines are the microtransducers with the highest theoretical power density, but owing to the extreme requirements that have to be fulfilled; volume production remains a long way ahead.

The major constraint of a microturbine involves the need to achieve rotational speeds of up to 1 million revolutions per minute with a rotor diameter of less than 20 mm. The mechanical stress under such extreme requirements can only be withstood by absolutely defect-free materials.

Temperature, too, poses a significant problem, since input temperature must reach at least 1500 °C in order to achieve acceptable efficiency. In large turbines the rotor blades are cooled by internal cooling channels, but this is not possible for rotor blades less than a few mm in size. In order to withstand high temperatures over long periods, the rotor blades are made of Si3N4-TiN ceramic composite. The temperature cycles from startup to stable operation also put considerable strain on the bearings, and so self-stabilising, aerodynamic gas bearings are used.

Microturbines only operate reliably if the surface forces are mastered by avoiding contact between moving parts during operation, or if high-precision plane-parallel silicon surfaces a few nm apart 'stick together' through van der Waals forces.

► **Microthermophotovoltaics.** Thermophotovoltaic energy conversion involves coupling the black body radiation from a hot emitter with low band gap photovoltaic cells, in other words, with high infrared sensitivity. The process is the same as in solar photovoltaics, the only difference being that the sun's surface temperature of 5,500 K cannot be replicated in a thermophotovoltaic system (TPV) and therefore colder radiation in the range of about 1,200 K is used. The distance of the radiation source is only a few centimetre (instead of approx. 150 million km), however, and in special cases is even under one µm, producing a power density of over 1 W/cm^2 of photovoltaically active surface. The challenge in realising a TPV system lies in matching the emissions spectrum of a burner to the spectral sensitivity of the photovoltaic cell. This is achieved by means of intermediary infrared optics in the form of wavelength-selective emitters or optical filters, which ensure that only photons with an energy level above the semiconductor-specific band edge reach the photovoltaic cell. A system with a radial configuration

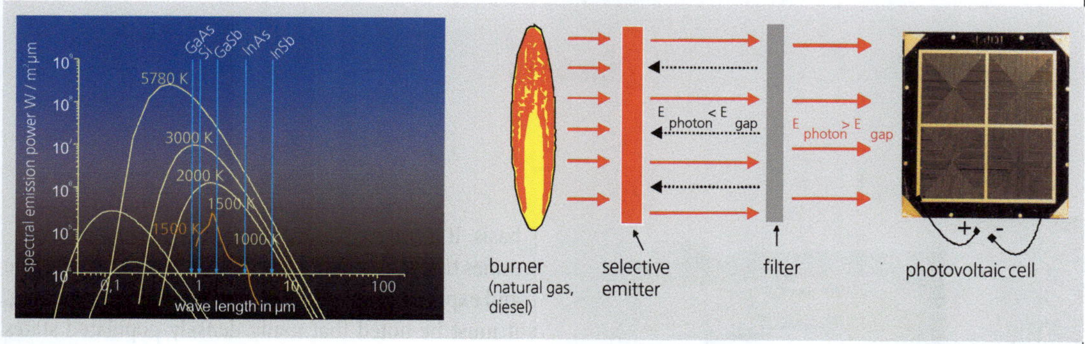

Thermophotovoltaic conversion. Left: Spectral emission ranges of black body radiation at different temperatures. Most photovoltaic materials are well matched to the spectrum of the sun at 5780 K. For radiators emitting lower temperatures, under 1500 K, low band gap semiconductors, such as GaSb or InSb, of high infrared sensitivity must be used for electrical conversion. The emitted spectrum can, however, be strongly suppressed (red spectrum) using selective emitters such as nanostructured tungsten in the long-wave part of the spectrum, outside the semiconductor-specific band gap. To increase the efficiency of the system further, spectral filters can be used in addition or as an alternative to selective emitters, in order to ensure that only those photons that can contribute to the generation of an electron-hole pair in the cell reach the photovoltaic cell (right). Thermophotovoltaic transducers can achieve power densities of over 1 W/cm² of photovoltaically active surface. The electrical efficiency based on to the energy density of the fuel used ideally is in the order of 10%. Source: Fraunhofer ISE

therefore consists of a centrally placed combustion chamber into which the fuel gas compound is injected, a spectral selective emitter configuration surrounding the chamber, a cooling system or transparent thermal insulation, appropriate reflective filters, and finally the rear-cooled photovoltaic cells. Advantages of this configuration are its variability with regard to the energy sources that can be used, as well as the absence of noise and moving parts – which would indicate a robust system with a long lifespan.

There are, however, some fundamental problems which still need to be solved: even with an emissions spectrum of 1,200 K, adjustment to the spectral sensitivity of low band gap photovoltaic cells is still very poor. As a result, surface-modified selective emitters, such as structured tungsten emitters, or tungsten photonic crystals, are needed to suppress the proportion of emitted infrared outside of the band gap. Current TPV systems therefore have a low efficiency of just a few percent, which can be attributed to the complexity of the system, the high temperature gradient and unstable selective emitters.

Micro-TPVs are cutting a different path in which the space between the selective emitter and the photovoltaic cell is held in a vacuum under a wavelength of around 1 µm by means of MEMS manufacturing methods in order to increase the transfer of radiative heat in the cell by a factor of 10–100. It is interesting to note that a constant temperature gradient of around 1,000 K has to be maintained over a distance of less than 1 µm during operation, but so far this has not been achieved.

Prospects

- Microenergy conversion technologies have a market potential given the inadequate energy densities of batteries for many electronic applications. Electronic systems which harvest energy from their local environment are about to be developed. They will be supplied with energy for their entire lifetime making it possible to provide real 'mount and forget' solutions. This is an advantage for sensor networks with thousands of communicating sensors.
- Amongst the technologies mentioned, device-integrated solar cells and micro fuel cell systems have the highest marketability, because they are being developed in parallel for large-scale use and because they feature unique characteristics which are highly attractive for many applications.
- Thermoelectric applications still need to make significant progress in terms of system costs if they are to achieve market introduction on a wide scale. Microturbines and microthermophotovoltaics are still in the research stage and so it is not yet possible to assess their long-term marketability.

DR. CHRISTOPHER HEBLING
Fraunhofer Institute for Solar Energy Systems ISE, Freiburg

Internet

- www.powermems.org
- www.iop.org/EJ/journal/JMM
- www.fraunhofer.ru/pics/FIT_MET_Portrait_06_e%5B1%5D.pdf

9

ENVIRONMENT AND NATURE

Climate change, the increasing scarcity of raw materials and global networking are present-day megatrends, calling for a change in former perceptions which saw commercial success and environmental protection as a contradiction. The principle of sustainability is now widely accepted, giving equal consideration to social, economic and ecological criteria. Ecological responsibility, but also economical rationality, indicate the need to improve efficiency worldwide in the use of energy and raw materials.

In contrast to the 1970s and 1980s when the emphasis was on downstream environmental engineering – such as filter systems to keep air and water clean – "life cycle" is now the dominant theme. The entire life-cycle of a product, from its manufacture, to its use, through to its reuse or recycling, are taken into account as the basis for further development of the product. With this approach, natural resources are to be used efficiently, and harmful effects on the environment minimised, throughout a product's entire lifecycle.

The expected demand-driven global intensification of environmental protection efforts, are likely to cause an increasing internationalisation of markets, which until now have largely operated on a national basis. It is indeed the highly developed national economies that must answer the call to develop transferable and exportable technologies and products. In so doing, it must be noted that many densely populated states such as India and China still have an enormous need to catch up on relevant technical implementation. Downstream end-of-pipe technologies still make up a large part of the industry in order to fulfil minimum standards in environmental protection. The future, however, clearly resides in integrated technologies, which can transform waste products into reusable materials. Zero emission technologies are ultimately expected to make waste-free production a possibility.

Climate change presents a major global challenge. It is not only a matter of protecting and preserving ecosystems, but also of protecting people from injury and illness. High value must be placed on preserving an environment in which people not only survive, but also feel comfortable – where elements like the soil and landscape, air and climate, as well as water, are treated as vital resources.

As the principle of sustainability is now well rooted in industry, politics and society, the engineering of environmental protection can no longer be seen as a collection of specific technologies; designers today can hardly afford to ignore the need to develop their products on sustainable principles, as "Green IT" has more recently demonstrated. Environmental protection technologies can be defined as follows:

- Traditional end-of-pipe technologies, e.g. air filters, sewage or waste recovery units
- Environmental monitoring technologies, e.g. satellite observation, measurement of fine particulates in inner cities
- Concepts and technologies for reuse or recycling, e.g. new joining processes that enable a product to be dismantled into individual parts at a later stage
- Concepts and technologies for saving resources, or to conserve both materials and energy when producing and operating products, e.g. minimisation of friction in movable engine parts to save fuel

As far as climate change is concerned, it remains unclear under the present circumstances of energy use, whether all technical measures for increasing efficiency will be enough to sufficiently reduce CO_2 concentration. This has prompted discussions on geo-engineering measures, including such exotic alternatives as fertilising the sea with iron to further plankton growth, or spraying aerosols into the atmosphere to damp the harmful effects of sunlight. A measure currently being pursued is to separate the CO_2 directly at power stations and to store it in underground reservoirs.

► **The topics.** *Environmental monitoring* is enabling scientists to gradually understand the earth as a system, comprising a multitude of interconnected components. Satellite images enable us, for example, to witness the short-, medium- and long-term changes human interventions have on our living environment, ranging from an oil slick spreading across the sea to observing melting ice at the poles.

Soil protection is in the first instance regulated indirectly by provisions on air pollution, on waste removal and on the use of fertilisers and pesticides in agriculture. Where soil pollution exists, environmental biotechnology processes can be employed to purify it. The soil beneath former industrial and military sites, in particular, is often very heavily polluted.

Good air quality in large parts of the industrial world can be attributed to the development of production processes which emit diminishing amounts of pollutants. Nevertheless, provisions to separate pollutants often have to be installed downstream of certain processes, as is the case for waste incineration plants.

The concept of "*product life cycles*" significantly contributes to sustainability. This means avoiding waste and regarding every product used as a source of valuable recyclable material. Decisions are made at the design stage about a product's use or processing after its service life – whereby the "thermal" option, or incineration, is seen only as the solution of last resort. While separate collection of individual materials, such as paper and glass, is now widely established, the dismantling and reuse or recycling of complex product components such as cars still poses great challenges.

Water purification technologies ensure the steady supply of drinking water and water treated for use by industry. At the same time, existing infrastructure networks present new challenges. The frequently very old and centrally organised sewage systems no longer meet present-day requirements. While water consumption in private households is falling, large amounts of water from heavy rainstorms must be dealt with. New, decentralised utilisation concepts integrate rainwater as a resource for industrial water.

Agriculture is no longer the labour-intensive sector that it was in the middle of the 20th century. With the aid of modern machinery, an individual farmer can now maintain large stocks of animals or work large expanses of land. Information and communication technology also helps the farmer to control processes, which may determine the individual feed composition for each individual cow, based on milk values, or the amount of fertiliser required on each individual square metre of the field, based on prior measurement of vegetation.

Measures to reduce greenhouse gases can be divided into prevention, reduction, and also storage. The last approach involves carbon dioxide capture. CO_2 is separated before or after the incineration process in power stations and pumped into underground reservoirs. ■

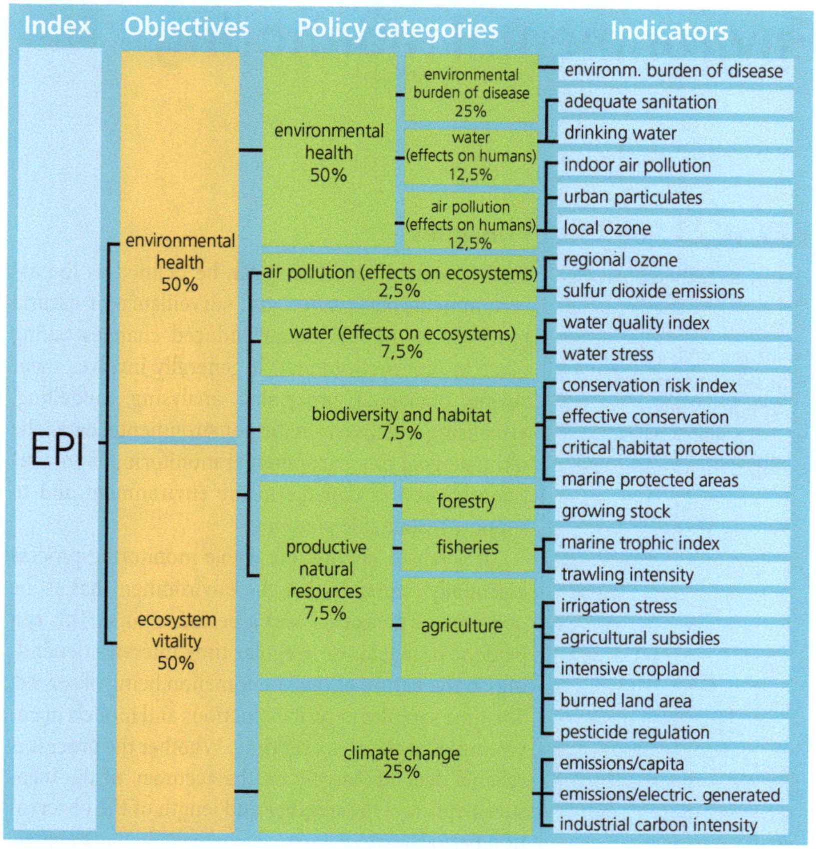

Environmental Performance Index (EPI) 2008: The EPI was formally released in Davos, Switzerland, at the annual meeting of the World Economic Forum 2008. The value focuses on two overarching objectives, reducing environmental stresses on human health and promoting ecosystem vitality and sound natural resource management. These broad goals reflect the policy priorities of environmental authorities around the world as well as the environmental dimension of the Millennium Development Goals. Success in meeting these objectives is gauged using 25 indicators of on-the-ground results tracked in six well-established policy categories. The 2008 EPI deploys a proximity-to-target methodology that quantitatively measures country-scale performance on a core set of environmental policy goals for which every government can be held accountable. A ranking of 149 states is available. Source: Yale University

Environmental monitoring

Related topics

- Optical technologies
- Laser
- Sensor systems
- Measuring techniques
- Image evaluation and interpretation
- Precaution against disasters

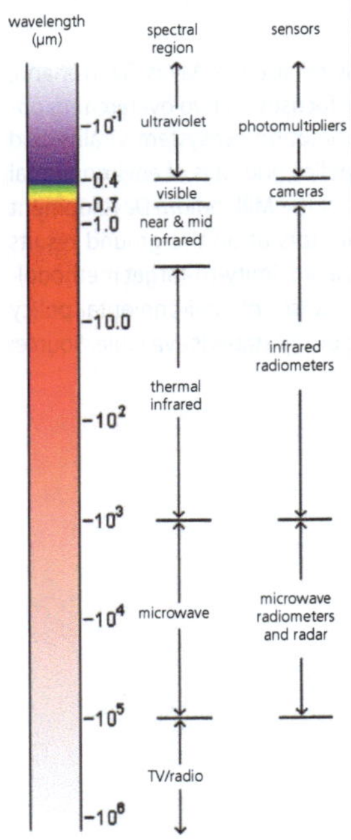

Electromagnetic spectrum: Each sensor responds to a small part of the spectrum only, so multiple sensors are often fitted to a remote sensing scanner. Source: NASA

Principles

Environmental monitoring can be defined as the ascertainment, observation and surveillance of natural processes or of the human-induced changes taking place in our environment. It generally involves measuring, managing, processing, analysing, modelling, visualising and disseminating environmental data. The ultimate goal of environmental monitoring is to consciously prevent damage to the environment and to support sustainable planning.

It is fair to say that the whole monitoring process essentially demands that the environment has to be observed or measured over a period of time. This can be done in regular or irregular time intervals depending on the nature of the phenomenon being observed. The time samples permit deductions and models of environmental changes over time. Whether the process is reliable depends largely on the accuracy of the measurements and the number and length of the observation periods.

The methods and processing techniques used for making measurements are by and large standardised. The recorded data can be compared directly or indirectly and then integrated. Direct methods allow specific measurement of environmental parameters, such as temperature or precipitation. Indirect methods, on the other hand, do not deliver immediate results for a specific parameter or set of parameters. The information has to be derived from additional analysis.

▸ **Optical methods.** Nowadays, there is a distinct preference for indirect methods. One such method, *spectrometry*, identifies substances by their electromagnetic radiation such as heat or light. The energy stimulates molecular and atomic electrons and is partially absorbed. Each chemical element has a unique number and formation of electrons, hence each element will absorb a different amount of radiated energy. Several types of spectrometer have been developed. They can be used for identifying the presence of contaminants such as heavy metals in soil or water samples (environmental analytics).The *atomic absorption spectrometer (AAS)*, for example, measures the wavelengths absorbed by the samples being analysed. The flame emission spectrometer, on the other hand, measures the radiation emitted when a substance resumes its normal energy state after having been artificially stimulated. Specific data plus the concentration of the absorbed or emitted radiation provide information about composition and quantity of the elements contained in the analysed substances.

▸ **Remote sensing.** Remote sensing is the science and technology of acquiring information about remote objects. It basically works in the following manner: electromagnetic radiation is reflected, absorbed, transmitted or emitted from a sample. The electromagnetic radiation can come from natural sources, such as the sun, or from the sample itself. It can even be created artificially, so here again, one differentiates between passive and active remote sensing techniques.

The fact that elements have a characteristic interaction with electromagnetic radiation is used to build remote sensors. Every *sensor* (such as a *CCD* chip in a *digital camera*) is made using one or more chemical elements that are sensitive to a specific range of electromagnetic wavelengths. The intensity of the radiation is then recorded by a calibrated sensor or set of sensors. The registered electromagnetic radiation also depends on the material composition of the objects from which it is reflected or emitted. Remote sensing can be used to identify and analyse forested areas or bodies of water. Many environmental programmes, such as the European Union Global Environmental Monitoring for Security (GMES) programme, make use of this technique for a number of reasons:

- Large areas can be monitored simultaneously.
- Sensors are calibrated to provide standardised comparisons.
- Repeat measurements in brief intervals are possible.
- The data are recorded in digital format.
- The acquired information can be used in many different applications.
- Data acquisition is by and large automated.

▸ **Resolution.** The quality and level of detail of the results and the possibility for exact differentiation and delineation of the monitored phenomena depend largely on the accuracy with which the original environmental data are acquired. For optical sensors, for example, this accuracy hinges on the resolution of the instrument used. There are four different types of resolution:

Platform	LANDSAT4, 5 (7)	TERRA (EOS-1)		ENVISAT
sensor system	TM (thematic mapper)	ASTER (Advanced Spaceborne Thermal Emission and Reflection Radiometer)	MODIS (Moderate Resolution Imaging Spectroradiometer)	SCIAMACHY (Scanning Imaging Absorption Spectrometer for Atmospheric Cartography)
spatial ground resolution	30 m (band 6:120 m)	15 m 30 m 90 m	band blue–green: 250 m band red–infrared: 500 m band: 1000 m	limb vertical 3 x 132 km, nadir horiz. 32 x 215 km
spectral channels	1: blue–green 2: green 3: red 4: near infrared 5: infrared 6: thermal infrared	visible–near IR infrared thermal infrared	red–near infrared visible–infrared visible–near IR thermal infrared	1-2: ultraviolet–blue 3-7: visible–near infrared 8-19: near IR–infrared
characteristics and field of use	series of earth-observing satellites – unique resource for climate change research and multiple applications	serves as a 'zoom' lens for the other Terra instrument; used for change detection and land surface studies	global measurements of large-scale atmospherical dynamics, processes occurring in the oceans, on land, and in the lower atmosphere	global monitoring of trace gases in the atmosphere by image spectrometers

Basic features of common remote sensing systems used in environmental monitoring.

- Spectral: *CCD* cameras use an array of sensors that register radiation at different wavelengths. Only a limited number of sensor types can be integrated on one chip owing to the finite space on a CCD chip. In other words, radiation can only be recorded in a number of discrete wavelength intervals, while the recording of a continuous spectrum is not possible.
- Spatial: The smaller the sensor elements, the more detail can be recorded. However, the elements have to be sufficiently large to accumulate enough electromagnetic energy to provide a satisfactory signal-to-noise ratio (SNR). Highest spatial resolution can therefore be delivered by panchromatic sensors, which record electromagnetic radiation over a single band covering a wide range of wavelengths, usually from the blue to the near infrared domain. So high spatial resolution is provided at the expense of spectral resolution and vice versa.
- Radiometric: This is the maximum number of levels that can be differentiated in a signal picked up by a sensor. The human eye is capable of differentiating between 60 levels of grey or different colours. Current remote sensors produce signals with up to 8192 levels of grey. The data are mostly recorded digitally, so the radiometric resolution is usually measured in bits (8192 levels = 13 bits).
- Temporal: This parameter constitutes the minimum time between two recordings of the same area. Monitoring dynamic events requires sensors that have a high temporal resolution. Weather observation and forecasting, for example, are done by geostationary satellites. These satellites are in orbits over the equator and move at the same speed

The shrinking of the Aral Sea (located between Kazakhstan and Uzbekistan in Central Asia): Shot of 3 different dates 1977, 1989, 2006. Once the fourth largest lake in the world, the Aral Sea is now less than half of its original size. Source: USGS

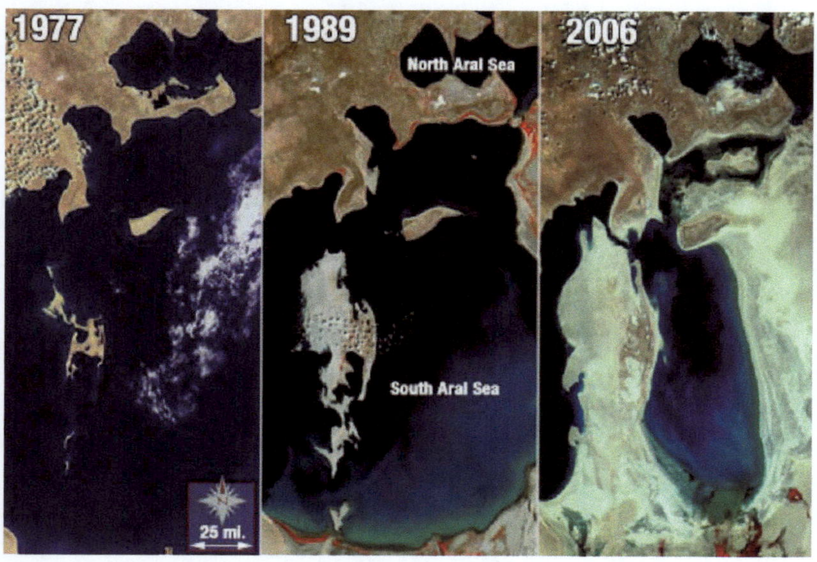

June 19, 2005

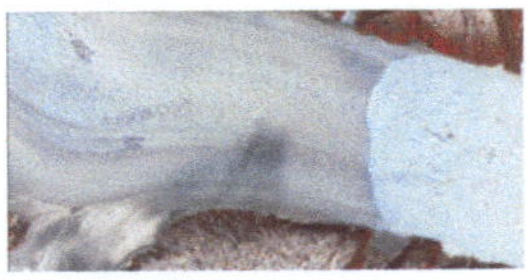
June 07, 2003

May 21, 2001

Retreat of the Helheim Glacier (East Greenland) 2001–2005: Infrared images show bare ground in brown or tan colour, vegetation is red. The glacier is on the left side of the images, the icebergs in the fjord are on the right. Source: NASA

as Earth, so they are always focused on the same region. Because the orbits are 36,000 km from Earth, the spatial resolution has to be very coarse for enough energy to reach the sensors. Therefore, temporal resolution comes at the expense of spatial resolution and vice versa.

Applications

Remote sensing high-resolution imagery from satellite sensors and aerial photography has become an important tool in environmental impact studies. New and expanded opportunities for data integration, analysis, modelling and cartographic representation have emerged when it is combined with *geographic information systems* (GIS). As the world population grows, natural resources are being increasingly exploited and countries, in turn, are boosting their economies, which is gradually impacting the environment. So governments and the general public are relying more and more on information about Earth and its resources. The following examples illustrate the different areas of application for remote sensing.

▶ **Landscape change.** Even relatively simple methods can be used to demonstrate substantial changes in the environment. A time series of colour composite images from the US Landsat satellite revealed the alarming shrinking of the Aral Sea between Kazakhstan and Uzbekistan. The spatial resolution of the satellite data had to be sufficient to observe the phenomenon in order to analyse it. The Landsat Thematic Mapper (TM) sensor provides multispectral images with a spatial resolution of 30 m which allows accurate delineation of the changes in the shoreline.

▶ **Climate change.** *Satellite* images with moderate and high resolutions (30 m–0.6 m) have facilitated scientific efforts tracking environmental change. The extent of the changes observed has prompted concern about the rise in the average surface temperature of the planet and the possible effects of this global warming on the earth's ecosystems. Satellite images from the Advanced Spaceborne Reflection Radiometer (ASTER) reveal the dramatic retreat the eastern Greenland's Helheim glacier from 2001 to 2005. The glacier's margin, where the icebergs "calve", has retreated inland by about 7.5 km. The glacier has become up to 40 m thinner and is flowing faster. Overall, the margins of the entire Greenland Ice Sheet have been retreating over the last decade. An imagery system with multispectral bands providing spatial resolutions of 30 m and higher are required to detect these changes.

▶ **Pollution.** Remote sensing also allows the synoptic observation and determination of air and water pollution. The European Environmental Satellite ENVISAT has an instrument on board called SCIAMACHY (SCanning Imaging Absorption spectroMeter for Atmospheric Cartography), which detects the spread of pollutants in the atmosphere. The satellite employs several hundred different highly-sensitive sensors to record the sun's radiation as it is reflected from earth's surface and from air particles and transmitted through the atmosphere. The sensors can measure a great range of wavelengths (UV – thermal infrared), and are thus able to monitor a host of chemical pollutants, plus ozone and greenhouse gases.

Observing earth's atmosphere using different line of sight geometries is possible by measuring different vertical profiles of the atmosphere and alternating between limb and nadir scans. Large-scale tropospheric profiles can be created by combining slope profiles with a low spatial resolution to the scans of vertical columns with a high spatial resolution (up to $30 \times 60\ km^2$, depending on wavelength and solar elevation).

▶ **Meteorology.** One of the first tasks of remote sensing satellites orbiting Earth has been to monitor and forecast the weather. The TIROS-1 satellite, launched in 1960, was the first. Today, meteorological satellites see more than just cloud distribution. Their

SCIAMACHY sensor, nadir and limb scanning: Nadir measurements focus on the atmospheric column directly beneath the satellite (horizontal track size 960 km, maximum resolution $6 \times 15\ km^2$). Limb measurements cover the edges of the atmosphere in 3-km steps with tangent height range of 0–100 km (horizontal track size 960 km, geometrical vertical resolution approx. 2.6 km). Source: Institute of Environmental Physics (IUP), University of Bremen

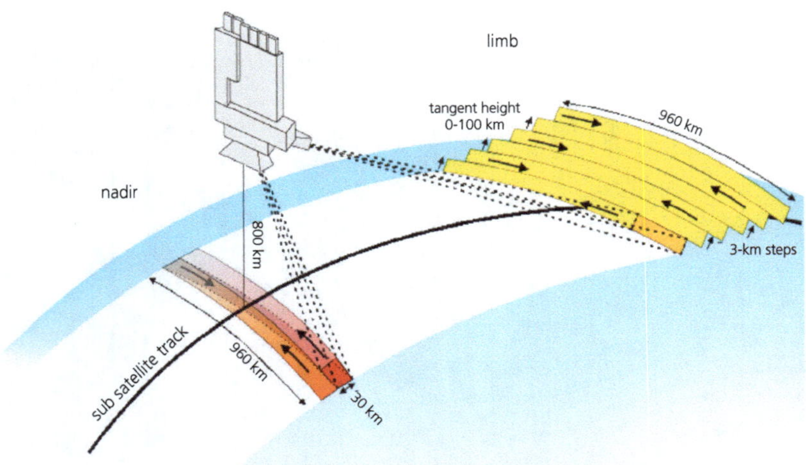

passive sensors detect visible light and infrared radiation, which can be used to determine cloud height and temperature. They can also be used to monitor city lights, fires, auroras, sand storms, snow and ice cover, ocean currents and waves, and lots more. In addition, passive radar sensors at ground stations can detect the density of clouds and the amount of precipitation.

Remote sensing imagery can also be used in meteorology for monitoring dust plumes. Each year, hundreds of millions of tonnes of fine Sahara sand are transported westward across the Atlantic towards the Americas. Hot air currents can lift this dust as high as 5,000 m. On March 10, 2007, a huge blanket of dust covered parts of the Atlantic Ocean engulfing the Canary Islands. It was one of the most severe sand storms ever recorded. The scene was captured by the 250-m resolution MODIS sensor (Moderate Resolution Imaging Spectroradiometer) aboard NASA's Terra satellite. The turbulences around the bigger islands' volcanic peaks churned up the dust plumes and spread them out. These dust storms have an impact on the ecosystems of the Caribbean and South American coast, but in the summer of 2006, they may also have hampered hurricane development by shading incoming sunlight and reducing the ocean's surface temperatures. The MODIS sensors capture data in 36 spectral bands ranging in a wide bandwith (wavelength from 0.4 μm–14.4 μm) at varying spatial resolutions. Their mission objective is to survey large-scale global processes occurring in the oceans, on land, and in the lower atmosphere.

► **Natural hazard management**. Remote sensing techniques are crucial when it comes to natural disasters. Satellite imagery is an important source of data needed to make predictions of catastrophic events, to determine hazardous areas, support rescue operations and facilitate damage estimation for events such as hurricanes and cyclones around the globe. High resolution data from Quickbird, IKONOS, or SPOT help determine where severe storms are likely to occur and predict them. After the storm has passed, remote sensing data is used to map the extent of landscape changes and the progress of recovery. Disaster mapping and rescue management of the tropical cyclone "Nargis" shows the importance of advanced image processing and analysis. The cyclone made landfall in Burma (Myanmar) on May 2, 2008, causing catastrophic destruction and at least 146,000 fatalities with tens of thousands more people still missing. It was the worst natural disaster in the recorded history of the country. According to the United Nations more than 1.5 million people were "severely affected" by this catastrophe.

Predicting the weather is especially important in the United States: in August 2005, hurricane Katrina caused devastating floods along the coast of Louisiana. The news was filled with images of New Orleans under water. A variety of remote sensing sensors – including weather satellites and systems for observing land – registered the continuous cyclical movement of the air masses around the eye of the storm, which kept heading towards the coast. These movements allowed scientists to analyse and forecast wind speeds and trajectories and hence the strength of the hurricane. Thanks to additional sensors that were dropped into the menacing air masses from special aeroplanes, additional and more detailed data on the structure of, and weather conditions within, the hurricane could be gathered (vertical air pressure distribution, for instance, and temperature). Today, New Orleans, which was founded

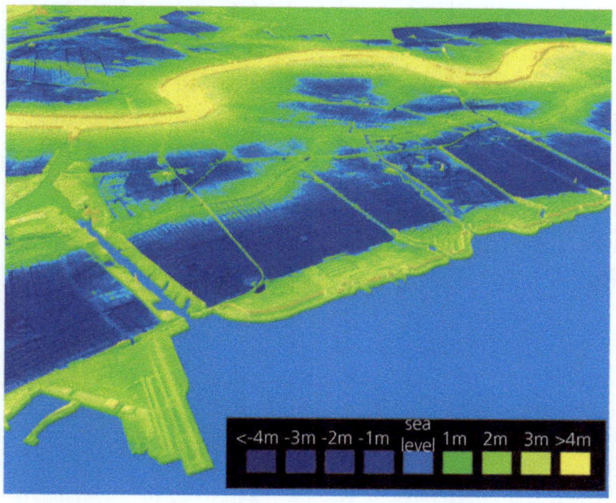

Digital elevation model of New Orleans: The model was generated with the help of airborne laser scanners. It clearly reveals that large parts of the city lie below sea level and are particularly at risk to storm flooding. At the same time, the raised embankments of the Mississippi have brought its water level to above that of the sea, posing a further threat in the event of flooding. Source: German Aerospace Center (DLR)

MODIS recording of the southwestern coast of Burma (Myanmar) before and after cyclone Nargis flooded the region: The images recorded by NASA's MODIS sensor use a combination of visible and infrared light to depict floodwater coverage. Top: Coastlines and freshwater can be clearly distinguished from vegetation and agricultural areas. Bottom: Coastal areas are entirely flooded and the agricultural areas have also been severely affected by the waters. Images taken with the DMC's (multi-satellite Disaster Monitoring Constellation) 32-m resolution camera. The standard three-band multispectral DMC sensor provides mid-resolution image information of the earth's surface and is mainly used for disaster relief. Source: NASA

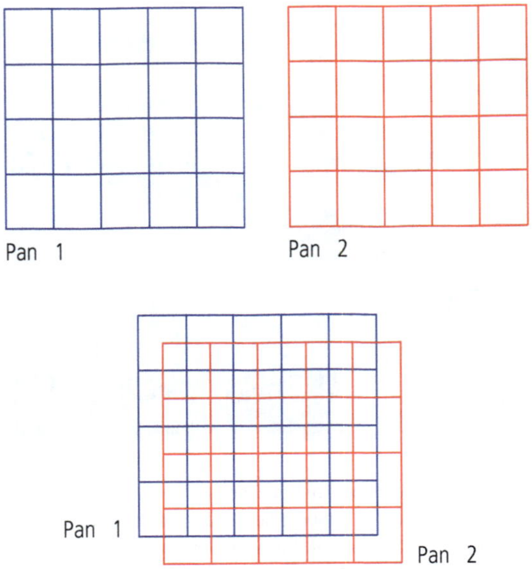

Functional principle of supermode (SPOT 5 satellites): The combination of two panchromatic sensors (5-m spatial resolution) merges overlapping edges of the pixels to create a new image with a spatial resolution of 2.5 m. Source: G. Bohmann (IGF)

on a rise surrounded by marshland, lies mostly between two and three metres below sea level. To simulate floods, data on the elevation and characteristics of the surface are important in order to reveal especially risky areas. This requires computing *Digital Ground Models* (DGM) that can be generated by sampling the earth's surface by means of active methods. These include radar (Radio Detecting And Ranging) and, of more recent vintage, *lidar* (Light Detecting And Ranging). Radar emits beams in the microwave range as separate bursts, whereas lidar uses infrared light as laser beams. Each respective impulse reaches the earth's surface and is then reflected in part back to the sensor. The distance between the sensor and earth's surface can be determined by the time difference between the emission of the impulse and the moment the reflected signal is measured. Since microwaves tend not to interact with the molecules in the atmosphere, radar sensors are usually deployed in satellite systems. Laser systems (lidar) are mostly installed in aeroplanes. They have the advantage of producing higher spatial resolution owing to the lower altitude of the sensor platform. Lidar DGM are therefore more detailed and more precise and are thus preferred over radar DGM, in particular for hydrological simulations. Remote sensing data can also contribute to rooting out the cause of other catastrophes as well, and containing damage. In 2004 and 2005, for example, thermal imaging was applied to differentiate the forest fires in Portugal according to the various temperatures among the sources of the fires. To gather these images, the thermal sensors of the satellite-based remote sensing systems were ample, since the spatial resolution was between 60–90 m. This allowed observers to record and assess large areas of Portugal simultaneously, and evaluate the risk from the individual fires and the forces needed to combat them.

MODIS recording of a dust storm over the Canary Islands on March 10, 2007: The satellite image clearly shows where the desert sand from the Sahara can be blown to. The plume of sand that appears on the image highlights the warm air currents which typically move northwards ahead of cold fronts. Source: NASA

Trends

Resolution. The current research on remote sensing is focused on increasing the resolution of the various sensors. Spatial and spectral resolution are considered the most important because they define the limits of object recognition and classification.

At the current time, the sensors on the Worldview 1 satellite are offering the highest spatial resolution – albeit panchromatic only – with about 50-cm ground sampling distance (GSD). GEOEYE-1, which was launched in 2008, generates panchromatic and multispectral images with a nominal resolution of 41 cm and 1.64 m, respectively. GEOEYE-1's spectral characteristics resemble those of the existing high-resolution satellites Quickbird and Ikonos. WorldView 2, on the other hand, which is to be launched in 2009, will offer four additional spectral bands for coastal monitoring (coastal and yellow) as well as a red edge and a second nearinfrared band. The German TerraSAR-X was launched in 2007. Its sensors operate at a wavelength that is almost unaffected by atmospheric conditions, so it can continue monitoring Earth almost continuously even in bad weather or at night.

Research activities are concentrated on improving temporal resolution. In 2008 a set of five multispectral satellites with a spatial resolution of 6.5 m are launched. The coordinated orbits of these satellites mesh to provide daily coverage of a large area, weather permitting.

Improvements in optical technology and the combination of several sensors has increased spatial resolution. The French satellite SPOT 5, for example, employs two identical 5-m resolution panchromatic sensors aligned in such a way that the combination of two images yields an effective resolution of 2.5 m.

Another way to improve spatial resolution is to place a recording platform closer to earth's surface. Remote sensing images taken from aircraft, helicopters, drones, balloons or zeppelins provide resolutions far better than anything recorded by satellite-borne sensors. These missions do not depend on orbits or earth's rotation, so the monitoring sensor can be positioned at the required temporal interval over the areas under surveillance. Often, the combination of large area satellite coverage and very high resolution airborne imagery allows for better calibration of the information obtained from satellites.

Improvement of spectral resolution, on the other hand, has physical limitations. The space on the CCD chips cannot accommodate enough sensors, which need to be of a minimum size to accumulate enough energy. Sensors used in remote sensing differ in size and arrangement to the ones on the usual CCD chips but they face the same restrictions with respect to the limitation of the recording area. The solution is to use optical tools such as prisms, grates and mirrors to split the incoming radiation and redirect the resulting wavelengths to spatially separated sensor chips. This technique can also be applied to generating remote sensing images of high spectral resolution. The sensor elements employed are usually sensitive to a broad range of wavelengths, so the incoming radiation has to be filtered for the required narrow spectral bands. Transmission filters can absorb certain parts of the electromagnetic spectrum, allowing only certain wavelengths to reach the sensor. *Interference filters* use a specific non-conducting material, for example glass. If the wavelength of the incoming radiation is equivalent to the thickness of the medium then the wave will be extinguished. To obtain very narrow spectral band coverage, many filter layers are required. This procedure, however, also lowers effective incoming energy, so sensors have to be extremely sensitive or exposed to the incoming radiation for a long period of time. This is crucial for satellite sensors because of the great distance separating them from the features to be observed on the ground. Atmospheric scattering and absorption are additional factors that weaken the radiated energy.

So far, longer dwelling or recording times are only possible for sensor platforms that offer flexible flying speeds such as aircraft or helicopters. The continuous movement of the sensor platforms on satellites, for instance, creates a so-called forward motion blur at certain resolutions. Aircraft, on the other hand, have sophisticated methods to compensate for forward motion. The only option for satellites at the present time is amplifiers such as photomultipliers that have to be used to enhance the signal before it can be recorded.

A panchromatic Quickbird image of the pyramids in Giza, Egypt with a spatial resolution of 0.60 m (left) is fused with a 2.4-m resolution multispectral image from the same sensor (centre): The image is resampled to the pixel size of the panchromatic image, which makes it look blurred. Advanced fusion techniques are used to inject the high resolution panchromatic image into the lower-resolution multispectral image to form a high resolution multispectral fused data set (right). Source: Digital Globe

Other suitable methods for the improvement of spatial resolution include data fusion techniques, which involve "injecting" high-resolution panchromatic images into lower resolution multispectral images. This procedure is often used to produce multispectral images with high spatial resolution.

Prospects

- Optical laboratory methods are the most accurate analysis techniques at the elementary level.
- Satellite and airborne remote sensing provides standardised methods for monitoring large areas and is especially well-suited for regional and international programmes.
- Current trends are focusing on the improvement of resolution, which is necessary to obtain detailed information for spatio-temporal analyses and visualisations.
- Improvement of spatial and temporal resolution can be achieved with airborne platforms. New developments (data fusion, multiple satellite constellations) also foster space-borne solutions.
- Improvement of spectral resolution requires specific optical tools.
- Improvement of radiometric resolution depends on methods for signal amplification.
- Use of multiple sensors may help overcome the shortcomings of a single-sensor monitoring strategy.

PROF. DR. MANFRED EHLERS
THOMAS KASTLER
Institute for Geoinformatics and Remote Sensing of the University of Osnabrück

Internet

- http://asterweb.jpl.nasa.gov
- http://envisat.esa.int
- http://eospso.gsfc.nasa.gov
- www.star.nesdis.noaa.gov/star/emb_index.php
- www.wdc.dlr.de/sensors/sciamachy

Environmental biotechnology

Related topics
- Industrial biotechnology
- Water treatment
- Measuring techniques
- Plant biotechnology
- Precaution against disasters

Principles

Many new technologies have been developed in recent years based on physical, chemical and biological processes capable of mitigating or remediating environmental damage. These include environmental biotechnology, defined as the development, use and regulation of biological systems to remediate contaminated environments (land, air, water). Nowadays this definition includes the development of environmentally friendly processes such as green manufacturing technologies and sustainable development.

Biological systems designed to eliminate contaminants are mainly based on the biological oxidation of organic compounds (electron acceptors: mainly oxygen, but also nitrate, sulphate, iron (III) and others, and biological reduction (in oxygen-free conditions, e. g. halocarbons).

Additionally, biological reactions can convert contaminants into insoluble, encapsulated or chemically bonded compounds, making them non-hazardous.

Biotechnological processes are often a preferred method, but are by no means a universal solution. Factors restricting their use include the chemical structure of the contaminant and its local concentration, and also bioavailability, which can be limited by its solubility in water or by incorporation or absorption in fixed matrices. Biological processes are usually slower than physical or chemical reactions, but they are more cost-effective.

The contaminants for which biological methods are unsuitable include:
- many metals (processes used: *leaching*, precipitation, fixing)
- complex high-molecular polycyclic hydrocarbons (process used: *incineration*)
- chlorinated hydrocarbons in high concentrations (process used: mainly *vaporisation* followed by chemical/catalytic incineration, partial recycling)

Selection of products representing potential sources of contamination, the contaminants, and preferred methods to eliminate them (BTEX aromatics: benzene – toluene – ethylbenzene – xylene – derivatives)

Source of pollution/ contamination (selection)	Products/potentially contaminating waste	Contaminants	Preferred treatment
gas stations, road accidents (soil, groundwater)	gasoline, diesel	petroleum hydrocarbons	biological
chemicals industry (soil, groundwater)	paints, polymers, pesticides, insecticides, coke	BTEX aromatics, polycyclic aromatic hydrocarbons, heterocyclic compounds, haloaromatics, tar	biological, biological physical/chemical (catalytic oxidation, sorption) incineration
agriculture (soil, groundwater, surface water)	leachates, agrochemicals	pesticides, nitrogen compounds, animal pharmaceuticals	biological
metal processing (groundwater, soil)	leachates	heavy metals	chemical, physical
explosives, munitions (soil, groundwater)	production lines, wastewater	nitro-aromatic compounds, heavy metals, chlorinated aromatic compounds	chemical / physical

► **Environmental contaminants.** The many organic compounds that contaminate former industrial sites even today, especially gas works, steel mills, chemical plants and recycling centres, and landfill waste sites, have not been broken down by natural processes since their formation. The chemical structure of these pollutants is a significant factor, affecting not only their mobility, and thus their dispersion, but also more generally their resistance to degradation (persistence). Degradation normally involves the effect of light and oxygen breaking down the molecules internal bonds. Biological degradation by various types of microorganisms can work in a similar way to such nonliving (abiotic) reactions. Generally, substances are seen as persistent if their half-life (the time it takes for their measured concentration to drop by half) is longer than 4 months in the environment.

Simple relationships between structure and persistency can be summed up as follows:
- Saturated hydrocarbons (alkanes) are more persistent than unsaturated compounds (alkenes, alkynes, carbonyl compounds, etc.).
- Aromatic compounds are more persistent than saturated hydrocarbons.
- OH groups in the molecule reduce persistence.
- Halogens (chlorine, bromine) in a molecule, increase persistence.

► **Biological principles.** Environmental biotechnology uses the metabolic abilities of microorganisms (reproducing or resting cells of bacteria, yeasts, fungi) or plants. Decontamination is usually not only achieved by a single organism, but only occurs within an entire biological system (biocoenosis).

Microorganisms are single-celled life forms that reproduce very quickly and can adapt to the most extreme habitats. All they require is a source of organic carbon compound, atmospheric oxygen and support by nutrients. They are equally capable of "digesting" carbon compounds that do not occur in nature – either in the same way as the natural compounds or by transforming them in a cleavage or oxidation reaction. Hence, in principle, non-natural microorganisms can be used to clean up foreign substances in the natural environment (xenobiotics) originating from chemical synthesis processes and subsequent waste streams. However, certain technical manipulations are needed to support the biological reaction, to ensure that in addition to carbon and oxygen the cells are always adequately supplied with other nutrients they require for reproduction, such as nitrogen, phosphorus and trace elements. A constant temperature in the range of 20–35 °C must also be maintained. All living cells require sufficient water to dissolve and absorb nutrients, and on the other hand to dissolve and dilute metabolites. Most organisms do not tolerate excessively acidic or alkaline conditions; such conditions have to be altered to within the life-supporting neutral range.

However, biological decontamination does not necessarily call for sophisticated technology. Within certain limits, nature can employ its own means to rid itself of foreign substances in groundwater, soils and watercourses. This natural attenuation can be enhanced (accelerated) using simple technological methods. This is often markedly cheaper and requires less energy, although it often takes longer to achieve the desired results. Comprehensive site analysis is required, and the process can take several years depending on the specific conditions.

Long-term transformation in soils can be on a similar scale, as with the explosive TNT. In this particular case, as with certain pesticides, the changes in the soil are exceptionally complex and can be difficult to ascribe to individual reaction steps. Several reactions may take place simultaneously: biological reactions in the absence of oxygen, with specially adapted bacteria, chemical reactions with the products of this biological transformation, physical adsorption to soil particles, and reactions with humic matter.

Contaminants entering the soil may adsorb to the surface of the particles, be partly dissolved in water or penetrate into the fine capillaries. On the surface, they are accessible to the bacteria, which populate the surface of the particles, where they are also supplied with air. The bacteria can use the pollutant as a source of carbon for their growth. Inside the pores, which the bacteria cannot penetrate due to their size, the contaminant is hidden and not bioavailable.

One obvious answer is to support microbial activity in contaminated soils or water by introducing production strains of bacteria (sometimes genetically engineered) that have been grown in suspension in bioreactors (bioaugmentation). Leaving aside the legal issues raised by genetically modified organisms and the possibility that certain strains might present a health risk, the results so far have not always been convincing. Even apparently positive results must be carefully interpreted, given that the addition of nutrients to support the introduced bacteria might also be supporting existing microorganisms. It is rare for a production strain brought in from the outside to become so firmly established that it has an overall positive effect on events as a whole and does not destabilise the complex interactions – thus improving performance.

Applications

► **Soil decontamination**. Every gramme of "soil" contains millions of different kinds of bacteria and fungi. These always include some that can recycle contaminants in the soil or adapt themselves to recycling them. The main aim of any human intervention must be to improve and promote their survival and growth. This can take place "in situ", at the site where the foreign substance enters the soil, or "ex situ", in which case all the contaminated soil is dug up and transported to a soil decontamination unit.

The in situ option requires aeration of the soil, the infiltration of nutrient solutions, and the introduction of steam to raise the temperature. The first results (e.g. a lower concentration of hydrocarbons) are normally detectable within a few weeks. However, long-term analysis has often revealed that initial successes were not "sustainable". Certain contaminants enter the humic matter by penetrating the fine pores of the soil matrix. Here they are no longer bioavailable, and sometimes not even detectable by analysis instruments. Over time, they diffuse out of the soil particles – making it necessary to introduce new remediation measures.

The simplest and safest ex situ treatment method involves removing the contaminated soil to a kind of compost heap or "windrow" on a nearby site. To increase the concentration of microorganisms, the pile is mixed, kept moist and supplemented with nutrients such as phosphate or nitrogen fertilisers. It is mechanically turned at intervals to ensure even mixing and aeration. Such methods permit hydrocarbons from gas station forecourts, road accidents or pipe ruptures to be cleaned to within legally stipulated limits for reuse

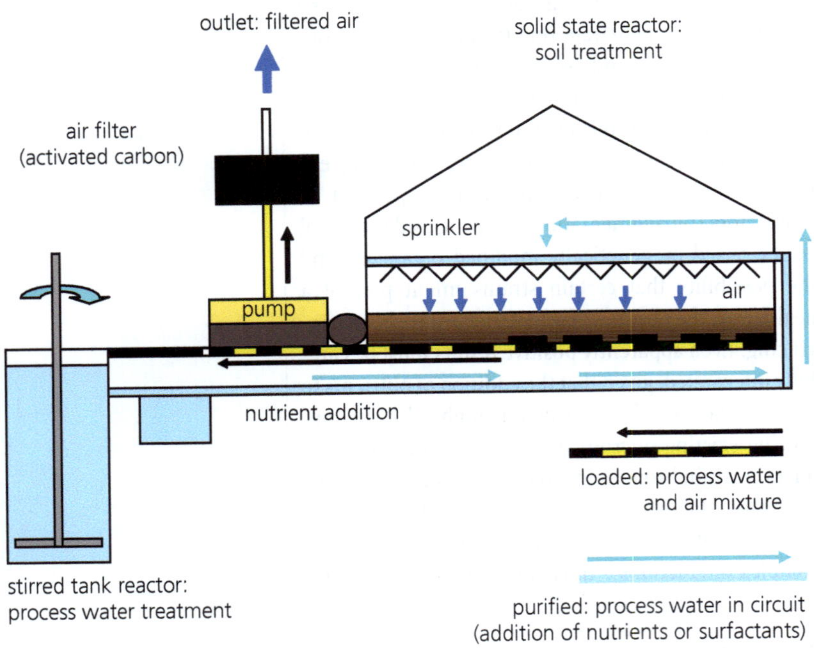

Diagram of a soil decontamination facility: The treatment basin (solid-state bioreactor) is fitted with a drainage and soil vapour extraction system. The vapour is extracted from the soil inside the reactor and channeled through an air filter. A mixing vessel holds and treats the process water, which is routed around a circulation system. If necessary, the process water can be supplemented with various additives or essential nutrients. The water can be cleaned during the circulation process. (Usual area of the soil treatment basin: 525 m^2, filling height 2 m, volume of the stirred tank reactor: 23 m^3, average duration of treatment 6 months). The rate of the reaction can be monitored by measuring oxygen levels or soil moisture and temperature, and can be controlled by varying the amount of nutrients added. Online sensors measure the nutrient level and nutrients are added via a computer-controlled dosing station. Source: Helmholtz Centre for Environmental Research UFZ

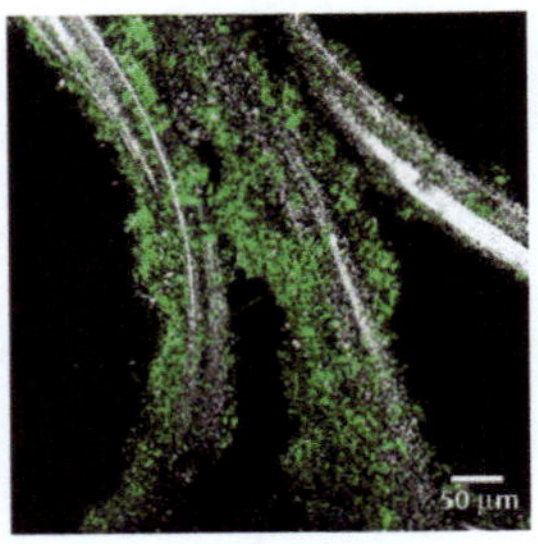

The polymer mats used as a substrate in groundwater biofilters serve as a matrix for the colonisation of biofilms: Bacteria lay down a solid film of biopolymers on the synthetic substrate. This film is populated in turn by many different types of microorganisms, which together degrade the pollutant substances dissolved in the groundwater as it flows through the filter. The image shows bacteria stained with nucleic acid (green) and the reflection of the plastic matrix (white). Source: C. Vogt, S. Jechalke, T.R. Neu, Helmholtz Centre for Environmental Research UFZ

within one or two years at most. The piles are placed on a solid foundation to prevent water seepage and are enclosed in tents. Escaping volatile hydrocarbons are captured inside the tent and cleaned in filters.

If the soil contains higher concentrations of contaminants or less easily degradable compounds, simple windrow composting systems have little effect on biological processes. Such cases call for specially equipped soil decontamination facilities, which allow conditions to be optimised for the action of soil bacteria. The higher transportation costs are balanced by a much shorter treatment time.

► **Groundwater decontamination.** Contaminants entering the soil almost always end up in the groundwater – either dissolved or suspended as a separate phase. A distinction based on density is made between LNAPLs (light non-aqueous phase liquids) and DNAPLs (dense non-aqueous phase liquids). Substances particularly resistant to natural degradation processes include chlorinated hydrocarbons, often originating from dry-cleaning establishments, and organic fuel additives used as anti-knock agents and oxygen carriers.

Depending on geological features of the site, the contamination moves through the soil at different rates – from just a few centimetres a year up to several metres a day – and may ultimately cause the groundwater to become unsafe for human use.

Groundwater contaminants can be degraded by natural biotic and abiotic processes in the aquifer and in some cases may no longer be detectable just a few kilometres away from the original site ("natural attenuation"). In active remediation processes, the groundwater is pumped to the surface for cleaning in external filter reactors. These are either sorption filters or biofilters. In the latter, contaminant-degrading microorganisms form colonies on a substrate such as pumice stone or coke. Here, they create a natural coating (biofilm) where the required decontamination takes place. In practice, the formation of biofilms has even been observed on sorption filters filled with activated charcoal after a certain length of time. Consequently, the cleaning effect increases during use, as now not only adsorption is taking place but also degradation.

► **Seawater purification.** The worst cases of pollution at sea are caused by mineral oil (5–10 million t a year from spillages, tanker accidents and pipeline ruptures). Roughly 25% of the escaped crude oil evaporates. The original amount is reduced to about 15–30% over the course of the first few months, due to the natural processes of photooxidation and biodegradation. The rest can only be removed mechanically (shorelines). Otherwise it eventually sinks down to the seabed.

To minimise the impact on marine flora and fauna, the applied environmental biotechnology methods focus on providing targeted support to natural processes. These closely interrelated processes are dispersion (physical: forming a mixture of two immiscible or nearly immiscible liquid phases), solubilisation (biochemical: changing substances into soluble substances by means of enzyme reactions), assimilation and mineralisation (metabolic: growing organisms and recovering energy with carbon dioxide as an end product).

When tanker accidents occur in cold waters, the naturally occurring microorganisms require additional support. This is done by isolating oil-degrading bacteria from the site of the accident after they been given enough time to develop in sufficient quantities (only a few weeks). The bacteria harvested from the seawater are cultured in *bioreactors* in the laboratory. The process is then scaled up to produce the required quantities amounting to several cubic meters. This takes an estimated two to three weeks. The bacteria suspension is sprayed onto the oil slick from planes or ships (bioaugmentation). In warmer climates, the microorganisms are already more abundant, but can still be stimulated to multiply more rapidly and thus increase their activity by spraying nutrient solutions onto the oil slick (biostimulation). The costs are considerably lower than for mechanical treatment: After the Exxon Valdez accident in 1989, the cost of biological treatment for 120 km of coastline was equivalent to one day of manual work by volunteers. Nevertheless, it is true that there are limits to these applications, and that such forms of bioremediation are only capable of fully degrading around 90% of the mineral oil substances. On the other hand, they do so in roughly half the time it takes for the unsupported, natural process to achieve the same result (usually several months).

► **Reactive barriers.** Pumping groundwater to the surface requires heavy financial investments. Hence the development of technological solutions that force the groundwater to flow through a narrow channel cut crosswise through the aquifer and filled with a substrate with low flow resistance, such as gravel. Here, too, a natural biofilm forms over time, which degrades the contaminants. This "reactive barrier" is suitable for the biological degradation of low concentrations of organic contaminants. Another more sophisticated process involves building a non-permeable barrier across the path of the aquifer in the shape of a funnel, leaving open a gate through which the groundwater can flow. This "funnel and gate" technology allows the addition of a biofilter filled with activated charcoal or chemically oxidising substances.

Trends

For quite a long time, environmental biotechnology processes were limited to treating soil, water and groundwater; they successfully implemented principles from technological microbiology and process engineering. Microbiological processes reach their full potential when combined with chemical or physical methods.

► **Catalysing reactions.** Biological reactions are generally slow. Their progress depends on the availability of the reactant, which is oxygen in the case of aerobic processes. To ensure a long-term supply of oxygen, special oxygen-releasing compounds (e.g. magnesium hydroperoxide) have been developed. These are mixed into the contaminated soil and gradually decompose in the presence of moisture in the soil, releas-

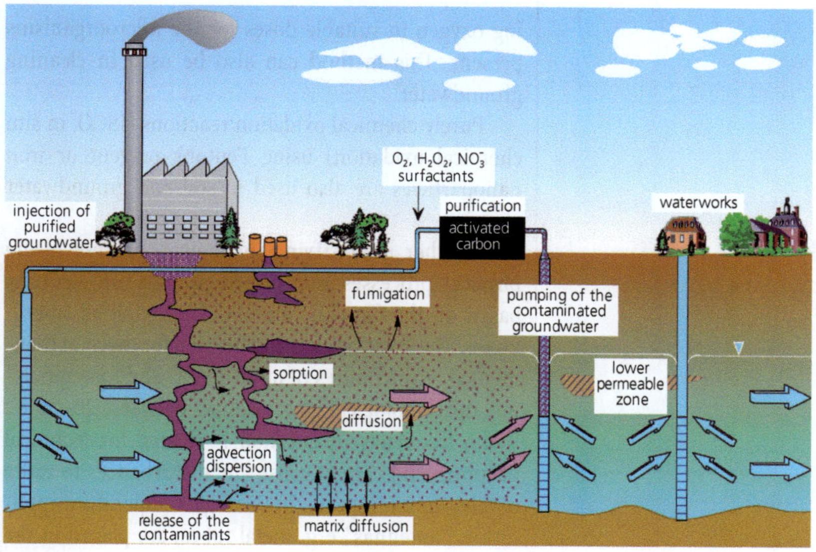

Pump-and-treat process with source remediation for contamination scenario caused by organic contaminants in the subsoil: Some of the contaminants which have entered the soil from an industrial plant sorb to the soil matrix, disperse or are dissolved, or are diffused through the heterogeneous subsoil composition. In the pump-and-treat process, the contaminated groundwater is pumped off as it flows away and is channeled through active charcoal filters, where the contaminant is bonded. The purified water can be funneled away for further use, or be saturated with air and furnished with nutrients or surfactants to support microbial decontamination processes in soil and groundwater. Then the water is returned to the aquifer. Source: K. Mackenzie, Helmholtz Centre for Environmental Research UFZ

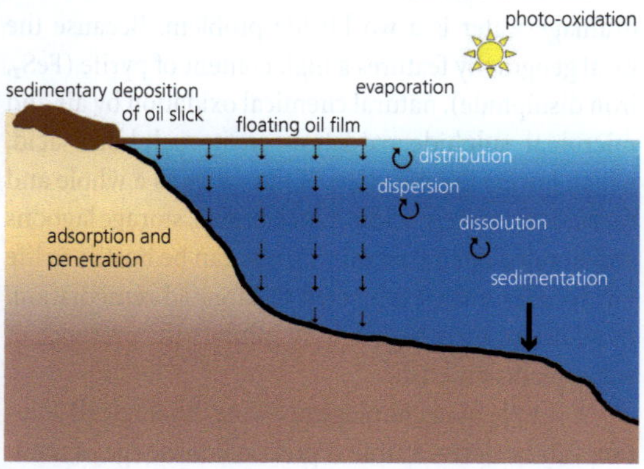

Paths taken by a mineral oil spill in a coastal zone: An oil slick partly evaporates, is dispersed in the seawater or partly dissolves. The heavy components sink down to the seabed. Certain aromatic compounds (polycyclic aromatic hydrocarbons PAH) decay when exposed to sunlight. The parts of the oil slick that are washed onto the beach adsorb onto soil and stone particles. Biological reactions are most frequent in places easily reached by the oxygen in the air. However, all life forms in the water are involved in the degradation, including plants, algae, etc. Many fat-soluble and often also carcinogenic components (PAHs) accumulate in fish.

ing oxygen in suitable doses for the microorganisms present. This method can also be used in cleaning groundwater.

Purely chemical oxidation reactions (ISCO, in situ chemical oxidation) using Fenton's reagent or iron nanoparticles are also used in soil and groundwater cleaning.

Another intensifying biological process using physical processes is *radio-frequency soil heating*, based on the familiar household principle of microwaves. The same process can be used for thermal treatment at temperatures of up to 300 °C and for supporting microbiological processes at between 20–30 °C. Radio frequency in question is cleared for technical purposes and cannot cause any disturbances to radio or television reception. In the high-temperature application, "antennas", e. g. metal tubes, are placed in the ground to the required depth, and the radio frequency is applied. The soil is heated to 300 °C, the contaminants vaporise and are extracted and destroyed. In the low-temperature application, the soil is only warmed up to a maximum of 40 °C and – if necessary – a nutrient solution is added. Microorganisms in the soil are activated to degrade contaminants at their ideal reproduction temperature. This method is particularly useful in winter or in cold climate zones.

► **De-acidification of surface water.** Acidic mine drainage water is a worldwide problem. Because the local geography features a high content of pyrite (FeS_2, iron disulphide), natural chemical oxidation by air and microbial sulphide oxidation create sulphuric acid, which lowers the pH value of the water as a whole and makes it extremely acidic. Catch basins, storage lagoons and flooded open-cast mining pits can be hostile to life for decades unless subjected to targeted remediation. In addition, their high acidity inhibits the presence of water plants and fish.

A simple biotechnological (ecotechnological) solution can be derived from a precise knowledge of natural chemical and biological processes and how they can be influenced. The sulphuric acid must be reduced back to sulfide, and this must be prevented from re-oxidising. To achieve this, naturally occurring sulphate-reducing bacteria must be provided with an ideal habitat. This can be done by triggering targeted putrefaction processes in organic substances (straw, organic waste) on the bed of the acidic body of water. In the oxygen-free area of the decaying zone, acid does not reform and the sulphide precipitates as the poorly soluble iron sulphide, which is deposited in an insoluble state – the acid-forming reaction is reversed. This simple technical measure can cause the pH level of the water to rise up to the neutral zone within a few years, creating a natural, living body of water with living conditions for organisms.

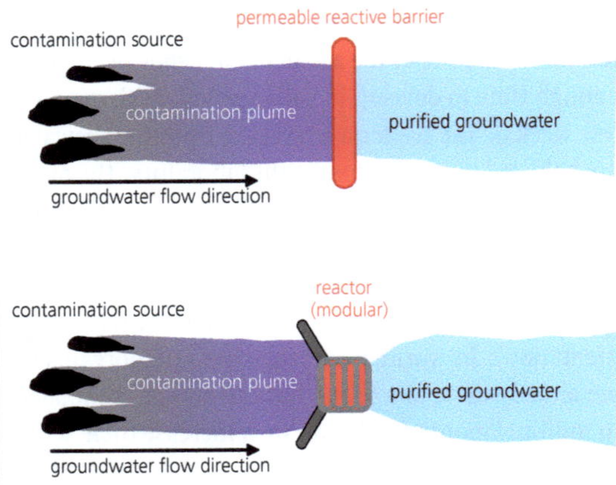

Principles of groundwater remediation. Top: Passive flow remediation using a permeable reactive barrier. A semi-permeable vertical layer of a reactive compound is inserted in the flow path of the contaminated groundwater to absorb and enable microbiological reactions. Bottom: Passive flow remediation using the "funnel and gate" concept. The reaction zone comprises a gate through which water channeled between cutoff walls (funnel), in a similar way to the permeable reactive barrier. Oxygen-releasing compounds (ORGs®) in the reactive barrier have proven effective, as have hydrogen-releasing compounds (HRCs® = polylactides). Source: K. Mackenzie and R. Köhler, Helmholtz Centre for Environmental Research UFZ

► **Sediment leaching.** Regulation and maintenance work creates large quantities of contaminated excavated river sediment. Disposal of these sediments is a worldwide problem. They contain contaminants (heavy metals, organic material) as the rivers are used as drainage points. Safely dumping the dredgings is expensive. The toxic heavy metals are fixed in the sediments as insoluble sulphides. They can only be removed at high cost by reducing the pH value using strong mineral acids. Biotechnological processes are a far more economical alternative. They make use of the ability of microorganisms (Thiobacillii sp.) to convert poorly soluble heavy metal sulphides into soluble sulphates, and to oxidise elemental sulphur, turning it into sulphuric acid by means of a complex biochemical

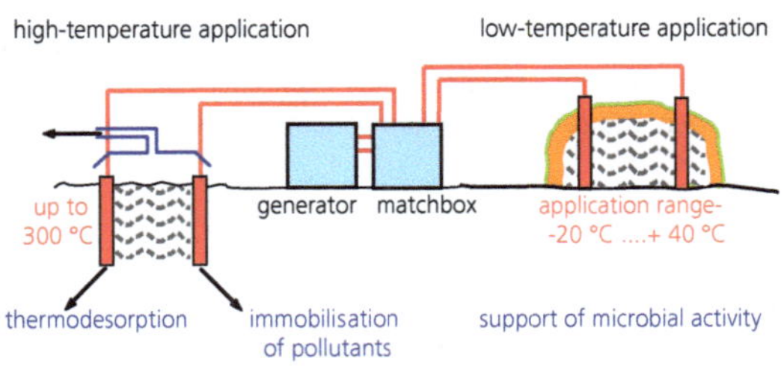

Radio-frequency soil heating for in situ and biological ex situ applications: Similarly to the household microwave, the soil is either heated (up to 300 °C) so that contaminants are vaporised, or only warmed up to 25–30 °C so that the soil microorganisms find their ideal reaction temperature. Source: K. Mackenzie, Helmholtz Centre for Environmental Research UFZ reaction. In nature, similar, yet uncontrolled and unwanted reactions take place when the sediments are dumped and when acidic mine drainage takes place. The biotechnological process combines two principles:

- *Phytoremediation*: Bioconditioning using plants. Freshly excavated slurry river sediment containing heavy metals (e.g. Cd, Cr, Co, Cu, Mn, Ni, Pb, Zn) is planted with grasses. Reeds (e.g. Phalaris arundinacea, Phragmites australis) tolerate both waterlogging and the heavy metals. The sludgy sediment, still containing all the heavy metals, has become a "soil" and can now be sieved and easily transported.
- *Bioleaching*: Biological acidification in percolation bioreactors

Percolation reactors (premium steel vessels with a sprinkler facility and extraction at the base) activate the microbial oxidation of sulphides and elemental sulphur, converting them into sulphuric acid. The solution emerging at the base of the vessel is extremely acidic (pH 2) and contains the soluble heavy metals in the form of sulphates. The biological oxidation process causes the temperature in the reactor to rise as high as 40 °C.

After only 20 days, it is possible to reach the maximum overall degree of heavy metal solubilisation; about 65% of the heavy metals can be removed from the solid matter. For Zn, Cd, Ni, Co and Mn, leaching levels of up to 80% can be achieved; about 30% of the Cu can be removed. Pb and As can not be solubilised. The heavy metals are precipitated out of the leachate with limestone as carbonates and/or hydroxides.

▸ Phytoconditioning: Interaction between plants and sediment. In one vegetation period (about 6 months), the roots of selected grasses can penetrate what was once a slimy, paste-like river sediment, forming a soil-like structure which can be added in further processing steps (heavy metal leaching). Roots grow through the sludge to a depth of roughly 1 m. Water is withdrawn, and chemical and microbial oxidation by air takes place in the root area. Only a small proportion of the heavy metals can be accumulated in the roots. Source: Helmholtz Centre for Environmental Research UFZ

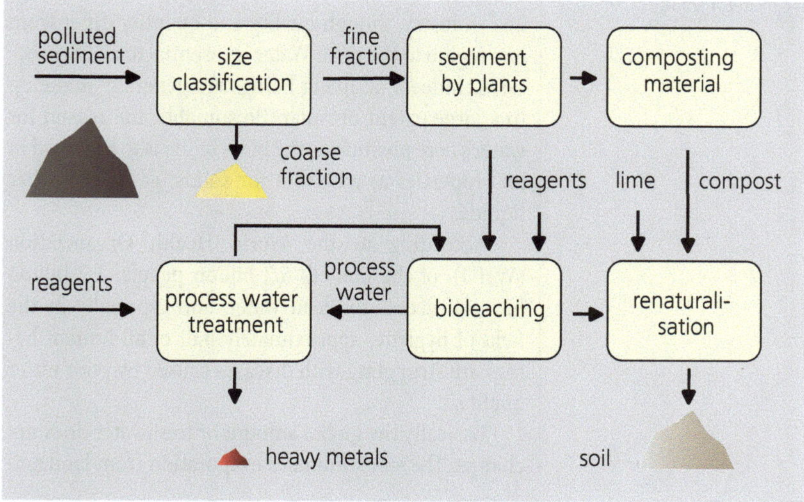

▴ Sediment treatment with plant conditioning and bio-leaching: Polluted sediment is classified by size. The excavated sediment is sieved (if it is sandy, the coarse fraction is removed) or, if it is paste-like or slimy, it is conditioned using plants. The conditioned sediment is fed into the bioleaching system (sulphur added up to 2% by weight). The result of the process as a whole is acidic process water ($pH < 2.5$) containing the dissolved heavy metals (precipitated using limestone suspension). The plant biomass from the conditioning is composted and mixed with the bioleaching product. The "soil" obtained can be used, for example, for restoration measures in mining (UFZ BIOLEA process). Source: Helmholtz Centre for Environmental Research UFZ

Prospects

- Environmental biotechnology will continue to play an important role as a means of removing contaminants from soils, groundwater, landfill waste sites and surface water, complementing established physical and chemical technologies.
- In most cases, an approach based on environmental biotechnology is less expensive than other alternatives. However, the longer timescale needs to be taken into account.
- There is scope for extending the range of applications and improving effectiveness by developing new combinations of microbiological processes with chemical and physical methods.
- Recent findings, especially in the field of microbiology, can provide useful contributions to the development of simple, effective technologies and support decision-making on remediation strategies. Specific topics include:
  soil / microorganism interactions (targeted activation of specific groups of microorganisms), reactions in biofilms (biofilters), plant root zone / microorganism / soil interactions (phytoremediation).
- Instead of focusing on "end-of-the-pipe" treatment, future technological developments will take an increasingly holistic view, promoting ecofriendly processes involving green manufacturing technologies and sustainable development.

PROF. DR. ULRICH STOTTMEIER
Helmholtz Centre for Environmental Research, Leipzig

Internet

- www.ebcrc.com.au
- www.nrc-cnrc.gc.ca/randd/areas/biotechnology_e.html
- www.epa.gov/oilspill/oilsites.htm
- www.environmental-expert.com/technology/regenesis/regenesis.htm

Water treatment

Related topics

- Sustainable building
- Industrial biotechnology
- Environmental biotechnology

Principles

Humankind has 1.4 trillion litres of water at its disposal on Earth, of which only 3% is freshwater. Around 60–70% of fresh water resources are used for irrigation in agriculture, the rest is consumed by human beings and industry, though quality and quantity differ from one region to the next. Water is essential to all life processes. Indeed, all living beings are generally made up to a large extent of water. Presumably, the reason for water's core position as the basis of life is to be found in its properties as a solvent for solids, gases and other liquids.

According to the World Health Organization (WHO), of the planet's 6.2 billion people, 1.2 billion have no access to clean water, and especially in the field of hygiene, approximately half of all human beings are struggling with diseases caused by poor water quality.

Basically the global amount of freshwater does not change. The sun stimulates evaporation from land and water surfaces. The vapour condenses to form clouds in the atmosphere, which are carried along by the wind. Given certain conditions, the water precipitates again and is returned to the earth's surface in the form of rain, snow or hail. A portion of the precipitated water seeps into the ground. Surface run-off water and parts of groundwater collects in rivers which return it to the sea.

Along this freshwater circuit the amount of collectable water does not change if atmospheric temperature is constant. But each 1 K rise in atmospheric temperature brought about by climate change will increase the air's vapour content by 8%. This additional mass stored physically in the atmosphere will reduce the water available to human beings.

► **Water demand.** Freshwater is used in households and industry and is collected after usage in sewer systems, where it is treated before being released into rivers, oceans or other bodies of water. Water demand from industry and urban areas has been decreasing steadily over the past several years, not only because of the rising cost of freshwater, but also because of the improved water efficiency of household appliances and sanitation equipment in use today.

Global water cycle: The cycle begins with rain or snow precipitation. On the ground the precipitation either infiltrates and is added to groundwater, or it is collected along surface run-offs to form rivers and lakes. Solar radiation evaporates water from the ocean (80%) and ground surfaces (20%). In order to close the circuit, vapour transported from the ocean to land surfaces must be equivalent to that transferred from land to the ocean.

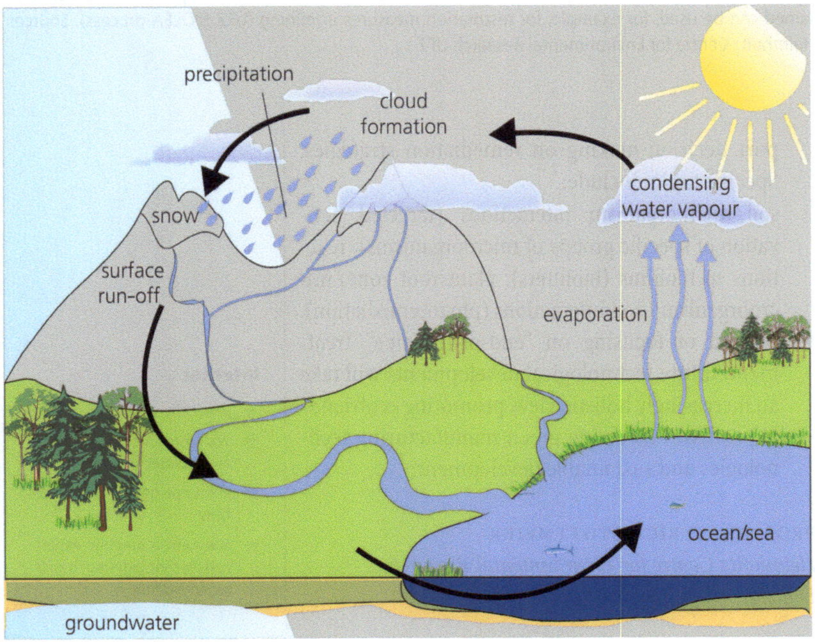

Applications

► **Drinking water.** In Europe both ground and surface waters serve as sources of drinking water. In most cases, freshwater treatment is carried out mechanically, employing sand filtration to remove solids. 45% of the water is also disinfected to guarantee a perfectly hygienic supply. In 1995, chlorine was still used for this purpose 90% of the time, although there is now an awareness of problems relating to the byproducts of this type of disinfection. In 6% of waterworks, disinfecting is performed using ultra-violet (UV) radiation, and in 1% using ozone.

The disinfecting effect of chlorine results from the element's ability to oxidise living microorganisms or organic impurities alike. However, it also activates the resistance mechanisms in the living microorganisms, which can generate halogenated compounds such as trihalomethane, which is carcinogenic.

UV radiation affects the reproductive ability of microorganisms, and occasionally destroys the genetic database – the *DNA* – of bacterial germs. Ozone, for

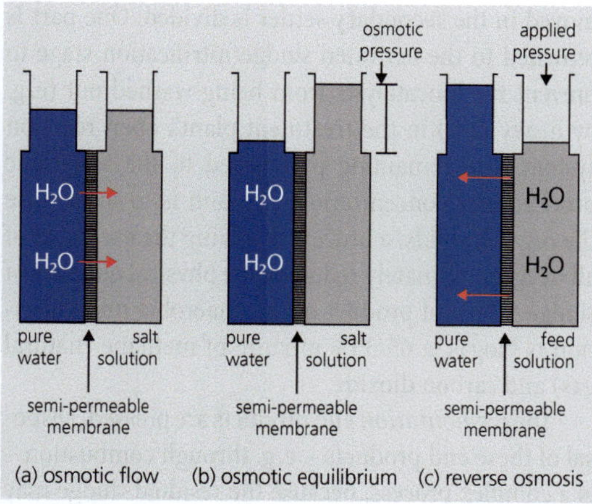

Osmosis and reverse osmosis. Left: If two salt solutions of different concentrations are separated by a membrane, the water molecules will flow towards the more concentrated solution (osmosis). Salt molecules can not pass through the membrane due to their size. Centre: Pressure in the right chamber increases and counteracts the incoming flow of water. Osmotic balance is achieved when the flow of water through the membrane is halted by the hydraulic counter-pressure exerted by the right chamber. Right: If additional external pressure is placed upon the container holding the higher salt concentration, water will nevertheless flow into the left container against the natural osmotic direction dictated by differences in concentration (reverse osmosis).

its part, forms OH radicals in the water. They are very reactive and oxidise the organic substance, i.e. they also attack living cells. Ozonification of drinking water is increasingly seen as an important method for treating water.

Activated carbon filters can also be employed to neutralise unwanted tastes that might be present in the water. Since Europe is located in a climatic zone with plenty of rain, the collection of rainwater from decentralised urban collectors, as well as partially or fully desalinated water produced by reverse osmosis or distillation, do not play any part in the production of drinking water. However, climate change will increase regional demand for water with low-salt content, especially for sustainable irrigation technology in agriculture.

► **Water desalination**. Seawater desalination is commonly used in countries where freshwater is not available in sufficient quantity, and where cheap energy can be used to operate reverse osmosis plants.

Osmosis is a natural process which occurs when a diluted solution is separated from a more concentrated one by means of a semi-permeable membrane. The driving force of osmosis is created through a difference in concentration between the two solutions. This causes water to flow through the membrane to the concentrated solution. This process continues until the concentrated solution is diluted and pressure in the opposite direction prevents a flow through the membrane – a state known as osmotic balance. If pressure is exerted on the side of the membrane containing the concentrated solution, and providing this pressure exceeds that caused by osmosis, the direction of the flow is reversed. Pure water flows from the concentrated solution through the membrane and is separated from impurities in a process called reverse osmosis.

Depending on its salt concentration, seawater has an osmotic pressure of around 30 bar. So a high pressure of around 60 or even 80 bar is required to desalinate ocean water. The process produces not only salt-free water, but also a concentrated salt solution, which has to be disposed of. Using solar radiation for centralised or decentralised desalination plants would be desirable, since the process does require a fair amount of energy.

Ultra- and micro-filtration techniques with porous membranes (from 0.2 μm–60 nm) are used to produce germ-free or pathogen-free drinking water in disaster relief operations, in areas afflicted by floods, earthquakes, or war. In the future, this technique will play a more important role in centralised freshwater production plants.

► **Wastewater purification**. From the standpoint of civil engineering, no two wastewater purification plants look the same, though most of them function according to the same basic principles. The wastewater is first of all channelled through a catcher, which separates out large solids, and then a sand and fat collector – the two groups of substances are collected separately. The next step is preliminary sedimentation, simply using the force of gravity to precipitate remaining organic and inorganic solids. The dissolved substances then remaining in the water flow are degraded by microorganisms. This normally takes place under aerobic conditions, i.e. by adding oxygen, which has to be fed into the wastewater. The dissolved organic compounds are ingested by the microorganisms and metabolised for use in the organisms' own reproduction (activated sludge stage/nitrification). Dissolved organic pollut-

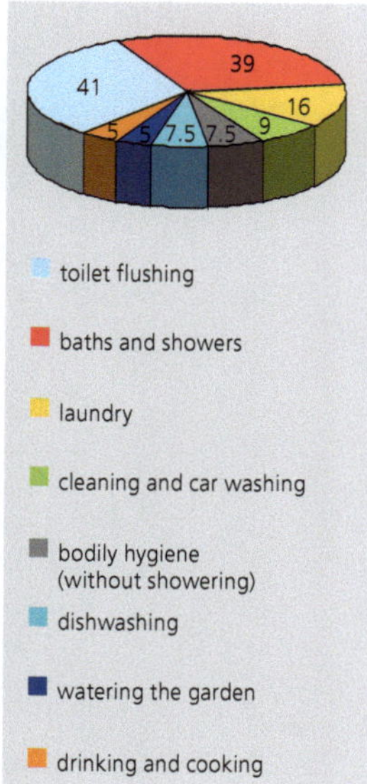

A variety of specific purposes contribute towards the total household water consumption in industrialised countries. The largest share of freshwater is used for hygiene and well-being. Drinking and cooking only use up to 5 l per person per day.

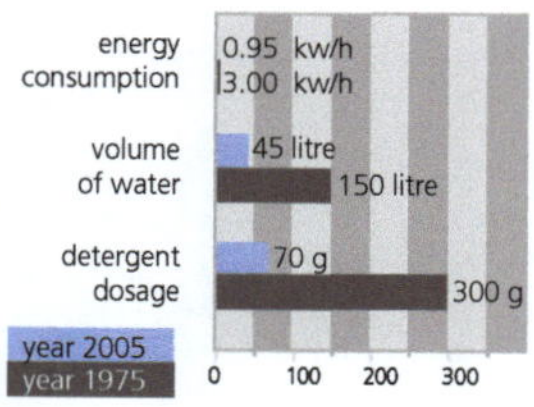

Improved efficiency of modern washing machines since 1975 (per 5 kg wash). Source: Technology Review

Drinking water recovery: Raw water is collected from the lake and pumped to a raw water reservoir. The first step involves cleansing water from this source, using a micro-strainer with a porosity of 15 μm. Residual organics of the raw water are then degraded in a second step by ozonification followed by final rapid sand filtration. The water has to be chlorinated before being pumped back into the extended drinking water grid. This ensures that the hygienic status of the source point is maintained.

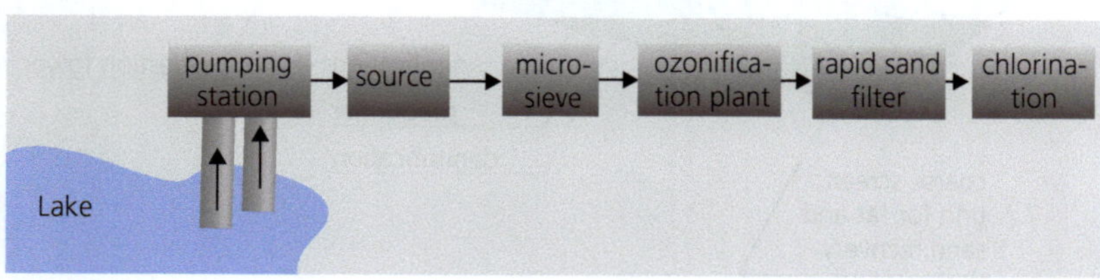

ants thus become bacterial mass – solids that once again have to be separated from the wastewater. This is done in a second settler, again using gravitational sedimentation. This produces excess sludge. Water freed of microorganism flocculi leaves the treatment plant and is discharged into an aqueous recipient.

While the metabolic process to form biomass (activated sludge stage/nitrification) is underway, organic pollutants containing nitrogen are oxidised to form nitrate. In dissolved form, this nitrate would actually make it through the treatment plant in spite of secondary settling. To prevent this, the wastewater is recirculated several times. It is thereafter redirected to a separate basin to which no oxygen may be added (advanced *denitrification*). Here, organic impurities are metabolised by aerobic microorganisms. These break down the nitrate (NO_3) contained in the water to oxygen, which serves as an electron acceptor, and molecular nitrogen, which escapes into the atmosphere. The separated solids from preliminary sedimentation together with the precipitated fats are then subjected to an anaerobic (oxygen-free) bioprocess. The sludge removed in the secondary settler is divided. One part is returned to the activated sludge/nitrification stage to prevent the biocatalysts from being washed out (e.g. by heavy rain) in the treatment plant's open reaction system. The remaining part is fed to the anaerobic process after concentration. The aim is to mineralise the organic solids, a process involving the exclusion of all oxygen, ultimately reducing the physical quantity of sludge. The end product of the anaerobic mineralisation is biogas, a 65/35% mixture of methane (natural gas) and carbon dioxide.

The *fermentation* end products are pressed. Disposal of these end products – e.g. through combustion – is a complex process, because the residual sludge may contain environmental hazards (absorbent pharmaceuticals, multi-resistance-bearing bacteria, endocrines). Phosphorus must be eliminated in treatment plants as it would otherwise result in "eutrophication" in the recipients: the mass growth of algae attributable to over-fertilisation. Phosphorus removal is done by a precipitation reaction using iron or aluminium salts. The precipitates are removed in the secondary settler at the latest.

▶ **Challenges of water treatment.** In the traditionally highly industrialised regions of Europe, water quality has improved dramatically in recent years as a result of comprehensive, state-of-the-art wastewater purification. In spite of these efforts, however, the run-off from the treatment plants presents several challenges:

- When purified water from the treatment plant is fed to the outfall, the hygienic quality of the receiving water may be diminished by germs that originated in wastewater discharged by hospitals into the sewer system. Such germs often become resistant to antibiotics through genetic transfer.
- Chemical decomposition products from households and industry that exhibit hormone-like effects (endocrine substances) are not removed by treatment plants and have an impact on the aquatic environment. They can distort the gender balance of the offspring of amphibians and fish.
- Persistent compounds, including numerous pharmaceuticals, pass through the treatment plants and accumulate in downstream ecosystems, though concentration levels are never so high as to impact on human beings. However, the compounds can be detected in the early stages of treating water which is suitable for drinking.
- Each time drinking water passes through a treatment plant the concentration of certain salts increases in the wastewater. If this wastewater has to be used again directly for obtaining drinking water, in the absence of dilution by rainwater in the river, the levels of salt concentration rise continually.

Operation of a sewage plant: Raw wastewater enters the treatment plant and passes through a coarse screen often combined with sand and fat recovery. The water then continues through a sedimentation unit (pre-clarification) and a non-aerated denitrification reactor where nitrate is reduced to molecular nitrogen. The next treatment step is nitrification, where soluble impurities are degraded and ammonia is oxidised to nitrate. The surplus biomass and degraded impurities produced during nitrification are collected in the final clarifier by sedimentation. Clarified water is discharged into a river or other body of water. Surplus sludge sediment from the final clarifier is in part recycled to the nitrification unit to ensure permanent inoculation. The remaining part is collected together with the primary sludge and the recovered fat to be digested in an anaerobic reactor. The end product of digestion is residual sludge, the non-convertible remains of raw waste. This material is dewatered and finally discharged to an incinerator or to be used as fertiliser.

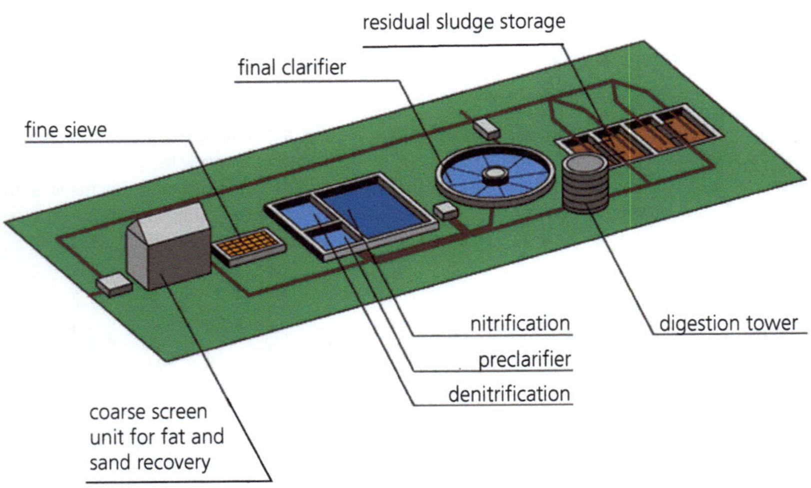

Trends

▶ **Use of rainwater.** Roof rainwater has a special quality when compared with drinking water. On the one hand, it corresponds almost to drinking water with regards to the degree of impurities it contains, yet it is comparable to desalinated water in terms of salt content. Using this grade of water in households could be very advantageous economically. It would avoid the problems of calcium build-up in systems in which water has to be heated, like dishwashers, water boilers, coffee-makers and washing machines.

An effective water supply can be implemented using two different systems. In addition to the centrally supplied drinking water, a second distribution system independently treats rainwater for use as drinking water.

Rainwater must correspond to drinking water both from the standpoint of hygiene and with regards to the concentration of various substances specified by the WHO (with the exception of the salt content). It must therefore be treated with a ceramic ultrafiltration membrane, because ceramic membranes guarantee higher effluent quality compared to polymer membranes in spite of deep filtration characteristics. If this primary technology proves insufficient, the quality standard can be achieved by using additional technologies (activated carbon biology, wet oxidation, etc.).

▶ **Vacuum systems.** These are today employed for toilets on aircraft, passenger ships and high-speed trains. A partial vacuum is created in a collection container linked to the entire network of pipes, so that a vaccum exists at all collection points in the network. A vacuum toilet functions as follows: faecal matter is sucked out of a vacuum toilet installed at the beginning of the network by briefly opening a valve. The means of transport here is air, which is sucked in along with the water containing faeces and transported to the centralised storage tank, where the air is then sucked off again in order to maintain the partial vacuum. The system has several inherent advantages. First, the faeces are drawn quickly through the pipes thus minimising the occurrence of unintentional random biological transformations in the faeces. Second, while the size of vacuum systems is limited – because theoretically a difference of only 1 bar pressure is available to operate the system – the system can easily be scaled up. The biggest advantages over gravity sewers, however, are the use of smaller bore piping and the plain fact that an installed vacuum system requires less digging.

▶ **Anaerobic treatment.** Anaerobic processing of wastewater has a lot of advantages compared to aerobic treatment: organic pollutants are converted to carbon dioxide (35%) and methane (65%), which can be used as a chemical raw material or as a renewable source of energy. Mineralisation efficiency is about 90% compared to about 50% for aerobic treatment, meaning that only small amounts of surplus sludge are produced.

When the wastewater purification is carried out anaerobically in a high-performance membrane biology process the first section of a sewage treatment facility must be a separation unit that concentrates all solids – including the inflowing bacteria and germs – into a concentrate stream of up to 100 g of dry volatile solids per litre. No auxiliary substances are added. The concentrate stream (containing solids) and the filtrate stream are transformed into biogas in a dual-stage, high-load anaerobic reactor with integrated microfiltration. This filtration allows the retention time of the solids and liquids to be dealt with separately, which improves the rate of anaerobic decomposition.

The filtrate water is then subjected to two processes. First comes MAP precipitation (N/P fertiliser). The second step involves the recovery of ammonia, which produces ammonia water or an ammonium salt, i. e. nitrogen fertiliser. The wastewater impurities are thus sustainably transformed into products that can be used again:

- Organic carbon compounds into biogas, a regenerative energy carrier
- Organic nitrogen and phosphorus compounds into fertiliser salts that can replace synthetic fertiliser

The treatment plant run-off is free of bacteria thanks to integrated microfiltration, and the organic trace substances (pharmaceuticals/endocrines contained in the run-off) are substantially reduced when compared to the run-off of conventional plants.

Prospects

- National and international research in the field of water management aims at inventing technologies that will contribute more to achieving the Millenium Development Goals than the traditional state of the art. These technologies include anaerobic wastewater purification in the low temperature range, the implementation of membrane separation technology in the treatment processes and the development of robust, adaptable, multi-purpose infrastructures.
- Reduction of costs is important, but there is also a need for a systemic paradigm change from centralised systems to smaller, robust and more flexible ones.

PROF. DR. WALTER TRÖSCH
Fraunhofer-Institute for Interfacial Engineering and Biotechnology IGB, Stuttgart

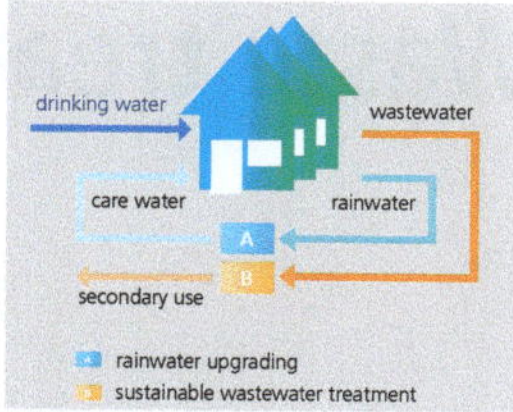

Rainwater separation: Sustainable urban water management is based on a strict separation of rainwater collection and reuse from the wastewater flow. Rainwater (blue) from the roof is mostly collected and stored in a reservoir near the buildings. This fresh (rain) water is processed to obtain drinking water quality by ultra-filtration (to remove micro-organisms) and is redistributed as care water to the households. Wastewater (orange) is collected by a vacuum-sewer system and channelled to an anaerobic wastewater treatment plant located near the buildings. Products of the cleaning process are biogas and minerals. The latter can be used as artificial fertiliser. Source: Fraunhofer IGB

Internet

- www.unesco.org
- www.germanwaterpartnership.de
- www.who.int
- www.water-treatment.org.uk

Waste treatment

Related topics

- Product life cycles
- Logistics
- Image evaluation and interpretation
- Water treatment
- Air purification technologies
- Bioenergy
- Measurement techniques

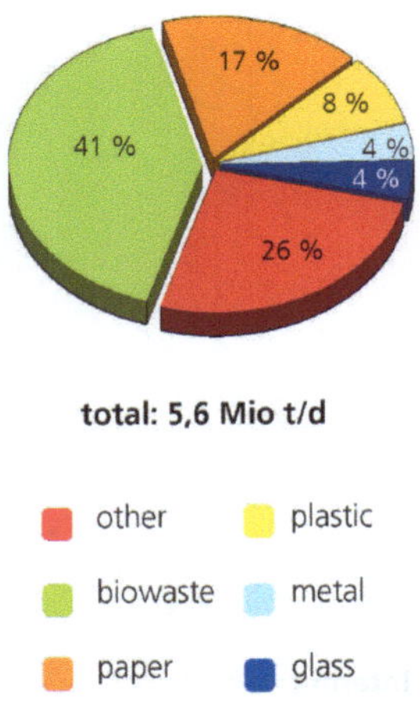

Worldwide generation and composition of municipal solid waste in t/d. Source: Kranert 2007

Principles

Climate protection and the efficient use of resources are major challenges facing modern society. There is a need for waste management solutions based on efficient recycling and treatment strategies to replace end-of-pipe solutions. Depending on the economic development of a country, its inhabitants produce from 70–800 kg of municipal solid waste (MSW) per year. The amount of waste generated is often linked directly to income level and lifestyle. Higher incomes lead to increased consumption and thus more waste – a problem also faced by industrialised economies, which have to find ways to avoid, minimise or recycle the waste they produce.

The worldwide output of MSW amounts to about 5-6 million t/d. If a single day's output was packed into 20-foot containers and piled up over a surface area of 10,000 m², it would form a tower stretching to a height of 1,500 m.

Most of this waste pollutes the environment, is buried in dumps or disposed of to a greater or lesser extent in sanitary landfills. Decomposition of the biodegradable fraction of this waste results in methane emissions which account for an estimated 10% of anthropogenic methane emissions worldwide. In many places waste is simply burned outdoors, thus representing a further source of air pollution. Only a small fraction of waste is burned in modern incinerators equipped with state-of-the-art air filters.

The recovery of recyclable materials from unsorted waste plays an important role in waste management. Municipal solid waste contains about 60–75% potentially recyclable material. Based on average daily accumulation rates, this corresponds to approximately 1.8 million t of paper, plastics, glass and metal and about 2.3 million t of biowaste. Appropriate collection systems would enable a substantial proportion of this waste to be recycled for reuse as secondary raw materials or refuse-derived fuels.

In addition to reducing the consumption of primary raw materials, recycling also helps to combat climate change. By using scrap metal in the production of crude steel, it is possible to achieve energy savings of up to 5,000 kWh/t. For every one percent of cullet added during the glassmaking process, the energy required to melt the raw materials drops by 0.3%. Pulp and paper manufacturing activities typically discharge wastewater at a rate of 250 m³/t. This can be reduced to 5 m³/t by using recovered paper. The corresponding energy saving amounts to about 6,000 kWh/t.

Biowaste is the separately collected biodegradable fraction of municipal solid waste and includes food waste, garden waste and kitchen waste. Biowaste can be transformed into a valuable fertiliser or soil conditioner. The fermentation of kitchen waste produces around 100 Nm³ of biogas per tonne of feedstock, with an energy content of approximately 6 kWh/m³.

Applications

▶ **Collection and transport.** Domestic waste and recyclables are collected in different containers or bags at the source, e. g. at the house (curbside collection), or decentralised in containers (e. g. bottle banks) or recycling stations (communal collection points). The containers are emptied into special trucks or refuse collection vehicles equipped with packer blades to compress the waste (front loader, rear loader or side loader). The waste is taken directly to treatment or recycling plants. However, if the distance, e. g. to the landfill, is greater than approximately 30 km, the waste is taken to a transfer station where it is transshipped into larger units for its further transport. The collection of recyclables is carried out in different ways. Different systems can be used depending on the recycling process:

- Mixed collection of *recyclables*: In this system, the dry recyclables, paper, packaging made from plastic, metal and beverage cartons are collected together and then sorted.
- Single collection of recyclables: If certain recyclables are not allowed to be mixed up with others, they are collected separately in a single container or bag. Biowaste used for the production of soil conditioner or fertilizer should not be mixed up with other substances. Mixing increases the risk of contamination with pollutants (e. g. heavy metals) and the products are no longer marketable. Another example is the collection of glass. If glass is collected together with other recyclables, the latter may be contaminated with glass splinters. This leads, e. g. to problems during the paper recycling process.

▶ **Material recovery facility.** Certain materials, such as paper and cardboard, must be collected separately and processed prior to recycling. This is necessary firstly because the commodity market demands non-mixed secondary raw materials, and secondly, because they can become so heavily contaminated with household waste that they are no longer directly marketable. The cleaner the state in which recyclable materials can be sold, the greater the financial returns. To ensure that these quality requirements are met, the recyclables must be processed in a material recovery facility. Besides manual sorting, more and more automatic systems are being used for the processing of recyclables.

- *Near-infrared spectroscopy*: This technology is used for sorting packaging waste. Infrared light is shone on the mixed recyclable waste from above and reflected back to sensors located above it. The spectrum of the reflected light varies as a function of the material characteristics. By comparing the result with reference spectra, specific materials (such as various types of plastic or beverage cartons) can be identified and selectively removed from the material stream by means of blower units.
- *Air classifiers*: They are used for the separation of light materials (such as foils). Different materials are sorted out in a stream of air according to their characteristic settling rates. The degree of separation depends on the air velocity and the weight and composition of the materials.
- *Magnetic separator*: Magnetic metals (such as tin cans) can be separated by magnets, mostly electromagnets.
- *Eddy-current separators*: These are used for the separation of non-ferrous metals (such as aluminium cans). When electrically conductive metals are exposed to a magnetic field, eddy currents are generated in them. The eddy currents create an opposing magnetic field. The resulting repulsive force deflects the conductive metals from the material stream. Eddy-current separators are installed downstream of the magnetic separators.

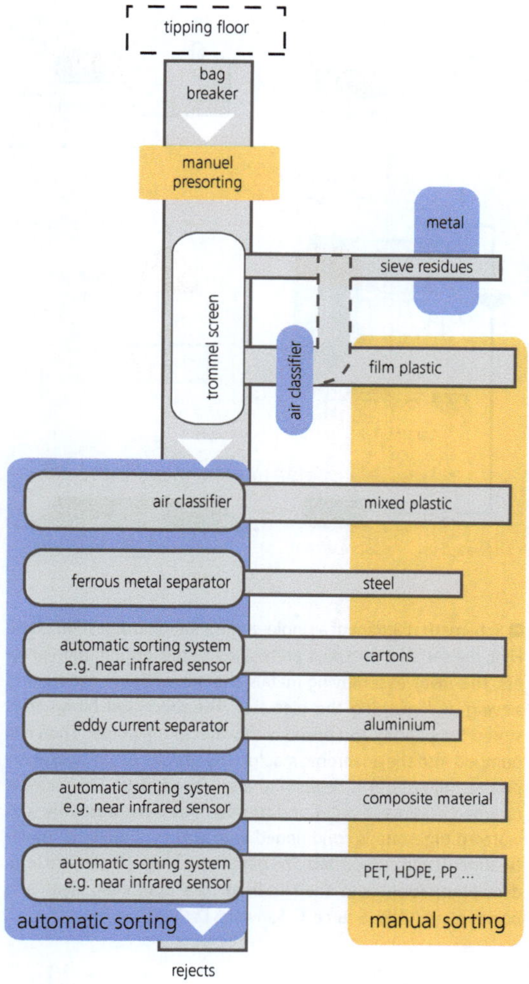

Schematic diagram of a material recovery facility for packaging waste: After delivery, the packaging waste is fed onto a conveyor belt. As a first step, closed bags are opened automatically (bag breaker). Before screening (trommel screen), large items of waste are sorted out manually (manual pre-sorting). Films and metals are separated from the sieve residues. In a first step, the screen overflow is pre-sorted automatically. Air classifiers, metal separators and near infrared sensors, for example, make it possible to sort out beverage cartons or specific types of plastic such as PET with a high purity grade. Remaining contaminants are sorted out manually. Source: D. Clauss

▶ **Biological treatment of biowaste.** Composting is the degradation and/or conversion of biowaste by microorganisms with atmospheric oxygen (aerobic). The main end products of this process are compost, carbon dioxide and water. The biowaste is turned frequently to ensure optimal mixing. In addition, the material is aerated and if necessary it can also be irrigated. This promotes the decomposition of the biowaste and prevents it from rotting. Composting is usually preceded by mechanical processing of the material (sieving, removal of foreign matter, crushing). The resulting compost can be used in agriculture, for example.

The degradation and/or conversion of biomass during the fermentation process (anaerobic) takes place in the absence of free oxygen. The organic substrate is converted through various intermediate stages into *biogas* (methane and carbon dioxide). Non-degraded substrate constitutes the fermentation residue. The fermentation of waste has a long tradition in agriculture and in the treatment of sewage sludge. New technologies are being developed to permit the utilisation of biowaste collected from homes and the food processing industry. Different process combinations are used. A distinction is made between mesophilic (30–40 °C) and thermophilic (55–65 °C) processes, high solids digestion (> 15% DS) or low solids digestion (<8% DS), single or multi-stage as well as between continuous and discontinuous facilities. The choice of process depends on the nature of the feedstock and economic considerations.

▶ **Pretreatment of municipal solid waste.** The mechanical biological treatment of MSW is designed to prepare it for disposal on landfill sites. It is not intended as a means of producing compost from mixed mu-

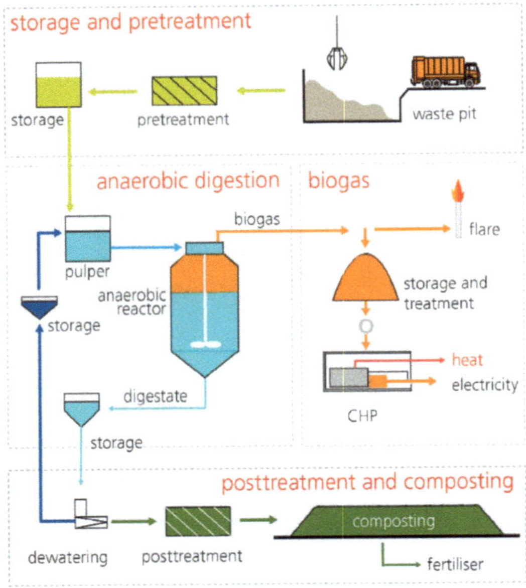

Schematic diagram of a single-stage anaerobic digestion plant: First, the sorted biowaste is pretreated for the fermentation process. This involves removing metals and other foreign matter, and sieving and crushing the biowaste. The processed biowaste is stored temporarily and mixed with water in the pulper. Then it is pumped into the anaerobic reactor. The residue of the hydrolysis process (digestate) is dewatered after 20–40 days. The water is returned as process water or disposed of as wastewater. The dewatered digestate is conditioned with structural material (posttreatment) and composted. The generated biogas is cleaned (e.g. desulphurisation) and then utilised in a combined heat and power plant (CHP). Source: C. Sanwald, D. Clauss

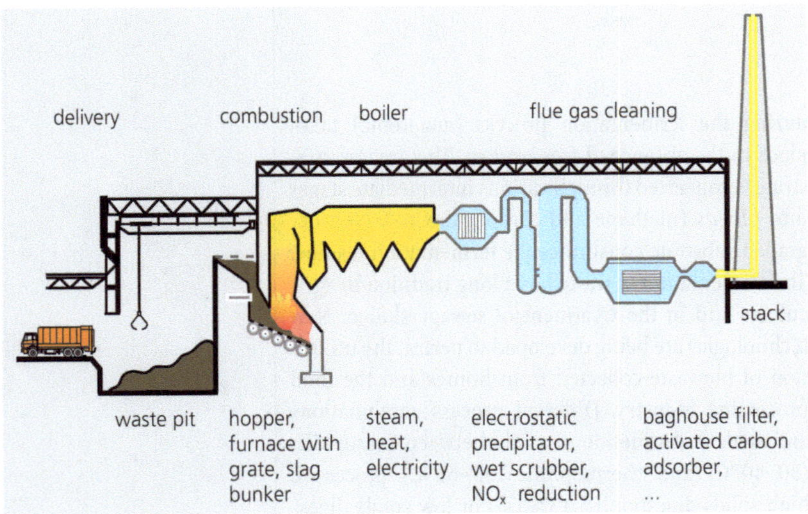

Schematic diagram of an incineration plant for municipal solid waste: Waste is delivered by means of trucks or trains to the waste pit and stored temporarily and blended. At the hopper the waste is fed onto the incineration grate and burned with primary air. The combustion residues are removed on the deslagging unit. The exhaust gas is burned in the afterburning chamber at 850 °C and conducted through the boiler area – use of the energy content – to the exhaust gas cleaning unit. Before the exhaust air is cleaned by scrubbers, particles are removed by means of electrostatic precipitators. The fabric filter with activated carbon removes the remaining pollutants. Emission measurements are carried out to verify the purity of the exhaust air before it leaves the incineration plant through the stack. Source: C. Sanwald, D. Clauss

nicipal waste. The aim of this biological treatment is to stabilise the waste as far as possible by technically assisted biodegradation. The idea is to reduce emissions of landfill gas as far as possible and to minimise the organic load (BOD) in the leachate. In addition, recyclables (e.g. metals) and high-calorific-value fractions (e.g. plastics) are extracted for use as secondary raw materials and refuse-derived fuels respectively. Secondary fuels are used for instance in cement kilns. The municipal waste is separated in several process steps into the following material flows: residues for material recycling, energy recovery and materials for biological treatment. The biodegradable fraction is taken for aerobic stabilisation or anaerobic digestion with aerobic post-treatment. The residence time (retention time) varies between 8–16 weeks, depending on the process. The stabilised waste is disposed of in a sanitary landfill.

▶ **Incineration of municipal solid waste**. During thermal treatment, waste is incinerated under optimum conditions. The incineration plant consists of five main areas:

- Waste pit: interim storage and homogenisation;
- Combustion line: incineration of the waste on moving grates at around 1,000 °C
- Afterburning chamber: burning of the resulting combustion gases at 850 °C
- Steam boiler: use of the energy content of the gases for steam generation (electricity, heat recovery)
- Flue gas cleaning: removal of pollutants that could not be destroyed by incineration or that were generated during the process. The technically most complex part of a waste incineration plant is the exhaust gas cleaning

The first step is the removal of particles from the exhaust air (fabric filters, electrostatic precipitators). In multistage scrubbers, the gaseous pollutants and heavy metals are absorbed by means of water, caustic soda or lime milk (SO_2, HCl, HF). The nitrogen oxides are removed by selective catalytic or non-catalytic reduction. At the end of the exhaust gas cleaning stage, the exhaust air is conducted through an activated carbon filter, which adsorbs the remaining contaminants (such as mercury or dioxins). The slag generated by the thermal treatment process is reused, for example in road construction or for filling disused mine shafts. It can also be disposed of in landfills. Due to their high pollutant content, flue ash and other residues of exhaust air cleaning must be disposed of in special underground salt mines. Other techniques in use around the world, in addition to moving grate incinerators, include rotary kilns (especially for hazardous waste) and fluidised bed processes (notably in Japan).

▶ **Sanitary landfill.** Landfill emissions can be reduced by installing vertical and/or horizontal gas wells for the active collection of landfill gas. The collected landfill gas is burned at high temperatures in a flare (1,200 °C) or, after cleaning, utilised in a combined heat and power plant (CHP). The polluted leachate is collected in a drainage system running underneath the landfill and treated on site or in a central wastewater treatment plant. On-site processes consist of nitrification and *de-nitrification* in combination with filtration processes (*ultra- and nano-filtration, reverse osmosis*). To prevent groundwater pollution, besides choosing a site with suitable hydro-geological conditions, a base liner must be placed below the drainage system. Controlled landfill operations supplemented by environmental monitoring help to contain dust, reduce the development of odours, and avoid windblown litter.

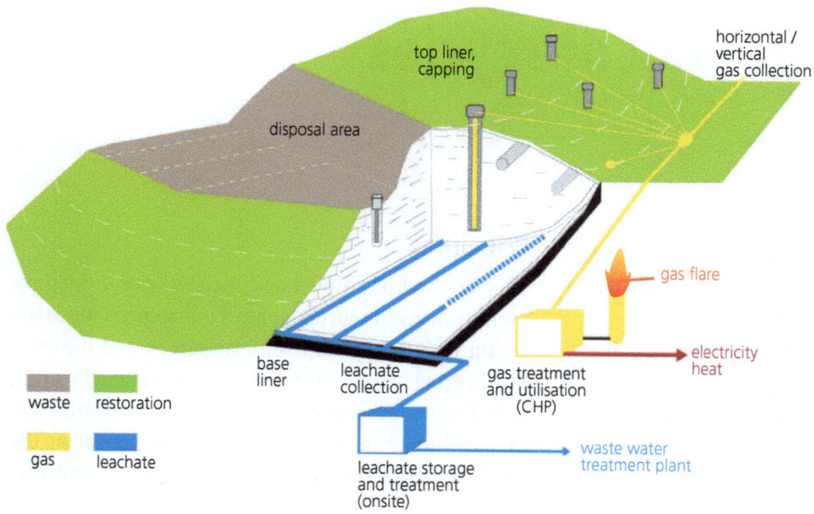

Model of a sanitary landfill: A landfill for municipal waste is divided into two main areas; the waste site itself and the technical infrastructure designed to minimise emissions. The base of the landfill consists of an appropriate hydro-geological underground and a technical base liner system. The liner is covered with a layer of gravel containing a drainage system to collect all of the generated leachate. The collected leachate is then treated either onsite or offsite (wastewater treatment plant). The waste is spread out and compacted on the disposal area. Once the final filling height is reached, the landfill is closed with a technical top liner and a capping (restoration). The landfill gas (mainly CH_4 and CO_2) generated in the landfill is actively collected through vertical or horizontal gas wells. The gas can be recovered in a combined heat and power plant (CHP) or must be burned in a flare. Source: D. Clauss

Trends

▶ **Recycling.** The increased worldwide demand for raw materials and the corresponding increase in commodity prices has led to a growing interest in the recovery of secondary raw materials from municipal solid wastes (buzzword: *urban mining*). In certain high-income countries, more than 50% of MSW is already being recovered (material and energy recovery). Many countries are only just beginning to implement systems for sorting and recovering waste, while others are already reaping the benefit of such initiatives. Brazil, for example, collects and recycles more than 90% of discarded aluminium beverage cans. In many cases, however, such quotas are only achieved as a result of the gray economy (scavengers etc.). Especially in the larger metropolitan cities, an attempt is being made to integrate this segment of the population in the public waste management system.

▶ **Climate protection.** The Kyoto Protocol defines the Clean Development Mechanism (CDM) as one means of reaching the set reduction targets for climate protection. Active landfill gas collection and recovery is a recognised method for reducing greenhouse gas emissions under the terms of the CDM. The reductions achieved in this way (CO_2-equivalents) are entitled to be traded as certified emission reductions (CERs). A substantial proportion of the costs incurred for upgrading landfill sites can be financed in this way. A typical example of a CDM project is a landfill of the city São Paulo (Brazil). Approximately 5,000 t of waste is dumped each day at the Bandeirantes landfill. Before the site was upgraded, landfill gas was passively recovered and burned directly at the gas well or in a simple flare on the gas well. However, this method only allows recovery of a part of the landfill gas, and the combustion temperatures are not high enough to guarantee complete destruction of the landfill gas. The new system permits the landfill gas to be actively collected and recovered in gas engines. Modern high-temperature flares have also been built to burn off excess gas and as a security measure in the event of emergencies.

Waste emissions can be further reduced by mechanical and biological pretreatment.

Prospects

- A wide diversity of both low-tech and high-tech processes are in use throughout the world for the treatment and recovery of municipal solid waste.
- The technologies and organisational structures employed in waste management have evolved to such a high level that they are capable of making a significant contribution to climate protection and the preservation of natural resources. There is a distinct trend towards material flow concepts that will no doubt replace end-of-pipe solutions in the near future.

PROF. DR. MARTIN KRANERT
DETLEF CLAUSS
Institute for Sanitary Engeering, Water Quality and Solid Waste Management, Stuttgart

Internet

- http://ec.europa.eu/environment/waste
- http://waste.eionet.europa.eu
- http://europa.eu/scadplus/leg/en/s15002.htm
- www.worldbank.org/html/extdr/thematic.htm
- www.oecd.org
- http://maps.grida.no/theme

Product life cycles

Related topics

- Waste treatment
- Metals
- Polymers
- Logistics
- Sensor systems
- Joining and production technologies
- Automobiles

Principles

The adoption of product and material closed-loop recycling forms the cornerstone of our responsible interaction with the environment, resources, and life cycle engineering. Apart from costs, life cycle engineering aims first and foremost to minimise the environmental impact throughout a product's life cycle, i. e. during the three phases of production, product usage and disposal.

The use of raw materials, auxiliary materials, automotive fluids and energy, not to mention waste and effluent disposal represent substantial cost factors for manufacturing companies. Most companies can achieve potential savings equivalent to return on sales by utilising resources more efficiently. Today, measures involving production-integrated environmental protection are being specifically leveraged to increase production efficiency. Thanks to *closed-loop recycling*, waste generated during production can be sorted by type into separate materials and fed back into production. This closed-loop recycling includes raw materials as well as finished and incorrectly classified components and products. Furthermore, the auxiliary materials, automotive fluids (e.g. cooling lubricants, oils, solvents) are reprocessed and reused as part of auxiliary materials closed-loop recycling. As these substances tend to be harmful to the environment, the cost of disposal is high so that recycling opens up huge potential savings for companies.

Applications

▶ **Disposal of end-of-life vehicles.** Modern-day automobiles are far from homogeneous, comprising around 10,000 individual components and over 40 different materials. Electronics, such as on-board computers, navigation devices or automatically controlled windscreen wipers, are increasingly finding their way into vehicles. The electronics market is highly dynamic and characterised by short innovation cycles. Meantime, supply obligations mean the auto industry that fits these electronics to its vehicles needs to guarantee that customers have access to spare parts with the same functional capabilities. To guarantee that the automotive industry has the spare parts it needs beyond the initial production period of new models, one strategy involves recycling and reprocessing used components from dismantled end-of-life vehicles. End-of-life vehicles are recovered in five key stages: draining the vehicle, dismantling, sorting and processing reusable components as well as shredding the remaining body.

Once the exterior and the engine compartment of the end-of-life vehicle have been cleaned and the tyres removed, the vehicle is drained by removing the automotive fluids and the components containing fluids (e.g. fuel, engine oil and engine, brake fluid, coolant and radiator as well as battery acid). Depending on the specification of the end-of-life vehicle, additional major assemblies such as ABS, automatic transmission, rear-axle and centre differential or air conditioning systems are included. Draining is essential, especially due to the risk of possible (water) contamination as fluids run out on the storage area, along with the risk of subsequent soiling of the residual waste.

Once drainage is complete, the components that are easily accessible from the outside are dismantled, including body components that can be recovered (e.g. hood and tailgate), bumpers, plastic fuel tank, auxiliary assemblies (e.g. alternator, carburettor) or windows. A start is then made on dismantling the interior with the removal of the side trim, steering wheel, seats and dashboard, etc. Once the trunk panelling has been removed, the power cable harnesses fitted in the vehicle can be easily accessed and pulled out of the end-of-life vehicle at a suitable point.

As part of the *dismantling process*, a decision needs to be made about where the components will ultimately end up. Some electronic components such as electric motors are virtually wear-free and therefore suitable for reuse or continued usage in similar vehicle models. The servo and actuating motors for the electric central locking, power windows, cruise control system,

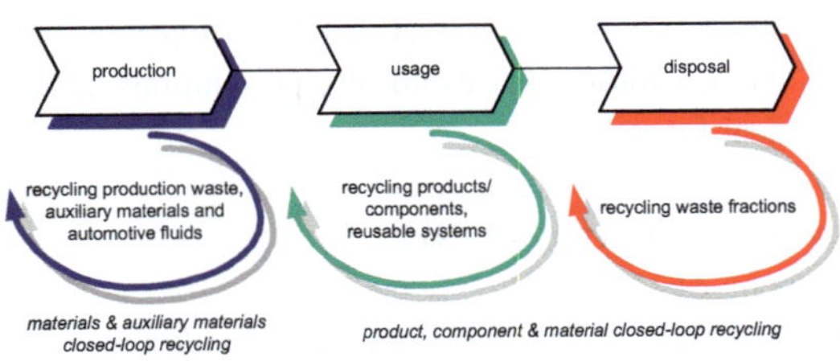

▣ The environmental impact of a product can be reduced through closed-loop recycling in all phases of the product life cycle.

and much more besides, are sorted, tested to see they work and, where appropriate, overhauled. Faulty parts are replaced and areas of wear reworked. Cable harnesses by contrast are only suitable to a limited extent for reuse in other vehicles since they are very difficult to remove without being damaged. They can generally be recovered to obtain the raw materials as part of copper recycling. *Plastics* that have been separated by type following dismantling are then passed on for plastic recycling. The remaining body is shredded, and the steel and nonferrous metals subsequently recycled.

▶ **PET drinks bottles.** Polyethylene terephthalate (PET) is a material used worldwide to manufacture packaging (e.g. bottles for carbonated drinks) and fibres for clothing. PET can be completely recycled. Once at the recycling plant, the used PET drinks bottles, which have previously been crushed into bales, are initially broken apart in separation drums. Coarse impurities such as metal, stones as well as coloured PET bottles are sorted and the loose PET bottles ground and washed in a wet-grinding mill to create uniform flakes. These flakes are a mixture of PET from the bottles and polyolefines from the labels and closures. Since polyolefines float on water, while PET is heavier than water and will sink in a tank filled with water, both material flows can be separated in a float-sink process by virtue of their different densities. The polyolefines are drained off from the surface of the water and fed into a separate recycling process once dry.

Moistening the PET flakes with caustic soda and the subsequent heating in a rotary kiln forms the core URRC process during which classic mechanical recycling is combined with chemical recycling. The top PET layer reacts with the caustic soda to create sodium salt of terephthalic acid and separates off, similar to peeling an onion. In this way, remaining adherent and migrated impurities can be removed from the PET flakes. The impurities are bonded to the salt; other plastics form lumps or change colour (PVC goes amber to black, for instance).

Fine and oversized grains are sieved, metal removed and finally colour impurities are eliminated during subsequent aftertreatment. The subsequent usage of the recycled PET demands a high level of colour purity since consumers tend to leave 'yellow' bottles on the shelves. For this reason, the entire material flow is managed by a grinding stock sorter which scans the grinding stock flake by flake with a multilinear colour camera. The recorded spectrum of the PET flakes is compared electronically against a precisely defined sample and reject flakes are blown out using pneumatic nozzles. The dried PET flakes, which are once again suitable for food use, can be used to produce PET bottles.

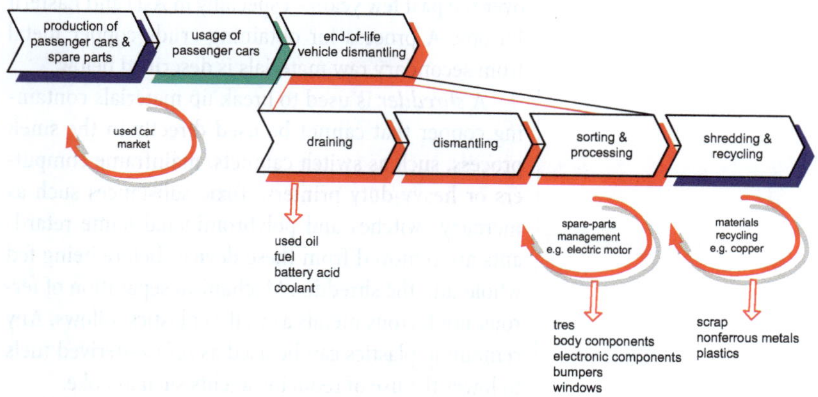

End-of-life vehicle dismantling includes various stages: draining, dismantling, sorting, processing and shredding, and offers a wide range of options for recycling components and the raw materials in automotive components.

End-of-life vehicle dismantling. Top left: Modern-day vehicles are far from homogeneous, comprising around 10,000 individual components and over 40 different materials. From bottom left to top right: All the automotive fluids (e.g. fuel) and the components containing fluids (e.g. fuel tank, fuel lines) are first removed (draining) from the end-of-life vehicle. Easily accessible components are then dismantled, e.g. the instrument panels or the doors. Dismantled end-of-life-vehicle components are reprocessed, labelled and stored for subsequent use as spare parts. End-of-life vehicle components that cannot be reprocessed are collected for subsequent recycling: e.g. copper in cables (bottom right). The remaining body is crushed for further storage and transport. In this way, up to 40 crushed bodies can be transported on a single tractor/trailer unit. Source: BMW AG, AWG Wuppertal

▶ **Copper recycling.** Copper is used in a wide range of applications thanks to its excellent heat and electrical conductivity as well as its durability: air-conditioning units, heat exchangers and brake pads contain copper, as do electric motors, cables, microchips or sanitary installations, piping and coins. As a result, the demand for copper has been constantly increasing over the past few years – especially in Asia and Eastern Europe. A process for obtaining crude copper metal from secondary raw materials is described below.

A *shredder* is used to break up materials containing copper that cannot be used directly in the smelt process, such as switch cabinets, mainframe computers or heavy-duty printers. Toxic substances such as mercury switches and polybrominated flame retardants are removed from these devices before being fed whole into the shredder. Mechanical separation of ferrous/nonferrous metals as well as plastics follows. Any remaining plastics can be used as refuse-derived fuels to lower the use of reducing agents such as coke.

Pyrometallurgical and wet metallurgical processes are used to recover what tend to be pre-sorted secondary raw materials containing copper. The *pyrometallurgical process* begins with the smelting and conversion stages, which are implemented consecutively in a single unit. The key feature of the process is the use of a 15-m long submerged combustion lance, which is immersed into the furnace from above and provides heat using oil while adding air and oxygen. Conversion involves blowing air and oxygen through the liquid crude copper whereby the crude copper is contaminated with copper oxide, turning into what is known as converter copper. The crude copper obtained after this process stage has an average copper content of 95%. The converter copper, still in a liquid state, is fed via a holding furnace to the anode furnace. Here a large amount of copper scrap containing high levels of copper is used in addition to the liquid crude copper and refined to produce a copper content of 98.5–99 %. The refined copper is then cast into copper sheets which are used as anodes in subsequent *electrolytic refining*.

A DC current is passed through two electrodes immersed in a conducting fluid (electrolyte) to facilitate the electrodissolution of metals in, or the deposition of metals from, an aqueous medium. Reaction products from the materials contained in the electrolyte are formed on the electrodes. The electrolyte contains positively and negatively charged ions; once a voltage is induced, the positively charged cations move to the negatively charged cathode where they acquire one or more electrons and are reduced in the process. The opposite process takes place at the anode. Electrolytic refining takes place in electrolysis cells lined with lead; each individual cell is fitted with several cathodes and anodes. The copper moves to the anode under the influence of the DC current in the solution and is deposited on the adjacent stainless-steel cathode. After a week, 99.99% pure cathode copper can be mechanically separated from the stainless steel sheet in a stripping machine.

Bales of crushed PET drinks bottles. Source: Günter Menzl

Trends

▸ **Aluminium recycling.** Aluminium is the third most abundant element in the earth's crust and the most abundant metal therein. Due to its high propensity to react with non-metals, especially oxygen, it does not occur naturally in a purely metallic form. Bauxite provides the raw material for producing aluminium. This ore can be turned into aluminium using a process that consumes large amounts of energy.

The recycling concept is nothing new in the aluminium industry since the secondary aluminium industry has been a successful part of the market since the 1920s. Waste containing aluminium is used as the feed material for aluminium recycling. Sources can include material that can be sorted by type, i. e. out of scrap whose composition is exactly known, or as aluminium scrap consisting of mixed scrap made up of various constituents. Scrap that can be sorted by type includes production waste from foundries, end-cuts, spoiled castings and rejects; mixed scrap covers old scrap and waste metal. The latter is included when smelting and resmelting aluminium and alloys, and must be 'scraped off' from the smelt surface.

Processing equipment for waste or scrap containing aluminium includes shredders that crush the mixed scrap and break down the compound material as far as possible. This is followed by airstream sorting that separates the components that will fly (the so-called shredder light fraction, SLF), such as plastic, wood and textiles, from the material stream. In the subsequent magnetic separation, ferromagnetic materials such as steel, nickel and iron fragments are separated from the non-magnetisable metals. The metals (such as aluminium) are then separated from the non-metals during eddy current separation.

In an eddy current separator, a rapidly revolving magnetic rotor is positioned inside a non-metallic drum which rotates at a much slower speed. This produces flux variations at the surface of the drum, which is the drive pulley for the conveyor belt carrying the stream of mixed materials. As the conducting metallic particles are carried by the conveyor over the drum, the magnetic field passing through the particles induces currents in them. Because the waste is of random shape, the current cannot flow within them in an orderly way. The currents tend to swirl around – or eddy – within them – hence the name "eddy current". The effect of such eddy currents is to induce a secondary magnetic field around the nonferrous particle. This field reacts with the magnetic field of the rotor, resulting in a combined driving and repelling force which literally ejects the conducting particle from the stream of mixed materials. This repulsion force in combination with the conveyor belt speed and vibration provides the means for an effective separation. Without prior magnetic separation, the magnetic attraction induced in ferrous metals in a mixed stream is stronger than the eddy effect. Hence, items containing steel stay magnetically attached to the belt, and the separation can still be effective. However, because this can cause excessive vibration and belt wear, magnetic separation of ferrous materials is best done before the mix reaches the eddy current stage. This can be achieved by putting strong magnets above the conveyer belt.

This is followed by static float-sink separation which separates the material flows into various density classes. The alloy composition of the aluminium fraction produced in the separation machinery is unknown. The resulting mixed fraction is used to produce secondary cast aluminium. This entails a downcycling for the wrought alloys contained in the mixture.

Depending on the level of contamination of the aluminium scrap used, a suitable metallurgy smelt process is selected. In general, scrap can be smelted with or without salt. Smelting equipment includes salt-based rotary drum furnaces that are fixed or can be tipped, as well as furnaces operated without salt (closed-well furnaces, side-well furnaces) or induction furnaces. Organically contaminated but low-oxide aluminium scrap is melted in the closed-well furnace operated without molten salt. In the side-well furnaces, bulky scrap, among other things, is flooded with liquid metal in a smelting basin (side-well). A rotary drum furnace is used if the fines and/or the oxide component of the aluminium scrap increases, such as with waste metal, mixed turnings and with a more frequent change of alloy. Induction furnaces are only used separately in secondary smelting plants due to the need for clean, practically oxide-free scrap. Foundries represent the main application for induction furnaces.

▸ **Recovery of minute amounts.** Phosphorous compounds are essential for the survival of plants and living creatures. Phosphorous, for instance, is one of the most important nutrients for plants and is added via organic fertilisers (e.g. liquid manure) as well as industrial mineral fertilisers. Phosphorous and its compounds are used industrially in the production of flame retardants, additives, softeners, pesticides or even matches. Limited phosphorous resources, increasing demand and the dwindling quality of mined ores have led to an increased effort to obtain phosphorous from secondary raw materials.

Most technology research projects to obtain phosphorous from secondary raw materials involve the chemical/thermal aftertreatment of sewage-sludge ash. As part of these processes, carcass meal or sewage sludge is burned at high temperature in mono-incineration plants. Organic substances, such as pathogens, hormones and drug residues are destroyed, leaving phosphorous behind as an oxidic compound. In subsequent process steps, heavy metals are removed from the ash and the phosphorous is chemically converted into a form usable for plants.

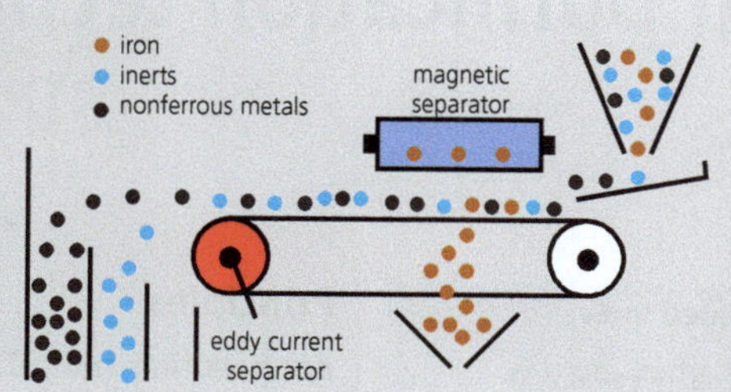

Eddy current separator (ECS) – a conveyor belt used to separate aluminium. It features a particular magnetic field in the head, which is generated by a high frequency polar wheel: when the nonferrous metals approach the magnetic field, they are lifted and "expulsed" to the appropriate collecting channel, while the inert materials drop freely into another container. Even the tiniest traces of iron, which differ from the nonferrous metals, are retained by the magnetic rotor and dropped into the proper container using the eddy current separator.

The refined copper is cast into sheets. Source: Norddeutsche Affinerie AG

Prospects

- Resources are limited: lower ore concentrations in raw material deposits have made obtaining high-quality raw materials more expensive. Meanwhile, the increasing influence of speculative investors in commodity markets (roving capital) is also making raw materials artificially scarce.
- Ever shorter product lifecycles increase the demand for closed-loop product recycling, the requisite takeback structures as well as flexible dismantling concepts for recycling product groups and/or materials.
- A paradigm shift is underway in international and European waste and resource legislation: Instead of the purely product-related regulations in the form of takeback obligations for individual products/product groups (e.g. WEEE, batteries), materials-related targets relating to the substances and materials contained in products are increasingly becoming the focus of international waste monitoring and economic cycles.

KERSTIN DOBERS
DR. KATHRIN HESSE
Fraunhofer Institute for Material Flow and Logistics IML, Dortmund

Internet

- www.tms.org/pubs/journals/JOM/0408/Gesing-0408.html
- www.pcimag.com/CDA/Articles/Feature_Article/
- www.copperinfo.com
- www.aluminium.org

Air purification technologies

Related topics

- Carbon capture and storage
- Nanomaterials
- Defence against hazardous materials
- Environmental monitoring
- Automobiles
- Product life cycles

Principles

Air pollution control includes all measures taken to avoid or reduce the release into the atmosphere of any potentially harmful substances caused by human activity. Emissions are attributed to specific sources. Technical facilities such as furnaces, power plants or stationary engines, which emit via stovepipes, smoke stacks, or tailpipes, are known as point sources. These can easily be analysed, quantified and controlled using standardised measurement techniques. Mobile sources of emission such as ships, cars, and trucks are known as line sources. In contrast, fugitive emissions from stockpiles, dumps or tips caused by wind erosion and dispersing dust are defined as area sources and are difficult to control. Fugitive emissions are also produced by ventilation through doors, windows, and the ridge turrets of halls.

Data of emissions from human activities all over the world are scant and unreliable. Where industrial activities are concerned, CO_2, NO_x and dust emissions are caused primarily by power generation from fossil fuels along with the use of electricity. Melting and metalworking processes and cement production also have high specific emissions, i. e. high emissions per unit produced. Every effort is being made to reduce emissions by cleaning the off-gas. However, as production rates increase, emissions will also continue to rise. There is reason to be concerned about the explosion of industrial emissions in developing countries.

Emissions are carried away – usually upwards – by air circulation, mingle with the air mass, spread out, and are steadily diluted. The remaining portion of the emission that returns to the earth's surface and interacts with people, the ecology and buildings is called ambient air quality. Federal air quality directives define emission limit values for air pollutants from point and fugitive sources in order to improve ambient air quality.

Emissions and ambient air quality can only be controlled and evaluated with the aid of suitable measurement techniques and methods. These are directly incorporated in the production process and thus inherently linked to regulated emissions. It is becoming increasingly important to monitor ambient air quality, interpret the respective data and understand ambient processes. It is now evident that mathematical, statistical, and analytical methods must be combined in order to correctly understand the processes underlying the changing contributions from different sources under different meteorological conditions.

If emissions cannot be avoided, suitable techniques must be applied to reduce them. In terms of air pollution control, this means implementing methods for the detection and removal of solid and liquid particles as well as specific gaseous compounds, including odour.

Emissions of solid particles	Emissions of liquid particles	Emissions of gases
• ash from combustion	• oil fog	• carbon dioxide combustion
• soot combustion cars	• finest droplets overspray condensing gas	• ammonia stock farming
• cement production processing	• finest droplets with dissolved substances	• solvents paints • fuel vapour fueling cars
• metal dust processing welding	• aerosols solvents	• odours wood fire

Classification of typical emissions by composition and sources

Applications

► **Filters for industrial facilities.** Filters separate harmful substances from the exhaust emitted during technical processes. It is important to differentiate between large flue gas streams such as those from power plants at 100,000 m³/h, and small ones such as those from coffee roasting facilities at 1,000 m³/h. With regard to the diversity of substances that have to be separated, the greatest effort is made at waste incineration plants. Besides particulate matter, heavy metals such as mercury or arsenic, and gases such as hydrogen chloride and sulphur dioxide, must be separated from the exhaust trace and dissipated as harmless compounds, e. g. by converting nitrogen oxides into nitrates. Another option is to destroy the compounds. The various methods are connected in series. The chain usually begins with a dust or electrostatic filter for particulate matter. When gas flows through a dust filter, particles are held back according to their size or adhere to the fibres of the filter. In the case of an electrostatic filter, particles are charged by a high-voltage source and then pass through the electric field to the electrode. If the gas flow is to be cleaned more extensively, gas separation is followed by absorption filters that dissolve gas components in a solvent or detergent. If the compounds are soluble, the simplest method is dilution in water.

To separate the sulphur dioxide created by the combustion of fossil fuels, a limestone suspension – a suspension of fine limestone particles in water – is preferred. The SO_2 reacts with the limestone to form anhydrite plaster, which is discharged from the process and used in numerous products such as gypsum plaster board.

Catalytic methods are another way of separating gas compounds from exhaust gas. *Catalysts* used for air pollution control often look like combs through which the gas passes. The operating temperature is often over 200 °C. To decrease the amount of nitrogen oxides produced by combustion, a small quantity of ammonia is mixed with the gas flow upstream from the catalyst. It reacts with the catalyst and the nitrogen oxides to form nitrate and water vapor. A secondary effect of this technique is that organic trace gases from incomplete combustion are broken down into CO_2 and water by the residual dioxygen in the exhaust gas. This method is also used in the catalytic converters of vehicles.

A particularly suitable method for reducing gas components of a lower concentration is adsorption. This method operates with highly porous solids that have a huge inner surface due to their microscopically fine pores. One gram of activated carbon, for example, can have a surface area of up to 1500 m². The inner surfaces can adsorb gas molecules. These universal sedimentation properties are used in waste incineration plants. A filter with many tonnes of activated carbon acts as a "police filter", reducing many harmful substances to a low level. It thus ensures that the emission limit values are observed even if operations are interrupted. The active carbon is incinerated; recovery of the activated carbon by thermal or of treatment and removal of the harmful treatment is worthwhile in small amounts.

► **Filters for indoor rooms.** Air pollution control in enclosed spaces and production facilities or for process air is easier than at industrial facilities. Fibre filters with a paper-like structure can be used to prevent air pollution in indoor rooms. The air is pressed or drawn through the filter medium; the particles that it contains impact the fibres, adhere to their surface and are thus removed. The fine fibres are close together, with larger gaps here and there. Nevertheless, particles that are much smaller than the gaps in this grid of fibre layers are still unable to pass through it. Filtration rates of over 99% are quite realistic. Because of their inertia, larger particles cannot follow the flow lines and there-

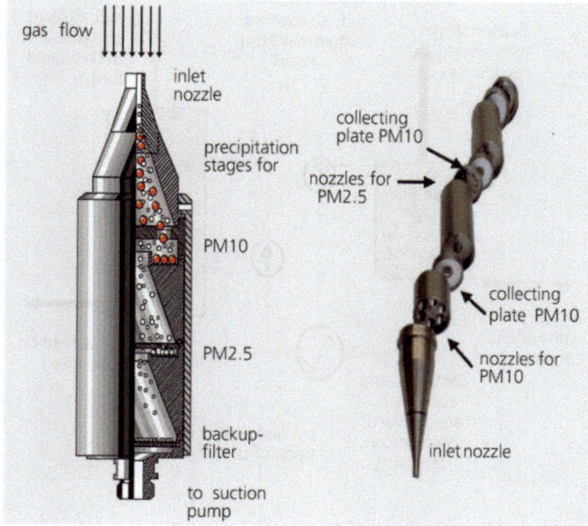

A cascade impactor as a measuring device to collect and classify fractions of particle matter according to particle size: The particle-loaded gas is drawn through the impactor by a suction pump. The gas passes from top to bottom. The impactor consists of an assembly of several (in this case 3) stages, each with an impaction plate where the particles are precipitated on a filter paper or membrane. Because of the special fluidic construction and the physical attributes such as size, mass and inertia of the particles, every stage preferentially intercepts one size of particles – for example, particles with a size of about 10 µm (PM10) are separated from particles of about 2.5 µm (PM2.5). For analysis purposes, the filters are removed and weighed in a laboratory. During the sampling period the flow at the inlet is adjusted to the flow velocity of the gas to be measured. This avoids disturbance of the gas flow and loss of particles. The device is made of stainless steel. The diameter is 70 mm, length 350 mm Source: IUTA e.V.

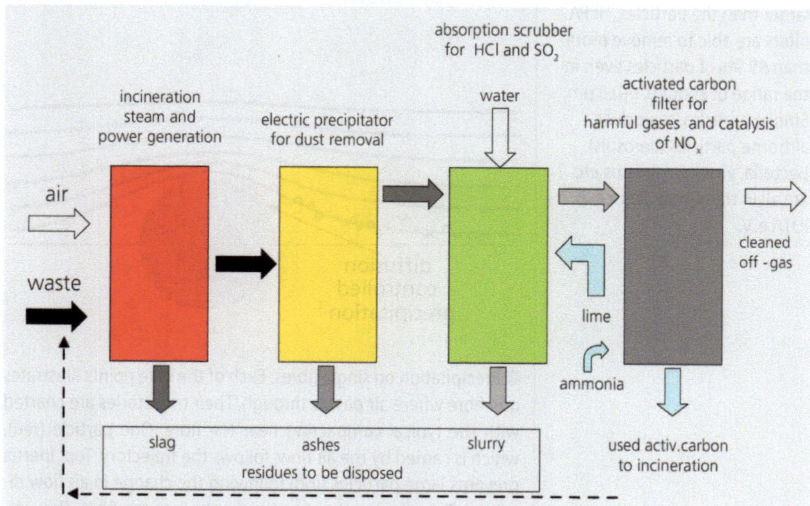

Equipment for cleaning at waste incineration plants: Particulate matter is deposited in the first step, followed by absorption of hydrogen chloride and sulphur dioxide and a multilevel adsorption filter with activated carbon to separate low-concentration gaseous substances, e.g. dioxins and furans. Furthermore, a catalytic stage at the filter converts nitrogen oxides into nitrate and water. All residues are disposed of separately. Source: IUTA e.V.

fore impact the fibres. Particles that are smaller than the gaps between the flow lines and the fibres can partially pass through. Even smaller particles are increasingly subject to the influence of Brownian motion and thus to diffusion; this means that they no longer move in a straight line and are very likely to impact a fibre. In terms of particle deposition, these filters have a characteristic deposition curve for the typical minimum passage. Cigarette smoke, with a particle size of 0.1 µm, lies in the range of the filter's minimum deposition size and therefore does not impact very easily. This implies a low filter efficiency.

Trends

▶ **Particulate matter.** Many studies have shown that the mortality rate in areas polluted with particulate matter is significantly higher than in unpolluted areas. There are numerous theories as to the mechanisms at work, but they are not yet very reliable. The problem lies in the emmission of particulate matter with a diameter of less than 10 µm. According to an EU directive, this must not exceed an annual mean of about 40 µm/m³, and a daily mean of 50 µm/m³ must not be exceeded on more than 35 days a year. Due to their small diameter, the mass of the particles is also low. Such particles therefore remain in the air for a long time before being deposited or washed out; they may also cover great distances. This is why dust from the Sahara is sometimes found in Western Europe.

Measurements have proven that severe background pollution already exists. This is the basic contamination in an area that is not directly affected by a pollutant source. Particles can be generated in many ways, e.g. by agricultural activities with dust and gaseous emissions (e.g. ammonia), natural dust eddies, or volcanic activity. Sea salt, too, contributes to background pollution. Up to 50% of the contamination in cities is due to background pollution. 10 to 20% can be attributed to natural causes and the rest to anthropogenic sources. This background contamination is augmented by dust from exhaust gases, tyre abrasion and resuspension, and industrial facilities. The burning of domestic fuels, especially wood, also contributes to background pollution. In order to burn wood at low emission rates in future, it will be necessary to use filters even in small-scale heating systems and furnaces; e.g. catalytic converters in stovepipes that oxidise CO to CO_2 and convert the residual hydrocarbons into CO_2 and water. This reduces odour emissions and helps filter the dust that impacts the catalyst. These filters require occasional cleaning.

▶ **Odours.** The human nose is able to recognise specific substances from just a few molecules. It is consid-

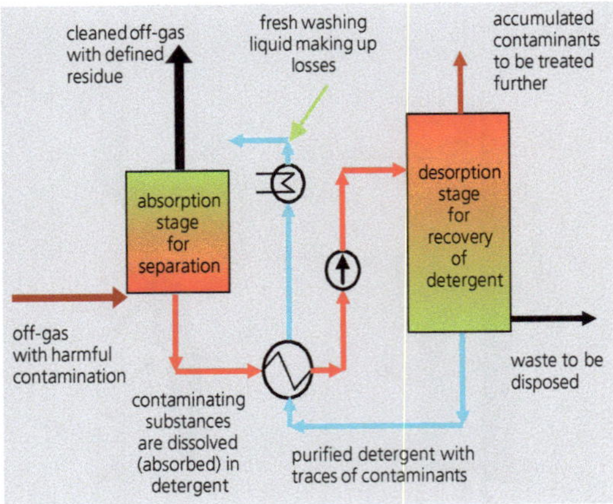

▶ Elution of gas components from an exhaust gas: In the actual cleaning stage (absorber), harmful substances are dissolved with a liquid detergent. The detergent is purified and traced back in a desorption stage. If the material can be used economically, it is purified and re-used; otherwise it is disposed of as waste and must be replaced. Source: IUTA e.V.

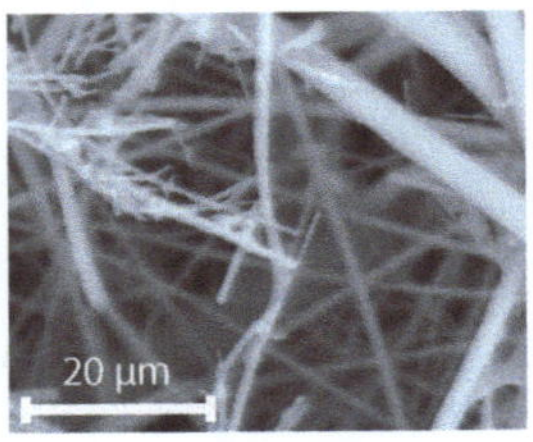

▲ HEPA (high-efficiency particulate air) filter in two magnifications by a scanning reflection electron microscope (REM): The images show the structure as an irregular composition of single fibres in multiple layers. Despite gaps between the fibres that are larger than the particles, HEPA filters are able to remove more than 99.9% of particles even in the range of about 0.1–0.3 µm. Smoke particles, toxic dust, airborne particles (aerosols), bacteria, viruses, asbestos etc. are all in this range. Source: IUTA e.V.

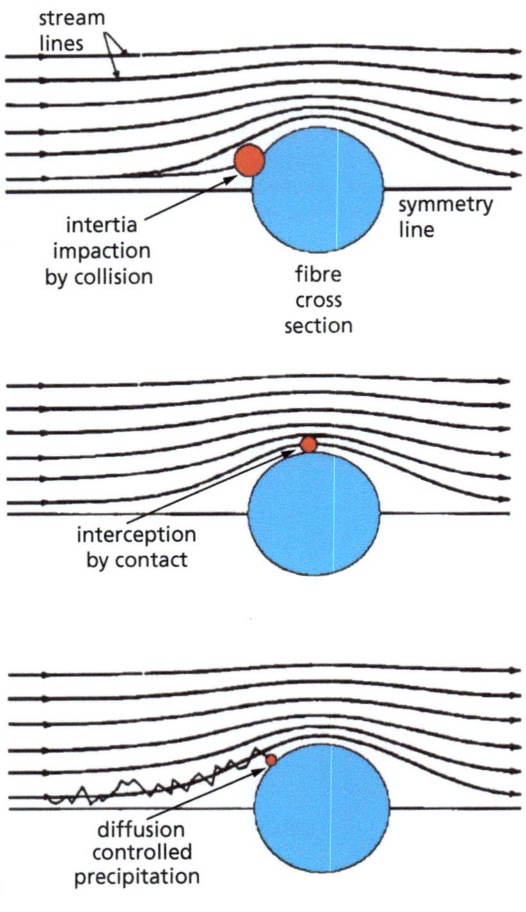

▲ Precipitation on single fibres: Each of the blue points illustrates one fibre where air passes through. Their trajectories are charted with the typical compacting near the fibre. One particle (red), which is carried by the air flow, follows the trajectory. Top: Inertia prevents large particles from following the change in air flow direction. They fly straight ahead and adhere to the fibre. They impact the fibre if the distance from the fibre during flight is less than the particle radius. Otherwise, they pass by and are not precipitated (barricade effect). Bottom: Small particles set in motion by the temperature no longer move in a straight line and follow the mass transport by diffusion. This increases the probability of impacting and adhering to the fibres. Source: IUTA e.V.

erably more sensitive than (almost) any other measuring instrument. However, there is no connection between the odour and the pollutant potential of chemical compounds (e.g. hydrogen sulfide, solvent, benzol). Strongly smelling substances may be harmless, while toxic ones might not be noticeable at all. Filters for odour precipitation must be tailored to each individual application. *Biofilters* in which bacteria "devour" the odorous substances are often used to reduce odours. Corresponding bacterial strains, for example, are settled on damp brushwood through which the exhaust gas to be cleaned is made to flow, e.g. at industrial livestock stables or in the food industry. Filters for solid matter that bind or disperse odours are easier to design, and their efficiency can be verified in preliminary tests. Care must be taken to preserve the right milieu for the bacteria, which need a constantly humid environment with a correspondingly optimised temperature. To ensure their success, these filters require relatively complex pilot experiments.

In developing new filters, scientists are attempting to enhance the precipitation or adhesion capability of the fibres and to harness supplementary effects that will drive the particles towards the fibres. Simultaneous precipitation of gases is only possible with fibres capable of adsorbing certain noxious gases. In straightforward cases, this can be achieved with activated carbon fibres that adsorb specific gas molecules on their microstructured inner surfaces. A simple filter medium of this kind thus possesses additional characteristics: besides the usual particle precipitation, it also binds gases or odours. The same effect is achieved by coating normal fibre filters for particle precipitation with finely dispersed activated carbon. Another modification might be to construct the filter from microscopic globules of activated carbon. The gaps between these globules form channels through which air can pass. Recent developments also involve catalytic effects: finely dispersed microscopic titanium dioxide TiO_2 is incorporated in the outer surface coating of the filter (the substance contained as a whitener in almost all white products from paint to toothpaste). Constant radiation with ultraviolet light causes TiO_2 to develop catalytic properties enabling organic gas molecules to be oxidised. In future, these filters will also have a bactericidals effect.

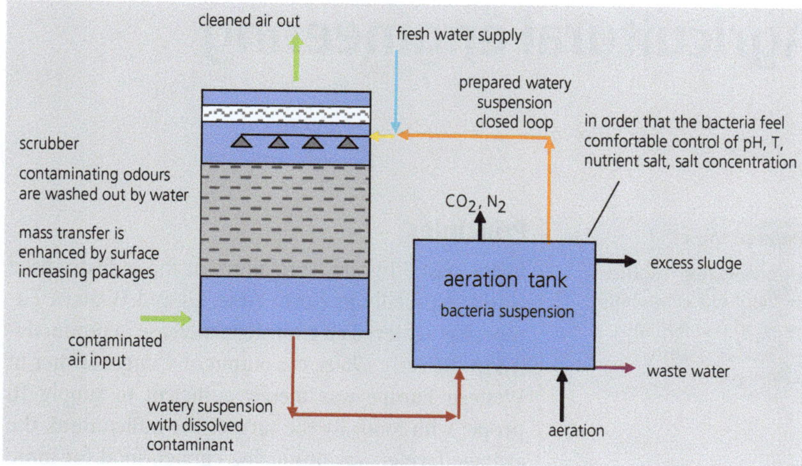

Bioscrubber: A bioscrubber is a scrubber which is operated with a recirculating watery suspension containing bacteria. The air to be treated passes the scrubber from the bottom to the top of the tank through a package of high porosity and surface area. The suspension countercurrent flows from the top to the bottom. Soluble contaminants are dissolved in the water. They are intercepted in both the scrubber and the aeration tank by bacteria that metabolise the absorbed contaminants. The reaction products are CO_2 and biomass, i.e. more bacteria. Excess bacteria must be removed from the system. Controlled operating conditions are important in order to maintain the biological activity of the bacteria. Source: IUTA e.V.

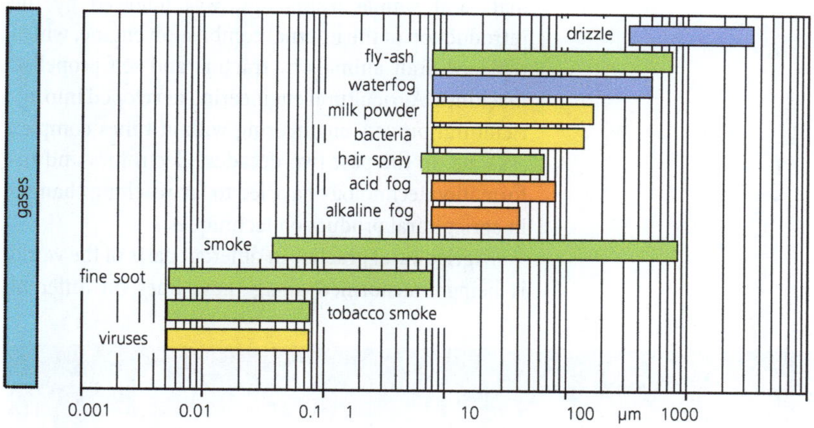

Examples and size ranges of particles

Prospects

- A wide diversity of filter systems is in use throughout the world to ensure air quality.
- Air pollution control and filtration has reached a high technical standard and can therefore make a major contribution to air quality. Filter systems now under development include integrated multifunction filter systems; high volume filter systems for power plants; filter systems with a reduced pressure drop, higher efficiency and reduced cost; filter systems to lower indoor pollution; process and cabin air filtration for highly sensitive production processes; and filter systems to capture or separate particles in the nano range.

PROF. DR. KLAUS GERHARD SCHMIDT
Institute of Energy and Environmental Technology e.V., University Duisburg-Essen

Internet

- www.umweltbundesamt.de/luft-e/index.htm
- www.air-purifier-power.com/clean-air-purifier.html
- www.airpurifiers.com/technology

Agricultural engineering

Related topics

- Renewable resources
- Plant biotechnology
- Food technology
- Sensor systems
- Robotics

Principles

In the past 50 years, agriculture in the industrialised world, especially in North America and Western Europe, has achieved an enormous increase in productivity. In the early 1950s, the output of a single farmer in Western Europe was merely sufficient to supply 10 people with food. By the turn of the millennium, the average farmer was producing enough food for more than ten times as many persons. The reasons for this increase in productivity are not only progress in plant breeding, plant nutrition and the control of plant diseases and pests, but primarily also technical developments.

Agriculture has always been an early testing ground for new technologies. The first wave of mechanisation, during which manual labor became more efficient and less onerous thanks to the use of draft animals and simple technology, was followed by the introduction of the internal combustion engine, which replaced draft animals by tractors and self-propelled machines. Agricultural engineering developed into agricultural process engineering with at times complex systems. In the past two decades, electronics and information technology have led to far-reaching changes in agricultural production techniques.

Agricultural practice in different parts of the world is being transformed along these lines on different time scales, depending on the conditions at the outset and the rate of progress.

The demand for agricultural products is set to increase in future as the world population grows and more and more biomass is needed for energy production. Agricultural engineering will therefore gain in importance as a means of producing enough food crops and animal foodstuffs for more people with less available arable land.

Stage of development	Characteristics	Example
manual labor	- subsistence farming	- manual tillage (hoe)
mechanisation	- reduction of heavy physical work increased productivity	- implements drawn by animals
motorisation	- increased productivity - intensive farming - release of land formerly used for growing forage for draft animals	- tractor - self propelled combine harvester
automation	- increased productivity and efficiency - higher quality of work - reduced workload - reduction of emissions	- automatic guidance systems - automatic milking systems - on-demand automatic feeding systems

Stages of development in agricultural engineering

Applications

► **Tractor**. The tractor is the key machine in agriculture. It has to meet a growing number of requirements related to changes in agricultural practice and tighter regulations concerning especially engine emissions. Modern tractors are designed to improve efficiency and optimise the control of and interaction with implements. Different subsystems of tractors contribute to these aims.

Typical agricultural tractors are powered by state-of-the-art diesel engines. In the past 50 years, the average engine power of new tractors grew from 12 kW (16 hp) to 88 kW (120 hp) in Western Europe. To fulfil emission regulations for off-road vehicles applied in Europe and North America, electronically controlled high-pressure fuel-injection systems combined with exhaust gas recirculation are used. Higher combustion temperatures are a challenge for cooling systems although the space for integrating larger radiating surfaces is very limited.

Different working speeds (from 0.5–50 km/h), which often require full engine power, are characteristic of tractor use in agriculture. Manufacturers have made an effort to meet these demands by offering manual multiple-gear transmissions (up to 40 forward and reverse gears). The ideal transmission should allow speed to be adjusted continuously at any engine output and engine speed depending on the specific requirements of use, to maximise working speed (and output) on changing soils and hill gradients or to minimise fuel consumption. This is possible today with the aid of hydraulic-mechanical powersplit transmissions. The power generated by the engine is spilt into a hydraulic and a mechanical branch. The hydraulic branch allows for continuously variable adjustment of the transmission ratio. The mechanical branch transmits power with very low losses. The two transmission branches are merged in a merging transmission. Improved hydraulic units and modern control electronics allowed powersplit transmission concepts to be realised. Communication with the electronic control systems of modern diesel engines enables optimised driving strategies to be applied (e.g. consumption or power-optimised strategies).

The hydraulics account for approximately 15% of the production costs of a tractor. Such systems are used within the tractor to support operation (hydraulic steering, hydraulic brakes, hydraulic clutch) but their major application is in interaction with implements. Front and rear three-point linkages, which are controlled hydraulically, connect implements to the tractor to create optimised combinations.

Besides the simple coupling of implements to the tractor, many agricultural machines used together with the tractor need power to run tools and functions. The mechanical way to transfer almost full engine power is the power take-off (PTO), a splined driveshaft running down the centre-line of the tractor. PTO speeds and shaft details are standardised, the speeds have fixed ratios to the engine speed.

More and more implements require hydraulic power from the tractor to control action by moving ram cylinders or to power hydraulic motors. To reduce power losses especially without or with little load, load-sensing hydraulic systems are used in medium and large-size tractors. The hydraulic power output of tractors range from 5–70 kW.

Newly developed engine-coupled generators supply electrical power to implements that require more than the 1–3 kW typically provided by the established 12-volt onboard electrical systems. Electrical power is easier to control than mechanical or hydraulic systems. Applications based on a 480-volt supply are capable of delivering up to 20 kW. Future generations of agricultural equipment will feature more electrical drives.

Many tractors are equipped with sophisticated electronic vehicle management systems. These systems coordinate the operating hydraulics, all-wheel drive shifting (electric, electro-hydraulic, or electronic), steering and brakes, as well as the power take-off PTO, especially during the repeated complex turning maneuvers at the headland.

► **Sensor-based distribution of fertiliser.** Nitrogen is one of the most important plant nutrients. It is very mobile in the soil, and its natural availability is highly dependent on the soil and the weather. For this reason, the available quantity of nitrogen fluctuates very significantly within fields and over the course of the year. High, reliable plant yields requires the additional supply of plant nutrients containing nitrogen. These fertilisers must be applied at the right time and in neither insufficient nor excessive quantities because failure to do so can have a negative influence on the quantity and quality of the harvest. In addition, nutrients not absorbed by the plants can have negative environmental effects.

To ensure that crops receive precisely the amount of fertiliser that they need to meet their nitrogen requirements, sensor systems have been developed to

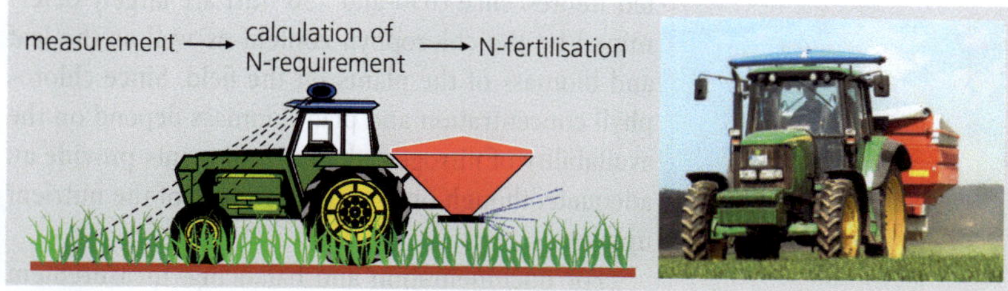

Sensor-based application of nitrogen fertiliser. Left: Measuring principle is reflection measurement. Right: Sensor at farm work. The measurement values are determined once per second (approximately once every 3 m at a working speed of 12 km/h). Depending on the fertilising algorithms, they are immediately converted into a fertilising recommendation and directly transmitted to the electronic control system of the fertiliser spreader. Source: Yara GmbH & Co. KG

Left: Section drawing of a modern tractor with six cylinder turbo-charged diesel engine, drive train with a power split continuously variable transmission (CVT), mechanical four wheel drive and additional front PTO and front linkage. Right: Rear end of a tractor. 1) lower links and top link of the hydraulically lifted three-point linkage. 2) hitch to couple trailers. 3) PTO shaft to transfer mechanical power to implements. 4) break away couplers to connect hydraulic hoses to transfer hydraulic power to implements. Source: AGCO Feudt determine the plants' uptake of nitrogen while the tractor passes the crop stand, enabling applications of mineral nitrogen fertiliser to be dosed accordingly.

In the illustrated spectral reflection measurement system, the reflected radiation is measured at two points each on the left and the right side of the tractor. This permits an integral analysis of a crop stand covering an area of 3×3 m². The arrangement of the sensors can compensate for differences in reflection caused by the position of the sun.

In addition to the light reflected by the plants, the fluorescence emitted by the plant material can also be measured and analysed. The spectral composition of the reflected light (e.g. 680 and 800 nm) and the emitted fluorescence (690 and 730 nm) are largely determined by the chlorophyll content as well as the leaf and biomass of the plants on the field. Since chlorophyll concentration and plant biomass depend on the availability of nitrogen, the measurements provide an adequate, though only indirect picture of the nutrient uptake.

For documentation and balancing, measurement values and fertiliser quantities are recorded together with positioning information provided by a GPS receiver. On a typical winter wheat field, the quantity of nitrogen dispensed can vary by up to 25%. The targeted distribution of nitrogen fertiliser enables the total quantity applied to be reduced by 5–20%. Furthermore, yields are slightly higher and quality (protein content) is increased and stabilised (less variation in protein content).

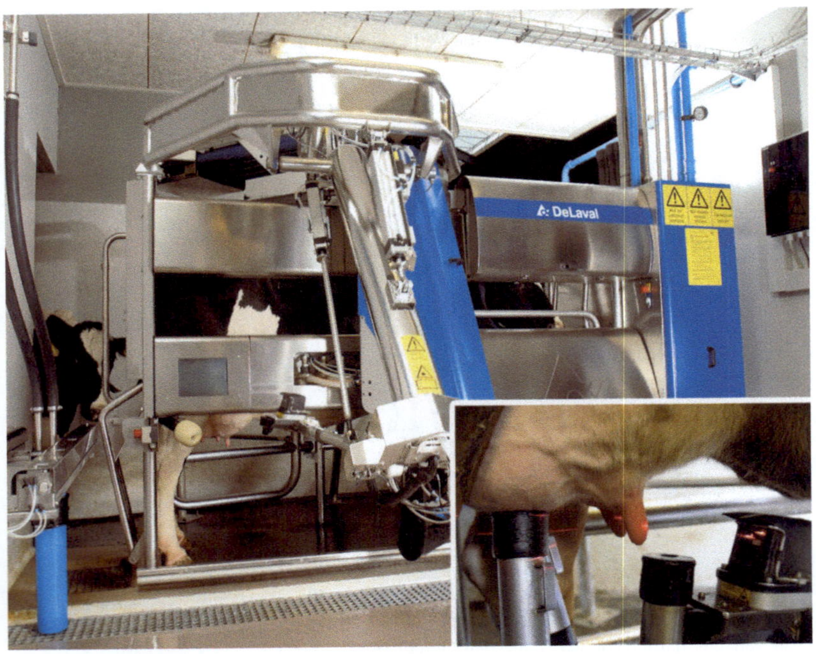

Automatic milking system ("milking robot"): The entire milking process takes place automatically without the presence of personnel. Inset: Automatic attachment of the teat cups. A laser-assisted image processing system locates the position of the teat. Source: DeLaval GmbH, Sweden

► **Automatic milking systems.** The milking of dairy cows is the most time-consuming work in animal husbandry. This work must be carried out by qualified personnel at fixed times twice or three times a day, 365 days a year.

When cows are milked conventionally in stanchion barns or loose houses, the milker goes to the cow, or the cows are driven by the milker to the milking parlor. In automatic milking, by contrast, the cows make their own way to the milking box, attracted by the feed concentrate available there. When a cow enters the milking box, its udder and the teats are cleaned with the aid of a cleaning brush or a cleaning cup. Afterwards, the individual teat is localised using appropriate techniques (laser measuring system, image processing, or ultrasonic sensors), and the milking cup is attached automatically. During the milking process, the milk flow is recorded together with various milk parameters (e.g. electrical conductivity, color and somatic cell count) that allow the quality of the milk to be assessed. After the milk flow stops, the milking cups are removed, and the cow leaves the milking box. All milking work is handled by the machine. The presence of humans during the milking process is unnecessary, and the cows can go to the milking system following their own rhythm.

► **Automatic fattening pig feeding systems.** The distribution of feed to animals is a time-consuming task involving hard physical labor. A cow, for example, needs more than 10 t of foodstuff per year. Therefore, animal feeding is highly mechanised in agriculture, and, where possible, it is also automated. In pig fattening, feeding technology is very far advanced. In the group husbandry of pigs, fully automatic, computer-controlled dry and liquid feeding systems are state-of-the-art. These systems automatically measure and mix the individual components of a feed mixture depending on the nutritional requirements of animals of different sizes. In pig fattening, it is standard practice to dispense the feed mixture to compartments or animal groups, whereas cows and breeding sows are fed individually.

Newer systems feature sensors that register the filling height of the feed troughs and thus integrate the eating behavior of the animals into the control process. In order to reduce the environmental impact of nitrogen and phosphorus excreted in feces and urine, nutrient-adapted phase feeding strategies have become widely established. Instead of dispensing one uniform ration to the animals over the entire fattening period, the feed components are adapted to the animals' nutritional requirements at different stages of growth.

Trends

Trends in agricultural engineering differ according to the development level of different regions. The most advanced stage is "precision agriculture", an approach made feasible by key technologies like satellite positioning and electronic animal identification.

► **Satellite positioning.** Global navigation satellite systems (GNNSs) make it possible to locate and navigate agricultural machines in the field and enable all collected information and data to be allocated to specific areas in a field. Differential global positioning systems (DGPSs) allow the position of the machine to be determined with a precision of a few centimetres, i. e. to guide tractors or harvesting machines. In DGPS, a fixed reference station is used that receives the GPS signal from the GPS satellites. The exact knowledge of the geographic position of the reference station enables GPS positioning errors to be detected and corrected to achieve great precision.

► **Electronic animal identification.** Today, every individual animal can be identified electronically using a reliable, automatic and contactless technique. In the housing systems that were common in the past (e. g. stanchion barns for cows), the animals generally were tied up so that each one could be fed and cared for individually. The use of electronic animal identification in combination with computer-based techniques, however, allows the individual animal to move freely in the group. With the aid of the transponder, each animal can nevertheless be identified and supplied with feed or cared for individually at the different stations (feeding station, weighing scales, milking stall, etc.). Thus, electronic animal identification allows for animal-friendly group housing with individual care and monitoring of the individual animal. The identification technique used is *RFID* technology in the form of electronic ear tags, neck collars, injectable transponders, and bolus transponders. The use of electronic animal identification systems is likely to become increasingly widespread in future, not least because it also improves traceability.

► **Sensors and sensor systems.** Sensors are another key technology for precision agriculture. They are used to measure physical, chemical, or biological parameters of the production process in order to improve open-loop and closed-loop control, monitor the production process and provide better documentation and traceability. Sensors must be inexpensive, sufficiently precise, reliable, robust and stable over long periods. In addition, they must be suitable for use under the particular conditions common in agriculture (heat, cold, dust, aggressive and corrosive gases, shocks and vibrations).

The establishment of control loops for yield and environmentally optimised sowing, fertilising, and crop protection as well as species-appropriate and environmentally compatible animal husbandry and performance-oriented animal nutrition requires online sensors for the determination of the following types of parameters:

- Physical and chemical properties of the soil (e. g. density, pore size distribution, soil moisture, nutrient contents)
- Health and nutritional status of plants and animals, crop growth, animal behavior (e. g. nutrient supply, diseases, activity, stress)

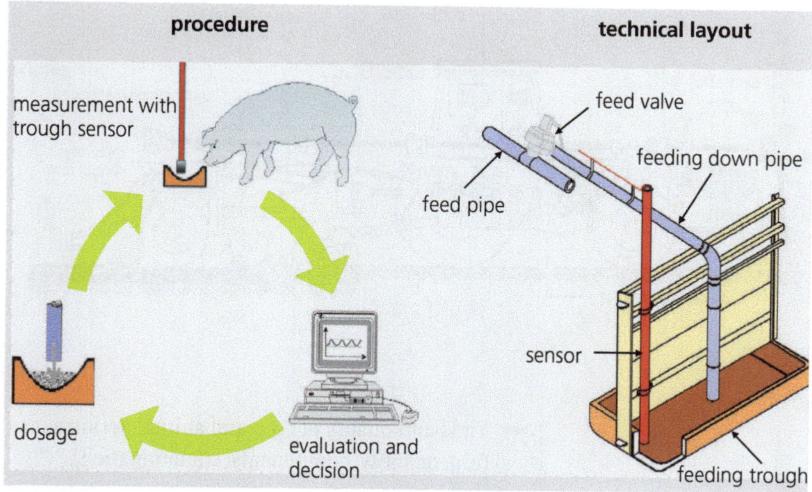

Principle of sensor feeding in fattening pig husbandry: The individual feeding behaviour of the animals is measured and evaluated, and the feed quantity is adapted accordingly. Source: G. Wendl according to Raschow, Bavarian State Research Centre for Agriculture

Electronic animal identification. Top: Ear tags are attached to the animals externally. Centre: Bolus transponders are swallowed by ruminants and remain in the fore-stomach. Bottom: Injectable transponders are placed subcutaneously or intramuscularly Source: G. Wendl, Bavarian State Research Centre for Agriculture

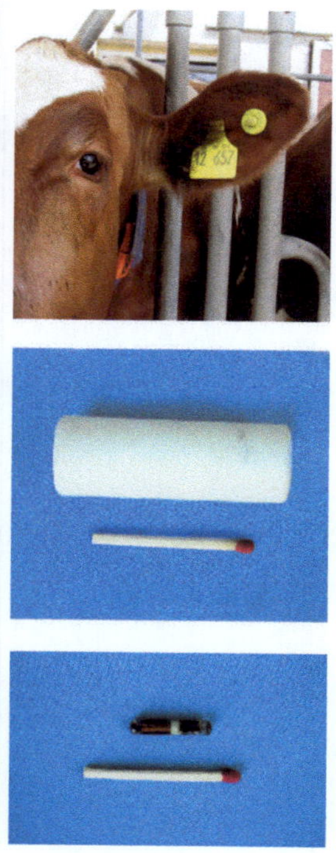

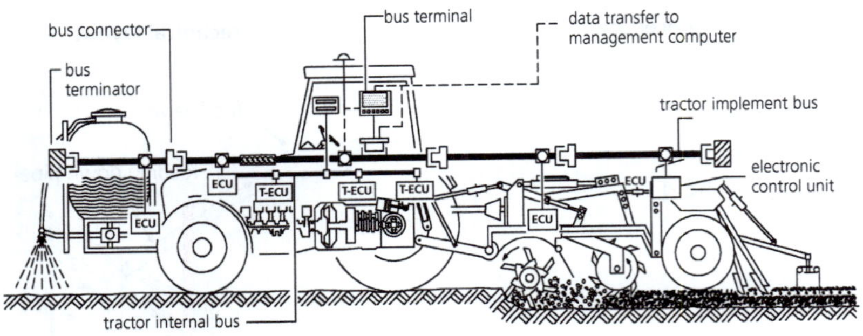

Agricultural BUS system: The tractor-implement BUS is used for implement operation with the aid of the user station and connection between the implement computers/implement control units (Electronic Control Unit, ECU) of the sprayer, the tillage implement and the sowing machine. The system is coupled to the tractor-internal BUS system (T-ECU) for the control of the engine, the transmission and the hydraulics. The connections to the satellite positioning system and data transmission to the farm computer also run via the tractor-implement BUS and the user station (hatched). Source: Auernhammer, Technical University Munich

- Yield and quality of crop and animal performance (e. g. quantity, components, contamination)

Simpler sensors (e. g. yield sensors, milk meters) are already available and in use. Many complex sensor systems, however, are still in the development and testing phase.

▶ **"Precision crop farming"**. Automatic data collection and documentation based on sensor information combined with the associated geographical coordinates ("*geo-referencing*") obtained by means of satellite position detection provide important information for both farm management (site-specific determination of operating resource consumption, work time requirements, yields) and public administration (land area, land use). Such information can also be used in quality management and to provide product traceability from the store counter to the field (tracing of every single product lot back to the relevant date and place of sowing, all fertilising and crop protection measures, and the harvest), as required by modern consumers. In addition, it provides the basic information for site-specific cultivation (soil properties and yield potential of different field areas) and simplifies the realisation of experiments on farms ("on-farm research", e. g. the reaction of different plant varieties to different soil characteristics within one field or several fields on one farm).

Site-specific arable farming modifies crop management to take account of differences (heterogeneity) within the fields, which are determined using technologies and methods of automatic data collection. It allows for site-specific tillage, sowing, fertilising (e. g. sensor-controlled nitrogen fertilising), irrigation, and plant protection with the goal of using the potential of the individual areas sustainably.

For these complex work procedures, it is necessary to operate, control, or regulate different implements (e. g. a sprayer, a tillage implement, and a sowing machine) simultaneously. Reliable information exchange between the tractor and implements is realised with the aid of agricultural *BUS systems*, which today define the physical basis and the communication protocols. Whereas, in the past, each implement deployed by the tractor had its own separate system, these systems are now networked and able to communicate with one another. Thus, the fertiliser sensor, for example, can directly transmit fertilising recommendations to the spreading equipment.

Agricultural tractors and self-propelled machines have a separate vehicle-internal data communication network for the engine, transmission, safety and other electronic circuits, which can output data, but does not permit any feedback control. So the agricultural BUS system for the control and regulation of the different mounted and drawn implements must work parallel to vehicle-internal control. It connects the operating terminal in the tractor to the implement control systems and guarantees that mounted and drawn implements from different manufacturers work together without problems and can exchange information. The agricultural BUS system uses a gateway to receive information from tractor electronics, such as engine or PTO speed. The sensors for positioning and navigation (GPS) as well as the condition of the soil and the crop stands (reflection measurement) are also integrated in the agricultural BUS system. All necessary information for farm and plant cultivation management, e. g. fertiliser or harvest quantities or parameters of tillage

Autonomous rice planter: An unmanned vehicle with automatic guidance and control is combined with an automated rice seedling transplanter. Source: Terao BRAIN, Japan

work, is transmitted to the farm computer via the BUS system.

Automatic vehicle control allows work processes of large agricultural machinery to be optimised by avoiding overlap due to complex cultivation structures or the inexperience of workers. It also facilitates the work of the driver during long working hours and under poor visibility caused by dust, fog, and darkness, for example. Since 1980, harvesting machines have been equipped with automatic steering systems using feelers that work along stable plant rows (maize stalks). For machine guidance along the cutting edge of a grain stand, these systems were extended to include laser scanners in combines 10 years ago. Attempts to steer tractors automatically during soil cultivation and tillage were only successful when satellite positioning was used. While the automatic steering of agricultural machinery with the aid of differential satellite positioning (DGPS) is available and the combination of a manned guiding vehicle and unmanned following vehicles is already being studied, unresolved safety problems prohibit the use of entirely unmanned vehicles on the field at present.

▶ **"Precision livestock farming".** Computer-controlled techniques, such as concentrate dispensers, have been used in livestock farming for quite some time in order to feed and monitor animals individually. But a sustainable approach capable of reconciling the demands of profitability with the different requirements of animal, environmental and consumer protection requires the use of integrated systems. The goal is to use the individual animal rather than the average as the basis for the control and monitoring of animal-based production. With the aid of sensors, different parameters like individual feed intake, animal behaviour, animal performance, product quality, and environmental conditions are continuously registered in real time and used to adapt the production process.

Pig fattening can serve as an illustrative example of this technique. Currently, an individual animal's daily weight gain is rarely or insufficiently taken into account in daily feed dispensing (feed quantity and composition). If, however, animal weight is measured automatically, daily weight gain can be incorporated in feeding and management decisions. Therefore, new techniques measure body weight using an electronic scale or indirectly with the aid of an image processing system and sort the animals into different feeding areas with different feed mixtures. Image processing even allows the weight of the value-determining cuts (ham, loin, and belly) to be estimated, thus enabling the optimal marketing time to be determined.

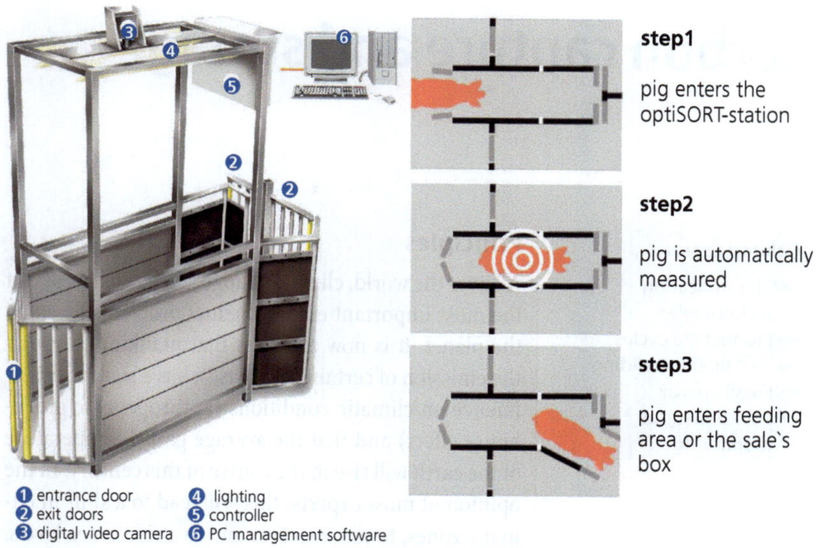

Pig fattening in large groups with a sorting station: On the way from the lying to the feeding area, an image processing system classifies fattening pigs and sorts them into the different feeding areas and sorts them into the stalling-out (sale) pen at the end of the fattening period based on their weight. Source: Hölscher and Leuschner

Prospects

- Progress in agricultural engineering strongly contributes to the optimisation of agricultural production with regard to economic, ecological and social requirements.
- The development stages of agricultural engineering depend on the conditions at the outset and the rate of progress.
- While many developing countries are in the process of mechanising agricultural work, newly industrialising countries like India or China are forcing motorisation by introducing tractors, self-propelled combine harvesters and transport vehicles adapted to their specific conditions.
- In industrialised countries, electronic control and communication technologies that allow optimised mechatronic applications to be realised together with mechanical, hydraulic, and electric systems are the main focus of agricultural engineering developments

DR. MARKUS DEMMEL
DR. GEORG WENDL
Bavarian State Research Center for Agriculture, Freising

Internet
- www.fao.org
- www.cigr.org
- www.eurageng.net
- www.clubofbologna.org
- http://cabi.org

Carbon capture and storage

Related topics
- Air purification technologies
- Product life cycles
- Sustainable building
- Fossil energy
- Environmental monitoring

Principles

All over the world, climate change is regarded as one of the most important environmental problems faced by the planet. It is now accepted that mankind, through the emission of certain trace gases, has a significant influence on climatic conditions (anthropogenic greenhouse effect) and that the average global temperature of the earth will rise in the course of this century. In the opinion of most experts, this will lead to a shift in climatic zones, to more storms and to a rise in sea levels. The most important trace gases, which intensify the greenhouse effect, are water vapor, carbon dioxide (CO_2), methane (CH_4), nitrous oxide (N_2O) and ozone (O_3). Of all these emissions, CO_2 is making the greatest contribution to global warming. For this reason, most of the measures for combating climate change target the reduction of CO_2 emissions into the atmosphere.

Most of the CO_2 is released by the combustion of fossil energy sources and by changes in land use. Power and heat generation is responsible, globally, for the largest emissions–followed by industry, traffic and agriculture.

The measures for reducing anthropogenic CO_2 emissions can be divided into three areas: avoidance, reduction and storage. CO_2 storage technologies basically involve two different processes:

- Technical CO_2 storage: CO_2 is removed from the gas flow before or after the combustion process and is stored in suitable geological or marine reservoirs. This process is primarily of interest for point sources of CO_2 (power plants, gas production plants).
- Biogenic CO_2 storage: CO_2 is extracted from the air by plant photosynthesis and is stored in biomass or soil humus.

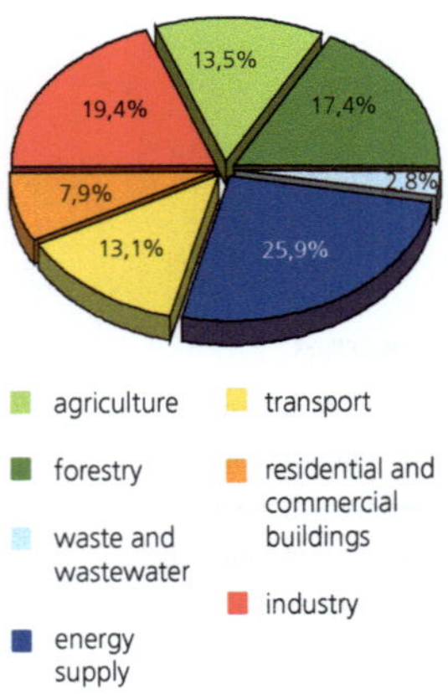

Breakdown of global anthropogenic greenhouse gas emissions in 2004 (converted to CO_2 equivalents): In order to account for other greenhouse gases, such as CH_4 (methane), N_2O (nitrous oxide), HFC (semi and full fluorinated hydrocarbons), and SF_6 (sulphur hexafluoride) a correspondingly uniform measurement principle (CO_2 equivalent) is applied, by which means the potential of other gases to precipitate global warming is seen in relation to the climatic effect of CO_2. Sources: IPCC, 4th Assessment 2007

Applications

▶ **Separation.** CO_2 can be captured at large point sources such as major fossil fuel or biomass energy facilities, or industries with major CO_2 emissions. Broadly speaking, three different types of technologies exist:

- Pre-combustion: In this process, the fossil energy source (e. g. natural gas) is converted in two steps into a synthetic gas which essentially consists of hydrogen (H_2) and carbon dioxide (CO_2). The CO_2 is removed from the gas flow by a liquid or solid absorber; the hydrogen is transferred to the combustion turbine, where it is burnt to form water vapor.
- Post-combustion: In this process, the fossil energy source is burnt in a turbine and the CO_2 is separated out by means of a downstream flue gas purifier. This takes place with the help of solvents, solid absorption materials or membranes. Pressure and temperature changes in the tank then separate the CO_2 from the solvent.
- *Oxyfuel process*: In this process, the fossil fuel is not burnt in air (80% nitrogen, 20% oxygen), but in pure oxygen. The concentrations of CO_2 in the flue gas are higher than in normal combustion. The separation of highly concentrated CO_2 is more efficient, and therefore cheaper, than the post-combustion process.

CO_2 separation is an energy-intensive process that leads to considerable efficiency losses in relation to a plant's overall efficiency. This is because the elaborate separation technology requires additional energy (e. g. to regenerate the absorber).

▶ **Transport.** As CO_2 is produced at point sources and there are normally no suitable storage facilities in the immediate vicinity, the separated CO_2 has to be transported over longer distances. CO_2 is transported overland through pipelines and in ships at sea (for storage at sea). The gas is generally compressed to a pressure of 80 bar for pipeline transport. Transport by ship is still not widely used.

Direct re-injection of CO_2 into deposits during the extraction of oil (Enhanced Oil Recovery – EOR) or natural gas is a separate case entirely. Natural gas typically contains about 4% CO_2, which can be separated from the remaining gas during recovery and pumped back into the deposit. This is the most highly developed CO_2 storage technique, and is already in routine use at some facilities, such as the Sleipner gas field in the North Sea and in Weyburn, Canada.

► **Storage in geological formations.** Separated CO_2 has to be stored in suitable storage media for it to remain "gas tight" on a long-term basis. This permanently prevents the emission of CO_2 into the earth's atmosphere. One possible solution is the storage of CO_2 underground.

One established technique is the compression of CO_2 into active or former gas and oil deposits. Gas or water vapour is already pumped into nearly depleted deposits (tertiary gas and oil recovery) in order to recover the last residues of fossil raw material by increasing the pressure in the strata holding the oil or gas. The gas (in this case CO_2) largely remains in the exhausted deposit.

A further option is the use of deep saline aquifers and ground water reservoirs. Deep saline aquifers are geological formations frequently found in the vicinity of large marine and terrestrial sediment basins. Deep saline aquifers have the potential to become the largest possible storage medium for CO_2. CO_2 is pressed into deep water-bearing strata, where it dissolves in the water. This leads to the formation of hot, aggressive carbonic acid solutions, which put great strain on the corrosion stability of the pumping technology. In some cases, when this technique is used, the geothermal use of hot water reserves can be self-defeating, as it makes little sense to pump water enriched with CO_2 up to the surface, where the greenhouse gas can escape back into the atmosphere. In Europe, CO_2 storage in a saline aquifer is currently being tested in situ in Ketzin, near Berlin. A sandstone reservoir at a depth of 700 m, which is covered by cap rocks comprising gypsum and clay, is the target geological structure for CO_2 injection and testing. The research project involves intensive monitoring of the fate of the injected CO_2 using a wide range of geophysical and geochemical techniques, as well as the development of models and risk assessment strategies. Approximately 30,000 t of CO_2 per year will be injected into the reservoir for a period of up to 3 years. CO_2 injection started in 2008. High-purity CO_2 is being used for the project – rather than gas from a power plant.

Unminable coal seams (at depths of over 1,500 m) can be used to store CO_2 because the coal surface absorbs CO_2. However, the technical feasibility of this solution depends on the permeability of the coal bed. In the process of absorption, the coal releases previously absorbed methane, and the methane can be recovered (enhanced coal bed methane recovery). As in the case of deep ground water storage, however, future use of these coal deposits will be excluded to prevent the release of absorbed CO_2.

► **Storage in the ocean.** CO_2 dissolves in water where it exists in the form of carbonates. The oceans

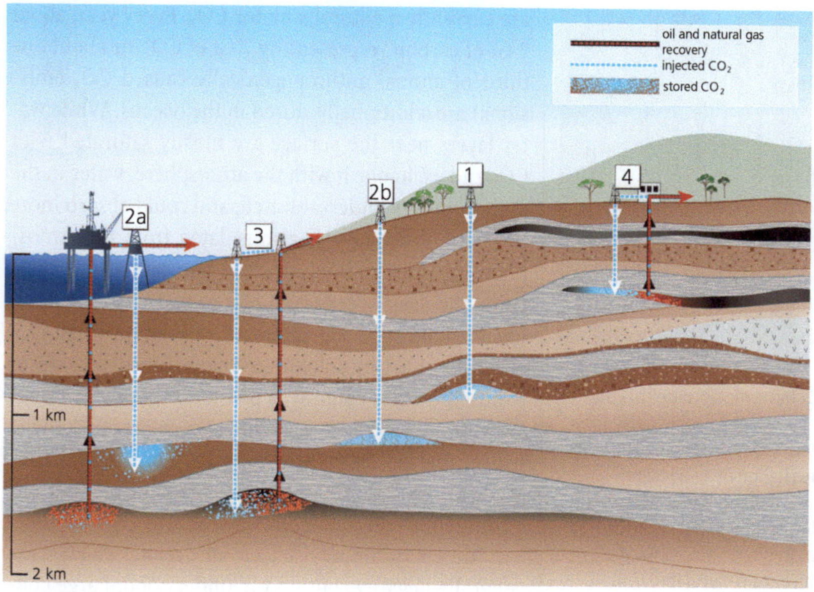

Geological storage options for CO_2: Exhausted oil and natural gas fields (1) and deep salt formations in the sea (2a) and on land (2b) can be used for storage. CO_2 is also used for tertiary oil and natural gas recovery (3) from depleting deposits. A further option is the adsorption of CO_2 in deep coal deposits (4). Source: IPCC Special Report CCS

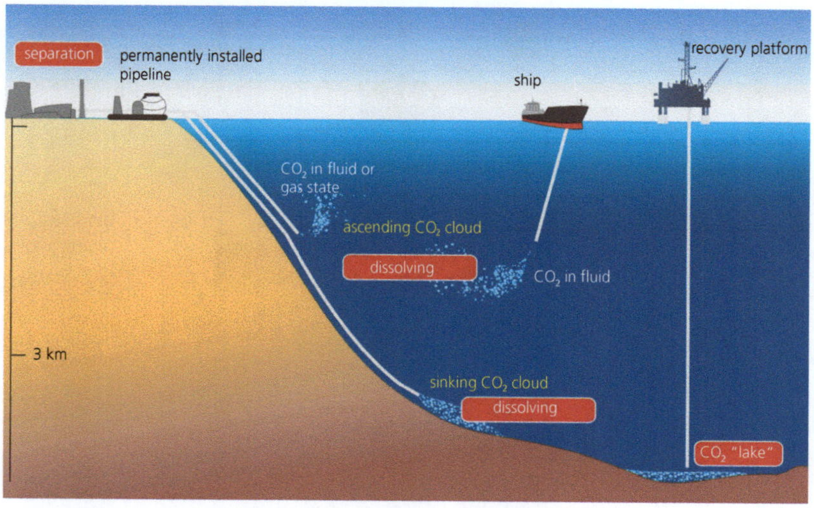

CO_2 storage options in the ocean: 3000 m below sea level, CO_2 is denser and becomes heavier than seawater. The liquid or solid CO_2 (in the form of dry ice, for example) is pumped or conveyed into deep water layers or on to the sea floor from ships, pipelines or platforms – it either dissolves in the water or forms pools of liquid CO_2 on the seabed. Source: IPCC Special Report CCS

Micrometeorological eddy covariance tower in Italy (viewed from the air): The exchange of air packets between the forest and the lower atmosphere is measured at frequent intervals and a balance is calculated. This determines whether a forest is actually storing the quantity of carbon specified as the target value by a forestation project. Source: Forest Department of the Autonomous Province Bolzano/Bozen Minerbi

are currently a natural sink for CO_2. Every year, about 2 Gt of carbon (representing 7 Gt of CO_2, or about one third of annual anthropogenically caused CO_2 emissions) are additionally stored in the oceans. While water layers near the surface are mostly saturated with CO_2 and exchange it with the atmosphere, water in the deep oceans is undersaturated, and could absorb more CO_2. The water here is also isolated from the atmosphere for long periods, so that like geological reservoirs, the deep oceans could be considered as a medium for long-term storage of CO_2. The aim of all CO_2 ocean storage technologies is to inject the gas as deeply into the water column as possible. These techniques are still only used on a small scale or are in the trial phase; market-ready processes are not yet available. Studies are predominantly focusing on the possible environmental consequences.

The environmental effects of oceanic storage are generally negative, but poorly understood. Large concentrations of CO_2 kill ocean organisms; another problem is that dissolved CO_2 would eventually equilibrate with the atmosphere, so storage would not be permanent. Furthermore, as part of the CO_2 would react with the water to form carbonic acid, H_2CO_3, the acidity of the ocean water would increase.

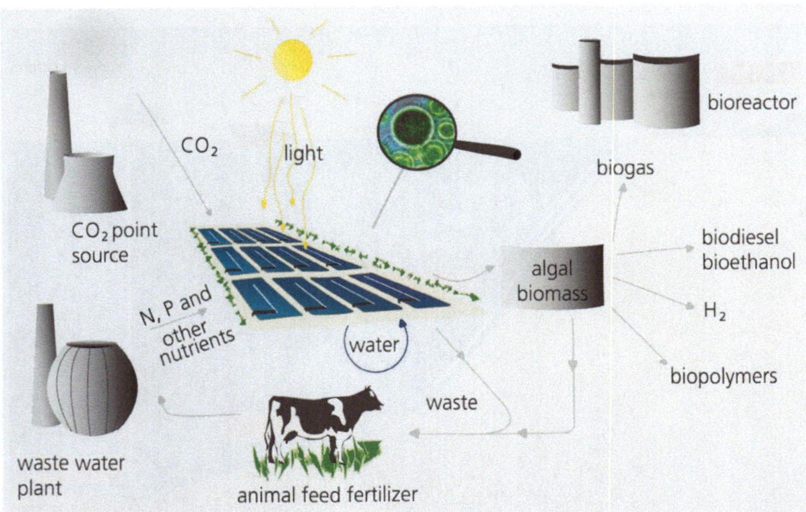

Overview of biofixation with microalgae in an open raceway pond: The CO_2 supply for these fast-growing algae must come from point sources. Sumps or diffusers transport the CO_2 into the pond. Point sources could be wastewater treatment plants, landfills or power plants. Other nutrients like N, P or minor elements essential for algal growth are added, ideally from municipal, agricultural or industrial wastewaters. Algal biomass is mainly harvested through flocculation: chemical flocculation or bio-flocculation. The challenge for these techniques is to produce biomass slurry with sufficient density to allow for further treatment. The last step of the biofixation process includes the processing of the algal biomass into biofuels and other products. To obtain these products, additional processes for acquiring ethanol are needed. These processes range from anaerobic digestion to biogas or yeast fermentation. H_2 could be produced enzymatically or through steam reformation. The former process is still, however, in the early stages of research.

▶ **Biogenic CO_2 storage**. The biosphere, where carbon is stored in plants and soil, is the earth's second largest natural carbon sink – after the oceans. Through the process of photosynthesis, green plants convert CO_2 into vegetable biomass with the help of sunlight. When the biomass dies, it is decomposed again by microorganisms. Part of the biomass is released back into the atmosphere in the form of CO_2, and part is transformed into longer term organic carbon reservoirs such as soil humus. Forest ecosystems, wetlands and grasslands represent the most important carbon sinks, while agricultural land or deserts store relatively little carbon.

The aim of biogenic carbon storage is to store additional carbon in terrestrial ecosystems. This involves human intervention in the biological carbon cycle. A series of measures in the agricultural and forestry sectors are currently under discussion:

- Protection of natural forest ecosystems against deforestation
- Protection and restoration of wetlands
- Forestation and reforestation of former forest areas
- Forestry measures, such as changes in rotation times or minimisation of humus and nutrient losses by carefully combining agricultural crops and trees
- Reduced tillage techniques. These techniques involve less frequent ploughing, which reduces the availability of oxygen in the soil and slows down the decomposition of humus by microorganisms. This decomposition releases CO_2. Reduced tillage, however, often leads to increased use of herbicides, as weeds grow vigorously when left undisturbed by the plough.

The technological challenge is to establish a suitable system for accurately measuring net carbon storage in an area set aside for sequestration. Until now, a simple inventory roughly estimated the biomass above ground level and the carbon content of the soil.

Newer techniques directly measure the exchange of CO_2 between ecosystem and atmosphere and produce a medium-term net carbon balance for the ecosystem. Eddy covariance towers are used to measure at frequent intervals the exchange of air packets between the area of forest concerned and the lower atmosphere. The balance is then calculated. This determines whether a forest is actually storing the quantity of carbon specified as the target value by a forestation project.

Biogenic processes offer another option for CO_2 sequestration through the biofixation of the gas by microalgae. The technology is based on these microscopic plants' ability to convert water, CO_2 and sunlight into biomass. Due to their small size, their high surface-to-volume ratio allows for rapid nutrient uptake and high growth rates if nutrients like CO_2, nitrogen and phosphorous are plentiful.

The cultivation of microalgae in open raceway ponds, which are constantly turned by paddlewheels, is the basic technology for this form of greenhouse gas abatement. The algal pond should ideally be located near point sources, like wastewater treatment plants, landfills or power plants, since CO_2 transport would increase costs and reduce any CO_2 abatement gained by the system. Ponds are usually one hectare or larger and approximately 30 cm deep. Typical productivity rates for commercial systems using Spirulina algae are currently 50 t/(ha · a). Research aims to increase this level of productivity.

The greenhouse gas abatement potential for microalgal biofixation is estimated at 100 t/(ha · a) in the short- to medium-term. To reach this goal, genetic improvement of algal strains is required to optimise their adaptability to pond conditions.

An alternative to open ponds are closed systems, where the algae are grown in glass tubes or plastic layers stacked or piled up in a greenhouse or similar structure. Closed systems offer higher productivity, better control of environmental conditions and CO_2 release, continuous production, and optimised use of land. The release of genetically modified organisms is also much less probable than it would be in open systems. The challenges are to optimise the availability of light and to prevent the coagulation of microalgae in the tubes.

▶ **Risks of CO_2 storage.** In itself, CO_2 is a harmless gas, but in higher concentrations it can cause asphyxiation. CO_2 is heavier than air. If CO_2 leaks from storage it will collect in depressions in the land, endangering the population there. For this reason – and because the aim of CO_2 storage is the long-term isolation of the gas from the earth's atmosphere – research into safe, long-term storage solutions is important to avert risks. Ocean storage dominates discussions on negative environmental consequences. A danger of contaminating deep water with pollutants transported in the gas flow is a risk in saline aquifers. The additional energy required to operate the separation technology is another aspect that has to be taken into account when making an overall energetic evaluation of the system.

Production of microalgae in a closed system (photobioreactor): The algal slurry in these tubes is constantly mixed so that all algal cells can be briefly exposed to high light intensities. This also prevents coagulation. Source: Fraunhofer IGB

Trends

Publicly-funded and industrial projects are conducting research into the technical storage of CO_2. Reducing the cost of the technology components, especially for the separation technologies, and increasing the energy efficiency of the separation process are crucial issues. Applications for smaller (biomass-fed) plants and for larger plants with an output of several hundred megawatt need to be investigated to broaden the possibilities for future use. The long-term safety and impermeability of geological reservoirs is another important issue dominating research. The ability of many of the geological structures (considered for CCS) to store gases over long periods of time has been proven (as in natural gas reservoirs). Once the structure is made permeable during extraction, its safety must then be examined carefully. This involves modelling and simulation techniques to reproduce the underground dispersion of CO_2 (especially along geological faults). Monitoring technologies also need to be developed. In addition to safety aspects, the issue of leakage is also important if CCS is to feature in emission trading systems.

Prospects

Combining CO_2 avoidance, reduction and separation technologies will be the only way of stabilising the emission of greenhouse gases into the atmosphere at a level that will mitigate the consequences of climate change.

- Because many of the CO_2 storage technologies are still under development, their current contribution to climate protection is rather small. Climate protection scenarios presented by the Intergovernmental Panel on Climate Change (IPCC) assess their relative contribution, over the next 100 years, to be between 15–55%. These are modells based assumptions.
- These technologies are generally regarded as a medium-term intermediate solution, in order to gain time for the development of a hydrogen-based energy economy.

CHRISTIANE PLOETZ

VDI Technologiezentrum GmbH, Düsseldorf

Internet

- www.ieagreen.org.uk
- www.ipcc.ch
- www.unfccc.int
- www.co2captureproject.org
- www.jsg.utexas.edu/carboncapture

10 BUILDING AND LIVING

One of the essential human needs, besides food, is housing to protect us against the elements.

The construction industry is one of the largest economic sectors in Europe, and it is closely linked to many other industries as well. Technological developments focus mainly on quality and – as in all other applications – above all on sustainable building.

In comparison to consumer products or other investment goods, like machines, buildings have a far longer useful life. A building is used on average for 50 to 100 years. This means it often has to be upgraded to incorporate the latest innovations for saving resources in its operation, and this requires corresponding investments (new heating systems, insulation, etc.). Whatever renovations are undertaken, they must be carried out with minimal impact, so that residents can remain in their apartments while the work is being completed, for example.

The existing inventory of buildings offers significant potential to reduce CO_2 emissions and energy use, because heating is what accounts for three-quarters of private energy consumption. Renovations carried out by specialists and modern building technology can economise a significant amount of energy. That is why many countries now subsidise the renovation of older buildings. New low-energy houses that use 85 kWh/m²a for heating are currently state-of-the-art, as opposed to 220 kWh/m²a for conventional buildings.

The aim of sustainable building, when it comes to new projects, is to minimise the use of energy and resources in constructing and operating the building. All life cycle phases of a building are taken into account in the process: from the primary materials and construction, to the dismantling of the building. This type of closed-loop management system for building materials and components – as is already established in the automotive industry – is still in its fledging stages in the building sector.

The following factors must be considered sustainable building practices:

- Lowering of energy requirements
- Lowering consumption of operating resources
- Use of recyclable building materials and components
- Avoidance of transportation costs (for the building materials and components used)
- Safe return of the materials used into the natural material cycle
- Possibility of subsequent uses
- Avoiding damage to the natural surroundings
- Space-saving construction

In addition to sustainability, reliability is an indispensable principle in construction. Once they have been completed, buildings must be safe to use, bridges must not give way and roofs must be able to support heavy snowfall without collapsing. Wear and tear and fatigue have to be controllable over a long period of use. This is where sensors are increasingly being deployed: a system of signal-processing sensors that can be put in place to perform what is referred to as "structural health monitoring". The system sounds an alarm when critical movements in the structure are registered, in time to prevent the building from collapsing.

The art of construction is an excellent reflection of the human progress toward increasing performance in terms of "faster, higher, deeper". Tunnels are becoming longer, bridges cross broader valleys, and skyscrapers by now truly touch the clouds. These new achievements require not only new materials (high-performance concrete), but also new construction concepts (e. g. special shuttering) and new machines (e. g. to dig tunnels).

Ultimately, a building – especially a residential one – must offer people protection and comfort. Thermal, acoustical and visual comfort have an impact on the building and its inhabitants, as does the use of low-emission construction products. This reciprocity between the residential environment and the feeling of wellbeing is increasingly becoming a part of interdisciplinary research projects carried out by construction physicists, engineers, psychologists, medical doctors and ergonomists.

Buildings are very long-lasting investment goods that take a great deal of effort to alter to suit new purposes. There is a need for modular constructions whose walls and infrastructures (water and data connections) can be rapidly adapted to changing requirements; however this is still just a future vision.

► **The topics.** The basis of each building project is first and foremost the choice of *building materials*. New technological developments with regard to the building's service life, material properties and manufacturing processes are constantly being made with ecological and economical considerations in mind. Each construction material demands a special type of construction. A conscious choice of construction materials can reduce transportation costs, for example, or lower consumption of resources, or allow for efficient dismantling, through to recyclability.

Building technology covers the technical and construction aspects of civil engineering. It includes a large number of special areas, such as structural design, construction physics, building biology, structure planning, construction engineering (skyscrapers), hydraulic engineering (dams), or traffic route engineering (bridges, tunnels). Of particular importance for developing transportation and supply infrastructures are modern engineering and construction methods for bridges, tunnel systems, skyscrapers or dams.

The aim of sustainable building is to perform construction in an economically efficient and environmentally friendly manner and with an eye to sparing resources. The life-cycle perspective on a building is important, and so is efficient sharing of information among those involved on the construction site. Paying attention to sustainable planning ideas early, one can considerably improve the overall cost-effectiveness of buildings (costs for construction, operation, use, environment and health, and non-monetary values). The possibilities of influencing the costs of a measure are greatest at the beginning of the measure. Decisions with the greatest impact on expenses are already made while defining the programme and during the initial conception phase.

Indoor climate is a significant aspect when considering a building's life cycle and the need for quality living and comfort. What is meant by indoor climate is the micro-climate in a room of a given building. It is influenced by a number of factors, principally temperature, humidity and air movements, but also light, odour and the ability of the building materials to breathe. ■

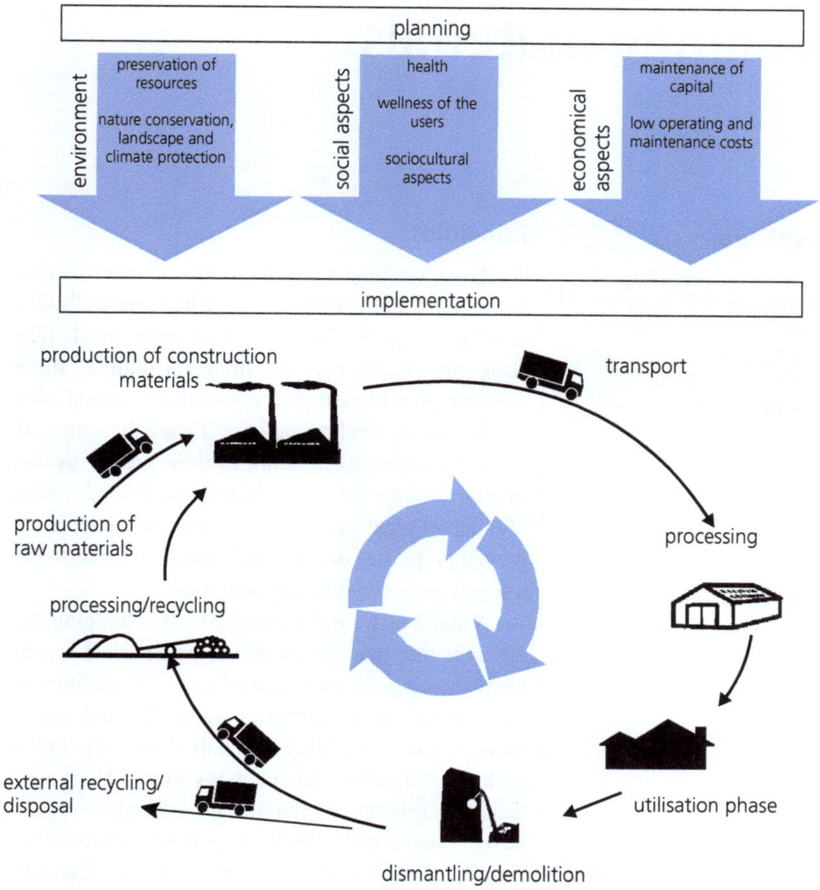

▲ Life cycle considerations: The possibilities of influencing the costs and the environmental footprint of buildings are highest in the planning phase and then decline sharply as planning advances. Life cycle considerations aim at preserving material resources, or using them efficiently, reducing environmental impact through emissions, and minimizing waste. The ecological balance sheet of any product should take the following points into account: extraction and provision of raw materials, manufacturing and processing of raw materials, transportation and distribution, care and re-use, recycling and waste processing

Building materials

Related topics

- Structural engineering
- Sustainable building
- Composite materials
- Metals
- Nanomaterials
- Intelligent materials

Principles

The term "building materials" refers to materials used in the construction trade, and which are generally classified in categories based on their type; metals (e.g. steel, aluminium, copper), minerals (natural stone, concrete, glass) and organic materials (e.g. wood, plastic, bitumen). Modern building materials cannot always be readily placed in one of these groups, as they may be developed through the systematic combination of different types to form composites that provide improved properties over the individual materials themselves (*composite building materials*).

A major factor in the development of building materials is that new structures are being asked to perform increasingly multifaceted tasks. In addition to their traditional load-bearing capacities and use as room partitions, building materials also need to fulfil a multitude of additional functions today. Along with technical criteria, economic and ecological criteria have become increasingly important factors when choosing and developing building materials. Materials with the smallest possible environmental impact (such as low levels of toxic emissions or required primary energy) are considered sustainable and suitable for use in the future.

In the planning phase for buildings and construction, the most important aspects include technical features (e.g. mechanical stability, deformation capacity, and fire protection) and the materials' physical and structural characteristics.

The technology of building materials has developed into a true science; studies are performed on the structural characteristics of materials using research methods common to both materials science and solid state physics. Based on this information, specific building material characteristics can be optimised for specific applications by performing targeted structural alterations (down to the nano scale). Building materials such as these are often referred to as "tailor-made building materials".

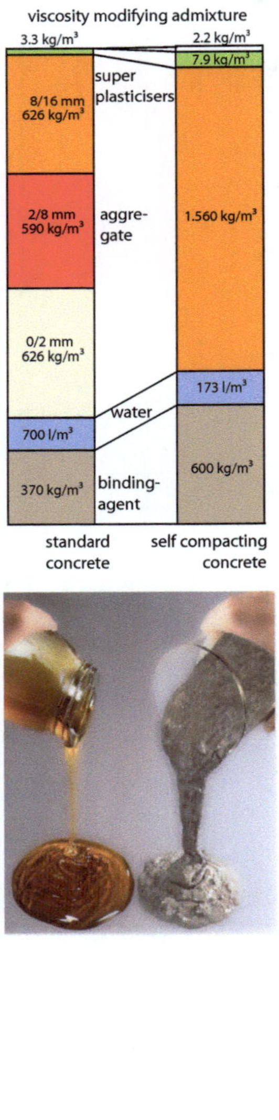

Self-compacting concrete (SCC): Comparison of the composition of conventional vibrated concrete and SCC (top), specifically for a tunnel construction, depiction of the flow capacity of SCC compared to honey (bottom). By altering the composition of concrete (increasing the binding agent content, using plasticiser and viscosity modifying agents), so-called self-compacting concretes can be produced which have a flowability like honey and thus enable construction components to be manufactured for which mechanical compaction is difficult or impossible to achieve. Source: Dyckerhoff Zement AG

▸ **Concrete**. For a variety of reasons, concrete is the most frequently used building material in the world. The required raw materials are usually readily available and inexpensive. Due to its rapidly-obtained plastic properties after mixing, concrete can be used for an almost unlimited range of designs. The finished product (in its hardened state) is highly stable and offers good long-term durability.

Concrete principally consists of five raw materials: an aggregate filler, cement (which reacts hydraulically with water, the other main ingredient), and the so-called additives and admixtures, which improve the physical and chemical properties of fresh and hardened concrete. The additives are mostly reactive or inert powder materials, such as pulverized stone or industrial by-products (pulverised fly ash, granulated blast furnace slag). Admixtures are usually liquids developed through chemical construction processes (polymer-based plasticisers are one example).

Cements are binding agents which solidify both in air and under water, and which harden to become stone-like and water-resistant. Compared to other hydraulic binding agents such as hydraulic lime, cements offer much higher stability and strength. Lime and clay – or their naturally occurring compound (marl lime) - are mainly used as raw materials for manufacturing cement. The hardening process of cement-bound materials after mixing with water is known as hydration. Due to complex gel and crystal formation processes, aqueous reaction products are formed in the cement clinker stage, or hydrate phases. In the early phase of hydration, gel formation is predominant with respect to crystal formation. Gel formation causes the cement paste to set. The weak bonds between the dispersed particles can be broken however by mechanical disturbances, resulting in gel reliquefaction. This phenomena, known as thixotropic behaviour, is used to produce concrete construction elements with improved compaction (vibration compaction). Vibrators are employed to oscillate the concrete and obtain mechanical compaction.

Applications

► **Self-compacting concrete**. Through the use of high-capacity plasticisers, the viscosity of the cement-bound materials can be altered so that no exogenous compacting work is required to obtain optimised compaction (self-compacting concrete, SCC).

SCC has a honey-like consistency and compacts under its own weight. SCC applications may be used for dense, reinforced construction components; for complicated, geometrical construction components; as well as for manufacturing exposed concrete surfaces that are almost entirely nonporous.

► **Ultra-high performance concrete**. By reducing the water content and the diameter of the coarse aggregate grain, and by using reactive (e. g. pulverised fly ash, silica fume) and non-reactive (e. g. stone powder) additives, the packing density of a cement-bound material can be increased to its maximum level. The maximum particle diameter of the aggregate in ordinary concrete is lowered from 16–32 mm to between 1–0.063 mm for ultra-high performance concrete (UHPC). The interstices between the aggregates and the cement-stone matrix are filled with the above-mentioned ingredients. In addition to the physical filling, the by-products which accumulate in the cement hydration (such as portlandite – $Ca(OH)_2$ are transformed into mineral phases (calcium silicate hydrate phases – CSH-phases) which increase strength and durability. The water needs the dry mixture to increase in reverse proportion to the particle size (aggregate, stone powder). In order to ensure good workability, liquefying polymer-based admixtures are added.

By reducing the recognised weak points in the texture (such as micro cracks, shrinkage cracks and cavity pores), t and hydrogen technology he maximum possible load-bearing capacity of the building material is increased. What's more, the higher density increases resistance to chemical and physical exposure. Brittleness, which increases along with the compressive strength, is reduced by adding suitable types of fibres. These improved materials with increased durability can help reduce the load-bearing cross-section of a structural member, making it possible to build constructions which up to now were unimaginable for bridges, high-rise buildings or factories.

► **Composite building materials**. Composites consist of two or more materials which are positively linked. By combining several different material characteristics, the advantages of the individual materials can be used, while the disadvantages are discarded. Several systems exist, depending on the type of composite:

- particle reinforced composites
- laminated composites
- fibre reinforced composites

► **Textile concrete**. For the manufacture of textile concretes, textile structures made of high-performance fibres are used. For reinforced concrete structures, a minimum coverage is required between the steel reinforcement and the surfaces which are exposed or otherwise stressed due to risks of corrosion. Since there is no risk of corrosion with alkali-resistant textile fibres, they are particularly well-suited for filigree and thin-walled construction components, and are now used for manufacturing façade components and, more recently, for pedestrian bridges and shell structures as well.

► **Fibre concrete**. The primary use of fibres is to improve the marginal tensile strength of solid building materials and the associated risks of cracking. Cement-bound building materials can only withstand low tensile forces, and their brittle material behaviour is problematic when the maximum load is exceeded. But, by combining materials with a high tensile strength, composite materials with improved material qualities can be manufactured. Different types of fibre materials, quantities and geometries are added to the concrete depending on the field of application. Due to their high tensile strength, steel fibres are often used. Since steel fibres tend to be vulnerable to corrosion, alternative materials such as stainless steel, glass, plastic and carbon fibres are used in certain applications.

Concrete behaviour in fires and the risk of shrinkage cracks can be improved by adding plastic fibres (polypropylene, polyethylene). In a fire, the temperature can rise to over 1,000 °C within a few minutes. When concrete heats at such a rapid rate, the pore water in the concrete component evaporates.

Spalling is caused by the formation of a quasi-saturated layer as a result of water vapour condensation flowing in from colder zones. The lower the permeability of the concrete, the faster these gastight layers occur, and the higher the resulting vapour pressures. The water barrier which is formed ("moisture clog") is impermeable to any further water vapor. If the water vapour pressure exceeds the tensile strength of the concrete, spalling will occur.

The positive effect of PP-fibres on the high-temperature behaviour of concrete is based on the improved permeability before and during exposure to fire. Additional micropores are formed when fibres are added. The resulting transition zones between the fibres and the cement-stone matrix consist of additional pores and weak cement hydration products (ettringite, portlandite). The transition zones and the contact surfaces of the aggregate form a permeable system even before exposure to high temperatures. As temperature increases, the PP-fibres melt and burn up, creating more capillary pores in the hardened concrete. In this way, the entire system obtains higher filtration capaci-

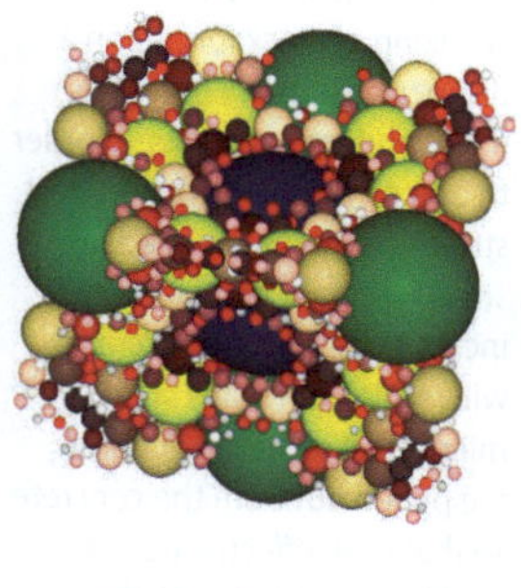

Modelling of packing density: The targeted selection of various superfine materials and aggregates enables a high packing density. The improved homogeneity and density of the matrix increase the material strength and durability at the same time. Source: Schmidt / Geisenhanslüke, University of Kassel

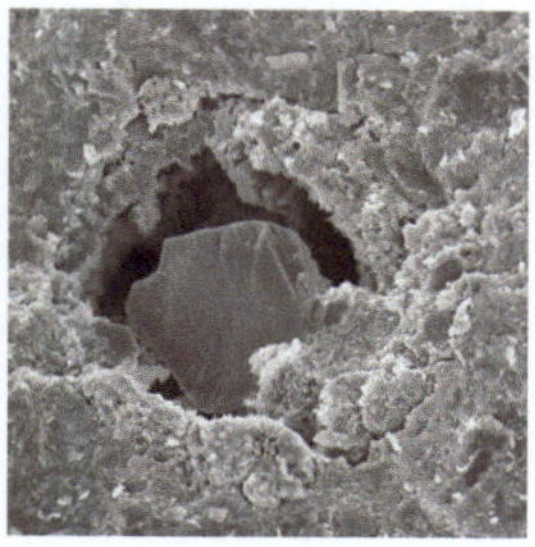

Cavities of melted plastic fibres: Scanning electron microscopy images of cavities which are created in the concrete structure by the melting of plastic fibres at approx. 170 °C. The image was made using a concrete sample which was exposed to 1000 °C. These cavities enable water vapour which forms in the interior to escape outwards. Source: MFPA Leipzig GmbH

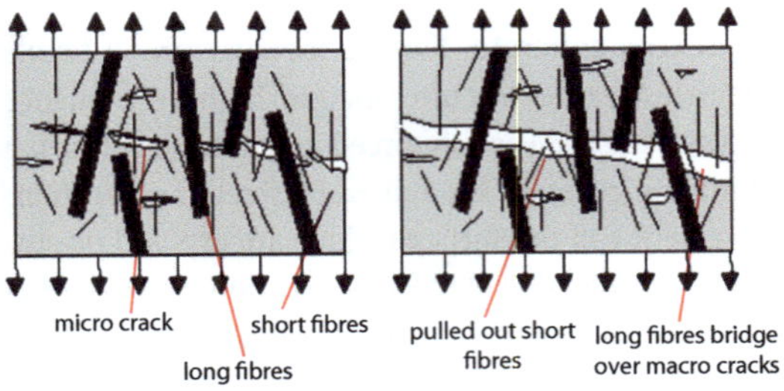

Fibre concrete: When speaking of crack formation a distinction is made between micro and macro cracks. In order to overcome micro cracks, short straight fibres are applied in small intervals. If the load is increased, larger macro cracks will result from many small micro cracks. The small fibres are pulled out from the concrete so that their effectiveness is starkly reduced, should the width of the cracks increase. In order to effectively bridge over the newly resulting macro cracks, long fibres are added. They bridge over wide cracks effectively and thus increase the ductility and load-bearing capacity. Source: Markovic et al. Technical University Delft

Polypropylene fibres in concrete: Prefabricated concrete components after exposure to fire. During fires in tunnels, the temperature can rise within a few minutes to over 1,000 °C. Conventional types of concrete would spall in an explosive manner during these rapid increases in temperature, since the physically bound water within the interior of the components would evaporate very quickly and thus create a very high vapour pressure. In order to prevent an explosive spalling of the concrete in the event of fire, plastic fibres are added to the concrete during manufacture. These melt at approx. 170 °C and thus create a capillary network through which the water vapour can be released. Left: Tunnel concrete with plastic fibres after fire testing; Right: tunnel concrete without plastic fibres after fire testing. Source: MFPA Leipzig GmbH

ty and allows the increasing vapour pressure in the interior to escape.

▶ **Wood-Polymer-Compound.** WPC is a three-material system, which mainly consists of wood and thermoplastic materials, and a smaller proportion of additives. For economic reasons, pine wood in various forms and quantities is added to the plastic matrix to manufacture WPC. The wood-powder filler content is between 50% and 75%. Since the natural wood material is damaged when the temperature exceeds 200 °C and the mechanical qualities of the final product are irretrievably degraded, plastics must be used which have a melting and processing temperature under 200 °C. For this reason, conventional mass-produced plastics such as polyethylene (PE), polyvinyl chloride (PVC) or polypropylene (PP) are increasingly being used for WPC. Additives markedly change the material properties of the processed material and the end product. In order to modify processing qualities, conventional additives such as plasticisers and lubricants are used. Plasticisers increase the elasticity of the plastics, and lubricants decrease the interior and exterior friction of the plastic melt. The material qualities of the final product are altered by introducing additives such as stabilisers, dyes, fillers and reinforcing materials. Flame-retarding additives are also used. Adhesion promoter is the decisive additive for the interaction of polar wood fibres and non-polar plastics. The adhesion promoter improves the bond, which improves the mechanical properties. At the same time, water needs the wood particles to decrease because they have been completely surrounded by polymer matrix.

Before carrying out the different manufacturing techniques, the raw materials are dried and mixed to form a homogeneous mixture. As an alternative to energy-intensive drying processes, water binding materials such as starch can also be used. For today's applications, WPC is mainly used for linear profiles using extrusion technology. In extrusion technology, the solid mass of the raw materials is heated to 200 °C to form a homogeneous, plasticised melt, and is then pressed out by an extrusion tool at 100–300 bar. Various units are located in front of and behind the extrusion line, which perform cooling, finishing, shaping, cutting to size, and curing.

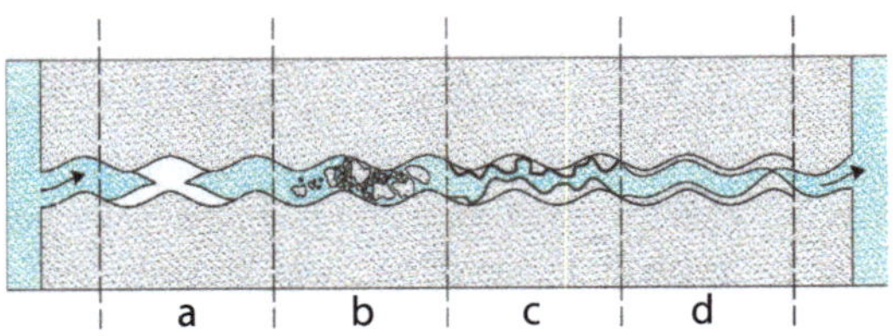

Self-healing of water-permeated tear-off cracks in cement-bound materials: a) formation of calcium carbonate: The calcium hydroxide in the cement stone reacts with the carbonic acid of the water to form calcium carbonate (lime), b) clogging due to sedimentation of loose particles, c) re-hydration of cement particles: when the cement hardens (hydration), not all cement particles are completely transformed; these react when the crack is permeated with water, which increaseses the volume of the particles, d) swelling of the cement stone: the moisture penetration of the zone at the edge of the crack triggers an expansion of the material.

Trends

▶ **Nanotechnology in construction.** Since the mid-1990s, the use of photo-catalytically activated titanium dioxide nano particles for building materials has been studied. This semi-conductive material can be used to produce surfaces which exhibit extraordinary features when exposed to UV light; oxygen-rich radicals e.g., which kill micro-organisms form on titanium dioxide surfaces (antibacterial effect). Furthermore, it has been proven that poisonous nitric oxide (NO, NOx) in the air is dissipated by photocatalytic oxidation (air purification effect). This "burning at room temperature" can be systematically controlled, since it is only activated by exposure to UV light and can thus be stopped by "switching off" the light. Coatings with titanium dioxide nano particles also create super-hydrophilic (strongly hygroscopic) surfaces with self-cleaning characteristics. The exposed surfaces of buildings are one area in particular where such usage has high potential.

▶ **Simulation of material behaviour.** The modelling of construction behaviour which describe complex material properties and their behaviour at the micro and nano scales are being used more and more often. The developed models can simulate experiments and help increase the understanding of material mechanisms with less time and effort. In addition, simulations make it possible to predict chronological sequences of physical-chemical damaging processes.

▶ **Self-healing.** Building materials are being developed which not only exhibit long-term durability with respect to environmental conditions, but which also react autonomously to possible damage. The self-healing of macro-cracks on water-impermeable concrete constructions has been studied intensively. In this process, tear-off cracks are autonomously sealed. Self-healing of cracks in cement-bound building materials is a process in which the mechanical, chemical and physical processes take place simultaneously. Self-healing is possible for cracks due to the interaction of mechanical processes in combination with various growth and expansion processes, which vary according to temperature, water pressure and the concentration of the materials used. This so-called passive self-healing is not a targeted design feature but rather an inherent quality of cement-bound building materials. In targeted self-healing (self-healing by design), materials are developed which possess an "inherent" capacity to repair structural damage autonomously or with minimal external help. The cause of the initial material damage is often the smallest micro-cracks. In self-healing materials, the appearance of these cracks is "recognised" and the self-healing potential of certain material components is activated to initiate the repair of the damaged areas. These processes are triggered "automatically" by the material damage (appearance of cracks) itself.

▶ **Self-cleaning.** For the development of self-cleaning building material surfaces such as ceramic tiles, roofing tiles, concrete, natural stone and glass façade elements, there are two main approaches: developing highly water-resistant (super hydrophobic) or water-attracting (super hydrophilic) surfaces. In theory, self-cleaning is mainly influenced by the wettability of a surface, and therefore by the interactions on the boundary surfaces between the gaseous, liquid and solid phases. The degree of wettability can be described by the so-called wetting angle. Very large wetting angles can be observed for example on hydrophobic surfaces (adoption of the so-called lotus effect of the natural plant). By exposing photocatalytically active metal-oxide coated surfaces to UV light, very small wetting angles ($< 1°$) can be created (super hydrophilicity).

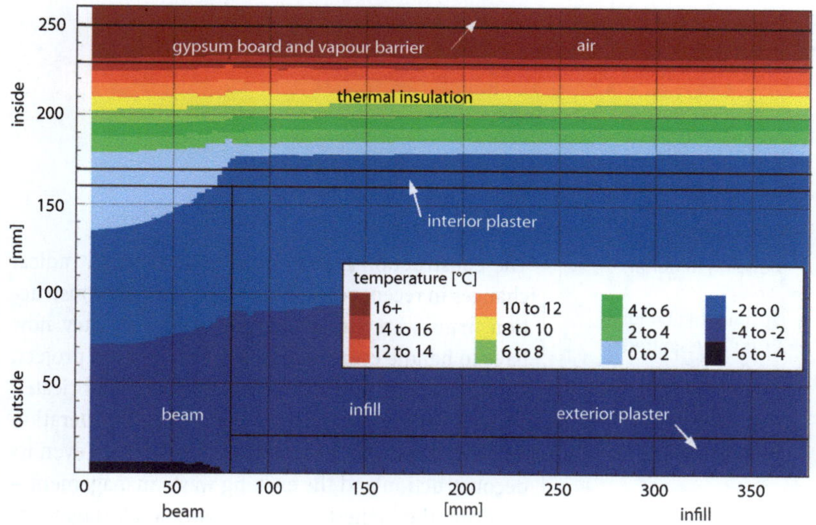

Hygrothermal simulation of a half-timbered wall, insulated on the interior: The illustration shows the temperature profile in a wall. The heat transfer coefficient (U-value) for the infill is worse than for the exposed beam, e.g. the beam has a lower outside surface temperature. The low thermal conductivity of the insulation can be easily seen by its large thermal gradient which amounts to approx. 14 K in the infill sections. Source: MFPA Leipzig GmbH

Prospects

- Sustainable building materials that have the smallest possible impact on the environment at all stages of their service life are finding growing use. This also applies to secondary raw materials, which are used directly as building materials or indirectly for manufacturing building materials (cement production, recycled aggregates in concrete construction, etc.).
- Natural building materials have become increasingly important, and are combined with conventional materials (wood-concrete, wood-steel, and wood-plastic) to form composite materials and composite constructions.
- The application of micro analytical methods has led to the development of a whole new generation of autonomously reacting "smart materials". They can perform new tasks, in addition to their conventional structural function, such as self-healing, self-cleaning and energy and pollutant transformation.

PROF. DR. FRANK DEHN
ANDREAS KÖNIG
KLAUS PISTOL
Leipzig Institut for Materials Research and Testing

Internet

- www.concretenetwork.com/glass-fiber-reinforced-concrete
- www.dcnonl.com/article/id28401
- www.zkg-online.info/en
- www.tech-wood.nl
- www.ecobusinesslinks.com/links/sustainable_building_supplies.htm
- www.cwc.ca
- www.ectp.org

Structural engineering

Related topics

- Building materials
- Bionics
- Sustainable building
- Composite materials
- Testing of materials and structures
- Environmental monitoring

Principles

The construction industry has gone through radical changes in recent years. Whereas in the past construction firms were simply required to build, they now have to handle every aspect of a construction project, from planning and manufacturing of prefabricated components to the finished structure, its operation and maintenance, repairs, and in some cases even its deconstruction and the ensuing waste management – ensuring the highest quality of work at all stages. Innovations in structural engineering extend from the development of new materials and construction techniques to new financing and operating models.

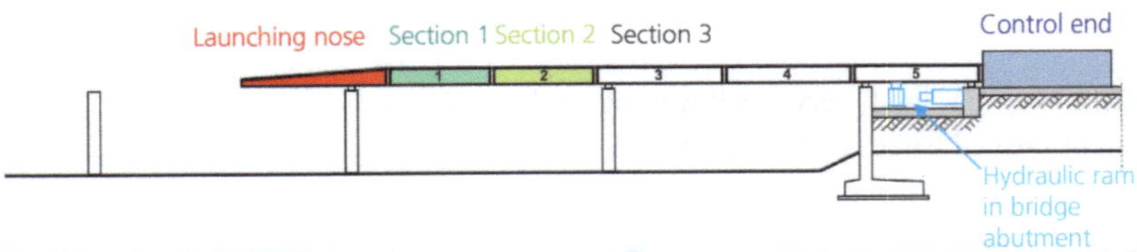

Incremental launching of a bridge made of pre-stressed concrete: The first segment is reinforced and cast in concrete at the casting site. Once the concrete has sufficiently set, stress is applied to the tendons. The completed segment is pushed out along the longitudinal axis of the bridge in gradual stages by a hydraulic ram. A launching nose – for instance made of steel – shortens the distance to the next pillar or, in the case of very wide spans, to a midway temporary support. Then the next segment is cast and pushed out behind the first. Source: Silke Scheerer, CJBU in Wikipedia

► **Bridges**. Man has been building bridges for thousands of years, in forms varying from simple wooden walkways to roads spanning several kilometres across a strait. They are built of wood, ropes and cord, brickwork, concrete, steel, modern composites and all imaginable combinations of materials. Each style of bridge and each type of material demands its own specific construction techniques:

- When building certain types of bridge, a temporary structure known as falsework or formwork is used to hold the superstructure in place until it is capable of supporting its own weight. Falsework is employed in connection with many types of building material and different types of bridge.
- The technique of incremental launching is economical for bridges with a span of 100 m or more, up to lengths of over 1 km. It requires that the bridge is of equal cross-section throughout, and that the pillars are spaced at the same or a similar distance apart.
- Cantilever structures are a suitable option for bridges consisting of girders, arches or segments spanning medium-to-long distances, i. e. in cases where it is not possible to erect a falsework. The conventional technique involves casting successive sections of the bridge in concrete using a traveling platform that moves out from a pillar in alternate directions or symmetrically, either until it reaches the next pillar, or until it meets the end of the opposite cantilever arm.

► **Tunnel construction.** The first tunnels were built in ancient times as a means of transporting water or as part of defence works. Some of the oldest known tunnels were dug almost 5,000 years ago in what is now Iran. The very first tunnel boring machine was developed in 1846 for the construction of the Mont-Cenis tunnel linking France and Italy. Methods of excavation are:

- Open-face excavation: The tunnel is built in an open pit, a method that calls for special safety measures to shore up the side walls of the construction site. At first, the bed of the tunnel is cast in concrete on a level foundation course. The walls and roof are then cast using a pivoted, travelling shuttering system, which can often extend to over 20 m in length. Expansion joints are inserted at regular intervals between adjoining units to compensate for restrained stress. The space above the finished tunnel is then filled with earth and levelled and either restored to its original state or landscaped and developed.

- Mechanised tunnelling: This involves underground mining techniques using tunnel boring or shield tunnelling machines. Precise up-front planning is a vital necessity in this case, because each geological formation calls for different machines and cutting tools. Allowance also has to be made in the planning phase for the on-site assembly and installation of the excavation machines, which can take several months. To amortise the high investment costs, tunnels built by this method need to be at least 2 km long. There is little scope for variation in the geometry of the tunnel. Advantages are a safer working environment, higher efficiency in spite of fewer operators, and less intense shock waves while excavating.
- Caisson methods: These are employed, e. g. when a tunnel passes below a body of water. Using the floating caisson method, portions of a tunnel can be fabricated on land, then towed to the construction site, ballasted to sink them in place and joined using watertight seals. The rest of the infrastructure is constructed afterwards. Submerged caissons are box-like tunnel sections measuring up to 40 m in length. They are placed on firm ground, then the soil or sand below the caisson is dug away until the structure is located at the required depth and can be anchored there.

▶ **High-rise construction**. Like bridges in civil engineering, high-rise buildings represent the epitome of the structural engineer's art. The first *skyscrapers* only became possible thanks to the invention of elevators, which allowed rapid vertical access to high buildings, and thanks to novel materials and the development of advanced computation and construction methods. The steel-framed multistory buildings that dominated the cityscape until the middle of the 20th century have meanwhile given way to completely new concepts of load-bearing structures. Vertical loads are supported down to the foundations through ceilings, floor pillars and walls. Solutions capable of resisting the enormous horizontal forces created by winds and earthquakes are far more problematic. Nowadays it is possible to simulate the effects of wind under near-realistic conditions in wind tunnels and define the related parameters. The distribution of wind speeds and turbulence at great height over built-up areas is simulated using reduced-scale models. One way of withstanding horizontal loads is to have a rigid central core that limits

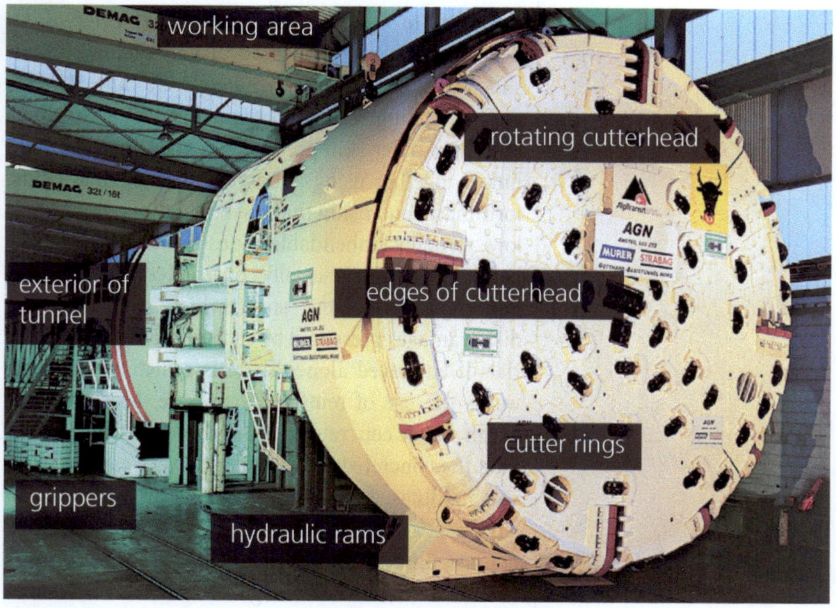

Tunnel boring machines (TBM): Often several hundred metres in length, the preferred tools for drilling through solid rock. This almost 10-m high gripper TBM drives the Gotthard Base Tunnel through granite in Switzerland. The rotating cutterhead (1) equipped with cutter rings (2) is advanced into the rock face by means of hydraulic rams (3). The detached rock chips are transported through the vents on the outer edges of the cutterhead (4) via chutes and conveyor belts through to the other extremity of the machine, where a hopper expedites them to the exterior of the tunnel (5). While in operation, the machine's radially configured grippers (6) provide the necessary support against the tunnel wall. The thrust cylinder permits a specific bore displacement. The machine itself then has to be moved forward. Source: Herrenknecht AG

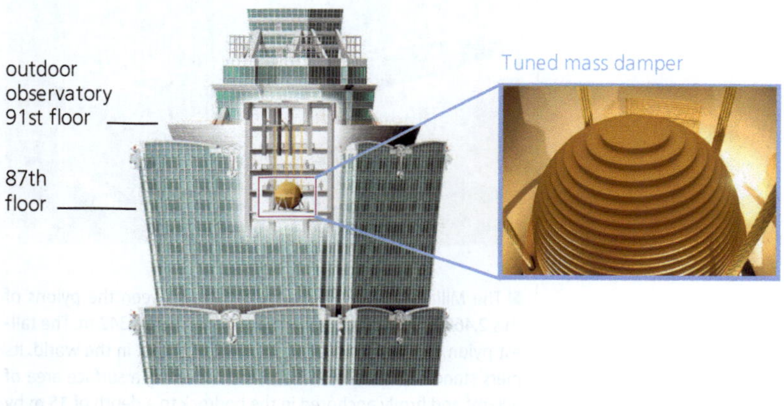

Tuned mass damper near the top of Taipei 101: Hurricane-force winds blowing at over 200 km/h often occur in Taipei. A gilded steel sphere weighing 660 t helps to keep the 508-m high skyscraper Taipei 101 perpendicular. The spherical pendulum swings to counterbalance any movement of the tower caused by earthquake tremors or high winds. With a diameter of 5.5 m, it is the largest tuned mass damper in the world. Source: Guillaume Paumier / Wikimedia Commons the extent to which the building can be shifted or deflected sideways from its vertical axis. Hollow elevator shafts built of solid reinforced concrete can serve this purpose. Using the tube system, the elements of the building's outer skin are made of rigid materials and linked together using almost totally inflexible joints, forming an outer skin that resembles the walls of a hollow, virtually unbendable pipe. Nowadays the load-bearing structure is usually constructed in a combination of concrete and steel, which overcomes the disadvantages presented by each of the building materials if employed alone. There is a distinct trend towards the use of reinforced concrete. Modern high-performance concrete mixes are up to 10 times firmer than the concrete used to build the very earliest high-rise buildings. At the same time they are very stiff and have a positive vibration-damping effect. A further, not insignificant advantage is that rein-forced concrete can resist fire for much longer than steel, which starts to yield and become unstable at temperatures of 600 °C and over.

Installation of deformable steel inserts: The advance workings in the heavily folded Tavetsch intermediate massif were a particularly challenging task when constructing the Gotthard Base Tunnel. In this region, hard rock strata repeatedly alternate with layers of soft, easily deformable rock. A system of flexible steel arches was devised to stabilise the soft zones. These allow a certain degree of deformation that attenuates the pressure exerted by the rock. The arches are embedded in shotcrete after they have been assembled. Source: AlpTransit Gotthard AG

Applications

▶ **The Millau Viaduct.** The tallest road bridge in the world at the present date was inaugurated in 2004. The road surface is supported by prefabricated steel box girders measuring 32 m in width and up to 351 m in length. Components subjected to particularly heavy loads are made of high-strength, fine-grained steel. Instead of using the cantilever technique normally employed when constructing cable-stayed bridges, a variant of the incremental launching method was used. At assembly sites at each bridge abutment, two of the central masts were pre-assembled on the corresponding road segments complete with cables. These two sections were then pushed out simultaneously from each end of the bridge in 60-cm stages towards the centre of the valley. Temporary support pylons with a height of up to 164 m halved the span. When these first sections of the road deck had bridged the two boundary fields, the next sections were assembled, joined and pushed out. During the rollout operation, the structure was monitored by GPS to within an accuracy of ± 0.3 mm. Last of all, the remaining five masts were slid horizontally over the completed deck to their final position and erected, the cables were attached and tensioned, and the temporary supports were removed.

The Millau Viaduct: The longest span between the pylons of this 2,460-m long, cable-stayed bridge measures 342 m. The tallest pylon is 343 m high, and is thus the highest in the world. Its piers stand on foundation slabs each covering a surface area of 200 m² and firmly anchored in the bedrock to a depth of 15 m by 4 piles. The reinforced concrete piers stretching to a height of up to 245 m were constructed using specially designed shuttering frames to accommodate a tapering cross-section that narrows towards the top. The stays are attached to steel masts rising to a height of 89 m above the road deck. The deck, which is only 4.20 m deep, is capable of withstanding wind speeds of up to 180 km/h. Source: PERI GmbH

▶ **Gotthard Base Tunnel.** Once completed, this will be the longest tunnel in the world. It consists of two separate single-track rail tunnels, each with a length of 57 km and linked by connecting galleries. Two multi-function stations sited between the tunnels accommodate technical services, safety and ventilation plant, evacuation routes, emergency halts and rail-switching facilities. Comprising a total of 142 km of tunnels and shafts, this important transalpine rail link traverses every major type of rock formation to be found in the Central Alps. The variety of geological conditions requires the use of a correspondingly wide range of excavation techniques, from blasting to the use of tunnel boring machines up to 400 m in length. As a result of the geological studies conducted prior to starting work, the cutterhead tools used to excavate the tunnel underneath the St. Gotthard massif were optimised for a very high contact pressure and extremely solid rock strata lying up to 2.3 km below the surface. It had to be ensured that worn tools could be replaced rapidly, so as not to slow down the drilling progress. The roof of the tunnel is secured with anchoring devices and shotcrete directly behind the cutterhead. In solid rock, the reinforced concrete coating is 40–60 cm thick, rising to 2 m in places where the pressure is exceptionally high. The ring of concrete inserted inside this coating is 1.2 m thick. The excavated material is separated into different fractions when it reaches the outside of the tunnel. Around one fifth of this material is reused underground for producing the concrete. An extensive geodetic monitoring system is installed on the surface to register any movement of the rock mass. Particular attention is paid to the region of influence of three large reservoirs under which the tunnel passes. The system, which utilises GPS data, is capable of measuring movements in the flanks of the valley to a tolerance of ± 4 mm and assessing the risk of damage by subsidence to the dam walls

► **Burj Dubai**. Since the summer of 2007, Burj Dubai has been the tallest building in the world. It is supported by 850 concrete piles driven up to a maximum depth of 50 m into the ground. The key reinforcing element is a torsionally rigid core of reinforced concrete. It is tied to the floors and partition walls in the separate wings to form an exceptionally rigid honeycomb structure. The exterior of the building features convex surfaces of glass, stainless steel and polished aluminium designed to limit the impact of the desert winds. The steplike shape of the tower was optimised using scale models in a wind tunnel. It repels circulatory vortex effects and prevents the buildup of wind pressure. The 585-m high reinforced concrete structure also damps vibrations. A steel skeleton is now being built on top of it. Special extra-high-pressure pumps had to be developed to pump concrete up to this exceptional height. Vital parameters such as hydraulic pressure, control signals, oil temperature, etc. are remotely monitored. The pumps deliver the concrete through a system of pipes and concrete distribution booms attached to the self-climbing formwork and pour it into the wall formwork that gives the skyscraper its distinctive profile. Despite a throughput rate of 30 m^3/h, the high-strength concrete often remains for up to 35 minutes in the pipelines at air temperatures exceeding 40 °C. To prevent it from setting prematurely, liquefying agents and ice are incorporated in the concrete mix. No less than 54 elevators will be installed to serve the apartments, hotels, leisure centres and shopping malls in the building – including the world's fastest elevator travelling at a speed of 18 m/s. The ultimate height of Burj Dubai when it has been completed is expected by 818 m. It is uncertain how long this skyscraper will be able to hold this record, given the number of construction projects competing for the title of the tallest building in the world.

▲ Self-climbing formwork system employed in the construction of Burj Dubai. The formwork system is based on a modular design that enables it to be installed progressively by a hydraulic system without the use of a crane. The climbing scaffold is always anchored in the concrete. The more than 200 automatic climbing formwork machines are capable of withstanding wind speeds of over 200 km/h. This type of self-climbing formwork system can be adapted to a wide range of structures of differing shapes and heights. Source: Doka/Amstetten

► **Toshka project.** By building the Aswan High Dam, Egypt created an enormous reservoir to regulate the annual flooding of the Nile valley. But since the 1990s, too much water has been flowing into this reservoir, known as Lake Nasser, and it threatens to overflow. The Toshka project aims to remedy this situation

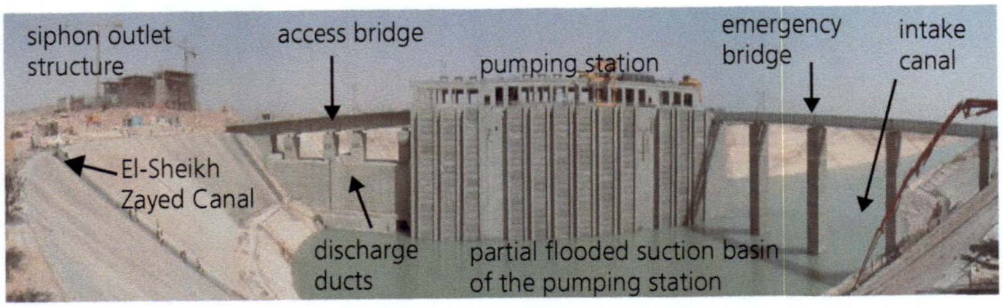

▲ Mubarak Pumping Station: The 140 x 45 x 60 m^3 concrete structure is constructed almost entirely without expansion joints and took 5 years to complete. The 140,000 m^3 of reinforced concrete contains 20,000 t of reinforcement and is watertight at pressures up to 5 bar. The pumping station houses 21 pump units with a total pump capacity of approx. 350 m^3/s. The discharge ducts have a cross-section of 2.7 x 2.4 m^2. Today, 18 continuously operating pumps transfer around 25 million m^3 of water daily from Lake Nasser to the Sheikh Zayed Canal. Source: Lahmeyer International

◄ Reinforcement of a 45-year-old reinforced concrete shell: The multi-curved shell spans an area of 38 x 39 m^2 and is only 8 cm thick at the centre. The load-bearing structure has to be conserved because it is classed as a historic building, but it no longer meets present-day safety standards in terms of load-bearing capacity. Textile reinforced concrete was the only technology capable of reinforcing the curvilinear roof. After removing the original, damaged roof covering, an initial 3–5 mm layer of special fine-grained concrete was sprayed onto the prepared undersurface. The textile carbon fabric was cut to size, laid on top of this initial layer, and tamped into the concrete. In this particular case, a 1.5-cm thick reinforcement layer consisting of three successive applications of textile fabric was sufficient to restore the decayed structure and bring it up to date with the latest safety directives on load-bearing capacity. Source: Ulrich van Stipriaan and Silvio Weiland

▼ Examples of bionic constructions. Left and centre: Lyon-Saint Exupéry airport station. The buildings and bridges designed by architect and construction engineer Santiago Calatrava often incorporate stylised animal and plant forms. The dynamic force lines are clearly visible in the load-bearing structures, which seem to be alive. The more closely the supporting systems reflect this interplay of forces, the more effectively materials can be employed, giving the building a lighter and more airy appearance. Right: The resemblance to the spreading branches of a tree is unmistakable in the 18 pillars that support the roof of Terminal 3 at Stuttgart Airport. The supports have a light and airy appearance. The lines along which the weight of the roof is carried down to the foundations are immediately evident, even to someone with no notion of structural engineering. Supports of this type have to withstand a combination of high compressive and tensile stresses. They are often constructed of steel tubes linked together by bolts, welded joints or cast-iron fittings. The fact that the roof is supported at a multitude of points means that lighter-weight materials can be used in the roof construction than in conventional designs. Source: Manfred Curbach and Jürgen Schmidt

through the construction of the Mubarak Pumping Station and the Sheikh Zayed Canal. The surplus water now flows through a 4.5-km intake channel to the pumping station's suction basin. Ten million cubic metres of rock had to be hewn out to build this channel. The project's centerpiece is sited in the middle of the suction basin – the world's largest pumping station, capable of lifting a volume of water equivalent to an average-sized river to a height 54 m above the desert plains. The siphon outlet structure leads into the main irrigation canal. The canal runs 50 km westwards and then splits into four side branches. It is lined with concrete, polymer sheeting and a protective coat to prevent the water from seeping away before it reaches its destination. When the first phase of the Toshka project is completed in 2015, the water will be used to transform over 500,000 acres of desert sand into intensively farmed agricultural land. Thanks to this irrigation system, the area of inhabitable land in Egypt is expected to increase significantly. However, not everybody is in favour of the ambitious project. Critics fear that irrigation will increase the salinity of the soil due to evaporation, and that the high level of water consumption could lead to disputes with the other countries that share the Nile basin.

Trends

► **Rehabilitation with textile reinforced concrete.** Buildings determine our lives. The building infrastructure is one of the most important location factors in a country's economy. It is hard to ascribe a monetary value to a country's total building infrastructure. Demolition and redevelopment is an expensive way of replacing structures that no longer fulfil their designated purpose. And it is not the only answer. For many reasons it can often be more expedient to preserve existing buildings and rehabilitate them. A completely new method of rehabilitating concrete structures involves the use of textile reinforced concrete. Textile multi-axial fabrics made of glass or carbon fibres embedded in high-strength, fine-grained concrete can give new stability to old building structures. A layer of textile concrete only a few millimetres thick can significantly improve the strength of building components. Textile concrete offers numerous advantages over conventional reinforcement methods. Technical textiles are resistant to corrosion. Consequently, a thinner layer is required than when using shotcrete, and the additional weight on the load-bearing structure is extremely low. The new material can be formed into almost any shape during its manufacture, allowing it to be moulded to fit profiled cross-sections or narrow radii, unlike adhesive-bonded sheets. Reinforcement strips of steel or other materials can be inserted to support single-axis loads. Suitably designed textiles are capable of supporting high forces over a wide area in many load directions. This innovative composite is more heat-resistant than materials such as carbon sheets. Moreover, old concrete and fine-grained concrete are similarly structured and can be joined almost perfectly. And in contrast to rigid, heavy steel straps or reinforced concrete, textile reinforced concrete can be mostly processed by hand. It is therefore predestined for use in confined spaces and for construction work on existing structures.

► **Bionics.** Progress in structural engineering is mirrored in changing architectural styles. New technologies permit architects and engineers to realise increasingly unusual designs. One of the current trends, as in other disciplines, is that of bionics – learning from the structures found in nature. Load-bearing systems are optimised in terms of load-bearing behaviour, the distribution of forces, and specific material properties. The result is broken-up elevations, unconventional forms, and hybrid systems.

► **Monitoring.** Bridges are monitored to obtain data on real-life load cycles due to traffic or the effect of wind and temperature. When a dam is being built across a valley, data are collected on the water level of underground and surface watercourses and the flow characteristics of the groundwater. The construction of tunnels can cause faults to develop in rock formations several kilometres away. Skyscrapers sway when confronted with hurricanes. Monitoring enables engineers to observe the effect of construction projects on the natural environment and surrounding buildings. By recording, monitoring and assessing the condition of manmade structures or the aging of materials, measures can be taken at the appropriate time to maintain or reinforce their stability. Building subsidence is detected using survey benchmarks aligned with trigonometric points outside the perimeter of the building itself. GPS is used to observe earth movements in connection with major projects such as the construction of tunnels or dams. Thermal *sensors* record changes in the temperature of different parts of engineered structures. Lines of sensors embedded at intervals throughout the length and width of a building component supply data that can be used to compile vertical and horizontal temperature profiles. Deformations inside building components can be detected with fibre-optic sensors. Acoustic sensors can be used to locate areas of corrosion, and ultrasound can be used to locate cracks in concrete or breaks in cables or tendons. These various types of on-site data can be transmitted to online processors to provide a basis for verifying impact models or producing lifetime predictions.

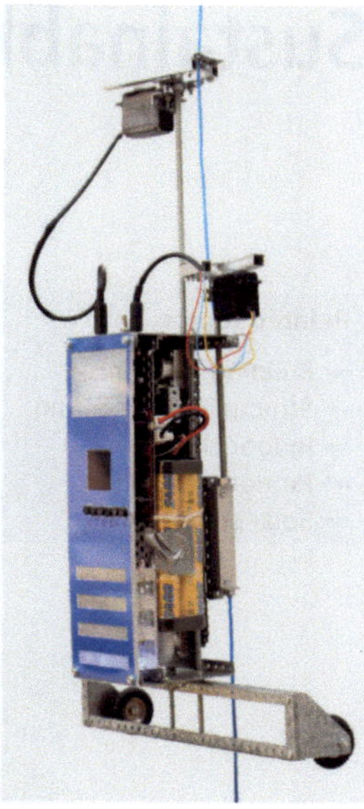

Climbing robot: This remote-controlled climbing robot can ascend cables under its own power. Its grippers can be adapted and optimised to meet the needs of different applications. The robot can be equipped to carry a wide selection of measuring instruments or cameras. It was developed for use in connection with the long-term monitoring of difficult-to-access building structures such as church spires. Source: Silke Scheerer

Prospects

- The development of new and enhanced materials and composites improves efficiency and reduces the cost of many applications. There are specialised building materials for almost every imaginable purpose.
- There is considerable potential for optimising the design of load-bearing structures.
- Rehabilitation, conversion and revitalisation projects are likely to grow in significance as an alternative to constructing new buildings from scratch.
- Construction projects can be simplified and accelerated through the use of mechanisation, computed-assisted design tools, and three-dimensional planning.

PROF. DR. MANFRED CURBACH
SILKE SCHEERER
Technical University Dresden

Internet

- sfb528.tu-dresden.de/index.en.html
- en.structurae.de/index.cfm
- www.icivilengineer.com
- www.emporis.com/en
- www.allaboutskyscrapers.com
- www.ectp.org

Sustainable building

Related topics

- Building materials
- Structural engineering
- Indoor climate
- Product life cycles
- Solar energy

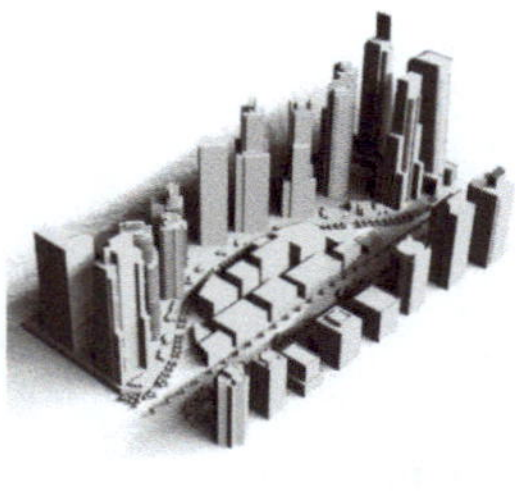

Design analysis: At the beginning of a planning process, design variations are used for shading analyses, to investigate the basic problems of solar radiation in streets and buildings. Source: WSGreen-Technologies

Principles

The aim of sustainability in building is to satisfy present needs and to preserve a healthy living environment for future generations. This means creating living conditions that are ecologically compatible, economically acceptable and which give users' needs top priority. Sustainable building has an effect on ecology and the economy, as well as socio-cultural factors such as the health and comfort of users. Sustainable building takes a holistic approach, looking at the ecological, economical and social effects of the built environment on human beings as well as on the natural environment.

In the past few years, significant innovations have been made in the field of efficient energy use, particularly in Central Europe, where stricter legal conditions are in force. However, many other areas of sustainability have not yet been developed in the same way. Efforts still need to be made to reduce emissions. The building sector has only just started to introduce recycling – it urgently needs to develop methods of construction with higher levels of recycling.

Sustainable building covers all the life cycle phases of a building, from planning to construction and utilisation, through to deconstruction ("end of life"). Designing, constructing and managing a building requires integral planning, incorporating all the relevant trades from the start.

- Design: A building's sustainability can be worked out most efficiently and cost-effectively during its design phase. Climate and location analysis form an integral part of the design, ensuring optimal planning. This analysis also ensures the building envelope, orientation and technical equipment are correctly adapted to the in situ conditions. Deconstruction should also be considered when selecting construction types and materials.
- Construction: Optimising the design process and content results in a faster, more cost-effective production process, offering higher quality. Maintaining quality assurance during the construction process is also important for meeting sustainability targets. This involves constant supervision, thermal images, etc.
- Building Management: If energy and water consumption are not optimised during the planning process, this will result in considerable, but avoidable, additional costs in the building's utilisation phase. In addition to accurate planning, building automation plays a significant role in efficient utilisation. The user profile and systems engineering must be optimised within this automation. Socio-cultural aspects of sustainability (such as a comfortable indoor climate) also contribute to user satisfaction and in turn affect the building's performance in terms of economic efficiency.
- Conversion: If a building has to be converted or renovated, intelligent and flexible design and construction are required. An intelligent building concept will incorporate conversion and renovation measures right from the start. This concept ensures that individual buildings, or the entire building, can be easily deconstructed, taking recyclability and emissions into account.
- End of Life: In most cases, the deconstruction process is not considered sufficiently when the building is designed and constructed. "End of life management" must, however, be an integral part of planning to ensure effective sustainable building. If this is not sufficiently considered, building deconstruction and removal become extremely complicated, and recycling larger items of building materials becomes impossible. Intelligent management of material flows and accurate recycling planning ensure that the building components can be deconstructed easily at the end of the building's life.

Modern Design Tools include:

- Microclimatic *simulation*: An important prerequisite for minimising energy requirements and optimising comfort is to adapt a building to the site and local climate. Microclimatic simulation is an external adaptation tool that takes parameters such as solar radiation, air temperature and humidity, wind direction and velocity into account. Microclimatic simulation investigates the effects of architecture and urban design on the transfer and flow of heat characteristics for the urban area - for example, the choice of materials, shading, and the orientation of a building.
- Thermal simulation: Thermal simulations reveal specific information about the energetic behaviour of a building, depending on weather conditions,

technical equipment, control mechanisms and user profiles. These simulations can predict spatial mean values over short periods (e. g. to assess peak temperatures in summer) or longer periods. It is possible to calculate the mean radiation temperature, in addition to the mean air temperature and the surface temperature of enclosed buildings. Thermal comfort in interior spaces can be predicted precisely.

- Daylight simulation: Daylight simulation can show the interior space in a photo-realistic and physically accurate way. This is used to design a facade in the best possible way, developing energetic annual balances and reducing artificial lighting to a minimum. Daylight simulation examines glare protection, shading mechanisms, and makes the best use of daylight.
- *Computational fluid dynamics (CFD)*: Air flow simulations are closely linked to thermal simulations, which are frequently used for complex building structures with natural ventilation or for designing double-skin facades. This type of simulation can be used to calculate spatial distribution within a zone. This is especially important for problems relating to local air flow velocity (e. g. air drafts, cold air drop, etc.) or local air temperature (e. g. thermal comfort). Due to the complex calculation methods, fluid dynamic simulations are usually only carried out for subzones.
- Building component simulation: Different types of building component simulations can be used for different purposes. An example is the 2D or 3D calculation of thermal bridges for problems of condensation and/or mould growth, and for the distribution of temperature and humidity within the building component (hygrothermal FEM simulation). Building component simulations are mainly applied when standard specifications are not yet available.
- Building equipment simulation: System simulations are used in advanced planning phases to harmonise the technical system components, in preparation for the operational phase. Building equipment simulations take the building envelope and occupancy into account. Sensor technologies, such as earth heat probes, require calibration. For example, the building's thermal output or input must be adapted to the soil before initial operation, to avoid extensive cooling or heating at a later stage. System simulation can also be used to optimise the control strategies for a building.
- Life cycle analysis: A total life cycle analysis (life cycle assessment, LCA) can be carried out for individual materials, building components or the whole building. The life cycle analysis provides parameters related to essential ecological problems, such as greenhouse effect, summer smog, acidification of the air and soils or scarcity of resources. Predictions about the effects of various alternatives on the eco-balance can be made in the early planning phases.

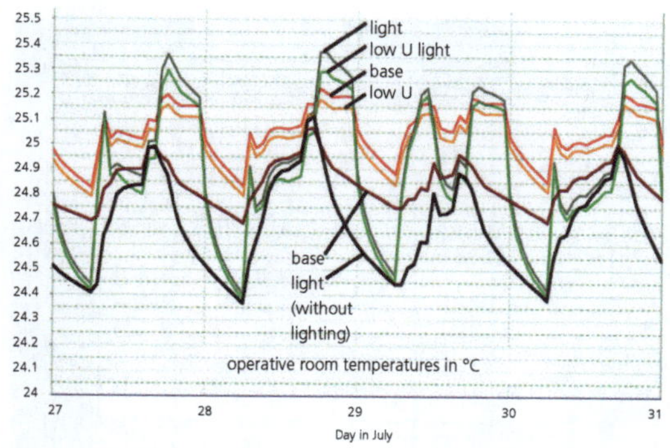

Thermal simulation: Example of a thermal building simulation, showing user thermal comfort. The diagram shows the operative room temperature for building materials of different weights and U-values, and for different lighting appliances. The simulation emphasises the impact of lighting on interior temperatures. It also shows how lightweight structures perform well in hot climates: the low thermal mass results in lower cooling loads for mechanical ventilation. Source: Ingenieurbüro Sick

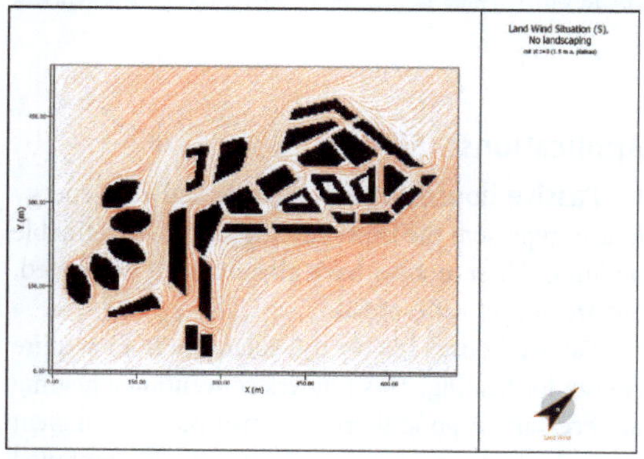

Computational fluid dynamics (CFD): Investigation of the air flow within an urban setting, using computational fluid dynamics simulation. This tool simulates factors such as wind speed, air flow direction and the resulting air temperature at any time, on any day of the year. Planning data for an urban layout, for example, can be adapted to any specific location at an early planning stage. Source: WSGreenTechnologies

Daylight simulation: Daylight simulation is used to check internal lighting conditions. It offers a realistic image showing the luminance of surrounding surfaces. Source: Hans Jürgen Schmitz, e[2] Energieberatung

▶ Completely dismantable and recyclable-R128: The structure consists of a screwed steel frame with high-quality insulating glazing acting as a curtain wall. Source: Roland Halbe

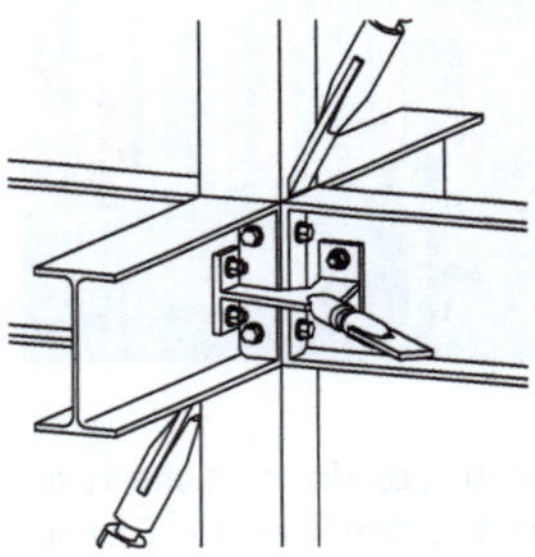

▲ R128 constructional joint: The whole structure of the building consists of a screwed steel frame (steel skeleton) with diagonal buttresses. It can be easily constructed, re-constructed and removed, as the structure does not need welded joints. Source: Werner Sobek Stuttgart

Applications

▶ **Passive houses.** Passive houses and zero-energy houses represent the latest development in sustainable building. These houses have already been developed, but are not yet widespread.

Passive houses largely use solar power to acquire energy for heating. This renders conventional heating unnecessary. A good thermal transmission coefficient for the building envelope must be generated, combined with a construction without thermal bridges and a high degree of air tightness. Ventilation systems with efficient heat recovery provide hygienic air conditions and avoid heat loss through window ventilation. Heating a passive house requires 90% less energy in comparison to a conventional building. In terms of fuel oil, the passive house uses less than 1.5 l/m^2 a year.

▶ **Zero-energy houses.** A zero-energy house uses no external energy (electricity, gas, oil, etc.) in its annual balance. Solar systems produce the annual energy requirement for heating (hot water supply, etc.). However, the energy required to construct the house is not taken into consideration. If a house produces more energy than it consumes on average annually, it is referred to as a plus-energy house.

Apart from the advantages of passive and zero-energy houses, a Triple Zero® Concept offers:

- Zero Energy: On average, the amount of energy consumed must not exceed the amount of energy produced annually.
- Zero Emission: CO_2 neutral (in operation), no combustion, use of emission-free materials
- Zero Waste: All components must be completely removable and recyclable.

The R128, an emission-free zero-energy house, is an example of the Triple Zero® Concept. It not only differs from previous buildings in that it is completely transparent – its modular construction and use of pre-fabricated building components means that it can also be quickly erected, and completely and easily recycled. A planning team has been closely involved from the start, implementing design, technology and climate concepts.

The R128 structure (steel frame structure with joints) is minimal in weight; it can be quickly constructed, reconstructed and disassembled and does not require composite materials. A recently developed climate concept regulates indoor temperature. Water-filled ceiling components absorb the *solar energy* radiated through the façade into the building. These components then deliver the solar energy to a heat storage facility, which heats the building in winter by reversing the heat exchange process. The ceiling components function as a radiant heat source and additional heating is not necessary. Photovoltaic cells on the roof generate electrical power, which is supplied to the national power grid. The national power supply system is used as an intermediate storage medium; the same amount of energy is consumed as is produced on average annually.

▶ **Renovation of existing buildings.** Sustainable renovation of existing buildings can also generate noticeable savings in terms of energy. This method is also much more environmentally friendly. A comparison of residential buildings throughout Europe shows that the utilisation phase's impact (energy use for heating and cooling) on the greenhouse effect (equiv. kg of CO_2) is four times greater than that of the construction phase. If we are to reduce the overall impact, in as short a time as possible, top priority must be given to the renovation of existing buildings (especially the quality of the building envelopes and building services engineering).

Optimised areas are:

- energetic quality
- thermal comfort in indoor environments (indoor air temperature, indoor air quality and acoustic characteristics)
- reduction of the electrical power required for artificial lighting and visual comfort
- rhermal insulation / shading in summer
- optimised ventilation

A primary school (built in 1936, renovated in 1997) provides an example of how this kind of renovation can use energy more efficiently. Measures implemented include insulation of the building envelope, installation of thermal insulating glazing and replacement of the heat generator. The original glazing had U-values of 2.5–3.5 and was replaced by thermal insulating glazing with a Low-e coating of 1.4 W/m²K. Additional thermal insulation improved the U-value of the walls from 1.7–0.26 W/m² K. These measures reduced energy use for heating from 200–49 kWh/m² a, i. e. a reduction of 75%. Controller units were also installed in each room, providing additional energy savings of 8%. Apart from the obvious reduction in total energy consumption, excessive overheating in summer was reduced to zero, improving the thermal comfort of the building considerably.

Trends

▶ **Life Cycle Assessment.** Life cycle assessment in building design is not yet the norm. However, increased certification requirements for sustainability in buildings will require life cycle assessments. Their use will increase in the future. Investigations can be carried out in the early stages of design, to determine which variations have the lowest environmental impact during building construction, operation and removal. Life cycle assessment includes all energy and material flows during the complete life cycle of a material, component or the whole building. This assessment considers energy and land use during removal of raw materials, emissions during production, effects on the energetic balance during operation, and recycling potential when the building is disassembled.

▶ **Adjustment to extreme climatic zones.** The Dubai Crystal project in the United Arab Emirates is a valuable example of how important it is to optimise the micro-climate in foreign climatic zones. As temperatures in summer exceed 40 °C, exterior comfort is extremely significant. Microclimatic *simulation* was applied in its entirety to optimise the urban design – i. e. the orientation and density of individual buildings - in terms of ventilation and solar radiation. Microclimatic simulation also optimises the use of reflecting materials, planting, shading equipment and water elements, which cool by evaporation. These measures considerably improve conditions for people outside in comparison with local construction methods in Dubai. Dense building development and the corresponding reduction in solar radiation also cut energy use for cooling by up to 12%. Optimised building envelopes and efficient building services, engineering and equipment are helping to reduce energy consumption further. The total energy consumption of the project will be reduced by between a third and a half in comparison with local construction standards in Dubai.

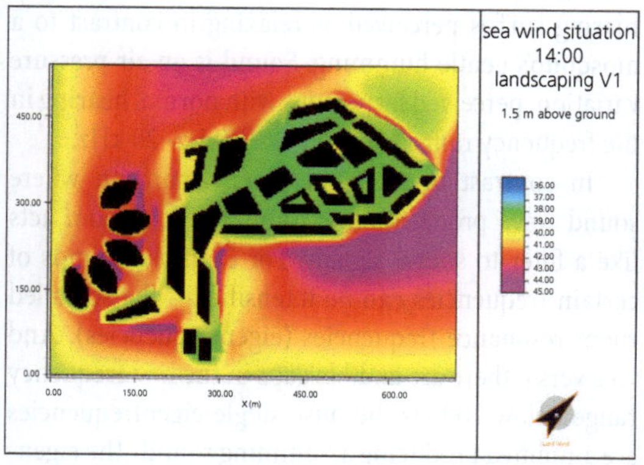

Microclimatic simulation for Dubai Crystal: Examination of the air temperature in an urban planning area in Dubai shows a clear reduction of temperatures within the urban area. Dense building development reduces solar radiation considerably, improving thermal comfort. Source: WSGreenTechnologies

Prospects

- Climate change and the scarcity of resources will increase the need for sustainability in building and urban development design. Transparency and interdisciplinary cooperation between all the partners involved in building design will become increasingly important in the future.
- Sustainability certification requirements will become more widespread. This is especially important for investors, guaranteeing long-term rates of return and profitability.
- Planning processes are becoming more and more integral and iterative; plans for managing and disassembling buildings are developed at the design stage.

Sustainable building offers advantages not only in terms of energy and finance (especially during the operation phase), it also improves user productivity and performance (as people work in a more comfortable environment).Technological developments will make sustainable building easier to implement.

PROF. DR. WERNER SOBEK
University of Stuttgart, Stuttgart

PROF. DR. KLAUS SEDLBAUER
Fraunhofer Institute for Building Physics IBP, Stuttgart

DR. HEIDE SCHUSTER
WSGreenTechnologies, Stuttgart

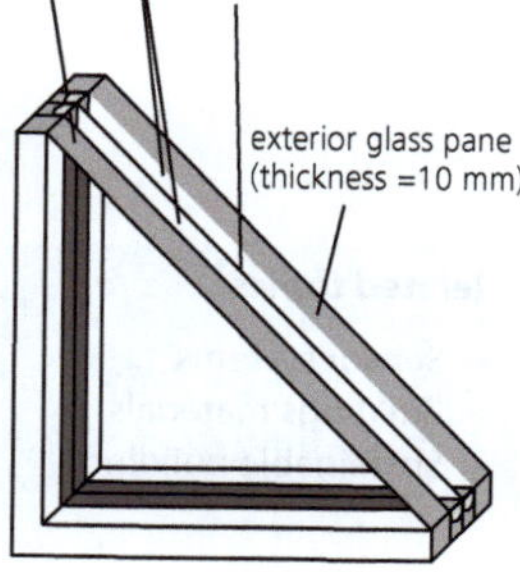

R128 cross section (heat, reflection and glazing: The glazing has three layers and consists of an external and internal pane and a foil with a vapor-deposited metal coating, which fits tightly in the gap between the panes. The space between the foil and glass is filled with the inert gas Argon. These specially designed panes allow extremely low thermal conductivity through the panes. Source: Werner Sobek Stuttgart

Internet

- www.dgnb.de
- www.wsgreentechnologies.com
- www.lowex.net
- www.ibp.fraunhofer.de/index_e.html
- www.ectp.org/fa_quality_of_life.asp

Indoor climate

Related topics

- Sensor systems
- Buildings materials
- Sustainable building

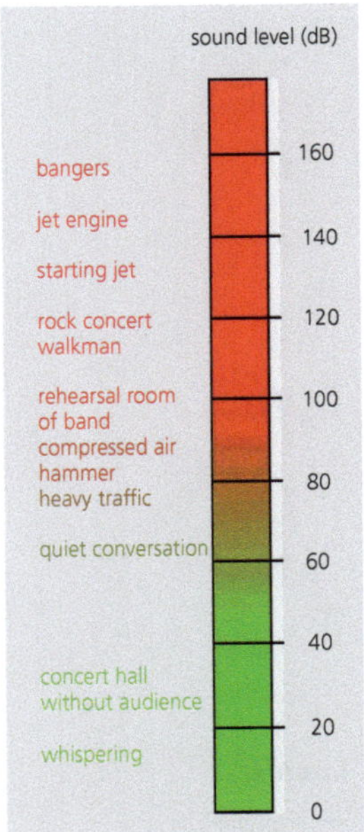

Sound level of sound sources in everyday life: The risk of hearing impairment (red range) already occurs at a load of 80 dB (A).

Principles

Human beings spend around 90% of their lives in closed environments such as apartments, offices, stores or cars. Well-being, health and performance are influenced by various perceived factors inside these enclosed spaces. The sum of these physical impacts is described as indoor climate.

Indoor climate comprises the following main aspects:

- Thermal comfort: This is primarily influenced by the location of the building, its utilisation, the method of construction, the type of heating and ventilation. The parameters are air temperature, radiant temperature, relative humidity and air velocity.
- Room acoustics: This is determined by external noise sources (e.g. road traffic), internal noise sources such as equipment and people, and the acoustic design (e.g. sound insulation of building components). Important factors are not only loudness and frequency components, but also the information content of the sound source (background conversations or dripping taps).
- Visual comfort: This is defined by levels of natural daylight (the proportion of windows in the façade and solar control) and artificial *lighting* (illumination, luminance). Comfort is primarily influenced by the distribution of the light sources in the room, by the colours of wall and ceiling surfaces and glare effects.
- Air quality: Indoor air quality is determined by the carbon dioxide content and by air pollution caused by volatile organic compounds (VOCs), particles or ozone, which may enter from the exterior or be generated indoors (emissions from materials, photocopiers, laser printers).
- Odour: The well-being of human beings in interiors is also influenced by unpleasant odours, e.g. kitchen smells, but also odours from materials.
- Microbial air quality (indoor air hygiene): This is determined by the content of dissemination units in the indoor air (e.g. spores) and excretions/emissions from microorganisms (e.g. mycelium fragments, endotoxins, MVOC, etc.). It is influenced by sources of humidity, temperature and the chemical (potential food supply) as well as physical (surface texture; hygrothermics) properties of the materials and construction.

▶ **Thermal comfort.** Thermal comfort is influenced not only by the indoor air temperature but also by the temperatures of surrounding surfaces, i.e. walls and windows, because the body not only feels the ambient air temperature, but also the radiant properties of the surrounding surfaces of walls or windows. Thus, a cold window has a "radiating effect" even on persons some distance away in the same room. This means that low surface temperatures must be compensated by higher air temperatures to achieve thermal comfort. This is a typical problem in older buildings: a higher air temperature is required in winter to obtain thermal comfort due to the inadequate thermal insulation of external building components and the resulting low temperatures of surrounding surfaces. The perceived temperature is the same in a room with warm, thermally insulated external walls and a low air temperature as in a room with cold external walls and a higher air temperature. In summer, the situation is reversed. The higher the surface temperature, the lower the air temperature needs to be to obtain the same degree of comfort.

▶ **Room acoustics.** Whether a particular sound is judged to be a nuisance or pleasurable depends on the loudness, content, pitch strength, duration and last but not least on the context and the surroundings. Thus roaring surf is perceived as relaxing in contrast to a mosquito's gentle humming. Sound is an air pressure variation, perceived by persons with normal hearing in the frequency range from approx. 30 Hz–20 kHz.

In contrast to the outdoor environment, where sound wave propagation is unhindered, a room acts like a filter to sound signals. For example, sounds of certain frequencies can be intensified – the so-called room resonance frequencies (eigenfrequencies). And vice versa, there are audible gaps in the low-frequency range below 100 Hz, because single eigenfrequencies are amplified producing a humming sound. The eigenfrequencies are dependent on the room geometry and the sound absorption coefficient of the total surface of the room. This also applies to the sound field at average and high frequencies, whereby a so-called diffuse sound field occurs in rooms at this frequency range.

The first acoustic impression of a room is usually determined by reverberation. The larger the room and the higher the reflectivity of the surfaces, the longer the reverberation time. This applies especially to modern buildings featuring wide expanses of non-absorbent concrete and glass. Whereas a long reverberation time is positively desirable when listening to an organ concert in a church, it is more of a nuisance in an open-plan office

▶ **Visual comfort.** 80–90% of the information we take in from our surroundings is visual. The human eye has a huge capacity to adapt to different levels of brightness. Luminance may vary in a dynamic range from a few photons on dark nights to the glaring brightness of snow fields at the limit of absolute dazzling. The way we perceive the brightness of an object is essentially determined by the surrounding environment. The face of a person seated with their back to a window with glare protection, lit by a constant level of light, appears to be bright. If the same window is not fitted with glare protection (i. e. bright background), the face appears to be dark. Thus brightness is a relative perception.

Applications

▶ **Individual air-conditioning.** Conventional air-conditioning systems evacuate heat by replacing a relatively high volume of warm air with cool air supplied through technically complex systems of conduits and fans. Since this can cause discomfort in the case of high thermal loads (due to high air velocities), systems were developed which dissipate a large part of the thermal load by cooling the surfaces. So-called chilled ceilings provide a greater feeling of comfort. This system allows surface temperatures to be reduced to below the indoor air temperature in summer, resulting in the climate control of individual workplaces due to "radiant cooling". The person sitting at the next desk can choose to set a higher temperature than their neighbour. In winter, by contrast, a rolling stream of warm air, which can be focussed on the individual workplace, produces the individual climate. This airflow is generated by a ceiling ventilator that directs the air through air deflector plates past a convector.

▶ **Thermally activated building components.** In the case of thermally activated building components, the entire building component (floor or wall structure) is used as a heat transfer surface by installing tubes carrying air or water in the component itself. This provides a very large surface area for heating and cooling, possibly allowing system temperatures to be reduced. The required space heating or cooling performance can be obtained using relatively low flow temperatures for heating and relatively high temperatures for cooling. This permits the use of alternative, renewable energy sources for heating and cooling, such as geothermal energy from the soil or groundwater. The total thermal mass of the construction is actively used as well. Load peaks can be buffered due to the thermal storage capacity of the solid floors and walls, and ambient temperature can be kept constant by this effect to a large extent.

▶ **Room acoustics.** The acoustic working conditions in an open-plan office should ideally allow good speech intelligibility at the same time as a sense of privacy. The individual instruments in an orchestra have to compete against the sound produced by other instruments, with the result that, in an orchestra pit without acoustic panels, the musicians are unable to assess their own playing and consequently play louder than required. The solution is called transparent accustics. This is the designation for the acoustic conditioning of a room that allows the occupants to communicate at an appropriate loudness level. Thus, broadband absorbing acoustic barriers are used, which absorb frequencies lower than 100 Hz with a low thickness of material. Conventional porous absorbers usually absorb frequencies at wavelengths not exceeding one quarter of the depth of the absorber. A 100 Hz sound has a wavelength of 3.4 m, i. e. the necessary wall panel would have a thickness of 80 cm. To avoid this kind of construction, acoustic resonance systems and porous foams are combined to allow high absorption at both low and high frequencies. These constructions are usually made of sheet metal and foamed plastics: the vibrating sheet absorbs the low frequency, the foamed plastic attenuates the vibration of the sheet, and due to subtle design even reduces the sound at medium and high frequencies in the room.

Micro-perforated absorbers (e. g. made of acrylic glass) produce acoustic and optic transparency. These are sheets or foils, and their absorption spectrum is determined exclusively by the dimensions and quantity of minute holes and by the thickness of the sheet and the distance to a wall. The acoustic energy, i. e. kinetic energy, is converted into heat by friction. The compressed air molecules rub against the walls as they pass through the holes and warm up the sheet and foil material. This process reduces the "acoustic" energy. The diameter of the holes is < 1 mm, making an unhindered air passage within the holes impossible, but representing the typical boundary layer for laminar flow.

▶ **Visual comfort.** Glare due to an excessive contrast between brightness levels may result in the physical re-

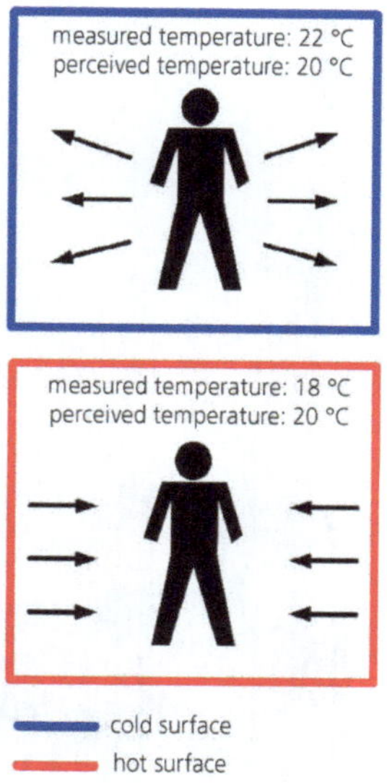

Thermal comfort: Difference between perceived and real temperature in a room with cold (top) and hot (bottom) surfaces. The same thermal comfort is perceived in both cases, as the different air temperatures are compensated by the radiant effect of the walls.

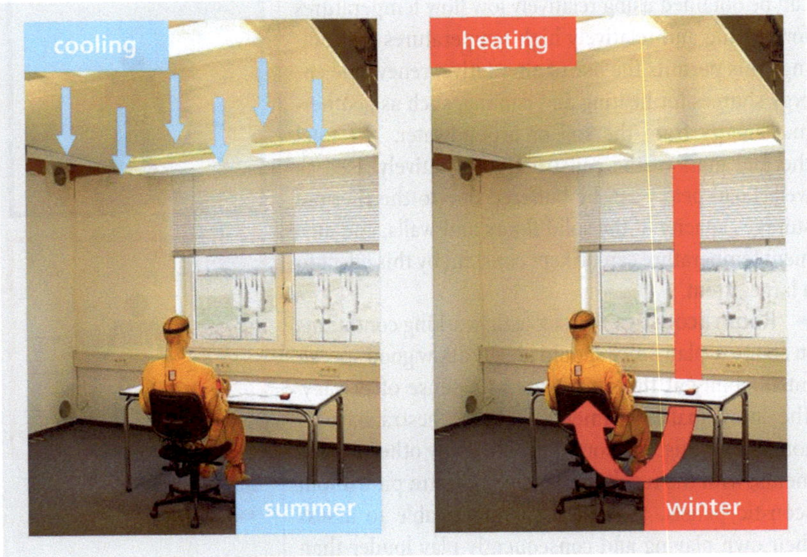

Chilled ceilings. Left: The individual cooling of a small-sized area is guaranteed by radiant heat transfer in summer. Right: The chilled ceiling generates a rolling stream of warm air in winter. Source: Fraunhofer IBP

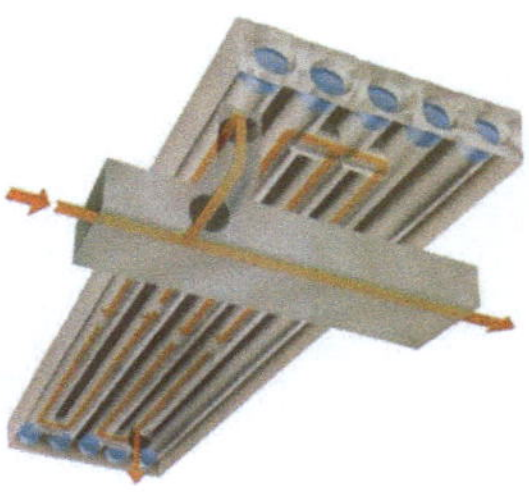

Thermally activated building components: A hot or cold medium (air or water) is pumped through a serpentine array of tubes integrated in the building component, heating or cooling the room according to the selected operating mode. The high thermal mass of the concrete ensures a more or less constant ambient temperature, even when no air or water is being pumped through it. Source: Fraunhofer IBP

duction of visual acuity. Moreover, certain variations in brightness may be perceived as unpleasant. The massive invasion of our places of work by computer workstations makes higher demands on lighting design, an issue that has to be repeatedly reassessed whenever tasks are reassigned in the wake of (frequent) reorganisation. Unlike paper, displays have their own background light source. Instead of improving the contrast, as is the case with paper, increasing the intensity of light falling on the screen in fact reduces contrast. Artificial lighting systems can be adapted to these new requirements through the use of louvred light fittings designed to reduce glare or new types of lamps designed specifically for use with inclined flat screens.

In rooms lit from one side, the external façade must be equipped with adequate glare protection at eye level. If windows are placed towards the top of an outside wall, special daylighting systems can be used to reflect natural light off the ceiling in more remote parts of the room, thus reducing the need for artificial lighting and reducing the consumption of energy.

Trends

► **Adaptive indoor climate control.** By combining selected room-specific data with forecasting models, the indoor climate of a room can be adapted to its utilisation. Intelligent systems are capable of learning how to react to changing numbers of occupants – e. g. in conference rooms – or how to maintain a stable indoor climate under erratic weather conditions or control the stream of visitors wishing to view works of art, which are highly sensitive to changes in their surroundings. Global climate change presents new challenges with respect to the preservation of cultural assets. Local extremes such as increased driving rain loads or rapidly changing weather conditions require foresighted and adaptive control technology for the different climatic zones in Europe and throughout the world. Preventive conservation aims to avoid damage to works of art and other cultural assets by ensuring that they are displayed or stored under the appropriate conditions, rather than having to restore them at repeated intervals. Works of art and historic buildings can be durably protected by programmable systems that control indoor environmental conditions as a function of changing outdoor weather conditions, based on knowledge of the physical processes taking place in the materials obtained through dynamic simulations. Similarly, the installation of sensors to monitor the presence of people in a conference room can help to optimise working conditions in terms of thermal comfort and performance, if combined with the adaptive control of thermally active building components and air-conditioning systems. Adaptive systems, e. g. neuro-adaptive regulations, respond automatically to unexpected events and allow the occupants of the room to concentrate on their work without disturbance.

► **Air conditioning in vehicles.** The results of research on indoor climate in buildings is transferable to other enclosed spaces, particularly all forms of vehicles designed to carry passengers. The objective here is to improve the physical comfort of passengers, taking into account the specific factors that may have a phy-

Daylighting systems. Adjustable slatted blind systems with specially designed reflectors direct the light deep into the room at a controlled angle. The slats often have a special profile enabling them to absorb the diffuse daylight at a large opening angle, focus it and direct it into the interior at a narrower angle of reflection. Source: Fraunhofer IBP

Flight Test Facility: The physical parameters of an aircraft cabin during flight can be realistically simulated on the ground. The test vessel is equipped to generate the low pressures typically encountered by a widebody aircraft during flight. During normal operation, the cabin pressure in an aircraft can fall to as low as 753 hPa. Normal air pressure at sea level is around 1000 hPa. At standard cruising altitudes, the air pressure outside the aircraft cabin can be as low as 180 hPa. Each of these scenarios can be simulated in the test facility. Source: Fraunhofer IBP

siological effect on the health of certain persons. A typical case in point is that of cabin pressurisation in aircraft. Aircraft passengers are also exposed to low levels of humidity, a frequent source of physical discomfort. However, whatever measures are taken to improve thermal comfort in an aircraft must be balanced against technical safety considerations, given that electrical components are sensitive to any condensation forming on the cold outer surface of the fuselage. Operators of other forms of transportation carrying passengers over short distances, such as tramways, have to contend with a different type of challenge. In this case, embarking passengers often have to deal with the transition from extremely cold or hot environments outside to a heated or air-conditioned interior. Thermal comfort must be ensured even during a 20-minute ride after a prolonged wait at the tramstop. Efforts to design the optimum indoor climate for passenger vehicles must include direct measurements of each of these distinct environments.

▶ **Multifunctional indoor building components.** An unfortunate consequence of improving the airtightness of buildings in the interests of energy efficiency is that this frequently results in an increase in the concentration of air pollutants in the interior, and associated health problems (fatigue, breathing difficulties, headaches, etc.). One way of counteracting high concentrations of VOCs in interiors involves equipping indoor building components with materials capable of adsorbing toxic substances (activated charcoal, zeolites) or degrading them by photooxidisation (titanium dioxide in the anatase configuration, effective in the UV range or in combination with antenna pigments in the visible wavelength range). *Photooxidative degradation* of chemical substances ideally results in the generation of carbon dioxide and water, but in the case of higher molecular and ramified hydrocarbons it is possible that imperfectly oxidised intermediates emerge, maybe even new and odour active substances. Besides the adsorptive or catalytic degradation of chemical air pollutants, this kind of equipment can also have an antimicrobial effect (possibly by adding "biocide" agents). Additional functions such as sound absorption (sound field conditioning) or the conditioning of temperature and light, can be integrated in newly developed building components, resulting in multifunctional building components that provide added value in terms of the quality of the indoor climate.

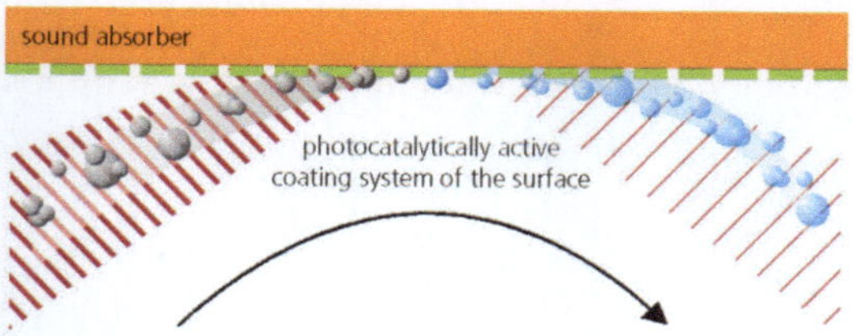

Conversion of odours and pollutants by means of catalysts in the coating of room acoustically optimised sound absorber sheets.

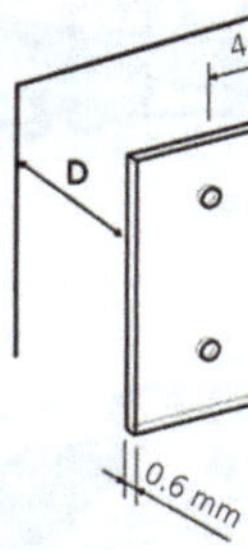

Diagram of a micro-perforated absorber in front of a wall: The maximum absorption of 75% for a frequency of e.g. 500 Hz occurs at a distance (D) of 10 cm from the wall.

Prospects

- The term "indoor climate" refers to the impact of heat, humidity, air movement, light and sound on the occupants of an enclosed space. In the past, optimisation measures aiming to improve thermal comfort have focused on individual parameters.
- The future goal, by contrast, is to develop "efficient" environments. Room acoustics is a determining factor in people's ability to work efficiently in a communication-intensive environment. But hygrothermal, thermal, lighting, olfactory and organisational parameters also determine human performance. The interaction of these different factors and their cumulative effect on performance are not yet elucidated. But the results of initial tests appear to indicate the existence of intensifying or compensatory effects, whereby room temperature and different-coloured surfaces can affect the perceived acoustic load.

PROF. DR. KLAUS SEDLBAUER
Fraunhofer Institute for Building Physics IBP, Stuttgart

Internet

- www.iea.org
- www.ibpsa.org
- www.ie.dtu.dk
- www.ibp.fraunhofer.de

LIFESTYLE AND LEISURE

Work fills a large part of our time, yet it is what we do when we are not working that defines our attitudes, behavioural patterns and lifestyle habits, more than anything else. At the beginning of the 21st century our understanding of leisure fundamentally changed – nowadays, the meaning of leisure is reflected in aspects such as self-fulfilment, enjoyment and maintaining health and fitness, rather than in recovery from the (physical) demands of the job. At the same time, our leisure budget and ability to consume is growing, to which a booming leisure market is responding with an abundance of leisure-time pursuits.

Basically, the "free time" at our disposal can be divided into three core areas:

- Domestic activities, which apart from active pursuits (with family or friends maybe) include quieter activities, such as contemplation, relaxation, or caring for our health and body
- The media sector: It has undergone rapid expansion since the beginning of the nineties and notably now appears to complement leisure activities rather than replacing them
- Leisure outside of the home: This is active time, when we engage in activities with other people, take part in sport, attend cultural events or get involved socially

It is almost impossible, however, to fit the modern consumer into a definite target group with fixed behavioural patterns. Today's consumers change their preferences continuously and want to keep as many options open as possible in terms of how and what leisure activities they pursue. The trend is towards a "multi-option society", characterized by individualisation and quickly changing value patterns. These trends make it impossible to evaluate what people enjoy doing with their leisure time – leisure behaviour patterns are becoming shorter and shorter, while long-term, enduring social structures in our society are on the decline. The leisure market is reacting to this increasing discontinuity of lifestyle and our passion for acting out our own interests by expanding and individualising what is on offer – which is often the unusual. Information and communication technologies have increasingly refined networking and accessibility, further obscuring, meanwhile, the already weak distinction between work and leisure. This is also exemplified by models such as teleworking, and the unlimited availability of people to work independently of constraints on time and place.

The term "leisure" is also determined by other factors – our society is becoming more elderly and fewer couples choose to have children. An important principle applies to all sectors in the future – anyone who ignores the older generation will probably end up with the short end of the stick. Older people are the "best agers", who are not only destined to represent the majority of European citizens in terms of numbers, but who also have the financial means and the time to shape tomorrow's leisure market.

Major changes can be seen in the intensity with which leisure activities are pursued. The saying "one thing at a time" no longer applies. The motto is now "do more in the same time". Whereas the hallmarks of free time in the past were unrestraint, unconcern and timelessness, genuinely free time is now becoming increasingly rare in people's lives. Leisure opportunities have had to become more and more time-efficient. Activities, which take more than two hours - whether with partners, children or alone – are stagnating or are decreasing.

It appears likely that *Information and Communication* technology will have the greatest influence on the leisure sector in the future. These technologies are increasing average media consumption and have a considerable impact on leisure behaviour. Traditional video games have now developed into *Virtual Worlds*. An "always on" society needs terminal units which are mobile, networked and self-sufficient in power supply. Developments in materials are required in all areas of leisure, whether this involves high-performance *sports equipment*, functional clothing or inductive heating cooking utensils.

The elderly population must not be banned to the fringes of social life. *Assistive technologies* and *ambient technologies* provide them with mobility and support in their domestic environment. These technologies significantly contribute to increasing the independence and quality of life of older people and the disabled. In addition to their importance for the clothing industry, *textiles* and materials now fulfil important functions in various other areas of technology, such as the *automotive industry*, *structural engineering* and medicine. The borders between *pharmaceutical products* and *cosmetics* are merging.

▸ **The topics.** *Textiles* not only serve as protection for the body and as a fashion accessory; for some time now they have assumed a host of other uses in different areas of industry and technology. New spinning and dyeing technologies, odour inhibiting qualities in the clothes we wear, luminous fabrics – when combined with related technologies, textiles can become exciting products for the private consumer, simultaneously fulfilling a growing number of functions.

In a similar way to textiles, *domestic appliances* have continued to develop from being purely functional items to being utensils of leisure design (e.g. cooking, do-it-yourself). New products are supposed to be fun to use and to save time. Both of these criteria are finding their way into the development of appliances, increasingly reducing the amount of work they involve. Low consumption and more efficient technologies in the design of future household appliances also respond to the concept of sustainability through the efficient use of resources. Still inside the house, health and body care – wellness – is booming. Women, and men also, now describe regular health and body care as a leisure pursuit, and related products extend far beyond the area of cosmetics. It is increasingly difficult to delineate care products from pharmaceutical products. The skin is both the focus of beauty (as the first visual impression of a person) and of medicine (as an important organ of the body). Research supports both areas, as it develops alternative methods to replace animal testing experiments for new cosmetics, for example, or to establish non-invasive methods for examing the skin.

Sport plays an important role in maintaining personal health and fitness – it is a major pursuit in outdoor leisure activities. Further developments in sport technology can be found not only in the professional sphere, but also in popular everyday sports. Sports products are becoming "smarter and smarter", safer and more effective – as exemplified by tennis rackets that absorb vibrations, or running shoes with integrated microprocessors that can detect the condition of the ground and alter the firmness of the soles accordingly.

Another popular leisure area outside the home is public concerts. The enjoyment of live music, such as a rock or pop concert, is a passion shared by many. Electro-acoustic technologies have made music events in bigger and bigger arenas for bigger and bigger audiences an unforgettable visual and acoustic experience. Loudspeaker technologies are consistently expanding the quality of sound and the possibilities in live performance. ■

	Organisation times	Individual use of time
Acceleration	- faster pace of social life - quicker global communication - more and faster cars	- less sleep and rest - less time for meals - more people have the feeling of time pressure - speed on motorways
Compaction	- in professional development: shorter training periods, faster passage through career milestones - in work: fewer breaks	- more simultaneous activities - closer merging of paid and unpaid work
Continuity	- production, trade, transport, leisure time, 24-hour media consumption	- more time later in the evening - typical weekend activities also on weekdays
Deregulation	- disbandment of normal working hours and closing times	- more flexible working hours - fewer communal meals
Desynchronisation	- disbandment of normal working hours	- less collective leisure time
Individualisation	- disbandment of class-specific leisure cultures	- differentiated leisure activities - domestication of leisure time
Time Management	- temporary and multiple paid employment	- more time expenditure on management of household, family and external relationships

Trends in time structure: In all leisure activities within and outside our living environment, technological developments and innovations have led to different user attitudes, or a different attitude to new technological developments. The ever-increasing pace of life, individualisation, as well as mobility and internationalisation, characterise the world we live in and dictate developments in leisure technologies. Source: F-CON GmbH

Sports technologies

Related topics

- Nanomaterials
- Textiles
- Sensor systems
- Intelligent materials
- Laser
- Automobiles
- Space technologies

Principles

In the 21st century, more people than ever before are engaging in sporting activities. Sport has evolved into a global business that puts both athletes and coaches into more and more complex systems. It relies heavily on advanced technologies. However, sports technology is not only applied to meet the needs of elite athletes and professional competitors – the recreational and occasional athlete also plays a major role in the use of technology-based sporting goods. This has resulted in ongoing research into and development of new sport techniques and equipment for athletes of all kinds.

Sports technologies cover a broad range of interests from leisure sports, training and practice to serious competition. "Performance enhancement" is the key factor in competitive sports. Aesthetics, lifestyle, fun and adventure are additional considerations when playing and participating. Sports-related technologies are used in sports equipment, footwear, apparel, sports facilities and even playing surfaces, and need to be adapted for both novice and elite athletes, for youngsters, adults and – more and more frequently – the elderly, for the able-bodied and the handicapped, for male and female athletes. The transfer and integration of knowledge from a wide range of disciplines and industries has provided a catalyst for rapid technological change in the modern-day sports business.

▶ **Apparatus and equipment**. Sport as we know it today can hardly be separated from technology. New technologies that were initially developed and trialled for the aerospace, automotive and defence industries in particular have had a profound effect on sporting goods and techniques, e. g. new materials have made sporting equipment stronger, lighter and more rigid; digital video and laser technologies have paved the way for new hardware and software for measuring, monitoring and analysis in sports. These technologies have made sports faster, more powerful and enjoyable in many ways. Over the years, research has focused on understanding the consequences of increasingly complex sports technologies and on developing new technologies and techniques that can improve not only performance and enjoyment, but also the safety and overall wellbeing of athletes at all performance levels. Sports apparatus, sports equipment and sporting apparel generally focus on enhancing performance, preventing injury, and increasing both comfort and enjoyment.

Racing bike: High-performance bicycles are designed to minimise the energy loss due to air drag or turbulence. They allow the rider a body position encountering a minimum of air resistance. Lightweight technology generally minimises the athlete's energy expenditure by using lightweight materials in a design that enhances stiffness. Heavyweight metals are replaced by plastics combined with glass, carbon fibres, nylon, Teflon or new materials like PU (polyurethane) or EVA (ethylene vinyl acetate).

▶ **Lightweight technology**. Sports equipment in elite sport is optimised to provide better performance by minimising the energy loss due to the apparatus or the environment (e. g. air drag or turbulence) and to allow the storage of elastic energy and the reutilisation of the energy stored by the athlete's body. High-performance bicycles designed so that the user rides in a position encountering a minimum of air resistance are an excellent illustration of how the energy loss due to air flow or air resistance can be minimised. The weight of the equipment is minimised at the same time by using lightweight materials in a design that enhances rigidity. Nowadays, these principles are increasingly being applied to equipment for the recreational user. Lightweight technology is not only required in motor sports (e. g. Formula One racing cars) but also in many other applications such as bicycles, ice-hockey sticks, alpine or cross-country skiing sticks. Consequently, heavyweight metals are being replaced by plastics combined with glass, carbon fibres, nylon or Teflon fibres in a wide range of sporting equipment. New plastic materials such as lightweight PU (polyurethane), which has about the same density as EVA (ethylene vinyl acetate), better damping properties and higher durability, are frequently used in sports equipment and footwear.

▶ **Nanotubes**. The requirement of low weight combined with high rigidity has opened the way for a variety of nanotechnology applications in sporting goods. Nanotubes with 16% of the weight of steel and a strength 50 to 100 times higher than that of steel are the weight-bearing components used not only in racing cars, but also in the frames of tennis rackets. Carbon nanotubes are large molecules of pure carbon that are long, thin and shaped like tubes, about 1-3 nm in diameter, and hundreds to thousands of nanometres long. A 13-m fishing pole made with carbon nanotubes weighs only about 840 g. High-performance ice hockey sticks, golf clubs and even baseball bats contain nanotubes to decrease weight and increase rigidity. Nanotube-based components are becoming more and more prevalent because they are almost fully resistant to shock and impact and extremely lightweight, which makes them the ideal material for sports equipment manufacturers.

▶ **Piezoelectric materials**. Smart or "*intelligent*" materials are used more and more frequently in sporting equipment. Smart materials are materials that have one or more properties that can be significantly changed in a controlled fashion by external stimuli such as stress, temperature, moisture, pH, electric or magnetic fields. Piezoelectric materials, for example, produce a voltage when stress is applied. Since this effect also applies in reverse, a voltage across the sample will produce stress within the sample. Suitably designed structures that bend, expand or contract when a voltage is applied can therefore be made from these materials. Using intelligent materials in a ski or tennis racket, for instance, can enable the torsional stiffness or the damping to be increased or decreased in relation to the mechanical stress applied. Mechanical energy from the impact of the tennis ball is converted into electrical energy in less than 1 ms. This leads to a 20% reduction in vibration due to the fact that, unlike mechanical energy, electrical energy does not produce vibration. The result is a more comfortable racket. The electrical response to the piezoelectric fibre creates changes in the fibre shape and increases torsional stability by up to 42%. This results in increased racket stiffness for ultimate power. A similar application is the electronic ski management system. The piezoelectric fibre is integrated in the body of the ski in front of the binding at an angle of 45°. The mechanical stress to the fibre when cruising on snow is converted into electrical energy. The electrical response to the piezoelectric fibre changes the shape of the fibres and increases torsional stability. The increased torsional stiffness gives the ski a better grip and improves the skier's performance.

Applications

▶ **Intelligent footwear.** Footwear is used in the majority of sports. The greatest amount of public interest is undoubtedly devoted to the technical developments of the running shoe. New concepts to cushion the impact force and control the motion of the rear foot have been developed in response to the hypothesis that running-related injuries are caused by excessive foot pronation and the impact forces incurred when the foot strikes the ground during running. Shoe designs and materials have been chosen to mitigate the impact forces when the foot hits the ground and to control foot pronation during the early and mid-ground contact phase. Technologies and materials such as air pads, capsule gels, EVA (ethylene vinyl acetate) or PU (polyurethane) in different densities are used. An "intelligent" running shoe has been developed to enable the stiffness of the shoe's mid-sole to adapt to the surface the subject is running on. The shoe provides "intelligent cushioning" by automatically and continuously adjusting itself. A deformation-sensitive sensor embedded in the mid-sole registers the peak mid-sole deformation when the foot hits the ground. A microprocessor then calculates whether the cushioning is too soft or too firm, and adapts it with a motor-driven cable system to provide the correct cushioning for the specific purpose at that time. The sensor works by measuring the distance from a small magnet at the bottom of the shoe, takes 1,000 readings a second, and is accurate to within a tenth of a millimetre. Its "brain" is a small microprocessor capable of making five million calculations per second, and it adapts the cushioning with a motor-driven cable system that spins at 6,000 rpm.

▶ **Swift-spin body suit.** Two types of air drag determine the overall aero performance of a cycling suit. Pressure drag is overwhelmingly dominant on regions of the body that approximate cylinders in cross-flow, such as the upper arms and the thighs. To optimise the suit design, fabric textures were chosen that minimise

Vibration damping in tennis rackets: Piezoelectric fibres are embedded in the carbon fibre reinforced frame of the racket while the electronics is hidden in the handle. Mechanical energy from the tennis ball impact is converted into electrical energy, resulting in reduced vibration. The electrical response to the piezoelectric fibre changes the shape of the fibres and increases torsional stability. Left: The electronics used for converting the energy from the piezo fibres so as to damp vibrations is fully integrated in the racket handle and thus has no impact on the racket's behaviour during play. Right: Embedded in the structure of the racket, the piezo fibres (light brown) rest between several layers of carbon fiber composites (CFC). When the ball hits the racket, the fibre modules generate electrical energy by the deformation of the fibres (longitudinal elongation). This energy is processed via the built-in electronics and passed back to the fibres out of phase, thus damping the vibration of the racket head. Source: HEAD Germany GmbH

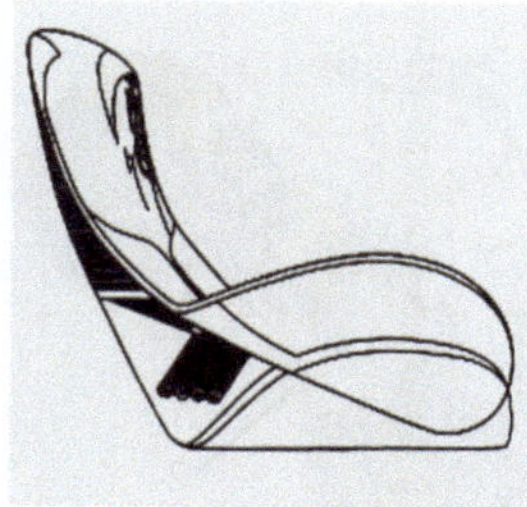

Managing the torsional stiffness of an alpine ski: The mechanical stress to the piezoelectric fibre in the body of the ski in front of the binding is converted into electrical energy when cruising on snow. The electrical response changes the shape of the fibres and increases the torsional stiffness of the ski. Source: HEAD Germany GmbH

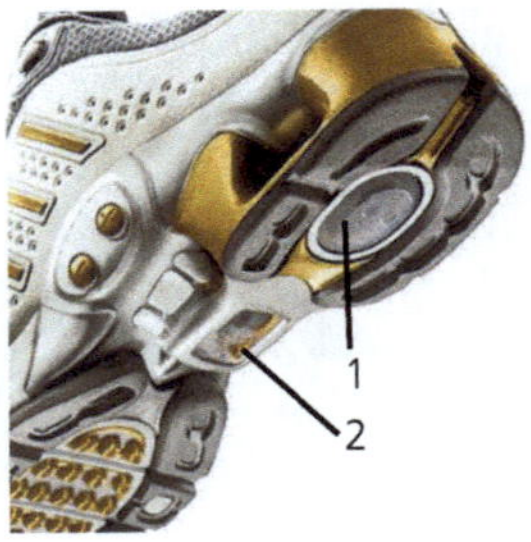

Intelligent running shoe: The running shoe provides "intelligent cushioning" by automatically and continuously adjusting itself. A deformation-sensitive sensor (1) embedded in the mid-sole registers the peak mid-sole deformation when the foot hits the ground, then a microprocessor (2) checks the cushioning level and adapts the mid-sole stiffness with a motor-driven cable system. Source: adidas Germany

The swift-spin body suit: The racing suit provides (1) "zoned aerodynamics" treating the surfaces of certain body parts with proprietary textiles to reduce turbulence, (2) a strategic seam placement minimising drag, (3) an articulated fit to create a suit that is cut specially for the riding position, and (4) strategic ventilation balancing the results from thermal and aerodynamic studies.

either pressure drag or friction drag, depending on the flow regime that exists on the particular body segment. It was shown in wind tunnel tests that a rougher woven fabric was superior to smooth coated fabrics and stretch knit fabrics. The swift-spin cycling suit incorporates design features which control the boundary layer through the careful placement of appropriately textured fabrics on zones of the shoulders, arms, torso and thighs. The combination of rough and smooth textures permits boundary layer control while minimising both pressure and frictional drag. On some portions of the body, such as the hands and feet, where limb geometry precludes boundary layer control, a very smooth, coated fabric is utilised to reduce frictional drag. To provide ventilation and cooling, panels of fine, breathable mesh have been placed in areas of separated flow, such as behind the upper arms and at the back of the neck. Also, the number of seams crossing the airflow has been minimised, with the majority of seams now oriented parallel to the airflow. More than 50 textiles were investigated and tested for several key qualities, including wind resistance, elasticity and breathability. Six selected fabrics were placed on certain body locations to work strategically and harmoniously with the athlete's unique motion in relation to the airflow. Through the combination of these design features, the swift-spin suit (first used for the Tour de France in 2002) provides over 2 N less drag at 50 km/h in a time trial than the second-to-last-generation suit. Mathematical modelling suggests that the swift-spin suit would yield a time saving of 79 s in a 55 km time trial.

▶ **Artificial turf**. Artificial turf surfaces manage various performance characteristics – ball roll behaviour, ball rebound, resilience, sliding behaviour, traction/foot safety, and abrasiveness – so that they correspond as far as possible to the structural model of natural turf surfaces. The latest (third) generation of artificial turf systems generally consists of a base layer and the synthetic turf itself, which is filled with silica sand and synthetic granules (turf infill). The turf infill improves the technical properties of the turf in terms of playing performance, e.g. ball roll behaviour or reduced sliding friction. The third generation of artificial turf is more readily accepted by soccer players because it no longer involves a risk of carpet burns to bare skin and because these surfaces have a noticeable resilience. The more comfortable sliding behaviour is caused by a reduction of the pile fibre density compared with the former "hockey" type surfaces. Conversely, the characteristics of the pile fibres provide less sliding resistance and more softness due to their greater length (up to 70 mm). Another way of reducing sliding resistance is

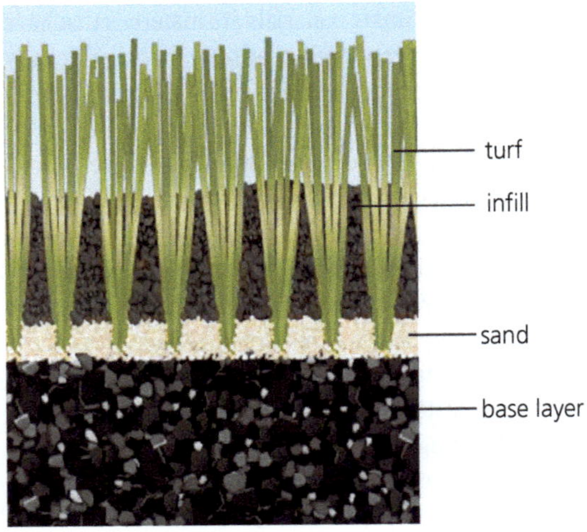

Artificial turf: The third generation of artificial turf is now widely accepted as a playing surface for sports such as football or hockey. It has significantly improved the technical properties of the turf in terms of playing performance, e.g. ball roll behaviour and reduced sliding friction. State-of-the-art artificial turf uses modelled and environmentally safe infill material and reduces the build-up of heat through its light colouring and through the addition of special mineral fillers for evapotranspiration and special pigments to increase solar reflection. Third-generation turf has longer piles, softer fibres (typically polyethylene), a sand-plus-rubber blend or rubber only as its infill materials, and an elastic base layer.

by covering the pile fibres with a wax-like coating. Two types of pile fibres are combined: textured and long, mostly straight fibers. The textured fibres are located between the long ones; they are not visible on the surface and act as "support" fibres, whereas the long fibres are the "play" fibres. The purpose of the sand infill is to weigh down the turf carpet to stabilise it against horizontal stress. The technical challenge is to preserve the characteristics for as many years as possible when exposed to elements such as high stress and UV radiation. Most synthetic turf systems essentially consist of the following components:

- Base layer: The structure directly affects the quantity of infill granules. If an elastic layer is used as a base, for example, shock absorption will be largely provided by the base layer and the quantity of infill granules can be reduced.
- Synthetic turf: The pile fibres of the synthetic turf mainly consist of polyethylene, polypropylene or a blend of the two materials.
- Silica sand: Silica sand is used to weigh down and hold in place the loosely laid artificial turf.
- Infill granules: Four different infill materials have become established; granules made from recycled tyres (SBR), recycled SBR granules coated with

pigmented polyurethane, rubber granules, and thermoplastic elastomers (TPE).

Trends

► **Optical motion analysis.** Optical 3D motion analysis technology is widely applied in elite and recreational sports as well as for rehabilitation purposes. Optical systems utilise data captured from image sensors to triangulate the 3D position of a subject from the position of one or more calibrated cameras. Data are acquired by using special markers attached to an athlete or the patient. Further cameras can be added in order to track a large number of performers or to expand the capture area. Passive optical systems use markers coated with a retro-reflective material to reflect the light that is generated near the camera lens. The camera's threshold can be adjusted so only the bright reflective markers will be sampled, ignoring skin and fabric. The centroid of the marker is estimated as a position within the two-dimensional image that is captured. Typically, a system will consist of around 6–24 cameras. Extra cameras are required for full coverage around the capture subject and multiple subjects. The markers are usually attached directly to the skin, or they are glued to a performer wearing a full body suit specifically designed for motion capture. Integrated software is used for modelling body segments and associated muscles. Real-time features make it possible to monitor dynamic changes in muscle length as well as muscle moment arms, as the subject is physically active. The latest trends include reconstruction of skeletal alignment or the movement of joints.

► **Inertial measurement systems.** The use of small inertial sensors such as *gyroscopes* and *accelerometers* has become common practice in motion tracking for a variety of applications. Typically, angular orientation of a body segment is determined by integrating the output from the angular rate sensors strapped on the segment. Accelerometers measure the vector sum of sensor acceleration and gravitational acceleration. In most cases of human movement sensing, the acceleration is dominant, thus providing inclination information that can be used to correct the drifted orientation estimate from the gyroscopes. A Kalman filter operates on the principle of combining gyroscopes and accelerometers to measure the orientation of a moving body segment. The magnetometer is sensitive to the earth's magnetic field. It gives information about the heading direction in order to correct the drift of the gyroscope about the vertical axis. An innovative fusion motion capture (FMC) technique has recently been used to capture the 3D kinematics of alpine ski racing. The FMC merges an IMU (inertial measurement unit) with GPS (global positioning system) data. IMUs may contain accelerometers, gyroscopes and magnetometers to track the local orientation and acceleration of each limb segment of interest. GPS data are merged with local acceleration data to track the athlete's global trajectory.

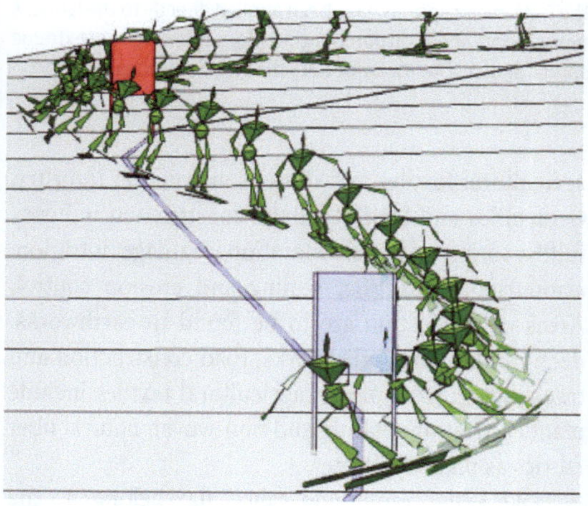

Fusion motion capture (FMC): 3D reconstruction of an alpine ski race. Fusion motion capture technology is able to capture the 3D kinematics of alpine ski racing. The crucial point of the FMC is the fusion of an IMU (inertial measurement unit) containing accelerometers, gyroscopes and magnetometers, and GPS (global positioning system) data. IMUs track the local orientation and acceleration of each limb segment of interest. GPS data are fused with local acceleration data to track the athlete's global trajectory. Fusion motion capture (FMC) technology makes it possible to survey the movements of athletes over a large object space and a long period of time. Source: Bodie, Walmsley & Page

Prospects

- Sports engineering will continue to focus on a range of sciences and technologies (e. g. smart materials or nanotechnology) in order to enhance the efficacy of sports equipment and its impact on individual performance.
- Sports technology will have an increasing impact not only on sporting performance, but also on the safety of athletes and recreational users through special sports equipment, footwear and apparel.
- Future sports products and technology will be designed in a more environmentally responsible way and focused on easy recovery and reuse of materials, recyclable materials, reduction of chemical emissions by products, and reduction of noise.

PROF. DR. GERT-PETER BRÜGGEMANN
German Sport University Cologne

Optical motion analysis: This 3D gait analysis utilises data captured from image sensors to triangulate the 3D position of a subject. Data are acquired using special markers attached to an athlete or the patient. Integrated software is used for modelling body segments and for reconstruction of skeletal alignment (shown here). Typically, a system will consist of around 6–24 cameras. The markers are usually attached directly to the skin, or they are glued to a performer wearing a full body suit specifically designed for motion capture. The integration of ground reaction force measurements permits inverse dynamic calculation of net joint kinetics, e. g. of the knee joint. Optical motion capture allows unrestricted movement without hindrance to the athlete or patient.

Internet

- www.sportandtechnology.com
- www3.interscience.wiley.com/journal/117899685/home
- www.design-technology.org/sportsshoes1.htm
- http://dvice.com/archives/2008/08/10_new_technolo.php
- www.motionanalysis.com/html/movement/sports.html
- www.vicon.com/applications/biomechanical.htm

Textiles

Related topics
- Polymers
- Renewable resources
- Nanomaterials
- Surface and coating technologies
- Intelligent materials
- Building materials
- Sport technologies
- Implants and protheses

Principles

Textiles play a role in numerous walks of life. While clothing still leads the field with a 43.5% share of worldwide textile production, this figure lay at 50% in the mid-1990s. The share of household textiles – and especially that of technical textiles – has risen simultaneously over the last few years. Household textiles include curtains, towels, carpets, table and bed linen. Technical textiles are found not only in cars, but also for example as filters or in the agricultural and medical sectors. There are a number of reasons for the production increase in technical textiles in recent years:

- Increasing use of textiles in the automotive sector
- New processing technologies, e. g. in the use of natural fibres for *composite materials*
- New areas of application for textiles, e. g. in medicine
- Food surplus in Europe, which led to an active search for alternative field crops (such as hemp, kenaf, calico)

Textiles are made of natural fibres, synthetic fibres or a mixture of the two. The proportion of synthetic fibres has been increasing constantly. Nowadays, natural fibres only account for 45% of the overall textile market. More and more production processes are being outsourced to second and third world countries. The domestic textile industry therefore has to concentrate on high-quality products.

▶ **Natural fibres.** Cotton outweighs all other natural fibres – despite the fact that cotton plants require special care. Monocultures and cultivation in regions with an unsuitable climate have made them particularly susceptible to pests. Cotton plants therefore need to be protected by wide-scale use of herbicides and insecticides. It is becoming increasingly common to grow genetically modified cotton, such as BT cotton (transgenic cotton). BT cotton contains a gene from Bacillus thuringiensis, a soil bacterium; the transgene triggers the production of toxins. These toxins kill certain butterfly species that damage the cotton. The proportion of genetically modified cotton has risen to more than 43% in recent years.

Besides cotton, other natural fibres such as flax, kenaf, ramie and calico are finding more frequent use, particularly in the area of technical textiles used in car manufacturing, for insulating materials, for geo- and agricultural textiles and in the construction industry. Geotextiles and textiles for the construction industry fulfil a variety of tasks: separation, drainage, filtration, armouring, protection, sealing and erosion control. Areas of application are to be found in earthworks, deep mining, hydraulic works, road construction and track laying. Examples of agricultural textiles include plant cultivation matting and non-woven natural fiber fabrics as plant substrates.

Like kenaf, ramie and calico, flax belongs to the group of bast or stalk fibres, i. e. the fibre material is obtained from the stalk. After harvesting the stalks are retted causing them to ferment. In the frequently used dew retting method, the stalks are spread out on the ground. Dew and soil humidity foster the decomposing effect brought about by fungi and bacteria. The pectin, also known as vegetable glue, is disintegrated. After retting, the straw can be separated from the fibres (a process called scutching). Since the fibres are present in the stalk as meshes instead of single fibres, a great deal of experience is required to obtain these nat-

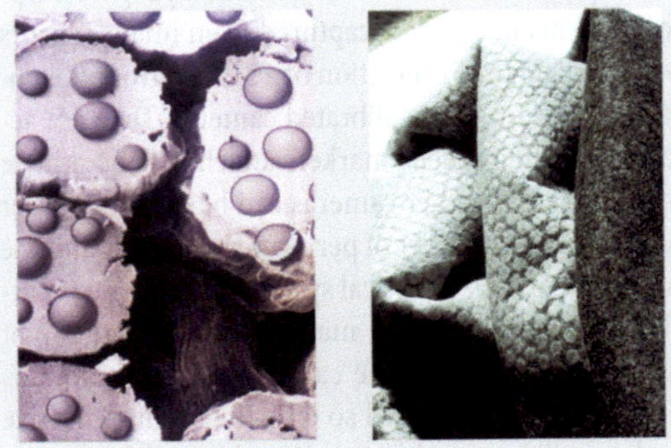

Warm clothing thanks to phase change materials: If a latent heat repository is heated to its melting point, it undergoes a phase change from solid to liquid. PCM are capable of compensating for extreme temperatures, for example in skiing outfits. The encapsulated globules incorporated in or on the textile surface become liquid or solid depending on the temperature, thus ensuring a constant body temperature. Left: Close-up of an Outlast® polyacrylic fibre. The microcapsules are incorporated in the fibers, which can be further spun to create insulating garments. Right: Woolen fabric with a new punctiform PCM coating. In this case the PCM microcapsules have been applied directly to the fabric. A typical application is in outdoor clothing, e. g. for active or extreme sports. Source: Outlast Europe & Schoeller, Switzerland

SEM images of textiles. Top: Knitted fabric with single stitches. Individual fibres, from which the thread was spun, are clearly visible. Bottom: Woven fabric with warp and weft. Source: Empa, Switzerland

ural fibres. Unlike cotton, the quality and properties of bast fibres are changed by each step of the specific production processes and conditions.

► **Synthetic fibres.** The worldwide production figures for fibres clearly demonstrate that synthetic fibres dominate our lives. Synthetic fibres are produced by dissolving or melting a starting material, pressing it through a spinning nozzle, and then hardening it. The main raw material is petroleum. Fibres that have long become established are polyester, polypropylene and polyamide.

► **Spinning.** Natural fibres and cut chemical fibers are mechanically twisted to form staple fibre yarns; they are spun. Silk is produced by silkworms, and chemical fibres are spun by chemical and technical processes. The word "spinning" thus describes different procedures. The thread (yarn) first has to be created before a woven or knitted fabric can be produced. The fabric of a shirt, for example, is created by crossing two right-angled thread systems (warp and weft). Knitted fabrics (such as ladies' hosiery) or knitted and crocheted goods are made up of stitches (loops of thread), which are in turn hooked into other stitches and thus gain their stability. These fabrics are more elastic than woven fabrics.

► **Textile finishing processes.** Textiles undergo various stages of "refinement" or treatment including pre-treatment, dyeing, printing and permanent finishing. In pre-treatment, undesired substances are removed from the fibres. The fabrics are coloured by means of dyeing and/or printing. Dyes are organic pigments that dissolve in water or other solvents. During the dyeing process, the dye moves out of the solution (dip) and into the fibres. Since not all types of fibre have the same properties, different types of dyes are used. In the case of cotton, these are often reactive and vat dyes. One very well-known cotton dye is indigo blue as used in "blue jeans". Polyester, on the other hand, is usually dyed with dispersed dyes. The point in the textile production process, at which colouration takes place, depends on the type of fibre, the article and the prescribed standard of quality. *Printing* is another way of colouring textiles. Printing produces only local colouration. The dyes or pigments are printed on the fibre in paste form. In contrast to dyes, pigments do not bond with the fibre, which means that they require a bonding agent to make them adhere to the fibres. After printing, the paste is fixed (e.g. by hot steam) and washed. In permanent finishing, the textile is given properties that it does not normally possess: examples include water-repellent or fire-resistant cotton articles, crease-resistant treatment, or improved comfort for the wearer by reducing odour from perspiration.

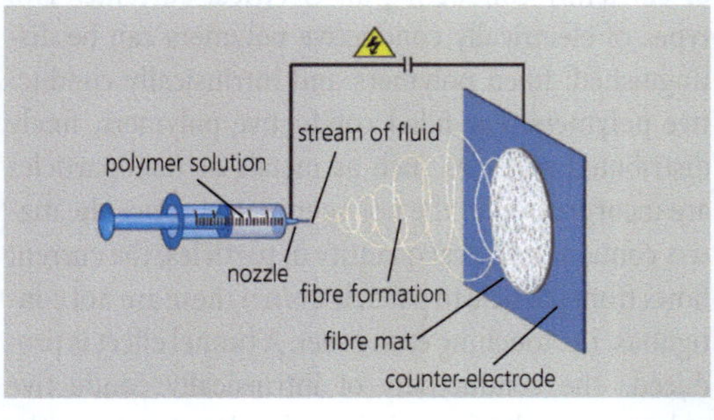

Spinning of artificial threads: Diagram of the electric spinning device. An electric field is applied to the polymer solution in the syringe. The resulting surface charge on the polymer drop creates a stream of air to the counter-electrode. The solvent in the polymer solution evaporates, creating a thread that is deposited on the counter-electrode, for instance in the randomly oriented state depicted. Source: Empa

Applications

► **Odour protection.** Today, the release of odour from perspiration can be reduced by special textile finishing. For instance, cyclodextrin molecules can be fixed permanently on the textile surface.Cyclodextrins are macrocyclic molecules containing a hydrophobic (water-repellent) cavity. Molecules that are also hydrophobic or at least possess a hydrophobic portion can be stored in this cavity. Cyclodextrin molecules fixed on a textile surface can thus enclose organic components of perspiration upon skin contact. The odorous molecules are removed from the cyclodextrin "cages" every time the garment is washed. This greatly reduces the formation of the typical sweat odour. In recent years, textiles such as men's suits are increasingly being treated with cyclodextrins. Another possibility is to use photocatalytically active polyester fibres. This type of fibre destroys adsorbed odour molecules, thereby reducing unpleasant odour or even preventing it altogether.

► **Humidity and heat management.** Specially constructed fibres with moisture transporting profiles, hollow fibres or *phase change materials* (PCM), and of course hydrophilic/hydrophobic treatment all find applications in the production of garments that offer their wearers a well regulated climate. PCM are latent heat repositories that undergo a phase change from solid to liquid as soon as they reach their melting point. The absorbed heat is "trapped", while the material's temperature itself remains constant. Because of the changes in their aggregate state, PCM occur in both solid and liquid form and therefore have to be encapsulated for use. Ceramic globules or paraffins encased in tiny plastic globules are used for applications in and on textiles. Their heat storage ability ensures a consistently pleasant body climate despite extreme tempera-

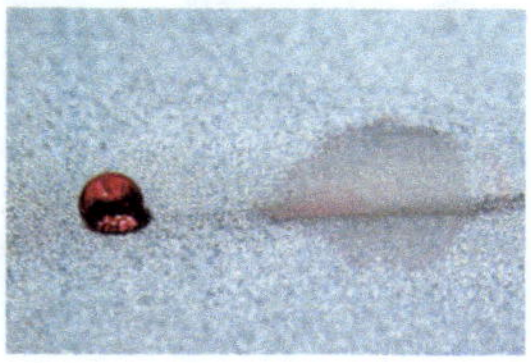

Silk with a silicon coating, partially treated by plasma technology: An extremely thin layer, less than 100 nm in thickness, is applied to the material. Wetting qualities can then be selectively adjusted. Furthermore, functional groups can be added that allow chemical bonding with other molecules possessing special functions, e.g. colour.

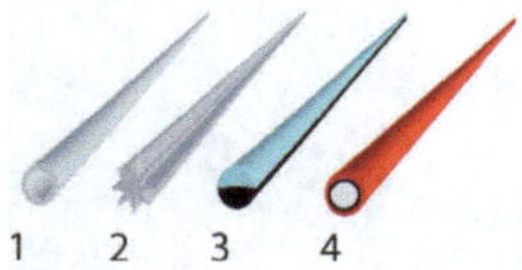

Cross-sections and profiles of spinnable fibres. From left: (1) Hollow fibre. These fibres can provide better insulation than a pure fibre due to their lower weight and greater content of entrapped air. (2) Star-shaped fibres carry the moisture away from the body to the surface of the fabric where it can quickly evaporate. (3) Side-by-side fibres, e.g. adhesive coating fibres, and (4) cladded core fibres, e.g. for crimped fibres. Source: Empa, Switzerland

Two-component adhesive yarn: Stabilised edge of a piece of lace. After heat treatment, the coating of the two-component fibre (red) is melted and seals the edge of the textile with adhesive. Source: EMS-GRILTECH

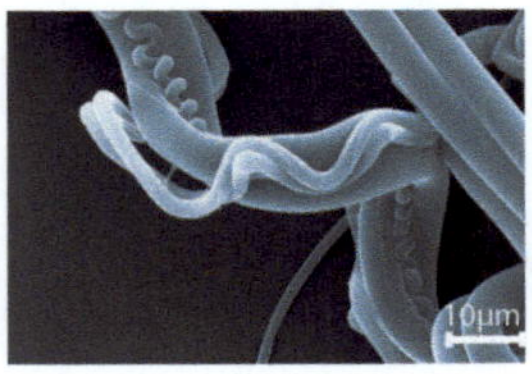

Spider silk: Electron microscope image of natural spider silk from the web of a garden spider (Araneus diadematus): It shows silk threads spun by different glands (major and minor ampullate glands, flagelliform gland). The stable lifeline is at the top right of the image, and a flexible capture thread can be seen in the centre. Source: University of Bayreuth

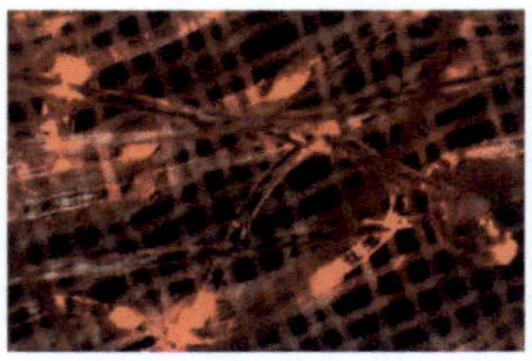

Luminous fabrics: Textiles with integrated optical fibres. These are embroidered onto a carrier fabric. The light is emitted through curvature of the optically conductive fibres, at certain points in the area of the curve. This enables a predefined surface to be irradiated with a strength of several mW/cm². The fabric is used for decorative purposes or in applications such as medicine to treat melanoma of the skin by photodynamic therapy with red light. Here: Carrier fabric embroidered with optical filaments. Selected emission of light within the curvature of the fibre. Source: Empa

ture fluctuations, e.g. in spacesuits, ski suits or firefighting equipment. The textile is capable of absorbing excess body heat when too much is created, and releasing it at a later stage. PCM are also used in other areas: in building technology, for instance, construction materials with integrated micro-encapsulated paraffins help to increase the thermal inertia of a building and thus save energy.

▶ **Electro-spinning**. *Nano-scale* structures in the form of fibres or tubes with diameters of only a few tens to a few hundred of nanometres cannot be produced using conventional processes – quantum physics effects have to be taken into account. Electro-spinning works by applying a strong electric field in the kilovolt range to a drop of a polymer solution or melt. The surface charge on the drop causes gravitation towards the counter-electrode as soon as the electric field has overcome the surface tension of the solution or melt. The solvent evaporates, the fibre is elongated and can be deposited on the counter-electrode with a suitable geometry. Due to their special properties such as a large specific surface and a selectively adjustable diameter, these fibres are used in applications such as filter systems and for tissue engineering in medicine. The most advanced method is capable of spinning two-component fibres. This method combines two different polymers, enabling the positive properties of the two materials to complement each other. The use of differing cross-sections and profiles for the two-component fibres can produce unusual fibre properties. The most common types are side-by-side fibres and cladded core fibres. It is possible to produce adhesive yarns, yarns with different shrinking properties and conductive fibres. Each single filament of an adhesive yarn consists of a standard fibre raw material (core) with a high melting point and a copolymer (cladding) with a low melting point. It is, therefore, a cladded core fibre. The targeted application of heat causes the cladding to melt, sticking the surrounding textile fibres together. A two-component adhesive yarn of this kind can be used to stabilise fabrics such as the edges of lace, e.g. for ladies' underwear. If two polymers with different shrinking properties are used as side-by-side fibres, a durable crimping effect is produced.

Trends

▶ **Spider silk**. Spider silk has a tensile strength about 10 times higher than that of aramid fibres. It is finer than a human hair, visco-elastic, biologically degradable and suitable for recycling. Since spiders cannot be bred like silkworms – they devour each other – scientists are trying to produce the fibre using genetic engineering. To do this, the spider silk genes are isolated and transferred to other genomes. Bacterial cells, tobacco and potato plants, and even cows, goats and silkworms have been used more or less successfully as a host for spider silk genes. In goats and cows, the silk proteins are produced with the milk. One litre of goat milk contains approximately five grams of silk proteins. These are precipitated and cleansed. However, fibres obtained in this way only possess about 30% of the strength of natural spider silk. In tobacco or potato plants, the spider silk protein is extracted from the leaves.

▶ **Luminous embroidery**. Textiles with integrated *optical fibres* are capable of glowing when connected to a light source. These are best known as woven products for fashion wear and as knitted luminous nets for decorative purposes. Optical fibres can also be embroidered onto a carrier fabric for medical applications. The light is then emitted from the steep curvature of the optically conductive fibres in the area of the curve. This makes it possible to irradiate a defined surface area with a strength of several mW/cm². Melanoma and other types of superficial cancer can thus be selectively treated with red light in what is known as photodynamic therapy. The two-dimensional distribution of light makes new methods of treatment possible. In contrast to conventional treatment methods, the textile is in direct contact with the body surface to be treated and can easily be formed and draped. This new way of generating a certain dose of light is more flexible than conventional systems such as objective lenses (a kind of tiny torch that casts a lustre cone on a surface to be irradiated) or rigid films.

▶ **Conductive fibres**. Nearly all pure polymers have a low electrical conductivity at low to moderate temperatures. Either mobile ions or electrons can be used as charge carriers to develop polymers that are capable of efficiently conducting an electrical currente. Two types of electrically conductive polymers can be distinguished: filled polymers and intrinsically conductive polymers. For filled conductive polymers, finely distributed particles such as metals or nanoparticles are incorporated in the polymer matrix. Once the matrix contains a critical quantity of particles, the current flows from particle to particle even if these are not contiguous, i.e. touching each other. A tunnel effect is produced. The conductivity of intrinsically conductive polymers results from their molecular structure. The conductivity is caused by distinctive conjugated double bonds. Conductive polymers are insoluble and cannot be melted, but they can be incorporated in low-melting polymers such as polypropylene, providing conductivity in these materials.

► **Dyeing**. Large amounts of water and energy are consumed by the dyeing process, so that resource efficient approaches are currently being investigated. Colouration by means of supercritical carbon dioxide is a novel approach. In this method, scientists make use of the specific property of certain substances, in which it is no longer possible to distinguish between their gas and liquid phases after exceeding certain temperatures and pressure values (critical point). In this state, known as "supercritical fluid", carbon dioxide exhibits the properties of both a gas and a fluid as well as possessing a high capacity for dissolving water-repellent substances, including dispersed dyes that are used in the colouration of polyester. As a supercritical fluid, CO_2 absorbs and dissolves the dispersed dye. The colouration process can now take place. After colouration, CO_2 returns to its gaseous state, while the remaining unused dispersed dye is left in powder form. This is therefore a water-free colouration process that requires no subsequent washing.

In the field of nanotechnology, attempts are being made to imitate the wing colouration of the butterfly species Morpho rhetenor. Its wings are covered with countless transparent scales. The nanoscopic surface structures of these scales create colour effects by reflecting, refracting and scattering ambient light. The luminous blue colour is thus produced by the physical effect of refracting the light. This colour effect has already been successfully employed in the Australian ten-dollar note, making the note forgery-proof at present.

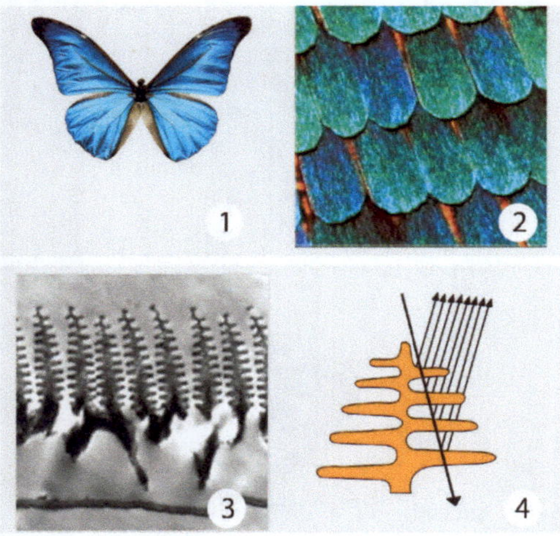

South American butterfly Morpho rhetenor (1). Each of the roughly 0.1 mm wide transparent wing scales conveys a luminous blue colour (2). The cross-section of a scale as seen under the electron microscope (3) reveals the surface structure, which causes about 75% of the blue light but only 3% of the red light to be reflected. Each "tree" (4) is roughly 1 µm high and the "branches" are 80–90 nm thick. A biomimetic conception of this microscopic analysis has led to completely new colouring techniques for textiles by modifying the surface structure of fibres. By precisely controlling the thickness of extremely thin polyester or nylon films (70 nm thickness, laminated in up to 61 layers), exciting structural colours can be created. Source: P. Vukusic, University of Exeter

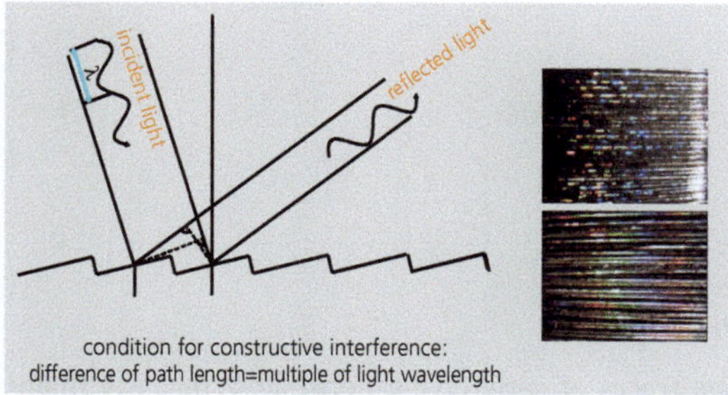

Dyeing of textiles by embossing the fibre surface using the physical principle of constructive interference: Differences in path length of the reflected light cause certain constructive interferences, i.e. cancellation effects, on a periodic structure. Physical calculation can produce completely new colour effects depending on the fibre geometry, the structural properties and the surface properties. In the examples illustrated, the lithographic structures were produced by embossing stamps which exist in the form of a brass plate. These structures are embossed on the fibres by a continuous process. Right: Visible colour imprint of a monostructural bundle of fibres (top, individual glittering spots of colour) as compared with a polystructural bundle of fibres (bottom, rainbow-like progression of colours). Source: P. Zolliker, Empa

Prospects

- The technical textile sector has experienced the strongest growth in recent years. Textiles with tailored properties are increasingly being applied in sectors such as medicine and automobile manufacturing.
- New spinning processes (two-component and electro-spinning) have made it possible to create completely new types of fibre with novel properties.
- The use of nanotechnology in textiles is undergoing intensive research. There is hardly any part of the textile production chain in which various forms of nanotechnology are not already applied. The integration of electrical and optical fibres for "smart textiles" and "wearable electronics" is the subject of intensive research.

PETRA KRALICEK
Empa – Swiss Federal Laboratories for Materials Testing and Research, St. Gallen

Internet

- www.emergingtextiles.com
- www.ntcresearch.org
- http://textile-platform.eu/textile-platform
- http://textileinnovation.com
- www.textileweb.com
- www.swisstextiles.ch

Cosmetics

Related topics

- Pharmaceutical research
- Intelligent materials
- Systems biology
- Polymers
- Industrial biotechnology
- Laser

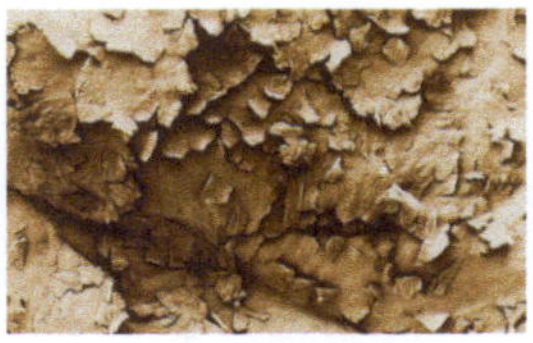

The surface of dry skin viewed under a scanning electron microscope: Loose skin flakes are clearly visible, explaining the uncomfortable feeling of dry skin and also the likelihood of the skin cracking and being invaded by external contaminants. Source: Beiersdorf

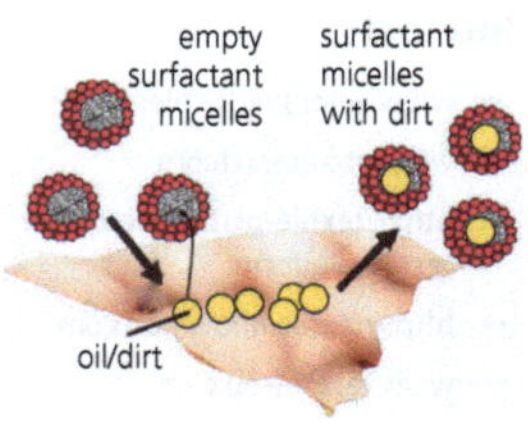

Skin cleansing: Active cleansing products remove lipophilic dirt from the skin. A single tensid monomer consists of a water-soluble head (red) and a lipophilic tail (grey). Together, they build micelles, small globes with a lipophilic nucleus, which enclose the dirt and enable rinsing and cleansing of greasy skin with water.

Principles

The traces of early civilisations show that externally applied cosmetics were an integral part of daily life even then. While this topical mode of application forms the basis of some international regulations, the European Cosmetics Directive defines cosmetics by their intended use, not by their mode of action. Altogether, this legal framework permits a wide spectrum of technologies. Cosmetics can be divided into two broad groups: personal hygiene articles, and products to care for and improve the appearance. Although in physicochemical terms the majority of cosmetics are dispersions, there are also many solutions, some of them colloidal, and mixtures of solids. The cosmetic chemist always chooses the physicochemical state that can best achieve the intended effect.

► **Skin**. The skin is a complex organ – each square centimetre of skin is on average 3 mm thick and contains 10 hair follicles, 100 sweat glands, and up to 2,500 sensory cells, not to mention 3 m of lymphatic and blood capillaries, 12 m of nerve fibres, and much more. With an area of approximately $2m^2$, the skin is the largest organ of the human body. The outermost layer is the stratum corneum, which regulates transepidermal water loss and prevents the intrusion of many harmful substances and microorganisms. The epidermis renews itself approximately every four weeks in a process called desquamation, in which surface cells are sloughed off and continually replaced by new cells. Cosmetics play an important role in maintaining the skin's function:

- Cleansing products remove not only dirt when rinsed off, but also excess sebum, microorganisms and loose cell debris.
- Skincare products moisturise the skin and strengthen the skin barrier.
- Deodorants and antiperspirants combat odour-causing bacteria and reduce perspiration.
- Cosmetics containing active ingredients stimulate natural skin regeneration and cell metabolism.

Applications

► **Cleansing products**. Pure water is often not enough to clean our skin – fat-soluble (lipophilic) dirt cannot be removed by water alone, but needs the help of a surfactant to detach it and wash it off. Amphiphilic surface-active molecules entrap the dirt, make it hy-

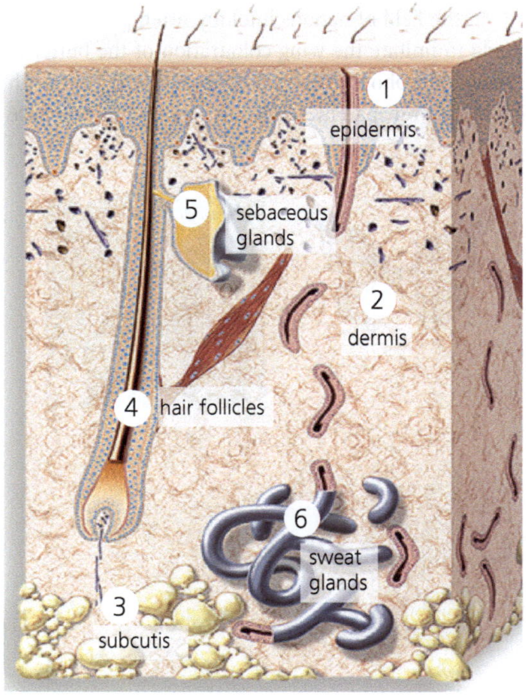

Diagram of skin structure consisting of the subcutis (3), the dermis (2) and the epidermis (1). The epidermis serves as the major skin barrier and is translucent (it allows light to pass partially through it). The second layer, the dermis, accounts for skin firmness with collagen and elastin. It contains blood capillaries, hair follicles (4), sweat glands (6) and sebaceous glands (5). Below the epidermis and the dermis lies a layer of subcutaneous fat. This is connected to muscles and bones, to which the whole skin structure is attached by connective tissues. The attachment is quite loose, so the skin can move fairly freely. Source: Eucerin/Beiersdorf drophilic and thus enable it to be rinsed off with water.

▶ **Skincare products.** By supporting the barrier function and smoothing the rough surface of the skin, skincare products improve the symptoms of dry skin by increasing its moisture content. Emulsions of lipids, water and additives are normally used as skincare products because only emulsions allow a skin-friendly combination of the required hydrophilic and lipophilic components. The hydrophilicity of the chosen emulsifier system determines whether water-in-oil (W/O) or the much more common oil-in-water (O/W) emulsions are obtained. The water in the emulsion evaporates within a few minutes after product application. The lipid phase then forms a film that reduces water loss from the skin by strengthening the skin barrier. The skincare effect comes not only from this moisture-increasing barrier effect but also from moisture-retaining substances (humectants such as urea or polyols such as glycerin) that penetrate the skin. Among the many different emulsifiers available, the derivatives of fatty acids and fatty alcohols with various hydrophilic head groups yield particularly good results. Skincare products usually also contain preservatives as a protection against microbial contamination. These require special approval in the EU and are recorded in a "positive list".

▶ **Skin ageing.** Beyond these basic skin caring ingredients, particularly face care products increasingly tend to contain bioactive substances, primarily to ameliorate the signs of skin ageing, especially skin wrinkling. To protect against premature skin ageing, face care products are formulated with UV filters and special antioxidants that counteract the degenerative oxidative processes induced primarily by ultraviolet solar radiation. Flavonoids and vitamin C are proven to be especially potent antioxidants. Some of the damage incurred by aged skin can even be repaired by increasing skin regeneration and by stimulating the skin metabolism. Vitamin A and its derivatives, fruit acids, energy metabolites such as ubiquinone Q10, creatine, and various botanicals are the most prominent representatives of the multitude of active ingredients in anti-ageing cosmetics. They specifically address the key deficits of ageing skin. For instance, ubiquinone Q10 and creatine supplement the declining energy metabolism in ageing skin cells by enhancing their ATP production and storage capacity.

▶ **Sun protection products.** Generally recognised as a key measure to maintain health, sun protection products strongly reduce the impact of dangerous ultraviolet (UV) sunlight. UV rays penetrate deeply into the skin, causing not only the acute inflammation known as sunburn but also long-term damage manifested as sun-induced skin ageing and skin cancer.

The sun protection factor (SPF) has for decades been used to indicate the protective efficacy of products against UVB radiation, the highest energy part of the solar spectrum that reaches the surface of the earth. Since the less energy-rich UVA radiation, with its ability to penetrate more deeply into the skin, has also been identified as a relevant pathological factor of sun-induced skin damage, more recent regulations have also addressed the required level of UVA protection. In terms of their formulation, sunscreen products are skincare emulsions with a high UV filter content. The basic formulation is adjusted for easy application, both to achieve maximum compliance and because only a sufficiently uniform film ensures adequate protection. Sunscreen sprays, which are low-viscosity

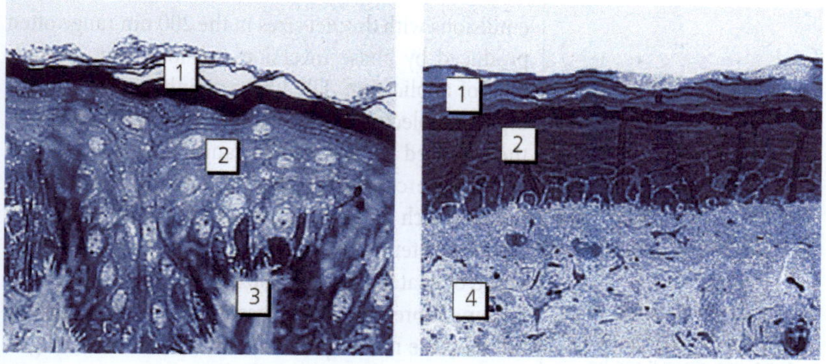

▲ Histological appearance of skin ageing. Left: Young skin shows regular interdigitation of the epidermis (2) and the dermis (3) with papillae. Capillaries ranging into the papillae efficiently supply the epidermis with nutrients. Right: Old skin shows a flattened junction between the dark-stained epidermis (2) and the dermis (4), and a thickened stratum corneum (1). Source: Eucerin/Beiersdorf

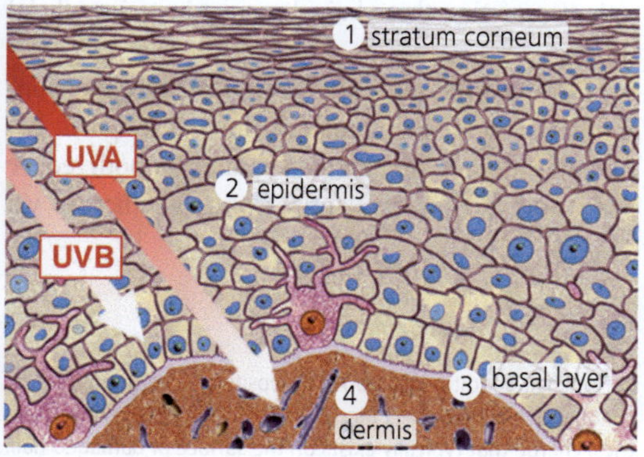

◀ Penetration depth of ultraviolet radiation: The energy-rich UVB rays (B: burn) interact mostly with the epidermis, damage especially the DNA of the keratinocytes, and cause sunburn. The less active UVA rays (A: ageing) partially pass the epidermis and account for the damage to the connective tissue in the dermis. Source: Eucerin/Beiersdorf

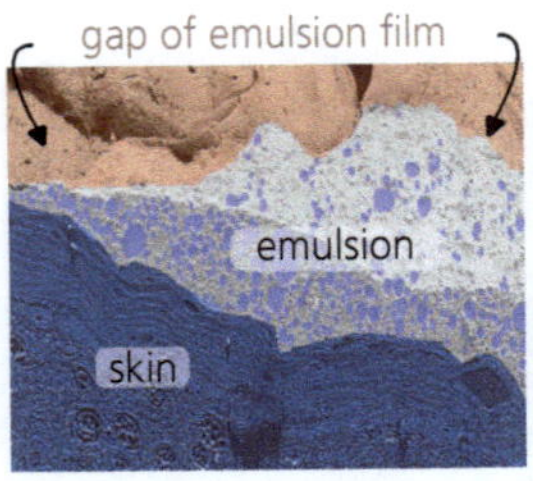

Cyro scanning electron microscopy of an emulsion film on the skin surface: It can be seen that the film clearly only partially covers the surface. This explains why sunscreen emulsions cannot completely shield the skin from damaging irradiation.

emulsions with droplet sizes in the 200 nm range often produced by phase inversion methods, are one new type of application. UV filters mainly comprise large organic molecules with chromophores tailor-made for the intended UV absorption range. Common UV absorbers are conjugated, carbonylated aromatic compounds such as cinnamic acid esters and, to an increasing extent, triazines. In addition, nano-pigments made of coated titanium dioxide and zinc oxide have become more popular in recent years. Unlike organic filters, these not only absorb but also reflect the light. The coatings often consist of siloxanes to allow a uniform distribution of pigments and to prevent the photocatalytic reactions known from uncoated metal oxides.

▶ **Hair products.** In addition to regular washing agents, hair cleansing products usually contain conditioning substances such as protein derivatives or cationic amphiphiles that neutralise negatively charged sites on damaged hair surfaces. Hair styling products are based on amphiphilic polymers, often copolymers of vinyl pyrrolidone, vinyl acetate and acrylates. These products are applied as ethanolic solutions in spray, aerosol or liquid form. Upon application, the solvent rapidly evaporates, leaving the dried polymers as minute droplets that fix hair fibres in the form of a three-dimensional network.

Hair colour products can be classified into three categories:

- Bleaches destroy the melanin pigment of the hair, usually by oxidation with hydrogen peroxide.
- Direct dyes act via electrostatic bonding of the dye to the hair.
- Permanent hair colours are often oxidation dyes that can penetrate deeply into the hair, where they are subsequently oxidised to produce the desired hair colour.

Solutions and emulsions are normally used as product forms for hair colourants. As is the case with all decorative cosmetics, these hair dyes require prior official listing.

▶ **Deodorants and antiperspirants.** Some consumers in moderate climates cannot distinguish between deodorants and antiperspirants, for both reduce body odour. However, antiperspirants function via partial, reversible obstruction of the sweat gland duct. They rely on partially hydrolysed aluminium chloride as the active principle, which precipitates as an insoluble aluminium hydroxide complex during its diffusion into the sweat gland duct. By "blocking" the sweat gland duct, it reduces perspiration by as much as 20–60% and deprives the odour-forming bacteria of the skin flora of essential nutrients. In contrast, deodorants directly control the microbial count on the axilla skin with their antimicrobial active ingredients.

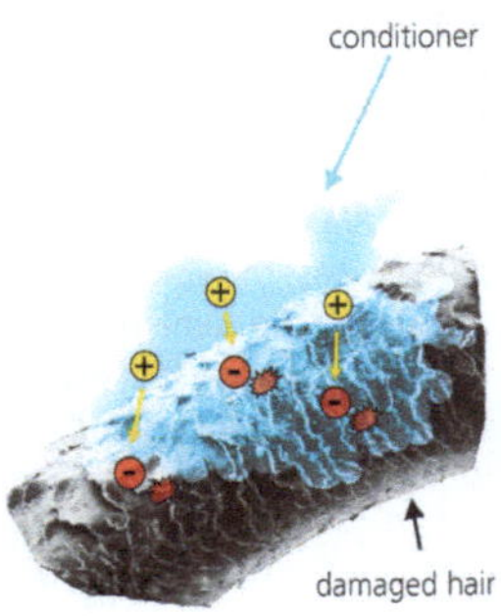

How hair conditioners work: The positively charged polymers contained in conditioners bind and neutralise the negative electrical charge that typically exists on the surface of damaged hair. Source: Beiersdorf

Trends

▶ **Improved application properties.** There is a clear trend towards enhancing the effectiveness of cosmetic products while at the same time maintaining high skin compatibility. We can expect to see new products with improved application properties, such as light sunscreens that provide high levels of protection yet exhibit little of the undesired greasiness. There will also be more products affording multiple effects, such as deodorants with whitening effects in Asia, and hair softening moisturisers that facilitate or reduce the need for shaving. Another recent example of multiple effects are self-tanning products for daily use. These are popular due to their ability to moisturise and deliver a light tan at the same time. Self-tanning products are mostly based on dihydroxy acetone (DHA). DHA penetrates the outmost layers of the stratum corneum, where it rapidly undergoes a chemical reaction with skin proteins. Known as the Maillard reaction, this process is well known from the browning of sugar in caramel or in fruit.

▶ **Skin-identical metabolites.** Scientists have gained a significantly better understanding of the metabolic processes in the skin in recent years, and this has been a major stimulus to enhancing product efficacy. The cellular metabolism can now be selectively influenced. Cosmetic chemists are seeking novel active substances from natural sources such as plants, algae and bacteria, and also to an increasing extent from biotechnological sources. Biotechnology is expected to provide access to skin-identical metabolites synthesized by fermentation. The focus will remain on preventing and treating the symptoms of skin ageing. The search for active substances will employ many of the proven methods of the pharmaceutical industry. This may include the adoption of in silico screening for potential actives, receptor binding studies, and bioavailability screening. These computer-based tools help predict specific properties of potential ingredients based on underlying specific knowledge databases and algorithms, and are increasingly useful once the biological targets have been elucidated. Knowledge derived from pharmaceutical formulation and encapsulation technology will permit the targeted penetration of an active substance into the respective site of action in the skin or into the follicle or epidermis, or, alternatively, the controlled release of substances. Encapsulation or entrapment of ingredients is facilitated by cyclodextrins, polymers or amphiphiles, to mention a few. This method is currently being investigated in relation to fragrances with prolonged efficacy, which would be particularly useful for deodorants. Fragrance research has also bridged into neuroscience. There is clear evidence that certain perfume components positively influence the wearer's mood, an effect that can not only be detected subjectively but can also be demonstrated in lower levels of the endogenous stress hormone cortisol.

▶ **Skin surface measurement.** Product technology will also advance due to new biophysical skin measuring techniques that can detect, with increasing precision, even the small improvements common in iterative product development. These techniques include skin surface measurements (skin topography) by fringe projection and computer-aided area matching, and confocal laser microscopy techniques. Using special laser microscopy in a non-invasive study with volunteers, it has been shown that deep skin layers can be accessed and visualised at high resolution and can thus partially replace the traditional skin morphology analysis methods of biopsy and histology.

▶ **Alternatives to animal testing.** In the area of skin compatibility and product safety, in vitro alternatives to animal testing will become increasingly important because, since 2004, cosmetics may no longer be tested in animal experiments in the EU. The 7th Amendment to the EU Cosmetic Directive prohibits all animal testing of dedicated cosmetic raw materials as of 2013. This European movement has influenced policy makers worldwide, leading to similar initiatives in the U.S. The development and validation of alternative methods will require more cross-sector basic and applied research. Even though there is a general consensus among scientists that it will be impossible to replace all of the proven obligatory animal toxicological tests in the short term, several novel toxicology approaches based on alternative in vitro test systems are under development. As an example, major advancements have taken place and will continue to do so in the development of non-animal skin sensitisation assays that rule out potentially allergenic materials. A promising set of projects utilises several dendritic cell types to simulate how skin-borne immune cells respond to potential sensitisers. Scientists hope that carefully assessing the behaviour of these cells upon exposure to test materials will allow reliable screening and prediction of the sensitisation potential of novel ingredients.

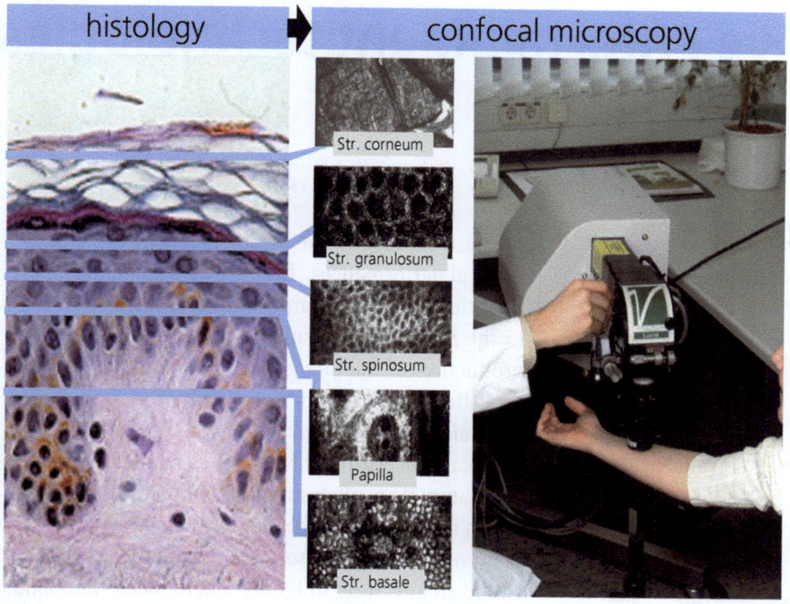

Confocal laser scanning microscopy: This new biophysical skin measuring technique allows non-invasive monitoring of the skin state and the metabolism throughout the epidermis to the papillaries, which feature the nutrient-supplying blood capillaries. Using special laser microscopy, deep skin layers can be accessed and visualised at high resolution. This can thus partially replace the traditional skin morphology analysis methods of biopsy and histology. Source: Beiersdorf

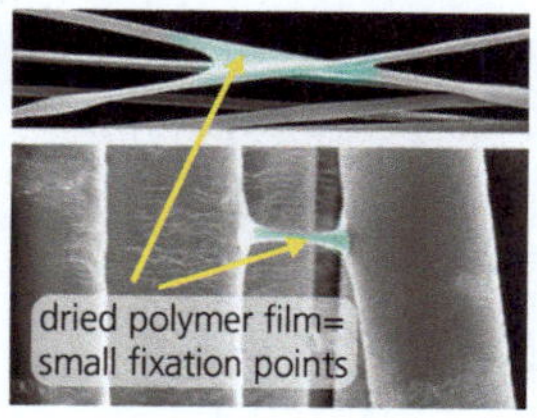

Function of hairspray shown as image of a polymer on hair fibres: Following evaporation of solvents the deposited polymers dry on the hair surface in the form of small fixation points that glue hair fibres together, thus leading to a rigid three-dimensional network. Source: Beiersdorf

Prospects

- A huge variety of product technologies has evolved from rapid development in the biological and material sciences to meet new demands from emerging markets and the needs of an ageing population (50+).
- Modern cosmetics combine excellent skin compatibility with a broad range of multiple effects.
- Skin-active ingredients with tailor-made biological effects and suitable forms of application represent an important trend.

DR. HORST WENCK
PROF. DR. KLAUS-PETER WITTERN
Research and Development, Beiersdorf AG, Hamburg

Internet
- www.colipa.eu
- www.cosmeticsinfo.org
- www.derm.net
- www.cosmeticsandtoiletries.com/research
- www.skin-care.health-cares.net
- www.pg.com/science/skincare/Skin_tws_toc.htm

Live entertainment technologies

Related topics

- Digital infotainment
- Human-computer cooperation
- Ambient intelligence
- Software
- Domestic appliances
- Sensor systems
- Image evaluation and interpretation

Electric guitar. Top: 3 rows of magnetic single coil pickups are mounted at different locations between the bridge (bottom) and the neck (top) of the body. The switch (right) lets the musician select various combinations of the three pickup devices. 3 rotary knobs drive active filter/amplifier circuits to change the volume and timbre of the sound. Bottom: Magnetic guitar pickup. An alternating current is induced in the coil by the magnetic field of a vibrating steel string. Source: Fender® Musical Instruments

Media technologies are an integral part of most live musical performances. They are the source of components for musical instruments and the technology used in the recording, transmission, sound design and reproduction of audio signals. Electronic and electro-acoustic devices are not only involved in the live performance of popular music (pop, rock and jazz). Also classical concerts (featuring orchestral music or opera, for example) are often supported by media technology, which is at times invisible, for instance when the reverberation of an opera house is enhanced by an "electronic architecture" of hidden microphones and loudspeakers for the performance of a symphony. For contemporary composers, electric and electronic instruments are the natural extension of their sonic repertoire. The live performance of music in large indoor or open-air venues with audiences of more than 10,000 people would be virtually impossible without electro-acoustic technology.

Principles

▶ **Musical instruments.** Musical instruments can be routed to a sound reinforcement (SR) system via several different paths. In the case of acoustic instruments, the airborne sound is picked up by microphones. Electric instruments have integrated devices to convert the mechanically produced sound into electrical signals, whereas electronic instruments are purely electronic sound generators and are controlled by a separate interface. Electric instruments include the electric guitar, the electric bass, and electromechanical pianos such as the Rhodes piano, the Wurlitzer piano and the Hammond organ. Their sound is often closely associated with musical styles of a certain period. Electric guitars have a solid body (as opposed to acoustic or semi-acoustic instruments) with 6 or 12 strings. Their vibration is picked up by a coil of very fine wire wound around a permanent magnet (magnetic pickup). The latter magnetises the steel strings mounted above it, whose vibration induces an alternating current in the coils. This, in turn, drives an amplifier/speaker system. As with all electric instruments, the sound quality of an electric guitar is the result of its mechanical properties, which are determined by the materials and design used in the body and strings as well as the electrical circuitry. Typically, an electric guitar consists of various magnetic pickups, with different guitar effects that shape the tonal balance of the electrical signal. Finally, it has a guitar amplifier with a pre-amplifier, built-in effects, a power amplifier and speakers. Characteristic sound effects such as distortion are produced electronically by driving parts of the signal chain into saturation. This produces additional harmonics, with a particular sound quality.

▶ **Musical instrument digital interface.** The most common protocol for musical control data is *MIDI (Musical Instrument Digital Interface)*, introduced in 1983 and still an industry standard for electronic instruments. It requires a physical interface (5-pin DIN connectors) and a protocol for messages to control electronic instruments such as synthesizers, sample players, drum machines or computer sound cards, using messages such as "note on", "pitch", and "velocity". Several instruments can be addressed simultaneously on different logical channels, and complex arrangements can be saved as standard MIDI files (SMF). Quite a number of acoustic instruments can be equipped with audio-to-midi converters. Even guitar players, therefore, can control synthesized sounds. The Open Sound Control (OSC) protocol, introduced in 1997, provides freely definable parameters, finer numerical resolution and faster transmission over standard network protocols. Although it has not superseded MIDI for the control of musical instruments, OSC is now widely used in electronic music and sound art.

▶ **Electro-acoustic transducers.** Microphones convert sound into electrical signals by exposing a thin and lightweight membrane to varying sound pressure. The resulting vibration is transformed into an electrical signal. In stage applications, most *microphones* use electromagnetic transducers (also known as dynamic microphones): a small moving coil attached to the membrane moves in the field of a permanent magnet, thus generating a varying current by way of electromagnetic induction. *Loudspeakers* use the same principle in reverse. A varying current causes a coil to vibrate in the field of a permanent magnet, while a diaphragm attached to the moving coil (voice coil) generates varying sound pressure in the surrounding air volume. A linear transmission of the whole audible frequency range is usually achieved by loudspeaker systems with two, three or four drivers of different sizes which are known as woofers, mid-range speakers and tweeters. Most consumer loudspeaker systems are still passive devices, but there is a general trend towards active systems with integrated digital signal processing and power amplification. Better frequency control and time response can thus be achieved using digital filters and linear phase crossover networks. Programmable filters even make it possible to equalise loudspeaker systems for specific installations to accommodate the interaction between loudspeakers and room acoustics.

► **Mixing and sound processing.** The central node for the distribution, control, effect processing and balancing of all audio signals is the mixing console. It is usually located in the middle of the audience, an area called Front of House (FOH), which is the part of a theatre or live music venue that is accessible to the audience. All signals from the stage converge here. Different mixes are then assigned to the reproduction channels feeding the loudspeaker system, to the monitoring system and to recording/broadcasting. Large-scale music productions use separate mixing consoles and engineers for each purpose. The stage monitor mixing console is dedicated to creating mixes for the performers' on-stage monitors (loudspeakers or in-ear-monitoring). It is usually situated at the side of the stage so that the engineer can communicate with the performers on stage. The surface of a mixing console is organised into vertical channel strips assigned to individual audio signals. One might be the output of a singer's microphone, another the electric signal of a keyboard output. The channel strips can perform individual audio processing, including equalizers that shape the spectral envelope of the incoming sounds or dynamic range processors such as compressors, used to increase the clarity and perceived loudness of voices and instruments. Digital consoles offer a wide variety of processing options, though external devices are still needed to create specific effects such as artificial reverberation, which is used to enhance the "dry" sound of microphones close to the sound source and to substitute for the lack of room response in open-air concerts. In analogue mixing consoles, the electronic signal processing circuitry is connected directly to the control panel. Digital consoles, on the other hand, serve as user interfaces connected to a digital signal processing (DSP) hardware unit, which is frequently separate. In a studio recording environment, digital audio workstations are often operated with standard computer interfaces (keyboard, mouse). In live sound reinforcement situations, engineers generally prefer the traditional mixing console because it allows immediate access to the complete signal flow.

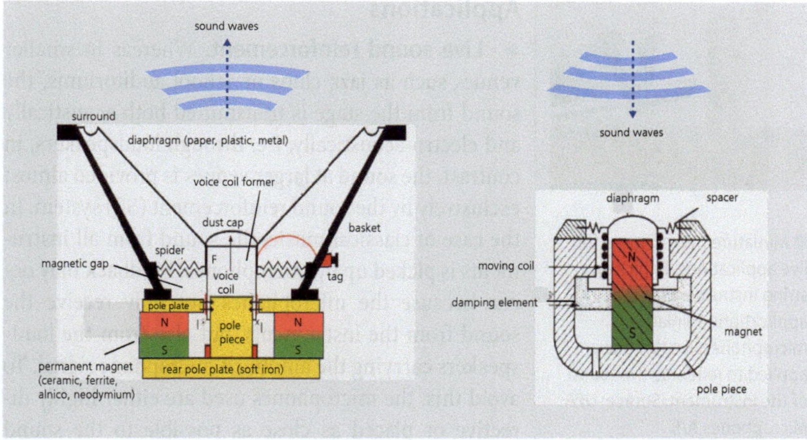

Electro-acoustic transducers: Two electro-acoustic transducers using the same physical principle in reverse direction. Left: The electro-dynamic loudspeaker causes a coil to vibrate in an alternating magnetic field, thus producing sound through the attached diaphragm. Right: The electro-dynamic microphone picks up the adjacent sound pressure, which causes a diaphragm to oscillate and induces an alternating current in the attached moving coil.

Sound reinforcement: An open-air symphonic concert from the mixing engineer's perspective (Berlin, Waldbühne). On either side of the stage, two loudspeaker line arrays are suspended in the air ("flown"). The digital mixing console is organised into vertical channel strips consisting of faders and control dials for the processing of individual audio signals such as the outputs of all microphones on stage. Volume meters are located at the top of each channel, while the middle section gives control over signal routing, monitoring options, and basic configurations such as the digital sampling rate. Source: Holger Schwark

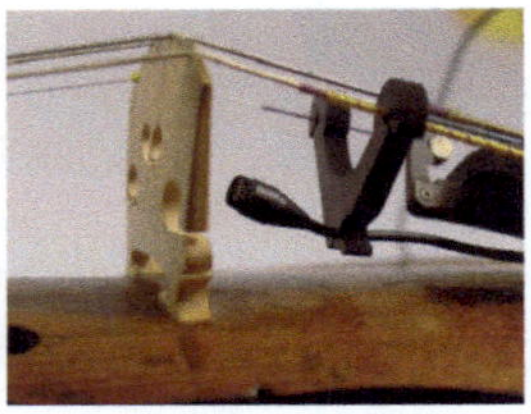

Miniature microphones for live applications: To pick up string instruments in live applications, miniature microphones like this one are applied to reinforce the sound of the instrument. Source: DPA Microphones A/S

Applications

▶ **Live sound reinforcement.** Whereas in smaller venues such as jazz clubs or school auditoriums, the sound from the stage is transmitted both acoustically and electro-acoustically, i. e. through loudspeakers, in contrast, the sound at larger venues is provided almost exclusively by the sound reinforcement (SR) system. In the case of classical music, the sound from all instruments is picked up by microphones. Feedback may occur because the microphones not only receive the sound from the instruments, but also from the loudspeakers carrying the amplified microphone signal. To avoid this, the microphones used are either highly directive or placed as close as possible to the sound source, which lowers the amplification ratio. Special microphones have been developed for near-field applications. Most popular music genres involve a mix of acoustic sound sources (singer, drum set, piano, acoustic guitars, bass), electric instruments and electronic sound sources. Some of these electronic instruments, such as digital sampling keyboards or instruments with an audio-to-MIDI converter, are operated like traditional instruments. Others are controlled by precomposed MIDI sequences, such as a symphonic sample library or a drum machine running synchronously with the "human" drummer. To synchronise man and machine, the drummer has to follow a MIDI-click signal played over headphones or in-ear monitors.

▶ **Live interactive sound art.** Some musical genres explore the creative potential of media technologies by making use of sounds or techniques of sonic design that are not available to traditional composition and traditional musical instruments. Among these are art forms such as electro-acoustic music, electronic music, computer music or audiovisual media art. Compositional techniques include the electronic synthesis and transformation of sounds, the structuring of sounds in space using loudspeaker-based spatialisation techniques, and the electronically mediated interaction of live performers, pre-recorded sounds and real-time sound synthesis by computers. New developments involve the creation of new algorithmic concepts in sound synthesis:

- Granular synthesis is based on very short sound grains that are stringed together, processed, modulated and layered, forming complex and fluctuating soundscapes.
- Concatenative synthesis is based on large databases of sound segments assembled according to unit selection algorithms in order to match a desired target sound. Successfully used in speech synthesis systems, this approach can be applied to music synthesis by using arbitrary input data (targets) and sound material, including voice-driven synthesizers.
- Physical modelling algorithms simulate the behaviour of natural sound sources using mathematical models, for example, differential equations for structure-borne sound and its acoustic emission under certain boundary conditions that can be accessed by the composer/performer.

Another important domain is the introduction of new sensor technologies to establish musical interaction between electronics and performers/composers. Multi-touch displays operated like musical instruments and connected to sound synthesizers via open sound control (OSC) interfaces are just one example.

Wave field synthesis loudspeaker: Loudspeaker array around the audience area at the open-air opera stage in Bregenz, Austria (red arrows). The system simulates the acoustical reflection pattern of a concert hall using wave field synthesis technology. Horizontal loudspeaker arrays around the audience area can simulate the sound of virtual sound sources or recreate the complex wave fields of acoustic environments.

Trends

▶ **Loudspeaker array technology.** Arrays of closely spaced loudspeakers provide much better control over the acoustic wave front than single, discrete loudspeakers. Vertical arrangements of loudspeaker modules, known as line arrays, with integrated digital signal processing are widely used in large sound reinforcement installations. A sound beam with largely frequency-independent directivity can be formed and electronically directed towards specific areas of the auditorium by applying band pass filtering and signal delay to the individual speaker modules. Curving the array will extend the area covered by the sound beam. *Wave field synthesis* (WFS) technology uses horizontal loudspeaker arrays arranged around the audience area to simulate the sound field of virtual sound sources or to recreate complex wave fields of acoustic environments. Loudspeaker arrays are operated with an object-oriented approach, as opposed to the channel-oriented approach of traditional multi-channel reproduction. Rendering software makes real-time calculations of individual loudspeaker signals interfering with each other, allowing wave fronts of virtual sound sources to be formed. The sound engineer can control the positions of those virtual sources online or offline and attribute a source signal such as a musical track to them. The technology has been used mainly in laboratories and small reproduction rooms such as cinemas and studios, because installing these systems in a larger space would be highly complex. The loudspeakers are usually positioned between 10–30 cm apart. Recent WFS installations for large-scale sound reinforcement applications – open-air concerts, for example – provide almost 1,000 audio channels. They are both used for public acoustical demonstrations and at outdoor venues to simulate the acoustic behaviour of a concert hall with its complex pattern of wall reflections. New approaches to avoid errors resulting from greater distances between the speakers are expected to boost the popularity of WFS for live reinforcement installations.

▶ **Simulation of acoustical environments**. Computer simulations are becoming increasingly important in the design of acoustic and electro-acoustic environments. They make it possible to predict the acoustic quality of performance spaces and loudspeaker installations. Such simulations are based on computer models representing the geometry and the sound absorption of walls and objects inside a given environment. With sound sources of a defined level, frequency distribution and directivity, it is possible to predict the acoustical behaviour at certain listener positions with the aid of an algorithmic sound propagation model.

Acoustical simulation programmes are based on ray tracing or mirror imaging techniques, which are also used to render photorealistic images of objects constructed in 3D computer graphics. The algorithm traces the path of an acoustic sound wave through space until it hits a given listener position. Tracing a high number of rays (typically 10^5–10^7) will produce a pattern of room reflections with realistic intensity and delay at the given listener position. Because each ray is computed independently, acoustic simulation greatly benefits from current multiprocessor architectures or other parallelization strategies. The straight ray is by and large the correct model for light waves. In the case of sound waves, however, it fails to take into account effects such as diffraction, diffusion or standing waves, which become important at low frequencies (long wavelengths). Current research is focused on the refinement of algorithms to account for those wave propagation phenomena. By the same token, models for the time-variant spectral and directional characteristics of natural sound sources such as singers or musical instruments are also being developed. These are important in order to simulate not only the deterministic behaviour of loudspeaker configurations, but also the sound of a human performer, a choir or an orchestra. The resulting reflection pattern, a so-called impulse response, can be used to auralize the sound source in the simulated environment, in other words, to convey a plausible sound experience as if one were personally present.

Computer simulation of acoustic environments: Distribution of sound energy on the interior surfaces of a factory building used for a rock concert, simulated using a computer model for the room and loudspeakers. Different colours indicate different sound pressure levels and illustrate how well the sound will be focused in the audience area (central plane). Source: Anselm Goertz

Prospects

- Technologies based on innovations in electronics, electro-acoustics, digital signal processing, sensor devices and algorithmic models of musical and acoustic phenomena will play an increasingly important role in public performances of music.
- Digital signal processing and transmission will provide more precise control over all components of the audio signal path.
- Due to progress in the modelling of acoustic phenomena, simulation techniques will be increasingly realistic and at the same time less "obvious" to the audience. This is true for the simulation of musical instruments as well as the simulation of room acoustic environments.
- Acousticians will be able to predict the behaviour of acoustical or electro-acoustical systems with increasing precision in the planning and design stage. Auralisations of the projected environment will provide an authentic impression of its sound quality even for non-expert listeners.

PROF. DR. STEFAN WEINZIERL
Technical University of Berlin

Human-computer interfaces used for musical interaction: The „reactable" is a translucent table with a video projector and a video camera beneath it. Different objects, called „tangibles", represent different sound synthesis devices. These can be moved by one or several performers at the same time. The movements are analysed and translated into musical processes via open sound control (OSC) messages. Source: Xavier Sivecas

Internet

- www.aes.org
- http://cnmat.berkeley.edu
- http://ccrma.stanford.edu
- www.ircam.fr
- www.nime.org

Domestic appliances

Related topics

- Sensor systems
- Surface and coating technologies
- Laser
- Ceramics
- Water treatment
- Ambient intelligence

Principles

The domestic appliances industry focuses on resource-saving technologies with the following aims:

- Saving resources: Energy (e. g. economic compressors in refrigerators), water (e. g. water detection and dirt sensors in dishwashers and washing machines), consumables (e. g. optimised technology for washing machines – dosage of detergent according to the quantity of laundry)
- Improved performance (e. g. vacuum-cleaner motors), time savings (e. g. induction cooking), "Quantum Speed" (combination of several heating types, better cleaning performance, e. g. variable water pressure in dishwashers)
- Comfort: Noise reduction, easy operation (e. g. automatic programmes, sensor pushbuttons, plain text dialogue), ergonomics (e. g. new appliance concept with optimised access), design (e. g. premium materials, large displays)

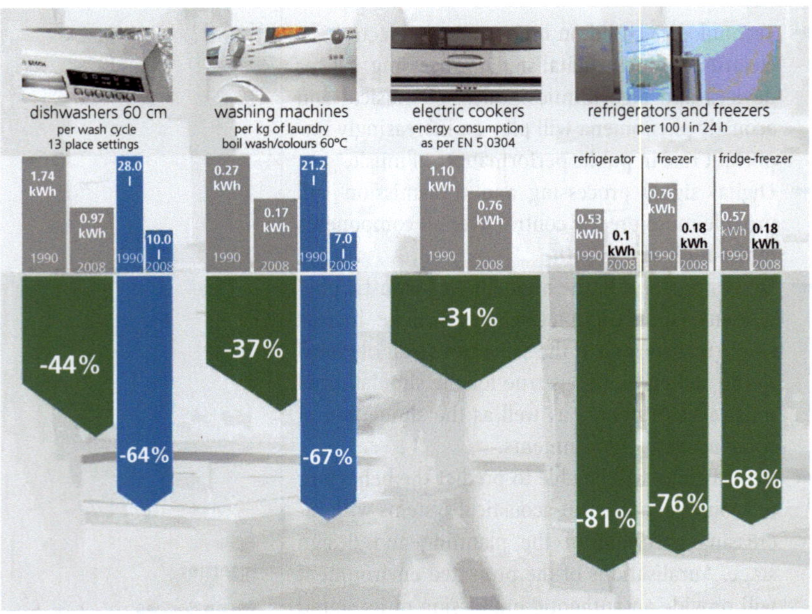

Development of energy and water consumption: All home appliances have become significantly more frugal, as evidenced by a comparison between average consumption values for each category of product in 1990 and the current figures for the most energy-efficient appliances. The diagram shows the results of this comparison for each product category (dishwasher, washing machine, electric cooker, refrigerator, freezer and fridge-freezer). Source: BSH Home Appliances Corporation

▶ **Induction**. An electric current can be generated through the movement or change of strength of a magnetic field. In the first case, the electric current is produced by moving a metal loop (conductor) through a magnetic field created by a magnet. Alternatively, the magnetic field can be created by passing a direct current (DC) through a coil (induction coil). The electric current is induced by the movement of the magnetic field relative to the conductor. In the second case, the conductor and the coil, and thus also the magnetic field, remain fixed in their positions and instead the static magnetic field is made to fluctuate by passing an alternating current (AC) through the coil. This causes an induced electric current to flow through the conductor (the same principle on which electrical transformers work). If, instead of the metal loop, a relatively large piece of metal is held in the changing magnetic field, eddy currents are induced in it, which in turn generate heat in the metal. This is the principle used in induction cooking.

▶ **Self-cleaning oven**. The most convenient way of cleaning an oven is to use a technology that does the work itself. Organic matter can be oxidised by heating and thereby be evacuated from the oven. In conventional ovens with a catalytic enamel lining, liquid dirt particles (mainly oil or fat) adhere to the porous surface structure and penetrate its pores. These organic substances are decomposed by an oxidation process in which they react with the surrounding oxygen in the presence of a catalyst incorporated in the oven lining (e. g. manganese dioxide). The waste products are water vapour and carbon dioxide, which are expelled in gaseous form through the heating process. A more advanced technique involves augmenting the supply of oxygen by using a special porous surface coating capable of storing oxygen in its pores. The liquid dirt particles penetrate this structure and become trapped between the microscopic bubbles of oxygen. This additional supply of oxygen accelerates the oxidative decomposition of the dirt particles. After they have decomposed, a fresh supply of oxygen fills up the pores and regenerates the system. A crucial property of this structured coating is that the pores should be of a size that allows only oxygen and not the dirt to penetrate inside them. This guarantees that the self-cleaning surface will remain effective throughout the life of the oven. The cleaning cycle necessitates a temperature of close to 500 °C. This and the duration of the cycle are controlled electronically. For safety reasons, the oven door locks automatically. Above 450 °C, the organic substances adhering to the walls of the oven oxidise, leaving behind flakes of ash that can be simply wiped off using a damp cloth.

► **Compressor.** In a refrigerator, this component compresses the coolant gas. The heat generated by this process is released to the surrounding air by the condenser. The working medium then streams through a throttle that lowers the pressure in the evaporator inside the refrigerator. As the cooling agent evaporates, it extracts heat from the interior of the refrigerator and then flows back - as gas - to the compressor. The energy efficiency of the compressor and the performance of the insulation material are the two main factors influencing the overall energy efficiency of the appliance.

► **Energy saving compressors.** A more efficient motor, reduced friction and an optimised flow can reduce the energy consumption of the compressor. The use of an improved lubrication system (e. g. lower viscosity oil, smaller contact surfaces) reduces friction and hence the energy loss of the bearings. Efficiency can also be reduced by pressure losses within the compressor occurring between the intake connection and cylinder. Vibrations caused by the compressor must be isolated from the evaporator, where they are likely to set up resonant frequencies in the housing of the refrigerator. New sound absorbers in the intake area help to increase efficiency as well as reduce noise. Further improvements can be achieved within the valve system by optimising the valve geometry and the opening and closing timing. A newly developed special type of reciprocating compressor is the linear compressor. Here, the oscillating piston movement is not generated by a crank drive (no transformation of energy), but instead directly by a linear oscillating drive. Linear compressors are particularly efficient. A feedback system linked to the compressor's electronic power control circuits offers additional potential to save energy, by varying output as a function of load and ambient temperature. This allows the compressor to operate at the minimum necessary level of cooling performance. Conventional compressors always operate at maximum output in order to ensure that the contents of the refrigerator are sufficiently cooled, even under the worst conditions. It is possible to reduce energy consumption by more than 40% merely by improving the performance of the compressor. Thanks to additional improvements to the cooling circuit and the use of more effective insulation, the energy consumption of refrigerators has been reduced by up to 81% over the last 15 years.

► **Open sorption systems.** A considerable amount of energy can be saved through the appropriate use of open sorption systems, which operate like heat pumps or thermal energy storage systems. The heat transfer effect of such systems is based on the adsorption of water molecules on microporous materials (e. g. zeolites). The adsorbed water is expelled from the adsorbent (e. g. zeolite) by evaporation when heat is applied. Any water vapour that subsequently comes into contact with the adsorbent is re-adsorbed, and the generated heat is stored.

Damp air is blown through a supply of zeolite and is both heated and dried. The moisture is almost completely absorbed by the zeolite. The exhaust air from this process, which has been heated and dried, can be reused for drying. Desorption then takes place during the rinse water heating process. Hot air streams through the zeolite and the resultant humid air is conducted to the dishwasher interior. There, the desorbed water vapour condenses and heats the dishes and the water. The condensation heat of the water vapour is completely used. During adsorption, again, humid air from the dishwasher interior is channelled into the desorbed adsorbent. The hot and dry air is then re-conducted into the dishwasher interior and, as a result, the dishes are dried in a very efficient manner. The advantage is that no additional heat source is required to dry the dishes. Dishwashers with an integrated sorption system consume up to 25% less energy than conventional appliances.

Applications

► **Limescale sensors in dishwashers.** The accumulation of limescale in dishwashers can be avoided if it is detected in time, e. g. by special sensors. This helps to

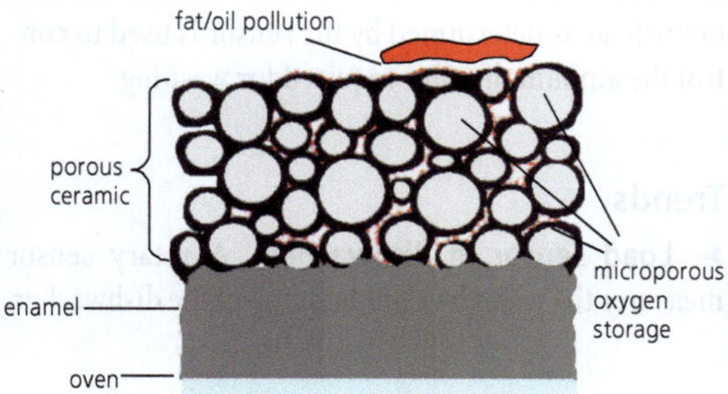

Self-cleaning oven: The optimised structure of the highly porous ceramic coating provides a large surface area on which the particles of dirt can collect. During the high-temperature cleaning cycle, these particles react with the oxygen stored in the pores of the ceramic coating as well as with the oxygen in the surrounding air. Through this oxidisation process, the dirt particles are transformed into water vapour and carbon dioxide. These gases volatilise, allowing the porous ceramic to fill with oxygen again and preserving the oven's self-cleaning ability throughout its entire life. Source: BSH Home Appliances Corporation

"Liftmatic": A wall-mounted oven utilising a lift system that allows access to the food from three sides. The oven closes automatically when it starts to heat by gently raising the lower half. Because heat rises, it cannot escape even when the oven is open, thus saving energy. Source: BSH Home Appliances Corporation

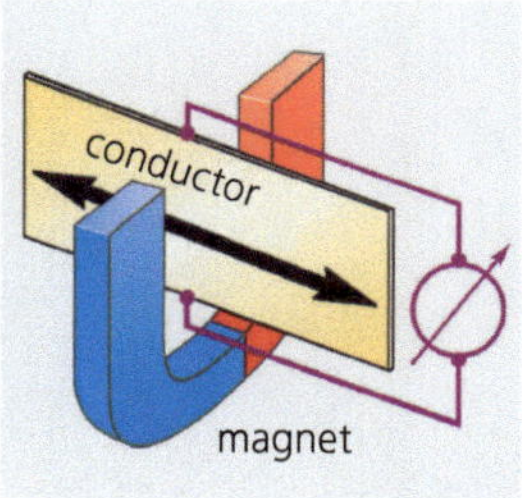

Induction process: An electric current is generated in a metal loop (conductor) when it is moved within a magnetic field. Induction cooking is based on this electrophysical principle.

protect glassware, which would otherwise become scratched or cloudy. All dishwashers have a built-in water softening unit (ion exchanger), which works by exchanging Ca- and Mg- ions from the water with Na- ions in the salt (regeneration). With the aid of a sensor, the regeneration process runs fully automatically as soon as too much lime is detected. The sensor takes the form of an optical fibre, which is immersed in the water in the dishwasher. Infrared light conducted through the fibre during the washing process is totally reflected at the interface between the wall of the optical fibre and the surrounding water, if the light impinges on the interface at a sufficiently oblique angle. The angle at which this total internal reflection occurs (critical angle) depends on the refractive indices of the two media, i. e. glass fibre and water. If a film of limescale forms on the outside of the *optical fibre*, the critical angle is reduced, because the interface is now between the optical fibre and the limescale. As a consequence, part of the light is no longer reflected, leading to a diminution of the intensity of the light emitted at the end of the fibre. In response, a detector activates a control lamp indicating the accumulation of limescale.

▶ **Induction cooking.** Induction cooking works on the principle of static magnetic fields. Alternating current flowing though an induction coil underneath the ceramic cooktop generates a magnetic field with changing polarities. This alternating field "induces" *eddy currents* that are trapped within the steel material that forms the base of the pan and transfer heat to the food it contains. Their dimensions depend on the specific electric resistance. The base of the pan must be perfectly flat and made of a ferromagnetic material to allow heat to be generated by the eddy current induced there. A large proportion of the cookware on sale today is suitable for use with induction cookers. The particularity of induction technology is that the base of the pan itself functions as the actual heat source rather than the ceramic cooktop (hob). This results in lower heat loss and a safer environment for children. Shorter heating and cooling times save time. Different heating levels can be set by regulating the current. The more current flows through the induction coils, the greater the induction. The frequency of the change in the magnetic field can be accelerated, too. Water boils in half the time on an induction hob than on a conventional cooktop.

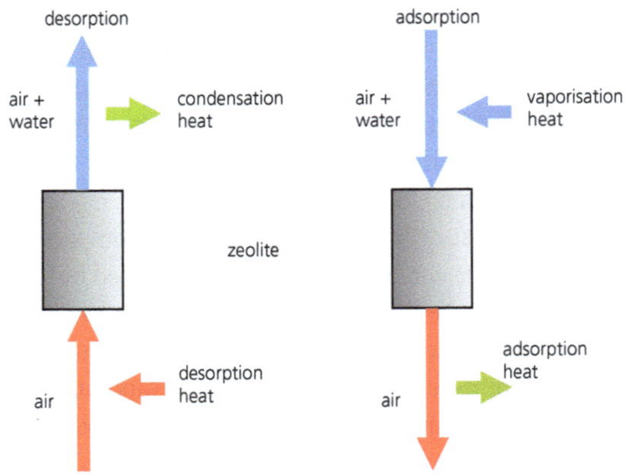

Open sorption system: Diagram of an open sorption process utilising zeolites (a microporous material) with the corresponding air and heat exchange processes. Such systems can be integrated in the washing process of dishwashers, thereby helping to save energy. Right: The moisture in the air is absorbed by the zeolite, and the dried air is heated by adsorption. Left: The zeolite is regenerated by being heated by an external source.

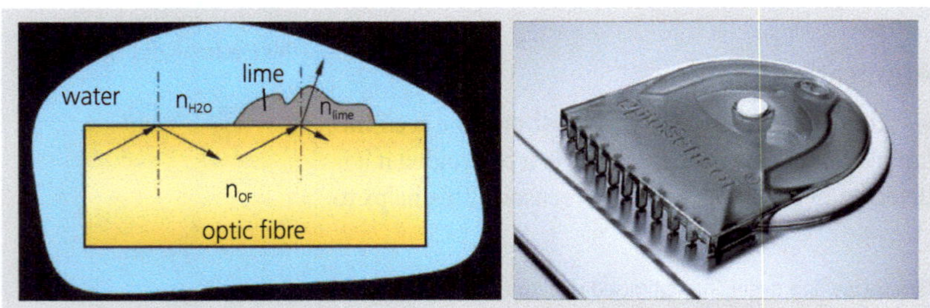

Limescale sensor (OptoSensor) in a dishwasher. Left: If limescale accumulates on the optical fibre, the proportion of reflected light is reduced due to a new interface between the conductor and the lime (before: conductor and the water) and the brightness of the light emitted by the fibre diminishes. Right: Original sensor. Source: BSH Home Appliances Corporation

▶ **Load sensor in washing machines.** A washing machine is not necessarily fully loaded when it is switched on. Water consumption can be reduced by automatic measurement of the load and the use of an appropriate amount of rinsing water. The loading level can be detected via distance measurement. The drum and outer tub form a unit, which is suspended within the washing machine housing with the aid of springs and supported by shock absorbers. The actual distance measurement is gauged with the aid of a Hall-effect sensor. This detects the position of a small magnet attached to the washing drum, which sinks down as a function of the weight of laundry it contains. The size of the load as determined by the sensor is used to control the amount of water required for washing.

Trends

▶ **Load sensor in dishwashers.** A rotary sensor measures the water level in the sump of the dishwasher.

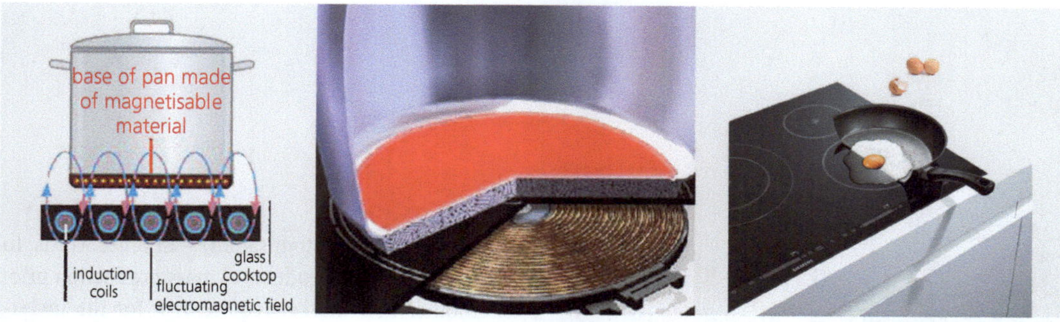

Induction cooking: The bottom of the pan is heated by induced eddy currents. Left: Induction coils below the ceramic hob generate heat in the base of the pan without directly heating the cooking area. Right: Although the element is switched on, the egg only cooks where heat has been transferred to it by the pan. Source: BSH Home Appliances Corporation

The greater the quantity of crockery in the machine, the more water is displaced onto the dishes rather than flowing into the sump. In a fully loaded dishwasher, the water level in the sump initially drops when the cleaning cycle begins. At this stage, a fixed amount of water is used regardless of the load. But as the water level drops in the sump, so does the available water pressure at the intake to the circulating pump, forcing it to work harder. As a result, the rotary speed exceeds a defined range. If water is added to refill the sump, the operating speed of the circulating pump returns to normal: this indicates the correct filling volume. The water flow to the machine can be adjusted in several steps, thus constantly adapting the volume of water to the actual load level.

► **Cooktop-integrated temperature sensors.** The temperature of the pan is a crucial factor when preparing fried food. A sensor underneath the surface of the cooktop detects the temperature of the ceramic element. This in turn is used to calculate the temperature of the pan on the basis of a mathematical model. The cook merely needs to select the required temperature range. An electronic processor performs the necessary calculations to regulate the heat of the cooking surface. This also helps to reduce energy consumption. An added advantage is that it eliminates the risk of burning or fat fires, which can occur if a pan is inadvertently left for too long on the range.

► **Infrared sensors.** Another way to constantly monitor and control the temperature and energy supply during cooking is the use of infrared sensors integrated in the cooktop. The user selects the required temperature profile. The sensor detects the temperature of the pan, compares it with the preset profile, and passes these data to a circuit that regulates the cooking temperature by adjusting the input of heating energy. The system precisely controls the input of energy to within the limits required to heat the food to the desired temperature. The food cannot be overcooked. The types of cookware compatible with this system include pans made of high-quality steel incorporating a special enamel strip, enamel pans, or alternatively any normal household high-quality steel pan to which a special tag has been affixed that can be detected by the IR sensor.

► **Automatic water pressure regulation.** This system adapts the water pressure to the degree of soiling in dishwashers. A water sensor detects the turbidity of the water by measuring the transmittance of a ray of infrared light reflected by a sample of water. A high concentration of dirt particles causes a diminution of the light emitted through the sample of water. The water pressure emerging from the spray arm is varied as a function of the soiling of the water.

Prospects

- Development cycles are growing increasingly shorter and the application of technologies more multifaceted. Consumers are demanding more efficient domestic appliances offering greater efficiency and lower energy/water consumption, while at the same time expecting them to offer the same high quality in terms of daily use and durability.
- In response to the market's growing ecological awareness, in addition to the challenges of rising energy and material costs, manufacturers of domestic appliances are seeking new ways of making more efficient use of energy, water and raw materials. Key trends are new materials (like zeolites) or the development of new sensors to reduce energy and water consumption.

DR. CHRISTOPH THIM
BSH Home Appliances Corporation

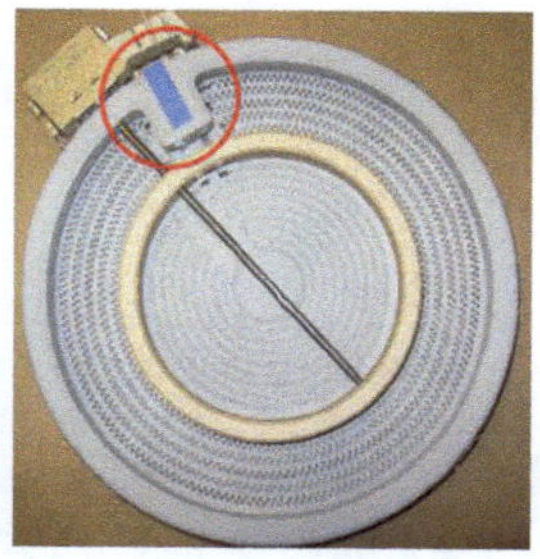

Two-zone heating element with temperature sensor: A sensor integrated in the ceramic cooktop below the element measures the temperature of the ceramic glass and regulates the temperature of the pan – in accordance with the temperature range set by the user. This eliminates the risk of overheating and burning. Source: BSH Home Appliances Corporation

Automatic cooking with infrared sensors: An infrared sensor detects the temperature of the cookware, compares it with the preset profile, processes the data and regulates the temperature by adjusting the supply of energy to the pan. The input of heating energy is precisely matched to the level needed to reach and maintain a constant temperature in the cooking vessel. Source: BSH Home Appliances Corporation

Internet

- www.intappliance.com
- www.ceced.org
- www.eais.info/TREAM_UK/pag1_eng.htm
- http://fcs.tamu.edu/housing/efficient_housing/equipment_and_appliances

12 PRODUCTION AND ENTERPRISES

The far-reaching and long-lasting global economic structural change currently taking place is creating world-wide interdependencies that are having a considerable impact on industrial production in high-wage countries. The principles, practices and strategies on industrial innovation and product that held sway for decades no longer suffice to meet the challenges arising from this transformation. The close integration of global economic goods, information and capital flows means that it is far easier today to organise distributed production of goods world-wide than it was a few years ago. New, aggressive competitors are showing up in the markets, which is putting far greater cost pressures on the manufacturing of products.

The globalisation aspect covers both competition and new markets. Globalisation is leading to an accelerated competitive situation with short production cycles, product development times and delivery times, as well as a sharp increase in competitive pressures between high- and low-wage countries. Markets are increasingly demanding that products be adapted to individual customer wishes. One response is to modularise products and create a large number of product variations while at the same time accelerating mass production (mass customisation) – in other words, to generate individual products at mass production prices. Another new market requirement for the investment goods industry is service that supports the product and is adapted to the individual customer. These range from classical maintenance to training, new leasing business models and even tele-service.

On the one hand, forecasts do suggest that globalisation will continue to ramp up competition. On the other hand, new markets are being tapped which, owing to cultural differences, make varying demands on products.

Faster development cycles, closer contact with the customer and faster response to change in the markets, coupled with their cultural differentiation, is generating greater change within companies, adjustments in their organisation and the expertise they offer. Active innovation management, therefore, is the most sought after competence of the future. The development of information and communication technology (*e-services*) and the growing significance of resource knowledge play a key role, while new types of products and an increasingly scarce workforce will demand modified forms of organisation.

The drivers of technological change are information and communication technologies, miniaturisation –with microelectronic components making up an ever larger share of all products – as well as new materials and *lightweight construction*. An important trend in production that urgently calls for new technologies, is the flexibilisation and acceleration of production with the aim of reducing throughput times. This comes in response to markets that are changing faster. One approach seen as very promising is *rapid prototyping* and manufacturing, which is of key importance to the implementation of flexible processes. *Laser* technology has also made broad inroads. It is used among other things for the integration of measurement and testing procedures during production. *Sensor* technology and intelligent controlling concepts are also indispensable for the implementation of robust processes. Mechatronic systems will enjoy growing significance, as will *human-machine interfaces* adapted to human use. Finally, simulation technology will support both production organisation and product development.

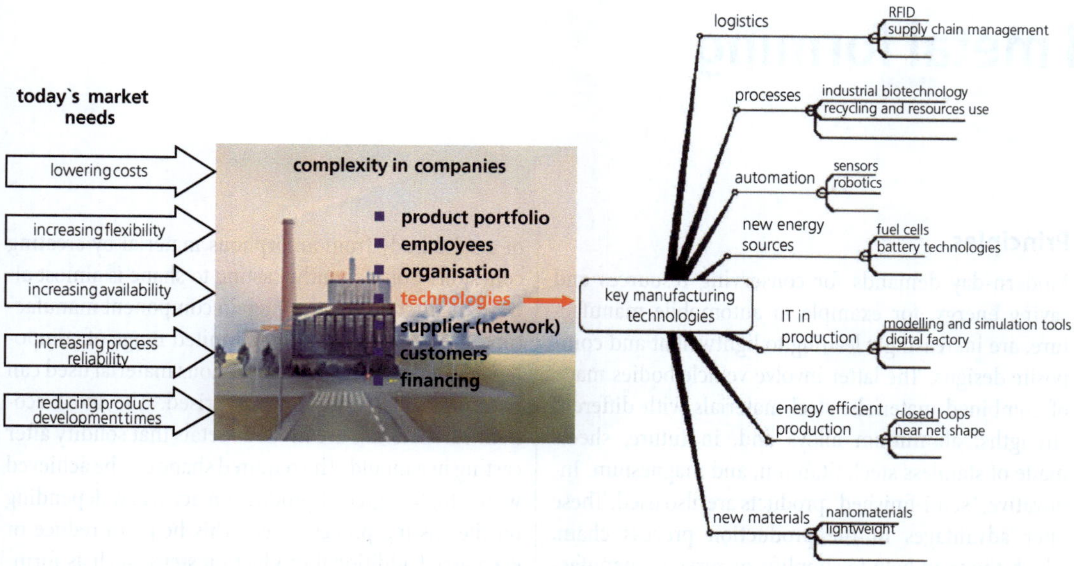

The demands made on manufacturing companies today are diverse. In addition to the need to build new networks with customers, suppliers and producers along the supply chain, other changes in society, politics and the financial world have to be taken into consideration. Quick technical developments and constantly changing trends are just one aspect of the challenge companies face. Source: Fraunhofer

Extensive work on collaboration in production networks is urgently needed. Greater customer orientation requires the development of individualised, customer-adapted solutions in the design of product networks. Resource availability and the careful use of limited resources is an important topic, especially within the context of the greater demand placed on them through global growth. Concentration on core competencies requires that processes be outsourced, and this will lead to a new forms of workforce organisation. New cooperation models offer new possibilities for cooperation across the entire supply chains.

▶ **The topics.** New concepts in *joining and production, primary shaping and reforming technologies* are clearly in evidence in automobile production. The processing of new materials, such as light metals, new joining technologies like bonding, or simulation to determine production parameters, present almost routine challenges for all automobile makers. That is why the automobile is globally a major innovation driver for a great many technologies. Information and communication technologies play a role not only in the simulation of production processes, but also in the planning of an entire production line. In the case of digital production, a future factory will be depicted first as a model before the real factory is built.

Alongside *digital products*, virtual models make it possible to test production processes, machine deployment, building technology or personnel resources. The use of *robots* in industrial production is known mainly from the welding robots in automobile manufacturing. But robots have meanwhile also established themselves in new applications. A distinction can be drawn between machines that lack artificial intelligence – such as welding robots in production – and intelligent robots of the future that recognise things in context, move freely, sense the environment, communicate with people and use flexible manipulators. These range from the inspection of critical environments, to assistant robots that perform heavy work, or service robots that help in the home.

Logistics represent a significant competitive factor in production. Procurement logistics, internal logistics to control material flows and distribution logistics to distribute the finished product are all essential elements of a comprehensive supply chain that must be holistically optimised. Synchronisation of material and information flows along the entire supply chain play a decisive role. Meanwhile, the entire logistics chain – from the supplier (of components) to the customer (who receives the final product) – has to be seen as an integrated process. One example is food chain management, in which the entire value chain of a food product is taken into account – from the production of agricultural raw materials to transport, processing and finally the end product on the consumer's table.

Whereas in production and forming processes, pre-fabricated, bulk-shaped material is worked, reformed or bonded, *process technology* turns bulk solids, liquid or gaseous substances into new production materials. The manufacturing of plastics for tote bags, of processed foods or of new lubricants for engines takes place in reactors of various types and sizes and under varying reaction conditions. One particular trend that has endured and will continue to prosper in the future is the development of catalysts. In this area, too, the extended possibilities offered by nanotechnology have opened the door to new applications. ■

Casting and metal forming

Related topics

- Metals
- Laser
- Materials simulation
- Joining and production technologies
- Automobiles

Car crankcase ventilation box, manufactured by aluminium sand casting. Source: GF Automotive

Principles

Modern-day demands for conserving resources and saving energy, for example in automobile manufacture, are increasingly leading to lightweight and composite designs. The latter involve vehicle bodies made of combined materials: steel materials with different strengths, aluminium alloys and, in future, sheets made of stainless steel, titanium, and magnesium. Innovative, "semi-finished" products are also used. These offer advantages in the production process chain, which they can help to simplify, or serve to manufacture components that have multifunctional properties and meet additional requirements such as the ability to damp noise. The challenges involved can only be addressed successfully if the entire product creation process is examined in detail. The production processes available at present and those set to be introduced in the future can be divided into six main categories according to two criteria: those that change material cohesion and those that change material properties.

- Casting to shape: Initial creation of a shape from the molten, gaseous, or amorphous solid state. A typical example of these production processes is casting.
- Forming: Production processes in which the given shape of a workpiece is transformed into a different, geometrically defined shape. Examples of typical forming processes in bulk forming and sheet metal forming are drop forging and deep drawing respectively.
- Separating: Machining or erosion to remove material, i.e. reduction of material cohesion. Typical processes include chip-type machining processes (milling, turning, etc.) and cutting processes.
- *Joining*: Connecting individual workpieces to create assemblies. Examples of typical processes are welding, gluing, and folding.
- Coating: Application of thin layers to workpieces, e.g. by electroplating or paint spraying.
- Modifying material properties: Deliberate modification of material properties, e.g. by heat treatment. A typical process is workpiece hardening performed after chip-type machining.

► **Casting to shape**. The production process referred to as casting to shape involves "the manufacture of a solid body from amorphous material by creating cohesion". Consequently, casting to shape is almost always the first production step in component manufacture. It permits a virtually unlimited range of component design options. The amorphous material used can be molten, pulverised, or vaporised. The most economical materials are molten metals that solidify after casting in a mould. The required shape can be achieved with a high degree of production accuracy, depending on the casting process used. This helps to reduce or even avoid additional production steps, such as forming, separating, or joining. Casting to shape makes it possible to manufacture semi-finished products – for example blocks, strips, pipes, or sections – as individual components. Semi-finished products are usually subjected to further processing by forming or chip-type machining. However, in the casting to shape of individual components, a large number of process steps can be dispensed with on account of the near-final contour design. A near-final contour design is one that scarcely requires any further machining steps to become a finished product. The main processes for manufacturing castings using investment moulds include sand casting and precision casting. "Investment mould" in this context means that the casting mould has to be destroyed after just one casting process in order to remove the component. The sand or ceramics moulding material can be reduced to a certain extent. If permanent moulds are used, i.e. steel moulds that make several components in succession, die-casting and permanent-mould casting processes for large quantities play a vital role. Castings have a wide range of applications, from mechanical and automotive engineering to medical equipment technology.

► **Aluminium sand casting**. Before the design data can be turned into a finished cast aluminium component, it is necessary to create a split model of the component. Model plates made of plastic, into which the positive mould of the component is milled, enable the upper and lower component contours to be reproduced. A moulding box is placed over each of these model plates. Oil-bonded sand is now introduced into the box and compacted. This procedure makes it possible to reproduce the component contour in the moulding material, i.e. the oil-bonded sand. The model plates are removed and the upper and lower parts of

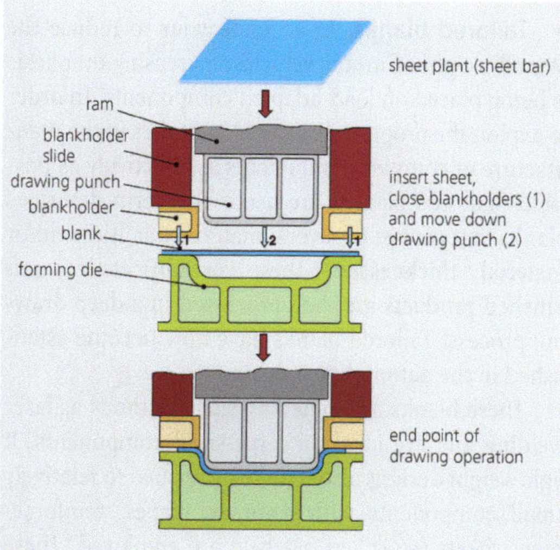

Drawing a component from a flat sheet blank: A flat sheet blank is inserted into a press. When the drawing punch is moved down, the sheet is deformed, whereby smaller sheet thicknesses are produced at the bend radii and the lateral surfaces. Source: Daimler AG

the moulding box are positioned relative to one another so as to create a hollow space that has the negative geometric shape of the subsequent component. A channel is kept free for the runner in the upper part of the moulding box. The sand mould is filled with molten aluminium through this port. When the aluminium has solidified, the mould can be opened. This is done by destroying the compacted sand mould, which is why sand casting ranks among the investment-mould casting processes. The raw cast component now has to be cleaned. Casting material and any casting flash is removed. After this step, drilling and milling surfaces can be machined and the component is then ready for fitting.

► **Forming processes.** Materials forming is an important production technology. As a production process, forming is used to systematically change the shape, surface, and material properties of a usually metal workpiece, while retaining its mass and material cohesion. A distinction is made between sheet metal forming and bulk forming, and – with regard to process temperatures – between cold, warm, and hot forming. Cold and warm forming are performed below recrystallisation temperature and hot forming takes place above that temperature. Forming processes can be further divided into 1st and 2nd production step processes. 1st production step processes are ones that are chiefly used to manufacture semi-finished products. A typical 1st production step forming process is the manufacture of metal sheets by rolling.

► **Rolling.** In the forming process referred to as rolling, the material is formed between two or more rotating tools known as rolls, and the cross-section of the workpiece is reduced or formed by section rolling to the required component geometry. The rolling process can be split up into three main categories: longitudinal rolling (flat longitudinal rolling and shape longitudinal rolling), cross rolling (flat cross rolling and shape cross rolling), and skew rolling.

Applications

► **Internal high-pressure forming (IHF).** The principle of internal high-pressure forming is that hollow metal bodies are made by the action of a fluid exposed to high pressure. The pressure fluid inside the hollow body acts along the inner wall of the tool. Effective fluid pressure is applied with pumps. The advantage of internal high-pressure forming is that a wide variety of workpiece geometries can be made, e. g. pipe branches. The disadvantage of this technology is the high cost involved. Examples of applications are chassis components, e. g. camshafts and vehicle exhaust systems, or component calibration for aluminium space frame parts.

► **Hot forming metal sheets.** For some years now, materials with high strengths and/or low density have been used increasingly as structural materials in vehicle body construction and in some cases subjected to hot forming. Examples are:

- The heat treatment of metal sheets made of aluminium alloys, which takes place before the forming process proper, aims to temporarily increase the material's formability for the forming process and raise its level of strength after the subsequent hardening process.

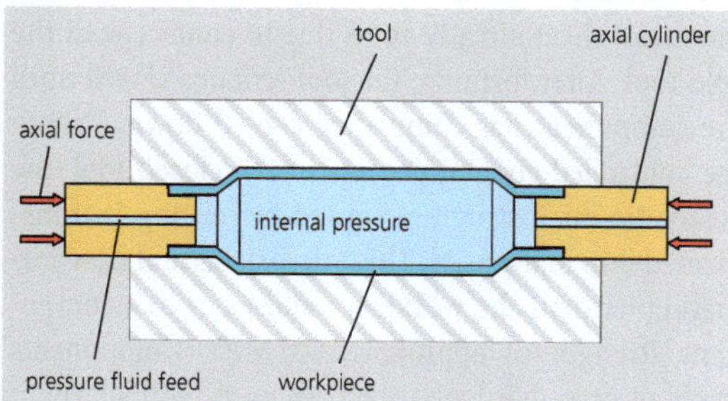

The principle of internal high-pressure forming: In this process, the workpiece geometry is shaped as a result of high pressure inside the tool, which presses the hollow body of the workpiece against the tool's inside wall. The axial ram presses the material inside and this axial feed makes it possible to control the forming process accurately.

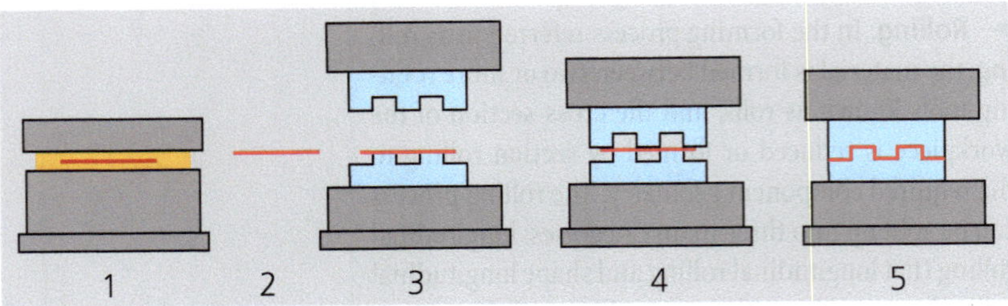

The process of hot forming boron-alloyed materials (press hardening). 1: The workpiece is heated to 950 °C in a furnace. In doing so, the microstructure is transformed into austenite. 2: Then the workpiece is transferred to the tool. Heat is lost due to radiation and convection, and the temperature field remains largely homogeneous. 3: The workpiece is positioned in the cold tool and the blank cools down inhomogeneously due to contact with the tool. 4, 5: The workpiece is hot formed and then cooled down in the closed tool. At a cooling rate higher than 29 K/s, the microstructure is transformed from austenite into martensite to harden the workpiece.

- The warm/hot forming of magnesium alloys aims to increase forming capability in the cold state and thus particularly to exploit the low density of the material as a positive structural property.
- The forming of metal sheets made of boron-alloyed heat-treatable steel in the temperature range from 900–950 °C in a cold tool. Martensitic hardening of the material occurs, coupled with a significant increase in strength. This process is called press hardening.

The *press hardening processes* used at present can be split up into direct and indirect processes: in direct processes, the blank is heated to 950 °C in a furnace and the microstructure of the material is chiefly transformed into austenite. Then the blank, which has a temperature of 950 °C, is inserted into the press tool. The hot blank is formed in a single step, during which the metal sheet already cools due to contact with the cold tool. After forming, the tool remains closed until the component has completely cooled down. Due to the controlled cooling process in the closed tool, the austenitic microstructure caused by heating is transformed into martensite. This microstructural transformation hardens the workpiece and leads to a component strength of approx. 1,500 MPa. Components made of press-hardened steel are used particularly in crash-relevant areas of the car. In indirect press hardening, the component geometry is predrawn as far as possible in a conventional cold forming stage. Then the components are austenitised in a furnace and, in a further process step, hot formed and hardened.

▶ **Tailored blanks.** In an endeavour to reduce the overall weight of motor vehicles, increasing emphasis is being placed on load-adapted components. In order to exploit the properties of metal materials in the manufacture of complex components as effectively as possible, “tailored blanks” are used. This term describes blanks that consist of several material qualities and/or material thicknesses. These prefabricated semi-finished products are then processed in a deep drawing process. Tailored blanks have now become established in the automotive industry.

These blanks are joined by such methods as laser welding and are ideal for large-sized components. If lightweight designs also have to be applied to relatively small components with complex shapes, reinforcement sheets (so-called patches) are employed. These patches are then formed in the same tool as the base sheet. Patchwork sheets, in which the patches are spot-welded, have been used in industrial series production for some time now. Bonded blanks, on the other hand, where the reinforcement sheets are glued on, are still at the development stage. They might compensate for the application limits of welded patchwork blanks, such as inhomogeneous force flow because there is no surface connection between the blanks, only a spot connection, and crevice corrosion in the surface due to possible penetration of liquids. In addition, there are other functional benefits such as noise and vibration damping due to the intermediate layer of glue and no need for seals or sealing compounds.

▶ **Honeycomb-structured products.** Load-oriented designs based on the strength-enhanced materials already mentioned are being increasingly used, as are innovative semi-finished products such as metal sheets provided with macrostructures, or metal sheets with

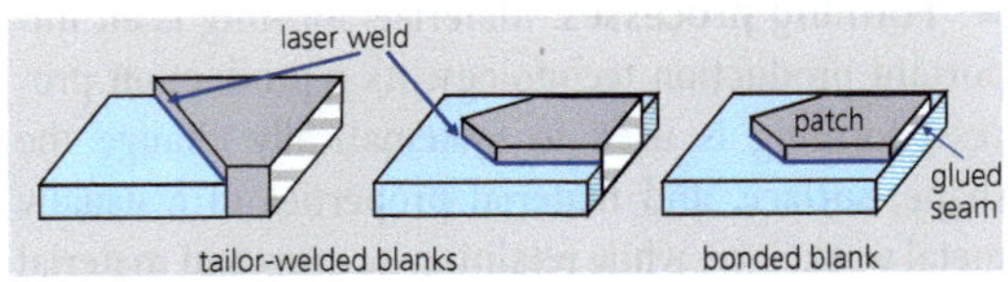

Schematic diagram of tailored blanks. Left and centre: Laser-welded blanks as a starting product (semi-finished product) for the drawing process. Right: Bonded blank (a patch blank glued to the base blank). Tailored blanks are used in particular to implement locally varying properties in a single component. Different sheet materials can be combined with the same wall thickness, the same materials can be combined with different wall thicknesses, or different materials can be combined with different wall thicknesses. Nowadays, tailored blanks are chiefly made by welding processes and glued tailored blanks are also being used increasingly.

Example of an application of prestructured sheets: By reducing material thicknesses, it has been possible to save up to 30% on weight compared to flush sheets and still maintain rigidity. Left: Tailboard made of honeycomb-structured starting sheet. Right: Position of the tailboard in the vehicle. Source: Daimler AG

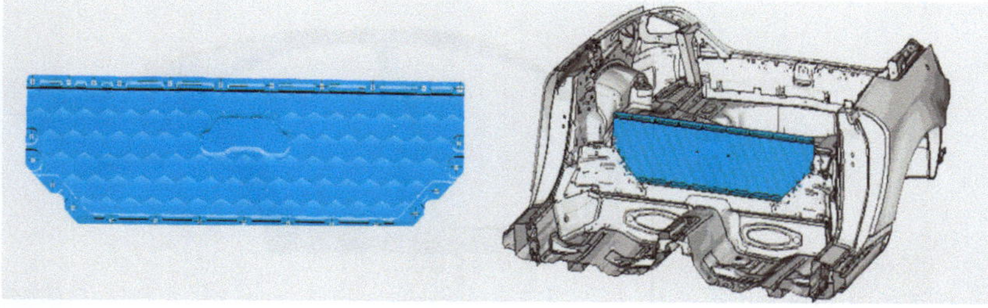

intermediate polymer layers (so-called sandwich sheets). Sandwich sheets can handle the function of noise damping in a vehicle to a certain extent, which obviates the need for additional noise-reducing damping mats. However, some rules have to be observed when forming such sheets in order to prevent damage to the material and thus to avoid impairing its function.

Honeycomb-structured tailboards serve as a local design measure in lightweight shaping, chiefly due to the following benefits:

- Increased homogeneous rigidity over the entire area of the tailboard. The hexagonal structures reduce the direction dependence of properties compared to conventionally beaded components. The sheet thickness can be reduced to a certain extent, thus saving weight while maintaining component rigidity.
- Cost savings in the tooling area, one of the reasons being a shorter manufacturing process chain.

In order to form structured sheets, however, it is necessary to make several modifications to the forming tool and the press parameters. In the area of the blankholder, for example, measures are required to increase the retaining forces that regulate sheet feed to the tool.

► **Use of magnesium sheets**. Magnesium has a high potential for lightweight design because it only has 70% of the density of aluminium (2.7 g/cm³) and 25% of the density of steel (7.85 g/cm³). The disadvantage of magnesium, however, is its low forming capability at room temperature so it has to be heated to approx. 230 °C. The blank is a sheet that serves as a flush "unmachined part" for forming. The technique used for large tools in sheet forming at high temperature, however, is not yet state-of-the-art. The main requirements for this are:

- Virtually uniform temperature distribution in the various tool components
- Blank heating above tool temperature
- Error-free continuous operation for several hours

A drawn part was made in an experimental tool, and the next forming steps took place at room temperature in the production tool for the aluminium roof.

Trends

► **Simulation of sheet forming processes.** The simulation of production workflows and individual processes is becoming increasingly important. Referred to by the terms "virtual manufacturing" or "digital factory", this technology makes it possible to simulate the effects of a material concept, for example, and the associated production technology for the various workflows at a very early stage of product development. It also enables them to be visualised in three dimensions. In this way, weak points in production can be revealed and influenced at the initial stage of product development.

Until a few years ago, workpieces could usually only be formed after carrying out preliminary tests and/or by drawing on expert knowledge. The reason why preliminary tests were needed is that forming processes cannot be described by closed analytical relationships. Over the past 30 years, processes have been developed that permit close-to-real "prefabrication" of the workpiece on a computer and thus eliminate or drastically reduce the amount of cost-intensive testing required. This procedure, in which process development is aided by process simulation software, is also referred to as

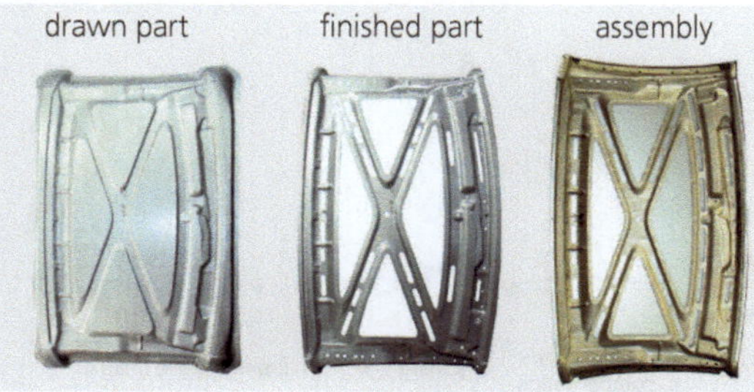

Experimental process chain with series geometry, taking the folding roof reinforcement as an example. Left: The magnesium interior roof made by hot forming. Centre: The finished interior part made by further operations (cutting). Right: The finished series-produced roof with aluminium exterior and magnesium interior. A roof made in this way is more than 1 kg lighter than a roof made with an aluminium interior and exterior. Source: Daimler AG

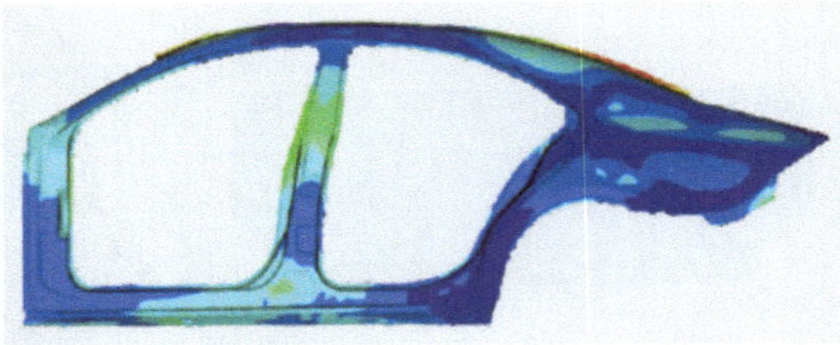

Simulation of the springback of a side wall after deep drawing and trimming: After trimming, the part deviates by more than 5 mm from the specified geometry due to springback (areas with a red color scale). This information is used to alter the geometry of the tool in such a way that the workpiece springs back to the correct shape after trimming. This process, referred to as springback compensation, is chiefly performed on a computer. The results of springback simulation can also be used directly in CAD systems in order to generate compensated tools. Source: Daimler AG

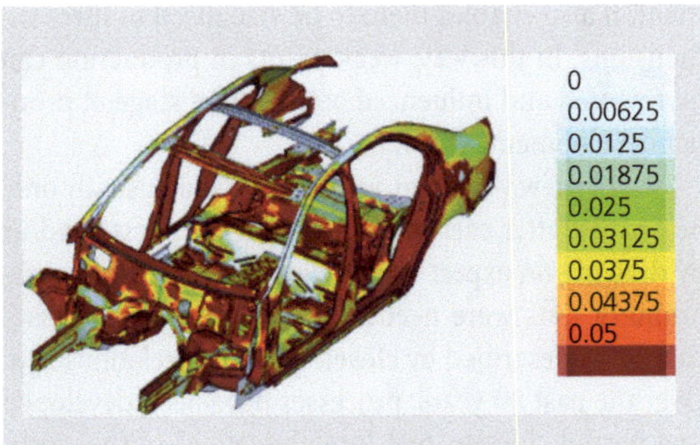

Plastic-equivalent elongation (in %) of sheet metal parts of a car structure made by forming. Prior plastic deformation of the structural parts by forming has an influence on crash behaviour. If the components' properties after forming have been taken into consideration, the crash simulation is more realistic than one in which they have not. Source: Daimler AG

virtual forming and digital production. Virtual forming will be accorded a high level of priority in our information and knowledge society in the future and will become increasingly important. The development of new forming processes or the application of known processes to new materials will no longer take place without the use of virtual forming technology.

► **Simulation of component properties.** Up to now, simulation analyses of the various processes have usually been kept separate, and any influence exerted by a preceding step of the process chain has not been taken into consideration. Depending on the material being investigated, however, the deformation history of the component is crucial to obtaining realistic simulation results. For this reason, efforts are currently being made to close the digital process chain of sheet components production in a holistic manner.

The digital process chain of sheet components production can be roughly divided into the following simulation packages: casting, rolling, annealing, forming, joining, lifetime analysis, and crash. *Finite element simulations* that reproduce the crash behaviour of components, for example, are nowadays chiefly made up of CAD records that solely contain the geometric information of the respective component assembly or of the respective component, but not their forming history. Recent years have seen the development of "mapping algorithms", which make it possible to apply certain parameters (element thickness, elongation) from the results of a forming simulation to the initial crash simulation mesh. The challenge is that finite element meshes for forming simulation, which are computed for each component, are much finer than those for crash simulation, which are computed for each vehicle. A useful mapping algorithm must therefore be able to summarise and apply the result parameters of a large number of fine forming elements to a coarser element of the crash mesh.

► **Rapid prototyping and tooling.** In almost every branch of industry, the development of new products is subject to rapid changes in market requirements. This calls for increasingly reduced development times and short technology phases within each project. In

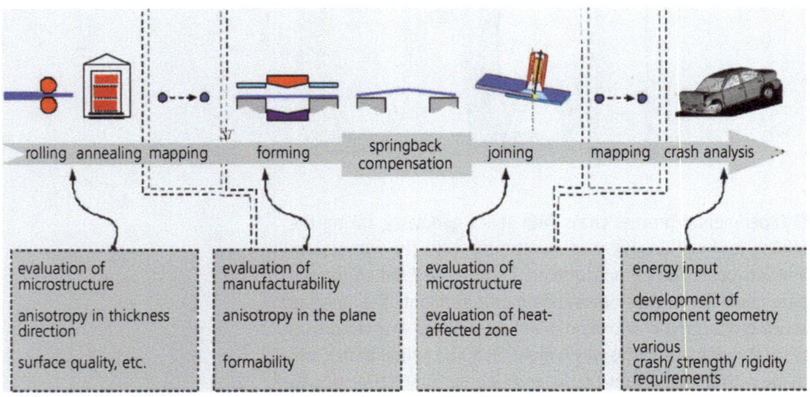

Simulation of component properties along the process chain of sheet component production: In the first step, the process for making the metal sheet is simulated, starting with rolling. The material specifications determined are used as input data in the subsequent forming and joining simulations. The component properties obtained after forming and the downstream joining processes are used as input data for any subsequent crash or lifetime analysis.

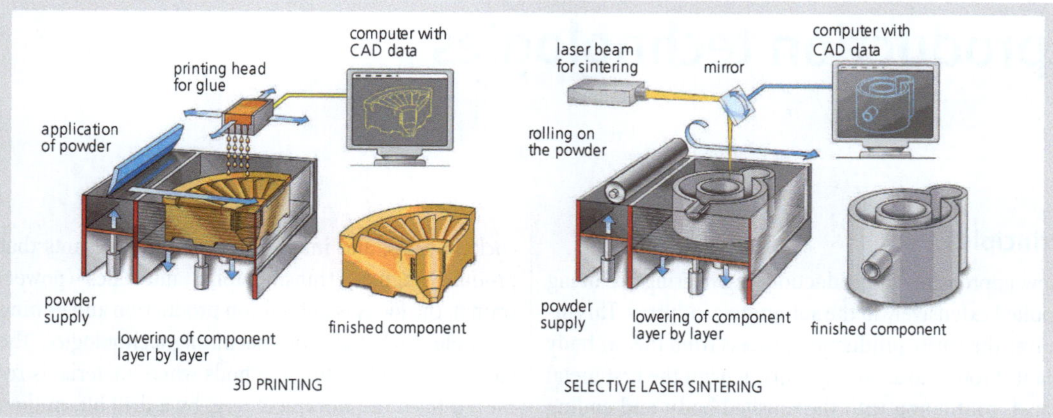

Rapid prototyping. Left: 3D printing. As with a conventional printer, powder is applied in layers and glued at points specified by the CAD model on the computer. In this way, a plastic component, for example, is built up layer by layer. Right: Selective laser sintering. If metal powder is used, it can be fused with a laser and a metal component can be produced. Source: Daimler AG

these ever briefer development periods, components have to be designed, manufactured and tested. One option is to use rapid prototyping (RP) parts, which have the following benefits:

- fast component availability
- supply availability of small quantities of components
- low total cost (cost of manufacture and materials)
- considerable flexibility in component design (shaping) and further development

The three major processes in rapid prototyping technology are:

- *laser sintering*
- stereo lithography
- layer laminate manufacturing (LLM)

These methods are based, for example, on 3D printing processes: as with a conventional printer, a layer is applied to a substrate and then fixed, e. g. with glue. Alternatively, a whole-surface metal powder layer can also be applied with a roller and then fused to the layer below at the desired points using a laser (laser sintering). Light-curing resin (e. g. epoxy resin) can also be used. The workpiece grows by being dipped deeper and deeper, layer by layer, into a bath of light-curing resin, whereby the top film is systematically hardened by means of a laser (stereo lithography).

Another method is to place thin layers of paper, plastic film, or thin metal sheets on top of one another and to glue or weld them together in layers (LLM).

Prospects

- The further development of processes for casting thin-walled parts will make it possible to manufacture lightweight components. As a result, assemblies that are nowadays made by forming and subsequent joining of several sheet metal components will be produced in a casting process in the future.
- Increasing importance is being attached to the forming of innovative semi-finished products such as tailored blanks, honeycomb-structured sheets, and hybrid materials including composite sheets (metal sheets with sandwich structures). These can be steel-plastic-steel sandwich sheets or hybrid steel-plastic composites, for example.
- In the forming of metal sheets made of high-strength and ultra-high-strength steel there are new, partially still unresolved, challenges regarding the design of highly stressed forming tools.
- Increasing diversification in niche models will call for a high level of process flexibility in future. Forming technologies oriented towards mass production will also need to be able to produce small quantities economically. In view of the requirements in terms of greater process flexibility, process reliability, quality assurance, and cost reduction there is considerable demand for close-to-real simulation methods, functionalised surfaces, and flexible tooling concepts that can be implemented for short periods.
- In the field of production process simulation, work is being increasingly performed on the "digital sheet components production" process chain. The objective is to calculate as accurately as possible the component properties resulting from the various production processes and thus make improved statements regarding dimensional accuracy, crash behaviour, and lifetime.

PROF. DR. KARL ROLL
DR. DIETER STEEGMÜLLER
Daimler AG

Internet

- www.wzl.rwth-aachen.de/en/629c52491e476b86c1256f580026aef2/mtii_e2.pdf
- www.wildefea.co.uk/download/case/6/wildefea010_CBM2001.pdf
- www.metalcastingzone.com/art-casting

Joining and production technologies

Related topics
- Metals
- Laser
- Robotics
- Sensor sytems
- Digital production
- Automobiles

Principles

New approaches to production engineering are being applied extensively in the automotive industry. This involves the entire production process from the car body plant through to assembly, starting with the first metal panel, continuing with the finished body, and ending with the completed vehicle. After the individual body components have been made in the press shop, they are then assembled in the body shop to create more complex vehicle components such as the side wall, floor, and flaps (trunk lid, engine hood, and doors). In order to connect them permanently, various joining processes are used, such as welding, gluing, clinching, and riveting. In the body shop, the entire vehicle body is built up step by step. In the surface treatment shop the car body is protected against corrosion and painted. This is followed by the assembly line, the first step being what is known as a "marriage", i.e. fitting the powertrain, which comprises the engine and transmission, into the body; the second step involves fitting the cockpit, seats, and interior equipment. At plants that produce engines, transmissions, and axles (powertrains), the focus is not only on production and joining technologies but also on machining technologies. The latter consist of cutting methods where material is removed from the workpiece, e.g. by a drill bit, milling tool, reaming tool, grinding tool, or turning tool. Today's rapid pace of development in the field of lightweight design in conjunction with a continuous rise in crash resistance requirements and the need for increasingly flexible yet efficient work practices, are having a major influence on the process chain as a whole.

▶ **Mechanical joining processes.** *Lightweight* design requirements in the automotive industry nowadays can be realised by combining different types of materials. This so-called composite design uses various grades of steel and aluminium, including high-strength and ultra-high-strength steel, as well as *fibre-reinforced plastics* or suitably coated panels. The objective is to reduce the weight of the body-in-white and simultaneously increase rigidity. Clinching and self-pierce riveting are proven, economical joining processes that are becoming increasingly important in composite design. *Clinching* uses a tool set, consisting of a punch and a die cavity, to make a positive and non-positive connection between components. In self-pierce riveting the inseparable connections are made by means of additional auxiliary joining parts, i.e. rivets. These reforming joining processes are ideal for connecting differently coated materials. Since the materials to be joined, and hence the microstructure of the connection (in contrast to welding), are not subjected to thermal influence, the resulting dynamic connection strengths are high. In addition, the processes are easy to automate, have a simple process flow, and require minimal capital investment.

▶ **Chemical joining processes (gluing).** Another joining process is surface gluing with high-strength adhesives, which increase the rigidity, crash performance, and durability of the body. Since ongoing demands in these areas have risen significantly, while lightweight design and composite design are also on the increase, surface joining with high-strength adhesives has become highly important. Combinations with other joining processes, such as spot welding,

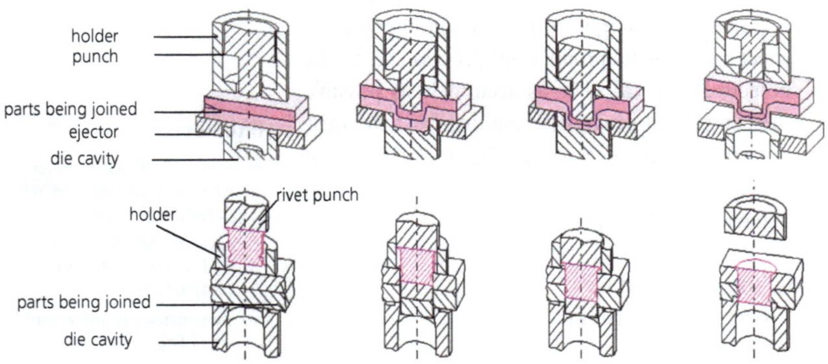

Clinching and self-pierce riveting. Top: Clinching – mechanical joining without auxiliary joining parts. Components are kept in place by the holder. At a high joining force the punch presses the parts being joined into the die cavity. The special design of the die cavity causes extrusion, which joins the metal panels to one another to create a positive and non-positive connection. Bottom: Self-pierce riveting – joining by introducing a rivet into the joint. The parts being joined are preloaded by a setting unit. Now the rivet punch presses the rivet element through the parts being joined, thus cutting a circular piece out of the metal panels (slug). The joining material on the die cavity side flows into the radial groove of the rivet shank. This creates a positive and non-positive connection. Source: LWF, University of Paderborn

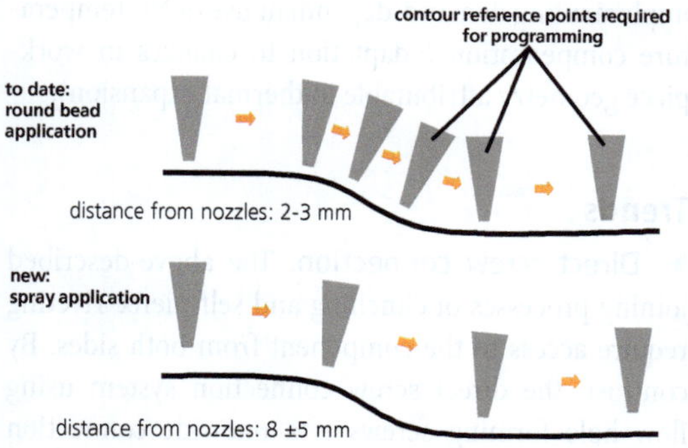

Nozzle application of glue. Top: Conventionally, highly viscous glue is applied to the component in the form of a round bead. The distance between the nozzle and the component is approximately the same as the height of the applied bead, so the robot can only roughly follow the three-dimensional component contours, some of which are extremely complex, and only at medium speed. Bottom: Spray application of high-strength body-in-white glue is twice as fast as conventional round bead application. Since high-strength glue is less viscous, the distance from the nozzles to the component can be greater (with no need to travel along the component with precision). This also simplifies robot programming because far fewer contour reference points are required. Source: Daimler AG

clinching, and self-pierce riveting, have proved particularly successful in terms of higher rigidity and crash resistance. Gluing is also being used to an increasing extent with growing modularisation in final assembly, involving the fitting of supplier-prefabricated components, such as the entire panoramic glass roof or the cockpit of micro-cars. In total, up to 150 m of glued seams are used in bodies-in-white; the glue (one-component epoxy resin glue) is applied to the components by spraying. Quality control is performed online throughout the process using a synchronised sensor. In another step, the components are then connected by spot welding or clinching. The glued joint acquires its ultimate crash resistance in the drying oven of the surface treatment shop. The joining properties of a high-strength surface-glued joint are much better than that achieved by spot joining processes.

▶ **Thermal joining processes (welding).** Thermal joining processes are widely used in automobile production, both for body-in-white structures and for engine, transmission and chassis components in the powertrain. Arc, resistance, friction, and laser processes are the most common. The components to be joined are heated locally (by supplying energy) and connected to one another in an integral joint, which provides benefits in terms of mass, material input (weight), and robustness. Process-related heating and cooling can lead to undesirable changes in material properties, especially in the case of composite design. This presents a major challenge in thermal joining processes, resulting in the ongoing development of techniques that make it possible to weld steel, for example, to aluminium, or cast materials and steel materials. Consequently, arc and laser brazing processes are in widespread use. The additional call for increases in productivity can be met with laser techniques since the high energy density generated permits faster welding speeds.

Applications

▶ **Laser hybrid welding.** Arc welding and laser welding have been combined in a modern process that unites the benefits of laser welding – low heat input and high welding speed – and the excellent gap bridgeability and high process stability of the metal-inert gas (MIG) method. The result is an integrated process that is characterised by high process stability, reliability, and productivity. In contrast to pure laser welding, the laser hybrid process is more robust with regard to component tolerances, such as the gaps that typically occur with pressed parts. The process can be used for both steel and aluminium. Inert or active gases are employed, depending on what materials are being joined. Welding speeds of up to 6 m/min are achieved. The welding head integrates every function, including the monitoring of all introduced media (inert gas, cooling

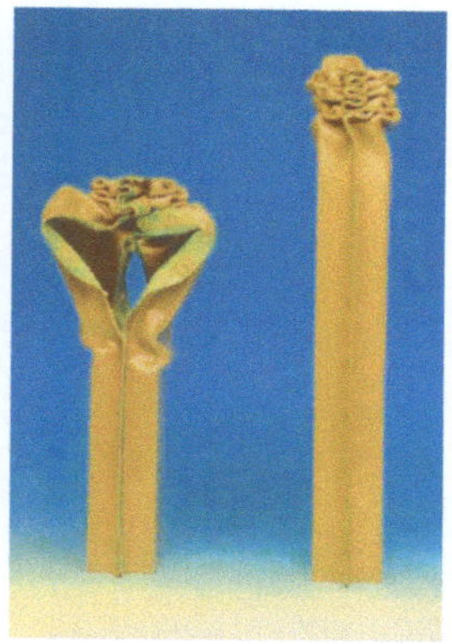

Comparison of rigidity and crash performance: If the components (double top hat sections) have only been joined by clinching (left), deformation by a crash load is far greater than if the join was made by clinch gluing (right). Source: Daimler AG

Laser hybrid welding: Combination of laser welding and arc welding in an inert-gas atmosphere. The heat sources for welding are an electric arc that burns between an electrode, which also serves as the welding rod, and the workpiece, plus a laser beam. As the welding rod is moved along the seam, an inert gas, usually argon or an argon mixture, prevents oxidation of the liquid metal under the arc, which would weaken the seam. Thanks to the laser's focused introduction of energy into the material, thermal deformation of the component is kept to a minimum, making it possible to bridge gaps. Source: Daimler AG

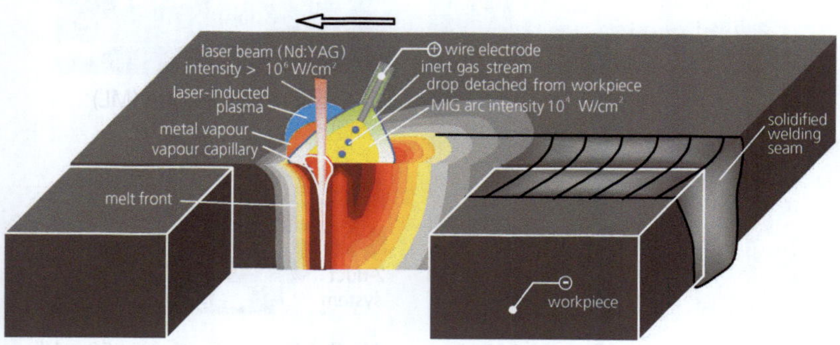

water and compressed air), as well as the seam guidance and quality assurance systems.

▶ **Dry machining.** One new technology to supersede conventional "wet" methods in the field of workpiece machining (e.g. drilling and milling) is dry machining, involving micro-lubrication applied directly to the point of friction in the form of an oil mist (approx. 5–50 ml per process hour, depending on the method of machining). This method has many advantages over conventional flood *lubrication* with oil or emulsion. Not only is it possible to substantially reduce or dispense with a cooling lubricant treatment system and supply line, the more compact dimensions of dry machining systems reduce costs further, and a simpler infrastructure permits greater flexibility in the factory layout. Dispensing with cooling lubricant does however cause a rise in component temperature since part of the process power generated is adsorbed by the component in the form of heat, in the absence of cooling. This represents a special challenge when it comes to compliance with dimensional tolerances, especially in the machining of aluminium components (the coefficient of linear expansion of aluminium is more than twice that of steel). Nowadays, these problems can be controlled by optimising the tooling and process parameters, adapting the machining sequence, and implementing demand-dependent use of NC temperature compensation (adaptation to changes in workpiece geometry attributable to thermal expansion).

Trends

▶ **Direct screw connection.** The above-described joining processes of clinching and self-pierce riveting require access to the component from both sides. By contrast, the direct screw connection system using flow hole forming screws is a one-side connection technology for metal materials. A pilot hole operation in the screw-in part is no longer required. Direct screw connection is based on a combination of "flow hole forming" and "thread forming". In the screwing-in operation, the tip of the screw penetrates the screw-in part, which has no pilot hole, and a metric thread forms automatically without any metal being removed. The screwing-in operation has a number of phases: the screw is first placed on the part being joined at a speed and axial force appropriate to the material in question. Frictional heat plasticises the material of the part being joined, allowing it to be penetrated by the screw to form an aperture. A thread is made by chipless tapping in the cylindrical aperture. When the screw is then inserted up to the in-built washer, it is tightened at a defined torque.

▶ **Remote laser welding.** Conventional welding tongs are heavy, require access to the component from both sides, and have a fixed welding spot geometry. Remote laser welding, on the other hand, only requires access from one side, so it is suitable for any seam geometry, each with different strength characteristics and load properties. Seams can be placed and designed in such a way that optimal strength is achieved, also in terms of the crash performance of the component ("design-for-crash seam"). A remote laser welding system consists of three components: an industrial *robot*,

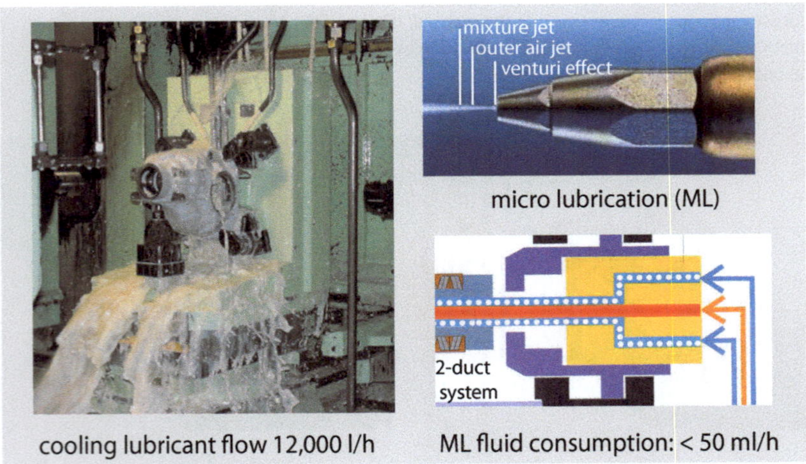

Wet and dry machining. Left: In wet machining, the cooling lubricants handle not only lubrication of the machining process and cooling of the tool and workpiece but also the withdrawal of chips by flooding with large quantities of fluid. Right: In a "2-duct system" the lubricating oil (red) and compressed air (blue, white) are fed separately to the machining tool, where they are atomised. Micro-lubrication uses only very small amounts of oil, chiefly for lubricating the cutting edge; the machining workspace, workpiece clamping fixture, and tools must be designed in such a way that the resulting chips are directly withdrawn to the chip conveyor. Source: Daimler AG

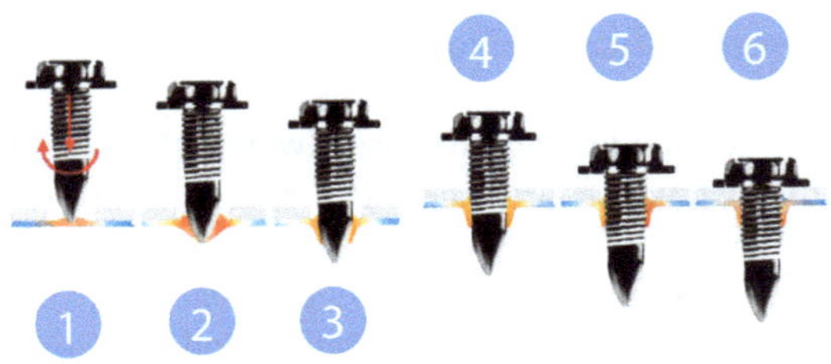

Flow drill screw process: Plasticising a metal panel by axial force at high speed (approx. 2,500 to 5,000 rpm). From left to right: Penetration of the material with a tapered screw tip; forming of the cylindrical aperture and chipless tapping of a metric thread; screwing through and tightening the screw at the correct torque. Source: EJOT GmbH & Co. KG a laser welding head with scanner and laser source, and a jig. The laser welding head with its highly dynamic focusing lens is attached to the head of the industrial robot. The laser beam is guided to the welding head by means of a fibre optic cable. The laser beam is precisely directed at the required weld location by a movable scanner mirror on the welding head. Motion of the robot arm is superimposed with that of the scanner mirror, thus achieving high welding speeds. Positioning times are reduced to a minimum, which permits a virtually 100% utilisation of the laser welding system's capacity. In addition, it is also possible to evaluate the quality of the weld during the process. Seam progress and seam quality are monitored directly during the welding operation. The process is monitored by a unit integrated into the welding head, consisting of an image acquisition and evaluation system.

► **"Best-fit" assembly**. The best-fit process allows sensor-based, gap and transition-oriented assembly. On the one hand, tolerances have to be intentionally permitted in the manufacture of components for reasons of cost efficiency, while on the other hand it is essential to meet increased demands in terms of the dimensional accuracy of the vehicle body, and hence the high standards expected of the skin of the bodies of premium vehicles. Initial applications of the process were implemented in the assembly of add-on parts: side doors, trunk lids, fenders, and engine hoods. Due to new methodology in gap and transition measurement, genuine relational measurements between the component and the production environment are carried out at selected quality-determining positions. For this purpose, the sensor system is actually located on the grab tool and thus performs extremely accurate online measurements at short intervals throughout the assembly process. On conventional automated add-on lines, sensors on stationary gantries only detect the position of the vehicle body at each station, so the robot position is adjusted with less precision. Now, rapid communication between the online measuring system and the robot via a dedicated interface makes it possible to perform direct sensor-guided robot positioning;

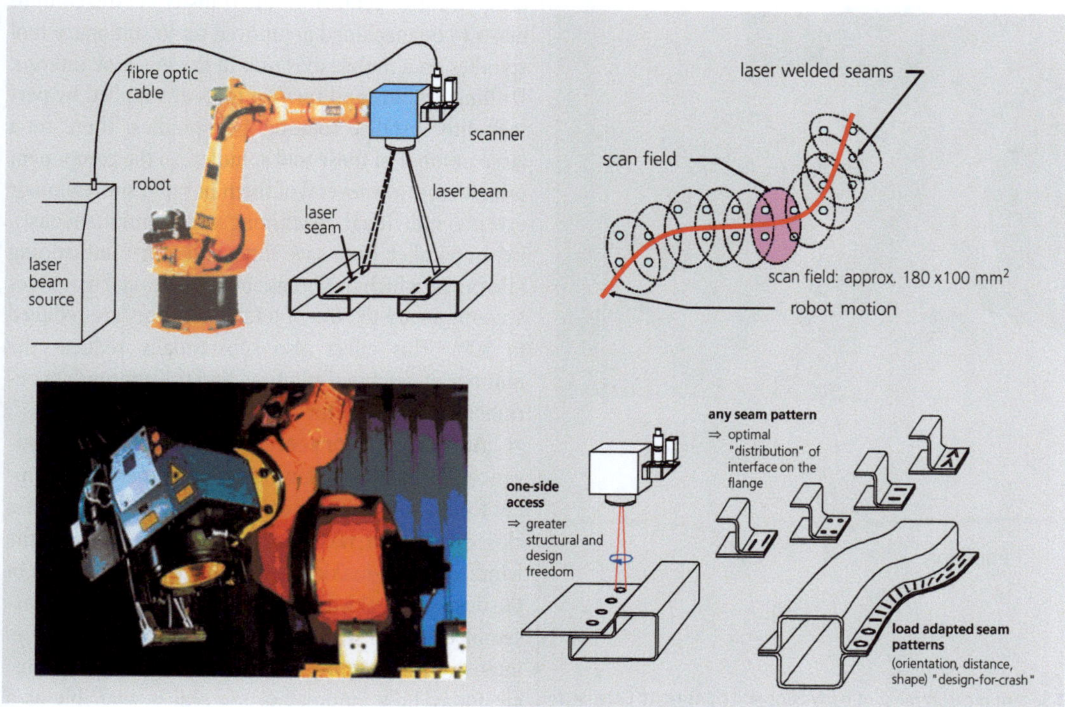

Remote laser welding. Top left: The dynamic scanner mirror makes it possible to accomplish variously positioned welding seams during the motion of the robot. Bottom left: The RobScan machining head for high-speed laser machining combines the speed and position of a laser welding head with the flexibility of a robot and thus allows optimal process speeds and seam qualities with quality control throughout the welding operation (in-process control). Top right: In remote laser welding, the seams are made by superimposing robot and scanner motion. Scanner motion allows deviations of robot motion within a certain radius (scan field) when placing the seam. Bottom right: As opposed to the conventional technique using a fixed welding spot profile, remote laser welding makes it possible to weld seam patterns with different strength and load properties, which, when suitably combined, optimise overall behaviour in a crash. Source: Daimler AG

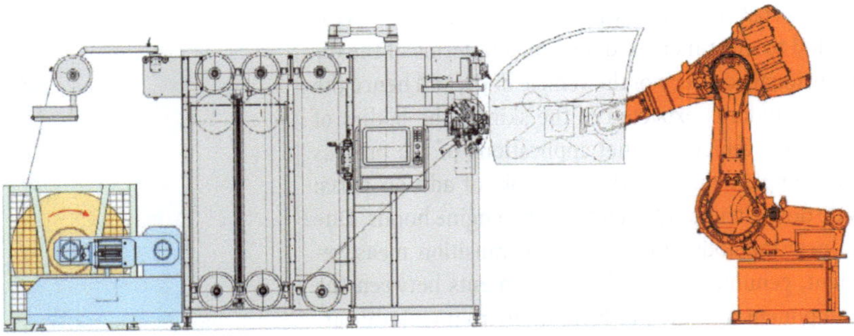

Door seal application: The door seal runs off a reel (left) to a stationary application head via a buffer (centre). This head cleans and applies the door seal to the robot-guided door. Source: Grohmann Engineering

component tolerances can thus be optimally compensated and communicated, allowing add-on parts to be fitted in an optimal position ("best-fit"). Design allowances are exploited in order to compensate for tolerance. For example, a fender can be ideally adjusted to

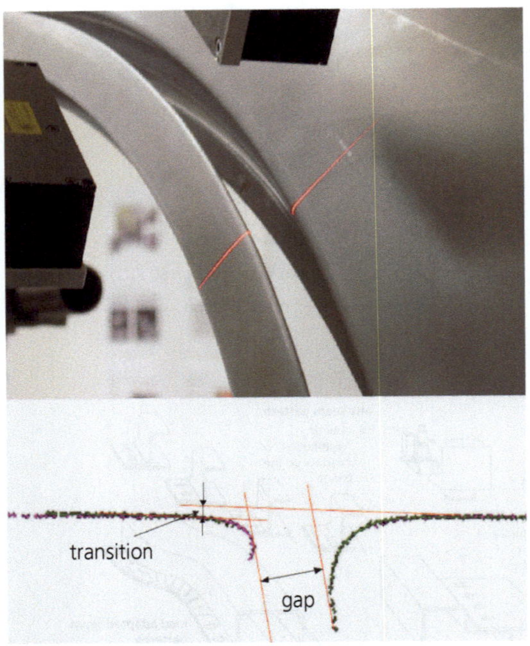

Best-fit measurement of body gaps and transitions: Laser sensors measure the gap and transition between the vehicle door to be positioned (left) and the body (right). In a control loop consisting of measurement and movement of the component with the assembly robot, the optimal installation position is attained with consistently narrow gaps and small transitions. Directly after installation, the result of fitting is checked with the same laser measuring system. This reduces labour in the form of manual readjustment, and the optimal fit lends the vehicle a top-quality appearance and ensures a high-quality body-in-white. Source: Daimler AG

the engine hood because its assembly holes were made large enough for some play, not allowing a positive fit with the fastening screw. Directly after assembly, the sensor system employed is also used to detect the achieved gap and transition dimensions and feed them into quality control loops for monitoring purposes. Measuring systems that until now had to be placed at the end of production lines are no longer required because the assembly process itself includes quality measurement. This directly available test data makes the process transparent. It is a prerequisite for achieving high quality standards, especially in the case of new launches.

► **Transfer centre.** One aspect of modern production systems is that they are becoming increasingly flexible and consequently have a high level of non-productive idle time. This idle time is due to the frequent tool changes necessary in order to match the relevant component machining required with the appropriate tooling. This has led to the development of a machine concept with minimal idle time to boost machining productivity by metal removal on a sustained basis (transfer centre). In the transfer centre, components to be machined are moved up to stationary tool spindles on a mobile axis unit in the machine interior. Drilling, milling and turning tools are applied by permanently installed rotating tool spindles. There are a large number of these tool spindles, so the component can either move several of them in rapid succession or execute individual machining steps simultaneously, i.e. in parallel. Both save time, and non-value-adding idle times, which in the case of conventional machines account for up to 70% of total cycle time, are reduced to 30%. This effect also substantially reduces the number of machines required and the appropriate peripheral equipment for automation.

► **Automatic door seal application.** Enormous advances have been achieved in development of assembly for door seals. This seal once took the form of a closed ring that was fixed in a channel around the window frame, making a positive fit. At the bottom of the door, the seal was attached with elaborately preassembled clip-on strips. These manual and hence time-intensive preassembly and final assembly operations, are today fully automated: the seal is available as a continuous strip on 1,000-m reels. With a stationary application head, the door seal is "unrolled" at a controlled force and speed, during which an industrial robot grabs the door and guides it past the application head. The start of the unrolled door seal is detected by the application head and the end is cut off with precision; the gap between the two seal ends is reliably prevented from becoming too large. In a final step, the two ends at the join are stuck together (sealed) by means of hot-melt adhesive foil. Since the door seal is delivered on a reel, all operations previously performed in order to manufacture special seal lengths for individual door types are no longer required. Cargo volume has been halved in relation to previous deliveries of closed rubber rings. The new method makes it possible to process all door types of current and future model series by simply customising the robot programmes.

▶ **"Assembly 21".** There is little reminiscent of former assembly processes in this new assembly concept, which is based on a holistic approach to human-robot interaction. There are no conveyors or workpiece carriers, and workplace-specific clamping, turning and tilting jigs have been abolished because workpiece positioning is handled by *robots*. There is no longer any dual material supply on the left and right along the conveyor, and the still existing logistical and material supply volumes are clearly separated from the value-adding assembly activities. A complete rear axle is assembled by over 45 cooperating robots (i. e. interlinked systems operating hand in hand). Up to 6 robots work in the robot team and demonstrate the substantial benefit of corporate capabilities by making it possible to machine a workpiece while it is in motion. Workpieces move from one cell to the next on the fly, eliminating non-productive transfer times and raising output rate and added value. Material is always supplied from a central point, the so-called supermarket. Inventories within the assembly line, and hence transit times, have been reduced to a minimum. All the assembly data for each individual axle are stored in a quality database in order to ensure seamless documentation and traceability.

Only at the end is the axle handed over to a turntable for completion, comprising 12 manual workplaces to fit a wide variety of add-on parts.

Prospects

In vehicle manufacture, there are now a large number of metal materials available that are used in their pure forms or as a material mix for lightweight design. High-strength/ultra-high-strength steels and aluminium offer enormous potential in this context, with the associated challenges in terms of production and joining technologies.

- The level of automation on the production lines of powertrain and body plants will continue to increase, in order to make optimal use of non-productive times in particular (such as component transit times) by means of robots.
- Simulation technology, which has already reached a high level of importance, will also continue to improve: product simulation is all about vehicle properties such as durability, crash resistance, and aerodynamics. In process simulation, robust methods of calculation have to be generated for comprehensive processes (e. g. painting and coating processes, material flow).
- One future point of emphasis will be modular vehicle design, the aim being to reduce the complexity and cost of development and production. Modular architecture uses a building-block system. This contains the preferred vehicle solutions in terms of function, cost, weight, and quality, which then represent the basis for utilisation in a number of model series and their successors.

DR. DIETER STEEGMÜLLER
Daimler AG

▲ Strut assembly: The attachment of five struts to a wheel carrier using nut-and-bolt connections, where strut position relative to a wheel spider is precisely defined, is reliably and reproducibly performed by four interlinked robots in a complex combination of robot grabs, robot screwdrivers, and workpieces. The robot team only splits up temporarily in order to get more struts, bolts, or nuts via innovative grab or pick-up concepts. The bolts are joined, the nuts are threaded on, and the bolt assemblies are tightened to the required torque/angles through the wheel carrier holes and strut holes positioned relative to each other. Specially developed cooperative search and troubleshooting algorithms, and even entire repetition strategies, ensure a high level of process stability and availability even with component, grab and positioning tolerances. A complete wheel spider is made in less than 60 seconds and goes for "marriage" with the rear axle carrier on the fly. Source: Daimler AG

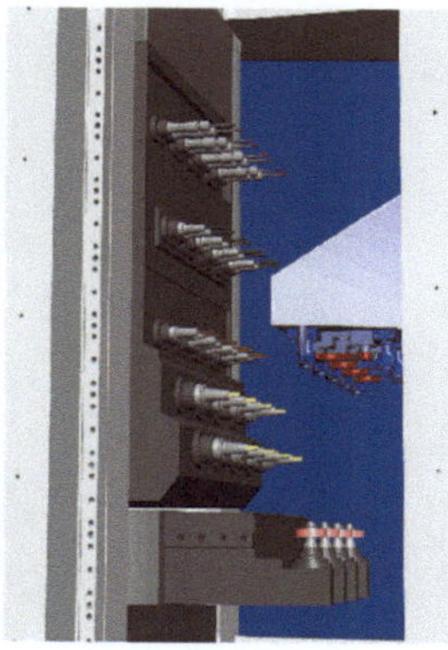

▶ Transfer centre: The components to be machined are clamped to a mobile access unit and moved up to the stationary tool spindles. Due to the large number of spindles in the machine interior (multi-spindle capability), tool positioning and changing times are reduced; in addition, parallel machining processes increase productivity. Source: Daimler AG

Internet

- www.vdw.de
- www.dvs-ev.de
- www.kunststoffe.de
- www.khake.com/page89.html

Process technologies

Related topics

- Plant safety
- Polymers
- Water treatment

Principles

The chemical industry transforms the earth's raw materials into many important basic materials in use every day. More than one quarter of the petroleum pumped worldwide ends up in films, fibres, paints, coatings, flavourings, fertilisers and crop protection agents, as well as body-care products and medicines. Annual production volumes cover a vast range: from 150 g of a haemophilia drug to more than 48 mio. t of the commodity plastic polyethylene. For each new product, the chemical synthesis must be developed in a laboratory, the process created on a small scale and finally brought up to production scale – including the preparation, isolation and often formulation or design of the product. Process technology provides the necessary plant and equipment for each process step. There is a continuous need to modify existing plants or to design new, tailored facilities.

Plant components in a production process can take on impressive dimensions. Stirred vessels for mixing liquids boast volumes of up to 1,000 m^3; some distillation columns are 5 m in diameter and 50 m high. The chemical plant itself is typically only a small part of a larger chemical site.

Flammable liquids, explosive gases and toxic solids are frequently used as raw materials, and the intermediate products can also be hazardous. Because the starting material is often not available in the proper size and shape, in the required concentration or at the desired temperature, processes such as comminution, dissolution, cooling, evaporation, mixing and transport have to be conducted to ensure an optimum reaction. The reaction – the step in which one material is converted into another – should ideally convert all starting materials into the finished product. In reality, however, what we find at the outlet of the reactor is a mixture of starting materials, auxiliaries, products and unwanted byproducts. A broad variety of reactor types can be found in chemical plants. Stirred tank reactors, reactor cascades and tubular reactors represent the basic reactor types, which can be subdivided into different reactor concepts specifically developed for different types of reaction. For example fluidised bed reactors are used to bring solids in contact with a gaseous feed component, bubble columns are used to process gas-liquid reaction mixtures and tubes with internal static mixers are used to redisperse liquid-liquid mixtures.

In addition to processing the reaction mixture, other important tasks include the recovery of auxiliaries, product purification and environmental disposal of waste products. Basic downstream operations such as filtration for solid-liquid separation, extraction for separation of a component from a liquid stream by a second liquid, distillation for separation of low and high boiling components, evaporation for concentration and drying for product formulation are standard process steps and are performed in specialised equipment. Because the requirements are different for each product, these are usually custom-built one-off solutions.

The operation of chemical plants is likewise a complex affair. The process equipment must be started up, monitored and maintained. To ensure safe operation during cleaning or in the event of faults, additional piping, standby tanks and monitoring instruments must be installed. Production plants are largely automated and controlled centrally from control rooms.

▣ A chemical site typically comprises multiple plants: For example the 220-hectare Nanjing production complex brought on stream in 2005 is home to several plants, which together produce 1.7 mio. t of fine chemicals and plastics from petroleum each year. Each plant itself comprises several units of chemical process equipment for reaction and separation. Source: BASF

Applications

► **Catalysts.** In process technology the development of new catalysts generally represents the first step towards a new process. Between 70–80 % of all chemical products are produced with the help of catalysts, i.e.

compounds which accelerate a reaction without being changed themselves.

Some reactions, such as the production of gasoline from biological waste, could not occur without the use of catalysing substances. Although a large number of materials are available for catalysis, suitable varieties have yet to be found for numerous thermodynamic reactions. *Nanotechnology* is opening up new possibilities for the development of catalysts with specific properties, the reactivity of supported catalysts is, for example, often influenced by the specific surface of the catalytically active metal cluster deposited on the inner surface of the supporting material. To increase reactivity, the size of these metal clusters is reduced.

The targeted production of such tiny particles presents a number of challenges for process technology, however. Processes that allow such particles to be freely tailor-made have now been developed.

Today a catalyst's mode of action during a reaction can be monitored at the molecular level. Chemists can now discover which parts of the catalyst surface are actually active during a reaction and which are not, how the reaction products move on the surface of the catalyst and when undesired byproducts arise. Copper zinc oxide catalysts, for example, are used in the industrial synthesis of methane. This suitable catalyst was identified by chance – the zinc was added to prevent the oxidation of the copper catalyst. Contrary to prior assumptions, we now know that it is not just the copper that is catalytically active; the tiny zinc oxide particles that migrate to the copper particles also significantly increase the activity of the catalytic centre.

Mesoporous materials have a strictly ordered matrix with pores ranging between 2–20 nm in size. The matrix frequently comprises amorphous silicon dioxide, but matrices of carbon-bearing components have also been produced. Like zeolites, the chemically complex minerals frequently used to date, mesoporous materials serve to immobilise the catalysts. Compared with the zeolites and other materials, however, they have an even larger inner surface and are also extremely versatile. The walls of the matrix and their composition can vary greatly: they can be amorphous or crystalline; they can be composed of inorganic materials or a mixture of organic and inorganic materials. If necessary, the walls of the matrix can be produced in such a way as to be hydrophobic. This prevents the catalyst from dissolving in water; the material remains filterable and can be separated off after the reaction. The tiny pores offer myriad possibilities for influencing reactions. The size and shape of the pores define the inner surface and thus control the selectivity of the reaction, for example determining the size of molecules in a mixture of multiple starting materials which will be converted to the products. Thus the pore structure geometry of the catalyst plays a major role in the processing of petroleum, for example. If the pores are too large, the primary reaction is accompanied by the undesired conversion of substances that can damage the catalyst. If the pore openings are too small, however, it is difficult for the educts and products to reach or exit the catalytically active centres inside the pore system, resulting in low utilisation of the catalyst.

Catalysis is making increasing use of nanoparticular materials such as silicon carbide and carbon nanotubes. These tiny particles possess an array of extraordinary properties (specific surface area, electrical conductivity, thermal conductivity, oxidation stability, etc.), making them candidates for use as substrates for catalytically active materials.

Another peculiarity of nanoparticles is that they often possess entirely different properties from larger particles of the same substance. They therefore often accelerate reactions more strongly than larger particles of the same material. The noble metal gold, for example, is normally a material with very low reactivity. In the form of particles less than 8 nm in size, however, it

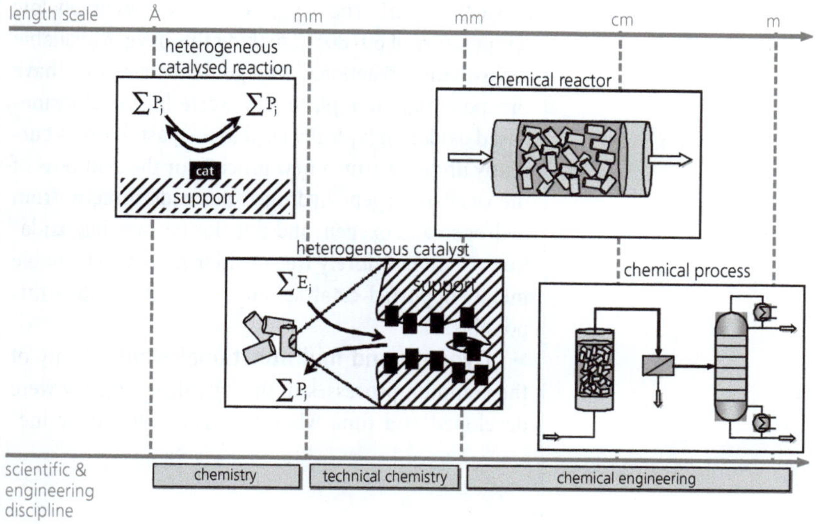

From catalyst to process: The development of heterogeneous catalysts is one of the major goals in the design of new and more efficient processes. Once a catalyst is developed by chemists and its reaction kinetics is known, the basic mechanism has to be implemented in a chemical reactor using inert highly porous material. The activated supporting particles (e.g. cylinders, spheres) are installed in the chemical reactor as a fixed bed offering the highest density of particles per volume. The reactor itself is connected to other downstream modules such as membrane separators or distillation columns by pipes in the plant. This is where the chemical engineer gets involved.

Catalysts accelerate chemical reactions through specific affinities to the reactants (squares and triangles) without being changed themselves. The result of such strong affinities is a highly selective transformation of reactants into the desired products.

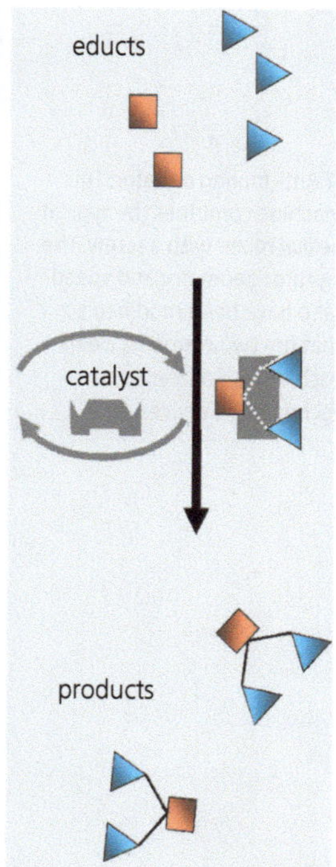

can activate airborne oxygen even at the relatively low temperature of 60–80 °C, making this oxygen available for oxidation reactions. Fine gold particles thus have the potential to replace the waste-laden, chlorine-based oxidation typically used in the past. Work is currently under way on a new process for the synthesis of the oxidising agent hydrogen peroxide directly from hydrogen and oxygen, and thus for performing oxidation reactions entirely free of chlorine. NxCAT™ noble metal supported catalysts are expected to make this possible.

▶ **Dynamic and multifunctional plants.** Many of the common processes in the chemical industry were developed at a time when the scope offered by mechanical engineering was limited and the range of engineering materials available for machine construction was restricted. Frequently the only process solution was to use solvents which enabled complex material systems to be processed in simple, robust equipment. Chemical process equipment to a large extent comprises simple, static plant units with no moving parts, such as distillation columns. Machines, on the other hand, are complex and dynamic plant components.

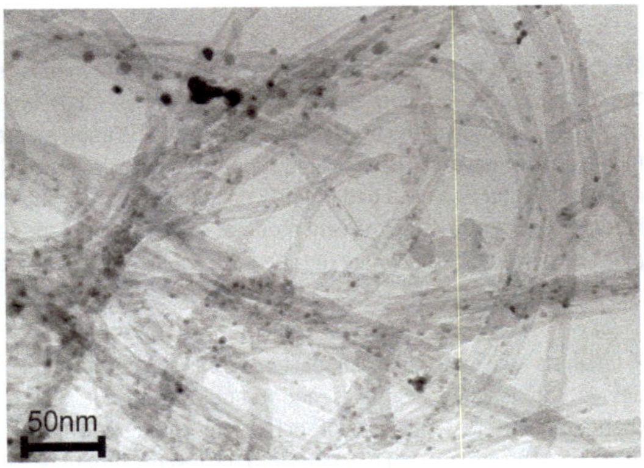

TEM images of noble metals (dark-gray dots) on carbon nanotubes (light-gray tubes): The noble metal particles only a few nanometres in size on the carbon nanotubes catalyse oxidation reactions. Source: Bayer Technology Services

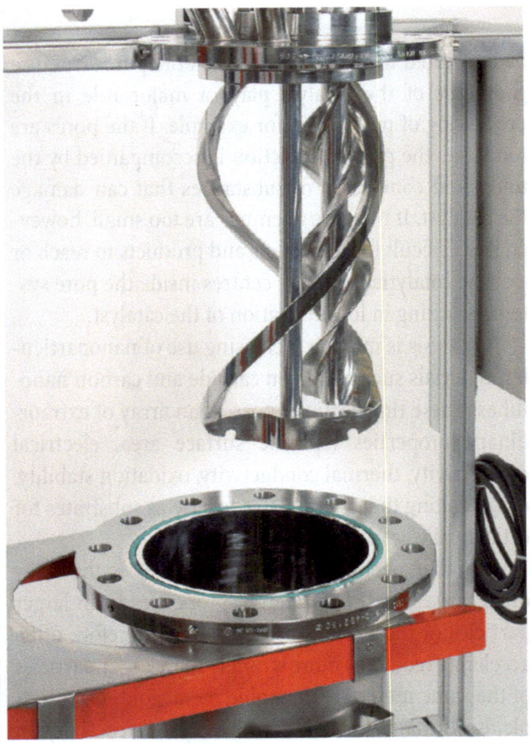

Anti-fouling agitator: This machine combines the typical helical mixer with a screw. The agitator geometry and speed ratio have been modified so that the two elements clean each other. Source: Bayer Technology Services

Process intensification is currently focused on transferring complex reactions from plant simple units to high-performance machines in which difficult process conditions can be easily managed. A self-cleaning stirred tank, for example, is suitable for mixing two viscous liquids. This machine resembles a food processor, but comprises two shafts whose surfaces are shaped so that each removes the viscous reaction mass from the other. Laboratory tests show that these machines mix more effectively than a standard helical mixer and can shorten mixing times by a factor of 20.

Another self-cleaning machine is the high-viscosity reactor, in which the individual components for plastics are quickly mixed, react to form long chains, are cooled and degassed. Two moving shafts inside the reactor scrape the viscous reaction product from all of the surfaces, ensuring effective mixing.

A unidirectional, twin-screw extruder has long played an important role in the production of plastics such as polycarbonates, which are used in the manufacture of CDs and toys, for example. In recent years, such double-shafted extruders have been greatly improved by the use of new materials to the point that various processes such as granule sintering, additive mixing, reaction and degassing can take place within 20 seconds.

▶ **Multifunctional reactors.** Also of interest are multifunctional reactors, such as membrane reactors or reactive rectification columns. The basic idea of such reactor types is to separate the products from the reactants inside the reactor simultaneously. This method can be used to increase the yield of equilibrium-limited reactions, for example. The limiting reaction products are removed directly where the reaction takes place. Significantly greater yields can thus be achieved in a single distillation column with integrated reaction than with a conventional process in which the reaction and multiple distillation steps occur sequentially. This is particularly true for esterification and etherification processes. Furthermore, it is often possible to use the heat of reaction released for distillation, thus reducing the energy requirements of the process. Such integration concepts can significantly reduce operating costs.

Trends

▸ **Ionic liquids.** The molecules of two substances must be brought as close as possible to the catalyst and be in close contact with each other to enable catalysed reactions between the substances to occur. This is easiest to do when all the substances are available in liquid form. Solvents are frequently required to convert all reaction participants to the liquid phase. In many reactions, in particular in the production of plastics, high-viscosity substances occur as an intermediate stage and are thinned with solvents to make them easier to handle. In chemical synthesis, the quantity of solvent is often ten times greater than the product produced. The solvent cycle is complex and energy-intensive. First the solvent is mixed with the starting materials. After the reaction, it is separated from the resulting product and purified. Finally, the product has to be dried.

One way of achieving process intensification is to replace the conventional organic solvents with ionic liquids. These are salts which are liquid at temperatures below 100 °C. Common salts, such as table salt, are made up of two small ions – a positively charged sodium ion and a negatively charged chlorine ion – which combine to form a solid crystal matrix because of their charge. The melting point of table salt is 800 °C. With ionic liquids, however, at least one of the two ions is so bulky that it is difficult to arrange in a crystal matrix, hence the low melting point.

Advantages of ionic liquids:

- Many substances can be easily dissolved in ionic liquids. This facilitates chemical reactions.
- They are versatile: whereas only approximately 600 different organic solvents are used in the chemical industry, millions of ionic fluids can be produced by combining various organic ions.
- They can be adapted optimally to a reaction. They can dramatically increase the rate and yield of the reaction.
- The reaction product can often be separated very easily from the ionic liquid, for example if the two are not miscible and separate spontaneously like oil and water. This is often the case owing to the high polarity of the ionic liquids. Complex distillations such as those generally required for the removal of organic solvents are then eliminated.
- Ionic liquids have no vapour pressure. Unlike organic solvents, they do not emit any harmful vapors. They can be toxic or highly corrosive, however.

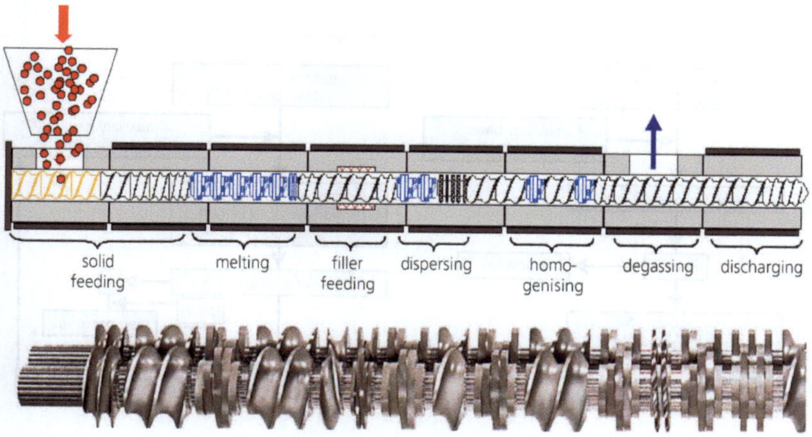

A unidirectional, twin-screw extruder consists of different sections: The polymer is fed in, plasticised, mixed with additives, separated from traces of monomers and discharged at the extruder outlet.

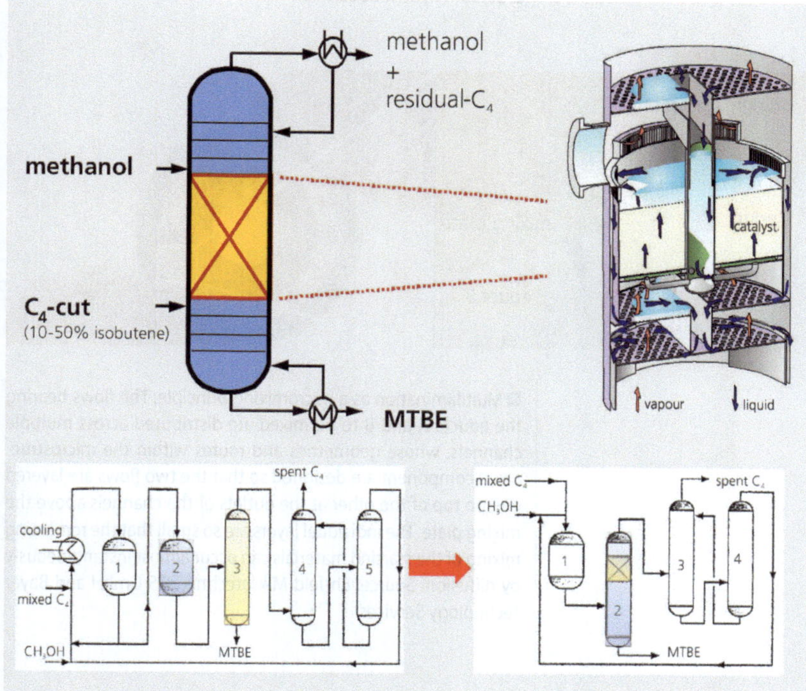

Reactive distillation. Top left: Methanol and C4 (e.g. isobutene) are converted in the catalytic packing (yellow part of column). The reaction takes place on the catalyst surface in the fixed bed of catalyst particles. The product MTBE is separated from the reactants in the lower (blue) part of the column, which comprises different stages of separated sieve plates. Top right: On the sieve plates the liquid is in contact with the gas phase generated by evaporation of the liquid in the bottom of the column which flows through the holes of the sieve to the upper stage. The liquid flows to the lower stage through downcomers. This countercurrent flow of liquid and gas phase results in the separation of components at every stage. Bottom: This type of reactor incorporates the functions of two separate reactors (reaction and separation). Source: J.L. Nocca, J.A. Chodorge

The production of alkoxyphenylphosphines, which play an important role in the manufacture of UV-resistant coatings, was increased by a factor of 10,000 using ionic liquids. In the original process an auxiliary base was used to remove the hydrochloric acid produced during the reaction from the reaction mixture.

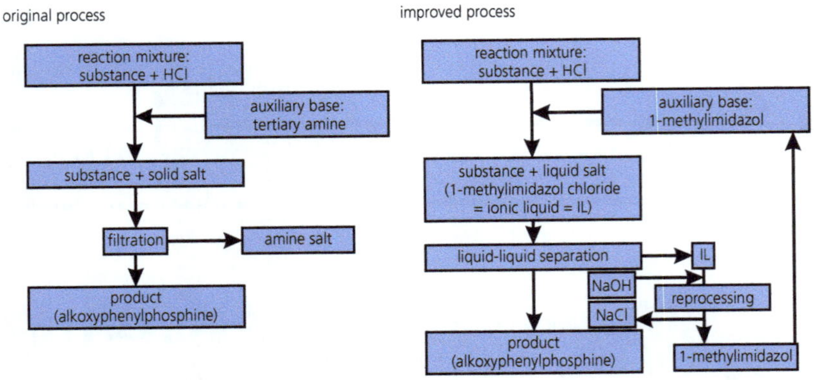

BASIL process: Using ionic liquids BASF achieved a 10,000-fold increase in the production productivity of certain chemicals. The ionic liquid 1-methylimidazol chloride is formed by the reaction of the base 1-methylimidazol with HCl. To recover the 1-methylimidazol, the ionic liquid can be reacted with a strong base such as sodium hydroxide solution.

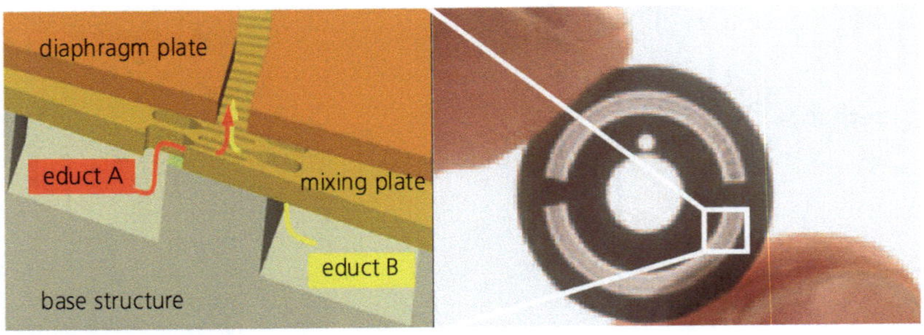

Multilamination as a micromixing principle: The flows bearing the educts A and B to be mixed are distributed across multiple channels, whose geometries and routes within the microstructured component are designed so that the two flows are layered one on top of the other at the outlets of the channels above the mixing plate. The individual layers are so small that the remaining mixing of the starting materials can occur almost instantaneously by diffusion. Source: Ehrfeld Mikrotechnik BTS GmbH and Bayer Technology Services

The solid salt produced by the acid/base reaction then had to be removed in a complex filtering process. In the improved process, 1-methylimidazol is used as the auxiliary base. 1-methylimidazol chloride, the salt arising from this auxiliary base used as an acid scavenger, is an ionic liquid that can be removed by simple liquid-liquid separation. The base also serves as a catalyst and is recovered at the end of the process.

▸ **Microreactors.** While microelectronic and micromechanical components have been around for decades, process engineering did not venture into the realm of miniaturisation until the mid-1990s. Microplant modules suitable for making products in quantities ranging from a few kilogrammes to a few tonnes are the size of a thimble or at most a pack of cigarettes. The structures inside the modules, such as the slots in a mixer, measure only a few micrometres. As in the nanocosmos, the ratio of surface area to volume plays an important role in the microworld. Here, however, it has less to do with the reactivity of individual molecules than with physical processes, in particular the exchange of heat and the mixing of matter.

With its modified physics, microtechnology offers numerous advantages:

- Processes such as mixing, heating and cooling are significantly faster than in conventional plants thanks to the favourable surface-area-to-volume ratio in micromodules.
- New, particularly effective catalysts, which for reasons of safety cannot be used in conventional plants, make microplants especially attractive.
- Process parameters such as pressure, temperature and mixing ratios can be much more precisely defined than in a large tank. The yield can be increased if reaction conditions are precisely defined. Waste volumes decrease.
- Reactions are safer. Thanks to the rapid heat transfer properties, pressure and temperature cannot move into ranges where the reaction gets out of control. In particular, highly exothermal reactions which are too dangerous for conventional plants can be controlled in microreactors with little safety effort. What's more, even in the event of a fault the potential hazard represented by the maximum escape of materials is low given the small volume of the reactor.
- Microtechnology also facilitates screening to find the optimum settings for a process: a reaction can be quickly repeated with different pressures, temperatures and mixing ratios. The time-to-market of a product can be shortened.

Two strategies are applied in the design of microplants. Some suppliers develop complete microplants tailored for different reactions; others supply different modules for the individual reaction steps that can be assembled like building blocks to form a complete plant. For example, various mixers, reactors and heat exchangers exist which are suitable for a wide range of reactions, including those in which solid particles are produced or processed. An array of different modules is already available for the process steps of mixing and heat exchange as well as for various types of reactions.

Microplants are currently used primarily in laboratories, mostly as tools for the development of new products or processes. Only in a few isolated cases can micromodules be found in production plants. Precursors for the production of liquid crystals for example are made in a microreactor.

▶ **Modular plant concepts.** Microtechnology offers the opportunity to develop a new plant philosophy. This point is illustrated by the traditional scale-up problem in the development of new chemicals. Scale-up involves transferring a process that functions on a laboratory scale to a pilot plant and then taking it up to the technical service lab scale. A large plant subsequently has to be built for production. This whole process is therefore very time-consuming.

Microtechnololgy offers the possibility of significantly reducing the time between the discovery of an active substance and its introduction on the market. The transition from laboratory to pilot scale can, for example, be avoided by simply setting up a large number of the same microplants in parallel (numbering up). The production volume can be easily increased in this way. This is particularly interesting for pharmaceutical active substances, where the continued development of the process technology virtually ceases during clinical trials.

But numbering up is not enough to produce several thousand tonnes of a substance per year; the modules have to be enlarged to contain more microstructures (equalling up). Currently there are only a handful of these enlarged modules which can convert quantities of up to one thousand litres per hour.

Propylene is one example of how microreaction technology can be used in production. The hydrocarbon propene is oxidised in a microreactor with the help of hydrogen peroxide vapour to form propylene oxide, a key raw material for the production of polyurethane. The solid catalyst titanium silicate is applied to the walls of the microreactor. The reaction with hydrogen peroxide gas, which is not possible in conventional reactors, obviates the need for solvents.

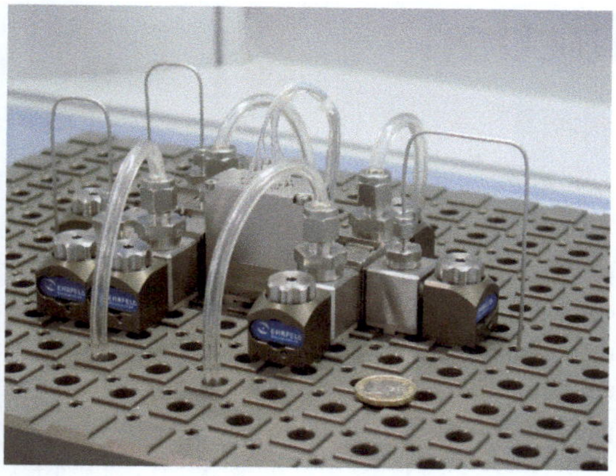

A modular microtechnology plant: Microtechnology plants generally comprise individual modules (mixer, heat exchanger) into which the microstructures are integrated. Besides equipment such as mixers, reactors and heat exchangers, modules for measuring temperature, pressure, volumetric flow and concentration can also be integrated on the base plate. The individual modules are supplied with specific volumes of heat exchange medium via external hose or tube connections for temperature control (heating, cooling). Source: Ehrfeld Mikrotechnik BTS GmbH

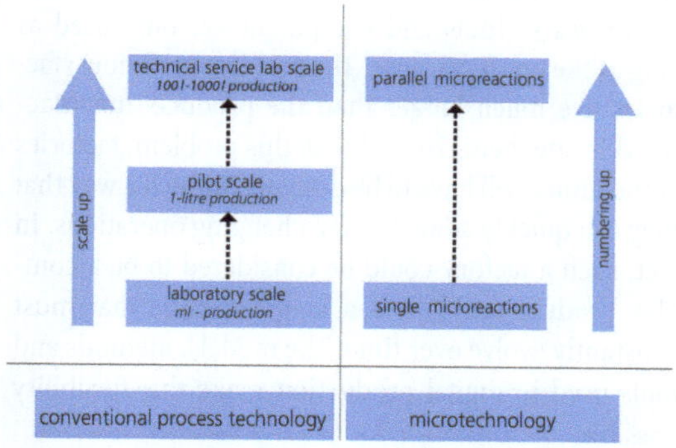

Microreaction technology has the potential to significantly shorten the time-to-market of an active substance. The transition from laboratory to pilot scale can be avoided by simply setting up a large number of the same microplants in parallel.

Prospects

- Superior processes cannot be realised by means of integrated, general solutions unless full advantage is taken of new reaction agents such as ionic liquids in microstructured equipment offering optimum energy and material properties. Dynamic machines open up new possibilities for process intensification, in particular for more viscous materials.
- Instead of conventional chemical plants, the future could bring intelligent, decentralised factories. The prospects for the European chemical industry in global competition lie in these future system solutions built around intensified process technologies.

DR. HELMUT MOTHES
DR. MARCUS GRÜNEWALD
Bayer Technology Services GmbH, Leverkusen

Internet

- www.rsc.org
- www.bayertechnology.com/en/home.html
- www.achema.de/en/ACHEMA.html
- www.basf.com/group/innovations/index
- www.chemguide.co.uk/physical/basicrates/catalyst.html

Digital production

Related topics

- Virtual and augmented reality
- Logistics
- Human-computer cooperation
- Information and knowledge management

▶ **The factory as product.** The mobile phone is a short-lived consumer good. New models appear on the market every six to twelve months, and the necessary adaptations then have to be made in factories to keep pace: adjustments must be made to machines, new plants must be acquired, and the required amounts of space and personnel may change as well. Even the automobile industry, which has traditionally produced more long-lasting products, now has to adapt: the life span of a new generation of cars may be as short as four years. Both major and minor product reassessments may arise just as often, and as the product changes, the related production means must change as well.

Unlike mobile phones or automobiles, buildings and infrastructures may have service lives of over 30 years. The lifetime of machinery is normally between 10–15 years. Tools and equipment are only used as long as the products "live". The problem is obvious: factories live much longer than the products manufactured inside them. To deal with this problem, factories of the future will have to be designed in such a way that they can quickly adapt to ever-changing operations. In fact, such a factory could be considered to be a complex product itself, with a long life cycle that must constantly evolve over time. The models, methods and tools used in digital production make this flexibility possible.

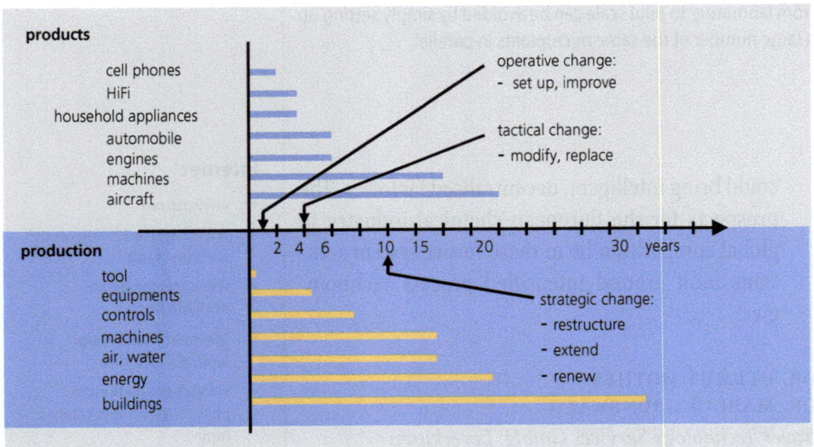

The factory as product: Different lifetimes of products and production. Products can sometimes have very short life cycles, and factories have to be adapted to the ever-changing nature of products. Machines must be quickly set up (operational), production lines must be reorganised over the medium term (tactical) and production infrastructures may be changed over the long term (strategic). Source: Fraunhofer IPA

▶ **Aspects of digital production.** The goal of digital production is to ensure that the planning, ramp-up of a new product for production, and the required adaptations in a running production system are implemented more rapidly, efficiently and economically. These adaptations include: scheduling of machines, allocation of labour resources to production processes, and the dimensioning of lot sizes or assigned work spaces. Digital production is a digital representation of the complete production process. It includes three main components: the digital factory, the virtual factory and the corresponding data management.

▶ **The digital and virtual factory.** The digital factory is a static representation of the current factory in which all the real and relevant aspects are modelled. It models the products, the buildings and infrastructure, the machines and equipment, the manufacturing processes, and human resources. Digital factory modelling takes into consideration feedback gained from the former factory set-up. This feedback data is valuable for determining the proper reaction to unexpected situations, such as how to quickly react in the event of machine failure based on similar cases in the past.

The projection of the factory into the future, the so-called virtual factory, uses simulation methods and tools. Compared with the digital factory, which is always "static", the virtual factory represents "dynamic" states over a defined period by using the time factor. The simulation predicts the factory behaviour and manufacturing processes over time (material flows, ergonomics, workplace, layout, etc.) using input parameters such as the configured production orders. The results show different future states of the factory "as it may be", based on key performance indicators: the cycle times, workload and the capacity of resources.

The simulation results can be visualised using virtual reality and augmented reality. This state-of-the-art technology is a valuable tool for decision makers from all different areas of expertise to continuously adapt and optimise the current factory based on the simulation results.

▶ **Data management system.** The main component of the digital and virtual factory is the product and factory data management system, which stores all the necessary information used by the digital manufacturing tools and applications. Data from autonomous and heterogeneous applications are integrated and exchanged. There are several main applications: investment planning, site and building planning, infrastructure and media planning, process and material flow planning, *logistics* planning, layout and workplace planning, tool management, and human resources management. The management system has to control the data distribution so that planners from different areas can access the updated or approved data and display their planning results transparently. The different planning object views (production area, machinery, tools, transportation equipment, workers, etc.) and the corresponding features and characteristics must be coordinated. In this way, a machine can be viewed from a factory layout perspective (how much area is required?) as well as from a logistics perspective (what is the highest capacity of a work station?).

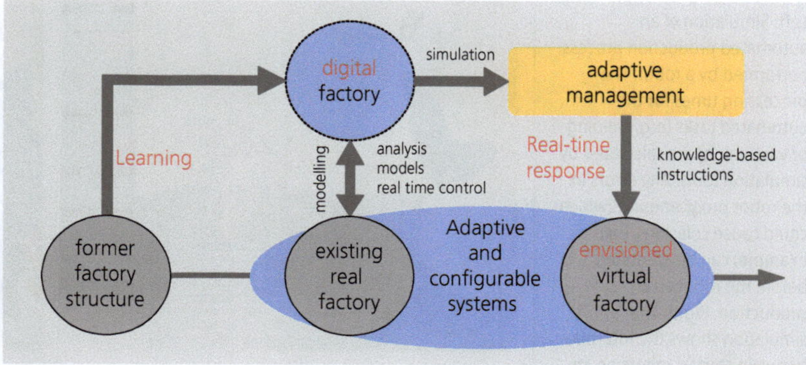

The digital factory includes the history of the factory (past experiences) and its current state (as it is today). The real factory and manufactured products are modelled using tools such as CAD systems and object-oriented modelling systems. The core component of the entire approach is a central database that manages all relevant manufacturing data (product, processes and resources) to help planners to continuously adapt the current state of the factory. Special virtual factory simulation tools are used. These tools calculate how the current factory conditions (cycle times, workload resources) can be projected into the future. The virtual factory data can be exchanged with the digital factory tools to ensure permanent adaptation via a common database. Source: IFF, Stuttgart University

Applications

▸ **Digital product.** The core application of digital production is production and process planning, acting as a go-between for product development and the actual manufacturing of products. A 3D model of a product (digital product) results in an enhanced representation, and serves as the basis for many different planning tasks within digital production. An example of an intermediate phase could be an employee involved in *logistics* planning, who can decide very quickly if the chosen container is appropriate for the intended purpose. The 3D product data model can also be used as the basis for planning the required welding spots. A blown-up drawing can also be created based on a 3D data model, to ensure transparency for all the specialists involved in planning activities.

All production planning in the digital factory is based on the digital product. Specialists involved in planning activities may come from several different areas of expertise, including not only engineers working in industrial and process planning (industrial engineers), but also managers. All these specialists must use the same data, but they work with different applications for process design, work time analysis, line balancing and factory planning. It is therefore extremely important to provide them with a centralised, consistent data management system.

New products undergo a number of changes before they go into series production. If, for example, the size or quantity of a screw is changed or the location of a drilling hole is modified, logistics experts need the new product information to adapt the material flows.

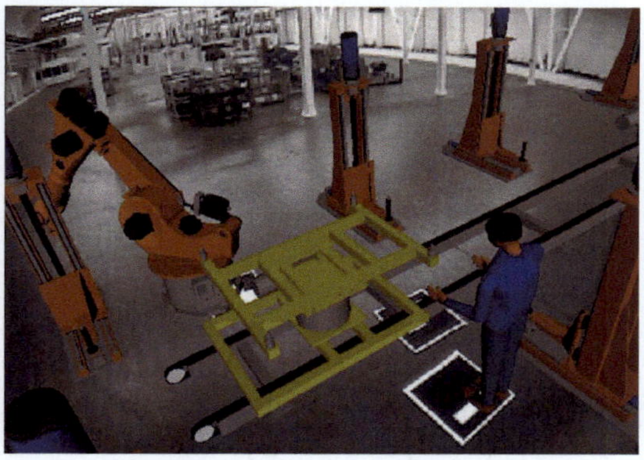

Virtual reality (VR) and augmented reality (AR): The picture shows a realistic representation of a production line in a virtual reality environment. Using this technology, an interdisciplinary team involved in many different levels of factory planning can analyse the structure and processes of the factory, such as the positioning of machines in the layout or the handling activities performed by robots on automated assembly lines. All these activities can be performed in a digital, virtual environment before the factory and its processes are developed and the factory begins operations. In this way, situations such as collisions between robots can be analysed, and processes such as operational flows for manual activities can be optimised in the early phases of factory and process planning. Source: Delmia GmbH

Managers need this information for cost calculations, and resource planners to adapt equipment assembly. The 3D representation of the product, production area

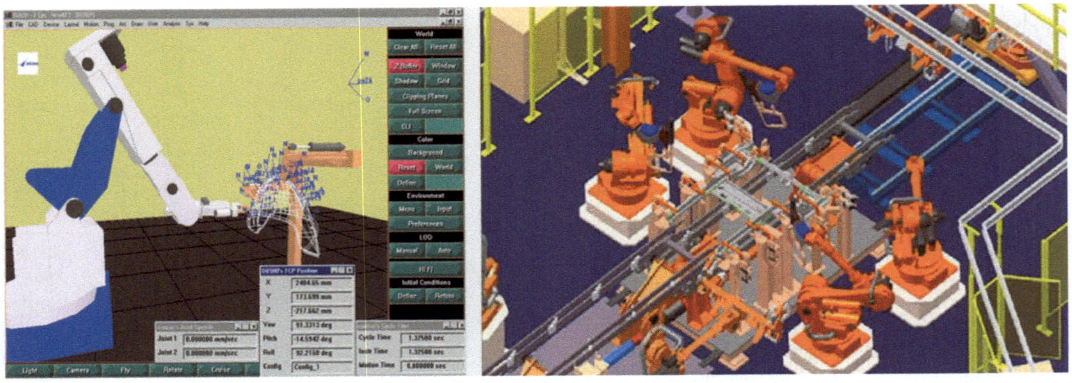

Production process planning. Left: Simulation of an automated production process performed by a robot. The processing times for the automated tasks (e. g. welding or varnishing) are calculated by simulation tools. Any errors in the robot programming (which could cause collisions, for example) can be identified before the robots start production. Right: The simulation shows the interplay between several robots on an automated assembly line. The correct positioning of the robots and any possible collisions can be identified. Source: Delmia GmbH

and processes makes communication between these professionals easier and helps prevent planning errors.

▶ **Production process planning.** The product design proposals for the entire system provide a representation of the work input data and indicate to planners how the manufacturing processes should be carried out. The work sequence must be a set procedure – and the same is true for the production machinery and plants, and the duration of the respective work processes. In digital production this planning is calculated by a computer.

Not only automated activities but also manual work performed by humans can be simulated. To analyse the ergonomic design of a work space, digital human models are developed. These on-going studies are mainly concerned with the effective ranges of the work space. The required data are provided by object libraries. The effective ranges help determine which production means (tools, parts) are easy or difficult to reach in a work space. Analyses can also be performed of field of view or handling capabilities. Determining the optimal position of production means and work tools ensures long-term safety, security and other ergonomic aspects for workers on the shop floor.

Even more aspects have to be considered to balance production lines for manual work stations; the required tools, equipment and space must be available at the work station. In addition, material flows between production stations should run in the same direction. To ensure planning for all these activities is optimised from the outset, close cooperation is required between assembly process planners and factory planners. The factory planner is responsible for positioning the machinery and equipment in the production area, for establishing the infrastructure and media, for determining access points and traffic flows, and for planning the preparation and shipping area – in other words, planning for the entire factory layout.

Trends

▶ **Real-time capability.** This is the timely transfer of data from the real factory shop floor to the digital factory so that the real-time state of the factory can be represented in the digital model. In the future, it will be possible to quickly input actual data from production and order management into distributed simulation models that calculate indicators to continuously adapt the manufacturing processes.

The "as-is" status and the simulation forecasts have to be accessible, with displays of different data levels (the entire production, production segments, individual production units and individual machines).

Based on this real-time capability, decision makers at different management levels in the factory (overall managers, technical managers, workers) can navigate between and display many different views of the factory and its manufacturing processes, as well as key indicators for decision-making. The systems which provide such functionalities are called *"factory cockpits"*.

The challenges for developing factory cockpits are many: integrating factory and logistics planning, determining capacity and occupancy schedules, developing factory controls, monitoring processes, and real-time updating of the actual factory with simulation results. In the future, a real-time factory cockpit will help the user identify trouble spots and assist in taking the appropriate measures.

▶ **Multi-scale simulation.** Currently, the applied simulation technology concentrates on particular structural levels: material logistics inside and outside the factory, simulation of machinery and robots, and the simulation of individual technological processes. The simulation technology solutions of the future will be used to construct complete factory models encompassing all structural levels by taking into consideration space and time dimensions as well. This simulation technology will provide engineers with a very flexible approach that includes each structural level and point of view from the factory, production areas, machinery and technical processes.

The term "multi-scalability" refers to both space and time within individual technological processes and the different scales in an operating factory, as well as in the model design itself. Currently, the technical and organisational processes in a factory are simulated for each scale – from the most detailed to the most general – in a stand-alone, heterogeneous set-up. The main goal is then to find a means of integrating these different simulation models which have different scales. The differences may include the migration of digital process simulation models and material models to discrete simulation for logistics simulation. The results of manufacturing process simulations must then be taken into consideration for logistical simulations, so that the duration of processes can be more precisely calculated and used to establish more realistic simulations of the throughput time for the factory.

The work becomes quite complex when modelling must be performed for horizontal scales on the entire factory, beginning with the technical manufacturing processes and including equipment, robotics, production systems, segments and networks. The following aspects must be considered:

- All levels of factory structures: Manufacturing processes, machinery, systems and networks
- Different disciplines and concepts: Mathematics, physics, chemistry, engineering, economic sciences
- Different simulation methods: Molecular dynamics, finite elements, event-oriented simulation
- Spatial expansion: Atomic, microscopic, mesoscopic, macroscopic
- Temporal expansion: Nanoseconds, seconds, hours, days, weeks, years

The following example shows how the concepts listed above can be applied. When simulating the factory at the lowest scale (manufacturing processes such as coating processes in the automotive industry), concepts from mathematics, physics and chemistry are used. Simulation methods for molecular dynamics and finite elements in the coating process are also employed. The duration of these processes is measured in nanoseconds and seconds, and also draws on atomic and microscopic spatial dimensions. For logistical planning, which is the next scale, engineering and mathematical concepts are used to simulate the flow of material on the shop floor through event-oriented methods at the mesoscopic level with durations in hours, days and weeks. The highest scale for the entire factory establishes plans for production sites and production areas using economic methods (e.g. predictions and trends, key performance indicators) and event-oriented simulation at the macroscopic level with the duration in years.

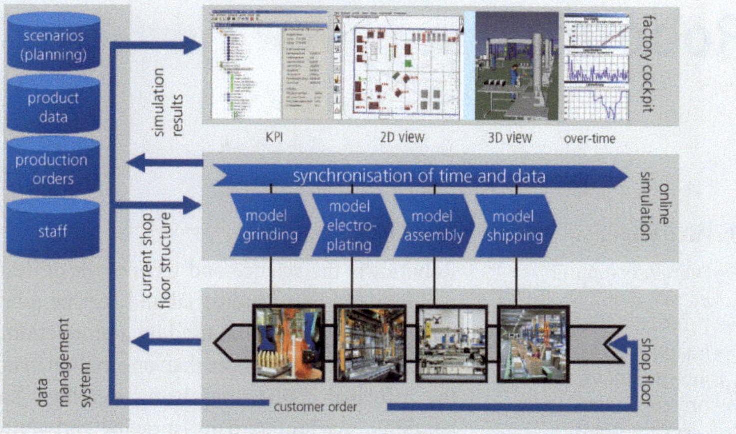

Factory cockpit: This instrument is used to monitor the factory's current operating conditions, but also to plan tasks. The view is based on an actual digital factory model. Each actual resource has an equivalent in the digital factory cockpit, indicating the current conditions and using "traffic light" control logic. The core component is the data management system (left), which maintains all relevant information on product data, production orders, staff and scenario planning. The operational data coming from the shop floor is continuously updated in the common data management system. The simulation environment (central rectangle) can access this data from the shop floor by accessing the database. It predicts possible states of the factory at several different moments in the future. These results are transferred back from the simulation to the database. The factory cockpit displays the actual data from the shop floor (bottom) and the predicted data from the simulation (key performance indicator). Source: IFF, Stuttgart University

Prospects

- Digital production integrates state-of-the-art technological developments such as: NC programming, PPC and CAx systems, virtual production design and development, virtual reality, and simulation.
- Digital production is a comprehensive and integrated representation on a computer in digital format of the entire factory and its production activities, components and systems. It includes the digital factory, the virtual factory and data management.
- Digital production improves efficiency, effectiveness, safety and reproducibility of planning processes. Data collected or produced in a planning phase can be transferred to other systems and be reused.
- Digital production integrates product design, production process planning, and planning for machinery, equipment and tools. Displays and simulations help pinpoint any mistakes that may already exist in the planning phases.

PROF. DR. ENGELBERT WESTKÄMPER
Fraunhofer-Institut for Manufacturing Engineering and Automation IPA, Stuttgart

Internet

- www.delmia.de
- www.ugs.com/products/tecnomatix
- www.ipa.fraunhofer.de

Robotics

Related topics

- Human-computer cooperation
- Artificial intelligence
- Image analysis and interpretation
- Ambient intelligence
- Sensor systems
- Logistics
- Casting and metal forming
- Disaster response

Principles

Robotics is the science and technology of designing, building and using industrial and service robots. Industrial robots were first used on a production line in 1962. Today, they are more than one million in number, the principal applications being welding, assembly, handling, machining, paint-spraying and gluing. The 1980s witnessed the first uses of robots outside the area of industrial production, particularly in applications in which the respective task could not be executed manually because it was unacceptable or too dangerous for a human to perform or because the required accuracy and/or force could not be achieved by manual means. Robots of this type are known as service robots. Nowadays, service robots are at work cleaning buildings and roads or monitoring public areas, such as museums. Ever more robots are being used to carry out hazardous maintenance and inspection operations in industry, local authorities and the energy sector. At the personal and domestic level, robots are increasingly finding applications as robot vacuum cleaners, automated lawnmowers and toy robots. These personal robots may be early examples of future systems which will do useful jobs and assist humans in everyday environments.

► **Kinematics of industrial robots.** Basically, the job of an industrial robot is the programmed manipulation of tools or grippers in a workspace. This is reflected in the typical construction of a robot: input devices are used to create and modify a program, which is executed by a controller. Motion and switching commands are executed by the robot kinematics and its end-effector, which is its gripper or tool. Robot kinematics consist of a chain of usually four to seven driven joints. In the special case of parallel kinematics, the joints consist of three to six parallel struts which are arranged side-by-side, connecting a base plate to a freely positionable platform. *Sensors* are used to measure object geometries, forces or process parameters and to feed this information back to the controller.

The workspace, dexterity and load-carrying capacity of an industrial robot are essentially influenced by the robot's kinematic construction. Some basic forms of construction have become standard. The majority of present-day industrial robots for handling and production tasks feature 6-axis vertical articulated-arm kinematics. The 4-axis SCARA robot (Selective Compliance Assembly Robot Arm) is predominant in the handling of small parts as well as in assembly operations as its design supports the highest speeds and combines rigidity in the robot's vertical axis and compliance in the horizontal axis.

► **Object localisation.** Workpieces can be handled automatically if their position and orientation are known. When the workpieces are randomly arranged in a container, it is necessary to use suitable sensors for detecting the location of an object. Just under 10% of

Cartesian	SCARA	Articulated	Parallel kinematics
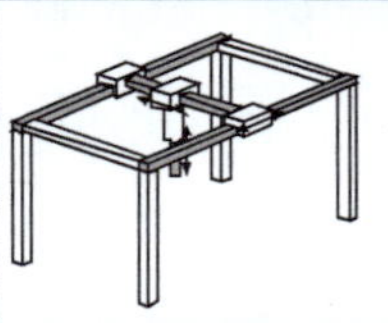	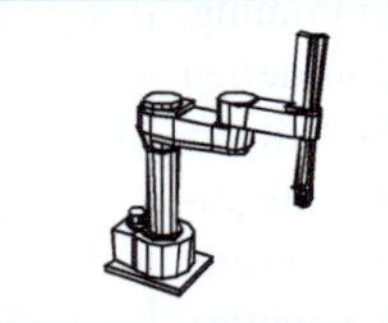		
Shape of the robot work space			
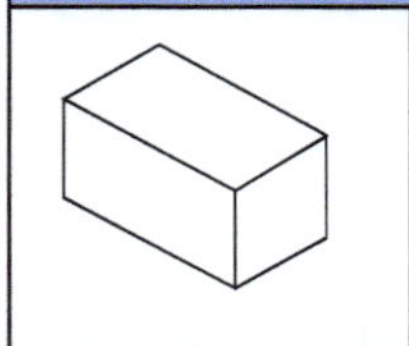	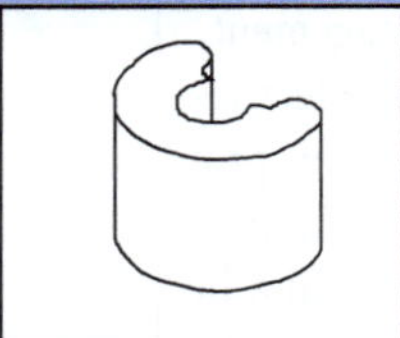	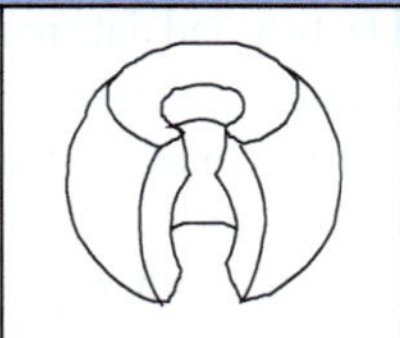	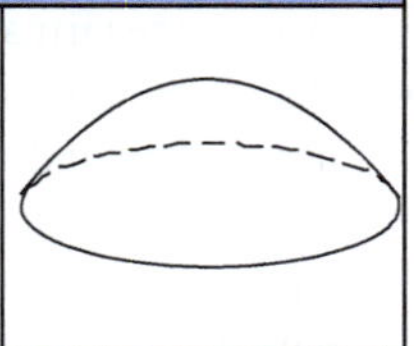
Robot kinematic characteristics			
• Translational or prismatic axes • Block-shaped work space	• Three rotational, one prismatic axes • Fast motions prismatic axes	• 5-7 degrees of freedom (DoF) • Predominant kinematics	• Parallel combination of prismatic axes • Stiff, low-weight, low inertia structure
Typical applications			
• Palletising • Machine tending	• Light assembly • Sorting of objects	• Welding, machining • Handling	• Positioning • Machining

Diagrammatic sketch and workspace geometry of typical robot kinematics: A Cartesian coordinate robot is typically referred to as a gantry. The SCARA (Selective Compliance Assembly Robot Arm) is mostly used in planar handling and assembly, it allows fast and compliant arm motions. The articulated robot (all joints are rotary and placed in series after each other) is predominant in applications such as welding, assembly, handling, machining, paint-spraying and gluing. In the special case of parallel kinematics, the motion axes are arranged between the base plate and the freely positioned platform. Source: Fraunhofer IPA today's industrial robots are equipped with geometry-measuring sensors such as 2D cameras or laser scanners for workpiece identification and localisation, but there is an ever growing demand for "bin-picking", i. e. picking out roughly or randomly arranged workpieces from a bin.

Modern localisation processes use a sensor-generated 3D point cloud of the surface of the bin containing the randomly arranged workpieces. This data can be generated, for example, by a *laser* stripe sensor or a rotating laser scanner. First, characteristic distributions (histograms) of the height contours and surface norms are generated from rotating views of the CAD workpiece model; these are stored in a database. These characteristic patterns are compared with the measured point cloud and the workpiece model is fitted virtually into the point cloud in a coincident orientation. State-of-the-art graphics processing units are particularly suited for this matching process so that a typical localisation process is carried out in a fraction of a second. The gripping point as well as collision-free gripping movements are generated from the detected object location. While one workpiece is being gripped and moved, a new measuring cycle is started until the bin has been emptied.

► **Navigation of mobile robots.** Most service robots are mobile, wheel-driven systems. Navigation is a key technology for mobile service robots and includes the secondary functions of environment mapping and localisation, motion planning, motion control and obstacle avoidance.

Employed navigation methods differ depending on the extensiveness, previous knowledge and preparation of the environment as well as the use of sensors for mapping. The map of the environment can be generated in two different ways. In the simplest case, it can be provided by the user in the form of simple 2D *CAD data* of the ground plan. Alternatively, the robot can generate its own map while actively exploring an initially unknown environment. This method, known as Simultaneous Localisation and Mapping (SLAM), describes the step-by-step generation of an intrinsically consistent environment map by a mobile robot as it charts its environment.

Since the advent of high-performance laser scanners (measuring angle +/– 140°, range 15 m, accuracy approx. 1 cm, with a measuring frequency of typically 50 Hz per scan) mobile robots have been able to navigate safely in busy public environments. Using an environment map, the robot can compute its whereabouts by constantly scanning its environment while at the same time detecting obstacles and object geometries.

By contrast, present-day robot vacuum cleaners and lawnmowers employ a much simpler form of nav-

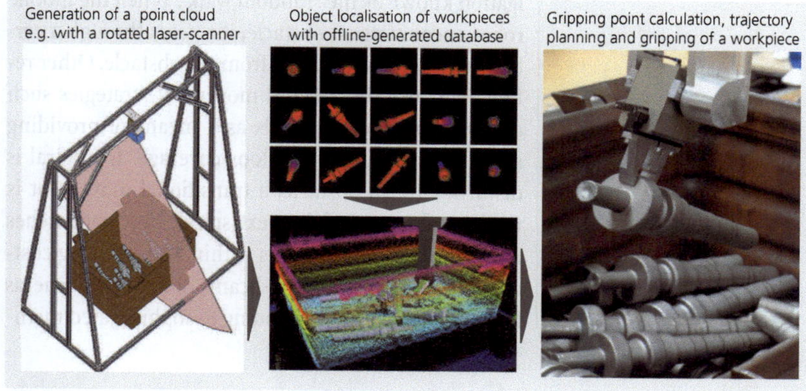

▲ Process for object localisation: Views of a rotated workpiece are generated offline on the basis of surface norms and depth histograms. The workpiece models are fitted into the 3D point cloud of an image of the bin containing the workpieces. The best fit provides the location of the workpiece. Gripping points and robot movements are computed. Source: Fraunhofer IPA

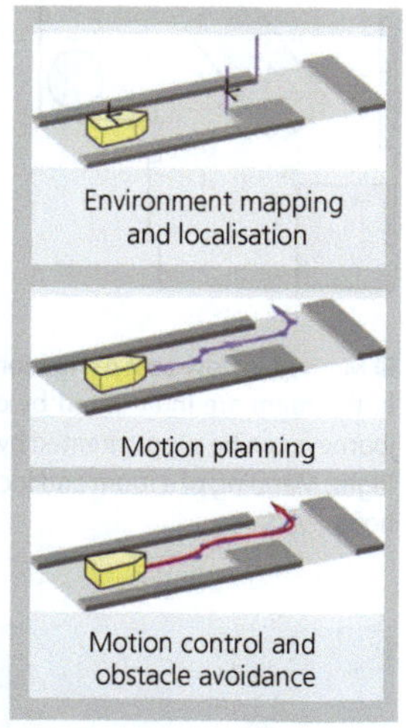

► Main stages in the navigation of mobile robots: The sensor based registration of the environment is referred to as mapping. Localisation determines the position of the mobile platform within the environment. The motion-planning stage makes a preliminary calculation of the nominal path of the mobile platform to a defined destination. During the mission, the nominal motion is continuously adapted as a result of suddenly occurring obstacles or other movement restrictions so that the destination is safely reached. Source: Fraunhofer IPA

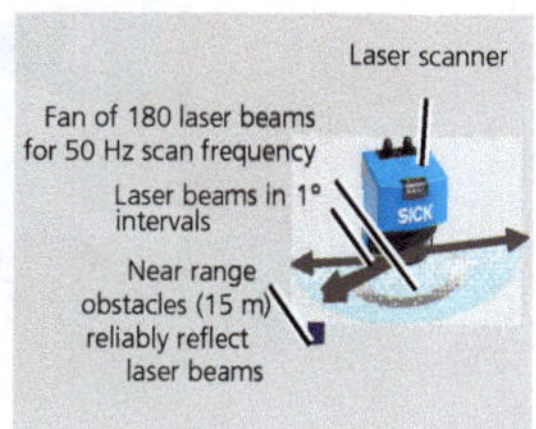

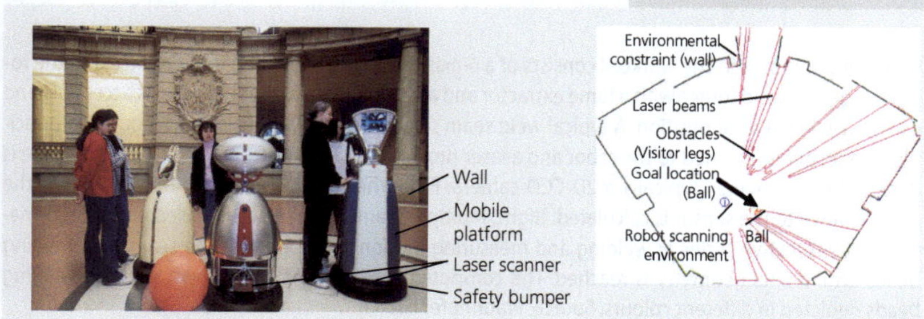

▼ Robots in the Museum für Kommunikation, Berlin. Left: The robots greet visitors, invite them to play ball games and escort them through the museum. They navigate by means of laser scanners in the atrium of the museum. The map of the atrium is in the form of a simple two-dimensional polygonal series of lines. Laser beams emitted by the robot strike the surrounding walls, the legs of people or the ball and are reliably reflected back to the scanner within some 15 m. The ball is identified on the basis of its characteristic round shape and size. The free area, marked by the connected boundaries of the emitted laser beams and their reflection points, is used for planned movements or for chasing after the ball. Source: Fraunhofer IPA igation known as the "random walk". When the mobile robot encounters an obstacle, it rotates through a certain angle and moves away from the obstacle. Other refinements use simple local movement strategies such as the generation of spirals as a means of providing some form of systematic floor coverage. If a spiral is disturbed by an obstacle, a transitional movement is made to the start of the next spiral. New approaches for localising mobile robots within a space using cost-effective sensors such as webcams or low-cost cameras from mobile phones are leading to sophisticated methods for the automatic detection of characteristic environmental features such as distinctive points, contours or colour patterns. The feature map generated is used for space-filling movement planning. Obstacles are detected by near-field sensors, such as ultrasound or infrared sensors, and are reactively circumnavigated.

Method for low-cost navigation. Left: Starting out from a journey round the room, random journeys in the room are interrupted by obstacles, triggering a new random trajectory. Centre: The random journeys can be supplemented by simple movement patterns, such as spirals, for better floor coverage. Right: Mapping of a room and localisation of the robot allows complete floor coverage. Source: Fraunhofer IPA

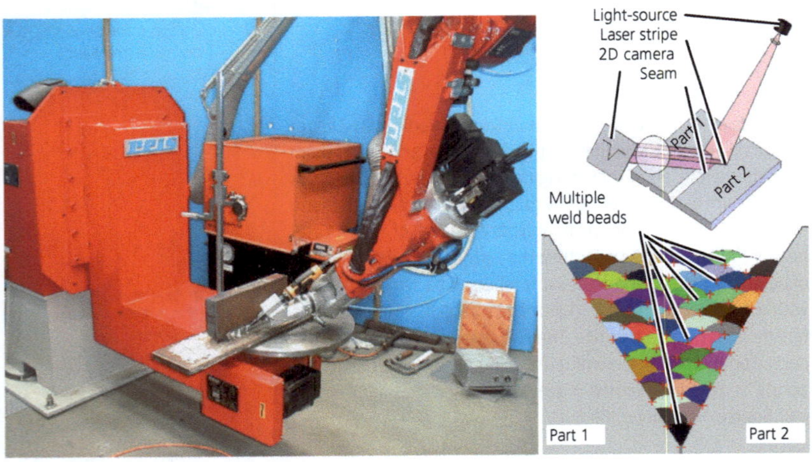

Welding robot. Left: The workcell consists of a 6-axis robot with a welding torch attached to the robot flange, the weld source with a fume extractor and a turning table so that the seam is accessible and can be welded in a flat position. A typical weld seam sensor is based on the laser triangulation principle. The sensor is attached to the robot and a laser projects a sharp stripe onto the seam. The stripe is detected by an imager, typically a 2D-CCD camera. From the extracted contour the position of the seam relative to the sensor is calculated. Right: Multiple seams are generated based on this information. In consecutive cycles of welding and measuring motions the robot fills the seam with welding beads until a final geometry is reached. The computed seam geometry is shown with the welding beads depicted in different colours. Source: Fraunhofer IPA

Applications

▶ **Welding.** Welding ranks among the most important joining processes in manufacturing and is the predominant application of industrial robotics today. Manual welding requires highly skilled workers, as small imperfections in the weld can lead to severe consequences. Modern welding robots have the following characteristics:

- Computer controls allow the flexible programming of task sequences, robot motions and sensors, as well as communication to external devices
- Free definition and parameterisation of robot positions/orientations, reference frames, and trajectories
- High repeatability and positioning accuracy of trajectories. Typically, repeatability is some ± 0.1 mm and positioning accuracy is around ± 1.0 mm
- High speeds of up to 5 m/s
- Payloads typically between 6–100 kg
- Interfacing to high-level factory control through industrial computer network protocols (field-buses or Ethernet)

The automatic arc welding process is based on a consumable wire electrode and a shielding gas which are fed through a welding gun. Electric current sources, torches and peripheral devices for automatically cleaning and maintaining the torch (anti-splatter, wire-cutting, tool changer, etc.) are offered by specialised companies. Often sensors are used to track welding gaps and measure the weld seams either before or synchronously with the welding process, adapting the robot's trajectory in the presence of workpiece variation and distortion. Also, cooperating robots have been introduced, where one robot fixes and moves the workpiece in synchronisation with another robot carrying a welding torch, so that the weld can be performed with the pool of molten metal in a flat position.

Another robot task is multilayer welding. At regular intervals the robot measures the profile of the weld seam connecting the two steel parts and adaptively generates subsequent paths for the insertion of successive weld beads until the final required geometry is reached. In the example shown up to 70 welding beads can be produced by the robot to weld extremely thick metal plates, as found on hydroelectric turbines.

► **Machining.** In addition to applications such as assembly, handling and paint-spraying, industrial robots are being used increasingly for the mechanical working of materials. An example that takes advantage of both the kinematic mobility and also the simple programmability of a robot is the hammer-forming of sheet metals, in which metal parts are hammered into shape by an industrial robot directly from CAD data without the use of a die. This metal forming process permits the production of sheet-metal parts in small batch sizes with low capital investment costs. The fact that there is no die means that changes to the geometry of the sheet-metal part can be realised within an extremely short space of time, because all that needs to be done is to generate a new CAD model. Areas of application include auto body panels, stainless-steel wash basins and aircraft panels.

► **Inspection and maintenance.** MIMROex (Mobile Inspection and Manipulation RObot experimental) represents a novel mobile robot used for remote and autonomous inspection and maintenance in hazardous locations, especially the offshore oil and gas environment. The mobile platform hosts a robotic arm which guides a camera for visual inspection, along with a variety of sensors such as microphones, gas and fire sensors, and laser scanners. The robot can also physically interact with process equipment, taking samples, turning valves or clearing minor obstructions, for example.

The robot can be set to remote or autonomous mode. The latter enables the robot to travel safely through industrial plants, stopping or changing direction when people walk in its way or any fixed or moving obstacle appears in front of it. The robot can also plan its path and travel autonomously between intervention points. Missions can last up to eight hours before the robot automatically docks onto a recharger unit.

Trends

► **Human-robot cooperation.** Today, mobile robots can be safely operated in public environments as sensors for monitoring their surroundings according to standardised safety regulations have become available. Robot arm motions are more complex and dynamic so that adequate safety sensors which monitor the robot's workplace are not commonly available. However, considerable gains in flexibility could be achieved through cooperative work-sharing between worker and robot, with the robot making use of its strengths in the areas of accuracy, endurance and power while the human contributes experience allied to unequalled perceptive and cognitive abilities.

The three minimum technical requirements for safe interaction between a human worker and a robot assistant are:

- The gripper must be controlled at reduced speed (typically approx. 25 mm/s).
- The robot may only move at reduced speed in the area near to the worker. Position and speed must be safely monitored.
- While in free movement, the robot must maintain a safe distance from the worker, depending on the speed of the robot.

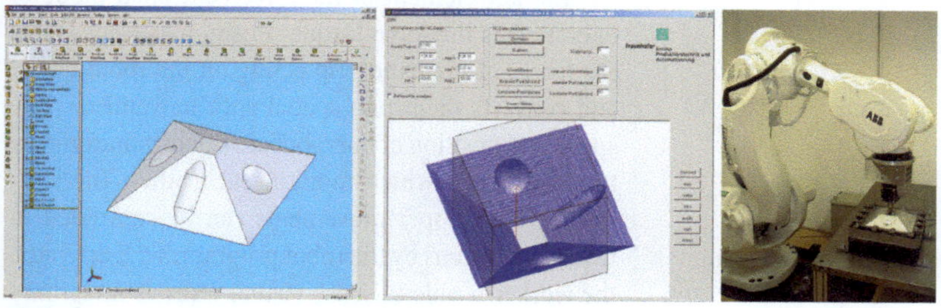

The hammer-forming of sheet metals is based on an oscillating stamp (amplitude 1 mm, 50 Hz frequency) which locally plastifies the metal in incremental steps. From the CAD model (left) the robot's trajectories are calculated on the basis of specific material models. Source: Fraunhofer IPA

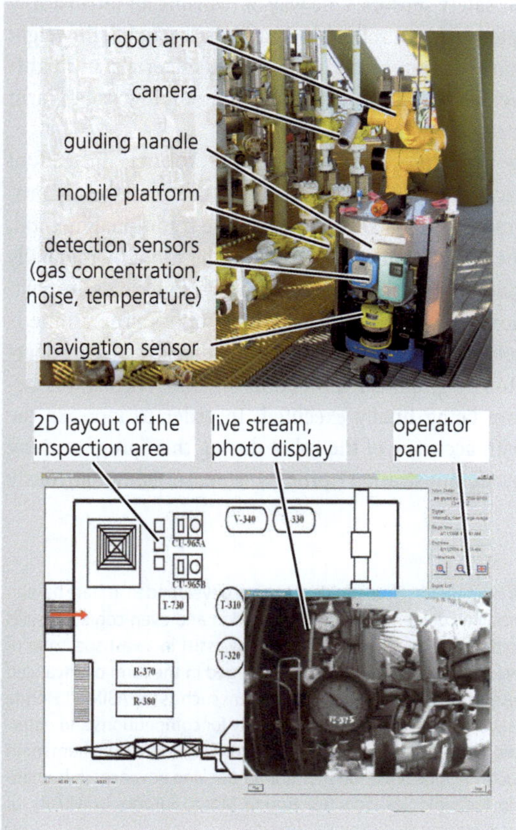

Mobile robot for offshore inspection: The explosion-proof MIMROex is a prototype robot system for reading gauges, checking for leaks, taking samples, etc. on oil rigs or in chemical plants. Data are transmitted to a central control room where the robot can also be tele-operated and programmed. Source: Fraunhofer IPA

A robot assistant of this type could be used, for example, to pick up a passenger-car gearbox from a pallet and move it, in suspended condition, to a worker at an assembly station for careful positioning and installation. The robot is controlled by the worker on the hand grip. The mode of operation switches from a simple path-following mode to a force-controlled mode, in which the robot and the handled load are weight-compensated, with small forces on the hand grip being sufficient to sensitively control the hovering movements of the robot.

The key element of the force-controlled mode of operation is the force/torque sensor, which measures the forces of the hand grip acting on the robot flange and converts them into spatial path movements. The speed and position of the robot are safely monitored by two controllers which monitor each other. The stiffness of the gripper movements can be controlled in three dimensions by the robot programme. If activated the worker is able to sense virtual walls as imaginary limits on the area of movement as well as the inertia generated by the workpiece.

An alternative principle for achieving safe movement of the robot is to ensure, by means of design measures, that the forces and energies attainable by the robot in the event of a malfunction are maintained in the non-critical range, thereby removing the need for safety sensors. One way of realising an intrinsically safe system of robot kinematics is through lightweight construction, in particular using materials and lightweight drives which contribute to low robot arm masses and inertia.

▶ **Intuitive programming of robots.** Instead of each individual movement of the robot being programmed on a keyboard as in the traditional method, the operator intuitively teaches the robot by manually guiding its kinematics and also using voice instructions and graphical inputs. The robot's programme is generated from these inputs and displayed to the operator by graphical simulation for final corrections before being finally executed. In order to increase the path accuracy of the robot during this "programming by demonstration" process a sensor simultaneously records characteristic workpiece geometries such as line or surface contours. In a subsequent step the manually recorded trajectory is corrected by the sensor data.

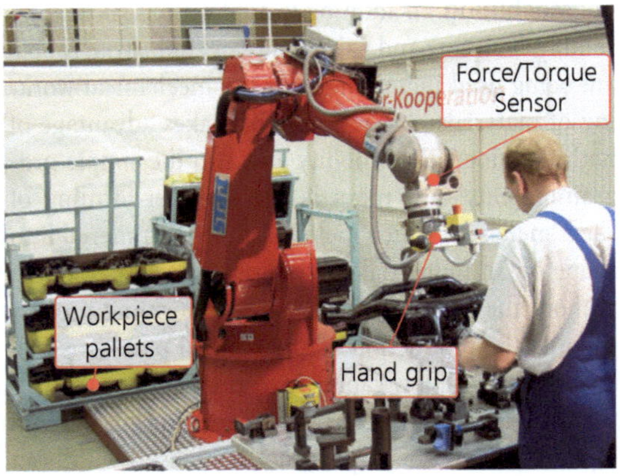

Robot assistant for work-sharing in assembly and handling: Gearboxes are picked up automatically from pallets. At the assembly station the robot switches to a movement speed of 25 mm/s and is sensitively controlled on the hand grip by the operator to fit the gearbox into the welding frame. Source: Fraunhofer IPA

Further developments include the activation of a dialogue with the user to automatically disambiguate or complete programmes which have been generated from available CAD data or from past experience. Advanced technologies from Artificial Intelligence such as planning and problem solving, natural language processing and machine learning have meanwhile become a most useful vehicle for making the robot more intuitive to use.

▶ **Personal robots.** Numerous developments are aimed at the application of robots, e. g. as assistants or butlers for performing useful jobs in everyday environments. Such robots are required to carry out bring-and-collect operations (drinks, books, etc.), to perform simple tasks in public or home environments (picking up objects off the floor, watering plants, operating and arranging furniture, heating up food, etc.) or to provide access to information on demand.

The physical layout of our everyday environment is organised with a view to human comfort and not to the use of robots. Most robots have a problem negotiating stairs and opening/closing doors. "Humanoid robots" are being designed to imitate human mobility, dexterity, sensory and cognitive skills, and intensive research is being conducted in key technologies, increasingly including the social sciences. Particularly in Japan, humanoids have been the subject of development work for years, and advanced humanoid designs such as the ASIMO™ at Honda, the partner robots at Toyota and

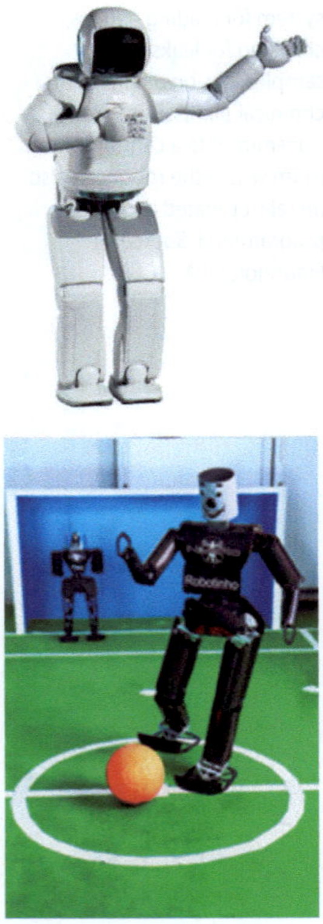

Humanoids: Humanoids are being developed to imitate human mobility, dexterity, sensory perception and even cognitive skills so that one day these machines may assist in a vast spectrum of everyday tasks. Products have emerged in the form of advanced research and demonstration platforms such as the ASIMO™ Honda as well as toy-like humanoids mostly for competitions and entertainment (fighting, soccer, dancing). The RoboCup™ humanoid soccer league is a good monitor for tracking advances in humanoid technologies. Sources: Honda Motor Europe, University of Freiburg the HRP-3 have received wide recognition from the general public and the scientific community. Given the versatility of these machines and the potential mass markets that exist, their long-term relevance for an extremely wide range of tasks in manufacturing, services and personal use is obvious. The international RoboCup™ project is a platform for fostering intelligent robotics research by having teams of robots compete at soccer. It is claimed by many robotics experts that by 2050 a team of fully autonomous humanoid robots will win against the human world soccer champions.

A different approach to designing robot assistants without emphasising a humanoid appearance has been suggested by the Care-O-bot® 3 concept. Systematic analysis of different types of work identified a spectrum of household chores which could be mastered by the robot using today's technologies, including fetch-and-carry jobs, finding and lifting an object from the floor, cleaning up a table or operating simple devices such as a microwave. The system consists of a mobile platform, a gripping arm, a tray and a movable body for the positioning and orientation of sensors, such as a 3D camera for body-pose and object recognition and a 2D camera for facial recognition. Alongside the aspects of cost, the key challenges include intuitive operation, reliable execution of tasks even in unstructured and changeable environments, as well as safe detection and gripping of at least 100 everyday objects.

Care-O-bot® 3 features a hybrid architecture consisting of planning and reactive components. Interactive instruction is provided by means of voiced, graphical or gestural user inputs. The input commands are compressed according to stored position data, object names or activities and forwarded to the symbolic planner. These inputs are used to generate a list of actions for executing the task and the list is then passed on to the sequence controller, which manages the data streams to the component controllers and the sensor computer and accesses stored or learned environment, object or personal data from the database. Knowledge of the environment is constantly updated by the sensor data. The status of the system is indicated graphically or is communicated by speech synthesis to the user, who is thus able to supervise the correct execution of the task and to intervene if necessary.

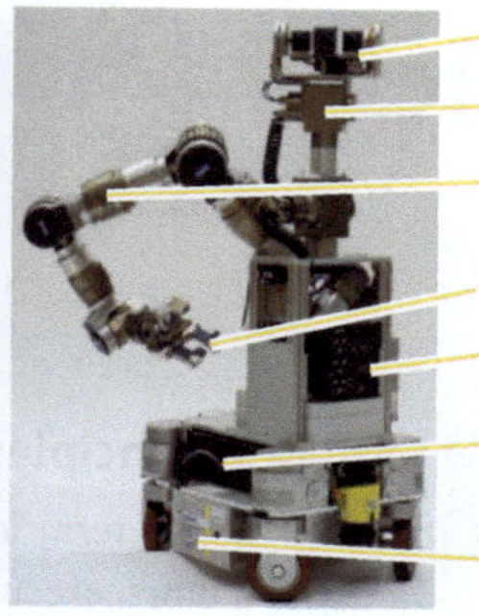

Care-O-bot® 3 service robot: The robot brings and collects objects. The front serves as the presentation side, the back as the working side. In addition to the gripping arm, objects are presented to the user on the tray. Force/Torque sensors as well as tactile sensors are integrated in the gripper. A 3D sensor alongside a stereo camera system serves the purpose of detecting faces, body poses, objects and obstacles. The Man-Machine Interface (MMI) comprises voice, graphics and gestures. Source: Fraunhofer IPA

Prospects

Alongside industrial robotics, which continues to represent the economically dominant pillar of robotics, it has been possible in recent years for new categories of robot systems to establish themselves in areas of application outside of production, offering the prospect of considerable growth potential for robots in the home as well as in the entertainment sector. The following trends in robotics can be identified:

- In future, robot assistants and service robots will be dependent on information about the outside world which will require large data repositories to be constantly updated by means of sensors or user commands. The challenges in terms of cognition relate in particular to the availability of reliable high-resolution 3D sensors for detection and recognition of environments, objects and persons given everyday environmental conditions in terms of lighting or object reflectivity.
- Future robotic arms and grippers will be light, compact, mobile and sensitive, not only on service robots. Solutions will combine the use of new materials with novel, safety-conformable actuators featuring integrated sensors for force and contact measuring.
- The aim is that service robots should also be capable of intuitive operation and suitable for use by untrained operators through several simultaneous information channels such as voice, gesture, graphics, haptics, etc.
- Home robots, such as vacuum cleaners, lawnmowers, mobile media terminals and surveillance robots, will evolve from generation to generation in terms of appearance and functionality, opening up a broad commercial market. There will be fluid boundaries between home robots, intelligent home automation and entertainment electronics.

MARTIN HÄGELE
Fraunhofer Institute for Manufacturing Engineering and Automation IPA, Stuttgart

Internet

- www.ifrstat.org
- www.eu-nited-robotics.net
- www.service-robots.org
- www.everydayrobots.com
- www.robotory.com
- www.care-o-bot.de
- www.smerobot.org
- www.androidworld.com
- www.robotics-platform.eu
- www.robots.net

Logistics

Related topics

- Ships
- Rail traffic
- E-services
- Sensor systems

Principles

The term logistics can be used to describe the planning, execution, design and control of all material and information flows between a company and its suppliers, all in-house flows, and the flows between a company and its customers. Put simply, logistics ensures that the right quantity of the right goods, with the right level of quality, arrive at the right location at the right time–and at the lowest possible cost. Logistics not only covers the control of transport, transshipment and warehousing processes, but also services such as customised packaging, assembly, data storage and information management that provide value added to the logistics process. As outsourcing is becoming increasingly common in certain sectors, the use of value-added chains is growing all over the world. The supply chain has therefore developed into an extremely complex delivery network requiring a high level of transparency, planning, tracking and control.

The following example clearly shows this level of complexity: a sub-supplier sells a range of 500 products which it buys from 50 manufacturers. The company also supplies 12 major customers with 50 shops each. In this example, 15 million combinations of products, links and sources have to be differentiated to clearly establish their origin. If the sub-supplier delivers a damaged product, it has to be recalled from every shop. An effective logistics information system is therefore the only way for the sub-supplier to identify those shops that have been supplied with the damaged products. Without such a system, the sub-supplier would be faced with an enormous task. It would also be more difficult to recall the products in time and prevent any harm from being done to the final customers.

Standardised container units. Top: One of the biggest container ships in the world XIN SHANGHAI with a max. load of 9500 containers. Centre: A Port with enough container bridges will be able to load and unload about 4,000 containers from ships in 24 hours. The containers will be loaded on trucks or railway cars. Bottom: The containers´ standard dimensions ensure smooth loading onto trucks for onward transport. Standardised, modular container units come in two different sizes, which can be combined and handled easily. These robust containers can be stacked one on top of the other. The exterior dimensions of the basic container module are 6.1 x 2.4 x 2.6 m^3 with an approximate capacity of 33 m^3 or a maximum loading capacity of 22 t. Daimler, for example, transports the components for six of its Sprinter models in one container from Germany to Argentina where the cars are assembled.

	Advantages	Disvantages
Road	- flexible - fast	- reduced bulk transport capacity - high levels of environmental pollution
Rail	- efficient energy consumption - superior bulk transport capacity	- inflexible - infrastructure not fully developed
Ship	- high energy consumption - optimum bulk transport capacity	- slow - high transhipment costs
Aircraft	- very fast - low capital binding	- expensive -high levels of environmental pollution

In traffic logistics, the expansion of the global exchange of goods is leading to gigantic means of transport. Enormous aircraft like the Airbus A380 and large, modern container vessels will have a considerable impact on existing infrastructures at airports and container ports. While the majority of goods are transported in containers by ship, more time-critical and capital-intensive goods – such as high-quality electronic articles – are transported by plane. In the future, two thirds of all goods will be packed in containers and loaded onboard a ship.

Logistics is divided into sub-systems such as procurement logistics (from the supplier to the receiving warehouse), distribution logistics (from the distribution warehouse to the customer) and intralogistics (in-house processes).

► **Traffic logistics.** Traffic logistics can be seen as the link between a company's purchasing and sales markets. Traffic logistics is the movement of goods or people to their destinations by road, rail, air and water. This essentially covers the planning and optimisation of traffic systems with regard to the design of transport chains, roads and distribution structures for freight traffic, and the study of public transport systems for passengers. A good example is the new distribution structure for goods where a central warehouse (hub) supplies several regional warehouses (spokes).

Logistics service providers are seeing their costs increasingly squeezed by the rising price of crude oil and road tolls. But traffic logistics provides the following services, which can improve their competitive edge:

- Efficient inter-company transportation (e.g. just-in-sequence (JIS) delivery of complete modules to the automotive industry)

- Provision of additional services ("value-added services" like the labelling of clothes in the picking area of a warehouse)
- Processing and provision of relevant information like precise forecasting of traffic volumes on motorways

Within intermodal transport chains – i. e. transportation using different means of transport – bodyshells, for example, are transported by rail from a low-wage country to the assembly plant in a high-wage country. This "principal" means of transport is enhanced by pre- and post-transportation – generally speaking, transportation by truck – to combine the advantages of both means of transport. The truck offers greater flexibility (bundling of goods) while rail significantly increases bulk transport capacity.

▶ **Order-picking**. In the past, goods were generally picked from pick-lists, i. e. order lists for each picking process, which told the picker how many products were needed for a specific consignment. The inherent capacity for pickers to read the lists incorrectly made this procedure extremely prone to error. Today, the quantities and locations of small parts are specified, for example, by light signals (pick-by-light) or acoustic signals (pick-by-voice) and, in comparison with the traditional method, these techniques have significantly improved the quality of the picking process. In the pick-by-voice procedure, the picker wears headphones and is informed how many of which goods he or she has to retrieve from which aisle and shelf. Once completed, the process is confirmed manually and the system guides the picker to the next retrieval point.

▶ **E-logistics**. Extremely efficient picking processes are required in the e-shopping sector where products are usually delivered within 24 hours of ordering. Route-optimised picking, where software indicates the shortest route from bin A to bin B, is generally used. Automatic picking of single items is also commonly used in e-logistics. Picking robots and automated warehouse systems transport the goods quickly on tracks to the picker who consolidates the consignment for delivery. Software also enhances the transport packaging selection process to ensure optimum use is made of the available space during shipment. This is done to reduce shipping costs, which compared with other distribution channels, are very high. In conventional commerce, a pallet with 60 shipping units is sent to a central warehouse for picking. In *e-commerce*, the pallet is split up into 600 customer units ("atomisation of shipments") which parcel service providers have to deliver to the final customer's door.

▶ **Radio Frequency Identification (RFID)**. An RFID tag is a "transponder", a device that "transmits" and "responds". The tag is scanned by a reader and provides information about the product to which it is attached. The reader's antenna transmits the data by transferring power to the tag's coil via a high frequency field. Every transponder is equipped with a coil. A chip then uses this power to transmit or save the data, i. e. data are sent to the reader or new data are saved on the tag. A passive transponder consists of a micro-chip with different memory capacities, an antenna, and a carrier or body. An active transponder, on the other hand, has a built-in power supply (battery). This battery allows the tags to send radio signals intuitively and span distances of up to 100 m.

Unlike the barcode, RFID is ideal if data has to be transmitted over long distances and/or goods cannot be accessed easily. The transportation of goods is no longer centrally controlled from a single point; smaller, decentralised networks have taken over this role. In this context, decentralisation means that individual components, such as junctions with gates on a conveyor belt at an airport, are equipped with "smart" tags. This means they can then be integrated in logistical processes. RFID helps to decentralise the information and make it accessible wherever it is needed.

▶ **Tracking and tracing**. Identification technologies like RFID allow the parcel processing and shipping cycle to be recorded and analysed over the complete supply chain. This is known as tracking and tracing. Tracking and tracing may come in useful, for example, when a recipient claims a consignment has been lost. RFID continuously tracks the transfer of goods. This establishes clear legal responsibility and is therefore extremely important in logistics where many sub-suppliers are involved.

▶ **Monitoring**. External or internal conditions have to be supervised, monitored and controlled for sensi-

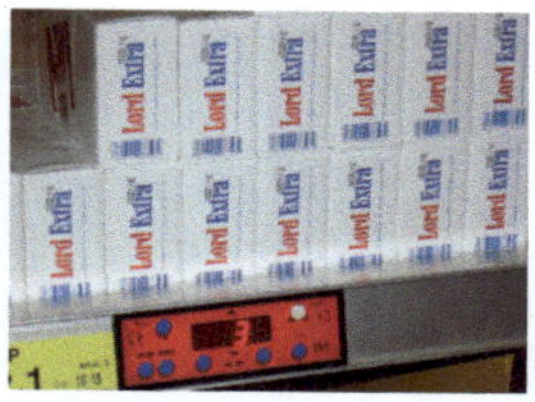

Pick-by-light system: A display shows the picker the quantities to be retrieved for a specific consignment.

In the future, everyday products will be fitted with a tag like this.

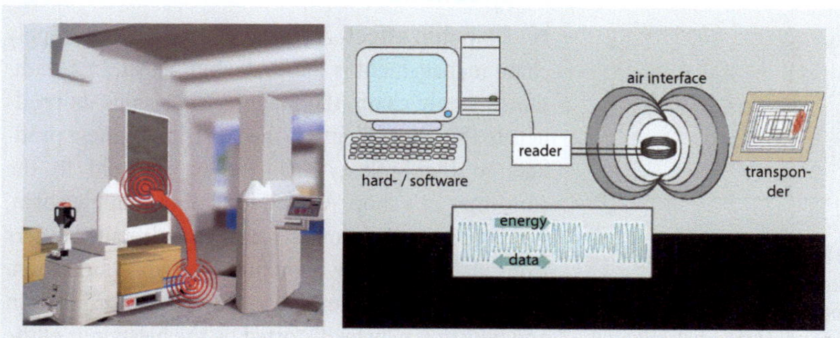

RFID transponders allow contact-free communication with the antenna. Left: Operation of an RFID system. A transponder containing important information is attached to a product or pallet. Air interface: The antenna emits an electro-magnetic field and starts exchanging data with the transponder. Reader: Receives and transmits data to a PC (hardware). Software: Filters the desired data and saves them in the system.

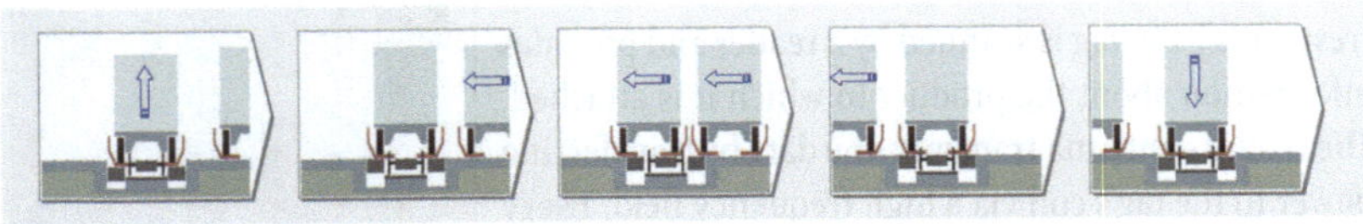

CargoBeamer. From left to right: Trucks drive onto a platform and their trailers are raised and moved sideways onto the wagons of the train. At the same time, the incoming trailers are simultaneously taken off the train. This system is compatible with any type of articulated trailer and reduces the length of time it takes to load and unload a train composed of 30 wagons from 3 hours to 10 minutes. Source: CargoBeamer AG

tive goods. Active transponders use their built-in energy supply to save data, measured by sensors, such as temperature or acceleration. This information can be saved for up to several days. The data can then be read and analysed in order to monitor, for example, the quality of temperature-sensitive drugs or food.

▶ **Supply Chain Management (SCM).** This principle addresses the planning and management of goods, information and cash flows within a network of suppliers, manufacturers, retailers and final customers. In the past, companies only tended to optimise processes at their premises (industrial companies optimised production processes while retailers improved the stock at their shops. Both, however, kept safety stocks). Today, organisations put more focus on improving the overall situation by perfecting inter-company relationships along the complete supply chain. The "pull" principle reduces a company's stocks considerably and creates a lean organisation that allows the company to respond much more flexibly to the market. Unlike the "push" production system of the past, which required large stocks, the pull principle focuses on the final customer. At the end of the chain, the customer places an order triggering production downstream.

Applications

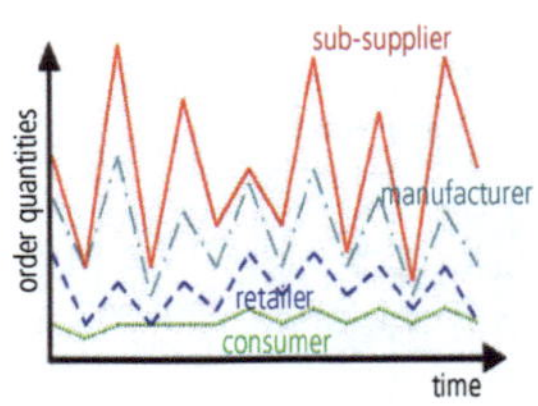

Bullwhip effect: The slightly fluctuating demand from the final customer causes stock to fluctuate at the wholesale level. The poor flow of information then causes the stock curve to rise even higher for the manufacturer, which has a negative impact on stock levels at the supplier's warehouse.

▶ **Bullwhip effect.** A typical phenomenon in supply chain management is the "bullwhip" effect. In a traditional supply chain, the manufacturer ships its products to the wholesaler who then distributes these products to the broker. Finally, they are sold to the consumer in a shop. At each level of the supply chain, goods are ordered downstream, e. g. a retailer orders from a broker. The demand-focused information, which the manufacturer uses to plan its production and logistics, may be distorted by the behaviour of the multiple players in the supply chain. Demand for nappies in a shop, for example, remains relatively stable. The trading company's central dispatching department collects individual orders from several shops, consolidates them into an overall order and sends this to the broker with no further information about how many shops are ordering which quantities of products. The broker receives a large order, but no information about the quantities ordered by each shop, and therefore assumes that demand is at a peak. In order to maintain its own safety stocks, the broker then orders a large quantity from the wholesaler. This process triggers a domino effect: the manufacturer receives a large order from the wholesaler and orders an even larger quantity of semi-finished products to adjust its safety stocks. Studies have shown that from the first level (the retail trade) upwards, orders no longer reflect the original demand for nappies and increase from level to level. This is the bullwhip effect, which has a negative impact on the availability of goods and costs along the supply chain. Nowadays, central dispatching departments have software solutions that reduce this problem. Suppliers have access to data like stocks, sales, receipt of goods and reservations made by upstream and downstream players so that they can, in turn, optimise their orders. This helps to reduce or adapt stocks and minimises delivery times.

▶ **RFID.** Trading companies are increasingly using RFID for logistical purposes. The aim is to fit transponders to shipments and entire containers within the global value-added chain to optimise the receipt and issue of goods and ensure overall traceability. Pilot projects have already integrated sub-suppliers in these processes. During the picking process, RFID transponders are encrypted with the internationally recognised *Electronic Product Code (EPC)*. These transponders are attached to packaging, cartons or pallets to ensure goods are clearly identified. Unlike other digital ID systems, which only indicate the product category, EPC clearly identifies each individual final product (e. g. TV set, type no. X 150 manufactured by Müller Inc., serial no. 4711). The network-based collection and evaluation of all the necessary data therefore ensures continuous tracking and tracing. The relevant information about the picked goods is saved in the warehouse management system while the goods are brought to the goods issue point for onwards distribution. When the goods are loaded, the RFID readers scan the transponder EPC and send the relevant data to the warehouse management system. The goods are then booked out and an electronic delivery note informs the recipient that the truck transporting its goods is on its way. As soon as the truck reaches its destination, RFID antennas automatically scan the consignment EPC and the data is sent to the recipient's EDP system. The data is compared with the delivery note sent by the manufacturer, and wrong and incomplete deliveries can be identified immediately (with no need for manual checking).

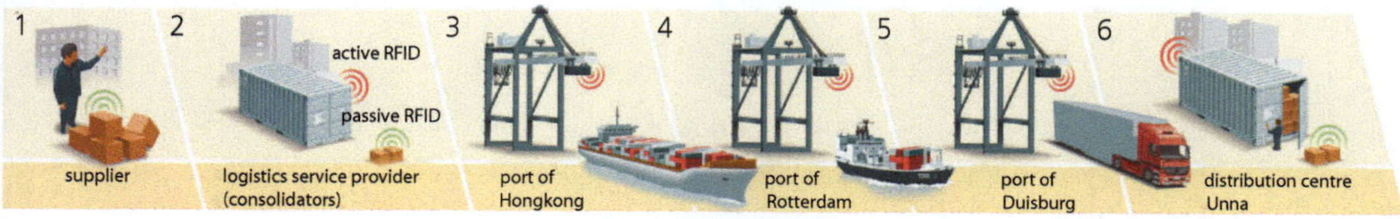

RFID helps to make freight traffic more efficient and transparent by means of active and passive transponder.

1. Suppliers can choose to equip their consignments with RFID transponders or to outsource this activity to logistics service providers.
2. The deliveries (export parcels) are equipped with a passive RFID transponder, which is scanned by an RFID reader at the goods issue point. The deliveries are then loaded into transport containers. These containers are equipped with active RFID transponders so that they can be read over longer distances.
3. The containers are scanned by an RFID reader during loading.
4. The containers are scanned by an RFID reader during transshipment.
5. The containers are scanned by an RFID reader during loading.
6. The containers and their contents are scanned by an RFID reader at the goods receipt point of the distribution centre and are automatically checked for completeness.

Source: Metro group

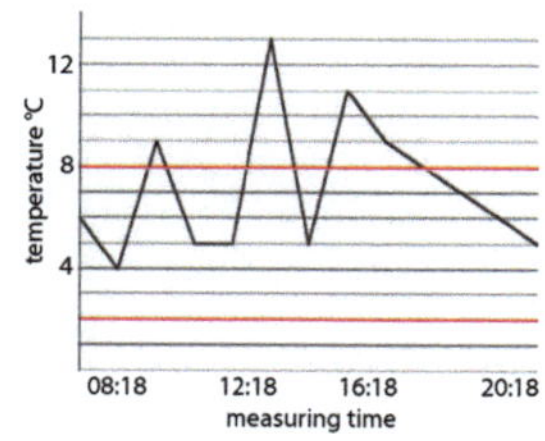

Continuous temperature tracking: A data logger saves the temperature values for a perishable good. This data shows whether the product has been transported within the specified range of 2–8 °C. If this were not the case, the average threshold values for the cooling chain would have been exceeded. This normally occurs when products are loaded from one means of transport onto another.

Trends

▶ **The Internet of Things.** RFID technology decentralises intralogistics systems and makes them more flexible. In the same way that emails always find the free or the shortest route, logistics objects will also be able to find their own way. In the future, junctions with gates on a conveyor belt, for example, will be equipped with "smart" tags so that a suitcase will be able to find its own way to the aircraft. Luggage is currently controlled from a central point. However, in the future, pieces of luggage will be able to select their own route. Each piece of luggage at an airport will be equipped with a transponder containing target information such as flight number and gate. On its journey, the suitcase passes junctions on a conveyor belt, which are equipped with RFID antennas and programmed with decision algorithms. When the antenna scans the suitcase target information, the system can decide whether the shortest route to the target gate is available or overloaded. If the shortest route is available, the junction releases the suitcase. In the event of failure or overload, the suitcase is directed to the next junction, where data on all the possible routes are also stored. The suitcase transponder sends its target signal again, and the decision making process is repeated. The suitcase and the "smart" junction separate themselves from a superimposed controlling instance and make their own routing decisions according to the given frame conditions. This kind of application can also be used for a variety of logistical systems. Furthermore, it is planned to incorporate automatically guided transport vehicles into the static conveyor belt system.

▶ **Food Chain Management.** Food Chain Management sees the entire food chain – from production, processing and trade, up to the consumer – as a uniform process where the importance of food safety and quality, as well as traceability, is paramount. E.g. when a Danish retailer receives strawberries from Spain, the wholesaler or fruit company does not provide any data in the transport information about the use of pesticides, although this information is still available. Trading companies set maximum pesticide levels for fruit and vegetables but, for time and money reasons, quality checks are only carried out randomly to ascertain whether these levels have been exceeded. Avoiding excessive levels of pesticides therefore remains an unsolved problem. This issue might be resolved by quick analyses or information provided on the Internet that lists the quantities of pesticide used by the producer for each respective batch. Retailers and customers could then access this information by entering the batch number. Temperature during the transportation from Spain, and during the handling of goods at the distribution centre, is another factor that has an influence on the product lifecycle. If the temperature is measured regularly and a transponder saves this information, the data can be checked when the goods are received and the transponder is scanned. If the thresholds have been exceeded, this data may trigger a quality check. In Europe, up to 25% of all goods are currently thrown away because they have perished.

Prospects

- Modern logistics uses advanced information and communication technologies to design, control and optimise logistical processes.
- From 2000 to 2020, freight volume in Europe will increase by 50%. Trucks are not environmentally acceptable because they consume large amounts of fossil fuels, and emit high levels of carbon monoxide and dioxide, as well as nitric oxide. Freight traffic must become more efficient and sustainable.

DR. VOLKER LANGE
CHRISTIANE AUFFERMANN
Fraunhofer-Institute for Material Flow and Logistics IML, Dortmund

Internet

- www.logisticsmgmt.com
- www.logistics.about.com
- www.ec.europa.eu/transport/logistics/index_en.htm

13 SECURITY AND SAFETY

The term security and safety covers the broad range of issues involved in combating individual and collective risks. A distinction is made between security in the sense of taking preventive measures against attack, and safety in the sense of reducing possible dangers from the operation of systems, machinery and the like. The focus in risk assessment is changing over the course of time. Threats from natural disasters and wars have always existed. During the second half of the 20th century the military threat emanating from the two dominant military alliances was perceived as being particularly dangerous, but since the turn of the century the debate in the political arena and in the media has concentrated increasingly on internal security following terrorist attacks on certain centres of the Western world. Despite this perception, most deaths are caused by natural disasters. 2008 was one of the worst years for disasters in recent history. Around the world more than 220,000 people died as a result of natural catastrophies such as hurricanes, floods and earthquakes. In particular, the increasing number of hurricanes attributed to climate change has prompted an urgent demand for technological solutions to reduce carbon dioxide emissions.

Our complex, tightly interconnected infrastructures (traffic, water, energy) are nowadays very vulnerable to disruptions of all kinds. Major cities around the world are growing in number and size and they depend very strongly on smoothly functioning infrastructures. Even small faults can affect the system significantly. This is particularly the case for communication infrastructures, where it takes little effort to cause major damage worldwide. Technical IT security therefore plays a major role.

Threats also exist in everyday life. Here the main focus is on the safety of products, systems, buildings and entire infrastructures. People need to be certain that the goods they buy are safe (e. g. food and electrical appliances), that chemical plants do not emit any toxic substances, that bridges will not collapse and that the electricity network will not break down. The complexity of modern technology is in part the reason why it sometimes fails. Equipment has to be controllable to the extent that people can understand and confidently operate it (with technical support). Sensors and measuring devices are important here because they enable dangers to be detected at an early stage. Communication and information technology is necessary for collecting and rapidly processing data so that dangers can be averted. But even if they are not complex, components and structures sometimes fail. Roofs collapse, pressure vessels burst and trains derail. The technological challenge here is to test the reliability of safety-relevant components before and during use by means of nondestructive techniques.

The term security and safety therefore encompasses:

- external security against a military threat
- internal security against terrorist attacks
- measures to deal with natural disasters (protection, warning and emergency relief)
- reliability of products, systems, buildings and infrastructures
- human failure in the operation of technical systems.

Accidents can happen because of human error or technical defects. As technical defects for their part are, however, always directly or indirectly due to human error (inadequate maintenance, inadequate quality control, design faults), all accidents are strictly speaking due to human error, apart from unavoidable natural events. Accidents with major consequences such as the Exxon Valdez or Chernobyl remain engraved in the memory. Technical systems therefore have to be developed which support people and as far as possible rule out mistakes, e. g. by providing intelligent control technology. Driver-assistance systems in the automobile industry, for example, help to avoid driving errors.

► **The topics.** Throughout history, the main threat to humankind has come from war. Since the end of the Cold War, however, the *weapons and military systems* deployed have changed considerably. In today's peacekeeping missions the focus is on well protected mobile special units rather than on heavy artillery. Guerrilla and civil conflicts are more prevalent than conventional warfare in which there is a well defined front. Technology has to respond to these changes by, for example, providing defence against hazardous substances used in military or terrorist attacks. Chemical, biological and radioactive substances and explosives must be reliably detected in order to avoid proliferation, i. e. the propagation of weapons of mass destruction.

In addition to the collective threat to the population, individuals are also at risk from criminal acts. Technology attempts to provide an initial deterrent, as seen in the increasing use of closed-circuit TV monitoring in cities and intelligent image analysis. If the deterrent does not work, *forensic science* has become a much stronger weapon in recent years, and is able to identify criminals and provide evidence for their conviction from the tiniest traces at the scene of the crime.

Crimes, including industrial espionage, are prevented by access control and surveillance. While it used to be keys and passwords which opened doors, today access to security-relevant areas is being increasingly controlled by the use of biometric features to identify authorized persons. Over the past few years, fingerprint identification has already gained widespread use for simple applications. It will not be long before automatic face recognition is introduced.

An "arms race" is in full swing between those who want to protect our information and communication systems and those who want to hack in and disrupt them. In our strongly networked global world *information security* is essential for healthy economic activity. The systems used for electronic communication need to be absolutely secure, since sectors such as e-services can only grow with the full confidence of consumers.

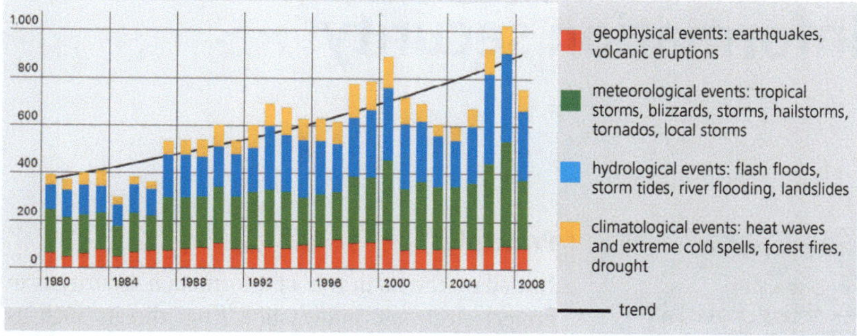

Number of natural disasters worldwide 1980–2008: While geophysical events (red) remain largely constant, the ten-year average for disasters is increasing significantly as a result of floods (blue), storms (green) and forest fires (orange). This trend can only be brought to a halt by stopping the rise in atmospheric temperature caused by the increase in greenhouse gases. Otherwise, technology can only help to deal with the effects of disasters. Source: München Rück Topics Geo 2007

While active defence measures can be taken against terrorism and war, technology cannot do much to counteract forces of nature such as hurricanes, floods, earthquakes and tsunamis. However extreme weather conditions (e. g. hurricanes) can now be accurately predicted, it is still very difficult to make medium-term prognoses of natural disasters such as tsunamis or earthquakes. At best, these can be rapidly detected as and when they occur. At present, the only way to reduce the effects of natural catastrophes in advance is to implement a policy of *precaution against disasters* so that the population can be warned at an early stage and evacuated in an orderly fashion if necessary. After a disaster has happened, effective emergency aid must respond rapidly. Assistance not only has to be available on call after natural disasters, but following accidents of all kinds, including car crashes, ship collisions, plane crashes and house fires. The aim must be to rescue people and protect the environment (e. g. after an oil spill) and to rapidly restabilise the infrastructure (e. g. after train crashes). This is where stable organisational structures and a robust communication infrastructure are particularly important.

For major facilities such as chemical and power plants, redundant safety systems are now available which sharply reduce the risk of failure and increase *plant safety*. Even if faults occur, they are quickly detected and a safety system deals with the effects of the fault, keeping the risk to people as low as possible. ■

Information security

Related topics

- E-services
- Access control and surveillance
- Internet technologies
- Ambient intelligence

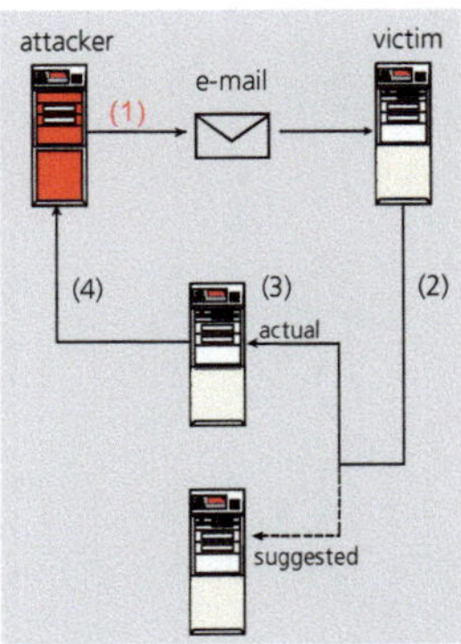

Phishing attack: The attacker sends out e-mails inviting people to visit a particular web page (1). Victims clicking on the Internet address (2) think they have reached the server of a reputable company but have actually been directed to the attacker's server (3). Any sensitive information entered, e.g. access data for online banking, falls into the hands of the attacker (4). Source: Fraunhofer INT

Principles

Nowadays the main role of information security is to protect electronic information from threats such as loss, manipulation and espionage. As information technology systems are used in numerous sectors, e.g. in power generation and distribution, telecommunications and transport infrastructure, their protection is essential for modern life. This is accomplished by the following characteristics:

- Authenticity: identifying persons and proving the origin of information beyond doubt
- Confidentiality: avoiding the unauthorised collection of information
- Integrity: preventing the unauthorised and unnoticed modification of information
- Availability: ensuring that resources and services are available whenever required by authorised parties

The most important branch of information security is IT security, which is primarily dedicated to protecting IT systems and information against deliberate IT attacks.

▶ **Malware.** This term relates to computer programmes specifically developed to do harm. A distinction is made between computer viruses, computer worms, and Trojan horses. Viruses spread by writing copies of themselves into computer programmes, electronic documents or data storage devices, whereas worms spread directly through computer networks such as the Internet and infiltrate other IT systems. If a useful programme contains a hidden malicious component it is referred to as a Trojan horse. Malware can cause different types of damage, including file manipulation, the installation of a programme enabling the attacker to access the system and manipulation of the computer's security equipment. In the last case it is possible for the administrator to completely lose control over the system. The Loveletter worm is an example of the damage malware can cause. In 2000, this worm infected about 20% of all computers worldwide and caused approximately 9 billion dollars worth of damage.

▶ **Phishing and pharming.** These new types of attack are among the biggest threats on the Internet. Phishing uses e-mails to induce the victim to visit the copy of an existing web page, e.g. the web page of an online bank, and reveal sensitive access information. Pharming uses, Trojan horses to manipulate the connection between the domain name and computer address (IP address) of the online bank so that victims are directed to the attacker's computer when entering the correct Internet address.

▶ **Denial-of-service attacks.** By saturating a system with a flood of senseless requests, denial-of-service attacks (DoS) try to overload the system to such an extent that it is no longer able to carry out its intended tasks. DoS attacks are particularly effective if a large number of computers – the 'zombies' or 'bots' that an attacker has brought under control e.g. by using a Trojan horse – conduct them simultaneously (distributed denial-of-service attack, DDoS).

▶ **E-mail spam.** Unsolicited bulk e-mails containing advertisements or malware are not only a nuisance, they also take up transfer and storage space as well as valuable working time to sort through and delete. E-mail spam makes up 60–90% of all e-mails, i.e. more than 10 billion e-mails per day worldwide.

There are a number of technical security measures that can be taken to address the diversity and complexity of the threats outlined above.

▶ **Antivirus software.** The use of antivirus software is a widely accepted and essential security measure to protect computers against malware. Malware is detected either from a bit sequence typical of the malicious programme concerned (signature detection) or from suspicious activities on the computer, such as the unauthorised deletion of files (behaviour detection). Whereas signature detection can only detect known malware, behaviour detection can detect unknown malicious programmes as well. In 2004, the MyDoom Internet worm caused 15 billion US dollars worth of damage. Within 6 h it had spread to its maximum extent, approximately 2 h after that the first signatures had been developed, and another 10 h later downloads of new signatures had protected 90% of the computers.

▶ **Firewalls.** These systems, which may contain hardware as well as software components, are situated between the IT system to be protected and the potentially insecure network (mostly the Internet) and check the data traffic between the two sub-networks. Based on a

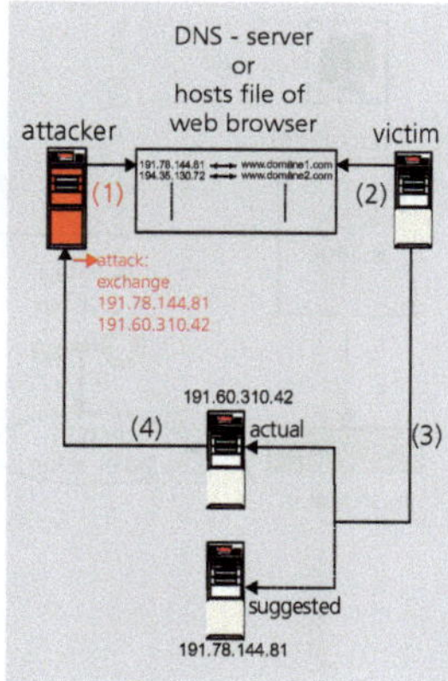

▶ Pharming attack: The attacker manipulates the connection between the domain names, e.g. www.domäne1.com, and the IP addresses, e.g. 191.78.144.81, of different web pages (1). On the Internet this data is stored on DNS servers (DNS = Domain Name Server) and is also found in the hosts file of every computer for frequently used web pages. In the illustration the attacker replaces the IP address 191.78.144.81 of www.domäne1.com with the new IP address 191.60.310.42 which belongs to a computer controlled by the attacker. On entering the domain name www.domäne1.com (2) the victim is directed to a computer controlled by the attacker (3), who is now able to acquire and misuse any data entered by the victim (4). Source: Fraunhofer INT

set of individual filter rules, basic implementations inspect every incoming data packet (packet filter). They reject or accept data packets on the basis of the sending computer's IP address. Application filters check the data traffic on the layer of individual applications or services. They inspect e-mails sent via SMTP (Simple Mail Transfer Protocol) for suspicious content such as viruses or other unwanted content.

▶ **Encryption procedures.** These are important for the protection of data confidentiality. If both communication partners possess the same key, the encryption is symmetric. Symmetric encryption requires the secure exchange of the shared key before the communication starts. Asymmetric encryption requires the generation of a pair of keys for each subscriber. One key of each pair is published, while the other remains secret, i. e. with the subscriber. The responsibility for generating and distributing the key pairs lies with a Trusted Third Party. Mathematical principles forming the basis of asymmetric encryption make it impossible to decrypt a text that has been encrypted with a public key by using that same key and allow it to be decrypted only with the secret key. Thus, if the sender wants to send the addressee a secret message, the sender encrypts it with the addressee's public key, which can be downloaded from the addressee's web page. The message encrypted in this way can be decrypted only with the addressee's secret key, which ensures that only the addressee is able to decrypt the message. Unlike symmetric procedures, asymmetric encryption procedures have the advantage that the key required for decryption does not need to be exchanged between the partners involved, but remains with the addressee all the time. The Trusted Third Party responsible for key generation confirms that the public key and the person associated with it go together by issuing a certificate.

▶ **Authentication procedures.** These are used to ensure that only authorised persons can access IT systems. The most common methods are based on the entry of a password or a personal identification number (PIN). In conjunction with asymmetric encryption these procedures make it possible to infer the addressee's identity from a certificate appended to the public key. Challenge-response procedures involve encryption routines and their execution is usually based on chip cards. They require the chip card to be authenticated to encrypt a random number generated by the system. A user is granted access to the system if the decryption of the value transmitted by the chip card produces the original random number on the system side. This procedure is used e. g. when a subscriber in a mobile communication network (GSM) authenticates to the network. Biometric procedures identifying a person on the basis of endogenous characteristics are another possibility and offer great authentication potential.

Applications

Co-contractors need to be able to authenticate each other and to have a trusted link between them so that they can securely execute electronic business transactions (e-business).

▶ **Trusted links.** HTTPS (Hypertext Transfer Protocol over Secure Sockets Layer), the most widely used security protocol on the Internet, is usually integrated in Internet browsers and establishes an encrypted network connection between the business partners. It combines symmetric with asymmetric encryption. To

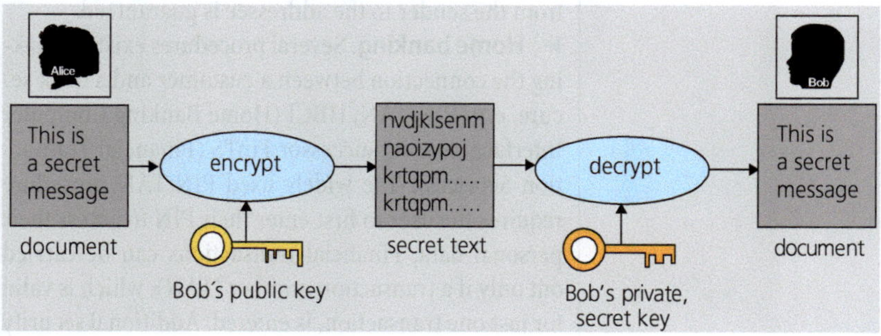

▶ Asymmetric encryption: If Alice wants to send Bob an encrypted message, she encrypts it with Bob's public key. It can be decrypted only with Bob's private and secret key. This ensures that only Bob can decrypt the message sent by Alice. Source: Fraunhofer INT

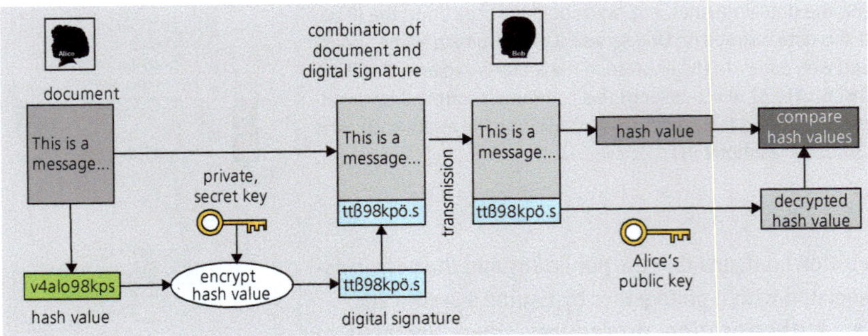

Digital signature: A commonly known mathematical algorithm is used to compress the document, which may be of any length, to a fixed-length sequence (hash value). Any change to the document text results in a change to the hash value. The secret key of the person generating the signature is used to encrypt the hash value, which is converted into a digital signature that can be identified as truly belonging to the holder of the secret key. Then Bob receives a document with a digital signature from Alice and uses Alice's publicly available key to the signature, decrypting the digital signature and generating the hash value from the signature. If both hash values match, the integrity and authenticity of the document are guaranteed. Source: Fraunhofer INT

establish a secure connection, the provider first transmits its public key to the customer. The enclosed certificate enables the customer to verify the provider's identity. The customer then uses the provider's public key to encrypt the message containing the symmetric key for communication. The customer has to be confident that the provider's system is free of security leaks.

▶ **Digital signature.** Digital signatures are important for ensuring a secure and legally binding exchange of data and are used e. g. for contracts and banking transactions, as well as for communication between citizens and public bodies (*e-government*). They are based on asymmetric encryption procedures and are designed to guarantee both the integrity and the authenticity of electronic documents. The first step towards signing a document digitally is to calculate a checksum, the so-called hash value. This hash value, which represents a fixed-length sequence, results from the application of a known mathematical algorithm to the document text, which may be of any length. If only one character in the document text changes, the hash value also changes. Using the secret key of the person creating the signature, the signature is then generated on the basis of the hash value and appended to the document. Unlike the hash value, only the holder of the secret key can generate this particular signature. The addressee can use the sender's public key to reconstruct the hash value from the signature and also use the known mathematical algorithm to generate a hash value of their own from the document. If both values match, the unaltered transmission of the document from the sender to the addressee is guaranteed.

▶ **Home banking.** Several procedures exist for making the connection between a customer and a bank secure, e. g. PIN/TAN, HBCI (Home Banking Computer Interface) and its successor FinTS (Financial Transaction Services). The widely used PIN/TAN procedure requires the user to first enter their PIN to access their personal data. Financial transactions can be carried out only if a transaction number (TAN), which is valid for just one transaction, is entered. Additional security results from the fact that the PIN and TAN are not transmitted in unencrypted form, but after having been encrypted via https. In contrast, HBCI uses asymmetric encryption procedures, and its execution is usually based on chip cards. The bank customer's chip card, which can be read out on the home computer, contains two digital keys, namely the customer's secret key – to generate a digital signature – and the bank's public key – to encrypt the data transmitted. Interesting features of FinTS include an increase in key length and the definition of particular file formats.

▶ **Security in wireless networks.** Wireless Internet or Intranet access via *WLAN* (Wireless Local Area Network) hot spots such as at airports and hotels can be risky for mobile devices. The radio link is susceptible to eavesdropping, so it is essential to use encryption procedures, like the Wired Equivalent Privacy (WEP). For the purpose of encrypting, the user's device first generates a bit sequence, the initialisation vector, which is different for every single data packet of the message to be transmitted. This initialisation vector as well as a static WLAN key provided by and valid for the whole WLAN have to be input into a number generator, which generates a key sequence that is mathematically combined with the message, thus encoding it. What is transmitted in the end is the encrypted message along with the initialisation vector prefixed to the message. The addressee uses the transmitted initialisation vector and the known static WLAN key to generate the same key sequence as originally used by the sender for encryption. The mathematical algorithm already used by the sender then combines the encrypted message with the key sequence, thus reproducing the original message. WEP encryption, however, no longer provides sufficient security for wireless networks. Its weaknesses range from relatively short initialisation vectors and WLAN keys to the fact that the WLAN key is static and valid for the whole wireless network. Another problem is that the user, but not the base station, is required to authenticate for the connection – which is how attackers reproducing the access side of the hot spots are able to get hold of sensitive data. Increased security is provided by the successor procedures WPA (WiFi Protected Access) and WPA2, which not only use enhanced encryption and authentication techniques but also an occasionally changing network key.

▶ **VPN tunneling**. VPN (Virtual Private Network) technology makes it possible to securely connect mobile terminals to an intranet via the Internet. To implement VPNs the IP Security (IPsec) transmission protocol can be used. It encrypts and encapsulates the entire IP packet, i. e. provides it with a new message header (IP header). The purpose of this header is to address the tunnel ends, while the addresses of the actual communication endpoints are inside the inner IP header. VPN technology can also be used to make Internet use on mobile terminals secure. E. g., whenever a terminal connects to the Internet, special software can be used to automatically establish a VPN tunnel to the home intranet that will handle the further Internet traffic. This makes all the IT security resources of the Intranet available to the user whenever the Internet is used.

Trends

▶ **Trusted computing**. Trusted IT platforms represent an alternative approach to software-only IT security solutions. Their main component is a TPM (Trusted Platform Module) hardware module soldered to the motherboard, which provides manipulation-proof directories. Data encrypted with the TPM's public key can be decrypted only with the TPM's private key and under specified conditions. The provider is able to prevent the transfer of data by stipulating e.g that the data transmitted must not be copied or read by e-mail programmes. This is especially important for content subject to copyright or data privacy protection as well as for licensed products, and facilitates *Digital Rights Management* (DRM) e. g. by the film and music industries. The platform owner can also encrypt (seal) data which can then be decrypted only under specified conditions for the TPM, ensuring that all users of the system, including the system administrator, follow certain rules. Before TPM supporting systems can be used in

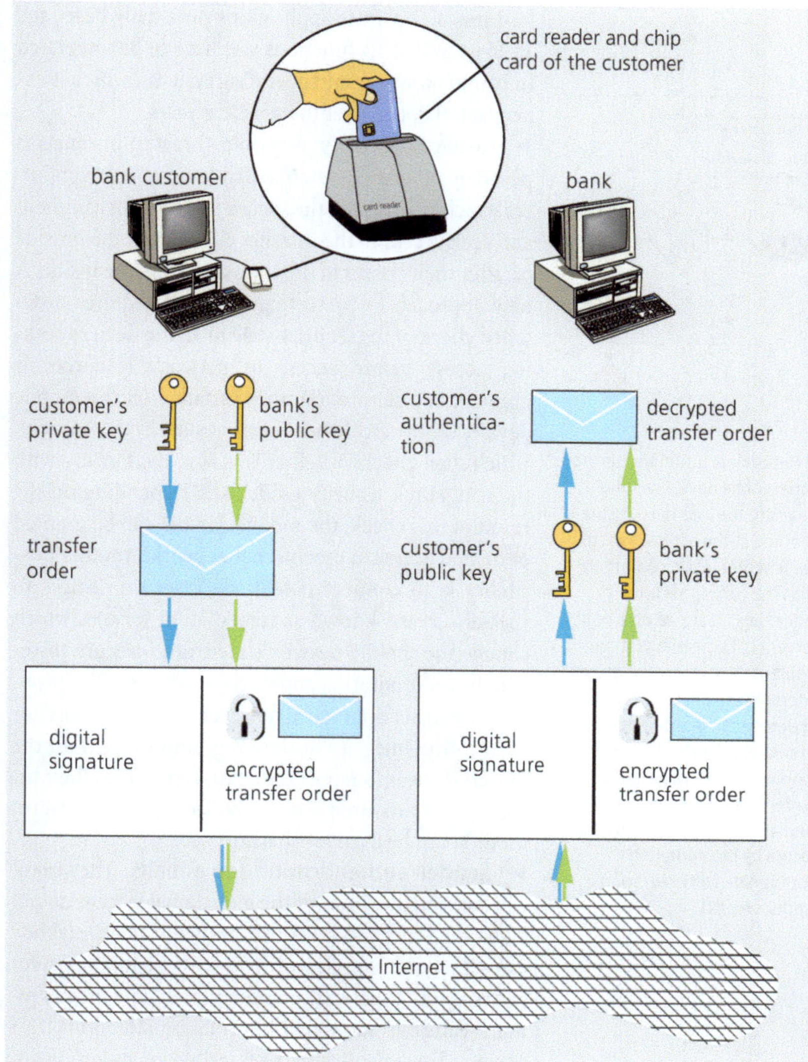

HBCI: This online banking procedure uses asymmetric encryption procedures. The bank and the customer possess a pair of keys. The customer has a chip card containing the private key to generate a signature and the bank's public encryption key. A card reader connected to the home computer reads them out and uses them in a transaction. The customer uses the private key to generate a digital signature which the bank uses for authentication, using the customer's public key (blue arrows). The customer uses the bank's public key to encrypt data, ensuring the confidentiality of the transfer order. The bank uses its private key to reconstruct the encrypted data (green arrows). Source: Fraunhofer INT

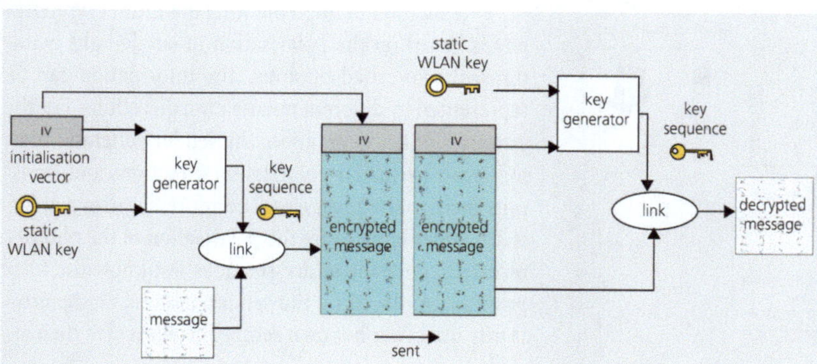

Message encryption in WLANs with WEP: For every message to be encrypted the user selects a bit sequence, the initialisation vector, which has to be input into a key generator along with the static WLAN key valid for the whole wireless network. The key generator produces a key sequence that is mathematically combined with the message to be encrypted, thus producing the encrypted message. The message and the initialisation vector are transmitted. Message decryption: The initialisation vector prefixed to the incoming encrypted message and the WLAN key valid for the whole wireless network are input into a key generator, which produces the same key sequence as the one used by the sender to encrypt the message. This key sequence is then combined with the encrypted message to reproduce the original message. Source: Fraunhofer INT

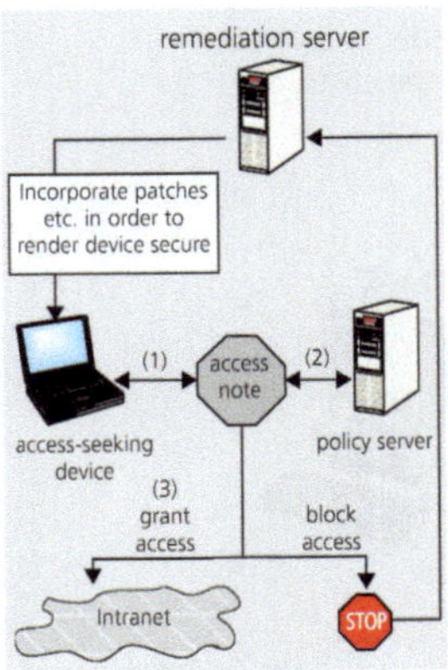

Network security: The state of each mobile device seeking access to the internal network is checked at the access node to see whether it is in conformity with the current security guidelines (1) stored on a policy server (2). Depending on the result of the check, access to the internal network can be granted or denied (3). Another option is to connect the possibly insecure mobile device to a remediation server (4) which changes the device's state from insecure to secure by incorporating appropriate software. Source: Fraunhofer INT

business and private applications on a daily basis, the TPM as well as its functions will have to be integrated fully into hardware and operating systems, which is expected to happen over the next few years.

▶ **Network security.** A serious threat to Intranets is posed by mobile IT systems such as notebook computers which run outside the secure network environment and seek access to the intranet either from the outside or, after their return to the company, from the inside. A new approach to protecting intranets requires a detailed check of the security state of all the devices seeking access before access to network resources is granted. Special programmes installed on the mobile devices report the devices current state to the intranet, which then checks whether this is in conformity with the company's security guidelines. Depending on the result of this check, the mobile devices can be granted or denied access to the internal network. Another possibility is to connect potentially insecure devices to special servers, known as remediation servers, which change the mobile device's state from insecure to secure by incorporating appropriate software. The problem with this approach, however, is that it calls for trust in the integrity of the programmes used on the mobile devices to report the status or requires their integrity to be ensured separately. The Trusted Platform Module can be used for this purpose.

▶ **Sender authentication for e-mails.** The protocols forming the basis of the global mail system do not allow the sender to be clearly authenticated. Anybody can send e-mails under any name. Procedures exist for verifying the trustworthiness of e-mails, and their common feature is additional information stored on DNS servers. These servers are responsible for assigning domain names to the IP addresses associated with this domain. The DomainKeys procedure provides for all e-mails to be digitally signed by the sending domain's mail server with the private key of that domain. The addressee retrieves the sending domain's public key from

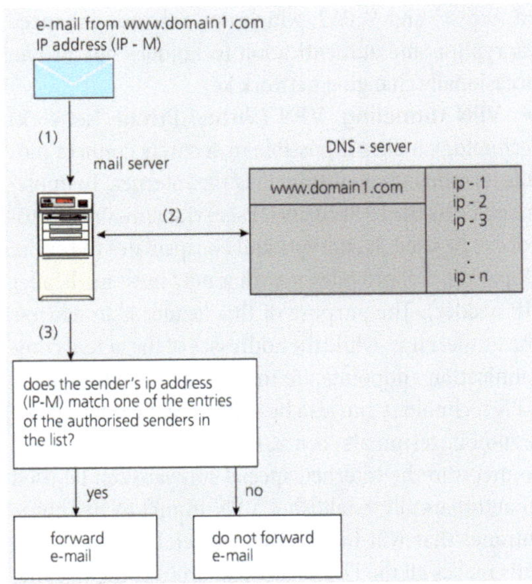

SenderID procedure: To authenticate e-mail senders, SenderID stores a list of IP addresses allowed to send e-mails from the domains on the DNS server. Every incoming e-mail (1) is checked to see whether the sender's IP address is listed (2). If the incoming mail's IP address is not listed, the e-mail can be prevented from being forwarded to the addressee (3). Source: Fraunhofer INT

a DNS server and verifies the signature. SenderID and Sender Policy Framework (SPF) procedures store the IP addresses of the computers allowed to send e-mails from that domain on the DNS server. If an e-mail is sent from a computer that is not registered there it can be prevented from being forwarded to the addressee.

▶ **Quantum cryptography.** This involves the generation of a reliable key by quantum physics methods. The important and unique property is the ability of the two communicating users to detect the presence of any third party trying to gain knowledge of the key. This results from the fact that the process of measuring a quantum system disturbs the system. A third party trying to eavesdrop on the key must in some way measure it, thus introducing detectable anomalies.

One method of implementing quantum cryptography is based on the polarisation of single light quantums, the so-called photons. The information can be represented by different polarisation directions. For the purposes of key generation, the sender generates photons with particular polarisation directions and transmits them to the addressee. The addressee uses a polarisation filter to measure the polarisation of the photons received. Since the addressee does not know the filter position employed by the sender, uses he or she arbitrarily use his or her own setup. The addressee then as-

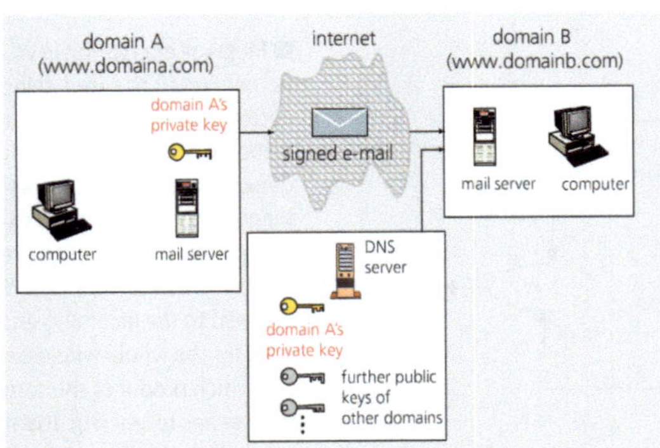

DomainKeys procedure: Every e-mail sent from domain A is digitally signed with the private and secret domain key before it is transmitted via the Internet. Having received the e-mail, the addressee, in this case domain B, uses the public key of domain A to verify the e-mail's digital signature. This key is publicly available on the respective DNS servers. Source: Fraunhofer INT signs binary zeros and ones to each single measurement, as agreed beforehand. When the sender and addressee filters are identical, they possess the same bits. These measurements can thus be used as a key, whereas the others are discarded. The sender and addressee can compare the filter positions selected (and only these positions) e. g. by phone. Since they only talk about the filter positions selected and not about single results, an attacker eavesdropping on this conversation cannot make assumptions about the actual key.

If the key's bits are altered in such a way that a certain threshold is exceeded, the key cannot be used to encrypt the message. In this case, it must be assumed that a third person has intercepted. In practice it is difficult to send light pulses containing exactly one photon. This enables Eve to launch an attack in which she branches off some photons from each pulse, which contains several photons. When Alice and Bob exchange their filter positions via a public channel, Eve can repeat Bob's measurement using her photons and get hold of the information. As a countermeasure, Alice can add additional photons (bait photons) to her pulses, which change their intensity in such a way that it will be noticed if photons are branched off.

Initial commercial products combine quantum cryptography as a means of securely generating keys with both symmetric and asymmetric encryption procedures. These products are based on optical fibre connections. It has also been possible, however, to demonstrate that quantum cryptography is able to wirelessly bridge distances of about 150 km, matching the maximum range of fibre-based transmission. This could permit the secure exchange of keys between satellites and ground stations.

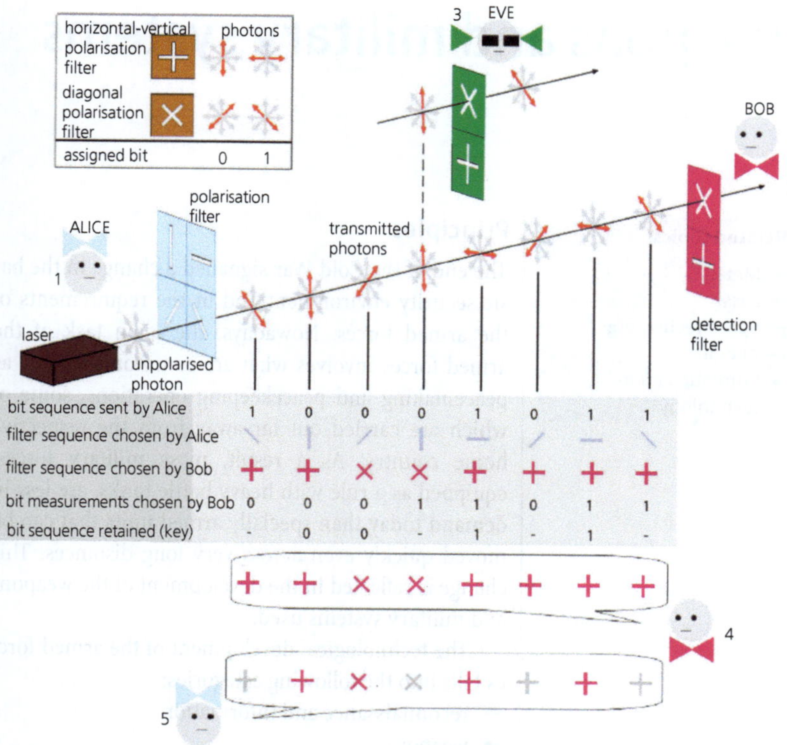

Quantum cryptography encryption: To generate a key, Alice sends single photons through the "0" or "1" slots of the horizontal/vertical or diagonal polarisation filters. She records the different directions (1). Bob randomly selects one of his two filters for every incoming photon and records both the filter position and the bit value measured (2). Attacker Eve, who wants to spy on the photon sequence, also has to opt for a filter, but if she makes the wrong choice the photon's polarisation will change (3). After transmission of all the photons, Bob and Alice exchange information on the filter selection via a public channel. The bits read out when the right filter selection is made provide the key that can be used to encrypt a message (4, 5). Source: VDI Technologiezentrum GmbH

Prospects

- It will remain relatively easy to conduct IT attacks, but, as the efficiency of IT security mechanisms steadily increases, along with users' awareness of IT security problems, the effectiveness of such attacks will decline significantly.
- IT security is especially important in e-business. Companies cannot afford to lose their customers' trust as a result of security leaks. IT security will therefore become an essential part of a comprehensive company security policy.
- IT security will play a major role in the design of future information technology systems. One approach is to integrate a hardware module (TPM) guaranteeing specific security characteristics.
- New ways of securing networks will take root. They will provide a detailed security check of all terminals seeking access before access is granted to the network or network resources.
- Since many attacks are based on faking existing e-mail addresses, future efforts will focus on the establishment and further development of procedures to authenticate e-mail senders with the aim of reducing phishing, malware-infected e-mails and e-mail spam.
- As the technical implementation of quantum cryptography improves, commercial utilisation will become a reality.

THOMAS EUTING
DR. BIRGIT WEIMERT
Fraunhofer Institute for Technological Trend Analysis INT, Euskirchen

Internet

- www.enisa.europa.eu
- www.cert.org
- www.bsi.bund.de/english/index.htm
- www.heise-online.co.uk/security
- www.iwar.org.uk/comsec

Weapons and military systems

Related topics

- Metals
- Laser
- Space technology
- Aircraft
- Communication technology

Principles

The end of the Cold War signalled a change in the basic security environment and in the requirements of the armed forces. Nowadays, the main task of the armed forces involves what are commonly known as peacemaking and peacekeeping operations, some of which are carried out far away from the respective home country. As a result, mass military forces, equipped as a rule with heavy battle tanks, are less in demand today than specially armed units that can be moved quickly even across very long distances. This change is reflected in the development of the weapons and military systems used.

The technological development of the armed forces falls into the following categories:

- reconnaissance and information
- weapons
- mobile land-, air-, sea- or space-based delivery systems
- protection

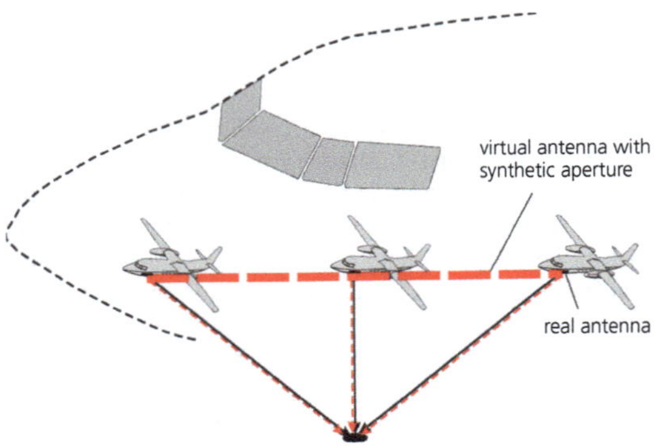

SAR side-looking radar: The different records made during the flight are added together to generate a picture similar to that provided by a far larger virtual antenna (red) that is installed on a far larger airplane (black dotted line), which would complete the recording in one step. The position of the real SAR antenna relative to the point being observed must be known precisely at all times. Source: Fraunhofer INT

▶ **Synthetic-Aperture Radar (SAR).** SAR components belong to reconnaissance and information. They are used from airborne platforms and are pointed sideways to deliver two-dimensional pictures of the terrain sectors under observation. They combine the advantages of radar, in particular that of deployment in nearly all weather conditions, with an especially high ground resolution (down to decimetre range) that radar *sensors* normally cannot reach in such a set-up. The length of the antenna that can be used on the platform usually limits a side-looking radar's ground resolution along a flight track. The visible antenna surface is commonly referred to as an aperture. The larger the aperture, the higher the ground resolution, i.e. the narrower the distance allowed between two points that the sensor will recognise as two separate objects. SAR improves ground resolution along the flight track, so the antenna's length along the flight track is of particular importance.

Common parabolic reflector antennas can be used as SAR sensors. They virtually extend the aperture's length by recording the target area from different positions during flight. Extensive computing power is then required to collate the data and generate a picture similar to that provided by a large antenna recording everything at once. Another advantage is an especially high ground resolution, independent of the antenna's distance from the target area. The larger distance allows the same point in the target area to be recorded over a greater flight distance and a longer period of time, even with a non-rotatable antenna on the platform. But while a real antenna's resolution deteriorates with increasing distance, the time it takes to collate a picture in SAR mode must also be considered.

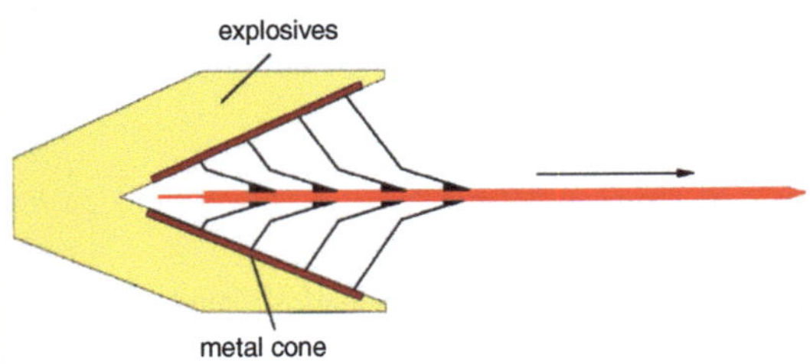

Warheads: After the detonation of a shaped charge's high explosive, the metal cone is converted into a jet that consists of metallic particles that penetrate the target. The metals selected are easy to shape (ductile) and as dense as possible (copper is often used). Source: Fraunhofer INT

► **Warheads**. The warhead represents a system's armament. Conventional warheads range from basic charges with pure blast and pressure effects, to kinetic energy, fragmentation and thermal warheads, and even shaped charges. The latter are designed to penetrate armour. They consist of a cone-shaped metallic liner filled with an explosive charge. The blast triggered at a specific distance to the target creates a metallic jet that leaves the cone at a very high speed (up to over 10,000 m/s) and stretches during flight. When this jet strikes the target, it produces such extreme high pressure that the armour behaves according to the principles of hydrodynamics – like a fluid – and can be penetrated.

► **Nuclear weapons**. Nuclear explosions have various destructive impacts. There is the blast wave, then thermal radiation, neutron and gamma radiation and radioactive fallout. Furthermore, every nuclear explosion causes an electromagnetic pulse in its proximity as a result of the intensive gamma radiation triggered by the so-called Compton effect in the atmosphere. This pulse couples into electric cables and electronic networks and damages or destroys them. New nuclear weapons are less focused on pure nuclear fission and fusion weapons, but rather on so-called 3rd and 4th generation weapons. 3rd generation weapons can selectively enhance or suppress any of the destructive effects mentioned above, generating high-level neutron radiation, for example. In terms of explosive power, 4th generation nuclear weapons fall between conventional bombs and traditional nuclear weapons. One example is the bunker buster, or earth-penetrating weapon, which is said to be able to destroy bunkers buried up to 10 m underground. In principle, they have the same effects as conventional bombs but their blast wave is considerably stronger.

► **Composite armours**. In addition to the materials used, particular importance is assigned to the design of armour protection. Multi-layer armours consist of two or more plates or layers made of similar or different materials that are fused together. If different types of materials are combined, the armour is referred to as composite armour. The most basic design involves an outer layer made of a very hard material that is intended to break, deform or erode armour-piercing projectiles upon impact, generating smaller elements of lower energy. The task of the inner layer, which is usually made of a more ductile material, is to prevent fractions from the first layer from moving inward, as well as catch fragments and avoid interior spalling.

► **Stealth technology**. Stealth technology is applied as a special precaution to hide systems from enemy radar reconnaissance. The characteristic traces emitted by the system, or the signature that is detectable within the radar's range, is referred to as the *radar cross section* (RCS). It is described as the surface of an ideal virtual reflector, i. e. reflecting the same amount of radar rays back to the receiving antenna as those reflected by the system being observed.

The outer shape and the materials used in the design of the system under observation are of particular importance to the radar cross section. *Radar absorbent materials* (RAM) prevent the reflection of incident electromagnetic radar waves by absorbing and converting them into heat. They often absorb within a relatively broad wavelength band. Modern *airplanes*

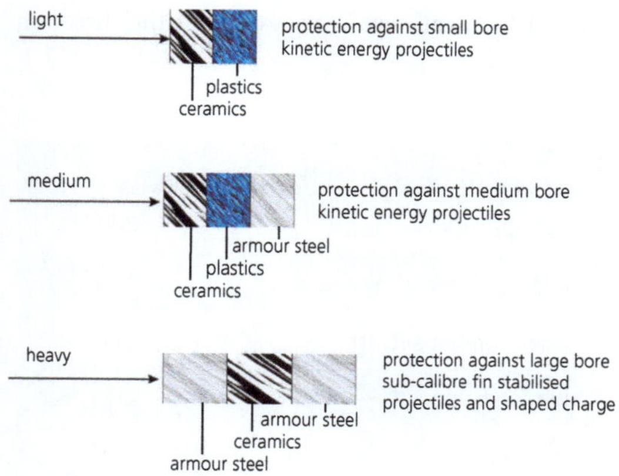

Composite armours: Material combinations typical of modern composite armours are ceramics-plastics for light armours, ceramics-plastics-armour steel for medium armours and armour steel-ceramics-armour steel for heavy armours. The sequence in which the respective materials are mentioned characterises the plates' sequence from the outside inwards. Source: Fraunhofer INT

Radar Cross Section (RCS): "F" corresponds to the surface of an ideal virtual reflector that would reflect the same signal back to the radar receiver as the system being observed. The radar sender and receiver are usually in the same place. Source: Fraunhofer INT

mostly use a design based on aramid fibres that already incorporate absorbent material. So-called resonant absorbers have effects similar to RAM. They are based on the overlapping and mutual extinction of radar rays, which are reflected at the material's rear and front sides and have very narrowband effects on their own. Attempts to reduce the RCS through the shape given to the system under observation, focus on designing and positioning its reflecting surfaces in such a way that most of the radar's rays are reflected in directions other than the direction from which the rays came, since the sender would then intercept and measure them. One approach is to avoid so-called corner reflectors. These are corners and edges reflecting electromagnetic rays back to the sender – the way light is reflected by a cat's eye – regardless of the direction from which they came. In this case, external superstructural parts must be avoided altogether, or covered with radar absorbent materials.

▸ Typical radar cross sections: Some data are estimated (since some of the exact values are confidential). Source: Fraunhofer INT

system	RCS[m²]
pickup-truck	200
large civilian aeroplane	100
small combat aircraft	2-3
human being	1
small bird	0.01
F-117 stealth combat aircraft	0.1
B2 stealth bomber aircraft	0.01

▴ US-American B2 stealth bomber aircraft: This aircraft leads the field in radar camouflage. From the radar's point of view, it is no larger than a small bird. In addition to the use of radar absorbent composite materials, the aircraft features a surface that conceals highly reflective metallic components in structural areas situated beneath the skin: the power plants, landing gears or loaded bombs. The smooth surfaces, in particular those at the bomber aircraft's front, are mostly curved. They hardly ever reflect incident radar rays back to the sender. This is also true for the flat underside, which can be detected only by a radar station situated immediately beneath the aircraft. Source: www.af.mil

Applications

▸ **Armoured vehicles.** The German Bundeswehr's new PUMA infantry fighting vehicle represents the latest generation in medium-armoured vehicles. It features a modular protection system combining air transportability on board the future A400M Airbus with maximum on-site protection. It also boasts highly effective level-A (air-transportable) protection against mines. So-called add-on additional armours, which can be transported separately, provide maximum level-C (combat) protection when installed.

The armoured vehicle also uses a 360-degree periscope for surveillance, independent of the position of the turret. It transmits images to the interior of the vehicle via a vertical fibre-optical channel (similar to a submarine periscope). These can be viewed by the commander through an eyepiece. A built-in camera also transfers the images to a monitor, which can also display images captured by a daytime camera or a nighttime thermal imaging device.

Not least because of weight limitations, armoured vehicles will in future be equipped with active mechanical protection systems. The PUMA will also be designed with the corresponding interfaces to such future systems. These systems are designed to detect incoming projectiles or missiles and damage or destroy them before they hit the vehicle. Such protective shields can be based on fragments or projectiles, as well as on flying plates discharged by a proactive blast to meet the attacking penetrator. The use of high-explosive grenades with powerful blast effects, which cause the penetrator to sway off target before it strikes the vehicle, is also being analysed.

Implementation of active mechanical protection systems depends essentially on the progress achieved in the field of sensor systems. Their efficacy against relatively large and slow (up to approximately 800 m/s) missiles has been reliably demonstrated. However, even if the development of active protection systems progresses favourably, it will remain impossible to forgo the use of basic conventional armour protection.

▸ **The soldier as a system.** This approach assumes that the individual soldier can also be regarded as a system equipped with all required technical components. Initial international developments in that field focus on the infantryman's equipment. The German Bundeswehr has already introduced a system comprising an adapted radio set and satellite navigation system GPS (Global Positioning System), as well as a PDA (Personal Digital Assistant) equipped with a digital map and a digital camera. Maps and compass have served their time.

Supporting soldiers by mechanically effective auxiliary systems is is being given extensive attention, Exoskeletons, which are fitted to the soldier's body like a combat uniform to augment the force behind his mechanical bodily functions, exemplify such developments. Other considerations pertain to partly autonomous robots. These accompany the soldier as heavy load bearers and may even independently transport him back to the base in case of injury.

▶ **Guided missiles.** Nowadays, large-calibre guns such as tank guns are competing with guided missiles, which are becoming more and more powerful: their flight path can be controlled and they usually have a propulsion system of their own. In addition to their potential to prevent collateral damage, interest is also focused on their specific suitability for use as standoff weapons, which allow engaging a target from a safe distance. So-called cruise missiles are of particular importance when engaging land targets in peacemaking operations. These guided missiles cover ranges of up to some thousand kilometres in extreme cases, flying for the most part under their own propulsion. Cruise missiles can thus also be referred to as unmanned aerial vehicles. They can be launched from airborne or waterborne platforms as well as from the ground and are able to manoeuvre during the entire flight. Equipped with a suitable data link, they can be reprogrammed during flight to approach a new target and are highly precise. Modern subsonic cruise missiles are capable of terrain flight, escaping detection at very low altitudes (10–200 m). In doing so, they feature extremely low signatures (characteristic detectable reflections emitted by the system).

Trends

▶ **Missile defence.** Since potential enemies are increasingly using guided missiles of different sizes and ranges, it has become increasingly important in recent years to protect against such threats. This development has led to the tasks of traditional air defence, originally limited to engaging airplanes, being incorporated into a so-called extended air defence. The size of the areas to be protected varies: from single sites or ships, to entire stationing areas or combat fields, through to to entire continents. Depending on the scenario, defence systems based on the use of lasers or high power microwaves are destined to play an important role in the medium term.

In essence, current defence strategies are based on the use of defensive missiles, which may strike the attacking missile directly and destroy it. At best, that has to take place as far as possible from one's own territory, or at the highest possible altitude to minimise the effects of the warhead, which might be equipped with weapons of mass destruction. To this end, the interceptor missile receives initial guidance from a ground based radar system, until its own sensors pick up the attacker. The interceptor's *sensors* can be based on ra-

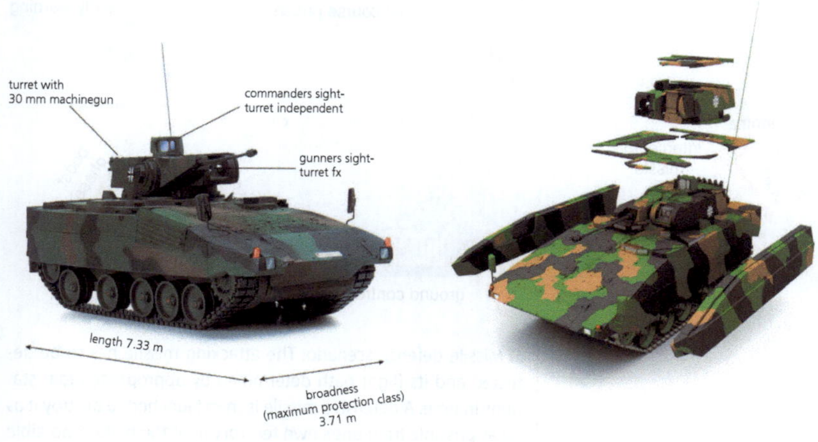

PUMA armoured infantry fighting vehicle. Right: Modular protective components, which can be installed if needed. Source: Rheinmetall AG

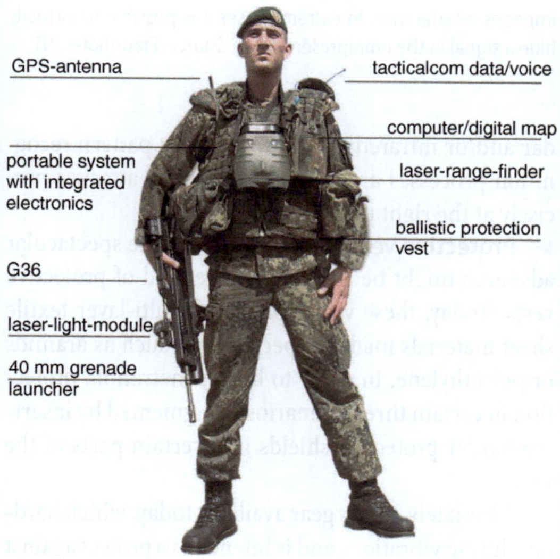

Infantryman; equipment used by the German Bundeswehr. Source: EADS

The Exoskeleton: Augmenting the force of mechanical bodily functions. Source: Kazerooni, University of California at Berkeley

Tomahawk cruise missile in action: The major components of this cruise missile are (from front to back) detector head, navigation system, warhead, fuel tank, navigational computer and jet engine. Source: Raytheon

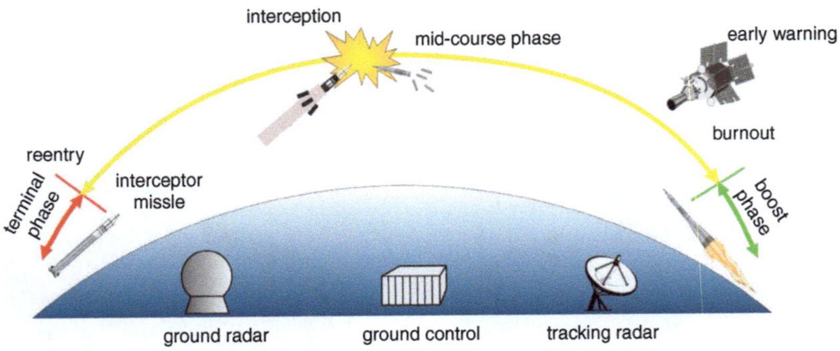

▲ Missile defence scenario: The attacking missile has to be detected and its flight path determined by appropriate radar stations in time. A defensive missile is then launched to destroy it as far as possible from one's own territory or at the highest possible altitude. Source: Fraunhofer INT

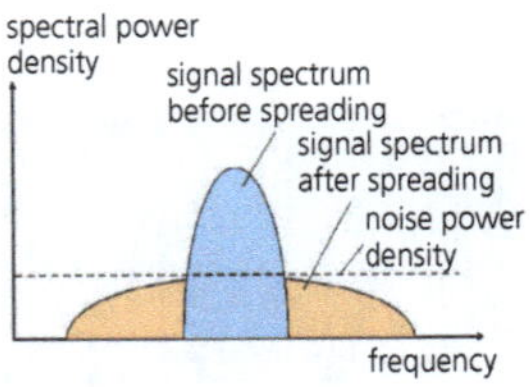

◄ Spread spectrum procedure: The low power density makes it difficult to detect the multipath signal and the broad bandwidth impedes interference. In extreme cases it is possible to virtually hide a signal in the omnipresent noise. Source: Fraunhofer INT

dar and/or infrared cameras. Complex pattern recognition processes are required to hit the attacker precisely at the right part of the warhead.

▸ **Protective vests.** In the long run, the spectacular advances might be achieved in the field of protective vests. Today, these vests consist of multi-layer textile sheet materials made of special fibres such as aramids or polyethylene. In order to limit penetration, protection in certain threat scenarios is augmented by inserting harder protective shields into certain parts of the vest.

Ultimately *sports* gear available today, which hardens during vibrations and is intended to protect against injuries from falling, might serve as a model for the further development of protective vests. Its material consists of flexible chains of molecules, which only become entangled under repeated rapid movement. Whether such materials will ever be suitable for armour protection purposes in military applications remains to be seen.

Regardless of the fibre type, soaking the material with a so-called shear-thickening fluid (STF) might enhance protective performance. The constituent particles in the nanoscale suspensions of such liquid armour agglomerate when subjected to rapid shear force, significantly enhancing the fluid's viscosity. Soaking the material with a *magnetorheological fluid*, the viscosity of which enhances when a magnetic field is applied, might be another way to implement active protection (instant armour). Such highly flexible materials would serve to protect soldiers' body parts without limiting their mobility, but there is still a long way to go in achieving any practical application of such developments.

▸ **Low probability of intercept procedures.** Communications networks form the backbone of network centric warfare. Since they must be mobile, they will continue to rely primarily on wireless links. Measures to protect these systems against detection and interference are of decisive importance. So-called LPI (Low Probability of Intercept) procedures serve this aim, and in particular multipath transmission procedures. Based on exactly defined rules, these spread the signal to be transmitted over a wide frequency range and simultaneously reduce its spectral power density. It thus becomes more or less possible to hide a signal in the omnipresent noise. Since the power density is low, it is hard to detect the multipath signal. The broad bandwidth simultaneously increases the immunity to interference. Only the authorised addressee knows the multipath procedure used and is able to restore the original signal.

The most important multipath procedures are frequency hopping and direct sequence procedures. Frequency hopping constantly changes the transmission frequency in a way that seems arbitrary to possible intruders. Thus, it is difficult for them to follow the quick frequency changes and interfere effectively. Direct sequence is based on adding a bit sequence featuring a much higher rate to the bit sequence to be transmitted, thereby spreading the signal spectrally as well. All in all, direct sequence provides a far better protection against detection or interference, which is why it will be used increasingly in future.

▸ **Laser weapons.** Depending on the intended use of laser weapons, a distinction is made between different power classes. So-called low-energy lasers (medium output less than 1 kW) are meant to cause in-band damage to sensors, i.e. interfere with how they work within their respective wavelength range. Such lasers blind the sensor by overpowering it with a much higher signal density. This makes it possible to knock out sensors in the detector heads of attacking guided missiles, for example. Medium-energy lasers (medium output of up to some 100 kW) are aimed at destroying optical and electro-optical devices (so-called out-of-band damage). Their radiation is absorbed, thereby heating the target material, such as that used for the sensor window, which softens or even evaporates. High-energy lasers feature far higher outputs in the megawatt range and are intended to destroy (relatively light) structures, e.g. missiles.

The major disadvantage of laser weapons is that the atmosphere limits their effectiveness. The relatively low wavelengths of their radiation allows it to be absorbed, e. g. during bad weather conditions on its way to the target (their wavelength is smaller than the diameter of existing floating particles and droplets). Hence, lasers aimed at destroying structures will primarily play a role in defence scenarios against missiles within upper layers of the atmosphere or from space.

▶ **Microwave weapons.** In principle, so-called high-power microwave (HPM) weapons are also ready for deployment today. They radiate within wavelengths exceeding 1 mm, and so the atmosphere does not absorb their radiation. Their microwave field couples into electric circuits, where they generate voltages and currents high enough to interfere with or destroy electronic components. In principle, they are meant to disrupt systems that rely heavily on electronics, these being inadequately protected against such radiation. There is a wide range of possible targets for high-power microwave weapons today, and the list will continue to grow, from "switching off" electronically initiated mines or booby traps hidden at the side of the road, to stopping enemy vehicles.

For conventional transmitter antennas, field intensity decreases proportionately to the growing distance to the target. Hence, one scenario for the use of HPM weapons envisions the one-time firing of an emitter near the target. Reusable systems, which remain effective from a distance, are also being developed. They are presently effective up to a distance of 1 km.

▶ **Non-lethal weapons**. Non-lethal weapons are intended to deliver a measured but adequate response to opponents who are, in principle, inferior. Such weapons are meant to incapacitate persons without causing them permanent harm, if possible, or to render their equipment useless, for example, by "deactivating" electronic functions with HPM weapons. Pistol-like electroshock devices, which transmit electric pulses by firing hook-like electrodes, are an example of non-lethal weapons used already against human beings in a non-military context.

Prospects

- The information factor is becoming increasingly dominant in the military sector, too, along with related technologies to collect, transmit and process data. This leads to new threats and types of conflicts, now referred to as information operations.
- This development is accompanied by an increase in technologies that can be used for both civilian and military applications. In this context, the civilian sector is increasingly taking the lead, which is attributable to the economic importance of information and communication technologies.
- Development in the field of communication media ultimately results in any information being available at any place and any time. This enables the comprehensive networking of sensors, command and control and weapon systems. The implementation of such technology to increase the performance of armed forces is referred to as network-centric warfare.
- The increasing availability of intelligent systems, as well as substantial improvement in the performance of sensors and navigation systems, facilitate a far more efficient use of weapons.
- The entire field of protection is increasingly characterised by automated or intelligent procedures. The growing deployment of so-called standoff weapons, in particular of guided missiles, is important for the protection of soldiers and avoiding the risks of direct force-on-force situations. The use of unmanned systems is also intended to reduce human exposure to dangerous situations.

◂ Mobile high-power microwave weapon: The large transmitter antenna, which is mounted to the vehicle's roof, is the most prominent feature. Source: www.de.afrl.af.mil

▸ Electroshock device: This "pistol" fires two hook-like electrodes, which receive electric pulses via thin wires. These pulses cause a human target to lose physical coordination. The disposable cartridge at the front contains the hook-like electrodes and the two transmission wires. For military use, their range is intended to increase beyond the typical current range of approximately 7 m. Source: TASER International

JUERGEN KOHLHOFF

Fraunhofer Institute for Technological Trend Analysis INT, Euskirchen

Internet

- www.globalsecurity.org/military/systems/index.html
- www.fas.org/man/dod-101/sys/index.html
- www.eda.europe.eu

Defence against hazardous materials

Related topics
- Weapons and military systems
- Plant safety
- Environmental monitoring
- Sensor systems

Principles

Hazardous materials are matter, substances, mixtures and solutions with harmful characteristics (e.g. toxic, flammable, mutagenic). Normally a distinction is made between the four main categories of chemical agents, contagious agents, radioactive substances and explosives when considering possible protective measures. Technological progress has put mankind and the natural world at greater risk of accidents arising from the industrial use, manufacture or transport of such hazardous materials. There is also the ever-present threat of warlike or terrorist acts. Threat analyses are attaching more and more importance to risks posed to the population by terrorism.

▶ **Ion mobility spectrometry**. Nowadays ion mobility spectrometry (IMS) or mass spectrometry (MS) is frequently used to chemically analyse substances in sample form. Using either IMS or MS, the sample to be analysed must first be converted into a gas. The molecules in the sample are then ionised. Using IMS the ions subsequently move through a drift gas under the influence of an electric field. The resulting drift velocity and therefore also the measured drift time of the ions in the gas depend on the strength of the electric field and on ion mobility. Ion mobility is a characteristic value of the relevant ions and is used as a basis for identifying the substances that need to be analysed. It is determined by, amongst other things, the mass, charge, size and shape of the ions. Using the MS method, analysis of the ions takes place in a vacuum, unlike IMS. The ions are identified on the basis of their mass-to-charge ratio. This ratio can be determined, for example, by means of a time-of-flight mass spectrometer. In this case, the ions are accelerated by an electric field. Then they travel a fixed flight distance in a field-free vacuum. During this process the flight time required by each ion is measured. This flight time depends on the mass-to-charge ratio of the ions. MS is typically a much more accurate method of identifying substances than IMS.

▶ **Immunological methods of detection**. Immunological methods of detection are used to detect contagious agents and are based on so-called antigen-antibody reactions. Antibodies are produced by the immune system as a reaction to foreign substances (antigens) that have invaded the body. Antibodies always combine with the antigens belonging to them. This characteristic is exploited by immunological detection methods. Antibodies that bind specifically with a contagious agent are used for this purpose. These

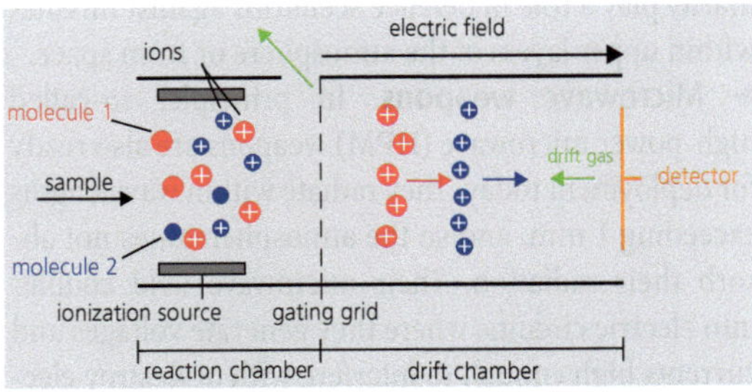

Ion mobility spectrometer: The molecules in a sample are ionised in the reaction chamber. The generated ions are then sent into the drift chamber via the gating grid. Here they move through a counter-flowing drift gas under the influence of an electric field until they reach the detector. The drift time of the different ions depends on, amongst other things, their respective ion mobility, which is a characteristic feature. The resulting ion mobilities are then used to identify the substances present in the sample. The drift gas moves from the detector to the gating grid and cleanses the drift chamber of unwanted neutral impurities. Source: Fraunhofer INT

The main classes of hazardous materials and some examples.

hazardous materials			
radioactive substances	**contagious agents**	**explosives**	**chemical agents**
fissile materials: - plutonium-239 - uranium-235 other radioactive substances: - caesium-137 - cobalt-60	bacteria: - bacillus anthracis (anthrax) - yersinia pestis (plague) viruses: - variola virus (smallpox) - yellow fever virus	warfare agents: - soman (nerve agent) - sulphur mustard (s-lost) - (blister agent) chlorpicrin (choking agent) toxic industrial chemicals (TIC): - sulphur dioxide - cyanogen chloride	military explosives: - TNT (trinitrotoluene) - RDX (hexogen, e.g. in plastic explosives) commercial explosives: - dynamite - ammonit 3

antibodies are attached to the surface of a carrier. When a sample is washed over the carrier, any hazardous substances that may be in the sample bind with the antibodies attached to the carrier. A check is then done to see if there has been any binding activity. This can be done, for example, by washing antibodies over the carrier that bind with any hazardous substance present, but are also marked with a fluorescent dye. These antibodies then stick to the hazardous substance now attached to the carrier. Any sign of fluorescence is therefore an indication that the hazardous material is present. This method can be used to detect various contagious agents at the same time. This is done by applying the specific antibodies to the carrier section by section. An antibody-based biochip is then produced.

▶ **Nuclear radiation detectors.** Nuclear radiation takes the form of alpha, beta, gamma or neutron radiation, for instance. This radiation can be detected by gas-filled detectors, *semiconductor* detectors or scintillation detectors. An ionisation chamber is an example of a gas-filled detector. The detector can be a parallel-plate capacitor filled with an easily ionisable detector gas, e.g. argon. Incident nuclear radiation then ionises the detector gas as it flows through the ionisation chamber, creating charge carriers. Capacitor plates collect these charge carriers, depending on their charge. This generates an electric signal. Semiconductor detectors work in a similar way to ionisation chambers. However, a suitable semiconductor, such as germanium, is used instead of a gas. Nuclear radiation creates charge carriers made up of electron-hole pairs as it passes through the semiconductor. These charge carriers are then, as in the ionisation chamber, collected and measured using an electric field. Scintillation detectors use scintillation materials such as sodium iodide as a detector. These materials produce flashes of light (scintillations) when they interact with nuclear radiation. The scintillation material also transmits this light to a detector able to transform it into an electric signal.

Applications

Systems used to detect hazardous materials can generally be used in the following areas:

- To monitor the concentration of hazardous materials in the environment (hazardous substance monitoring in workplaces and environmental monitoring to assess the effects of any hazardous materials released after there has been an accident or if weapons have been used)
- To identify local hazards if any hazardous materials have been released (e.g. by first-aiders or military NBC defence unit personnel)
- To detect hazardous materials to stop them being distributed illegally (safety checks, e.g. at airports or national borders)

Any objects found to be contaminated with chemical or contagious agents or radioactive substances can then be decontaminated.

▶ **Detection of hazardous chemical agents.** With regard to hazardous chemical agents, a fundamental distinction is made between remote detection methods and detection methods based on the analysis of samples. Remote detection methods are used to identify hazardous chemical agents in aerosol clouds at a distance. One such method is to measure the exact wavelengths absorbed by the aerosol cloud in the infrared range of the electromagnetic spectrum. These wavelengths then indicate if any hazardous chemical agents are present.

Analysis of samples permits, in principle, on-site identification of any chemical agent that could be released at the scene of an emergency. Mass spectrometry methods are ideally used in this scenario. However, mass spectrometers are very expensive and non-portable. Also, specialist personnel have to be trained to use them as they are difficult to operate and evaluating the measured data is a complex process. So mass spectrometry only has a limited range of applications at present. This particularly applies to mobile first-aiders and NBC defence units of the armed forces. They have to have reasonably priced, portable devices. Ion mobility spectrometers are often used in these areas of application. No universal on-site detection solution is envisaged as yet because such a solution has to take a multitude of operational scenarios and hazardous substances into consideration.

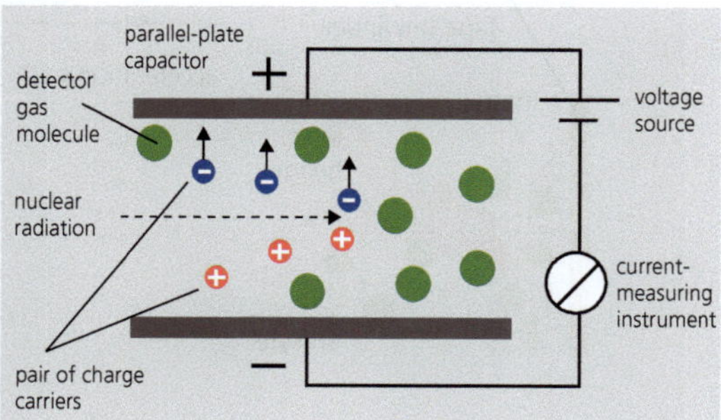

Ionisation chamber: Nuclear radiation ionises molecules of the detector gas (green), creating pairs of charge carriers (red or blue). These move to the positive or negative capacitor plates, where they generate a measurable electric signal. Source: Fraunhofer INT

Portable explosive trace detector: In this case, an ion mobility spectrometer is used for detection purposes. Source: General Electrics

Arriving at the scene of an emergency, it may be difficult to determine the hazards present. This technology, installed in miniature chemical and biological detection systems, puts the power of an analytical laboratory in the palm of your hand. As they can rapidly detect and identify a range of chemical and biological weapons, they could help the authorities to quickly ascertain if a chemical or biological attack has taken place – and to respond accordingly. Source: Sandia National Laboratories

▶ **Detection of contagious agents.** Devices for detecting infectious substances exhibit a lower level of technological maturity compared with those for detecting hazardous chemical substances. This ist due to, among other reasons, infectious substances being able to exert their effect in smaller quantities than hazardous chemical substances. Therefore more sensitive instruments have to be used to detect them. It is also much more difficult to distinguish between infectious and harmless substances that are naturally present in the environment, compared with hazardous chemical substances. Most of the biological methods used to detect contagious agents are based on exploiting biomolecular binding events, such as the antigen-antibody reactions in immunological detection methods. However, a chemical method of detection also exists in the form of mass spectrometry. It is used to search for certain characteristic building blocks, or so-called biomarkers, of contagious agents. Certain proteins, for example, can be *biomarkers*. Agents are identified on the basis of these biomarkers. In addition to detection devices used to analyse samples, detectors also remotely scan aerosol clouds containing biological substances which could have a detrimental effect on health. These detectors work in a similar way to detectors used in hazardous chemical identification systems.

▶ **Detection of radioactive substances.** A pencil dosimeter, used as a personal dosimeter, is an example of a gas-filled detector. In addition to these detectors which can be read on the spot, many systems used for personal dose monitoring are read at a later point in time. These systems are based, for example, on nuclear radiation blackening a photographic film (film badge).

In some areas of application it is important that radioactive substances be identified beyond doubt. During border checks, for example, to distinguish between legitimate radiation sources, such as those used for medical purposes, and illegal ones. Often substances are identified by analysing the energy spectrum of the gamma radiation emitted by radioactive substances. To do this, the radiation detector in use must be able to resolve energy to a level high enough to be able to correctly identify the characteristics of the energy spectrum emitted by a particular substance. Semiconductor

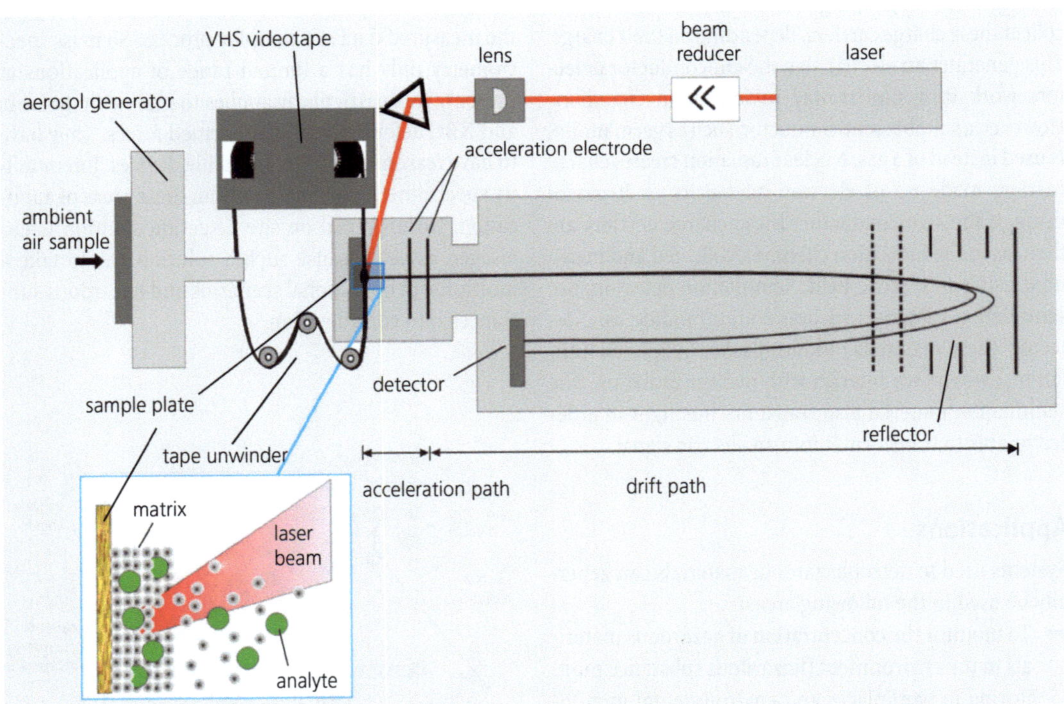

Analysis of bioaerosols using portable "Tiny-TOF" (Time-Of-Flight) mass spectrometer: The device is the size of a briefcase and is designed for automatic analysis of contagious agents in aerosol samples. An aerosol collector takes a concentrated aerosol sample from the ambient air. This aerosol sample is then deposited onto the tape of a commercial VHS video cassette. The sample is conveyed on this tape to the actual mass spectrometer, onto the aerosolised sample spot and moved to the focal point of the laser. The laser is fired, causing the desorbing of the matrix and sample, and a drifting of the charged molecular ions towards the detector, where possible hazardous substances are identified. Source: Applied Physics Laboratory, John Hopkins University detectors permit a high level of energy resolution and are therefore especially suited to this sort of analysis.

Plutonium and, in particular, highly enriched uranium, only emit a relatively low level of radiation, which can easily be shielded. It is therefore difficult to detect such substances in freight containers, for example, when using normal passive detection methods. Consequently, methods involving active stimulation of these materials by neutron or gamma radiation to bring about nuclear fission are currently being considered. The radiation resulting from nuclear fission is more penetrating and is therefore more easily detected than the natural radiation emitted by these substances.

▶ **Explosives detection.** There are basically two different approaches to detecting explosives:

- Bulk detection is the detection of large quantities of explosives by imaging X-ray devices, e. g.
- Trace detection is used to detect traces the actual explosive leaves behind. These traces can appear as vapors or particles and can be detected using ion mobility spectrometry and mass spectrometry. However, trained sniffer dogs are still the most effective way of detecting vapors given off by explosives.

Security screening, especially at airports, is an important area of application for explosives detection. Imaging X-ray devices are mainly used to examine passenger baggage. These devices harness the ability of X-rays to penetrate objects to some extent. This makes it possible to see what is inside the items of baggage. X-ray devices typically operate by positioning an X-ray source on one side of an item to be screened. The X-rays pass through the object and a detector on the other side of the item converts them into an (transmission) X-ray image. The radiation is weakened when the X-rays interact with the item and its contents. This reduction in strength depends on the density of the materials in the item being examined. Therefore potentially explosive materials can be identified not only by their appearance but also by their material density. However, legitimate materials with densities similar to explosives cannot be distinguished from real explosives and therefore trigger false alarms. Monitoring equipment can also measure penetration results using different X-ray energies. This process is used to estimate the atomic number of a material. This additional information is used in the X-ray image to assign different artificial colours to materials with different atomic numbers. These systems make it easier to identify explosives, but they cannot be used to uniquely identify them. Modern X-ray devices used to screen luggage are often based on medical technology, specifically the principle of computed tomography. Computed tomography helps to detect explosives by generating a 3D image of the item in question. To this end, several cross-sectional 2D images of an object are taken at different angles and then combined to produce a 3D image. Other devices can be used to produce an (backscatter) X-ray image from the radiation that is reflected back to the X-ray source. *Backscatter images* are more readily used to detect organic materials such as plastic explosives than transmission images. Explosive trace detectors can be used to screen people on an individual basis. Explosive detectors are available in portable form, amongst other things, and also as personnel portals for high throughput screening. They can also be used to detect legacy military ordnance, such as mines. These detectors can be put to good use searching for the explosive contained in mines as metal detectors cannot easily identify modern mines.

▶ **Decontamination.** Decontamination is the process of removing contamination caused by hazardous materials from people, objects and areas. The aim is to protect contaminated persons from any further injury and to stop the harmful substances being spread into areas as yet uncontaminated. It is also meant to make affected objects and areas usable without having to take any protective measures. The decontamination measures used depend not only on what the hazardous substance is, but also on the object that needs to be cleansed. It should not be damaged during this process. The following decontamination methods, which can also be combined, are used:

- Physical methods: The hazardous substances are physically removed from the contaminated object using, for example, water or hot air. Sometimes washing down with water can decontaminate people contaminated by radioactive substances. Hot air is also used to remove hazardous substances from surfaces such as those of vehicles. Air temperatures of between approx. 100–200 °C can also destroy chemical warfare agents.
- Chemical methods: These are used, amongst other things, to facilitate the process of removing hazardous materials to be decontaminated by making them soluble. They are also used to neutralise chemical or contagious agents with the help of suitable oxidizing agents and strong bases, for example.
- Biological methods: Special enzymes degrade hazardous biological and chemical agents in an environmentally friendly way. If applicable, usually much smaller quantities of enzyme are needed than of most chemical decontaminants that come into use.

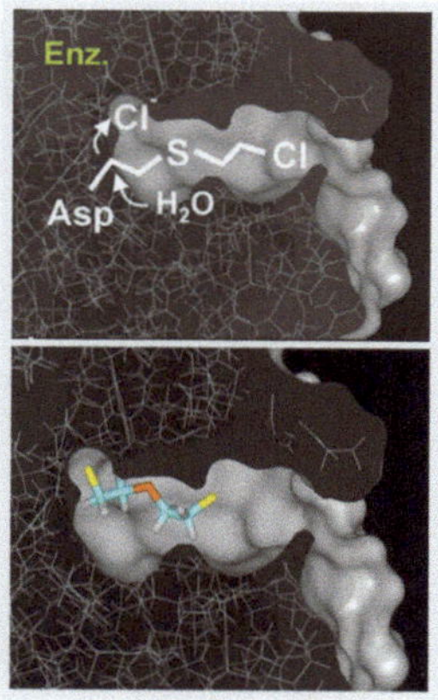

▣ Enzyme-based decontamination: Haloalkane dehalogenase, an engineered enzyme that uses a hydrolytic mechanism to convert toxic sulphur mustard into nontoxic thiodiglycol. The binding site for the enzyme and the chemical agent sulphur mustard is shown above. The molecule is split when the enzyme catalyses the chemical reaction by reacting with water. Source: Dr. Zbynek Prokop, University Brno

Any end products of the decontamination process may need to be appropriately disposed of, depending on the type of contamination and the decontamination method used. For example, many of the chemical agents used for decontamination purposes are also environmentally unfriendly themselves.

Trends

More efficient detection systems will help to protect us against hazardous substances in the future. This improvement in efficiency could be achieved by integrating different detection principles into one combined system, thereby utilising the advantages of various detection methods. Detection systems that trigger fewer false alarms should therefore become feasible. New types of detection methods, such as *surface acoustic wave (SAW)* detectors or nuclear *quadrupole resonance (NQR)* systems, will provide the basis for further positive developments in this field.

▶ **Surface acoustic wave.** SAW detectors are coming more and more into use and are mainly suited to detecting traces of explosives and hazardous chemicals in samples. An acoustic wave is generated at one end of the surface when an electric field is applied to a piezoelectric substrate such as a quartz crystal. This wave then propagates on the surface and is received as an electric signal by a receiver at the other end. If a gaseous substance becomes attached to the surface, the propagation conditions change, as do the surface acoustic wave's characteristics, such as frequency. The receiver then uses these changes to detect the attached substance. This makes it possible to, for example, concurrently determine the increase in mass caused by the substance. To ensure that the detector responds purely selectively to certain substances, the surface can be covered with a suitable coating that allows only certain substances to attach themselves. Several, differently coated SAW detectors can be used in a detector array to detect a large range of different substances. SAW detectors can be compact, portable devices.

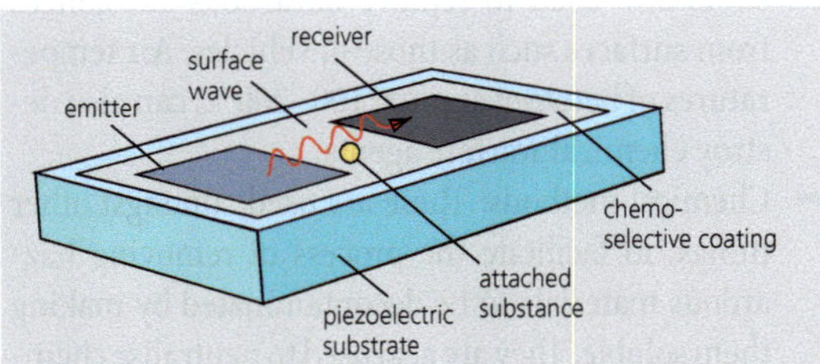

Surface acoustic wave detector: An emitter electrically generates an acoustic wave on the surface of a piezoelectric substrate. This wave travels along the surface in the direction of the receiver, which then detects it. The wave's characteristics, such as its frequency, are changed when the surface acoustic wave interacts with the substances attached to the surface. The receiver senses these changes and the substance is detected. A chemoselective coating on the surface stops substances other than those needing to be identified becoming attached to the surface. Source: Fraunhofer INT

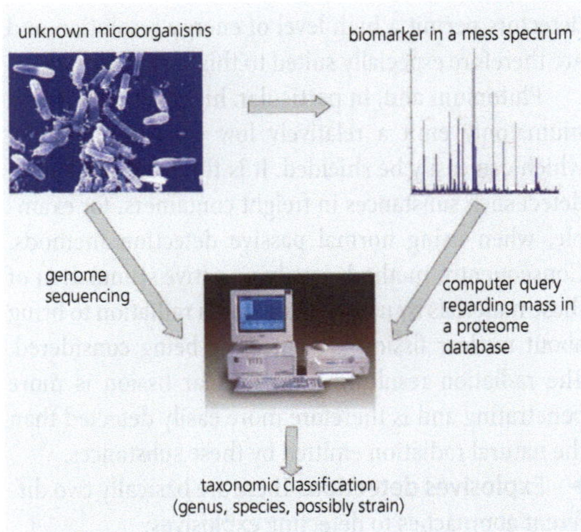

Bioinformatics for identification of microorganisms: Mass spectrometry is used to identify unknown microorganisms. The analysis data is subsequently compared with the masses of protein biomarkers of known microorganisms to classify the microorganism present. The masses of the different biomarkers are not determined beforehand by way of experimentation, but are calculated using computers. This process uses data from the genome sequencing of different microorganisms, which allows conclusions to be drawn about all the proteins of a microorganism. Source: Fraunhofer INT

▶ **Nuclear quadrupole resonance.** NQR systems are more able than X-ray devices currently in use to specifically identify explosives during bulk detection of explosives. The technique is somewhat similar to the nuclear magnetic resonance method used in medical science, but, in contrast, no external magnetic field is required. NQR explosive detectors are based on the principle that almost all explosives have certain constituents, such as the ^{14}N isotope of nitrogen, whose atomic nuclei have a nuclear quadrupole moment. This nuclear quadrupole moment is produced by a non-spherical distribution of electric charge in the nucleus. When an explosive is exposed to suitable electromagnetic waves, these waves interact with the atomic nuclei and an electromagnetic return signal is emitted. Every explosive has a characteristic signal, which makes unambiguous identification possible. As the NQR method does not use hazardous radiation, it can be used to examine both objects and people.

▶ **Use of bioinformatics.** Bioinformatics uses information technology methods to find answers to biological problems. It is increasingly being used in mass spectrometry to detect contagious agents. The resulting analysis data is compared with reference data for known contagious agents to identify the substance present in the sample. Reference data for a large number of hazardous substances is therefore required if a multitude of possible contagious agents are to be identified. Such data must also be available for different conditions. For example, when harmless biological substances are also present in the environment. This is because collectively present substances form a collective spectrum. The conventional approach of collecting this reference data by experimentation therefore quickly reaches its limits in practice. This can be overcome by using bioinformatics methods, where the reference data is no longer collected by way of experimentation but is calculated using computers. Proteins, amongst other things, are used as biomarkers in this connection. It is therefore possible to use existing and steadily growing databases for data about the protein content of different biological substances. Biomarkers that still retain their validity under variable conditions are then selected on the basis of these databases. The masses of the biomarkers are then calculated and compared with the collected analysis data.

▶ **Cell-based sensor systems**. Most of the procedures for detecting hazardous biological and chemical agents are currently based on molecular analysis methods. However, cell-based bioanalytical systems could possibly be used to obtain information about chemical or biological agents. This is based on the biological activity patterns induced by a certain substance in the cellular sensor element. Designing and developing suitable systems is a major technological challenge. However, if it can be done, they will provide access to data on substance characteristics that are more closely correlated with the reaction of the human organism to substances to be analysed in a given dosage. Cell-based sensor systems potentially offer the capability to detect a large number of possible hazardous substances in environmental samples simultaneously, as well as to identify unknown, modified or unexpected agents. A wide range of different transformation and evaluation techniques are used to classify the cell signals of cultivated cells and tissues. Cellular signal events are transformed using optical techniques, particularly staining, as well as conventional, non-invasive extracellular recording of the electrical activity of nerve and myocardial cells using microelectrode arrays. Since the introduction of cell engineering methods, which make it possible to integrate optical or bright reporter elements such as green fluorescent protein (GFP), specific *biomakers* have also been employed.

Developing and using living cells or cell tissues in biological-chemical analysis devices remains an extremely complex technical issue. The difficulties range from production of physiologically representative cell cultures and their preservation in a technical environment to suitable sample preparation and exposure and, lastly, recording of the often weak cell signals and correct interpretation of the complex activity patterns of cells reacting to changes in their environmental conditions. Furthermore, cells or tissues must be integrated with synthetic materials to create a suitable, functional biosensor or a diagnostic device. Cell-based sensors are currently almost exclusively laboratory devices. Most of them are prototypes, although some are already commercially available on the market.

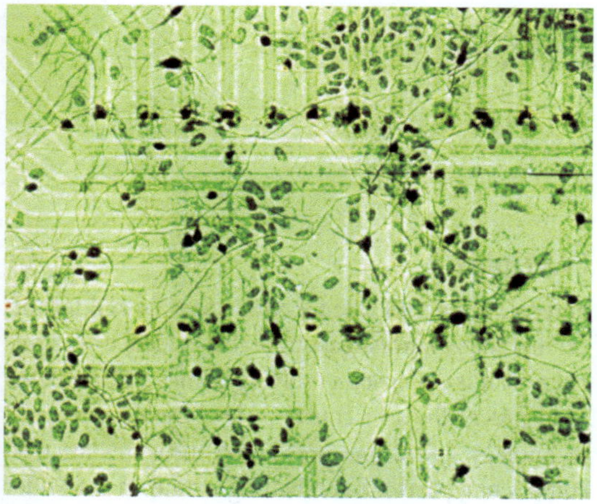

▲ Network of nerve cells from the spinal cord of a mouse on a 64-electrode microarray. Nerve cells react sensitively to even small doses of toxic substances. The electrical activity of the cells, which is correlated with their physiological condition at the time, is picked up by the microelectrodes and a signal amplifier and evaluated by a processing unit. If the cells are exposed to a hazardous substance, e.g. nerve gas, a certain pattern of electrical activity emerges as a cellular reaction to this exposure. Source: University of North Texas

Prospects

- A number of new detection systems are being developed, which, for example, are more sensitive than conventional systems or are less likely to trigger a false alarm.
- Most of the technologies used to detect contagious agents are not as technologically advanced as those used to detect other hazardous substances.
- Cheaper and more portable detectors will allow more widespread use in the future.

DR. ROMAN KERNCHEN
DR. KLAUS RUHLIG
Fraunhofer Institute for Technological Trend Analysis INT, Euskirchen

Internet

- www.eudem.vub.ac.be
- www.iaea.org
- www.sandia.gov
- www.opcw.org

Forensic science

Related topics

- Optical technologies
- Molecular diagnostics
- Measuring techniques
- Testing of materials and structures

Principles

Forensic science in essence uses scientific and technical methods to investigate traceable evidence of criminal acts. It is employed in solving crimes and in related legal proceedings in courts of law. Developments in technology and crime prevention measures are also component parts of combatting crime in its wider sense.

Forensic science is particularly challenging as it necessarily involves combining a number of different disciplines in natural and engineering sciences, as well as the humanities and empirical science. Physics, chemistry and biology are needed as well as handwriting analyses, linguistic text analyses and investigations of hardware and software. The findings must be easily understood and comprehensible if they are to stand up in court.

As the range of scientific and technical methods available has grown, so too have the demands being placed on forensic science. Material evidence is increasingly being used to confirm, supplement or refute witness statements precisely because it is objective, can be verified and is highly accurate. In the field of crime prevention, and to an even greater extent for investigations of trace evidence, importance is attached to the use of the most up-to-date methods and to the highest reliability of the findings.

The impact on society increases as technology is used in more and more areas. These social concerns can restrict use of existing technologies or even rule out their use all together. For example, methods of analysis associated with data protection (e.g. *DNA analysis* and other biometric tests) or preventive measures which can intrude on people's privacy (e.g. CCTV, *face recognition* and screening through passengers' garments at airport security control points).

Applications

► **Fingerprinting**. Detection and development of fingerprints has long been a traditional forensic science task. Although increasingly powerful computer systems can help when comparing fingerprints, it is ultimately down to a trained fingerprint expert and not technology to determine if two prints come from the same person.

In contrast, the methods used to find fingerprints and make them visible go far beyond the tried and tested method of using powder. Various technical methods are used, depending mainly on the material and the structure of the background.

Fluorescence detection is often used on irregular (e.g. printed) backgrounds. The fingerprint is treated with chemical compounds (liquid or powder form), which emit light of a specific different wavelength (i.e. a different colour) when exposed to light of another wavelength. If a suitable light source and filter to observe (or to take a photo of) the fingerprint are used, only the fluorescence light can be seen and the distracting background is dark.

Often many standard methods of developing fingerprints cannot be used on other surfaces, such as flexible plastic film used in packaging, leather, fabrics and certain types of banknotes. The *vacuum metal deposition* (VMD) method takes advantage of the fact that certain metals used to coat a surface stick to the surface itself but not to the area contaminated by a fingerprint. Often even the smallest traces can be made visible if the right metals are used and applied in the right order. *Autoradiography* involves treating fingerprints with radioactive tracers, which only bond fin-

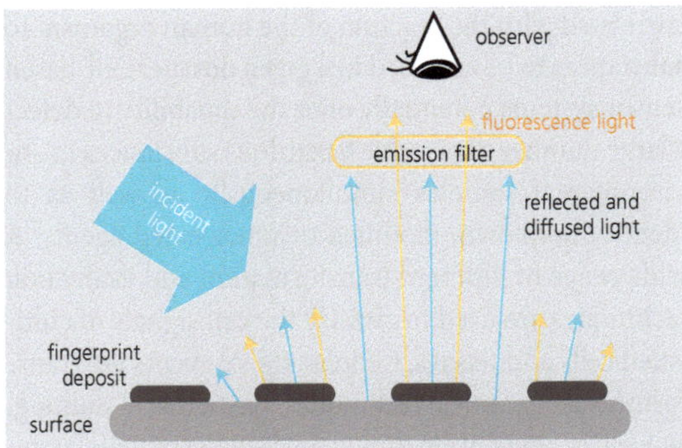

Fluorescence detection of fingerprints: After fingerprints have been treated with a suitable fluorescent substance, incident light of a given colour (i.e. wavelength range) illuminates the surface inducing fluorescence in the fingerprint deposits. The emission filter absorbs the reflected incident light (from substrate and deposits) and transmits only the fluorescent wavelength (from the deposits) to the observer. Only the fingerprint deposits appear illuminated when observed through the filter.

gerprint residues and not the background. Even for the smallest traces it is possible to capture the radioactive radiation emitted by the fingerprints on photographic film. This is because even less background radiation is emitted by the surface than in the case of fluorescence light.

Another method has come into use, drawing on biology. It is mainly used for valuable and sensitive objects, such as works of art. If more conventional methods (powder, solutions, VMD, etc.) cannot be used because the objects might be damaged, then it is possible to coat the surface with a certain type of bacteria found on human skin. As these bacteria (Acinetobacter calcoaceticus) feed on substances in fingerprints, the prints become visible once the bacteria have sufficiently multiplied.

Inducing fluorescent light in the fingerprint deposits by scanning the surface with X-rays is one of a variety of methods currently being tested. In the "*Fourier transform infrared (FTIR) chemical imaging*" method, infrared light is used to identify the chemical substances at various points on a surface, so that the particular chemical constituents of the fingerprints make them visible if the level of spatial resolution is sufficiently high. Detecting fingerprints on skin and being able to determine the age of a print are other particularly challenging concepts.

► **DNA analysis.** Modern forensic DNA analysis is based on short tandem repeats (STR) of DNA, which are present in the genetic makeup of all higher evolved beings. These repeat sequences cannot be used to draw any conclusions about an individual's appearance (unlike genes, which determine the physiological characteristics of a living being).

However, the number of repeats in a sequence can vary greatly, so the sequence in question is very likely to be a different length in two different people. To determine its length and therefore the number of repeats, specific enzymes are first used to extract these short tandem repeats of DNA. These are then specifically amplified using the *polymerase chain reaction*, PCR. This is a cyclical process, whereby an exact copy of each DNA sequence is made during each cycle. Consequently the number of sequences is doubled at each stage. After 30 cycles the amount of DNA material to be analysed has been increased by a factor in excess of a billion (2^{30}), making it possible to analyse even the smallest traces (right down to a single hair that has fallen out). The amplified DNA sequences are then separated according to their size using electrophoresis. An electric field is used to force the charged DNA molecules to migrate through a suitable medium. Shorter and therefore smaller sequences "move" more quickly.

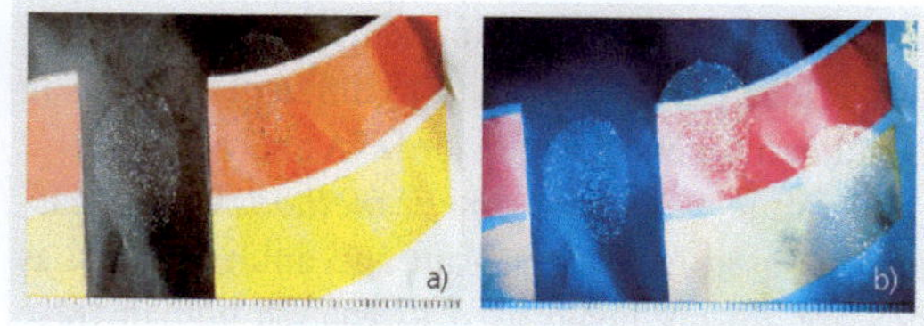

▣ Fluorescence: Whilst fingerprints on a plastic bag could hardly be seen after standard treatment (a), follow-up treatment using a fluorescent compound improved image quality (b), making the prints on the dark and light background visible. Source: Federal Criminal Police Office, Germany

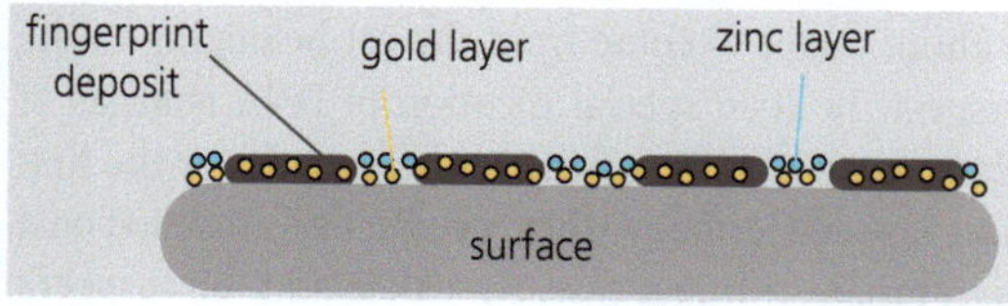

▣ Vacuum metal deposition: A surface with fingerprints on it is first covered with a thin layer of gold, then a zinc layer is deposited on top of the gold layer. Whilst the invisible gold layer penetrates the fingerprint deposits and covers the whole surface, the visible zinc layer does not penetrate the deposits, but instead forms preferentially on the gold between the deposits. The zinc deposits produce a visible negative image of the fingerprint.

Fluorescent markers already used during the amplification process, for example, can then be used to make the DNA fragments visible. The relative lengths of the individual DNA fragments are subsequently determined on the basis of how far they have moved. To avoid any confusion between two different people, several different STR sequences from various DNA loca-

▣ Genetic fingerprint: The polymerase chain reaction (PCR) can be used to amplify the DNA to a level where even the smallest traces can be analysed. The results can be compared with an analysis made of the DNA taken from a suspect. Source: Birte Schlund

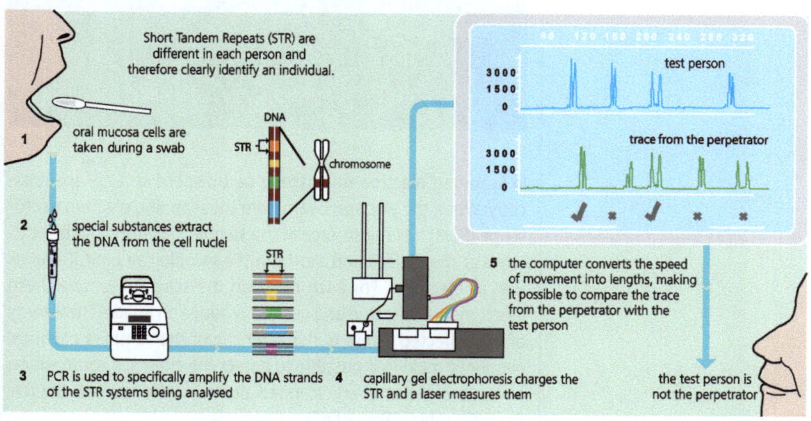

tions (loci) are analysed. Nowadays countries usually take STR sequences from 8 to 15 different locations in the human genetic makeup for DNA analysis.

In future, *molecular genetic methods* will increasingly be used to analyse plant and animal traces, as they are already being used to analyse human trace material. As is the case with people, a trace can effectively be assigned to a single perpetrator. For example, a dog hair from a getaway car can clearly be assigned to a dog's master suspected of having committed a crime. In 1994 analysis of animal DNA led to a conviction in a murder case for the very first time. This was in Canada, where cat hair at the scene of the crime was shown without doubt to come from the cat belonging to the suspect. In 2006 special mention of DNA analysis of plant material for examination was made for the first time ever in Germany when a verdict was reached on a high profile criminal case. An oak leaf in the suspect's vehicle was shown to have come from a certain tree in the area where the body was found.

▶ **Scanning electron microscope**. Most forensic science tests are based on a principle formulated a hundred years ago by Edmond Locard, a pioneer in forensic science: "Every contact leaves a trace". The smaller the trace, the more difficult it is for the perpetrator to avoid or destroy it. This is why microscopes are some of the most useful tools of the trade in forensic science. Scanning Electron Microscopes (SEM) are particularly useful because they allow higher magnifications than light microscopes and can be combined with many other methods. The comparison SEM, specially developed to meet the challenges posed by forensic science, makes it possible to compare two surfaces in questions, say details of the striation pattern of a bullet from the scene of a crime and those of a similar bullet fired by a suspect's weapon, simultaneously and under the same conditions. Accordingly, the comparison SEM is able to not only record any similarities between the two objects very accurately, but also very clearly. The microscope has therefore become indispensable when carrying out comparative analyses of bullets, cartridge cases, fragments of tools and tiny tool marks and imprint traces.

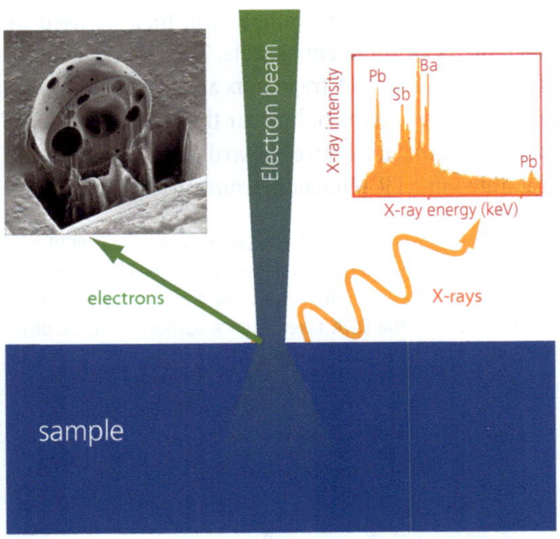

Scanning electron microscope and dispersive X-ray spectroscopy: When the electron beam from the scanning electron microscope (SEM) hits the surface of the sample, electrons can be scattered at the surface and atoms in the sample can emit X-rays as well as electrons. For each point on the scanned sample these electrons are detected and used to produce an image. The energy of the reflected X-rays is characteristic of the sample's chemical elements, making it possible to detect the chemical composition of the sample at every scanned point on the surface. Source: Federal Criminal Police Office, Germany

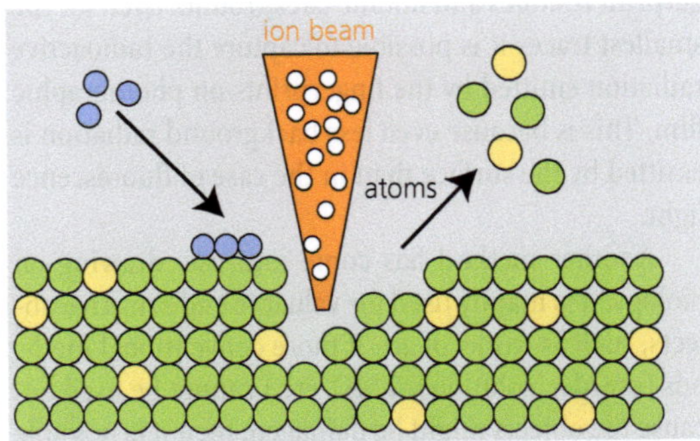

Focused ion beam (FIB): Using a FIB, it is possible to produce structures less than 100 nm thick by removing atoms from a sample where the ion beam (which can be less than 10 nm wide) hits the surface. A FIB can also be used to deposit material on the surface, e. g. when modifying microprocessor electrical circuits, if a suitable gas is present (blue). Source: Federal Criminal Police Office, Germany

A SEM can also be used to determine the chemical composition of a sample with a high level of spatial resolution. This is particularly important for very small samples or samples with marked localised differences in their chemical composition.

Energy dispersive X-ray spectroscopy (EDX) measures the energy of the X-rays emitted by the sample when an electron beam hits the surface of the sample. The emitted energy is characteristic of the sample's chemical elements. An electron microscope and EDX can be combined to provide very reliable verification, for example, of gun shot residues that can be left behind on a gunman's hand and clothing when the gun is fired. This is because both the shape and size, as well as a special combination of elements can be verified.

The SEM, combined with Focused Ion Beam technology (FIB), can be used in many new applications. If

Gun shot residues: Gun shot residues left behind by new types of ammunition cannot always be clearly verified on the basis of chemical composition and appearance alone. However, in many cases the internal structure of particles "cut open" using a FIB can be used to distinguish gun shot residues from other types of traces. The structures (on the left, a particle from old, conventional ammunition and, on the right, from more modern ammunition) can even allow conclusions to be drawn about the type of ammunition in question. Source: Federal Criminal Police Office, Germany

Car paint. Left: Some new car paints are made up of tiny pigments, which cannot be analysed using conventional preparation and microscopy methods and identified as being made by a particular manufacturer or from a particular make of car. Right: Using their internal structure and the chemical composition of their layers, cross-sections made by a FIB through particles that are just a few micrometres in size and often only a few hundred nanometres thick can be used to identify the paint, even from minute flakes of paint, after a hit and run accident, for example. Source: Federal Criminal Police Office, Germany

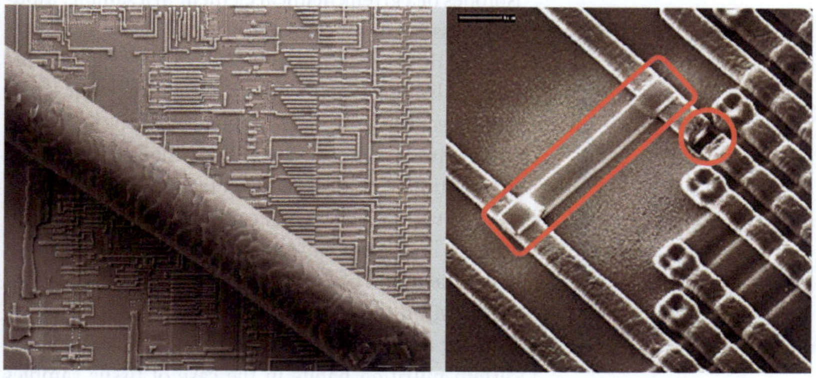

Microprocessors. Left: A human hair laid over the surface of a typical conductor circuit gives an indication of the size of the structures being worked on. Right: A FIB is used to break (red circle), as well as make (red rectangle) conductors in microprocessors and is an important tool in the analysis of manipulated microprocessors, in credit cards, for example. Source: Federal Criminal Police Office, Germany

a FIB system is integrated into a SEM, an ion beam just a few nanometres wide (less than a ten thousandth of a human hair) acts like a miniature sand jet to erode tiny areas of the sample and cut microscopic objects open. It is therefore now possible for the SEM to image not only the outside of samples, but also the inside. This can be critical for gun shot residues left behind by modern ammunition because its chemical composition varies. It can be equally important when analysing flakes of paint from hit and run accidents, from ignition devices on explosive charges, or microprocessors embedded in credit cards, for example. This method has until now only been used in isolated cases, precisely because it is so complex: however, in future it is likely to be used in many other areas.

Trends

► **Isotope analysis.** Forensic science is particularly challenged by problems posed by materials and material traces which can be assigned to a certain geographic area or are shown to be synthetic or natural in origin. The finish and chemical composition of technical mass-produced articles, in particular, is such that they often fall into uniform groups and it is therefore very difficult to distinguish between them. In this connection, it can be helpful to determine the isotope ratios of the material or individual component parts.

In fraud investigations, incorrect declarations regarding origin of goods can be refuted, such as when importing butter. In certain cases, adhesive tapes and adhesives made and formulated in the same way – and used to package or secure drug shipments and explosive devices – can be traced back to specific batches.

Isotopes are atoms of the same chemical element which have different numbers of neutrons in the nuclei and therefore different mass numbers. Most naturally occurring elements have one or a few stable isotopes, whereas the other isotopes are radioactive and decay in time. For practical reasons, the stable isotopes are of most interest to forensic scientists. Isotopes of a given element exhibit nearly identical chemical behaviour. However, for light elements such as hydrogen, carbon, nitrogen, oxygen or sulphur the relatively big differences in mass between the isotopes (up to 12%) mean that some physical and chemical processes, such as diffusion or phase transition (change in aggregate state), can cause the isotope ratios to deviate from the average value (isotope fractionation). For heavy elements, such as lead or strontium, similar discrepancies can be found in various samples – despite negligible differences in mass – as, for example, lead deposits in different locations around the world have different isotope ratios.

Isotope ratio analyses measure the abundance of isotopes of a given chemical element in a sample. Modern mass spectrometers allow very precise measurement of this isotope mixture. SNIF-NMR (Site-Specific Natural Isotope Fractionation Nuclear Magnetic Resonance) is especially used in the analysis of hydrogen. In this special process, it is even possible to determine the location of an isotope in a molecule.

The various effects that can cause isotope fractionation in light elements allow conclusions to be drawn about the origin of the sample (e.g. geographical classification on the basis of continental and altitude effects for oxygen isotopes in the water cycle) or about certain nutrition habits in the case of tissue samples (e.g. dependency of the N isotope ratio on food chains and biogeochemical cycles). For heavy elements, geographical classification is made using information about local isotope ratios in mineral samples or those brought about by environmental pollution.

In a specific case where explosive devices did not detonate, hardened epoxy resin adhesive among other traces was subjected to forensic examinations. Stable isotope ratio analyses were carried out as part of the comprehensive process in which traces of adhesive with the same composition were compared. The ratios of the nitrogen isotopes in the hardener component of the adhesive mixture of two components provided valuable information about which batches the adhesive came from.

In other cases counterfeit medical drugs and pharmaceutical products must be verified, as must chemically or biotechnologically synthesised vanilla flavouring, which is sold to the consumer as a product made using natural ingredients. If there is any suspicion of performance enhancing drugs being used, isotope analysis can differentiate very reliably between synthetic steroids and more pronounced natural production of testosterone by the body. In some cases, the carbon and nitrogen isotope ratios, in particular, have been used to delimit the geographical origin of heroin and cocaine samples. Overall, isotope analysis encompasses: Authentification; verifying an identity (A), identification (I), origin classification or limitation (O), indications of synthesis (S), batch differentiation (D), quality assurance (Q), information about eating habits and lifestyles (E), information about individuals' migration behaviour (M), labelling/product protection (P). General applications go far beyond forensic applications:

- Medicines (A, I, D, Q, P)
- Foodstuffs (A, O): Fruit juice, wine, honey, vegetable oils, flavourings, meat, fodder, dairy products
- Archaeological finds (O, E, M)
- Technical products and pre-products (O, D): Glass, steel, adhesive, adhesive tape, bullets, solvents, fuels, fuel additives, fire accelerants, explosives, paints
- Illegal substances (O, S, D): Cannabis, heroin, cocaine, synthetic drugs
- Illegal substances in body fluid (I): Doping agents
- Human residues (O, E, M): Traces of blood, bone, teeth, hair, nails

X-rayed lorry: Goods smuggled in lorries and transport containers are often hidden behind several tonnes of camouflage goods. A customs officer without an X-ray image at his or her disposal is very unlikely to have found the 4 million cigarettes in the lorry, which can be clearly seen on the image. Source: Smiths-Heimann

▸ **Baggage scanning.** If weapons, explosives and smuggled goods in passenger baggage, freight containers or packages are to be clearly identified on an X-ray, it must be possible to see what shape the objects are, as well as to identify what they are made of, or at least to narrow the field. Explosives or drugs are mainly made

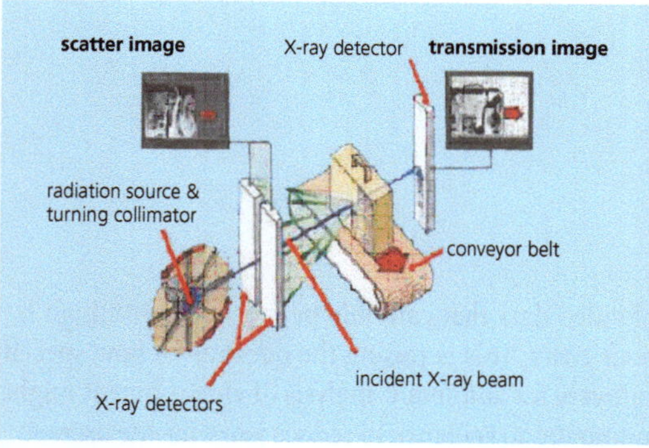

Backscatter principle: When an X-ray beam scans a piece of luggage, the transmitted beam is used to produce a standard X-ray image. Backscattered X-ray beams are used to produce quite a different X-ray image, which clearly shows organic materials. This is important if explosives and drugs are to be identified.

up of light, organic elements, whilst weapons can typically be expected to contain heavier metallic materials. Simple x-raying methods, as used in medicine, only help to a limited extent as darkness of a shadow in an image does not allow to conclude if the object is very thick or dense or is even made of metal.

However, this information can be obtained by using X-ray beams with two different energies (if visible light were used, this would correspond to two different colours). With low energy beams, the level of absorption greatly depends on the chemical elements in the object being x-rayed. For example, lead, which has a high atomic number (the number of protons in the nucleus of an atom), absorbs about 500,000 times more radiation than carbon. In contrast, there is almost no difference between absorption of different chemical elements when higher energies are used. The two different energies can be combined to determine both the density or thickness of the object (indicated in an image by the object's brightness), as well as if it is organic in nature or a metal element (usually indicated by colour).

Instead of using the method of capturing the radiation that passes through an object (transmission scanner), another method is used, based on a thin X-ray beam to scan the object and then to measure the intensity of the radiation that comes back from the target (*backscatter scanner*). Using this method, image resolution is not as high and the x-raying process is longer than when the transmission method is used. However, if the backscatter method is used, objects only have to be accessed from one side and organic materials can be very easily identified on the images thanks to their ability to backscatter X-rays.

These methods are used to X-ray letters and cases, as well as loaded ship containers and lorries. For some time now baggage X-ray systems based on *Nuclear Quadrupole Resonance (NQR)* have been in development, with the particular aim to identify explosives, in particular, more readily. NQR is a technique related to the *Nuclear Magnetic Resonance Method (NMR)*, which is very widely used for medical diagnostic purposes. However, NQR does not need a strong magnetic field, so the systems can be more compact. NQR can only identify a limited number of materials as it is only sensitive to the atomic nuclei of certain elements, although it readily detects nitrogen, an essential component of many explosives.

Prospects

- In addition to developing its own special lab procedures, forensic science is benefitting greatly from adapting and combining methods from other areas of technology.
- Forensic science will continue to benefit from the advances being made in the development of material analysis and microscopy methods. Higher levels of accuracy and lower detection limits of existing technologies will become increasingly important, as will completely new analysis methods used to prevent and investigate crimes.
- The increase in analyses made possible through technology, accompanied by rising demands for forensic examinations, makes it necessary to prioritise and initiate intelligent rationalisation measures or to increase expenditure (or both).
- Miniaturisation and automation, especially of sensors and analysis systems, can provide new opportunities, as exemplified by mobile fingerprint scanners, current DNA analysis systems or 3D scanners. However, these concepts must also be applied intelligently.
- New and further developed image-generation methods, in combination with automated analyses, will increasingly be used in the field of crime prevention.
- Forensic science will be challenged anew by increasing "digitalisation" in public and private life. New digital tools can help to solve as well as commit crimes. Therefore forensic science is increasingly adressing the area of digital evidence.

DR. PATRICK VOSS-DE HAAN
DR. ULRICH SIMMROSS
Federal Criminal Police Office, Wiesbaden

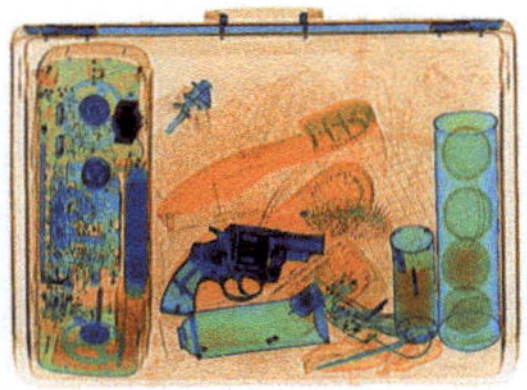

X-rayed case: When X-ray beams of two different energies are used, it is easy to distinguish between organic material (shoes, tennis racket; orange in the picture) and metal objects (portable radio, spray cans, firearm; green and blue in the picture).These different energy X-ray beams have to be used together if any conclusions are to be drawn about an object's thickness and composition. Source: Smiths-Heimann

Internet

- www.fbi.gov/hq/lab/lab-home.htm
- www.forensic.gov.uk/html/services/
- www.crimeandclues.com/
- www.all-about-forensic-science.com/
- www.forensic.tk/www.ispub.com/journal/the_internet_journal_of_forensic_science.html

Access control and surveillance

Related topics

- Measuring techniques
- Sensor systems
- Human-computer cooperation
- Image evaluation and interpretation
- Information security
- Optics and information technology

Principles

In recent years, the growing threat of international terrorism and the increasing networking of organised crime have led to a worldwide boom in surveillance technology. In Great Britain alone, where video control has been pioneered since the eighties, more than one million security cameras are in use today in the private and public spheres. In London alone, several thousand cameras are deployed in underground stations, buses, football stadiums and other public places. In other countries, too, surveillance technology is making progress. Application areas are predominantly the monitoring of objects, i. e. buildings, public spaces, stations, public transport and stadiums, as well as traffic control. The increasing use of security cameras also poses new challenges for operators. Although the installation costs of the camera systems have become rather low by now, using them is labour-intensive. Due to the great number of existing cameras there is a flood of video data that can only be mastered with high labour costs. This is raising the question of how special software for automatic analysis of video content might automate surveillance to reduce work for the users.

Furthermore, access to special areas or information is a matter of control. In order to give authorised persons admission to particular areas, two principles have always been used for *authentication*: either a key (or chip card or the like) or by controlling knowledge (PIN or password). While using a key or a general password only can prevent unauthorised access, information technology allows personalisation of access. By assigning a specific string of numbers or characters to an individual, the person actually using the key can be identified. Different technologies can be applied:

- Knowledge: The string, i. e. the PIN or the password, must be communicated to the system through an input device, be it a keyboard or touch screen. We know this from logging in to our computers or from withdrawing cash from an ATM.
- Possession: The access code is represented by, say, a key, a chip or magnetic card, or even an *RFID* chip. The RFID chip has a built-in reader that continuously emits electromagnetic radio waves. The signals, sent through a fine coil or an antenna, are strong enough to activate a chip within a range of a few centimetres up to more than a mile, and initiate the stored data to be sent to the reader. This system can also be used for contactless access control.

	Ease of use	Precision	Availability	Costs
Passive Criteria				
Iris recognition	********	*********	********	********
Retina recognition	******	********	*****	*******
Fingerprinting	*******	*******	****	***
Facial geometry	*********	***	*******	*****
Hand geometry	******	*****	******	*****
DNA	*	*******	*********	*********
Active criteria				
Voice	****	**	***	**
Writing dynamics	***	****	*****	****
Touch dynamics	****	*	**	*
Comparison: Password	*****	**	********	*

Biometric methods: Characteristics that have met the demands for universality, uniqueness, invariability over time and measurability. When considering ease of use, precision, availability and cost of the individual methods, it becomes obvious that actually choosing a single type of biometry is quite difficult and they should be used in combination depending on the application (green star: good; red star: unfavourable; for example, iris recognition is very precise, but it is also very costly). Source: Fraunhofer IGD, according to Bromba GmbH

In addition to the "key" or "knowledge" used, the readers must receive a correct code and convey it to the background system for verification. The individualisation of the key allows categorical control of who is getting access at what time.

However, these methods of acquisition or knowledge representation have one major drawback: they cannot guarantee that the owner of the key or password is in fact the person who should be given access. Keys get lost, can be found or stolen by others, and can also be given to third parties. The same applies to passwords or PINs: often, people write these down somewhere in case they should forget them, and that runs the risk that some unauthorised person might get hold of those keys.

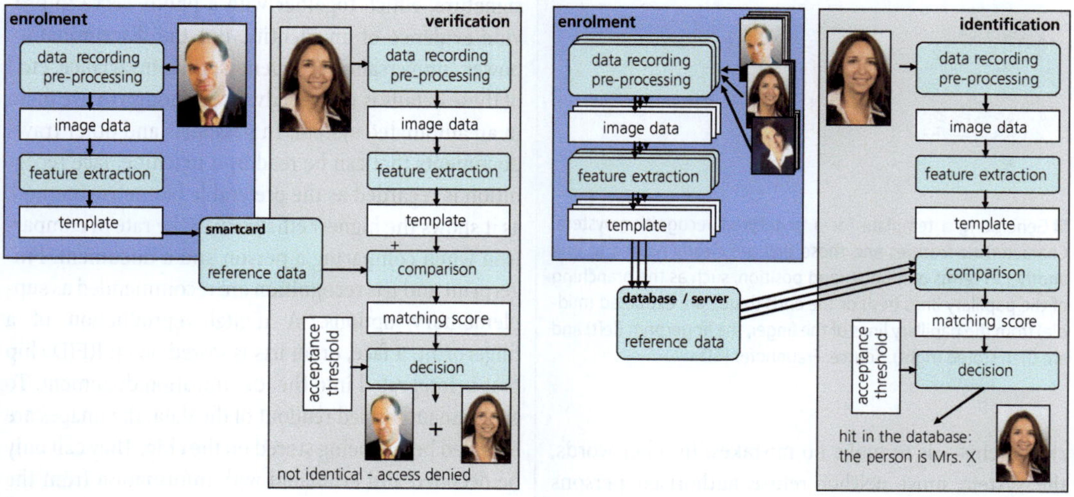

Biometric access controls. Left: Verification. The recorded feature is merely compared with a specific template that was previously recorded by the user. Right: Identification. The biometric feature is collected and compared with a large number of templates that have been deposited previously (in some cases in large databases), in order to identify the recorded persons. Source: Fraunhofer

Biometrics could solve this problem. The advantage is evident, since analysis is based on physical characteristics, making the key inseparable from the person wanting access. This type of key cannot be forgotten or misplaced, and certainly not easily copied.

▶ **Biometrics.** A biometric feature apt to be used for access control must meet the following requirements:

- Universality: It must be universal, i. e. nearly every person should have it. It must be ensured that the characteristic is unique to each person. If at all possible, even monozygotic twins should be different when it comes to this feature.
- Robustness: The criterion must be permanent, or invariable over time. The ageing of a person, a change of hairstyle, or a beard should not have a fundamental impact on the biometric feature.
- Measurability: The feature must be recordable and measurable if an automated, comparative approach is to be used.
- Acceptability: The selected attribute must be acceptable to the user and offer no risk of physical impairment.

The system is no different, really, to a human guardian: It must first get to know the individual before granting access to an off-limits area. So the person seeking access must go through a registration process called enrolment. The first step, then, is to record and digitise the relevant characteristic, e. g. the fingerprint, iris or face. Disturbing factors are removed and contrasts are improved. A normalisation process allows comparability. The different data fragments are now bundled into a reference dataset and stored. This dataset is also called a reference template, and all new acquisitions of characteristics are compared with it.

The actual authentication can now be executed in two different ways. The identification method records the biometric attribute and compares it with a multitude of stored templates. This method is mostly used in surveillance and will find persons previously enrolled, such as known criminals, within a group of observed persons. If there is a similarity to a reference dataset within specified limits the person can be considered as identified (anti-crime measure).

Speaking of verification, however, the registered attribute is only compared to the template stored for the user requesting access to a protected area. Somebody claiming to be Mr X will be checked to see if he is really Mr X. This method has an advantage, because the reference dataset does not need to be located in the system itself, but can also be stored on a smart card or a personal document to be presented to the system for authentication. Furthermore, as the comparison is done against one template, this method is relatively fast compared to the identification method.

In contrast to passwords or PINs, biometric attributes are never the same. The system only can calculate a probability suggesting that an acquired biometric characteristic is the same as the enrolled one. Therefore the objective – and the challenge of biomet-

Generating a template for a fingerprint recognition system: Characteristic features and those that are clearly related to one another in terms of number and position, such as the branching of the papillary lines (red) or the ends (blue), are extracted (middle) from the papillary lines of the finger, the fingerprint (left) and are then stored (right). Source: Fraunhofer IGD

ric systems – is to make no mistakes. In other words, the system must neither refuse authorised persons (false reject), nor give access to unknown persons (false acceptance). Depending on the situation, the tolerance of the identification method might be higher than that of the verification method (security). Individual settings are also an option: lower tolerance for a bank vault (higher security), and higher for the public library (more comfort).

The process of identification based on a certain number of stored attributes raises the issue of data protection because of two major reasons:

- prevention of misuse of person-related data
- identity theft prevention

In the past, we learned to assign different passwords for different systems. This is to inhibit access to all other areas once one of them has been compromised. In such cases the password can be changed easily. This is unfortunately not possible with biometric systems. If a fingerprint has been misused once, that finger can never again be used in a biometric system if we want to stay on the safe side. However, we only have 10 fingers, two eyes and one face. Therefore measurements are needed to securely store biometric reference data and allow for renewability and revocation of the biometric template. The underlying technology is referred to as template protection. It ensures enhanced privacy, since the original biometric characteristic cannot be deduced from the stored template, and improves security, because now information in one database can no longer be retrieved by using the template from another system (cross-matching).

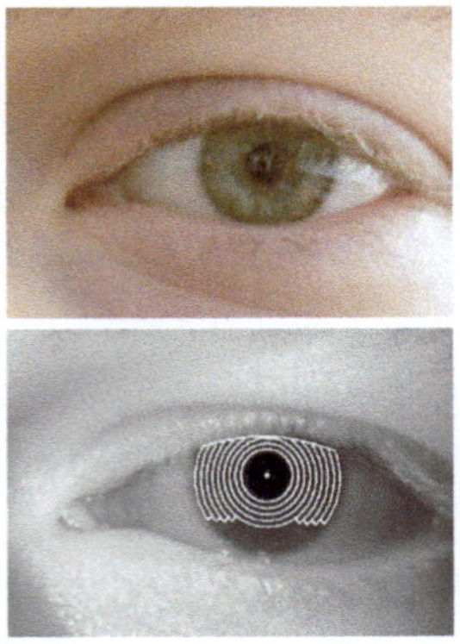

Iris recognition: The iris is localised and segmented by the scanner. The separate areas are analysed and an iris code is produced and deposited in the database. Source: Fraunhofer IGD

Applications

► **Identification cards.** Identification cards contain specific details, such as height, eye colour, or a person's signature, which together with a photo, serve to provide evidence of an identity. But the discriminatory power (universality, uniqueness, time invariance, etc.) of these details is not yet advanced enough to use them in an automated system. In passports and other travel documents that can be read by a machine, face recognition is regarded as the preferable biometric method, as it shows the highest efficiency in the rate of comparison when comparing a person and a document. Fingerprint and iris recognition are recommended as supplementary options. A digital reproduction of a fingerprint, a face, or an iris is stored on an RFID chip that is integrated into the identification document. To avoid unauthorised readout of the data, the images are encoded before being stored on the chip. They can only be decoded and evaluated with information from the machine readable zone of the passport. Passports with these specifications have been stipulated by the International Civil Aviation Organisation (ICAO) and have been mandatory in the European Union ever since mid-2006.

► **Access control by iris recognition.** The airports Schiphol in Amsterdam, Netherlands and Fraport in Frankfurt, Germany, have relied on iris recognition for some time now to give frequent travellers the opportunity to sidestep conventional passport controls, thus avoiding long waiting times. To take advantage of this privilege, the interested person must first take the time to become familiar with the system. Usually, a monochrome picture of the iris is taken using the same system that will be making the comparison later on. These systems work with wavelengths in the near-infrared range, so even an iris with dark pigments will produce useable templates. The scanned picture will then have to be processed in several steps. First, the outer iris rims are localised and then the iris is segmented into different areas. Characteristic features are analysed and their position in relationship to each other is displayed in vectors. Finally, an iris code is stored in the system. At the passport check, the traveller now has the option of bypassing the queue and going into a separate room where he or she can have another picture taken with the iris recognition system. The new pattern is compared to the one stored. In the case of congruence the traveller may continue his or her journey.

► **Video control.** Progressing beyond initial technological innovations such as IP-based data transfer, modern control systems are increasingly using methods that automatically evaluate video data, making the existing systems more intelligent and efficient. Depending on the application, the transmission of video data is carried out using Motion-JPEG or MPEG-4

Enrolment at the airport for iris recognition. Source: Bosch

compression, which is already integrated into the camera system. This reduces the amount of data to be transferred via the network. Furthermore, most systems can create a digital record, which permits an evaluation of material taken during surveillance or following a criminal incident. The primary goal of intelligent surveillance is to detect possible offences as early as possible so that appropriate countermeasures can be taken immediately. This extends beyond the possibility of using records merely as a means of solving crimes. Applications might include an intelligent camera system to control car number plates in road traffic and the automatic recognition of suspicious activities close to public buildings. When combined with biometrics, video control can be used to detect the presence of persons wanted by the police at major events, for example. Number plate recognition, which is technically well developed, is already in widespread use. Image analysis systems are used to detect the number plate of the vehicle. Afterwards an *OCR technique* (Optical Character Recognition) is used to identify the number plate. The scene is illuminated using infrared light, which means that the retro-reflective properties of the plate's surface allow successful recognition even under bad lighting conditions.

Once the plate has been recognised the vehicle and the driver can easily be identified by means of a database. This can automate speed monitoring or general traffic supervision. Another application could be the highway toll system. Football clubs have started a pilot project with a German IT service provider to use a face recognition system to identify spectators who are under a stadium ban. The system has been developed jointly with the stadium's operator, the police, public prosecutors and the municipality and is based on the existing infrastructure of the stadium. The security cameras of the stadium send pictures to the face recognition software, which compares their facial biometrics to images of stadium attendees in an existing database.

Trends

▸ **Improved sensor systems.** The sensors employed in existing access control systems can sometimes be easily deceived. Currently almost all fingerprint identification systems can be circumvented easily by applying, for example, a silicone finger. In many cases, face recognition systems can be fooled by showing a printed photo. Hence, a biometric sensor should not only be able to identify persons but also to detect fakes. This method is called liveness detection. It not only performs feature detection, but also looks for vital signs. Fingerprint identification systems might be supplemented by blood flow or pulse recognition. The system will measure the distance between papillary lines using a laser. Significant differences between the measurements suggest a pulse, meaning that the finger is alive. Other options include recording the surface of the skin of a finger instead of the papillary lines and applying suitable methods to measure the depth of the underlying skin. Or to detect motion, such as the twinkling of an eye, using a sequence of images. Checking pupil dilation is yet another option.

License plate recognition: Image analysis methods are used to detect the license plate in the recordings of passing vehicles and to identify them using optical character recognition (OCR). Source: Fraunhofer IGD

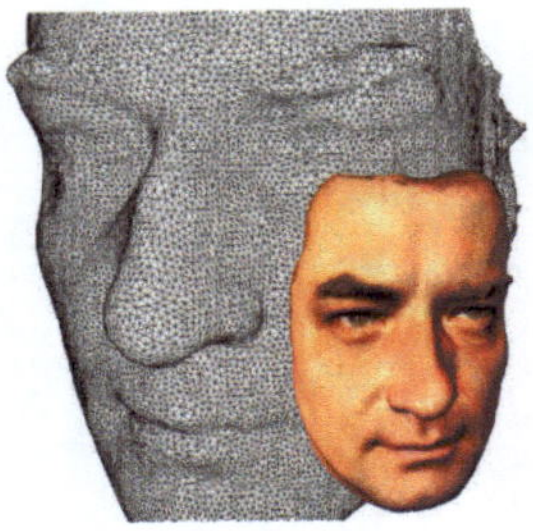

Recording of the 3D form of a face: The polygon model of the face allows computation of the relationship between features. The composition of the surface is also relevant for 3D facial recognition. Recognising a face requires only limited resolution. But the higher the resolution, the greater the recognition rate and lower the rate of error. Source: Fraunhofer IGD

▶ **3D face recognition.** Performance in face recognition – and hence resistance to attacks – will be substantially improved by recording the three-dimensional form of the face. Once the 3D geometry and the corresponding texture information have been recorded, the face is rotated in the computer to the frontal view and then compared to previously stored 3D faces. The subject is only authenticated and given access to a protected area if there is sufficient conformity in the geometries. The basic advantage of this method is that the 3D model enables a correction of the face image in case of head rotations or unfavourable camera angles. Another characteristic of the 3D record is its scale-invariance. With 2D recordings, the unknown distance between the person and the camera results in pictures of different sizes. The 3D models, on the other hand, are always metrically correct and free of perspective and radial distortion. Obviously, the basic proportions of the head, like the distance between the eyes, are retained; they do not change when converted to a uniform picture size (including uniform eye distance).

3D face recognition can also be carried out by a stripe projection scanner. After recording the 3D geometry and the corresponding texture information, the face is rotated to a front view and then compared to previously stored 3D faces. Acceptable agreement will give the subject authorisation to proceed.

The increasing number of monitoring systems and the corresponding increase in video data pose new challenges for research particularly in the field of computer vision. The large number of cameras available already makes it difficult today to evaluate the pictures taken by certain types of camera. Manual evaluation of the monitoring camera does not make much sense because of the different systems deployed. Methods for automatic evaluation of video data streams are therefore becoming more and more important in the field of security. Current research activities are focused on the following areas:

- Robust traceability of individual objects across a network of monitoring cameras by tracking methods
- Classification of individual objects in a picture: differentiation of persons and vehicles, detection of suspicious activities by means of motion and gesture recognition
- Biometric recognition of particular persons within a group
- Advancement of the methodology in the field of video mining to improve evaluation of stored video data

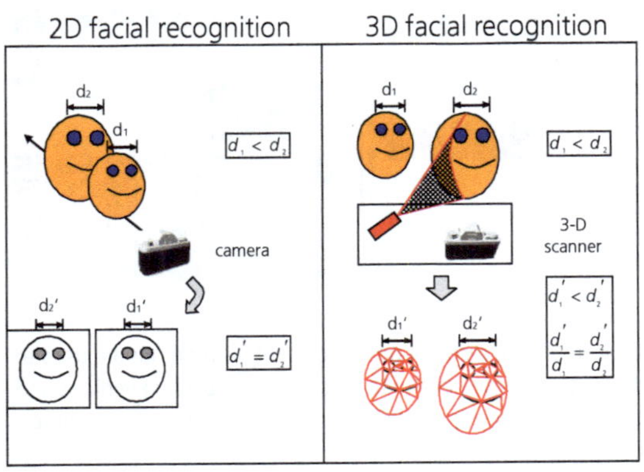

The advantages of 3D facial recognition over 2D recognition: If the images of faces of different sizes are standardised (d → d'), 2D recordings will no longer provide any information on dimensions. 3D facial recognition also records relationships, such as the relative distance between the eyes, and these can be applied towards recognising the original dimensions of the face. Source: Fraunhofer IGD

▶ **Smart camera networks.** Advances in computing, communication and sensor technology have triggered the development of high-performance integrated processing platforms. Smart cameras are one example of innovation in this field, and they already integrate complex and powerful heterogeneous hardware platforms, including general purpose processors, digital signal processing units and a so-called field programmable gate array (FPGA). Today's *surveillance cameras* with onboard processing capabilities can only perform a set of low-level image processing operations on captured frames to improve video compression, for example. But new processing technology allows the implementation of smart cameras, which are capable of performing complex video analysis tasks directly and effectively. These cameras integrate video sensing, video processing and communication in a single embedded device. They act as flexible processing nodes with capabilities for self-reconfiguration, self-monitoring and self-diagnosis. With the dawn of these cameras, the system architecture of surveillance camera networks is shifting from a central to a distributed network design. One advantage of this shift is the increase in the surveillance system's functionality, availability, autonomy and scalability. In a central network design, modifying or reconfiguring an assignment during operations is difficult, a distributed design of smart cameras is far more flexible. Smart cameras are the key components within these new architectures because they provide the means for distributed onboard video processing and control. Surveillance systems that employ smart sensors can react more rapidly and more autonomously to changes in the system's environment while detecting events of interest in the monitored scenes. One field where smart camera networks can be deployed is traffic surveillance, where typical video analysis tasks include the detection of stationary vehicles and wrong-way drivers, as well as the computation of traffic statistics such as average speed, lane occupancy and vehicle classification. Smart camera networks generate many views, so that by using a good selection of processing tasks, they can also support more complex and challenging applications such as crowd monitoring for sporting events, concerts, and so on. Knowing the density of a crowd, for example, or its flow, or even the identity of single individuals within it can be of interest to observers. The results of this evaluation can be used to prevent panic or to determine whether the capacity of a certain location is being fully utilised.

▶ **Tracking methods.** A person or, more generally, an object, must be identified before it can be tracked automatically. The process therefore begins with face detection (for persons) or number plate detection (for vehicles). In a process called anchoring, certain critical visual features of the object to be tracked are stored as a symbolic representation, appropriate to the comparable differentiation of features. The system then searches for the same constellation of features in the pictures following the first sequence to identify the object that has now moved to another position. Optical flow is often used for this purpose because it displays the movements of particular picture elements. Assumptions about the possible movements of the object (for instance, direction and maximum speed) can facilitate detection in the subsequent images. More complex stochastic models, like the sequential Monte Carlo method, are used for the support of object tracking.

Panorama video technology is an ideal platform for a number of applications in the field of object tracking. The system's 360° view allows use of one single camera for the surveillance of whole rooms or spaces.

Intelligent video analysis software: New filtering options improve guard-supported surveillance. New filters can set off an alarm if a subject follows a defined path, crosses certain defined lines, has a particular colour, or changes shape. In the images, the detected subjects are yellow prior to the alarm, but bordered in red when the alarm is set off and afterwards. Source: Bosch

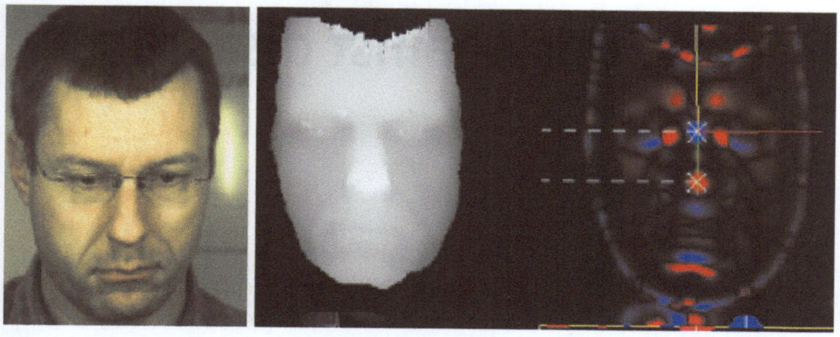

Results of 3D facial recognition. Left: The person is photographed from various angles. Right: The images are subjected to analytical processes and standardised in order to enable comparability. A depth image is produced, which identifies noticeable features, whereby the lighter areas are closer to the camera than the darker ones. A curvature image is used to examine the curves in a face (right). Outward curves (red) allow identification of the tip of the nose in this example, inward curves (blue) allow the root of the nose to be identified. The information generated – i.e. the relationship of characteristic parts of the face to each other – tell the system to either allow or deny access to the user. Source: Fraunhofer IGD

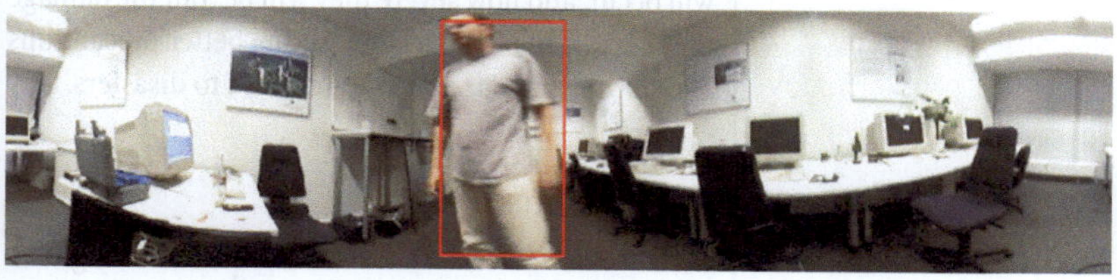

Images from a panoramic video camera depicting the tracking of a subject. Source: Fraunhofer IGD

The comprehensive view of the monitored area is especially suitable to the automatic tracking of objects.

Prospects

- Access control by means of biometric methods is simple for users.
- The combination of knowledge- and acquisition-based access methods with biometric systems can improve levels of security.
- The combination of panorama video technology and object monitoring makes it possible to automate the surveillance of room/space.

ALEXANDER NOUAK
VOLKER HAHN
Fraunhofer Institute for Computer Graphics Research IGD, Darmstadt

Internet

- www.biometrics.org
- http://pagesperso-orange.fr/fingerchip/biometrics/biometrics.htm
- www.research.ibm.com/peoplevision
- www.securityinfowatch.com

Precautions against disasters

Related topics

- Environmental monitoring
- Sensor systems
- Communication technologies
- Image evaluation and interpretation

Principles

A catastrophe is a natural occurrence, a technical disaster, a terrorist attack, or some similar event which affects a large number of people or a large region, with the result that standard emergency measures fail to provide everyone with the required assistance in time. Recent examples include the Elbe River floods of 2002, the East Asian tsunami of 2004, earthquakes in Pakistan, forest fires in Portugal, and most recently hurricanes in the Caribbean in 2008, all of which caused devastating damage.

The dimensions of these natural disasters have noticeably increased, not least due to our higher vulnerability compared with previous decades. At present it is usually not possible to predict when and where they will occur, and how severe they will be. But the damage they cause can be reduced by taking better precautions and optimising our ability to respond to disasters.

Applications

▶ **Satellite-based crisis information**. The global upsurge in the number of natural disasters, along with the resulting humanitarian emergencies and threats to the civilian population, have led to a need for more, up-to-date information about existing conditions. The experience of the past few years has confirmed the need for timely, comprehensive and wide-coverage earth observation information for a variety of civilian crisis situations. Besides its immediate usefulness when reacting to and assessing a crisis, the generation of spatial information is especially important for reconstruction work and crisis prevention.

Generation of spatially-related crisis information products for the Elbe floods of 2002 (Dessau, Germany) from IKONOS imagery and topographic data. Left: In order to assess the extent of flooding, satellite images from before and after flooding are visually combined to facilitate change detection. Right: Converting the complex content of a satellite image into readily comprehensible crisis information requires the generation of situation and damage maps, which is accomplished by combining the post-flood satellite data with a topographic map (centre) to show the extent of flooding. Source: European Space Imaging

▶ **Monitoring earthquakes**. The world was informed about the 2004 level 9.3 seaquake in the Indian Ocean just 12 minutes after the event. Earthquakes are detected with the help of a network of globally distributed seismometers, which are highly sensitive vibration recorders. This measurement technique is based on powerful instruments which continually record any movement of the earth's surface and relay the information to the appropriate situation centre if it surpasses a critical intensity.

A *seismograph* consists of a measuring device, the seismometer, and a data recording system. Most seismometers measure ground movement with respect to a fixed reference point by means of an inertial mass suspended as a pendulum. Attached to the pendulum is an induction coil which moves through the field of a permanent magnet rigidly fastened to the frame. An electrical voltage is induced by the pendulum's movement and digitally recorded. In such seismographs, the suspended mass can only move in one direction or rotate around a vertical axis. The three motion components must therefore be measured using three independent seismographs. The suspended masses can weigh anywhere from a few grams to several kilograms, depending on the design.

Recorded information about the amplitude and arrival time of earthquake waves at separate seismometers can be used to determine the geographical location, the depth and the strength of an earthquake.

Spatially restricted, regional, or very dense seismometer networks are used for special tasks such as monitoring high-risk zones, or for time-critical early warning systems. Well-known examples include the earthquake monitoring system in Japan (with about 800 stations) or the recently established Indonesian tsunami early warning system (with some 200 stations). In Japan, critical infrastructures such as high speed trains are brought to a halt immediately following any earthquake that exceeds a certain magnitude threshold. As for the tsunami early warning system, it is essential to determine, in the shortest possible space of time (ideally within 2–3 minutes), the exact location of any powerful earthquake that could lead to a tsunami.

In the case of the tsunami warning system, seismometers are also installed on the ocean floor at depths of about 3–5 km and send their data as sound waves via a special acoustic modem to a buoy. The buoy then sends the information to an assessment centre via satellite communication links. Other oceanographic systems use fibreglass cables on the ocean floor for the data transfer.

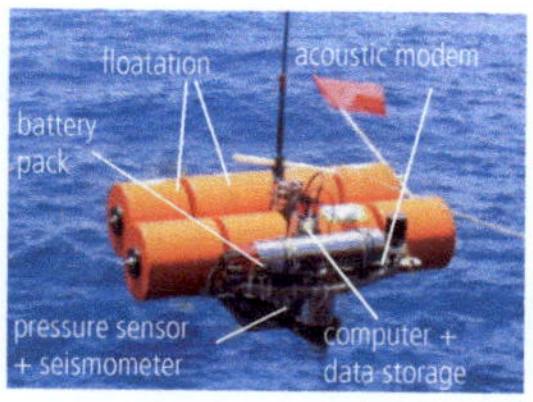

Ocean bottom unit (OBU): The main components are a pressure sensor to record changes in ocean level (the resolution at 5 km depth is about 2–3 mm), a seismometer to record the earthquake waves on the ocean floor, a computer with a data recording unit, an acoustic modem to relay data to the surface, and a battery unit as the power supply. The OBU is equipped with floatation devices. When these are released from their anchoring weight by a message from the acoustic modem, the OBU rises to the surface so that it can be serviced. It is then equipped with a new anchoring device and deployed again. Servicing takes place every 6–12 months, depending on how much power is consumed. The maximum operating depth for such sensors is 6 km. The water column over the pressure sensor acts like a filter. Frequent changes to the ocean surface caused by surface waves are not registered as pressure changes at the depth of the sensor; only long-wave phenomena such as tides or tsunamis are recorded. Source: GFZ

Trends

► **Satellite altimetry.** Using this approach, changes in ocean height and waves can be measured to centimetre accuracy. Altimeters are usually satellite-borne radar instruments which send out brief microwave impulses to earth at regular intervals. Each impulse is reflected by the sea surface and returns to the satellite, where it is detected and recorded. If the time difference between emission and reception is known, the distance between the satellite and the sea surface can be determined. If the satellite orbit is known precisely enough, the sea surface height can be determined to an accuracy of a few centimetres. Satellite altimeters typically orbit the earth in 90 minute cycles. While the earth rotates underneath at right angles to the orbit, measurements are recorded along a track. It takes several days or weeks before the satellite flies over the same region again and can record the same track. To achieve continuous monitoring with constant coverage, as is required for the early detection of tsunamis, it is necessary to operate a larger number of satellites moving in different orbits.

► **Radar information from satellite systems.** Radar (wavelength 3–25 cm) enables images to be recorded day and night and irrespective of cloud cover. This method is therefore well suited to "real time" reception, particularly during bad weather. In the case of an *SAR (synthetic aperture radar)* system, the radar instrument obliquely "illuminates" the region to be imaged as it flies past. The pulses are scattered in different directions by the objects on the earth or sea surface, and some of them return to the radar. The returning echo is also affected by the roughness and dielectric constants of the objects, which depend on their water content. With this approach it is possible, for example, to detect oil spills because oil films flatten the sea surface. It has also been demonstrated that extreme waves, sometimes called monster waves, can be identified by SAR radar.

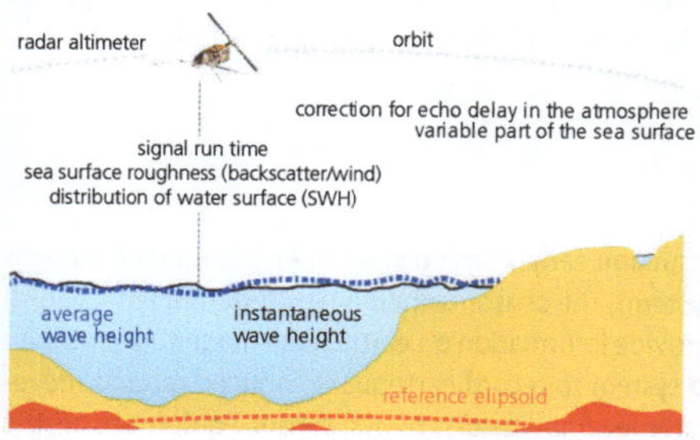

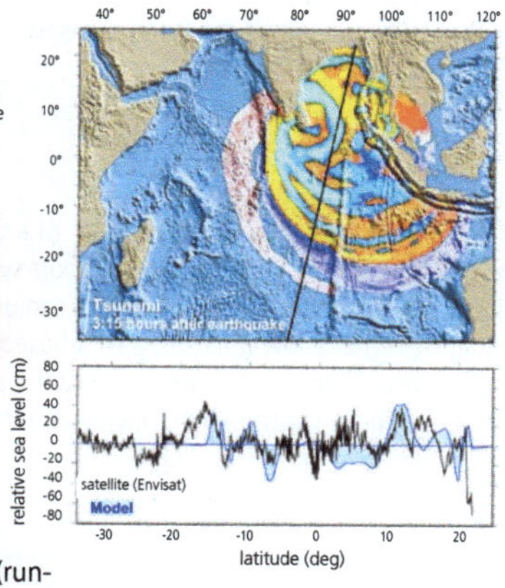

Satellite altimetry. Left: Radar altimeter measurements of sea surface topography. The round trip time (runtime) of a radar pulse, surface roughness and wave distribution are measured (SWH: significant wave height). Top right: Propagation of a tsunami wave after two hours, determined from satellite altimeter measurements. Sea surface changes and wave activity can be measured to centimetre accuracy. The black line shows the orbit of the TOPEX/Poseidon altimeter. The coloured wave pattern shows the propagation of the tsunami wave (model calculation). Lower right: Propagation and height of the tsunami wave along the altimeter orbit with the tsunami wave at 5° southern latitude. Source: GFZ and NOAA

SAR images not only consist of pixel brightness values but also contain phase information. If two or more images are recorded from slightly different viewing angles, a phase difference can be calculated to obtain what is known as an interferogram. It can be used to generate a digital elevation model, which is a valuable tool for a variety of applications, including disaster response.

If the two interferometric SAR images are not recorded at the same time but with a time delay of anything between one day and several years, it is possible to determine whether the earth's surface or any objects on it have moved during the interval. This differential

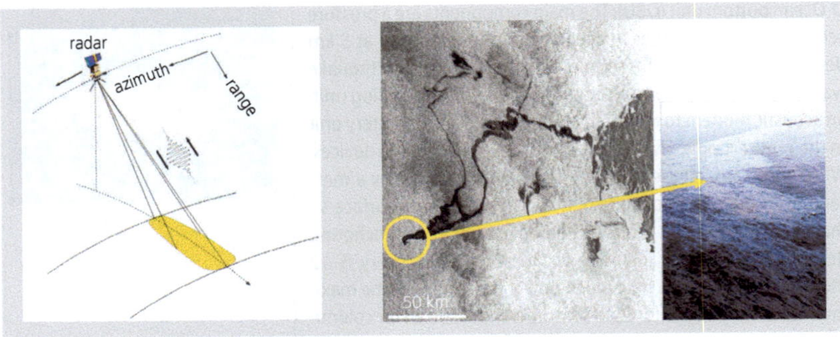

Synthetic aperture radar (SAR). Left: Illustration of SAR geometry. Synthetic aperture radar yields a two-dimensional representation of a portion of the ground by scanning the earth's surface with electromagnetic waves. At regular intervals the instrument sends out microwave pulses which are reflected from the earth's surface and which return to the sensor at different times, depending on the distance between the object and the SAR system. From the strength of the returned signal it is possible to draw conclusions about the roughness and dielectric constants of the ground surface in question. Centre: SAR image of the Galician coast during the shipwreck of the Prestige tanker, showing an oil film. Right: Photograph of the Prestige disaster taken from a helicopter. Source: ESA, DLR.

The modular and expandable concept of a tsunami early warning system: All information streams come together in the DSS (decision support system). The sensor technologies (top from left to right) include: the seismological system, which can provide information on earthquake magnitude, location and depth within a few minutes, a GPS-based system to record earthquake-induced ground movements measured by seismometers, coastal water level indicators, buoys, sensor equipment on the ocean floor ("ocean bottom units"), and data from earth observation satellites. Source: DLR

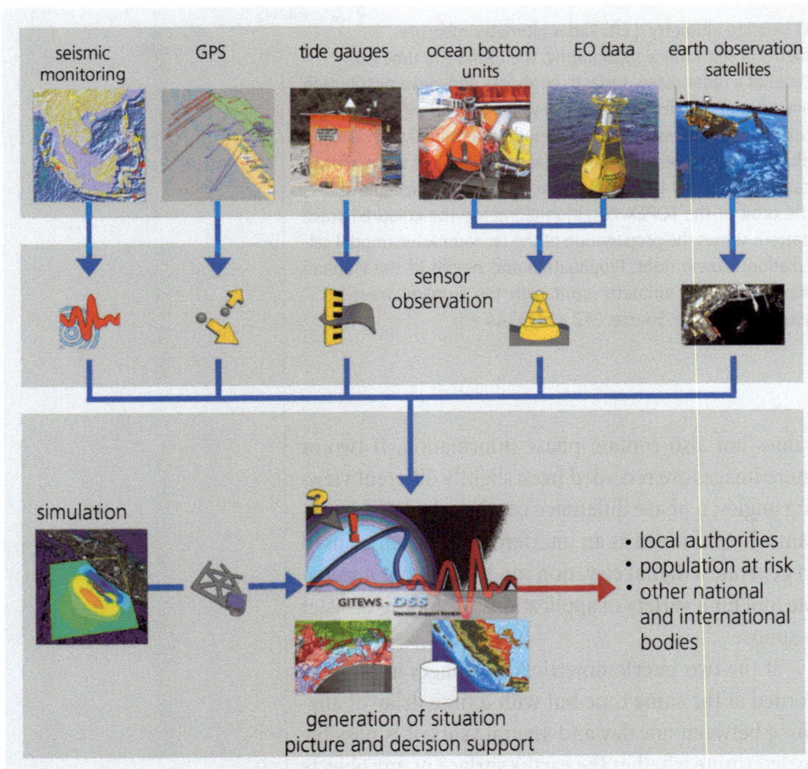

interferometry procedure makes it possible to detect and map surface glacier flow, slope collapse, volcanic activity and deformations caused by earthquakes, as well as vertical land elevation or subsidence, to within an accuracy of a millimetre per year.

► **Optical spectrometric remote sensing**. Geodynamic processes and deposits of fossil hydrocarbons can cause outgassing at the earth's surface. Locating and monitoring these gas flows can contribute to understanding the underlying geophysical processes. Changes in gas quantity and/or temperature may indicate impending earthquakes or volcanic eruptions. The measuring principle is based on the fact that the escaping gas has a constant temperature over the medium term. If it is colder than the surface temperature, which depends on the weather, then it will cool down the area near the outflow point. If it is warmer, it will in turn heat up this region. In other words, the temperature near the gas exit point is only partially influenced by surface temperature changes caused by the weather. Infrared cameras (7.5–13 μm wavelengths) can be used to take spatially identical thermal images at regular intervals of time, which can then be analysed to find pixels showing minimal temperature differences. These are the locations of the gas exit points. Changes in the amount and/or temperature of the gas alter the intensity of this thermal effect and can be identified in the time series.

► **Monitoring the ionosphere**. The surface wave spreading radially from the epicentre of an earthquake (Rayleigh wave) excites the air layers above it like a loudspeaker membrane. The acoustic waves stimulated by an earthquake on land reach the ionosphere in about 10 minutes, whereas tsunami-induced atmospheric gravity waves require a longer time, about 90 minutes. Atmospheric waves spread up to the level of the ionosphere at dramatically increasing amplitude because the air density decreases exponentially with altitude while the wave energy is conserved. A 1-mm rise of the ground causes amplitudes of more than 100 m in the atmosphere. The ionosphere is an electrically conductive layer of the earth's atmosphere and is located at an altitude of 100 – 1,000 km. In this very thin atmospheric layer, gas molecules are ionised – in other words, split into ions and free electrons – by energy-rich incoming solar radiance. The atmospheric gas thus enters a plasma phase. The earthquake-induced atmospheric waves alter the plasma, and this can be detected.

As was recently demonstrated, transionospheric GPS signals are sensitive enough to detect earthquake-induced surface waves as density changes in the plasma of the ionosphere. These signatures, found with the help of GPS (GALILEO), are processed together with seismic and hydrographic data coming from pressure sensors located on the ocean floor and from the GPS buoys. Together with highly accurate position information from the buoys, the recorded data is forwarded via communication satellites to the early warning centres. This procedure could provide supplementary information that would improve the accuracy of tsunami detection and warning messages.

► **Decision support.** During many types of natural disaster, decision makers only have a few minutes in which to evaluate a situation and decide whether to send out a warning message or begin an evacuation. One example is a *tsunami early warning*: if a potential tsunami wave is due to hit the coast in less than 30 minutes, any decision as to whether to issue a warning will inevitably be uncertain, despite the numerous sensor systems employed for tsunami detection. A new concept is now being implemented as part of the German-Indonesian Tsunami Early Warning project (GITEWS) to aid such decisions.

The measurement values obtained on the basis of this multisensor approach are collected in a *decision support system* (DDS), analysed, and condensed with the help of a simulation system to a constantly updated and geographically differentiated situation picture. Building on this information, the DSS generates recommendations for decision makers. As soon as a decision has been made, the system automatically triggers all subsequent processes, for example the generation of a warning message.

Prospects

- Innovative technologies improve the flow of information during natural disasters and help decision makers to react more rapidly, efficiently and purposefully.
- Rapid wide-coverage mapping of areas damaged by natural disasters can only be accomplished with the help of remote sensing data. A combination of different optical sensors and the latest radar technologies (both aerial and satellite) assures efficient emergency mapping.
- Early warning systems which combine different sensor systems and data streams and integrate their output into a complete situation picture with the help of data fusion, numeric models, simulations and decision support systems give decision makers a head start in terms of timing and quality.

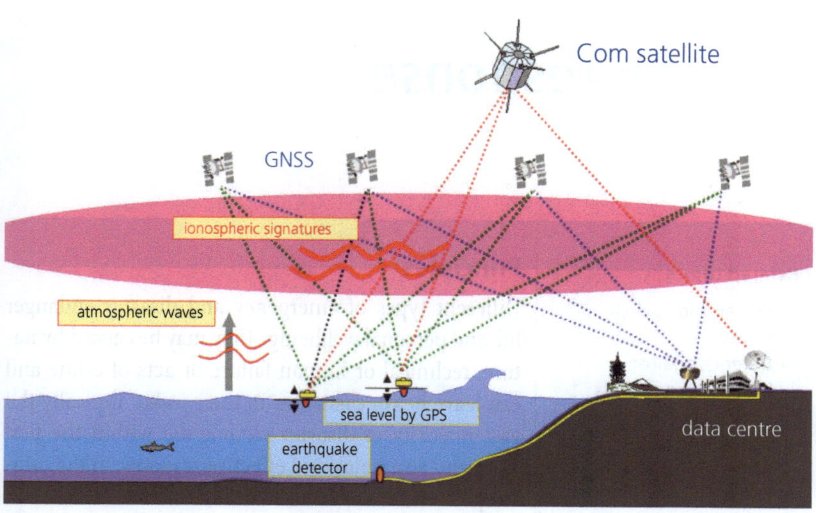

Monitoring the ionosphere: Illustration of a tsunami early warning and monitoring system which includes ionospheric monitoring data derived from measurements of wave motion by GNSS (Global Navigation Satellite System) buoys installed for this purpose. The GNSS data are relayed via a telecommunication link (in this case a communication satellite) to a data centre, where it is possible to correlate this information with other tsunami indicators such as data from pressure sensors located on the ocean floor in order to assure reliable early warning. Green: GNSS measurements from the buoys; blue: GNSS measurements to reference stations; red: measurement data transfer to the data centre. Source: DLR

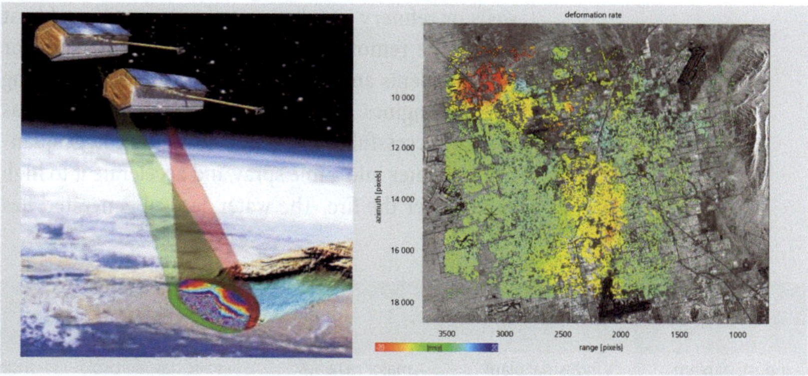

SAR interferometry. Left: This procedure makes it possible to detect plate tectonics, slope collapse, and terrain elevation and subsidence to within an accuracy of 1 mm/a. Source: ASTRIUM. Right: Map of ground elevation and subsidence in the Las Vegas urban area based on SAR interferometric data. Red corresponds to subsidence of up to 20 mm/a, green to stable terrain and blue to uplift. This movement is caused by groundwater extraction.

DR. WOLFGANG METT
German Aerospace Center DLR, Cologne

Internet

- www.zki.dlr.de
- www.gfz-potsdam.de
- http://helmholtz-eos.dlr.de
- www.infoterra.de/terrasar-x.html
- www.gitews.de

Disaster response

Related topics

- Precaution against disaster
- Defence against hazardous materials
- Plant safety
- Environmental biotechnology
- Environmental monitoring

Principles

Different types of emergency and disaster endanger life and personal wellbeing. They may be caused by nature, technical or human failure or acts of crime and can leave heavy casualties in their wake. Typically the vast majority of disaster victims die within the first 72 h, and so a quick and effective response to the catastrophic event is crucial.

Depending on the scale of the emergency, various agencies at different levels of public administration, including governmental and non-governmental organisations, have to act and cooperate on emergency relief. In a coordinated effort, initial responders are supplemented by secondary emergency services, eventually from farther away. Their capabilities are increasingly being improved by a range of technological solutions and innovations.

Applications

▶ **Firefighting**. To extinguish fires at oil and gas wells or refinery installations, great quantities of heat have to be removed over a short period of time. For such incidents an aerosol extinguishing method using aircraft engines mounted on a carrier vehicle has proven successful. The jet engines vaporise huge quantities of water into a fine spray and accelerate it to high speeds over the fire. The water aerosol generated has an extremely large droplet surface and as the water evaporates it removes energy from the fire very quickly. The water vapour that forms also cuts off the oxygen supply.

In this aerosol firefighting system water, extinguishing foam or extinguishing powder is injected into the exhaust jet. Vehicles of this type are suitable for fighting refinery and chemical plant fires, extinguishing burning oil wells and putting out forest fires. They are able to cool fire-endangered chemical installations or tanker cars effectively, disperse noxious gas clouds and dilute inflammable gas clouds. In tunnel fires, the jet engines are used to blow the dangerous smoke from the tunnel (the exhaust jet entrains roughly fifteen times the quantity of air) and cool the area around the fire. This allows the firefighters to advance more quickly to the seat of the fire and start the extinguishing process. The system is not suitable for use at airports, where extinguishing operations have to get underway within seconds whereas the jet engines need some time to start up.

▶ **Emergency rescue**. People endangered by fire must be removed from the danger area as quickly as possible, but the escape routes from elevated locations may be blocked. Evacuation chutes are therefore being increasingly installed in addition to escape routes, stairs and emergency slides. These chutes decelerate the descent by friction and allow evacuation from high

major fire	explosion	major accident	ecological disaster	natural disaster	epidemic/ pandemic	radioactivity
- industrial plant - hazardous materials transportation - ship - residential area - oil well, coal mine - forest, moorland	- industrial plant - hazardous materials transportation - coal mine - pyrotechnic items, ammuniation - bomb blast	- major railway/ highway accident - aircraft emergency landing - plane crash - shipping accident	- toxic gas, aerosol smoke cloud - water contamination - soil contamination	- thunderstorm - tsunami - avalanche, landslide - earthquake - flooding, flash flood - volcanic eruption - drought - heavy snowfall	- human disease - epizootic disease - epidemic plant disease	- nuclear power plant accident - nuclear materials transportation - radiation sources release from industrial or medical radiation sources - radiological attack

Overview of various types of large-scale emergency and disaster: Some emergencies arise quite suddenly, while others may develop over a prolonged period of time. Source: Fraunhofer INT buildings as well as from ships, offshore oil platforms and elevated means of transport. They offer the psychological advantage that the people seeking to escape do not see the height they have to descend.

There are two variants for decelerating the descent in such a chute: a closely surrounding chute that brakes the sliding motion solely by friction and a chute with an internal spiral in which the user is not so tightly enclosed. In most cases the chute is of multilayer design. The outer layer is a temperature-resistant protective skin made of fibreglass material or aramide fibres which can withstand temperatures of several hundred degrees. At least one further layer is similarly made of aramide fibres in order to provide the load capacity required for several people to use the chute at the same time. The length manufactured depends on the intended place of use and may vary between 3–80 m.

► **Maritime rescue.** Rescues at sea require personal rescue equipment and suitable boats or life-rafts. It is essential to prevent or, by suitable design, render insignificant any capsizing or overturning of the boats or rafts in the sea and to build them in such a way that even exhausted or injured persons can gain access. Since the vessels from which people in danger have to escape are getting bigger and bigger, jumping directly into the water is hardly survivable. Emergency slides, evacuation chutes, quick-launching lifeboats, etc. are therefore gaining in importance. For large / high ships and offshore platforms the use of free-fall lifeboats has proven successful. The occupants have to be carefully strapped-in so as to survive without injury the impact on the sea's surface from a height of up to 40 m.

► **Personal protective equipment.** Rescuers need personal protective equipment to be able to move into incident areas under hazardous conditions without running excessive risk. The type of equipment is dictated by the operational requirements. For firefighting it must be as light as possible in order to avoid exhaustion and heat accumulation. For head protection against falling debris a helmet made of heat-insulating thermoset fibre composite is worn. It is designed for the attachment of supplementary equipment such as

The aerosol system firefighting vehicle in service with the BASF plant fire department can atomise water, foam or extinguishing powder. The control panel and control electronics are housed in the driver's cab. The second vehicle is used to supply water. The two Alphajet engines used by the aerosol generator have a combined thrust of 1.3 t (picture detail). Four multipurpose nozzles are installed on top to inject up to 8,000 l of water (or water/foam mixture) per minute into the exhaust jets. The unit can be swivelled through 180° and tilted. The maximum range of the water vapour is 150 m. Source: Zikun Fahrzeugbau GmbH

cameras, etc. The clothing also has to withstand cuts and tears and be heat-insulating and fire-retardant. Aramide fibres are used as a fire-retardant material.

In future this basic equipment will be complemented by helmet visors onto which information will be projected, as already used in combat aircraft. The helmets will be fitted with an integrated headset for communication purposes, and in order to be able to see in smoke-filled rooms, a thermal imaging camera whose images are available to both the wearer and (by radio data transmission) to the operational commander. Firemen already wear a respirator which, depending on the oxygen content of the ambient air, is designed either as a mask respirator fitted with a filter or as an independent respirator (with a compressed air cylinder) so that they can work in areas which are full of smoke or toxic fumes. Efforts are being made to improve the comfort of protective clothing (dissipation of

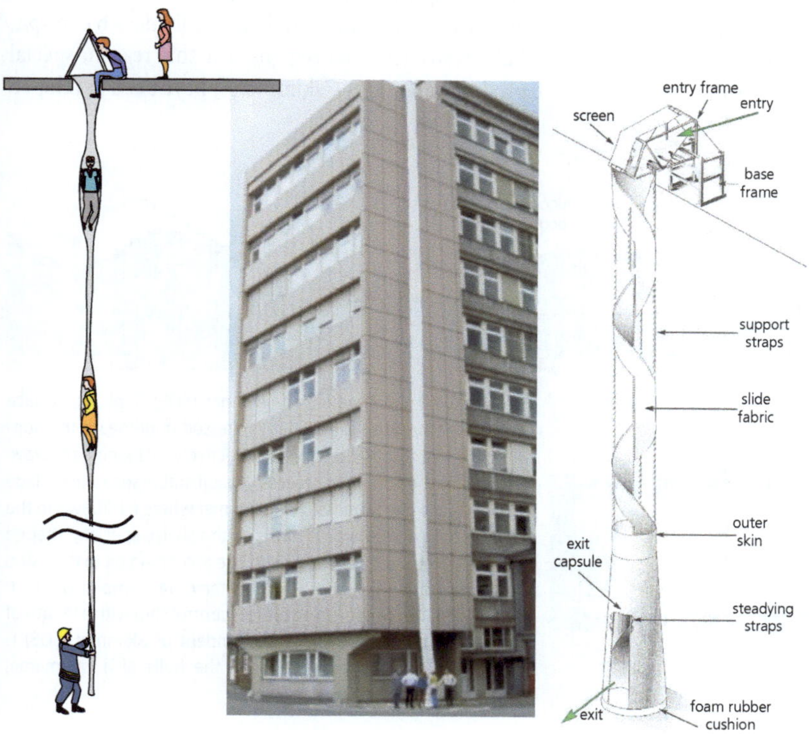

Evacuation chute: The evacuation chute is fastened securely at the point of entry by means of its storage container and is unfolded downwards in an emergency. The person being evacuated sits down on the edge of the opening with their legs dangling in the chute and then drops down. The chutes are designed to expand and then to constrict again as people slide through. People can reduce the sliding speed by stretching out their elbows and knees. A rescuer standing at the lower end of the chute keeps the point of exit free for the next person descending. Depending on the design it is possible to evacuate up to 350 people within 15 minutes using one chute. Source: Fraunhofer INT, b.i.o. BRANDSCHUTZ GmbH

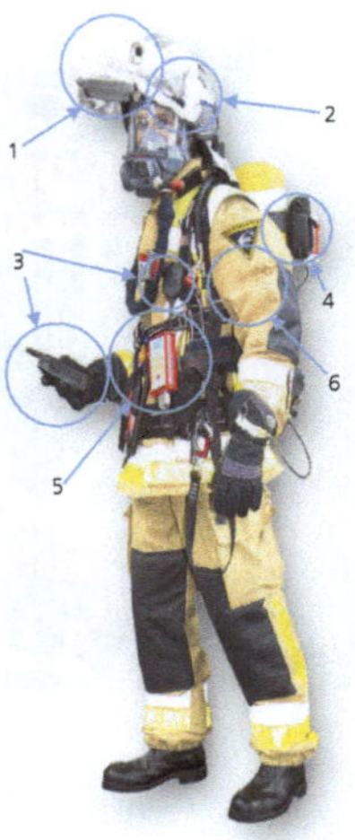

Blueprint for the "fireman of the future": The features complementing the basic equipment are designed to leave the hands largely free: (1) Helmet-mounted thermal imaging camera with head-up display and connection to the digital radio set for better orientation at the incident location, keeping the operational commander informed and for documentation of the operation. (2) Helmet-integrated headset. (3) Digital radio set with various functions which can be operated by gloved hands. (4) Transmitter as location aid, with connection to the digital radio set for locating the wearer in buildings where GPS does not function. (5) Respirator monitoring device with pressure fittings and remote monitoring. (6) Enhanced protective suit offering improved comfort and, envisaged for the future, integrated monitoring of body functions. Source: EADS / Fraunhofer ICT

Free-fall lifeboats are now the standard evacuation method for the crews of cargo ships out at sea: They provide a quick escape even from very high superstructures. Source: Helmut Seger

heat and sweat). Other developments include the monitoring of body temperature and functions so that the operational commander can give instructions for withdrawal if levels become critical. This requires the integration of an advanced communication system (digital radio) in the equipment.

► **Oil spill control.** Catastrophic oil spills count among the gravest ecological disasters, particularly when open bodies of water are contaminated. It is possible to fight crude oil spills by means of bioengineering methods, but this takes time. The use of chemical agents to dissolve oil slicks is controversial, because they shift rather than reduce the damage to the environment. Skimming off the oil mechanically before it reaches the particularly sensitive coastal biotopes still plays an important role, but the use of booms to prevent the spread of oil contamination at sea takes a relatively long time and is only limitedly possible in spatially restricted tidal regions. For this reason, special oil skimming ships ("skimmers") have been developed which in some cases use special booms or are designed as catamarans or split-hull vessels.

The disadvantage of such designs is that they require a relatively calm sea with wave heights of less than 1.5 m. Even the use of a blade that follows the wave motion and separates off a surface layer in a predefined thickness loses its efficiency if the sea becomes rough (MPOSS – Multi-Purpose Oil-Skimming System). A "sea swell-independent oil skimmer" (SOS) is being developed for combating spilled oil even in rough sea. The special design of the hull cushions the wave motion, and a vortex is generated specifically to siphon off the oil film. The advantages of the new oil skimming system will include a high travel speed to the location of the oil spill, the ability to operate in rough sea and high oil-skimming efficiency. There are plans to build a prototype for testing under real conditions.

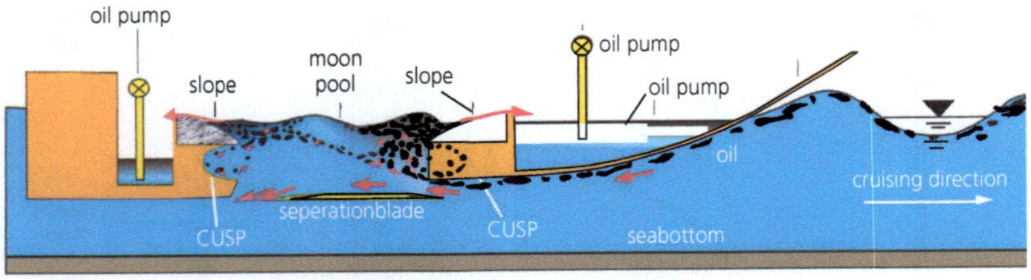

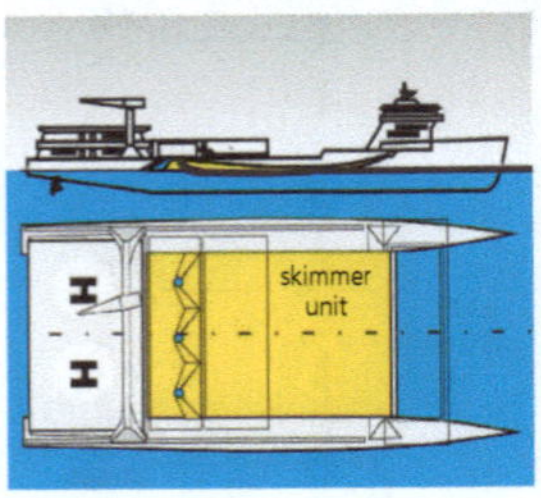

Sea swell-independent oil skimmer (SOS). Top: The surface water together with the oil film is pressed underneath the pontoon without turbulence, thereby effectively cushioning the wave motion for the skimming process. An adjustable separation blade conducts the oily water into the skimmer where it is drawn to the surface by a vortex which forms at the sharp, specially shaped trailing edge of the bow segment. The ascending oil masses slop into a collection tank, from where they are pumped off. Left: Sketch of a new ocean-going oil spill control ship with a length of approx. 80 m. The sea swell-independent oil skimmer (SOS) is suspended to float freely between the hulls of a catamaran. Source: TU Berlin

Trends

► **Command and control equipment.** Effective communication and optimum operational planning are essential for the swift and coordinated control of disasters and their aftermath. To allow rapid assessment of the broader as well as the local situation a multitude of different data and information sources, including satellite data and geographic information systems (GIS), have to be combined in a joint information system.

In a first step, the globally used digital trunked radio standard TETRA ("terrestrial trunked radio") will enable information to be passed between all rescue agencies and services. Communication systems are also being developed which combine multimedia communication (using various mobile radio standards) and satellite navigation with a compilation of the situation picture and with background data.

In a proposed future European rescue service management system all relevant information will be broadcast continuously to all participating services, using existing as well as spontaneously built up infrastructure. The picture is constantly updated as each network user can feed new data into the system ("push-to-share"). The data can be displayed on different media, e. g. a tablet PC or PDA, with additional information levels available behind symbols.

So that the system can be used hands-free during operations, speech control, speech input and graphic output on displays will be integrated in the personal equipment.

► **Search-and-rescue robots.** When a major incident occurs it is often necessary to work at many different search locations simultaneously in order to find the casualties as quickly as possible. As only a limited number of search specialists are available in most cases and sniffer dogs get tired, robots might support search specialists increasingly in future. This would minimise the risk to rescuers during the search-and-rescue work and help to localise secondary hazards (e.g. escape of gas or pollutants).

For search robots working on the ground or underneath wreckage, the size and shape of the openings between the debris through which they have to move are crucial and may measure less than 30 cm. Very small robots, however, have problems in negotiating obstacles, cannot carry much of an energy supply and are limited in terms of the sensors and communication devices they can accommodate. For small, winding openings and for work in sewage systems snake-like robots comprising a chain of functional elements have therefore been developed.

Each link of the chain has its own drive system, which can incorporate a miniature chain drive as well as a wheeled chassis or a running gear/walking support system. As it is not foreseeable where the snake robot will touch the opening or the obstacle, the drive elements are mounted on all four longitudinal sides of the individual elements so that advance movement is ensured on any side that is in contact with the surroundings. The individual elements are linked by means of actively adjustable joints, which permit both lateral and vertical bending. Here, too, a great number of technical solutions are possible in which advancing and bending mechanisms allow the robot to move like a snake through narrow openings.

▶ **Sensors.** Search devices and sensors for detecting casualties with the aid of robots should be suitable for use on various platforms and also be small enough to attach to data transmission equipment. The data processing should take place prior to transmission so as not to place too great a strain on difficult data transmission routes (e.g. from cavities). While sensors which are small enough are already available, some still lack the operative capability or use too much energy. The plan is to use robots to search in small swarms, incorporating cost-effective sensors.

Thermal imaging cameras provide an option for searching in darkness underneath rubble. Sensors using the coded-light method might be used where temperature differences of a living body in relation to the environment are negligible, or where searches for other objects are carried out. This method makes use of a laser to scan a sector of space linearly. The reflected laser light is captured by a camera. A shadow of the three-dimensional object is registered as a function of the distance between the laser and the camera. By processing the camera data with fuzzy logic software this shadow is used to obtain a three-dimensional image with what is actually a two-dimensional method.

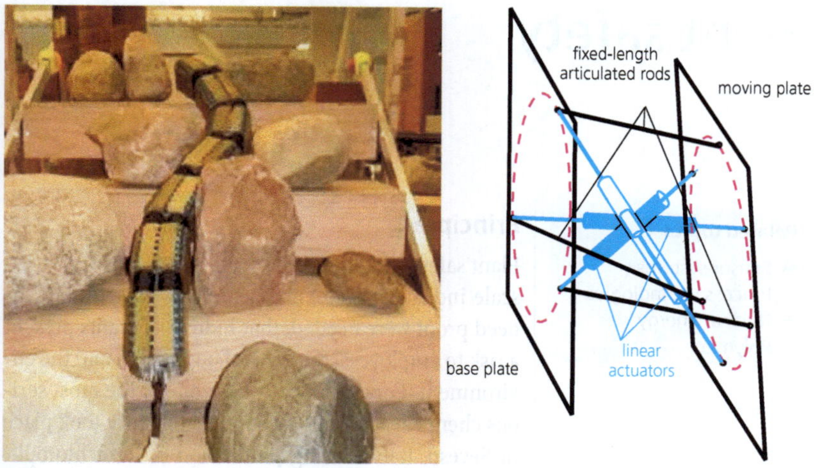

Snake robot. Left: This robot consists of several elements with track-like drive belts on all four longitudinal sides. The individual segments are linked by articulated joints, permitting controlled lateral and vertical bending. Right: Moving joint unit of the snake robot, comprising three actuators and three fixed-length articulated rods. Source: Johann Borenstein, University of Michigan

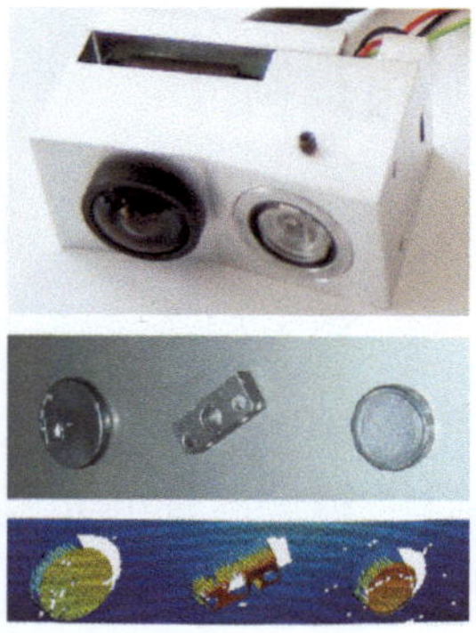

3D coded-light sensor. Top: CCD camera and scanning laser. Because the camera is located to the side of the laser, the image appears together with a shadow, making it possible to extract 3D information. Centre: Test objects. Bottom: Images of the test objects as generated by the 3D coded-light sensor. Source: Fraunhofer IITB

Prospects

- Different types of disaster and emergency call for a diverse range of technologies to facilitate situation assessment, the search for casualties, short- and medium-term countermeasures and relief efforts.
- Numerous technical innovations are improving the scope for providing help and conducting rescue activities, particularly in special emergency and hazard situations.
- Robot systems are increasingly being used for situational fact-finding, casualty localisation and risk neutralisation. They enable a better knowledge of the situation to be gained and speed up the search-and-rescue measures.
- The management and coordination of rescue services, as well as communication between them, are crucial, particularly when they are dealing with major incidents and disasters. The use of new technologies can be expected to increase their effectiveness substantially in the near future.

DR. MATTHIAS GRÜNE
WOLFGANG NÄTZKER
Fraunhofer Institute for Technological Trend Analysis INT, Euskirchen

Internet

- www.thw.bund.de
- www.bbk.bund.de
- www.civilprotection.net
- www.dhs.gov/xres
- www.cordis.europa.eu/fp7/cooperation/security_en.html
- http://cool-palimpsest.stanford.edu/bytopic/disasters

Plant safety

Related topics

- Sensor systems
- Process technologies
- Nuclear energy
- Polymers

Principles

Plant safety should be non-negotiable. However, large-scale industrial accidents in the past are proof, if indeed proof were needed, that industrial plants present a risk to employees, the general population and the environment. For example, in 1976 one of the most serious chemical accidents in European history took place in Seveso, Italy. During production of hexachlorophene, a medical disinfectant, there was an uncontrolled reaction and about 2 kg of this toxic substance was released into the atmosphere through a safety valve. An 18 km^2 area was contaminated and many people were poisoned. The accident in Bhopal, India in 1984 was even worse. Temperature and pressure increased dramatically when a tank containing methyl isocyanate was flooded with water. 24 t of the toxic substance subsequently escaped through a safety valve, killing large numbers of people.

Analyses of accidents have shown there is a strong correlation between technology and organisational factors. Accidents were generally caused not only by a technical failure, but also by mistakes made by personnel, organisational mistakes, incorrect assumptions about safety systems and because procedure was ignored.

Numerous national and international bodies are now working on safety and the risks associated with industrial plants. Defined objectives governing safe operation of industrial plants are laid down in regulations, guidelines and manuals. To guarantee a minimum level of plant safety, the technology, organisation and human factor, which directly affect plant safety, must all be seen as being part of the same thing. A large number of technical facilities and systems are used to reduce the risks and to guarantee industrial plant safety. In the area of organisation, safety management systems have been introduced, which have improved plant safety. The human factor plays a decisive role when it comes to safe operation of industrial plants.

Avoiding technical failure always takes precedence over damage limitation. Failure prevention measures are used to keep the likelihood of serious technical failure at an acceptable, low level.

Because the industrial applications are so diverse, emphasis in the field of plant safety is placed on developing and optimising individual safety components. Particularly favoured are measures and technologies that identify technical failures early on and then immediately deal with the reasons for the failure in order to rectify the situation or limit the effects to a manageable level.

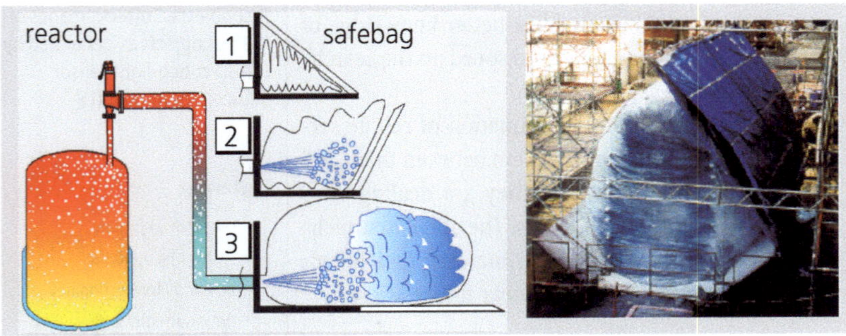

Pressure is released in a controlled way through safety valves into safety bags to prevent any substance leakage. The bags are made of chemical-resistant, coated material and are used, amongst other things, to render boilers or distillation columns safe. They take up almost no space when empty. Left: An empty safebag (1), at the beginning (2) and towards the end (3) of a pressure release cycle. Right: A safebag after pressure has been released. Source: Siemens Axiva GmbH

Applications

▶ **Closed systems.** Chemical production processes are used to produce a large number of substances. Special processes, such as recrystallisation or refining, are used to produce other substances from naturally occurring resources. A plant can only operate safely if enough is known about the properties of the substances it is producing. For example, the substances produced or reaction mixtures must not be allowed to escape into the atmosphere if the environment is to be protected. Instead, these substances must be produced in reliably safe closed systems. One way of doing this is to use safebags.

Many chemical reactions are exothermal. This means that heat is generated during the reaction. The potential risk posed by exothermal reactions can be largely attributed to the fact that chemical reactions take place far more quickly as the temperature rises. If more heat is released during the reaction than the safety systems can deal with, the result will be a self-accelerating increase in temperature. This, in turn, can produce a thermal explosion (spontaneous release of the accumulated reaction potential). Cooling and other control systems (regulating dosage speed, for example) are therefore of critical importance in the field of plant safety.

Processes must be monitored as accurately and reliably as possible if substances are to be prevented from leaking because of uncontrolled, exothermal reactions. If the results of system failure analyses are properly taken into account, serious failures can be avoided by detecting potentially critical reactor states early on, by intelligent signal processing and taking remedial action at an early stage. Depending on the type of reaction, critical reactor states can be identified, for example, by detecting the pressure and temperature parameters. Measured against "good" curve diagrams prestored in the process control system, critical deviations can quickly be identified and the process control system can initiate countermeasures, such as increasing the level of cooling, stopping dosage of reactants, adding phlegmatisation agents or reaction stoppers, stopping the reaction and putting the system into safe state.

► **Emergency condenser**. The emergency condensers feed residual heat into the core flooding pools from the reactor pressure vessel. There is therefore no need for high-pressure injection systems. The condensers also serve as an alternative means of depressurising the system.

The emergency condensers are heat exchangers composed of a number of parallel horizontal U-tubes between two headers. A pipe runs from the top header to the reactor pressure vessel steam plenum, whilst the bottom header is connected to the reactor vessel below water level inside the reactor. The heat exchangers are situated in a side pool filled with cold water. The emergency condensers and the reactor vessel are thus interconnected by a system of pipes. When the water level inside the reactor is normal, the emergency condensers are flooded with cold water, but if the water level drops, the heat-exchanging surfaces inside the tubes are gradually exposed and the incoming steam condenses on the cold surfaces. The condensate is fed back into the reactor vessel.

► **Passive pressure pulse transmitter**. Passive pressure pulse transmitters are installed in boiling water reactors as a safety feature. They operate independently of an electric power supply, external media or actuation via instrumentation and control signals. These transmitters initiate reactor scram, isolate main steam lines and automatically depressurise the reactor pressure vessel. The actual valve operation is accomplished using system fluids and valves with stored actuation energy. The passive pressure pulse transmitters initiate the above operations when the water level inside the reactor pressure vessel falls.

The passive pressure pulse transmitters consist of small heat exchangers connected to the reactor by an unisolable pipe. When the water level inside the reactor is normal, the main side of the heat exchanger is filled with water and no heat is transferred. However, when the water level inside the reactor pressure vessel drops below a certain level, the main side of the passive pressure pulse transmitter is filled with steam, which con-

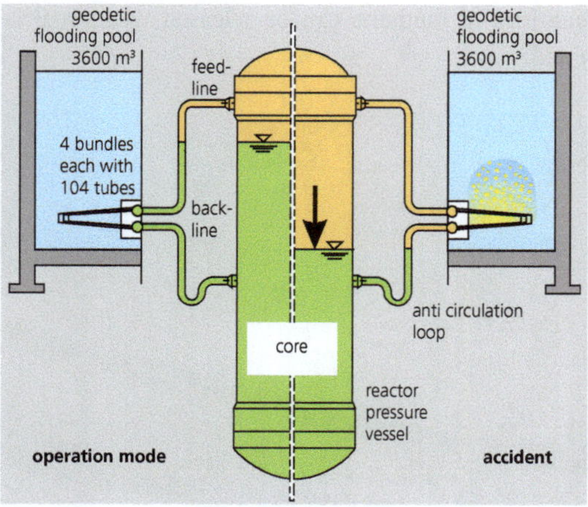

Emergency condenser operating mode. Left: During normal operation the emergency condenser is filled with water and heat losses are minimised by an anti-circulation loop. Right: When there is an accident, steam is sucked into the heat exchanger pipes as the water level inside the reactor pressure vessel falls. The steam condenses and the condensate only flows back into the reactor pressure vessel under the force of gravity.

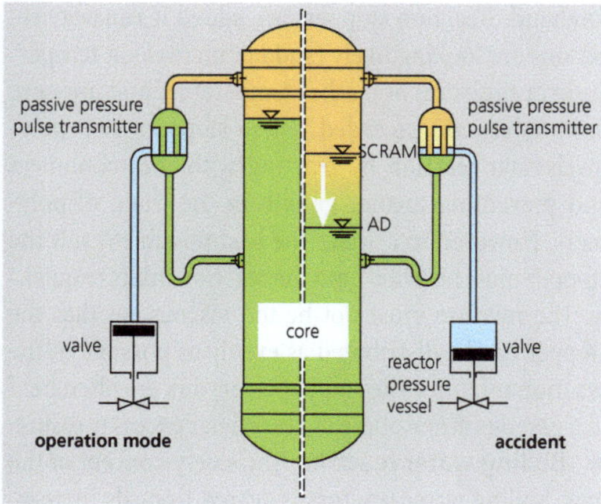

Passive pressure pulse transmitter operating mode. Left: During normal operation the main side of the passive pressure pulse transmitter is filled with water and no heat is transferred to the secondary side. Right: When there is an accident, steam is sucked into the pressure pulse transmitter, where the condensation heat heats up the water stored on the secondary side. The water partly evaporates. The rise in pressure triggers the safety functions via pilot valves.

denses and drains back into the reactor pressure vessel. During this process the water stored on the secondary side is heated and partly evaporates. This leads to a rapid increase in pressure. The rise in pressure triggers the safety functions via diaphragm pilot valves.

These devices function entirely independently and are therefore a good addition to instrumentation and control equipment. They are integrated into the plant's system in a "2 out of 2 configuration" so safety functions are not triggered if a single pressure pulse transmitter is actuated, but, by the same token, failure of one pressure pulse transmitter does not mean that safety functions cannot be triggered. It took less than 10 seconds to reach the 6 bar actuation pressure.

▶ **Inhibitors to polymerisation reactions.** Depending on the *polymer* chain growth process, a distinction is usually made between monomer and polymer combinations. A huge amount of heat is released during polymer reactions, which are based on a combination of monomers. The idea is to control the dosage of monomers in order to keep this heat generation in check. The key parameters to monitor here are temperature and pressure. Furthermore, the increase in polymer concentration makes the reaction mixture more viscous. This, in turn, makes it more difficult for the released heat to dissipate and the released energy is more likely to accumulate.

Temperature and pressure monitoring devices control dosage of monomers to make sure that there is no runaway reaction. The reactant dosing process is stopped if critical values are exceeded. Each value has been determined during experiments carried out beforehand. Reaction stoppers are added if runaway reactions are looking likely (sudden increase in temperature or pressure) or if other temperature and pressure limit values are exceeded. These stoppers stop polymerisation reaction by reacting with the monomers and preventing further growth in the chain of polymers. However, to achieve the best possible result the stopper must be "shot" into the reaction mixture quickly. The mixture must not be too viscous, so that the stopper can be distributed as evenly as possible in the reaction mixture. The stopper reservoirs are often held in a gaseous atmosphere of several bar excess pressure.

▶ **Boiling water reactors.** The safety concept of the new, boiling water reactors is geared towards increasing safety margins and giving operating personnel more time to act when there is an accident by:

- Increasing the amount of water inside the pressure vessel
- Creating large flooding pools inside (core flooding pool) and outside the pressure vessel (dryer-separator storage pool)
- Simplifying the (emergency) cooling systems
- Further decreasing the possibility of safety systems failing

Active safety systems are therefore replaced by passive ones or are combined with passive safety systems if the active systems cannot be replaced. Passive systems are actuated by natural laws such as gravity, natural convection and/or evaporation. No input is required from operating personnel. Consequently, passive safety systems are less likely to fail and there is also far less chance of a serious accident taking place.

▶ **Fire and explosion prevention.** A tailored explosion prevention concept is developed if a risk assessment shows that dangerous amounts of combustible gases, steam or dust can occur in a potentially explosive mixture. The concept can include primary explosion prevention measures (preventing a dangerous, potentially explosive atmosphere from forming or restricting its formation), as well as measures to prevent an ignition source from forming or designing the containment to be explosion proof. In practice, the ideas of preventing an ignition source from forming and explosion-proofing the containment structure are linked.

▶ **Industrial ventilation system.** Suitable industrial ventilation systems can be used to slow down development of potentially explosive gas/air mixtures or to stop them from forming at all. However, the systems must be monitored to ensure that they remain effective enough to do the job. Such systems are used, for example, in mine ventilation, where they have long been used to ensure good "ventilation" in coal mines. Large quantities of methane can be released when coal is mined.

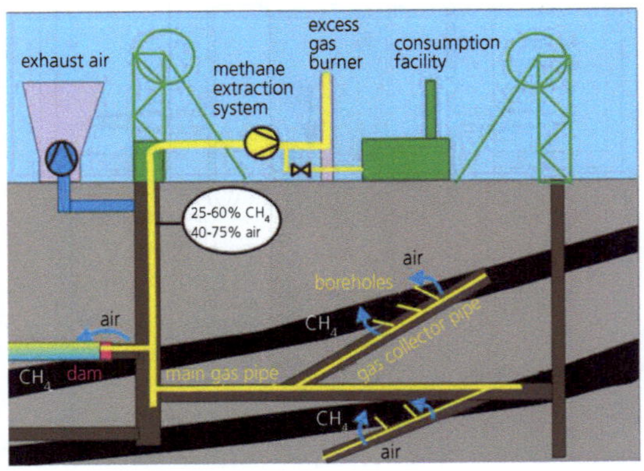

Extraction of methane from working mines: There is a separate gas collector pipe, which has vents in areas where large quantities of methane are released, for example, in the vicinity of boreholes. The methane extracted by the methane extraction system is fed to the excess gas burner or to an energy recovery unit (consumption facility).

The ventilation system has at least one hole in the ground to feed through fresh air and at least one hole in the ground for spent air ("exhaust air") discharge. Fans are used for ventilation purposes. They can be up to 5 m in diameter. A mine ventilation system can be used to dilute small amounts of methane to below 25% of the lower explosive limit. The methane/air mixture produced is sent above ground.

If large quantities of methane are released, the process of safely diluting the gas and extracting it via the ventilation system is very complex. Therefore the methane is extracted through specific holes in the seam. If the methane/air mixture is concentrated at about 150% of the upper explosive limit, it is fed to the methane extraction system through separate gas collector pipes. Here it can be diverted to the energy recovery unit. Above ground this method of directly extracting combustible substances at or near the point of release or discharge is known as "point extraction".

▶ **Intelligent automatic fire detectors.** Nowadays multi-parameter or multi-*sensor* detectors are often used as automatic fire detectors in the commercial/industrial sector. The indications of impending fire depend on what is burning. Detectors which react to indicators such as temperature, light scatter caused by smoke particles, or concentration of carbon monoxide, can be combined to detect a smokeless ethyl alcohol fire as well as a smouldering fire giving off quite a lot of smoke and not much heat. In smoke detectors using the light scatter principle an infrared light emitting diode (*LED*) in a measuring chamber transmits a pulsed beam of light. Normally this beam of light does not touch the photodiode connected to the evaluation unit. However, the beam of light is scattered by the smoke particles if smoke gets into the measuring chamber. Once the smoke reaches a certain level of concentration some of the scattered light reaches the photodiode and the alarm goes off. Fire detectors triggered by smoke can be used to detect fires when they are still smouldering. However, they are also sensitive to other environmental criteria, such as the presence of dust, steam, diesel engine exhaust fumes or tobacco smoke. This means that the alarm can sometimes go off for no reason. Heat detectors can be triggered by operational changes in temperature. Sensors that measure the concentration of various gases typical of fires, especially carbon monoxide, are gaining importance. Intelligent signal evaluation of the various fire indicators is being used to make fire detectors more sensitive to fire smoke whilst maintaining a high level of resistance to the alarm going off for no reason (e.g. algorithm evaluation of the signals from the optical, thermal and chemical sensors).

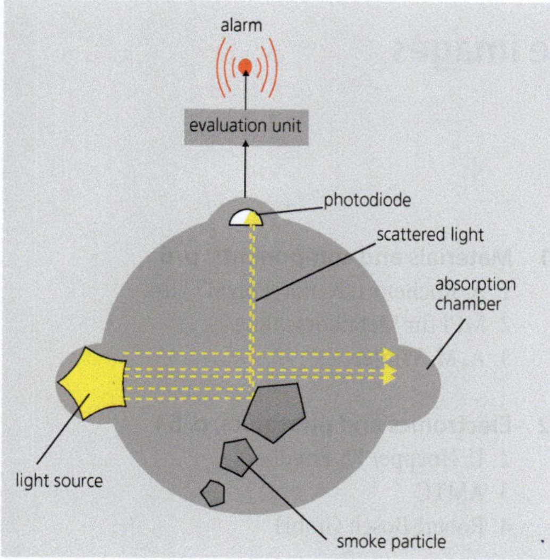

Optical smoke detector: Any particles of smoke that get into the measuring chamber diffusely scatter beams of light transmitted by the light source, which can then come into contact with the photodiode and the alarm goes off.

Trends

▶ **Reducing the concentration of oxygen.** For some years now the principle of permanently reducing the concentration of oxygen in the air has been applied in areas in buildings not constantly used by people and which are already sealed off from adjacent areas (e.g. dangerous materials stores). The odds of a fire starting are slashed simply by reducing the concentration from 21% (ambient air) to about 17% per volume by adding an inert gas (normally nitrogen) in a controlled way. Consequently, flammable materials become far less flammable. Areas protected in this way can also continue to be used by people.

Prospects

- Developments in the field of plant safety are being geared towards far more effective control and monitoring of safety processes using process control systems than in the past.
- In future, safety-related processes will be more automated, making any action by personnel largely unnecessary.

VOLKER KLOSOWSKI
TÜV NORD AG, Hannover

DR. ANDREAS SCHAFFRATH
TÜV NORD SysTec GmbH and Co. KG, Hamburg

DR. ULRICH SEIFERT
Fraunhofer Institute for Environmental, Safety and Energy Technology IUSE, Oberhausen

Internet

- www.csb.gov/index.cfm
- www.umweltbundesamt.de/anlagen/eng/index.html
- www.nuclearsafetyhistory.org
- www.fike.com/
- www.grecon.de/en/news/index.htm
- www.wagner.de/home/index.html?L=2

Sources of collage images

1 Materials and components, p. 6
1 A. Rochefort, Nano@PolyMTL.ca
2 MPI für Metallforschung
3 A. M. Winter

2 Electronics and photonics, p. 64
2 U. Hoeppe/ Fh Friedberg
3 AMTC
4 Robert Bosch GmbH

3 Information and communication, p. 120
2 K. Armendariz/ NRAO/AUI

4 Life Sciences and biotechnology, p. 156
1 Max-Planck-Institut für Biochemie
2 Schweizerische Brotinformation SBI
3 W. Barthlott, Nees-Institut
4 National Human Genome Research Institute

5 Health and nutrition, p. 184
1 Pacific Parkinson's Research Centre (UBC)

6 Communication and knowledge, p. 236
1 Fraunhofer-Verbund Mikroelektronik VµE
2 Nokia GmbH
3 Hokkaido Sapporo Toun High School
4 Fraunhofer-Gesellschaft

7 Mobility and transport, p. 282
1 Hapag-Loyd AG
2 D. Schollbach, ICE-fanpage.de
3 E.Ramseier/ETH Life, ETH Zürich
4 Transrapid International GmbH & Co. KG

8 Energy and resources, p. 316
2 H.-P. Strauß/Eon
3 Vestas Deutschland GmbH
4 Royal Dutch Shell plc

9 Environment and nature, p. 380
1 Renggli AG

10 Building and living, p. 420
1 HP Velotechnik
2 Gaggenau Hausgeräte GmbH
4 Igedo Company GmbH & Co. KG

11 Lifestyle and leisure, p. 440
1 DaimlerChrysler AG
2 Fraunhofer-IPT
3 Fraunhofer-IPA
4 DaimlerChrysler AG

12 Production and enterprise, p. 462
2 NOAA
3 Iridian Technologies, Inc.
4 K. Tokunaga

13 Security and safety, p. 496
1 AVN Abfallverwertung Niederösterreich Ges.m.b.H
2 Envisat (bearbeitet), European Space Agency/ Denmann Productions
4 City of Delaware, Ohio, USA

Subject index

W

X